Birkhäuser

**Operator Theory: Advances and Applications**
**Volume 221**

**Founded in 1979 by Israel Gohberg**

Wolfgang Arendt
Joseph A. Ball
Jussi Behrndt
Karl-Heinz Förster
Volker Mehrmann
Carsten Trunk
Editors

# Spectral Theory, Mathematical System Theory, Evolution Equations, Differential and Difference Equations

21st International Workshop on Operator Theory and Applications, Berlin, July 2010

Birkhäuser

*Editors*
Wolfgang Arendt
Abt. Mathematik V
Universität Ulm
Ulm, Germany

Joseph A. Ball
Department of Mathematics
Virginia Polytechnic Institute
Blacksburg, VA, USA

Jussi Behrndt
Institut für Mathematik
TU Berlin
Berlin, Germany

Karl-Heinz Förster
Institut für Mathematik
TU Berlin
Germany

Volker Mehrmann
Institut für Mathematik
TU Berlin
Berlin, Germany

Carsten Trunk
Institut für Mathematik
TU Ilmenau
Ilmenau
Germany

ISBN 978-3-0348-0296-3 ISBN 978-3-0348-0297-0 (eBook)
DOI 10.1007/978-3-0348-0297-0
Springer Basel Heidelberg New York Dordrecht London

Library of Congress Control Number: 2012941137



Printed on acid-free paper

Springer Basel AG is part of Springer Science+Business Media (www.birkhauser-science.com)

# Contents

# Preface

The present book displays recent advances in modern operator theory and contains a collection of original research papers written by participants of the $21^{st}$ *International Workshop on Operator Theory and Applications* (IWOTA) at the Technische Universität Berlin, Germany, July 12 to 16, 2010.

IWOTA dates back to the first meeting in Santa Monica, USA in 1981, and continued as a satellite meeting of the *International Symposium on the Mathematical Theory of Networks and Systems*. Nowadays, the IWOTA meetings are among the largest conferences in operator theory worldwide. The 2010 meeting in Berlin attracted more than 350 participants from 50 countries.

The articles collected in this volume contain new results for

- boundary value problems, inverse problems, spectral problems for linear operators,
- spectral theory for operators in indefinite inner product spaces, invariant subspaces,
- problems from operator ideals and moment problems,
- evolution equations and semigroups,
- differential and integral operators, Schrödinger operators, maximal $L^p$-regularity,
- Jacobi and Toeplitz operators,
- problems from system theory and mathematical physics.

It is a pleasure to acknowledge substantial financial support for the $21^{st}$ *International Workshop on Operator Theory and Applications* received from the Deutsche Forschungsgemeinschaft (Germany), the National Science Foundation (USA), the Institute of Mathematics of the Technische Universität Berlin (Germany) and Birkhäuser.

September 2011, Berlin-Blacksburg-Graz-Ilmenau-Ulm

The Editors

Operator Theory:
Advances and Applications, Vol. 221, 1-10


# About the Gradient of the Conformal Radius

A.N. Akhmetova and L.A. Aksentyev

**Abstract.** Let $D$ be a simply connected domain in $\overline{\mathbb{C}}$ and $R(D,z)$ the conformal radius of $D$ at the point $z \in D/\{\infty\}$. We discuss the function $\nabla R(D,z)$. In particular, we prove that $\nabla R(D,z)$ is a quasi-conformal mapping of $D$ for different types of domains.

**Mathematics Subject Classification (2000).** Primary 30C35; Secondary 30C62.

**Keywords.** Conformal radius, hyperbolic radius, gradient of the conformal radius, quasi-conformal mapping.

## 1. Introduction

Let $D$ be a simply connected domain in $\overline{\mathbb{C}}$. According to the Riemann mapping theorem, there exists the conformal map $F: D \to E = \{\omega : |\omega| < 1\}$ such that $F(z) = 0$ ($z$ is fix point into $D$) whose inverse map $f: E \to D$ satisfies $f\left(\frac{\omega+\zeta}{1+\bar{\zeta}\omega}\right)\Big|_{\omega=0} = z,\ \zeta \in E$.

The quantity

$$R(D,z) = \frac{1}{|F'(z)|} = |f'(\zeta)|(1-|\zeta|^2), \zeta \in E, \tag{1.1}$$

is called the conformal radius of $D$ at the point $z = f(\zeta)$.

We study the gradient of $R$ defined as

$$\nabla R(D,z) = \frac{\partial R(D,z)}{\partial x} + i\frac{\partial R(D,z)}{\partial y} = 2R_{\bar{z}},\ \ z = x + iy \in D. \tag{1.2}$$

F.G. Avkhadiev and K.-J. Wirths [1], [2] proved that $\nabla R(D,z)$ is a diffeomorphism of $D$ onto a certain domain $G$ (the type of $G$ depends on the type of domain $D$).

In the papers [3], [4] we showed that the gradient of the conformal radius is a conformal map if and only if $D$ is the unit disk $E$.

This work adds to a recent paper [1] and extends results in [3], [4] from the point of view of quasi-conformal map.

**Definition 1.1.** The map $\nabla R(D,z)$ is called the $K(k)$-quasi-conformal map, if $\nabla R(D,z)$ satisfies Beltrami equation

$$(\nabla R)_{\bar{z}} = \mu(z,\bar{z})(\nabla R)_z, \tag{1.3}$$

for which

$$\sup|\mu(z,\bar{z})|,\ z \in D = \frac{K-1}{K+1} < 1\ (1 \leq K < \infty).$$

## 2. Auxiliary results

To prove main results, we use two special cases of the principle of the hyperbolic metrics ([5], p. 326).

**Lemma 1.** *If a function $\varphi$ is regular in the exterior $E^-$ of the unit disk $E$ and $\varphi(E^-) \subset E^-$, then the derivative of this function satisfies*

$$|\varphi'(\zeta)| \leq \frac{|\varphi|^2 - 1}{|\zeta|^2 - 1},\ \zeta \in E^-.$$

*An equality is attained for the function*

$$\varphi(\zeta) = e^{i\alpha}\frac{1+\bar{a}\zeta}{\zeta + a},\ a \in E^-,$$

*at every point $\zeta \in E^-$.*

**Lemma 2.** *If the function $\varphi$ is regular in semi-plane $P = \{\zeta : \operatorname{Re}\zeta > 0\}$ and $\varphi(P) \subset P$, then for the derivative $\varphi'$ the estimate*

$$|\varphi'(\zeta)| \leq \frac{\operatorname{Re}\varphi}{\operatorname{Re}\zeta},\ \zeta \in P,$$

*is valid. The equality is attained for the function*

$$\varphi(\zeta) = \frac{ai\zeta + b}{-c\zeta + id},\ a,b,c,d \in \mathbb{R},\ ad - bc > 0,$$

*at every point $\zeta \in P$.*

## 3. Main results

Let $D_\alpha = \{z : |\arg z| < \alpha\pi/2\}$, $\alpha \in (0,1]$, and $D_0 = \{z : |\operatorname{Im} z| < \pi/2\}$.

**Theorem 3.1.** *For any compact subset of a convex domain $D = f(E)$, except $D_\alpha$, $\alpha \in [0,1]$, the gradient (1.2) of the conformal radius (1.1) is a quasi-conformal mapping.*

*For the angular domain $D_\alpha$, $\alpha \in (0,1]$, and the strip $D_0$ the identity $\left|\frac{R_{\bar{z}\bar{z}}(D,z)}{R_{\bar{z}z}(D,z)}\right| \equiv 1$ is valid, and for any compact subset of $D_\alpha$ and $D_0$ (1.2) is a degenerate map.*

*In the class $S^0$ of the convex functions $f(r\zeta)$, $\zeta \in E$, $0 < r < 1$, there is exact estimator for the coefficient of the quasi-conformality in the form*

$$\sup_{f(\zeta)\in S^0} K(f(r\zeta)) = \frac{1+r^2}{1-r^2}.$$

*Proof.* As known ([5], p. 166), a necessary and sufficient condition of convexity of the domain $D = f(E)$ is

$$\operatorname{Re}\zeta\frac{f''(\zeta)}{f'(\zeta)} \geq -1,\ \zeta \in E. \tag{3.1}$$

If $\Phi_0(\zeta)$ is a univalent map of the disk $E$ onto the semi-plane $\{\operatorname{Re}\Phi_0 > -1\}$, then the function $\Phi_0^{-1}\left(\zeta\frac{f''(\zeta)}{f'(\zeta)}\right) = \tilde{\varphi}(\zeta)$ satisfies the conditions of the Schwarz lemma ([5], p. 319) by (3.1). It means that $|\tilde{\varphi}(\zeta)| \leq 1$ and $\tilde{\varphi}(0) = 0$. Therefore $\tilde{\varphi}(\zeta) = \zeta\varphi(\zeta)$, where $\varphi(\zeta)$ is regular function in $E$, and the following inequalities are valid

$$|\tilde{\varphi}| \leq |\zeta| \implies \frac{|\tilde{\varphi}|}{|\zeta|} \leq 1 \implies |\varphi| \leq 1.$$

Thus, the function $\varphi$ satisfies the conditions of the extended Schwarz lemma.

Let $\Phi_0(\zeta) = \frac{2\zeta}{1-\zeta}$. Then $\zeta\frac{f''(\zeta)}{f'(\zeta)} = \Phi_0(\tilde{\varphi}(\zeta)) = 2\frac{\zeta\varphi(\zeta)}{1-\zeta\varphi(\zeta)}$ and hence

$$\frac{f''(\zeta)}{f'(\zeta)} = 2\frac{\varphi(\zeta)}{1-\zeta\varphi(\zeta)}. \tag{3.2}$$

We calculate the derivatives

$$R_{\bar{z}\bar{z}} = \frac{1}{f'}\frac{1-|\zeta|^2}{2}\{\bar{f},\bar{\zeta}\},$$

$$R_{\bar{z}z} = \frac{1}{|f'|(1-|\zeta|^2)}\left(\left|\frac{f''}{f'}\frac{1-|\zeta|^2}{2} - \bar{\zeta}\right|^2 \quad 1\right).$$

Due to (3.2) we write

$$R\,|R_{\bar{z}\bar{z}}| = \frac{(1-|\zeta|^2)^2}{2}|\{f,\zeta\}| = \frac{(1-|\zeta|^2)^2}{2}\left|\left(\frac{f''}{f'}\right)' - \frac{1}{2}\left(\frac{f''}{f'}\right)^2\right| = \frac{|\varphi'|(1-|\zeta|^2)^2}{|1-\zeta\varphi|^2},$$

$$R\,|R_{\bar{z}z}| = 1 - \left|\frac{f''}{f'}\frac{1-|\zeta|^2}{2} - \bar{\zeta}\right|^2 = 1 - \left|\frac{2\varphi}{1-\zeta\varphi}\frac{1-|\zeta|^2}{2} - \bar{\zeta}\right|^2 = \frac{(1-|\zeta|^2)(1-|\varphi|^2)}{|1-\zeta\varphi|^2}.$$

Finally, we get

$$\left|\frac{R_{\bar{z}\bar{z}}}{R_{\bar{z}z}}\right| = \frac{1-|\zeta|^2}{1-|\varphi|^2}|\varphi'|.$$

By the extended Schwarz lemma, the equality in $|\varphi'| \leq \frac{1-|\varphi|^2}{1-|\zeta|^2}$ is attained for the extremal function $\varphi(\zeta) = e^{i\alpha}\frac{\zeta+a}{1+\bar{a}\zeta}$, $|a| < 1$. In our case this function satisfies the identity $|\frac{R_{\bar{z}\bar{z}}}{R_{\bar{z}z}}| \equiv 1$.

Let us find a function $f$ for which this identity is valid. We replace the extremal function $\varphi$ from the Schwarz lemma in (3.2) and decompose the right-hand side of last equality into partial fractions. We have

$$\frac{f''}{f'} = \frac{2(\zeta+a)e^{i\alpha}}{1+(\bar{a}-ae^{i\alpha})\zeta - e^{i\alpha}\zeta^2} = -e^{i\alpha/2}\frac{2t+2ae^{i\alpha/2}}{(t-t_1)(t-t_2)}$$
$$= e^{i\alpha/2}\left(\frac{A}{t-t_1}+\frac{B}{t-t_2}\right),$$

where $t = e^{i\alpha/2}\zeta$ and

$$t_1 = -i\,\mathrm{Im}\{ae^{i\alpha/2}\} + \sqrt{1-\mathrm{Im}^2\{ae^{i\alpha/2}\}} \in \partial E,$$
$$t_2 = -i\,\mathrm{Im}\{ae^{i\alpha/2}\} - \sqrt{1-\mathrm{Im}^2\{ae^{i\alpha/2}\}} \in \partial E,$$

$t_1 \neq t_2$, as $\mathrm{Im}^2\{ae^{i\alpha/2}\} \neq 1$, if $|a|<1$.

Solving the system

$$\begin{cases} A+B=-2, \\ At_2+Bt_1 = 2ae^{i\alpha/2}, \end{cases}$$

we find

$$A = \frac{2t_1+2ae^{i\alpha/2}}{t_1-t_2} = -1-\beta, \quad B=-2-A=-1+\beta,$$
$$\beta = \frac{\mathrm{Re}\{ae^{i\alpha/2}\}}{\sqrt{1-\mathrm{Im}^2\{ae^{i\alpha/2}\}}} \in \mathbb{R}.$$

Then integrating the formula

$$\frac{f''}{f'} = e^{i\alpha/2}\left(\frac{-1-\beta}{\zeta e^{i\alpha/2}-t_1}+\frac{-1+\beta}{\zeta e^{i\alpha/2}-t_2}\right),$$

we get

$$\ln f' = \ln\left(C\left(\frac{\zeta e^{i\alpha/2}-t_2}{\zeta e^{i\alpha/2}-t_1}\right)^{\beta-1}\frac{1}{(\zeta e^{i\alpha/2}-t_1)^2}\right),$$

and therefore

$$f(\zeta) = \begin{cases} C_1\left(\frac{\zeta e^{i\alpha/2}-t_2}{\zeta e^{i\alpha/2}-t_1}\right)^{\beta} + C_2, \text{ if } \beta\neq 0, \\ C_3 \ln\frac{\zeta e^{i\alpha/2}-t_2}{\zeta e^{i\alpha/2}-t_1} + C_4, \quad \text{if } \beta = 0. \end{cases} \tag{3.3}$$

The obtained function is a map of the unit disk onto the angular domain with opening $\pi\beta$ when $\beta\neq 0$, and onto a horizontal strip otherwise.

For noted functions the equal-sign is reached in the estimator for $|\varphi'(\zeta)|$, and the quasi-conformal map degenerates. Therefore we name these functions exclusive and designate a class of convex functions without functions of an exclusive kind (3.3) as $\tilde{S}^0$.

As for any function $f(\zeta)\in\tilde{S}^0$ the relator

$$\max_{|\zeta|\le r}\left[\frac{1-|\zeta|^2}{1-|\varphi(\zeta)|^2}|\varphi'(\zeta)|\right] < 1$$

is valid (where the function $\varphi$ is defined from (3.2)) we get in the domain $D_r = f(E_r)$, $E_r = \{\zeta : |\zeta| < r\}$ that

$$k_1(f(\zeta), E_r) = \sup_{z \in D_r} \left| \frac{R_{\bar{z}\bar{z}}}{R_{\bar{z}z}} \right| < 1.$$

It means the gradient (1.2) of the conformal radius (1.1) is a quasi-conformal map for any compact subset of the domain $D_r$. For the class $\tilde{S}^0$ we have

$$\sup_{f \in \tilde{S}^0} k_1(f(\zeta), E_r) = 1,$$

where $k_1(f(\zeta), E_r) = 1$ for exclusive functions which have not entered in $\tilde{S}^0$.

The first part of the theorem is proved.

To proof the second part we move on to $r$-equipotential line and rewrite (3.2) as

$$r \frac{f''(r\zeta)}{f'(r\zeta)} = 2r \frac{\varphi(r\zeta)}{1 - r\zeta\varphi(r\zeta)}.$$

Since

$$R|R_{\bar{z}\bar{z}}| = \frac{r^2|\varphi'(r\zeta)|(1 - |\zeta|^2)^2}{|1 - r\zeta\varphi(r\zeta)|^2}, \qquad R|R_{\bar{z}z}| = \frac{(1 - |\zeta|^2)(1 - r^2|\varphi(r\zeta)|^2)}{|1 - r\zeta\varphi(r\zeta)|^2},$$

we have

$$\left| \frac{R_{\bar{z}\bar{z}}}{R_{\bar{z}z}} \right| = \frac{1 - |\zeta|^2}{1 - r^2|\varphi(r\zeta)|^2} r^2 |\varphi'(r\zeta)| \le r^2 \frac{1 - |\zeta|^2}{1 - r^2|\varphi(r\zeta)|^2} \frac{1 - |\varphi(r\zeta)|^2}{1 - r^2|\zeta|^2} \le r^2.$$

It is clear we get the sign-equal only if $\varphi(\zeta) = e^{i\alpha}\zeta$ for $\zeta = 0$. Therefore we input the coefficient

$$k_2(f(r\zeta), E) = r^2 \max_{|\zeta| \le 1} \left[ \frac{1 - |\zeta|^2}{1 - r^2|\varphi(r\zeta)|^2} |\varphi'(r\zeta)| \right] \le r^2,$$

hence the gradient $\nabla R(D_r, z)$ is a quasi conformal map. Note that

$$\max_{f \in S^0} k_2(f(r\zeta), E) = r^2 < 1,$$

as for $z = \ln \frac{1+r\zeta}{1-r\zeta}$ we have

$$R(D_r, z) = 2r \frac{1 - |\zeta|^2}{|1 - r^2\zeta^2|}$$

$$|\mu(\zeta, \bar{\zeta})| = \frac{r^2(1 - |\zeta|^2)}{1 - r^4|\zeta|^2} \le |\mu(0, 0)| = r^2 = k_2 \left( \ln \frac{1 + r\zeta}{1 - r\zeta}, E \right). \qquad \square$$

We consider the following example in which the coefficient of quasi-conformality $\frac{1+k_2(f(r\zeta),E)}{1-k_2(f(r\zeta),E)}$ is less than $\frac{1+r^2}{1-r^2}$.

*Example.* Let domain $D$ be the square with the mapping function ([5], p. 77)

$$z = f(\zeta) = \int_0^\zeta \frac{d\zeta}{(1-\zeta^4)^{1/2}}.$$

The conformal radius for the domain

$$z = f(r\zeta) = \int_0^{r\zeta} \frac{d\zeta}{(1-r^4\zeta^4)^{1/2}}$$

has the view

$$R(\zeta) = \frac{1-|\zeta|^2}{|1-r^4\zeta^4|^{1/2}},$$

therefore

$$R_{\bar z} = \frac{r^4\bar\zeta^4 - \zeta}{\sqrt[4]{1-r^4\zeta^4}\sqrt[4]{(1-r^4\bar\zeta^4)^3}}, \qquad R_{\bar z\bar z} = \frac{3}{\sqrt[4]{1-r^4\zeta^4}} \frac{r^4\bar\zeta^2(1-|\zeta|^2)}{\sqrt[4]{(1-r^4\bar\zeta^4)^5}},$$

$$R_{\bar z z} = -\frac{1-r^8|\zeta|^6}{\sqrt[4]{(1-r^4\bar\zeta^4)^3}\sqrt[4]{(1-r^4\zeta^4)^3}}.$$

Finally

$$\left|\frac{(R_{\bar z})_{\bar z}}{(R_{\bar z})_z}\right| = \frac{3r^4|\zeta|^2(1-|\zeta|^2)}{1-r^8|\zeta|^6}.$$

Then

$$\tilde k_2(r) = \sup_{0\le|\zeta|\le 1}\left|\frac{R_{\bar z\bar z}}{R_{\bar z z}}\right| = r^4 \max_{0\le u\le 1}\frac{3u(1-u)}{1-r^8u^3} = r^4\frac{3u_0(1-u_0)}{1-r^8u_0^3} < r^4,$$

$$0<r<1,\ 0<u_0<1.$$

Note that

$$\sup_{0\le|\zeta|\le 1}\left|\frac{R_{\bar z\bar z}}{R_{\bar z z}}\right|\Bigg|_{r=1} = \max_{0\le|\zeta|\le 1}\frac{3|\zeta|^2(1-|\zeta|^2)}{1-|\zeta|^6} = \frac{3}{\frac{1}{|\zeta|^2}+1+|\zeta|^2}\Bigg|_{|\zeta|=1} = 1.$$

On the other hand,

$$\tilde{k_1}(\rho) = \left|\frac{R_{\bar z\bar z}}{R_{\bar z z}}\right|\Bigg|_{r=1,|\zeta|\le\rho} = \frac{3|\zeta|^2(1-|\zeta|^2)}{1-|\zeta|^6}\Bigg|_{|\zeta|\le\rho} \le \frac{3\rho^2}{\rho^4+\rho^2+1},$$

and $\tilde k_2(r) < \tilde{k_1}(r)$, as

$$r^4 < \frac{3r^2}{r^4+r^2+1}.$$

Apparently, we will have $k_2(f(r\zeta),E) < k_1(f(\zeta),E_r)$, $0<r<1$, for any convex domain $D = f(E)$.

Let us turn now to a class $\Sigma^0$ of functions mapping $E^- = \{|\zeta|>1\}$ onto domains $D^-$, $\infty\in D^-$, which represent an exterior of some closed convex curve and for which the surface of the conformal radius is convex downwards. Then it holds that

**Theorem 3.2.** *For any compact subset of the domain $D^- = f(E^-)$ with finite convex complement the gradient* (1.2) *of the conformal radius $R(f(E^-), f(\zeta))$ is a quasi-conformal map.*

*There are exact estimators of analogs of quantities for two kinds of quasi-conformal maps.*

*Proof.* The necessary and sufficient condition of the convexity of the boundary of the domain $D^- = f(E^-)$ is (3.1). Arguing as earlier, we get $\zeta\frac{f''(\zeta)}{f'(\zeta)} \prec \Phi_0 = \frac{2}{\zeta-1}$. Then $\Phi_0^{-1}\left(\zeta\frac{f''(\zeta)}{f'(\zeta)}\right) = \varphi(\zeta) \ : \infty \mapsto \infty$. Repeating the calculations from the proof of the previous theorem, we have

$$\zeta\frac{f''(\zeta)}{f'(\zeta)} = \Phi_0(\varphi(\zeta)) = \frac{2}{\varphi(\zeta)-1}, \tag{3.4}$$

and hence

$$\frac{f''(\zeta)}{f'(\zeta)} = \frac{2}{\zeta(\varphi-1)}.$$

Since the expression for the conformal radius is

$$R(D^-, z) = |f'(\zeta)|(|\zeta|^2-1), \ |\zeta|>1, \ z = f(\zeta), \tag{3.5}$$

we estimate second derivatives. Since

$$\frac{1}{2}\frac{f''}{f'}(|\zeta|^2-1)+\bar{\zeta} = \frac{|\zeta|^2-1}{\zeta(\varphi-1)}+\bar{\zeta} = \frac{|\zeta|^2\varphi-1}{\zeta(\varphi-1)},$$

we use

$$R\,|R_{\bar{z}z}| = 1-\left|\frac{f''}{f'}\frac{|\zeta|^2-1}{2}+\bar{\zeta}\right|^2 = 1-\frac{|\varphi|\zeta|^2-1|^2}{|\zeta|^2|\varphi-1|^2}$$
$$= \frac{(|\zeta|^2-1)(|\zeta\varphi|^2-1)}{|\zeta|^2|\varphi-1|^2}.$$

Further

$$\{f,\zeta\} = -\frac{2}{\zeta^2(\varphi-1)} - 2\frac{\varphi'}{\zeta(\varphi-1)^2} - 2\frac{1}{\zeta^2(\varphi-1)^2}$$
$$= -2\frac{\varphi+\zeta\varphi'}{\zeta^2(\varphi-1)^2} = -2\frac{(\varphi\zeta)'}{\zeta^2(\varphi-1)^2},$$

therefore

$$\left|\frac{(R_{\bar{z}})_{\bar{z}}}{(R_{\bar{z}})_z}\right| = \frac{|(\varphi\zeta)'|(|\zeta|^2-1)}{|\zeta\varphi|^2-1}.$$

As $|\varphi\zeta| \geq |\varphi| > 1$, we use Lemma 1 for the function $\varphi\zeta$, to deduce the relation

$$|(R_{\bar{z}})_{\bar{z}}/(R_{\bar{z}})_z| \leq 1.$$

We know (Lemma 1) that the equality is attained for the extremal function $\zeta\varphi(\zeta) = e^{i\alpha}\frac{1+\bar{a}\zeta}{\zeta+a}$, where $\varphi(\zeta)$ is a function with a simple zero at $\infty$. But in expression (3.4) function $1/\varphi(\zeta)$ has to have a zero of the second order. Therefore for the class $\Sigma^0$, it holds that $|(R_{\bar z})_{\bar z}/(R_{\bar z})_z| \not\equiv 1$ and

$$k_1(f(\zeta), E_r^-) = \max_{|\zeta|\geq r} \frac{|(\varphi\zeta)'|(|\zeta|^2-1)}{|\varphi\zeta|^2-1} < 1.$$

For any compact subset of a domain $D_r^- = f(E_r^-)$, $E_r^- = \{\zeta : |\zeta| > r > 1\}$, the gradient of the conformal radius is a quasi-conformal map, and

$$\sup_{f\in\Sigma^0} k_1(f(\zeta), E_r^-) < 1,\ 1 < r < \infty, \quad \sup_{f\in\Sigma^0} k_1(f(\zeta), E^-) = 1.$$

For $r$-equipotential line we have

$$r\frac{f''(r\zeta)}{f'(r\zeta)} = \frac{2}{\zeta(\varphi(r\zeta)-1)},$$

which yields

$$1 - \left|\frac{f''}{f'}\frac{|\zeta|^2-1}{2} + \bar\zeta\right|^2 = \frac{(|\zeta|^2-1)(|\zeta\varphi(r\zeta)|^2-1)}{|\zeta|^2|\varphi(r\zeta)-1|^2},$$

$$\{f(r\zeta),\zeta\} = -2\frac{(\varphi(r\zeta)r\zeta)'_{r\zeta}}{\zeta^2(\varphi(r\zeta)-1)^2}.$$

From Lemma 1 we get

$$\left|\frac{(R_{\bar z})_{\bar z}}{(R_{\bar z})_z}\right| = \frac{|(\varphi(r\zeta)r\zeta)'_{r\zeta}|(|\zeta|^2-1)}{|\zeta|^2|\varphi(r\zeta)|^2-1} < 1.$$

Equal-sign achievement in last inequality is impossible and therefore

$$k_2(f(r\zeta), E^-) = \max_{|\zeta|\geq 1} \frac{|(\varphi(r\zeta)r\zeta)'_{r\zeta}|(|\zeta|^2-1)}{|\zeta|^2|\varphi(r\zeta)|^2-1} < 1,$$

where $\sup_{f\in\Sigma^0} k_2(f(r\zeta), E^-) = 1$ for $r \geq 1$. □

Now we consider domains $D$, $\infty \in \partial D$. Such domains require additional analysis. To this end, we further split them into two separate classes. The first class is characterized by the convexity of the equipotential lines similar to the boundary curve, and the other by non-convexity of such lines. The first case is, in fact, shown in Theorem 2.1. In the second case we replace the equipotential lines defined by the circles $|\zeta| = r$, by the equipotential lines defined by the circles $|\zeta + r| = 1 - r$. The following result reflects the effect of such a change.

**Theorem 3.3.** *For any closed finite domain from $D = \tilde f(E)$, $\tilde f(1) = \infty$ ($\mathbb{C}\backslash D$ – a convex domain) the gradient* (1.2) *of the conformal radius is a quasi-conformal mapping.*

*In particular, this statement is valid for the conformal radius $R(\tilde f(\tilde E_r), \tilde f(\omega))$ for a domain $\tilde f(\tilde E_r)$, $\tilde E_r = \{\omega : |\omega + r| \leq 1 - r\}$, $0 < r < 1$.*

As $\{\omega : |\omega + r| \leq 1 - r\} \Leftrightarrow \{\zeta : \operatorname{Re}\zeta \geq \rho\}$, $\rho = \frac{1-r}{r}$, $\zeta = \frac{1-\omega}{1+\omega}$, it is more convenient to speak about a surface of conformal radius

$$R(f(P), f(\zeta)) = 2\operatorname{Re}\zeta|f'(\zeta)|,\ f(\zeta) = \tilde f\left(\frac{1-\zeta}{1+\zeta}\right), \tag{3.6}$$

which is constructed over the semi-plane $P = \{\zeta : \operatorname{Re}\zeta > 0\}$ (corresponding mapping functions $f(\zeta)$ form a class $\Sigma^0(P)$).

*Proof.* The necessary and sufficient condition of the convexity of a domain in this case has the form

$$\operatorname{Re}\left(\frac{f''(i\eta)}{f'(i\eta)}\right) \geq 0. \tag{3.7}$$

Therefore, it clearly holds

$$\frac{f''}{f'} = \varphi, \tag{3.8}$$

where $\varphi(\zeta)$ is a function satisfying the conditions of Lemma 2.

Next, we compute the derivatives of the conformal radius (3.6). We have

$$R_{\bar{z}} = \frac{\sqrt{f'}}{f'}\left(\sqrt{\bar{f}'}(\zeta+\bar{\zeta})\right)'_{\bar{\zeta}} = \frac{\sqrt{f'}}{\sqrt{\bar{f}'}}\left(\frac{\bar{f}''}{\bar{f}'}\operatorname{Re}\zeta + 1\right),$$

$$\begin{aligned} R_{\bar{z}\bar{z}} &= \frac{\sqrt{f'}}{\bar{f}'}\left(\frac{1}{\sqrt{\bar{f}'}}\left(\frac{\bar{f}''}{\bar{f}'}\operatorname{Re}\zeta + 1\right)\right)'_{\bar{\zeta}} \\ &= \frac{\sqrt{f'}}{\sqrt{\bar{f}'^3}}\left(\left(\frac{\bar{f}''}{\bar{f}'}\right)' - \frac{1}{2}\left(\frac{\bar{f}''}{\bar{f}'}\right)^2\right)\operatorname{Re}\zeta = \frac{\sqrt{f'}}{\sqrt{\bar{f}'^3}}\operatorname{Re}\zeta\{\bar{f},\bar{\zeta}\} \end{aligned}$$

and consequently

$$R|R_{\bar{z}\bar{z}}| = 2\operatorname{Re}^2\zeta|\{f,\zeta\}|.$$

Due to (3.8) we rewrite the last expression in the form

$$R|R_{\bar{z}\bar{z}}| = 2\operatorname{Re}^2\zeta|\varphi' - \frac{\varphi^2}{2}|.$$

Since

$$R_{\bar{z}z} = \frac{1}{f'\sqrt{\bar{f}'}}\left(\sqrt{f'}\left(\frac{\bar{f}''}{\bar{f}'}\operatorname{Re}\zeta + 1\right)\right)'_{\zeta} = \frac{1}{2|f'|}\left(\left|\frac{f''}{f'}\right|^2\operatorname{Re}\zeta + \frac{f''}{f'} + \frac{\bar{f}''}{\bar{f}'}\right),$$

it holds

$$R\,R_{\bar{z}z} = \left|\frac{f''}{f'}\operatorname{Re}\zeta + 1\right|^2 - 1.$$

By (3.8) we have

$$R\,R_{\bar{z}z} = |\varphi\operatorname{Re}\zeta + 1|^2 - 1 = |\varphi|^2\operatorname{Re}^2\zeta + 2\operatorname{Re}\zeta\operatorname{Re}\varphi.$$

Finally we get

$$\left|\frac{(R_{\bar{z}})_{\bar{z}}}{(R_{\bar{z}})_z}\right| = \frac{2\operatorname{Re}\zeta|\varphi' - \frac{\varphi^2}{2}|}{|\varphi|^2\operatorname{Re}\zeta + 2\operatorname{Re}\varphi} \leq \frac{\operatorname{Re}\zeta(2|\varphi'| + |\varphi|^2)}{|\varphi|^2\operatorname{Re}\zeta + 2\operatorname{Re}\varphi} \leq \frac{\operatorname{Re}\zeta\left(2\frac{\operatorname{Re}\varphi}{\operatorname{Re}\zeta} + |\varphi|^2\right)}{|\varphi|^2\operatorname{Re}\zeta + 2\operatorname{Re}\varphi} = 1.$$

Let us find the function $f$ for which the equality $|(R_{\bar{z}})_{\bar{z}}/(R_{\bar{z}})_z| \equiv 1$ is valid. Note that the equality in the inequality

$$|\varphi' - \varphi^2/2| \leq |\varphi'| + |\varphi|^2/2,$$

is attained, when functions $\varphi'$ and $-\varphi^2/2$ have an identical arguments. Let

$$\varphi' = R_1 e^{i\theta} \quad \text{and} \quad -\varphi^2/2 = R_2 e^{i\theta}.$$

Then the function $\frac{\varphi'}{-\varphi^2/2} = \frac{R_1}{R_2}$ has to be a real-analytic function positive in the half-plane. This is possible if and only if $\frac{R_1}{R_2} \equiv 2\alpha > 0$. The solution of the differential equation $\frac{\varphi'}{-\varphi^2/2} = 2\alpha$ has the form $\frac{1}{\varphi} = \alpha\zeta + b$. Among all such functions we choose the one that maps the right half-plane on itself, namely $\varphi = \frac{1}{\alpha\zeta + i\beta}$, $\alpha, \beta \in \mathbb{R}$, $\alpha > 0$. Due to (3.8) we get $f = C_1(\alpha\zeta + i\beta)^{\frac{1+\alpha}{\alpha}} + C_2$.

For any closed domain $\bar{G}_\rho \subset \bar{D}$, which is an image of the strip

$$\{\zeta : \ 0 \le \operatorname{Re}\zeta \le \rho\} = f^{-1}(\bar{G}_\rho),$$

we will have

$$k_1(f(\zeta), f^{-1}(G)) = \max_{\zeta \in f^{-1}(G)} \frac{2\operatorname{Re}\zeta |\varphi' - \frac{\varphi^2}{2}|}{|\varphi|^2 \operatorname{Re}\zeta + 2\operatorname{Re}\varphi} \equiv \psi(\rho, f) \equiv \psi(\rho, f) < 1, \quad (3.9)$$

but

$$\sup_{f \in \Sigma^0(P) \setminus \{f(\alpha,\zeta)\}} \psi(\rho, f) = 1.$$

We can concretize an estimation (3.9), having included $G$ in $\tilde{f}(\tilde{r}E)$ for some $\tilde{r} \in (0,1)$.

There is not a finite exhaustion for domain $f(P)$, $f(\zeta) \in \Sigma^0(P)$, by finite domains with convex boundaries. Therefore there is not the analog of the coefficient $k_2$ from proofs of previous theorems. □

## References


[1] F.G. Avkhadiev, K.-J. Wirths, *The conformal radius as a function and its gradient image.* Israel J. of Mathematics **145** (2005), 349–374.

[2] F.G. Avkhadiev, K.-J. Wirths, *Schwarz-Pick type inequalities.* Birkhäuser Verlag, 2009.

[3] L.A. Aksentyev, A.N. Akhmetova, *On mappings related to the gradient of the conformal radius.* Izv. VUZov. Mathematics **6** (2009), 60–64 (the short message).

[4] L.A. Aksentyev, A.N. Akhmetova, *On mappings related to the gradient of the conformal radius.* Mathematical Notes **1** (2010), 3–12.

[5] G.M. Goluzin, *Geometric theory of the function of complex variables.* Nauka, 1966.



A.N. Akhmetova
Kazan State Technical University, Kazan 420111, Russia
e-mail: achmetowa@inbox.ru

L.A. Aksentyev
Kazan (Volga region), Federal University, Kazan 420008, Russia
e-mail: Leonid.Aksentev@ksu.ru

Operator Theory:
Advances and Applications, Vol. 221, 11-24


# The Influence of the Tunnel Effect on $L^\infty$-time Decay

F. Ali Mehmeti, R. Haller-Dintelmann and V. Régnier

**Abstract.** We consider the Klein-Gordon equation on a star-shaped network composed of $n$ half-axes connected at their origins. We add a potential that is constant but different on each branch. Exploiting a spectral theoretic solution formula from a previous paper, we study the $L^\infty$-time decay via Hörmander's version of the stationary phase method. We analyze the coefficient $c$ of the leading term $c \cdot t^{-1/2}$ of the asymptotic expansion of the solution with respect to time. For two branches we prove that for an initial condition in an energy band above the threshold of tunnel effect, this coefficient tends to zero on the branch with the higher potential, as the potential difference tends to infinity. At the same time the incline to the $t$-axis and the aperture of the cone of $t^{-1/2}$-decay in the $(t, x)$-plane tend to zero.

**Mathematics Subject Classification (2000).** Primary 34B45; Secondary 47A70, 35B40.

**Keywords.** Networks, Klein-Gordon equation, stationary phase method, $L^\infty$-time decay.

## 1. Introduction

In this paper we study the $L^\infty$-time decay of waves in a star shaped network of one-dimensional semi-infinite media having different dispersion properties. Results in experimental physics [10, 11], theoretical physics [9] and functional analysis [5, 8] describe phenomena created in this situation by the dynamics of the tunnel effect: the delayed reflection and advanced transmission near nodes issuing two branches. Our purpose is to describe the influence of the height of a potential step on the $L^\infty$-time decay of wave packets above the threshold of tunnel effect, which sheds a new light on its dynamics.

---

Parts of this work were done, while the second author visited the University of Valenciennes. He wishes to express his gratitude to F. Ali Mehmeti and the LAMAV for their hospitality.

In this proceedings contribution we state results for a special choice of initial conditions. The proofs in a more general context will be the core of another paper.

The dynamical problem can be described as follows:
Let $N_1, \ldots, N_n$ be $n$ disjoint copies of $(0, +\infty)$ with $n \geq 2$. Consider numbers $a_k, c_k$ satisfying $0 < c_k$, for $k = 1, \ldots, n$ and $0 \leq a_1 \leq a_2 \leq \cdots \leq a_n < +\infty$. Find a vector $(u_1, \ldots, u_n)$ of functions $u_k : [0, +\infty) \times \overline{N_k} \to \mathbb{C}$ satisfying the Klein-Gordon equations

$$[\partial_t^2 - c_k \partial_x^2 + a_k] u_k(t, x) = 0, \ k = 1, \ldots, n,$$

on $N_1, \ldots, N_n$ coupled at zero by usual Kirchhoff conditions and complemented with initial conditions for the functions $u_k$ and their derivatives.

Reformulating this as an abstract Cauchy problem, one is confronted with the self-adjoint operator $A = (-c_k \cdot \partial_x^2 + a_k)_{k=1,\ldots,n}$ in $\prod_{k=1}^n L^2(N_k)$, with a domain that incorporates the Kirchhoff transmission conditions at zero. For an exact definition of $A$, we refer to Section 2.

Invoking functional calculus for this operator, the solution can be given in terms of

$$e^{\pm i\sqrt{A}t} u_0 \text{ and } e^{\pm i\sqrt{A}t} v_0.$$

In a previous paper ([4], see also [3]) we construct explicitly a spectral representation of $\prod_{k=1}^n L^2(N_k)$ with respect to $A$ involving $n$ families of generalized eigenfunctions. The $k$th family is defined on $[a_k, \infty)$ which reflects that $\sigma(A) = [a_1, \infty)$ and that the multiplicity of the spectrum is $j$ in $[a_j, a_{j+1}), j = 1, \ldots, n$, where $a_{n+1} = +\infty$. In this band $(a_j, a_{j+1})$ the generalized eigenfunctions exhibit exponential decay on the branches $N_{j+1}, \ldots, N_n$, a fact called "multiple tunnel effect" in [4].

In Section 2 we recall the solution formula proved in [4]. In Section 3 we use Hörmander's version of the stationary phase method to derive the leading term of the asymptotic expansion of the solution on certain branches and for initial conditions in a compact energy band included in $(a_j, a_{j+1})$. We obtain $c \cdot t^{-1/2}$ in cones in the $(t, x)$-space delimited by the group velocities of the limit energies and the dependence of $c$ on the coefficients of the operator is indicated. One can prove that outside these cones the $L^\infty$-norm decays at least as $t^{-1}$. The complete analysis will be carried out in a more detailed paper.

For the case of two branches and wave packets having a compact energy band included in $(a_2, \infty)$, we show in Section 4 that $c$ tends to zero on the side of the higher potential, if $a_1$ stays fixed and $a_2$ tends to infinity. We observe further that the exact $t^{-1/2}$-decay takes place in a cone in the $(t, x)$-plane whose aperture and incline to the $t$-axis tend to zero as $a_2$ tends to infinity. Physically the model corresponds to a relativistic particle, more precisely a pion, in a one-dimensional world with a potential step of amount $a_2 - a_1$ in $x = 0$. Our result represents thus a dynamical feature for phenomena close to tunnel effect, which might be confirmed by physical experiments.

Our results are designed to serve as tools in some pertinent applications as the study of more general networks of wave guides (for example microwave networks [17]) and the treatment of coupled transmission conditions [7].

For the Klein-Gordon equation in $\mathbb{R}^n$ with constant coefficients the $L^\infty$-time decay $c{\cdot}t^{-1/2}$ has been proved in [15]. Adapting their method to a spectral theoretic solution formula for two branches, it has been shown in [1, 2] that the $L^\infty$-norm decays at least as $c \cdot t^{-1/4}$.

In [14] and several related articles, the author studies the $L^\infty$-time decay for crystal optics using similar methods.

In [13], the authors consider general networks with semi-infinite ends. They give a construction to compute some generalized eigenfunctions but no attempt is made to construct explicit inversion formulas. In [6] the relation of the eigenvalues of the Laplacian in an $L^\infty$-setting on infinite, locally finite networks to the adjacency operator of the network is studied.

## 2. A solution formula

The aim of this section is to recall the tools we used in [4] as well as the solution formula of the same paper for a special initial condition and to adapt this formula for the use of the stationary phase method in the next section.

**Definition 2.1 (Functional analytic framework).**

i) Let $n \geq 2$ and $N_1, \dots, N_n$ be $n$ disjoint sets identified with $(0, +\infty)$. Put $N := \bigcup_{k=1}^{n} \overline{N_k}$, identifying the endpoints 0.
For the notation of functions two viewpoints are useful:
  - functions $f$ on the object $N$ and $f_k$ is the restriction of $f$ to $N_k$.
  - $n$-tuples of functions on the branches $N_k$; then sometimes we write $f = (f_1, \dots, f_n)$.

ii) Two transmission conditions are introduced:

$$(T_0)\colon\ (u_k)_{k=1,\dots,n} \in \prod_{k=1}^{n} C(\overline{N_k}) \text{ satisfies } u_i(0) = u_k(0),\ i,k \in \{1, \dots, n\}.$$

This condition in particular implies that $(u_k)_{k=1,\dots,n}$ may be viewed as a well-defined function on $N$.

$$(T_1)\colon\ (u_k)_{k=1,\dots,n} \in \prod_{k=1}^{n} C^1(\overline{N_k}) \text{ satisfies } \sum_{k=1}^{n} c_k \cdot \partial_x u_k(0^+) = 0.$$

iii) Define the Hilbert space $H = \prod_{k=1}^{n} L^2(N_k)$ with scalar product

$$(u, v)_H = \sum_{k=1}^{n} (u_k, v_k)_{L^2(N_k)}$$

and the operator $A : D(A) \longrightarrow H$ by

$$D(A) = \Big\{(u_k)_{k=1,\dots,n} \in \prod_{k=1}^{n} H^2(N_k) : (u_k)_{k=1,\dots,n} \text{ satisfies } (T_0), (T_1)\Big\},$$
$$A((u_k)_{k=1,\dots,n}) = (A_k u_k)_{k=1,\dots,n} = (-c_k \cdot \partial_x^2 u_k + a_k u_k)_{k=1,\dots,n}.$$

Note that, if $c_k = 1$ and $a_k = 0$ for every $k \in \{1, \dots, n\}$, $A$ is the Laplacian in the sense of the existing literature, cf. [6, 13].

**Definition 2.2 (Fourier-type transform $V$).**

i) For $k \in \{1, \dots, n\}$ and $\lambda \in \mathbb{C}$ let

$$\xi_k(\lambda) := \sqrt{\frac{\lambda - a_k}{c_k}} \quad \text{and} \quad s_k := -\frac{\sum_{l \neq k} c_l \xi_l(\lambda)}{c_k \xi_k(\lambda)}.$$

Here, and in all what follows, the complex square root is chosen in such a way that $\sqrt{r \cdot e^{i\phi}} = \sqrt{r} e^{i\phi/2}$ with $r > 0$ and $\phi \in [-\pi, \pi)$.

ii) For $\lambda \in \mathbb{C}$ and $j, k \in \{1, \dots, n\}$, we define generalized eigenfunctions $F_\lambda^{\pm,j} : N \to \mathbb{C}$ of $A$ by $F_\lambda^{\pm,j}(x) := F_{\lambda,k}^{\pm,j}(x)$ with

$$\begin{cases} F_{\lambda,k}^{\pm,j}(x) = \cos(\xi_j(\lambda)x) \pm i s_j(\lambda) \sin(\xi_j(\lambda)x), & \text{for } k = j, \\ F_{\lambda,k}^{\pm,j}(x) = \exp(\pm i \xi_k(\lambda)x), & \text{for } k \neq j. \end{cases}$$

for $x \in \overline{N_k}$.

iii) For $l = 1, \dots, n$ let

$$q_l(\lambda) := \begin{cases} 0, & \text{if } \lambda < a_l, \\ \frac{c_l \xi_l(\lambda)}{|\sum_{j=1}^{n} c_j \xi_j(\lambda)|^2}, & \text{if } a_l < \lambda. \end{cases}$$

iv) Considering for every $k = 1, \dots, n$ the weighted space $L^2((a_k, +\infty), q_k)$, we set $L_q^2 := \prod_{k=1}^{n} L^2((a_k, +\infty), q_k)$. The corresponding scalar product is

$$(F, G)_q := \sum_{k=1}^{n} \int_{(a_k, +\infty)} q_k(\lambda) F_k(\lambda) \overline{G_k(\lambda)} \, d\lambda$$

and its associated norm $|F|_q := (F, F)_q^{1/2}$.

v) For all $f \in L^1(N, \mathbb{C})$ we define $Vf : \prod_{k=1}^{n} [a_k, +\infty) \to \mathbb{C}$ by

$$(Vf)_k(\lambda) := \int_N f(x) \overline{(F_\lambda^{-,k})}(x) \, dx, \ k = 1, \dots, n.$$

In [4], we show that $V$ diagonalizes $A$ and we determine a metric setting in which it is an isometry. Let us recall these useful properties of $V$ as well as the fact that the property $u \in D(A^j)$ can be characterized in terms of the decay rate of the components of $Vu$.

**Theorem 2.3.** *Endow* $\prod_{k=1}^n C_c^\infty(N_k)$ *with the norm of* $H = \prod_{k=1}^n L^2(N_k)$. *Then*

i) $V : \prod_{k=1}^n C_c^\infty(N_k) \to L_q^2$ *is isometric and can be extended to an isometry* $\tilde{V} : H \to L_q^2$, *which we shall again denote by* $V$ *in the following.*
ii) $V : H \to L_q^2$ *is a spectral representation of* $H$ *with respect to* $A$. *In particular,* $V$ *is surjective.*
iii) *The spectrum of the operator* $A$ *is* $\sigma(A) = [a_1, +\infty)$.
iv) *For* $l \in \mathbb{N}$ *the following statements are equivalent:*
   (a) $u \in D(A^l)$,
   (b) $\lambda \mapsto \lambda^l (Vu)(\lambda) \in L_q^2$,
   (c) $\lambda \mapsto \lambda^l (Vu)_k(\lambda) \in L^2((a_k, +\infty), q_k),\ k = 1, \dots, n.$

Denoting $F_\lambda(x) := (F_\lambda^{-,1}(x), \dots, F_\lambda^{-,n}(x))^T$ and $P_j = \left(\begin{array}{c|c} I_j & 0 \\ \hline 0 & 0 \end{array}\right)$, where $I_j$ is the $j \times j$ identity matrix, for $\lambda \in (a_j, a_{j+1})$ it holds: $F_\lambda^T q(\lambda) F_\lambda = (P_j F_\lambda)^T q(\lambda)(P_j F_\lambda)$ and

$$P_j F_\lambda = \begin{pmatrix} \left(+, *, *, \dots, *, e^{-|\xi_{j+1}|x}, \dots, e^{-|\xi_n|x}\right) \\ \left(*, +, *, \dots, *, e^{-|\xi_{j+1}|x}, \dots, e^{-|\xi_n|x}\right) \\ \left(*, *, +, \dots, *, e^{-|\xi_{j+1}|x}, \dots, e^{-|\xi_n|x}\right) \\ \vdots \\ \left(*, *, \dots, *, +, e^{-|\xi_{j+1}|x}, \dots, e^{-|\xi_n|x}\right) \\ 0 \\ \vdots \\ 0 \end{pmatrix}. \tag{1}$$

Here $*$ means $e^{-i\xi_k(\lambda)}$ and $+$ means $\cos(\xi_k(\lambda)x) - i s_k(\lambda)\sin(\xi_k(\lambda)x)$ in the $k$th column for $k = 1, \dots, j$. This can be interpreted as a multiple tunnel effect (tunnel effect in the last $(n-j)$ branches with different exponential decay rates). For $\lambda$ near $a_{j+1}$, the exponential decay of the function $x \mapsto e^{-|\xi_{j+1}|x}$ is slow. The tunnel effect is weaker on the other branches since the exponential decay is quicker.

We are now interested in the Abstract Cauchy Problem

$$(\text{ACP}) : u_{tt}(t) + Au(t) = 0,\ t > 0,\ \text{with } u(0) = u_0,\ u_t(0) = 0.$$

Here, the zero initial condition for the velocity is just chosen for simplicity, as we will not deal with the general case in this contribution.

By the surjectivity of $V$ (cf. Theorem 2.3 (ii)) for every $j, k \in \{1, \dots, n\}$ with $k \leq j$ there exists an initial condition $u_0 \in H$ satisfying

**Condition** $(A_{j,k})$: $(Vu_0)_l \equiv 0,\ l \neq k$, and $(Vu_0)_k \in C_c^2((a_j, a_{j+1}))$.

*Remark* 2.4.

i) We use the convention $a_{n+1} = +\infty$.
ii) For $u_0$ satisfying $(A_{j,k})$ there exist $a_j < \lambda_{\min} < \lambda_{\max} < a_{j+1}$ such that

$$\text{supp}(Vu_0)_k \subset [\lambda_{\min}, \lambda_{\max}].$$

iii) If $u_0 \in H$ satisfies $(A_{j,k})$, then $u_0 \in D(A^\infty) = \bigcap_{l\geq 0} D(A^l)$, due to Theorem 2.3 (iv), since $\lambda \mapsto \lambda^l (Vu)_m(\lambda) \in L^2((a_m,+\infty),q_m)$, $m = 1,\ldots,n$ for all $l \in \mathbb{N}$ by the compactness of $\operatorname{supp}(Vu_0)_m$.

**Theorem 2.5 (Solution formula of (ACP) in a special case).** *Fix $j,k \in \{1,\ldots,n\}$ with $k \leq j$. Suppose that $u_0$ satisfies Condition $(A_{j,k})$. Then there exists a unique solution $u$ of (ACP) with $u \in C^l([0,+\infty), D(A^{m/2}))$ for all $l,m \in \mathbb{N}$. For $x \in N_r$ with $r \leq j$ such that $r \neq k$ and $t \geq 0$, we have the representation*

$$u(t,x) = \frac{1}{2}\big(u_+(t,x) + u_-(t,x)\big)$$

*with*

$$u_\pm(t,x) := \int_{\lambda_{\min}}^{\lambda_{\max}} e^{\pm i\sqrt{\lambda}t} q_k(\lambda) e^{-i\xi_r(\lambda)x} (Vu_0)_k(\lambda)\, d\lambda. \tag{2}$$

*Proof.* Since $v_0 = u_t(0) = 0$, we have for the solution of (ACP) the representation

$$u(t) = V^{-1}\cos(\sqrt{\lambda}t)Vu_0$$

(cf. for example [1, Theorem 5.1]). The expression for $V^{-1}$ given in [4] yields the formula for $u_\pm$. □

*Remark* 2.6. Expression (2) comes from a term of the type $*$ in $F_\lambda$ (see (1)) via the representation of $V^{-1}$. A solution formula for arbitrary initial conditions which is valid on all branches is available in [4]. This general expression is not needed in the following.

## 3. $L^\infty$-time decay

The time asymptotics of the $L^\infty$-norm of the solution of hyperbolic problems is an important qualitative feature, for example in view of the study of nonlinear perturbations.

In [16] the author derives the spectral theory for the 3D-wave equation with different propagation speeds in two adjacent wedges. Further he attempts to give the $L^\infty$-time decay which he reduces to a 1D-Klein-Gordon problem with potential step (with a frequency parameter). He uses interesting tools, but his argument is technically incomplete: the backsubstitution (see the proof of Theorem 3.2 below) has not been carried out, and thus his results cannot be reliable. Nevertheless, we have been inspired by some of his techniques.

The main problem to determine the $L^\infty$-norm is the oscillatory nature of the integrands in the solution formula (2). The stationary phase formula as given by L. Hörmander in Theorem 7.7.5 of [12] provides a powerful tool to treat this situation.

In the following theorem we formulate a special case of this result relevant for us.

**Theorem 3.1 (Stationary phase method).** *Let $K$ be a compact interval in $\mathbb{R}$, $X$ an open neighborhood of $K$. Let $U \in C_0^2(K)$, $\Psi \in C^4(X)$ and $\mathrm{Im}\Psi \geq 0$ in $X$. If there exists $p_0 \in X$ such that $\frac{\partial}{\partial p}\Psi(p_0) = 0$, $\frac{\partial^2}{\partial p^2}\Psi(p_0) \neq 0$, and $\mathrm{Im}\Psi(p_0) = 0$, $\frac{\partial}{\partial p}\Psi(p) \neq 0$, $p \in K \setminus \{p_0\}$, then*

$$\Big| \int_K U(p) e^{i\omega\Psi(p)}\,dp \;-\; e^{i\omega\Psi(p_0)} \left[\frac{\omega}{2\pi i}\frac{\partial^2}{\partial p^2}\Psi(p_0)\right]^{-1/2} U(p_0)\Big| \leq C(K)\,\|U\|_{C^2(K)}\,\omega^{-1}$$

*for all $\omega > 0$. Moreover $C(K)$ is bounded when $\Psi$ stays in a bounded set in $C^4(X)$.*

**Theorem 3.2 (Time-decay of the solution of (ACP) in a special case).** *Fix $j,k \in \{1,\dots,n\}$ with $k \leq j$. Suppose that $u_0$ satisfies Condition $(A_{j,k})$ and choose $\lambda_{\min}, \lambda_{\max} \in (a_j, a_{j+1})$ such that*

$$\mathrm{supp}(Vu_0)_k \subset [\lambda_{\min}, \lambda_{\max}] \subset (a_j, a_{j+1}).$$

*Then for all $x \in N_r$ with $r \leq j$ and $r \neq k$ and all $t \in \mathbb{R}^+$ such that $(t,x)$ lies in the cone described by*

$$\sqrt{\frac{\lambda_{\max}}{c_r(\lambda_{\max} - a_r)}} \leq \frac{t}{x} \leq \sqrt{\frac{\lambda_{\min}}{c_r(\lambda_{\min} - a_r)}}\,, \tag{3}$$

*there exists $H(t,x,u_0) \in \mathbb{C}$ and a constant $c(u_0)$ satisfying*

$$\left|u_+(t,x) - H(t,x,u_0)t^{-1/2}\right| \leq c(u_0)\cdot t^{-1}\,, \tag{4}$$

*where $u_+$ is defined in Theorem 2.5, with*

$$|H(t,x,u_0)| \leq \left(\frac{2\pi c_k}{c_r}\right)^{1/2} \lambda_{\max}^{3/4} \cdot \frac{\max_{v\in[v_{\min},v_{\max}]}\sqrt{|(a_r - a_k)v + a_r|}}{\left(\sum_{l\leq r}\sqrt{c_l}\sqrt{(a_r - a_l)v_{\min} + a_r}\right)^2} \cdot \|(Vu_0)_k\|_\infty,$$

*where $v_{\min} := \dfrac{a_r}{\lambda_{\max} - a_r}$ and $v_{\max} := \dfrac{a_r}{\lambda_{\min} - a_r}$.*

*Remark* 3.3. i) Note that (3) is equivalent to

$$v_{\min} \leq v(t,x) := c_r(t/x)^2 - 1 \leq v_{\max}\,. \tag{5}$$

ii) The hypotheses of Theorem 3.2 imply that $j \geq 2$.
iii) An explicit expression for $H(t,x,u_0)$ is given at the end of the proof in (8).
iv) We have chosen to investigate only $u_+$ in this proceedings article, since the expression for $u_-$ does not possess a stationary point in its phase. Hence, one can prove that its contribution will decay at least as $ct^{-1}$. A detailed analysis will follow in a forthcoming paper.

*Proof.* We divide the proof in five steps.

*First step: Substitution.* Realizing the substitution $p := \xi_r(\lambda) = \sqrt{\frac{\lambda - a_r}{c_r}}$ in the expression for $u_+$ given in Theorem 2.5 leads to:

$$u_+(t,x) = 2c_r \int_{p_{\min}}^{p_{\max}} e^{i\sqrt{a_r + c_r p^2}t} q_k(a_r + c_r p^2) e^{-ipx} (Vu_0)_k(a_r + c_r p^2) p \, dp$$

with $p_{\min} := \xi_r(\lambda_{\min})$ and $p_{\max} := \xi_r(\lambda_{\max})$.

*Second step: Change of the parameters* $(t,x)$. In order to get bounded parameters, we change $(t,x)$ into $(\tau,\chi)$ defined by

$$\tau = \frac{t}{\omega} \qquad \text{and} \qquad \chi = \frac{x}{\omega} \qquad \text{with} \qquad \omega = \sqrt{t^2 + x^2},$$

following an argument from [16]. Thus the argument of the exponential in the integral defining $u_+$ becomes:

$$i\omega(\sqrt{a_r + c_r p^2}\tau - p\chi) =: i\omega\varphi(p,\tau,\chi).$$

Note that $\tau, \chi \in [0,1]$ for $t, x \in [0,\infty)$.

*Third step: Application of the stationary phase method.* Now we want to apply Theorem 3.1 to $u_+$ with the amplitude $U$ and the phase $\Psi$ defined by:

$$U(p) := q_k(a_r + c_r p^2)(Vu_0)_k(a_r + c_r p^2)p, \quad \Psi(p) := \varphi(p,\tau,\chi), \quad p \in [p_{\min}, p_{\max}].$$

The functions $U$ and $\Psi$ satisfy the regularity conditions on the compact interval $K := [p_{\min}, p_{\max}]$ and $\Psi$ is a real-valued function. One easily verifies, that for $\tau \neq 0$

$$\Psi'(p) = \frac{c_r p}{\sqrt{a_r + c_r p^2}}\tau - \chi = 0 \iff p = p_0 := \sqrt{\frac{a_r \chi^2}{c_r(c_r\tau^2 - \chi^2)}} = \sqrt{\frac{a_r x^2}{c_r(c_r t^2 - x^2)}},$$

and that this stationary point $p_0$ belongs to the interval of integration $[p_{\min}, p_{\max}]$, if and only if $(t,x)$ lies in the cone defined by (3). Furthermore, for $p \in \mathbb{R}$

$$\frac{\partial^2 \Psi}{\partial p^2}(p) = \frac{\partial^2 \varphi}{\partial p^2}(p,\tau,\chi) = \tau \frac{c_r a_r}{(a_r + c_r p^2)^{3/2}} \neq 0 \,.$$

Thus, Theorem 3.1 implies that for all $(t,x)$ satisfying (3) there exists a constant $C(K,\tau,\chi) > 0$ such that

$$\Bigg| u_+(t,x) - \underbrace{e^{-i\omega\varphi(p_0,\tau,\chi)} \left( \frac{\omega}{2i\pi} \frac{\partial^2 \varphi}{\partial p^2}(p_0,\tau,\chi) \right)^{-1/2} U(p_0)}_{(*)} \Bigg| \leq C(K,\tau,\chi) \|U\|_{C^2(K)}\, \omega^{-1}$$

for all $\omega > 0$.

*Fourth step: Backsubstitution.* We must now control the dependence of $C(K,\tau,\chi)$ on the parameters $\tau,\chi$. To this end, one has to assure that $\Psi=\varphi(\cdot,\tau,\chi)$ stays in a bounded set in $C^4(X)$, if $\tau$ and $\chi$ vary in $[0,1]$, where we choose $X=(p_m,p_M)$ such that $0<p_m<p_{\min}<p_{\max}<p_M<\infty$. This follows using the above expressions for $\frac{\partial\varphi}{\partial p}, \frac{\partial^2\varphi}{\partial p^2}$ and

$$\frac{\partial^3\varphi}{\partial p^3}(p,\tau,\chi)=-\frac{3a_rc_r^2p}{(a_r+c_rp^2)^{5/2}}\,\tau\,,\quad \frac{\partial^4\varphi}{\partial p^4}(p,\tau,\chi)=-\frac{3a_rc_r^2(a_r-4p^2c_r)}{(a_r+c_rp^2)^{7/2}}\,\tau$$

for $p\in X$. Thus Theorem 3.1 implies that there exists a constant $C(K)>0$ such that $C(K,\tau,\chi)\le C(K)$ for all $\tau,\chi\in[0,1]$.
To evaluate $(*)$ we observe that $p_0=\frac{1}{\sqrt{c_r}}\sqrt{\frac{a_r}{c_r(t/x)^2-1}}$. This implies

$$\xi_l(a_r+c_rp_0^2)=\sqrt{\frac{(a_r+c_rp_0^2)-a_l}{c_l}}=\frac{1}{\sqrt{c_l}}\frac{\sqrt{(a_r-a_l)\,(c_r(t/x)^2-1)+a_r}}{\sqrt{c_r(t/x)^2-1}}$$

and thus

$$\begin{aligned}q_k(a_r+c_rp_0^2)&=\frac{c_k\xi_k(a_r+c_rp_0^2)}{|\sum_{l=1}^n c_l\xi_l(a_r+c_rp_0^2)|^2}\\&=\sqrt{c_k}\sqrt{c_r(t/x)^2-1}\frac{\sqrt{(a_r-a_k)\,(c_r(t/x)^2-1)+a_r}}{\left|\sum_{l=1}^n\sqrt{c_l}\sqrt{(a_r-a_l)\,(c_r(t/x)^2-1)+a_r}\right|^2}\,.\end{aligned}$$

Finally,

$$\begin{aligned}\frac{\partial^2\varphi}{\partial p^2}(p_0,\tau,\chi)-\tau\frac{c_ra_r}{(a_r+c_rp_0^2)^{3/2}}&=\tau\frac{\left(c_r\tau^2-\chi^2\right)^{3/2}}{(a_rc_r)^{1/2}\tau^2}\\&=\tau(c_ra_r)^{-1/2}\left(\frac{c_r(t/x)^2-1}{(t/x)^2}\right)^{3/2}.\end{aligned}$$

Combining these results and using $\omega\tau=t$ we find

$$\begin{aligned}(*)&=\left(\frac{\omega}{2i\pi}\,\frac{\partial^2\varphi}{\partial p^2}(p_0,\tau,\chi)\right)^{-1/2}\quad q_k(a_r+c_rp_0^2)\;p_0\;(Vu_0)_k(a_r+c_rp_0^2)\\&=(2i\pi)^{1/2}t^{-1/2}(c_ra_r)^{1/4}\left(\frac{(t/x)^2}{c_r\,(t/x)^2-1}\right)^{3/4}\\&\quad\times\sqrt{c_k}\sqrt{c_r(t/x)^2-1}\frac{\sqrt{(a_r-a_k)\,(c_r(t/x)^2-1)+a_r}}{\left|\sum_l\sqrt{c_l}\sqrt{(a_r-a_l)\,(c_r(t/x)^2-1)+a_r}\right|^2}\\&\quad\times\frac{1}{\sqrt{c_r}}\sqrt{\frac{a_r}{c_r(t/x)^2-1}}\;(Vu_0)_k(a_r+c_rp_0^2)\\&=(2i\pi)^{1/2}a_r^{3/4}c_r^{1/4}\;h_1(t,x)\;h_2(t,x)\;(Vu_0)_k(a_r+c_rp_0^2)\;t^{-1/2}\end{aligned}$$

with

$$h_1(t,x) := \left(\frac{\left(\frac{t}{x}\right)^2}{c_r\left(\frac{t}{x}\right)^2 - 1}\right)^{3/4}, \quad h_2(t,x) := \frac{\sqrt{(a_r - a_k)\left(c_r(\frac{t}{x})^2 - 1\right) + a_r}}{\left|\sum_l \sqrt{c_l}\sqrt{(a_r - a_l)\left(c_r(\frac{t}{x})^2 - 1\right) + a_r}\right|^2}.$$

*Fifth step: Uniform estimates.* It remains to estimate $h_1(t,x)h_2(t,x)(Vu_0)_k(a_r + c_r p_0^2)$ uniformly in $t$ and $x$, if $(t,x)$ satisfies (3). To this end we note that the function $b \mapsto \dfrac{b}{c_r b - 1}$ is a decreasing function on $(1/c_r, +\infty)$. Thus the maximum of $h_1$ for $(t,x)$ satisfying (3) is attained at $\frac{t}{x} = \sqrt{\frac{\lambda_{\max}}{c_r(\lambda_{\max} - a_r)}}$. This implies

$$h_1(t,x) \le \left(\frac{1}{c_r a_r}\lambda_{\max}\right)^{3/4} \text{ for } (t,x) \text{ satisfying (3).} \tag{6}$$

Let us now estimate $h_2$. For a fixed $v$ in $[v_{\min}, v_{\max}]$, we denote by $I_1$ the set of indices $l$ such that $(a_r - a_l)v + a_r \ge 0$ and $I_2 = \{1,\dots,n\} \setminus I_1$.
Then, for any $v$ in $[v_{\min}, v_{\max}]$ we have $\{1,\dots,r\} \subset I_1$ (since $v_{\min} > 0$) and

$$\begin{aligned}
&\Big|\sum_l \sqrt{c_l}\sqrt{(a_r - a_l)v + a_r}\Big|^2 \\
&\quad = \Big(\sum_{l\in I_1} \sqrt{c_l}\sqrt{(a_r - a_l)v + a_r}\Big)^2 + \Big|\sum_{l\in I_2} \sqrt{c_l}\sqrt{(a_r - a_l)v + a_r}\Big|^2 \\
&\quad \ge \Big(\sum_{l\le r} \sqrt{c_l}\sqrt{(a_r - a_l)v + a_r}\Big)^2 \\
&\quad \ge \Big(\sum_{l\le r} \sqrt{c_l}\sqrt{(a_r - a_l)v_{\min} + a_r}\Big)^2 .
\end{aligned}$$

Thus, (5) implies

$$|h_2(t,x)| \le \frac{\max_{v\in[v_{\min}, v_{\max}]} \sqrt{|(a_r - a_k)v + a_r|}}{\left(\sum_{l\le r}\sqrt{c_l}\sqrt{(a_r - a_l)v_{\min} + a_r}\right)^2} . \tag{7}$$

Putting everything together, the assertion of the theorem is valid for

$$H(t,x,u_0) := e^{-i\varphi(p_0,t,x)}(2i\pi)^{1/2} a_r^{3/4} c_r^{1/4} c_k^{1/2}\; h_1(t,x)\; h_2(t,x)\; (Vu_0)_k(a_r + c_r p_0^2) . \tag{8}$$

Finally the right-hand side of estimate (4) is derived from the inequality

$$\begin{aligned}
&C(K,\tau,\chi)\|U\|_{C^2(K)}\ \omega^{-1} \\
&\qquad \le C(K)\big\|p \mapsto q_k(a_r + c_r p^2)(Vu_0)_k(a_r + c_r p^2)p\big\|_{C^2([p_{\min},p_{\max}])}\ t^{-1}.
\end{aligned} \tag{9}$$

The $C^2$-norm is finite, since the involved functions are regular on the compact set $[p_{\min}, p_{\max}]$. □

## 4. Growing potential step

For this section we specialize to the case of two branches $N_1$ and $N_2$ and, for the sake of simplicity, we also set $c_1 = c_2 = 1$. We show that, choosing a generic initial condition $u_0$ in a compact energy band included in $(a_2, \infty)$, the coefficient $H(t, x, u_0)$ in the asymptotic expansion of Theorem 3.2 tends to zero, if the potential step $a_2 - a_1$ tends to infinity. Simultaneously the cone of the exact $t^{-1/2}$-decay shrinks and inclines toward the $t$-axis.

**Theorem 4.1.** *Let $0 < \alpha < \beta < 1$ and $\psi \in C_c^2((\alpha, \beta))$ with $\|\psi\|_\infty = 1$ be given. Setting $\tilde{\psi}(\lambda) := \psi(\lambda - a_2)$, we choose the initial condition $u_0 \in H$ satisfying $(Vu_0)_2 \equiv 0$ and $(Vu_0)_1 = \tilde{\psi}$. Furthermore, let $u_+$ be defined as in Theorem 2.5.*

*Then there is a constant $C(\psi, \alpha, \beta)$ independent of $a_1$ and $a_2$, such that for all $t \in \mathbb{R}^+$ and all $x \in N_2$ with*

$$\sqrt{\frac{a_2 + \beta}{\beta}} \leq \frac{t}{x} \leq \sqrt{\frac{a_2 + \alpha}{\alpha}}$$

*the value $H(t, x, u_0)$ given in (8) satisfies*

$$\left|u_+(t, x) - H(t, x, u_0) \cdot t^{-1/2}\right| \leq C(\psi, \alpha, \beta) \cdot t^{-1}$$

*and*

$$\left|H(t, x, u_0)\right| \leq \sqrt{2\pi} \frac{\sqrt{\beta}(a_2 + \beta)^{3/4}}{\sqrt{a_2}\sqrt{a_2 - a_1 + \beta}}.$$

*Proof.* Note that it is always possible to choose the initial condition in the indicated way, thanks to the surjectivity of $V$, cf. Theorem 2.3 ii).

The constant $C(\psi, \alpha, \beta)$ has been already calculated in Theorem 3.2. It remains to make sure that it is independent of $a_1$ and $a_2$ and to prove the estimate for $|H(t, x, u_0)|$.

We start with the latter and carry out a refined analysis of the proof of Theorem 3.2 for our special situation. Using the notation of this proof, (8) yields

$$\left|H(t, x, u_0)\right| = \sqrt{2\pi} a_2^{3/4} h_1(t, x) \left|h_2(t, x)\right| \cdot \|(Vu_0)_1\|_\infty.$$

By (6) and $\lambda_{\max} = a_2 + \beta$ we find

$$h_1(t, x) \leq \frac{(a_2 + \beta)^{3/4}}{a_2^{3/4}}$$

and, investing the definition of $h_2$ together with (5), we have

$$\begin{aligned}\left|h_2(t, x)\right| &= \left|\frac{\sqrt{(a_2 - a_1)((t/x)^2 - 1) + a_2}}{\left(\sqrt{(a_2 - a_1)((t/x)^2 - 1) + a_2} + \sqrt{a_2}\right)^2}\right| \\ &\leq \frac{1}{\sqrt{(a_2 - a_1)((t/x)^2 - 1) + a_2}} \leq \frac{1}{\sqrt{(a_2 - a_1)v_{\min} + a_2}}.\end{aligned}$$

Putting in the definitions of $v_{\min}$ and afterwords $\lambda_{\max}$ and rearranging terms, this leads to

$$\big|h_2(t,x)\big| \le \frac{\sqrt{\beta}}{\sqrt{a_2}\sqrt{a_2 - a_1 + \beta}}.$$

Since $\|(Vu_0)_1\|_\infty = \|\tilde{\psi}\|_\infty = \|\psi\|_\infty$ was set to 1, we arrive at the estimate

$$\big|H(t,x,u_0)\big| \le \sqrt{2\pi}(a_2+\beta)^{3/4}\frac{\sqrt{\beta}}{\sqrt{a_2}\sqrt{a_2 - a_1 + \beta}}.$$

Going again back to Theorem 3.2 for the constant $C$ we have by (9)

$$C = C(K)\|U(p)\|_{C^2(K)},$$

where

$$U(p) = pq_1(a_2+p^2)(Vu_0)_1(a_2+p^2), \quad p \in K,$$

and

$$K = [p_{\min}, p_{\max}] = [\xi_2(a_2+\alpha), \xi_2(a_2+\beta)] = [\sqrt{\alpha}, \sqrt{\beta}].$$

Thus, the constant $C(K)$ is independent of $a_1$ and $a_2$ and we can start to estimate the $C^2$-norm of $U$:

$$U(p) = p\tilde{\psi}(a_2+p^2)\frac{\xi_1(a_2+p^2)}{|\xi_1(a_2+p^2)+\xi_2(a_2+p^2)|^2} = p\psi(p^2)\frac{\sqrt{a_2-a_1+p^2}}{\big(\sqrt{a_2-a_1+p^2}+p\big)^2}$$
$$= p\psi(p^2)\frac{f(p)}{(f(p)+p)^2}$$

where $f(p) := \sqrt{a_2-a_1+p^2}$. For the function $U$ itself we find

$$|U(p)| \le \sqrt{\beta}\|\psi\|_\infty\frac{1}{f(p)} = \frac{\sqrt{\beta}}{\sqrt{a_2-a_1+p^2}} \le \frac{\sqrt{\beta}}{\sqrt{a_2-a_1+\alpha}} \le \frac{\sqrt{\beta}}{\sqrt{\alpha}}.$$

Calculating the derivatives is lengthy, but using $f'(p)f(p) = p$, one finds constants $C_1$ and $C_2$ depending only on $\psi$, $\alpha$ and $\beta$ with

$$|U'(p)| = \Big|\big(p\psi(p^2)\big)'\frac{f(p)}{(f(p)+p)^2} + p\psi(p^2)\frac{p^2 - pf(p) - 2f(p)^2}{f(p)(f(p)+p)^3}\Big|$$
$$\le C_1\Big(\frac{1}{f(p)} + \frac{2p^2+4pf(p)+2f(p)^2}{f(p)(f(p)+p)^3}\Big) \le C_1\Big(\frac{1}{f(p)} + \frac{2}{f(p)^2}\Big)$$
$$\le C_1\Big(\frac{1}{\sqrt{\alpha}} + \frac{2}{\alpha}\Big)$$

and in a similar manner

$$|U''(p)| = \Big|\big(p\psi(p^2)\big)''\frac{f(p)}{(f(p)+p)^2} + 2\big(p\psi(p^2)\big)'\frac{p^2-pf(p)-2f(p)^2}{f(p)(f(p)+p)^3}$$
$$+ p\psi(p^2)\frac{5f(p)^4+8pf(p)^3-4p^3f(p)-p^4}{f(p)^3(f(p)+p)^4}\Big|$$
$$\le C_2\Big(\frac{1}{\sqrt{\alpha}} + \frac{4}{\alpha} + \frac{5}{\alpha^{3/2}}\Big).$$

□

*Remark* 4.2.

i) In the situation of Theorem 4.1, we have

$$\begin{aligned}\left|H(t,x,u_0)\right| &\le \sqrt{2\pi}(a_2+\beta)^{3/4}\frac{\sqrt{\beta}}{\sqrt{a_2}\sqrt{a_2-a_1+\beta}}\\ &\sim \sqrt{2\pi\beta}\, a_2^{-1/4} \quad \text{as } a_2 \to +\infty.\end{aligned}$$

ii) Suppose that $\psi(\mu) \ge m > 0$ for $\mu \in [\alpha', \beta']$ with $\alpha < \alpha' < \beta' < \beta$. Then one can show that

$$\left|H(t,x,u_0)\right| \ge \sqrt{2\pi\alpha}\, a_2^{-1/4} m.$$

for $(t, x)$ satisfying

$$\sqrt{\frac{a_2+\beta'}{\beta'}} \le \frac{t}{x} \le \sqrt{\frac{a_2+\alpha'}{\alpha'}}$$

if $a_2$ is sufficiently large. Thus the coefficient of $t^{-1/2}$ behaves exactly as $\text{const} \cdot a_2^{-1/4}$ (in particular it tends to zero) as $a_2 \to +\infty$.

iii) The cone in the $(t, x)$-plane, where $u_+$ decays as $\text{const} \cdot t^{-1/2}$ is given by

$$\sqrt{\frac{\beta}{a_2+\beta}} \le \frac{x}{t} \le \sqrt{\frac{\alpha}{a_2+\alpha}}.$$

Clearly it shrinks and inclines toward the $t$-axis as $a_2 \to +\infty$. One can prove that outside this cone, $u_+$ decays at least as $t^{-1}$. This exact asymptotic behavior of the $L^\infty$-norm might be experimentally verified.

iv) Note that (4) also implies that

$$\begin{aligned}\left|u_+(t,x)\right| &\le \left|u_+(t,x) - H(t,x,u_0)t^{-1/2} + H(t,x,u_0)t^{-1/2}\right|\\ &\le C(\psi,\alpha,\beta)t^{-1} + \left|H(t,x,u_0)\right|t^{-1/2}\\ &\le D(\psi,\beta,a_1,a_2)t^{-1/2}\end{aligned}$$

for $(t, x)$ in the cone indicated there, if $t$ is sufficiently large.

**Acknowledgement**

The authors thank Otto Liess for useful remarks.

## References

[1] F. Ali Mehmeti, *Spectral Theory and $L^\infty$-time Decay Estimates for Klein-Gordon Equations on Two Half Axes with Transmission: the Tunnel Effect.* Math. Methods Appl. Sci. **17** (1994), 697–752.

[2] F. Ali Mehmeti, *Transient Waves in Semi-Infinite Structures: the Tunnel Effect and the Sommerfeld Problem.* Mathematical Research, vol. 91, Akademie Verlag, Berlin, 1996.

[3] F. Ali Mehmeti, R. Haller-Dintelmann, V. Régnier, *Dispersive Waves with multiple tunnel effect on a star-shaped network;* to appear in: "Proceedings of the Conference on Evolution Equations and Mathematical Models in the Applied Sciences (EEMMAS)", Taranto 2009.

[4] F. Ali Mehmeti, R. Haller-Dintelmann, V. Régnier, *Multiple tunnel effect for dispersive waves on a star-shaped network: an explicit formula for the spectral representation.* arXiv:1012.3068v1 [math.AP], preprint 2010.

[5] F. Ali Mehmeti, V. Régnier, *Delayed reflection of the energy flow at a potential step for dispersive wave packets.* Math. Methods Appl. Sci. **27** (2004), 1145–1195.

[6] J. von Below, J.A. Lubary, *The eigenvalues of the Laplacian on locally finite networks.* Results Math. **47** (2005), no. 3-4, 199–225.

[7] S. Cardanobile and D. Mugnolo. *Parabolic systems with coupled boundary conditions.* J. Differential Equations **247** (2009), no. 4, 1229–1248.

[8] Y. Daikh, *Temps de passage de paquets d'ondes de basses fréquences ou limités en bandes de fréquences par une barrière de potentiel.* Thèse de doctorat, Valenciennes, France, 2004.

[9] J.M. Deutch, F.E. Low, *Barrier Penetration and Superluminal Velocity.* Annals of Physics **228** (1993), 184–202.

[10] A. Enders, G. Nimtz, *On superluminal barrier traversal.* J. Phys. I France **2** (1992), 1693–1698.

[11] A. Haibel, G. Nimtz, *Universal relationship of time and frequency in photonic tunnelling.* Ann. Physik (Leipzig) **10** (2001), 707–712.

[12] L. Hörmander, *The Analysis of Linear Partial Differential Operators I.* Springer, 1984.

[13] V. Kostrykin, R. Schrader, *The inverse scattering problem for metric graphs and the travelling salesman problem.* Preprint, 2006 (www.arXiv.org:math.AP/0603010).

[14] O. Liess, *Decay estimates for the solutions of the system of crystal optics.* Asymptotic Analysis **4** (1991), 61–95.

[15] B. Marshall, W. Strauss, S. Wainger, *$L^p$-$L^q$ Estimates for the Klein-Gordon Equation.* J. Math. Pures et Appl. **59** (1980), 417–440.

[16] K. Mihalincic, *Time decay estimates for the wave equation with transmission and boundary conditions.* Dissertation. Technische Universität Darmstadt, Germany, 1998.

[17] M. Pozar, *Microwave Engineering.* Addison-Wesley, New York, 1990.

F. Ali Mehmeti and V. Régnier
Univ Lille Nord de France, F-59000 Lille, France
UVHC, LAMAV, FR CNRS 2956, F-59313 Valenciennes, France
e-mail: felix.ali-mehmeti@univ-valenciennes.fr
Virginie.Regnier@univ-valenciennes.fr

R. Haller-Dintelmann
TU Darmstadt, Fachbereich Mathematik
Schloßgartenstraße 7, D-64289 Darmstadt, Germany
e-mail: haller@mathematik.tu-darmstadt.de

Operator Theory:
Advances and Applications, Vol. 221, 25-46


# Some Classes of Operators on Partial Inner Product Spaces

Jean-Pierre Antoine and Camillo Trapani

**Abstract.** Many families of function spaces play a central role in analysis, such as $L^p$ spaces, Besov spaces, amalgam spaces or modulation spaces. In all such cases, the parameter indexing the family measures the behavior (regularity, decay properties) of particular functions or operators. Actually all these space families are, or contain, scales or lattices of Banach spaces, which are special cases of *partial inner product spaces* (PIP-*spaces*). In this paper, we shall give an overview of PIP-spaces and operators on them, defined globally. We will discuss a number of operator classes, such as morphisms, projections or certain integral operators. We also explain how a PIP-space can be generated from a *-algebra of operators on a Hilbert space and we prove that, under natural conditions, every lattice of Hilbert spaces is obtained in this way.

**Mathematics Subject Classification (2000).** 46C50, 47A70, 47B37, 47B38.

**Keywords.** Partial inner product spaces, function spaces, operators, homomorphisms.

## 1. Introduction

When dealing with singular functions, one generally turns to distributions, most often to tempered distributions. In the latter case, one is in fact working in the triplet (Rigged Hilbert space or RHS)

$$\mathcal{S}(\mathbb{R}) \subset L^2(\mathbb{R}, dx) \subset \mathcal{S}^\times(\mathbb{R}), \tag{1.1}$$

where $\mathcal{S}(\mathbb{R})$ is the Schwartz space of smooth functions of fast decay and $\mathcal{S}^\times(\mathbb{R})$ is the space of tempered distributions, taken as antilinear continuous functionals over $\mathcal{S}(\mathbb{R})$, so that the embeddings in (1.1) are linear (we restrict ourselves to one dimension, for simplicity, but the argument is general).

The problem with the triplet (1.1) is that, besides the Hilbert space vectors, it contains only two types of elements, "very good" functions in $\mathcal{S}$ and "very bad" ones in $\mathcal{S}^\times$. If one wants a fine control on the behavior of individual elements, one has to interpolate somehow between the two extreme spaces. In the case of the Schwartz triplet (1.1) a well-known solution is given by a chain of Hilbert spaces, the so-called Hermite representation of tempered distributions [14].

In fact, this is not at all an isolated case. Indeed many function spaces that play a central role in analysis come in the form of families, indexed by one or several parameters that characterize the behavior of functions (smoothness, behavior at infinity, ...). The typical structure is a *chain of Hilbert or (reflexive) Banach spaces*. Let us give two familiar examples.

(i) The Lebesgue $L^p$ spaces on a finite interval, e.g., $\mathcal{I} = \{L^p([0,1],dx), 1 \leqslant p \leqslant \infty\}$:

$$L^\infty \subset \cdots \subset L^{\overline{q}} \subset L^{\overline{r}} \subset \cdots \subset L^2 \subset \cdots \subset L^r \subset L^q \subset \cdots \subset L^1, \quad (1.2)$$

where $1 < q < r < 2$. Here $L^q$ and $L^{\overline{q}}$ are dual to each other ($1/q + 1/\overline{q} = 1$), and similarly $L^r$ and $L^{\overline{r}}$ ($1/r + 1/\overline{r} = 1$). By the Hölder inequality, the ($L^2$) inner product

$$\langle f|g\rangle = \int_0^1 \overline{f(x)}\, g(x)\, dx \quad (1.3)$$

is well defined if $f \in L^q$, $g \in L^{\overline{q}}$. However, it is *not* well defined for two arbitrary functions $f, g \in L^1$. Take for instance, $f(x) = g(x) = x^{-1/2} : f \in L^1$, but $fg = f^2 \notin L^1$. Thus, on $L^1$, (1.3) defines only a *partial* inner product. The same result holds for any compact subset of $\mathbb{R}$ instead of [0,1].

(ii) The scale of Hilbert spaces[1] built on the powers of a positive self-adjoint operator $A \geqslant 1$ in a Hilbert space $\mathcal{H}_0$. Let $\mathcal{H}_n$ be $D(A^n)$, the domain of $A^n$, equipped with the graph norm $\|f\|_n = \|A^n f\|$, $f \in D(A^n)$, for $n \in \mathbb{N}$ or $n \in \mathbb{R}^+$, and $\mathcal{H}_{-n} = \mathcal{H}_n^\times$ (conjugate dual):

$$\mathcal{H}_\infty(A) := \bigcap_n \mathcal{H}_n \subset \cdots \subset \mathcal{H}_2 \subset \mathcal{H}_1 \subset \mathcal{H}_0 \subset \cdots$$
$$\cdots \subset \mathcal{H}_{-1} \subset \mathcal{H}_{-2} \cdots \subset \mathcal{H}_{-\infty}(A) := \bigcup_n \mathcal{H}_n. \quad (1.4)$$

Note that here the index $n$ could also be taken as real, the link between the two cases being established by the spectral theorem for self-adjoint operators. In this case also, the inner product of $\mathcal{H}_0$ extends to each pair $\mathcal{H}_n, \mathcal{H}_{-n}$, but on $\mathcal{H}_{-\infty}(A)$ it yields only a partial inner product. The following examples are standard:

- $(A_{\rm p} f)(x) = (1 + x^2) f(x)$ in $L^2(\mathbb{R}, dx)$.
- $(A_{\rm m} f)(x) = (1 - \frac{d^2}{dx^2}) f(x)$ in $L^2(\mathbb{R}, dx)$: Sobolev spaces $H^s(\mathbb{R})$, $s \in \mathbb{Z}$ or $\mathbb{R}$.
- $(A_{\rm osc} f)(x) = (1 + x^2 - \frac{d^2}{dx^2}) f(x)$ in $L^2(\mathbb{R}, dx)$.

[1] A discrete chain of Hilbert spaces $\{\mathcal{H}_n\}_{n\in\mathbb{Z}}$ is called a *scale* if there exists a self-adjoint operator $B \geqslant 1$ such that $\mathcal{H}_n = D(B^n)$, $\forall n \in \mathbb{Z}$, with the graph norm $\|f\|_n = \|B^n f\|$. A similar definition holds for a continuous chain $\{\mathcal{H}_\alpha\}_{\alpha\in\mathbb{R}}$.

(The notation is suggested by the operators of position, momentum and harmonic oscillator energy in quantum mechanics, respectively.) Note that both $\mathcal{H}_\infty(A_{\rm p}) \cap \mathcal{H}_\infty(A_{\rm m})$ and $\mathcal{H}_\infty(A_{\rm osc})$ coincide with the Schwartz space $\mathcal{S}(\mathbb{R})$and $\mathcal{H}_{-\infty}(A_{\rm osc})$ with the space $\mathcal{S}^\times(\mathbb{R})$ of tempered distributions.

However, a moment's reflection shows that the total order relation inherent in a chain is in fact an unnecessary restriction, partially ordered structures are sufficient, and indeed necessary in practice. For instance, in order to get a better control on the behavior of individual functions, one may consider the lattice built on the powers of $A_{\rm p}$ and $A_{\rm m}$ simultaneously. Then the extreme spaces are still $\mathcal{S}(\mathbb{R})$ and $\mathcal{S}^\times(\mathbb{R})$. Similarly, in the case of several variables, controlling the behavior of a function in each variable separately requires a nonordered set of spaces. This is in fact a statement about tensor products (remember that $L^2(X \times Y) \simeq L^2(X) \otimes L^2(Y)$). Indeed the tensor product of two chains of Hilbert spaces, $\{\mathcal{H}_n\} \otimes \{\mathcal{K}_m\}$ is naturally a lattice $\{\mathcal{H}_n \otimes \mathcal{K}_m\}$ of Hilbert spaces. For instance, in the example above, for two variables $x, y$, that would mean considering intermediate Hilbert spaces corresponding to the product of two operators, $\big(A_{\rm m}(x)\big)^n \big(A_{\rm m}(y)\big)^m$.

Thus the structure to analyze is that of *lattices of Hilbert or Banach spaces*, interpolating between the extreme spaces of a RHS, as in (1.1). Many examples can be given, for instance the lattice generated by the spaces $L^p(\mathbb{R}, dx)$, the amalgam spaces $W(L^p, \ell^q)$, the mixed norm spaces $L^{p,q}_m(\mathbb{R}, dx)$, and many more. In all these cases, which contain most families of function spaces of interest in analysis and in signal processing, a common structure emerges for the "large" space $V$, defined as the union of all individual spaces. There is a lattice of Hilbert or reflexive Banach spaces $V_r$, with an (order-reversing) involution $V_r \leftrightarrow V_{\overline{r}}$, where $V_{\overline{r}} = V_r^\times$ (the space of continuous conjugate linear functionals on $V_r$), a central Hilbert space $V_o \simeq V_{\overline{o}}$, and a partial inner product on $V$ that extends the inner product of $V_o$ to pairs of dual spaces $V_r, V_{\overline{r}}$.

Moreover, many operators should be considered globally, for the whole scale or lattice, instead of on individual spaces. In the case of the spaces $L^p(\mathbb{R})$, such are, for instance, operators implementing translations ($x \mapsto x - y$) or dilations ($x \mapsto x/a$), convolution operators, Fourier transform, etc. In the same spirit, it is often useful to have a *common* basis for the whole family of spaces, such as the Haar basis for the spaces $L^p(\mathbb{R})$, $1 < p < \infty$. Thus we need a notion of operator and basis defined globally for the scale or lattice itself.

This state of affairs prompted A. Grossmann and one of us (JPA), some time ago, to systematize this approach, and this led to the concept of *partial inner product space* or PIP-*space* [1, 2, 3]. However, the topic has found a new interest in recent years, in particular through developments of signal processing and the underlying mathematics. Thus the time had come for reviewing the whole subject, which we achieved in the recent monograph [5], to which we refer for further information. The aim of this paper is to present briefly this formalism of PIP-spaces. In a first part, the structure of PIP-space is derived systematically from the abstract notion of compatibility and then particularized to some examples.

In a second part, operators on PIP-spaces are introduced. In particular, several classes of operators, such as morphisms, projections or certain integral operators, are discussed at some length. Most of the contents is based on our monograph [5], but some new results are included.

## 2. Partial inner product spaces

### 2.1. Basic definitions

**Definition 2.1.** A *linear compatibility relation* on a vector space $V$ is a symmetric binary relation $f\#g$ which preserves linearity:

$$\begin{aligned} f\#g &\iff g\#f,\ \forall\, f,g \in V, \\ f\#g, f\#h &\implies f\#(\alpha g + \beta h),\ \forall\, f,g,h \in V,\ \forall\, \alpha,\beta \in \mathbb{C}. \end{aligned}$$

As a consequence, for every subset $S \subset V$, the set $S^{\#} = \{g \in V : g\#f, \forall\, f \in S\}$ is a vector subspace of $V$ and one has

$$S^{\#\#} = (S^{\#})^{\#} \supseteq S,\ S^{\#\#\#} = S^{\#}.$$

Thus one gets the following equivalences:

$$\begin{aligned} f\#g &\iff f \in \{g\}^{\#} &&\iff \{f\}^{\#\#} \subseteq \{g\}^{\#} \\ &\iff g \in \{f\}^{\#} &&\iff \{g\}^{\#\#} \subseteq \{f\}^{\#}. \end{aligned} \tag{2.1}$$

From now on, we will call *assaying subspace* of $V$ a subspace $S$ such that $S^{\#\#} = S$ and denote by $\mathcal{F}(V,\#)$ the family of all assaying subsets of $V$, ordered by inclusion. Let $F$ be the isomorphy class of $\mathcal{F}$, that is, $\mathcal{F}$ considered as an abstract partially ordered set. Elements of $F$ will be denoted by $r, q, \ldots$, and the corresponding assaying subsets $V_r, V_q, \ldots$. By definition, $q \leqslant r$ if and only if $V_q \subseteq V_r$. We also write $V_{\overline{r}} = V_r^{\#}$, $r \in F$. Thus the relations (2.1) mean that $f\#g$ if and only if there is an index $r \in F$ such that $f \in V_r$, $g \in V_{\overline{r}}$. In other words, vectors should not be considered individually, but only in terms of assaying subspaces, which are the building blocks of the whole structure.

It is easy to see that the map $S \mapsto S^{\#\#}$ is a closure, in the sense of universal algebra, so that the assaying subspaces are precisely the "closed" subsets. Therefore one has the following standard result.

**Theorem 2.2.** *The family $\mathcal{F}(V,\#) \equiv \{V_r, r \in F\}$, ordered by inclusion, is a complete involutive lattice, i.e., it is stable under the following operations, arbitrarily iterated:*

- *involution:* $V_r \leftrightarrow V_{\overline{r}} = (V_r)^{\#}$,
- *infimum:* $V_{p\wedge q} \equiv V_p \wedge V_q = V_p \cap V_q$, $\quad (p,q,r \in F)$
- *supremum:* $V_{p\vee q} \equiv V_p \vee V_q = (V_p + V_q)^{\#\#}$.

The smallest element of $\mathcal{F}(V,\#)$ is $V^{\#} = \bigcap_r V_r$ and the greatest element is $V = \bigcup_r V_r$. By definition, the index set $F$ is also a complete involutive lattice; for instance,

$$(V_{p\wedge q})^{\#} = V_{\overline{p\wedge q}} = V_{\overline{p}\vee\overline{q}} = V_{\overline{p}} \vee V_{\overline{q}}.$$

**Definition 2.3.** A *partial inner product* on $(V, \#)$ is a Hermitian form $\langle\cdot|\cdot\rangle$ defined exactly on compatible pairs of vectors. A *partial inner product space* (PIP-space) is a vector space $V$ equipped with a linear compatibility and a partial inner product.

Note that the partial inner product is not required to be positive definite.

The partial inner product clearly defines a notion of *orthogonality*: $f \perp g$ if and only if $f\#g$ and $\langle f|g\rangle = 0$.

**Definition 2.4.** The PIP-space $(V, \#, \langle\cdot|\cdot\rangle)$ is *nondegenerate* if $(V^{\#})^{\perp} = \{0\}$, that is, if $\langle f|g\rangle = 0$ for all $f \in V^{\#}$ implies $g = 0$.

We will assume henceforth that our PIP-space $(V, \#, \langle\cdot|\cdot\rangle)$ is nondegenerate. As a consequence, $(V^{\#}, V)$ and every couple $(V_r, V_{\overline{r}})$, $r \in F$, are dual pairs in the sense of topological vector spaces [9]. We also assume that the partial inner product is positive definite.

Now one wants the topological structure to match the algebraic structure, in particular, the topology $\tau_r$ on $V_r$ should be such that its conjugate dual be $V_{\overline{r}}$: $(V_r[\tau_r])^{\times} = V_{\overline{r}}$, $\forall r \in F$. This implies that the topology $\tau_r$ must be finer than the weak topology $\sigma(V_r, V_{\overline{r}})$ and coarser than the Mackey topology $\tau(V_r, V_{\overline{r}})$:

$$\sigma(V_r, V_{\overline{r}}) \preceq \tau_r \preceq \tau(V_r, V_{\overline{r}}).$$

From here on, we will assume that every $V_r$ carries its Mackey topology $\tau(V_r, V_{\overline{r}})$. This choice has two interesting consequences. First, if $V_r[\tau_r]$ is a Hilbert space or a reflexive Banach space, then $\tau(V_r, V_{\overline{r}})$ coincides with the norm topology. Next, $r < s$ implies $V_r \subset V_s$, and the embedding operator $E_{sr} : V_r \to V_s$ is continuous and has dense range. In particular, $V^{\#}$ is dense in every $V_r$.

From the previous examples, we learn that $\mathcal{F}(V, \#)$ is a huge lattice (it is complete!) and that assaying subspaces may be complicated, such as Fréchet spaces, nonmetrizable spaces, etc. This situation suggests to choose an involutive sublattice $\mathcal{I} \subset \mathcal{F}$, indexed by $I$, such that

(i) $\mathcal{I}$ is generating:

$$f\#g \Leftrightarrow \exists\, r \in I \text{ such that } f \in V_r, g \in V_{\overline{r}}; \tag{2.2}$$

(ii) every $V_r, r \in I$, is a Hilbert space or a reflexive Banach space;

(iii) there is a unique self-dual assaying subspace $V_o = V_{\overline{o}}$, which is a Hilbert space.

In that case, the structure $V_I := (V, \mathcal{I}, \langle\cdot|\cdot\rangle)$ is called, respectively, a *lattice of Hilbert spaces* (LHS) or a *lattice of Banach spaces* (LBS). Both types are particular cases of the so-called indexed PIP-spaces [5]. By this, one means the structure obtained from a PIP-space by imposing only condition (i) above. Actually, the two concepts are closely related. Given an indexed PIP-space $V_I$, it defines a unique PIP-space, namely, $(V, \#_{\mathcal{I}}, \langle\cdot|\cdot\rangle)$, with $\mathcal{F}(V, \#_{\mathcal{I}})$ the lattice completion of $\mathcal{I}$ (the converse is not true). And a PIP-space is a particular indexed PIP-space for which $\mathcal{I}$ happens to be a *complete* involutive lattice.

In this context, the behavior of a given vector $f \in V$ is characterized by the set $\mathrm{j}(f) := \{r \in I : f \in V_r\}$. Then the relation (2.2) means that $f\#g$ if and only if $\overline{\mathrm{j}(f)} \cap \mathrm{j}(g) \neq \emptyset$, where $\overline{\mathrm{j}(f)} := \{\overline{r} : r \in \mathrm{j}(f)\}$. Hence we have

$$\{g\}^{\#} = \bigcup \{V_r : r \in \overline{\mathrm{j}(g)}\} \text{ and } \{g\}^{\#\#} = \bigcap \{V_r : r \in \mathrm{j}(g)\}.$$

Note that $V^{\#}, V$ themselves usually do *not* belong to the family $\{V_r,\, r \in I\}$, but they can be recovered as

$$V^{\#} = \bigcap_{r \in I} V_r, \quad V = \sum_{r \in I} V_r.$$

In the LBS case, the lattice structure takes the following form

- $V_{p \wedge q} = V_p \cap V_q$, with the *projective* norm
$$\|f\|_{p \wedge q} = \|f\|_p + \|f\|_q \,;$$
- $V_{p \vee q} = V_p + V_q$, with the *inductive* norm
$$\|f\|_{p \vee q} = \inf_{f=g+h} \left(\|g\|_p + \|h\|_q\right), g \in V_p,\, h \in V_q.$$

These norms are usual in interpolation theory [8]. In the LHS case, one takes similar definitions with squared norms, in order to get Hilbert norms throughout.

### 2.2. Examples

**2.2.1. Sequence spaces.** Let $V$ be the space $\omega$ of *all* complex sequences $x = (x_n)$ and define on it:

- a compatibility relation by $x\#y \iff \sum_{n=1}^{\infty} |x_n\, y_n| < \infty$;
- a partial inner product $\langle x|y\rangle = \sum_{n=1}^{\infty} \overline{x_n}\, y_n$.

Then $\omega^{\#} = \varphi$, the space of finite sequences, and the complete lattice $\mathcal{F}(\omega, \#)$ consists of all Köthe's perfect sequence spaces [9]. They include all $\ell^p$-spaces ($1 \leqslant p \leqslant \infty$),

$$\ell^1 \subset \cdots \subset \ell^p \cdots \subset \ell^2 \cdots \subset \ell^{\overline{p}} \cdots \subset \ell^{\infty} \quad (1 < p < 2). \tag{2.3}$$

Of course, duality reads, as usual, $\ell^{p\#} = \ell^{\overline{p}}$ where $1/p + 1/\overline{p} = 1$. This PIP-space also contains the LHS of weighted Hilbert spaces $\ell^2(r)$,

$$\ell^2(r) = \left\{(x_n) \in \omega : (x_n/r_n) \in \ell^2, \text{ i.e., } \sum_{n=1}^{\infty} |x_n|^2\, r_n^{-2} < \infty\right\}, \tag{2.4}$$

where $r = (r_n)$, $r_n > 0$, is a sequence of positive numbers. The family possesses an involution, namely, $\ell^2(r) \leftrightarrow \ell^2(\overline{r}) = \ell^2(r)^{\times}$ where $\overline{r}_n = 1/r_n$ and the central, self-dual Hilbert space is $\ell^2$. These assaying subspaces of $\omega$ constitute a lattice, and indeed a generating, noncomplete sublattice of $\mathcal{F}(\omega, \#)$.

**2.2.2. Lebesgue spaces.** Take the chain of Lebesgue $L^p$ spaces on a finite interval $\Lambda \subset \mathbb{R}$, $\mathcal{I} = \{L^p(\Lambda, dx),\ 1 \leqslant p \leqslant \infty\}$, already described in the introduction:

$$L^\infty \subset \cdots \subset L^{\overline{p}} \subset \cdots \subset L^2 \subset \cdots \subset L^p \subset \cdots \subset L^1,\ 1 < p < 2. \tag{2.5}$$

This is a chain of Banach spaces, reflexive for $1 < p < \infty$.

On the other hand, the spaces $L^p(\mathbb{R})$ no longer form a chain, no two of them being comparable. We have only $L^p \cap L^q \subset L^s$, for all $s$ such that $p < s < q$. Hence we have to take the lattice generated by $\mathcal{I} = \{L^p(\mathbb{R}, dx), 1 \leqslant p \leqslant \infty\}$, by intersection and direct sum, with projective, resp. inductive norms. In this way, one gets a nontrivial LBS. See [5, Sec. 4.1.2] for a thorough analysis.

**2.2.3. Spaces of locally integrable functions.** Consider $V = L^1_{\rm loc}(X, d\mu)$, the space of all measurable, locally integrable functions on a measure space $(X, \mu)$. Define on it

- a compatibility relation by $f\#g \iff \int_X |f(x)g(x)|\, d\mu < \infty$
- a partial inner product $\langle f|g\rangle = \int_X \overline{f(x)}g(x)\, d\mu$.

Then $V^\# = L^\infty_c(X, d\mu)$ consists of all essentially bounded measurable functions of compact support. The complete lattice $\mathcal{F}(L^1_{\rm loc}, \#)$ consists of all Köthe function spaces [5, Sec. 4.4]). Here again, typical assaying subspaces are weighted Hilbert spaces

$$L^2(r) := \Big\{ f \in L^1_{\rm loc}(X, d\mu) : \int_X |f|^2 r^{-2} d\mu < \infty, \\ \text{with } r^{\pm 1} \in L^2_{\rm loc}(X, d\mu), r > 0 \text{ a.e..} \Big\}$$

Duality reads $[L^2(r)]^\# = L^2(\overline{r})$, with $\overline{r}(x) = r^{-1}(x)$. These spaces form a generating, involutive, noncomplete sublattice of $\mathcal{F}(L^1_{\rm loc}, \#)$.

## 3. Operators in pip-spaces

### 3.1. Definitions

**Definition 3.1.** Let $V_I$ and $Y_K$ be two nondegenerate indexed PIP-spaces (in particular, two LHSs or LBSs). Then an *operator* from $V_I$ to $Y_K$ is a map from a subset $\mathcal{D}(A) \subset V$ into $Y$, such that

(i) $\mathcal{D}(A) = \bigcup_{q \in \mathsf{d}(A)} V_q$, where $\mathsf{d}(A)$ is a nonempty subset of $I$;
(ii) For every $r \in \mathsf{d}(A)$, there exists $u \in K$ such that the restriction of $A$ to $V_r$ is a continuous linear map into $Y_u$ (we denote this restriction by $A_{ur}$);
(iii) $A$ has no proper extension satisfying (i) and (ii).

We denote by $\mathrm{Op}(V_I, Y_K)$ the set of all operators from $V_I$ to $Y_K$ and, in particular, $\mathrm{Op}(V_I) := \mathrm{Op}(V_I, V_I)$. The continuous linear operator $A_{ur} : V_r \to Y_u$ is called a *representative* of $A$. In terms of the latter, the operator $A$ may be characterized by the set $\mathsf{j}(A) = \{(r, u) \in I \times K : A_{ur} \text{ exists}\}$. Indeed, if $(r, u) \in \mathsf{j}(A)$, then the representative $A_{ur}$ is uniquely defined. Conversely, if $A_{ur}$ is any continuous linear map from $V_r$ to $Y_u$, then there exists a unique $A \in \mathrm{Op}(V, Y)$ having $A_{ur}$ as $\{r, u\}$-representative. This $A$ can be defined by considering $A_{ur}$ as a map from $V^{\#}$ to $Y$ and then extending it to its natural domain. Thus the operator $A$ may be identified with the collection of its representatives,

$$A \simeq \{A_{ur} : V_r \to Y_u : (r, u) \in \mathsf{j}(A)\}.$$

By condition (ii), the set $\mathsf{d}(A)$ is obtained by projecting $\mathsf{j}(A)$ on the "first coordinate" axis. The projection $\mathsf{i}(A)$ on the "second coordinate" axis plays, in a sense, the role of the range of $A$. More precisely,

$$\mathsf{d}(A) = \{r \in I : \text{there is a } u \text{ such that } A_{ur} \text{ exists}\},$$
$$\mathsf{i}(A) = \{u \in K : \text{there is a } r \text{ such that } A_{ur} \text{ exists}\}.$$

The following properties are immediate:

- $\mathsf{d}(A)$ is an initial subset of $I$: if $r \in \mathsf{d}(A)$ and $r' < r$, then $r \in \mathsf{d}(A)$, and $A_{ur'} = A_{ur}E_{rr'}$, where $E_{rr'}$ is a representative of the unit operator.
- $\mathsf{i}(A)$ is a final subset of $K$: if $u \in \mathsf{i}(A)$ and $u' > u$, then $u' \in \mathsf{i}(A)$ and $A_{u'r} = E_{u'u}A_{ur}$.
- $\mathsf{j}(A) \subset \mathsf{d}(A) \times \mathsf{i}(A)$, with strict inclusion in general.

Operators may be defined in the same way on a PIP-space (or between two PIP-spaces), simply replacing the index set $I$ by the complete lattice $F(V)$.

Since $V^{\#}$ is dense in $V_r$, for every $r \in I$, an operator may be identified with a separately continuous sesquilinear form on $V^{\#} \times V^{\#}$. Indeed, the restriction of any representative $A_{pq}$ to $V^{\#} \times V^{\#}$ is such a form, and all these restrictions coincide. Equivalently, an operator may be identified with a continuous linear map from $V^{\#}$ into $V$ (continuity with respect to the respective Mackey topologies).

But the idea behind the notion of operator is to keep also the *algebraic operations* on operators, namely:

(i) *Adjoint:* every $A \in \mathrm{Op}(V_I, Y_K)$ has a unique adjoint $A^{\times} \in \mathrm{Op}(Y_K, V_I)$, defined by the relation

$$\langle A^{\times}x|y\rangle = \langle x|Ay\rangle, \text{ for } y \in V_r,\ r \in \mathsf{d}(A), \text{ and } x \in V_{\overline{s}},\ s \in \mathsf{i}(A),$$

that is, $(A^{\times})_{\overline{r}\overline{s}} = (A_{sr})^*$ (usual Hilbert/Banach space adjoint).

It follows that $A^{\times\times} = A$, for every $A \in \mathrm{Op}(V_I, Y_K)$: no extension is allowed, by the maximality condition (iii) of Definition 3.1.

(ii) *Partial multiplication:* Let $V_I$, $W_L$, and $Y_K$ be nondegenerate indexed PIP-spaces (some, or all, may coincide). Let $A \in \mathrm{Op}(V_I, W_L)$ and $B \in \mathrm{Op}(W_L, Y_K)$. We say that the product $BA$ is defined if and only if there exist $r \in I, t \in L$, $u \in K$ such that $(r, t) \in \mathsf{j}(A)$ and $(t, u) \in \mathsf{j}(B)$. Then $B_{ut}A_{tr}$ is a continuous map from $V_r$ into $Y_u$. It is the $\{r, u\}$-representative of a unique element of $BA \in \mathrm{Op}(V_I, Y_K)$, called the product of $A$ and $B$. In other words, $BA$ is defined if and only if there is a $t \in \mathsf{i}(A) \cap \mathsf{d}(B)$, that is, if and only if there is continuous factorization through some $W_t$:

$$V_r \overset{A}{\to} W_t \overset{B}{\to} Y_u, \quad \text{i.e.} \quad (BA)_{ur} = B_{ut}A_{tr}. \tag{3.1}$$

If $BA$ is defined, then $A^\times B^\times$ is also defined, and equal to $(BA)^\times \in \mathrm{Op}(Y_K, V_I)$. It is worth noting that, for a LHS/LBS, the domain $\mathcal{D}(A)$ is always a vector subspace of $V$ (this is not true for a general PIP-space). Therefore, in that case, $\mathrm{Op}(V_I, Y_K)$ is a vector space and $\mathrm{Op}(V_I)$ is a *partial *-algebra* [4].

### 3.2. Regular operators

**Definition 3.2.** An operator $A \in \mathrm{Op}(V_I, Y_K)$ is called *regular* if $\mathsf{d}(A) = I$ and $\mathsf{i}(A) = K$ or, equivalently, if $A : V^\# \to Y^\#$ and $A : V \to Y$ continuously for the respective Mackey topologies.

This notion depends only on the pairs $(V^\#, V)$ and $(Y^\#, Y)$, *not* on the particular compatibilities on them. Accordingly, the set of all regular operators from $V$ to $Y$ is denoted by $\mathrm{Reg}(V, Y)$. Thus a regular operator may be multiplied both on the left and on the right by an arbitrary operator.

Of particular interest is the case $V = Y$, then we write simply $\mathrm{Reg}(V)$ for the set of all regular operators of $V$ onto itself. In this case $A$ is regular if, and only if, $A^\times$ is regular. Clearly the set $\mathrm{Reg}(V)$ is a *-algebra.

### 3.3. Morphisms

**Definition 3.3.** An operator $A \in \mathrm{Op}(V_I, Y_K)$ is called a *homomorphism* if

(i) for every $r \in I$ there exists $u \in K$ such that both $A_{ur}$ and $A_{\overline{ur}}$ exist;
(ii) for every $u \in K$ there exists $r \in I$ such that both $A_{ur}$ and $A_{\overline{ur}}$ exist.

Equivalently, for every $r \in I$, there exists $u \in K$ such that $(r, u) \in \mathsf{j}(A)$ and $(\overline{r}, \overline{u}) \in \mathsf{j}(A)$, and for every $u \in K$, there exists $r \in I$ with the same property.

The definition may also be rephrased as follows: $A : V_I \to Y_K$ is a homomorphism if

$$\mathrm{pr}_1(\mathsf{j}(A) \cap \overline{\mathsf{j}(A)}) = I \quad \text{and} \quad \mathrm{pr}_2(\mathsf{j}(A) \cap \overline{\mathsf{j}(A)}) = K, \tag{3.2}$$

where $\overline{\mathsf{j}(A)} = \{(\overline{r}, \overline{u}) : (r, u) \in \mathsf{j}(A)\}$ and $\mathrm{pr}_1, \mathrm{pr}_2$ denote the projection on the first, resp. the second component.[2]

We denote by $\mathrm{Hom}(V_I, Y_K)$ the set of all homomorphisms from $V_I$ into $Y_K$ and by $\mathrm{Hom}(V_I)$ those from $V_I$ into itself.

**Proposition 3.4.** *Let $A \in \mathrm{Hom}(V_I, Y_K)$. Then, $f \,\#_I\, g$ implies $Af \,\#_K Ag$.*

*Proof.* By the assumption, there exists $r \in I$ such that $f \in V_r$, $g \in V_{\overline{r}}$. Let $u \in K$ be such that $(r, u) \in \mathsf{j}(A)$ and $(\overline{r}, \overline{u}) \in \mathsf{j}(A)$. Then $Af \in Y_v$ and $Ag \in Y_{\overline{u}}$. Hence $Af$ and $Ag$ are compatible. □

[2]Contrary to what is stated in [5, Def. 3.3.4], the condition (3.2), which is the correct one, does *not* imply $\mathsf{j}(A) = I \times K$ and $\mathsf{j}(A^\times) = K \times I$.

The following properties are immediate.

**Proposition 3.5.** *Let $V_I, Y_K, \dots$ be indexed* PIP*-spaces. Then:*

(i) *Every homomorphism is regular.*
(ii) $A \in \mathrm{Hom}(V_I, Y_K)$ *if and only if* $A^\times \in \mathrm{Hom}(Y_K, V_I)$.
(iii) *The product of any number of homomorphisms (between successive* PIP*-spaces) is defined and is a homomorphism.*
(iv) *If $B$ is an arbitrary operator and $A$ is a homomorphism, then the two products $BA$ and $AB$ are defined.*
(v) *If $A \in \mathrm{Hom}(V_I, Y_K)$, then $\mathsf{j}(A^\times A)$ contains the diagonal of $I \times I$ and $\mathsf{j}(AA^\times)$ contains the diagonal of $K \times K$.*

A scalar multiple of a homomorphism is a homomorphism. However, the sum of two homomorphisms may fail to be one. Of course, if $A_1$ and $A_2$ in $\mathrm{Hom}(V_I)$ are such that both $\mathsf{j}(A_1)$ and $\mathsf{j}(A_2)$ contain the diagonal of $I \times I$, then their sum does, too; consequently it belongs to $\mathrm{Hom}(V_I)$. This happens, in particular, for projections, that we shall study in Section 3.3.4.

The definition of homomorphisms just given is tailored in such a way that one may consider the category **PIP** of all indexed PIP-spaces, with the homomorphisms as morphisms (arrows) [6, 11]. This language is useful for defining particular classes of morphisms, such as monomorphisms, epimorphisms and isomorphisms.

**3.3.1. Monomorphisms.** We define monomorphisms according to the language of general categories, namely,

**Definition 3.6.** Let $M \in \mathrm{Hom}(W_L, Y_K)$. Then $M$ is called a *monomorphism* if $MA = MB$ implies $A = B$, for any two elements of $A, B \in \mathrm{Hom}(V_I, W_L)$, where $V_I$ is any PIP-space.

Then we have the following

**Proposition 3.7.** [5, Prop. 3.3.9] *If every representative of $M \in \mathrm{Hom}(W_L, Y_K)$ is injective, then $M$ is a monomorphism.*

Clearly, if $M$ is an injective monomorphism, then every representative is injective. However we don't know whether every monomorphism in the category **PIP** is injective. In other words, the converse of Proposition 3.7 is open: Does there exist monomorphisms with at least one non-injective representative?

Typical examples of monomorphisms are the inclusion maps resulting from the restriction of a support. Take, for instance, $L^1_{\mathrm{loc}}(X, d\mu)$, the space of locally integrable functions on a measure space $(X, \mu)$(Example 2.2.3). Let $\Omega$ be a measurable subset of $X$ and $\Omega'$ its complement, both of nonzero measure, and construct the space $L^1_{\mathrm{loc}}(\Omega, d\mu)$. Given $f \in L^1_{\mathrm{loc}}(X, d\mu)$, define $f^{(\Omega)} = f\chi_\Omega$, where $\chi_\Omega$ is the characteristic function of $\chi_\Omega$. Then we obtain an injection monomorphism $M^{(\Omega)} : L^1_{\mathrm{loc}}(\Omega, d\mu) \to L^1_{\mathrm{loc}}(X, d\mu)$ as follows:

$$(M^{(\Omega)} f^{(\Omega)})(x) = \begin{cases} f^{(\Omega)}(x), & \text{if } x \in \Omega, \\ 0, & \text{if } x \notin \Omega, \end{cases} \quad f^{(\Omega)} \in L^1_{\mathrm{loc}}(\Omega, d\mu).$$

If we consider the lattice of weighted Hilbert spaces $\{L^2(r)\}$ in this PIP-space, then the correspondence $r \leftrightarrow r^{(\Omega)} = r\chi_\Omega$ is a bijection between the corresponding involutive lattices.

**3.3.2. Epimorphisms.** This is the notion dual to monomorphisms.

**Definition 3.8.** Let $N \in \text{Hom}(W_L, Y_K)$. Then $N$ is called an *epimorphism* if $AN = BN$ implies $A = B$, for any two elements $A, B \in \text{Hom}(Y_K, V_I)$, where $V_I$ is any PIP-space.

Then we have the following result, dual to Proposition 3.7

**Proposition 3.9.** *If every representative of $N \in \text{Hom}(W_L, Y_K)$ is surjective, then $N$ is an epimorphism.*

*Proof.* With the same notation as above, we have $AN - BN = 0$, thus $(A - B)N$ is well defined, i.e., there exist $r \in I, u \in K, l \in L$ such that $(u, r) \in \text{j}(A - B)$ and $(l, u) \in \text{j}(N)$. Hence $\big((A - B)N\big)_{rl} = (A - B)_{ru}N_{ul} = 0$. Since $N_{ul} : W_l \to Y_u$ is surjective, it follows that $(A - B)_{ru}f = 0$, $\forall f \in Y_u$. Thus $A = B$, since the restriction to $Y_u$ determines completely the operator $A - B$. □

Here too, we may ask whether there exist epimorphisms with at least one non-surjective representative.

Again, typical examples of epimorphisms are the restriction maps to a subset of the support. Take the same example as above, $V = L^1_{\text{loc}}(X, d\mu)$. Then the projection map $P^{(\Omega)} : L^1_{\text{loc}}(X, d\mu) \to L^1_{\text{loc}}(\Omega, d\mu)$ defined by $P^{(\Omega)}f = f\chi_\Omega$, $f \in L^1_{\text{loc}}(X, d\mu)$ is an epimorphism. Notice that $P^{(\Omega)}M^{(\Omega)} = 1_\Omega$, but $M^{(\Omega)}P^{(\Omega)} \neq 1_X$.

**3.3.3. Isomorphisms.** In a Hilbert space, a unitary operator is the same thing as an (isometric) isomorphism (i.e., a bijection that, together with its adjoint, preserves all inner products), but the two notions differ for a general PIP-space.

We say that an operator $U \in \text{Op}(V_I, Y_K)$ is *unitary* if $U^\times U$ and $UU^\times$ are defined and $U^\times U = 1_V, UU^\times = 1_Y$, the identity operators on $V, Y$, respectively. We emphasize that a unitary operator need *not* be a homomorphism, in fact it is a rather weak notion, and it is insufficient for group representations. Indeed, given a group $G$ and a PIP-space $V_I$, a unitary representation of $G$ into $V_I$ should be a homomorphism $g \mapsto U(g)$ from $G$ into some class of unitary operators on $V_I$, that is, one should have $U(g)U(g') = U(gg')$ and $U(g)^\times = U(g^{-1})$ for all $g, g' \in G$. In addition, we expect that the operators $U(g)$ always preserve the structure of the representation space. Therefore, in the present case, $U(g)$ should map compatible vectors into compatible vectors, and this leads us to the notion of isomorphism.

**Definition 3.10.** An operator $A \in \text{Op}(V_I,, Y_K)$ is an *isomorphism* if $A \in \text{Hom}(V_I, Y_K)$ and there is a homomorphism $B \in \text{Hom}(Y_K, V_I)$ such that $BA = 1_V$, $AB = 1_Y$, the identity operators on $V, Y$, respectively.

Of course, every isomorphism is a monomorphism as well.

To give an example, take $V = L^2_{\rm loc}(\mathbb{R}^n, dx)$, the space of locally square integrable functions, with $n \geqslant 2$. Let $R$ be the map $(Rf)(x) = f(\rho^{-1}x)$, where $\rho \in {\rm SO}(n)$ is an orthogonal transformation of $\mathbb{R}^n$. Then $R$ is an isomorphism of $L^2_{\rm loc}(\mathbb{R}^n, dx)$ onto itself, $R^\times$ is one, too, but $\mathsf{j}(R)$ does not contain the diagonal of $I \times I$. For instance, the assaying subspace $V_r = L^2(r)$, $r^{\pm 1} \in L^\infty$, is invariant under $R$ only if the weight function $r$ is rotation invariant. Note that, in addition, $R$ is unitary, in the sense that both $R^\times R$ and $RR^\times$ are defined and equal $1_V$.

Combining the two notions, we arrive at a good definition of a unitary group representation.

**Definition 3.11.** Let $G$ be a group and $V_I$ a PIP-space. A *unitary representation* of $G$ into $V_I$ is a homomorphism of $G$ into the unitary isomorphisms of $V_I$.

It is worth emphasizing that the two notions of monomorphism and epimorphism are *not* sufficient for defining isomorphisms. Instead, we need that of (co)retraction. Let $A \in {\rm Hom}(V_I, Y_K)$. Then $A$ is called a *coretraction* if there is a homomorphism $B \in {\rm Hom}(Y_K, V_I)$ such that $BA = 1_V$. Dually, $A$ is called a *retraction* if there is a $B \in {\rm Hom}(Y_K, V_I)$ such that $AB = 1_Y$. The interest of these notions lies in the following result [11].

**Proposition 3.12.**

(i) *If $A \in {\rm Hom}(V_I, Y_K)$ is a coretraction and is also an epimorphism, then it is an isomorphism.*

(ii) *If $A \in {\rm Hom}(V_I, Y_K)$ is a retraction and is also a monomorphism, then it is an isomorphism.*

(iii) *If $A$ is both a retraction and a coretraction, then it is an isomorphism.*

**3.3.4. Orthogonal projections.** In the case of a Hilbert space, projection operators play a fundamental role. Thanks to the bijection between orthogonal projections and orthocomplemented subspaces, they provide the correct notion of "subobjects", namely closed (Hilbert) subspaces. As such, they are essential for the study of group representations and operator algebras (von Neumann algebras). A similar situation may be achieved in PIP-spaces. In the general case [5], the partial inner product is *not* required to be positive definite, but it is supposed to be nondegenerate. Here we will restrict ourselves to the case of a LHS/LBS.

**Definition 3.13.** An *orthogonal projection* on a nondegenerate indexed PIP-space $V_I$ is a homomorphism $P \in {\rm Hom}(V_I)$ such that $P^2 = P^\times = P$. A similar definition applies in the case of a LBS or a LHS $V_I$.

It follows immediately from the definition that an orthogonal projection $\mathsf{j}(P)$ contains the diagonal $I \times I$, or still that $P$ leaves every assaying subspace invariant. Equivalently, $P$ is an orthogonal projection if $P$ is an idempotent operator (that is, $P^2 = P$) such that $\{Pf\}^\# \supseteq \{f\}^\#$ for every $f \in V$ and $\langle g|Pf\rangle = \langle Pg|f\rangle$ whenever $f \# g$.

On the other hand, assume we have a direct sum decomposition of $V$ into two subspaces, $V = W \oplus Z$, meaning $W \cap Z = \{0\}, W + Z = V$. For any $f \in V$, write its (unique) decomposition as $f = f_W + f_Z$, $f_W \in W, f_Z \in Z$. Then we say that $W$ is an *orthocomplemented* subspace if there exists a vector subspace $Z \subseteq V$ such that $V = W \oplus Z$ and

(i) $\{f\}^{\#} = \{f_W\}^{\#} \cap \{f_Z\}^{\#}$ for every $f \in V$;
(ii) if $f \in W, g \in Z$ and $f\#g$, then $\langle f|g\rangle = 0$.

Condition (i) means that the compatibility # can be recovered from its restriction to $W$ and $Z$.

Armed with these two notions, we may now state the fundamental result.

**Proposition 3.14.** *A vector subspace $W$ of the nondegenerate* PIP*-space $V$ is orthocomplemented if and only if it is the range of an orthogonal projection:*

$$W = PV \text{ and } V = W \oplus W^{\perp} = PV \oplus (1-P)V.$$

In fact, this proposition (which applies also in the nonpositive case) may be reformulated in topological terms ([5, Sec. 3.4.2]). Things simplify in the case of a LHS/LBS $V_I = \{V_r\}$. For each $r \in I$, write $W_r = W \cap V_r$ and $W_{\overline{r}} = W \cap V_{\overline{r}}$. Then, if $W$ is orthocomplemented, it follows that, for each $r \in I$, $W_r = P_{rr}V_r$ is a closed subspace of $V_r$ with dual $W_{\overline{r}}$ and $V_r = W_r \oplus W_r^{\perp}$, but vectors of $W_r$ need *not* be compatible with those of $W_r^{\perp}$. If they are, they are mutually orthogonal.

The following result is remarkable.

**Proposition 3.15.** *A finite-dimensional vector subspace $W$ of the nondegenerate* PIP*-space $V$ is orthocomplemented if and only if $W \cap W^{\perp} = \{0\}$ and $W \subset V^{\#}$.*

Of course, if the partial inner product is definite (i.e., $f\#f$ and $\langle f|f\rangle = 0$ imply $f = 0$), the condition $W \cap W^{\perp} = \{0\}$ is superfluous.

We conclude this section by an example, actually the same as before, in $V = L^1_{\rm loc}(X, d\mu)$. Take again any partition of $X$ into two measurable subsets, of nonzero measure, $X = \Omega \cup \Omega'$. Then $V$ is decomposed in two orthocomplemented subspaces:

$$V = L^1_{\rm loc}(\Omega,\, d\mu_\Omega) \oplus L^1_{\rm loc}(\Omega', d\mu_{\Omega'}),$$

where $\mu_\Omega, \mu_{\Omega'}$ are the restrictions of $\mu$ to $\Omega$, resp. $\Omega'$. The orthogonal projection $P_\Omega$ is the operator of multiplication by the characteristic function $\chi_\Omega$ of $\Omega$. Similarly, $P_{\Omega'} = 1 - P_\Omega$ is the operator of multiplication by $\chi_{\Omega'}$. We notice that the projection $P_\Omega$ essentially coincides with the epimorphism $P^{(\Omega)}$ introduced above. In this particular case, of course, we have that $P_\Omega f \# P_{\Omega'} f$ for every $f \in L^1_{\rm loc}(X, d\mu)$.

The same discussion can be made about the LHS of weighted Hilbert spaces $L^1_{\rm loc}(X, d\mu) = \{L^2(r)\}$.

## 4. Examples of operators

### 4.1. General operators

**4.1.1. Regular linear functionals.** Let $V$ be arbitrary. Notice that $\mathbb{C}$, with the obvious inner product $\langle \xi|\eta\rangle = \overline{\xi}\eta$, is a Hilbert space. Hence both $\mathrm{Op}(\mathbb{C}, \mathrm{V})$ and $\mathrm{Op}(\mathrm{V}, \mathbb{C})$ are well defined and anti-isomorphic to each other. Elements of $\mathrm{Op}(\mathrm{V}, \mathbb{C})$

are called *regular linear functionals* on $V$. They are given precisely by the functionals

$$\langle g| : f \mapsto \langle g|f\rangle, \quad f \in \{g\}^{\#} = \bigcup\{V_r : r \in \overline{\mathsf{j}(g)}\}.$$

Indeed, for every $f \in V$, define a map $|f\rangle : \mathbb{C} \to V$ by $|f\rangle : \xi \mapsto \xi f$. The correspondence $f \mapsto |f\rangle$ is a linear bijection between $V$ and $\mathrm{Op}(\mathbb{C}, V)$. Clearly, $\mathsf{j}(|f\rangle) = \{e\} \times \mathsf{j}(f)$, where $e$ denotes the unique element of the index set of the PIP-space $\mathbb{C}$. Then, the adjoint of $|f\rangle$ is $\langle f|$ as defined above. Hence $\mathsf{j}(\langle f|) = \overline{\mathsf{j}(f)} \times \{e\}$

**4.1.2. Dyadics.** Using the fact that $\langle g| \in \mathrm{Op}(V, \mathbb{C})$ is defined exactly on $\{g\}^{\#}$, one verifies that:

- The product $(\langle g|)(|f\rangle)$ is defined if and only if $g\#f$, and equals $\langle g|f\rangle$. This follows also from the condition (3.1) on the partial multiplication, since $\mathsf{i}(|f\rangle) \cap \mathsf{d}(\langle g|) = \mathsf{j}(f) \cap \overline{\mathsf{j}(g)}$.
- The product $P_{fg} := |f\rangle\langle g|$ is always defined; it is an element of $\mathrm{Op}(V)$, called a *dyadic*. Its action is given by:

$$|f\rangle\langle g|\,(h) = \langle g|h\rangle f, \quad h \in \{g\}^{\#}.$$

The adjoint of $|f\rangle\langle g|$ is $|g\rangle\langle f|$. One constructs in the same way operators between different spaces and finite linear combinations of dyadics.

Concerning the domain of a dyadic, we have $\mathsf{j}(P_{fg}) = \{(r, u) : g \in V_{\overline{r}}\} = \overline{\mathsf{j}(g)} \times I$. Thus $\overline{\mathsf{j}(P_{fg})} = \mathsf{j}(g) \times I$ and $\overline{\mathsf{j}(P_{fg})} \cap \mathsf{j}(P_{fg}) = \big(\mathsf{j}(g) \cap \overline{\mathsf{j}(g)}\big) \times I$. From this we conclude immediately that $P_{fg}$ is a homomorphism if and only if $\overline{\mathsf{j}(g)} \cap \mathsf{j}(g) = I$, that is, $f$ and $g$ belong to $V^{\#}$.

Now, taking $f = g$, we obtains the projector $P_f$ on the one-dimensional subspace generated by $f$. This corroborates Proposition 3.15 and the fact that the orthocomplemented subspaces, that is, the range of orthogonal projections, are pip-subspaces ("subobjects" in the category **PIP**). The converse is true in the case of a LBS/LHS but, in a general indexed pip-space, there might be subobjects which are not orthocomplemented [6].

**4.1.3. Matrix elements.** One can continue in the same vein and define matrix elements of an operator $A \in \mathrm{Op}(V)$: The matrix element $\langle f|A|h\rangle$ is defined whenever $\big(\mathsf{j}(f) \times \overline{\mathsf{j}(h)}\big) \cap \mathsf{j}(A)$ is nonempty. However, one has to be careful, there might be pairs of vectors $h, f$ such that $\langle f|Ah\rangle$ is defined, but $\langle f|A|h\rangle$ is not. This may happen, for instance, if $Ah$ is defined and $Ah = 0$. We see here at work the definition of the product of three operators.

### 4.2. Integral operators on $L^p$ spaces

In this section, we discuss some properties of integral operators of the form

$$(A_K f)(x) = \int_0^1 K(x, y) f(y)\, dy$$

acting on the LBS (1.2) of $L^p$-spaces on the interval [0,1]. We let the kernel $K$ belong to different spaces of functions on the square $Q = [0,1] \times [0,1]$. The nature of the operator $A_K$ clearly depends on the behavior of the kernel $K$.

The simplest situation occurs when $K$ is an essentially bounded function on $Q$. Assume indeed that $\operatorname{ess\,sup}_{(x,y)\in Q} |K(x,u)| = M < \infty$. Then, if $f \in L^p([0,1])$, $|(A_K f)(x)| \leqslant M\|f\|_p$ a.e., which means that $A_K$ maps every $L^p$ into $L^\infty([0,1])$. $A_K$ is totally regular, that is, it maps every $L^p$-space continuously into itself.

Let us now suppose that $K$ has the property

$$\|K\|_{(s,r)} := \left(\int_0^1 \left(\int_0^1 |K(x,y)|^r dy\right)^{s/r} dx\right)^{1/s} < \infty$$

for some $r, s \geqslant 1$.

This condition is satisfied, in particular, if $K(x,y) \in L^m(Q)$, $m \geqslant 1$, since by Fubini's theorem the integral $\int_0^1 |K(x,y)|^m dy$ exists for almost every $x \in [0,1]$ and

$$\int_0^1 \left[\int_0^1 |K(x,y)|^m dy\right] dx = \int_Q |K(x,y)|^m dx dy.$$

Coming back to the general case, define

$$k(x) = \left[\int_0^1 |K(x,y)|^r dy\right]^{1/r}.$$

Then $k(x)$ is finite for almost every $x \in [0,1]$, $k(\cdot) \in L^s([0,1])$ and $\|k\|_s = \|K\|_{(s,r)}$. Let now $f \in L^{\overline{r}}([0,1])$. The integral

$$\int_0^1 K(x,y) f(y) dy$$

is defined for all $x$ such that $k(x)$ is finite. We check that the function

$$g(x) = \int_0^1 K(x,y) f(y) dy$$

belongs to $L^s([0,1])$. Indeed, by the Hölder inequality we have

$$\left|\int_0^1 K(x,y) f(y) dy\right|^s \leqslant \left(\int_0^1 |K(x,y)|^r dy\right)^{s/r} \cdot \left(\int_0^1 |f(y)|^{\overline{r}} dy\right)^{s/\overline{r}} = k(x)^s \|f\|_{\overline{r}}^s$$

and

$$\|g\|_s^s = \int_0^1 \left|\int_0^1 K(x,y) f(y) dy\right|^s dx \leqslant \int_0^1 k^s(x) dx \cdot \|f\|_{\overline{r}}^s.$$

In conclusion, the operator

$$(A_K f)(x) = \int_0^1 K(x,y) f(y) dy, \qquad f \in L^{\overline{r}}([0,1]), r \geqslant \overline{m},$$

is a bounded linear operator from $L^{\overline{r}}([0,1])$ into $L^s([0,1])$ and, *a fortiori*, from $L^m([0,1])$ into $L^q([0,1])$, whenever $m \geqslant \overline{r}$ and $q \leqslant s$. Let now

$$\gamma(K) := \{r \geqslant 1;\ \|K\|_{(s,r)} < \infty \text{ for some } s \geqslant 1\} \quad\text{and}\quad r_o := \sup \gamma(K).$$

Then

$$\bigcup_{q > \overline{r}_o} L^q([0,1]) \subseteq D(A_K).$$

If $r_o \in \gamma(K)$, then $L^{r_o}([0,1]) \subset D(A_K)$ too.

Assume that $r_o = \infty$ with respect to $y$ for almost every $x \in [0,1]$ (this means that $K$ belongs to the Arens algebra $L^\omega([0,1])$ with respect to $x$). In this case, $D(A_K) = L^1([0,1]) = \bigcup_{p \geqslant 1} L^p([0,1])$, so that $A_K$ is everywhere defined, but it need not be a homomorphism.

However, the following elementary example shows that, in general, $A_K$ is not everywhere defined on $L^1([0,1])$. Take $K(x,y) = x^{-1/2}y^{-1/3}, (x,y) \in Q, xy \neq 0$. Then $\|K\|_{(s,r)} < \infty$ for $1 \leqslant r < 3$ and $1 \leqslant s < 2$. Now consider the function $f(x) = x^{-2/3}$. Then $f \in L^1([0,1])$, but

$$(A_K f)(x) = \int_0^1 \frac{1}{x^{1/2}y^{1/3}} f(y)dy = \int_0^1 \frac{1}{x^{1/2}y} dy = \infty, \ \forall x \in (0,1].$$

Finally, assume that $K \in L^\omega(Q)$. In this case $K \in L^m(Q)$, for every $m \geqslant 1$. From the previous discussion, it follows that $A_K$ is a bounded linear operator from $L^p([0,1])$ into $L^m([0,1])$ for every $m$ such that $\overline{m} \leqslant p$. Since this is true for every $m \geqslant 1$, it follows that $A_K$ is totally regular and a homomorphism.

A complete characterization of the domain of $A_K$, in the general case, goes beyond the framework of this paper and we hope to discuss it again in a future work.

## 5. LHS generated by O*-algebras

Up to now, we have studied operators on a PIP-space. Now we want to invert the perspective and explain how a PIP-space can be generated by a *-algebra of (unbounded) operators on a Hilbert space.

Let $\mathcal{H}$ be a complex Hilbert space and $\mathcal{D}$ a dense subspace of $\mathcal{H}$. We denote by $\mathcal{L}^\dagger(\mathcal{D},\mathcal{H})$ the set of all (closable) linear operators $A$ such that $D(A) = \mathcal{D},\ D(A^*) \supseteq \mathcal{D}$. The map $A \mapsto A^\dagger = A^* \upharpoonright \mathcal{D}$ defines an involution on $\mathcal{L}^\dagger(\mathcal{D},\mathcal{H})$, which can be made into a partial *-algebra with respect to the so-called weak multiplication [4]; however, this fact will not be used in this paper. A subset $\mathcal{O}$ of $\mathcal{L}^\dagger(\mathcal{D},\mathcal{H})$ is called an O-family (and an O*-family, if it is stable under involution).

Let $\mathcal{L}^\dagger(\mathcal{D})$ be the subspace of $\mathcal{L}^\dagger(\mathcal{D},\mathcal{H})$ consisting of all elements which leave, together with their adjoints, the domain $\mathcal{D}$ invariant. Then $\mathcal{L}^\dagger(\mathcal{D})$ is a *-algebra with respect to the usual operations. A *-subalgebra $\mathfrak{M}$ of $\mathcal{L}^\dagger(\mathcal{D})$ is called an O*-algebra.

Let $\mathcal{O}$ be an O*-family in $\mathcal{L}^\dagger(\mathcal{D}, \mathcal{H})$. The *graph topology* $t_{\mathcal{O}}$ on $\mathcal{D}$ is the locally convex topology defined by the family $\{\|\cdot\|, \|\cdot\|_A;\ A \in \mathcal{O}\}$ of seminorms, where $\|\xi\|_A := \|A\xi\|$, $\xi \in \mathcal{D}$. If the locally convex space $\mathcal{D}[t_{\mathcal{O}}]$ is complete, then $\mathcal{O}$ is said to be *closed.* See our monograph [4, Sec. 2.2] or [13] for a detailed discussion.

An O*-algebra, or even an O-family $\mathcal{O}$, generates also a RHS having $\mathcal{D}$ as smallest space. More precisely, let $\mathcal{O}$ be an O-family on $\mathcal{D}$. For any $A \in \mathcal{O}$, we write $R_A = 1 + A^*\overline{A}$, where $\overline{A}$ is the closure of $A$. Each $R_A$ is a self-adjoint, invertible operator, with bounded inverse. The graph topology $\mathsf{t}_{\mathcal{O}}$ on $\mathcal{D}$ can also be defined by the family of norms

$$f \in \mathcal{D} \mapsto \|(1 + A^*\overline{A})^{1/2} f\| = \|R_A^{1/2} f\|, \quad A \in \mathcal{O}.$$

Let $\mathcal{D}^\times$ be the conjugate dual of $\mathcal{D}[\mathsf{t}_{\mathcal{O}}]$, endowed with the strong dual topology $\mathsf{t}_{\mathcal{O}}^\times$. The RHS

$$\mathcal{D}[\mathsf{t}_{\mathcal{O}}] \hookrightarrow \mathcal{H} \hookrightarrow \mathcal{D}^\times[\mathsf{t}_{\mathcal{O}}^\times],$$

where $\hookrightarrow$ denotes a continuous embedding with dense range, will be called the RHS *associated to* $\mathcal{O}$.

As is customary, the domain $D(\overline{A})$ of the closure of $A$ can be made into a Hilbert space, to be denoted by $\mathcal{H}(R_A)$, when it is endowed with the graph norm $\|f\|_{R_A} := \|R_A^{1/2} f\|$. Then $D(\overline{A}) = D(R_A^{1/2}) = Q(R_A)$, the form domain of $R_A$.

The conjugate dual of $\mathcal{H}(R_A)$, with respect to the inner product of $\mathcal{H}$, is (isomorphic to) the completion of $\mathcal{H}$ in the norm $\|R_A^{-1/2} \cdot \|$; we denote it by $\mathcal{H}(R_A^{-1})$. Thus we have

$$\mathcal{H}(R_A) \hookrightarrow \mathcal{H} \hookrightarrow \mathcal{H}(R_A^{-1}). \tag{5.1}$$

The operator $R_A^{1/2}$ is unitary from $\mathcal{H}(R_A)$ onto $\mathcal{H}$, and from $\mathcal{H}$ onto $\mathcal{H}(R_A^{-1})$. Hence $R_A$ is the Riesz unitary operator mapping $\mathcal{H}(R_A)$ onto its conjugate dual $\mathcal{H}(R_A^{-1})$, and similarly $R_A^{-1}$ from $\mathcal{H}(R_A^{-1})$ onto $\mathcal{H}(R_A)$.

From (5.1) it follows that for every $A \in \mathcal{O}$, $\mathcal{H}(R_A)$ and $\mathcal{H}(R_A^{-1})$ are interspaces in the sense of Definition 5.4.4 of [5].

First of all, we show that, to any such family $(\mathcal{O}, \mathcal{D})$, there corresponds a canonical LHS. In a standard fashion, the spaces $\mathcal{H}(R_A)$ generate, by set inclusion and vector sum, a lattice of Hilbert spaces, all dense in $\mathcal{H}$. For any $A, B \in \mathcal{O}$, let us define:

$$\begin{aligned} R_{A\wedge B} &:= R_A \dotplus R_B, \\ R_{A\vee B} &:= (R_A^{-1} \dotplus R_B^{-1})^{-1}, \end{aligned} \tag{5.2}$$

where $\dotplus$ denotes the form sum, so that the operators on the left-hand side are indeed self-adjoint.

*Remark* 5.1. The left-hand side of the preceding equations should be read as symbols for denoting the right-hand side: indeed, this does not mean that there exist operators $A \wedge B, A \vee B$ defined on $\mathcal{D}$ for which the required equalities hold.

For the corresponding Hilbert spaces, one has:

$$\begin{aligned} \mathcal{H}(R_{A\wedge B}) &= \mathcal{H}(R_A) \cap \mathcal{H}(R_B), \\ \mathcal{H}(R_{A\vee B}) &= \mathcal{H}(R_A) + \mathcal{H}(R_B), \end{aligned} \tag{5.3}$$

where the first space carries the projective norm, the second the inductive norm. In addition, the norms corresponding to $R_A$ and $R_B$ are consistent on $\mathcal{H}(R_{A\wedge B})$, since the operators $R_A, R_B$ are closed.

Doing the same with the dual spaces $\mathcal{H}(R_A^{-1})$, one gets another lattice, dual to the first one. The conjugate duals of the spaces (5.3) are, respectively:

$$\begin{aligned} \mathcal{H}(R_{A\wedge B}^{-1}) &= \mathcal{H}(R_A^{-1}) + \mathcal{H}(R_B^{-1}), \\ \mathcal{H}(R_{A\vee B}^{-1}) &= \mathcal{H}(R_A^{-1}) \cap \mathcal{H}(R_A^{-1}). \end{aligned} \tag{5.4}$$

We will denote by $\mathcal{R}$ the set of all positive self-adjoint operators $R_A^{\pm 1}$:

$$\mathcal{R} = \mathcal{R}(\mathcal{O}) := \{R_A^{\pm 1} : A \in \mathcal{O}\}.$$

**Definition 5.2.** Given the O-family $\mathcal{O}$ on $\mathcal{D}$ and the corresponding set of (Riesz) operators $\mathcal{R} = \mathcal{R}(\mathcal{O})$, we define $\Sigma_{\mathcal{R}}$ as the minimal set of self-adjoint operators containing $\mathcal{R}$ and satisfying the following conditions:

(c1) for every $R \in \Sigma_{\mathcal{R}}$, $R^{-1} \in \Sigma_{\mathcal{R}}$;

(c2) for every $R, S \in \Sigma_{\mathcal{R}}$, $R \dotplus S \in \Sigma_{\mathcal{R}}$.

Then the set $\Sigma_{\mathcal{R}}$ is said to be an *admissible cone of self-adjoint operators* if, in addition,

(c3) $\mathcal{D}$ is dense in every $\mathcal{H}(R), R \in \Sigma_{\mathcal{R}}$.

In particular, all the operators $R_{A\wedge B}^{\pm 1}, R_{A\vee B}^{\pm 1}$ belong to $\Sigma_{\mathcal{R}}$, so that every element $R \in \Sigma_{\mathcal{R}}$ is the Riesz operator of the dual pair of Hilbert spaces $\mathcal{H}(R), \mathcal{H}(R^{-1})$. Note also that the norms corresponding to any $R, S \in \Sigma_{\mathcal{R}}$ are consistent, since all operators in $\Sigma_{\mathcal{R}}$ are closed.

Then the family $\Sigma_{\mathcal{R}}$ obtained in this way generates an involutive lattice of Hilbert spaces $\mathcal{I}(\Sigma_{\mathcal{R}})$ indexed by self-adjoint operators. We have the following picture:

$$\mathcal{D} \subseteq V^{\#} = \bigcap_{R\in\Sigma_{\mathcal{R}}} \mathcal{H}(R) = \bigcap_{A\in\mathcal{O}} \mathcal{H}(R_A) \subset \Big\langle \mathcal{H}(R_A), A \in \mathcal{O} \Big\rangle \subset \mathcal{H} \subset \cdots \tag{5.5}$$

$$\cdots \subset \Big\langle \mathcal{H}(R_A^{-1}), A \in \mathcal{O} \Big\rangle \subset V := \sum_{A\in\mathcal{O}} \mathcal{H}(R_A^{-1}) = \sum_{R\in\Sigma_{\mathcal{R}}} \mathcal{H}(R),$$

where $\langle \mathcal{H}(R_A), A \in \mathcal{O} \rangle$ denotes the lattice generated by the operators $R_A$ according to the rules (5.3), and similarly for the other one. Actually, this lattice is peculiar, in the sense that each space $\mathcal{H}(R_A)$ is contained in $\mathcal{H}$ and each space $\mathcal{H}(R_A^{-1})$ contains $\mathcal{H}$.

*Remark* 5.3. The space $\mathcal{H}(R_A)$ does not determine the operator $R_A$, or $A$, uniquely. Indeed one sees easily that $\mathcal{H}(R_A) = \mathcal{H}(R_B)$ wherever $R_A^{1/2} R_B^{-1/2}$ is bounded with bounded inverse.

Following the definition given in Section 2.1, the lattice $\mathcal{I}(\Sigma_{\mathcal{R}}) = \{\mathcal{H}(R), R \in \Sigma_{\mathcal{R}}\}$ is a LHS with central Hilbert space $\mathcal{H}$ and total space $V = \sum_{A \in \mathcal{O}} \mathcal{H}(R_A^{-1})$. Thus we may state

**Theorem 5.4.** *Let $\mathcal{O}$ be a family of closable linear operators on a Hilbert space $\mathcal{H}$, with common dense domain $\mathcal{D}$. Assume that the corresponding set $\Sigma_{\mathcal{R}}$ is an admissible cone. Let $\mathcal{I}(\Sigma_{\mathcal{R}})$ be the lattice of Hilbert spaces generated by $\mathcal{O}$, as in Eq.* (5.5). *Then*

(i) *$\mathcal{O}$ generates a* PIP*-space, with central Hilbert space $\mathcal{H}$ and total space $V = \sum_{A \in \mathcal{O}} \mathcal{H}(R_A^{-1})$, where $\mathcal{H}(R_A^{-1})$ is the completion of $\mathcal{H}$ with respect to the norm $\|(1 + A^*\overline{A})^{-1/2} \cdot \|$. The compatibility is*

$$f\#g \iff \exists R \in \Sigma_{\mathcal{R}} \ \text{ such that } f \in \mathcal{H}(R),\ g \in \mathcal{H}(R^{-1}) \tag{5.6}$$

*and the partial inner product is*

$$\langle R^{1/2} f | R^{-1/2} g \rangle_{\mathcal{H}}. \tag{5.7}$$

(ii) *The lattice $\mathcal{I}(\Sigma_{\mathcal{R}})$ itself is a LHS, with central Hilbert space $\mathcal{H}$, with respect to the compatibility* (5.6)*and the partial inner product* (5.7) *inherited from the* PIP*-space of (i).*

(iii) *One has $V^\# = \bigcap_{A \in \mathcal{O}} \mathcal{H}(R_A)$, where $\mathcal{H}(R_A)$ is $\mathcal{D}(\overline{A})$ with the graph norm $\|(1 + A^*\overline{A})^{1/2} \cdot \|$, and $\mathcal{D} \subseteq V^\#$.*

By this construction, the space $V^\#$ acquires a natural topology $\mathrm{t}_{\mathcal{R}}$, as the projective limit of all the spaces $\mathcal{H}(R), R \in \mathcal{R}$ (or, equivalently, $R \in \Sigma_{\mathcal{R}}$). With this topology, $V^\#$ is complete and semi-reflexive, with dual $V$. However the Mackey topology $\tau(V^\#, V)$ may be strictly finer than the projective topology. On the dual $V$, on the contrary, the topology of the inductive limit of all the $\mathcal{H}(R), R \in \mathcal{R}$, coincides with both $\tau(V, V^\#)$ and $\beta(V, V^\#)$, i.e., $V$ is barreled.

If the family $\mathcal{O} \equiv \{A_i\}$ is finite, then $V^\#$ is the Hilbert space $\mathcal{H}(R_C)$ with $R_C = 1 + \sum_i A_i^* \overline{A_i}$. If $\mathcal{O}$ is countable, $V^\#$ is a reflexive Fréchet space (and then $\tau(V^\#, V)$ coincides with $t_R$). Otherwise $V^\#$ is nonmetrizable.

The most interesting case arises when we start with a $^*$-*invariant* family $\mathcal{O}$ of closable operators, with a common dense *invariant* domain $\mathcal{D}$. For then $\mathcal{O}$ generates a $^*$-algebra $\mathfrak{M}$ of operators on $\mathcal{D}$, i.e., an O$^*$-algebra. We equip $\mathcal{D}$ with the graph topology $\mathrm{t}_{\mathfrak{M}}$ defined by $\mathfrak{M}$. Then all the operators $A \in \mathfrak{M}$ are continuous from $\mathcal{D}$ into $\mathcal{D}$. The domain $\mathcal{D}$ need not be complete in the topology $\mathrm{t}_{\mathfrak{M}}$. In any case, its completion $\widetilde{\mathcal{D}}(\mathfrak{M})$ coincides with the full closure $\widehat{\mathcal{D}}(\mathfrak{M}) := \bigcap_{A \in \mathfrak{M}} D(\overline{A})$, and we have $V^\# = \widehat{\mathcal{D}}(\mathfrak{M})$. In other words, we can assume from the beginning that the $^*$-algebra $\mathfrak{M}$ is fully closed, i.e., $\mathcal{D} = \widehat{\mathcal{D}}(\mathfrak{M})$.

**Theorem 5.5.** *Let $\mathfrak{M}$ be an $O^*$-algebra on the dense domain $\mathcal{D} \subset \mathcal{H}$. Then $\mathfrak{M}$ generates a* PIP*-space structure and a LHS structure on $V = \sum_{A \in \mathfrak{M}} \mathcal{H}(R_A^{-1})$. The subspace $V^\# = \bigcap_{A \in \mathfrak{M}} \mathcal{H}(R_A)$ is the completion $\widehat{\mathcal{D}}(\mathfrak{M})$ of $\mathcal{D}$ in the $\mathfrak{M}$-topology and*

*$\widehat{\mathfrak{M}} \subseteq \mathrm{Reg}(V)$, where $\widehat{\mathfrak{M}}$ is the full closure of $\mathfrak{M}$ and $\mathrm{Reg}(V)$ denotes the set of regular operators on $V$.*

Now, we ask the following question: Given a LHS $V_I := (V, \mathcal{I}, \langle\cdot|\cdot\rangle)$, under which conditions does there exist an O-family $\mathcal{O}$ on $V^\#$ such that $V_I := (V, \mathcal{I}, \langle\cdot|\cdot\rangle)$ coincides with the LHS generated by $\mathcal{O}$? The following statement generalizes a result by Bellomonte and one of us (CT) for inductive limits of contractive families of Hilbert spaces [7].

**Theorem 5.6.** *Let $V_I := \{V_r\}$ be a LHS such that every $r$ is comparable with $o$. Then the following statements hold true.*

(i) *For every $r \in I$, there exists a linear operator $A_r$ with domain $V^\#$, closable in $V_o$ (the central Hilbert space) such that $V_r$ is the completion $\mathcal{H}(R_{A_r})$ of $V^\#$ with respect to the norm $\|\xi\|_{A_r} = \|(I + A_r^*\overline{A}_r)^{1/2}\xi\|_o$, $\xi \in V^\#$.*

(ii) *The family $\mathcal{O} = \{A_r, r \in I, r \geqslant 0\}$ is directed upward by $I$ (i.e., $o \leqslant r \leqslant s \Leftrightarrow A_r \subseteq A_s$).*

(iii) *$V^\# = \bigcap_{r\in\mathcal{I}} \mathcal{H}(R_{A_r})$ and $V = \bigcup_{r\in I} R_{A_r}\mathcal{H}(R_{A_r})$ is the conjugate dual of $V^\#$ for the graph topology $\mathsf{t}_{\mathcal{O}}$. The inductive topology on $V$ coincides with the Mackey topology $\tau(V, V^\#)$.*

*Proof.* (i): Since, for $r \geqslant o$, $V_r \subset V_o$, the inner product $\langle\cdot|\cdot\rangle_r$ of $V_r$ can be viewed as a closed positive sesquilinear form defined on $V_r \times V_r \subset V_o \times V_o$ (up to an isomorphism) which, as a quadratic form on $V_r$, has 1 as greatest lower bound. Then there exists a selfadjoint operator $B_r$ with dense domain $D(B_r)$ in $V_o$, with $B_r \geqslant 1$, such that $D(B_r) = V_r$ and

$$\langle\xi|\eta\rangle_r = \langle B_r\xi|B_r\eta\rangle_o, \quad \forall \xi, \eta \in V_r.$$

Since $V^\#$ is dense in $V_r$, $V^\#$ is a core for $B_r$ and, hence, also for the operator $(B_r^2 - 1)^{1/2}$. We define $A_r = (B_r^2 - 1)^{1/2} \restriction \mathcal{D}$. The proofs of (ii) and (iii) follow then from simple considerations. In particular, the fact that the inductive topology of $V$ coincides with the Mackey topology $\tau(V, V^\#)$ is well known (see [12, Ch. IV, Sec. 4.4] or [5, Sec. 2.3]). □

## 6. Conclusion

Most families of function spaces used in analysis and in signal processing come in scales or lattices and in fact are, or contain, PIP-spaces. The (lattice) indices defining the (partial) order characterize the properties of the corresponding functions or distributions: smoothness, local integrability, decay at infinity, etc. Thus it seems natural to formulate the properties of various operators globally, using the theory of PIP-space operators, in particular the set $\mathsf{j}(A)$ of an operator encodes its properties in a very convenient and visual fashion. In addition, it is often possible to determine uniquely whether a function belongs to one of those spaces simply by estimating the (asymptotic) behavior of its Gabor or wavelet coefficients, a real breakthrough in functional analysis [10].

A legitimate question is whether there are instances where a PIP-space is really *needed*, or a RHS could suffice. The answer is that there are plenty of examples, among the applications enumerated in Chapters 7 and 8 of our monograph [5]. We may therefore expect that the PIP-space formalism will play a significant role in Gabor/wavelet analysis, as well as in mathematical physics.

Concerning the applications in mathematical physics, in almost all cases, the relevant structure is a scale or a chain of Hilbert spaces, which allows a finer control on the behavior of operators. For instance: (i) The description of singular interactions in quantum mechanics; (ii) The formulation of the Weinberg-van Winter approach to quantum scattering theory; (iii) Various aspects of quantum field theory, such as the energy bounds or Nelson's approach to Euclidean field theory. Details and references may be found in [5, Chap. 7].

Coming back to quantum mechanics, the notation for operators introduced in Section 4.1 coincides precisely with the familiar Dirac notation. It allows a rigorous formulation of the latter, more natural that the RHS formulation, and its extension to quantum field theory. In addition, it provides an elegant way of describing very singular operators, thus providing a far-reaching generalization of bounded operators. It allows indeed to treat on the same footing all kinds of operators, from bounded ones to very singular ones. By this, we mean the following, loosely speaking. Take

$$V_r \subset V_o \simeq V_{\overline{o}} \subset V_s \quad (V_o = \text{Hilbert space}).$$

Three cases may arise:

- if $A_{oo}$ exists, then $A$ corresponds to a bounded operator $V_o \to V_o$;
- if $A_{oo}$ does not exist, but only $A_{or} : V_r \to V_o$, with $r < o$, then $A$ corresponds to an unbounded operator, with domain $D(A) \supset V_r$;
- if no $A_{or}$ exists, but only $A_{sr} : V_r \to V_s$, with $r < o < s$, then $A$ corresponds to a singular operator, with Hilbert space domain possibly reduced to $\{0\}$.

The singular interactions ($\delta$ or $\delta'$ potentials) are a beautiful example [5, Sec. 7.1.3].

As for the applications in signal processing, all families of spaces routinely used are, or contain, chains of Banach spaces, which are needed for a fine tuning of elements (usually, distributions) and operators on them. Such are, for instance, $L^p$ spaces, amalgam spaces, modulation spaces, Besov spaces or coorbit spaces. Here again, a RHS is clearly not sufficient. See [5, Chap. 8] for further details.

## References

[1] J.-P. Antoine and A. Grossmann, *Partial inner product spaces I. General properties*, J. Funct. Anal. **23** (1976), 369–378; *II. Operators*, ibid. **23** (1976), 379–391.

[2] J.-P. Antoine and A. Grossmann, *Orthocomplemented subspaces of nondegenerate partial inner product spaces*, J. Math. Phys. **19** (1978), 329–335.

[3] J.-P. Antoine, *Partial inner product spaces III. Compability relations revisited*, J. Math. Physics **21** (1980), 268–279; *IV. Topological considerations*, ibid. **21** (1980), 2067–2079.

[4] J.-P. Antoine, A. Inoue, and C. Trapani, *Partial *-Algebras and Their Operator Realizations*, Kluwer, Dordrecht, 2002.

[5] J.-P. Antoine and C. Trapani, *Partial Inner Product Spaces – Theory and Applications*, Lecture Notes in Mathematics, vol. 1986, Springer-Verlag, Berlin, Heidelberg, 2009.

[6] J.-P. Antoine, D. Lambert and C. Trapani, *Partial inner product spaces: Some categorical aspects*, Adv. in Math. Phys., vol. 2011, art. 957592.

[7] G. Bellomonte and C. Trapani, *Rigged Hilbert spaces and contractive families of Hilbert spaces*, Monatshefte f. Math. Monatshefte f. Math., 164 (2011), 271–285.

[8] J. Bergh and J. Löfström, *Interpolation Spaces*, Springer-Verlag, Berlin, 1976.

[9] G. Köthe, *Topological Vector Spaces, Vols. I, II*, Springer-Verlag, Berlin, 1969, 1979.

[10] Y. Meyer, *Ondelettes et Opérateurs. I*, Hermann, Paris, 1990.

[11] B. Mitchell, *Theory of Categories*, Academic Press, New York, 1965.

[12] H.H. Schaefer, *Topological Vector Spaces*, Springer-Verlag, Berlin, 1971.

[13] K. Schmüdgen, *Unbounded Operator Algebras and Representation Theory*, Akademie-Verlag, Berlin, 1990.

[14] B. Simon, *Distributions and their Hermite expansions*, J. Math. Phys., **12** (1971), 140–148.

Jean-Pierre Antoine
Institut de Recherche en Mathématique et Physique (IRMP)
Université catholique de Louvain
B-1348 Louvain-la-Neuve, Belgium
e-mail: `jean-pierre.antoine@uclouvain.be`

Camillo Trapani
Dipartimento di Matematica e Informatica
Università di Palermo
I-90123 Palermo, Italia
e-mail: `trapani@unipa.it`

Operator Theory:
Advances and Applications, Vol. 221, 47-69


# From Forms to Semigroups

Wolfgang Arendt and A.F.M. ter Elst

**Abstract.** We present a review and some new results on form methods for generating holomorphic semigroups on Hilbert spaces. In particular, we explain how the notion of closability can be avoided. As examples we include the Stokes operator, the Black–Scholes equation, degenerate differential equations and the Dirichlet-to-Neumann operator.

**Mathematics Subject Classification (2000).** Primary 47A07; Secondary 47D06.

**Keywords.** Sectorial form, semigroup.

## Introduction

Form methods give a very efficient tool to solve evolutionary problems on Hilbert space. They were developed by T. Kato [Kat] and, in a slightly different language by J.L. Lions. In this article we give an introduction based on [AE2]. The main point in our approach is that the notion of closability is not needed anymore. In the language of Kato the form merely needs to be sectorial. Alternatively, in the setting of Lions the Hilbert space $V$, on which the form is defined, no longer needs to be continuously embedded in the Hilbert space $H$, on which the semigroup acts. Instead one merely needs a continuous linear map from $V$ into $H$ with dense range.

The new setting is particularly efficient for degenerate equations, since then the sectoriality condition is obvious, whilst the form is not closable, in general, or closability might be hard to verify. The Dirichlet-to-Neumann operator is normally defined on smooth domains, that is, domains with at least a Lipschitz boundary. The new form method allows us to consider the Dirichlet-to-Neumann operator on rough domains. Besides this we give several other examples.

This presentation starts by an introduction to holomorphic semigroups. Instead of the contour argument found in the literature, we give a more direct argument based on the Hille–Yosida Theorem.

## 1. The Hille–Yosida Theorem

A $C_0$*-semigroup* on a Banach space $X$ is a mapping $T\colon (0,\infty) \to \mathcal{L}(X)$ satisfying

$$\begin{aligned} &T(t+s) = T(t)T(s) && (t,s>0) \\ &\lim_{t\downarrow 0} T(t)x = x && (x \in X) \,. \end{aligned}$$

The *generator* $A$ of such a $C_0$-semigroup is defined by

$$D(A) := \{x \in X : \lim_{t\downarrow 0} \frac{T(t)x - x}{t} \text{ exists}\}$$

$$Ax := \lim_{t\downarrow 0} \frac{T(t)x - x}{t} \qquad (x \in D(A)) \,.$$

Thus the domain $D(A)$ of $A$ is a subspace of $X$ and $A\colon D(A) \to X$ is linear. One can show that $D(A)$ is dense in $X$. The main interest in semigroups lies in the associated Cauchy problem

$$\text{(CP)} \qquad \left\{ \begin{aligned} \dot u(t) &= Au(t) \qquad (t>0) \\ u(0) &= x \,. \end{aligned} \right.$$

Indeed, if $A$ is the generator of a $C_0$-semigroup, then given $x \in X$, the function $u(t) := T(t)x$ is the unique *mild* solution of (CP); i.e.,

$$u \in C([0,\infty);X) \,, \quad \int_0^t u(s)\,ds \in D(A)$$

for all $t > 0$ and

$$\begin{aligned} u(t) &= x + A\int_0^t u(s)\,ds \\ u(0) &= x \,. \end{aligned}$$

If $x \in D(A)$, then $u$ is a *classical solution*; i.e., $u \in C^1([0,\infty);X)$, $u(t) \in D(A)$ for all $t \geq 0$ and $\dot u(t) = Au(t)$ for all $t > 0$. Conversely, if for each $x \in X$ there exists a unique mild solution of (CP), then $A$ generates a $C_0$-semigroup [ABHN, Theorem 3.1.12]. In view of this characterization of well-posedness, it is of big interest to decide whether a given operator generates a $C_0$-semigroup. A positive answer is given by the famous Hille–Yosida Theorem.

**Theorem 1.1 (Hille–Yosida (1948)).** *Let $A$ be an operator on $X$. The following are equivalent.*

(i) *$A$ generates a contractive $C_0$-semigroup;*
(ii) *the domain of $A$ is dense, $\lambda - A$ is invertible for all $\lambda > 0$ and $\|\lambda(\lambda - A)^{-1}\| \leq 1$.*

Here we call a semigroup $T$ *contractive* if $\|T(t)\| \leq 1$ for all $t > 0$. By $\lambda - A$ we mean the operator with domain $D(A)$ given by $(\lambda - A)x := \lambda x - Ax \quad (x \in D(A))$. So the condition in (ii) means that $\lambda - A\colon D(A) \to X$ is bijective and $\|\lambda(\lambda - A)^{-1}x\| \leq \|x\|$ for all $\lambda > 0$ and $x \in X$. If $X$ is reflexive, then this existence of the *resolvent* $(\lambda - A)^{-1}$ and the contractivity $\|\lambda(\lambda - A)^{-1}\| \leq 1$ imply already that the domain is dense [ABHN, Theorem 3.3.8].

Yosida's proof is based on the Yosida-approximation: Assuming (ii), one easily sees that

$$\lim_{\lambda \to \infty} \lambda(\lambda - A)^{-1}x = x \qquad (x \in D(A)) ,$$

i.e., $\lambda(\lambda - A)^{-1}$ converges strongly to the identity as $\lambda \to \infty$. This implies that

$$A_\lambda := \lambda A(\lambda - A)^{-1} = \lambda^2(\lambda - A)^{-1} - \lambda$$

approximates $A$ as $\lambda \to \infty$ in the sense that

$$\lim_{\lambda \to \infty} A_\lambda x = Ax \qquad (x \in D(A)) .$$

The operator $A_\lambda$ is bounded, so one may define

$$e^{tA_\lambda} := \sum_{n=0}^{\infty} \frac{t^n}{n!} A_\lambda^n$$

by the power series. Note that $\|\lambda^2(\lambda - A)^{-1}\| \leq \lambda$. Since

$$e^{tA_\lambda} = e^{-\lambda t} e^{t\lambda^2(\lambda - A)^{-1}} ,$$

it follows that

$$\|e^{tA_\lambda}\| \leq e^{-\lambda t} e^{t\|\lambda^2(\lambda - A)^{-1}\|} \leq 1 .$$

The key element in Yosida's proof consists in showing that for all $x \in X$ the family $(e^{tA_\lambda}x)_{\lambda > 0}$ is a Cauchy net as $\lambda \to \infty$. Then the $C_0$-semigroup generated by $A$ is given by

$$T(t)x := \lim_{\lambda \to \infty} e^{tA_\lambda}x \qquad (t > 0)$$

for all $x \in X$. We will come back to this formula when we talk about holomorphic semigroups.

*Remark* 1.2. Hille's independent proof is based on Euler's formula for the exponential function. Note that putting $t = \frac{1}{\lambda}$ one has

$$\lambda(\lambda - A)^{-1} = (I - tA)^{-1} .$$

Hille showed that

$$T(t)x := \lim_{n \to \infty} \left(I - \frac{t}{n}A\right)^{-n} x$$

exists for all $x \in X$, see [Kat, Section IX.1.2].

## 2. Holomorphic semigroups

A $C_0$-semigroup is defined on the real half-line $(0,\infty)$ with values in $\mathcal{L}(X)$. It is useful to study when extensions to a sector

$$\Sigma_\theta := \{re^{i\alpha} : r > 0,\ |\alpha| < \theta\}$$

for some $\theta \in (0, \pi/2]$ exist. In this section $X$ is a complex Banach space.

**Definition 2.1.** A $C_0$-semigroup $T$ is called *holomorphic* if there exist $\theta \in (0, \pi/2]$ and a holomorphic extension

$$\widetilde{T}\colon \Sigma_\theta \to \mathcal{L}(X)$$

of $T$ which is locally bounded; i.e.,

$$\sup_{\substack{z\in\Sigma_\theta \\ |z|\le 1}} \|\widetilde{T}(z)\| < \infty .$$

If $\|\widetilde{T}(z)\| \le 1$ for all $z \in \Sigma_\theta$, then we call $T$ a *sectorially contractive holomorphic* $C_0$-semigroup (*of angle* $\theta$, if we want to make precise the angle).

The holomorphic extension $\widetilde{T}$ automatically has the semigroup property

$$\widetilde{T}(z_1 + z_2) = \widetilde{T}(z_1)\widetilde{T}(z_2) \qquad (z_1, z_2 \in \Sigma_\theta) .$$

Because of the boundedness assumption it follows that

$$\lim_{\substack{z\to 0 \\ z\in\Sigma_\theta}} \widetilde{T}(z)x = x \qquad (x \in X) .$$

These properties are easy to see. Moreover, $\widetilde{T}$ can be extended continuously (for the strong operator topology) to the closure of $\Sigma_\theta$, keeping these two properties. In fact, if $x = T(t)y$ for some $t > 0$ and some $y \in X$, then

$$\lim_{w\to z} T(w)x = \lim_{w\to z} T(w + t)y = T(z + t)y$$

exists. Since the set $\{T(t)y : t \in (0,\infty),\ y \in X\}$ is dense the claim follows. In the sequel we will omit the tilde and denote the extension $\widetilde{T}$ simply by $T$. We should add a remark on vector-valued holomorphic functions.

*Remark* 2.2. If $Y$ is a Banach space and $\Omega \subset \mathbb{C}$ open, then a function $f\colon \Omega \to Y$ is called *holomorphic* if

$$f'(z) = \lim_{h\to 0} \frac{f(z+h) - f(z)}{h}$$

exists in the norm of $Y$ for all $z \in \Omega$ and $f'\colon \Omega \to Y$ is continuous. It follows as in the scalar case that $f$ is analytic. It is remarkable that holomorphy is the same as weak holomorphy (first observed by Grothendieck): A function $f\colon \Omega \to Y$ is holomorphic if and only if

$$y' \circ f \colon \Omega \to \mathbb{C}$$

is holomorphic for all $y' \in Y'$. In our context the space $Y$ is $\mathcal{L}(X)$, the space of all bounded linear operators on $X$ with the operator norm. If the function $f$

is bounded it suffices to test holomorphy with fewer functionals. We say that a subspace $W \subset Y'$ *separates points* if for all $x \in Y$,

$$\langle y', x \rangle = 0 \text{ for all } y' \in W \text{ implies } x = 0 \, .$$

Assume that $f\colon \Omega \to Y$ is bounded such that $y' \circ f$ is holomorphic for all $y' \in W$ where $W$ is a separating subspace of $Y'$. Then $f$ is holomorphic. This result is due to [AN], see also [ABHN, Theorem A7]. In particular, if $Y = \mathcal{L}(X)$, then a bounded function $f\colon \Omega \to \mathcal{L}(X)$ is holomorphic if and only if $\langle x', f(\cdot)x \rangle$ is holomorphic for all $x$ in a dense subspace of $X$ and all $x'$ in a separating subspace of $X'$.

We recall a special form of Vitali's Theorem (see [AN], [ABHN, Theorem A5]).

**Theorem 2.3 (Vitali).** *Suppose $\Omega \subset \mathbb{C}$ is connected. For all $n \in \mathbb{N}$ let $f_n\colon \Omega \to \mathcal{L}(X)$ be holomorphic, let $M \in \mathbb{R}$ and suppose that*

a) $\|f_n(z)\| \le M$ *for all $z \in \Omega$ and $n \in \mathbb{N}$, and;*

b) $\Omega_0 := \{z \in \Omega : \lim_{n\to\infty} f_n(z)x$ *exists for all* $x \in X\}$ *has a limit point in $\Omega$, i.e., there exist a sequence $(z_k)_{k\in\mathbb{N}}$ in $\Omega_0$ and $z_0 \in \Omega$ such that $z_k \neq z_0$ for all $k \in \mathbb{N}$ and $\lim\limits_{k\to\infty} z_k = z_0$.*

*Then*

$$f(z)x := \lim_{n\to\infty} f_n(z)x$$

*exists for all $x \in X$ and $z \in \Omega$, and $f\colon \Omega \to \mathcal{L}(X)$ is holomorphic.*

Now we want to give a simple characterization of holomorphic sectorially contractive semigroups. Assume that $A$ is a densely defined operator on $X$ such that $(\lambda - A)^{-1}$ exists and

$$\|\lambda(\lambda - A)^{-1}\| \le 1 \qquad (\lambda \in \Sigma_\theta) \, ,$$

where $0 < \theta \le \pi/2$. Let $z \in \Sigma_\theta$. Then for all $\lambda > 0$,

$$(zA)_\lambda = zA_{\frac{\lambda}{z}}$$

is holomorphic in $z$. For each $z \in \Sigma_\theta$, the operator $zA$ satisfies Condition (ii) of Theorem 1.1. By the Hille–Yosida Theorem

$$T(z)x := \lim_{\lambda\to\infty} e^{(zA)_\lambda} x$$

exists for all $x \in X$ and $z \in \Sigma_\theta$. Since $z \mapsto e^{(zA)_\lambda} = e^{zA_{\lambda/z}}$ is holomorphic, $T\colon \Sigma_\theta \to \mathcal{L}(X)$ is holomorphic by Vitali's Theorem. If $t > 0$, then

$$T(t) = \lim_{\lambda\to\infty} e^{tA_{\lambda/t}} = T_A(t)$$

where $T_A$ is the semigroup generated by $A$. Since $T_A(t+s) = T_A(t)T_A(s)$, it follows from analytic continuation that

$$T(z_1 + z_2) = T(z_1)T(z_2) \qquad (z_1, z_2 \in \Sigma_\theta) \, .$$

Thus $A$ generates a sectorially contractive holomorphic $C_0$-semigroup of angle $\theta$ on $X$. One sees as above that

$$T_{zA}(t) = T(zt)$$

for all $t > 0$ and $z \in \Sigma_\theta$. We have shown the following.

**Theorem 2.4.** *Let $A$ be a densely defined operator on $X$ and $\theta \in (0, \pi/2]$. The following are equivalent.*

(i) *$A$ generates a sectorially contractive holomorphic $C_0$-semigroup of angle $\theta$;*
(ii) *$(\lambda - A)^{-1}$ exists for all $\lambda \in \Sigma_\theta$ and*

$$\|\lambda(\lambda - A)^{-1}\| \leq 1 \qquad (\lambda \in \Sigma_\theta) \ .$$

We refer to [AEH] for a similar approach to possibly noncontractive holomorphic semigroups.

## 3. The Lumer–Phillips Theorem

Let $H$ be a Hilbert space over $\mathbb{K} = \mathbb{R}$ or $\mathbb{C}$. An operator $A$ on $H$ is called *accretive* or *monotone* if

$$\mathrm{Re}(Ax|x) \geq 0 \qquad (x \in D(A)) \ .$$

Based on this notion the following very convenient characterization is an easy consequence of the Hille–Yosida Theorem.

**Theorem 3.1 (Lumer–Phillips).** *Let $A$ be an operator on $H$. The following are equivalent.*

(i) *$-A$ generates a contraction semigroup;*
(ii) *$A$ is accretive and $I + A$ is surjective.*

For a proof, see [ABHN, Theorem 3.4.5]. Accretivity of $A$ can be reformulated by the condition

$$\|(\lambda + A)x\| \geq \|\lambda x\| \qquad (\lambda > 0,\ x \in D(A)) \ .$$

Thus if $\lambda + A$ is surjective, then $\lambda + A$ is invertible and $\|\lambda(\lambda + A)^{-1}\| \leq 1$. We also say that $A$ is *$m$-accretive* if Condition (ii) is satisfied. If $A$ is $m$-accretive and $\mathbb{K} = \mathbb{C}$, then one can easily see that $\lambda + A$ is invertible for all $\lambda \in \mathbb{C}$ satisfying $\mathrm{Re}\,\lambda > 0$ and

$$\|(\lambda + A)^{-1}\| \leq \frac{1}{\mathrm{Re}\,\lambda} \ .$$

Due to the reflexivity of Hilbert spaces, each $m$-accretive operator $A$ is densely defined (see [ABHN, Proposition 3.3.8]). Now we want to reformulate the Lumer–Phillips Theorem for generators of semigroups which are contractive on a sector.

**Theorem 3.2 (Generators of sectorially contractive semigroups).** *Let $A$ be an operator on a complex Hilbert space $H$ and let $\theta \in (0, \frac{\pi}{2})$. The following are equivalent.*

(i) *$-A$ generates a holomorphic $C_0$-semigroup which is contractive on the sector $\Sigma_\theta$;*
(ii) *$e^{\pm i\theta} A$ is accretive and $I + A$ is surjective.*

*Proof.* (ii)⇒(i). Since $e^{\pm i\theta}A$ is accretive the operator $zA$ is accretive for all $z \in \Sigma_\theta$. Since $(I+A)$ is surjective, the operator $A$ is $m$-accretive. Thus $(\lambda + A)$ is invertible whenever $\operatorname{Re}\lambda > 0$. Consequently $(I+zA) = z(z^{-1}+A)$ is invertible for all $z \in \Sigma_\theta$. Thus $zA$ is $m$-accretive for all $z \in \Sigma_\theta$. Now (i) follows from Theorem 2.4.

(i)⇒(ii). If $-A$ generates a holomorphic semigroup which is contractive on $\Sigma_\theta$, then $e^{i\alpha}A$ generates a contraction semigroup for all $\alpha$ with $|\alpha| \le \theta$. Hence $e^{i\alpha}A$ is $m$-accretive whenever $|\alpha| \le \theta$. □

If $A$ is self-adjoint, then both conditions of Theorem 3.2 are valid for all $\theta \in (0, \frac{\pi}{2})$ and the semigroup is holomorphic on $\Sigma_{\frac{\pi}{2}}$.

## 4. Forms: the complete case

We recall one of our most efficient tools to solve equations, the Lax–Milgram lemma, which is just a non-symmetric generalization of the Riesz–Fréchet representation theorem from 1905.

**Lemma 4.1 (Lax–Milgram (1954)).** *Let $V$ be a Hilbert space over $\mathbb{K}$, where $\mathbb{K} = \mathbb{R}$ or $\mathbb{K} = \mathbb{C}$, and let $a\colon V \times V \to \mathbb{K}$ be sesquilinear, continuous and coercive, i.e.,*

$$\operatorname{Re} a(u) \ge \alpha \|u\|_V^2 \qquad (u \in V)$$

*for some $\alpha > 0$. Let $\varphi\colon V \to \mathbb{K}$ be a continuous anti-linear form, i.e., $\varphi$ is continuous and satisfies $\varphi(u+v) = \varphi(u) + \varphi(v)$ and $\varphi(\lambda u) = \overline{\lambda}\varphi(u)$ for all $u, v \in V$ and $\lambda \in \mathbb{K}$. Then there is a unique $u \in V$ such that*

$$a(u,v) = \varphi(v) \qquad (v \in V)\ .$$

Of course, to say that $a$ is continuous means that

$$|a(u,v)| \le M\|u\|_V\|v\|_V \qquad (u, v \in V)$$

for some constant $M$. We let $a(u) := a(u,u)$ for all $u \in V$.

In general, the range condition in the Hille–Yosida Theorem is difficult to prove. However, if we look at operators associated with a form, the Lax–Milgram Lemma implies automatically the range condition. We describe now our general setting in the complete case. Given is a Hilbert space $V$ over $\mathbb{K}$ with $\mathbb{K} = \mathbb{R}$ or $\mathbb{K} = \mathbb{C}$, and a continuous, coercive sesquilinear form

$$a\colon V \times V \to \mathbb{K}\ .$$

Moreover, we assume that $H$ is another Hilbert space over $\mathbb{K}$ and $j\colon V \to H$ is a continuous linear mapping with dense image. Now we associate an operator $A$ on $H$ with the pair $(a, j)$ in the following way. Given $x, y \in H$ we say that $x \in D(A)$ and $Ax = y$ if there exists a $u \in V$ such that $j(u) = x$ and

$$a(u,v) = (y|j(v))_H \qquad \text{for all } v \in V\ .$$

We first show that $A$ is well defined. Assume that there exist $u_1, u_2 \in V$ and $y_1, y_2 \in H$ such that

$$j(u_1) = j(u_2) \ ,$$

$$a(u_1, v) = (y_1 | j(v))_H \qquad (v \in V), \text{ and,}$$

$$a(u_2, v) = (y_2 | j(v))_H \qquad (v \in V) \ .$$

Then $a(u_1 - u_2, v) = (y_1 - y_2 | j(v))_H$ for all $v \in V$. Since $j(u_1 - u_2) = 0$, taking $v := u_1 - u_2$ gives $a(u_1 - u_2, u_1 - u_2) = 0$. Since $a$ is coercive, it follows that $u_1 = u_2$. It follows that $(y_1 | j(v))_H = (y_2 | j(v))_H$ for all $v \in V$. Since $j$ has dense image, it follows that $y_1 = y_2$.

It is clear from the definition that $A\colon D(A) \to H$ is linear. Our main result is the following generation theorem. We first assume that $\mathbb{K} = \mathbb{C}$.

**Theorem 4.2 (Generation theorem in the complete case).** *The operator $-A$ generates a sectorially contractive holomorphic $C_0$-semigroup $T$. If $a$ is symmetric, then $A$ is selfadjoint.*

*Proof.* Let $M \geq 0$ be the constant of continuity and $\alpha > 0$ the constant of coerciveness as before. Then

$$\frac{|\operatorname{Im} a(v)|}{\operatorname{Re} a(v)} \leq \frac{M \|v\|_V^2}{\alpha \|v\|_V^2} = \frac{M}{\alpha}$$

for all $v \in V \setminus \{0\}$. Thus there exists a $\theta' \in (0, \frac{\pi}{2})$ such that

$$a(v) \in \overline{\Sigma_{\theta'}} \qquad (v \in V) \ .$$

Let $x \in D(A)$. There exists a $u \in V$ such that $x = j(u)$ and $a(u, v) = (Ax | j(v))_H$ for all $v \in V$. In particular, $(Ax|x)_H = a(u) \in \overline{\Sigma_{\theta'}}$. It follows that $e^{\pm i\theta} A$ is accretive where $\theta = \frac{\pi}{2} - \theta'$. In order to prove the range condition, consider the form $b\colon V \times V \to \mathbb{C}$ given by

$$b(u, v) = a(u, v) + (j(u) | j(v))_H \ .$$

Then $b$ is continuous and coercive. Let $y \in H$. Then $\varphi(v) := (y | j(v))_H$ defines a continuous anti-linear form $\varphi$ on $V$. By the Lax–Milgram Lemma 4.1 there exists a unique $u \in V$ such that

$$b(u, v) = \varphi(v) \qquad (v \in V) \ .$$

Hence $(y|j(v))_H = a(u, v) + (j(u)|j(v))_H$; i.e., $a(u, v) = (y - j(u)|j(v))_H$ for all $v \in V$. This means that $x := j(u) \in D(A)$ and $Ax = y - x$. □

The result is also valid in real Banach spaces. If $T$ is a $C_0$-semigroup on a real Banach space $X$, then the $\mathbb{C}$-linear extension $T_{\mathbb{C}}$ of $T$ on the complexification $X_{\mathbb{C}} := X \oplus iX$ of $X$ is a $C_0$-semigroup given by $T_{\mathbb{C}}(t)(x + iy) := T(t)x + iT(t)y$. We call $T$ *holomorphic* if $T_{\mathbb{C}}$ is holomorphic. The generation theorem above remains true on real Hilbert spaces.

In order to formulate a final result we want also allow a rescaling. Let $X$ be a Banach space over $\mathbb{K}$ and $T$ be a $C_0$-semigroup on $X$ with generator $A$. Then for all $\omega \in \mathbb{K}$ and $t > 0$ define

$$T_\omega(t) := e^{\omega t} T(t) \ .$$

Then $T_\omega$ is a $C_0$-semigroup whose generator is $A + \omega$. Using this we obtain now the following general generation theorem in the complete case.

Let $V, H$ be Hilbert spaces over $\mathbb{K}$ and $j\colon V \to H$ continuous linear with dense image. Let $a\colon V \times V \to \mathbb{K}$ be sesquilinear and continuous. We call the form $a$ *j-elliptic* if there exist $\omega \in \mathbb{R}$ and $\alpha > 0$ such that

$$\operatorname{Re} a(u) + \omega \|j(u)\|_H^2 \geq \alpha \|u\|_V^2 \qquad (u \in V) . \tag{4.1}$$

Then we define the operator $A$ associated with $(a, j)$ as follows. Given $x, y \in H$ we say that $x \in D(A)$ and $Ax = y$ if there exists a $u \in V$ such that $j(u) = x$ and

$$a(u, v) = (y | j(v))_H \qquad \text{for all } v \in V \ .$$

**Theorem 4.3.** *The operator defined in this way is well defined. Moreover, $-A$ generates a holomorphic $C_0$-semigroup on $H$.*

*Remark* 4.4. The form $a$ satisfies Condition (4.1) if and only if the form $a_\omega$ given by

$$a_\omega(u, v) = a(u, v) + \omega (j(u) | j(v))_H$$

is coercive. If $T_\omega$ denotes the semigroup associated with $(a_\omega, j)$ and $T$ the semigroup associated with $(a, j)$, then

$$T_\omega(t) = e^{-\omega t} T(t) \qquad (t > 0)$$

as is easy to see.

## 5. The Stokes operator

In this section we show as an example that the Stokes operator is selfadjoint and generates a holomorphic $C_0$-semigroup. The following approach is due to Monniaux [Mon]. Let $\Omega \subset \mathbb{R}^d$ be a bounded open set. We first discuss the Dirichlet Laplacian.

**Theorem 5.1 (Dirichlet Laplacian).** *Let $H = L^2(\Omega)$ and define the operator $\Delta^D$ on $L^2(\Omega)$ by*

$$D(\Delta^D) = \{u \in H_0^1(\Omega) : \Delta u \in L^2(\Omega)\}$$

$$\Delta^D u := \Delta u \ .$$

*Then $\Delta^D$ is selfadjoint and generates a holomorphic $C_0$-semigroup on $L^2(\Omega)$.*

*Proof.* Define $a\colon H_0^1(\Omega) \times H_0^1(\Omega) \to \mathbb{R}$ by $a(u, v) = \int_\Omega \nabla u \nabla v$. Then $a$ is clearly continuous. Poincaré's inequality says that $a$ is coercive. Consider the injection $j$ of $H_0^1(\Omega)$ into $L^2(\Omega)$. Let $A$ be the operator associated with $(a, j)$. We show that $A = -\Delta^D$. In fact, let $u \in D(A)$ and write $f = Au$. Then $\int_\Omega \nabla u \nabla v = \int_\Omega f v$ for all

$v \in H_0^1(\Omega)$. Taking in particular $v \in C_c^\infty(\Omega)$ we see that $-\Delta u = f$. Conversely, let $u \in H_0^1(\Omega)$ be such that $f := -\Delta u \in L^2(\Omega)$. Then $\int_\Omega f\varphi = \int_\Omega \nabla u \nabla \varphi = a(u,\varphi)$ for all $\varphi \in C_c^\infty(\Omega)$. This is just the definition of the weak partial derivatives in $H^1(\Omega)$. Since $C_c^\infty(\Omega)$ is dense in $H_0^1(\Omega)$, it follows that $\int_\Omega fv = a(u,v)$ for all $v \in H_0^1(\Omega)$. Thus $u \in D(A)$ and $Au = f$. Now the theorem follows from Theorem 4.2. □

For our treatment of the Stokes operator it will be useful to consider the Dirichlet Laplacian also in $L^2(\Omega)^d = L^2(\Omega) \oplus \cdots \oplus L^2(\Omega)$.

**Theorem 5.2.** *Define the symmetric form* $a\colon H_0^1(\Omega)^d \times H_0^1(\Omega)^d \to \mathbb{R}$ *by*

$$a(u,v) = \int_\Omega \nabla u \nabla v := \sum_{j=1}^d \int_\Omega \nabla u_j \nabla v_j \ ,$$

*where* $u = (u_1, \ldots, u_d)$. *Moreover, let* $j\colon H_0^1(\Omega)^d \to L^2(\Omega)^d$ *be the inclusion. Then* $a$ *is continuous and coercive. The operator* $A$ *associated with* $(a,j)$ *on* $L^2(\Omega)^d$ *is given by*

$$D(A) = \{u \in H_0^1(\Omega)^d : \Delta u_j \in L^2(\Omega) \text{ for all } j \in \{1,\ldots,d\}\} \ ,$$

$$Au = (-\Delta u_1, \ldots, -\Delta u_d) =: -\Delta u \ .$$

*We call* $\Delta^D := -A$ *the Dirichlet Laplacian on* $L^2(\Omega)^d$.

In order to define the Stokes operator we need some preparation. Let $\mathcal{D}(\Omega) := C_c^\infty(\Omega)^d$ and let $\mathcal{D}_0(\Omega) := \{\varphi \in \mathcal{D}(\Omega) : \operatorname{div}\varphi = 0\}$, where $\operatorname{div}\varphi = \partial_1\varphi_1 + \cdots + \partial_d\varphi_d$ and $\varphi = (\varphi_1, \ldots, \varphi_d)$. By $\mathcal{D}(\Omega)'$ we denote the dual space of $\mathcal{D}(\Omega)$ (with the usual topology). Each element $S$ of $\mathcal{D}(\Omega)'$ can be written in a unique way as $S = (S_1, \ldots, S_d)$ with $S_j \in C_c^\infty(\Omega)'$ so that

$$\langle S, \varphi\rangle = \sum_{j=1}^d \langle S_j, \varphi_j\rangle$$

for all $\varphi = (\varphi_1, \ldots, \varphi_d) \in \mathcal{D}(\Omega)$.

We say that $S \in H^{-1}(\Omega)$ if there exists a constant $c \geq 0$ such that

$$|\langle S, \varphi\rangle| \leq c\Big(\int |\nabla\varphi|^2\Big)^{\frac12} \qquad (\varphi \in \mathcal{D}(\Omega))$$

where $|\nabla\varphi|^2 = |\nabla\varphi_1|^2 + \cdots + |\nabla\varphi_d|^2$. For the remainder of this section we assume that $\Omega$ has Lipschitz boundary. We need the following result (see [Tem, Remark 1.9, p. 14]).

**Theorem 5.3.** *Let* $T \in H^{-1}(\Omega)$. *The following are equivalent.*

(i) $\langle T, \varphi\rangle = 0$ *for all* $\varphi \in \mathcal{D}_0(\Omega)$;
(ii) *there exists a* $p \in L^2(\Omega)$ *such that* $T = \nabla p$.

Note that Condition (ii) means that

$$\langle T, \varphi\rangle = \sum_{j=1}^{d} \langle \partial_j p, \varphi_j\rangle = -\sum_{j=1}^{d} \langle p, \partial_j \varphi_j\rangle = -\langle p, \operatorname{div} \varphi\rangle \ .$$

Now the implication (ii)⇒(i) is obvious. We omit the other implication.

Consider the real Hilbert space $L^2(\Omega)^d$ with scalar product

$$(f|g) = \sum_{j=1}^{d} (f_j|g_j)_{L^2(\Omega)} = \sum_{j=1}^{d} \int_\Omega f_j g_j \ .$$

We denote by

$$H := \mathcal{D}_0(\Omega)^{\perp\perp} = \overline{\mathcal{D}_0(\Omega)}$$

the closure of $\mathcal{D}_0(\Omega)$ in $L^2(\Omega)^d$. We call $H$ the space of all *divergence free vectors* in $L^2(\Omega)^d$. The orthogonal projection $P$ from $L^2(\Omega)^d$ onto $H$ is called the *Helmholtz projection.* Now let $V$ be the closure of $\mathcal{D}_0(\Omega)$ in $H^1(\Omega)^d$. Thus $V \subset H_0^1(\Omega)^d$ and $\operatorname{div} u = 0$ for all $u \in V$. One can actually show that

$$V = \{u \in H_0^1(\Omega)^d : \operatorname{div} v = 0\} \ .$$

We define the form $a\colon V \times V \to \mathbb{R}$ by

$$a(u,v) = \sum_{j=1}^{d} (\nabla u_j|\nabla v_j)_{L^2(\Omega)} \quad (u = (u_1, \ldots, u_d), v - (v_1, \ldots, v_d) \in V) \ .$$

Then $a$ is continuous and coercive. The space $V$ is dense in $H$ since it contains $\mathcal{D}_0(\Omega)$. We consider the inclusion $j\colon V \to H$. Let $A$ be the operator associated with $(a, j)$. Then $A$ is selfadjoint and $-A$ generates a holomorphic $C_0$-semigroup. The operator can be described as follows.

**Theorem 5.4.** *The operator $A$ has the domain*

$$D(A) = \{u \in V : \exists \pi \in L^2(\Omega) \text{ such that } -\Delta u + \nabla \pi \in H\}$$

*and is given by*

$$Au = -\Delta u + \nabla \pi \ ,$$

*where $\pi \in L^2(\Omega)$ is such that $-\Delta u + \nabla \pi \in H$.*

If $u \in H_0^1(\Omega)^d$, then $\Delta u \in H^{-1}(\Omega)$. In fact, for all $\varphi \in \mathcal{D}(\Omega)$,

$$|\langle -\Delta u, \varphi\rangle| = |-\langle u, \Delta\varphi\rangle| = \left|\sum_{j=1}^{d} \int_\Omega \nabla u_j \nabla \varphi_j\right| \le \|u\|_{H_0^1(\Omega)^d} \|\varphi\|_{H_0^1(\Omega)^d} \ .$$

*Proof of Theorem* 5.4. Let $u \in D(A)$ and write $f = Au$. Then $f \in H$, $u \in V$ and $a(u,v) = (f|v)_H$ for all $v \in V$. Thus, the distribution $-\Delta u \in H^{-1}(\Omega)$ coincides with $f$ on $\mathcal{D}_0(\Omega)$. By Theorem 5.3 there exists a $\pi \in L^2(\Omega)$ such that $-\Delta u + \nabla \pi =$

$f$. Conversely, let $u \in V$, $f \in H$, $\pi \in L^2(\Omega)$ and suppose that $-\Delta u + \nabla \pi = f$ in $\mathcal{D}(\Omega)'$. Then for all $\varphi \in \mathcal{D}_0(\Omega)$,

$$a(u, \varphi) = \int_\Omega \nabla u \nabla \varphi = \int_\Omega \nabla u \nabla \varphi + \langle \nabla \pi, \varphi \rangle = (f|\varphi)_{L^2(\Omega)^d} .$$

Since $\mathcal{D}_0(\Omega)$ is dense in $V$, it follows that $a(u, \varphi) = (f|\varphi)_{L^2(\Omega)^d}$ for all $\varphi \in V$. Thus, $u \in D(A)$ and $Au = f$. □

The operator $A$ is called the *Stokes operator*. We refer to [Mon] for this approach and further results on the Navier–Stokes equation. We conclude this section by giving an example where $j$ is not injective. Further examples will be seen in the sequel.

**Proposition 5.5.** *Let $\widetilde{H}$ be a Hilbert space and $H \subset \widetilde{H}$ a closed subspace. Denote by $P$ the orthogonal projection onto $H$. Let $\widetilde{V}$ be a Hilbert space which is continuously and densely embedded into $\widetilde{H}$ and let $a \colon \widetilde{V} \times \widetilde{V} \to \mathbb{R}$ be a continuous, coercive form. Denote by $A$ the operator on $\widetilde{H}$ associated with $(a, j)$ where $j$ is the injection of $\widetilde{V}$ into $\widetilde{H}$ and let $B$ be the operator on $H$ associated with $(a, P \circ j)$. Then*

$$D(B) = \{Pw : w \in D(A) \text{ and } Aw \in H\} ,$$

$$BPw = Aw \qquad (w \in D(A),\ Aw \in H) .$$

This is easy to see. In the context considered in this section we obtain the following example.

**Example 5.6.** Let $\widetilde{H} = L^2(\Omega)^d$, $H = \overline{\mathcal{D}_0(\Omega)}$ and $\widetilde{V} := H_0^1(\Omega)^d$. Define $a \colon \widetilde{V} \times \widetilde{V} \to \mathbb{R}$ by

$$a(u, v) = \int_\Omega \nabla u \nabla v .$$

Moreover, define $j \colon \widetilde{V} \to \widetilde{H}$ by $j(u) = u$. Then the operator associated with $(a, j)$ is $A = -\Delta^D$ as we have seen in Theorem 5.2. Now let $P$ be the Helmholtz projection and $B$ the operator associated with $(a, P \circ j)$. Then

$$D(B) = \{u \in H : \exists \pi \in L^2(\Omega) \text{ such that}$$
$$u + \nabla \pi \in D(\Delta^D) \text{ and } \Delta(u + \nabla \pi) \in H\}$$

and

$$Bu = -\Delta(u + \nabla \pi) ,$$

if $\pi \in L^2(\Omega)$ is such that $u + \nabla \pi \in D(\Delta^D)$ and $\Delta(u + \nabla \pi) \in H$. This follows directly from Proposition 5.5 and Theorem 5.3. The operator $B$ is selfadjoint and generates a holomorphic semigroup.

## 6. From forms to semigroups: the incomplete case

In the preceding sections we considered forms which were defined on a Hilbert space $V$. Now we want to study a purely algebraic condition considering forms whose domains are arbitrary vector spaces. At first we consider the complex case. Let $H$ be a complex Hilbert space. A *sectorial form* on $H$ is a sesquilinear form

$$a\colon D(a) \times D(a) \to \mathbb{C} ,$$

where $D(a)$ is a vector space, together with a linear mapping $j\colon D(a) \to H$ with dense image such that there exist $\omega \geq 0$ and $\theta \in (0, \pi/2)$ such that

$$a(u) + \omega\|j(u)\|_H^2 \in \overline{\Sigma_\theta} \qquad (u \in D(a)) .$$

If $\omega = 0$, then we call the form 0-*sectorial.* To a sectorial form, we associate an operator $A$ on $H$ by defining for all $x, y \in H$ that $x \in D(A)$ and $Ax = y :\Leftrightarrow$ there exists a sequence $(u_n)_{n\in\mathbb{N}}$ in $D(a)$ such that

a) $\lim_{n\to\infty} j(u_n) = x$ in $H$; b) $\sup_{n\in\mathbb{N}} \operatorname{Re} a(u_n) < \infty$; and

c) $\lim_{n\to\infty} a(u_n, v) = (y|j(v))_H$ for all $v \in D(a)$.

It is part of the next theorem that the operator $A$ is well defined (i.e., that $y$ depends only on $x$ and not on the choice of the sequence satisfying a), b) and c)). We only consider single-valued operators in this article.

**Theorem 6.1.** *The operator $A$ associated with a sectorial form $(a, j)$ is well defined and $-A$ generates a holomorphic $C_0$ semigroup on $H$.*

The proof of the theorem consists in a reduction to the complete case by considering an appropriate completion of $D(a)$. Here it is important that in Theorem 4.2 a non-injective mapping $j$ is allowed. For a proof we refer to [AE2, Theorem 3.2].

If $C \subset H$ is a closed convex set, we say that $C$ is *invariant* under a semigroup $T$ if

$$T(t)C \subset C \qquad (t > 0) .$$

Invariant sets are important to study positivity, $L^\infty$-contractivity, and many more properties. If the semigroup is associated with a form, then the following criterion, [AE2, Proposition 3.9], is convenient.

**Theorem 6.2 (Invariance).** *Let $C \subset H$ be a closed convex set and let $P$ be the orthogonal projection onto $C$. Then the semigroup $T$ associated with a sectorial form $(a, j)$ on $H$ leaves $C$ invariant if and only if for each $u \in D(a)$ there exists a sequence $(w_n)_{n\in\mathbb{N}}$ in $D(a)$ such that*

a) $\lim_{n\to\infty} j(w_n) = Pj(u)$ *in* $H$;

b) $\limsup_{n\to\infty} \operatorname{Re} a(w_n, u - w_n) \geq 0$; *and*

c) $\sup_{n\in\mathbb{N}} \operatorname{Re} a(w_n) < \infty$.

**Corollary 6.3.** *Let $C \subset H$ be a closed convex set and let $P$ be the orthogonal projection onto $C$. Assume that for each $u \in D(a)$, there exists a $w \in D(a)$ such that*

$$j(w) = Pj(u) \qquad \text{and} \qquad \operatorname{Re} a(w, u - w) \geq 0 \; .$$

*Then $T(t)C \subset C$ for all $t > 0$.*

In this section we want to use the invariance criterion to prove a generation theorem in the incomplete case which is valid in real Hilbert spaces. Let $H$ be a real Hilbert space. A *sectorial* form on $H$ is a bilinear mapping

$$a \colon D(a) \times D(a) \to \mathbb{R} \; ,$$

where $D(a)$ is a real vector space, together with a linear mapping $j \colon D(a) \to H$ with dense image such that there are $\alpha, \omega \geq 0$ such that

$$|a(u,v) - a(v,u)| \leq \alpha(a(u) + a(v)) + \omega(\|j(u)\|_H^2 + \|j(v)\|_H^2) \qquad (u, v \in D(a)) \; .$$

It is easy to see that the form $a$ is sectorial on the real space $H$ if and only if the sesquilinear extension $a_{\mathbb{C}}$ of $a$ to the complexification of $D(a)$ together with the $\mathbb{C}$-linear extension of $j$ is sectorial in the sense formulated in the beginning of this section.

To such a sectorial form $(a, j)$ we associate an operator $A$ on $H$ by defining for all $x, y \in H$ that $x \in D(A)$ and $Ax = y :\Leftrightarrow$ there exists a sequence $(u_n)$ in $D(a)$ satisfying

a) $\lim_{n \to \infty} j(u_n) = x$ in $H$;

b) $\sup_{n \in \mathbb{N}} a(u_n) < \infty$; and

c) $\lim_{n \to \infty} a(u_n, v) = (y | j(v))_H$ for all $v \in D(a)$.

Then the following holds.

**Theorem 6.4.** *The operator $A$ is well defined and $-A$ generates a holomorphic $C_0$-semigroup on $H$.*

*Proof.* Consider the complexifications $H_{\mathbb{C}} = H \oplus iH$ and $D(a_{\mathbb{C}}) := D(a) + iD(a)$. Let

$$a_{\mathbb{C}}(u,v) := a(\operatorname{Re} u, \operatorname{Re} v) + a(\operatorname{Im} u, \operatorname{Im} v) + i(a(\operatorname{Re} u, \operatorname{Im} v) + a(\operatorname{Im} u, \operatorname{Re} v))$$

for all $u = \operatorname{Re} u + i \operatorname{Im} u, v = \operatorname{Re} v + i \operatorname{Im} v \in D(a_{\mathbb{C}})$. Then $a_{\mathbb{C}}$ is a sesquilinear form. Let $J \colon D(a_{\mathbb{C}}) \to H_{\mathbb{C}}$ be the $\mathbb{C}$-linear extension of $j$. Let

$$b(u,v) = a_{\mathbb{C}}(u,v) + \omega(J(u)|J(v))_{H_{\mathbb{C}}} \qquad (u, v \in D(a_{\mathbb{C}})) \; .$$

Then

$$\operatorname{Im} b(u) = a(\operatorname{Im} u, \operatorname{Re} u) - a(\operatorname{Re} u, \operatorname{Im} u),$$
$$\operatorname{Re} b(u) = a(\operatorname{Re} u) + a(\operatorname{Im} u) + \omega(\|j(\operatorname{Re} u)\|_H^2 + \|j(\operatorname{Im} u)\|_H^2) \; .$$

The assumptions imply that there is a $c > 0$ such that $|\operatorname{Im} b(u)| \le c \operatorname{Re} b(u)$ for all $u \in D(a_{\mathbb{C}})$. Consequently, $b(u) \in \overline{\Sigma_\theta}$, where $\theta = \arctan c$. Thus the operator $B$ associated with $b$ generates a $C_0$-semigroup $S_{\mathbb{C}}$ on $H_{\mathbb{C}}$. It follows from Corollary 6.3 that $H$ is invariant. The part $A_\omega$ of $B$ in $H$ is the generator of $S$, where $S(t) := S_{\mathbb{C}}(t)_{|H}$. It is easy to see that $A_\omega - \omega = A$. □

*Remark* 6.5. It is remarkable, and important for some applications, that Condition b) in Theorem 6.1 as well as in Theorem 6.4 may be replaced by

b′)
$$\lim_{n,m\to\infty} a(u_n - u_m) = 0\ .$$

For later purposes we carry over the invariance criterion Corollary 6.3 to the real case.

**Corollary 6.6.** *Let $H$ be a real Hilbert space and $(a, j)$ a sectorial form on $H$ with associated semigroup $T$. Let $C \subset H$ be a closed convex set and $P$ the orthogonal projection onto $C$. Assume that for each $u \in D(a)$ there exists a $w \in D(a)$ such that*

$$j(w) = Pj(u) \qquad \text{and} \qquad a(w, u - w) \ge 0\ .$$

*Then $T(t)C \subset C$ for all $t > 0$.*

We want to formulate a special case of invariance. An operator $S$ on a space $L^p(\Omega)$ is called

$$\textit{positive} \text{ if } \Big(f \ge 0 \text{ a.e. implies } Sf \ge 0 \text{ a.e.}\Big) \text{ and}$$
$$\textit{submarkovian} \text{ if } \Big(f \le \mathbb{1} \text{ a.e. implies } Sf \le \mathbb{1} \text{ a.e.}\Big).$$

Thus, an operator $S$ is submarkovian if and only if it is positive and $\|Sf\|_\infty \le \|f\|_\infty$ for all $f \in L^p \cap L^\infty$. A semigroup $T$ is called *submarkovian* if $T(t)$ is submarkovian for all $t > 0$.

**Proposition 6.7.** *Consider the real space $H = L^2(\Omega)$ and a sectorial form $a$ on $H$. Assume that for each $u \in D(a)$ one has $u \wedge \mathbb{1} \in D(a)$ and*

$$a(u \wedge \mathbb{1}, (u - \mathbb{1})^+) \ge 0\ .$$

*Then the semigroup $T$ associated with $a$ is submarkovian.*

Recall that $u \wedge v := \min(u, v)$ and $v^+ = \max(v, 0)$.

*Proof.* The set $C := \{u \in L^2(\Omega) : u \le \mathbb{1} \text{ a.e.}\}$ is closed and convex. The orthogonal projection $P$ onto $C$ is given by $Pu = u \wedge \mathbb{1}$. Thus $u - Pu = (u - \mathbb{1})^+$ and the result follows from Corollary 6.6. □

We conclude this section with a remark concerning closable forms.

*Remark* 6.8 (Forget closability). In many text books, for example [Dav], [Kat], [MR], [Ouh], [Tan] one finds the notion of a sectorial form $a$ on a complex Hilbert space $H$. By this one understands a sesquilinear form $a\colon D(a) \times D(a) \to \mathbb{C}$ where

$D(a)$ is a dense subspace of $H$ such that there are $\theta \in (0, \pi/2)$ and $\omega \geq 0$ such that $a(u) + \omega \|u\|_H^2 \in \overline{\Sigma_\theta}$ for all $u \in D(a)$. Then

$$\|u\|_a := (\operatorname{Re} a(u) + (\omega + 1)\|u\|_H^2)^{1/2}$$

defines a norm on $D(a)$. The form is called *closed* if $D(a)$ is complete for this norm. This corresponds to our complete case with $V = D(a)$ and $j$ the inclusion. If the form is not closed, then one may consider the completion $V$ of $D(a)$. Since the injection $D(a) \to H$ is continuous for the norm $\| \ \|_a$, it has a continuous extension $j: V \to H$. This extension may be injective or not. The form is called *closable* if $j$ is injective. In the literature only for closable forms generation theorems are given, see [AE2] for precise references. The results above show that the notion of closability is not needed.

In this special setting it is easy to give the proof of Theorem 6.1. There exists a unique continuous sesquilinear form $\tilde{a}: V \times V \to \mathbb{C}$ such that $\tilde{a}(u, v) = a(u, v)$ for all $u, v \in D(a)$. Since the form $a$ is sectorial, it follows that $\tilde{a}$ is $j$-elliptic (see (4.1)). Let $\widetilde{A}$ be the operator associated with $(\tilde{a}, j)$ from Theorem 4.3. Let $x, y \in H$ and suppose that $x \in D(A)$ and $Ax = y$. By assumption there exists a sequence $(u_n)_{n \in \mathbb{N}}$ in $D(a)$ such that $\lim u_n = x$ in $H$, $\sup \operatorname{Re} a(u_n) < \infty$ and $\lim a(u_n, v) = (y|v)_H$ for all $v \in D(a)$. Then $(u_n)_{n \in \mathbb{N}}$ is bounded in $V$, so passing to a subsequence if necessary, it is weakly convergent, say to $u \in V$. Then $\tilde{a}(u, v) = \lim \tilde{a}(u_n, v) = (y|v)_H$ for all $v \in D(a)$. Hence by density, $\tilde{a}(u, v) = (y|j(v))_H$ for all $v \in V$. So $x = j(u) \in D(\widetilde{A})$ and $\widetilde{A}x = y$. Therefore $A$ is well defined and $\widetilde{A}$ is an extension of $A$. It is easy to show that also $A$ is an extension of $\widetilde{A}$. So $A = \widetilde{A}$ and $-A$ generates a holomorphic semigroup.

It is clear that one needs to consider approximating sequences in the definition of the operator $A$ in the incomplete case. Just consider the trivial form $a = 0$ with $D(a)$ a proper dense subspace of $H$. Then the associated operator is the zero operator.

There is a unique correspondence between sectorially quasi contractive holomorphic semigroups and closed sectorial forms (see [Kat, Theorem VI.2.7]). One looses uniqueness if one considers forms which are merely closable or in our general setting if one allows arbitrary maps $j: D(a) \to H$ with dense image. However, examples show that in many cases a natural operator is obtained by this general framework.

## 7. Degenerate diffusion

In this section we use our tools to show that degenerate elliptic operators generate holomorphic semigroups on the real space $L^2(\Omega)$. We start with a 1-dimensional example.

**Example 7.1 (Degenerate diffusion in dimension 1).** Consider the real Hilbert space $H = L^2(a, b)$, where $-\infty \leq a < b \leq \infty$, and let $\alpha, \beta, \gamma \in L^\infty_{\text{loc}}(a, b)$ be real coefficients. We assume that there is a $c_1 \geq 0$ such that

$$\gamma^- := \max(-\gamma, 0) \in L^\infty(a,b) \text{ and } \beta^2(x) \leq c_1 \cdot \alpha(x) \qquad (x \in (a,b)) \ .$$

We define the bilinear form $a$ on $L^2(a,b)$ by

$$a(u,v) = \int_a^b \Big(\alpha(x)u'(x)v'(x) + \beta(x)u'(x)v(x) + \gamma(x)u(x)v(x)\Big)\, dx$$

with domain

$$D(a) = H^1_c(a,b) = \{u \in H^1(a,b) : \operatorname{supp} u \text{ is compact in } (a,b)\} \ .$$

We choose $j\colon H^1_c(a,b) \to L^2(a,b)$ to be the inclusion map. We next show that the form $a$ is *sectorial*, i.e., there exist constants $c, \omega \geq 0$, such that

$$|a(u,v) - a(v,u)| \leq c(a(u) + a(v)) + \omega(\|u\|^2_{L^2} + \|v\|^2_{L^2}) \tag{7.1}$$

$$(u, v \in D(a)) \ .$$

For the proof of (7.1) we use Young's inequality

$$|xy| \leq \varepsilon x^2 + \frac{1}{4\varepsilon} y^2$$

twice. Let $u, v \in D(a)$. On one hand we have for all $\delta > 0$,

$$|a(u,v) \quad a(v,u)| = |\int_a^b \beta(u'v - uv')| \leq \int_u^b \delta\beta^2(u'^2 + v'^2) + \frac{1}{4\delta}(u^2 + v^2) \ .$$

On the other hand, for all $c, \omega, \varepsilon > 0$ one has

$$\begin{aligned} & c(a(u) + a(v)) + \omega(\|u\|^2_H + \|v\|^2_H) \\ &= \int_a^b c\alpha(u'^2 + v'^2) + c\beta(u'u + v'v) + (c\gamma + \omega)(u^2 + v^2) \\ &\geq \int_a^b (c\alpha - \varepsilon\beta^2)(u'^2 + v'^2) - c^2 \frac{1}{4\varepsilon}(u^2 + v^2) + (c\gamma + \omega)(u^2 + v^2) \\ &\geq \int_a^b (c\alpha - \varepsilon\beta^2)(u'^2 + v'^2) + (\omega - c\|\gamma^-\|_{L^\infty} - \frac{c^2}{4\varepsilon})(u^2 + v^2) \ . \end{aligned}$$

Therefore (7.1) is valid if $(c\alpha - \varepsilon\beta^2) \geq \delta\beta^2$ and $(\omega - c\|\gamma^-\|_{L^\infty} - \frac{c^2}{4\varepsilon}) \geq \frac{1}{4\delta}$. Since $\beta^2 \leq c_1\alpha$ one can find $\delta, \varepsilon, c, \omega$ such that the conditions are satisfied.

Thus $a$ is sectorial. As a consequence, if $A$ is the operator associated with $(a, j)$, then it follows from Theorem 6.4 that $-A$ generates a holomorphic $C_0$-semigroup $T$ on $L^2(\Omega)$. Moreover, $T$ is submarkovian by Proposition 6.7.

The condition $\beta^2 \leq c_1\alpha$ shows in particular that $\{x \in (a,b) : \alpha(x) = 0\} \subset \{x \in (a,b) : \beta(x) = 0\}$. This inclusion is a natural hypothesis, since in general an operator of the form $\beta u'$ does not generate a holomorphic semigroup.

A special case is the *Black–Scholes Equation*

$$u_t + \frac{\sigma^2}{2} x^2 u_{xx} + r x u_x - r u = 0 ,$$

with $\sigma \in \mathbb{R}$ and $r \in L^\infty(\mathbb{R})$, together with the condition that $r = 0$ if $\sigma = 0$. This one obtains by choosing $H = L^2(0,\infty)$,

$$a(u,v) = \int_0^\infty \left( \frac{\sigma^2}{2} x^2 u' v' + (\sigma^2 - r) x u' v + r u v \right)$$

and $D(a) = H^1_c(0,\infty)$.

It is not difficult to extend the example above to higher dimensions.

**Example 7.2.** Let $\Omega \subset \mathbb{R}^d$ be open and for all $i,j \in \{1,\ldots,d\}$ let $a_{ij}, b_j, c \in L^\infty_{\mathrm{loc}}(\Omega)$ be real coefficients. Assume $c^- \in L^\infty(\Omega)$, $a_{ij} = a_{ji}$ and there exists a $c_1 > 0$ such that

$$c_1 A(x) - B^2(x) \text{ is positive semidefinite}$$

for almost all $x \in \Omega$, where

$$A(x) = (a_{ij}(x)) \text{ and } B(x) = \mathrm{diag}(b_1(x), \ldots, b_d(x)) .$$

Define the form $a$ on $L^2(\Omega)$ by

$$a(u,v) = \int_\Omega \Big( \sum_{i,j=1}^d a_{ij} (\partial_i u)(\partial_j v) + \sum_{j=1}^d b_j (\partial_j u) v + c u v \Big)$$

with domain

$$D(a) = H^1_c(\Omega) .$$

Then $a$ is sectorial. The associated semigroup $T$ on $L^2(\Omega)$ is submarkovian.

This and the previous example incorporate Dirichlet boundary conditions. In the next one we consider a degenerate elliptic operator with Neumann boundary conditions.

**Example 7.3.** Let $\Omega \subset \mathbb{R}^d$ be an open, possibly unbounded subset of $\mathbb{R}^d$. For all $i,j \in \{1,\ldots,d\}$ let $a_{ij} \in L^\infty(\Omega)$ be *real* coefficients and assume that there exists a $\theta \in (0,\pi/2)$ such that

$$\sum_{i,j=1}^d a_{ij}(x) \xi_i \overline{\xi_j} \in \overline{\Sigma_\theta} \qquad (\xi \in \mathbb{C}^d,\ x \in \Omega) .$$

Consider the form $a$ on $L^2(\Omega)$ given by

$$a(u,v) = \int_\Omega \sum_{i,j=1}^d a_{ij} (\partial_i u)(\partial_j v)$$

with domain $D(a) = H^1(\Omega)$. Then $a$ is sectorial. Let $T$ be the associated semigroup. Our criteria show right away that $T$ is submarkovian. Therefore $T$ extends consistently to a semigroup $T_p$ on $L^p(\Omega)$ for all $p \in [1, \infty]$, the semigroup $T_p$ is strongly continuous for all $p < \infty$ and $T_\infty$ is the adjoint of a strongly continuous semigroup on $L^1(\Omega)$. It is remarkable that even

$$T_\infty(t)\mathbb{1}_\Omega = \mathbb{1}_\Omega \qquad (t > 0) .$$

For bounded $\Omega$ this is easy to prove, but otherwise more sophisticated tools are needed (see [AE2, Corollary 4.9]).

We want to mention an abstract result which shows that our solutions are some kind of *viscosity solutions*. This is illustrated particularly well in the situation of Example 7.3.

**Proposition 7.4 ([AE2, Corollary 3.9]).** *Let $V, H$ be real Hilbert spaces such that $V \underset{d}{\hookrightarrow} H$. Let $j \colon V \to H$ be the inclusion map. Let $a \colon V \times V \to \mathbb{R}$ be continuous and sectorial. Assume that $a(u) \geq 0$ for all $u \in V$. Let $b \colon V \times V \to \mathbb{R}$ be continuous and coercive. Then for each $n \in \mathbb{N}$ the form*

$$a + \frac{1}{n} b \colon V \times V \to \mathbb{R}$$

*is continuous and coercive. Let $A_n$ be the operator associated with $(a + \frac{1}{n} b, j)$ and $A$ with $(a, j)$. Then*

$$\lim_{n\to\infty} (A_n + \lambda)^{-1} f = (A + \lambda)^{-1} f \text{ in } H$$

*for all $f \in H$ and $\lambda > 0$. Moreover, denoting by $T_n$ and $T$ the semigroup generated by $-A_n$ and $-A$ one has*

$$\lim_{n\to\infty} T_n(t) f = T(t) f \text{ in } H$$

*for all $f \in H$.*

The essence in the result is that the form $a$ is merely sectorial and may be degenerate. For instance, in Example 7.3 $a_{ij}(x) = 0$ is allowed. If we perturb by the Laplacian, we obtain a coercive form

$$a_n \colon H^1(\Omega) \times H^1(\Omega) \to \mathbb{R}$$

given by

$$a_n(u, v) = a(u, v) + \frac{1}{n} \int_\Omega \nabla u \nabla v .$$

Then Proposition 7.4 says that in the situation of Example 7.3 for this perturbation one has $\lim_{n\to\infty} (A_n + \lambda)^{-1} f = (A + \lambda)^{-1} f$ in $L^2(\Omega)$ for all $f \in L^2(\Omega)$.

## 8. The Dirichlet-to-Neumann operator

The following example shows how the general setting involving non-injective $j$ can be used. It is taken from [AE1] where also the interplay between trace properties and the semigroup generated by the Dirichlet-to-Neumann operator is studied. Let $\Omega \subset \mathbb{R}^d$ be a bounded open set with boundary $\partial\Omega$. Our point is that we do not need any regularity assumption on $\Omega$, except that we assume that $\partial\Omega$ has a finite $(d-1)$-dimensional Hausdorff measure. Still we are able to define the Dirichlet-to-Neumann operator on $L^2(\partial\Omega)$ and to show that it is selfadjoint and generates a submarkovian semigroup on $L^2(\Omega)$. Formally, the Dirichlet-to-Neumann operator $D_0$ is defined as follows. Given $\varphi \in L^2(\partial\Omega)$, one solves the Dirichlet problem

$$\begin{cases} \Delta u = 0 \text{ in } \Omega \\ u_{|\partial\Omega} = \varphi \end{cases}$$

and defines $D_0\varphi = \frac{\partial u}{\partial \nu}$. We will give a precise definition using weak derivatives. We consider the space $L^2(\partial\Omega) := L^2(\partial\Omega, \mathcal{H}^{d-1})$ with the $(d-1)$-dimensional Hausdorff measure $\mathcal{H}^{d-1}$. Integrals over $\partial\Omega$ are always taken with respect to $\mathcal{H}^{d-1}$, those over $\Omega$ always with respect to the Lebesgue measure. Throughout this section we only assume that $\mathcal{H}^{d-1}(\partial\Omega) < \infty$ and that $\Omega$ is bounded.

**Definition 8.1 (Normal derivative).** Let $u \in H^1(\Omega)$ be such that $\Delta u \in L^2(\Omega)$. We say that

$$\frac{\partial u}{\partial \nu} \in L^2(\partial\Omega)$$

if there exists a $g \in L^2(\partial\Omega)$ such that

$$\int_\Omega (\Delta u)v + \int_\Omega \nabla u \nabla v = \int_{\partial\Omega} gv$$

for all $v \in H^1(\Omega) \cap C(\overline{\Omega})$. This determines $g$ uniquely and we let $\frac{\partial u}{\partial \nu} := g$.

Recall that for all $u \in L^1_{\text{loc}}(\Omega)$ the Laplacian $\Delta u$ is defined in the sense of distributions. If $\Delta u = 0$, then $u \in C^\infty(\Omega)$ by elliptic regularity. Next we define traces of a function $u \in H^1(\Omega)$.

**Definition 8.2 (Traces).** Let $u \in H^1(\Omega)$. We let

$$\begin{aligned} \operatorname{tr}(u) = \Big\{ g \in L^2(\partial\Omega) : \exists\, (u_n)_{n\in\mathbb{N}} \text{ in } H^1(\Omega) \cap C(\overline{\Omega}) \text{ such that} \\ \lim_{n\to\infty} u_n = u \text{ in } H^1(\Omega) \text{ and} \\ \lim_{n\to\infty} u_{n|\partial\Omega} = g \text{ in } L^2(\partial\Omega) \Big\}. \end{aligned}$$

For arbitrary open sets and $u \in H^1(\Omega)$ the set $\operatorname{tr}(u)$ might be empty, or contain more than one element. However, if $\Omega$ is a Lipschitz domain, then for each $u \in H^1(\Omega)$ the set $\mathrm{tr}(u)$ contains precisely one element, which we denote by $u_{|\partial\Omega} \in L^2(\partial\Omega)$. Now we are in the position to define the Dirichlet-to-Neumann operator $D_0$. Its domain is given by

$$D(D_0) := \Big\{\varphi \in L^2(\partial\Omega) : \exists u \in H^1(\Omega) \text{ such that}$$
$$\Delta u = 0,\ \varphi \in \mathrm{tr}(u) \text{ and } \frac{\partial u}{\partial \nu} \in L^2(\partial\Omega)\Big\}$$

and we define

$$D_0\varphi = \frac{\partial u}{\partial \nu}$$

where $u \in H^1(\Omega)$ is such that $\Delta u = 0$, $\frac{\partial u}{\partial \nu} \in L^2(\partial\Omega)$ and $\varphi \in \mathrm{tr}(u)$. It is part of our result that this operator is well defined.

**Theorem 8.3.** *The operator $D_0$ is selfadjoint and $-D_0$ generates a submarkovian semigroup on $L^2(\partial\Omega)$.*

In the proof we use Theorem 6.4. Here a non-injective mapping $j$ is needed. We also need Maz'ya's inequality. Let $q = \frac{2d}{d-1}$. There exists a constant $c_M > 0$ such that

$$\Big(\int_\Omega |u|^q\Big)^{2/q} \leq c_M\Big(\int_\Omega |\nabla u|^2 + \int_{\partial\Omega} |u|^2\Big)$$

for all $u \in H^1(\Omega) \cap C(\overline{\Omega})$. (See [Maz, Example 3.6.2/1 and Theorem 3.6.3] and [AW, (19)].)

*Proof of Theorem* 8.3. We consider real spaces. Our Hilbert space is $L^2(\partial\Omega)$. Let $D(a) = H^1(\Omega) \cap C(\overline{\Omega})$, $a(u,v) = \int_\Omega \nabla u \nabla v$ and define $j\colon D(a) \to L^2(\partial\Omega)$ by $j(u) = u_{|\partial\Omega} \in L^2(\partial\Omega)$. Then $a$ is symmetric and $a(u) \geq 0$ for all $u \in D(a)$. Thus the sectoriality condition before Theorem 6.4 is trivially satisfied. Denote by $A$ the operator on $L^2(\partial\Omega)$ associated with $(a,j)$. Let $\varphi, \psi \in L^2(\partial\Omega)$. Then $\varphi \in D(A)$ and $A\varphi = \psi$ if and only if there exists a sequence $(u_n)_{n\in\mathbb{N}}$ in $H^1(\Omega) \cap C(\overline{\Omega})$ such that $\lim_{n\to\infty} u_{n|\partial\Omega} = \varphi$ in $L^2(\partial\Omega)$, $\lim_{n\to\infty} a(u_n, v) = \int_{\partial\Omega} \psi v_{|\partial\Omega}$ for all $v \in D(a)$ and $\lim_{n,m\to\infty} \int_\Omega |\nabla(u_n - u_m)|^2 = 0$ (here we use Remark 6.5). Now Maz'ya's inequality implies that $(u_n)_{n\in\mathbb{N}}$ is a Cauchy sequence in $H^1(\Omega)$. Thus $\lim_{n\to\infty} u_n = u$ exists in $H^1(\Omega)$, and so $\varphi \in \mathrm{tr}(u)$. Moreover $\int_{\partial\Omega} \psi v = \lim_{n\to\infty} \int_\Omega \nabla u_n \nabla v = \int_\Omega \nabla u \nabla v$ for all $v \in H^1(\Omega) \cap C(\overline{\Omega})$. Taking as $v$ test functions, we see that $\Delta u = 0$. Thus

$$\int_\Omega \nabla u \nabla v + \int_\Omega (\Delta u) v = \int_{\partial\Omega} \psi v$$

for all $v \in H^1(\Omega)$. Consequently, $\frac{\partial u}{\partial \nu} = \psi$. We have shown that $A \subset D_0$.

Conversely, let $\varphi \in D(D_0), D_0\varphi = \psi$. Then there exists a $u \in H^1(\Omega)$ such that $\Delta u = 0$, $\varphi \in \operatorname{tr}(u)$ and $\frac{\partial u}{\partial \nu} = \psi$. Since $\varphi \in \operatorname{tr}(u)$ there exists a sequence $(u_n)_{n\in\mathbb{N}}$ in $H^1(\Omega) \cap C(\overline{\Omega})$ such that $u_n \to u$ in $H^1(\Omega)$ and $u_{n|\partial\Omega} \to \varphi$ in $L^2(\partial\Omega)$. It follows that $j(u_n) = u_{n|\partial\Omega} \to \varphi$ in $L^2(\partial\Omega)$, the sequence $(a(u_n))_{n\in\mathbb{N}}$ is bounded and

$$a(u_n, v) = \int_\Omega \nabla u_n \nabla v \to \int_\Omega \nabla u \nabla v = \int_\Omega \nabla u \nabla v + \int_\Omega (\Delta u) v = \int_{\partial\Omega} \psi v$$

for all $v \in H^1(\Omega) \cap C(\overline{\Omega})$. Thus, $\varphi \in D(A)$ and $A\varphi = \psi$ by the definition of the associated operator. Since $a$ is symmetric, the operator $A$ is selfadjoint. Now the claim follows from Theorem 6.4.

Our criteria easily apply and show that semigroup generated by $-D_0$ is submarkovian. □

## References


[ABHN] Arendt, W., Batty, C., Hieber, M. and Neubrander, F., *Vector-valued Laplace transforms and Cauchy problems*, vol. 96 of Monographs in Mathematics. Birkhäuser, Basel, 2001.

[AEH] Arendt, W., El-Mennaoui, O. and Hieber, M., Boundary values of holomorphic semigroups. *Proc. Amer. Math. Soc.* **125** (1997), 635–647.

[AE1] Arendt, W. and Elst, A.F.M. ter, The Dirichlet-to-Neumann operator on rough domains *J. Differential Equations* **251** (2011), 2100–2124.

[AE2] ——, Sectorial forms and degenerate differential operators. *J. Operator Theory* **67** (2011), 33–72.

[AN] Arendt, W. and Nikolski, N., Vector-valued holomorphic functions revisited. *Math. Z.* **234** (2000), 777–805.

[AW] Arendt, W. and Warma, M., The Laplacian with Robin boundary conditions on arbitrary domains. *Potential Anal.* **19** (2003), 341–363.

[Dav] Davies, E.B., *Heat kernels and spectral theory.* Cambridge Tracts in Mathematics 92. Cambridge University Press, Cambridge etc., 1989.

[Kat] Kato, T., *Perturbation theory for linear operators.* Second edition, Grundlehren der mathematischen Wissenschaften 132. Springer-Verlag, Berlin etc., 1980.

[MR] Ma, Z.M. and Röckner, M., *Introduction to the theory of (non-symmetric) Dirichlet Forms.* Universitext. Springer-Verlag, Berlin etc., 1992.

[Maz] Maz'ja, V.G., *Sobolev spaces.* Springer Series in Soviet Mathematics. Springer-Verlag, Berlin etc., 1985.

[Mon] Monniaux, S., Navier–Stokes equations in arbitrary domains: the Fujita–Kato scheme. *Math. Res. Lett.* **13** (2006), 455–461.

[Ouh] Ouhabaz, E.-M., *Analysis of heat equations on domains*, vol. 31 of London Mathematical Society Monographs Series. Princeton University Press, Princeton, NJ, 2005.

[Tan] Tanabe, H., *Equations of evolution.* Monographs and Studies in Mathematics 6. Pitman, London etc., 1979.

[Tem] TEMAM, R., *Navier–Stokes equations. Theory and numerical analysis. Reprint of the* 1984 *edition.* AMS Chelsea Publishing, Providence, RI, 2001.

Wolfgang Arendt
Institute of Applied Analysis
University of Ulm
D-89069 Ulm, Germany
e-mail: wolfgang.arendt@uni-ulm.de

A.F.M. ter Elst
Department of Mathematics
University of Auckland
Private Bag 92019
Auckland 1142, New Zealand
e-mail: terelst@math.auckland.ac.nz

Operator Theory:
Advances and Applications, Vol. 221, 71-108


# Accretive (∗)-extensions and Realization Problems

Yury Arlinskiĭ, Sergey Belyi and Eduard Tsekanovskiĭ

*Dedicated to Heinz Langer on the occasion of his $75^{th}$ birthday*

**Abstract.** We present a solution of the extended Phillips-Kato extension problem about existence and parametrization of all accretive (∗)-extensions (with the exit into triplets of rigged Hilbert spaces) of a densely defined non-negative operator. In particular, the analogs of the von Neumann and Friedrichs theorems for existence of non-negative self-adjoint (∗)-extensions are obtained. Relying on these results we introduce the extremal classes of Stieltjes and inverse Stieltjes functions and show that each function from these classes can be realized as the impedance function of an L-system. It is proved that in this case the realizing L-system contains an accretive operator and, in case of Stieltjes functions, an accretive (∗)-extension. Moreover, we establish the connection between the above-mentioned classes and the Friedrichs and Kreĭn-von Neumann extremal non-negative extensions.

**Mathematics Subject Classification (2000).** Primary 47A10, 47B44; Secondary 46E20, 46F05.

**Keywords.** Accretive operator, L-system, realization.

## 1. Introduction

We recall that an operator-valued function $V(z)$ acting on a finite-dimensional Hilbert space $E$ belongs to the class of operator-valued Herglotz-Nevanlinna functions if it is holomorphic on $\mathbb{C}\setminus\mathbb{R}$, if it is symmetric with respect to the real axis, i.e., $V(z)^* = V(\bar z)$, $z\in\mathbb{C}\setminus\mathbb{R}$, and if it satisfies the positivity condition

$$\operatorname{Im} V(z) \geq 0, \quad z\in\mathbb{C}_+.$$

It is well known (see, e.g., [14], [16]) that operator-valued Herglotz-Nevanlinna functions admit the following integral representation:

$$V(z) = Q + Lz + \int_{\mathbb{R}} \left(\frac{1}{t-z} - \frac{t}{1+t^2}\right) dG(t), \quad z\in\mathbb{C}\setminus\mathbb{R}, \tag{1.1}$$

where $Q = Q^*$, $L \geq 0$, and $G(t)$ is a nondecreasing operator-valued function on $\mathbb{R}$ with values in the class of nonnegative operators in $E$ such that

$$\int_{\mathbb{R}} \frac{(dG(t)x, x)_E}{1+t^2} < \infty, \quad x \in E. \tag{1.2}$$

The realization of a selected class of Herglotz-Nevanlinna functions is provided by a system $\Theta$ of the form

$$\begin{cases} (\mathbb{A} - zI)x = KJ\varphi_- \\ \varphi_+ = \varphi_- - 2iK^*x \end{cases} \tag{1.3}$$

or

$$\Theta = \begin{pmatrix} \mathbb{A} & K & J \\ \mathcal{H}_+ \subset \mathcal{H} \subset \mathcal{H}_- & & E \end{pmatrix}. \tag{1.4}$$

In this system $\mathbb{A}$, the *state-space operator* of the system, is a so-called $(*)$-extension, which is a bounded linear operator from $\mathcal{H}_+$ into $\mathcal{H}_-$ extending a symmetric operator $A$ in $\mathcal{H}$, where $\mathcal{H}_+ \subset \mathcal{H} \subset \mathcal{H}_-$ is a rigged Hilbert space. Moreover, $K$ is a bounded linear operator from the finite-dimensional Hilbert space $E$ into $\mathcal{H}_-$, while $J = J^* = J^{-1}$ is acting on $E$, are such that $\operatorname{Im}\mathbb{A} = KJK^*$. Also, $\varphi_- \in E$ is an input vector, $\varphi_+ \in E$ is an output vector, and $x \in \mathcal{H}_+$ is a vector of the state space of the system $\Theta$. The system described by (1.3)-(1.4) is called an *L-system.* The operator-valued function

$$W_\Theta(z) = I - 2iK^*(\mathbb{A} - zI)^{-1}KJ \tag{1.5}$$

is a transfer function of the system $\Theta$. It was shown in [14] that an operator-valued function $V(z)$ acting on a Hilbert space $E$ of the form (1.1) can be represented and realized in the form

$$V(z) = i[W_\Theta(z) + I]^{-1}[W_\Theta(z) - I] = K^*(\operatorname{Re}\mathbb{A} - zI)^{-1}K, \tag{1.6}$$

where $W_\Theta(z)$ is a transfer function of some canonical scattering ($J = I$) system $\Theta$, and where the "*real part*" $\operatorname{Re}\mathbb{A} = \frac{1}{2}(\mathbb{A} + \mathbb{A}^*)$ of $\mathbb{A}$ satisfies $\operatorname{Re}\mathbb{A} \supset \hat{A} = \hat{A}^* \supset \dot{A}$ if and only if the function $V(z)$ in (1.1) satisfies the following two conditions:

$$\begin{cases} L = 0, \\ Qx = \int_{\mathbb{R}} \frac{t}{1+t^2}\, dG(t)x \quad \text{when} \quad \int_{\mathbb{R}} (dG(t)x, x)_E < \infty. \end{cases} \tag{1.7}$$

The class of all realizable Herglotz-Nevanlinna functions with conditions (1.7) is denoted by $N(R)$ (see [14]).

In the first part of this paper we present a solution of the extended Phillips-Kato extension problem. We show the existence and parameterize all accretive $(*)$-extensions (with the exit into triplets of rigged Hilbert spaces) of a densely defined non-negative symmetric operator. Moreover, the analogs of the von Neumann and Friedrichs theorems for existence of non-negative self-adjoint $(*)$-extensions are obtained. In the remaining part of the paper we focus on the introduced extremal classes of Stieltjes and inverse Stieltjes functions. We show that any function belonging to these classes can be realized as the impedance function of an L-system with special properties. In the end we establish the connection between the above-mentioned classes and the Friedrichs and Kreĭn-von Neumann extremal non-negative extensions.

The complete proofs of some parts of the material from [6], [20], [30] are presented here for the first time.

## 2. Preliminaries

For a pair of Hilbert spaces $\mathcal{H}_1$, $\mathcal{H}_2$ we denote by $[\mathcal{H}_1, \mathcal{H}_2]$ the set of all bounded linear operators from $\mathcal{H}_1$ to $\mathcal{H}_2$. Let $\dot{A}$ be a closed, densely defined, symmetric operator in a Hilbert space $\mathcal{H}$ with inner product $(f,g), f,g \in \mathcal{H}$. Any operator $T$ in $\mathcal{H}$ such that

$$\dot{A} \subset T \subset \dot{A}^*$$

is called a *quasi-self-adjoint extension* of $\dot{A}$.

Consider the rigged Hilbert space (see [14]) $\mathcal{H}_+ \subset \mathcal{H} \subset \mathcal{H}_-$, where $\mathcal{H}_+ = \mathrm{Dom}(\dot{A}^*)$ and

$$(f,g)_+ = (f,g) + (\dot{A}^* f, \dot{A}^* g), \quad f,g \in \mathrm{Dom}(A^*).$$

Let $\mathcal{R}$ be the *Riesz-Berezansky operator* $\mathcal{R}$ (see [14]) which maps $\mathcal{H}_-$ onto $\mathcal{H}_+$ such that $(f,g) = (f, \mathcal{R}g)_+$ $(\forall f \in \mathcal{H}_+,\ g \in \mathcal{H}_-)$ and $\|\mathcal{R}g\|_+ = \|g\|_-$. Note that identifying the space conjugate to $\mathcal{H}_\pm$ with $\mathcal{H}_\mp$, we get that if $\mathbb{A} \in [\mathcal{H}_+, \mathcal{H}_-]$ then $\mathbb{A}^* \in [\mathcal{H}_+, \mathcal{H}_-]$.

**Definition 2.1.** An operator $\mathbb{A} \in [\mathcal{H}_+, \mathcal{H}_-]$ is called a self-adjoint bi-extension of a symmetric operator $\dot{A}$ if $\mathbb{A} = \mathbb{A}^*$ and $\mathbb{A} \supset \dot{A}$.

Let $\mathbb{A}$ be a self-adjoint bi-extension of $\dot{A}$ and let the operator $\widehat{A}$ in $\mathcal{H}$ be defined as follows:

$$\mathrm{Dom}(\widehat{A}) = \{f \in \mathcal{H}_+ : \widehat{A}f \in \mathcal{H}\}, \ \widehat{A} = \mathbb{A} \upharpoonright \mathrm{Dom}(\widehat{A}).$$

The operator $\widehat{A}$ is called a *quasi-kernel* of a self-adjoint bi-extension $\mathbb{A}$ (see [35]). We say that a self-adjoint bi-extension $\mathbb{A}$ of $\dot{A}$ is *twice-self-adjoint* or *t-self-adjoint* if its quasi-kernel $\widehat{A}$ is a self-adjoint operator in $\mathcal{H}$.

**Definition 2.2.** Let $T$ be a quasi-self-adjoint extension of $\dot{A}$ with nonempty resolvent set $\rho(T)$. An operator $\mathbb{A} \in [\mathcal{H}_+, \mathcal{H}_-]$ is called a $(*)$-extension (or correct bi-extension) of an operator $T$ if

1. $\mathbb{A} \supset T \supset \dot{A}, \quad \mathbb{A}^* \supset T^* \supset \dot{A}$,
2. the quasi-kernel of self-adjoint bi-extension $\mathrm{Re}\,\mathbb{A} = \frac{1}{2}(\mathbb{A} + \mathbb{A}^*)$ is a self-adjoint extension of $\dot{A}$.

The existence, description, and analog of von Neumann's formulas for self-adjoint bi-extensions and $(*)$-extensions were discussed in [35] (see also [5], [9], [14]). In what follows we suppose that $\dot{A}$ has equal deficiency indices and will say that a quasi-self-adjoint extension $T$ of $\dot{A}$ belongs to the **class** $\Lambda(\dot{A})$ if $\rho(T) \neq \emptyset$, $\mathrm{Dom}(\dot{A}) = \mathrm{Dom}(T) \cap \mathrm{Dom}(T^*)$, and $T$ admits $(*)$-extensions.

Recall that two quasi-self-adjoint extensions $T_1$ and $T_2$ of $\dot{A}$ are called **disjoint** if

$$\mathrm{Dom}(T_1) \cap \mathrm{Dom}(T_2) = \mathrm{Dom}(\dot{A})$$

and **transversal** if, in addition,

$$\mathrm{Dom}(T_1) + \mathrm{Dom}(T_2) = \mathrm{Dom}(\dot{A}^*).$$

Note that from von Neumann formulas immediately follows that two transversal self-adjoint extensions are automatically disjoint.

Let $\dot{A}$ be a closed densely defined symmetric operator and let $T \in \Lambda(\dot{A})$. It has been shown in [4] that $T \in \Lambda(\dot{A})$ if and only if there exists a self-adjoint extension $\tilde{A}$ of $\dot{A}$ transversal to $T$, and, moreover, the formulas

$$\mathbb{A} = \dot{A}^* - \mathcal{R}^{-1}\dot{A}^*(I - \mathcal{P}_{T\tilde{A}}), \quad \mathbb{A}^* = \dot{A}^* - \mathcal{R}^{-1}\dot{A}^*(I - \mathcal{P}_{T^*\tilde{A}}), \tag{2.1}$$

set a bijection between the set of all $(*)$-extensions of $T \in \Lambda(\dot{A})$ and their adjoint and the set of all all self-adjoint extensions $\tilde{A}$ of the operator $\dot{A}$ that are transversal to $T$. Here $\mathcal{P}_{T\tilde{A}}$ and $\mathcal{P}_{T^*\tilde{A}}$ are the projectors in $\mathcal{H}_+$ onto $\mathrm{Dom}(T)$ and $\mathrm{Dom}(T^*)$, corresponding to the direct decompositions

$$\mathcal{H}_+ = \mathrm{Dom}(T)\dot{+}\mathfrak{M}_{\tilde{A}}, \quad \mathcal{H}_+ = \mathrm{Dom}(T^*)\dot{+}\mathfrak{M}_{\tilde{A}}, \tag{2.2}$$

where $\mathfrak{M}_{\tilde{A}} = \mathrm{Dom}(\tilde{A}) \ominus \mathrm{Dom}(\dot{A})$. If a $(*)$-extension $\mathbb{A}$ of $T$ takes the form (2.1), we say that $\mathbb{A}$ is **generated** by $\tilde{A}$.

It is shown in [4] that if deficiency indices of $\dot{A}$ are finite and equal, then for each quasi-self-adjoint extension $T$ of $\dot{A}$ with $\rho(T) \neq \emptyset$ there exists a self-adjoint extension of $\dot{A}$ transversal to $T$. The latter is also true if there is $z \in \mathbb{C}$ such that $z, \bar{z} \in \rho(T)$ even for the case of infinite deficiency indices of $\dot{A}$.

Recall that a linear operator $T$ in a Hilbert space $\mathfrak{H}$ is called **accretive** [24] if $\mathrm{Re}\,(Tf, f) \geq 0$ for all $f \in \mathrm{Dom}(T)$ and **maximal accretive** ($m$-accretive) if it is accretive and has no accretive extensions in $\mathfrak{H}$. The following statements are equivalent [31]:

(i) the operator $T$ is $m$-accretive;
(ii) the operator $T$ is accretive and its resolvent set contains points from the left half-plane;
(iii) the operators $T$ and $T^*$ are accretive.

The resolvent set $\rho(T)$ of $m$-accretive operator contains the open left half-plane $\Pi_-$ and

$$||(T - zI)^{-1}|| \leq \frac{1}{|\mathrm{Re}\, z|}, \quad \mathrm{Re}\, z < 0.$$

Let $\mathcal{A}$ and $\mathcal{B}$ be two densely defined closed accretive operators such that

$$(\mathcal{A}f, g) = (f, \mathcal{B}g), \ f \in \mathrm{Dom}(\mathcal{A}), \ g \in \mathrm{Dom}(\mathcal{B}).$$

It was proved in [31] that there exists a maximal accretive operator $T$ such that

$$T \supset \mathcal{A} \quad \text{and} \quad T^* \supset \mathcal{B}.$$

In particular, it follows that if $\dot{A}$ is nonnegative symmetric operator, then there exist maximal accretive quasi-self-adjoint extensions of $\dot{A}$.

Let $T$ be a quasi-self-adjoint maximal accretive extension of a nonnegative operator $\dot{A}$. A (*)-extension $\mathbb{A}$ of $T$ is called accretive if $\mathrm{Re}\,(\mathbb{A}f, f) \geq 0$ for all $f \in \mathcal{H}_+$. This is equivalent to that the real part $\mathrm{Re}\,\mathbb{A} = (\mathbb{A} + \mathbb{A}^*)/2$ is nonnegative self-adjoint bi-extension of $\dot{A}$.

**Definition 2.3.** Let $\dot{A}$ have finite equal deficiency indices. A system of equations

$$\begin{cases} (\mathbb{A} - zI)x = KJ\varphi_- \\ \varphi_+ = \varphi_- - 2iK^*x \end{cases},$$

or an array

$$\Theta = \begin{pmatrix} \mathbb{A} & K & J \\ \mathcal{H}_+ \subset \mathcal{H} \subset \mathcal{H}_- & & E \end{pmatrix} \tag{2.3}$$

is called an **L-system** if:

(1) $\mathbb{A}$ is a (*)-extension of an operator $T$ of the class $\Lambda(\dot{A})$;
(2) $J = J^* = J^{-1} \in [E, E]$, $\quad \dim E < \infty$;
(3) $\mathrm{Im}\,\mathbb{A} = KJK^*$, where $K \in [E, \mathcal{H}_-]$, $K^* \in [\mathcal{H}_+, E]$, and

$$\mathrm{Ran}(K) = \mathrm{Ran}(\mathrm{Im}\,\mathbb{A}). \tag{2.4}$$

In the definition above $\varphi_- \in E$ stands for an input vector, $\varphi_+ \in E$ is an output vector, and $x$ is a state space vector in $\mathcal{H}$. An operator $\mathbb{A}$ is called a *state-space operator* of the system $\Theta$, $J$ is a *direction operator*, and $K$ is a *channel operator*. A system $\Theta$ of the form (2.3) is called an *accretive system* [17] if its main operator $\mathbb{A}$ is accretive and *accumulative system* [18] if its main operator $\mathbb{A}$ is accumulative, i.e., satisfies

$$(\mathrm{Re}\,\mathbb{A}f, f) \leq (\dot{A}^*f, f) + (f, \dot{A}^*f), \quad f \in \mathcal{H}_+. \tag{2.5}$$

We associate with an L-system $\Theta$ the operator-valued function

$$W_\Theta(z) = I - 2iK^*(\mathbb{A} - zI)^{-1}KJ, \quad z \in \rho(T), \tag{2.6}$$

which is called a **transfer operator-valued function** of the L-system $\Theta$. We also consider the operator-valued function

$$V_\Theta(z) = K^*(\mathrm{Re}\,\mathbb{A} - zI)^{-1}K. \tag{2.7}$$

It was shown in [14] that both (2.6) and (2.7) are well defined. The transfer operator-function $W_\Theta(z)$ of the system $\Theta$ and an operator-function $V_\Theta(z)$ of the form (2.7) are connected by the relations valid for $\mathrm{Im}\,z \neq 0$, $z \in \rho(T)$,

$$\begin{aligned} V_\Theta(z) &= i[W_\Theta(z) + I]^{-1}[W_\Theta(z) - I]J, \\ W_\Theta(z) &= (I + iV_\Theta(z)J)^{-1}(I - iV_\Theta(z)J). \end{aligned} \tag{2.8}$$

The function $V_\Theta(z)$ defined by (2.7) is called the **impedance function** of an L-system $\Theta$ of the form (2.3). It was shown in [14] that the class $N(R)$ of all Herglotz-Nevanlinna functions in a finite-dimensional Hilbert space $E$ that can be realized as impedance functions of an L-system is described by conditions (1.7). In particular, the following theorem [3], [14] takes place.

**Theorem 2.4.** *Let $\Theta$ be an L-system of the form* (2.3)*. Then the impedance function $V_\Theta(z)$ of the form* (2.7) *belongs to the class $N(R)$.*

*Conversely, let an operator-valued function $V(z)$ belong to the class $N(R)$. Then $V(z)$ can be realized as the impedance function of an L-system $\Theta$ of the form* (2.3) *with a preassigned direction operator $J$ for which $I+iV(-i)J$ is invertible.*

We will heavily rely on Theorem 2.4 in the last two sections of the present paper.

## 3. The Friedrichs and Kreĭn-von Neumann extensions

We recall that a symmetric operator $\dot{B}$ is called **non-negative** if

$$(\dot{B}f,f)\geq 0,\quad \forall f\in \mathrm{Dom}(\dot{B}).$$

Let $\dot{B}$ be a closed densely defined non-negative operator a Hilbert space $\mathcal{H}$ and let $\dot{B}^*$ be its adjoint. Consider the sesquilinear form $\tau_{\dot{B}}[f,g]=(\dot{B}f,g)$, $f,g\in\mathrm{Dom}(\dot{B})$. A sequence $\{f_n\}\subset\mathrm{Dom}(\dot{B})$ is called $\tau_{\dot{B}}$**-converging** to the vector $u\in\mathcal{H}$ if

$$\lim_{n\to\infty} f_n=u \quad\text{and}\quad \lim_{n,m\to\infty}\tau_{\dot{B}}[f_n-f_m]=0.$$

The form $\tau_{\dot{B}}$ is **closable** [24], i.e., there exists a minimal closed extension (the closure) of $\tau_{\dot{B}}$. Following the M. Kreĭn notations we denote by $\dot{B}[\cdot,\cdot]$ the closure of $\tau_{\dot{B}}$ and by $\mathcal{D}[\dot{B}]$ its domain. By definition $\dot{B}[u]=\dot{B}[u,u]$ for all $u\in\mathcal{D}[\dot{B}]$. Because $\dot{B}[u,v]$ is closed, it possesses the property: if

$$\lim_{n\to\infty}u_n=u \quad\text{and}\quad \lim_{n,m\to\infty}\dot{B}[u_n-u_m]=0,$$

then $\lim\limits_{n\to\infty}\dot{B}[u-u_n]=0$. The **Friedrichs extension** $B_F$ of $\dot{B}$ is defined as a non-negative self-adjoint operator associated with the form $\dot{B}[\cdot,\cdot]$ by the First Representation Theorem [24]:

$$(B_F u,v)=\dot{B}[u,v]\quad\text{for all}\quad u\in\mathrm{Dom}(B_F)\quad\text{and for all}\quad v\in\mathcal{D}[\dot{B}].$$

It follows that

$$\mathrm{Dom}(B_F)=\mathcal{D}[\dot{B}]\cap\mathrm{Dom}(\dot{B}^*),\ B_F=\dot{B}^*\restriction\mathrm{Dom}(B_F).$$

The Friedrichs extension $B_F$ is a unique non-negative self-adjoint extension having the domain in $\mathcal{D}[\dot{B}]$. Notice that by the Second Representation Theorem [24] one has

$$\mathcal{D}[\dot{B}]=\mathcal{D}[B_F]=\mathrm{Dom}(B_F^{1/2}),\ \dot{B}[u,v]=(B_F^{1/2}u,B_F^{1/2}v),\ u,v\in\mathcal{D}[\dot{B}].$$

Let $\dot{B}$ be a non-negative closed densely defined symmetric operator. Consider the family of symmetric contractions

$$\dot{A}^{(a)}=(aI-\dot{B})(aI+\dot{B})^{-1},\ a>0,$$

defined on $\mathrm{Dom}(\dot{A}^{(a)}) = (aI+\dot{B})\mathrm{Dom}(\dot{B})$. Notice that the orthogonal complement $\mathfrak{N}^{(a)} = \mathcal{H} \ominus \mathrm{Dom}(\dot{A}^{(a)})$ coincides with the defect subspace $\mathfrak{N}_{-a}$ of the operator $\dot{B}$. Let $\dot{A} = \dot{A}^{(1)}$ and let $b = (1-a)(a+1)^{-1}$. Then $b \in (-1,1)$ and

$$\dot{A}^{(a)} = (\dot{A} - bI_{\mathcal{H}})(I - b\dot{A})^{-1}.$$

Clearly, there is a one-one correspondence given by the Cayley transform

$$B = a(I - A^{(a)})(I + A^{(a)})^{-1}, \quad A^{(a)} = (aI - B)(aI + B)^{-1},$$

between all non-negative self-adjoint extensions $B$ of the operator $\dot{B}$ and all self-adjoint contractive ($sc$) extensions $A^{(a)}$ of $\dot{A}^{(a)}$. As was established by M. Kreĭn in [25], [26] the set of all $sc$-extensions of $\dot{A}$ forms an operator interval $[A_\mu, A_M]$. Following M. Kreĭn's notations we call the extreme contractive self-adjoint extensions $A_\mu$ and $A_M$ of a symmetric contraction $\dot{A}$ by the **rigid** and the **soft** extensions, respectively. The next result describe the sesquilinear form $B[u,v]$ by means the fractional-linear transformation $A = (I-B)(I+B)^{-1}$.

**Proposition 3.1.**

(1) *Let $B$ be a non-negative self-adjoint operator and let $A = (I-B)(I+B)^{-1}$ be its Cayley transform. Then*

$$\begin{aligned} &\mathcal{D}[B] = \mathrm{Ran}((I+A)^{1/2}), \\ &B[u,v] = -(u,v) + 2\left((I+A)^{-1/2}u, (I+A)^{-1/2}v\right), \quad u,v \in \mathcal{D}[B]. \end{aligned} \tag{3.1}$$

(2) *Let $\dot{B}$ be a closed densely defined non-negative symmetric operator and let $B$ be its non-negative self-adjoint extension. If $\dot{A} = (I-\dot{B})(I+\dot{B})^{-1}$, $A = (I-B)(I+B)^{-1}$, then*

$$\mathcal{D}[B] = \mathrm{Ran}(I+A_\mu)^{1/2} \dotplus \mathrm{Ran}(A-A_\mu)^{1/2}. \tag{3.2}$$

*Proof.* (1). Since $B = (I-A)(I+A)^{-1}$, one obtains with $f = (I+A)h$,

$$\begin{aligned} B[f] = ((I-A)h,(I+A)h) &= -\|(I+A)h\|^2 + 2\|(I+A)^{1/2}h\|^2 \\ &= -\|f\|^2 + 2\|(I+A)^{-1/2}f\|^2. \end{aligned}$$

Now the closure procedure leads to (3.1).

(2) Since $A$ is a $sc$-extension of $\dot{A}$, we get $A_\mu \le A \le A_M$. Hence $I + A = I + A_\mu + (A - A_\mu)$. Because $I + A_\mu$ and $A - A_\mu$ are non-negative self-adjoint operators, we get the equality [21]:

$$\mathrm{Ran}((I+A)^{1/2}) = \mathrm{Ran}((I+A_\mu)^{1/2}) + \mathrm{Ran}((A-A_\mu)^{1/2}).$$

Since $\mathrm{Ran}((I+A_\mu)^{1/2}) \cap \mathfrak{N} = \{0\}$, where $\mathfrak{N} = \mathcal{H} \ominus \mathrm{Dom}(\dot{A})$, and $\mathrm{Ran}(A - A_\mu) \subseteq \mathfrak{N}$, we get $\mathrm{Ran}((I+A_\mu)^{1/2}) \cap \mathrm{Ran}((A-A_\mu)^{1/2}) = \{0\}$. Then we arrive to (3.2). □

We note that $\mathrm{Ran}(\tilde{B}^{1/2}) = \mathrm{Ran}((I-\tilde{A})^{1/2})$. Now let $A_\mu$ and $A_M$ be the rigid and the soft extensions of $\dot{A}$. Then the operators

$$B_F = (I - A_\mu)(I + A_\mu)^{-1}, \tag{3.3}$$

and

$$B_K = (I - A_M)(I + A_M)^{-1}, \tag{3.4}$$

are non-negative self-adjoint extensions of $\dot{B}$. It also follows (see [25], [26]) that

$$B_F = a(I - A_\mu^{(a)})(I + A_\mu^{(a)})^{-1}, \quad B_K = a(I - A_M^{(a)})(I + A_M^{(a)})^{-1}.$$

Since, the operators $A_\mu^{(a)}$ and $A_M^{(a)}$ posses the properties

$$\operatorname{Ran}((I + A_\mu^{(a)})^{1/2}) \cap \mathfrak{N}_{-a} = \operatorname{Ran}((I - A_M^{(a)})^{1/2}) \cap \mathfrak{N}_{-a} = \{0\},$$

we get the following result [25].

**Proposition 3.2.** *Let $B$ be a non-negative self-adjoint extension of $\dot{B}$ and let $E(\lambda)$ be its resolution of identity. Then*

1. $B = B_F$ *if and only if at least for one $a > 0$ (then for all $a > 0$) the relation*

$$\int_0^\infty \lambda (dE(\lambda)\varphi, \varphi) = +\infty, \tag{3.5}$$

*holds for each $\varphi \in \mathfrak{N}_{-a} \setminus \{0\}$;*

2. $B = B_K$ *if and only if at least for one $a > 0$ (then for all $a > 0$) the relation*

$$\int_0^\infty \frac{(dE(\lambda)\varphi, \varphi)}{\lambda} = +\infty, \tag{3.6}$$

*holds for each $\varphi \in \mathfrak{N}_{-a} \setminus \{0\}$.*

The self-adjoint extension $B_F$ given by (3.3) coincides [25] with the *Friedrichs extension* of $\dot{B}$. In the sequel we will call the operator $B_K$ defined in (3.4) by the **Kreĭn-von Neumann extension** of $\dot{B}$.

## 4. Bi-extensions of non-negative symmetric operators

First we consider the case of bounded non-densely defined non-negative symmetric operator $\dot{B}$.

**Theorem 4.1.** *Let $\dot{B}$ be a bounded non-densely defined non-negative symmetric operator in a Hilbert space $\mathcal{H}$, $\operatorname{Dom}(\dot{B}) = \mathcal{H}_0$. Let $\dot{B}^* \in [\mathcal{H}, \mathcal{H}_0]$ be the adjoint of $\dot{B}$. Put $\dot{B}_0 = P_{\mathcal{H}_0}\dot{B}$, $\mathfrak{L} = \mathcal{H} \ominus \mathcal{H}_0$, where $P_{\mathcal{H}_0}$ is an orthogonal projection in $\mathcal{H}$ onto $\mathcal{H}_0$. Then the following statements are equivalent:*

(i) *$\dot{B}$ admits bounded non-negative self-adjoint extensions in $\mathcal{H}$;*

(ii) $\sup\limits_{f \in \mathcal{H}_0} \dfrac{||\dot{B}f||^2}{(\dot{B}f, f)} < \infty$*;*

(iii) $\dot{B}^*\mathfrak{L} \subseteq \operatorname{Ran}(\dot{B}_0^{1/2})$.

*Proof.* Since $(\dot{B}f, f) = ||\dot{B}_0^{1/2} f||^2$, $f \in \mathcal{H}_0$, and

$$\dot{B}^* = \dot{B}_0 P_{\mathcal{H}_0} + \dot{B}^* P_{\mathfrak{L}},$$

conditions (i) and (ii) are equivalent due to the Douglas Theorem [19]. Suppose $\dot{B}$ admits a bounded non-negative self-adjoint extension $B$. Then for $f \in \mathcal{H}_0$ one has

$$||\dot{B}f||^2 = ||Bf||^2 = ||B^{1/2}B^{1/2}f||^2 \leq ||B^{1/2}||^2||B^{1/2}f||^2$$
$$= ||B^{1/2}||^2(Bf,f) = ||B^{1/2}||^2(\dot{B}f,f) = ||B^{1/2}||^2||\dot{B}_0^{1/2}f||^2.$$

It follows that statement (ii) holds true.

Now suppose that (iii) is fulfilled. Then the operator $L_0 := \dot{B}_0^{[-1/2]}\dot{B}^*\upharpoonright\mathfrak{L}$ is bounded, where $\dot{B}_0^{[-1/2]}$ is the Moore-Penrose inverse to $\dot{B}_0^{1/2}$. Let $L_0^* \in [\mathcal{H}_0, \mathfrak{L}]$ be the adjoint to $L_0$. Set

$$\mathcal{B}_0 = \dot{B}P_{\mathcal{H}_0} + (\dot{B}^* + L_0^*L_0)P_{\mathfrak{L}}. \tag{4.1}$$

Then $\mathcal{B}_0$ is bounded extension of $\dot{B}$ in $\mathcal{H}$. Let $P_{\mathfrak{L}}$ be the orthogonal projection operator in $\mathcal{H}$ onto $\mathfrak{L}$. For $h \in \mathcal{H}$ we have

$$(\mathcal{B}_0h,h) = (\dot{B}P_{\mathcal{H}_0}h + (\dot{B}^* + L_0^*L_0)P_{\mathfrak{L}}h, P_{\mathcal{H}_0}h + P_{\mathfrak{L}}h)$$
$$= ||\dot{B}_0^{1/2}P_{\mathcal{H}_0}h||^2 + ||L_0P_{\mathfrak{L}}h||^2 + 2\mathrm{Re}\,(P_{\mathcal{H}_0}h, \dot{B}^*P_{\mathfrak{L}}h)$$
$$= ||\dot{B}_0^{1/2}P_{\mathcal{H}_0}h||^2 + ||L_0P_{\mathfrak{L}}h||^2 + 2\mathrm{Re}\,(\dot{B}_0^{1/2}P_{\mathcal{H}_0}h, \dot{B}_0^{[-1/2]}\dot{B}^*P_{\mathfrak{L}}h)$$
$$= ||\dot{B}_0^{1/2}P_{\mathcal{H}_0}h + L_0P_{\mathfrak{L}}h||^2.$$

Thus, $\mathcal{B}_0$ is non-negative bounded self-adjoint extension of $\dot{B}$. Therefore (i) is equivalent to (iii). □

*Remark* 4.2. It is easy to see that the conditions

1. $\sup\limits_{f \in \mathcal{H}_0} \dfrac{||\dot{B}f||^2}{(\dot{B}f,f)} < \infty$,
2. there exists $c > 0$ such that $|(\dot{B}f,g)|^2 \leq c(\dot{B}f,f)||g||^2$, $f \in \mathcal{H}_0, g \in \mathcal{H}$,
3. there exists $c > 0$ such that $|(\dot{B}f,g)|^2 \leq c(\dot{B}f,f)||g||^2$, $f \in \mathcal{H}_0, g \in \mathfrak{L}$

are equivalent.

Now we consider semi-bounded (in particular non-negative) symmetric densely defined operators $\dot{A}$,

$$(\dot{A}x,x) \geq m(x,x), \qquad x \in \mathrm{Dom}(\dot{A}).$$

According to the classical von Neumann's theorem there exists a self-adjoint extension $A$ of $\dot{A}$ with an arbitrary close to $m$ lower bound. It was shown later by Friedrichs that operator $\dot{A}$ actually admits a self-adjoint extension with the same lower bound. In this section we are going to show that for the case of a self-adjoint bi-extension of $\dot{A}$ the analogue of von Neumann's theorem is true while the analogue of the Friedrichs theorem, generally speaking, does not take place.

**Theorem 4.3.** *Let $\dot{A}$ be a semi-bounded operator with a lower bound $m$ and $\hat{A}$ be its symmetric extension with the same lower bound. Then $\dot{A}$ admits a self-adjoint* *bi-extension* $\mathbb{A}$ *with the same lower bound and containing* $\hat{A}$ ($\mathbb{A} \supset \hat{A}$) *if and only if there exists a number* $k > 0$ *such that*

$$\left|((\hat{A} - mI)f, h)\right|^2 \leq k((\hat{A} - mI)f, f)\,\|h\|_+^2, \tag{4.2}$$

*for all* $f \in \mathrm{Dom}(\hat{A})$, $h \in \mathcal{H}_+$.

*Proof.* Let $\mathcal{H}_+ \subseteq \mathcal{H} \subseteq \mathcal{H}_-$ be the rigged triplet generated by $\dot{A}$ and $\mathcal{R}$ be a Riesz-Berezansky operator corresponding to this triplet. In the Hilbert space $\mathcal{H}_+$ consider the operator

$$\dot{B} := \mathcal{R}(\hat{A} - mI), \quad \mathrm{Dom}(\dot{B}) = \mathrm{Dom}(\hat{A}).$$

Then $(\dot{B}f, f)_+ = ((\hat{A}f - mI)f, f) \geq 0$ for all $f \in \mathrm{Dom}(\dot{B})$. Observe that $\mathbb{A}$ is a self-adjoint bi-extension of $\dot{A}$ containing $\hat{A}$ if and only if the operator $B := \mathcal{R}\mathbb{A}$ is a $(+)$-bounded and $(+)$-self-adjoint extension of the operator $\dot{B}$ in $\mathcal{H}_+$. It follows from Theorem 4.1 and Remark 4.2 that the operator $\dot{B}$ admits $(+)$-non-negative bounded self-adjoint extension in $\mathcal{H}_+$ if and only if there exists $k > 0$ such that

$$|(\dot{B}f, h)_+|^2 \leq k(\dot{B}f, f)_+ ||h||_+^2, \; f \in \mathrm{Dom}(\dot{B}), h \in \mathcal{H}_+.$$

This is equivalent to (4.2) □

Remark 4.2 yields that if $\dot{A}$ has at least one self-adjoint bi-extension $\mathbb{A}$ containing $\hat{A}$ with the same lower bound, then it has infinitely many of such bi-extensions.

**Corollary 4.4.** *Inequalities* (4.2) *take place if and only if there exists a constant* $C > 0$ *such that*

$$|((\hat{A} - mI)f, \varphi_a)|^2 \leq C((\hat{A} - mI)f, f)\,\|\varphi_a\|_+^2, \tag{4.3}$$

*for all* $f \in \mathrm{Dom}(\hat{A})$ *and all* $\varphi_a$ *such that* $(\dot{A}^* - (m - a)I)\varphi_a = 0$, $(a > 0)$.

*Proof.* Suppose (4.2). Then for $h = \varphi_a \in \ker(\dot{A}^* - (m - a)I))$ we have (4.3). Now let us show that (4.2) follows from (4.3). It is known that there exists a self-adjoint extension $A$ of $\hat{A}$ (for instance, the Friedrichs extension of $\hat{A}$) with the lower bound $m$. If $\lambda$ is a regular point for $A$, then

$$\mathcal{H}_+ = \mathrm{Dom}(A) \dotplus \mathfrak{N}_\lambda. \tag{4.4}$$

Indeed, if $f \in \mathcal{H}_+ = \mathrm{Dom}(\dot{A}^*)$, then there exists an element $g \in \mathrm{Dom}(A)$ such that $(\dot{A}^* - \lambda I)f = (A - \lambda I)g$. This implies $(\dot{A}^* - \lambda I)(f - g) = 0$ and hence $(f - g) \in \mathfrak{N}_\lambda$ for any $f \in \mathcal{H}_+$ and $g \in \mathrm{Dom}(A)$, which confirms (4.4). Further, applying Cauchy-Schwartz inequality we obtain

$$\begin{aligned}|((\hat{A} - mI)f, g)|^2 &\leq ((\hat{A} - mI)f, f)((A - mI)g, g)\\ &\leq \hat{C}((\hat{A} - mI)f, f)\|g\|_+^2,\end{aligned} \tag{4.5}$$

for $f \in \mathrm{Dom}(\hat{A})$ and $g \in \mathrm{Dom}(A)$. Clearly, all the points of the form $(m - a)$, $(a > 0)$ are regular points for $A$ and the points of a regular type for $\dot{A}$. Thus (4.4) implies

$$\mathcal{H}_+ = \mathrm{Dom}(A) \dotplus \mathfrak{N}_{m-a}. \tag{4.6}$$

Let $h \in \mathcal{H}_+$ be an arbitrary vector. Applying (4.6) we get $h = g + \psi_a$, where $g \in \mathrm{Dom}(A)$ and $\psi_a \in \mathfrak{N}_{m-a}$. Adding up inequalities (4.3) and (4.5) and taking into account that the norms $\|\cdot\|$ and $\|\cdot\|_+$ are equivalent on $\mathfrak{N}_{m-a}$ we get (4.2). □

The following theorem is the analogue of the classical von Neumann's result.

**Theorem 4.5.** *Let $\varepsilon$ be an arbitrary small positive number and $\dot{A}$ be a semi-bounded operator with the lower bound $m$. Then there exist infinitely many semi-bounded self-adjoint bi-extensions with the lower bound $(m-\varepsilon)$.*

*Proof.* First we show that the inequality

$$|((\dot{A}-(m-\varepsilon)I)f,g)|^2 \le k((\dot{A}-(m-\varepsilon)I)f,f)\|g\|_+^2,$$

takes place for all $f \in \mathrm{Dom}(\dot{A})$, $g \in \mathfrak{M}$, and $k > 0$. Indeed,

$$\begin{aligned}
|((\dot{A}-(m-\varepsilon)I)f,g)| &= |(f,(\dot{A}^*-(m-\varepsilon I)g)| \\
&\le |(f,\dot{A}^*g)| + |m-\varepsilon|\cdot|(f,g)| \le \|f\|\cdot\|A^*g\| + |m-\varepsilon|\cdot\|f\|\cdot\|g\| \\
&\le \frac{1}{\sqrt{\varepsilon}}((\dot{A}-(m-\varepsilon)I)f,f)^{1/2}\|g\|_+ + \frac{|m-\varepsilon|}{\sqrt{\varepsilon}}((\dot{A}-(m-\varepsilon)I)f,f)^{1/2}\|g\|_+ \\
&= \frac{1+|m-\varepsilon|}{\sqrt{\varepsilon}}((\dot{A}-(m-\varepsilon)I)f,f)\|g\|_+.
\end{aligned}$$

The statement of the theorem follows from Theorem 4.3 and Remark 4.2. □

**Theorem 4.6.** *A non-negative densely-defined operator $\dot{A}$ admits a non-negative self-adjoint bi-extension if and only if the Friedrichs and Kreĭn-von Neumann extensions of $\dot{A}$ are transversal.*

*Proof.* It was shown in [28] (see also [12]) that the Friedrichs and Kreĭn-von Neumann extensions are transversal if and only if

$$\mathrm{Dom}(\dot{A}^*) \subseteq \mathcal{D}[A_K]. \tag{4.7}$$

Suppose that the Friedrichs extension $A_F$ and the Kreĭn-von Neumann extension $A_K$ of the operator $\dot{A}$ are transversal. Then the inclusion (4.7) holds. This means that $\mathcal{H}_+ \subseteq \mathrm{Dom}(A_K^{1/2})$. Since $||h||_+ \ge ||h||$ for all $h \in \mathcal{H}_+$, and $A_K^{1/2}$ is closed in $\mathcal{H}$, the closed graph theorem yields now that $A_K^{1/2} \in [\mathcal{H}_+, \mathcal{H}]$, i.e., there exists a number $c > 0$ such that

$$||A_K^{1/2}u||^2 = A_K[u] \le c||u||_+^2.$$

It follows that the sesquilinear form $A_K[u,v] = (A_K^{1/2}u, A_K^{1/2}v)$, $u,v \in \mathcal{H}_+$ is bounded on $\mathcal{H}_+$. Therefore, by Riesz theorem, there exists an operator $\mathbb{A}_K \in [\mathcal{H}_+, \mathcal{H}_-]$ such that

$$(\mathbb{A}_K u, v) = A_K[u,v], \quad u,v \in \mathcal{H}_+,\ u \in \mathcal{H}_+.$$

Due to $A_K[u] \geq 0$ for all $u \in \mathcal{D}[A_K]$, the operator $\mathbb{A}_K$ is non-negative. Since $(A_K u, v) = A_K[u, v]$ for all $u \in \mathrm{Dom}(A_K)$ and all $v \in \mathcal{D}[A_K]$, we get

$$(\mathbb{A}_K u, v) = (A_K u, v), \quad u \in \mathrm{Dom}(A_K),\ v \in \mathcal{H}_+.$$

Hence $\mathbb{A}_K \supset A_K$, i.e., $\mathbb{A}_K$ is t-self-adjoint bi-extension of $\dot{A}$ with quasi-kernel $A_K$.

Conversely, let $\dot{A}$ admits a non-negative self-adjoint bi-extension. Then from Theorem 4.3 we get the equality

$$|(\dot{A}f, h)|^2 \leq k(\dot{A}f, f)||h||_+^2,$$

for all $f \in \mathrm{Dom}(\dot{A})$ and all $h \in \mathcal{H}_+ = \mathrm{Dom}(\dot{A}^*)$, and some $k > 0$. Applying the theorem by T. Ando and K. Nishio [2] (see also [12]) we get that $\mathcal{H}_+ \subseteq \mathcal{D}[A_K]$. Now (4.7) yields that $A_F$ and $A_K$ are transversal. □

**Corollary 4.7.** *If a non-negative densely-defined symmetric operator $\dot{A}$ admits a non-negative self-adjoint bi-extension, then it also admits a non-negative self-adjoint bi-extension $\mathbb{A}$ with quasi-kernel $A_K$.*

It follows from Theorem 4.6 that if $A_K = A_F$, then the operator $\dot{A}$ does not admit non-negative self-adjoint bi-extensions. Consequently, in this case the analogue of the Friedrichs theorem is not true. The following theorem provides a criterion on when the analogue of the Friedrichs theorem does take place.

**Theorem 4.8.** *A non-negative densely-defined symmetric operator $\dot{A}$ admits a non-negative self-adjoint bi-extensions if and only if*

$$\int_0^{\infty} t\, d(E(t)h, h) < \infty \quad \text{for all} \quad h \in \mathfrak{N}_{-a},\ a > 0, \tag{4.8}$$

*where $E(t)$ is a spectral function of the Kreĭn-von Neumann extension $A_K$ of $\dot{A}$.*

*Proof.* The inequality (4.8) is equivalent to the inclusion

$$\mathfrak{N}_{-a} \subset \mathrm{Dom}(A_K^{1/2}) = \mathcal{D}[A_K].$$

Since $-a$ is a regular point of $A_K$, the direct decomposition

$$\mathrm{Dom}(\dot{A}^*) = \mathrm{Dom}(A_K) \dot{+} \mathfrak{N}_{-a},$$

holds. So, from (4.7) we get that (4.8) is equivalent to transversality of $A_F$ and $A_K$. The latter is equivalent to existence of non-negative self-adjoint bi-extension of $\dot{A}$ (see Theorem 4.6). □

Observe, that since $\mathfrak{N}_{-a}$ is a subspace in $\mathcal{H}$, $A_K$ is closed in $\mathcal{H}$, condition (4.8) is equivalent to the following: there exists a positive number $k > 0$, depending on $a$, such that

$$\int_0^{\infty} t\, d(E(t)h, h) < k||h||^2, \quad \forall h \in \mathfrak{N}_{-a},\ a > 0. \tag{4.9}$$

On the other hand, (4.8) is equivalent (see proof of Theorem 4.6) to the existence of $k > 0$ such that

$$\int_0^\infty t\, d(E(t)f, f) < k\|f\|_+^2, \quad \forall f \in \mathrm{Dom}(\dot{A}^*).$$

## 5. Accretive bi-extensions

Let $\dot{A}$ be a densely defined and closed non-negative symmetric operator. In this section we will study the existence of accretive (*)-extensions of a given maximal accretive operator $T \in \Lambda(\dot{A})$.

**Theorem 5.1.** *If $\mathbb{A}$ is a quasi-self-adjoint bi-extension of $T \in \Omega(\dot{A})$ generated by $\tilde{A}$ via (2.1), then for all $\phi = h + f \in \mathcal{H}_+$, $h \in \mathrm{Dom}(T)$, and $f \in \mathrm{Dom}(\tilde{A})$ we have*

$$(\mathbb{A}\phi, \phi) = (Th, h) + (\tilde{A}f, f) + 2\mathrm{Re}\,(Th, f). \tag{5.1}$$

*Proof.* Let $\mathbb{A} = \dot{A}^* - \mathcal{R}^{-1}\dot{A}^*(I - \mathcal{P}_{T\tilde{A}})$ according to (2.1). Note that for any $f \in \mathrm{Dom}(\tilde{A})$ we have that $P_{T\tilde{A}}f \in \mathrm{Dom}(\dot{A})$ and hence

$$(\mathcal{P}_{T\tilde{A}}f, Th) = (\dot{A}\mathcal{P}_{T\tilde{A}}f, h),$$

for any $h \in \mathrm{Dom}(T)$. Besides,

$$\left(\mathcal{R}^{-1}\dot{A}^*(I - \mathcal{P}_{T\tilde{A}})f, g\right) = 0, \qquad \forall f, g \in \mathrm{Dom}(\tilde{A}).$$

Indeed, since $\mathrm{Dom}(\tilde{A}) = \mathrm{Dom}(\dot{A}) \oplus (U + I)\mathfrak{N}_i$, where $U$ is a unitary operator from $\mathfrak{N}_i$ onto $\mathfrak{N}_{-i}$, we have $(I - \mathcal{P}_{T\tilde{A}})f = (I + U)\varphi$, for $\varphi \in \mathfrak{N}_i$, and

$$\dot{A}^*(I - \mathcal{P}_{T\tilde{A}})f = i(I - U)\varphi.$$

Moreover, from the (+)-orthogonality of $(U + I)\mathfrak{N}_i$ and $(U - I)\mathfrak{N}_i$ we obtain the desired equation. Further,

$$\begin{aligned}(\mathbb{A}\phi, \phi) &= (Th + \tilde{A}f - \mathcal{R}^{-1}\tilde{A}(I - \mathcal{P}_{T\tilde{A}})f, h + f)\\ &= (Th, h) + (\tilde{A}f, f) + (Th, f) - (\tilde{A}(I - \mathcal{P}_{T\tilde{A}})f, h)_+ + (\tilde{A}f, h),\end{aligned}$$

and

$$\begin{aligned}(\tilde{A}(I - \mathcal{P}_{T\tilde{A}})f, h)_+ &= (\tilde{A}(I - \mathcal{P}_{T\tilde{A}})f, h) - ((I - \mathcal{P}_{T\tilde{A}})f, Th)\\ &= (\tilde{A}f, h) - (\dot{A}\mathcal{P}_{T\tilde{A}}f, h) - (f, Th) + (\mathcal{P}_{T\tilde{A}}f, Th)\\ &= (\tilde{A}f, h) - (f, Th).\end{aligned}$$

Consequently, $(\mathbb{A}\phi, \phi) = (Th, h) + (\tilde{A}f, f) + 2\mathrm{Re}\,(Th, f)$. □

**Corollary 5.2.** *Let $T$ be a quasi-self-adjoint maximal accretive extension of $\dot{A}$. Assume that $\mathbb{A} \in [\mathcal{H}_+, \mathcal{H}_-]$ is given by (2.1) and generated by a self-adjoint extension $A$ transversal to $T$. Then $\mathbb{A}$ is accretive if and only if the form*

$$\mathrm{Re}\,(Th, h) + (Ag, g) + 2\,\mathrm{Re}\,(Th, g), \tag{5.2}$$

*is non-negative for all $h \in \mathrm{Dom}(T)$ and $g \in \mathrm{Dom}(A)$.*

By $\Xi(\dot{A})$ we denote the set of all maximal accretive quasi-self-adjoint extensions of the operator $\dot{A}$. In particular, the class $\Xi(\dot{A})$ contains all nonnegative self-adjoint extensions of $\dot{A}$. It follows from Lemma 5.2 that if $T \in \Xi(\dot{A})$ and if $\mathbb{A} \in [\mathcal{H}_+, \mathcal{H}_+]$ of the form (2.1) is accretive, then $A \in \Xi(\dot{A})$. On the class $\Xi(\dot{A})$ we define Cayley transform given by the formula

$$K(T) = (I - T)(I + T)^{-1}, \qquad T \in \Xi(\dot{A}). \tag{5.3}$$

This Cayley transform sets one-to-one correspondence between the class $\Xi(\dot{A})$ and the set of quasi-self-adjoint contractive (qsc) extensions of a symmetric contraction

$$\dot{S} = (I - \dot{A})(I + \dot{A})^{-1},$$

defined on a subspace $\mathrm{Dom}(\dot{S}) = (I + \dot{A})\mathrm{Dom}(\dot{A})$, i.e., both $Q$ and $Q^*$ are extensions of $S$ and $||Q|| \leq 1$. Put

$$\mathfrak{N} = \mathcal{H} \ominus \mathrm{Dom}(\dot{S}). \tag{5.4}$$

Notice that $\mathfrak{N} = \mathfrak{N}_{-1} = \ker(\dot{A}^* + I)$ (the deficiency subspace of $\dot{A}$).

Let $S_\mu = K(A_F)$ and $S_M = K(A_K)$. It was shown in [7], [8], [10] that $Q \in [\mathcal{H}, \mathcal{H}]$ is a qsc-extension of a symmetric contraction $\dot{S}$ if and only if it can be represented in the form

$$Q = \frac{1}{2}(S_M + S_\mu) + \frac{1}{2}(S_M - S_\mu)^{1/2} X (S_M - S_\mu)^{1/2}, \tag{5.5}$$

where $X$ is a contraction in the subspace $\overline{\mathrm{Ran}(S_M - S_\mu)} \subseteq \mathfrak{N}$.

Clearly, if $X$ is a self-adjoint contraction, then (5.5) provides a description of all sc-extensions of a symmetric contraction $S$.

**Lemma 5.3.**

1) *The class $\Xi(\dot{A})$ contains mutually transversal operators if and only if $A_F$ and $A_K$ are mutually transversal.*
2) *Let $T_1$ and $T_2$ belong to $\Xi(\dot{A})$. Then $T_1$ and $T_2$ are mutually transversal if and only if*

$$(K(T_1) - K(T_2))\mathfrak{N} = \mathfrak{N}.$$

*Proof.* It follows from (5.5) that

$$K(T_1) - K(T_2) = \frac{1}{2}(S_M - S_\mu)^{1/2}(X_1 - X_2)(S_M - S_\mu)^{1/2},$$

where $X_l$, $(l = 1, 2)$ are the corresponding to $T_l$ contractions in $\overline{\mathrm{Ran}(S_M - S_\mu)}$. Relation (5.3) yields

$$K(T_1) - K(T_2) = 2\left((I + T_1)^{-1} - (I + T_2)^{-1}\right).$$

Thus

$$(I + T_1)^{-1} - (I + T_2)^{-1} = \frac{1}{4}(S_M - S_\mu)^{1/2}(X_1 - X_2)(S_M - S_\mu)^{1/2}.$$

Furthermore, using [35] we get that

$$\begin{aligned}&\left((I+T_1)^{-1}-(I+T_2)^{-1}\right)\mathfrak{N}_{-1}=\mathfrak{N}_{-1}\\ \iff &\begin{cases}\overline{\operatorname{Ran}(S_M-S_\mu)}=\operatorname{Ran}(S_M-S_\mu)=\mathfrak{N}=\mathfrak{N}_{-1},\\ \qquad\qquad\qquad\qquad \operatorname{Ran}(X_1-X_2)\mathfrak{N}=\mathfrak{N}.\end{cases}\end{aligned}$$ □

In what follows we assume that $A_K$ and $A_F$ are mutually transversal. Let $A_1$ and $A_2$ be two mutually transversal operators from $\Xi(\dot{A})$. Consider a form defined on $\operatorname{Dom}(A_1)\times\operatorname{Dom}(A_2)$ as follows

$$B(f_1,f_2)=(A_1f_1,f_1)+(A_2f_2,f_2)+2\operatorname{Re}(A_1f_1,f_2), \tag{5.6}$$

where $f_l\in\operatorname{Dom}(A_l)$, $(l=1,2)$. Let

$$\phi_l=\frac{1}{2}(I+A_l)f_l,\qquad S_l\phi_l=\frac{1}{2}(I-A_l)f_l,$$

be the Cayley transform of $A_l$ for $l=1,2$. Then

$$f_l=(I+S_l)\phi_l,\qquad A_lf_l=(I-S_l)\phi_l,\quad (l=1,2). \tag{5.7}$$

Substituting (5.7) into (5.6) we obtain a form defined on $\mathcal{H}\times\mathcal{H}$

$$\tilde{B}(\phi_1,\phi_2)=\|\phi_1+\phi_2\|^2-\|S_1\phi_1+S_2\phi_2\|^2-2\operatorname{Re}\left((S_1-S_2)\phi_1,\phi_2\right).$$

Let us set

$$F=\frac{1}{2}(S_1-S_2),\quad G=\frac{1}{2}(S_1+S_2),\quad u=\frac{1}{2}(\phi_1+\phi_2),\quad v=\frac{1}{2}(\phi_1-\phi_2). \tag{5.8}$$

Then $\tilde{B}(\phi_1,\phi_2)=4H(u,v)$ where

$$H(u,v)=\|u\|^2+(Fv,v)-(Fu,u)-\|Fv+Gu\|^2. \tag{5.9}$$

Moreover, $F\pm G$ are contractive operators. From the above reasoning we conclude that non-negativity of the form $B(f_1,f_2)$ on $\operatorname{Dom}(A_1)\times\operatorname{Dom}(A_2)$ is equivalent to non-negativity of the form $H(u,v)$ on $\mathcal{H}\times\mathcal{H}$.

**Lemma 5.4.** *The form $H(u,v)$ in (5.9) is non-negative for all $u,v\in\mathcal{H}$ if and only if operator $F$ defined in (5.8) is non-negative.*

*Proof.* If $H(u,v)\geq 0$ for all $u,v\in\mathcal{H}$ then $H(0,v)\geq 0$ for all $v\in\mathcal{H}$. Hence $(Fv,v)\geq\|Fv\|^2\geq 0$, i.e., $F\geq 0$.

Conversely, let $F\geq 0$. Since both operators $F\pm G$ are self-adjoint contractions, then $-I\leq F+G\leq I$ and $-I\leq F-G\leq I$. This implies $-(I-F)\leq G\leq I-F$, and thus

$$G=(I-F)^{1/2}X(I-F)^{1/2}, \tag{5.10}$$

where $X$ is a self-adjoint contraction. Then (5.10) yields that for all $u,v\in\mathcal{H}$

$$\begin{aligned}\|Fv+Gu\|&=\|Fv\|^2+\|Gu\|^2+2\operatorname{Re}(Fv,Gu)\\ &\leq\|Fv\|^2+\left((I-F)X(I-F)^{1/2}u,X(I-F)^{1/2}u\right)\\ &\quad+2\left|\left(F(I-F)^{1/2}X(I-F)^{1/2}u,v\right)\right|\end{aligned}$$

$$\begin{aligned}
&= \|Fv\|^2 + \|X(I-F)^{1/2}u\|^2 - (FX(I-F)^{1/2}u, X(I-F)^{1/2}u)\\
&\quad + 2\left|\left(FX(I-F)^{1/2}u, (I-F)^{1/2}v\right)\right|\\
&\leq \|Fv\|^2 + \|X(I-F)^{1/2}u\|^2 - (FX(I-F)^{1/2}u, X(I-F)^{1/2}u)\\
&+ (FX(I-F)^{1/2}u, X(I-F)^{1/2}u) + (F(I-F)^{1/2}v, (I-F)^{1/2}v)\\
&= \|Fv\|^2 + \|X(I-F)^{1/2}u\|^2 + (Fv, v) - \|Fv\|^2\\
&\leq (Fv, v) + \|u\|^2 - (Fu, u).
\end{aligned}$$

Therefore, for all $u, v \in \mathcal{H}$

$$H(u,v) = \|u\|^2 - (Fu,u) + (Fv,v) - \|Fv + Gu\|^2 \geq 0.$$

The lemma is proved. □

**Theorem 5.5.** *Let $\mathbb{A} = \dot{A}^* - \mathcal{R}^{-1}\dot{A}^*(I - \mathcal{P}_{\hat{A}A})$ be a self-adjoint $(*)$-extension of a non-negative symmetric operator $\dot{A}$, with a self-adjoint quasi-kernel $\hat{A} \in \Xi(\dot{A})$, and generated (via (2.1)) by a self-adjoint extension $A$. Then the following statements are equivalent*

(i) *$\mathbb{A}$ is non-negative*
(ii) *$\left(K(\hat{A}) - K(A)\right) \upharpoonright \mathfrak{N}$, (where $\mathfrak{N}$ is defined by (5.4)) is positively defined,*
(iii) *$(\hat{A} + I)^{-1} \geq (A + I)^{-1}$, and $\hat{A}$ is transversal to $A$,*
(iv) *$\hat{A} \leq A$ and $\hat{A}$ is transversal to $A$.*

*Proof.* Let $(\mathbb{A}f, f) \geq 0$ for all $f \in \mathcal{H}_+$. Then due to (5.1) and Corollary 5.2 we have that the form

$$B(g,h) = (\hat{A}g, h) + (Ah, g) + 2\mathrm{Re}\,(\hat{A}g, h), \quad (g \in \mathrm{Dom}(\hat{A}),\, h \in \mathrm{Dom}(A)),$$

is non-negative on $\mathrm{Dom}(\hat{A}) \times \mathrm{Dom}(A)$. Consequently, the form $H(u,v)$ given by (5.9) is non-negative for all $u, v \in \mathcal{H}$ where

$$F = \frac{1}{2}\left(K(\hat{A}) - K(A)\right) \text{ and } G = \frac{1}{2}\left(K(\hat{A}) + K(A)\right).$$

Using Lemma 5.4 we conclude that $F \upharpoonright \mathfrak{N} \geq 0$ and applying Lemma 5.3 yields $F\mathfrak{N} = \mathfrak{N}$. This proves that (i) $\Rightarrow$ (ii). The implication (ii) $\Rightarrow$ (i) can be shown by reversing the argument. Since

$$K(\hat{A}) = -I + 2(\hat{A} + I)^{-1},\ K(A) = -I + 2(A + I)^{-1},$$

we get that (ii) $\Longleftrightarrow$ (iii). Applying inequalities from [24] yields that (iii) $\Longleftrightarrow$ (iv). □

**Theorem 5.6.** *A self-adjoint operator $\hat{A} \in \Xi(\dot{A})$ admits non-negative $(*)$-extensions if and only if $\hat{A}$ is transversal to $A_F$.*

*Proof.* If $\hat{A}$ is transversal to $A_F$, then $\Big(K(\hat{A}) - K(A_F)\Big) \upharpoonright \mathfrak{N}$ is positively defined. Applying Theorem 5.5 we obtain that

$$\mathbb{A} = \dot{A}^* - \mathcal{R}^{-1}\dot{A}^*(I - \mathcal{P}_{\hat{A}A_F}),$$

is a non-negative (*)-extension.

Conversely, if $\mathbb{A} = \dot{A}^* - \mathcal{R}^{-1}\dot{A}^*(I - \mathcal{P}_{\hat{A}A})$ is a (*)-extension of $\hat{A}$, then via Theorem 5.5 we get that $\Big(K(\hat{A}) - K(A)\Big) \upharpoonright \mathfrak{N}$ is positively defined. But then due to the chain of inequalities

$$K(\hat{A}) \geq K(A) \geq K(A_F),$$

the operator $\Big(K(\hat{A}) - K(A_F)\Big) \upharpoonright \mathfrak{N}$ is positively defined as well. According to Lemma 5.3 $\hat{A}$ is transversal $A_F$. □

We note that if $\hat{A}$ is a self-adjoint extension of $\dot{A}$, then all self-adjoint (*)-extensions of $\hat{A}$ coincide with t-self-adjoint bi-extensions of $\dot{A}$ with the quasi-kernel $\hat{A}$. Consequently, Theorem 5.6 gives the criterion of the existence of a non-negative t-self-adjoint bi-extension of $\dot{A}$ and hence provides the conditions when Friedrichs theorem for t-self-adjoint bi-extensions is true.

Now we focus on non-self-adjoint accretive (*)-extensions of operator $T \in \Xi(\dot{A})$.

**Lemma 5.7.** *Let $\mathbb{A}$ be a (*)-extensions of operator $T \in \Xi(\dot{A})$ generated by an operator $A \in \Xi(\dot{A})$. Then the quasi-kernel $\hat{A}$ of the operator* $\operatorname{Re}\mathbb{A}$ *is defined by the formula*

$$\begin{aligned} f &= (Q + I)g + \frac{1}{2}(S + I)(Q^* - S)^{-1}(Q - Q^*)g, \\ \hat{A}f &= (I - Q)g + \frac{1}{2}(I - S)(Q^* - S)^{-1}(Q - Q^*)g, \end{aligned} \tag{5.11}$$

*where $g \in \mathcal{H}$, $Q = K(T)$, $Q^* = K(T^*)$, and $S = K(A)$.*

*Proof.* Let $\mathbb{A} = \dot{A}^* - \mathcal{R}^{-1}\dot{A}^*(I - \mathcal{P}_{TA})$ (of the form (2.1)) be a (*)-extensions of operator $T$ generated by a self-adjoint extension $A$. Let

$$\operatorname{Dom}(A) = \operatorname{Dom}(\dot{A}) \oplus (\mathcal{U} + I)\mathfrak{N}_i,$$

where $\mathcal{U} \in [\mathfrak{N}_i, \mathfrak{N}_{-i}]$ is a unitary mapping. Suppose $f \in \operatorname{Dom}(\hat{A})$, where $\hat{A}$ is a quasi-kernel of $\operatorname{Re}\mathbb{A}$. Due to the transversality of $T^*$ and $A$ and $T$ and $A$ we have

$$f = u + (\mathcal{U} + I)\varphi, \qquad f = v + (\mathcal{U} + I)\psi,$$

where $u \in \operatorname{Dom}(T)$, $v \in \operatorname{Dom}(T^*)$, and $\varphi, \psi \in \mathfrak{N}_i$. Also

$$\begin{aligned} \mathbb{A}f &= Tu + \dot{A}^*(\mathcal{U} + I)\varphi - i\mathcal{R}^{-1}(I - \mathcal{U})\varphi, \\ \mathbb{A}^*f &= T^*v + \dot{A}^*(\mathcal{U} + I)\psi - i\mathcal{R}^{-1}(I - \mathcal{U})\psi, \end{aligned}$$

and

$$\hat{A}f = \frac{1}{2}(\mathbb{A}f + \mathbb{A}^* f)$$
$$= \frac{1}{2}\left(Tu + T^* v + \dot{A}^*(\mathcal{U}+I)\varphi + \dot{A}^*(\mathcal{U}+I)\psi - i\mathcal{R}^{-1}(I-\mathcal{U})(\varphi+\psi)\right).$$

Since $\hat{A}f \in \mathcal{H}$, then $\varphi = -\psi$ and hence any vector $f \in \mathrm{Dom}(\hat{A})$ is uniquely represented in the form

$$f = u + \phi, \quad u \in \mathrm{Dom}(T),\ \phi \in (\mathcal{U}+I)\mathfrak{N}_i,$$

or in the form $f = v - \phi$, $v \in \mathrm{Dom}(T^*)$. By (2.1) (see also [11]) $\hat{A}$ is transversal to $A$ and

$$\mathrm{Re}\,\mathbb{A} = \dot{A}^* - \mathcal{R}^{-1}(I - P_{\hat{A}A}).$$

Thus $\mathfrak{M}_{\hat{A}} \dotplus (\mathcal{U}+I)\mathfrak{N}_i = \mathfrak{M}$, where $\mathfrak{M}_{\hat{A}} = \mathrm{Dom}(\hat{A}) \ominus \mathrm{Dom}(\dot{A})$. It follows from

$$\mathfrak{M}_T \dotplus (\mathcal{U}+I)\mathfrak{N}_i = \mathfrak{M}, \quad \text{where} \quad \mathfrak{M}_T = \mathrm{Dom}(T) \ominus \mathrm{Dom}(\dot{A}),$$

that $\mathcal{P}_{\hat{A}A}\mathfrak{M}_T = \mathfrak{M}_{\hat{A}}$ and hence for any $u \in \mathrm{Dom}(T)$ there exists a $\phi \in (\mathcal{U}+I)\mathfrak{N}_i$ and $f \in \mathrm{Dom}(\hat{A})$ such that $f = u + \phi$. Similarly, for any $v \in \mathrm{Dom}(T)$ there exists a $\phi \in (\mathcal{U}+I)\mathfrak{N}_i$ and $f \in \mathrm{Dom}(\hat{A})$ such that $f = u - \phi$. Since

$$\mathrm{Dom}(T) = (I+Q)\mathcal{H},\ \mathrm{Dom}(T^*) = (I+Q^*)\mathcal{H},\ \mathrm{Dom}(A) = (I+S)\mathcal{H},$$

and

$$Q\!\upharpoonright\!\mathrm{Dom}(\dot{S}) = Q^*\!\upharpoonright\!\mathrm{Dom}(\dot{S}) = S\!\upharpoonright\!\mathrm{Dom}(\dot{S}),$$

we conclude that for any $f \in \mathrm{Dom}(\hat{A})$ there are uniquely defined $g, g_* \in \mathcal{H}$ and $h \in \mathfrak{N}$ such that

$$f = (Q+I)g + (S+I)h, \quad f = (Q^*+I)g_* - (S+I)h. \tag{5.12}$$

Conversely, for every $g \in \mathcal{H}$ (respectively, $g_* \in \mathcal{H}$) there are $g_* \in \mathcal{H}$ (respectively, $g \in \mathcal{H}$) and $h \in \mathfrak{N}$, such that (5.12) holds with $f \in \mathrm{Dom}(\hat{A})$. Since $\dot{A}^*(Q+I)g = (I-Q)g$, $\dot{A}^*(I+Q^*)g_* = (I-Q^*)g_*$, and $\dot{A}^*(I+S)h = (I-S)h$, then

$$\hat{A}f = (I-Q)g + (I-S)h, \quad \hat{A}f = (I-Q^*)g_* - (I-S)h. \tag{5.13}$$

From (5.12) and (5.13) we have $2h = g_* - g$ and $2Sh = Q^* g_* - g$, which implies

$$2(Q^* - S)h = (Q - Q^*)g. \tag{5.14}$$

Since $T^*$ and $A$ are mutually transversal, according to Lemma 5.3 $(Q^* - S)\!\upharpoonright\!\mathfrak{N}$ is an isomorphism on $\mathfrak{N}$. Then (5.14) implies

$$h = \frac{1}{2}(Q^* - S)^{-1}(Q - Q^*)g. \tag{5.15}$$

Substituting (5.15) into (5.12) and (5.13) we obtain (5.11). □

**Lemma 5.8.** *Let $T \in \Xi(\dot{A})$ and $A \in \Xi(\dot{A})$ be a transversal to $T$ self-adjoint operator. If the operator*

$$[K(T) + K(T^*) - 2K(A)]\upharpoonright\mathfrak{N},$$

*is an isomorphism of the space $\mathfrak{N}$ (defined in (5.4)), then the quasi-kernel $\hat{A}$ of the real part of the operator $\mathbb{A} = \dot{A}^* - \mathcal{R}^{-1}\dot{A}^*(I - \mathcal{P}_{TA})$ is a Cayley transform of the operator*

$$\hat{S} = S + (Q - S)(\mathrm{Re}\, Q - S)^{-1}(Q^* - S),$$

*where $S = K(A)$ and $\mathrm{Re}\, Q = (1/2)[K(T) + K(T^*)]$.*

*Proof.* Let $\mathbb{A} = \dot{A}^* - \mathcal{R}^{-1}\dot{A}^*(I - \mathcal{P}_{TA})$. Then by the virtue of Lemma 5.7, formula (5.11) defines the quasi-kernel $\hat{A}$ of the operator $\mathrm{Re}\,\mathbb{A}$. It also follows from (5.11) that

$$f + \hat{A}f = 2g + (Q^* - S)^{-1}(Q - Q^*)g.$$

Let $P_{\mathfrak{N}}$ and $P_{\dot{S}}$ denote the orthoprojection operators in $\mathcal{H}$ according to (5.4) onto $\mathfrak{N}$ and $\mathrm{Dom}(\dot{S})$, respectively. Then

$$\begin{aligned} 2g + (Q^* - S)^{-1}(Q - Q^*)g &= 2P_{\dot{S}}g + (Q^* - S)^{-1}(2Q^* - 2S + Q - Q^*)P_{\mathfrak{N}}g \\ &= 2P_{\dot{S}}g + 2(Q^* - S)^{-1}(\mathrm{Re}\, Q - S)P_{\mathfrak{N}}g, \end{aligned}$$

and

$$(I + \hat{A})f = 2P_{\dot{S}}g + 2(Q^* - S)^{-1}(\mathrm{Re}\, Q - S)P_{\mathfrak{N}}g. \tag{5.16}$$

From the statement of the lemma we have that $(\mathrm{Re}\, Q - S)\upharpoonright\mathfrak{N}$ is an isomorphism of the space $\mathfrak{N}$. Hence, (5.16) $\mathrm{Ran}(I + \hat{A}) = \mathcal{H}$ and the Cayley transform is well defined for $\hat{A}$. Let

$$\hat{S} = (I - \hat{A})(I + \hat{A})^{-1}.$$

It follows from (5.11) that

$$\begin{aligned} (\hat{S} + I)\phi &= (Q + I)g + \tfrac{1}{2}(S + I)(Q^* - S)^{-1}(Q - Q^*)g, \\ (I - \hat{S})\phi &= (I - Q)g + \tfrac{1}{2}(I - S)(Q^* - S)^{-1}(Q - Q^*)g. \end{aligned}$$

Therefore,

$$\begin{aligned} \phi &= g + \tfrac{1}{2}(Q^* - S)^{-1}(Q - Q^*)g, \\ \hat{S}\phi &= Qg + \tfrac{1}{2}S(Q^* - S)^{-1}(Q - Q^*)g. \end{aligned}$$

and hence

$$\begin{aligned} \hat{S}\phi &= S\phi + (Q - S)P_{\mathfrak{N}}g, \\ \phi &= P_{\dot{S}}g + (Q^* - S)^{-1}(\mathrm{Re}\, Q - S)P_{\mathfrak{N}}g. \end{aligned} \tag{5.17}$$

Using the second half of (5.17) we have

$$P_{\mathfrak{N}}g = (\mathrm{Re}\, Q - S)^{-1}(Q^* - S)P_{\mathfrak{N}}\phi. \tag{5.18}$$

Substituting, (5.18) into the first part of (5.17) we obtain

$$\hat{S}\phi = S\phi + (Q - S)(\mathrm{Re}\, Q - S)^{-1}(Q^* - S)\phi,$$

which proves the lemma. □

Let $T \in \Xi(\dot{A})$. By the **class** $\Xi_{AT}$ we denote the set of all non-negative self-adjoint operators $A \supset \dot{A}$ satisfying the following conditions:

1. $[K(T) + K(T^*) - 2K(A)]\upharpoonright \mathfrak{N}$ is a non-negative operator in $\mathfrak{N}$, where $\mathfrak{N}$ is defined in (5.4);
2. $K(A)+2[K(T)-K(A)][K(T)+K(T^*)-2K(A)]^{-1}[K(T^*)-K(A)] \leq K(A_K)$.[1]

**Theorem 5.9.** *A* $(*)$*-extension of operator* $T \in \Xi(\dot{A})$

$$\mathbb{A} = \dot{A}^* - \mathcal{R}^{-1}\dot{A}^*(I - \mathcal{P}_{TA}),$$

*generated by a self-adjoint operator* $A \supset \dot{A}$ *is accretive if and only if* $A \in \Xi_{AT}$.

*Proof.* We prove the necessity part first. Let $\mathbb{A} = \dot{A}^* - \mathcal{R}^{-1}\dot{A}^*(I - \mathcal{P}_{TA})$ be an accretive $(*)$-extension, then $\operatorname{Re}\mathbb{A}$ is a non-negative $(*)$-extension of the quasi-kernel $\hat{A} \in \Xi(\dot{A})$. But according to Lemma 5.7 $\hat{A}$ is defined by formulas (5.11). Since $(-1)$ is a regular point of operator $\hat{A}$, then (5.16) implies that the operator

$$(\operatorname{Re} Q - S)\upharpoonright \mathfrak{N} = \frac{1}{2}[K(T) + K(T^*) - 2K(A)]\upharpoonright \mathfrak{N},$$

is an isomorphism of the space $\mathfrak{N}$. According to Lemma 5.8 we have

$$K(\hat{A}) = K(A) + 2[K(T) - K(A)][K(T) + K(T^*) - 2K(A)]^{-1}[K(T^*) - K(A)].$$

Since $K(\hat{A})$ is a self-adjoint contractive extension of $\dot{S}$, then $K(\hat{A}) \leq K(A_K)$. Also, since $\operatorname{Re}\mathbb{A}$ is generated by $A$ and $\operatorname{Re}\mathbb{A} \geq 0$, then by Theorem 5.5 the operator $[K(\hat{A}) - K(A)]\upharpoonright \mathfrak{N}$ is non-negative. Consequently, the operator $[K(T) + K(T^*) - 2K(A)\upharpoonright \mathfrak{N}$ is non-negative as well and we conclude that $A \in \Xi_{AT}$.

Now we prove sufficiency. Let $A \in \Xi_{AT}$, then by Lemma 5.8, $\hat{A}$ is a Cayley transform of a self-adjoint extension $\hat{S}$ of the operator $\dot{S}$. Since

$$[\hat{S} - S]\upharpoonright \mathfrak{N} = [K(\hat{A}) - K(A)]\upharpoonright \mathfrak{N},$$

is a non-negative operator, then due to Theorem 5.5 the operator $\operatorname{Re}\mathbb{A}$ is a non-negative $(*)$-extension of $\hat{A}$. That is why $\mathbb{A}$ is an accretive $(*)$-extension of operator $T$. □

**Theorem 5.10.** *An operator* $T \in \Xi(\dot{A})$ *admits accretive* $(*)$*-extensions if and only if* $T$ *is transversal to* $A_F$.

*Proof.* If $T$ admits accretive $(*)$-extensions, then the class $\Xi_{AT}$ is non-empty, i.e., there exists a self-adjoint operator $A \in \Xi(\dot{A})$ such that the operator

$$[(K(T) + K(T^*) - 2K(A)]\upharpoonright \mathfrak{N},$$

is non-negative. But then $[(K(T) + K(T^*) - 2K(A_F)]\upharpoonright \mathfrak{N}$ is non-negative as well. This yields that $[K(T) - K(A_F)]\upharpoonright \mathfrak{N}$ is an isomorphism of $\mathfrak{N}$. Then by Lemma 5.3 $T$ and $A_F$ are mutually transversal. This proves the necessity.

[1]When we write $[K(T) + K(T^*) - 2K(A)]^{-1}$ we mean the operator inverse to $[K(T) + K(T^*) - 2K(A)]\upharpoonright \mathfrak{N}$.

Now let us assume that $T$ and $A_F$ are mutually transversal. We will show that in this case $A_F \in \Xi_{AT}$. By Lemma 5.3, $[K(T)-K(A_F)]\upharpoonright\mathfrak{N}$ is an isomorphism of the space $\mathfrak{N}$. Then using formula (5.5) we have

$$K(T) = Q = \frac{1}{2}(S_M + S_\mu) + \frac{1}{2}(S_M - S_\mu)^{1/2}X(S_M - S_\mu)^{1/2},$$

where $X \in [\mathfrak{N}, \mathfrak{N}]$ is a contraction. Furthermore,

$$(Q - S_\mu)\upharpoonright\mathfrak{N} = \frac{1}{2}(S_M - S_\mu)^{1/2}(X + I)(S_M - S_\mu)^{1/2}\upharpoonright\mathfrak{N}.$$

Thus, $X + I$ is an isomorphism of the space $\mathfrak{N}$. Moreover, $\operatorname{Re} X + I \geq 0$ and for every $f \in \mathfrak{N}$

$$((\operatorname{Re} X + I)f, f) = \frac{1}{2}\left(\|f\|^2 - \|Xf\|^2 + \|(X + I)f\|^2\right). \tag{5.19}$$

But since $\|(X + I)f\|^2 \geq a\|f\|^2$, where $a > 0$, $f \in \mathfrak{N}$, we have

$$((\operatorname{Re} X + I)f, f) \geq b\|f\|^2, \qquad (b > 0).$$

Hence, $\operatorname{Re} X + I$ is a non-negative operator implying that

$$[(1/2)(K(T) + K(T^*) - K(A)]\upharpoonright\mathfrak{N} = \frac{1}{2}(S_M - S_\mu)^{1/2}(\operatorname{Re} X + I)(S_M - S_\mu)^{1/2}\upharpoonright\mathfrak{N},$$

is non-negative too. Also (5.19) implies

$$\operatorname{Re} X + I \geq \frac{1}{2}(X^* + I)(X + I).$$

It is easy to see then that $(\operatorname{Re} X + I)^{-1} \leq 2(X + I)^{-1}(X^* + I)^{-1}$. Therefore,

$$\frac{1}{2}(X + I)(\operatorname{Re} X + I)^{-1}(X^* + I) \leq I. \tag{5.20}$$

Now, since

$$\begin{aligned} K(A_F) &+ 2[K(T) - K(A)][K(T) + K(T^*) - 2K(A)]^{-1}[K(T^*) - K(A)] \\ &= S_\mu + \frac{1}{2}(S_M - S_\mu)^{1/2}(X + I)(\operatorname{Re} X + I)^{-1}(X^* + I)(S_M - S_\mu)^{1/2}, \end{aligned}$$

then applying (5.20) we obtain

$$K(A_F) + 2[K(T) - K(A)][K(T) + K(T^*) - 2K(A)]^{-1}[K(T^*) - K(A)] \leq K(A_K).$$

Thus, $A_F$ belongs to the class $\Xi_{AT}$ and applying theorem (5.9) we conclude that $\mathbb{A} = \dot{A}^* - \mathcal{R}^{-1}\dot{A}^*(I - \mathcal{P}_{AT})$ is an accretive (*)-extension of $T$. □

A qsc-extension

$$Q = \frac{1}{2}(S_M + S_\mu) + \frac{1}{2}(S_M - S_\mu)^{1/2}X(S_M - S_\mu)^{1/2}$$

is called **extremal** if $X$ is isometry.

**Theorem 5.11.** *Let $T \in \Xi(\dot{A})$ be transversal to $A_F$. Then the accretive (*)-extension $\mathbb{A}$ of $T$ generated by $A_F$ has a property that $\operatorname{Re}\mathbb{A} \supset A_K$ if and only if $T$ and $T^*$ are extremal extensions of $\dot{A}$.*

*Proof.* Suppose $\mathrm{Re}\,\mathbb{A} \supset A_K$ and $\mathrm{Re}\,\mathbb{A} = \dot{A}^* - \mathcal{R}^{-1}\dot{A}^*(I - \mathcal{P}_{TA_F})$. Then by Lemma 5.8 we have

$$S_M = S_\mu + (Q - S_\mu)(\mathrm{Re}\,Q - S_\mu)^{-1}(Q^* - S_\mu).$$

Thus,

$$(X + I)(\mathrm{Re}\,X + I)^{-1}(X^* + I) = 2I, \tag{5.21}$$

where

$$Q = K(T) = \frac{1}{2}(S_M + S_\mu) + \frac{1}{2}(S_M - S_\mu)^{1/2}X(S_M - S_\mu)^{1/2}.$$

It is easy to see that

$$(X^* + I)(\mathrm{Re}\,X + I)^{-1}(X + I) = (X + I)(\mathrm{Re}\,X + I)^{-1}(X^* + I). \tag{5.22}$$

Then it follows from (5.21) and (5.22) that

$$X^*X = XX^* = I,$$

i.e., $X$ is a unitary operator in $\mathfrak{N}$. Then both operators $T$ and $T^*$ are extremal $m$-accretive extensions of $\dot{A}$.

The second part of the theorem is proved by reversing the argument. □

## 6. Realization of Stieltjes functions

**Definition 6.1.** An operator-valued Herglotz-Nevanlinna function $V(z)$ in a finite-dimensional Hilbert space $E$ is called a **Stieltjes function** if $V(z)$ is holomorphic in $\mathrm{Ext}[0, +\infty)$ and

$$\frac{\mathrm{Im}[zV(z)]}{\mathrm{Im}\ z} \geq 0. \tag{6.1}$$

Consequently, an operator-valued Herglotz-Nevanlinna function $V(z)$ is Stieltjes if $zV(z)$ is also a Herglotz-Nevanlinna function. Applying the integral representation (1.1) (see also [17]) for this case we get that

$$\sum_{k,l=1}^{n} \left( \frac{z_k V(z_k) - \bar{z}_l V(\bar{z}_l)}{z_k - \bar{z}_l} h_k, h_l \right)_E \geq 0, \tag{6.2}$$

for an arbitrary sequence $\{z_k\}$ $(k = 1, \ldots, n)$ of $(\mathrm{Im}\, z_k > 0)$ complex numbers and a sequence of vectors $\{h_k\}$ in $E$.

Similar to (1.1) formula holds true for the case of a Stieltjes function. Indeed, if $V(z)$ is a Stieltjes operator-valued function, then

$$V(z) = \gamma + \int_0^\infty \frac{dG(t)}{t - z}, \tag{6.3}$$

where $\gamma \geq 0$ and $G(t)$ is a non-decreasing on $[0, +\infty)$ operator-valued function such that

$$\int_0^\infty \frac{(dG(t)h, h)_E}{1 + t} < \infty, \quad h \in E. \tag{6.4}$$

**Theorem 6.2.** *Let $\Theta$ be an L-system of the form* (2.3) *with a densely defined non-negative symmetric operator $\dot{A}$. Then the impedance function $V_\Theta(z)$ defined by* (2.7) *is a Stieltjes function if and only if the operator $\mathbb{A}$ of the L-system $\Theta$ is accretive.*

*Proof.* Let us assume first that $\mathbb{A}$ is an accretive operator, i.e., $(\mathrm{Re}\,\mathbb{A}f, f) \geq 0$, for all $f \in \mathcal{H}_+$. Let $\{z_k\}$ $(k = 1, \dots, n)$ be a sequence of $(\mathrm{Im}\, z_k > 0)$ complex numbers and $h_k$ be a sequence of vectors in $E$. Let us denote

$$Kh_k = \delta_k, \quad g_k = (\mathrm{Re}\,\mathbb{A} - z_k I)^{-1}\delta_k, \quad g = \sum_{k=1}^{n} g_k. \tag{6.5}$$

Since $(\mathrm{Re}\,\mathbb{A}g, g) \geq 0$, we have

$$\sum_{k,l=1}^{n} (\mathrm{Re}\,\mathbb{A}g_k, g_l) \geq 0. \tag{6.6}$$

By formal calculations one can have $(\mathrm{Re}\,\mathbb{A})g_k = \delta_k + z_k(\mathrm{Re}\,\mathbb{A} - z_k I)^{-1}\delta_k$, and

$$\begin{aligned}\sum_{k,l=1}^{n} (\mathrm{Re}\,\mathbb{A}g_k, g_l) = \sum_{k,l=1}^{n} &\big[(\delta_k, (\mathrm{Re}\,\mathbb{A} - z_l I)^{-1}\delta_l) \\ &+ (z_k(\mathrm{Re}\,\mathbb{A} - z_k I)^{-1}\delta_k, (\mathrm{Re}\,\mathbb{A} - z_k I)^{-1}\delta_l)\big].\end{aligned}$$

Using obvious equalities

$$\big((\mathrm{Re}\,\mathbb{A} - z_k I)^{-1}Kh_k, Kh_l\big) = \big(V_\Theta(z_k)h_k, h_l\big)_E,$$

and

$$\big((\mathrm{Re}\,\mathbb{A} - \bar{z}_l I)^{-1}(\mathrm{Re}\,\mathbb{A} - z_k I)^{-1}Kh_k, Kh_l\big) = \left(\frac{V_\Theta(z_k) - V_\Theta(\bar{z}_l)}{z_k - \bar{z}_l}h_k, h_l\right)_E,$$

we obtain

$$\sum_{k,l=1}^{n} ((\mathrm{Re}\,\mathbb{A})g_k, g_l) = \sum_{k,l=1}^{n} \left(\frac{z_k V_\Theta(z_k) - \bar{z}_l V_\Theta(\bar{z}_l)}{z_k - \bar{z}_l}h_k, h_l\right)_E \geq 0, \tag{6.7}$$

which implies that $V_\Theta(z)$ is a Stieltjes function.

Now we prove necessity. First we assume that $\dot{A}$ is a prime operator[2]. Then the equivalence of (6.7) and (6.6) implies that $(\mathrm{Re}\,\mathbb{A}g, g) \geq 0$ for any $g$ from c.l.s.$\{\mathfrak{N}_z\}$. It was shown in [11] that a symmetric operator $\dot{A}$ with the equal deficiency indices is prime if and only if

$$\underset{z \neq \bar{z}}{c.l.s.}\, \mathfrak{N}_z = \mathcal{H}. \tag{6.8}$$

Thus $(\mathrm{Re}\,\mathbb{A}g, g) \geq 0$ for any $g \in \mathcal{H}_+$ and therefore $\mathbb{A}$ is an accretive operator.

Now let us assume that $\dot{A}$ is not a prime operator. Then there exists a subspace $\mathcal{H}^1 \subset \mathcal{H}$ on which $\dot{A}$ generates a self-adjoint operator $A_1$. Let us denote by

[2]We call a closed linear operator in a Hilbert space $\mathcal{H}$ a **prime operator** if there is no non-trivial reducing invariant subspace of $\mathcal{H}$ on which it induces a self-adjoint operator.

$\dot{A}_0$ an operator induced by $\dot{A}$ on $\mathcal{H}^0 = \mathcal{H} \ominus \mathcal{H}^1$. As it was shown shown in the proof of Theorem 12 of [14] the decomposition

$$\mathcal{H}_+ = \mathcal{H}_+^0 \oplus \mathcal{H}_+^1, \quad \mathcal{H}_+^0 = \mathrm{Dom}(\dot{A}_0^*),\ \mathcal{H}_+^1 = \mathrm{Dom}(A_1), \tag{6.9}$$

is (+)-orthogonal. Since $\dot{A}$ is a non-negative operator, then

$$(\mathrm{Re}\,\mathbb{A}g, g) = (A_1 g, g) = (\dot{A}g, g) \geq 0, \qquad \forall g \in \mathcal{H}_+^1 = \mathrm{Dom}(A_1).$$

On the other hand operator $\dot{A}_0$ is prime in $\mathcal{H}^0$ and hence $\underset{z\neq\bar{z}}{c.l.s.}\,\mathfrak{N}_z^0 = \mathcal{H}^0$, where $\mathfrak{N}_z^0$ are the deficiency subspaces of the symmetric operator $\dot{A}_0$ in $\mathcal{H}^0$. Then the equivalence of (6.7) and (6.6) again implies that $(\mathrm{Re}\,\mathbb{A}g, g) \geq 0$ for any $g \in \mathcal{H}_+^0$. Taking into account decomposition (6.9) we conclude that $\mathrm{Re}\,(\mathbb{A}g, g) \geq 0$ holds for all $g \in \mathcal{H}_+$ and hence $\mathbb{A}$ is accretive. □

Now we define a class of realizable Stieltjes functions. At this point we need to note that since Stieltjes functions form a subset of Herglotz-Nevanlinna functions, then according to (1.7) and realization Theorems 8 and 9 of [14], we have that the class of all realizable Stieltjes functions is a subclass of $N(R)$. To see the specifications of this class we recall that aside of the integral representation (6.3), any Stieltjes function admits a representation (1.1). According to (1.7) a Herglotz-Nevanlinna operator-function can be realized if and only if in the representation (1.1) $L = 0$ and

$$Qh = \int_{-\infty}^{+\infty} \frac{t}{1+t^2}\, dG(t)h, \tag{6.10}$$

for all $h \in E$ such that

$$\int_{-\infty}^{\infty} (dG(t)h, h)_E < \infty. \tag{6.11}$$

holds. Considering this we obtain

$$Q = \frac{1}{2}\left[V(-i) + V^*(-i)\right] = \gamma + \int_0^{+\infty} \frac{t}{1+t^2}\, dG(t). \tag{6.12}$$

Combining (6.10) and (6.12) we conclude that $\gamma h = 0$ for all $h \in E$ such that (6.11) holds.

**Definition 6.3.** An operator-valued Stieltjes function $V(z)$ in a finite-dimensional Hilbert space $E$ belongs to the **class** $S(R)$ if in the representation (6.3)

$$\gamma h = 0$$

for all $h \in E$ such that

$$\int_0^{\infty} (dG(t)h, h)_E < \infty. \tag{6.13}$$

We are going to focus though on the subclass $S_0(R)$ of $S(R)$ whose definition is the following.

**Definition 6.4.** An operator-valued Stieltjes function $V(z)$ in a finite-dimensional Hilbert space $E$ belongs to the **class** $S_0(R)$ if in the representation (6.3) we have

$$\int_0^\infty (dG(t)h,h)_E = \infty, \tag{6.14}$$

for all non-zero $h \in E$.

An L-system $\Theta$ of the form (2.3) is called an **accretive L-system** if its operator $\mathbb{A}$ is accretive. The following theorem is the direct realization theorem for the functions of the class $S_0(R)$.

**Theorem 6.5.** *Let $\Theta$ be an accretive L-system of the form* (2.3) *with an invertible channel operator $K$ and a densely defined symmetric operator $\dot{A}$. Then its impedance function $V_\Theta(z)$ of the form* (2.7) *belongs to the class $S_0(R)$.*

*Proof.* Since our L-system $\Theta$ is accretive, then by Theorem 6.2, $V_\Theta(z)$ is a Stieltjes function. Now let us show that $V_\Theta(z)$ belongs to $S_0(R)$. It follows from Theorem 7 of [14] that $E_1 = K^{-1}\mathfrak{L}$, where $\mathfrak{L} = \mathcal{H} \ominus \overline{\mathrm{Dom}(\dot{A})}$ and

$$E_1 = \left\{ h \in E : \int_0^{+\infty} (dG(t)h,h)_E < \infty \right\}.$$

But $\overline{\mathrm{Dom}(\dot{A})} = \mathcal{H}$ and consequently $\mathfrak{L} = \{0\}$. Next, $E_1 = \{0\}$,

$$\int_0^\infty (dG(t)h,h)_E = \infty,$$

for all non-zero $h \in E$, and therefore $V_\Theta(z) \in S_0(R)$. □

We can also state and prove the following inverse realization theorem for the classes $S_0(R)$.

**Theorem 6.6.** *Let an operator-valued function $V(z)$ belong to the class $S_0(R)$. Then $V(z)$ can be realized as an impedance function of a minimal accretive L-system $\Theta$ of the form* (2.3) *with an invertible channel operator $K$, a densely defined non-negative symmetric operator $\dot{A}$,* $\mathrm{Dom}(T) \neq \mathrm{Dom}(T^*)$, *and a preassigned direction operator $J$ for which $I + iV(-i)J$ is invertible.*[3]

*Proof.* We have already noted that the class of Stieltjes function lies inside the wider class of all Herglotz-Nevanlinna functions. Thus all we actually have to show is that $S_0(R) \subset N_0(R)$, where the subclass $N_0(R)$ was defined in [16], and that the realizing L-system in the proof of Theorem 11 of [16] appears to be an accretive L-system. The former is rather obvious and follows directly from the definition of the class $S_0(R)$. To see that the realizing L-system is accretive we need to recall that the model L-system $\Theta$ was constructed in the proof of Theorem 11 of [16] and

[3] It was shown in [14] that if $J = I$ this invertibility condition is satisfied automatically.

then it was shown that $V_\Theta(z) = V(z)$. But $V(z)$ is a Stieltjes function and hence so is $V_\Theta(z)$. Applying Theorem 6.2 yields the desired result. □

Let us define a subclass of the class $S_0(R)$.

**Definition 6.7.** An operator-valued Stieltjes function $V(z)$ of the class $S_0(R)$ is said to be a member of the **class** $S_0^K(R)$ if

$$\int_0^\infty \frac{(dG(t)h,h)_E}{t} = \infty, \tag{6.15}$$

for all non-zero $h \in E$.

Below we state and prove direct and inverse realization theorem for this subclass.

**Theorem 6.8.** *Let $\Theta$ be an accretive L-system of the form* (2.3) *with an invertible channel operator $K$ and a densely defined symmetric operator $\dot{A}$. If the Kreĭn-von Neumann extension $A_K$ is a quasi-kernel for* $\mathrm{Re}\,\mathbb{A}$, *then the impedance function $V_\Theta(z)$ of the form* (2.7) *belongs to the class $S_0^K(R)$.*

*Conversely, if $V(z) \in S_0^K(R)$, then it can be realized as the impedance function of an accretive L-system $\Theta$ of the form* (2.3) *with* $\mathrm{Re}\,\mathbb{A}$ *containing $A_K$ as a quasi-kernel and a preassigned direction operator $J$ for which $I + iV(-i)J$ is invertible.*

*Proof.* We begin with the proof of the second part. First we use realization Theorem 2.4 and Theorem 6.6 to construct a minimal model L-system $\Theta$ whose impedance function is $V(z)$. Then we will show that (6.15) is equivalent to the fact that self-adjoint operator $A$ introduced in the proof of Theorem 2.4 (see [14]) and constructed to be a quasi-kernel for $\mathrm{Re}\,\mathbb{A}$, coincides with $A_K$, that is the Kreĭn-von Neumann extension of the model symmetric operator $\dot{A}$ of multiplication by an independent variable (see [14]). Let $L^2_G(E)$ be a model space constructed in the proof of Theorem (2.4) (see [14]). Let also $E(s)$ be the orthoprojection operator in $L^2_G(E)$ defined by

$$E(s)f(t) = \begin{cases} f(t), & 0 \le t \le s \\ 0, & t > s \end{cases} \tag{6.16}$$

where $f(t) \in C_{00}(E,[0,+\infty))$. Here be $C_{00}(E,[0,+\infty))$ is the set of continuous compactly supported functions $f(t)$, $([0 < t < +\infty))$ with values in $E$. Then for the operator $A$, that is the operator of multiplication by independent variable defined in the proof of Theorem 2.4 (see [14]), we have

$$A = \int_0^\infty s\,dE(s), \tag{6.17}$$

and $E(s)$ is the resolution of identity of the operator $A$. By construction provided in the proof of Theorem 2.4, the operator $A$ is the quasi-kernel of $\mathrm{Re}\,\mathbb{A}$, where $\mathbb{A}$

is an accretive (*)-extension of the model system. Let us calculate $(E(s)f(t),f(t))$ and $(Af(t),f(t))$ (here we use $L^2_G(E)$ scalar product).

$$(E(s)f(t),f(t)) = \int_0^\infty (dG(t)E(s)f(t),f(t))_E = \int_0^s (dG(t)f(t),f(t))_E, \qquad (6.18)$$

$$(Af(t),f(t)) = \int_0^\infty s\,d\left\{\int_0^s (dG(t)f(t),f(t))_E\right\} = \int_0^\infty s\,d(G(s)x(s),x(s))_E. \quad (6.19)$$

The equality $A = A_K$ holds (see Proposition 3.2) if for all $\varphi \in \mathfrak{N}_{-a}$, $\varphi \neq 0$

$$\int_0^\infty \frac{(dE(t)\varphi,\varphi)}{t} = \infty, \qquad (6.20)$$

where $\mathfrak{N}_{-a}$ is the deficiency subspace of the operator $\dot{A}$ corresponding to the point $(-a)$, $(a > 0)$. But according to Theorem 2.4 we have

$$\mathfrak{N}_z = \left\{\frac{h}{t-z} \in L^2_G(E) \mid h \in E\right\},$$

and hence

$$\mathfrak{N}_{-a} = \left\{\frac{h}{t+a} \in L^2_G(E) \mid h \in E\right\}. \qquad (6.21)$$

Taking into account (6.15) we have for all $h \in E$

$$\int_0^\infty \frac{(dE(s)\varphi,\varphi)_{L^2_G(E)}}{s} = \int_0^\infty \frac{(dE(s)\frac{h}{t+a},\frac{h}{t+a})_{L^2_G(E)}}{s} = \int_0^\infty \frac{(dG(s)h,h)_E}{s(s+a)^2}.$$

Hence the operator $A = A_K$ iff

$$\int_0^\infty \frac{(dG(t)h,h)_E}{t(t+a)^2} = \infty, \quad \forall h \in E,\ h \neq 0. \qquad (6.22)$$

Let us transform (6.15)

$$\begin{aligned}\int_0^\infty \frac{(dG(t)h,h)_E}{t} &= \int_0^\infty \frac{(t+a)^2}{t}\left(dG(t)\frac{h}{t+a},\frac{h}{t+a}\right)_E \\ &= \int_0^\infty t\left(dG(t)\frac{h}{t+a},\frac{h}{t+a}\right)_E + 2a\int_0^\infty \left(dG(t)\frac{h}{t+a},\frac{h}{t+a}\right)_E \qquad (6.23)\\ &\quad + a^2\int_0^\infty \frac{(dG(t)h,h)_E}{t(t+a)^2}.\end{aligned}$$

Since $\mathrm{Re}\,\mathbb{A}$ is a non-negative self-adjoint bi-extension of $\dot{A}$ in the model system, then we can apply Theorem 4.8 to get (4.9). Then first two integrals in (6.23) converge for a fixed $a$ because of (4.9) and equality

$$\int_0^\infty \left(dG(t)\frac{h}{t+a},\frac{h}{t+a}\right)_E = \int_0^\infty d\left(E(t)\varphi,\varphi\right), \quad \varphi \in \mathfrak{N}_{-a}.$$

Therefore the divergence of integral in (6.15) completely depends on divergence of the last integral in (6.23).

Now we can prove the first part of the theorem. Let $\Theta$ be our L-system with $A_K$ that is a quasi-kernel for $\mathrm{Re}\,\mathbb{A}$, and the impedance function $V_\Theta(z)$. Without loss of generality we can consider $\Theta$ as a minimal system, otherwise we would take the principal part of $\Theta$ that is minimal and has the same impedance function (see [14]). Furthermore, $V_\Theta(z)$ can be realized as an impedance function of the model L-system $\Theta_1$ constructed in the proof of Theorem 2.4. Some of the elements of $\Theta_1$ were already described above during the proof of the second part of the theorem. If the L-system $\Theta_1$ is not minimal, we consider its principal part $\Theta_{1,0}$ that is described in Theorem 12 of [14] and has the same impedance function as $\Theta_1$. Since both $\Theta$ and $\Theta_{1,0}$ share the same impedance function $V_\Theta(z)$ they also have the same transfer function $W_\Theta(z)$ and thus we can apply the theorem on bi-unitary equivalence of [11]. According to this theorem the quasi-kernel operator $A_0$ of $\Theta_{1,0}$ is unitary equivalent to the quasi-kernel $A_K$ in $\Theta$. Consequently, property (6.20) of $A_K$ gets transferred by the unitary equivalence mapping to the corresponding property of $A_0$ making it, by Proposition 3.2, the Kreĭn-von Neumann self-adjoint extension of the corresponding symmetric operator $\dot{A}_0$ of $\Theta_{1,0}$. But this implies that the quasi-kernel operator $A$ of $\Theta_1$ (defined by (6.17)) is also the Kreĭn-von Neumann self-adjoint extension and hence has property (6.20) that causes (6.22). Using (6.22) in conjunction with (6.23) we obtain (6.15). That proves the theorem. □

## 7. Realization of inverse Stieltjes functions

**Definition 7.1.** We will call an operator-valued Herglotz-Nevanlinna function $V(z)$ in a finite-dimensional Hilbert space $E$ by an **inverse Stieltjes** if $V(z)$ it is holomorphic in $\mathrm{Ext}[0,+\infty)$ and

$$\frac{\mathrm{Im}[V(z)/z]}{\mathrm{Im}\ z} \geq 0. \tag{7.1}$$

Combining (7.1) with (1.1) we obtain (see [18])

$$\sum_{k,l=1}^{n}\left(\frac{V(z_k)/z_k - V(\bar{z}_l)/\bar{z}_l}{z_k-\bar{z}_l}h_k, h_l\right)_E \geq 0,$$

for an arbitrary sequence $\{z_k\}$ $(k=1,\dots,n)$ of $(\mathrm{Im}\,z_k>0)$ complex numbers and a sequence of vectors $\{h_k\}$ in $E$. It can be shown (see [23]) that every inverse Stieltjes function $V(z)$ in a finite-dimensional Hilbert space $E$ admits the following integral representation

$$V(z) = \alpha + z\beta + \int_0^\infty \left(\frac{1}{t-z} - \frac{1}{t}\right) dG(t), \tag{7.2}$$

where $\alpha \leq 0$, $\beta \geq 0$, and $G(t)$ is a non-decreasing on $[0,+\infty)$ operator-valued function such that

$$\int_0^\infty \frac{(dG(t)h,h)}{t+t^2} < \infty, \quad \forall h \in E.$$

The following definition provides the description of all realizable inverse Stieltjes operator-valued functions.

**Definition 7.2.** An operator-valued inverse Stieltjes function $V(z)$ in a finite-dimensional Hilbert space $E$ is a member of the **class** $S^{-1}(R)$ if in the representation (7.2) we have

$$\text{i)} \qquad \beta = 0,$$

$$\text{ii)} \qquad \alpha h = 0,$$

for all $h \in E$ with

$$\int_0^\infty (dG(t)h,h)_E < \infty.$$

In what follows we will, however, be mostly interested in the following subclass of $S^{-1}(R)$.

**Definition 7.3.** An inverse Stieltjes function $V(z) \in S^{-1}(R)$ is a member of the **class** $S_0^{-1}(R)$ if

$$\int_0^\infty (dG(t)h,h)_E = \infty,$$

for all $h \in E$, $h \neq 0$.

We recall that an L-system $\Theta$ of the form (2.3) is called **accumulative** if its state-space operator $\mathbb{A}$ is accumulative, i.e., satisfies (2.5). It is easy to see that if an L-system is accumulative, then (2.5) implies that the operator $\dot{A}$ of the system is non-negative and both operators $T$ and $T^*$ are accretive.

The following statement is the direct realization theorem for the functions of the class $S_0^{-1}(R)$.

**Theorem 7.4.** *Let $\Theta$ be an accumulative L-system of the form* (2.3) *with an invertible channel operator $K$ and $\overline{\mathrm{Dom}(\dot{A})} = \mathcal{H}$. Then its impedance function $V_\Theta(z)$ of the form* (2.7) *belongs to the class $S_0^{-1}(R)$.*

*Proof.* First we will show that $V_\Theta(z)$ is an inverse Stieltjes function. Let $\{z_k\}$ $(k = 1,\dots,n)$ is a sequence of non-real $(z_k \neq \bar{z}_k)$ complex numbers and $\varphi_k$ $(z_k \neq \bar{z}_k)$ is a sequence of elements of $\mathfrak{N}_{z_k}$, the defect subspace of the operator $\dot{A}$. Then for every $k$ there exists $h_k \in E$ such that

$$\varphi_k = z_k(\mathrm{Re}\,\mathbb{A} - z_k I)^{-1} K h_k, \qquad (k = 1,\dots,n). \tag{7.3}$$

Taking into account that $\dot{A}^*\varphi_k = z_k\varphi_k$, formula (7.3), and letting $\varphi = \sum_{k=1}^n \varphi_k$ we get

$$\begin{aligned}
&(\dot{A}^*\varphi,\varphi) + (\varphi,\dot{A}^*\varphi) - (\mathrm{Re}\,\mathbb{A}\varphi,\varphi)\\
&= \sum_{k,l=1}^n \Big[(\dot{A}^*\varphi_k,\varphi_l) + (\varphi_k,\dot{A}^*\varphi_l) - (\mathrm{Re}\,\mathbb{A}\varphi_k,\varphi_l)\Big]\\
&= \sum_{k,l=1}^n ([-\mathrm{Re}\,\mathbb{A} + z_k + \bar{z}_l]\varphi_k,\varphi_l)
\end{aligned}$$

$$
\begin{aligned}
&= \sum_{k,l=1}^{n} \Bigg( \frac{(\operatorname{Re}\mathbb{A} - \bar z_l I)^{-1}(\bar z_l(\operatorname{Re}\mathbb{A} - \bar z_l I) - z_k(\operatorname{Re}\mathbb{A} - z_k I))(\operatorname{Re}\mathbb{A} - z_k I)^{-1}}{z_k \bar z_l (z_k - \bar z_l)} \\
&\qquad \times K h_k, K h_l \Bigg) \\
&= \sum_{k,l=1}^{n} \left( \frac{\bar z_l K^*(\operatorname{Re}\mathbb{A} - z_k I)^{-1}K - z_k K^*(\operatorname{Re}\mathbb{A} - z_l I)^{-1}K}{z_k \bar z_l (z_k - \bar z_l)} h_k, h_l \right) \\
&= \sum_{k,l=1}^{n} \left( \frac{\bar z_l V_\Theta(z_k) - z_k V_\Theta(\bar z_l)}{z_k z_l (z_k - \bar z_l)} h_k, h_l \right) \geq 0.
\end{aligned}
$$

The last line can be re-written as follows

$$
\sum_{k,l=1}^{n} \left( \frac{V_\Theta(z_k)/z_k - V_\Theta(\bar z_l)/\bar z_l}{z_k - \bar z_l} h_k, h_l \right) \geq 0. \tag{7.4}
$$

Letting in (7.4) $n = 1$, $z_1 = z$, and $h_1 = h$ we get

$$
\left( \frac{V_\Theta(z)/z - V_\Theta(\bar z)/\bar z}{z - \bar z} h, h \right) \geq 0, \tag{7.5}
$$

which means

$$
\frac{\operatorname{Im}(V_\Theta(z)/z)}{\operatorname{Im} z} \geq 0,
$$

and therefore $V_\Theta(z)/z$ is a Herglotz-Nevanlinna function. In Theorem 8 of [14] we have shown that $V_\Theta(z) \in N(R)$. Applying (7.1) we conclude that $V_\Theta(z)$ is an inverse Stieltjes function.

Now we will show that $V_\Theta(z)$ belongs to $S^{-1}(R)$. As any inverse Stieltjes function $V_\Theta(z)$ has its integral representation (7.2) where $\alpha \leq 0$, $\beta \geq 0$, and

$$
\int_0^\infty \frac{(dG(t)h, h)}{t + t^2} < \infty, \quad \forall h \in E.
$$

In a neighborhood of zero the expression $(t + t^2)$ is equivalent to the $(t + t^3)$ and in a neighborhood of the point at infinity

$$
\frac{1}{t + t^3} < \frac{1}{t + t^2}.
$$

Hence,

$$
\int_0^\infty \frac{(dG(t)h, h)}{t + t^3} < \infty, \quad \forall h \in E.
$$

Furthermore,

$$
\begin{aligned}
V_\Theta(z) &= \alpha + z\beta + \int_0^\infty \left( \frac{1}{t - z} - \frac{t}{1 + t^2} + \frac{t}{1 + t^2} - \frac{1}{t} \right) dG(t) \\
&= \left( \alpha - \int_0^\infty \frac{dG(t)}{t + t^3} \right) + z\beta + \int_0^\infty \left( \frac{1}{t - z} - \frac{t}{1 + t^2} \right) dG(t).
\end{aligned}
$$

On the other hand, as it was shown in [14], a Herglotz-Nevanlinna function can be realized if and only if it belongs to the class $N(R)$ and hence in representation (1.1) condition (1.7) holds. Considering this and the uniqueness of the function $G(t)$ we obtain

$$\left(\alpha - \int_0^{\infty} \frac{dG(t)}{t+t^3}\right) f = \int_0^{+\infty} \frac{t}{1+t^2} dG(t)f, \tag{7.6}$$

for all $f \in E$ such that $\int_{-\infty}^{+\infty} (dG(t)f, f)_E < \infty$. Solving (7.6) for $\alpha$ we get

$$\alpha f = \int_0^{\infty} \frac{1}{t}\, dG(t)f, \tag{7.7}$$

for the same selection of $f$. The left-hand side of (7.7) is non-positive but the right-hand side is non-negative. This means that $\alpha = 0$ and $V_\Theta(z) \in S^{-1}(R)$. The proof of the fact that $V_\Theta(z) \in S_0^{-1}(R)$ is similar to the proof of Theorem 6.5. □

The inverse realization theorem can be stated and proved for the class $S_0^{-1}(R)$ as follows.

**Theorem 7.5.** *Let an operator-valued function $V(z)$ belong to the class $S_0^{-1}(R)$. Then $V(z)$ can be realized as an impedance function of an accumulative minimal L-system $\Theta$ of the form (2.3) with an invertible channel operator $K$, a non-negative densely defined symmetric operator $\dot{A}$ and $J = I$.*

*Proof.* The class $S_0^{-1}(R)$ is a subclass of $N_0(R)$ and hence it is realizable by a minimal L-system $\Theta$ with a densely defined symmetric operator $\dot{A}$ and $J = I$. Thus all we have to show is that the L-system $\Theta$ we have constructed in the proof of Theorem 11 of [16] is an accumulative L-system, i.e., satisfying the condition (2.5).

Since the L-system $\Theta$ is minimal then the operator $\dot{A}$ is prime. Applying (6.8) yields

$$\underset{z \neq \bar{z}}{c.l.s.}\, \mathfrak{N}_z = \mathcal{H}, \quad z \neq \bar{z}. \tag{7.8}$$

In the proof of Theorem 7.4 we have shown that

$$(\mathrm{Re}\, \mathbb{A}\varphi, \varphi) \leq (\dot{A}^*\varphi, \varphi) + (\varphi, \dot{A}^*\varphi), \quad \varphi = \sum_{k=1}^{n} \varphi_k, \quad \varphi_k \in \mathfrak{N}_{z_k}, \tag{7.9}$$

is equivalent to (7.4), where $z_k$ are defined by (7.3). Combining (7.8) and (7.9) we get property (2.5) and conclude that $\Theta$ is an accumulative L-system. □

It is not hard to see that members of the classes $S_0(R)$ and $S_0^{-1}(R)$ are the Kreĭn-Langer $Q$-functions [27] corresponding to a self-adjoint extensions of a densely defined symmetric operator.

Now we define a subclass of the class $S_0^{-1}(R)$.

**Definition 7.6.** An operator-valued Stieltjes function $V(z)$ of the class $S_0^{-1}(R)$ is said to be a member of the **class** $S_{0,F}^{-1}(R)$ if

$$\int_0^\infty \frac{t}{t^2+1}\,(dG(t)h,h)_E = \infty, \tag{7.10}$$

for all non-zero $h \in E$.

**Theorem 7.7.** *Let $\Theta$ be an accumulative L-system of the form (2.3) with an invertible channel operator $K$ and a symmetric densely defined operator $\dot{A}$. If Friedrichs extension $A_F$ is a quasi-kernel for* $\mathrm{Re}\,\mathbb{A}$, *then the impedance $V_\Theta(z)$ of the form (2.7) belongs to the class $S_{0,F}^{-1}(R)$.*

*Conversely, if $V(z) \in S_{0,F}^{-1}(R)$, then it can be realized as an impedance of an accumulative L-system $\Theta$ of the form (2.3) with* $\mathrm{Re}\,\mathbb{A}$ *containing $A_F$ as a quasi-kernel and a preassigned direction operator $J$ for which $I + iV(-i)J$ is invertible.*

*Proof.* Following the framework of the proof of Theorem 6.8, we begin with the proof of the second part. First we use the realization Theorem 2.4 and Theorem 7.5 to construct a minimal model L-system $\Theta$ whose impedance function is $V(z)$. Then we will show that (6.15) is equivalent to the fact that self-adjoint operator $A$ introduced in the proof of Theorem (2.4) (see [14]) and constructed to be a quasi-kernel for $\mathrm{Re}\,\mathbb{A}$, coincides with $A_F$, that is the Friedrichs extension of the symmetric operator $\dot{A}$ of multiplication by an independent variable (see [14]). Let $L_G^2(E)$ be a model space constructed in the proof or of Theorem (2.4). Let also $E(s)$ be the orthoprojection operator in $L_G^2(E)$ defined by (6.16). Then for the operator $A$ defined in the proof of Theorem 2.4 (see [14]) we have

$$A = \int_0^\infty t\,dE(t),$$

and $E(t)$ is the spectral function of operator $A$. As we have shown in the proof of Theorem 6.8 the relations (6.18) and (6.19) take place. The equality $A = A_F$ holds (see Proposition 3.2) if for all $\varphi \in \mathfrak{N}_{-a}$

$$\int_0^\infty t\,(dE(t)\varphi,\varphi)_E = \infty, \tag{7.11}$$

where $\mathfrak{N}_{-a}$ is the deficiency subspace of the operator $\dot{A}$ corresponding to the point $(-a)$, $(a > 0)$. But according to Theorem 2.4 we have $\mathfrak{N}_{-a}$ described by (6.21). Taking into account (7.10) we have for all $h \in E$

$$\int_0^\infty s(dE(s)\varphi,\varphi)_{L_G^2(E)} = \int_0^\infty s d\left(E(s)\frac{h}{t+a},\frac{h}{t+a}\right)_{L_G^2(E)} = \int_0^\infty \frac{s\,(dG(s)h,h)_E}{(s+a)^2}.$$

Hence the operator $A = A_F$ iff

$$\int_0^\infty \frac{t\,(dG(t)h,h)_E}{(t+a)^2} = \infty, \quad \forall h \in E,\ h \neq 0. \tag{7.12}$$

Let us transform (7.10)

$$\begin{aligned}\int_0^\infty \frac{t}{t^2+1}(dG(t)h,h)_E &= \int_0^\infty \frac{t(t+a)^2}{t^2+1}\left(dG(t)\frac{h}{t+a},\frac{h}{t+a}\right)_E\\ &= \int_0^\infty \frac{t}{(t+a)^2}\cdot\frac{t^2}{t^2+1}\left(dG(t)\frac{h}{t+a},\frac{h}{t+a}\right)_E\\ &+ 2a\int_0^\infty \frac{t^2}{(t+a)^2(t^2+1)}\left(dG(t)\frac{h}{t+a},\frac{h}{t+a}\right)_E\\ &+ a^2\int_0^\infty \frac{1}{t^2+1}\cdot\frac{t\,(dG(t)h,h)_E}{(t+a)^2}.\end{aligned}\tag{7.13}$$

Consider the following obvious inequality

$$\frac{t^2}{(t+a)^2(t^2+1)}-\frac{1}{t^2+1}=\frac{t^2-(t+a)^2}{(t+a)^2(t^2+1)}=\frac{(2t+a)(-a)}{(t+a)^2(t^2+1)}<0.$$

Taking into account this inequality and the fact that the integral

$$\int_0^\infty \frac{(dG(t)h,h)_E}{t^2+1},$$

converges for all $h\in E$, we conclude that the second integral in (7.13) is convergent. Let us denote this integral as $Q$. Then using (7.13) and obvious estimates we obtain

$$\begin{aligned}\int_0^\infty \frac{t}{t^2+1}(dG(t)h,h)_E \le &\int_0^\infty \frac{t}{(t+a)^2}\left(dG(t)\frac{h}{t+a},\frac{h}{t+a}\right)_E\\ &+2aQ+a^2\int_0^\infty \frac{t\,(dG(t)h,h)_E}{(t+a)^2},\end{aligned}$$

or

$$\int_0^\infty \frac{t}{t^2+1}(dG(t)h,h)_E \le (a^2+1)\int_0^\infty \frac{t}{(t+a)^2}\left(dG(t)\frac{h}{t+a},\frac{h}{t+a}\right)_E+2aQ.$$

Since $V(z)\in S_0^{-1}(R)$, then (7.7) holds and the integral on the left diverges causing the integral on the right side diverge as well. Thus $A=A_F$.

Now we can prove the first part of the theorem. Let $\Theta$ be our L-system with $A_F$ that is a quasi-kernel for $\mathrm{Re}\,\mathbb{A}$, and the impedance function $V_\Theta(z)$. Then $V_\Theta(z)$ can be realized as an impedance function of the model L-system $\Theta_1$ constructed in the proof of Theorem 2.4. Repeating the argument of the second part of the proof of Theorem 6.8 with $A_K$ replaced by $A_F$ we conclude that the quasi-kernel operator $A$ of $\Theta_1$ is the Friedrichs self-adjoint extension and hence has property (7.11) that in turn causes (7.12) for any $a>0$. Let $a=1$, then by (7.12)

$$\infty=\int_0^\infty \frac{t\,(dG(t)h,h)_E}{(t+1)^2}\le\int_0^\infty \frac{t\,(dG(t)h,h)_E}{t^2+1},\quad \forall h\in E,\ h\neq 0,$$

and hence the integral on the right diverges and (7.10) holds. This completes the proof. □

## 8. Examples

Let $\mathcal{H} = L_2[a,+\infty)$ and $l(y) = -y'' + q(x)y$ where $q$ is a real locally summable function. Suppose that the symmetric operator

$$\begin{cases} Ay = -y'' + q(x)y \\ y(a) = y'(a) = 0 \end{cases} \tag{8.1}$$

has deficiency indices (1,1). Let $D^*$ be the set of functions locally absolutely continuous together with their first derivatives such that $l(y) \in L_2[a,+\infty)$. Consider $\mathcal{H}_+ = D(A^*) = D^*$ with the scalar product

$$(y,z)_+ = \int_a^\infty \left( y(x)\overline{z(x)} + l(y)\overline{l(z)} \right) dx, \quad y,\ z \in D^*.$$

Let $\mathcal{H}_+ \subset L_2[a,+\infty) \subset \mathcal{H}_-$ be the corresponding triplet of Hilbert spaces. Consider operators

$$\begin{cases} T_h y = l(y) = -y'' + q(x)y \\ hy(a) = y'(a) \end{cases}, \quad \begin{cases} T_h^* y = l(y) = -y'' + q(x)y \\ \overline{h}y(a) = y'(a) \end{cases}, \tag{8.2}$$

$$\begin{cases} \widehat{A}y = l(y) = -y'' + q(x)y \\ \mu y(a) = y'(a) \end{cases}, \quad \operatorname{Im}\mu = 0.$$

It is well known [1] that $\widehat{A} = \widehat{A^*}$. The following theorem was proved in [11].

**Theorem 8.1.** *The set of all $(*)$-extensions of a non-self-adjoint Schrödinger operator $T_h$ of the form* (8.2) *in $L_2[a,+\infty)$ can be represented in the form*

$$\begin{aligned} \mathbb{A}y &= -y'' + q(x)y - \frac{1}{\mu - h}\,[y'(a) - hy(a)]\,[\mu\delta(x-a) + \delta'(x-a)], \\ \mathbb{A}^*y &= -y'' + q(x)y - \frac{1}{\mu - \overline{h}}\,[y'(a) - \overline{h}y(a)]\,[\mu\delta(x-a) + \delta'(x-a)]. \end{aligned} \tag{8.3}$$

*In addition, the formulas* (8.3) *establish a one-to-one correspondence between the set of all $(*)$-extensions of a Schrödinger operator $T_h$ of the form* (8.2) *and all real numbers $\mu \in [-\infty,+\infty]$.*

Suppose that the symmetric operator $A$ of the form (8.1) with deficiency indices (1,1) is nonnegative, i.e., $(Af,f) \geq 0$ for all $f \in D(A)$). It was shown in [34] that the Schrödinger operator $T_h$ of the form (8.2) is accretive if and only if

$$\operatorname{Re} h \geq -m_\infty(-0), \tag{8.4}$$

where $m_\infty(\lambda)$ is the Weyl-Titchmarsh function [1]. For real $h$ such that $h \geq -m_\infty(-0)$ we get a description of all nonnegative self-adjoint extensions of an operator $A$. For $h = -m_\infty(-0)$ the corresponding operator

$$\begin{cases} A_K\, y = -y'' + q(x)y \\ y'(a) + m_\infty(-0)y(a) = 0 \end{cases} \tag{8.5}$$

is the Kreĭn-von Neumann extension of $A$ and for $h = +\infty$ the corresponding operator

$$\begin{cases} A_F\, y = -y'' + q(x)y \\ y(a) = 0 \end{cases} \tag{8.6}$$

is the Friedrichs extension of $A$ (see [34], [11]).

We conclude this paper with two simple illustrations for Theorems 6.8 and 7.7.

*Example.* Consider a function

$$V(z) = \frac{i}{\sqrt{z}}. \tag{8.7}$$

A direct check confirms that $V(z)$ in (8.7) is a Stieltjes function. It was shown in [29] that the inversion formula

$$G(t) = C + \lim_{y\to 0} \frac{1}{\pi} \int_0^t \operatorname{Im}\left(\frac{i}{\sqrt{x+iy}}\right) dx \tag{8.8}$$

describes the measure $G(t)$ in the representation (6.3). By direct calculations one can confirm that

$$V(z) = \int_0^\infty \frac{dG(t)}{t-z} = \frac{i}{\sqrt{z}}, \quad \text{and that} \quad \int_0^\infty \frac{dG(t)}{t} = \int_0^\infty \frac{dt}{\pi t^{3/2}} = \infty.$$

Thus we can conclude that $V(z) \in S_0^K(R)$. It was shown in [17] that $V(z)$ can be realized as the impedance function of the L-system

$$\Theta = \begin{pmatrix} & \mathbb{A} & & K & 1 \\ \mathcal{H}_+ \subset L_2[a, +\infty) \subset \mathcal{H}_- & & & & \mathbb{C} \end{pmatrix},$$

where

$$\mathbb{A}\, y = -y'' + [iy(0) - y'(0)]\delta(x).$$

The operator $T_h$ in this case is

$$\begin{cases} T_h y = -y'' \\ y'(0) = iy(0), \end{cases} \tag{8.9}$$

and channel operator $Kc = cg$, $g = \delta(x)$, $(c \in \mathbb{C})$ with

$$K^* y = (y, g) = y(0).$$

The real part of $\mathbb{A}$

$$\operatorname{Re}\mathbb{A}\, y = -y'' - y'(0)\delta(x)$$

contains the self-adjoint quasi-kernel

$$\begin{cases} \widehat{A} y = -y'' \\ y'(0) = 0. \end{cases}$$

Clearly, $\widehat{A} = A_K$, where $A_K$ is given by (8.5).

*Example.* Consider a function

$$V(z) = i\sqrt{z}. \tag{8.10}$$

A direct check confirms that $V(z)$ in (8.10) is an inverse Stieltjes function. Applying the inversion formula similar to (8.8) we obtain

$$G(t) = C + \lim_{y\to 0}\frac{1}{\pi}\int_0^t \operatorname{Im}\left(i\sqrt{x+iy}\right)\,dx,$$

where $G(t)$ is the function in the representation (7.2). By direct calculations one can confirm that

$$V(z) = \int_0^\infty \left(\frac{1}{t-z}-\frac{1}{t}\right)dG(t) = i\sqrt{z},$$

and that

$$\int_0^\infty \frac{t}{t^2+1}\,dG(t) = \int_0^\infty \frac{dG(t)}{t} = \int_0^\infty \frac{dt}{\pi\sqrt{t}} = \infty.$$

Thus we can conclude that $V(z)\in S_{0,F}^{-1}(R)$. It was shown in [18] that $V(z)$ can be realized as the impedance function of the L-system

$$\Theta = \begin{pmatrix} \mathbb{A} & & K & 1\\ \mathcal{H}_+\subset L_2[a,+\infty)\subset\mathcal{H}_- & & & \mathbb{C}\end{pmatrix},$$

where

$$\mathbb{A}\,y = -y'' - [iy'(0)+y'(0)]\delta'(x).$$

The operator $T_h$ in this case is again given by (8.9) and channel operator $Kc = cg$, $g=\delta'(x)$, $(c\in\mathbb{C})$ with

$$K^*y = (y,g) = -y'(0).$$

The real part of $\mathbb{A}$

$$\operatorname{Re}\mathbb{A}\,y = -y'' - y(0)\delta'(x)$$

contains the self-adjoint quasi-kernel

$$\begin{cases} \widehat{A}y = -y''\\ y(0)=0.\end{cases}$$

Clearly, $\widehat{A} = A_F$, where $A_F$ is given by (8.6).

**Acknowledgment.** We would like to thank the referee for valuable remarks.

## References

[1] Akhiezer, N.I., Glazman, I.M.: Theory of Linear Operators in Hilbert Space. Dover, New York, (1993)

[2] Ando, T., Nishio, K.: Positive Selfadjoint Extensions of Positive Symmetric Operators. Tohóku Math. J., **22**, 65–75 (1970)

[3] Arlinskiĭ, Yu.M.: On inverse problem of the theory of characteristic functions of unbounded operator colligations. Dopovidi Akad. Nauk Ukrain. RSR, Ser. A, No. 2, 105–109 (1976)

[4] Arlinskiĭ, Yu.M.: Regular (∗)-extensions of quasi-Hermitian operators in rigged Hilbert spaces. (Russian), Izv. Akad. Nauk Armyan. SSR, Ser.Mat., **14**, No. 4, 297–312 (1979)

[5] Arlinskiĭ, Yu.M.: On regular (∗)-extensions and characteristic matrix-valued functions of ordinary differential operators. Boundary value problems for differential operators, Kiev, 3–13, (1980)

[6] Arlinskiĭ, Yu.M.: On accretive (*)-extensions of a positive symmetric operator. (Russian), Dokl. Akad. Nauk. Ukraine, Ser. A, No. 11, 3–5 (1980)

[7] Arlinskiĭ, Yu.M., Tsekanovskiĭ, E.R.: Nonselfadjoint contractive extensions of a Hermitian contraction and theorems of Kreĭn. Russ. Math. Surv., **37:1**, 151–152 (1982)

[8] Arlinskiĭ, Yu.M., Tsekanovskiĭ, E.R.: Sectorial extensions of positive Hermitian operators and their resolvents. (Russian), Akad. Nauk. Armyan. SSR, Dokl., **79**, No.5, 199–203 (1984)

[9] Arlinskiĭ, Yu.M., Tsekanovskiĭ, E.R.: Regular (∗)-extension of unbounded operators, characteristic operator-functions and realization problems of transfer mappings of linear systems. Preprint, VINITI, -2867. -79, Dep.-72 (1979)

[10] Yu.M. Arlinskiĭ and E.R. Tsekanovskiĭ: Quasi-self-adjoint contractive extensions of Hermitian contractions, Teor. Funkts., Funkts. Anal. Prilozhen, **50**, 9–16 (1988) (Russian). English translation in J. Math. Sci. **49**, No. 6, 1241–1247 (1990).

[11] Arlinskiĭ, Yu.M., Tsekanovskiĭ, E.R.: Linear systems with Schrödinger operators and their transfer functions. Oper. Theory Adv. Appl., **149**, 47–77 (2004)

[12] Arlinskiĭ, Yu.M., Tsekanovskiĭ, E.R.: The von Neumann problem for nonnegative symmetric operators. Integral Equations Operator Theory, **51**, No. 3, 319-356 (2005)

[13] Belyi, S.V., Hassi, S., de Snoo, H.S.V., Tsekanovskiĭ, E.R.: On the realization of inverse of Stieltjes functions. Proceedings of MTNS-2002, University of Notre Dame, CD-ROM, 11p., (2002)

[14] Belyi, S.V., Tsekanovskiĭ, E.R.: Realization theorems for operator-valued $R$-functions. Operator theory: Advances and Applications, **98**, Birkhäuser Verlag Basel, 55–91 (1997)

[15] Belyi, S.V., Tsekanovskiĭ, E.R.: Multiplication Theorems for J-contractive operator-valued functions. Fields Institute Communications, **25**, 187–210 (2000)

[16] Belyi, S.V., Tsekanovskiĭ, E.R.: On classes of realizable operator-valued $R$-functions. Operator theory: Advances and Applications, **115**, Birkhäuser Verlag Basel, (2000), 85–112.

[17] Belyi, S.V., Tsekanovskiĭ, E.R.: Stieltjes like functions and inverse problems for systems with Schrödinger operator. Operators and Matrices, vol. 2, No. 2, 265–296 (2008)

[18] Belyi, S.V., Tsekanovskiĭ, E.R.: Inverse Stieltjes like functions and inverse problems for systems with Schrödinger operator. Operator Theory: Advances and Applications, vol. 197, 21–49 (2009)

[19] Douglas, R.G.: On majorization, factorization and range inclusion of operators in Hilbert space. Proc. Amer. Math. Soc. **17**, 413–416 (1966)

[20] Dovzhenko, I., Tsekanovskiĭ, E.R.: Classes of Stieltjes operator-valued functions and their conservative realizations. Soviet Math. Dokl., **41**, no. 2, 201–204 (1990)

[21] Fillmore, P.A., Williams, J.P.: On operator ranges. Advances in Math. **7**, 254–281 (1971)

[22] Gesztesy, F., Tsekanovskiĭ, E.R.: On Matrix-Valued Herglotz Functions. Math. Nachr. **218**, 61–138 (2000)

[23] Kac, I.S., Kreĭn, M.G.: $R$-functions – analytic functions mapping the upper half-plane into itself. Amer. Math. Soc. Transl., (2), **103**, 1–18 (1974)

[24] Kato, T.: Perturbation Theory for Linear Operators. Springer-Verlag, (1966)

[25] Kreĭn, M.G.: The Theory of Selfadjoint Extensions of Semibounded Hermitian Transformations and its Applications. I, (Russian) Mat. Sbornik **20**, No. 3, 431–495 (1947)

[26] Kreĭn, M.G.: The Theory of Selfadjoint Extensions of Semibounded Hermitian Transformations and its Applications, II, (Russian) Mat. Sbornik **21**, No. 3, 365–404 (1947)

[27] Kreĭn, M.G., Langer, H.: Über die Q-Function eines $\Pi$-Hermiteschen Operators im Raum $\Pi_\kappa$. Acta Sci. Math. Szeged, **34**, 191–230 (1973)

[28] Malamud, M.: On some classes of Hermitian operators with gaps. (Russian) Ukrainian Mat.J., **44**, No. 2, 215–234 (1992)

[29] Naimark, M.A.: Linear Differential Operators II. F. Ungar Publ., New York, (1968)

[30] Okunskiĭ, M.D., Tsekanovskiĭ, E.R.: On the theory of generalized selfadjoint extensions of semibounded operators. (Russian) Funkcional. Anal. i Prilozen., **7**, No. 3, 92–93 (1973)

[31] Phillips, R.: On dissipative operators, in "Lectures in Differential Equations", vol. II, Van Nostrand-Reinhold, New York, 65–113 (1965).

[32] Tsekanovskiĭ, E.R.: Non-self-adjoint accretive extensions of positive operators and theorems of Friedrichs-Kreĭn-Phillips. Funct. Anal. Appl. **14**, 156–157 (1980)

[33] Tsekanovskiĭ, E.R.: Friedrichs and Kreĭn extensions of positive operators and holomorphic contraction semigroups. Funct. Anal. Appl. **15**, 308–309 (1981)

[34] Tsekanovskiĭ, E.R.: Accretive Extensions and Problems on Stieltjes Operator-Valued Functions Relations. Operator Theory: Advan. and Appl., **59**, 328–347 (1992)

[35] Tsekanovskiĭ, E.R., Šmuljan, Yu.L.: The theory of bi-extensions of operators on rigged Hilbert spaces. Unbounded operator colligations and characteristic functions. Russ. Math. Surv., **32**, 73–131 (1977)

Yury Arlinskiĭ
Department of Mathematics
East Ukrainian National University
Kvartal Molodyozhny, 20-A
91034 Lugansk, Ukraine
e-mail: `yma@snu.edu.ua`

Sergey Belyi
Department of Mathematics
Troy University
Troy, AL 36082, USA
e-mail: `sbelyi@troy.edu`

Eduard Tsekanovskiĭ
Department of Mathematics
Niagara University
New York 14109, USA
e-mail: `tsekanov@niagara.edu`

Operator Theory:
Advances and Applications, Vol. 221, 109-119


# The Dynamical Problem for a Non Self-adjoint Hamiltonian

Fabio Bagarello and Miloslav Znojil

**Abstract.** After a compact overview of the standard mathematical presentations of the formalism of quantum mechanics using the language of C*-algebras and/or the language of Hilbert spaces we turn attention to the possible use of the language of Krein spaces. In the context of the so-called three-Hilbert-space scenario involving the so-called PT-symmetric or quasi-Hermitian quantum models a few recent results are reviewed from this point of view, with particular focus on the quantum dynamics in the Schrödinger and Heisenberg representations.

**Mathematics Subject Classification (2000).** Primary 47B50; Secondary 81Q65 47N50 81Q12 47B36 46C20.

**Keywords.** Metrics in Hilbert spaces, hermitizations of a Hamiltonian.

## 1. Introduction

In the analysis of the dynamics of a closed quantum system $\mathcal{S}$ a special role is played by the energy $H$, which is typically the self-adjoint operator defined by the sum of the kinetic energy of $\mathcal{S}$ and of the potential energy giving rise to the conservative forces acting on $\mathcal{S}$. The *most common* approaches in the description of $\mathcal{S}$ are the following:

**1.** *The algebraic description* ($AD$): in this approach the observables of $\mathcal{S}$ are elements of a C*-algebra $\mathfrak{A}$ (which coincides with $B(\mathcal{H})$ for some Hilbert space $\mathcal{H}$). This means, first of all, that $\mathfrak{A}$ is a vector space over $\mathbb{C}$ with a multiplication law such that $\forall A, B \in \mathfrak{A}$, $AB \in \mathfrak{A}$. Also, two such elements can be summed up and the following properties hold: $\forall A, B, C \in \mathfrak{A}$ and $\forall \alpha, \beta \in \mathbb{C}$ we have

$$A(BC) = (AB)C, \quad A(B+C) = AB + AC, \quad (\alpha A)(\beta B) = \alpha\beta(AB).$$

An involution is a map $* : \mathfrak{A} \to \mathfrak{A}$ such that

$$A^{**} = A, \quad (AB)^* = B^* A^*, \quad (\alpha A + \beta B)^* = \overline{\alpha}\, A^* + \overline{\beta}\, B^*$$

A *-algebra $\mathfrak{A}$ is an algebra with an involution *. $\mathfrak{A}$ is a *normed algebra* if there exists a map, *the norm of the algebra*, $\|.\| : \mathfrak{A} \to \mathbb{R}_+$, such that:

$$\|A\| \geq 0, \qquad \|A\| = 0 \Longleftrightarrow A = 0, \qquad \|\alpha A\| = |\alpha| \, \|A\|,$$
$$\|A + B\| \leq \|A\| + \|B\|, \qquad \|AB\| \leq \|A\| \, \|B\|.$$

If $\mathfrak{A}$ is complete wrt $\|.\|$, then it is called a *Banach algebra*, or a *Banach *-algebra* if $\|A^*\| = \|A\|$. If further $\|A^*A\| = \|A\|^2$ holds for all $A \in \mathfrak{A}$, then $\mathfrak{A}$ is a *$C^*$-algebra.*

The *states* are linear, positive and normalized functionals on $\mathfrak{A}$, which look like $\rho(\hat{A}) = tr(\hat{\rho}A)$, where $\mathfrak{A} = B(\mathcal{H})$, $\hat{\rho}$ is a trace-class operator and $tr$ is the trace on $\mathcal{H}$. This means in particular that

$$\rho(\alpha_1 A + \alpha_2 B) = \alpha_1 \rho(A) + \alpha_2 \rho(B)$$

and that, if $\mathfrak{A}$ has the identity $1\!\!1$,

$$\rho(A^*A) \geq 0; \qquad \rho(1\!\!1) = 1.$$

An immediate consequence of these assumptions, and in particular of the positivity of $\rho$, is that $\rho$ is also continuous, i.e., that $|\rho(A)| \leq \|A\|$ for all $A \in \mathfrak{A}$.

The dynamics in the Heisenberg representation for the closed quantum system $\mathcal{S}$ is given by the map

$$\mathfrak{A} \ni A \to \alpha^t(A) = U_t A U_t^\dagger \in \mathfrak{A}, \, \forall t$$

which defines a 1-parameter group of *-automorphisms of $\mathfrak{A}$ satisfying the following conditions

$$\alpha^t(\lambda A) = \lambda \alpha^t(A), \qquad \alpha^t(A + B) = \alpha^t(A) + \alpha^t(B),$$
$$\alpha^t(AB) = \alpha^t(A) \, \alpha^t(B), \qquad \|\alpha^t(A)\| = \|A\|, \text{ and } \alpha^{t+s} = \alpha^t \, \alpha^s.$$

In the Schrödinger representation the time evolution is the dual of the one above, i.e., it is the map between states defined by $\hat{\rho} \to \hat{\rho}_t = \alpha^{t^*} \hat{\rho}$.

**2.** *The Hilbert space description* (*HSD*): this is much simpler, at a first sight. We work in some fixed Hilbert space $\mathcal{H}$, somehow related to the system we are willing to describe, and we proceed as follows:

- each observable $A$ of the physical system corresponds to a self-adjoint operator $\hat{A}$ in $\mathcal{H}$;
- the pure states of the physical system corresponds to normalized vectors of $\mathcal{H}$;
- the expectation values of $A$ correspond to the following mean values: $\langle \psi, \hat{A}\psi \rangle = \rho_\psi(\hat{A}) = tr(P_\psi \, \hat{A})$, where we have also introduced a projector operator $P_\psi$ on $\psi$ and $tr$ is the trace on $\mathcal{H}$;
- the states which are not pure, i.e., the mixed states, correspond to convex linear combinations $\hat{\rho} = \sum_j w_j \, \rho_{\psi_n}$, with $\sum_j w_j = 1$ and $w_j \geq 0$ for all $j$;
- the dynamics (in the Schrödinger representation) is given by a unitary operator $U_t := e^{iHt/\hbar}$, where $H$ is the self-adjoint energy operator, as follows: $\hat{\rho} \to \hat{\rho}_t = U_t^\dagger \hat{\rho} U_t$. In the Heisenberg representation the states do not evolve in time while

the operators do, following the *dual* rule: $\hat{A} \to \hat{A}_t = U_t \hat{A} U_t^\dagger$, and the Heisenberg equation of motion is satisfied: $\frac{d}{dt}\hat{A}_t = \frac{i}{\hbar}[H, \hat{A}_t]$. It is very well known that these two different representations have the same physical content: indeed we have $\hat{\rho}(\hat{A}_t) = \hat{\rho}_t(\hat{A})$, which means that what we measure in experiments, that is the time evolution of the mean values of the observables of $\mathcal{S}$, do not depend on the representation chosen.

The $AD$ is especially useful when $\mathcal{S}$ has an infinite number of degrees of freedom, [4, 10, 30], while the $HSD$ is quite common for ordinary quantum mechanical systems, i.e., for those systems with a finite number of degrees of freedom. The reason why the algebraic approach to ordinary quantum mechanics is not very much used in this simpler case follows from the following von Neumann uniqueness theorem: *for finite quantum mechanical systems there exists only one irreducible representation.* This result is false for systems with infinite degrees of freedom (briefly, in $QM_\infty$), for which $AD$ proved to be useful, for instance in the description of phase-transitions [32].

As we have already said, in the most common applications of quantum theory the self-adjoint Hamiltonian is just the sum of a kinetic plus an interaction term, and the Hilbert space in which the model is described is usually $\mathcal{H} := \mathcal{L}^2(\mathbb{R}^D)$, for $D = 1, 2$ or 3. Of course any unitary map defined from $\mathcal{H}$ to any (in general different) other Hilbert space $\tilde{\mathcal{H}}$ does not change the physics which is contained in the model, but only provides a different way to extract the results from the model itself. In particular, the mean values of the different observables do not depend from the Hilbert space chosen, as far as the different representations are connected by unitary maps. These observables are other self-adjoint operators whose eigenvalues are interesting for us since they have some physical meaning. For a quantum particle moving in $D$-dimensional Euclidian space, for example, people usually work with the position operator $\mathfrak{q}$, whose eigenvalues are used to label the wave function $\psi(\vec{x}, t)$ of the system, which is clearly an element of $\mathcal{L}^2(\mathbb{R}^D)$. It might be worth reminding that we are talking here of *representations* from two different points of view: the Heisenberg and the Schrödinger representations are two (physical) equivalent ways to describe the dynamics of $\mathcal{S}$, while in the $AD$ a representation is a map from $\mathfrak{A}$ to $B(\mathcal{H})$ for a chosen $\mathcal{H}$ which preserve the algebraic structure of $\mathfrak{A}$. Not all these kind of representations are unitarily equivalent, and for this reason they can describe different physics (e.g., different phases of $\mathcal{S}$), [4, 10, 30].

Recently, Bender and Boettcher [8] emphasized that many Hamiltonians $H$ which look unphysical in $\mathcal{H}$ may still be correct and physical, provided only that the conservative textbook paradigm is replaced by a modification called **PT**-symmetric quantum mechanics (PTSQM, cf., e.g., reviews [9, 13, 15, 25, 35] for more details). Within the PTSQM formalism the operators **P** and **T** (which characterize a symmetry of the quantum system in question) are usually pre-selected as parity and time reversal, respectively.

## 2. Quantization recipes using non-unitary Dyson mappings $\Omega$

The main appeal of the PTSQM formalism lies in the permission of selecting, for phenomenological purposes, various new and nonstandard Hamiltonians exhibiting the manifest non-Hermiticity property $H \neq H^\dagger$ in $\mathcal{H}$. It is clear that these operators cannot generate unitary time evolution in $\mathcal{H}$ via exponentiation, [2, 3], but this does not exclude [29] the possibility of finding a unitary time evolution in a different Hilbert space, not necessarily uniquely determined [36], which, following the notation in [35], we indicate as $\mathcal{H}^{(P)}$, $P$ standing for *physical.* Of course, the descriptions of the dynamics in $\mathcal{H}$ and $\mathcal{H}^{(P)}$ cannot be connected by a unitary operator, but still other possibilities are allowed and, indeed, these different choices are those relevant for us here.

In one of the oldest applications of certain specific Hamiltonians with the property $\hat{H} \neq \hat{H}^\dagger$ in $\mathcal{H}$ in the so-called interacting boson models of nuclei [29] it has been emphasized that besides the above-mentioned fact that each such Hamiltonian admits many non-equivalent physical interpretations realized via mutually nonequivalent Hilbert spaces $\mathcal{H}^{(P)}$, one can also start from a *fixed* self-adjoint Hamiltonian $\mathfrak{h}$ defined in $\mathcal{H}^{(P)}$ and move towards *many* alternative isospectral images defined, in some (different) Hilbert space $\mathcal{H}$, by formula $\hat{H} = \Omega^{-1}\,\mathfrak{h}\,\Omega$. Here the so-called Dyson map $\Omega$ should be assumed nontrivial, i.e., non-unitary: $\Omega^\dagger\Omega = \Theta \neq \mathbb{1}$.

In phenomenology and practice, the only reason for preference and choice between $\hat{H}$ and $\mathfrak{h}$ is the feasibility of calculations and the constructive nature of experimental predictions. However, we also should be aware of the fact that $\Omega$ is rather often an unbounded operator, so that many mathematical subtle points usually arise when moving from $\mathfrak{h}$ to $\hat{H}$ in the way suggested above because of, among others, *domain details.* Examples of this kind of problems are discussed in [5, 6, 7] in connection with the so-called *pseudo-bosons.*

### 2.1. The three-space scenario

For certain complicated quantized systems (say, of tens or hundreds of fermions as occur, typically, in nuclear physics, or for many-body systems) the traditional theory forces us to work with an almost prohibitively complicated Hilbert space $\mathcal{H}^{(P)}$ which is not "friendly" at all. Typically, this space acquires the form of a multiple product $\bigotimes L^2(\mathbb{R}^3)$ or, even worse, of an antisymmetrized Fock space. In such a case, unfortunately, wave functions $\psi^{(P)}(t)$ in $\mathcal{H}^{(P)}$ become hardly accessible to explicit construction. The technical difficulties make Schrödinger's equation practically useless: no time evolution can be easily deduced.

A sophisticated way towards a constructive analysis of similar quantum systems has been described by Scholtz et al. [29]. They felt inspired by the encouraging practical experience with the so-called Dyson's mappings $\Omega^{(\mathrm{Dyson})}$ between bosons and fermions in nuclear physics. Still, their technique of an efficient simplification of the theory is independent of any particular implementation details. One only has to assume that the *overcomplicated* realistic Hilbert space $\mathcal{H}^{(P)}$ is being mapped on a *much simpler*, friendly intermediate space $\mathcal{H}^{(F)}$. The latter space remains just auxiliary and unphysical but it renders the calculations (e.g., of spectra) feasible. In particular, the complicated state vectors $\psi^{(P)}(t)$ are made friendlier via an invertible transition from $\mathcal{H}^{(P)}$ to $\mathcal{H}^{(F)}$,

$$\psi^{(P)}(t) = \Omega\,\psi(t) \;\in\; \mathcal{H}^{(P)}, \qquad \psi(t) \;\in\; \mathcal{H} = \mathcal{H}^{(F)}. \tag{2.1}$$

The introduction of the redundant superscript $^{(F)}$ underlines the maximal friendliness of the space (note, e.g., that in the above-mentioned nuclear-physics context of [29], the auxiliary Hilbert space $\mathcal{H}^{(F)}$ was a bosonic space). It is clear that, since $\Omega$ is not unitary, the inner product between two functions $\psi_1^{(P)}(t)$ and $\psi_2^{(P)}(t)$ (treated as elements of $\mathcal{H}^{(P)}$) differs from the one between $\psi_1(t)$ and $\psi_2(t)$,

$$\langle\psi_1^{(P)}, \psi_2^{(P)}\rangle_P = \langle\psi_1, \Omega^\dagger\,\Omega\,\psi_2\rangle_F \;\neq\; \langle\psi_1, \psi_2\rangle_F\,. \tag{2.2}$$

Here we use the suffixes $P$ and $F$ to stress the fact that the Hilbert spaces $\mathcal{H}^{(F)}$ and $\mathcal{H}^{(P)}$ are different and, consequently, they have different inner products in general. Under the assumption that $\ker\{\Omega\}$ only contains the zero vector, and assuming for the moment that $\Omega$ is bounded, $\Theta := \Omega^\dagger\,\Omega$ may be interpreted as producing another, alternative inner product between elements $\psi_1(t)$ and $\psi_2(t)$ of $\mathcal{H}^{(F)}$. This is because $\Theta$ is strictly positive. This suggests to introduce a third Hilbert space, $\mathcal{H}^{(S)}$, which coincides with $\mathcal{H}^{(F)}$ but for the inner product. The new product, $\langle.,.\rangle^{(S)}$, is defined by formula

$$\langle\psi_1, \psi_2\rangle^{(S)} :- \langle\psi_1, \Theta\psi_2\rangle_F \tag{2.3}$$

and, because of (2.2), exhibits the following unitary-equivalence property

$$\langle\psi_1, \psi_2\rangle^{(S)} \;\equiv\; \langle\psi_1^{(P)}, \psi_2^{(P)}\rangle_P\,. \tag{2.4}$$

More details on this point can be found in [35]. It is worth stressing that the fact that $\Omega$ is bounded makes it possible to have $\Theta$ everywhere defined in $\mathcal{H}^{(F)}$. Under the more general (and, we should say, more common) situation when $\Omega$ is not bounded, we should be careful about the possibility of introducing $\Theta$, since $\Omega^\dagger\Omega$ could be not well defined [19, 20, 21]: indeed, for some $f$ in the domain of $\Omega$, $f \in D(\Omega)$, we could have that $\Omega f \notin D(\Omega^\dagger)$. If this is the case, the best we can have is that the inner product $\langle.,.\rangle^{(S)}$ is defined on a dense subspace of $\mathcal{H}^{(F)}$.

### 2.2. A redefinition of the conjugation

The adjoint $X^\dagger$ of a given (bounded) operator $X$ acting on a certain Hilbert space $\mathcal{K}$, with inner product $(.,.)$, is defined by the following equality:

$$(X\varphi, \Psi) = (\varphi, X^\dagger\Psi).$$

Here $\varphi$ and $\Psi$ are arbitrary vectors in $\mathcal{K}$. It is clear that changing the inner product also produces a different adjoint. Hence the adjoint in $\mathcal{H}^{(S)}$ is different from that in $\mathcal{H}^{(F)}$, since their inner products are different. The technical simplifying assumption that $X$ is bounded is ensured, for instance, if we consider finite-dimensional Hilbert spaces. In this way we can avoid difficulties which could arise, e.g., due to the unboundedness of the metric operator $\Theta$. The choice of $\dim \mathcal{H}^{(P,F,S)} < \infty$ gives also the chance of getting analytical results which, otherwise, would be out of our reach.

Once we have, in $\mathcal{H}^{(P)}$, the physical, i.e., safely Hermitian and self-adjoint $\mathfrak{h} = \Omega\, \hat{H}\, \Omega^{-1} = \mathfrak{h}^{\dagger}$, we may easily deduce that

$$\hat{H} = \Theta^{-1} \hat{H}^{\dagger}\, \Theta := \hat{H}^{\ddagger} \tag{2.5}$$

where $^{\dagger}$ stands for the conjugation in either $\mathcal{H}^{(P)}$ or $\mathcal{H}^{(F)}$ while $^{\ddagger}$ may be treated as meaning the conjugation in $\mathcal{H}^{(S)}$ which is metric-mediated (sometimes also called "non-Dirac conjugation" in physics literature). Any operator $\hat{H}$ which satisfies Eq. (2.5) is said to be *quasi-Hermitian.* The similarity of superscripts $^{\dagger}$ and $^{\ddagger}$ emphasizes the formal parallels between the three Hilbert spaces $\mathcal{H}^{(P,F,S)}$.

Once we temporarily return to the point of view of physics we must emphasize that the use of the nontrivial metric $\Theta$ is strongly motivated by the contrast between the simplicity of $\hat{H}$ (acting in friendly $\mathcal{H}^{(F)}$ as well as in the non-equivalent but physical $\mathcal{H}^{(S)}$) and the practical intractability of its isospectral partner $\mathfrak{h}$ (defined as acting in a constructively inaccessible Dyson-image space $\mathcal{H}^{(P)}$). Naturally, once we make the selection of $\mathcal{H}^{(S)}$ (in the role of the space in which the quantum system in question is represented), all of the other operators of observables (say, $\hat{\Lambda}$) acting in $\mathcal{H}^{(S)}$ must be also self-adjoint in the same space, i.e., they must be quasi-Hermitian with respect to *the same* metric,

$$\hat{\Lambda} = \hat{\Lambda}^{\ddagger} := \Theta^{-1} \hat{\Lambda}^{\dagger}\, \Theta\,. \tag{2.6}$$

In opposite direction, once we start from a given Hamiltonian $\hat{H}$ and search for a metric $\Theta$ which would make it quasi-Hermitian (i.e., compatible with the requirement (2.5)), we reveal that there exist *many different* eligible metrics $\Theta = \Theta(\hat{H})$. In such a situation the simultaneous requirement of the quasi-Hermiticity of another operator imposes new constraints of the form $\Theta(\hat{H}) = \Theta(\hat{\Lambda})$ which restricts, in principle, the ambiguity of the metric [29]. Thus, a finite series of the quasi-Hermiticity constraints

$$\hat{\Lambda}_j = \hat{\Lambda}_j^{\ddagger}\,, \qquad j = 1, 2, \ldots, J \tag{2.7}$$

often leads to a unique physical metric $\Theta$ and to the unique, optimal Hilbert-space representation $\mathcal{H}^{(S)}$. This is what naturally extends the usual requirement of the ordinary textbook quantum mechanics which requires the set of the observables to be self-adjoint in a *pre-selected* Hilbert-space representation or very concrete realization $\mathcal{H}^{(F)}$.

Let us now briefly describe some consequences of (2.5) to the dynamical analysis of $\mathcal{S}$. As we have already seen relation (2.3) defines the *physical* inner product. We will now verify that this is the natural inner product to be used to find the expected unitary evolution generated by the quasi-Hermitian operator $\hat{H}$.

The first consequence of property $\hat{H} = \Theta^{-1} \hat{H}^{\dagger}\, \Theta$ is that $e^{i\hat{H}^{\dagger}t} = \Theta e^{i\hat{H}t} \Theta^{-1}$. The proof of this equality is trivial whenever the operators involved are bounded, condition which we will assume here for simplicity (as stated above, we could simply imagine that our Hilbert spaces are finite-dimensional). Condition (2.5) allows us to keep most of the standard approach to the dynamics of the quantum system sketched in the Introduction, even in presence of a manifest non-Hermiticity of $\hat{H}$ in friendly $\mathcal{H}^{(F)}$. Indeed, once we consider the Schrödinger evolution of a vector $\Psi(0)$, which we take to be $\Psi(t) = e^{-i\hat{H}t}\Psi(0)$, we may immediately turn attention to the time-dependence of the *physical* norm in $\mathcal{H}^{(S)}$,

$$\begin{aligned}\|\Psi(t)\|^2 &:= \langle\Psi(t),\Psi(t)\rangle^{(S)} = \langle e^{-i\hat{H}t}\Psi(0),\Theta e^{-i\hat{H}t}\Psi(0)\rangle_F \\ &= \langle\Psi(0),e^{i\hat{H}^\dagger t}\Theta e^{-i\hat{H}t}\Psi(0)\rangle_F = \langle\Psi(0),\Theta\Psi(0)\rangle_F = \|\Psi(0)\|^2,\end{aligned}$$

Thus, for all $\Psi \in \mathcal{H}^{(S)}$ the natural use of the inner product $\langle .,.\rangle^{(S)}$ and of the related norm gives a probability which is preserved in time. This observation has also an interesting (even if expected) consequence: the dual evolution, i.e., the time evolution of the observables in the Heisenberg representation, has the standard form, $X(t) = e^{i\hat{H}t}Xe^{-i\hat{H}t}$, so that $\dot{X}(t) = i[\hat{H},X(t)]$. This is true independently of the fact that $\hat{H}$ is self-adjoint or not, as in the present case. Indeed if we ask the mean value (in the product $(S)$) of the observables to be independent of the representation adopted, i.e., if we require that

$$\langle\varphi(t),X\Psi(t)\rangle^{(S)} = \langle e^{-i\hat{H}t}\varphi(0),\Theta Xe^{-i\hat{H}t}\Psi(0)\rangle = \langle\varphi(0),X(t)\Psi(0)\rangle^{(S)}$$

then we are forced to put $X(t) = e^{i\hat{H}t}Xe^{-i\hat{H}t}$ (rather than the maybe more natural $e^{i\hat{H}t}Xe^{-i\hat{H}^\dagger t}$). This means that a consistent approach to the dynamical problem can be settled up also when the energy operator of the system is not self-adjoint, paying the price by replacing the Hilbert space in which the model was first defined, and its inner product, with something slightly different. In this different Hilbert space the time evolution *does its job*, preserving probabilities and satisfying the usual differential equations, both in the Schrödinger and in the Heisenberg pictures.

### 2.3. PT-symmetric quantum mechanics

One of the most efficient suppressions of the ambiguity of the metric has been proposed within the PTSQM formalism where the role of the additional observable $\hat{\Lambda}$ is being assigned to another involution $\mathbf{C}$ [9]. The presence as well as a "hidden" mathematical meaning of the original involution $\mathbf{P}$ may be further clarified by introducing, together with our three Hilbert spaces $\mathcal{H}^{(P,F,S)}$, also another auxiliary space $\mathbf{K}$ with the structure of the Krein space endowed with an invertible indefinite metric equal to the parity operator as mentioned above, $\mathbf{P} = \mathbf{P}^\dagger$ [24]. One requires that the given Hamiltonian proves $\mathbf{P}$-self-adjoint in $\mathbf{K}$, i.e., that it satisfies equation

$$\hat{H}^\dagger\,\mathbf{P} = \mathbf{P}\,\hat{H}\,. \tag{2.8}$$

This is an intertwining relation between two non self-adjoint operators $\hat{H}$ and $\hat{H}^\dagger$, and $P$ is the *intertwining operator* (IO, [14, 33]). In the standard literature on IO, see [22, 23, 28] for instance, the operators intertwined by the IO are self-adjoint, so that their eigenvalues are real, coincident, and the associated eigenvectors are orthogonal (if the degeneracy of each eigenvalue is 1). Here, see below, these eigenvalues are not necessarily real. Nevertheless, many situations do exist in the literature in which the eigenvalues of some non self-adjoint operator can be computed and turn out to be real, [5, 6, 7, 17, 18, 34], even if $\hat{H}$ is not self-adjoint in $\mathcal{H}^{(F)}$.

Let us go back to the requirement (2.8). Multiple examples of its usefulness may be found scattered in the literature [1, 16]. Buslaev and Grecchi [11] were probably the first mathematical physicists who started calling property (2.8) of the Hamiltonian a "**PT**-symmetry". Let us now briefly explain its mathematical benefits under a not too essential additional assumption that the spectrum $\{E_n\}$ of $\hat{H}$ is discrete and non-degenerate (though, naturally, not necessarily real). Then we may solve the usual Schrödinger equation for the (right) eigenvectors of $\hat{H}$,

$$\hat{H}\,\psi_n = E_n\,\psi_n\,, \qquad n = 0, 1, \ldots \tag{2.9}$$

as well as its Schrödinger-like conjugate rearrangement

$$\hat{H}^\dagger\,\psi^n = E_n^*\,\psi^n\,, \qquad n = 0, 1, \ldots. \tag{2.10}$$

In the light of Krein-space rule (2.8) the latter equation acquires the equivalent form

$$\hat{H}\,\left(\mathbf{P}^{-1}\,\psi^n\right) = E_n^*\,\left(\mathbf{P}^{-1}\,\psi^n\right)\,, \qquad n = 0, 1, \ldots \tag{2.11}$$

so that we may conclude that

- either $E_n = E_n^*$ is real and the action of $\mathbf{P}^{-1}$ on $\psi^n$ gives merely a vector proportional to the $\psi_n$,
- or $E_n \neq E_n^*$ is not real. In this regime we have $E_n^* = E_m$ at some $m \neq n$. This means that the action of $\mathbf{P}^{-1}$ on the left eigenket $\psi^n$ produces a right eigenvector of $\hat{H}$ which is proportional to certain right eigenket $\psi_m$ of Eq. (2.9) at a *different* energy, $m \neq n$.

We arrived at a dichotomy: Once we take all of the $n$-superscripted left eigenvectors $\psi^n$ of $\hat{H}$ and premultiply them by the inverse pseudometric, we obtain a new set of ket vectors $\phi_n = \mathbf{P}^{-1}\,\psi^n$ which are either all proportional to their respective $n$-subscripted partners $\psi_n$ (while the whole spectrum is real) or not. This is a way to distinguish between two classes of Hamiltonians. Within the framework of quantum mechanics only those with the former property of having real eigenvalues may be considered physical (see [6] for a typical illustration). In parallel, examples with the latter property may be still found interesting beyond the limits of quantum mechanics, i.e., typically, in classical optics [12, 27]. The recent growth of interest in the latter models (exhibiting the so-called spontaneously broken **PT**-symmetry) may be well documented by their presentation via the dedicated webpages [39, 40].

The Hamiltonians $\hat{H}$ with the real and non-degenerate spectra may be declared acceptable in quantum mechanics. In the subsequent step our knowledge of the eigevectors $\psi^n$ of $\hat{H}$ enables us to define the positive-definite operator of the metric directly [26],

$$\Theta = \sum_{n=0}^{\infty} |\psi^n\rangle\,\langle\psi^n|\,. \tag{2.12}$$

As long as this formula defines different matrices for different normalizations of vectors $|\psi^n\rangle$ (cf. [37] for details) we may finally eliminate this ambiguity via the factorization ansatz

$$\Theta = \mathbf{PC} \tag{2.13}$$

followed by the *double* involutivity constraint [9]

$$\mathbf{P}^2 = I\,, \qquad \mathbf{C}^2 = I\,. \tag{2.14}$$

We should mention that the series in (2.12) could be just formal. This happens whenever the operator $\Theta$ is not bounded. This aspect was considered in many details in connection with pseudo-bosons, see [5] for instance, where one of us has proved that $\Theta$ being bounded is equivalent to $\{\Psi^n\}$ being a Riesz basis.

Skipping further technical details we may now summarize: The operators of the form (2.13) and (2.14) have to satisfy a number of additional mathematical conditions before they may be declared the admissible metric operators determining the inner product in $\mathcal{H}^{(S)}$. *Vice versa,* once we satisfy these conditions (cf., e.g., [29, 31] for their list) we may replace the unphysical and auxiliary Hilbert space $\mathcal{H}^{(F)}$ and the intermediate Krein space $\mathbf{K}$ (with its indefinite metric $\mathbf{P} = \mathbf{P}^\dagger$) by the ultimate physical Hilbert space $\mathcal{H}^{(S)}$ of the PTSQM theory. The input Hamiltonian $\hat{H}$ may be declared self-adjoint in $\mathcal{H}^{(S)}$. In this space it also generates the correct unitary time-evolution of the system in question, at least for the time-independent interactions, [38].

## 3. Summary

We have given here a brief review of some aspects of PTSQM with particular interest to the role of different inner products and their related conjugations. We have also discussed how the time evolution of a system with a non self-adjoint Hamiltonian can be analyzed within this settings, and the role of the different inner products is discussed.

Among other lines of research, we believe that the construction of algebras of unbounded operators associated to PTSQM, along the same lines as in [4], is an interesting task and we hope to be able to do that in the near future.

**Acknowledgment**

Work supported in part by the GAČR grant Nr. P203/11/1433, by the MŠMT "Doppler Institute" project Nr. LC06002 and by the Inst. Res. Plan AV0Z10480505 and in part by M.I.U.R.

## References

[1] A.A. Andrianov, C.M. Bender, H.F. Jones, A. Smilga and M. Znojil, eds., *Proceedings of the VIIth Workshop "Quantum Physics with Non-Hermitian Operators"*, SIGMA **5** (2009), items 001, 005, 007, 017, 018, 039, 043, 047, 053, 064 and 069;

[2] F. Bagarello, A. Inoue, C. Trapani, *Derivations of quasi ∗-algebras*, Int. Jour. Math. and Math. Sci., **21** (2004), 1077–1096.

[3] F. Bagarello, A. Inoue, C. Trapani, *Exponentiating derivations of quasi *-algebras: possible approaches and applications*, Int. Jour. Math. and Math. Sci., **17** (2005), 2805–2820.

[4] F. Bagarello, *Algebras of unbounded operators and physical applications: a survey*, Reviews in Math. Phys, **19** (2007), 231–272.

[5] F. Bagarello, *Pseudo-bosons, Riesz bases and coherent states*, J. Math. Phys., **50** (2010), 023531, 10 pages.

[6] F. Bagarello, *Examples of Pseudo-bosons in quantum mechanics*, Phys. Lett. A, **374** (2010), 3823–3827.

[7] F. Bagarello, (*Regular*) *pseudo-bosons versus bosons*, J. Phys. A, **44** (2011), 015205.

[8] C.M. Bender and S. Boettcher, *Real Spectra in Non-Hermitian Hamiltonians Having PT Symmetry.* Phys. Rev. Lett. **80** (1998), 5243–5246.

[9] C.M. Bender, *Making sense of non-hermitian Hamiltonians.* Rep. Prog. Phys. **70** (2007), 947–1018, hep-th/0703096.

[10] O. Bratteli and D.W. Robinson, *Operator algebras and Quantum statistical mechanics*, vols. 1 and 2, Springer-Verlag, New York, 1987.

[11] V. Buslaev and V. Grechi, *Equivalence of unstable inharmonic oscillators and double wells.* J. Phys. A: Math. Gen. **26** (1993), 5541–5549.

[12] Y.D. Chong, Li Ge, A. Douglas Stone, *PT-symmetry breaking and laser-absorber modes in optical scattering systems*, Phys. Rev. Lett. **106** (2011), 093902.

[13] E.B. Davies, *Linear operators and their spectra.* Cambridge University Press, 2007.

[14] J. Dieudonné, *Quasi-Hermitian operators,* in Proc. Int. Symp. Lin. Spaces, Pergamon, Oxford, 1961, pp. 115–122.

[15] P. Dorey, C. Dunning and R. Tateo, *The ODE/IM correspondence.* J. Phys. A: Math. Theor. **40** (2007), R205–R283, hep-th/0703066.

[16] A. Fring, H. Jones and M. Znojil, eds., *Pseudo-Hermitian Hamiltonians in Quantum Physics VI*, Journal of Physics A: Math. Theor. **41** (2008), items 240301–244027.

[17] S.R. Jain and Z. Ahmed, eds., *Special Issue on Non-Hermitian Hamiltonians in Quantum Physics*, Pramana, Journal of Physics **73** (2009), 215–416 (= part I).

[18] S.R. Jain and Z. Ahmed, eds., *Special Issue on Non-Hermitian Hamiltonians in Quantum Physics*, Pramana, Journal of Physics **73** (2009), 417–626 (= part II).

[19] R. Kretschmer and L. Szymanowski, *The Interpretation of Quantum-Mechanical Models with Non-Hermitian Hamiltonians and Real Spectra*, arXiv:quant-ph/0105054.

[20] R. Kretschmer and L. Szymanowski, *Quasi-Hermiticity in infinite-dimensional Hilbert spaces*, Phys. Lett. A **325** (2004), 112–115.

[21] R. Kretschmer and L. Szymanowski, *The Hilbert-Space Structure of Non-Hermitian Theories with Real Spectra*, Czech. J. Phys. **54** (2004), 71–75.

[22] S. Kuru, A. Tegmen, and A. Vercin, *Intertwined isospectral potentials in an arbitrary dimension*, J. Math. Phys. **42** (2001), 3344–3360.

[23] S. Kuru, B. Demircioglu, M. Onder, and A. Vercin, *Two families of superintegrable and isospectral potentials in two dimensions*, J. Math. Phys. **43** (2002), 2133–2150.

[24] H. Langer, and Ch. Tretter, *A Krein space approach to PT symmetry.* Czechosl. J. Phys. **70** (2004), 1113–1120.

[25] A. Mostafazadeh, *Pseudo-Hermitian Quantum Mechanics.* Int. J. Geom. Meth. Mod. Phys., **7** (2010), 1191–1306.

[26] A. Mostafazadeh, *Metric Operator in Pseudo-Hermitian Quantum Mechanics and the Imaginary Cubic Potential*, J. Phys. A: Math. Theor. **39** (2006), 10171–10188.

[27] A. Mostafazadeh, *Optical Spectral Singularities as Threshold Resonances*, Phys. Rev. A **83** (2011), 045801.

[28] K.A. Samani, and M. Zarei, *Intertwined Hamiltonians in two-dimensional curved spaces*, Ann. of Phys. **316** (2005), 466–482.

[29] F.G. Scholtz, H.B. Geyer and F.J.W. Hahne, *Quasi-Hermitian Operators in Quantum Mechanics and the Variational Principle.* Ann. Phys. (NY) **213** (1992), 74–108.

[30] G.L. Sewell, *Quantum Theory of Collective Phenomena*, Oxford University Press, Oxford, 1989.

[31] P. Siegl, *The non-equivalence of pseudo-Hermiticity and presence of antilinear symmetry.* PRAMANA-Journal of Physics **73** (2009), 279–287.

[32] W. Thirring, *Quantum mathematical physics*, Springer-Verlag, Berlin and Heidelberg, 2010.

[33] J.P. Williams, *Operators Similar to their Adjoints.* Proc. Amer. Math. Soc. **20** (1969), 121–123.

[34] J.-D. Wu and M. Znojil, eds., *Pseudo-Hermitian Hamiltonians in Quantum Physics IX*, Int. J. Theor. Phys. **50** (2011), special issue Nr. 4, pp. 953–1333.

[35] M. Znojil, *Three-Hilbert-space formulation of Quantum Mechanics.* SYMMETRY, INTEGRABILITY and GEOMETRY: METHODS and APPLICATIONS (SIGMA) **5** (2009), 001, 19 pages.

[36] M. Znojil, *Complete set of inner products for a discrete PT-symmetric square-well Hamiltonian.* J. Math. Phys. **50** (2009), 122105.

[37] M. Znojil, *On the Role of Normalization Factors and Pseudometric in Crypto-Hermitian Quantum Models.* SIGMA **4** (2008), p. 001, 9 pages (arXiv: 0710.4432).

[38] M. Znojil, *Time-dependent version of cryptohermitian quantum theory.* Phys. Rev. D **78** (2008), 085003.

[39] http://gemma.ujf.cas.cz/%7Eznojil/conf/proceedphhqp.html

[40] http://ptsymmetry.net

Fabio Bagarello
Dieetcam, Facoltà di Ingegneria
Università di Palermo
I-90128 Palermo, Italy
e-mail: `fabio.bagarello@unipa.it`
URL: `www.unipa.it\fabio.bagarello`

Miloslav Znojil
Nuclear Physics Institute ASCR
250 68 Řež, Czech Republic
e-mail: `znojil@ujf.cas.cz`

Operator Theory:
Advances and Applications, Vol. 221, 121-130


# Extension of the $\nu$-metric: the $H^\infty$ Case

Joseph A. Ball and Amol J. Sasane

**Abstract.** An abstract $\nu$-metric was introduced by Ball and Sasane, with a view towards extending the classical $\nu$-metric of Vinnicombe from the case of rational transfer functions to more general nonrational transfer function classes of infinite-dimensional linear control systems. In this short note, we give an additional concrete special instance of the abstract $\nu$-metric, by verifying all the assumptions demanded in the abstract set-up. This example links the abstract $\nu$-metric with the one proposed by Vinnicombe as a candidate for the $\nu$-metric for nonrational plants.

**Mathematics Subject Classification (2000).** Primary 93B36; Secondary 93D15, 46J15.

**Keywords.** $\nu$-metric, robust control, Hardy algebra, quasianalytic functions.

## 1. Introduction

We recall the general *stabilization problem* in control theory. Suppose that $R$ is a commutative integral domain with identity (thought of as the class of stable transfer functions) and let $\mathbb{F}(R)$ denote the field of fractions of $R$. The stabilization problem is:

Given $P \in (\mathbb{F}(R))^{p\times m}$ (an unstable plant transfer function),
find $C \in (\mathbb{F}(R))^{m\times p}$ (a stabilizing controller transfer function),
such that (the closed loop transfer function)

$$H(P,C) := \begin{bmatrix} P \\ I \end{bmatrix} (I - CP)^{-1} \begin{bmatrix} -C & I \end{bmatrix}$$

belongs to $R^{(p+m)\times(p+m)}$ (is stable).

In the *robust stabilization problem*, one goes a step further. One knows that the plant is just an approximation of reality, and so one would really like the controller $C$ to not only stabilize the *nominal* plant $P_0$, but also all sufficiently close plants $P$ to $P_0$. The question of what one means by "closeness" of plants thus arises naturally.

So one needs a function $d$ defined on pairs of stabilizable plants such that

1. $d$ is a metric on the set of all stabilizable plants,
2. $d$ is amenable to computation, and
3. stabilizability is a robust property of the plant with respect to this metric.

Such a desirable metric, was introduced by Glenn Vinnicombe in [8] and is called the *$\nu$-metric.* In that paper, essentially $R$ was taken to be the rational functions without poles in the closed unit disk or, more generally, the disk algebra, and the most important results were that the $\nu$-metric is indeed a metric on the set of stabilizable plants, and moreover, one has the inequality that if $P_0, P \in \mathbb{S}(R, p, m)$, then

$$\mu_{P,C} \geq \mu_{P_0,C} - d_\nu(P_0, P),$$

where $\mu_{P,C}$ denotes the *stability margin* of the pair $(P, C)$, defined by

$$\mu_{P,C} := \|H(P, C)\|_\infty^{-1}.$$

This implies in particular that stabilizability is a robust property of the plant $P$.

The problem of what happens when $R$ is some other ring of stable transfer functions of infinite-dimensional systems was left open in [8]. This problem of extending the $\nu$-metric from the rational case to transfer function classes of infinite-dimensional systems was addressed in [1]. There the starting point in the approach was abstract. It was assumed that $R$ is any commutative integral domain with identity which is a subset of a Banach algebra $S$ satisfying certain assumptions, labelled (A1)–(A4), which are recalled in Section 2. Then an "abstract" $\nu$-metric was defined in this setup, and it was shown in [1] that it does define a metric on the class of all stabilizable plants. It was also shown there that stabilizability is a robust property of the plant.

In [8], it was suggested that the $\nu$-metric in the case when $R = H^\infty$ might be defined as follows. (Here $H^\infty$ denotes the algebra of bounded and holomorphic functions in the unit disk $\{z \in \mathbb{C} : |z| < 1\}$.) Let $P_1, P_2$ be unstable plants with the normalized left/right coprime factorizations

$$\begin{aligned} P_1 &= N_1 D_1^{-1} = \widetilde{D}_1^{-1}\widetilde{N}_1, \\ P_2 &= N_2 D_2^{-1} = \widetilde{D}_2^{-1}\widetilde{N}_2, \end{aligned}$$

where $N_1, D_1, N_2, D_2, \widetilde{N}_1, \widetilde{D}_1, \widetilde{N}_2, \widetilde{D}_2$ are matrices with $H^\infty$ entries. Then

$$d_\nu(P_1, P_2) = \begin{cases} \|\widetilde{G}_2 G_1\|_\infty & \text{if } T_{G_1^* G_2} \text{ is Fredholm with Fredholm index } 0, \\ 1 & \text{otherwise.} \end{cases} \tag{1.1}$$

Here $\cdot^*$ has the usual meaning, namely: $G_1^*(\zeta)$ is the transpose of the matrix whose entries are complex conjugates of the entries of the matrix $G_1(\zeta)$, for $\zeta \in \mathbb{T}$, and $G_k, \widetilde{G}_k$ arise from $P_k$ ($k = 1, 2$) according to the notational conventions given in Subsection 2.5 below. Also in the above, for a matrix $M \in (L^\infty)^{p\times m}$, $T_M$ denotes the *Toeplitz operator* from $(H^2)^m$ to $(H^2)^p$, given by

$$T_M\varphi = P_{(H^2)^p}(M\varphi) \quad (\varphi \in (H^2)^m)$$

where $M\varphi$ is considered as an element of $(L^2)^p$ and $P_{(H^2)^p}$ denotes the canonical orthogonal projection from $(L^2)^p$ onto $(H^2)^p$.

Although we are unable to verify whether there is a metric $d_\nu$ such that the above holds in the case of $H^\infty$, we show that the above does work for the somewhat smaller case when $R$ is the class $QA$ of quasicontinuous functions analytic in the unit disk. We prove this by showing that this case is just a special instance of the abstract $\nu$-metric introduced in [1].

The paper is organized as follows:

1. In Section 2, we recall the general setup and assumptions and the abstract metric $d_\nu$ from [1].
2. In Section 3, we specialize $R$ to a concrete ring of stable transfer functions, and show that our abstract assumptions hold in this particular case.

## 2. Recap of the abstract $\nu$-metric

We recall the setup from [1]:

(A1) $R$ is commutative integral domain with identity.

(A2) $S$ is a unital commutative complex semisimple Banach algebra with an involution $\cdot^*$, such that $R \subset S$. We use inv $S$ to denote the invertible elements of $S$.

(A3) There exists a map $\iota$ : inv $S \to G$, where $(G, +)$ is an Abelian group with identity denoted by $\circ$, and $\iota$ satisfies

- (I1) $\iota(ab) = \iota(a) + \iota(b)$ $(a, b \in \text{inv } S)$.
- (I2) $\iota(a^*) = -\iota(a)$ $(a \in \text{inv } S)$.
- (I3) $\iota$ is locally constant, that is, $\iota$ is continuous when $G$ is equipped with the discrete topology.

(A4) $x \in R \cap (\text{inv } S)$ is invertible as an element of $R$ if and only if $\iota(x) = \circ$.

We recall the following standard definitions from the factorization approach to control theory.

### 2.1. The notation $\mathbb{F}(R)$:

$\mathbb{F}(R)$ denotes the field of fractions of $R$.

### 2.2. The notation $F^*$:

If $F \in R^{p\times m}$, then $F^* \in S^{m\times p}$ is the matrix with the entry in the $i$th row and $j$th column given by $F_{ji}^*$, for all $1 \le i \le p$, and all $1 \le j \le m$.

### 2.3. Right coprime/normalized coprime factorization:

Given a matrix $P \in (\mathbb{F}(R))^{p\times m}$, a factorization $P = ND^{-1}$, where $N, D$ are matrices with entries from $R$, is called a *right coprime factorization of* $P$ if there exist matrices $X, Y$ with entries from $R$ such that $XN + YD = I_m$. If moreover it holds that $N^*N + D^*D = I_m$, then the right coprime factorization is referred to as a *normalized* right coprime factorization of $P$.

**2.4. Left coprime/normalized coprime factorization:**

A factorization $P = \widetilde{D}^{-1}\widetilde{N}$, where $\widetilde{N}, \widetilde{D}$ are matrices with entries from $R$, is called a *left coprime factorization of* $P$ if there exist matrices $\widetilde{X}, \widetilde{Y}$ with entries from $R$ such that $\widetilde{N}\widetilde{X} + \widetilde{D}\widetilde{Y} = I_p$. If moreover it holds that $\widetilde{N}\widetilde{N}^* + \widetilde{D}\widetilde{D}^* = I_p$, then the left coprime factorization is referred to as a *normalized* left coprime factorization of $P$.

**2.5. The notation $G, \widetilde{G}, K, \widetilde{K}$:**

Given $P \in (\mathbb{F}(R))^{p\times m}$ with normalized right and left factorizations $P = ND^{-1}$ and $P = \widetilde{D}^{-1}\widetilde{N}$, respectively, we introduce the following matrices with entries from $R$:

$$G = \begin{bmatrix} N \\ D \end{bmatrix} \quad \text{and} \quad \widetilde{G} = \begin{bmatrix} -\widetilde{D} & \widetilde{N} \end{bmatrix}.$$

Similarly, given a $C \in (\mathbb{F}(R))^{m\times p}$ with normalized right and left factorizations $C = N_C D_C^{-1}$ and $C = \widetilde{D}_C^{-1}\widetilde{N}_C$, respectively, we introduce the following matrices with entries from $R$:

$$K = \begin{bmatrix} D_C \\ N_C \end{bmatrix} \quad \text{and} \quad \widetilde{K} = \begin{bmatrix} -\widetilde{N}_C & \widetilde{D}_C \end{bmatrix}.$$

**2.6. The notation $\mathbb{S}(R, p, m)$:**

We denote by $\mathbb{S}(R, p, m)$ the set of all elements $P \in (\mathbb{F}(R))^{p\times m}$ that possess a normalized right coprime factorization and a normalized left coprime factorization.

We now recall the definition of the metric $d_\nu$ on $\mathbb{S}(R, p, m)$. But first we specify the norm we use for matrices with entries from $S$.

**Definition 2.1** **($\|\cdot\|$).** Let $\mathfrak{M}$ denote the maximal ideal space of the Banach algebra $S$. For a matrix $M \in S^{p\times m}$, we set

$$\|M\| = \max_{\varphi\in\mathfrak{M}} |\mathbf{M}(\varphi)|. \tag{2.1}$$

Here $\mathbf{M}$ denotes the entry-wise Gelfand transform of $M$, and $|\cdot|$ denotes the induced operator norm from $\mathbb{C}^m$ to $\mathbb{C}^p$. For the sake of concreteness, we fix the standard Euclidean norms on the vector spaces $\mathbb{C}^m$ to $\mathbb{C}^p$.

The maximum in (2.1) exists since $\mathfrak{M}$ is a compact space when it is equipped with Gelfand topology, that is, the weak-$*$ topology induced from $\mathcal{L}(S;\mathbb{C})$. Since we have assumed $S$ to be semisimple, the Gelfand transform

$$\widehat{\cdot} : S \to \widehat{S} \ (\subset C(\mathfrak{M}, \mathbb{C}))$$

is an isomorphism. If $M \in S^{1\times 1} = S$, then we note that there are two norms available for $M$: the one as we have defined above, namely $\|M\|$, and the norm $\|\cdot\|_S$ of $M$ as an element of the Banach algebra $S$. But throughout this article, we will use the norm given by (2.1).

**Definition 2.2 (Abstract $\nu$-metric $d_\nu$).** For $P_1, P_2 \in \mathbb{S}(R, p, m)$, with the normalized left/right coprime factorizations

$$\begin{aligned} P_1 &= N_1 D_1^{-1} = \widetilde{D}_1^{-1}\widetilde{N}_1, \\ P_2 &= N_2 D_2^{-1} = \widetilde{D}_2^{-1}\widetilde{N}_2, \end{aligned}$$

we define

$$d_\nu(P_1, P_2) := \begin{cases} \|\widetilde{G}_2 G_1\| & \text{if } \det(G_1^* G_2) \in \text{inv } S \text{ and } \iota(\det(G_1^* G_2)) = \circ, \\ 1 & \text{otherwise,} \end{cases} \tag{2.2}$$

where the notation is as in Subsections 2.1–2.6.

The following was proved in [1]:

**Theorem 2.3.** *$d_\nu$ given by (2.2) is a metric on $\mathbb{S}(R, p, m)$.*

**Definition 2.4.** Given $P \in (\mathbb{F}(R))^{p\times m}$ and $C \in (\mathbb{F}(R))^{m\times p}$, the *stability margin* of the pair $(P, C)$ is defined by

$$\mu_{P,C} = \begin{cases} \|H(P,C)\|_\infty^{-1} & \text{if } P \text{ is stabilized by } C, \\ 0 & \text{otherwise.} \end{cases}$$

The number $\mu_{P,C}$ can be interpreted as a measure of the performance of the closed loop system comprising $P$ and $C$: larger values of $\mu_{P,C}$ correspond to better performance, with $\mu_{P,C} > 0$ if $C$ stabilizes $P$.

The following was proved in [1]:

**Theorem 2.5.** *If $P_0, P \in \mathbb{S}(R, p, m)$ and $C \in \mathbb{S}(R, m, p)$, then*

$$\mu_{P,C} \geq \mu_{P_0,C} - d_\nu(P_0, P).$$

The above result says that stabilizability is a robust property of the plant, since if $C$ stabilizes $P_0$ with a stability margin $\mu_{P,C} > m$, and $P$ is another plant which is close to $P_0$ in the sense that $d_\nu(P, P_0) \leq m$, then $C$ is also guaranteed to stabilize $P$.

## 3. The $\nu$-metric when $R = QA$

Let $H^\infty$ be the Hardy algebra, consisting of all bounded and holomorphic functions defined on the open unit disk $\mathbb{D} := \{z \in \mathbb{C} : |z| < 1\}$.

As was observed in the Introduction, it was suggested in [8] to use (1.1) to define a metric on the quotient ring of $H^\infty$. It is tempting to try to do this by using the general setup of [1] with $R = H^\infty$, $S = L^\infty$ and with $\iota$ equal to the Fredholm index of the associated Toeplitz operator. However at this level of generality there is no guarantee that $\varphi$ invertible in $L^\infty$ implies that $T_\varphi$ is Fredholm (and hence $\iota$ equal to the Fredholm index of the associated Toeplitz operator is not well defined on inv $S$ (condition (A3)), as illustrated by the following example.

**Example 3.1** **($\varphi \in \mathrm{inv}\, L^\infty \not\Rightarrow T_\varphi \in \mathrm{Fred}(\mathcal{L}(H^2))$).** There is a simple example of a function which is invertible in $L^\infty$ for which the associated Toeplitz operator is not Fredholm. We take a piecewise continuous function $\varphi$ in $L^\infty$ for which:

1. the closure of the essential range of $\varphi$ misses the origin (guaranteeing invertibility of $\varphi \in L^\infty$) and
2. the line segment connecting the right- and left-hand limits at a jump discontinuity goes through the origin (which by the Gohberg-Krupnik theory, guarantees that the Toeplitz operator $T_\varphi \in \mathcal{L}(H^2)$ is not Fredholm; see for example [5, Lemma 16.6, p. 116]).

For instance, we can take

$$\varphi(e^{i\theta}) := \begin{cases} 1 & \text{if } \theta \in [0, \pi), \\ -1 & \text{if } \theta \in [\pi, 2\pi) \end{cases}$$

which is clearly $\varphi$ is invertible in $L^\infty$ (since $\varphi^{-1} = \varphi$) and moreover, for each of its two discontinuities, corresponding to $\theta = 0$ and $\theta = \pi$, the line segment joining the left- and right-hand limits is the interval $[-1, 1]$, which contains 0. ◇

However a perusal of the extensive literature on Fredholm theory of Toeplitz operators from the 1970s leads to the choices $R$ equal to the class $QA$ of quasi-analytic and $S$ equal to the class $QC$ of quasicontinuous functions as conceivably the most general subalgebras of $H^\infty$ and $L^\infty$ which fit the setup of [1], as we now explain.

$QC$ is the $C^*$-subalgebra of $L^\infty(\mathbb{T})$ of *quasicontinuous* functions:

$$QC := (H^\infty + C(\mathbb{T})) \cap \overline{(H^\infty + C(\mathbb{T}))}.$$

An alternative characterization of $QC$ is $QC = L^\infty \cap VMO$, where $VMO$ is the set of vanishing mean oscillation functions [4, Theorem 2.3, p. 368].

The Banach algebra $QA$ of analytic quasicontinuous functions is

$$QA := H^\infty \cap QC.$$

For verifying (A4), we will also use the result given below; see [2, Theorem 7.36].

**Proposition 3.2.** *If $f \in H^\infty(\mathbb{D}) + C(\mathbb{T})$, then $T_f$ is Fredholm if and only if there exist $\delta, \epsilon > 0$ such that*

$$|F(re^{it})| \geq \epsilon \text{ for } 1 - \delta < r < 1,$$

*where $F$ is the harmonic extension of $f$ to $\mathbb{D}$. Moreover, in this case the index of $T_f$ is the negative of the winding number with respect to the origin of the curve $F(re^{it})$ for $1 - \delta < r < 1$.*

**Theorem 3.3.** *Let*

$$\begin{aligned} R &:= QA, \\ S &:= QC, \\ G &:= \mathbb{Z}, \\ \iota &:= \Big(\varphi(\in \text{inv } QC) \mapsto \text{Fredholm index of } T_\varphi(\in \mathbb{Z})\Big). \end{aligned}$$

*Then* (A1)–(A4) *are satisfied.*

*Proof.* Since $QA$ is a commutative integral domain with identity, (A1) holds.

The set $QC$ is a unital ($1 \in C(\mathbb{T}) \subset QC$), commutative, complex, semisimple Banach algebra with the involution

$$f^*(\zeta) = \overline{f(\zeta)} \quad (\zeta \in \mathbb{T}).$$

In fact, $QC$ is a $C^*$-subalgebra of $L^\infty(\mathbb{T})$. So (A2) holds as well.

[6, Corollary 139, p. 354] says that if $\varphi \in \text{inv } QC$, then $T_\varphi$ is a Fredholm operator. Thus it follows that the map $\iota : \text{inv } QC \to \mathbb{Z}$ given by

$$\iota(\varphi) := \text{Fredholm index of } T_\varphi \quad (\varphi \in \text{inv } QC)$$

is well defined. If $\varphi, \psi \in \text{inv } QC$, then in particular they are in $H^\infty + C(\mathbb{T})$, and so the semicommutator $T_{\phi\psi} - T_\phi T_\psi$ is compact [6, Lemma 133, p. 350]. Since the Fredholm index is invariant under compact perturbations [6, Part B, 2.5.2(h)], it follows that the Fredholm index of $T_{\varphi\psi}$ is the same as that of $T_\phi T_\psi$. Consequently (A3)(I1) holds.

Also, if $\varphi \in \text{inv } QC$, then we have that

$$\begin{aligned} \iota(\varphi^*) &= \iota(\overline{\varphi}) \\ &= \text{Fredholm index of } T_{\overline{\varphi}} \\ &= \text{Fredholm index of } (T_\varphi)^* \\ &= -(\text{Fredholm index of } T_\varphi) \\ &= -\iota(\varphi). \end{aligned}$$

Hence (A3)(I2) holds.

The map sending a Fredholm operator on a Hilbert space to its Fredholm index is locally constant; see for example [7, Part B, 2.5.1.(g)]. For $\varphi \in L^\infty(\mathbb{T})$, $\|T_\varphi\| \leq \|\varphi\|$, and so the map $\varphi \mapsto T_\varphi : \text{inv } QC \to \text{Fred}(H^2)$ is continuous. Consequently the map $\iota$ is continuous from inv $QC$ to $\mathbb{Z}$ (where $\mathbb{Z}$ has the discrete topology). Thus (A3)(I3) holds.

Finally, we will show that (A4) holds as well. Let $\varphi \in H^\infty \cap (\text{inv } QC)$ be invertible as an element of $H^\infty$. Then clearly $T_\varphi$ is invertible, and so has Fredholm index ind $T_\varphi$ equal to 0. Hence $\iota(\varphi) = 0$. This finishes the proof of the "only if" part in (A4).

Now suppose that $\varphi \in H^\infty \cap (\text{inv } QC)$ and that $\iota(\varphi) = 0$. In particular, $\varphi$ is invertible as an element of $H^\infty + C(\mathbb{T})$ and the Fredholm index ind $T_\varphi$ of $T_\varphi$ is equal to 0. By Proposition 3.2, it follows that there exist $\delta, \epsilon > 0$ such that

$|\Phi(re^{it})| \geq \epsilon$ for $1-\delta < r < 1$, where $\Phi$ is the harmonic extension of $\varphi$ to $\mathbb{D}$. But since $\varphi \in H^\infty$, its harmonic extension $\Phi$ is equal to $\varphi$. So $|\varphi(re^{it})| \geq \epsilon$ for $1-\delta < r < 1$. Also since $\iota(\varphi) = 0$, the winding number with respect to the origin of the curve $\varphi(re^{it})$ for $1-\delta < r < 1$ is equal to 0. By the Argument principle, it follows that $f$ cannot have any zeros inside $r\mathbb{T}$ for $1-\delta < r < 1$. In light of the above, we can now conclude that there is an $\epsilon' > 0$ such that $|\varphi(z)| > \epsilon'$ for all $z \in \mathbb{D}$. Thus $1/\varphi$ is in $H^\infty$ with $H^\infty$-norm at most $1/\epsilon'$ and we conclude that $\varphi$ is invertible as an element of $H^\infty$. Consequently (A4) holds. □

In the definition of the $\nu$-metric given in Definition 2.2 corresponding to Lemma 3.3, the $\|\cdot\|_\infty$ now means the usual $L^\infty(\mathbb{T})$ norm.

**Lemma 3.4.** *Let $A \in QC^{p\times m}$. Then*

$$\|A\| = \|A\|_\infty := \text{ess.}\sup_{\zeta\in\mathbb{T}} \boldsymbol{|} A(\zeta) \boldsymbol{|}.$$

*Proof.* We have that

$$\begin{aligned}\|A\|_\infty &= \text{ess.}\sup_{\zeta\in\mathbb{T}} \boldsymbol{|} A(\zeta) \boldsymbol{|} = \text{ess.}\sup_{\zeta\in\mathbb{T}} \sigma_{\max}\Big(A(\zeta)\Big) \\ &= \max_{\varphi\in M(L^\infty(\mathbb{T}))} \widehat{\sigma_{\max}(A)}(\varphi) = \max_{\varphi\in M(L^\infty(\mathbb{T}))} \sigma_{\max}\Big(\widehat{A}(\varphi)\Big) \\ &= \max_{\varphi\in M(QC)} \sigma_{\max}\Big(\widehat{A}(\varphi)\Big) = \max_{\varphi\in M(QC)} \boldsymbol{|}\widehat{A}(\varphi)\boldsymbol{|} = \|A\|.\end{aligned}$$

In the above, the notation $\sigma_{\max}(X)$, for a complex matrix $X \in \mathbb{C}^{p\times m}$, means its largest singular value, that is, the square root of the largest eigenvalue of $X^*X$ (or $XX^*$). We have also used the fact that for an $f \in QC \subset L^\infty(\mathbb{T})$, we have that

$$\max_{\varphi\in M(L^\infty(\mathbb{T}))} \widehat{f}(\varphi) = \|f\|_{L^\infty(\mathbb{T})} = \max_{\varphi\in M(QC)} \widehat{f}(\varphi).$$

Also, we have used the fact that if $\mu \in L^\infty(\mathbb{T})$ is such that

$$\det(\mu^2 I - A^*A) = 0,$$

then upon taking Gelfand transforms, we obtain

$$\det((\widehat{\mu}(\varphi))^2 I - (\widehat{A}(\varphi))^*\widehat{A}(\varphi)) = 0 \quad (\varphi \in M(L^\infty(\mathbb{T}))),$$

to see that $\widehat{\sigma_{\max}(A)}(\varphi) = \sigma_{\max}(\widehat{A}(\varphi))$, $\varphi \in M(L^\infty(\mathbb{T}))$. □

Finally, our scalar winding number condition

$$\det(G_1^*G_2) \in \text{inv } QC \text{ and Fredholm index of } T_{\det(G_1^*G_2))} = 0$$

is exactly the same as the condition

$$T_{G_1^*G_2} \text{ is Fredholm with Fredholm index } 0$$

in (1.1). This is an immediate consequence of the following result due to Douglas [3, p. 13, Theorem 6].

**Proposition 3.5.** *The matrix Toeplitz operator $T_\Phi$ with the matrix symbol*

$$\Phi = [\varphi_{ij}] \in (H^\infty + C(\mathbb{T}))^{n\times n}$$

*is Fredholm if and only if*

$$\inf_{\zeta\in\mathbb{T}} |\det(\varphi(\zeta))| > 0,$$

*and moreover the Fredholm index of $T_\Phi$ is the negative of the Fredholm index of* $\det\Phi$.

Thus our abstract metric reduces to the same metric given in (1.1), that is, for plants $P_1, P_2 \in \mathbb{S}(QA, p, m)$, with the normalized left/right coprime factorizations

$$P_1 = N_1 D_1^{-1} = \widetilde{D}_1^{-1}\widetilde{N}_1,$$
$$P_2 = N_2 D_2^{-1} = \widetilde{D}_2^{-1}\widetilde{N}_2,$$

define

$$d_\nu(P_1,P_2) := \begin{cases} \|\widetilde{G}_2 G_1\|_\infty & \text{if } \det(G_1^* G_2) \in \text{inv } QC \text{ and} \\ & \quad \text{Fredholm index of } T_{\det(G_1^* G_2)} = 0, \\ 1 & \text{otherwise.} \end{cases} \tag{3.1}$$

Summarizing, our main result is the following.

**Corollary 3.6.** *$d_\nu$ given by (3.1) is a metric on $\mathbb{S}(QA,p,m)$. Moreover, if $P_0$, $P$ belong to $\mathbb{S}(QA,p,m)$ and $C \in \mathbb{S}(QA,m,p)$, then*

$$\mu_{P,C} \geq \mu_{P_0,C} - d_\nu(P_0,P).$$

## References

[1] J.A. Ball and A.J. Sasane. Extension of the $\nu$-metric. *Complex Analysis and Operator Theory*, to appear.

[2] R.G. Douglas. *Banach algebra techniques in operator theory.* Second edition. Graduate Texts in Mathematics, 179, Springer-Verlag, New York, 1998.

[3] R.G. Douglas. *Banach algebra techniques in the theory of Toeplitz operators.* Expository Lectures from the CBMS Regional Conference held at the University of Georgia, Athens, Ga., June 12–16, 1972. Conference Board of the Mathematical Sciences Regional Conference Series in Mathematics, No. 15. American Mathematical Society, Providence, R.I., 1973.

[4] J.B. Garnett. *Bounded analytic functions.* Revised first edition. Graduate Texts in Mathematics, 236. Springer, New York, 2007.

[5] N.Ya. Krupnik. *Banach algebras with symbol and singular integral operators.* Translated from the Russian by A. Iacob. Operator Theory: Advances and Applications, 26. Birkhäuser Verlag, Basel, 1987.

[6] N.K. Nikolski. *Treatise on the shift operator. Spectral function theory. With an appendix by S.V. Khrushchëv and V.V. Peller.* Translated from the Russian by Jaak Peetre. Grundlehren der Mathematischen Wissenschaften, 273. Springer-Verlag, Berlin, 1986.

[7] N.K. Nikolski. *Operators, functions, and systems: an easy reading. Vol. 1. Hardy, Hankel, and Toeplitz.* Translated from the French by Andreas Hartmann. Mathematical Surveys and Monographs, 92. American Mathematical Society, Providence, RI, 2002.

[8] G. Vinnicombe. Frequency domain uncertainty and the graph topology. *IEEE Transactions on Automatic Control*, no. 9, 38:1371–1383, 1993.

Joseph A. Ball
Department of Mathematics
Virginia Tech.
Blacksburg, VA 24061, USA.
e-mail: joball@math.vt.edu

Amol J. Sasane
Department of Mathematics
Royal Institute of Technology
Stockholm, Sweden.
e-mail: sasane@math.kth.se

Operator Theory:
Advances and Applications, Vol. 221, 131-159


# Families of Homomorphisms in Non-commutative Gelfand Theory: Comparisons and Examples

Harm Bart, Torsten Ehrhardt and Bernd Silbermann

**Abstract.** In non-commutative Gelfand theory, families of Banach algebra homomorphisms, and particularly families of matrix representations, play an important role. Depending on the properties imposed on them, they are called sufficient, weakly sufficient, partially weakly sufficient, radical-separating or separating. In this paper these families are compared with one another. Conditions are given under which the defining properties amount to the same. Where applicable, examples are presented to show that they are genuinely different.

**Mathematics Subject Classification (2000).** Primary: 46H15; Secondary: 46H10.

**Keywords.** Banach algebra homomorphism, matrix representation, sufficient family, weakly sufficient family, partially weakly sufficient family, radical-separating family, separating family, polynomial identity algebra, spectral regularity.

## 1. Introduction and preliminaries

Standard Gelfand theory for commutative Banach algebras heavily employs multiplicative linear functionals. A central result in the theory is that an element $b$ in a commutative unital Banach algebra $\mathcal{B}$ is invertible if (and only if) its image $\phi(b)$ is nonzero for each nontrivial multiplicative linear functional $\phi : \mathcal{B} \to \mathbb{C}$.

In generalizations of the theory, dealing with non-commutative Banach algebras, the role of the multiplicative linear functionals is taken over by continuous unital Banach algebra homomorphisms, particularly matrix representations. More specifically, the relevant literature features families of such homomorphisms with properties pertinent to dealing with invertibility issues. So far, the relationships between the different concepts introduced this way have not been pointed out. It is the aim of this paper to fill this gap.

The Banach algebras that we consider are non-trivial, unital, and (as usual) equipped with a submultiplicative norm. Norms of unit elements are (therefore) necessarily at least one, but they are not required to be equal to one.

In order to describe the contents of this paper, it is necessary to recall some notions that at present are around in non-commutative Gelfand theory. Throughout $\mathcal{B}$ will be a unital Banach algebra. Its unit element is written as $e_{\mathcal{B}}$, its norm as $\|\cdot\|_{\mathcal{B}}$. Let $\{\phi_\omega : \mathcal{B} \to \mathcal{B}_\omega\}_{\omega\in\Omega}$ be a family of continuous unital Banach algebra homomorphisms (where to avoid trivialities it is assumed that the index set $\Omega$ is nonempty). The norm and unit element in $\mathcal{B}_\omega$ are denoted by $\|\cdot\|_\omega$ and $e_\omega$, respectively. The algebras $\mathcal{B}_\omega$ are referred to as the *target algebras*; for $\mathcal{B}$ sometimes the term *underlying algebra* will be used.

Next we present three notions that are concerned with (sufficient) conditions for invertibility of Banach algebra elements. Our terminology is in line with what has become customary in the literature.

The family $\{\phi_\omega : \mathcal{B} \to \mathcal{B}_\omega\}_{\omega\in\Omega}$ is called *sufficient* when $b \in \mathcal{B}$ is invertible in $\mathcal{B}$ if and only if $\phi_\omega(b)$ is invertible in $\mathcal{B}_\omega$ for all $\omega \in \Omega$. The 'only if part' in this definition is a triviality. Indeed, if $b \in \mathcal{B}$ is invertible in $\mathcal{B}$, then $\phi_\omega(b)$ is invertible in $\mathcal{B}_\omega$ with inverse $\phi_\omega(b^{-1})$. Sufficient families play a role in, e.g., [Kr], [HRS01], [Si], [RSS], [BES94], [BES04], [BES12], and Section 7.1 in [P].

The family $\{\phi_\omega : \mathcal{B} \to \mathcal{B}_\omega\}_{\omega\in\Omega}$ is said to be *weakly sufficient* when $b \in \mathcal{B}$ is invertible in $\mathcal{B}$ if and only if $\phi_\omega(b)$ is invertible in $B_\omega$ for all $\omega \in \Omega$ and, in addition, $\sup_{\omega\in\Omega} \|\phi_\omega(b)^{-1}\|_\omega < \infty$. Note that this implies the finiteness of $\sup_{\omega\in\Omega} \|e_\omega\|_\omega$. Weakly sufficient families feature in [HRS01], [Si] and [BES12].

The family $\{\phi_\omega : \mathcal{B} \to \mathcal{B}_\omega\}_{\omega\in\Omega}$ is called *partially weakly sufficient*, or *p.w. sufficient* for short, if

(i) $\sup_{\omega\in\Omega} \|e_\omega\|_\omega < \infty$, and

(ii) $b \in \mathcal{B}$ is invertible in $\mathcal{B}$ provided $\phi_\omega(b)$ is invertible in $B_\omega$ for all $\omega \in \Omega$ and $\sup_{\omega\in\Omega} \|\phi_\omega(b)^{-1}\|_\omega < \infty$.

Partially weakly sufficient families are employed in [BES12], and the definition also occurs in Section 2.2.5 of [RSS]. There, however, the term weakly sufficient is used instead of the name partially weakly sufficient that has been chosen in [BES12].

Before proceeding by recalling two more definitions, a few comments are in order. First, even when a sufficient family is known to exist, it is sometimes preferable to work with weakly or partially weakly sufficient families. Indeed, it may happen that not all members of the sufficient family can be concretely described while such an explicit description can be given for a (partially) weakly sufficient family. For an example, consider the Banach algebra $\ell_\infty$. (Being a commutative algebra, the collection of all (continuous) multiplicative linear functionals on $\ell_\infty$ is sufficient. Not all its members can be described explicitly, however. The coordinate functionals form a weakly sufficient family in this case.) There are also situations where weakly or partially weakly sufficient families can be identified but a sufficient family is not known or does not even exist. As a second comment, we note that in the first of the above definitions, the norms in the target algebras do not play a role. Their role in the second and third definition is essential, however. An example illustrating this will be given in Section 2.

The family $\{\phi_\omega : \mathcal{B} \to \mathcal{B}_\omega\}_{\omega\in\Omega}$ is said to be *separating* if it separates the points of $\mathcal{B}$ or, what amounts here (by linearity) to the same, $\bigcap_{\omega\in\Omega} \operatorname{Ker}\phi_\omega = \{0\}$. It is called *radical-separating* if it separates the points of $\mathcal{B}$ modulo the radical of $\mathcal{B}$, i.e., if $\bigcap_{\omega\in\Omega} \operatorname{Ker}\phi_\omega \subset \mathcal{R}(\mathcal{B})$, where $\mathcal{R}(\mathcal{B})$ stands for the radical of $\mathcal{B}$. In these two definitions, again the norms in the target algebras do not play a role. The notions of a separating family and that of a radical-separating family were introduced in [BES12]. To complete the references, we note that in [RSS], the introduction to Chapter 2, there is a remark to the effect that a sufficient family of Banach algebra homomorphisms is radical-separating. Also the proof of [RSS], Theorem 2.2.10, shows that the same conclusion holds when the family is partially weakly sufficient (or has the more restrictive property of being weakly sufficient).

The goal of the present paper is to make clear the relationships between the different types of families just described. This will be done on two levels. The first is that of the individual families. Here the Banach algebra $\mathcal{B}$ and a family of homomorphisms are given, and the issue is which property defined above implies the other. So in this context, the questions are of the sort 'is a sufficient family always weakly sufficient?'. The second level is – so to speak – that of the underlying Banach algebra. Thus only the Banach algebra $\mathcal{B}$ is given a priori, and the question is whether the existence of one type of families implies the existence of another. Here we allow the families to change. A typical question which comes up in this setting is: 'if $\mathcal{B}$ possesses a radical-separating family, does it also possess a sufficient one?'. As the change to another family may also involve the target algebras, a caveat is in order here. Indeed, for any unital Banach algebra, the singleton family consisting of the identity mapping on the algebra has all the properties of being sufficient, weakly sufficient, partially weakly sufficient, radical-separating and separating. So, in order to avoid trivialities, some restriction must be imposed. Here this restriction will take the form of requiring the Banach algebra homomorphisms to be matrix representations. As was indicated before, the restriction to matrix representations is not uncommon in the literature dealing with non-commutative Gelfand theory.

The problems on the second (underlying Banach algebra) level are more fundamental than those on the first. Not only because they are generally (much) harder to handle, but especially in view of the fact that in the literature, families of homomorphisms serve as tools in criteria for establishing certain properties of Banach algebras. Here is an example – actually the one that prompted the authors to write this paper. The Banach algebra $\mathcal{B}$ is said to be *spectrally regular* if – analogous to the situation in the scalar case – contour integrals of logarithmic residues of analytic $\mathcal{B}$-valued functions can only vanish (or more generally be quasinilpotent) when the function in question takes invertible values on the interior of the contour. It was observed in [BES94] that a Banach algebra $\mathcal{B}$ is spectrally regular if it possesses a sufficient family of matrix representations. In this criterion for establishing spectral regularity, sufficient families may be replaced by radical-separating families (see [BES12]). The question whether the two criteria amount to the same or are genuinely different comes down to the second level issue (already mentioned as an example in the preceding paragraph) whether a Banach algebra that possesses a radical-separating family of matrix representations also has one that is sufficient. This is not the case, as appears from an example given in Section 4. Note that in order to prove this, one needs 'control' on the collection of all matrix representations of the Banach algebra in the example question, a highly non-trivial matter as is already suggested by the relatively simple special case of the (commutative) Banach algebra $\ell_\infty$.

Apart from the introduction also containing preliminaries (Section 1), and the list of references, the paper consists of four sections. Section 2 discusses problems on the level of individual families. One of the main theorems is concerned with the $C^*$-case where the underlying algebra $\mathcal{B}$ as well as the target algebras $\mathcal{B}_\omega$ are $C^*$-algebras, and, in addition, all the homomorphisms $\phi_\omega$ are $C^*$-homomorphisms. The theorem in question states that in such a situation the properties of being weakly sufficient, p.w. sufficient, radical-separating and separating all amount to the same. Without the $C^*$-condition, the picture is less simple; both positive results and counterexamples are given. Section 3 deals with what above were called second level issues. As was already mentioned, a restriction is made to families of matrix representations. One of the theorems says that a Banach algebra possesses a radical-separating family of matrix representations if and only if it possesses a p.w. sufficient family of matrix representations. It is also proved that for families of matrix representations of finite order (i.e., with a finite upper bound on the size of the matrices involved) the properties of being sufficient, weakly sufficient, p.w. sufficient and radical-separating all come down to the same. Families of matrix representations of finite order feature abundantly in the literature on non-commutative Gelfand theory.

Sections 4 and 5 deal with two specific questions. The first is: 'if a Banach algebra possesses a radical-separating (or even separating) family of matrix representations, does it follow that it also possesses a sufficient family of matrix representations?'; the second: 'if a Banach algebra possesses a radical-separating (or even separating) family of matrix representations, does it follow that it also possesses a weakly sufficient family of matrix representations?'. Both questions have a negative answer; the corresponding examples are complicated. The authors think that the two Banach algebras involved (one of them a $C^*$-algebra) are interesting in their own right.

## 2. First observations

We begin with a couple of simple observations. For terminology and notation, see the introduction.

Separating families are clearly radical-separating, but the converse is not true. A counterexample is easily given: take for $\mathcal{B}$ the Banach algebra of upper triangular $2 \times 2$ matrices, let $\phi_1$ pick out the first diagonal element, and $\phi_2$ the second; then $\{\phi_1, \phi_2\}$ is radical-separating but not separating.

Weakly sufficient families are evidently p.w. sufficient. An example given later in this section will show that the converse is not true. Weakly sufficient families are not necessarily sufficient; for an example, consider the Banach algebra $\ell_\infty$ and the associated coordinate homomorphisms.

A sufficient family need not be p.w. sufficient (so a fortiori it need not be weakly sufficient). To see this, take any infinite sufficient family $\{\phi_\omega : \mathcal{B} \to \mathcal{B}_\omega\}_{\omega\in\Omega}$ and (if necessary) renorm the target algebras $\mathcal{B}_\omega$ by taking appropriate scalar multiples of the given norms so as to get $\sup_{\omega\in\Omega} \|e_\omega\|_\omega = \infty$. Such a renorming does not influence the sufficiency, the new norms being equivalent to the corresponding original ones. Note here that the property of being a submultiplicative norm is not violated by multiplication with a scalar larger than one.

Summing up: of the five notions concerning families of homomorphisms introduced in Section 1, only two imply another in a completely obvious manner. Indeed, the properties of being separating and weakly sufficient trivially entail those of being radical-separating and p.w. sufficient, respectively; the other connections are not so immediately transparent, however.

One more positive result, still on a very simple level, can be added here: modulo a simple change to equivalent norms in the target algebras, a sufficient family of unital Banach algebra homomorphisms is p.w. sufficient. Indeed, when $\{\phi_\omega : \mathcal{B} \to \mathcal{B}_\omega\}_{\omega\in\Omega}$ is a sufficient family, then, for $\omega \in \Omega$, choose an equivalent Banach algebra norm on $\mathcal{B}_\omega$ for which the unit element $e_\omega$ in $\mathcal{B}_\omega$ has norm one. It is a standard fact from Banach algebra theory that this can be done. Whether or not sufficiency also implies weak sufficiency if one allows for a change to equivalent norms, we do not know.

So far for the obvious connections; next we concern ourselves with those that are less or, in some cases, not at all straightforward. To facilitate the discussion in the remainder of the paper, one more item of terminology is introduced. We call the family $\{\phi_\omega : \mathcal{B} \to \mathcal{B}_\omega\}_{\omega\in\Omega}$ *norm-bounded* if $\sup_{\omega\in\Omega} \|\phi_\omega\|_\omega < \infty$. Here, by slight abuse of notation, $\|\phi_\omega\|_\omega$ stands for the norm of $\phi_\omega$ considered as a bounded linear operator from $\mathcal{B}$ into $\mathcal{B}_\omega$ (equipped with the norms $\|\cdot\|_\mathcal{B}$ and $\|\cdot\|_\omega$, respectively).

**Theorem 2.1.** *The family $\{\phi_\omega : \mathcal{B} \to \mathcal{B}_\omega\}_{\omega\in\Omega}$ is weakly sufficient if and only if it is p.w. sufficient and norm-bounded. If $\{\phi_\omega : \mathcal{B} \to \mathcal{B}_\omega\}_{\omega\in\Omega}$ is sufficient and norm-bounded, then it is weakly sufficient.*

*Proof.* If $\{\phi_\omega : \mathcal{B} \to \mathcal{B}_\omega\}_{\omega\in\Omega}$ is norm-bounded and sufficient or p.w. sufficient, then it is weakly sufficient. To prove this, use that when $b$ is invertible in $\mathcal{B}$, then $\phi_\omega(b)$ is invertible in $\mathcal{B}_\omega$ with inverse $\phi_\omega(b^{-1})$, so that $\|\phi_\omega(b)^{-1}\|_\omega \leq \|\phi_\omega\|_\omega \cdot \|b^{-1}\|_\mathcal{B}$. If the family $\{\phi_\omega : \mathcal{B} \to \mathcal{B}_\omega\}_{\omega\in\Omega}$ is weakly sufficient, then clearly it is p.w. sufficient. It remains to establish that $\{\phi_\omega : \mathcal{B} \to \mathcal{B}_\omega\}_{\omega\in\Omega}$ is then norm-bounded too. By the uniform boundedness principle (Banach-Steinhaus) it is sufficient to prove that $\sup_{\omega\in\Omega} \|\phi_\omega(b)\|_\omega < \infty$ for every $b$ in $\mathcal{B}$. Take $b \in \mathcal{B}$. Fix a positive real number $\mu$ (sufficiently small) so that $e_\mathcal{B} + \mu b$ is invertible in $\mathcal{B}$. The invertibility of

$(e_{\mathcal{B}} + \mu b)^{-1}$ combined with the weak sufficiency of the family $\{\phi_\omega : \mathcal{B} \to \mathcal{B}_\omega\}_{\omega\in\Omega}$ gives that $\sup_{\omega\in\Omega} \|e_\omega + \mu\phi_\omega(b)\|_\omega < \infty$. We also have $\sup_{\omega\in\Omega} \|e_\omega\|_\omega < \infty$ (specialize to $b = 0$). Now $\|\phi_\omega(b)\| \le \mu^{-1}(\|e_\omega + \mu\phi_\omega(b)\| + \|e_\omega\|)$, and it follows that $\sup_{\omega\in\Omega} \|\phi_\omega(b)\|_\omega < \infty$, as desired. □

It is convenient to adopt the following notation. Writing $\Phi$ for the family $\{\phi_\omega : \mathcal{B} \to \mathcal{B}_\omega\}_{\omega\in\Omega}$, we denote by $\mathcal{G}_\Phi(\mathcal{B})$ the set of all $b \in \mathcal{B}$ such that $\phi_\omega(b)$ is invertible in $\mathcal{B}_\omega$ for every $\omega \in \Omega$, while, in addition, $\sup_{\omega\in\Omega} \|\phi_\omega(b)^{-1}\|_\omega < \infty$. A few comments are in order. The zero element in $\mathcal{B}$ does not belong to $\mathcal{G}_\Phi(\mathcal{B})$. If $b \in \mathcal{G}_\Phi(\mathcal{B})$ and $\beta \neq 0$, then $\beta b \in \mathcal{G}_\Phi(\mathcal{B})$. The set $\mathcal{G}_\Phi(\mathcal{B})$ is closed under taking products. It may happen that $\mathcal{G}_\Phi(\mathcal{B}) = \emptyset$ (cf., the fourth paragraph of this section). Writing $\mathcal{G}(\mathcal{B})$ for the group of invertible elements in $\mathcal{B}$, we have that the family $\{\phi_\omega : \mathcal{B} \to \mathcal{B}_\omega\}_{\omega\in\Omega}$ is weakly sufficient if and only if $\mathcal{G}_\Phi(\mathcal{B})$ coincides with $\mathcal{G}(\mathcal{B})$. Also, it is p.w. sufficient if and only if $e_{\mathcal{B}} \in \mathcal{G}_\Phi(\mathcal{B}) \subset \mathcal{G}(\mathcal{B})$ or, what amounts to the same, $\{\lambda e_{\mathcal{B}} \mid 0 \neq \lambda \in \mathbb{C}\} \subset \mathcal{G}_\Phi(\mathcal{B}) \subset \mathcal{G}(\mathcal{B})$.

**Proposition 2.2.** *If $\emptyset \neq \mathcal{G}_\Phi(\mathcal{B}) \subset \mathcal{G}(\mathcal{B})$, the family $\Phi = \{\phi_\omega : \mathcal{B} \to \mathcal{B}_\omega\}_{\omega\in\Omega}$ is radical-separating. In case $\mathcal{G}_\Phi(\mathcal{B}) = \{\lambda g \mid 0 \neq \lambda \in \mathbb{C}\}$ for some $g \in \mathcal{G}(\mathcal{B})$, the element $g$ is a nonzero multiple of $e_{\mathcal{B}}$ (so in fact $\mathcal{G}_\Phi(\mathcal{B}) = \{\lambda e_{\mathcal{B}} \mid 0 \neq \lambda \in \mathbb{C}\}$) and the family $\{\phi_\omega : \mathcal{B} \to \mathcal{B}_\omega\}_{\omega\in\Omega}$ is separating.*

It can happen that $\mathcal{G}_\Phi(\mathcal{B}) = \{\lambda e_{\mathcal{B}} \mid 0 \neq \lambda \in \mathbb{C}\} \neq \mathcal{G}(\mathcal{B})$; see the example below. We do not know whether $\emptyset \neq \mathcal{G}_\Phi(\mathcal{B}) \subset \mathcal{G}(\mathcal{B})$ implies that $e_{\mathcal{B}} \in \mathcal{G}_\Phi(\mathcal{B})$.

*Proof.* Suppose $\emptyset \neq \mathcal{G}_\Phi(\mathcal{B}) \subset \mathcal{G}(\mathcal{B})$, and take $x \in \bigcap_{\omega\in\Omega} \operatorname{Ker} \phi_\omega$. We need to show that $x \in \mathcal{R}(\mathcal{B})$. This can be done by proving that $e_{\mathcal{B}} + bx$ and $e_{\mathcal{B}} + xb$ are invertible for every $b \in \mathcal{B}$. Fix $h \in \mathcal{G}_\Phi(\mathcal{B})$, so $h \in \mathcal{G}(\mathcal{B})$ in particular, and let $b \in \mathcal{B}$. Then $\phi_\omega(h + hbx) = \phi_\omega(h) \in \mathcal{G}(\mathcal{B}_\omega)$ and $\big(\phi_\omega(h + hbx)\big)^{-1} = \phi_\omega(h)^{-1}$. Hence

$$\sup_{\omega\in\Omega} \|\big(\phi_\omega(h + hbx)\big)^{-1}\|_\omega = \sup_{\omega\in\Omega} \|\phi_\omega(h)^{-1}\|_\omega < \infty,$$

and we may conclude that $h + hbx \in \mathcal{G}_\Phi(\mathcal{B})$. By assumption $\mathcal{G}_\Phi(\mathcal{B}) \subset \mathcal{G}(\mathcal{B})$, and we get $h + hbx \in \mathcal{G}(\mathcal{B})$. As $h \in \mathcal{G}(\mathcal{B})$ too, it follows that $e_{\mathcal{B}} + bx \in \mathcal{G}(\mathcal{B})$. Likewise we have $e_{\mathcal{B}} + xb \in \mathcal{G}(\mathcal{B})$, and the first part of the proposition is proved.

To deal with the second part, assume that $\mathcal{G}_\Phi(\mathcal{B}) = \{\lambda g \mid 0 \neq \lambda \in \mathbb{C}\}$ for some $g \in \mathcal{G}(\mathcal{B})$. As $\mathcal{G}_\Phi(\mathcal{B})$ is closed under taking products, we have that along with $g$, the square $g^2$ of $g$ belongs to $\mathcal{G}_\Phi(\mathcal{B})$ too. Hence $g^2 = \gamma g$ for some nonzero $\gamma \in \mathbb{C}$, and it follows that $g = \gamma e_{\mathcal{B}}$. Our assumption on $\mathcal{G}_\Phi(\mathcal{B})$ can now be restated as $\mathcal{G}_\Phi(\mathcal{B}) = \{\lambda e_{\mathcal{B}} \mid 0 \neq \lambda \in \mathbb{C}\}$. Let $x \in \bigcap_{\omega\in\Omega} \operatorname{Ker} \phi_\omega$. Taking $b = h = e_{\mathcal{B}}$ in the reasoning presented in the first paragraph of this proof, we get $e_{\mathcal{B}} + x \in \mathcal{G}_\Phi(\mathcal{B})$. But then $e_{\mathcal{B}} + x = \mu e_{\mathcal{B}}$ for some nonzero scalar $\mu$. Applying $\phi_\omega$ we arrive at $e_\omega = \mu e_\omega$, hence $\mu = 1$ and, consequently, $x = 0$ as desired. □

**Corollary 2.3.** *If the family $\{\phi_\omega : \mathcal{B} \to \mathcal{B}_\omega\}_{\omega\in\Omega}$ is p.w. sufficient or sufficient, then it is radical-separating.*

As weak sufficiency implies p.w. sufficiency, a weakly sufficient family is radical-separating too, and the same conclusion trivially holds for separating families. Thus radical-separateness is the encompassing notion. Corollary 2.3 features as Proposition 3.4 in [BES12]; cf., [RSS], the introduction to Chapter 2, which contains a remark to the effect that a sufficient family of Banach algebra homomorphisms is radical-separating; see also the proof of [RSS], Theorem 2.2.10 which shows that the same conclusion holds when the family is partially weakly sufficient (or has the more restrictive property of being weakly sufficient).

Next we present an example showing that a family of homomorphisms, while being both sufficient and p.w. sufficient, can fail to be weakly sufficient. The example also illustrates a fact already noted in the introduction, namely that in the definitions of weak sufficiency and p.w. sufficiency, the norms in the target algebras play an essential role. Finally it supports the claim made right after Proposition 2.2.

**Example.** Consider the situation where $\mathcal{B}$ is the $\ell_\infty$-direct product of two copies of $\mathbb{C}$. Thus $\mathcal{B}$ consists of all ordered pairs $(a,b)$ with $a,b \in \mathbb{C}$, the algebraic operations in $\mathcal{B}$ are defined pointwise, and the norm on $\mathcal{B}$ is given by $\|(a,b)\| = \max\{|a|,|b|\}$. Clearly $\mathcal{B}$ is unital and the unit element $(1,1)$ has norm one. For $n = 1,2,\ldots$, let $\mathcal{B}_n = \mathcal{B}$, also with norm $\|\cdot\|$. Then the family $\{\phi_n : \mathcal{B} \to \mathcal{B}_n\}_{n\in\mathbb{N}}$ consisting of identity mappings from $\mathcal{B}$ into $\mathcal{B}_n = \mathcal{B}$ has all the properties of being sufficient, weakly sufficient, p.w. sufficient, radical-separating and separating.

Now we are going to change the (coinciding) norms on the algebras $\mathcal{B}_n$. This will not affect the sufficiency of the family $\{\phi_n : \mathcal{B} \to \mathcal{B}_n\}$ of identity mappings; it will, however, destroy its weak sufficiency. As a first step, we introduce

$$|||(a,b)|||_n = \frac{1}{2}|a+b| + n|a-b|, \qquad a,b \in \mathbb{C},\ n = 1,2,\ldots.$$

Then $|||\cdot|||_n$ is a norm on $\mathcal{B}_n$. One easily verifies that

$$\|(a,b)\| \le |||(a,b)|||_n \le (2n+1)\|(a,b)\|,$$

which concretely exhibits that the norms $|||\cdot|||_n$ and $\|\cdot\|$ are equivalent. For the unit element $(1,1)$ of $\mathcal{A}$ we have $|||(1,1)|||_n = 1$.

To transform $|||\cdot|||_n$ into a submultiplicative norm, we employ a standard procedure and put

$$\|(a,b)\|_n = \sup_{|||(x,y)|||_n=1} |||(a,b)\cdot(x,y)|||_n, \qquad a,b \in \mathbb{C},\ n = 1,2,\ldots.$$

Then $\|(1,1)\|_n = 1$. Further the norm $\|\cdot\|_n$ is equivalent $|||\cdot|||_n$ (and therefore also to $\|\cdot\|$). In fact, $|||(a,b)|||_n \le \|(a,b)\|_n \le (2n+1)|||(a,b)|||_n$.

Modulo the norms $\|\cdot\|_n$, the family $\{\phi_n : \mathcal{B} \to \mathcal{B}_n\}_{n\in\mathbb{N}}$ is still sufficient. Indeed, this property is not affected by a change of norms. For weak sufficiency, however, the situation is different. Actually with respect to the norms $\|\cdot\|_n$, the family $\{\phi_n : \mathcal{B} \to \mathcal{B}_n\}_n \in \mathbb{N}$ is not weakly sufficient. To see this consider (for instance) the element $(1,-1) \in \mathcal{B}$. This element is invertible in $\mathcal{B}$ with inverse $(1,-1)$. However,

$$\|(\phi_n(1,-1))^{-1}\|_n = \|(1,-1)\|_n \geq |||(1,-1)|||_n = 2n,$$

and so $\sup_{n=1,2,3,\ldots} \|(\phi_n(1,-1))^{-1}\|_n = \infty$.

A straightforward argument shows that the set $\mathcal{G}_\Phi(\mathcal{B})$ of all $(a,b) \in \mathcal{B}$ such that $\phi_n(a,b)$ is invertible for every $n$ while, in addition,

$$\sup_{n=1,2,3,\ldots} \|(\phi_n(a,b))^{-1}\|_n < \infty,$$

coincides with the set $\{\lambda(1,1) \mid 0 \neq \lambda \in \mathbb{C}\}$. As $\mathcal{G}_\Phi(\mathcal{B})$ is contained in the set $\mathcal{G}(\mathcal{B})$ of all invertible elements in $\mathcal{B}$, the family $\{\phi_n : \mathcal{B} \to \mathcal{B}_n\}_{n\in\mathbb{N}}$ retains the property of being p.w. sufficient. The circumstance that $\mathcal{G}_\Phi(\mathcal{B})$ does not coincide with $\mathcal{G}(\mathcal{B})$ corroborates the earlier observation that $\{\phi_n : \mathcal{B} \to \mathcal{B}_n\}_{n\in\mathbb{N}}$ is not weakly sufficient with respect to the norms $\|\cdot\|_n$. □

We close this section by considering the important $C^*$-case. The family $\{\phi_\omega : \mathcal{B} \to \mathcal{B}_\omega\}_{\omega\in\Omega}$ is said to be a *$C^*$-family* if all the Banach algebras $\mathcal{B}$ and $\mathcal{B}_\omega$ are $C^*$-algebras, and all the homomorphisms $\phi_\omega$ are unital $C^*$-homomorphisms. As such homomorphisms are contractions, a $C^*$-family is norm-bounded.

**Theorem 2.4.** *For a $C^*$-family $\{\phi_\omega : \mathcal{B} \to \mathcal{B}_\omega\}_{\omega\in\Omega}$, the properties of being weakly sufficient, p.w. sufficient, radical-separating and separating all amount to the same.*

*Proof.* We have already observed that weak sufficiency (trivially) implies p.w. sufficiency. From Corollary 2.3, we know that p.w. sufficiency implies the property of being radical-separating. Since $C^*$-algebras are semi-simple, the family $\{\phi_\omega : \mathcal{B} \to \mathcal{B}_\omega\}_{\omega\in\Omega}$ is radical-separating if and only if it is separating. It remains to prove that if $\{\phi_\omega : \mathcal{B} \to \mathcal{B}_\omega\}_{\omega\in\Omega}$ is separating, then it is weakly sufficient. For this, the reasoning is as follows.

First there is this simple observation. Let $b \in \mathcal{B}$ be invertible, and take $\omega \in \Omega$. Then $\phi_\omega(b)$ is invertible in $\mathcal{B}_\omega$ and $\phi_\omega(b)^{-1} = \phi_\omega(b^{-1})$. Now $C^*$-homomorphisms are contractions, so $\|\phi_\omega(b)^{-1}\|_\omega \leq \|b^{-1}\|_\mathcal{B}$.

The next step is more involved. Write $\mathbf{B}$ for the family of unital Banach algebras $\{\mathcal{B}_\omega\}_{\omega\in\Omega}$, and let $\mathcal{A} = \ell_\infty^\mathbf{B}$ be the $\ell_\infty$-direct product of the family $\mathbf{B}$ (cf., [P], Subsection 1.3.1). Thus $\ell_\infty^\mathbf{B}$ consists of all $F$ in the Cartesian product $\prod_{\omega\in\Omega} \mathcal{B}_\omega$ such that $|||F||| = \sup_{\omega\in\Omega} \|F(\omega)\|_\omega < \infty$. With the pointwise operations and the sup-norm, $\ell_\infty^\mathbf{B}$ is a $C^*$-algebra. Define $\phi : \mathcal{B} \to \mathcal{A}$ by $\phi(b)(\omega) = \phi_\omega(b)$. Because $C^*$-homomorphisms are contractions, $\phi$ is well defined. Clearly $\phi$ is a continuous $C^*$-algebra homomorphism. As (by assumption) the family $\{\phi_\omega : \mathcal{B} \to \mathcal{B}_\omega\}_{\omega\in\Omega}$ is separating, the homomorphism $\phi : \mathcal{B} \to \mathcal{A}$ is injective. Hence it is an isometry and $\phi[\mathcal{B}]$ is a closed $C^*$-subalgebra of $\mathcal{A}$. But then, as is known from general $C^*$-theory, $\phi[\mathcal{B}]$ is inverse closed in $\mathcal{A}$. Now let $b$ be an element of $\mathcal{B}$ such that $\phi_\omega(b)$ is invertible for all $\omega \in \Omega$ while, in addition, $\sup_{\omega\in\Omega} \|\phi_\omega(b)^{-1}\|_\omega < \infty$. Then $\phi(b) \in \phi[\mathcal{B}]$ is invertible in $\mathcal{A}$. Since its inverse belongs to $\phi[\mathcal{B}]$, there exists $b_1 \in \mathcal{B}$ such that $\phi_\omega(b)\phi_\omega(b_1) = \phi_\omega(b_1)\phi_\omega(b) = e_\omega = \phi_\omega(e_\mathcal{B})$, $\omega \in \Omega$. It follows that both

$bb_1 - e_{\mathcal{B}}$ and $b_1 b - e_{\mathcal{B}}$ belong to $\bigcap_{\omega\in\Omega} \operatorname{Ker}\phi_\omega$. By assumption, this intersection is trivial, and we get that $b$ is invertible in $\mathcal{B}$ with inverse $b_1$. □

From Theorem 2.4 and the second part of Theorem 2.1, taking into account that $C^*$-families are norm-bounded, we see that a sufficient $C^*$-family has the (in this case coinciding) properties of being weakly sufficient, p.w. sufficient, radical-separating and separating. Theorem 4.1 below shows that the converse is not true.

Finally we observe that in the $C^*$-setting, the target algebras play a secondary role. The reason is the following. If $\phi_1 : \mathcal{B} \to \mathcal{A}_1$ and $\phi_2 : \mathcal{B} \to \mathcal{A}_2$ are two $C^*$-homomorphisms such that $\phi_1(b)$ is invertible in $\mathcal{A}_1$ if and only if $\phi_2(b)$ is invertible in $\mathcal{A}_2$ for every $b \in \mathcal{B}$, then the null spaces of $\phi_1$ and $\phi_2$ coincide (see Proposition 5.43 in [HRS01]). So, as far as sufficiency is concerned, one can replace each member $\phi : \mathcal{B} \to \mathcal{A}$ from a given family of $C^*$-homomorphisms by the canonical mapping from $\mathcal{B}$ onto $\mathcal{B}/\operatorname{Ker}\phi$. This observation also indicates that the properties of families of homomorphisms relevant in the present context are closely related to those of families of ideals in the underlying algebra. The proofs of Theorems 4.1 and Theorem 5.1 illustrate this fact.

## 3. Matrix representations

Until now we have been dealing with fixed given families of homomorphisms, at most allowing for a modification of the norms in the target spaces. This restriction will now be dropped and both the target spaces as well as the homomorphisms are allowed to change. On the other hand, for reasons explained in the introduction, we will require the Banach algebra homomorphisms that we consider to be matrix representations.

A *matrix representation* on a unital Banach algebra $\mathcal{B}$ is a continuous unital Banach algebra homomorphisms of the type $\phi : \mathcal{B} \to \mathbb{C}^{n\times n}$ with $n$ a positive integer. Note that the continuity condition does not depend on the norm used on $\mathbb{C}^{n\times n}$. By the *order* of a family $\{\phi_\omega : \mathcal{B} \to \mathbb{C}^{n_\omega\times n_\omega}\}_{\omega\in\Omega}$ of matrix representations, we mean the extended integer $\sup_{\omega\in\Omega} n_\omega$. Families of matrix representations of finite order play a major role in the literature (see, for instance, [Kr] and [RSS]).

**Theorem 3.1.** *If $\mathcal{B}$ possesses a sufficient family of matrix representations of order $n$ (possibly $\infty$), then $\mathcal{B}$ also possesses a family of contractive surjective matrix representations of order not exceeding $n$, and having the properties of being sufficient and weakly sufficient.*

*Proof.* We start with an auxiliary observation. Let $k$ be a positive integer, and let $\phi : \mathcal{B} \to \mathbb{C}^{k\times k}$ be a unital matrix representation. Then there exist a positive integer $m$, positive integers $k_1, \ldots, k_m$ not exceeding $k$, and surjective unital matrix representations

$$\phi_j : \mathcal{B} \to \mathbb{C}^{k_j\times k_j}, \qquad j = 1, \ldots, m,$$

such that, for $b \in \mathcal{B}$, the matrix $\phi(b)$ is invertible in $\mathbb{C}^{k\times k}$ if and only if $\phi_j(b)$ is invertible in $\mathbb{C}^{k_j\times k_j}$, $j = 1, \ldots, m$. This can be seen as follows. If the matrix representation $\phi$ itself is surjective, there is nothing to prove (case $m = 1$). Assume it is not, so $\phi[\mathcal{B}]$ is a proper subalgebra of $\mathbb{C}^{k\times k}$. Applying Burnside's Theorem (cf., [LR]), we see that $\phi[\mathcal{B}]$ has a nontrivial invariant subspace, i.e., there is a nontrivial subspace $V$ of $\mathbb{C}^{k\times k}$ such that $\phi(b)[V]$ is contained in $V$ for all $b$ in $\mathcal{B}$. But then there exist an invertible $k \times k$ matrix $S$, positive integers $k_-$ and $k_+$ with $k = k_- + k_+$ (so, in particular, $k_-, k_+ < k$), a unital matrix representation $\phi_- : \mathcal{B} \to \mathbb{C}^{k_-\times k_-}$ and a unital matrix representation $\phi_+ : \mathcal{B} \to \mathbb{C}^{k_+\times k_+}$ such that $\phi$ has the form

$$\phi(b) = S^{-1}\begin{bmatrix} \phi_-(b) & * \\ 0 & \phi_+(b) \end{bmatrix} S, \qquad b \in \mathcal{B}.$$

Clearly $\phi(b)$ is invertible in $\mathbb{C}^{k\times k}$ if and only if $\phi_-(b)$ is invertible in $\mathbb{C}^{k_-\times k_-}$ and $\phi_+(b)$ is invertible in $\mathbb{C}^{k_+\times k_+}$. If $\phi_-$ and $\phi_+$ are both surjective we are done (case $m = 2$); if not we can again apply Burnside's Theorem and decompose further. This process terminates after at most $k$ steps. (A completely rigorous argument can of course be given using induction.)

Assume $\mathcal{B}$ possesses a sufficient family of matrix representations, of order $n$ say (possibly $\infty$). Then, by the observation contained in the preceding paragraph, $\mathcal{B}$ also possesses a sufficient family which consists of surjective matrix representations and which has order not exceeding $n$. Let $\{\phi_\omega : \mathcal{B} \to \mathbb{C}^{n_\omega\times n_\omega}\}_{\omega\in\Omega}$ be such a family. We shall show that by a suitable change of norms in the target algebras $\mathbb{C}^{n_\omega\times n_\omega}$, the matrix representations can be made contractive. The weak sufficiency is then immediate from the second part of Theorem 2.1.

Let $\kappa_\omega$ be the canonical mapping from $\mathcal{B}$ onto $\mathcal{B}/\operatorname{Ker}\phi_\omega$. Then $\kappa_\omega$ is a contraction. Define $\psi_\omega : \mathcal{B}/\operatorname{Ker}\phi_\omega \to \mathbb{C}^{n_\omega\times n_\omega}$ to be the homomorphism induced by $\phi_\omega$, so $\phi_\omega = \psi_\omega \circ \kappa_\omega$. As $\phi_\omega$ is surjective, $\psi_\omega$ is a bijection. For $A \in \mathbb{C}^{n_\omega\times n_\omega}$, let $|||A|||_\omega$ be the norm of $\psi_\omega^{-1}(A)$ in the quotient algebra $\mathcal{B}/\operatorname{Ker}\phi_\omega$. Then $|||\cdot|||_\omega$ is a norm on $\mathbb{C}^{n_\omega\times n_\omega}$. With respect to this norm, $\phi_\omega$ is a contraction. □

**Theorem 3.2.** *The Banach algebra $\mathcal{B}$ possesses a radical-separating family of matrix representations if and only if $\mathcal{B}$ possesses a p.w. sufficient family of matrix representations.*

As will become clear in the proof, the theorem remains true when one restricts oneself to considering families of matrix representations of finite order. It is convenient to have at our disposal the following lemma.

**Lemma 3.3.** *Let $\mathcal{B}$ be a unital Banach algebra, and let $\{\phi_\omega : \mathcal{B} \to \mathbb{C}^{n_\omega\times n_\omega}\}_{\omega\in\Omega}$ be a family of matrix representations. Then there exists a family $\{\psi_\gamma : \mathcal{B} \to \mathbb{C}^{d_\gamma\times d_\gamma}\}_{\gamma\in\Gamma}$ of matrix representations such that*

$$\bigcap_{\gamma\in\Gamma} \operatorname{Ker}\psi_\gamma = \bigcap_{\omega\in\Omega} \operatorname{Ker}\phi_\omega, \tag{1}$$

*while, writing $\Psi$ for the family $\{\psi_\gamma : \mathcal{B} \to \mathbb{C}^{d_\gamma\times d_\gamma}\}_{\gamma\in\Gamma}$, the set $\mathcal{G}_\Psi(\mathcal{B})$ is given by*

$$\mathcal{G}_\Psi(\mathcal{B}) = \Big\{\lambda e_{\mathcal{B}} + r \mid 0 \neq \lambda \in \mathbb{C},\, r \in \bigcap_{\omega\in\Omega} \operatorname{Ker}\phi_\omega\Big\}. \tag{2}$$

Recall that the set $\mathcal{G}_\Psi(\mathcal{B})$ consists of all $b \in \mathcal{B}$ with $\psi_\gamma(b)$ invertible in $\mathbb{C}^{d_\gamma \times d_\gamma}$ for every $\gamma \in \Gamma$, and, in addition, $\sup_{\gamma\in\Gamma} \|\phi_\gamma(b)^{-1}\|_\gamma < \infty$. The positive integers $d_\gamma$ and the norms $\|\cdot\|_\gamma$ on the algebras $\mathbb{C}^{d_\gamma \times d_\gamma}$ will be specified in the proof. The norms in question depend on those given on the original target algebras $\mathbb{C}^{n_\omega \times n_\omega}$ for which there is no restriction except that they satisfy the usual submultiplicativity condition. The choices made in the proof imply that the unit element in $\mathbb{C}^{d_\gamma \times d_\gamma}$ has norm one. They also entail that the order of the family $\{\psi_\gamma : \mathcal{B} \to \mathbb{C}^{d_\gamma \times d_\gamma}\}_{\gamma\in\Gamma}$ is finite if the order of $\{\phi_\omega : \mathcal{B} \to \mathbb{C}^{n_\omega \times n_\omega}\}_{\omega\in\Omega}$ is.

*Proof.* We begin by fixing some notation. The norm that we assume to be given on $\mathbb{C}^{n_\omega \times n_\omega}$ will be denoted by $\|\cdot\|_\omega$. Further, let $I_\omega$ be the $n_\omega \times n_\omega$ identity matrix, i.e., the unit element in $\mathbb{C}^{n_\omega \times n_\omega}$. Finally, write $\Gamma$ for the set of all (ordered) triples $(\tau, \sigma, n)$ with $\tau, \sigma \in \Omega$ and $n \in \mathbb{N}$, and introduce $\mathcal{B}_{(\tau,\sigma,n)}$ as the algebra of block diagonal matrices having two blocks, the one in the upper left corner of size $n_\tau$, and the one in the lower right corner of size $n_\sigma$.

Algebraically $\mathcal{B}_{(\tau,\sigma,n)}$ does not depend on $n$. This, however, is different for the norm $\|\cdot\|_{(\tau,\sigma,n)}$ that we will use on $\mathcal{B}_{(\tau,\sigma,n)}$. The definition goes in two steps. First, for $(X_\tau, Y_\sigma) \in \mathcal{B}_{(\tau,\sigma,n)}$, we define $|||(X_\tau, Y_\sigma)|||_{(\tau,\sigma,n)}$ by the expression

$$|\mathrm{tr}(X_\tau)| + n\big(\|X_\tau - \mathrm{tr}(X_\tau)I_\tau\|_\tau + \|Y_\sigma - \mathrm{tr}(Y_\sigma)I_\sigma\|_\sigma + |\mathrm{tr}(X_\tau) - \mathrm{tr}(Y_\sigma)|\big).$$

Here, for $M$ a square matrix, $\mathrm{tr}(M)$ is the normalized trace of $M$, i.e., the usual trace of $M$ divided by the size of $M$. It is easy to check that $|||\cdot|||_{(\tau,\sigma,n)}$ is a norm on $\mathcal{B}_{(\tau,\sigma,n)}$, possibly not submultiplicative though. For the unit element $(I_\tau, I_\sigma)$ in $\mathcal{B}_{(\tau,\sigma,n)}$ we have $|||(I_\tau, I_\sigma)|||_{(\tau,\sigma,n)} = 1$. As a second step we remedy the possible lack of submultiplicativity. For this we use the standard procedure involving left regular representations. Write

$$\|(X_\tau, Y_\sigma)\|_{(\tau,\sigma,n)} = \sup_{|||(R_\tau,S_\sigma)|||_{(\tau,\sigma,n)}=1} |||(X_\tau, Y_\sigma)\cdot(R_\tau, S_\sigma)|||_{(\tau,\sigma,n)},$$

or, what amounts to the same,

$$\|(X_\tau, Y_\sigma)\|_{(\tau,\sigma,n)} = \sup_{|||(R_\tau,S_\sigma)|||_{(\tau,\sigma,n)}=1} |||(X_\tau R_\tau, Y_\sigma S_\sigma)|||_{(\tau,\sigma,n)}.$$

Then $\|\cdot\|_{(\tau,\sigma,n)}$ is a norm on $\mathcal{B}_{(\tau,\sigma,n)}$, well defined (because of finite dimensionality), and submultiplicative as desired. Also, since $|||(I_\tau, I_\sigma)|||_{(\tau,\sigma,n)}$ has the value one, $|||(X_\tau, X_\sigma)|||_{(\tau,\sigma,n)} \le \|(X_\tau, X_\sigma)\|_{(\tau,\sigma,n)}$, and finally, $\|(I_\tau, I_\sigma)\|_{(\tau,\sigma,n)} = 1$.

Next we introduce a family $\{\psi_{(\tau,\sigma,n)} : \mathcal{B} \to \mathcal{B}_{(\tau,\sigma,n)}\}_{(\tau,\sigma,n)\in\Gamma}$ of unital Banach algebra homomorphisms. The defining expression is

$$\psi_{(\tau,\sigma,n)}(b) = \big(\phi_\tau(b), \phi_\sigma(b)\big), \qquad b \in \mathcal{B}.$$

As a mapping $\psi_{(\tau,\sigma,n)} : \mathcal{B} \to \mathcal{B}_{(\tau,\sigma,n)}$ does not depend on $n$. Clearly

$$\bigcap_{(\tau,\sigma,n)\in\Gamma} \mathrm{Ker}\, \psi_{(\tau,\sigma,n)} = \bigcap_{\omega\in\Omega} \mathrm{Ker}\, \phi_\omega.$$

Also, with $\Psi$ standing for the family $\{\psi_{(\tau,\sigma,n)} : \mathcal{B} \to \mathcal{B}_{(\tau,\sigma,n)}\}_{(\tau,\sigma,n)\in\Gamma}$, the set $\mathcal{G}_\Psi(\mathcal{B})$ is given by (2), but for this the argument is more involved.

First note that

$$\left\{\lambda e + r \mid 0 \neq \lambda \in \mathbb{C},\, r \in \bigcap_{(\tau,\sigma,n)\in\Gamma} \operatorname{Ker}\psi_{(\tau,\sigma,n)}\right\} \subset \mathcal{G}_\Psi(\mathcal{B}). \tag{3}$$

To see this, let $\lambda$ be a nonzero complex number, let $r \in \bigcap_{(\tau,\sigma,n)\in\Gamma} \operatorname{Ker}\psi_{(\tau,\sigma,n)}$, and take $(\tau,\sigma,n)$ in $\Gamma$. Then $\psi_{(\tau,\sigma,n)}(\lambda e + r) = \lambda(I_\tau, I_\sigma)$ is a nonzero scalar multiple of the unit element $(I_\tau, I_\sigma)$ in $\mathcal{B}_{(\tau,\sigma,n)}$. Hence $\psi_{(\tau,\sigma,n)}(\lambda e + r)$ is invertible in $\mathcal{B}_{(\tau,\sigma,n)}$ with inverse $\lambda^{-1}(I_\tau, I_\sigma)$. Using that $\|(I_\tau, I_\sigma)\|_{(\tau,\sigma,n)} = 1$, we get

$$\|\big(\psi_{(\tau,\sigma,n)}(\lambda e + r)\big)^{-1}\|_{(\tau,\sigma,n)} = \|\lambda^{-1}(I_\tau, I_\sigma)\|_{(\tau,\sigma,n)} = |\lambda^{-1}|,$$

and so $\lambda e + r \in \mathcal{G}_\Psi(\mathcal{B})$.

Combining (3) and (1), we may conclude that

$$\left\{\lambda e + r \mid 0 \neq \lambda \in \mathbb{C},\, r \in \bigcap_{\omega\in\Omega} \operatorname{Ker}\phi_\omega\right\} \subset \mathcal{G}_\Psi(\mathcal{B}).$$

Next take $b$ in $\mathcal{G}_\Psi(\mathcal{B})$. Thus $\psi_{(\tau,\sigma,n)}(b)$ is invertible in $\mathcal{B}_{(\tau,\sigma,n)}$ for all $(\tau,\sigma,n) \in \Gamma$ and, moreover,

$$\sup_{(\tau,\sigma,n)\,\in\,\Gamma} \|\psi_{(\tau,\sigma,n)}(b)^{-1}\|_{(\tau,\sigma,n)} < \infty.$$

Now $\psi_{(\tau,\sigma,n)}(b) = \big(\phi_\tau(b), \phi_\sigma(b)\big)$ and $\psi_{(\tau,\sigma,n)}(b)^{-1} = \big(\phi_\tau(b)^{-1}, \phi_\sigma(b)^{-1}\big)$. Hence

$$\sup_{(\tau,\sigma,n)\,\in\,\Gamma} \|\big(\phi_\tau(b)^{-1}, \phi_\sigma(b)^{-1}\big)\|_{(\tau,\sigma,n)} < \infty.$$

But then $\sup_{(\tau,\sigma,n)\,\in\,\Gamma} |||\big(\phi_\tau(b)^{-1}, \phi_\sigma(b)^{-1}\big)|||_{(\tau,\sigma,n)} < \infty$ too. Invoking the definition of the norm $|||\cdot|||_{(\tau,\sigma,n)}$, we obtain

$$\phi_\tau(b)^{-1} = \operatorname{tr}\big(\phi_\tau(b)^{-1}\big)I_\tau, \qquad \phi_\sigma(b)^{-1} = \operatorname{tr}\big(\phi_\sigma(b)^{-1}\big)I_\sigma,$$

and $\operatorname{tr}\big(\phi_\tau(b)^{-1}\big) = \operatorname{tr}\big(\phi_\sigma(b)^{-1}\big)$. With $\mu_{\tau,\sigma} = \operatorname{tr}\big(\phi_\tau(b)^{-1}\big) = \operatorname{tr}\big(\phi_\sigma(b)^{-1}\big)$, this gives $\phi_\tau(b)^{-1} = \mu_{\tau,\sigma}I_\tau$ and $\phi_\sigma(b)^{-1} = \mu_{\tau,\sigma}I_\sigma$. Clearly $\mu_{\tau,\sigma} \neq 0$. For $t, s \in \Omega$, we have $\mu_{\tau,\sigma}I_\tau = \phi_\tau(b)^{-1} = \mu_{\tau,s}I_\tau$, hence $\mu_{\tau,\sigma} = \mu_{\tau,s}$, and $\mu_{\tau,s}I_s = \phi_s(b)^{-1} = \mu_{t,s}I_s$, hence $\mu_{\tau,s} = \mu_{t,s}$. We conclude that $\mu_{\tau,\sigma} = \mu_{t,s}$. Thus the scalars $\mu_{\tau,\sigma}$ are independent of $\tau$ and $\sigma$.

The upshot of all of this is that there exists a nonzero scalar $\mu$ such that $\phi_\omega(b)^{-1} = \mu I_\omega = \mu\phi_\omega(e)$ for all $\omega \in \Omega$. With $\lambda = \mu^{-1}$, this can be rewritten as $\phi_\omega(b) = \lambda\phi_\omega(e)$, $\omega \in \Omega$. In other words

$$b - \lambda e \in \bigcap_{\omega\in\Omega} \operatorname{Ker}\phi_\omega = \bigcap_{(\tau,\sigma,n)\in\Gamma} \operatorname{Ker}\psi_{(\tau,\sigma,n)},$$

as desired.

To finish the proof one more step has to be made. The Banach algebra $\mathcal{B}_{(\tau,\sigma,n)}$ is finite dimensional. In fact it has dimension $d_{(\tau,\sigma,n)} = n_\tau^2 + n_\sigma^2$. Using left regular representations, $\mathcal{B}_{(\tau,\sigma,n)}$ can be (isomorphically) identified with a Banach subalgebra of the Banach algebra of all (bounded) linear operators on $\mathcal{B}_{(\tau,\sigma,n)}$, which in turn can be identified with $\mathbb{C}^{d_{(\tau,\sigma,n)}\times d_{(\tau,\sigma,n)}}$. We can now identify $\psi_{(\tau,\sigma,n)}$ with a mapping into $\mathbb{C}^{d_{(\tau,\sigma,n)}\times d_{(\tau,\sigma,n)}}$. As the subalgebra mentioned above is inverse closed (finite dimensionality), this identification does not affect the set $\mathcal{G}_\Psi(\mathcal{B})$. Finally we note that $\sup_{(\tau,\sigma,n)\in\Gamma} d_{(\tau,\sigma,n)}$ is finite whenever $\sup_{\omega\in\Omega} n_\omega$ is. □

*Proof of Theorem* 3.2. Each p.w. sufficient family is radical-separating (see Corollary 2.3). So the 'if part' of the theorem is obvious, and this is also true when one restricts oneself to families of finite order. Thus we focus on the 'only if part' of the theorem. Here the situation is more involved and requires a change of family (see the comment at the very end of the paper).

Suppose $\{\phi_\omega : \mathcal{B} \to \mathbb{C}^{n_\omega\times n_\omega}\}_{\omega\in\Omega}$ is a radical-separating family of matrix representations. This means that $\bigcap_{\omega\in\Omega}\operatorname{Ker}\phi_\omega \subset \mathcal{R}(\mathcal{B})$. Applying Lemma 3.3 we obtain a family $\Psi = \{\psi_\gamma : \mathcal{B} \to \mathbb{C}^{d_\gamma\times d_\gamma}\}_{\gamma\in\Gamma}$ of matrix representations such that the set $\mathcal{G}_\Psi(\mathcal{B})$ of all $b \in \mathcal{B}$ with $\psi_\gamma(b)$ is invertible in $\mathbb{C}^{d_\gamma\times d_\gamma}$ for all $\gamma \in \Gamma$, while, in addition, $\sup_{\gamma\in\Gamma} \|\psi_\gamma(b)^{-1}\|_\gamma < \infty$, coincides with

$$\left\{\lambda e_{\mathcal{B}} + r \mid 0 \neq \lambda \in \mathbb{C},\, r \in \bigcap_{\omega\in\Omega}\operatorname{Ker}\phi_\omega\right\}.$$

Here $\|\cdot\|_\gamma$ stands for an appropriate norm on $\mathbb{C}^{d_\gamma\times d_\gamma}$ which we (can) choose in such a way that the unit element in $\mathbb{C}^{d_\gamma\times d_\gamma}$ has norm one. With the help of the inclusion $\bigcap_{\omega\in\Omega}\operatorname{Ker}\phi_\omega \subset \mathcal{R}(\mathcal{B})$, it follows that $e_{\mathcal{B}} \in \mathcal{G}_\Psi(\mathcal{B}) \subset \mathcal{G}(\mathcal{B})$. Hence $\{\psi_\gamma : \mathcal{B} \to \mathbb{C}^{d_\gamma\times d_\gamma}\}_{\gamma\in\Gamma}$ is p.w. sufficient (see the paragraph preceding Proposition 2.2). As $\{\psi_\gamma : \mathcal{B} \to \mathbb{C}^{d_\gamma\times d_\gamma}\}_{\gamma\in\Gamma}$ can be constructed in such a way that its order is finite whenever that of $\{\phi_\omega : \mathcal{B} \to \mathbb{C}^{n_\omega\times n_\omega}\}_{\omega\in\Omega}$ is, the 'only if part' of the theorem is also true when one restricts oneself to considering families of finite order. □

Elaborating on the material presented above (and using notations employed there) we mention the following result.

**Theorem 3.4.** *The unital Banach algebra $\mathcal{B}$ with unit element $e_{\mathcal{B}}$ possesses a separating family of matrix representations if and only if $\mathcal{B}$ possesses a family $\Psi = \{\psi_\gamma : \mathcal{B} \to \mathbb{C}^{d_\gamma\times d_\gamma}\}_{\gamma\in\Gamma}$ of matrix representations having the property that $\mathcal{G}_\Psi(\mathcal{B}) = \{\lambda e_{\mathcal{B}} \mid 0 \neq \lambda \in \mathbb{C}\}$.*

Recall from the paragraph preceding Proposition 2.2 that the identity $\mathcal{G}_\Psi(\mathcal{B}) = \{\lambda e_{\mathcal{B}} \mid 0 \neq \lambda \in \mathbb{C}\}$ implies p.w. sufficiency. Theorem 3.4 remains true when one restricts oneself to considering families of matrix representations of finite order.

*Proof.* Suppose $\Psi = \{\psi_\gamma : \mathcal{B} \to \mathbb{C}^{d_\gamma\times d_\gamma}\}_{\gamma\in\Gamma}$ is a family of matrix representations such that $\mathcal{G}_\Psi(\mathcal{B}) = \{\lambda e_{\mathcal{B}} \mid 0 \neq \lambda \in \mathbb{C}\}$. Then, by Proposition 2.2, this family itself is separating. This settles the 'if part' of the theorem. As to the 'only if part', apply Lemma 3.3. □

Next we confine ourselves to considering families of matrix representations of finite order. As we shall see, under this restriction the properties of possessing a family of matrix representations which is sufficient, weakly sufficient, p.w. sufficient or radical-separating all amount to the same. A Banach algebra $\mathcal{A}$ is said to be a *polynomial identity (Banach) algebra*, PI-algebra for short, if there exist a positive integer $k$ and a nontrivial polynomial $p(x_1, \ldots, x_k)$ in $k$ non-commuting variables $x_1, \ldots, x_k$ such that $p(a_1, \ldots, a_k) = 0$ for every choice of elements $a_1, \ldots, a_k$ in $\mathcal{A}$. Such a polynomial is then called an *annihilating polynomial* for $\mathcal{A}$. By $F_{2n}$ we mean the polynomial in $2n$ non-commuting variables given by

$$F_{2n}(x_1, \ldots x_{2n}) = \sum_{\sigma \in S_{2n}} (-1)^{\operatorname{sign} \sigma} x_{\sigma_1} x_{\sigma_2} \cdots x_{\sigma_{2n}}.$$

Here $S_{2n}$ stands for the collection of all permutations of $\{1, \ldots, 2n\}$, and $\operatorname{sign} \sigma$ for the signature of an element $\sigma$ in $S_{2n}$. The algebra $\mathcal{A}$ is termed an $F_{2n}$-algebra when it has $F_{2n}$ as an annihilating polynomial. By a celebrated result of Amitsur and Levitzky [AL], the matrix algebra $\mathbb{C}^{m \times m}$ is an $F_{2n}$-algebra whenever $m$ does not exceed $n$.

**Theorem 3.5.** *Let $\mathcal{B}$ be a unital Banach algebra, and let $\mathcal{R}(\mathcal{B})$ be its radical. The following five statements are equivalent:*

(1) *$\mathcal{B}/\mathcal{R}(\mathcal{B})$ is a PI-algebra;*

(2) *$\mathcal{B}$ possesses a family of matrix representations which is both sufficient and of finite order;*

(3) *$\mathcal{B}$ possesses a family of matrix representations which is both weakly sufficient and of finite order;*

(4) *$\mathcal{B}$ possesses a family of matrix representations which is both p.w. sufficient and of finite order;*

(5) *$\mathcal{B}$ possesses a family of matrix representations which is both radical-separating and of finite order.*

*Proof.* The implication (1) ⇒ (2) is known; see [Kr]. The implication (2) ⇒ (3) follows from Theorem 3.1. As weak sufficiency implies p.w. sufficiency, we have (3) ⇒ (4). The implication (4) ⇒ (5) is covered by Proposition 3.4 in [BES12] or, if one prefers, Corollary 2.3 above. It remains to show that (5) ⇒ (1).

Suppose $\mathcal{B}$ possesses a family of matrix representations $\{\phi_\omega : \mathcal{B} \to \mathbb{C}^{n_\omega \times n_\omega}\}$ which is both radical-separating and of finite order, $n$ say. We shall prove that $\mathcal{B}/\mathcal{R}(\mathcal{B})$ is an $F_{2n}$-algebra. Here is the argument. As the positive integers $n_\omega$ do not exceed $n$, the polynomial $F_{2n}$ is annihilating for all the algebras $\mathbb{C}^{n_\omega \times n_\omega}$. Thus $\phi_\omega(F_{2n}(a_1, \ldots, a_{2n})) = F_{2n}(\phi_\omega(a_1), \ldots, \phi_\omega(a_{2n})) = 0$ for all $\omega \in \Omega$ and all $a_1, \ldots, a_{2n} \in \mathcal{B}$. This implies that $F_{2n}(a_1, \ldots, a_{2n})$ is in $\mathcal{R}(\mathcal{B})$, and it follows that $F_{2n}$ is an annihilating polynomial for $\mathcal{B}/\mathcal{R}(\mathcal{B})$. □

The main results obtained so far for families of matrix representations (without restriction on the order) can be schematically summarized as follows:

$$\begin{array}{ccccccc} \text{sufficient} & \Rightarrow & \text{weakly sufficient} & \Rightarrow & \text{p.w. sufficient} & \Leftrightarrow & \text{radical-separating} \\ & & & & & & \Uparrow \\ & & & & & & \text{separating.} \end{array}$$

Here the implications have to be understood as holding on what in the introduction has been called the second level, i.e., the level concerned with the existence of families with the properties in question for a given underlying Banach algebra (in contrast to the first level which pertains to individual families).

Three questions still deserve our attention. The first is: 'if a Banach algebra possesses a radical-separating family of matrix representations, does it follow that it also possesses a separating family of matrix representations?'; the second: 'if a Banach algebra possesses a radical-separating (or even separating) family of matrix representations, does it follow that it also possesses a sufficient family of matrix representations?'; the third: 'if a Banach algebra possesses a radical-separating (or even separating) family of matrix representations, does it follow that it also possesses a weakly sufficient family of matrix representations?'.

The first question can be rephrased as follows: 'if $\mathcal{B}$ is a Banach algebra, and the intersection of the null spaces of all possible matrix representations on $\mathcal{B}$ is contained in the radical of $\mathcal{B}$, can one conclude that this intersection consists of the zero element only?'. We conjecture that the answer is negative. A counterexample might be the Calkin algebra of the Banach space $X$ constructed in Section 3 of [GM] – which has the property that the ideal of strictly singular operators on $X$ has codimension one in $\mathcal{B}(X)$ – but we have not yet been able to verify this.

The second and the third question have a negative answer, the second one even when one restricts oneself to the $C^*$-case where the properties of being weakly sufficient, p.w. sufficient, radical-separating and separating all amount to the same. The arguments involved are complicated and lengthy. Therefore we present them in two separate sections.

## 4. Separating versus sufficient families of matrix representations

Let $\mathbf{M} = \{\mathbb{C}^{n\times n}\}_{n\in\mathbb{N}}$, where $\mathbb{C}^{n\times n}$ is identified with the Banach algebra $\mathcal{B}(\mathbb{C}^n)$ of (bounded) linear operators on the Hilbert space $\mathbb{C}^n$, equipped with the standard Hilbert space norm. In particular, the unit element $I_n$ in $\mathbb{C}^{n\times n}$ has norm one. By slight abuse of notation, the norms on $\mathbb{C}^n$ and $\mathbb{C}^{n\times n}$ will both be denoted by $\|\cdot\|_n$. Write $\ell_\infty^{\mathbf{M}}$ for the $\ell_\infty$-direct product of the family $\mathbf{M}$. Thus $\ell_\infty^{\mathbf{M}}$ consists of all $M$ in the Cartesian product $\prod_{n\in\mathbb{N}}\mathbb{C}^{n\times n}$ such that

$$|||M||| = \sup_{n\in\mathbb{N}} \|M(n)\|_n < \infty.$$

With the operations of scalar multiplication, addition, multiplication and that of taking the adjoint all defined pointwise, and with $|||\cdot|||$ as norm, $\ell_\infty^{\mathbf{M}}$ is a unital Banach algebra – in fact a $C^*$-algebra.

With the next theorem we make good on a promise made at the end of Section 2. The result also shows that the new Gelfand type criteria for spectral regularity of Banach algebras developed in [BES12] genuinely go beyond the scope of Theorem 4.1 in [BES94].

**Theorem 4.1.** *The $C^*$-algebra $\ell_\infty^{\mathbf{M}}$ possesses a $C^*$-family of matrix representations which is weakly sufficient (hence p.w. sufficient) and separating; it does, however, not possess any sufficient family of matrix representations.*

Note that this includes ruling out the existence of sufficient families of matrix representations that are not of $C^*$-type.

*Proof.* The existence of a $C^*$-family of matrix representations which is weakly sufficient (so a fortiori p.w. sufficient) and separating is easily established: just consider the point evaluations $\ell_\infty^{\mathbf{M}} \ni M \mapsto M(n) \in \mathbb{C}^{n\times n}$, $n \in \mathbb{N}$. This covers the first part of the theorem.

For the proof of the second part, one needs some 'grasp' on the collection of all unital matrix representations of the Banach algebra $\ell_\infty^{\mathbf{M}}$; a nontrivial matter, as can already be seen from the situation for the relatively simple Banach algebra $\ell_\infty$. We shall overcome the difficulty by identifying a $C^*$-subalgebra $\ell_{\infty,c}^{\mathbf{M}}$ of $\ell_\infty^{\mathbf{M}}$ for which we are able to explicitly describe the set of surjective matrix representations and use that description to prove that $\ell_{\infty,c}^{\mathbf{M}}$ does not possess a sufficient family of matrix representations. The (well-known) fact that such a $C^*$-subalgebra is inverse closed then guarantees that $\ell_\infty^{\mathbf{M}}$ does not have a sufficient family of matrix representations either. Indeed, if it had one we could just take the restrictions to the subalgebra to obtain a contradiction.

The detailed argument proving the above theorem will now be given in a number of steps. The first two of these are concerned with the introduction of the subalgebra meant above. The definition of the subalgebra is inspired by the numerically oriented work of Hagen, Roch, and the third author of the present paper; see [HRS95].

*Step* 1. For $n$ a positive integer, define the bounded linear operators

$$\Pi_n^{\downarrow} : \ell_2 \to \mathbb{C}^n \quad \text{and} \quad \Pi_n^{\uparrow} : \mathbb{C}^n \to \ell_2$$

by

$$\Pi_n^{\downarrow}\begin{pmatrix} x_1 \\ x_2 \\ x_3 \\ \vdots \end{pmatrix} = \begin{pmatrix} x_1 \\ \vdots \\ x_n \end{pmatrix}, \qquad \Pi_n^{\uparrow}\begin{pmatrix} x_1 \\ \vdots \\ x_n \end{pmatrix} = \begin{pmatrix} x_1 \\ \vdots \\ x_n \\ 0 \\ 0 \\ \vdots \end{pmatrix}.$$

Then $\Pi_n^{\downarrow}\Pi_n^{\uparrow}$ is a contraction and $\Pi_n^{\uparrow}$ is an isometry. Also $\Pi_n^{\downarrow}\Pi_n^{\uparrow} = I_n$, the identity matrix of size $n$, while $\Pi_n^{\uparrow}\Pi_n^{\downarrow}$ is the projection $\Pi_n$ on $\ell_2$ with action

$$\Pi_n \begin{pmatrix} x_1 \\ x_2 \\ x_3 \\ \vdots \end{pmatrix} = \begin{pmatrix} x_1 \\ \vdots \\ x_n \\ 0 \\ 0 \\ \vdots \end{pmatrix}.$$

Note that $\Pi_n$ tends strongly to the identity operator $I$ on $\ell_2$ when $n$ goes to infinity, written $\text{s-}\lim_{n\to\infty} \Pi_n = I$. The mapping $\mathbb{C}^{n\times n} \ni A \mapsto \Pi_n^{\uparrow} A \Pi_n^{\downarrow} \in \mathcal{B}(\ell_2)$ is a (non-unital) Banach algebra isomorphism from $\mathbb{C}^{n\times n}$ into $\mathcal{B}(\ell_2)$, the latter equipped with the operator norm $\|\cdot\|$ induced by the standard Hilbert space norm on $\ell_2$.

Write $\mathcal{N}$ for the set of all $H \in \ell_\infty^{\mathbf{M}}$ such that $\lim_{n\to\infty} \|H(n)\|_n = 0$. Clearly $\mathcal{N}$ is a closed two-sided ideal in $\ell_\infty^{\mathbf{M}}$. Next, let $\ell_{\infty,c}^{\mathbf{M}}$ consist of all $M \in \ell_\infty^{\mathbf{M}}$ such that there exist $\lambda \in \mathbb{C}$, a compact operator $K : \ell_2 \to \ell_2$ and $H \in \mathcal{N}$ such that

$$M(n) = \lambda I_n + \Pi_n^{\downarrow} K \Pi_n^{\uparrow} + H(n), \qquad n = 1, 2, 3, \ldots . \tag{4}$$

Note that $\Pi_n^{\downarrow} K \Pi_n^{\uparrow} : \mathbb{C}^n \to \mathbb{C}^n$ is the $n$th finite section of the operator $K$. In the next step shall prove that $\ell_{\infty,c}^{\mathbf{M}}$ is indeed a $C^*$-subalgebra of $\ell_\infty^{\mathbf{M}}$ (hence inverse closed by general $C^*$-algebra theory). The type of argument employed for this is also used in [HRS01], but for the convenience of the reader we present the reasoning in detail.

*Step* 2. Clearly $\ell_{\infty,c}^{\mathbf{M}}$ is closed under addition and scalar multiplication. It is also closed under multiplication but the argument for this is more involved. Take $\widehat{M}, \widetilde{M}$ in $\ell_{\infty,c}^{\mathbf{M}}$ and put $M = \widehat{M}\widetilde{M}$. We need to show that $M \in \ell_{\infty,c}^{\mathbf{M}}$. Choose $\hat{\lambda}, \tilde{\lambda} \in \mathbb{C}$, compact operators $\widehat{K}, \widetilde{K} : \ell_2 \to \ell_2$ and $\widehat{H}, \widetilde{H} \in \mathcal{N}$ such that

$$\widehat{M}(n) = \hat{\lambda} I_n + \Pi_n^{\downarrow} \widehat{K} \Pi_n^{\uparrow} + \widehat{H}(n), \qquad n = 1, 2, 3, \ldots ,$$

$$\widetilde{M}(n) = \tilde{\lambda} I_n + \Pi_n^{\downarrow} \widetilde{K} \Pi_n^{\uparrow} + \widetilde{H}(n), \qquad n = 1, 2, 3, \ldots .$$

A straightforward computation gives

$$M(n) = \hat{\lambda}\tilde{\lambda} I_n + \Pi_n^{\downarrow} (\hat{\lambda}\widetilde{K} + \tilde{\lambda}\widehat{K} + \widehat{K}\widetilde{K}) \Pi_n^{\uparrow} + H(n),$$

where $H \in \ell_\infty^{\mathbf{M}}$ is given by

$$\begin{aligned} H(n) = \hat{\lambda}\widetilde{H}(n) + \tilde{\lambda}\widehat{H}(n) + \widehat{H}(n)\widetilde{H}(n) + \Pi_n^{\downarrow}\widehat{K}\Pi_n^{\uparrow}\widetilde{H}(n) + \widehat{H}(n)\Pi_n^{\downarrow}\widetilde{K}\Pi_n^{\uparrow} \\ - \Pi_n^{\downarrow}\widehat{K}(I - \Pi_n)\widetilde{K}\Pi_n^{\uparrow}. \end{aligned}$$

As has already been mentioned, $\text{s-}\lim_{n\to\infty} \Pi_n = I$. Also $\widetilde{K}$ is compact. Recalling that pointwise convergence on compact subsets is uniform, we may conclude that $\lim_{n\to\infty} \|(I - \Pi_n)\widetilde{K}\| = 0$. Thus $H \in \mathcal{N}$. As the operator $\hat{\lambda}\widetilde{K} + \tilde{\lambda}\widehat{K} + \widehat{K}\widetilde{K}$ is compact, it follows that $M \in \ell_{\infty,c}^{\mathbf{M}}$, as desired.

The unit element of $\ell_\infty^{\mathbf{M}}$ is contained in $\ell_{\infty,c}^{\mathbf{M}}$. It is also easy to see that $\ell_{\infty,c}^{\mathbf{M}}$ is closed under the $C^*$-operation. So the only thing left to prove is that $\ell_{\infty,c}^{\mathbf{M}}$ is a closed subset of $\ell_\infty^{\mathbf{M}}$.

We begin by showing that for $M \in \ell_{\infty,c}^{\mathbf{M}}$ given in the form (4), the following norm inequalities hold:

$$|||M||| \leq |\lambda| + \|K\| + |||H||| \leq 7|||M|||. \tag{5}$$

The first inequality in (5) is obvious; so we concentrate on the second. For the operator $\Pi_n^\uparrow M(n)\Pi_n^\downarrow : \ell_2 \to \ell_2$, we have

$$\Pi_n^\uparrow M(n)\Pi_n^\downarrow = \lambda\Pi_n + \Pi_n K\Pi_n + \Pi_n^\uparrow H(n)\Pi_n^\downarrow, \qquad n = 1,2,3,\ldots.$$

Taking the strong limit, we obtain

$$\text{s-}\lim_{n\to\infty} \Pi_n^\uparrow M(n)\Pi_n^\downarrow = \lambda I + K. \tag{6}$$

The fact that we need to work with strong limits here is due to the circumstance that we only have $\text{s-lim}_{n\to\infty} \Pi_n = I$ and not $\Pi_n \to I$ in norm; we do have convergence in norm $\Pi_n^\uparrow H(n)\Pi_n^\downarrow \to 0$ (clearly) and $\Pi_n K\Pi_n \to K$. The latter can be seen as follows. First note that $\Pi_n K\Pi_n = K - (I-\Pi_n)K - \Pi_n K(I-\Pi_n)$. Next observe that the compactness of $K$ together with $\text{s-lim}_{n\to\infty} \Pi_n = I$ implies that $(I-\Pi_n)K \to 0$ in norm. Finally $K(I-\Pi_n) \to 0$ in norm too, which can be seen by taking adjoints.

For $x \in \ell_2$, we have $(\lambda I + K)x = \lim_{n\to\infty} \Pi_n^\uparrow M(n)\Pi_n^\downarrow x$, and it follows that

$$\|\lambda I + K\| \leq \sup_{n=1,2,3,\ldots} \|M(n)\|_n = |||M|||. \tag{7}$$

Also $|\lambda| \leq \|\lambda I + K\|$. This is trivial for $\lambda = 0$. For $\lambda \neq 0$, the compact operator $\lambda^{-1}K$ cannot be invertible, and so we must have $\|I+\lambda^{-1}K\| \geq 1$. Thus $|\lambda| \leq |||M|||$ and $\|K\| \leq \|\lambda I + K\| + |\lambda| \leq 2|||M|||$. From $H(n) = M(n) - \lambda I_n - \Pi_n^\downarrow K\Pi_n^\uparrow$, we get $\|H(n)\|_n \leq \|M(n)\|_n + |\lambda| + \|K\| \leq 4|||M|||$. Thus $|||H||| \leq 4|||M|||$ and $|\lambda| + \|K\| + |||H||| \leq 7|||M|||$, as desired.

Let $M_1, M_2, M_3, \ldots$ be a sequence in $\ell_{\infty,c}^{\mathbf{M}}$ converging in $\ell_\infty^{\mathbf{M}}$ to $M$. We need to show that $M \in \ell_{\infty,c}^{\mathbf{M}}$. This goes as follows. For $k = 1,2,3,\ldots$, write

$$M_k(n) = \lambda_k I_n + \Pi_n^\downarrow K_k\Pi_n^\uparrow + H_k(n), \qquad n = 1,2,3,\ldots,$$

with $\lambda_k$ a scalar, $K_k : \ell_2 \to \ell_2$ a compact operator and $H_k \in \mathcal{N}$. By the second inequality in (5), the sequence $\lambda_1, \lambda_2, \lambda_3, \ldots$ is a Cauchy sequence in $\mathbb{C}$, the sequence $K_1, K_2, K_3, \ldots$ is a Cauchy sequence in $\mathcal{B}(\ell_2)$, and $H_1, H_2, H_3, \ldots$ is a Cauchy sequence in $\ell_\infty^{\mathbf{M}}$. But then there exist a scalar $\lambda$, a bounded linear operator $K : \ell_2 \to \ell_2$, and an element $H \in \ell_\infty^{\mathbf{M}}$ such that $\lambda_k \to \lambda$ in $\mathbb{C}$, $K_k \to K$ in $\mathcal{B}(\ell_2)$, and $H_k \to H$ in $\ell_\infty^{\mathbf{M}}$. The operator $K$, being a limit of compact operators, is compact too. Also $\mathcal{N}$ is closed in $\ell_\infty^{\mathbf{M}}$, hence $H \in \mathcal{N}$. Put

$$\widehat{M}(n) = \lambda I_n + \Pi_n^\downarrow K\Pi_n^\uparrow + H(n), \qquad n = 1,2,3,\ldots.$$

Then $\widehat{M}$ belongs to $\ell_{\infty,c}^{\mathbf{M}}$ and the first inequality in (5) gives that $\widehat{M} = \lim_{k\to\infty} M_k$ in $\ell_\infty^{\mathbf{M}}$. It follows that $M = \widehat{M} \in \ell_{\infty,c}^{\mathbf{M}}$.

*Step* 3. In this step we introduce certain Banach algebra homomorphisms on $\ell_{\infty,c}^{\mathbf{M}}$. For $n$ a positive integer, $\phi_n : \ell_{\infty,c}^{\mathbf{M}} \to \mathbb{C}^{n\times n}$ is just the point evaluation given by $\phi_n(M) = M(n)$. Clearly this is a surjective contractive matrix representation and, in fact, a $C^*$-homomorphism.

Two more homomorphisms are needed. Here is the definition. Given $M$ in $\ell_{\infty,c}^{\mathbf{M}}$, there exist a unique scalar $\lambda_M$, a unique compact operator $K_M : \ell_2 \to \ell_2$ and a unique element $H_M \in \mathcal{N}$ such that

$$M(n) = \lambda_M I_n + \Pi_n^{\downarrow} K_M \Pi_n^{\uparrow} + H_M(n), \qquad n = 1, 2, 3, \ldots. \tag{8}$$

The uniqueness is immediate from (5). We now put

$$\phi_0(M) = \lambda_M,$$

$$\phi_\infty(M) = \lambda_M I + K_M = \text{s-}\lim_{n\to\infty} \Pi_n^{\uparrow} M(n) \Pi_n^{\downarrow},$$

the second expression for $\phi_\infty(M)$ coming from (6). In this way we obtain mappings $\phi_0 : \ell_{\infty,c}^{\mathbf{M}} \to \mathbb{C}$ and $\phi_\infty : \ell_{\infty,c}^{\mathbf{M}} \to \mathcal{B}(\ell_2)$. Clearly these mappings are unital homomorphisms. They are continuous too and, as we see from the paragraph containing (7), in fact contractions. For completeness, we note that $\phi_0$ and $\phi_\infty$ are $C^*$-homomorphisms.

Along with $\phi_1, \phi_2, \phi_3, \ldots$, the homomorphism $\phi_0 : \ell_{\infty,c}^{\mathbf{M}} \to \mathbb{C}$ is surjective. The image of $\phi_\infty$ is the Banach subalgebra $\mathcal{B}_{\mathcal{K}}(\ell_2)$ of $\mathcal{B}(\ell_2)$ consisting of all operators of the form $\lambda I + K$ with $\lambda \in \mathbb{C}$ and $K : \ell_2 \to \ell_2$ a compact linear operator. So, whenever this is convenient, $\phi_\infty$ can be considered as a (surjective) homomorphism from $\ell_{\infty,c}^{\mathbf{M}}$ onto $\mathcal{B}_{\mathcal{K}}(\ell_2)$. Note that $\phi_\infty(M)$ is a Fredholm operator (on $\ell_2$) if and only if $\phi_0(M) \neq 0$. The null space of $\phi_\infty$ is easily seen to be the ideal $\mathcal{N}$.

*Step* 4. Next we prove that the maximal two-sided ideals in $\ell_{\infty,c}^{\mathbf{M}}$ are precisely the null spaces of the matrix representations $\phi_0, \phi_1, \phi_2, \ldots$. These are surjective matrix representations. Hence, as is well known (and easy to see using the simplicity of the full matrix algebras $\mathbb{C}^{n\times n}$), their null spaces are maximal two-sided ideals in $\ell_{\infty,c}^{\mathbf{M}}$. It remains to show that there are no others. So let $\mathcal{J}$ be a maximal ideal in $\ell_{\infty,c}^{\mathbf{M}}$ and suppose $\mathcal{J} \neq \operatorname{Ker} \phi_n$ for all positive integers $n$. We shall prove that $\mathcal{J} = \operatorname{Ker} \phi_0$.

For $k = 1, 2, 3, \ldots$, define $P_k \in \mathcal{N}$ by

$$P_k(n) = \delta_{kn} I_n, \qquad n = 1, 2, 3, \ldots.$$

Our first aim is to show that $P_1, P_2, P_3, \ldots$ all belong to $\mathcal{J}$. Fix $k \in \mathbb{N}$. As $\phi_k$ is surjective, $\operatorname{Ker} \phi_k$ is a proper ideal in $\ell_{\infty,c}^{\mathbf{M}}$. By maximality, $\mathcal{J}$ is not strictly contained in any proper ideal in $\ell_{\infty,c}^{\mathbf{M}}$. Hence $\mathcal{J}$ cannot be a subset of $\operatorname{Ker} \phi_k$. Now introduce $\mathcal{J}_k = \{M(k) \mid M \in \mathcal{J}\}$. Then $\mathcal{J}_k$ is a two-sided ideal in $\mathbb{C}^{k\times k}$. If $\mathcal{J}_k$ is the trivial ideal consisting of the zero element $0_k$ in $\mathbb{C}^{k\times k}$ only, then $\phi_k(M) = M(k) = 0_k$ for all $M \in \mathcal{J}$ or, what amounts to the same, $\mathcal{J} \subset \operatorname{Ker} \phi_k$. But this is not the case. Since the algebra $\mathbb{C}^{k\times k}$ is simple, we may conclude that $\mathcal{J}_k = \mathbb{C}^{k\times k}$. Now choose $M_k \in \mathcal{J}$ such that $M_k(k) = I_k$. Then $P_k M_k \in \mathcal{J}$. Observing that $P_k M_k = P_k$ we arrive at $P_k \in \mathcal{J}$.

Next we show that $\mathcal{N} \subset \mathcal{J}$. Let $\mathcal{N}_0$ be the set of all $H \in \mathcal{N}$ such that $H(n) = 0$ for all but a finite set of positive integers $n$. Clearly $\mathcal{N}_0$ is dense in $\mathcal{N}$. Also $\mathcal{J}$, being a maximal ideal, is closed. So it is sufficient to establish the inclusion $\mathcal{N}_0 \subset \mathcal{J}$. Take $H_0 \in \mathcal{N}_0$ and choose $m$ such that $H_0(n) = 0$ for $n > m$. Then $H_0 = P_1H_0 + \cdots + P_mH_0$ and $H_0$ belongs to $\mathcal{J}$ along with $P_1, \ldots, P_m$.

We now bring into play the homomorphism $\phi_\infty$, here considered as a mapping from $\ell^{\mathbf{M}}_{\infty,c}$ into $\mathcal{B}_\mathcal{K}(\ell_2)$. Actually $\phi_\infty : \ell^{\mathbf{M}}_{\infty,c} \to \mathcal{B}_\mathcal{K}(\ell_2)$ is surjective. Further the null space of $\phi_\infty$ is the ideal $\mathcal{N}$, and so $\operatorname{Ker}\phi_\infty$ is contained in the maximal ideal $\mathcal{J}$. As is well known (and easy to deduce) these facts imply that $\phi_\infty[\mathcal{J}]$ is a maximal ideal in $\mathcal{B}_\mathcal{K}(\ell_2)$. It is also common knowledge that the only non-trivial closed ideal in $\mathcal{B}_\mathcal{K}(\ell_2)$ is the set $\mathcal{K}(\ell_2)$ of all compact linear operators on $\ell_2$. Hence $\phi_\infty[\mathcal{J}] = \mathcal{K}(\ell_2)$.

After all these preparations we are finally ready to prove the desired identity $\mathcal{J} = \operatorname{Ker}\phi_0$. First, take $M$ in $\ell^{\mathbf{M}}_{\infty,c}$, and assume $\phi_0(M) = 0$. In terms of the expression (8) this gives $\lambda_M = 0$ and $\phi_\infty(M) = K_M \in \mathcal{K}(\ell_2) = \phi_\infty[\mathcal{J}]$. Choose $X \in \mathcal{J}$ such that $\phi_\infty(M) = \phi_\infty(X)$. Then $M - X \in \operatorname{Ker}\phi_\infty = \mathcal{N} \subset \mathcal{J}$, and we get $M = (M-X)+X \in \mathcal{J}$. Thus $\operatorname{Ker}\phi_0 \subset \mathcal{J}$. Conversely, suppose $M \in \mathcal{J}$. Then $\lambda_M I + K_M = \phi_\infty(M) \in \phi_\infty[\mathcal{J}] = \mathcal{K}(\ell_2)$, and it follows that $\lambda_M = 0$, which can be rewritten as $\phi_0(M) = 0$.

*Step* 5. Before being able to proceed with the main line of the argument, we need a simple general observation. Let $\Phi : \mathcal{A} \to \mathbb{C}^{n\times n}$ and $\Psi : \mathcal{A} \to \mathbb{C}^{m\times m}$ be two surjective matrix representations of a unital Banach algebra $\mathcal{A}$, and suppose $\operatorname{Ker}\Phi = \operatorname{Ker}\Psi$. Then $n = m$ and $\Psi$ is an *inner transform* of $\Phi$, i.e., there exists an invertible $n\times n$ matrix $S$ such that $\Psi(a) = S^{-1}\Phi(a)S,\ a \in \mathcal{A}$. The reasoning is as follows. Put $\mathcal{J} = \operatorname{Ker}\Phi = \operatorname{Ker}\Psi$, and let $\Phi_\mathcal{J} : \mathcal{A}/\mathcal{J} \to \mathbb{C}^{n\times n}$ and $\Psi_\mathcal{J} : \mathcal{A}/\mathcal{J} \to \mathbb{C}^{m\times m}$ be the homomorphisms induced by $\Phi$ and $\Psi$, respectively. Then $\Phi_\mathcal{J}$ and $\Psi_\mathcal{J}$ are bijections, and it is clear that $n$ and $m$ must be the same. Consider $\Psi_\mathcal{J}\circ\Phi_\mathcal{J}^{-1} : \mathbb{C}^{n\times n} \to \mathbb{C}^{n\times n}$. This is an automorphism of $\mathbb{C}^{n\times n}$, i.e., a bijective unital homomorphism. It is well known (cf., [Se]) that such an automorphism is inner, which means that there exists an invertible $n \times n$ matrix $S$ such that $\left(\Psi_\mathcal{J}\circ\Phi_\mathcal{J}^{-1}\right)(M) = S^{-1}MS$ for all $M \in \mathbb{C}^{n\times n}$. With $\kappa$ the canonical mapping from $\mathcal{A}$ onto $\mathcal{A}/\mathcal{J}$ and $a \in \mathcal{A}$, we now have $\Psi(a) = \Psi_\mathcal{J}\big(\kappa(a)\big) = \left(\Psi_\mathcal{J}\circ\Phi_\mathcal{J}^{-1}\right)\Phi_\mathcal{J}\big(\kappa(a)\big) = \left(\Psi_\mathcal{J}\circ\Phi_\mathcal{J}^{-1}\right)\Phi(a) = S^{-1}\Phi(a)S$, so $\Psi$ is an inner transform of $\Phi$ indeed.

*Step* 6. If $\psi$ is surjective matrix representation of $\ell^{\mathbf{M}}_{\infty,c}$, then either $\psi = \phi_0$, or $\psi = \phi_1$ (the first point evaluation), or $\psi$ is an inner transform of one of the (other) point evaluations $\phi_2, \phi_3, \phi_4, \ldots$ . Here are the details. The null space of $\psi$ is a maximal ideal in $\ell^{\mathbf{M}}_{\infty,c}$, hence $\operatorname{Ker}\psi = \operatorname{Ker}\phi_n$ for some $n$ among $0, 1, 2, \ldots$ (Step 4). In Step 5 we saw that this guarantees the existence of an invertible matrix $S$, of appropriate size depending on $n$, such that $\psi(M) = S^{-1}\phi_n(M)S$ for all $M \in \ell^{\mathbf{M}}_{\infty,c}$. For $n \geq 2$, the size of $S$ is equal to $n$; when $n = 0$ or $n = 1$, it is equal to one. So in the latter two cases, the matrix $S$ actually is a scalar, and hence $\psi = \phi_0$ or $\psi = \phi_1$ depending on $n$ being equal to zero or one.

*Step* 7. We finally have the ingredients to prove that $\ell^{\mathbf{M}}_{\infty,c}$ does not possess a sufficient family of matrix representations. Suppose it does. Then, by Theorem 3.1

(which is based on Burnside's Theorem), the collection of all surjective matrix representations of $\ell_{\infty,c}^{\mathbf{M}}$ is a sufficient family of matrix representations for $\ell_{\infty,c}^{\mathbf{M}}$ too. The result obtained in the previous step now gives that $\{\phi_0, \phi_1, \phi_2, \ldots\}$ is a sufficient family of representations. However, as we shall see, it is not.

We need to identify an element $G \in \ell_{\infty,c}^{\mathbf{M}}$ such that $G$ is not invertible in $\ell_{\infty,c}^{\mathbf{M}}$ while $\phi_k(G)$ is invertible for all $k$. This is not difficult. Consider $G \in \ell_{\infty,c}^{\mathbf{M}}$ given by $G(n) = I_n + \Pi_n^{\downarrow} K \Pi_n^{\uparrow} + H(n)$, where $K \in \mathcal{K}(\ell_2)$ is given by

$$K \begin{pmatrix} x_1 \\ x_2 \\ x_3 \\ \vdots \end{pmatrix} = - \begin{pmatrix} x_1 \\ 0 \\ 0 \\ \vdots \end{pmatrix},$$

and $H(n)$ is the $n \times n$ matrix with $n^{-1}$ in the (1,1)th position and zeros everywhere else. One verifies without difficulty that $\phi_n(G) = G(n)$ is invertible in $\mathbb{C}^{n\times n}$ for every positive integer $n$. Note that $\phi_0(G) = 1$ is invertible in $\mathbb{C}$ too. Assume that $G$ is an invertible element of $\ell_{\infty,c}^{\mathbf{M}}$. For its inverse $G^{-1}$ we then have

$$G^{-1}(n) = G(n)^{-1}, \qquad n = 1, 2, 3, \ldots .$$

As $G(n)^{-1}$ has $n$ in the (1,1)th position, it follows that $\sup_{n=1,2,3,\ldots} \|G^{-1}(n)\|_n$ is not finite, contrary to how $\ell_{\infty,c}^{\mathbf{M}}$ (or, if one so wants $\ell_{\infty}^{\mathbf{M}}$), has been defined. □

Amplifying on the last step in the above proof: here is an alternative way to see that $G$ is not invertible. Apply the homomorphism $\phi_\infty : \ell_{\infty,c}^{\mathbf{M}} \to \mathcal{B}_{\mathcal{K}}(\ell_2)$ to $G$, having in mind that invertibility in $\mathcal{B}(\ell_2)$ amounts to the same as invertibility in $B_{\mathcal{K}}(\ell_2)$. If $G$ were invertible in $\ell_{\infty,c}^{\mathbf{M}}$, the operator $\phi_\infty(G) : \ell_2 \to \ell_2$ would be invertible, hence bijective. It is clear, however, that $\phi_\infty(G) = I + K$, with $K$ as above, is neither injective not surjective. This reasoning suggests that the family $\phi_\infty, \phi_0, \phi_1, \phi_2, \ldots$ might be a sufficient family for $\ell_{\infty,c}^{\mathbf{M}}$. It is indeed, but the proof is rather technical and will not be given here. Note that $\phi_0$ and $\phi_\omega$ are not independent. Indeed, $\phi_\infty(M) = \phi_0(M)I + K_M$ so that $\phi_\infty(M)$ can only be invertible in $B_{\mathcal{K}}(\ell_2)$ if $\phi_0(M) \neq 0$. Hence the family $\{\phi_\infty, \phi_1, \phi_2, \ldots\}$ is sufficient for $\ell_{\infty,c}^{\mathbf{M}}$. The homomorphisms $\{\phi_1, \phi_2, \ldots\}$ are matrix representations but $\phi_\infty : \ell_{\infty,c}^{\mathbf{M}} \to \mathcal{B}_{\mathcal{K}}(\ell_2)$ is not. As $\mathcal{B}_{\mathcal{K}}(\ell_2)$ is spectrally regular (see [BES94] or [BES04]), the existence of the sufficient family $\{\phi_\infty, \phi_1, \phi_2, \ldots\}$ nevertheless implies that $\ell_{\infty,c}^{\mathbf{M}}$ is spectrally regular (cf., Corollary 3.5 in [BES12]). This result can also be obtained by combining Corollary 4.1 and Theorem 4.6 in [BES12], or by observing that the family $\{\phi_1, \phi_2, \phi_3, \ldots\}$ of point evaluations separates the points of $\ell_{\infty,c}^{\mathbf{M}}$ so that [BES12], Corollary 3.3 can be applied.

Still under the standing assumption $\mathbf{M} = \{\mathbb{C}^{n\times n}\}_{n\in\mathbb{N}}$, let $\ell_{\infty,0}^{\mathbf{M}}$ consist of all $M \in \ell_{\infty}^{\mathbf{M}}$ such that there exist $\lambda \in \mathbb{C}$ and $H \in \mathcal{N}$ such that

$$M(n) = \lambda I_n + H(n), \qquad n = 1, 2, 3, \ldots .$$

In other words $\ell_{\infty,0}^{\mathbf{M}}$ consists of those elements of $\ell_{\infty,c}^{\mathbf{M}}$ for which the corresponding compact operator on $\ell_2$ vanishes. So, with the notation employed above,

$$\{M \in \ell_{\infty,c}^{\mathbf{M}} \mid \phi_\infty(M) - \phi_0(M)I = 0\},$$

hence $\ell_{\infty,0}^{\mathbf{M}}$ is closed in $\ell_{\infty,c}^{\mathbf{M}}$. It is now clear that $\ell_{\infty,0}^{\mathbf{M}}$ is a $C^*$-subalgebra of $\ell_{\infty,c}^{\mathbf{M}}$ (and of course of $\ell_{\infty}^{\mathbf{M}}$ too).

**Proposition 4.2.** *The $C^*$-algebra $\ell_{\infty,0}^{\mathbf{M}}$ possesses a sufficient family of matrix representations, but it is not a polynomial identity algebra.*

*Proof.* The family consisting of the restrictions of $\phi_0, \phi_1, \phi_2, \dots$ to $\ell_{\infty,0}^{\mathbf{M}}$ is sufficient. However, $\ell_{\infty,0}^{\mathbf{M}}$ is not a polynomial identity algebra. This can be seen as follows. Assume it is, so it has a polynomial identity algebra with an annihilating polynomial in a finite number of (noncommuting) variables, $k$ say. This polynomial then clearly also annihilates all algebras $\mathbb{C}^{n\times n}$, $n \in \mathbb{N}$. But then, necessarily (see [Lev] or [Kr], Theorem 20.2), $k$ must be larger than or equal to $2n$ for all $n \in \mathbb{N}$, something which is obviously impossible. □

Proposition 4.2 should be appreciated in light of the fact that a polynomial identity Banach algebra possesses a sufficient family of matrix representations (even of finite order); see Section 22 in [Kr] and also Theorem 3.5 above. We are not aware of another example of this type besides $\ell_{\infty,0}^{\mathbf{M}}$.

## 5. Separating versus weakly sufficient families of matrix representations

Let $\mathbf{M}_\bullet = \{\mathbb{C}^{n\times n}\}_{(n,k)\,\in\,\mathbb{N}_\bullet}$, where $\mathbb{N}_\bullet = \{(n,k) \mid k = 1,\dots,n; n \in \mathbb{N}\}$, and where $\mathbb{C}^{n\times n}$ is identified with the Banach algebra $\mathcal{B}(\mathbb{C}^n)$ of (bounded) linear operators on the Hilbert space $\mathbb{C}^n$, equipped with the standard Hilbert space norm. In particular, the unit element $I_n$ in $\mathbb{C}^{n\times n}$ has norm one. By slight abuse of notation, the norms on $\mathbb{C}^n$ and $\mathbb{C}^{n\times n}$ will both be denoted by $\|\cdot\|_n$. Write $\ell_\infty^{\mathbf{M}_\bullet}$ for the $\ell_\infty$-direct product of the family $\mathbf{M}_\bullet$. Thus $\ell_\infty^{\mathbf{M}_\bullet}$ consists of all $M$ in the Cartesian product $\prod_{(n,k)\,\in\,\mathbb{N}_\bullet} \mathbb{C}^{n\times n}$ such that

$$|||M||| = \sup_{(n,k)\,\in\,\mathbb{N}_\bullet} \|M(n,k)\|_n < \infty.$$

With the operations of scalar multiplication, multiplication, addition, and that of taking the adjoint all defined pointwise, and with $|||\cdot|||$ as norm, $\ell_\infty^{\mathbf{M}_\bullet}$ is a unital Banach algebra – in fact a $C^*$-algebra.

For $n = 1, 2, 3, \dots$, let $\Pi_n^\downarrow : \ell_2 \to \mathbb{C}^n$ and $\Pi_n^\uparrow : \mathbb{C}^n \to \ell_2$ be as in the previous section. Recall that the mapping $\mathbb{C}^{n\times n} \ni A \mapsto \Pi_n^\uparrow A \Pi_n^\downarrow \in \mathcal{B}(\ell_2)$ is a (non-unital) Banach algebra isomorphism from $\mathbb{C}^{n\times n}$ into $\mathcal{B}(\ell_2)$, the latter equipped with the operator norm $\|\cdot\|$ induced by the standard Hilbert space norm on $\ell_2$.

Let $\ell_{\infty,s}^{\mathbf{M}_\bullet}$ consist of all $M \in \ell_\infty^{\mathbf{M}_\bullet}$ such that for each $k \in \mathbb{N}$ the strong limit

$$M_k = \text{s-}\lim_{n\to\infty} \Pi_n^\uparrow M(n,k) \Pi_n^\downarrow \tag{9}$$

exists in $\mathcal{B}(\ell_2)$. Then $\ell_{\infty,s}^{\mathbf{M}_\bullet}$ is closed under scalar multiplication, addition and multiplication. Clearly $\ell_{\infty,s}^{\mathbf{M}_\bullet}$ contains the unit element of $\ell_{\infty}^{\mathbf{M}_\bullet}$. One also checks without difficulty that $\ell_{\infty,s}^{\mathbf{M}_\bullet}$ is a closed subset of $\ell_{\infty}^{\mathbf{M}_\bullet}$. Thus $\ell_{\infty,s}^{\mathbf{M}_\bullet}$ is a Banach subalgebra of $\ell_{\infty}^{\mathbf{M}_\bullet}$ (not closed under the $C^*$-operation, however). Finally, the mapping $\Theta : \ell_{\infty,s}^{\mathbf{M}_\bullet} \ni M \mapsto (M_1, M_2, M_3, \ldots) \in \ell_\infty\big(\mathcal{B}(\ell_2)\big)$ with the $M_k$ given by (9) is a contractive unital homomorphism. Here $\ell_\infty\big(\mathcal{B}(\ell_2)\big)$ is the Banach algebra of bounded sequences in $\mathcal{B}(\ell_2)$.

Let $\mathbb{T}$ and $\mathbb{D}$ stand for the unit circle and the open unit disc in the complex plane, respectively. Further, let $A(\mathbb{D}, \mathcal{B}(\ell_2))$ denote the set of all continuous functions $F : \overline{\mathbb{D}} \to \mathcal{B}(\ell_2)$ which are analytic on $\mathbb{D}$. For the sup-norm $\|\cdot\|_{A(\mathbb{D},\mathcal{B}(\ell_2))}$ on $A(\mathbb{D}, \mathcal{B}(\ell_2))$ we have (by the maximum modulus principle for operator-valued functions)

$$\|F\|_{A(\mathbb{D},\mathcal{B}(\ell_2))} = \sup_{\lambda \in \overline{\mathbb{D}}} \|F(\lambda)\| = \max_{\lambda \in \mathbb{T}} \|F(\lambda)\|.$$

Now fix a countable dense subset $\{\lambda_1, \lambda_2, \lambda_3, \ldots\}$ of $\mathbb{T}$. Then

$$\|F\|_{A(\mathbb{D},\mathcal{B}(\ell_2))} = \sup_{k \in \mathbb{N}} \|F(\lambda_k)\|. \tag{10}$$

Hence, if $F$ and $G$ are in $A(\mathbb{D}, \mathcal{B}(\ell_2))$, then $F = G$ if and only if $F(\lambda_k) = G(\lambda_k)$ for all $k \in \mathbb{N}$.

Let $\ell_{\infty,\bullet}^{\mathbf{M}_\bullet}$ consist of all $M \in \ell_{\infty,s}^{\mathbf{M}_\bullet}$ for which there exists a (necessarily unique) function $F_M \in A(\mathbb{D}, \mathcal{B}(\ell_2))$ such that

$$M_k = \text{s-}\lim_{n\to\infty} \Pi_n^{\uparrow} M(n,k) \Pi_n^{\downarrow} = F_M(\lambda_k), \qquad k \in \mathbb{N}.$$

Then $\ell_{\infty,\bullet}^{\mathbf{M}_\bullet}$ is closed under the operations of scalar multiplication, addition and multiplication. Clearly $\ell_{\infty,\bullet}^{\mathbf{M}_\bullet}$ contains the unit element of $\ell_{\infty,s}^{\mathbf{M}_\bullet}$ (or, what amounts to the same, that of $\ell_{\infty}^{\mathbf{M}_\bullet}$). The function associated with it is the one with constant value the identity operator on $\ell_2$. Define $\Upsilon : A(\mathbb{D}, \mathcal{B}(\ell_2)) \to \ell_\infty\big(\mathcal{B}(\ell_2)\big)$ by $\Upsilon(F) = (F(\lambda_1), F(\lambda_2), F(\lambda_3), \ldots)$. From (10) we see that $\Upsilon$ is a unital Banach algebra isomorphism. Hence the image $\operatorname{Im}\Upsilon$ of $\Upsilon$ is a closed subalgebra of $\ell_\infty\big(\mathcal{B}(\ell_2)\big)$ containing the unit element of $\ell_\infty\big(\mathcal{B}(\ell_2)\big)$. Now note that $\ell_{\infty,\bullet}^{\mathbf{M}_\bullet}$ is the inverse image of $\operatorname{Im}\Upsilon$ under the homomorphism $\Theta : \ell_{\infty,s}^{\mathbf{M}_\bullet} \to \ell_\infty\big(\mathcal{B}(\ell_2)\big)$. As the latter is continuous and $\operatorname{Im}\Upsilon$ is closed in $\ell_\infty\big(\mathcal{B}(\ell_2)\big)$, we arrive at the conclusion that $\ell_{\infty,\bullet}^{\mathbf{M}_\bullet}$ is closed in $\ell_{\infty,s}^{\mathbf{M}_\bullet}$.

The upshot of all of this is that $\ell_{\infty,\bullet}^{\mathbf{M}_\bullet}$ is a Banach subalgebra of $\ell_{\infty,s}^{\mathbf{M}_\bullet}$, hence of $\ell_{\infty}^{\mathbf{M}_\bullet}$ too (not closed under the $C^*$-operation, though). For later use we also observe that the mapping $\ell_{\infty,\bullet}^{\mathbf{M}_\bullet} \ni M \mapsto F_M = \Upsilon^{-1}\big(\Theta(M)\big) \in \mathcal{A}(\mathbb{D}, \mathcal{B}(\ell_2))$ is a contractive unital homomorphism from $\ell_{\infty,\bullet}^{\mathbf{M}_\bullet}$ into $A(\mathbb{D}, \mathcal{B}(\ell_2))$.

**Theorem 5.1.** *The Banach algebra $\ell_{\infty,\bullet}^{\mathbf{M}_\bullet}$ possesses a separating family of matrix representations; it does, however, not possess a weakly sufficient family of matrix representations.*

*Proof.* The proof of the first part of the theorem is easy. Indeed, the existence of a separating family of matrix representations is clear: just consider the point evaluations $\ell_{\infty,\bullet}^{\mathbf{M}_\bullet} \ni M \mapsto M(n,k) \in \mathbb{C}^{n\times n}$, $(n,k) \in \mathbb{N}_\bullet$. For the proof of the fact that $\ell_{\infty,\bullet}^{\mathbf{M}_\bullet}$ does not possess any weakly sufficient family of matrix representations, one needs some 'control' on the collection of all unital matrix representations of the Banach algebra $\ell_{\infty,\bullet}^{\mathbf{M}_\bullet}$ – again, as for the Banach algebra $\ell_\infty^{\mathbf{M}}$ in Theorem 4.1, a non-trivial matter. Another serious complication one has to deal with is the freedom one has in choosing norms on the target (matrix) algebras. The reasoning will be split up into several steps.

*Step* 1. Given a subset $N$ of $\mathbb{N}_\bullet$, put

$$\mathcal{J}^N = \{M \in \ell_{\infty,\bullet}^{\mathbf{M}_\bullet} \mid M(n,k) = 0 \text{ for all } (n,k) \in N\}.$$

Then $\mathcal{J}^N$ is a closed two-sided ideal in $\ell_{\infty,\bullet}^{\mathbf{M}_\bullet}$. Clearly $\mathcal{J}^N \cap \mathcal{J}^{\mathbb{N}_\bullet\setminus N} = \{0\}$. In general it is not true, however, that $\mathcal{J}^N + \mathcal{J}^{\mathbb{N}_\bullet\setminus N} = \ell_{\infty,\bullet}^{\mathbf{M}_\bullet}$. (Indeed, when $N$ consists of the pairs $(n,1)$ with $n \in \mathbb{N}$, the unit element of $\ell_{\infty,\bullet}^{\mathbf{M}_\bullet}$ does not belong to $\mathcal{J}^N + \mathcal{J}^{\mathbb{N}_\bullet\setminus N}$). In case $N$ is finite (the situation to be encountered below), we do have the direct sum decomposition $\ell_{\infty,\bullet}^{\mathbf{M}_\bullet} = \mathcal{J}^N \dot{+} \mathcal{J}^{\mathbb{N}_\bullet\setminus N}$.

*Step* 2. Define $\mathcal{N}$ by

$$\mathcal{N} = \bigcup_{N\subset\mathbb{N}_\bullet,\ N \text{ finite}} \mathcal{J}^{\mathbb{N}_\bullet\setminus N}. \tag{11}$$

Then $\mathcal{N}$ is an ideal in $\ell_{\infty,\bullet}^{\mathbf{M}_\bullet}$ (possibly non-closed). The unit element of $\ell_{\infty,\bullet}^{\mathbf{M}_\bullet}$ does not belong to $\mathcal{N}$. Therefore the quotient algebra $\ell_{\infty,\bullet}^{\mathbf{M}_\bullet}/\mathcal{N}$ is unital. We shall prove that if $p$ is a nonnegative integer and $\psi : \ell_{\infty,\bullet}^{\mathbf{M}_\bullet}/\mathcal{N} \to \mathbb{C}^{p\times p}$ is an algebra homomorphism, not necessarily continuous or unital, then $\psi$ maps all of $\ell_{\infty,\bullet}^{\mathbf{M}_\bullet}/\mathcal{N}$ onto the zero element in $\mathbb{C}^{p\times p}$. (By the way, for $p = 0$ there is nothing to prove.)

Let $\mathcal{C}$ be the $\ell_\infty$-direct product of the $p+1$ matrix algebras

$$\mathbb{C}^{(p+1)\times(p+1)}, \ldots, \mathbb{C}^{(2p+1)\times(2p+1)},$$

and consider the mapping $\phi : \mathcal{C} \ni (A_{p+1}, \ldots, A_{2p+1}) \mapsto M \in \ell_\infty^{\mathbf{M}_\bullet}$ with

$$M(n,k) = \begin{cases} 0_n, & k = 1,\ldots,n,\ n = 1,\ldots,p, \\ \left(\oplus_{s=1}^{s_n-1} A_{p+1}\right) \oplus A_{p+1+t_n}, & k = 1,\ldots,n,\ n = p+1, p+2,\ldots. \end{cases}$$

Here $0_n$ denotes the $n\times n$ zero matrix, and $s_n$ and $t_n$ are the unique nonnegative integers such that $n = s_n(p+1) + t_n$ and $t_n \in \{0,\ldots,p\}$. For $n$ larger than $p$ and $k = 1,\ldots,n$, the matrix $M(n,k)$ is a block diagonal matrix with $s_n$ blocks, the first $s_n - 1$ blocks being copies of $A_{p+1}$, and the last (in the lower right-hand corner) being a copy of $A_{p+1+t_n}$. In particular $M(n,k)$ is an $n \times n$ matrix, as desired. Clearly $\phi$ maps $\mathcal{C}$ indeed into $\ell_\infty^{\mathbf{M}_\bullet}$. In fact $\phi : \mathcal{C} \to \ell_\infty^{\mathbf{M}_\bullet}$ is an isometric algebra homomorphism. Note, though, that it is non-unital.

For $M$ given by the above expression, and $k \in \mathbb{N}$, we have

$$M_k = \text{s-}\lim_{n\to\infty} \Pi_n^{\uparrow} M(n,k) \Pi_n^{\downarrow} = \oplus_{m=1}^{\infty} A_{p+1},$$

where the far right-hand side of this expression stands for the block diagonal bounded linear operator on $\ell_2$ with blocks $A_{p+1}$, independent of $k$. From this we see that $\phi$ maps $\mathcal{C}$ into $\ell_{\infty,s}^{\mathbf{M}_\bullet}$. In fact, it maps $\mathcal{C}$ into $\ell_{\infty,\bullet}^{\mathbf{M}_\bullet}$. Indeed, defining $F_M \in A(\mathbb{D}, \mathcal{B}(\ell_2))$ to be the constant function with value $\oplus_{m=1}^{\infty} A_{p+1} \in \mathcal{B}(\ell_2)$, we have that $M_k = F_M(\lambda_k)$ for all $k \in \mathbb{N}$. Thus we can view $\phi$ as a (non-unital) isometric algebra homomorphism from $\mathcal{C}$ into $\ell_{\infty,\bullet}^{\mathbf{M}_\bullet}$.

For $j = 1, \dots, p+1$, let $\phi_j : \mathbb{C}^{(p+j)\times(p+j)} \ni A \to \widehat{A} \in \mathcal{C}$ be the mapping given by $\widehat{A} = (0_{p+1}, \dots, 0_{p+j-1}, A, 0_{p+j+1}, \dots, 0_{2p+1})$. Then $\phi_j : \mathbb{C}^{(p+j)\times(p+j)} \to \mathcal{C}$ is an algebra isomorphism, again non-unital. By assumption, $\psi : \ell_{\infty,\bullet}^{\mathbf{M}_\bullet}/\mathcal{N} \to \mathbb{C}^{p\times p}$ is an algebra homomorphism, not necessarily continuous or unital. Write $\kappa$ for the canonical mapping from $\ell_{\infty,\bullet}^{\mathbf{M}_\bullet}$ onto $\ell_{\infty,\bullet}^{\mathbf{M}_\bullet}/\mathcal{N}$, and let $\zeta_j$ be the composition of the homomorphisms $\psi$, $\kappa$, $\phi$ and $\phi_j$ as indicated in the scheme

$$\mathbb{C}^{(p+j)\times(p+j)} \xrightarrow{\ \phi_j\ } \mathcal{C} \xrightarrow{\ \phi\ } \ell_{\infty,\bullet}^{\mathbf{M}_\bullet} \xrightarrow{\ \kappa\ } \ell_{\infty,\bullet}^{\mathbf{M}_\bullet}/\mathcal{N} \xrightarrow{\ \psi\ } \mathbb{C}^{p\times p},$$

so $\zeta_j = \psi \circ \kappa \circ \phi \circ \phi_j$. Then $\zeta_j$ is an algebra homomorphism from $\mathbb{C}^{(p+j)\times(p+j)}$ into $\mathbb{C}^{p\times p}$. Clearly it cannot be injective (dimension argument). Hence its kernel is a nontrivial two-sided ideal in $\mathbb{C}^{(p+j)\times(p+j)}$. But this algebra is simple, and so the ideal must be all of $\mathbb{C}^{(p+j)\times(p+j)}$. In other words, the homomorphism $\zeta_j$ is a zero mapping.

We shall now infer that $\psi$ is identically zero too. For this it is sufficient to establish that $\psi$ maps the unit element in $\ell_{\infty,\bullet}^{\mathbf{M}_\bullet}/\mathcal{N}$ onto $0_p$. As the unit element in $\ell_{\infty,\bullet}^{\mathbf{M}_\bullet}/\mathcal{N}$ is $\kappa(e_\bullet)$, were $e_\bullet$ is the unit element in $\ell_{\infty,\bullet}^{\mathbf{M}_\bullet}$, we need to show that $\psi$ maps $\kappa(e_\bullet)$ onto $0_p$. This goes as follows. Clearly $(I_{p+1}, \dots, I_{2p+1}) = \sum_{j=1}^{p+1} \phi_j(I_{p+j})$. Applying $\phi$ we get

$$\phi(I_{p+1}, \dots, I_{2p+1}) = \sum\nolimits_{j=1}^{p+1} (\phi \circ \phi_j)(I_{p+j}).$$

Now, for the element $\phi(I_{p+1}, \dots, I_{2p+1}) \in \ell_{\infty,\bullet}^{\mathbf{M}_\bullet}$ we have

$$\phi(I_{p+1}, \dots, I_{2p+1})(n,k) = I_n, \qquad k = 1, \dots, n,\ n = p+1, p+2, \dots .$$

Hence $e_\bullet - \phi(I_{p+1}, \dots, I_{2p+1}) \in \ell_{\infty,\bullet}^{\mathbf{M}_\bullet}$ satisfies

$$e_\bullet - \phi(I_{p+1}, \dots, I_{2p+1})(n,k) = 0_n, \qquad k = 1, \dots, n,\ n = p+1, p+2, \dots ,$$

and so $e_\bullet - \phi(I_{p+1}, \dots, I_{2p+1}) \in \mathcal{J}^{\mathbb{N}_\bullet \setminus N}$ where $N$ is the set of all pairs of positive integers $(s,t)$ with $t \le s \le n$. As $N$ is a finite subset of $\mathbb{N}_\bullet$, we may conclude that $e_\bullet - \phi(I_{p+1}, \dots, I_{2p+1}) \in \mathcal{N}$. But then

$$\kappa(e_\bullet) = \kappa\big(\phi(I_{p+1}, \dots, I_{2p+1})\big) = \sum\nolimits_{j=1}^{p+1} (\kappa \circ \phi \circ \phi_j)(I_{p+j}).$$

It follows that $\psi\big(\kappa(e_\bullet)\big) = \sum_{j=1}^{p+1} \zeta_j(I_{p+j}) = \sum_{j=1}^{p+1} 0_p = 0_p$, as desired.

*Step* 3. We shall now prove that all two-sided ideals of $\ell_{\infty,\bullet}^{\mathbf{M}_\bullet}$ having finite codimension in $\ell_{\infty,\bullet}^{\mathbf{M}_\bullet}$ are of the form $\mathcal{J}^N$ with $N$ a finite subset of $\mathbb{N}_\bullet$. (From this it follows that such ideals are necessarily closed; this is not assumed a priori, however.)

For $(n_0, k_0) \in \mathbb{N}_\bullet$, put $\mathcal{P}^{(n_0,k_0)} = \mathcal{J}^{\mathbb{N}_\bullet \setminus \{(n_0,k_0)\}}$. Thus $\mathcal{P}^{(n_0,k_0)}$ is the closed two-sided ideal in $\ell_{\infty,\bullet}^{\mathbf{M}_\bullet}$ consisting of all $M \in \ell_{\infty,\bullet}^{\mathbf{M}_\bullet}$ such that $M(n,k) = 0_n$ for all $(n,k) \in \mathbb{N}_\bullet$ different from $(n_0,k_0)$. Clearly $\dim \mathcal{P}^{(n_0,k_0)} = n_0^2$.

Next suppose $N$ is a finite subset of $\mathbb{N}_\bullet$. Then the closed two-sided ideal $\mathcal{J}^{\mathbb{N}_\bullet \setminus N}$ is the direct sum of the ideals $\mathcal{P}^{(n,k)}$ with $(n,k) \in N$. Hence

$$\dim \mathcal{J}^{\mathbb{N}_\bullet \setminus N} = \sum_{(n,k)\in N} n^2 < \infty.$$

Also, as is easily verified, $\ell_{\infty,\bullet}^{\mathbf{M}_\bullet} = \mathcal{J}^N \dot{+} \mathcal{J}^{\mathbb{N}_\bullet \setminus N}$ and $\operatorname{codim} \mathcal{J}^N = \dim \mathcal{J}^{\mathbb{N}_\bullet \setminus N} < \infty$.

Assume $\mathcal{J}$ is a two-sided ideal in $\ell_{\infty,\bullet}^{\mathbf{M}_\bullet}$ having finite codimension in $\ell_{\infty,\bullet}^{\mathbf{M}_\bullet}$, and write $N$ for the set of all $(n,k) \in \mathbb{N}_\bullet$ such that $M(n,k) = 0_n$ for all $M \in \mathcal{J}$. Then $\mathcal{J} \subset \mathcal{J}^N$ and $\mathcal{J}^N$ must have finite codimension in $\ell_{\infty,\bullet}^{\mathbf{M}_\bullet}$ not exceeding that of $\mathcal{J}$. Now $\mathcal{J}^N \cap \mathcal{J}^{\mathbb{N}_\bullet \setminus N} = \{0\}$, and we may conclude that $\dim \mathcal{J}^{\mathbb{N}_\bullet \setminus N} \le \operatorname{codim} \mathcal{J}$. For a finite subset $L$ of $N$, we have that the direct sum of all $\mathcal{P}^{(n,k)}$ with $(n,k) \in L$ is contained in $\mathcal{J}^{\mathbb{N}_\bullet \setminus N}$. Hence, for all finite subsets $L$ of $N$,

$$\sharp(L) \le \sum_{(n,k)\in L} n^2 \le \dim \mathcal{J}^{\mathbb{N}_\bullet \setminus N} \le \operatorname{codim} \mathcal{J},$$

where $\sharp(L)$ stands for the (finite) cardinality of $L$. But this can only be true when $N$ itself is finite. It remains to show that $\mathcal{J}^N = \mathcal{J}$.

As an auxiliary item, we first prove that $\mathcal{P}^{(n,k)} \subset \mathcal{J}$ whenever $(n,k) \in \mathbb{N}_\bullet \setminus N$. The argument is as follows. Take $(n,k) \in \mathbb{N}_\bullet \setminus N$, and consider the set

$$\mathcal{J}_{(n,k)} = \{A \in \mathbb{C}^{n\times n} \mid A = M(n,k) \text{ for some } M \in \mathcal{J}\}.$$

Then $\mathcal{J}_{(n,k)}$ is a two-sided ideal in $\mathbb{C}^{n\times n}$. Since $(n,k) \notin N$, there exists $H \in \mathcal{J}$ such that $H(n,k) \neq 0_n$. Now $H(n,k) \in \mathcal{J}_{(n,k)}$, and so $\mathcal{J}_{(n,k)} \neq \{0_n\}$. As $\mathbb{C}^{n\times n}$ is a simple algebra, it follows that $\mathcal{J}_{(n,k)} = \mathbb{C}^{n\times n}$. In other words, for each $A \in \mathbb{C}^{n\times n}$ there exists $M \in \mathcal{J}$ such that $M(n,k) = A$. Take $P \in \mathcal{P}^{(n,k)}$, and choose $\widehat{P} \in \mathcal{J}$ such that $\widehat{P}(n,k) = P(n,k)$. Let $H^{(n,k)}$ be the element in $\ell_\infty^{\mathbf{M}_\bullet}$ with the $n\times n$ identity matrix $I_n$ on the $(n,k)$th position and the appropriate zero matrix everywhere else. Clearly $H^{(n,k)} \in \ell_{\infty,\bullet}^{\mathbf{M}_\bullet}$, and so $\widehat{P}H^{(n,k)} \in \mathcal{J}$. As $P = \widehat{P}H^{(n,k)}$, we may conclude that $P \in \mathcal{J}$.

Introduce $\widehat{\mathcal{J}} = \mathcal{J} + \mathcal{J}^{\mathbb{N}_\bullet \setminus N}$. Then $\widehat{\mathcal{J}}$ is a two-sided ideal in $\ell_{\infty,\bullet}^{\mathbf{M}_\bullet}$ with finite codimension. Also $\mathcal{P}^{(n,k)} \subset \widehat{\mathcal{J}}$ for all $(n,k) \in \mathbb{N}_\bullet$. For $(n,k) \in \mathbb{N}_\bullet \setminus N$ this is immediate from what was established in the previous paragraph; in case $(n,k) \in N$ we have $\mathcal{P}^{(n,k)} \subset \mathcal{J}^{\mathbb{N}_\bullet \setminus N}$. With $\mathcal{N}$ as in (11), it follows that $\mathcal{N} \subset \widehat{\mathcal{J}}$. In the next paragraph, this will be used to prove that $\ell_{\infty,\bullet}^{\mathbf{M}_\bullet} = \widehat{\mathcal{J}} = \mathcal{J} + \mathcal{J}^{\mathbb{N}_\bullet \setminus N}$. Taking this for granted, we have $\ell_{\infty,\bullet}^{\mathbf{M}_\bullet} = \mathcal{J} + \mathcal{J}^{\mathbb{N}_\bullet \setminus N}$. But then, using the direct sum decomposition $\ell_{\infty,\bullet}^{\mathbf{M}_\bullet} = \mathcal{J}^N \dot{+} \mathcal{J}^{\mathbb{N}_\bullet \setminus N}$ (valid because $N$ is finite; see Step 1) and the inclusion $\mathcal{J} \subset \mathcal{J}^N$, one immediately obtains $\mathcal{J} = \mathcal{J}^N$, as desired.

We finish this third step in the proof by establishing the equality of $\widehat{\mathcal{J}}$ and $\ell^{\mathbf{M}_\bullet}_{\infty,\bullet}$. As $\mathcal{N} \subset \widehat{\mathcal{J}}$, the canonical mapping from $\ell^{\mathbf{M}_\bullet}_{\infty,\bullet}$ onto $\ell^{\mathbf{M}_\bullet}_{\infty,\bullet}/\widehat{\mathcal{J}}$ induces an algebra homomorphism $\psi_0$ from $\ell^{\mathbf{M}_\bullet}_{\infty,\bullet}/\mathcal{N}$ onto $\ell^{\mathbf{M}_\bullet}_{\infty,\bullet}/\widehat{\mathcal{J}}$. The latter algebra has finite dimension, $p$ say, and can (via the use of left regular representations) be identified with a subalgebra of $\mathbb{C}^{p\times p}$. But then $\psi_0$ can be viewed as an algebra homomorphism $\psi$ from $\ell^{\mathbf{M}_\bullet}_{\infty,\bullet}/\mathcal{N}$ into $\mathbb{C}^{p\times p}$. From the second step of the proof we know that $\psi$ maps all of $\ell^{\mathbf{M}_\bullet}_{\infty,\bullet}/\mathcal{N}$ onto $0_p$. But then $\psi_0$ maps all of $\ell^{\mathbf{M}_\bullet}_{\infty,\bullet}/\mathcal{N}$ onto the zero element of $\ell^{\mathbf{M}_\bullet}_{\infty,\bullet}/\widehat{\mathcal{J}}$. On the other hand $\psi_0 : \ell^{\mathbf{M}_\bullet}_{\infty,\bullet}/\mathcal{N} \to \ell^{\mathbf{M}_\bullet}_{\infty,\bullet}/\widehat{\mathcal{J}}$ is surjective. We conclude that $\dim \ell^{\mathbf{M}_\bullet}_{\infty,\bullet}/\widehat{\mathcal{J}} = 0$ which can be rewritten as $\widehat{\mathcal{J}} = \ell^{\mathbf{M}_\bullet}_{\infty,\bullet}$.

*Step* 4. Here we prove that $\ell^{\mathbf{M}_\bullet}_{\infty,\bullet}$ does not have any weakly sufficient family of matrix representations. So, assuming that a family $\{\phi_\omega : \ell^{\mathbf{M}_\bullet}_{\infty,\bullet} \to \mathbb{C}^{n_\omega\times n_\omega}\}_{\omega\in\Omega}$ of matrix representations is given, we shall demonstrate that it fails to be weakly sufficient, and that this is the case – we emphasize – regardless of the choice of the (submultiplicative) norms on the target algebras. The norm on the target algebra $\mathbb{C}^{n_\omega\times n_\omega}$ will be denoted by $|||\cdot|||_\omega$, and the same notation is used for the norm of $\phi_\omega : \ell^{\mathbf{M}_\bullet}_{\infty,\bullet} \to \mathbb{C}^{n_\omega\times n_\omega}$ (seen as a bounded linear operator) induced by the norms $|||\cdot|||$ on $\ell^{\mathbf{M}_\bullet}_{\infty,\bullet}$ and $|||\cdot|||_\omega$ on $\mathbb{C}^{n_\omega\times n_\omega}$. By the first part of Theorem 2.1, weak sufficiency amounts to the combination of p.w. sufficiency and norm-boundedness. So the family $\{\phi_\omega\}_{\omega\in\Omega}$ cannot be weakly sufficient if it is not norm-bounded. Therefore we shall assume that $\sup_{\omega\in\Omega} |||\phi_\omega|||_\omega < \infty$, and prove that this is not compatible with $\{\phi_\omega\}_{\omega\in\Omega}$ being p.w. sufficient.

Let $Y \in \ell^{\mathbf{M}_\bullet}_{\infty}$ be given by $Y(n,k) = \lambda_k I_n$, $(n,k) \in \mathbb{N}_\bullet$. Here $\lambda_1, \lambda_2, \lambda_3 \ldots$ are elements of $\mathbb{T}$ as introduced in the fourth paragraph of this section. Clearly,

$$\text{s-}\lim_{n\to\infty} \Pi_n^{\uparrow} Y(n,k) \Pi_n^{\downarrow} = \lambda_k I, \qquad k = 1,2,3,\ldots,$$

where $I$ is the identity operator on $\ell_2$. So $Y$ belongs to $\ell^{\mathbf{M}_\bullet}_{\infty,\bullet}$, and the function $F_Y \in \mathcal{A}(\mathbb{D}, \mathcal{B}(\ell_2))$ is given by $F_Y(\lambda) = \lambda I$. As $F_Y(0) = 0$, the function $F_Y$ is not an invertible element of $\mathcal{A}(\mathbb{D}, \mathcal{B}(\ell_2))$. But then $Y$ is not invertible in $\ell^{\mathbf{M}_\bullet}_{\infty,\bullet}$.

Let us now see what happens to $Y$ under the application of the matrix representations $\phi_\omega$. Take $\omega \in \Omega$. Then the null space of $\phi_\omega$ is a two-sided ideal in $\ell^{\mathbf{M}_\bullet}_{\infty,\bullet}$ having finite codimension. Therefore (as we have seen in the previous step) it is of the form $\mathcal{J}^{N_\omega}$ with $N_\omega$ a finite subset of $\mathbb{N}_\bullet$. With the help of this set $N_\omega$, we define $X^{(\omega)} \in \ell^{\mathbf{M}_\bullet}_{\infty}$ by

$$X^{(\omega)}(n,k) = \begin{cases} \lambda_k^{-1} I_n, & (n,k) \in N_\omega, \\ I_n, & (n,k) \notin N_\omega. \end{cases}$$

Since $N_\omega$ is finite, we actually have that $X^{(\omega)}$ belongs to $\ell^{\mathbf{M}_\bullet}_{\infty,\bullet}$ with corresponding function $F_{X^{(\omega)}}$ identically equal to $I$. Put $Z^{(\omega)} = YX^{(\omega)} - E$, where $E$ is the unit element in $\ell^{\mathbf{M}_\bullet}_{\infty,\bullet}$. Then $Z^{(\omega)}(n,k) = 0_n$ for all $(n,k) \in N_\omega$, and so $Z^{(\omega)} \in \mathcal{J}^{N_\omega}$. The latter is equal to the null space of $\phi_\omega$, and so $\phi_\omega(Y)\phi_\omega(X^{(\omega)}) = \phi_\omega(E) = e_\omega$, where the latter stands for the $n_\omega \times n_\omega$ identity matrix. Likewise $\phi_\omega(X^{(\omega)})\phi_\omega(Y) = e_\omega$.

We conclude that for each $\omega \in \Omega$, the image $\phi_\omega(Y)$ of $Y$ under $\phi_\omega$ is invertible in $\mathbb{C}^{n_\omega \times n_\omega}$ with inverse $\phi_\omega(X^{(\omega)})$. Now note that $|||X^{(\omega)}||| = 1$. Therefore, with $K = \sup_{\omega\in\Omega} |||\phi_\omega|||_\omega$ (assumed to be finite), we have

$$|||\phi_\omega(Y)^{-1}|||_\omega = |||\phi_\omega(X^{(\omega)})|||_\omega \le |||\phi_\omega|||_\omega \cdot |||X^{(\omega)}||| \le K,$$

so $\sup_{\omega\in\Omega} |||\phi_\omega(Y)^{-1}|||_\omega < \infty$. This, together with the non-invertibility of $Y$ proved above, gives that the family $\{\phi_\omega : \ell^{\mathbf{M}_\bullet}_{\infty,\bullet} \to \mathbb{C}^{n_\omega \times n_\omega}\}_{\omega\in\Omega}$ is not partially weakly sufficient. □

We close this section (and the paper) with three comments. The first is that combining Theorems 3.1 and 5.1, one sees that a Banach algebra may possess a separating family of matrix representations, while not having a sufficient family of matrix representations. In this sense, part of Theorem 4.1 can also be obtained from Theorem 5.1. However, the Banach algebra in the latter is not a $C^*$-algebra whereas the one in Theorem 4.1 is (and, in addition, possesses a weakly sufficient family of matrix representations). We do not know whether there is a $C^*$-algebra which can replace $\ell^{\mathbf{M}_\bullet}_{\infty,\bullet}$ in Theorem 5.1. If one restricts to a full $C^*$-setting by requiring the families of matrix representations to be of $C^*$-type too, there is none (see Theorem 2.4).

The second remark is that a Banach subalgebra of a Banach algebra with a weakly sufficient family of matrix representations need not have a such a family. This is clear from Theorem 4.1 upon noting that $\ell^{\mathbf{M}_\bullet}_{\infty,\bullet}$ is a Banach subalgebra of $\ell^{\mathbf{M}_\bullet}_{\infty}$ which has the point evaluations $\ell^{\mathbf{M}_\bullet}_{\infty} \ni M \mapsto M(n,k) \in \mathbb{C}^{n\times n}$, $(n,k) \in \mathbb{N}_\bullet$ as a weakly sufficient family. Of course an inverse closed Banach subalgebra of a Banach algebra possessing a weakly sufficient family of matrix representations has such a family too: just restrict the matrix representations to the subalgebra.

The third comment concerns the family of point evaluations on $\ell^{\mathbf{M}_\bullet}_{\infty,\bullet}$ featuring in the first paragraph of the proof of Theorem 5.1. As has been mentioned there, this family, which we will denote by $\Xi$, is separating. Obviously $\Xi$ is also norm-bounded. Combining Theorems 2.1, 5.1 and 3.1, one sees that $\Xi$ is neither partially weakly sufficient, nor weakly sufficient, nor sufficient. Thus the family $\Xi$ provides a counterexample to the converse of Corollary 2.3. It also exhibits that on the level of individual families, the property of being (radical-)separating does not imply that of being p.w. separating (cf., Theorem 3.2).

**Acknowledgment**

The second author (T.E.) was supported in part by NSF grant DMS-0901434.

## References

[AL] S.A. Amitsur, J. Levitzky, Minimal identities for algebras, *Proc. Amer. Math. Soc.* **1** (1950), 449–463.

[BES94] H. Bart, T. Ehrhardt, B. Silbermann, Logarithmic residues in Banach algebras, *Integral Equations and Operator Theory* **19** (1994), 135–152.

[BES04] H. Bart, T. Ehrhardt, B. Silbermann, Logarithmic residues in the Banach algebra generated by the compact operators and the identity, *Mathematische Nachrichten* **268** (2004), 3–30.

[BES12] H. Bart, T. Ehrhardt, B. Silbermann, Spectral regularity of Banach algebras and non-commutative Gelfand theory, in: Dym et al. (Eds.), The Israel Gohberg Memorial Volume, Operator Theory: Advances and Applications, Vol. 218, Birkhäuser 2012, 123–154.

[GM] W.T. Gowers, B. Maurey, The unconditional basic sequence problem, *Journal A.M.S.* **6** (1993), 851–874.

[HRS95] R. Hagen, S. Roch, B. Silbermann, *Spectral Theory of Approximation Methods for Convolution Equations*, Operator Theory: Advances and Applications, Vol. 74, Birkhäuser Verlag, Basel 1995.

[HRS01] R. Hagen, S. Roch, B. Silbermann, *$C^*$-Algebras and Numerical Analysis*, Marcel Dekker, New York, Basel 2001.

[Kr] N.Ya. Krupnik, *Banach Algebras with Symbol and Singular Integral Operators*, Operator Theory: Advances and Applications, Vol. 26, Birkhäuser Verlag, Basel 1987.

[Lev] J. Levitzky, A theorem on polynomial identities, *Proc. Amer. Math. Soc.* **1** (1950), 334–341.

[LR] V. Lomonosov, P. Rosenthal, The simplest proof of Burnside's Theorem on matrix algebras, *Linear Algebra Appl.* **383** (2004), 45–47.

[P] T.W. Palmer, *Banach Algebras and The General Theory of *-Algebras, Volume I: Algebras and Banach Algebras*, Cambridge University Press, Cambridge 1994.

[RSS] S. Roch, P.A. Santos, B. Silbermann, *Non-commutative Gelfand Theories*, Springer Verlag, London Dordrecht, Heidelberg, New York 2011.

[Se] P. Šemrl, Maps on matrix spaces, *Linear Algebra Appl.* **413** (2006), 364–393.

[Si] B. Silbermann, Symbol constructions and numerical analysis, In: *Integral Equations and Inverse Problems* (R. Lazarov, V. Petkov, eds.), Pitman Research Notes in Mathematics, Vol. 235, 1991, 241–252.

Harm Bart
Econometric Institute, Erasmus University Rotterdam
P.O. Box 1738, NL-3000 DR Rotterdam, The Netherlands
e-mail: bart@ese.eur.nl

Torsten Ehrhardt
Mathematics Department, University of California
Santa Cruz, CA 95064, USA
e-mail: tehrhard@ucsc.edu

Bernd Silbermann
Fakultät für Mathematik, Technische Universität Chemnitz
D-09107 Chemnitz, Germany
e-mail: silbermn.toeplitz@googlemail.com

Operator Theory:
Advances and Applications, Vol. 221, 161-171


# Operator Splitting with Spatial-temporal Discretization

András Bátkai, Petra Csomós, Bálint Farkas and Gregor Nickel

**Abstract.** Continuing earlier investigations, we analyze the convergence of operator splitting procedures combined with spatial discretization and rational approximations.

**Mathematics Subject Classification (2000).** 47D06, 47N40, 65J10.

**Keywords.** Operator splitting, spatial-temporal approximation, rational approximation, A-stable.

## 1. Introduction

Operator splitting procedures are special finite difference methods one uses to solve partial differential equations numerically. They are certain time-discretization methods which simplify or even make the numerical treatment of differential equations possible.

The idea is the following. Usually, a certain physical phenomenon is the combined effect of several processes. The behaviour of a physical quantity is described by a partial differential equation in which the local time derivative depends on the sum of the sub-operators corresponding to the different processes. These sub-operators are usually of different nature. For each sub-problem corresponding to each sub-operator there might be a fast numerical method providing accurate solutions. For the sum of these sub-operators, however, we usually cannot find an adequate method. Hence, application of operator splitting procedures means that instead of the sum we treat the sub-operators separately. The solution of the original problem is then obtained from the numerical solutions of the sub-problems. For a more detailed introduction and further references, see the monographs by Hairer et al. [8], Faragó and Havasi [6], Holden et al. [9], Hunsdorfer and Verwer [10].

Research partially supported by the Alexander von Humboldt-Stiftung.

Since operator splittings are time-discratization methods, the analysis of their convergence plays an important role. In our earlier investigations in Bátkai, Csomós, Nickel [2] we achieved theoretical convergence analysis of problems when operator splittings were applied together with some spatial approximation scheme. In the present paper we additionally treat temporal discretization methods of special form. Since rational approximations often occur in practice (consider, e.g., Euler and Runge–Kutta methods, or any linear multistep method), we will concentrate on them. Let us start by setting the abstract stage.

**Assumption 1.1.** Suppose that $X$ is a Banach space, $A$ and $B$ are closed, densely defined linear operators generating the strongly continuous operator semigroups $(T(t))_{t\geq 0}$ and $(S(t))_{t\geq 0}$, respectively. Further, we suppose that the closure $\overline{A+B}$ of $A+B$ with $D(\overline{A+B}) \supset D(A)\cap D(B)$ is also the generator of a strongly continuous semigroup $(U(t))_{t\geq 0}$.

For the terminology and notations about strongly continuous operator semigroups see the monographs by Arendt et al. [1] or Engel and Nagel [5]. Then we consider the following abstract Cauchy problem

$$\begin{cases} \dfrac{\mathrm{d}u(t)}{\mathrm{d}t} = (A+B)u(t), & t\geq 0,\\ u(0) = x\in X. \end{cases} \tag{1.1}$$

For the different splitting procedures the *exact split* solution of problem (1.1) at time $t\geq 0$ and for $n$ steps is given by

$u_n^{\mathrm{sq}}(t) := [S(t/n)T(t/n)]^n x$ (sequential),

$u_n^{\mathrm{St}}(t) := [T(t/2n)S(t/n)T(t/2n)]^n x$ (Strang),

$u_n^{\mathrm{w}}(t) := [\Theta S(t/n)T(t/n) + (1-\Theta)T(t/n)S(t/n)]^n x$ with $\Theta\in(0,1)$ (weighted).

However, in practice, we obtain the numerical solution of the problem (1.1) by

* applying a *splitting procedure* with operator $A$ and $B$,
* defining a mesh on which the split problems should be *discretized in space*, and
* using a certain *temporal approximation* to solve these (semi-)discretized equations.

Thus, the properties of this complex numerical scheme should be investigated. In order to work in an abstract framework, we introduce the following spaces and operators, see Ito and Kappel [11, Chapter 4].

**Assumption 1.2.** Let $X_m$, $m\in\mathbb{N}$ be Banach spaces and take operators

$$P_m : X\to X_m \qquad \text{and} \qquad J_m : X_m\to X$$

having the following properties:

(i) $P_mJ_m = I_m$ for all $m\in\mathbb{N}$, where $I_m$ is the identity operator in $X_m$,

(ii) $\lim\limits_{m\to\infty} J_mP_m x = x$ for all $x\in X$,

(iii) $\|J_m\|\leq K$ and $\|P_m\|\leq K$ for all $m\in\mathbb{N}$ and some given constant $K>0$.

The operators $P_m$, $m \in \mathbb{N}$ are usually some projections onto the spatial "mesh" $X_m$, while the operators $J_m$ correspond to the interpolation method resulting the solution in the space $X$, but also Fourier–Galerkin methods fit in this framework.

Let us recall the following definitions and results from [2]. First we assume that the exact solution $u$ of problem (1.1) is obtained by using only a splitting procedure and discretization in space. Although operators $A$ and $B$ were required to be generators in Assumptions 1.1 we formulate the following Assumption more general.

**Assumption 1.3.** Let $(A_m, D(A_m))$ and $(B_m, D(B_m))$, $m \in \mathbb{N}$, be operators on $X_m$ and let $(A, D(A))$ and $(B, D(B))$ be operators on $X$, that satisfy the following:

(i) *Stability:*
Suppose that there exist a constant $M \geq 0$ such that
a) $\|(\operatorname{Re}\lambda)R(\lambda, A)\| \leq M$ and $\|(\operatorname{Re}\lambda)R(\lambda, A_m)\| \leq M$,
b) $\|(\operatorname{Re}\lambda)R(\lambda, B)\| \leq M$ and $\|(\operatorname{Re}\lambda)R(\lambda, B_m)\| \leq M$
for all $\operatorname{Re}\lambda > 0$, $m \in \mathbb{N}$.

(ii) *Consistency:*
Suppose that $P_m D(A) \subset D(A_m)$, $P_m D(B) \subset D(B_m)$, and
a) $\lim_{m\to\infty} J_m A_m P_m x = Ax \qquad$ for all $x \in D(A)$,
b) $\lim_{m\to\infty} J_m B_m P_m x = Bx \qquad$ for all $x \in D(B)$.

*Remark* 1.4.

1. If Assumption 1.3 is satisfied for $M = 1$, then by the Hille–Yosida Theorem $A, B, A_m, B_m$ are all generators of contraction semigroups $(T(t))_{t\geq 0}$, $(S(t))_{t\geq 0}$, $(T_m(t))_{t\geq 0}$, $(S_m(t))_{t>0}$. Furthermore, from the Trotter–Kato Approximation Theorem (see Ito and Kappel [12, Theorem 2.1]) it follows that the approximating semigroups converge to the original semigroups locally uniformly, that is:
*Convergence:*
a) $\lim_{m\to\infty} J_m T_m(h) P_m x = T(h)x \qquad \forall x \in X$ and uniformly for $h \in [0, t_0]$,
b) $\lim_{m\to\infty} J_m S_m(h) P_m x = S(h)x \qquad \forall x \in X$ and uniformly for $h \in [0, t_0]$
for any $t_0 \geq 0$.
2. In turn, the resolvent estimates are satisfied if $A, B, A_m, B_m$ are all generators of bounded semigroups, with the same bound for all $m \in \mathbb{N}$. One may even assume that these semigroups have the same exponential estimate, that is
$$\|T(t)\|,\ \|T_m(t)\|,\ \|S(t)\|, \|S_m(t)\| \leq M\mathrm{e}^{\omega t} \quad \text{for all } t \geq 0.$$
In the following this would result in a simple rescaling that we want to spare for the sake of brevity.

In order to prove the convergence of the *splitting* procedures in this case, we need to formulate a modified version of Chernoff's Theorem being valid also for the spatial discretizations. Our main technical tool will be the following theorem, whose proof can be carried out along the same lines as Theorem 3.12 in [2]. Let us agree on the following terminology. We say that for a sequence $a_{m,n}$ the limit

$$\lim_{m,n\to\infty} a_{m,n} =: a$$

exists if for all $\varepsilon > 0$ there exists $N \in \mathbb{N}$ such that for all $n, m \geq N$ we have $\|a_{m,n} - a\| \leq \varepsilon$.

**Theorem 1.5 (Modified Chernoff–Theorem, [2, Theorem 3.12]).** *Consider a sequence of functions $F_m : \mathbb{R}^+ \to \mathscr{L}(X_m)$, $m \in \mathbb{N}$, satisfying*

$$F_m(0) = I_m \qquad \text{for all } m \in \mathbb{N}, \tag{1.2}$$

*(with $I_m$ being the identity on $X_m$), and that there exist constants $M \geq 1$, $\omega \in \mathbb{R}$, such that*

$$\|[F_m(t)]^k\|_{\mathscr{L}(X_m)} \leq M\mathrm{e}^{k\omega t} \qquad \text{for all } t \geq 0,\ m, k \in \mathbb{N}. \tag{1.3}$$

*Suppose further that*

$$\exists \lim_{h\to 0} \frac{J_m F_m(h) P_m x - J_m P_m x}{h} \tag{1.4}$$

*uniformly in $m \in \mathbb{N}$, and that there is a dense subspace $D \subseteq X$ such that $(\lambda_0 - G)D$ too is dense for some $\lambda_0 > \omega$, and*

$$Gx := \lim_{m\to\infty} \lim_{h\to 0} \frac{J_m F_m(h) P_m x - J_m P_m x}{h} \tag{1.5}$$

*exists for all $x \in D$. Then the closure $\overline{G}$ of $G$ generates a strongly continuous semigroup $(U(t))_{t\geq 0}$, which is given by*

$$U(t)x = \lim_{m,n\to\infty} J_m [F_m(\tfrac{t}{n})]^n P_m x$$

*for all $x \in X$ uniformly for $t$ in compact intervals.*

## 2. Rational approximations

Our aim is to show the convergence of various splitting methods when combined with both spatial and temporal disretizations. As temporal discretizations we consider finite difference methods, or more precisely, rational approximations of the exponential function. Throughout this section, we suppose that $r$ and $q$ will be rational functions **approximating the exponential function** at least of order one, that is we suppose

$$r(0) = r'(0) = 1 \quad \text{and} \quad q(0) = q'(0) = 1.$$

Further, we suppose that these functions are bounded on the closed left half-plane

$$\mathbb{C}_- := \{z \in \mathbb{C} : \operatorname{Re} z \leq 0\}.$$

Rational (e.g., A-stable) functions typically appearing in numerical analysis satisfy these conditions. An important consequence of the boundedness in the closed left half-plane is that the poles of $r$ have strictly positive real part, and thus lie in some sector

$$\Sigma_\theta := \{z : z \in \mathbb{C},\ |\arg(z)| < \theta\}$$

of opening half-angle $\theta \in [0, \frac{\pi}{2})$.

It is clear that for an application of the Modified Chernoff Theorem 1.5 uniform convergence (w.r.t. $m$ or $h$) plays a crucial role here (cf. [2]). Hence, the following lemma will be the main technical tool in our investigations. For a rational function

$$r(z) = \frac{(\lambda_1 - z)(\lambda_2 - z)\cdots(\lambda_k - z)}{(\mu_1 - z)(\mu_2 - z)\cdots(\mu_n - z)}$$

the function of an operator $A$ is defined by

$$r(A) = (\lambda_1 - A)(\lambda_2 - A)\cdots(\lambda_k - A)(\mu_1 - A)^{-1}(\mu_2 - A)^{-1}\cdots(\mu_n - A)^{-1}$$

(where $\mu_i \in \rho(A)$). See Haase [7] for further explanations, generalizations, and applications to rational approximation schemes.

**Lemma 2.1.** *Let $A, A_m, P_m, J_m$ be as in Assumptions 1.2 and 1.3. Let $r$ be a rational approximation of the exponential being bounded on the closed left half-plane $\mathbb{C}_-$.*

*Then we have the following.*

a) *There is an $M \geq 0$ such that $\|r(hA_m)\| \leq M$ for all $h \geq 0$, $m \in \mathbb{N}$.*
b) *For all $x \in D(A)$ we have*

$$\left\| \frac{J_m r(hA_m)P_m x - J_m P_m x}{h} - J_m A_m P_m x \right\| \to 0 \tag{2.1}$$

*uniformly in $m \in \mathbb{N}$ for $h \to 0$.*

The proof of this lemma is postponed to the end of this section. With its help, however, one can prove the next results: (1) on convergence of spatial-temporal discretization without splitting, (2) on convergence of the splitting procedures combined with spatial and temporal approximations.

**Theorem 2.2.** *Let $A, A_m, P_m, J_m$ be as in Assumptions 1.2 and 1.3 and let $A$ generate the semigroup $(T(t))_{t\geq 0}$. Suppose that $r$ is a rational function approximating the exponential function bounded on $\mathbb{C}_-$, and that there exist constants $M \geq 1$ and $\omega \in \mathbb{R}$ with*

$$\|[r(hA_m)]^k\| \leq M e^{k\omega h} \quad \text{for all } h \geq 0,\ k, m \in \mathbb{N}. \tag{2.2}$$

*Then*

$$\lim_{m,n\to\infty} J_m r(\tfrac{t}{n}A_m)^n P_m x = T(t)x,$$

*uniformly for $t \geq 0$ in compact intervals.*

*Proof.* We will apply Theorem 1.5 with $F_m(h) := r(hA_m)$. The stability criteria (1.2)–(1.3) follow directly from $r(0) = 1$ and assumption (2.2). For the consistency (1.5) we have to show the existence of the limit in (1.4) uniformly in $m \in \mathbb{N}$. But this is exactly the statement of Lemma 2.1. □

Here is the announced theorem on the convergence of the sequential splitting with spatial and rational temporal discretization.

**Theorem 2.3.** *Let $A, B, A_m, B_m, P_m, J_m$ be as in Assumption* 1.1*, Assumptions* 1.2 *and* 1.3*, and let $(U(t))_{t\geq 0}$ denote the semigroup generated by the closure of $A+B$. Suppose that the following stability condition is satisfied:*

$$\left\|[q(hB_m)r(hA_m)]^k\right\| \leq M\mathrm{e}^{kh\omega} \quad \textit{for all } h \geq 0,\ k, m \in \mathbb{N}.$$

*Then the sequential splitting is convergent, i.e.,*

$$\lim_{m,n\to\infty} [q(\tfrac{t}{n}B_m)r(\tfrac{t}{n}A_m)]^n x = U(t)x$$

*for all $x \in X$ uniformly for $t$ in compact intervals.*

*Proof.* We apply Theorem 1.5 with the choice $F_m(t) := q(tB_m)r(tA_m)$ for an arbitrarily fixed $t \geq 0$. Since stability is assumed, we only have to check the consistency. To do that, first we have to show that

$$\lim_{h\to 0} \frac{J_m q(hB_m)r(hA_m)P_m x - J_m P_m x}{h} = J_m(A_m + B_m)P_m x \tag{2.3}$$

for all $x \in D(A+B)$ and uniformly for $m \in \mathbb{N}$. The left-hand side of (2.3) can be written as:

$$\frac{J_m q(hB_m)r(hA_m)P_m x - J_m P_m x}{h}$$
$$= J_m q(hB_m)P_m \frac{J_m r(hA_m)P_m x - J_m P_m x}{h} + \frac{J_m q(hB_m)P_m x - J_m P_m x}{h}.$$

Since the topology of pointwise convergence on a dense subset of $X$ and the topology of uniform convergence on relatively compact subsets of $X$ coincide on bounded subsets of $\mathscr{L}(X)$ (see, e.g., Engel and Nagel [5, Proposition A.3]), it follows from Lemma 2.1 that the expression above converges uniformly to $J_m(A_m + B_m)P_m x$. □

**Theorem 2.4.** *Suppose that the conditions of Theorem* 2.3 *are satisfied, but replace the stability assumption with either*

$$\|[r(\tfrac{h}{2}A_m)q(hB_m)r(\tfrac{h}{2}A_m)]^k\| \leq M\mathrm{e}^{kh\omega} \quad \textit{for all } h \geq 0,\ k, m \in \mathbb{N}$$

*for the Strang splitting, or*

$$\|\,[\Theta q(hB_m)r(hA_m) + (1-\Theta)r(hA_m)q(hB_m)]^k\,\| \leq M e^{kh\omega}$$

*for a $\Theta \in [0,1]$ and for all $h \geq 0$, $k, m \in \mathbb{N}$ in case of the weighted splitting. Then the Strang and weighted splittings, respectively, are convergent, i.e.,*

$$\lim_{m,n\to\infty} \left[r(\tfrac{t}{2n}A_m)q(\tfrac{t}{n}B_m)r(\tfrac{t}{2n}A_m)\right]^n x = U(t)x \quad (\textit{Strang}),$$

$$\lim_{m,n\to\infty} \left[\Theta q(\tfrac{t}{n}B_m)r(\tfrac{t}{n}A_m) + (1-\Theta)r(\tfrac{t}{n}A_m)q(\tfrac{t}{n}B_m)\right]^n = U(t)x \quad (\textit{weighted})$$

*for all $x \in X$ uniformly for $t$ in compact intervals.*

*Proof.* The proof is very similar as it was in the case of the sequential splitting in Theorem 2.3. The only difference occurs in formula (2.3). In the case of Strang splitting we take

$$F_m(t) := r(\tfrac{t}{2n}A_m)q(\tfrac{t}{n}B_m)r(\tfrac{t}{2n}A_m)$$

and write

$$\begin{aligned}
&\frac{J_m r(\frac{h}{2}A_m)q(hB_m)r(\frac{h}{2}A_m)P_m x - J_m P_m x}{h}\\
&\quad = J_m r(\tfrac{h}{2}A_m)q(hB_m)P_m \frac{J_m r(\frac{h}{2}A_m)P_m x - J_m P_m x}{h}\\
&\quad + J_m r(\tfrac{h}{2}A_m)P_m \frac{J_m q(hB_m)P_m x - J_m P_m x}{h} + \frac{J_m r(\frac{h}{2}A_m)P_m x - J_m P_m x}{h}.
\end{aligned}$$

By Lemma 2.1 this converges uniformly to

$$J_m(\tfrac{1}{2}A_m + B_m + \tfrac{1}{2}A_m)P_m x = J_m(A_m + B_m)P_m x.$$

For the weighted splitting we choose

$$F_m(t) := \Theta q(tB_m)r(tA_m) + (1-\Theta)r(tA_m)q(tB_m),$$

which results in

$$\begin{aligned}
&\frac{J_m[\Theta q(tB_m)r(tA_m) + (1-\Theta)r(tA_m)q(tB_m)]P_m x - J_m P_m x}{h}\\
&\quad = \Theta\frac{J_m q(hB_m)r(hA_m)P_m x - J_m P_m x}{h}\\
&\quad + (1-\Theta)\frac{J_m r(hA_m)q(hB_m)P_m x - J_m P_m x}{h}.
\end{aligned}$$

By using the argumentation for sequential splitting, this converges uniformly (in $m$) to

$$\Theta J_m(A_m + B_m)P_m x + (1-\Theta)J_m(B_m + A_m)P_m x = J_m(A_m + B_m)P_m x$$

as $h \to 0$. □

## Proof of Lemma 2.1

The proof consists of three steps, the first being the case of the simplest possible rational approximation, which describes the backward Euler scheme. The next two steps generalize to more complicated cases. We prove (a) and (b) together.

*Step* 1. Consider first the rational function $r(z) = \frac{1}{1-z}$. Then $r(hA_m) = \frac{1}{h}R(\frac{1}{h}, A_m)$ for $h > 0$ sufficiently small. Then (a) follows from the stability Assumption 1.3. Further, the left-hand side of (2.1) takes the form

$$\begin{aligned}
&\left\|\tfrac{1}{h}\left(\tfrac{1}{h}J_m R(\tfrac{1}{h}, A_m)P_m x - J_m P_m x\right) - J_m A_m P_m x\right\|\\
&\quad = \left\|\tfrac{1}{h}\left(\tfrac{1}{h}J_m R(\tfrac{1}{h}, A_m)P_m x - J_m\left(\tfrac{1}{h} - A_m\right)R(\tfrac{1}{h}, A_m)P_m x\right) - J_m A_m P_m x\right\|\\
&\quad = \left\|J_m\left(\tfrac{1}{h}R(\tfrac{1}{h}, A_m) - I_m\right)A_m P_m x\right\|. \qquad (2.4)
\end{aligned}$$

Since for $x \in D(A)$, by the consistency in Assumption 1.3, the set

$$\{J_m A_m P_m x \,:\, m \in \mathbb{N}\} \cup \{Ax\}$$

is compact, for arbitrary $\varepsilon > 0$ there is $N \in \mathbb{N}$ such that the balls $B(J_i A_i P_i x, \varepsilon)$ for $i = 1, \ldots, N$ cover this compact set. Now let $m \in \mathbb{N}$ arbitrary and pick $i \leq N$ with $\|J_i A_i P_i x - J_m A_m P_m x\| \leq \varepsilon$. Then we can write

$$\begin{aligned}
&\left\|\tfrac{1}{h} J_m R(\tfrac{1}{h}, A_m) A_m P_m x - J_m A_m P_m x\right\| \\
&\quad \leq \left\|\tfrac{1}{h} J_m R(\tfrac{1}{h}, A_m) A_m P_m x - J_i A_i P_i x\right\| + \|J_m A_m P_m x - J_i A_i P_i x\| \\
&\quad \leq \left\|\tfrac{1}{h} J_m R(\tfrac{1}{h}, A_m) P_m (J_m A_m P_m x - J_i A_i P_i x)\right\| \\
&\qquad + \left\|(\tfrac{1}{h} J_m R(\tfrac{1}{h}, A_m) P_m - I_m) J_i A_i P_i x\right\| + \varepsilon \\
&\quad \leq C\varepsilon + \left\|(\tfrac{1}{h} J_m R(\tfrac{1}{h}, A_m) P_m - I_m) J_i A_i P_i x\right\| + \varepsilon,
\end{aligned}$$

with an absolute constant $C \geq 0$ being independent on $m \in \mathbb{N}$ for $h$ sufficiently small. The term in the middle can be estimated as follows. Take $x \in D(A)$. Then for $\lambda > 0$

$$\lambda J_m R(\lambda, A_m) P_m x = J_m R(\lambda, A_m) A_m P_m x + J_m P_m x.$$

Hence,

$$\|\lambda J_m R(\lambda, A_m) P_m x - J_m P_m x\| \leq \|J_m R(\lambda, A_m) P_m\| \cdot \|J_m A_m P_m x\|$$

follows. By the stability in Assumption 1.3,

$$\|J_m R(\lambda, A_m) P_m\| \leq \frac{K^2 M}{\lambda} \quad \text{holds for } \lambda > 0,\ m \in \mathbb{N}.$$

Further, by the consistency in Assumption 1.3 the sequence $J_m A_m P_m x$ is bounded. Therefore

$$\lambda J_m R(\lambda, A_m) P_m x \to J_m P_m x \tag{2.5}$$

as $\lambda \to \infty$ uniformly in $m \in \mathbb{N}$. Since by (2) $\|J_m \lambda R(\lambda, A_m) P_m\|$ is uniformly bounded in $m \in \mathbb{N}$, we obtain by the denseness of $P_m D(A) \subset X_m$ that (2.5) holds even for arbitrary $x \in X$, therefore, (2.4) converges to 0 as $h \to 0$ (choosing $\lambda = \frac{1}{h}$). This proves the validity of (2.1) for our particular choice of the rational function $r$.

*Step* 2. Next, let $k \in \mathbb{N}$ and $r(z) := \frac{1}{(1-z/k)^k}$. Then $r(0) = 1$, $r'(0) = 1$ and $r(hA_m) = [\frac{k}{h} R(\frac{k}{h}, A_m)]^k$. The validity of (a) follows again by the stability Assumption 1.3. For (b) we have to prove

$$\tfrac{1}{h}\left[J_m (\tfrac{k}{h} R(\tfrac{k}{h}, A_m))^k P_m x - J_m P_m x\right] - J_m A_m P_m x \to 0 \tag{2.6}$$

uniformly for $m \in \mathbb{N}$ as $h \to 0$. To achieve this we shall repeatedly use the following "trick": for $y \in D(A_m)$ we have $y = R(\frac{k}{h}, A_m)(\frac{k}{h} - A_m)y$. Hence we obtain

$$\begin{aligned}
J_m P_m x &= J_m \tfrac{k}{h} R(\tfrac{k}{h}, A_m) P_m x - J_m R(\tfrac{k}{h}, A_m) A_m P_m x \\
&= \cdots = J_m [\tfrac{k}{h} R(\tfrac{k}{h}, A_m)]^k P_m x - \sum_{j=0}^{k-1} J_m [\tfrac{k}{h} R(\tfrac{k}{h}, A_m)]^j R(\tfrac{k}{h}, A_m) A_m P_m x.
\end{aligned}$$

By inserting this into the left-hand side of (2.6) we get

$$\tfrac{1}{h}\Big[J_m(\tfrac{k}{h}R(\tfrac{k}{h},A_m))^k P_m x - J_m P_m x\Big] - J_m A_m P_m x$$
$$= \frac{1}{k}\sum_{j=1}^{k} J_m[\tfrac{k}{h}R(\tfrac{k}{h},A_m)]^j A_m P_m x - J_m A_m P_m x. \tag{2.7}$$

By what is proved in Step 1 we have

$$J_m[\tfrac{k}{h}R(\tfrac{k}{h},A_m)]^j P_m x \to J_m P_m x$$

uniformly in $m \in \mathbb{N}$ as $h \to 0$. An analogous compactness argument as in Step 1 shows that

$$J_m[\tfrac{k}{h}R(\tfrac{k}{h},A_m)]^j A_m P_m x \to J_m A_m P_m x \quad (h \to 0)$$

uniformly for $m \in \mathbb{N}$. This shows that the expression in (2.7) converges to 0 uniformly in $m \in \mathbb{N}$.

*Step* 3. To finish the proof for the case of a general rational function

$$r(z) = \frac{a_0 + a_1 z + \cdots + a_k z^k}{b_0 + b_1 z + \cdots + b_n z^n}$$

we use the partial fraction decomposition, i.e., we write

$$r(z) = \sum_{i=1}^{l}\sum_{j=1}^{\nu_i} \frac{C_{ij}}{(1 - z/\lambda_i)^j},$$

with some uniquely determined $C_{ij} \in \mathbb{C}$. Since, by assumption, $r(0) = 1$ and $r'(0) = 1$, we obtain

$$\sum_{i=1}^{l}\sum_{j=1}^{\nu_i} C_{ij} = 1, \quad \text{and} \quad \sum_{i=1}^{l}\sum_{j=1}^{\nu_i} \frac{j}{\lambda_i} C_{ij} = 1. \tag{2.8}$$

Since $r$ is bounded on the left half-plane, we have that the poles $\lambda_i$ of $r$ have positive real part, $\operatorname{Re}\lambda_i > 0$. For $j = 1, \dots, \nu_i$, $i = 1, \dots, l$ consider the rational functions

$$r_j(z) := \frac{1}{(1 - z/j)^j},$$

appearing in Step 2., then

$$r(z) = \sum_{i=1}^{l}\sum_{j=1}^{\nu_i} C_{ij} r_j(\tfrac{j}{\lambda_i} z).$$

Defining the operators $A_{ij,m} := \frac{j}{\lambda_i} A_m$ we shall apply Step 2 to these rational functions and to these operators. To do that we have to check if the required assumptions are satisfied. Obviously, $r_j$ have the properties needed. The consistency part of Assumption 1.3 is trivially satisfied for $A_{ij,m}$, $P_m$ and $J_m$, $m \in \mathbb{N}$. The uniform boundedness of $\lambda R(\lambda, A_{ij,m})$ follows from the stability Assumption 1.3.(a) (from which, in turn Lemma 2.1.(a) follows, too). Indeed, that condition implies

$$\|\lambda R(\lambda, A_m)\| \leq M_\phi \quad \text{for all } \lambda \in \Sigma_\phi,$$

where $\Sigma_\phi$ is any sector with opening half-angle $\phi \in [0, \frac{\pi}{2})$. If $\lambda_i \in \Sigma_\varphi$ is a pole of $r$ then $\frac{\lambda_i \lambda}{j} \in \Sigma_\theta$ for $\lambda > 0$. So have

$$\|\lambda R(\lambda, A_{ij,m})\| = \|\lambda R(\lambda, \tfrac{j}{\lambda_i} A_m)\| = \|\tfrac{\lambda\lambda_i}{j} R(\tfrac{\lambda\lambda_i}{j}, A_m)\| \leq M_\theta \quad \text{for all } \lambda > 0.$$

By Step 2, we have that

$$\tfrac{1}{h}\Big[J_m\big(\tfrac{j}{h}R(\tfrac{j}{h}, A_{ij,m})\big)^j P_m x - J_m P_m x\Big] - J_m A_{ij,m} P_m x \to 0$$

uniformly in $m \in \mathbb{N}$ as $h \to 0$. By taking also the equalities (2.8) into account this yields

$$\begin{aligned}
&\frac{1}{h}\Big[J_m r(hA_m)P_m x - J_m P_m x\Big] - J_m A_m P_m x \\
&\quad = \frac{1}{h}\Big[\sum_{i=1}^{l}\sum_{j=1}^{\nu_i} C_{ij}\Big(J_m r_j(h\tfrac{j}{\lambda_i}A_m)P_m x - J_m P_m x\Big)\Big] - \sum_{i=1}^{l}\sum_{j=1}^{\nu_i} C_{ij} J_m A_{ij,m} P_m x \\
&\quad = \sum_{i=1}^{l}\sum_{j=1}^{\nu_i} C_{ij}\Big[\frac{1}{h}\Big(J_m r_j(h\tfrac{j}{\lambda_i}A_m)P_m x - J_m P_m x\Big) - J_m A_{ij,m} P_m x\Big] \to 0
\end{aligned}$$

uniformly in $m \in \mathbb{N}$ as $h \to 0$. This finishes the proof. □

Finally we remark that in the present paper we only treated an autonomous evolution equation (1.1). In the case of time-dependent operators $A(t)$ and $B(t)$ we have already shown the convergence in [3] for numerical methods applying splitting and spatial discretization together. The extension of our present results concerning the application of an approximation in time as well, will be the subject of forthcoming work.

**Acknowledgment.** A. Bátkai was supported by the Alexander von Humboldt-Stiftung and by the OTKA grant Nr. K81403. The European Union and the European Social Fund have provided financial support to the project under the grant agreement no. TÁMOP-4.2.1/B-09/1/KMR-2010-0003.
B. Farkas was supported by the János Bolyai Research Scholarship of the Hungarian Academy of Sciences.

# References

[1] W. Arendt, C. Batty, M. Hieber, F. Neubrander, *Vector Valued Laplace Transforms and Cauchy Problems*, Birkhäuser Verlag, Monographs in Mathematics **96**, 2001.

[2] A. Bátkai, P. Csomós, G. Nickel, *Operator splittings and spatial approximations for evolution equations*, J. Evolution Equations **9** (2009), 613–636, DOI 10.1007/s00028-009-0026-6.

[3] A. Bátkai, P. Csomós, B. Farkas, G. Nickel, *Operator splitting for non-autonomous evolution equations*, Journal of Functional Analysis **260** (2011), 2163–2190.

[4] P. Csomós, G. Nickel, *Operator splittings for delay equations*, Computers and Mathematics with Applications **55** (2008), 2234–2246.

[5] K.-J. Engel, R. Nagel, *One-Parameter Semigroups for Linear Evolution Equations*, Springer-Verlag, Berlin, 2000.

[6] I. Faragó and Á. Havasi, *Operator splittings and their applications*, Mathematics Research Developments, Nova Science Publishers, New York, 2009.

[7] M. Haase, *The functional calculus for sectorial operators*, Birkhäuser Verlag, Basel, 2006.

[8] E. Hairer, Ch. Lubich, G. Wanner, *Geometric Numerical Integration. Structure-Preserving Algorithms for Ordinary Differential Equations*, Springer-Verlag, Berlin, 2006.

[9] H. Holden, K.H. Karlsen, K.-A. Lie, N.H. Risebro, *Splitting Methods for Partial Differential Equations with Rough Solutions*, European Mathematical Society, 2010.

[10] W. Hundsdorfer, J. Verwer, *Numerical solution of time-dependent advection-diffusion-reaction equations*, Springer Series in Computational Mathematics **33**, Springer-Verlag, Berlin, 2003.

[11] K. Ito, F. Kappel, *Evolution Equations and Approximations*, World Scientific, River Edge, N.J., 2002.

[12] K. Ito, F. Kappel, *The Trotter–Kato theorem and approximation of PDE's*, Mathematics of Computation **67** (1998), 21–44.

András Bátkai
Eötvös Loránd University, Institute of Mathematics
Pázmány P. sétány 1/C
H-1117, Budapest, Hungary.
e-mail: batka@cs.elte.hu

Petra Csomós
Leopold-Franzens-Universität Innsbruck, Institut für Mathematik
Technikerstraße 13,
A-6020, Innsbruck, Austria
e-mail: petra.csomos@uibk.ac.at

Bálint Farkas
Technische Universität Darmstadt, Fachbereich Mathematik
Schloßgartenstraße 7
D-64289 Darmstadt, Germany
e-mail: farkas@mathematik.tu-darmstadt.de

Gregor Nickel
Universität Siegen, FB 6 Mathematik
Walter-Flex-Straße 3
D-57068 Siegen, Germany.
e-mail: nickel@mathematik.uni-siegen.de

Operator Theory:
Advances and Applications, Vol. 221, 173-187


# On a Theorem of Karhunen and Related Moment Problems and Quadrature Formulae

Georg Berschneider and Zoltán Sasvári

*Dedicated to Heinz Langer on the occasion of his 75th birthday*

**Abstract.** In the present paper we give a refinement of a classical result by Karhunen concerning spectral representations of second-order random fields. We also investigate some related questions dealing with moment problems and quadrature formulae. Some of these questions are closely related to Heinz Langer's work.

**Mathematics Subject Classification (2000).** Primary 60G12, 42A70, 44A60, 65D32; Secondary 42A82, 60G10.

**Keywords.** Random field, spectral representation, trigonometric moment problem, power moment problem, quadrature formula.

## 1. Introduction

In [11] K. Karhunen showed that if the correlation function of a second-order random field $Z : V \to L^2(\Omega, \mathcal{A}, P)$ defined on a set $V$ has a certain integral representation with a positive measure then the field itself has an analogous representation with a random orthogonal measure $\zeta$ (more precise formulations will be given below). He also proved that the range of $\zeta$ is equal to the subspace of $L^2(\Omega, \mathcal{A}, P)$ spanned by all random variables $Z(t)$, $t \in V$. Karhunen's theorem contains as special case the spectral representation of stationary fields which was obtained earlier by A. Kolmogorov and H. Cramér, we refer to [24] for historical remarks. As another application we mention the spectral representation of random processes with stationary increments of order $n$, due to A.M. Yaglom and M.S. Pinsker [25].

To prove his result Karhunen assumes a condition which at first seems to be quite technical. In the subsequent paper [12] he gives a detailed proof of his integral representation without using this condition. In this case, the random orthogonal measure $\zeta$ has values in some possibly larger space $L^2(\tilde{\Omega}, \tilde{\mathcal{A}}, \tilde{P})$. This can be seen from Karhunen's proof but is not stated explicitly in his theorem and thus it might be a source for misunderstanding. Some of the monographs on stochastic processes, e.g., [2, 8], follow Karhunen's second paper [12], others, e.g., [9], prefer the first version, where the range of $\zeta$ is contained in the underlying space $L^2(\Omega, \mathcal{A}, P)$.

As starting point of this paper we tried to find conditions which assure that the range of $\zeta$ is in $L^2(\Omega, \mathcal{A}, P)$. Theorem 3.1 contains a necessary and sufficient condition in terms of the dimensions of certain subspaces. Our investigations led us to interesting and nontrivial questions dealing with classical moment problems. Especially, there are connections to extension problems studied by Heinz Langer in numerous papers. These investigations are the content of Section 4.

We had correspondence with Torben M. Bisgaard on some of our open questions. He communicated an unpublished existence result to us with the permission to include it into this paper. His result gives a positive quadrature formula and is the content of Section 5.

Next we give a precise formulation of Karhunen's theorem.

## 2. Karhunen's theorem

Throughout, $(\Omega, \mathcal{A}, P)$ denotes a probability space, and any mapping $Z : V \to L^2(\Omega, \mathcal{A}, P)$ on a nonempty set $V$ with values in the (complex) Hilbert space $L^2(\Omega, \mathcal{A}, P)$ is called *second-order (complex) random field.* Denoting the inner product in $L^2(\Omega, \mathcal{A}, P)$ with $\langle \cdot\,; \cdot \rangle$,

$$C(s,t) = \langle Z(s)\,; Z(t) \rangle, \quad s,t \in V,$$

defines the *correlation function* $C : V \times V \to \mathbb{C}$ of the random field $Z$. Karhunen's original representation can be stated as follows.

**Theorem A (Karhunen, [11]).** *Let $Z : V \to L^2(\Omega, \mathcal{A}, P)$ be a random field with correlation function $C$. Assume that $C$ has the representation*

$$C(s,t) = \int_W g(s,x)\overline{g(t,x)}\mathrm{d}\sigma(x), \quad s,t \in V, \tag{2.1}$$

*where*

(i) *$\sigma$ is a positive, $\sigma$-finite measure on the measurable space $(W, \mathcal{B})$;*
(ii) *$g : V \times W \to \mathbb{C}$ such that $g(t,\cdot) \in L^2(W, \mathcal{B}, \sigma)$ for all $t \in V$;*
(iii) *the linear space*
$$L_g := \operatorname{span}\{g(t,\cdot) :\ t \in V\}$$
*is dense in $L^2(W, \mathcal{B}, \sigma)$.*

*Then there exists a uniquely determined random orthogonal measure $\zeta$ on $\mathcal{B}_0 := \{B \in \mathcal{B} :\ \sigma(B) < \infty\}$ with values in $L^2(\Omega, \mathcal{A}, P)$ and structure function $\sigma$ such that*

$$Z(t) = \int_W g(t,x)\mathrm{d}\zeta(x), \quad t \in V. \tag{2.2}$$

*Additionally, setting* $H(Z) := \overline{\text{span}}\{Z(t) : t \in V\}$ *and* $H(\zeta) := \overline{\text{span}}\{\zeta(A) : A \in \mathcal{B}_0\}$, *one obtains* $H(Z) = H(\zeta)$.

Here, a mapping $\zeta : \mathcal{B}_0 \to L^2(\Omega, \mathcal{A}, P)$ is called *random orthogonal measure with structure function* $\sigma$ if

(i) $\zeta(A \cup B) = \zeta(A) + \zeta(B)$ for all $A, B \in \mathcal{B}_0$ with $A \cap B = \emptyset$;
(ii) $\langle \zeta(A) ; \zeta(B) \rangle = \sigma(A \cap B)$ for $A, B \in \mathcal{B}_0$.

The integral in equation (2.2) is constructed as follows. For a simple function $f = \sum_{i=1}^n c_i \mathbf{1}_{A_i}$, $c_i \in \mathbb{C}$, $A_i \in \mathcal{B}_0$, $1 \le i \le n$, $n \in \mathbb{N}$, one defines

$$\int_W f(x) \mathrm{d}\zeta(x) := \sum_{i=1}^n c_i \zeta(A_i).$$

Using density of the set of simple functions the integral extends to $L^2(W, \mathcal{B}, \sigma)$ and satisfies

$$\Big\langle \int_W f(x) \mathrm{d}\zeta(x) ; \int_W g(x) \mathrm{d}\zeta(x) \Big\rangle = \int_W f(x)\overline{g(x)} \mathrm{d}\sigma(x), \quad f, g \in L^2(W, \mathcal{B}, \sigma). \tag{2.3}$$

As we already mentioned Karhunen gives in [12] a detailed proof of the aforementioned result without using the density condition in (A.iii). In this case, besides losing the equality of the spaces $H(Z)$ and $H(\zeta)$, the representing random orthogonal measure has values in a possibly larger space.

Using the same notation as in Theorem A we can now formulate the problems we are dealing with in the present note.

**Problem 1.** *Under which conditions on the functions* $g : V \times W \to \mathbb{C}$ *does there exist a random orthogonal measure* $\zeta$ *on* $(W, \mathcal{B})$ *with structure function* $\sigma$ *fulfilling the condition* (2.2) *and having values in* $L^2(\Omega, \mathcal{A}, P)$*?*

For fixed functions $C$ and $g$ the representing measure $\sigma$ is in general not uniquely determined. Thus, it is natural to ask whether one can find a representation measure for the function $C$ such that the density condition (A.iii) is satisfied. To be precise:

**Problem 2.** *Let* $(W, \mathcal{B}, \sigma)$ *be a measure space and* $V \neq \emptyset$. *Given functions* $g : V \times W \to \mathbb{C}$ *such that* $g(t, \cdot) \in L^2(W, \mathcal{B}, \sigma)$ *for all* $t \in V$ *find a measure* $\tilde{\sigma}$ *on* $(W, \mathcal{B})$ *such that* $g(t, \cdot) \in L^2(W, \mathcal{B}, \tilde{\sigma})$,

$$\int_W g(s, x)\overline{g(t, x)} \mathrm{d}\tilde{\sigma}(x) = \int_W g(s, x)\overline{g(t, x)} \mathrm{d}\sigma(x), \quad s, t \in V, \tag{2.4}$$

*and* $L_g$ *is dense in* $L^2(W, \mathcal{B}, \tilde{\sigma})$.

Our next problem deals with a special case and leads us to a quadrature problem.

**Problem 3.** *Let* $(W, \mathcal{B}, \sigma)$ *be a measure space where* $\sigma$ *is positive and assume that* $L_g$ *is finite dimensional. Find a positive molecular measure* $\tilde{\sigma}$ *on* $(W, \mathcal{B})$ *satisfying* (2.4).

Note that a measure is called *molecular* if it has finite support.

## 3. A necessary and sufficient condition

A simple example shows that without posing the density assumption in (A.iii) we cannot hope for a random orthogonal measure on the same probability space to exist. In fact, let the set $W = \{w_1, \dots, w_n\}$ have $n > 1$ elements and let $\sigma$ be a finite measure on $W$ having support $W$. Let further $V = \{1, \dots, m\}$, where $m < n$ and $g : V \times W \to \mathbb{C}$. Setting

$$C(i,j) := \int_W g(i,x)\overline{g(j,x)}\mathrm{d}\sigma(x), \quad 1 \le i,j \le m,$$

we can choose vectors $z_1, \dots, z_m \in \mathbb{C}^m$ such that

$$C(i,j) = \langle z_i\,;z_j\rangle.$$

The $z_j$ can be considered as a random field $Z$ on a probability space $(\Omega, \mathcal{A}, P)$, where $\Omega$ has $m$ elements, $\mathcal{A}$ consists of all subsets of $\Omega$ and $P$ is uniformly distributed. This random field cannot have the desired integral representation (2.2) with a random orthogonal measure $\zeta : 2^W \to L^2(\Omega, \mathcal{A}, P)$. For $\sigma$ being the structure function of $\zeta$ we obtain

$$\langle \zeta(\{w_i\})\,;\zeta(\{w_j\})\rangle = \sigma\big(\{w_i\} \cap \{w_j\}\big) = \delta_{i,j}\sigma\big(\{w_i\}\big),$$

where $\delta_{i,j}$ denotes the Kronecker symbol of $i,j \in \{1, \dots, n\}$. Since $\sigma(\{w_i\}) > 0$, $\zeta(\{w_i\})$ must be an orthogonal system in $L^2(\Omega, \mathcal{A}, P)$, a contradiction.

In fact, this concept transfers to the general case. We will use the same notation as in Theorem A. Additionally, let $L_g(\sigma)^\perp$ denote the orthogonal complement of $L_g$ in $L^2(W, \mathcal{B}, \sigma)$.

**Theorem 3.1.** *The integral representation* (2.2) *with a random orthogonal measure $\zeta : \mathcal{B}_0 \to L^2(\Omega, \mathcal{A}, P)$ having structure function $\sigma$ is possible if and only if* $\dim L_g(\sigma)^\perp \le \dim H(Z)^\perp$.

*Proof.* For short we will write $L_g^\perp$ instead of $L_g(\sigma)^\perp$. Assume first that $\dim L_g^\perp \le \dim H(Z)^\perp$ and consider the orthogonal decomposition

$$L^2(W, \mathcal{B}, \sigma) = L_g \oplus L_g^\perp.$$

Choose an orthonormal basis $\{e_\iota : \iota \in I\}$ of $L_g^\perp$, where $I$ is a nonempty index set such that $V \cap I = \emptyset$. By assumption there exists an orthonormal system $\{Y_\iota : \iota \in I\}$ in $H(Z)^\perp$. Now consider the random field

$$\widetilde{X}(t) := \begin{cases} Z(t), & \text{if } t \in V, \\ Y_t, & \text{if } t \in I, \end{cases}$$

for all $t \in V \cup I$. Further define for $t \in V \cup I$

$$\widetilde{g}(t,\cdot) := \begin{cases} g(t,\cdot), & \text{if } t \in V. \\ e_t, & \text{if } t \in I. \end{cases}$$

Then, for all $t \in V \cup I$, $\widetilde{X}(t) \in L^2(\Omega, \mathcal{A}, P)$ and

$$\widetilde{C}(s,t) = \langle \widetilde{X}(s) ; \widetilde{X}(t) \rangle = \int_W \widetilde{g}(s,x)\overline{\widetilde{g}(t,x)} \mathrm{d}\sigma(x).$$

Since $L_{\widetilde{g}}$ is dense in $L^2(W, \mathcal{B}, \sigma)$, Karhunen's original statement (i.e., with the density condition) yields the existence of a uniquely determined random orthogonal measure $\zeta$ with structure function $\sigma$ and values in $L^2(\Omega, \mathcal{A}, P)$ such that

$$\widetilde{X}(t) = \int_W \widetilde{g}(t,x) \mathrm{d}\zeta(x).$$

In particular,

$$Z(t) = \int_W g(t,x) \mathrm{d}\zeta(x)$$

for all $t \in V$. For the converse direction assume that

$$\dim H(Z)^{\perp} < \dim L_g^{\perp}.$$

As before, choose an orthonormal basis $\{e_\iota : \iota \in I\}$ of $L_g^{\perp}$, where $I \neq \emptyset$ and $V \cap I = \emptyset$. Assume $\zeta : \mathcal{B}_0 \to L^2(\Omega, \mathcal{A}, P)$ to be a random orthogonal measure satisfying (2.2) and having structure function $\sigma$. Set for $\iota \in I$

$$Y_\iota := \int_W e_\iota(x) \mathrm{d}\zeta(x) \in L^2(\Omega, \mathcal{A}, P).$$

The properties of random orthogonal measures (cf. (2.3)) now allow to obtain for all $s \in V$ and $\iota \in I$

$$\begin{aligned}\langle Z(s) ; Y_\iota \rangle &= \left\langle \int_W g(s,x) \mathrm{d}\zeta(x) ; \int_W e_\iota(x) \mathrm{d}\zeta(x) \right\rangle \\ &= \int_W g(s,x) \overline{e_\iota(x)} \mathrm{d}\sigma(x) = 0,\end{aligned}$$

and for $\iota, \jmath \in I$

$$\begin{aligned}\langle Y_\iota ; Y_\jmath \rangle &= \left\langle \int_W e_\iota(x) \mathrm{d}\zeta(x) ; \int_W e_\jmath(x) \mathrm{d}\zeta(x) \right\rangle \\ &= \int_W e_\iota(x) \overline{e_\jmath(x)} \mathrm{d}\sigma(x) = \delta_{\iota,\jmath}.\end{aligned}$$

Thus, $\{Y_\iota : \iota \in I\}$ is an orthonormal system in $H(Z)^{\perp}$. Hence, the system can be extended to an orthonormal basis of $H(Z)^{\perp}$. In particular,

$$\dim L_g^{\perp} = \mathrm{card}(I) \leq \dim H(Z)^{\perp} < \dim L_g^{\perp},$$

a contradiction. Thus, no random orthogonal measure of this form exists, which concludes the proof of the theorem. □

## 4. The second problem

This problem contains several well-studied special cases where solutions exist. In the present section we review some results connected with the trigonometric and power moment problem and give two examples which answer the question in the negative.

### 4.1. The trigonometric moment problem

Let us first consider the case of $W = \mathbb{R}$ equipped with the usual Borel $\sigma$-field $\mathcal{B}$, $a > 0$, $V = [-a; a]$, and the functions

$$g : [-a; a] \times \mathbb{R} \to \mathbb{C}, \ (t, x) \mapsto g(t, x) = e^{itx}.$$

For any bounded positive measure $\sigma$ on the real line the equation

$$f(t) = \int_{\mathbb{R}} e^{itx} \mathrm{d}\sigma(x), \quad |t| \leq 2a,$$

defines a positive definite function $f : [-2a; 2a] \to \mathbb{C}$. Thus, Problem 2 can be reformulated in terms of continuations of (continuous) positive definite functions.

**Problem 4.** *Let $\sigma$ be a bounded Borel measure on the real line and $a > 0$. For the positive definite function $f : [-2a; 2a] \to \mathbb{C}$ given by*

$$f(t) := \int_{\mathbb{R}} e^{itx} \mathrm{d}\sigma(x), \quad |t| \leq 2a,$$

*find a positive definite continuation $\widetilde{f} : \mathbb{R} \to \mathbb{C}$ (and representing measure $\tilde{\sigma}$) such that $L_g = \operatorname{span}\{e^{it\bullet} : |t| \leq a\}$ is dense in $L^2(\mathbb{R}, \mathcal{B}, \tilde{\sigma})$.*

The existence of continuations of positive definite functions without necessarily fulfilling the density requirement was established by M.G. Kreĭn in [13]. Additionally, for the case that $f$ has more than one continuation, he gives a description of all possible continuations: In this case there exist four entire functions $w_{jk}$ such that the relation

$$\int_{\mathbb{R}} \frac{1}{t - z} \mathrm{d}\tilde{\sigma}(t) = i \int_0^{\infty} e^{izt} \overline{\widetilde{f}(t)} \mathrm{d}t = \frac{w_{11}(z)T(z) + w_{12}(z)}{w_{21}(z)T(z) + w_{22}(z)}, \quad \operatorname{Im} z > 0, \tag{4.1}$$

establishes a bijective correspondence between all continuations $\widetilde{f}$ (and their representing measures $\tilde{\sigma}$) and all Nevanlinna-Pick functions $T$. A function $T$ is hereby called *Nevanlinna-Pick function* if $T \equiv \infty$ or $T$ is holomorphic in $\mathbb{C} \setminus \mathbb{R}$, has nonnegative imaginary part there, and satisfies $f(\overline{z}) = \overline{f(z)}$. The latter have the form

$$T(z) = \alpha + \beta z + \int_{\mathbb{R}} \frac{tz + 1}{t - z} \mathrm{d}\mu(t), \quad z \in \mathbb{C} \setminus \mathbb{R}, \tag{4.2}$$

where $\alpha \in \mathbb{R}$, $\beta \geq 0$, and $\mu$ is a positive, finite measure on $\mathbb{R}$.

A continuation $\widetilde{f}$ of the given positive definite function $f$ is called *orthogonal* if it is either unique or in the representation (4.1) $T$ is a real constant or $T \equiv \infty$.

In a series of papers [14, 15, 16, 17, 18] M.G. Kreĭn and H. Langer extended these results to functions with finitely many negative squares. Finally, in the unpublished manuscript [19] Kreĭn and Langer give on page 32 a characterization of orthogonal continuations in the following form:

**Theorem 4.1 (Kreĭn, Langer, [19, §4.1]).** *Let $\widetilde{f}$ be a continuation of the positive definite function $f$ with representation measure $\tilde{\sigma}$. Then* $\operatorname{span}\{e^{it\bullet} : |t| \le a\}$ *is dense in $L^2(\mathbb{R}, \mathcal{B}, \tilde{\sigma})$ if and only if $\widetilde{f}$ is an orthogonal continuation.*

In the article [20] the authors computed the entire functions $w_{jk}$ for the positive definite function $f(t) = 1 - |t|$ on $[-2a; 2a]$, where $a \le 1$. The $4a$-periodic extension treated on page 49 corresponds to $T \equiv \infty$, hence it is orthogonal.

Thus, in the case of $g$ being trigonometric polynomials Problem 4 always has a solution given by the representation measures of the orthogonal continuations of the implicitly given positive definite correlation function $f(s-t) = C(s,t)$, $s, t \in V - V$.

### 4.2. The power moment problem

For this special case we stay on the real axis, but consider the functions

$$g : \mathbb{N}_0 \times \mathbb{R} \to \mathbb{R}, \ (n, x) \mapsto g(n, x) = x^n,$$

instead. Now Problem 2 can be formulated in terms of the Hamburger moment problem:

**Problem 5.** *Let $\sigma$ be a bounded positive measure on the real line having moments of all orders. Find a bounded positive measure $\mu$ having the same moments as $\sigma$ such that the set of all polynomials is dense in $L^2(\mathbb{R}, \mathcal{B}, \mu)$.*

For more information on the (power) moment problem we refer the reader to the monograph [1] by N.I. Akhiezer. For the question of density of the set of polynomials, we refer to the paper [4] by C. Berg and J.P.R. Christensen. Denote the set of all measures that give rise to the same moment sequence as $\sigma$ by $V_\sigma$, i.e.,

$$V_\sigma = \Big\{ \mu \in \mathcal{M}_+^b(\mathbb{R}) : \int_{\mathbb{R}} x^n \mathrm{d}\mu(x) = \int_{\mathbb{R}} x^n \mathrm{d}\sigma(x), \ n \in \mathbb{N}_0 \Big\}.$$

The moment problem is called *determinate* if $V_\sigma = \{\sigma\}$ and *indeterminate* in the other case.

A necessary and sufficient condition for the polynomials to be dense in $L^1(\mathbb{R}, \mathcal{B}, \mu)$ was given by M.A. Naĭmark in [21].

**Theorem B (Naĭmark).** *The space of all polynomials is dense in $L^1(\mathbb{R}, \mathcal{B}, \mu)$ if and only if $\mu$ is an extreme point of $V_\sigma$.*

For the indeterminate case, R. Nevanlinna established in [22] a description of all elements of $V_\sigma$. Similar to the trigonometric case, there exist four entire functions $w_{jk}$ such that

$$\int_{\mathbb{R}} \frac{1}{t-z} \mathrm{d}\mu(t) = -\frac{w_{11}(z)T(z) - w_{12}(z)}{w_{21}(z)T(z) - w_{22}(z)}, \quad z \in \mathbb{C} \setminus \mathbb{R}, \tag{4.3}$$

establishes a bijective correspondence between all measures $\mu \in V_\sigma$ and all functions $T$ belonging to the Nevanlinna-Pick class, cf. (4.2). The question of density of the set of all polynomials in $L^2(\mathbb{R}, \mathcal{B}, \mu)$ was solved by M. Riesz in [23] by giving the following condition.

**Theorem C (M. Riesz).** *The space of all polynomials is dense in* $L^2(\mathbb{R}, \mathcal{B}, \mu)$ *if and only if either* $\mu$ *is determinate or* $\mu$ *is* N-extremal, *i.e., the function* $T$ *in Nevanlinna's parameterization* (4.3) *is a real constant or* $T \equiv \infty$.

### 4.3. A non-elementary counterexample

While Problem 5 always has a solution on the real line, the statement of Theorem C fails in higher dimensions. C. Berg and M. Thill give in [6] an example of a rotation-invariant Radon measure $\sigma$ on $\mathbb{R}^2$ such that $V_\sigma = \{\sigma\}$ and the space of all polynomials in two variables is not dense in $L^2(\mathbb{R}^2, \mathcal{B}^2, \sigma)$. In this case Problem 2 clearly has no solution.

### 4.4. An elementary counterexample

At the end of this section we give a simple example such that Problem 2 has no solution. For this purpose, let $W = \{1, 2, 3\}$, and denote the Dirac measure at $i \in W$ with $\varepsilon_i$. Choose $\sigma_1, \sigma_2, \sigma_3 > 0$ such that

$$\sigma = \sigma_1\varepsilon_1 + \sigma_2\varepsilon_2 + \sigma_3\varepsilon_3$$

is a positive measure on $W$ and a function $g : \{1, 2\} \times W \to \mathbb{C}$ such that $h_1 := |g(1,\cdot)|^2$, $h_2 := g(1,\cdot)\overline{g(2,\cdot)}$ and $h_3 := |g(2,\cdot)|^2$ are linearly-independent functions on $W$. Any measure $\mu$ on $W$ has a representation

$$\mu = \mu_1\varepsilon_1 + \mu_2\varepsilon_2 + \mu_3\varepsilon_3,$$

where $\mu_1, \mu_2, \mu_3 \geq 0$. Therefore

$$\int_W h_i(x)\mathrm{d}\mu(x) = \int_W h_i(x)\mathrm{d}\sigma(x), \quad 1 \leq i \leq 3, \tag{4.4}$$

can be written as the linear system

$$\mathsf{H}\upmu = \mathsf{m},$$

where $\upmu = [\mu_i]_{i=1}^3$, $\mathsf{m} = \left[\int_W h_i \mathrm{d}\sigma\right]_{i=1}^3$, and $\mathsf{H} = \left[h_i(j)\right]_{i,j=1}^3$.

Since $\{h_1, h_2, h_3\}$ is linearly independent by assumption, the linear system has the unique solution $\upsigma = [\sigma_i]_{i=1}^3$. In particular, there exists no measure $\mu$ on $W$ satisfying (4.4) such that the two-dimensional space span$\{g_1, g_2\}$ is dense in the three-dimensional space $L^2(W, 2^W, \mu)$.

## 5. Molecular measures and a quadrature formula

Problem 3 is always solvable, this follows from Theorem 5.1 below. For the case of $L_g$ being equal to the space of polynomials up to a certain degree the first part of this theorem is also known as Tchakaloff's theorem. A similar result has also been proved by C. Bayer and J. Teichmann in [3].

**Theorem 5.1 (T.M. Bisgaard).** *Let $\sigma$ be a positive measure on $(X, \mathcal{F})$, where $\mathcal{F}$ is a $\sigma$-ring with union $X$. Then:*

(i) *For every finite-dimensional subspace $D$ of $\mathcal{L}^1(X, \mathcal{F}, \sigma)$, there is a positive molecular measure $\xi$ such that*

$$\int f \mathrm{d}\xi = \int f \mathrm{d}\sigma$$

*for each $f \in D$. The molecular measure can be chosen to have support of cardinality at most* $\dim D$.

(ii) *There is a net $(\xi_\lambda)_{\lambda \in \Lambda}$ of molecular measures such that not only do we have*

$$\int f \mathrm{d}\xi_\lambda \to \int f \mathrm{d}\sigma$$

*for each $f \in \mathcal{L}^1(X, \mathcal{F}, \sigma)$: we even have, for every such $f$,*

$$\int f \mathrm{d}\xi_\lambda = \int f \mathrm{d}\sigma$$

*for all sufficiently large $\lambda$.*

*Notes.*

(a) The notion of measurable function that we use is that of Halmos [10]: a real-valued function $f$ on $X$ is said to be measurable if $f^{-1}(B) \in \mathcal{F}$ for every Borel subset $B$ of $\mathbb{R} \setminus \{0\}$. Also, functions in $\mathcal{L}^1(X, \mathcal{F}, \sigma)$ must not assume any value in the set $\{-\infty, \infty\}$. The integral of a nonnegative measurable function $f$ (and, afterwards, of a measurable function $f$ such that $\int |f| \mathrm{d}\sigma < \infty$) may be defined by

$$\int f \mathrm{d}\sigma := \int_{f^{-1}(\mathbb{R} \setminus \{0\})} f \mathrm{d}\sigma,$$

the point being that the right member is an integral with respect to a measure on a $\sigma$-*field*, viz, the set $\{A \in \mathcal{F} : A \subseteq f^{-1}(\mathbb{R} \setminus \{0\})\}$.

(b) A net may be indexed by any nonempty set equipped with an upward filtering transitive reflexive relation: we do not require anti-symmetry. (This will allow us to avoid using the Axiom of Choice.)

In the proof of the theorem, we need five lemmas.

**Lemma 5.2.** *Let $K$ be a convex set in a finite-dimensional real affine space $E$. If $K$ is dense in $E$ then $K = E$.*

*Proof.* Let $d$ be the dimension of $E$. Let $y \in E$. Choose an affinely independent $(d+1)$-tuple $s = (s_0, \dots, s_d)$ of points in $E$ such that $y$ is in the interior of the convex hull of $\{s_0, \dots, s_d\}$. For every affinely independent $(d+1)$-tuple $t = (t_0, \dots, t_d)$ of points in $E$, let $\eta(t) = (\eta_0(t), \dots, \eta_d(t))$ be the $(d+1)$-tuple of barycentric coordinates of $y$ with respect to $t$, i.e., the unique $(d+1)$-tuple of real numbers such that $\sum_{i=0}^d \eta_i(t) = 1$ and $y = \sum_{i=0}^d \eta_i(t) t_i$. If $K$ is dense in $E$ then we can choose, for each $i \in \{0, \dots, d\}$, a sequence $\left(t_i^{(n)}\right)_{n \in \mathbb{N}}$ of points in $K$

converging to $s_i$. Since $s$ is affinely independent, so is $t^{(n)} := \left(t_0^{(n)}, \dots, t_d^{(n)}\right)$ for all sufficiently large $n$, hence without loss of generality for all $n$. By the choice of $s$ we have $\eta_i(s) > 0$, $0 \le i \le d$. By the continuity of the function $\eta$ it follows that for all sufficiently large $n$ we have $\eta_i\left(t^{(n)}\right) > 0$, $0 \le i \le d$. Choosing any such $n$, we see that $y$ is in the convex hull of $\left\{t_0^{(n)}, \dots, t_d^{(n)}\right\}$, hence in $K$. Since $y$ was arbitrary, this completes the proof. □

We stress that by a *cone* in a real vector space we understand one with apex 0.

**Lemma 5.3.** *Let $E$ be a finite-dimensional real vector space and let $G$ be an open subset of $E$ such that the set $K := G \cup \{0\}$ is a convex cone. Let $y \in E \setminus K$. Then there is a linear form $\varphi$ on $E$ such that $\varphi(y) \le 0$ and $\varphi(z) > 0$ for each $z \in G$.*

*Proof.* Without loss of generality, assume $G \ne \emptyset$. Since $y \in E \setminus K$, it follows from Lemma 5.2 that $K$ is not dense in $E$. By the bipolar theorem [5, 1.3.6] it follows that there is a nonzero linear form $\psi$ on $E$ such that $\psi(z) \ge 0$ for each $z \in K$. It follows that $\psi(z) > 0$ for each $z \in G$.[1] If $\psi(y) \le 0$, we may take $\varphi = \psi$.

Assume $\psi(y) > 0$. Then the line $L := \{\lambda y : \lambda \in \mathbb{R}\}$ is disjoint from $G$. To see this, assume $\lambda y = z$ for some $\lambda \in \mathbb{R}$ and some $z \in G$. If $\lambda > 0$ then ($K$ being a cone) the point $y = \lambda^{-1} z$ is in $K$, a contradiction. If $\lambda \le 0$, we find $\psi(z) = \lambda \psi(y) \le 0$, again a contradiction.

The line $L$ being disjoint from $G$, by [5, 1.2.1] there is a hyperplane $H$ in $E$ which contains $L$ and is disjoint from $G$. Now $G$, being convex (as is easily seen), is contained in one of the two open half-spaces bounded by $H$, and there is therefore a linear form $\varphi$ on $E$ such that $H = \ker \varphi$ and $\varphi(z) > 0$ for each $z \in G$. Since $y \in L \subset H$, we have $\varphi(y) = 0$, completing the proof. □

For every topological space $Y$, let $\mathcal{B}(Y)$ be its Borel $\sigma$-field. If $\nu$ is a positive measure, let $\nu_*$ be the associated inner measure.

**Lemma 5.4.** *Let $E$ be a finite-dimensional real vector space and let $\nu$ be a positive measure on $\mathcal{B}(E)$. Assume that the dual space of $E$ is contained in $\mathcal{L}^1(E, \mathcal{B}(E), \nu)$, and let $y$ be the* resultant *of $\nu$, i.e., the unique point in $E$ such that $\varphi(y) = \int \varphi \mathrm{d}\nu$ for each $\varphi$ in the dual. Let $A$ be a convex cone in $E$, containing* 0 *and such that $\nu_*(E \setminus A) = 0$. Then $y \in A$.*

*Proof.* We proceed by induction on the dimension of $E$. The case $\dim E = 0$ being trivial, assume $\dim E \ge 1$ and suppose the statement is true for all lower dimensions. We may assume $E = A - A$: otherwise, define a subspace $E_1$ of $E$ by $E_1 := A - A$ and apply the induction hypothesis to $E_1$ and $\nu_1 := \nu_{|E_1}$ instead of $E$ and $\nu$.[2]

---

[1]If the kernel of $\psi$ intersected $G$ then the set $G$, being open, would intersect both of the open half-spaces bounded by $\ker \psi$. In particular, $G$ (hence $K$) would intersect that half-space where $\psi < 0$, a contradiction.

[2]For this argument, we need to show that $\nu$ is supported by $E_1$ and that $y \in E_1$. Since $E_1$ is a measurable set containing $A$, we have $\nu(E \setminus E_1) = 0$. Thus, if $\varphi$ is any linear form on $E$ which

We henceforth assume $E = A - A$. Now $A$, being a nonempty convex set, has a nonempty relative interior $G$ ([7, Thm. 3.1]), and $G$ is dense in $A$ [7, Thm. 3.4]. Since $E = A - A$, the set $G$ is just the interior of $A$, hence open. It is easily verified that the set $K := G \cup \{0\}$ is a convex cone. We may obviously assume $y \notin A$, hence $y \notin K$. By Lemma 5.3 there is a linear form $\varphi$ on $E$ such that $\varphi(y) \leq 0$ and $\varphi(z) \geq 0$ for each $z$ in the closure $\overline{A}$ of $A$. The measure $\nu$ being supported by $\overline{A}$, we conclude from

$$0 \geq \varphi(y) = \int \varphi \mathrm{d}\nu = \int_{\overline{A}} \varphi \mathrm{d}\nu$$

that $\nu$ is supported by the subspace $H := \ker \varphi$ and that $y \in H$. The linear form $\varphi$, being strictly positive on the nonempty set $G$, is nonzero. Thus $\dim H = \dim E - 1$. We can now apply the induction hypothesis to $H$, $\nu_{|H}$, and $A \cap H$ instead of $E$, $\nu$, and $A$, leading to the desired conclusion. □

**Lemma 5.5.** *Let $D$ be a finite-dimensional vector space of measurable real-valued functions on $X$ (recall that such functions are not allowed to take any value in $\{-\infty, \infty\}$). Let $E$ be the dual space of $D$ and define a mapping $x \mapsto \widehat{x}$ on $X$ into $E$ by $\widehat{x}(f) := f(x)$, $x \in X$, $f \in D$. Let $W := \{x \in X : \widehat{x} \neq 0\}$. Then the mapping $W \ni x \mapsto \widehat{x}$ is measurable with respect to $\mathcal{F}$ and $\mathcal{B}(E \setminus \{0\})$.*

*Proof.* Let $p$ be the restriction to $W$ of the mapping $x \mapsto \widehat{x}$. Clearly, the set

$$\mathcal{E} := \{B \in \mathcal{B}(E \setminus \{0\}) : \ p^{-1}(B) \in \mathcal{F}\}$$

is a $\sigma$-ring. We have to show that $\mathcal{E}$ is all of $\mathcal{B}(E \setminus \{0\})$. Since $E \setminus \{0\}$ is second countable, it suffices to show that $\mathcal{E}$ contains a base of the topology on $E \setminus \{0\}$. Since the set

$$\mathcal{H}_0 := \Big\{\{z \in E : \ z(f) \in I_f, \ f \in F\} : F \text{ is a finite subset of } D, \\ \text{and } I_f \text{ is an open interval in } \mathbb{R}, \ f \in F\Big\}$$

is a base of the topology on $E$, the set $\mathcal{H} := \{H \in \mathcal{H}_0 : \ 0 \notin H\}$ is a base of the topology on $E \setminus \{0\}$. It thus suffices to show $p^{-1}(\mathcal{H}) \subset \mathcal{F}$. So let $H \in \mathcal{H}$; we shall show $p^{-1}(H) \in \mathcal{F}$. Choose a finite subset $F$ of $D$ and open intervals $I_f$ in $\mathbb{R}$ $(f \in F)$ such that

$$H = \{z \in E : \ z(f) \in I_f, \ f \in F\}.$$

Since $0 \notin H$, we have

$$0 \notin I_f \text{ for at least one } f \in F. \tag{$*$}$$

vanishes on $E_1$, then

$$\varphi(y) = \int \varphi \mathrm{d}\nu = \int_{E_1} \varphi \mathrm{d}\nu + \int_{E \setminus E_1} \varphi \mathrm{d}\nu = 0.$$

This being so for every such $\varphi$, by the Hahn-Banach Theorem we get $y \in E_1$.

We compute

$$\begin{aligned}p^{-1}(H) &= \{x \in W : \widehat{x} \in H\} \\ &= \{x \in W : \widehat{x}(f) \in I_f,\ f \in F\} \\ &= \{x \in W : f(x) \in I_f,\ f \in F\} \\ &= \bigcap_{f \in F} f^{-1}(I_f). \end{aligned} \tag{$**$}$$

Let $F_+ := \{f \in F : 0 \notin I_f\}$ and $F_- := F \setminus F_+$. By $(*)$ we have $F_+ \neq \emptyset$. For $f \in F_-$, let $I_f^c := \mathbb{R} \setminus I_f$. Resuming the computation from $(**)$, we find

$$p^{-1}(H) = \Big( \bigcap_{f \in F_+} f^{-1}(I_f) \Big) \Big\backslash \Big( \bigcup_{f \in F_-} f^{-1}(I_f^c) \Big) \in \mathcal{F}$$

since $\mathcal{F}$ is, in particular, a ring, and since $f^{-1}(I_f)$ (resp. $f^{-1}(I_f^c)$) is in $\mathcal{F}$ whenever $f \in F_+$ (resp. $f \in F_-$). This completes the proof. □

**Lemma 5.6.** *Let $E$ be a real vector space and $M$ a subset of $E$. Then every element in the convex cone generated by $M$ is a linear combination, with positive coefficients, of the elements of some linearly independent subset of $M$.*

*Proof.* As the proof of [7, Thm. 2.3], mutatis mutandis. □

*Proof of Theorem* 5.1. : Let $\Delta$ be the set of all finite-dimensional subspaces of $\mathcal{L}^1(X, \mathcal{F}, \sigma)$.

(i) Let $D \in \Delta$. Let $\Xi(D)$ be the set of all $\xi$ as in the statement; we have to show $\Xi(D) \neq \emptyset$. Let $E$, $x \mapsto \widehat{x}$, and $W$ be as in Lemma 5.5. By that lemma, it makes sense to define a positive measure $\nu$ on $\mathcal{B}(E \setminus \{0\})$ as the image measure of $\sigma_{|W}$ under the mapping $p : W \to E \setminus \{0\}$ given by $p(x) := \widehat{x}$, $x \in W$. We extend $\nu$ to a measure on all of $\mathcal{B}(E)$, denoted by the same symbol, by the condition $\nu(\{0\}) = 0$. Let $f \mapsto \widehat{f}$ be the canonical isomorphism of $D$ onto the dual of $E$, i.e., $\widehat{f}(z) := z(f)$, $f \in D$, $z \in E$.

**Claim 1.** *Every linear form on $E$ is $\nu$-integrable.*

*Proof.* Let $\varphi$ be a linear form on $E$. Then $\varphi = \widehat{f}$ for some $f \in D$, and so

$$\begin{aligned}\int |\varphi| \mathrm{d}\nu &= \int |\widehat{f}| \mathrm{d}\nu = \int |\widehat{f}(z)| \mathrm{d}\nu(z) = \int |z(f)| \mathrm{d}\nu(z) \\ &= \int_W |p(x)(f)| \mathrm{d}\sigma(x) = \int_W |\widehat{x}(f)| \mathrm{d}\sigma(x) = \int_W |f(x)| \mathrm{d}\sigma(x).\end{aligned}$$

Since $f$ vanishes off $W$, it follows that

$$\int |\varphi| \mathrm{d}\nu = \int |f| \mathrm{d}\sigma < \infty. \qquad \square$$

**Claim 2.**

$$\int \widehat{f} \mathrm{d}\nu = \int f \mathrm{d}\sigma, \quad f \in D.$$

*Proof.* Repeat the argument in the proof of Claim 1, without taking absolute values. □

By Claim 1 it makes sense to define $y \in E$ (the resultant of $\nu$ in the sense of Lemma 5.4) by the condition that

$$\varphi(y) = \int \varphi \mathrm{d}\nu \tag{5.1}$$

for every linear form $\varphi$ on $E$.

**Claim 3.**

$$y(f) = \int f \mathrm{d}\sigma, \quad f \in D.$$

*Proof.* Using (5.1) and Claim 2, we find

$$y(f) = \widehat{f}(y) = \int \widehat{f} \mathrm{d}\nu = \int f \mathrm{d}\sigma.$$ □

Let $A$ be the convex cone (with zero) in $E$ generated by $p(W)$.

**Claim 4.** $\nu_*(E \setminus A) = 0$.

*Proof.* Given $B \in \mathcal{B}(E)$ with $B \subset E \setminus A$, we have to show $\nu(B) = 0$, that is, $\sigma(\{x \in W : \widehat{x} \in B\}) = 0$. But $\{x \in W : \widehat{x} \in B\} = \emptyset$ since $A \cap B = \emptyset$. □

By Lemma 5.4, we have $y \in A$, so we can choose $m \in \mathbb{N}_0$, $x_1, \ldots, x_m \in X$, and $a_1, \ldots, a_m > 0$ such that $y = \sum_{i=1}^m a_i \widehat{x}_i$. Define a molecular measure $\xi$ by

$$\xi := \sum_{i=1}^m a_i \varepsilon_{x_i}.$$

For $f \in D$ we find, using Claim 3:

$$\int f \mathrm{d}\xi = \sum_{i=1}^m a_i f(x_i) = \sum_{i=1}^m a_i \widehat{x}_i(f) = y(f) = \int f \mathrm{d}\sigma.$$

This proves that $\xi \in \Xi(D)$, and so $\Xi(D)$ is nonempty. The statement concerning the cardinality of the support of the molecular measure is a direct consequence of Lemma 5.6. This proves (i).

(ii) With $\Xi(D)$ (for $D \in \Delta$) as in the proof of (i), let $\Lambda := \{(D, \xi) : D \in \Delta, \xi \in \Xi(D)\}$. Introduce a transitive reflexive relation $\preceq$ in $\Lambda$ by

$$(D_1, \xi_1) \preceq (D_2, \xi_2) \Leftrightarrow D_1 \subset D_2.$$

To see that $(\Lambda, \preceq)$ is upwards filtering, let $(D_1, \xi_1)$, $(D_2, \xi_2) \in \Lambda$. Set $D := D_1 + D_2$ and choose $\xi \in \Xi(D)$ (possible by (i)). Then $(D, \xi) \in \Lambda$ and $(D_i, \xi_i) \preceq (D, \xi)$ $(i = 1, 2)$.

The mapping $\lambda \mapsto \xi_\lambda$ of the net shall be the mapping $\Lambda \ni (D, \xi) \mapsto \xi$. Given $f \in \mathcal{L}^1(X, \mathcal{F}, \sigma)$, choose $D_f \in \Delta$ such that $f \in D_f$ (e.g., $D_f := \{\alpha f : \alpha \in \mathbb{R}\}$). Choose $\xi_f \in \Xi(D_f)$ arbitrarily (possible by (i)) and set

$\lambda_f := (D_f, \xi_f) \in \Lambda$. For every $\lambda := (D, \xi) \in \Lambda$ satisfying $\lambda \succeq \lambda_f$ (i.e., $D \supset D_f$), we have $f \in D$, hence $\int f d\xi = \int f d\sigma$ (because $\xi \in \Xi(D)$). This proves (ii) and completes the proof of the theorem. □

**Acknowledgment**

The authors wish to thank Torben M. Bisgaard for providing the content of Section 5 and for calling our attention to the paper [6].

## References

[1] N.I. Akhiezer. *The classical moment problem.* Oliver & Boyd, Edinburgh-London, 1965.

[2] J. Anděl. *Statistische Analyse von Zeitreihen. Vom Autor autorisierte und ergänzte Übersetzung aus dem Tschechischen*, volume 35 of *Mathematische Lehrbücher und Monographien. I. Abteilung: Mathematische Lehrbücher.* Akademie Verlag, Berlin, 1984.

[3] C. Bayer and J. Teichmann. *The proof of Tchakaloff's theorem.* Proc. Amer. Math. Soc. **134**(2006), 3035–3040.

[4] C. Berg and J.P.R. Christensen. *Density questions in the classical theory of moments.* Ann. Inst. Fourier (Grenoble) **31**(1981), 99–114.

[5] C. Berg, J.P.R. Christensen, and P. Ressel. *Harmonic Analysis on Semigroups. Theory of Positive Definite and Related Functions*, volume 100 of *Graduate Texts in Mathematics.* Springer, New York, NY, 1984.

[6] C. Berg and M. Thill. *Rotation invariant moment problems.* Acta Math. **167**(1991), 207–227.

[7] A. Brøndsted. *An Introduction to Convex Polytopes*, volume 90 of *Graduate Texts in Mathematics.* Springer, New York-Heidelberg-Berlin, 1983.

[8] I.I. Gikhman and A.V. Skorokhod. *Introduction to the Theory of Random Processes. Translated from the* 1965 *Russian original. Reprint of the* 1969 *English translation.* Dover, Mineola, NY, 1996.

[9] I.I. Gikhman and A.V. Skorokhod. *The Theory of Stochastic Processes. I. Translated from the Russian by S. Kotz. Corrected printing of the first edition.* Classics in Mathematics. Springer, Berlin, 2004.

[10] P.R. Halmos. *Measure Theory*, volume 18 of *Graduate Texts in Mathematics.* Springer, New York-Heidelberg-Berlin, second edition, 1974.

[11] K. Karhunen. *Zur Spektraltheorie stochastischer Prozesse.* Ann. Acad. Sci. Fenn., Ser. I. A. Math. **34**(1946).

[12] K. Karhunen. *Über lineare Methoden in der Wahrscheinlichkeitsrechnung.* Ann. Acad. Sci. Fenn., Ser. I. A. Math. **37**(1947).

[13] M.G. Kreĭn. *Sur le problème du prolongement des fonctions hermitiennes positives et continues.* Dokl. Acad. Sci. URSS, n. Ser. **26**(1940), 17–22.

[14] M.G. Kreĭn and H. Langer. *Über einige Fortsetzungsprobleme, die eng mit der Theorie hermitescher Operatoren im Raume $\Pi_\kappa$ zusammenhängen. I: Einige Funktionenklassen und ihre Darstellungen.* Math. Nachr. **77**(1977), 187–236.

[15] M.G. Kreĭn and H. Langer. *Über einige Fortsetzungsprobleme, die eng mit der Theorie hermitescher Operatoren im Raume $\Pi_k$ zusammenhängen. II. Verallgemeinerte Resolventen, u-Resolventen und ganze Operatoren.* J. Funct. Anal. **30**(1978), 390–447.

[16] M.G. Kreĭn and H. Langer. *On some extension problems which are closely connected with the theory of hermitian operators in a space $\Pi_\kappa$. III: Indefinite analogues of the Hamburger and Stieltjes moment problems. I.* Beitr. Anal. **14**(1979), 25–40.

[17] M.G. Kreĭn and H. Langer. *On some extension problems which are closely connected with the theory of hermitian operators in a space $\Pi_\kappa$. III: Indefinite analogues of the Hamburger and Stieltjes moment problem. II.* Beitr. Anal. **15**(1981), 27–45.

[18] M.G. Kreĭn and H. Langer. *On some continuation problems which are closely related to the theory of operators in spaces $\Pi_\kappa$. IV: Continuous analogues of orthogonal polynomials on the unit circle with respect to an indefinite weight and related continuation problems for some classes of functions.* J. Operator Theory **13**(1985), 299–417.

[19] M.G. Kreĭn and H. Langer. *Continuation of Hermitian positive definite functions and related questions.* Unpublished manuscript.

[20] H. Langer, M. Langer, and Z. Sasvári. *Continuation of hermitian indefinite functions and corresponding canonical systems: an example.* Methods Funct. Anal. Topology **10**(2004), 39–53.

[21] M.A. Naĭmark. *On extremal spectral functions of a symmetric operator.* Dokl. Acad. Sci. URSS, n. Ser. **54**(1946), 7–9.

[22] R. Nevanlinna. *Asymptotische Entwicklungen beschränkter Funktionen und das Stieltjessche Momentenproblem.* Ann. Acad. Sci. Fenn., Ser. I. A. Math. **18**(1922).

[23] M. Riesz. *Sur le problème des moments et le théorème de Parseval correspondant.* Acta Litt. ac. Scient. Univ. Hung. **1**(1923), 209–225.

[24] A.M. Yaglom. *Correlation Theory of Stationary and Related Random Functions. Volume II.* Springer Series in Statistics. Springer, New York, NY, 1987.

[25] A.M. Yaglom and M.S. Pinsker. *Random processes with stationary increments of order n.* Dokl. Acad. Sci. URSS, n. Ser. **90**(1953), 731–734.

Georg Berschneider and Zoltán Sasvári
Technische Universität Dresden
Institut für Mathematische Stochastik
D-01062 Dresden, Germany
e-mail: `georg.berschneider@tu-dresden.de`
`zoltan.sasvari@tu-dresden.de`

Operator Theory:
Advances and Applications, Vol. 221, 189-217


# Explicit Error Estimates for Eigenvalues of Some Unbounded Jacobi Matrices

Anne Boutet de Monvel and Lech Zielinski

**Abstract.** We consider the self-adjoint operator $J$ defined by an infinite Jacobi matrix in the case when the diagonal entries $(d_k)_1^\infty$ form an increasing sequence $d_k \to \infty$ and the off-diagonal entries $(b_k)_1^\infty$ are small with respect to $(d_{k+1} - d_k)_1^\infty$. The main result of this paper is an explicit estimate of the difference between the $n$th eigenvalue of $J$ and the corresponding eigenvalue of a finite-dimensional block of the original Jacobi matrix.

**Mathematics Subject Classification (2000).** Primary 47B36; Secondary 47A10, 47A75, 15A42, 47A55.

**Keywords.** Jacobi matrices, eigenvalue estimates, error estimates, Rayleigh–Ritz method.

## 1. Introduction

Finite and infinite tridiagonal matrices are investigated in many branches of theoretical and applied mathematics, e.g., in the theory of orthogonal polynomials (cf. [9, 10, 16]), in spectral theory of differential operators (cf. [8, 23]), in numerical analysis (cf. [18, 24]), and in mathematical physics (cf. [1, 7, 20, 21]).

In particular the spectral analysis of Schrödinger operators and self-adjoint second-order differential operators acting in $L^2(\mathbb{R})$ can be compared with the analysis of infinite tridiagonal matrices

$$\begin{pmatrix} d_1 & \bar{b}_1 & 0 & 0 & 0 & \dots \\ b_1 & d_2 & \bar{b}_2 & 0 & 0 & \dots \\ 0 & b_2 & d_3 & \bar{b}_3 & 0 & \dots \\ 0 & 0 & b_3 & d_4 & \bar{b}_4 & \dots \\ \vdots & \vdots & \vdots & \vdots & \vdots & \ddots \end{pmatrix} \tag{1.1}$$

acting in the Hilbert space $l^2 = l^2(\mathbb{N}^*)$ of square summable complex-valued sequences $(x_j)_1^\infty$.

**Assumptions**

We assume that

(i) $(d_n)_1^\infty$ is a sequence of positive real numbers,

(ii) $(b_n)_1^\infty$ is a sequence of complex numbers satisfying

$$\lim_{n\to\infty} \frac{|b_{n-1}|+|b_n|}{d_n} = 0, \tag{1.2}$$

(iii) the diagonal satisfies

$$d_n \xrightarrow[n\to+\infty]{} +\infty. \tag{1.3}$$

It is well known (cf. [6]) that under the first two assumptions a self-adjoint operator $J$ with domain

$$\mathcal{D}(J) := \{(x_j)_1^\infty \in l^2 \mid (d_j x_j)_1^\infty \in l^2\} \tag{1.4}$$

can be defined by the matrix (1.1), i.e.,

$$Jx = (d_j x_j + \bar{b}_j x_{j+1} + b_{j-1} x_{j-1})_{j=1}^\infty \tag{1.5}$$

where by convention $x_0 = b_0 = 0$.

The third assumption ensures that $J$ has compact resolvent and there exists an orthonormal basis $(v_n)_1^\infty$ such that $Jv_n = \lambda_n(J)v_n$ with $(\lambda_n(J))_1^\infty$ nondecreasing:

$$\lambda_1(J) \le \cdots \le \lambda_n(J) \le \lambda_{n+1}(J) \le \cdots .$$

**Estimating the $n$th eigenvalue**

The aim of this paper is to present explicit estimates of the $n$th eigenvalue $\lambda_n(J)$ for a class of matrices for which the diagonal is increasing, i.e., $d_n < d_{n+1}$ and the off-diagonal entries are small with respect to the differences $d_{n+1} - d_n$.

The new form of our results allows us to join two aspects of the approximation of $\lambda_n(J)$ usually investigated for different reasons.

One of the aspects consists in a numerical evaluation of $\lambda_n(J)$ considered in concrete physical problems. As an example we mention the problem of finding eigenvalues $\lambda$ for which the Mathieu equation

$$u''(t) + (\lambda - 2h^2 \cos 2t)u(t) = 0$$

has non-trivial odd solutions of period $\pi$. It is possible to reformulate this problem in terms of the Jacobi matrix (1.1) with $d_n = 4n^2$ and $b_n = h^2$ and following [23] a good error estimate for $\lambda_n(J)$ can be obtained by the Rayleigh–Ritz method for small $h$ and $n \in \mathbb{N}^*$ fixed. Another problem described in [23] concerns eigenvalues of the spherical wave equations and the Rayleigh–Ritz method can be used in a similar way to estimate the $n$th eigenvalue of the infinite Jacobi matrix by means of suitably chosen finite submatrices.

Another aspect of estimating $\lambda_n(J)$ consists in the description of its asymptotic behaviour for large values of $n$ and is motivated, e.g., by the quantum mechanical theory of a molecule in a homogeneous magnetic field or the Jaynes–Cummings model related to the Hamiltonian of interactions with $N$-level atoms in quantum optics. The aspect of controlling the approximation of $\lambda_n(J)$ with respect to $n$ has not been investigated in [23] or other numerical investigations cited in the references of [23]. This type of models has been investigated by using the transformation operator method developed in [12] to obtain the asymptotics of large eigenvalues for Jacobi matrices with entries satisfying

$$d_n = n^2 + c_0 n + c_1 n^{-1} + c_2 n^{-2} + \mathrm{O}(n^{-3}), \tag{1.6a}$$

$$b_n = g_0 + g_1 n^{-1} + g_2 n^{-2} + \mathrm{O}(n^{-3}), \tag{1.6b}$$

where $c_0$, $c_1$, $c_2$ and $g_0$, $g_1$, $g_2$ are some constants, and by using the method of successive approximations in [3, 4] for entries of the form

$$d_n = n^2 + c_n, \tag{1.7a}$$

$$b_n = g_n n^{1/2}, \tag{1.7b}$$

where $(c_n)_1^\infty$, $(g_n)_1^\infty$ are periodic sequences of period $N$.

More recently the class of matrices satisfying

$$c_0(d_n - d_{n-1}) \geq n^\alpha, \tag{1.8a}$$

$$|b_n| \leq c_1 n^{\alpha-\kappa} \tag{1.8b}$$

for some constants $c_0$, $c_1$, $\kappa > 0$ and $\alpha \geq 0$ has been investigated in [15], where the following asymptotic formula

$$|\lambda_n(J) - \lambda_{n,l}| \leq C_l\, n^{\alpha - l\kappa} \tag{1.9}$$

is proved with $\lambda_{n,l}$ obtained by means of the characteristic polynomial of the $(2l+1)\times(2l+1)$ submatrix $J_{n,l}$ of $J$ centered at $(n,n)$, $n > l$, i.e.,

$$J_{n,l} := \begin{pmatrix} d_{n-l} & \bar{b}_{n-l} & 0 & \dots & 0 & 0 & 0 \\ b_{n-l} & d_{n-l+1} & \bar{b}_{n-l+1} & \dots & 0 & 0 & 0 \\ 0 & b_{n-l+1} & d_{n-l+2} & \dots & 0 & 0 & 0 \\ \vdots & \vdots & \vdots & \ddots & \vdots & \vdots & \vdots \\ 0 & 0 & 0 & \dots & d_{n+l-2} & \bar{b}_{n+l-2} & 0 \\ 0 & 0 & 0 & \dots & b_{n+l-2} & d_{n+l-1} & \bar{b}_{n+l-1} \\ 0 & 0 & 0 & \dots & 0 & b_{n+l-1} & d_{n+l} \end{pmatrix}. \tag{1.10}$$

In this paper we consider a larger class of Jacobi matrices and assuming that $\lambda_{n,l}$ is a suitable eigenvalue of $J_{n,l}$, $n > l$ we show how the difference $\lambda_n(J) - \lambda_{n,l}$ can be explicitly estimated in terms of $|b_k|/(d_j - d_i)$ with $n - l \leq i, j, k \leq n + l$, $i \neq j$. In particular we can deduce estimates (1.9) with an explicit constant $C_l$ if the entries satisfy (1.8) for some constants $c_0$, $c_1$, $\kappa > 0$ and $\alpha > -1$.

To complete our review of references we note that the problem of comparing $\lambda_n(J)$ with a suitable eigenvalue of finite blocks of $J$ is also investigated in [14, 23] and that the papers [5, 25] describe the asymptotic behaviour of $\lambda_n(J)$ for some special classes of Jacobi matrices when $b_n/d_n = \mathrm{O}(n^{-\kappa})$ holds with $\kappa \in (0, 1]$. This problem will be also investigated in a forthcoming paper [2] under additional conditions on the behaviour of $(b_n - b_{n-1})_{n\in\mathbb{N}^*}$ and $(d_n - d_{n-1})_{n\in\mathbb{N}^*}$. We finally cite [9, 10, 13, 11, 16], where the asymptotic behaviour of eigenvalues is considered for matrices which do not satisfy condition (1.8b).

Sections 2 and 7 are devoted to infinite Jacobi matrices. Section 2 contains the main results of this paper. Their proofs are given in Section 7, using auxiliary results on finite Jacobi matrices developed in Sections 3–6.

Section 3 contains preliminary results together with a first level approximation of the middle eigenvalue of a finite Jacobi matrix. Sections 4 and 5 give more refined first level approximations. Section 6 implement the method of successive approximations in order to obtain a higher level approximation.

## 2. Main results

We denote by $l^2$ the Hilbert space of complex-valued sequences $x = (x_j)_1^\infty$ satisfying

$$\|x\|_2^2 := \sum_{j=1}^{\infty} |x_j|^2 < \infty \tag{2.1}$$

and consider the operator

$$J\colon\ \mathcal{D}(J) \to l^2$$

defined by (1.4), (1.5) under assumptions (1.2), (1.3). Thus

(i) $J$ is self-adjoint,
(ii) its spectrum $\sigma(J)$ is discrete.

The eigenvalues $\{\lambda_n(J)\}_{n=1}^\infty$ of $J$ are enumerated in nondecreasing order and repeated according to their multiplicity:

$$\lambda_1(J) \le \cdots \le \lambda_n(J) \le \lambda_{n+1}(J) \le \cdots .$$

We introduce

$$\hat{d}_n^+ := \sup_{j\le n}(d_j + |b_j| + |b_{j-1}|), \tag{2.2a}$$

$$\hat{d}_n^- := \inf_{j\ge n}(d_j - |b_j| - |b_{j-1}|). \tag{2.2b}$$

Both sequences $\{\hat{d}_n^-\}_1^\infty$ and $\{\hat{d}_n^+\}_1^\infty$ are nondecreasing, and $\hat{d}_n^- \le d_n \le \hat{d}_n^+$.

**Theorem 2.1.** *Let $J$ be the Jacobi operator defined by (1.4), (1.5) under assumptions (1.2), (1.3). Let $\lambda_1(J) \le \cdots \le \lambda_n(J) \le \cdots$ be the eigenvalues of $J$ and let $\hat{d}_n^\pm$ be given by (2.2). Then*

$$\hat{d}_n^- \le \lambda_n(J) \le \hat{d}_n^+.$$

*Assume that*

- *$n$ is such that the following condition holds*

$$\hat{d}_{n-1}^+ \le d_n \le \hat{d}_{n+1}^-. \tag{2.3}$$

*Then one has the estimate*

$$|\lambda_n(J) - d_n| \le |b_n| + |b_{n-1}|. \tag{2.4}$$

*Proof.* See Subsection 7.2. □

*Remark* (on condition (2.3)). Condition (2.3) on $n$ implies that

$$d_k \le d_n \le d_m \text{ if } k < n < m.$$

Further on we consider the intervals

$$\Delta_n := [\hat{d}_n^-, \hat{d}_n^+], \tag{2.5}$$

where $\hat{d}_n^\pm$ are given by (2.2) and for $k \in \mathbb{N}^*$ we denote

$$\beta_{n,k} := \max_{n-k \le j < n+k} |b_j|. \tag{2.6}$$

Then the following result holds.

**Theorem 2.2.** *Let $J$ be as in Theorem 2.1 and let $n \ge 1$. Assume that*

- *$n$ is such that*

$$\Delta_n \cap (\Delta_{n-1} \cup \Delta_{n+1}) = \varnothing. \tag{2.7}$$

*Then $\lambda_n(J)$ is the only point of the spectrum $\sigma(J)$ belonging to $\Delta_n$:*

$$\sigma(J) \cap \Delta_n = \{\lambda_n(J)\}. \tag{2.8}$$

*Moreover, $\lambda_n(J)$ is an eigenvalue of multiplicity one and one has the estimate*

$$|\lambda_n(J) - d_n| \le 4\varepsilon_n \beta_{n,2}, \tag{2.9}$$

*where*

$$\varepsilon_n := \frac{|b_n| + |b_{n-1}|}{\min(d_{n+1} - d_n,\, d_n - d_{n-1})}. \tag{2.10}$$

*Proof.* See Subsections 7.4 and 7.3. □

*Remark* (on condition (2.7)). Obviously, condition (2.7) on $n$ means that

$$\hat{d}_{n-1}^+ < \hat{d}_n^- \le \hat{d}_n^+ < \hat{d}_{n+1}^-,$$

then it implies $d_{n-1} < d_n < d_{n+1}$.

Further on, for $n, k \in \mathbb{N}^*$, we denote

$$\gamma_{n,k} := \min_{n-k \le j < n+k} (d_{j+1} - d_j). \tag{2.11}$$

Under the *assumption* $\gamma_{n,k} > 0$, which means $d_{n-k} < d_{n-k+1} < \cdots < d_{n+k-1} < d_{n+k}$, we introduce

$$\varepsilon_{n,k} := \frac{\beta_{n,k}}{\gamma_{n,k}}. \tag{2.12}$$

Our next result is the following

**Theorem 2.3.** *Let $J$ be as in Theorem 2.1 and let $n \ge 1$. Assume that*

- $\varepsilon_{n,1} < \frac{1}{2}$,
- $\gamma_{n,2} > 0$,
- (2.7) *holds.*

*Then the estimate*

$$|\lambda_n(J) - d_n - r_n| \leq 48\varepsilon_{n,1}\varepsilon_{n,2}\beta_{n,3} \tag{2.13}$$

*holds with*

$$r_n := -\frac{|b_n|^2}{d_{n+1} - d_n} + \frac{|b_{n-1}|^2}{d_n - d_{n-1}}. \tag{2.14}$$

*Proof.* See Subsections 7.4 and 7.3. □

Moreover we will prove

**Theorem 2.4.** *Let $J$ be as in Theorem 2.1. Let also $n \geq 1$, $l \geq 3$ be given, and let $J_{n,l}$ denote the $(2l+1) \times (2l+1)$ submatrix of $J$ given by (1.10). Assume that*

- $18\beta_{n,l} < \gamma_{n,l}$,
- $\hat{d}^+_{n-2} \leq d_{n-1} < d_{n+1} \leq \hat{d}^-_{n+2}$.

*Then one has the estimate*

$$|\lambda_n(J) - \lambda_{n,l}| \leq 2\beta_{n,l}\varepsilon_n\, 6^{l-1} \prod_{k=2}^{l-1} \varepsilon_{n,k}, \tag{2.15}$$

*where $\lambda_{n,l}$ is the unique eigenvalue of $J_{n,l}$ belonging to $\Delta_n$.*

*Proof.* See Subsections 7.5 and 7.3. □

Moreover, following the indication at the end of Section 6 one can find an approximative value of $\lambda_{n,l}$ with the error given in the right-hand side of (2.15) by means of linear combinations and products of matrices (of size $2l+1$), the total number of these operations being of order $l^2$.

*Comments* (on the conditions involving $\hat{d}^\pm_n$). Under the *additional assumptions* that

(iv) there exists $n_0 \in \mathbb{N}^*$ such that $n > n_0 \implies d_{n-1} < d_n$,
(v) there exists $n_0' > n_0$ such that $n \geq n_0' \implies \varepsilon_n \leq 1$,

both sequences $(d_j \pm (|b_j| + |b_{j-1}|))_{j \geq n_0'}$ are nondecreasing, hence

$$k \geq n_0' \implies \hat{d}^-_k = d_k - |b_k| - |b_{k-1}|, \tag{2.16}$$

$$k \geq n_0' \implies \hat{d}^+_k = \max(d_k + |b_k| + |b_{k-1}|, \hat{d}^+_{n_0'-1}). \tag{2.17}$$

Let $n_0'$ be fixed so that (2.16), (2.17) hold. Then, due to assumption (1.3) we can choose $n_1 \geq n_0'$ such that

$$d_{n_1} \geq \hat{d}^+_{n_0'-1}, \tag{2.18}$$

then we obtain

- $\hat{d}^\pm_k = d_k \pm (|b_k| + |b_{k-1}|)$ for any $k > n_1$,
- $\hat{d}^+_{k-1} \leq d_k \leq \hat{d}^-_{k+1}$ for any $k \geq n_1$,
- $\hat{d}^+_k < \hat{d}^-_{k+1}$ for any $k \geq n_1$ such that $\varepsilon_k < 1/2$.

At the end of this section we discuss the case of entries satisfying (1.8) for $n \geq n_0$. It is easy to see that for $n \geq l + \max\{n_0, l\}$ one has

$$c_0\gamma_{n,l} \geq \min\{(n-l)^\alpha, (n+l)^\alpha\} \geq 2^{-|\alpha|}n^\alpha,$$
$$\beta_{n,l} \leq c_1 \max\{(n-l)^{\alpha-\kappa}, (n+l)^{\alpha-\kappa}\} \leq 2^{|\alpha-\kappa|}c_1 n^{\alpha-\kappa},$$

hence the estimate

$$\varepsilon_{n,l} \leq Cn^{-\kappa} \tag{2.19}$$

holds with

$$C := 2^{|\alpha|+|\alpha-\kappa|}c_0c_1. \tag{2.20}$$

Then we can use $n \geq (4C)^{1/\kappa} \implies Cn^{-\kappa} \leq \frac{1}{4}$ and $n \geq (18C)^{1/\kappa} \implies Cn^{-\kappa} \leq \frac{1}{18}$ to estimate the right-hand side of (2.19) and applying Theorems 2.1–2.4 we obtain

**Corollary 2.5.** *Let $J$ be as in Theorem 2.1. We assume that*

- *for some $c_0$, $c_1$, $\kappa > 0$, $\alpha > -1$, and $n_0 \in \mathbb{N}^*$ the entries of $J$ satisfy (1.8) for any $n \geq n_0$.*

*Let $n_1$ be chosen such that (2.18) holds with some $n_0' > \max\{n_0, (4C)^{1/\kappa}\}$, and let $C$ be given by (2.20).*

*Then for any $n > 3 + \max\{n_1, 3\}$ one has*

$$|\lambda_n(J) - d_n| \leq c_1 2^{|\alpha-\kappa|+1} n^{\alpha-\kappa},$$
$$|\lambda_n(J) - d_n| \leq 8c_1 2^{|\alpha-\kappa|} C n^{\alpha-2\kappa},$$
$$|\lambda_n(J) - d_n - r_n| \leq 48c_1 2^{|\alpha-\kappa|} C^2 n^{\alpha-3\kappa},$$

*where $r_n$ is given by (2.14).*

*If moreover $l \geq 3$ and $n > l + \max\{n_1, (18C)^{1/\kappa}, l\}$, then*

$$|\lambda_n(J) - \lambda_{n,l}| \leq c_1 2^{|\alpha-\kappa|+2}(6C)^{l-1} n^{\alpha-l\kappa}.$$

## 3. Finite Jacobi matrices

All results of Section 2 will be proved in Section 7 by using properties of finite Jacobi matrices developed in Sections 3–6. For this reason we give a presentation of our results on finite matrices in a special framework.

### 3.1. Notations

For $s, t \in \mathbb{R}$ we denote $[\![s, t]\!] := \{k \in \mathbb{Z} : s \leq k \leq t\}$. Let $l \in \mathbb{N}^*$ be fixed. We denote

$$\mathcal{J} = [\![-l, l]\!] \tag{3.1}$$

and define $V_{\mathcal{J}}$ as the $(2l+1)$-dimensional complex Hilbert space

$$V_{\mathcal{J}} := \mathbb{C}^{\mathcal{J}} = \{(x_j)_{-l}^{l} \mid x_j \in \mathbb{C}\} \tag{3.2}$$

with the scalar product

$$\langle x, y\rangle = \sum_{j\in\mathcal{J}} \bar{x}_j y_j \tag{3.3}$$

and the norm $\|x\|_2 = \langle x, x\rangle^{1/2}$. Let $(\mathrm{e}_k)_{k\in\mathcal{J}}$ be the canonical basis of $V_{\mathcal{J}}$, i.e., $\mathrm{e}_k = (\delta_{j,k})_{j\in\mathcal{J}}$ with $\delta_{k,k} = 1$ and $\delta_{j,k} = 0$ for $j \neq k$. A linear operator $T \in \mathrm{End}(V_{\mathcal{J}})$ is identified with its matrix with respect to the canonical basis,

$$T = (t_{i,j})_{i,\,j\in\mathcal{J}} = (\langle \mathrm{e}_i, T\,\mathrm{e}_j\rangle)_{i,\,j\in\mathcal{J}}. \tag{3.4}$$

The diagonal matrix with entries $(q_j)_{j\in\mathcal{J}} \in V_{\mathcal{J}}$ is denoted

$$\mathrm{diag}(q_j)_{j\in\mathcal{J}} = (q_j\delta_{i,j})_{i,\,j\in\mathcal{J}}.$$

For a self-adjoint operator $T \in \mathrm{End}(V_{\mathcal{J}})$ we denote $(\lambda_n(T))_{n=-l}^{l}$ the $2l+1$ eigenvalues of $T$ enumerated in nondecreasing order and repeated according to their multiplicity:

$$\lambda_{-l}(T) \le \cdots \le \lambda_0(T) \le \cdots \le \lambda_l(T).$$

For $T \in \mathrm{End}(V_{\mathcal{J}})$ we consider the two operator norms with respect to $\|\cdot\|_2$ and $\|x\|_\infty = \max_{j\in\mathcal{J}}|x_j|$, respectively:

$$\|T\|_2 := \sup_{\|x\|_2\le 1} \|Tx\|_2, \tag{3.5}$$

$$\|T\|_\infty := \sup_{\|x\|_\infty\le 1} \|Tx\|_\infty = \max_{i\in\mathcal{J}} \sum_{j\in\mathcal{J}} |t_{i,j}|. \tag{3.6}$$

These two norms $\|\cdot\|_p$, $p = 2, \infty$ satisfy

$$\|T_1T_2\|_p \le \|T_1\|_p\,\|T_2\|_p. \tag{3.7}$$

A discrete version of the Schur test implies

$$\|T\|_2 \le \|T\|_\infty^{1/2}\|T^*\|_\infty^{1/2}.$$

Thus, if $T$ is self-adjoint,

$$\|T\|_2 \le \|T\|_\infty. \tag{3.8}$$

In Sections 3–6 we investigate the $(2l+1)\times(2l+1)$ Jacobi matrix

$$J := \begin{pmatrix}
d_{-l} & \bar{b}_{-l} & 0 & 0 & \dots & 0 & 0 & 0 & 0\\
b_{-l} & d_{1-l} & \bar{b}_{1-l} & 0 & \dots & 0 & 0 & 0 & 0\\
0 & b_{1-l} & d_{2-l} & \bar{b}_{2-l} & \dots & 0 & 0 & 0 & 0\\
0 & 0 & b_{2-l} & d_{3-l} & \dots & 0 & 0 & 0 & 0\\
\vdots & \vdots & \vdots & \vdots & \ddots & \vdots & \vdots & \vdots & \vdots\\
0 & 0 & 0 & 0 & \dots & d_{l-3} & \bar{b}_{l-3} & 0 & 0\\
0 & 0 & 0 & 0 & \dots & b_{l-3} & d_{l-2} & \bar{b}_{l-2} & 0\\
0 & 0 & 0 & 0 & \dots & 0 & b_{l-2} & d_{l-1} & \bar{b}_{l-1}\\
0 & 0 & 0 & 0 & \dots & 0 & 0 & b_{l-1} & d_l
\end{pmatrix} \tag{3.9}$$

where $b_j \in \mathbb{C}$ for $j = -l, \dots, l-1$, and $(d_j)_{-l}^{l}$ is a sequence of real numbers. Let $D = \mathrm{diag}(d_j)_{j\in\mathcal{J}}$ and $B$ denote the diagonal and off-diagonal parts of $J$, respectively:

$$J = D + B. \tag{3.10}$$

With the convention that $b_{-l-1} = b_l = 0$ and $x_{\pm(l+1)} = 0$,

$$Bx = (b_{j-1}x_{j-1} + \bar{b}_j x_{j+1})_{j\in\mathcal{J}}. \tag{3.11}$$

### 3.2. Application of the min-max principle

Let $T \in \mathrm{End}(V_{\mathcal{J}})$ be self-adjoint, let $\lambda_{-l}(T) \leq \cdots \leq \lambda_l(T)$ be its eigenvalues and $m \in [\![1, 2l+1]\!]$. The min-max principle asserts that the $m$th eigenvalue of $T$ is given by

$$\lambda_{m-1-l}(T) = \inf_{\substack{V \subset V_{\mathcal{J}} \\ \dim V = m}} \sup_{\substack{x \in V \\ \|x\|_2 = 1}} \langle Tx, x \rangle, \tag{3.12a}$$

$$\lambda_{m-1-l}(T) = \sup_{\substack{V \subset V_{\mathcal{J}} \\ \dim V = m-1}} \inf_{\substack{x \in V^{\perp} \\ \|x\|_2 = 1}} \langle Tx, x \rangle, \tag{3.12b}$$

where $V \subset V_{\mathcal{J}}$ denotes an arbitrary linear subspace and $V^{\perp}$ its orthogonal. We write $T_1 \leq T_2$ if and only if $T_1$, $T_2$ are self-adjoint and $\langle T_1 x, x \rangle \leq \langle T_2 x, x \rangle$ holds for all $x \in V_{\mathcal{J}}$. Then by the min-max formula (3.12),

$$T_1 \leq T_2 \implies \lambda_n(T_1) \leq \lambda_n(T_2) \tag{3.13}$$

for any $n = -l, \ldots, l$. If $T_1$, $T_2$ are self-adjoint and $\eta = \|T_1 - T_2\|_2$, then

$$T_1 - \eta I \leq T_2 \leq T_1 + \eta I \implies \lambda_n(T_1) - \eta \leq \lambda_n(T_2) \leq \lambda_n(T_1) + \eta$$

for any $-l \leq n \leq l$, and we obtain the following useful estimate

$$|\lambda_n(T_1) - \lambda_n(T_2)| \leq \|T_1 - T_2\|_2. \tag{3.14}$$

**Lemma 3.1 (estimate of $\lambda_n(J)$).** *Let $J$ be the finite Jacobi matrix given by (3.9) and let $D_{\pm} \in \mathrm{End}(V_{\mathcal{J}})$ be defined by*

$$D_{\pm} := D \pm \mathrm{diag}(|b_j| + |b_{j-1}|)_{j \in \mathcal{J}}. \tag{3.15}$$

*Then*

(i) $D_- \leq J \leq D_+$,

(ii) $\hat{d}_n^- \leq \lambda_n(J) < \hat{d}_n^+$ *with*

$$\hat{d}_n^+ := \max_{-l \leq j \leq n} (d_j + |b_j| + |b_{j-1}|), \tag{3.16a}$$

$$\hat{d}_n^- := \min_{n \leq j \leq l} (d_j - |b_j| - |b_{j-1}|). \tag{3.16b}$$

*Proof.* To begin we observe that rewriting the expression

$$\langle Bx, x \rangle = \sum_j b_j \bar{x}_{j+1} x_j + \sum_k \bar{b}_{k-1} \bar{x}_{k-1} x_k \tag{3.17}$$

with $k = j+1$ we can estimate $|\langle Bx, x \rangle|$ by

$$\sum_j 2|b_j|\,|x_{j+1} x_j| \leq \sum_j |b_j|(|x_{j+1}|^2 + |x_j|^2). \tag{3.18}$$

Since the right-hand side of (3.18) can be written in the form

$$\sum_k |b_k|\,|x_{k+1}|^2 + \sum_j |b_j||x_j|^2 = \sum_j (|b_{j-1}| + |b_j|)|x_j|^2, \tag{3.19}$$

we obtain

$$\langle Dx, x\rangle + \langle Bx, x\rangle \leq \sum_j d_j|x_j|^2 + \sum_j (|b_j| + |b_{j-1}|)|x_j|^2, \tag{3.20}$$

i.e., $J \leq D_+$ and the inequality $J \geq D_-$ follows from (3.19) similarly.

Since (3.13) ensures $\lambda_n(D_-) \leq \lambda_n(J) \leq \lambda_n(D_+)$ in order to complete the proof it remains to observe that using (3.12) with $V_m$ generated by $\{\mathrm{e}_j\}_{j=-l}^{m-1-l}$ we obtain

$$\lambda_{m-1-l}(D_+) \leq \sup_{\substack{x\in V_m\\ \|x\|_2=1}} \langle D_+x, x\rangle = \hat{d}^+_{m-1-l},$$

$$\lambda_{m-1-l}(D_-) \geq \inf_{\substack{x\in V_{m-1}^\perp\\ \|x\|_2=1}} \langle D_-x, x\rangle = \hat{d}^-_{m-1-l}. \qquad \square$$

### 3.3. Estimate of the middle eigenvalue

Further on we are interested in operators $T = (t_{i,j})_{i,j\in\mathcal{J}}$ satisfying the condition

$$\max\{|i|,\, |j|\} > k \implies t_{i,j} = 0 \tag{3.21}$$

for a given $0 \leq k < l$. Condition (3.21) means that $T = P_kTP_k$, where $P_k \in \mathrm{End}(V_\mathcal{J})$ denotes the orthogonal projection on the linear subspace generated by $\{\mathrm{e}_j\}_{-k}^{k}$, i.e.,

$$P_kx := \sum_{-k\leq j\leq k} x_j\mathrm{e}_j. \tag{3.22}$$

We introduce the operator

$$B_1 := P_1BP_1. \tag{3.23}$$

Then $B_1x = y_{-1}\mathrm{e}_{-1} + y_0\mathrm{e}_0 + y_1\mathrm{e}_1$ holds with

$$\begin{pmatrix} y_{-1}\\ y_0\\ y_1\end{pmatrix} = \begin{pmatrix} 0 & \bar{b}_{-1} & 0\\ b_{-1} & 0 & \bar{b}_0\\ 0 & b_0 & 0\end{pmatrix}\begin{pmatrix} x_{-1}\\ x_0\\ x_1\end{pmatrix} \tag{3.24}$$

and we have

$$\|B_1\|_\infty = |b_0| + |b_{-1}|. \tag{3.25}$$

Next we introduce

$$\hat{J}_1 := J - B_1 = D + B - B_1. \tag{3.26}$$

By the min-max principle (3.14) and by (3.8) we obtain

$$|\lambda_0(J) - \lambda_0(\hat{J}_1)| \leq \|B_1\|_2 \leq \|B_1\|_\infty = |b_0| + |b_{-1}|. \tag{3.27}$$

Since

$$\hat{J}_1 = \begin{pmatrix} d_{-l} & \dots & 0 & 0 & 0 & 0 & 0 & \dots & 0 \\ \vdots & \ddots & \vdots & \vdots & \vdots & \vdots & \vdots & \vdots & \\ 0 & \dots & d_{-2} & \bar{b}_{-2} & 0 & 0 & 0 & \dots & 0 \\ 0 & \dots & b_{-2} & d_{-1} & \boxed{0} & 0 & 0 & \dots & 0 \\ 0 & \dots & 0 & \boxed{0} & d_0 & \boxed{0} & 0 & \dots & 0 \\ 0 & \dots & 0 & 0 & \boxed{0} & d_1 & \bar{b}_1 & \dots & 0 \\ 0 & \dots & 0 & 0 & 0 & b_1 & d_2 & \dots & 0 \\ \vdots & \vdots & \vdots & \vdots & \vdots & \vdots & \vdots & \ddots & \vdots \\ 0 & \dots & 0 & 0 & 0 & 0 & 0 & \dots & \hat{d}_l \end{pmatrix} \tag{3.28}$$

we have $\hat{J}_1 \, e_0 = d_0 \, e_0$, i.e., $d_0$ is an eigenvalue of $\hat{J}_1$.

**Lemma 3.2 (estimate of $\lambda_0(J)$).** *Let $J$ be the finite Jacobi matrix given by* (3.9), *and $\hat{J}_1$ given by* (3.28). *Assume that*

- $\hat{d}^+_{-1} \leq d_0 \leq \hat{d}^-_1$.

*Then*

(i) $\lambda_0(\hat{J}_1) = d_0$,
(ii) $|\lambda_0(J) - d_0| \leq |b_0| + |b_{-1}|$.

*Proof.* Applying Lemma 3.1 to $\hat{J}_1$ instead of $J$ we find

$$\min(d_0, \hat{d}^-_1) < \lambda_0(\hat{J}_1) \leq \max(\hat{d}^+_{-1}, d_0). \tag{3.29}$$

Hence (i) follows from our assumption $\hat{d}^+_{-1} \leq d_0 \leq \hat{d}^-_1$. Moreover (i) and (3.27) imply (ii). □

## 4. Middle eigenvalue estimate: a first improvement

### 4.1. Main result of this section

As before $J$ is the finite Jacobi matrix given by (3.9), $\hat{d}^{\pm}_j$ are given by (3.16) and for $k = 1, \dots, l$, we introduce

$$\beta_k := \max_{-k \leq j \leq k} |b_j|. \tag{4.1}$$

The purpose of this section is to prove the following

**Proposition 4.1 (estimate of $\lambda_0(J)$).** *Let $J$ be the finite Jacobi matrix given by* (3.9). *Assume that*

- $\hat{d}^+_{-1} \leq d_0 \leq \hat{d}^-_1$.

*Then the middle eigenvalue of $J$ satisfies*

$$|\lambda_0(J) - d_0| \leq 4\varepsilon\beta_2, \tag{4.2}$$

*with*

$$\varepsilon := \frac{|b_0|}{d_1 - d_0} + \frac{|b_{-1}|}{d_0 - d_{-1}}. \tag{4.3}$$

*Idea of proof.* The key idea consists in finding a self-adjoint operator $A_1$ such that the operator

$$J_1 = \mathrm{e}^{-\mathrm{i}A_1} J \mathrm{e}^{\mathrm{i}A_1} \tag{4.4}$$

is a good approximation of the operator $\hat{J}_1$ defined in (3.26).

Since $\lambda_0(J) = \lambda_0(J_1)$ (by (4.4)) and $d_0 = \lambda_0(\hat{J}_1)$ (by Lemma 3.2), the min-max principle (3.14) applied to $J_1$, $\hat{J}_1$, and $n = 0$ gives the estimate

$$|\lambda_0(J) - d_0| = |\lambda_0(J_1) - \lambda_0(\hat{J}_1)| \leq \|J_1 - \hat{J}_1\|_2. \tag{4.5}$$

Thus, it only remains to estimate $\|J_1 - \hat{J}_1\|_2$. □

Before starting the proof of Proposition 4.1 we describe

(i) a commutator equation which is used to construct $A_1$,

(ii) two integral representations which are used to estimate $\|J_1 - \hat{J}_1\|_2$.

### 4.2. Solution of a commutator equation

**Lemma 4.2.** *Let $k \in [\![1, l]\!]$ and let $D = \operatorname{diag}(d_j)_{j\in\mathcal{J}}$ be such that*

$$d_{-k} < d_{-k+1} < \cdots < d_{k-1} < d_k.$$

*Let $\hat{R} = (\hat{r}_{i,j})_{i,j\in\mathcal{J}} \in \operatorname{End}(V_{\mathcal{J}})$ be such that*

- *$\hat{r}_{j,j} = 0$ for all $j \in \mathcal{J}$,*
- *$\hat{R} = P_k \hat{R} P_k$ where $P_k$ is given by (3.22).*

*Then one can find $A \in \operatorname{End}(V_{\mathcal{J}})$ having the following properties:*

(i) *A is a solution of the commutator equation*

$$\mathrm{i}(AD - DA) = [\mathrm{i}A, D] = \hat{R}. \tag{4.6}$$

(ii) $A = P_k A P_k$.

(iii) *The estimate*

$$\|A\|_\infty \leq \frac{\|\hat{R}\|_\infty}{\gamma_k} \tag{4.7}$$

*holds with*

$$\gamma_k := \min_{-k\leq j<k} (d_{j+1} - d_j). \tag{4.8}$$

(iv) *A is self-adjoint if $\hat{R}$ is self-adjoint.*

*Proof.* Indeed, if we consider $A = (a_{i,j})_{i,j\in\mathcal{J}}$ defined by

$$a_{i,j} := \begin{cases} \mathrm{i}\,\dfrac{\hat{r}_{i,j}}{d_i - d_j} & \text{if } i \neq j, \\ 0 & \text{if } i = j, \end{cases} \tag{4.9}$$

then

$$\begin{aligned} \langle \mathrm{e}_i,\, [\mathrm{i}A, D]\,\mathrm{e}_j\rangle &= \mathrm{i}\langle \mathrm{e}_i,\, AD\,\mathrm{e}_j\rangle - \mathrm{i}\langle D\,\mathrm{e}_i,\, A\,\mathrm{e}_j\rangle \\ &= \mathrm{i}(d_j - d_i)\langle \mathrm{e}_i,\, A\,\mathrm{e}_j\rangle = \hat{r}_{i,j}, \end{aligned}$$

and it is easy to check that all additional properties are also satisfied. □

### 4.3. Auxiliary integral representations

**Lemma 4.3.** *Let $A$, $Q$, $D \in \mathrm{End}(V_{\mathcal{J}})$ and let*

$$F_A(Q) := \mathrm{e}^{-\mathrm{i}A} Q \mathrm{e}^{\mathrm{i}A} - Q, \tag{4.10}$$

$$G_A(D) := \mathrm{e}^{-\mathrm{i}A}(D - [D, \mathrm{i}A])\mathrm{e}^{\mathrm{i}A} - D. \tag{4.11}$$

(a) *We have the integral representations*

$$F_A(Q) = \int_0^1 \mathrm{e}^{-\mathrm{i}sA}[Q, \mathrm{i}A]\mathrm{e}^{\mathrm{i}sA}\,\mathrm{d}s = \int_0^1 F_{sA}\big([Q, \mathrm{i}A]\big)\,\mathrm{d}s + [Q, \mathrm{i}A], \tag{4.12}$$

$$G_A(D) = \int_0^1 \mathrm{e}^{-\mathrm{i}sA}\, s\, \mathrm{ad}_A^2(D)\mathrm{e}^{\mathrm{i}sA}\,\mathrm{d}s = \int_0^1 F_{sA}\big(s\, \mathrm{ad}_A^2(D)\big)\,\mathrm{d}s + \frac{1}{2}\,\mathrm{ad}_A^2(D), \tag{4.13}$$

*where* $\mathrm{ad}_A^2(D) := [[D, A], A]$.

(b) *If $A$ is self-adjoint, then*

$$\|F_A(Q)\|_2 \le \|[Q, A]\|_2 \le 2\|Q\|_2\|A\|_2, \tag{4.14}$$

$$\|G_A(D)\|_2 \le \int_0^1 s\,\|\mathrm{ad}_A^2(D)\|_2\,\mathrm{d}s = \frac{1}{2}\|\mathrm{ad}_A^2(D)\|_2. \tag{4.15}$$

*Proof.* The derivatives of $F_{sA}(Q)$ and $G_{sA}(D)$ with respect to $s \in \mathbb{R}$ are, respectively:

$$\begin{aligned}
\partial_s F_{sA}(Q) &= \mathrm{e}^{-\mathrm{i}sA}[Q, \mathrm{i}A]\mathrm{e}^{\mathrm{i}sA}, \\
\partial_s G_{sA}(D) &= \mathrm{e}^{-\mathrm{i}sA}([D - [D, \mathrm{i}sA], \mathrm{i}A] - [D, \mathrm{i}A])\mathrm{e}^{\mathrm{i}sA} \\
&= \mathrm{e}^{-\mathrm{i}sA}\, s\, [[D, A], A]\mathrm{e}^{\mathrm{i}sA} \\
&= \mathrm{e}^{-\mathrm{i}sA}\, s\, \mathrm{ad}_A^2(D)\mathrm{e}^{\mathrm{i}sA}.
\end{aligned}$$

Hence,

$$\begin{aligned}
F_A(Q) &= \int_0^1 \partial_s F_{sA}(Q)\,\mathrm{d}s = \int_0^1 \mathrm{e}^{-\mathrm{i}sA}[Q, \mathrm{i}A]\mathrm{e}^{\mathrm{i}sA}\,\mathrm{d}s, \\
G_A(D) &= \int_0^1 \partial_s G_{sA}(D)\,\mathrm{d}s = \int_0^1 \mathrm{e}^{-\mathrm{i}sA} s\, \mathrm{ad}_A^2(D)\mathrm{e}^{\mathrm{i}sA}\,\mathrm{d}s.
\end{aligned}$$

If $A$ is self-adjoint, then $\|\mathrm{e}^{\mathrm{i}sA}\|_2 = 1$ for every $s \in \mathbb{R}$ and the integral representations (4.12) and (4.13) lead, respectively, to estimates

$$\begin{aligned}
&\|F_A(Q)\|_2 \le \|[Q, A]\|_2 \le 2\|Q\|_2\|A\|_2, \\
&\|G_A(D)\|_2 \le \int_0^1 s\,\|\mathrm{ad}_A^2(D)\|_2\,\mathrm{d}s = \frac{1}{2}\|\mathrm{ad}_A^2(D)\|_2.
\end{aligned}$$

□

#### 4.4. Proof of Proposition 4.1

(i) *Construction of* $A_1$. We define $A_1 \in \mathrm{End}(V_{\mathcal{J}})$ by

$$A_1 x = \sum_{-1\le j\le 1} y_j\, \mathrm{e}_j \tag{4.16a}$$

with

$$\begin{pmatrix} y_{-1} \\ y_0 \\ y_1 \end{pmatrix} = \begin{pmatrix} 0 & -\mathrm{i}\frac{\bar{b}_{-1}}{d_0-d_{-1}} & 0 \\ \mathrm{i}\frac{b_{-1}}{d_0-d_{-1}} & 0 & -\mathrm{i}\frac{\bar{b}_0}{d_1-d_0} \\ 0 & \mathrm{i}\frac{b_0}{d_1-d_0} & 0 \end{pmatrix} \begin{pmatrix} x_{-1} \\ x_0 \\ x_1 \end{pmatrix}. \tag{4.16b}$$

Then $A_1 = A_1^*$ is self-adjoint and satisfies

$$[\mathrm{i}A_1, D] = B_1.$$

According to (3.8), (4.16) and (4.3) we have

$$\|A_1\|_2 \le \|A_1\|_\infty = \frac{|b_0|}{d_1-d_0} + \frac{|b_{-1}|}{d_0-d_{-1}} = \varepsilon. \tag{4.17}$$

(ii) *Estimate of* $\|J_1 - \hat{J}_1\|_2$. We introduce $B_2 = P_2BP_2$ and we observe that

$$(B-B_2)\left(\mathrm{e}^{\mathrm{i}sA_1} - I\right) = (B-B_2)(I-P_1)\left(\mathrm{e}^{\mathrm{i}sA_1} - I\right) = 0$$

implies $\mathrm{e}^{-\mathrm{i}A_1}(B-B_2)\mathrm{e}^{\mathrm{i}A_1} = B - B_2$. Therefore the operator $J_1$ given by (4.4) can be written in the form

$$J_1 = \mathrm{e}^{-\mathrm{i}A_1}(D+B_1)\mathrm{e}^{\mathrm{i}A_1} + \mathrm{e}^{-\mathrm{i}A_1}(B_2-B_1)\mathrm{e}^{\mathrm{i}A_1} + B - B_2.$$

Using $B_1 = -[D, \mathrm{i}A_1]$ and the notations (4.10), (4.11) we can write

$$J_1 - \hat{J}_1 = G_{A_1}(D) + F_{A_1}(B_2 - B_1). \tag{4.18}$$

However the estimate (4.14) ensures

$$\|F_{A_1}(B_2-B_1)\|_2 \le \|[B_2-B_1, A_1]\|_2 \le 2\|B_2-B_1\|_2\|A_1\|_2 \tag{4.19}$$

and the estimate (4.15) ensures

$$\|G_{A_1}(D)\|_2 \le \frac{1}{2}\|[B_1, A_1]\|_2 \le \|B_1\|_2\|A_1\|_2. \tag{4.20}$$

Thus combining (4.18) with (4.19) and (4.20) we obtain

$$\|J_1 - \hat{J}_1\|_2 \le (\|B_1\|_2 + 2\|B_2-B_1\|_2)\|A_1\|_2. \tag{4.21}$$

Next we observe that $(B_2 - B_1)x = y_{-2}\,\mathrm{e}_{-2} + y_{-1}\,\mathrm{e}_{-1} + y_1\,\mathrm{e}_1 + y_2\,\mathrm{e}_2$ holds with

$$\begin{pmatrix} y_k \\ y_{k+1} \end{pmatrix} = \begin{pmatrix} 0 & \bar{b}_k \\ b_k & 0 \end{pmatrix} \begin{pmatrix} x_k \\ x_{k+1} \end{pmatrix}$$

for $k = -2$ and $k = 1$, hence

$$\|B_2 - B_1\|_\infty = \max\{|b_{-2}|,\, |b_1|\} \le \beta_2. \tag{4.22}$$

Moreover, by (3.8)

$$\|B_1\|_2 \le \|B_1\|_\infty = |b_0| + |b_{-1}| \le 2\beta_1 \le 2\beta_2.$$

Then, using (4.17) we can estimate the right-hand side of (4.21) by $(2\beta_1+2\beta_2)\varepsilon \le 4\beta_2\varepsilon$, which gives

$$\|J_1 - \hat{J}_1\|_2 \le 4\beta_2\varepsilon.$$

(iii) *End of proof.* Combine this last estimate with (4.5). □

## 5. Middle eigenvalue estimate: a second improvement

### 5.1. Main result of this section

As before $J$ is given by (3.9). Let $\hat{d}_j^{\pm}$, $\beta_k$, $\gamma_k$ be given by (3.16), (4.1), and (4.8), respectively. If $\gamma_k > 0$, which means $d_{-k} < d_{-k+1} < \cdots < d_{k-1} < d_k$, then we set

$$\varepsilon_k := \frac{\beta_k}{\gamma_k}. \tag{5.1}$$

*Remark.* If $\gamma_k > 0$ then $\varepsilon_{k-1} \le \varepsilon_k$ follows from $\beta_{k-1} \le \beta_k$ and $\gamma_{k-1} \ge \gamma_k$. Moreover, $\varepsilon \le 2\varepsilon_1$, hence

$$\varepsilon \le 2\varepsilon_1 \le 2\varepsilon_2 \le \cdots \le 2\varepsilon_k.$$

**Proposition 5.1.** *Let $J$ be the finite Jacobi matrix given by (3.9). Assume that*

- $\varepsilon_1 \le \frac{1}{2}$,
- $\gamma_2 > 0$,
- $\hat{d}_{-1}^{+} \le \hat{d}_0^{-} \le \hat{d}_0^{+} \le \hat{d}_1^{-}$.

*Then*

$$|\lambda_0(J) - d_0 - r_{0,0}| \le 48\varepsilon_1\varepsilon_2\beta_3 \tag{5.2}$$

*where*

$$r_{0,0} := -\frac{|b_0|^2}{d_1 - d_0} + \frac{|b_{-1}|^2}{d_0 - d_{-1}}. \tag{5.3}$$

*Sketch of proof.* As before the starting point of our proof is the equality $\lambda_0(J) = \lambda_0(J_1)$ with $J_1$ given by (4.4). Our next step consists in defining $R \in \mathrm{End}(V_{\mathcal{J}})$ as a suitable correction term for $\hat{J}_1 = D + B - B_1$ so that $J_1' = \hat{J}_1 + R$ is a good approximation of $J_1$. In the final part of the proof we check that $d_0 + r_{0,0}$ is a good approximation of $\lambda_0(J_1')$. □

### 5.2. The operator $J_1' = \hat{J}_1 + R$

We consider $J_1$ defined by (4.4) with $A_1$ given by (4.16). As before $P_k$ is given by (3.22) and we denote

$$B_k := P_k B P_k. \tag{5.4}$$

**5.2.1.** We introduce a first correction term

$$R_1' := [B_2 - B_1,\, \mathrm{i}A_1]. \tag{5.5}$$

Applying (4.12) for $Q = B_2 - B_1$ and $A = A_1$ we get the relation

$$F_{A_1}(B_2 - B_1) - R_1' = \int_0^1 F_{sA_1}(R_1')\,\mathrm{d}s,$$

then the estimate

$$\|F_{A_1}(B_2 - B_1) - R_1'\|_2 \le \int_0^1 \|F_{sA_1}(R_1')\|_2\,\mathrm{d}s. \tag{5.6}$$

Applying (4.14) we get $\|F_{sA_1}(R_1')\|_2 \le \|[R_1',\, sA_1]\|_2$. Then we can estimate the right-hand side of (5.6) as follows:

$$\int_0^1 \|F_{sA_1}(R_1')\|_2\,\mathrm{d}s \le \int_0^1 \|[R_1',\, sA_1]\|_2\,\mathrm{d}s = \frac{1}{2}\|[R_1',\, A_1]\|_2 \le \|R_1'\|_2 \times \|A_1\|_2$$

and we conclude

$$\|F_{A_1}(B_2 - B_1) - R_1'\|_2 \le \|R_1'\|_2 \times \|A_1\|_2 \le 2\|B_2 - B_1\|_2 \times \|A_1\|_2^2. \tag{5.7}$$

**5.2.2.** We introduce a second correction term

$$R_1'' := [B_1,\, \mathrm{i}A_1] = [[\mathrm{i}A_1, D],\, \mathrm{i}A_1] = \mathrm{ad}^2_{A_1}(D). \tag{5.8}$$

Applying (4.13) for $A = A_1$, hence for $\mathrm{ad}^2_A(D) = R_1''$, we get the relation

$$G_{A_1}(D) - \frac{1}{2}R_1'' = \int_0^1 F_{sA_1}(sR_1'')\,\mathrm{d}s,$$

which gives the estimate

$$\|G_{A_1}(D) - \tfrac{1}{2}R_1''\|_2 \le \int_0^1 \|F_{sA_1}(sR_1'')\|_2\,\mathrm{d}s. \tag{5.9}$$

By (4.14) we get $\|F_{sA_1}(sR_1'')\|_2 \le 2s^2\|R_1''\|_2\|A_1\|_2$. Hence we can estimate the right-hand side of (5.9) as follows:

$$\int_0^1 \|F_{sA_1}(sR_1'')\|_2\,\mathrm{d}s \le \int_0^1 2s^2\|R_1''\|_2\|A_1\|_2\,\mathrm{d}s = \frac{2}{3}\,\|R_1''\|_2 \times \|A_1\|_2.$$

Finally:

$$\|G_{A_1}(D) - \tfrac{1}{2}R_1''\|_2 \le \frac{2}{3}\,\|R_1''\|_2 \times \|A_1\|_2 \le \frac{4}{3}\,\|B_1\|_2 \times \|A_1\|_2^2. \tag{5.10}$$

**5.2.3.** Let us now define the correction

$$R := R_1' + \frac{1}{2}R_1''. \tag{5.11}$$

We observe that (4.18) ensures

$$\|J_1 - \hat{J}_1 - R\|_2 \leq \|G_{A_1}(D) - \tfrac{1}{2}R_1''\|_2 + \|F_{A_1}(B_2 - B_1) - R_1'\|_2. \tag{5.12}$$

Combining (5.12) with (5.7) and (5.10), we estimate the right-hand side of (5.12) by

$$\big(\tfrac{4}{3}\|B_1\|_2 + 2\|B_2 - B_1\|_2\big)\|A_1\|_2^2 \leq \big(\tfrac{8}{3}\beta_1 + 2\beta_2\big)\varepsilon^2 \leq 20\,\varepsilon_1^2\beta_2, \tag{5.13}$$

using the estimates $\|A_1\|_2 \leq \varepsilon \leq 2\varepsilon_1$, $\|B_1\|_2 \leq 2\beta_1$, and (4.22). Thus we finally have

$$\|J_1 - \hat{J}_1 - R\|_2 \leq 20\,\varepsilon_1\varepsilon_2\beta_3. \tag{5.14}$$

### 5.3. Analysis of $R$

We observe that $R = [B', \mathrm{i}A_1]$ holds with $B' := B_2 - \frac{1}{2}B_1$. We have $B'\,\mathrm{e}_i = \bar{b}_{i-1}'\,\mathrm{e}_{i-1} + b_i'\,\mathrm{e}_{i+1}$ with

$$b_i' = \begin{cases} b_i & \text{for } i = 1,\,-2, \\ \frac{1}{2}\,b_i & \text{for } i = -1,\,0, \\ 0 & \text{otherwise.} \end{cases} \tag{5.15}$$

Moreover $\mathrm{i}A_1\,\mathrm{e}_j = -\bar{a}_{j-1}\,\mathrm{e}_{j-1} + a_j\,\mathrm{e}_{j+1}$ holds with

$$a_j = \begin{cases} \frac{b_j}{d_j - d_{j-1}} & \text{for } j = -1,\,0, \\ 0 & \text{otherwise,} \end{cases} \tag{5.16}$$

and we find that $R = (r_{i,j})_{i,j\in\mathcal{J}} = (2\,\mathrm{Re}\,z_{i,j})_{i,j\in\mathcal{J}}$ with

$$z_{i,j} = \langle \mathrm{i}B'\,\mathrm{e}_i, -\mathrm{i}A_1\,\mathrm{e}_j\rangle = (\bar{b}_{i-1}'a_{j-1} - b_i'\bar{a}_j)\delta_{i,j} + a_{j-1}\bar{b}_j'\delta_{i-2,j} + \bar{a}_j b_{i+1}'\delta_{i+2,j}.$$

Let $D' = \mathrm{diag}(d_j')_{j\in\mathcal{J}}$ be the diagonal part of $D + R$, i.e.,

$$d_j' = d_j + r_{j,j}, \tag{5.17}$$

we find that $r_{0,0}$ is given by (5.3) and

$$\|D - D'\|_2 = \max_{-1\leq j\leq 1} |r_{j,j}| \leq \varepsilon_1\beta_1. \tag{5.18}$$

Let now $\hat{R} = (\hat{r}_{i,j})_{i,j\in\mathcal{J}}$ be the off-diagonal part of $R$, i.e.,

$$\hat{r}_{i,j} := \begin{cases} r_{i,j} & \text{if } i \neq j, \\ 0 & \text{if } i = j. \end{cases} \tag{5.19}$$

We observe that $\hat{r}_{i,j} \neq 0 \implies |j - i| = 2$ and obtain the estimate

$$\|\hat{R}\|_\infty = \max_{i\in\mathcal{J}}(|r_{i,i+2}| + |r_{i,i-2}|) \leq 4\varepsilon_1\beta_2. \tag{5.20}$$

### 5.4. Proof of Proposition 5.1

Since $\hat{r}_{j,j} = 0$ for $j \in \mathcal{J}$, we can find $A' \in \operatorname{End}(V_{\mathcal{J}})$ such that $[\mathrm{i}A', D] = \hat{R}$. Moreover $\hat{R} = P_2\hat{R}P_2$ implies $A' = P_2A'P_2$ and

$$\|A'\|_\infty \le \frac{\|\hat{R}\|_\infty}{\gamma_2} \le \frac{4\varepsilon_1\beta_2}{\gamma_2} = 4\varepsilon_1\varepsilon_2. \tag{5.21}$$

Introducing $\hat{J}' := \mathrm{e}^{-\mathrm{i}A'}(\hat{J}_1 + R)\mathrm{e}^{\mathrm{i}A'}$ and using $D + R = D' + \hat{R}$ we can write

$$\hat{J}' = \mathrm{e}^{-\mathrm{i}A'}(D + \hat{R})\mathrm{e}^{\mathrm{i}A'} + \mathrm{e}^{-\mathrm{i}A'}(D' - D + B - B_1)\mathrm{e}^{\mathrm{i}A'}.$$

Since $\hat{R} = [\mathrm{i}A', D]$ and

$$(B - B_3)\big(\mathrm{e}^{\mathrm{i}sA'} - I\big) = (B - B_3)(I - P_2)\big(\mathrm{e}^{\mathrm{i}sA'} - I\big) = 0$$

implies $\mathrm{e}^{-\mathrm{i}A'}(B - B_3)\mathrm{e}^{\mathrm{i}A'} = B - B_3$, we can write

$$\hat{J}' = \mathrm{e}^{-\mathrm{i}A'}(D + [\mathrm{i}A', D])\mathrm{e}^{\mathrm{i}A'} + \mathrm{e}^{-\mathrm{i}A'}(D' - D + B_3 - B_1)\mathrm{e}^{\mathrm{i}A'} + B - B_3$$

and introducing $\hat{J}_1' := D' + B - B_1$ we find

$$\hat{J}' - \hat{J}_1' = G_{A'}(D) + F_{A'}(D' - D + B_3 - B_1).$$

Thus (4.14) and (4.15) allow us to estimate

$$\|\hat{J}' - \hat{J}_1'\|_2 \le (\|\hat{R}\|_2 + 2\|D' - D + B_3 - B_1\|_2)\|A'\|_2. \tag{5.22}$$

Using $\|B_3 - B_1\|_2 \le 2\beta_3$, (5.18), (5.20) and (5.21) to estimate the right-hand side of (5.22) we obtain

$$\|\hat{J}' - \hat{J}_1'\|_2 \le 4\varepsilon_1\varepsilon_2\,(6\varepsilon_1\beta_2 + 4\beta_3) \le 28\varepsilon_1\varepsilon_2\beta_3, \tag{5.23}$$

where we have used $\varepsilon_1 \le \frac{1}{2}$. Combining $\lambda_0(J) = \lambda_0(J_1)$ with $\lambda_0(\hat{J}_1 + R) = \lambda_0(\hat{J}')$ we estimate

$$|\lambda_0(J) - \lambda_0(\hat{J}_1')| \le |\lambda_0(J_1) - \lambda_0(\hat{J}_1 + R)| + |\lambda_0(\hat{J}') - \lambda_0(\hat{J}_1')| \tag{5.24}$$

and the min-max principle allows us to estimate the right-hand side of (5.24) by

$$\|J_1 - \hat{J}_1 - R\|_2 + \|\hat{J}' - \hat{J}_1'\|_2 \le 48\varepsilon_1\varepsilon_2\beta_3 \tag{5.25}$$

due to (5.14) and (5.23). Thus in order to complete the proof it remains to show that $\lambda_0(\hat{J}_1') = d_0 + r_{0,0} = d_0'$. For this purpose we introduce

$$\hat{d}_k'^{+} := \max_{j\le k}(d_j' + |b_j| + |b_{j-1}|), \tag{5.26a}$$

$$\hat{d}_k'^{-} := \min_{j\ge k}(d_j' - |b_j| - |b_{j-1}|) \tag{5.26b}$$

and we observe that (5.18) ensures $|\hat{d}_k'^{\pm} - \hat{d}_k^{\pm}| \le \max_{j\in\mathcal{J}}|d_j' - d_j| \le \beta_1\varepsilon_1 \le \frac{1}{2}\beta_1$. Thus $\hat{d}_1'^{-} - d_0' \ge \hat{d}_1^{-} - d_0 - \beta_1 \ge \hat{d}_1^{-} - \hat{d}_0^{+} \ge 0$. By similar considerations we obtain $d_0' - \hat{d}_{-1}'^{+} \le 0$, completing the proof of $\lambda_0(\hat{J}_1') = d_0'$ by Lemma 3.2 applied to $\hat{J}_1'$ instead of $\hat{J}_1$. □

## 6. Successive approximations

### 6.1. Main result of this section

In this section $J$ is always the $(2l+1)\times(2l+1)$ Jacobi matrix defined by (3.9). Let $1 \le m \le l$ and let $J_0 := J$. We construct by induction a sequence of unitary equivalent operators $J_1, J_2, \dots, J_m$ such that, for $k = 1, \dots, m$,

$$J_k = \mathrm{e}^{-\mathrm{i}A_k} J_{k-1} \mathrm{e}^{\mathrm{i}A_k} \tag{6.1}$$

for some self-adjoint operator $A_k \in \mathrm{End}(V_{\mathcal{J}})$ satisfying

$$A_k = P_k A_k P_k. \tag{6.2}$$

We already defined $\hat{J}_1 = D + B - B_1$ by (3.26) and $J_1 = \mathrm{e}^{-\mathrm{i}A_1} J_0 \mathrm{e}^{\mathrm{i}A_1}$ by (4.4) with $A_1$ given by (4.16). Then by (4.18)

$$R_1 := J_1 - \hat{J}_1 = G_{A_1}(D) + F_{A_1}(B_2 - B_1). \tag{6.3}$$

**Proposition 6.1.** *Let $J$ be the finite Jacobi matrix given by (3.9). Let $2 \le m \le l$ such that*

- $\varepsilon_m \le 1/18$, i.e.,

$$18\beta_m \le \gamma_m. \tag{6.4}$$

*Then for any $1 \le k \le m$ one can find a self-adjoint operator $A_k \in \mathrm{End}(V_{\mathcal{J}})$ satisfying (6.2) and such that (6.1) gives $J_k \in \mathrm{End}(V_{\mathcal{J}})$ of the form*

$$J_k = D_k + B - B_1 + R_k, \tag{6.5}$$

*where $D_k = \mathrm{diag}(d_j^{(k)})_{j\in\mathcal{J}}$ is a diagonal matrix whose entries satisfy*

$$d_j^{(k)} = d_j \quad \text{if } |j| \ge k+1, \tag{6.6}$$

$$\max_{j\in\mathcal{J}} \left| d_j^{(k)} - d_j^{(k-1)} \right| \le \|R_{k-1}\|_2 \quad \text{for } k \ge 2, \tag{6.7}$$

*and where $R_k$ satisfies*

$$R_k = P_{k+1} R_k P_{k+1}, \tag{6.8}$$

$$\|R_k\|_\infty \le 6\,\frac{\beta_{k+1}}{\gamma_k}\,\|R_{k-1}\|_\infty \quad \text{for } k \ge 2, \tag{6.9}$$

$$\|R_1\|_\infty \le \frac{9}{16}\,\beta_2. \tag{6.10}$$

*Remark.* Assumption (6.4) gives the estimates

$$\varepsilon \le 2\varepsilon_1 \le \cdots \le 2\varepsilon_m \le \frac{1}{9}, \tag{6.11}$$

and also, for any $2 \le k \le m$,

$$18\beta_2 \le 18\beta_k \le 18\beta_m \le \gamma_m \le \gamma_k \le \gamma_2. \tag{6.12}$$

### 6.2. First step ($k = 1$)

Let $D_1 := D$ and $A_1$ as in (4.16). Then (6.2) holds for $k = 1$ and (6.5) for $k = 1$ is equivalent to (6.3). Moreover $R_1 = P_2 R_1 P_2$ and

$$\|R_1\|_2 \le (\|B_1\|_2 + 2\|B_2 - B_1\|_2)\|A_1\|_2 \le 4\beta_2\varepsilon \tag{6.13}$$

due to (4.17), (4.21) and (4.22). We claim that we also have

$$\|R_1\|_\infty \le 4\beta_2\varepsilon \mathrm{e}^{2\varepsilon}. \tag{6.14}$$

Indeed, since $\|\mathrm{e}^{\mathrm{i}A}\|_\infty \le \sum_{n\in\mathbb{N}} \frac{1}{n!}\|A^n\|_\infty \le \mathrm{e}^{\|A\|_\infty}$, we can estimate the norm $\|\cdot\|_\infty$ of the expression (4.12) by

$$\|F_A(Q)\|_\infty \le \|[Q, A]\|_\infty \mathrm{e}^{2\|A\|_\infty} \le 2\|Q\|_\infty\|A\|_\infty \mathrm{e}^{2\|A\|_\infty}. \tag{6.15}$$

Similarly we can estimate the norm $\|\cdot\|_\infty$ of the expression (4.13) by

$$\|G_A(D)\|_\infty \le \frac{1}{2}\|\mathrm{ad}_A^2(D)\|_\infty \mathrm{e}^{2\|A\|_\infty}$$

and using $\|\mathrm{ad}_A^2(D)\|_\infty \le 2\|[D, A]\|_\infty\|A\|_\infty$, we obtain

$$\|G_A(D)\|_\infty \le \|[A, D]\|_\infty\|A\|_\infty \mathrm{e}^{2\|A\|_\infty}. \tag{6.16}$$

Thus combining (6.15) and (6.16) with $[\mathrm{i}A_1, D] = B_1$ to estimate (6.3) we find

$$\|R_1\|_\infty \le \big(\|B_1\|_\infty + 2\|B_2 - B_1\|_\infty\big)\,\|A_1\|_\infty\, \mathrm{e}^{2\|A\|_\infty} \le 4\beta_2\varepsilon \mathrm{e}^{2\varepsilon}. \tag{6.17}$$

Moreover $\|A_1\|_\infty \le \varepsilon \le 2\varepsilon_1 \le 2\varepsilon_m \le \frac{1}{9}$ ensures

$$\|R_1\|_\infty \le \frac{4}{9}\mathrm{e}^{2/9}\beta_2 \le \frac{9}{16}\beta_2. \tag{6.18}$$

### 6.3. Induction step ($k = 2, \dots, m$)

Let $D_1 = D$, $A_1$, $R_1$ be as before and fix $k \in [\![2, m]\!]$. We assume that we have already defined $D_{k-1}$ satisfying (6.6) and (6.7), and $R_{k-1}$ such that $R_{k-1} = P_k R_{k-1} P_k$.

#### 6.3.1. Construction of $D_k$.

We define

$$\hat{R}_k = (\hat{r}_{i,j}^{(k)})_{i,j\in\mathcal{J}}$$

with

$$\hat{r}_{i,j}^{(k)} := r_{i,j}^{(k-1)}(1 - \delta_{i,j}) \tag{6.19}$$

and we take $D_k := \mathrm{diag}(d_j^{(k)})_{j\in\mathcal{J}}$ as the diagonal part of $D_{k-1} + R_{k-1}$, i.e.,

$$d_j^{(k)} := d_j^{(k-1)} + r_{j,j}^{(k-1)}. \tag{6.20}$$

Therefore we have $\|\hat{R}_k\|_\infty \le \|R_{k-1}\|_\infty$ and

$$\max_{j\in\mathcal{J}} |d_j^{(k)} - d_j^{(k-1)}| = \max_{|j|\le k} |d_j^{(k)} - d_j^{(k-1)}| = \max_{|j|\le k} |r_{j,j}^{(k-1)}|, \tag{6.21}$$

hence (6.7) holds.

**6.3.2. Construction of $A_k$.** Since $\hat{r}^{(k)}_{j,j} = 0$ we can find $A_k$ self-adjoint such that

$$[\,\mathrm{i}A_k, D] = \hat{R}_k. \tag{6.22}$$

Since $\hat{R}_k = P_k\hat{R}_kP_k$ we can take $A_k$ satisfying $A_k = P_kA_kP_k$ and

$$\|A_k\|_\infty \le \frac{\|R_{k-1}\|_\infty}{\gamma_k}. \tag{6.23}$$

**6.3.3. Construction of $R_k$.** Since $D_{k-1}+R_{k-1} = D_k+\hat{R}_k = D+[\,\mathrm{i}A_k, D]+(D_k - D)$, introducing $J_k$ by (6.1) we obtain (6.5) with

$$R_k := G_{A_k}(D) + F_{A_k}(B - B_1 + D_k - D). \tag{6.24}$$

Since $A_k = P_kA_kP_k$ ensures $F_{A_k}(B - B_1) = F_{A_k}(B_{k+1} - B_1)$ we can use (6.15) and (6.16) to estimate

$$\|R_k\|_\infty \le \rho_k \mathrm{e}^{2\|A_k\|_\infty}\|A_k\|_\infty \tag{6.25}$$

with

$$\rho_k := \|R_{k-1}\|_\infty + 2\|B_{k+1} - B_1 + D_k - D\|_\infty. \tag{6.26}$$

It only remains to prove estimate (6.9) for $k = 2,\dots,m$.

### 6.4. Estimate of $R_2$

We check that (6.9) holds for $k = 2$. To begin we estimate

$$\|A_2\|_\infty \le \frac{\|R_1\|_\infty}{\gamma_2} \le \frac{9}{16}\frac{\beta_2}{\gamma_2} \le \frac{9}{16}\varepsilon_m \le \frac{1}{32} \tag{6.27}$$

using (6.18) and (6.4). Then $\|D_2 - D\|_2 \le \|R_1\|_2$ and $\|B_3 - B_1\|_\infty \le 2\beta_3$ allow us to estimate

$$\rho_2 \le \|R_1\|_\infty + 4\beta_3 + 2\|R_1\|_2 \le 4(\varepsilon\mathrm{e}^{2\varepsilon} + 1 + 2\varepsilon)\beta_3$$

and

$$\rho_2\mathrm{e}^{2\|A_2\|_\infty} \le 4\Big(\frac{1}{9}\mathrm{e}^{2/9} + 1 + \frac{2}{9}\Big)\mathrm{e}^{1/16}\beta_3 \le 6\beta_3. \tag{6.28}$$

Thus combining (6.25) with (6.28) and (6.27) we obtain

$$\|R_2\|_\infty \le \rho_2\mathrm{e}^{2\|A_2\|_\infty}\|A_2\|_\infty \le 6\beta_3\|A_2\|_\infty < 6\frac{\beta_3}{\gamma_2}\|R_1\|_\infty.$$

### 6.5. Proof of Proposition 6.1 (end)

It only remains to prove (6.9) for $k \ge 3$. For simplicity we can even assume $k = m$. By induction hypothesis we already know that (6.9) holds for $k = 2,\dots,m-1$. Our assumption (6.4) (see (6.12)) implies $18\beta_{k+1} \le \gamma_k$ for any $2 \le k \le m-1$. Using this estimate in (6.9) we obtain $\|R_k\|_\infty \le \frac{1}{3}\|R_{k-1}\|_\infty$, hence, using (6.18)

$$\|R_k\|_\infty \le 3^{1-k}\|R_1\|_\infty \le \frac{9}{16}\beta_2 \tag{6.29}$$

for any $2 \le k \le m-1$. By (6.4) we also have $18\beta_2 \le 18\beta_m \le \gamma_m$, hence, using (6.23)

$$\|A_m\|_\infty \le \frac{\|R_{m-1}\|_\infty}{\gamma_m} \le \frac{9}{16}\cdot\frac{\beta_2}{\gamma_m} \le \frac{9}{16}\cdot\frac{1}{18} = \frac{1}{32}\,. \tag{6.30}$$

Using $D_m - D = D_m - D_1 = \sum_{k=2}^{m}(D_k - D_{k-1})$ and (6.7) we obtain

$$\|D_m - D\|_\infty \le \sum_{k=1}^{m-1} \|R_k\|_2, \tag{6.31}$$

hence

$$\begin{aligned}\rho_m &= \|R_{m-1}\|_\infty + 2\|B_{m+1} - B_1 + D_m - D\|_\infty \\ &\le \|R_{m-1}\|_\infty + 2\|B_{m+1} - B_1\|_\infty + 2\sum_{k=1}^{m-1}\|R_k\|_2 \\ &\le 4\beta_{m+1} + 2\|R_1\|_2 + \|R_{m-1}\|_\infty + 2\sum_{k=2}^{m-1}\|R_k\|_\infty .\end{aligned}$$

However (6.29) allows us to estimate

$$\|R_{m-1}\|_\infty + \sum_{k=2}^{m-1} 2\|R_k\|_\infty \le \Big(3^{2-m} + 2\sum_{k=2}^{m-1} 3^{1-k}\Big)\|R_1\|_\infty = \|R_1\|_\infty$$

and using (6.14) and (6.17), we find

$$\rho_m \le 4\beta_{m+1} + 2\|R_1\|_2 + \|R_1\|_\infty \le 4\big(1 + 2\varepsilon + \varepsilon e^{2\varepsilon}\big)\beta_{m+1}.$$

Thus

$$\rho_m e^{2\|A_m\|_\infty} \le 4\big(1 + \tfrac{2}{9} + \tfrac{1}{9}e^{2/9}\big)e^{1/16}\beta_{m+1} \le 6\beta_{m+1} \tag{6.32}$$

holds due to $\varepsilon \le \frac{1}{9}$ and (6.30). Then (6.23), (6.25) with $k = m$ and (6.32) give

$$\|R_m\|_\infty \le \rho_m e^{2\|A_m\|_\infty} \frac{\|R_{m-1}\|_\infty}{\gamma_m} \le 6\,\frac{\beta_{m+1}}{\gamma_m}\|R_{m-1}\|_\infty,$$

i.e., (6.9) holds for $k = m$. □

### 6.6. Middle eigenvalue estimate: last improvement

**Corollary 6.2.** *Let $J$, $m \ge 2$, $J_1, \dots, J_m$, and $R_1, \dots, R_m$ be as in Proposition 6.1. Assume that*

- $18\beta_m \le \gamma_m$,
- $\hat{d}^+_{-2} \le d_{-1} \le d_1 \le \hat{d}^-_2$.

*Then*

$$|\lambda_0(J) - d_0^{(m)}| \le \|R_m\|_2 \le 6^m \beta_{m+1}\varepsilon \prod_{2\le k\le m} \varepsilon_k \tag{6.33}$$

*where $d_0^{(m)}$ is the middle diagonal entry of the $m$th approximation $J_m$.*

*Proof.* We first check that it is possible to replace $\hat{J}_1$ by $\hat{J}_m = D_m + B - B_1$ in Lemma 3.2. For this purpose we introduce

$$\hat{d}_k^{(m,+)} := \max_{j\le k}\{d_j^{(m)} + |b_j| + |b_{j-1}|\}, \tag{6.34a}$$

$$\hat{d}_k^{(m,-)} := \min_{j\ge k}\{d_j^{(m)} - |b_j| - |b_{j-1}|\}, \tag{6.34b}$$

hence

$$|\hat{d}_k^{(m,\pm)} - \hat{d}_k^{\pm}| \leq \max_{j \in \mathcal{J}} |d_j^{(m)} - d_j| \leq \|D_m - D\|_\infty.$$

However (6.31), (6.29) and (6.18) ensure

$$\|D_m - D\|_\infty \leq \sum_{k=1}^{m} 3^{1-k} \|R_1\|_\infty < \beta_2,$$

hence $\hat{d}_1^{(m,-)} - d_0^{(m)} \geq \hat{d}_1^- - d_0 - 2\beta_2 \geq \gamma_1 - 4\beta_2 \geq 0$, where we have used $\hat{d}_2^- \geq d_1 \implies \hat{d}_1^- = d_1 - |b_1| - |b_0|$. Similarly we check that $d_0^{(m)} - \hat{d}_{-1}^{(m,+)} \geq 0$, hence Lemma 3.2 applied to $\hat{J}_m$ ensures $\lambda_0(\hat{J}_m) = d_0^{(m)}$. But $R_m = J_m - \hat{J}_m$ gives

$$|\lambda_0(J) - d_0^{(m)}| = |\lambda_0(J_m) - \lambda_0(\hat{J}_m)| \leq \|R_m\|_2$$

due to the min-max principle. To complete the proof it remains to observe that

$$\|R_m\|_2 \leq 6^{m-1} \|R_1\|_\infty \prod_{2 \leq k \leq m} \frac{\beta_{k+1}}{\gamma_k}$$

and $\|R_1\|_\infty \leq 6\varepsilon\beta_2$. □

### 6.7. Computing an approximation of the middle eigenvalue

We indicate briefly how to compute an approximation of $\lambda_0(J)$ by means of linear combinations and products of matrices belonging to $\mathrm{End}(V_{\mathcal{J}})$. Using Corollary 6.2 we look for an approximation of $d_0^{(m)}$ with the error considered in the right-hand side of (6.24), i.e., it suffices to replace the formula (6.1) by an approximation using linear combinations and products of matrices. For this purpose we observe that

$$\partial_s^j F_{sA}(Q) = \mathrm{e}^{-\mathrm{i}sA}\, \mathrm{i}^j\, \mathrm{ad}_A^j(Q) \mathrm{e}^{\mathrm{i}sA},$$

where $\partial_s = \mathrm{d}/\mathrm{d}s$, $\mathrm{ad}_A^1(Q) = \mathrm{ad}_A(Q) = [Q, A]$ and $\mathrm{ad}_A^{j+1}(Q) = [\mathrm{ad}_A^j(Q), A]$. Thus the $N$th remainder in the Taylor's development of $s \to F_{sA}(Q)$ is

$$F_A(Q) - \sum_{1 \leq j \leq N} \frac{\mathrm{i}^j}{j!} \mathrm{ad}_A^j(Q) = \int_0^1 \frac{(1-s)^N}{N!} \mathrm{e}^{-\mathrm{i}sA}\, \mathrm{i}^{N+1}\, \mathrm{ad}_A^{N+1}(Q) \mathrm{e}^{\mathrm{i}sA} \mathrm{d}s. \quad (6.35)$$

Its $\|\cdot\|_\infty$ norm can be estimated by $2^{N+1}\|Q\|_\infty \|A\|_\infty^{N+1}/(N+1)!$. Thus it is clear that neglecting the remainders (6.35) with $N$ of order $m$ we can express $J_k$ and $D_k$ with the indicated errors.

## 7. Proofs of the main results

### 7.1. Notations

We denote by $\{\mathrm{e}_n\}_{n=1}^\infty$ the canonical basis of $l^2$, i.e., $\mathrm{e}_n = (\delta_{k,n})_{k=1}^\infty$ with $\delta_{n,n} = 1$ and $\delta_{k,n} = 0$ for $k \neq n$. If $q_j \in \mathbb{C}$ for $j \in \mathbb{N}^*$, then $Q = \mathrm{diag}(q_j)_{j \in \mathbb{N}^*}$ denotes the closed diagonal operator defined by $Q\,\mathrm{e}_j = q_j\,\mathrm{e}_j$ for $j \in \mathbb{N}^*$. In this section

$J\colon \mathcal{D}(J) \to l^2$ is the unbounded Jacobi operator defined by (1.4), (1.5) under the assumptions (1.2) and (1.3). We decompose

$$J = D + B, \tag{7.1}$$

where $D = \operatorname{diag}(d_j)_{j\in\mathbb{N}^*}$ is the diagonal part of $J$ and $B$ is the off-diagonal part of $J$, i.e.,

$$Bx = (\bar{b}_j x_{j+1} + b_{j-1}x_{j-1})_{j\in\mathbb{N}^*}$$

for $x \in \mathcal{D}(J)$ with the convention $x_0 = 0$ and $b_0 = 0$.

### 7.2. Proof of Theorem 2.1

We begin with the extension of Lemma 3.1 to infinite Jacobi matrices.

**Lemma 7.1.** *Let $J$ be as in Theorem 2.1. If*

$$\begin{aligned} D_\pm &:= \operatorname{diag}(d_j^\pm(J))_{j\in\mathbb{N}^*},\\ d_j^\pm(J) &:= d_j \pm (|b_j| + |b_{j-1}|), \end{aligned} \tag{7.2}$$

*then*

(i) $D_- \le J \le D_+$,
(ii) $\hat{d}_n^- \le \lambda_n(J) \le \hat{d}_n^+$ *with* $\hat{d}_n^\pm$ *as in* (2.2).

*Remark.* (i) means $\langle D_-x,\, x\rangle \le \langle (D+B)x,\, x\rangle \le \langle D_+x,\, x\rangle$ for any $x \in \mathcal{D}(J)$ and (ii) means $\lambda_n(J) \in \Delta_n$.

*Proof.* We obtain $D_- \le J \le D_+$ as in the proof of Lemma 3.1 and similarly we deduce $\lambda_n(D_-) \le \lambda_n(J) \le \lambda_n(D_+)$ by means of the min-max principle (cf. [17]), completing the proof due to $\hat{d}_n^- \le \lambda_n(D_-) \le \lambda_n(D_+) \le \hat{d}_n^+$ as before. □

*Proof of Theorem* 2.1. Let $R\colon l^2 \to l^2$ be a self-adjoint and bounded linear operator, then similarly as in Section 3.2 the min-max principle (cf. [17]) gives

$$|\lambda_n(J+R) - \lambda_n(J)| \le \|R\|_2 := \sup_{\substack{x\in l^2\\ \|x\|_2\le 1}} \|Rx\|_2. \tag{7.3}$$

Let $B^{n,1}\colon l^2 \to l^2$ be defined by $B^{n,1}x = y_{n-1}e_{n-1} + y_n e_n + y_{n+1}e_{n+1}$, where $x = (x_j)_1^\infty$ and

$$\begin{pmatrix} y_{n-1}\\ y_n\\ y_{n+1}\end{pmatrix} = \begin{pmatrix} 0 & \bar{b}_{n-1} & 0\\ b_{n-1} & 0 & \bar{b}_n\\ 0 & b_n & 0\end{pmatrix} \begin{pmatrix} x_{n-1}\\ x_n\\ x_{n+1}\end{pmatrix}. \tag{7.4}$$

If $J^{n,1} := D + B - B^{n,1}$, then

$$|\lambda_n(J) - \lambda_n(J^{n,1})| \le \|B^{n,1}\|_2 \le |b_n| + |b_{n-1}|. \tag{7.5}$$

Applying Lemma 7.1 to $J^{n,1}$ instead of $J$ we find

$$\min(d_n, \hat{d}_{n+1}^-) \le \lambda_n(J^{n,1}) \le \max(\hat{d}_{n-1}^+, d_n).$$

Therefore $\lambda_n(J^{n,1}) = d_n$ follows from assumption (2.3): $\hat{d}_{n-1}^+ \le d_n \le \hat{d}_{n+1}^-$. Then (7.5) becomes the estimate (2.4). □

### 7.3. Estimates by middle eigenvalues of finite Jacobi submatrices

Let $P^{n,l}$ denote the orthogonal projection on the subspace generated by $\{e_j\}_{j=n-l}^{n+l}$, $n > l$, i.e.,

$$P^{n,l}x = \sum_{j=n-l}^{n+l} x_j\, e_j, \tag{7.6}$$

and $B^{n,l} := P^{n,l}BP^{n,l}$. Then $B^{n,l}x = (\bar{b}_j^{n,l}x_{j+1} + b_{j-1}^{n,l}x_{j-1})_{j\in\mathbb{N}^*}$ holds with

$$b_j^{n,l} = \begin{cases} b_j & \text{if } n-l \le j < n+l, \\ 0 & \text{otherwise.} \end{cases}$$

We introduce

$$\hat{b}_j^{n,l} = b_j - b_j^{n,l} = \begin{cases} 0 & \text{if } n-l \le j < n+l, \\ b_j & \text{otherwise,} \end{cases}$$

and

$$D_\pm^{n,l} := D \pm \operatorname{diag}\big(|\hat{b}_j^{n,l}| + |\hat{b}_{j-1}^{n,l}|\big)_{j\in\mathbb{N}^*}. \tag{7.7}$$

Then replacing $B$ by $\hat{B}^{n,l} := B - B^{n,l}$ in Lemma 7.1 we obtain

$$D_-^{n,l} \le D + \hat{B}^{n,l} \le D_+^{n,l}, \tag{7.8}$$

hence $D_-^{n,l} + B^{n,l} \le D + \hat{B}^{n,l} + B^{n,l} \le D_+^{n,l} + B^{n,l}$. Since $D + \hat{B}^{n,l} + B^{n,l} = J$ we can rewrite the last inequality in the form

$$J_-^{n,l} \le J \le J_+^{n,l}, \tag{7.9}$$

where

$$J_\pm^{n,l} := D_\pm^{n,l} + B^{n,l} = \begin{pmatrix} \operatorname{diag}(d_j^\pm)_{j=1}^{n-l-1} & 0 & 0 \\ 0 & J_{n,l}^\pm & 0 \\ 0 & 0 & \operatorname{diag}(d_j^\pm)_{j=n+l+1}^\infty \end{pmatrix} \tag{7.10}$$

with

$$J_{n,l}^\pm = J_{n,l} + \operatorname{diag}(|b_{n-l-1}|, 0, \dots, 0, |b_{n+l}|), \tag{7.11}$$

i.e.,

$$J_{n,l}^\pm = \begin{pmatrix} d_{n-l} \pm |b_{n-l-1}| & \bar{b}_{n-l} & 0 & \dots & 0 & 0 & 0 \\ b_{n-l} & d_{n-l+1} & \bar{b}_{n-l+1} & \dots & 0 & 0 & 0 \\ 0 & b_{n-l+1} & d_{n-l+2} & \dots & 0 & 0 & 0 \\ \vdots & \vdots & \vdots & \ddots & \vdots & \vdots & \vdots \\ 0 & 0 & 0 & \dots & d_{n+l-2} & \bar{b}_{n+l-2} & 0 \\ 0 & 0 & 0 & \dots & b_{n+l-2} & d_{n+l-1} & \bar{b}_{n+l-1} \\ 0 & 0 & 0 & \dots & 0 & b_{n+l-1} & d_{n+l} \pm |b_{n+l}| \end{pmatrix}.$$

In the next lemma, the *finite* Jacobi matrices $J_{n,l}^\pm$ are indexed by $[\![-l, l]\!] \times [\![-l, l]\!]$, and the *infinite* Jacobi matrices $J$ and $J_\pm^{n,l}$ are indexed by $\mathbb{N}^* \times \mathbb{N}^*$.

**Lemma 7.2.** *Let $J$ be as in Theorem 2.1. Let $n, l \in \mathbb{N}^*$, $n > l$. Let $J_{\pm}^{n,l}$ and $J_{n,l}^{\pm}$ denote the Jacobi matrices defined by (7.10) and (7.11), respectively. Assume that*

- $\hat{d}_{n-1}^{+} < \hat{d}_n^{-} \leq \hat{d}_n^{+} < \hat{d}_{n+1}^{-}$.

*Then*

(i) $\lambda_n(J_{\pm}^{n,l}) = \lambda_0(J_{n,l}^{\pm})$,

(ii) *the nth eigenvalue of $J$ is estimated by the middle eigenvalues of $J_{n,l}^{\pm}$:*

$$\lambda_0(J_{n,l}^{-}) \leq \lambda_n(J) \leq \lambda_0(J_{n,l}^{+}). \tag{7.12}$$

*Proof.* Due to the min-max principle one has

$$J_{-}^{n,l} \leq J \leq J_{+}^{n,l} \implies \lambda_k(J_{-}^{n,l}) \leq \lambda_k(J) \leq \lambda_k(J_{+}^{n,l})$$

for all $k \in \mathbb{N}^*$, then it only remains to show

$$\lambda_n(J_{\pm}^{n,l}) = \lambda_0(J_{n,l}^{\pm}). \tag{7.13}$$

To begin we use Lemma 7.1 with $J_{\pm}^{n,l}$ instead of $J$ and we find that

$$d_j^{-}(J) = d_j^{-}(J_{-}^{n,l}) \leq d_j^{+}(J_{+}^{n,l}) = d_j^{+}(J)$$

implies $D_{-} \leq J_{-}^{n,l} \leq J_{+}^{n,l} \leq D_{+}$ and

$$\hat{d}_k^{-} \leq \lambda_k(J_{-}^{n,l}) \leq \lambda_k(J_{+}^{n,l}) \leq \hat{d}_k^{+}$$

for $k \in \mathbb{N}^*$. Thus the hypothesis $\hat{d}_{n-1}^{+} < \hat{d}_n^{-} \leq \hat{d}_n^{+} < \hat{d}_{n+1}^{-}$ ensures

$$\{\lambda_n(J_{\pm}^{n,l})\} = \sigma(J_{\pm}^{n,l}) \cap [\hat{d}_n^{-}, \hat{d}_n^{+}].$$

Reasoning similarly we can estimate eigenvalues of finite matrices

$$\hat{d}_k^{-}(J_{n,l}^{-}) \leq \lambda_k(J_{n,l}^{-}) \leq \lambda_k(J_{n,l}^{+}) \leq \hat{d}_k^{+}(J_{n,l}^{+})$$

with

$$\begin{aligned} \hat{d}_k^{-}(J_{n,l}^{-}) &= \inf_{k \leq j \leq l} d_{n+j}^{-}(J), \\ \hat{d}_k^{+}(J_{n,l}^{+}) &= \sup_{-l \leq j \leq k} d_{n+j}^{+}(J) \end{aligned}$$

for $k \in [\![-l, l]\!]$. Next we observe that

$$\begin{aligned} d_n < \hat{d}_{n+1}^{-} \leq \hat{d}_1^{-}(J_{n,l}^{-}) &\implies d_n - |b_n| - |b_{n-1}| = \hat{d}_n^{-} = \hat{d}_0^{-}(J_{n,l}^{-}), \\ d_n > \hat{d}_{n-1}^{+} \geq \hat{d}_{-1}^{+}(J_{n,l}^{+}) &\implies d_n + |b_n| + |b_{n-1}| = \hat{d}_n^{+} = \hat{d}_0^{+}(J_{n,l}^{+}) \end{aligned}$$

ensures

$$\hat{d}_n^{-} \leq \lambda_0(J_{n,l}^{-}) \leq \lambda_0(J_{n,l}^{+}) \leq \hat{d}_n^{+}.$$

Using moreover

$$\begin{aligned} &\lambda_{-1}(J_{n,l}^{-}) \leq \lambda_{-1}(J_{n,l}^{+}) \leq \hat{d}_{-1}^{+}(J_{n,l}^{+}) \leq \hat{d}_{n-1}^{+} < \hat{d}_n^{-}, \\ &\hat{d}_n^{+} < \hat{d}_{n+1}^{-} \leq \hat{d}_1^{-}(J_{n,l}^{-}) \leq \lambda_1(J_{n,l}^{-}) \leq \lambda_1(J_{n,l}^{+}), \end{aligned}$$

we deduce

$$\{\lambda_0(J^{\pm}_{n,l})\} = \sigma(J^{\pm}_{n,l}) \cap [\hat{d}^-_n, \hat{d}^+_n].$$

To complete the proof we observe that the block structure of (7.10) gives

$$\sigma(J^{n,l}_{\pm}) = \sigma(J^{\pm}_{n,l}) \cup \{d^{\pm}_k(J) : k \in \mathbb{N}^* \setminus [n-l,\, n+l]\}$$

and $k \notin [n-l; n+l] \implies d^{\pm}_k(J) \notin [\hat{d}^-_n, \hat{d}^+_n]$ ensures the equality

$$\sigma(J^{\pm}_{n,l}) \cap [\hat{d}^-_n, \hat{d}^+_n] = \sigma(J^{n,l}_{\pm}) \cap [\hat{d}^-_n, \hat{d}^+_n],$$

i.e., $\{\lambda_n(J^{n,l}_{\pm})\} = \{\lambda_0(J^{\pm}_{n,l})\}$. Thus, (7.13) is proved. □

### 7.4. Proofs of Theorems 2.2 and 2.3

*Proof.* Using Proposition 4.1 with $J = J^{\pm}_{n,2}$ and Theorem 5.1 with $J = J^{\pm}_{n,3}$ we can write for $l = 2$ or 3,

$$\eta^-_{n,l} \le \lambda_0(J^{\pm}_{n,l}) \le \eta^+_{n,l}$$

with

$$\begin{cases} \eta^{\pm}_{n,2} = d_n \pm 4\varepsilon_n\beta_{n,2}, \\ \eta^{\pm}_{n,3} = d_n + r_n \pm 48\varepsilon_{n,1}\varepsilon_{n,2}\beta_{n,3}. \end{cases}$$

Therefore, the estimate (7.12) from Lemma 7.2 gives

$$\eta^-_{n,l} \le \lambda_0(J^-_{n,l}) \le \lambda_n(J) \le \lambda_0(J^+_{n,l}) \le \eta^+_{n,l},$$

completing the proofs of Theorems 2.2 and 2.3. □

### 7.5. Proof of Theorem 2.4

*Proof.* For $k \in [\![1, l-1]\!]$ let $A_{n,l,k}$ be the self-adjoint operators described in Proposition 6.1 with $J = J_{n,l}$ and $m = l-1$. Then denoting

$$U_{n,l} = \exp(\mathrm{i}A_{n,l,1}) \dots \exp(\mathrm{i}A_{n,l,l-1})$$

we can write

$$U^{-1}_{n,l} J_{n,l} U_{n,l} = \mathrm{diag}(d^{(l-1)}_{n,l,j})_{j\in\mathcal{J}} + \hat{B}_{n,l} + R_{n,l},$$

where $\hat{B}_{n,l} = (\langle \mathrm{e}_{n+i}, \hat{B}^{n,1}\, \mathrm{e}_{n+j}\rangle)_{i,j\in\mathcal{J}}$ and Corollary 6.2 ensures

$$|\lambda_0(J_{n,l}) - d^{(l-1)}_{n,l,0}| \le \|R_{n,l}\|_2 \le \eta_{n,l} := 6^{l-1}\varepsilon_n\beta_{n,l}\prod_{k=2}^{l-1}\varepsilon_{n,k}. \tag{7.14}$$

Since $A_{n,l,k}\, \mathrm{e}_{n\pm l} = 0$ for $k \in [\![1, l-1]\!]$ ensures $U_{n,l}\, \mathrm{e}_{n\pm l} = \mathrm{e}_{n\pm l}$, we have

$$\pm\,\mathrm{diag}(|b_{n-l-1}|, 0, \dots, 0, |b_{n+l}|)_{-l\le j\le l} = J^{\pm}_{n,l} - J_{n,l} = U^{-1}_{n,l}(J^{\pm}_{n,l} - J_{n,l})U_{n,l}$$

due to (7.11). For $k \in [\![1, l-1]\!]$ the $k$th step of the procedure of Section 6 applied to $J^{\pm}_{n,l}$ instead of $J_{n,l}$ gives the same corrections on the subspace generated by $\{\mathrm{e}_j\}_{n-l<j<n+l}$. Thus we have the same $A_{n,l,k}$ and the same diagonal entry $d^{(l-1)}_{n,l,0}$ appears as the approximation of $\lambda_0(J^{\pm}_{n,l})$, i.e., we obtain

$$|\lambda_0(J^{\pm}_{n,l}) - d^{(l-1)}_{n,l,0}| \le \eta_{n,l}$$

as well. Combining the last estimate with (7.14) we obtain

$$|\lambda_0(J_{n,l}) - \lambda_0(J^{\pm}_{n,l})| \le 2\eta_{n,l}.$$

To complete the proof we use the estimate (7.12) from Lemma 7.2:

$$\lambda_0(J_{n,l}) - 2\eta_{n,l} \le \lambda_0(J^{-}_{n,l}) \le \lambda_n(J) \le \lambda_0(J^{+}_{n,l}) \le \lambda_0(J_{n,l}) + 2\eta_{n,l}.$$

It remains to explain why Lemma 7.2 applies. Both assumptions $\varepsilon_{n,l} < 1/18$ and $\hat{d}^{+}_{n-2} \le d_{n-1} < d_{n+1} \le \hat{d}^{-}_{n+2}$ of Theorem 2.4 imply

$$\hat{d}^{+}_{k} = d_k + |b_{k-1}| + |b_k| \quad \text{for} \quad k = n-1,\ n,\ n+1,$$
$$\hat{d}^{-}_{k} = d_k - |b_{k-1}| - |b_k| \quad \text{for} \quad k = n-1,\ n,\ n+1.$$

Then, using again the first condition $\varepsilon_{n,l} < 1/18$, we find that the assumption $\hat{d}^{+}_{n-1} \le \hat{d}^{-}_{n} \le \hat{d}^{+}_{n} \le \hat{d}^{-}_{n+1}$ of Lemma 7.2 is satisfied. □

## References


[1] M. Bonini, G.M. Cicuta, and E. Onofri, Fock space methods and large $N$, *J. Phys. A*, **40** (10) (2007), F229–F234.

[2] A. Boutet de Monvel, J. Janas, and L. Zielinski, Asymptotics of large eigenvalues for a class of band matrices, In preparation, 2011.

[3] A. Boutet de Monvel, S. Naboko, and L.O. Silva, Eigenvalue asymptotics of a modified Jaynes–Cummings model with periodic modulations, *C. R. Math. Acad. Sci. Paris*, **338** (1) (2004), 103–107.

[4] A. Boutet de Monvel, S. Naboko, and L.O. Silva, The asymptotic behavior of eigenvalues of a modified Jaynes–Cummings model, *Asymptot. Anal.*, **47** (3-4) (2006), 291–315.

[5] A. Boutet de Monvel and L. Zielinski, Eigenvalue asymptotics for Jaynes–Cummings type models without modulations, preprint 08-03-278, BiBoS, Bielefeld, 2008.

[6] P.A. Cojuhari and J. Janas, Discreteness of the spectrum for some unbounded Jacobi matrices, *Acta Sci. Math. (Szeged)*, **73** (3-4) (2007), 649–667.

[7] J.S. Dehesa, The eigenvalue density of rational Jacobi matrices. II, *Linear Algebra Appl.*, **33** (1980), 41–55.

[8] P. Djakov and B. Mityagin, Simple and double eigenvalues of the Hill operator with a two-term potential, *J. Approx. Theory*, **135** (1) (2005), 70–104.

[9] J. Dombrowski and S. Pedersen, Spectral measures and Jacobi matrices related to Laguerre-type systems of orthogonal polynomials, *Constr. Approx.*, **13** (3) (1997), 421–433.

[10] J. Edward, Spectra of Jacobi matrices, differential equations on the circle, and the su(1, 1) Lie algebra, *SIAM J. Math. Anal.*, **24** (3) (1993), 824–831.

[11] J. Janas and M. Malejki, Alternative approaches to asymptotic behaviour of eigenvalues of some unbounded Jacobi matrices, *J. Comput. Appl. Math.*, **200** (1) (2007), 342–356.

[12] J. Janas and S. Naboko, Infinite Jacobi matrices with unbounded entries: asymptotics of eigenvalues and the transformation operator approach, *SIAM J. Math. Anal.*, **36** (2) (2004), 643–658.

[13] R.V. Kozhan, Asymptotics of the eigenvalues of two-diagonal Jacobi matrices, *Mat. Zametki*, **77** (2) (2005), 313–316.

[14] M. Malejki, Approximation of eigenvalues of some unbounded self-adjoint discrete Jacobi matrices by eigenvalues of finite submatrices, *Opuscula Math.*, **27** (1) (2007), 37–49.

[15] M. Malejki, Asymptotics of large eigenvalues for some discrete unbounded Jacobi matrices, *Linear Algebra Appl.*, **431** (10) (2009), 1952–1970.

[16] D.R. Masson and J. Repka, Spectral theory of Jacobi matrices in $l^2(\mathbf{Z})$ and the su(1, 1) Lie algebra, *SIAM J. Math. Anal.*, **22** (4) (1991) 1131–1146.

[17] M. Reed and B. Simon, *Methods of modern mathematical physics. II. Fourier analysis, self-adjointness*, Academic Press [Harcourt Brace Jovanovich Publishers], New York, 1975.

[18] J. Sánchez-Dehesa, The asymptotical spectrum of Jacobi matrices, *J. Comput. Appl. Math.*, **3** (3) (1977), 167–171.

[19] P.N. Shivakumar, J.J. Williams, and N. Rudraiah, Eigenvalues for infinite matrices, *Linear Algebra Appl.*, **96** (1987), 35–63.

[20] G. Teschl, *Jacobi operators and completely integrable nonlinear lattices*, volume **72** of *Mathematical Surveys and Monographs*, American Mathematical Society, Providence, RI, 2000.

[21] E.A. Tur, Jaynes–Cummings model: solution without rotating wave approximation, *Optics and Spectroscopy*, **89** (4) (2000), 574–588.

[22] E.A. Tur, Asymptotics of eigenvalues for a class of Jacobi matrices with a limit point spectrum, *Mat. Zametki*, **74** (3) (2003), 449–462.

[23] H. Volkmer, Error estimates for Rayleigh-Ritz approximations of eigenvalues and eigenfunctions of the Mathieu and spheroidal wave equation, *Constr. Approx.*, **20** (1) (2004), 39–54.

[24] J.H. Wilkinson, Rigorous error bounds for computed eigensystems, *Comput. J.*, **4** (1961/1962), 230–241.

[25] L. Zielinski, Eigenvalue asymptotics for a class of Jacobi matrices, In *Hot Topics in Operator Theory: Conference Proceedings, Timişoara, June* 29*–July* 4, 2006, volume **9** of *Theta Ser. Adv. Math.*, pages 217–229, Theta, Bucharest, 2008.

Anne Boutet de Monvel
Institut de Mathématiques de Jussieu, Université Paris Diderot Paris 7
175 rue du Chevaleret
F-75013 Paris, France
e-mail: aboutet@math.jussieu.fr

Lech Zielinski
LMPA, Centre Universitaire de la Mi-Voix, Université du Littoral
50 rue F. Buisson, B.P. 699
F-62228 Calais Cedex, France
e-mail: Lech.Zielinski@lmpa.univ-littoral.fr

Operator Theory:
Advances and Applications, Vol. 221, 219-239


# Algebraic Reflexivity and Local Linear Dependence: Generic Aspects

Nir Cohen

**Abstract.** Both reflexivity and the LLD (local linear dependence) property for a space of linear operators are defined in terms of ("local") one-sided action on vectors. We survey the main developments concerning these two areas during the last decade, and observe a major difference between them. Whereas the LLD property is verified on generic sets of (so-called separating) vectors, reflexivity must be tested on the entire vector space. We duly introduce a modified notion of reflexivity (generic reflexivity) which is verified on generic sets of vectors, study its properties, and show that it is closer in spirit to LLD than the usual notion of reflexivity.

In passing, we simplify and complete some recent results on LLD spaces of low dimension, in the special case of matrix spaces. We answer an open problem on matrix pairs $(A, B)$ for which $AB$ belongs to the reflexive hull of $A$ and $B$. Our approach is based on determinant-based genericity constructions and the Kronecker-Weierstrass canonical form.

The study of reflexivity and the LLD property is restricted to operator spaces of very low dimension. We provide basic partial results which apply also in arbitrary dimension.

**Mathematics Subject Classification (2000).** Primary 46A25; Secondary 46B10, 47L05.

**Keywords.** Local linear independence, reflexivity, separating vector, genericity, singular matrix space, matrix pencil.

## 1. Introduction

Both the LLD (locally linear dependence) property (and with it the study of separating vectors) and algebraic reflexivity have their origin in the context of spectrum and invariant subspaces for operators and algebras, in the early 1970's. See for example [1], [5], [6], [10], [11], [12], [16], [17]. Judged by their definitions,

This work was supported by CNPq – Conselho Nacional de Pesquisa, Brazil.

one would expect algebraic reflexivity and the LLD property to be as close, on the local level, as span and linear dependence are on the global level. Nevertheless, in spite of evident similarities, the two theories have not been unified so far. The purpose of the present paper is to establish a simple link between the two theories, based on a critical examination of the recent advances in both areas, which include [4], [5], [7], [24], [25], [26], [27]. Our approach is so far limited to matrix spaces of finite dimension.

Our starting point is the following important difference between the two concepts. It has been observed [11], [15] that the set of separating vectors for a space $\mathbb{S} \in L(X,Y)$ is either "generic" or empty. In other words, to establish the LLD property one needs to check a generic set of vectors, rather than each and every vector. In contrast, checking reflexivity on a generic set is not good enough.

We therefore introduce a modified notion of reflexive hull, the *g-reflexive hull* $\mathbb{S}_{\mathrm{gref}}$, which is checked on generic sets, and which may be strictly larger than the classical reflexive hull, denoted here by $\mathbb{S}_{\mathrm{ref}}$. This naturally leads to a new notion of g-reflexivity, which is weaker than the classical one. Checking against the most complete source on reflexivity available to us, [17], it appears that this construction is new.

The relation between the two hulls is expressed by

$$\mathbb{S} \subset \mathbb{S}_{\mathrm{ref}} \subset \mathbb{S}_{\mathrm{gref}}. \tag{1.1}$$

On the other hand, our definitions will imply that a vector space $\mathbb{L}$ is LLD iff

$$\mathbb{S} \subset \mathbb{L} \subset \mathbb{S}_{\mathrm{gref}} \tag{1.2}$$

for some strict subspace $\mathbb{S}$ of $\mathbb{L}$. It is seen that (1.1) and (1.2) are of the same nature. In particular, we reach the following immediate conclusion: $\mathbb{S}$ is reflexive (resp. g-reflective) if $\mathbb{S}_{\mathrm{ref}}$ (resp. $\mathbb{S}_{\mathrm{gref}}$) is LLI.

In this paper we develop the basic results on generic reflexivity of low-dimensional spaces in finite dimension, in analogy to known results on classical reflexivity. Observing the perfect match of these results with their LLD counterparts, we deduce the latter from the former, using the relation (1.2). Our LLD results are somewhat simpler than those obtained by Breršar and Šemrl [5] and Chebotar and Šemrl [7], though they are restricted to the matrix case.

Along the way we solve an open problem, proposed by Hartwig et al. [18], concerning matrices $A, B$ so that $AB$ belongs to the reflexive hull of $A, B$.

## 2. Overview and results

### 2.1. Preliminaries

The notation introduced here will be used repeatedly in the paper. We fix an underlying field $F$ and linear spaces $X, Y$ over $F$ and consider $L(X,Y)$, the space of linear transformations. In the finite-dimensional case we replace $L(X,Y)$ by $M_{n,m}$ and set $M_n := M_{n,n}$. We denote by $S_k, T_k, U_k$ the spaces of skew-symmetric, upper Toeplitz and upper triangular matrices in $M_k$.

We use $X \to X'$ to denote the algebraic dual, and $A' \in L(Y', X')$ denotes the adjoint of $A \in L(X, Y)$ (or the adjoint matrix for $A \in M_{n,m}$). We shall use LI/LD for linearly independent/dependent (over $F$), and LLI/LLD for the local version. The F-linear span of a set $\mathcal{A}$ will be denoted by $[\mathcal{A}]$.

*Localization* in the space $L(X, Y)$ means the study of any property of a subset $\mathcal{A} \subset L(X, Y)$ in terms of any related property of the "localized sets" indexed by $X$,

$$\mathcal{A}x := \{Ax : A \in \mathcal{A}\} \subset Y \qquad (x \in X). \tag{2.1}$$

Our definition adopts the conventional, or *right-handed* orientation. A dual *left-handed* localization would rather use the sets $y'\mathcal{A} := \{y'A : A \in \mathcal{A}\}$ $(y \in Y)$.

We define the *dimension type* of a linear space $\mathbb{S} \in L(X, Y)$ as the pair $(d, r)$ where

$$d = \dim(\mathbb{S}), \qquad r = \operatorname{rank}(\mathbb{S}) := \dim[\mathbb{S}X]$$

(observe that normally $\mathbb{S}X$ is not a vector space). Up to strict equivalence ($\mathbb{S} \to S\mathbb{S}T$ with $S, T$ fixed and invertible) we may replace $\mathbb{S}$ by a space of matrices whose $j$th row vanishes for all $j > r$. Thus, the number of rows can always be reduced to satisfy $n = r$.

We also define the *local dimension* as

$$\bar{d} = \max_x d(x), \qquad d(x) = \dim(\mathbb{S}x).$$

Considering ranks of individual operators in $\mathbb{S}$, we call $\mathbb{S}$ :

- *k-regular* if $k$ is the rank of every non-zero operator in $\mathbb{S}$;
- *k-admitting* if $\mathbb{S}$ contains a non-zero operator of rank smaller than $k + 1$; and
- *k-forcing* if $\mathbb{S}$ does not contain operators of rank larger than $k$.

Clearly $\mathbb{S}$ is $r$-forcing.

### 2.2. Reflexivity

The (algebraic) reflexive hull $\mathbb{S}_{\text{ref}}$ of a linear subspace $\mathbb{S} \subset L(X, Y)$ consists of all the operators $Z \in L(X, Y)$ for which $Zx \in \mathbb{S}x$ for all $x \in X$. We say that $\mathbb{S}$ is reflexive if $\mathbb{S}_{\text{ref}} = \mathbb{S}$. It is known that $(\mathbb{S}')_{\text{ref}} = (\mathbb{S}_{\text{ref}})'$, hence the right- and left-handed notions of reflexivity coincide [24].

Results in the literature may be roughly divided into results which bound the ranks of matrices in $\mathbb{S}$, and results which characterize operator spaces $\mathbb{S}$ with a given (low) type $(d, r)$. We start with the former.

**Theorem 2.1.** *Assume that a space $\mathbb{S}$ of dimension $d$ is not reflexive.*

(i) *(Meshulam-Šemrl [26] Cor. 2.5). If $|F| \geq d+3$ then $\mathbb{S}$ is either $2d-3$-admitting or $2d-2$-regular.*

(ii) *(Meshulam-Šemrl [27]) If $F$ is algebraically closed then $\mathbb{S}$ is $d$-admitting.*

(In item (i) it follows that a non-reflexive space with $d \geq 3$ cannot be $d$-regular, although it may have k-regular subspaces.) We move to type-oriented results.

**Theorem 2.2.** *If $\min\{d, r\} = 1$ then $\mathbb{S}$ is reflexive.*

For $d = 1$ see [1],[10]; and in the case $r = 1$ we must have $\mathbb{S} = V \otimes U$ with ($U \subset X$, $V \subset Y'$, $\dim(U) = 1$) which can be studied directly. Next we consider $d = 2 \leq r$. Deddens and Fillmore [8], and Bračič and Kuzma [4], found the necessary and sufficient condition for reflexivity under the additional condition that $I \in \mathbb{S}$, using the Jordan form.

The following central result extends Theorem 2.2 to cases where $I \notin \mathbb{S}$. Our proof is quite complicated, and is based on the Kronecker-Weierstrass canonical form:

**Theorem 2.3.** *Assume that $F$ is algebraically closed and $\mathbb{S} \subset M_{n,m}$ is of type $(2, r)$ with $r \geq 2$. Then $\mathbb{S}$ is not reflexive iff $\mathbb{S}$ (resp. $\mathbb{S}_{\rm ref}$) is a copy of $T_2$ (upper Toeplitz), resp. $U_2$ (upper triangular).*

Here, and in the sequel, we say that $\mathbb{S}$ is a copy of $\mathbb{S}'$ if $S = K\mathbb{S}'N$ for some fixed left-invertible matrix $K$ and right-invertible matrix $N$. A very similar result, valid also in infinite dimension, has appeared recently as Theorem 3.10 of [4], using a completely different technique.

**Remarks. 1)** Theorem 2.3 should be compared with a result of Larson and Wogen [23] which exhibits a reflexive operator $T$ so that $T \oplus 0$ is not reflexive. In the matrix case topological reflexivity (used there) coincides with algebraic reflexivity (as used here), which can be treated using Theorem 2.2. **2)** It is easy to see that for Toeplitz matrices $(T_n)_{\rm ref} = U_n$ and $(T_n)_{\rm gref} = M_n$ for all $n$. This, plus Lemma 3.2, provides examples of triangular LLD spaces of arbitrarily large type.

Reflexivity has been studied extensively in the context of Banach algebra theory. We mention in particular the study of quasinilpotency of certain operators, related to a conjecture made by Brešar and Šemrl (see [5], [6], [25]). In these papers, the original quasinilpotency issue was reduced, in the final analysis, to the investigation of pairs $(A, B)$ for which $BA \in [A, B, I]_{\rm ref}$, and more specifically, pairs $(A, B)$ such that for all $x \in F^m$ there exist $a, b \in F$ with $(B - bI)(A - aI)x = 0$. While the original quasinilpotency issue was settled completely in [6], a full characterization of general pairs of these types is still unknown.

The following simpler problem was the initial motivation for the present paper: the classification of pairs $(A, B)$ for which $AB \in [A, B]_{\rm ref}$. For this problem, posed in 1997 by Hartwig et al. [18], we present the following solution:

**Theorem 2.4.** *Let $F$ be algebraically closed. If $A, B \in M_n$ and $AB \in [A, B]_{\rm ref}$ then either $AB \in [A, B]$ or $\{A, B\} = K\{I_2, J\}N$ where $J$ is a $2 \times 2$ Jordan block, $K \in M_{n,2}$ is left-invertible and $N \in M_{2,m}$ is right-invertible.*

We prove Theorem 2.4 by reduction to Theorem 2.3. The solution set displays two symmetries: the first symmetry $(A, B) \to (B', A')$ is justified by the observation made earlier that $(\mathbb{S}')_{\rm ref} = (\mathbb{S}_{\rm ref})'$. The second symmetry, $(A, B) \to (B, A)$, is somewhat surprising. Thus, if $AB$ belongs to $[A, B]_{\rm ref} \setminus [A, B]$ then so do $BA$ and the commutator $AB - BA$.

### 2.3. Local linear dependence

Assume that $d < \infty$. We denote by *sep*($\mathbb{S}$) the set of *separating vectors* for $\mathbb{S}$, i.e., vectors $x \in X$ for which $d(x) = d$. Properties of separating vectors are described in, e.g., [1], [11], [24]. A linear subspace $\mathbb{S} \subset L(X, Y)$ of finite dimension is called *LLI* (*locally linearly independent*) if $\bar{d} = d$, namely, if sep($\mathbb{S}$) is not empty. Otherwise, we say that $\mathbb{S}$ is LLD.

The following results concerning low individual ranks are, up to some cosmetics, due to Meshulam and Šemrl:

**Theorem 2.5.** *Let $\mathbb{S}$ be LLD of dimension $d$ and local dimension $\bar{d}$.*

(i) ([25] Theorem 2.2). *$\mathbb{S}$ is $\bar{d}$-admitting. This bound is tight.*
(ii) ([25]) *If $|F| \geq d+2$ then $\mathbb{S}$ is $d-2$-admitting or $d-1$-regular.*
(iii) ([26] Theorem 2.1) *If $|F| > \bar{d}$ then $\mathbb{S}$ contains a subspace of type $(d-\bar{d}, \bar{d})$ (which is therefore $\bar{d}$-forcing).*

Clearly, $\mathbb{S}$ is LLD if it has an LLD subspace. Thus, it makes sense to study minimal LLD spaces. For such spaces, Theorem 2.5(ii) can be improved (see [25] Theorem 22). We also have:

**Theorem 2.6 (Chebotar, Šemrl [7] Theorem 1.2).** *Assume that $d \geq 2$ and $|F| \geq d+2$. Let $\mathbb{S}$ be minimal LLD of type $(d, r)$. Then $d-1 \leq r \leq d(d-1)/2$. These bounds are tight.*

We move to low type results. The case $d = 1$ is trivial: $d = \bar{d} = 1$ hence $\mathbb{S}$ is automatically LLI. Another simple case concerns spaces of type $(d, 1)$ with $d > 1$. In this case, $\mathbb{S}$ must be of the form $V \otimes U$ ($U \in X$, $V \in Y'$, $\dim(U) = 1$), hence is LLD (though left-handed LLI!).

**Theorem 2.7 (Brešar and Šemrl, [5] Theorem 2.3).** *Assume that $F$ is infinite. If $\mathbb{S}$ is of type $(2, r)$ with $r \geq 2$ then $\mathbb{S}$ is LLI.*

For $d = 3$, again the case $r \leq 2$ is easily settled (see our proof of Theorem 2.10). Regarding $r = 3$, the space $S_3$ plays a central role (spaces of skew-symmetric matrices are mentioned in [13] as examples of singular spaces). The space $S_3$ is both right-handed LLD and left-handed LLD, but not reflexive. We start with two known results, conveniently rephrased.

**Theorem 2.8 (Brešar and Šemrl [5, Theorem 2.4]).** *Assume that $F$ is infinite and $ch(F) \neq 2$. Assume that the space $\mathbb{S}$ of type $(3, r)$ with $r \geq 3$ is LLD. Then one of the following holds:*

(i) *For some unit-rank idempotent $P \in L(Y)$,* $\operatorname{rank}(\mathcal{B}) = 1$ *where $\mathcal{B} = \{(I-P)A_i\}_{i=1}^3$;*
(ii) *$A_i x = \sum_{i=1}^3 [Q_1(Rx)Q_2]_{ki} u_k$ $(i = 1, 2, 3)$ for some $Q_1, Q_2 \in GL(3, F)$, LI vectors $u_1, u_2, u_3 \in Y$ and $R \in L(X, S_3)$.*

**Theorem 2.9 (Chebotar and Šemrl [7] Theorem 3.1)).** *Assume that $|F| \geq 5$. If the space $\mathbb{S}$ of type $(3, r)$ with $r \geq 3$ is LLD then one of the following holds:*

(i) $\mathbb{S}$ *contains a subspace of type* $(2,1)$;
(ii) $\mathbb{S}$ *is standard-three-dimensional.*

Both conditions in Theorem 2.8, and the definition of standard-three-dimensional spaces in Theorem 2.9, are complicated and implicit. The following result offers a fully explicit finite-dimensional simplification (compare with Theorem 2.3).

**Theorem 2.10.** *Assume that $F$ is algebraically closed. If the space $\mathbb{S} \subset M_{n,m}$ of type $(3,r)$ with $r \geq 3$ is LLD then one of the following holds:*
(i) $\mathbb{S}$ *contains a subspace of type* $(2,1)$;
(ii) $\mathbb{S}$ *is a copy of the skew symmetric space* $S_3$.

## 2.4. Generic reflexivity

The new generic reflexive hull is defined here, and its basic properties are studied. We base our genericity arguments on the following definition:

**Definition 2.11.**
(i) A set $\Omega \subset F^m$ is called generic if its complement $\Omega^C$ is contained in an algebraic set. Equivalently, if there exists a non-trivial polynomial $p(x)$ with coefficients in $F$ which vanishes on $\Omega^C$. We use "ae" as synonymous to "generically".
(ii) A subspace $\mathbb{S} \subset M_{n,m}$ is called GLD (resp. GLI) if $d(x) < d$ (resp. $d(x) = d$) ae.
(iii) We say that $Z \in \mathbb{S}_{\text{gref}}$ if $Zx \in [\mathbb{S}x]$ ae.
(iv) We call $\mathbb{S}$ *generically reflexive* (or g-reflexive) if $\mathbb{S} = \mathbb{S}_{\text{gref}}$.

The GLD property is redundant in view of the following observation, which is implicit in [12]:

**Lemma 2.12.** *Let $\mathbb{S} \subset M_{n,m}$ be a linear space.*
(i) *The set* sep($\mathbb{S}$) *is empty if $\mathbb{S}$ is LLD, generic otherwise.*
(ii) *Assume that $F$ is infinite. $\mathbb{S}$ is LLD (resp. LLI) iff it is GLD (resp. GLI).*

*Proof.* (i) Assume that $\mathbb{S}$ is LLI and $\mathcal{A}$ is a basis for $\mathbb{S}$. Then sep($\mathbb{S}$) is not empty. If $x_0 \in$ sep($\mathbb{S}$) then $\mathcal{A}x_0$ is LI, so the determinant $f(x)$ of a certain $k \times k$ minor of the $n \times k$ matrix $R(x)$ whose columns are $A_i x$ satisfies $f(x_0) \neq 0$. But then sep($\mathbb{S}$) contains the generic set defined by the inequality $f(x) \neq 0$.

(ii) If $\mathbb{S}$ is LLD, it is clearly GLD. And if it is LLI, (i) shows that it is GLI. □

Generalizing item (i), the set max($\mathbb{S}$) of vectors $x$ for which $d(x) = \bar{d}$ is always generic, and is equal to sep($\mathbb{S}$) if $\mathbb{S}$ is LLI.

In contrast to item (ii), the generic reflexive hull is not, in general, equal to its classical counterpart, and we merely have the obvious chain of inclusions

$$\mathbb{S} \subset \mathbb{S}_{\text{ref}} \subset \mathbb{S}_{\text{gref}} \subset M_{n,m}. \tag{2.2}$$

Below we present two low-type results for g-reflexivity. If $r = 1$ the situation is clear: $\mathbb{S} = \{uv' : v \in V\}$ for some $V \subset X$ and $u \in Y$, and $\mathbb{S}_{\text{gref}} = X' \otimes [u] = uM_{1,m}$.

If $t = \dim(V)$ then $\mathbb{S}$ is reflexive for all $t$, LLI for $t = 1$ only, and g-reflexive for $t = m$ only.

**Theorem 2.13.** *Assume that $F$ is infinite. Let the space $\mathbb{S} \subset M_{n,m}$ be of type $(1, r)$ with $r \geq 2$. Then $\mathbb{S}$ is g-reflexive.*

For $d = 2$, the case $r \leq 2$ is covered in Lemma 8.1 and the following central result covers the case $r \geq 3$.

**Theorem 2.14.** *Assume that $F$ is algebraically closed and the space $\mathbb{S} \subset M_{n,m}$ of type $(2, r)$ with $r \geq 3$ is not g-reflexive. then one of the following holds:*

(i) *$\mathbb{S}$ contains an essentially unique matrix $C = uv'$, and $\mathbb{S}_{\text{gref}} = \mathbb{S} + uM_{1,m}$;*
(ii) *$\mathbb{S}_{\text{gref}}$ is a copy of the skew symmetric space $S_3$.*

**Remark.** For the space $\mathbb{S} = S_k$ (skew-symmetric) of type (k(k-1)/2,k) the following are equivalent: (i) $\mathbb{S}$ is g-reflexive; (ii) $\mathbb{S}$ is reflexive; (iii) $k \leq 2$. Indeed, for skew-symmetric matrices with $k \geq 3$ we have $(S_k)_{\text{gref}} = M_n$ by Lemma 3.2(i). This, plus Lemma 2.15, gives many examples of non-triangular LLD spaces of larger type. Also, it is easy to see that $(S_k)_{\text{ref}}$ is the space of zero-diagonal matrices if $k \geq 4$.

### 2.5. The basic relation

Comparison of Theorems 2.7 versus 2.13, as well as 2.10 versus 2.14, shows that g-reflexivity results match nicely with their LLD counterparts; the same cannot be said about the corresponding (classical) reflexivity results. This is partly due to the following simple but fundamental relation.

**Lemma 2.15.** *A space $\mathbb{L} \subset L_{nm}$ is LLD iff $\mathbb{L} \subset \mathbb{S}_{\text{gref}}$ for some proper subspace $\mathbb{S}$ of $\mathbb{L}$.*

*Proof.* Let the space $\mathbb{L}$ be of dimension $d$ and local rank $\bar{d}$.

(i) Assume that $\mathbb{L}$ is LLD, i.e., $\bar{d} < d$. Choose $x \in \max(\mathbb{S})$ so that $d(x) = \bar{d}$. Choose a basis for $\mathbb{L}x$, which has the form $\mathcal{A}x$ for some $\mathcal{A} = \{A_1, \dots, A_{\bar{d}}\} \subset \mathbb{L}$, and set $\mathbb{S} := [\mathcal{A}]$. The inclusion $\mathbb{S} \subset \mathbb{L}$ is strict since $\dim(\mathbb{S}) = \bar{d} < d = \dim(\mathbb{L})$. On the other hand, for almost all $y \in F^m$ we have $\dim(\mathbb{S}y) = \bar{d} = \dim(\mathbb{L}y)$, hence $\mathbb{L}y \subset \mathbb{S}y$. Thus, by definition, $\mathbb{L} \subset \mathbb{S}_{\text{gref}}$.

Conversely, assume that $\mathbb{S} \subset \mathbb{L} \subset \mathbb{S}_{\text{gref}}$ for some strict subspace $\mathbb{S}$ of $\mathbb{L}$. By definition, for almost all $x$ $\mathbb{L}x \subset \mathbb{S}x$, hence $d(x) = \dim(\mathbb{L}x) \leq \dim(\mathbb{S}x) \leq \dim(\mathbb{S}) < \dim(\mathbb{L}) = d$, meaning that $\mathbb{L}$ is GLD, hence also LLD. □

We restate this result in different terms.

**Lemma 2.16.** *Let $\mathcal{A}$ be a finite set.*

(i) *If $Z \in [\mathcal{A}]$ then $[\mathcal{A} \cup \{Z\}]$ is LD. The converse holds if $\mathcal{A}$ is LI.*
(ii) *If $Z \in [\mathcal{A}]_{\text{ref}}$ then $[\mathcal{A} \cup \{Z\}]$ is LLD.*
(iii) *If $Z \in [\mathcal{A}]_{\text{gref}}$ then $[\mathcal{A} \cup \{Z\}]$ is GLD. The converse holds if $\mathcal{A}$ is GLI.*

**Remarks.**

1) In items 2–3, we use GLD and LLD synonymously, in view of Lemma 2.12.
2) The converse statement missing in item (ii) does not hold, as shown, e.g., by Theorem 2.14(i–ii).
3) Item (iii) is equivalent to Lemma 2.15.
4) Both inclusion $\mathbb{S} \subset \mathbb{S}_{\text{ref}} \subset \mathbb{S}_{\text{gref}}$ and $\mathbb{S} \subset \mathbb{L} \subset \mathbb{S}_{\text{gref}}$ can be used to derive low-rank results for sets which are not g-reflexive.

Our proof of Theorems 2.3 and 2.14 will occupy a big part of the paper. We shall jointly prove both results using exhaustive analysis of canonical forms. Theorem 2.10 is an almost immediate consequence of Theorem 2.14, using Lemma 2.15.

We comment that necessary and sufficient conditions for the LLD property (resp. any of the two reflexivity properties) are known only for $d \leq 3$ (resp. $d \leq 2$). The situation for larger dimensions is unclear, and probably much harder. Both the classification of singular subspaces of matrices, used by some authors (see comment in [5], [13]), or methods based on canonical forms, used here, are so far limited to small dimension and we seem to be on the edge between tame and wild cases (see, e.g., [9]).

It follows that qualitative partial results valid for $d$ arbitrary may be of considerable interest, for example, extensions of the results on small individual rank cited earlier. In Section 3 we mention more results of this type, which guarantee equality in one of the two inclusions

$$\mathbb{S} \subset \mathbb{S}_{\text{gref}} \subset M_{n,m}, \tag{2.3}$$

mentioned in (2.2). These, and further structural results mentioned in Section 5, are used heavily in our proofs.

## 2.6.. Total reflexivity

$\mathbb{S}$ is called totally reflexive if $\mathbb{S}$, as well as any of its subspaces, is reflexive; and elementary if the unit rank operators in $S_\perp$ span $L(X, Y)$. According to Proposition 2.10 of [1], $\mathbb{S}$ is totally reflexive iff it is both reflexive and elementary. This, plus Theorem 2.3, implies that $[A, B] \subset M_{nm}$ is totally reflexive iff the canonical form of $(A, B)$ (see Section 6) contains no $2 \times 2$ Jordan cell. An open conjecture is that a totally reflexive space in $M_n$ can have dimension $2n-1$. Azoff confirmed the conjecture for $n = 3$ (see [1] Question 9.4 and Proposition 9.11). These claims can be checked using canonical form.

An LLI space is elementary and 2-reflexive; if in addition it is reflexive then it is totally reflexive [1], [12]. The Toeplitz algebra $T_2$ featured in Theorem 2.3 was used as an example for an elementary, non-reflexive algebra [1], or an LLI non-reflexive algebra [12].

## 3. Equality in the inclusions $\mathbb{S} \subset \mathbb{S}_{\mathbf{gref}} \subset M_{n,m}$

Under Definition 2.11, a finite intersection of generic sets in $F^m$ is generic. When the field $F$ is infinite, it can be argued that a generic set is "fat", and its complement is "meager", since algebraic sets in $F^m$ have dimension $< m$. Our treatment of genericity will rely heavily on rank considerations. To this end we define two indices, the simple index $\bar{d}$ and the double index $\bar{d}_2$.

**Definition 3.1.** With every set $\mathcal{A} = \{A_1, \ldots, A_d\} \subset M_{n,m}$ we associate the matrix $R(x) \in M_{n,d}$ whose $j$th column is $A_j x$. Given $x, y \in F^m$, set

$$\begin{aligned} d(x) &= \text{rank } R(x), & \bar{d} &= \max_x d(x), \\ d_2(x,y) &= \text{rank } [R(x), R(y)], & \bar{d}_2 &= \max_{x,y} d_2(x,y). \end{aligned} \tag{3.1}$$

We say that $\mathbb{S} = [\mathcal{A}]$ is $2LI$ if $\bar{d}_2 = 2d$. We say that $\mathbb{S}$ has *full local rank* if $\bar{d} = n$.

We shall always assume that $\mathcal{A}$ is LI and $\mathbb{S} = [\mathcal{A}]$. In this case, $\bar{d}$ coincides with the local rank (also denoted $\bar{d}$) associated with $\mathbb{S}$ in Section 2. Also, according to Lemma 2.12 $\mathbb{S}$ (or $\mathcal{A}$) is LLI iff $\bar{d} = d$.

Definition 3.1 is relevant to the study of cases of equality in the chain of inclusions (2.3). In particular, the 2LI property means that $\mathbb{S}x$ and $\mathbb{S}y$ have trivial intersection for some separating vectors $x, y$. Ding [12], [10], [11] studied this property and showed that (assuming $|F| > d$) it implies reflexivity. We strengthen his result.

**Lemma 3.2.** *Consider the space* $\mathbb{S} \subset M_{n,m}$.

(i) $\mathbb{S}_{\text{gref}} = M_{n,m}$ *iff* $\mathbb{S}$ *has full local rank;*

(ii) $\mathbb{S}$ *is g-reflexive if* $\mathbb{S}$ *is* $2LI$.

*Proof.* (i) Let $x \in max(\mathbb{S})$, i.e., $d(x) = \bar{d}$. If $\bar{d} = n$ then $\mathbb{S}x = F^n$, hence $\mathbb{S}x$ spans $Zx$ for all $Z \in M_{n,m}$. Conversely, define

$$\Omega_i = \{x \in F^m : E_{i1}x \in \mathbb{S}x\}, \qquad \Delta = \{x \in F^m : x_1 \neq 0\}.$$

$\Delta$ is generic. Also, if $\mathbb{S}_{\text{gref}} = M_{nm}$ then $E_{i1} \in \mathbb{S}_{\text{gref}}$, hence $\Omega_i$ is generic. Thus, on the generic set $\cap_{i=1}^n \Omega_i \cap \Delta$ we have $E_{i1}x \in \mathbb{S}x$, hence $e_i \in \mathbb{S}x$ for all $i$, implying $\mathbb{S}x = F^n$ and $\bar{d} = n$.

(ii) For each $x$ define the set $\Omega(x) = \{y \in F^m : d_2(x,y) = 2d\}$. Clearly, $\Omega(x)$ is either empty or generic. If $\mathcal{A}$ is 2LI, we may find $\xi \in F^m$ such that $\Omega(\xi)$ is generic. It follows that $\Omega(x)$ is generic for all $x \in \Omega(\xi)$.

For each $Z \in \mathbb{S}_{\text{gref}}$ define the generic set $D = \{\zeta \in F^m : Z\zeta \in \mathbb{S}\xi\}$. Choose any $x \in D \cap \Omega(\xi)$ and any $y \in D \cap (D - x) \cap \Omega(x)$. Both choices are possible since both sets are generic. Defining $z = x+y$, we observe that $x, y, z \in D$. Therefore

$$Zx = \sum a_i A_i x, \qquad Zy \in \sum b_i A_i y, \qquad Zz = \sum c_i A_i z$$

for some $a_i, b_i, c_i \in F$. As $z = x + y$ we get $\sum (c_i - a_i) A_i x + \sum (c_i - b_i) A_i y = 0$. By assumption, the set $\{A_i x, A_i y\}_{i=1}^k$ is LI, so $a_i = c_i = b_i$. Namely, the linear combination $Zy = \sum a_i A_i y$ is unique, and independent on the (generic) choice of $y$. So, $Z - \sum a_i A_i$ vanishes ae, hence must be the zero matrix, i.e., $Z \in \mathbb{S}$ as required. □

(While item (ii) is valid over any field, it becomes meaningful mainly over infinite fields.) We see that the 2LI property implies both reflexivity and g-reflexivity. In comparison, when $F$ is algebraically closed, the LLI property implies g-reflexivity but not reflexivity. See Theorem 2.14.

The following result provides necessary, and sufficient, conditions for the 2LI property. Here, by a "direct sum decomposition $\mathbb{S} = \mathbb{S}_1 \oplus \mathbb{S}_2$" we understand the following:

$$\mathbb{S} = \operatorname{span}\{A_i \oplus B_i\}, \quad \mathbb{S}_1 = \operatorname{span}\{A_i\}, \quad \mathbb{S}_2 = \operatorname{span}\{B_i\}.$$

**Lemma 3.3.**

(i) *2LI implies LLI;*
(ii) *$\mathbb{S}$ is 2LI if $\mathbb{S}$ contains a 2LI minor;*
(iii) *$\mathbb{S}$ is 2LI if $\mathbb{S} = \mathbb{S}_1 \oplus \mathbb{S}_2$ is a "direct sum decomposition" where both $\mathbb{S}_1$ and $\mathbb{S}_2$ are LLI.*
(iv) *If $\mathbb{S}$ is 1-admitting then $\mathbb{S}$ is 2LD.*

*Proof.* (i) follows from the inequality $2d = \bar{d}_2 \leq 2\bar{d}$. (ii) Up to strict equivalence we may assume that $\mathbb{S} = \begin{pmatrix} \mathbb{S}_1 & * \\ * & * \end{pmatrix}$. If $\mathbb{S}_1$ is 2LI then $d_2(\xi, \eta) = 2\bar{d}$ for some vectors $\xi, \eta$. But then $\mathbb{S}$ is 2LI for $d_2((\xi', 0')', (\eta', 0')')$. (iii) Choosing $\xi \in \operatorname{sep}(\mathbb{S}_1)$ and $\eta \in \operatorname{sep}(\mathbb{S}_2)$, we see that $x = (\xi', 0')'$ and $y = (0', \eta')'$ are in $\operatorname{sep}(\mathbb{S})$ and $\mathbb{S}x \cap \mathbb{S}y = \{0\}$. (iv) Let $x, y$ be so that the subspace $V$ spanned by the vectors $\mathbb{S}x, \mathbb{S}y$ $(i = 1, \ldots, k)$ is of maximal dimension. Clearly, $V$ would remain unchanged if we replaced the basis $\mathcal{A}$ for $\mathbb{S}$. We may assume that the basis contains a unit-rank matrix $C$. But then $Cx, Cy$ are LD, hence $\dim(V) < 2d$, so that $\mathbb{S}$ is 2LD. □

To end this section, we make a comment on the chance of a k-tuple to generate a space with any of the properties discussed so far. We identify the collection of ordered k-tuples in $M_{n,m}$ with the vector space $F^{nmk}$, equipped with algebraic sets and a well-defined sense of genericity.

**Lemma 3.4.** *Let $k, n, m$ be fixed. For a k-tuple in $M_{n,m}$,*

(i) *the LI property is generic if $k \leq nm$ and void otherwise;*
(ii) *The full local rank property is generic if $k \geq n$ and void otherwise;*
(iii) *The LLI property is generic if $k \leq n$ and void otherwise;*
(iv) *The 2LI property is generic if $2k \leq n$ and void otherwise.*

*Proof.* Item (i) is well known; the remaining items are based on Lemma 2.11(i), if we choose a maximal rank minor of $R(x)$ or $[R(x), R(y)]$. □

## 4. Proofs of Theorems 2.7 and 2.13

Lemma 3.2(ii) is critical in proving these results. The following lemma subsumes Theorem 2.7 in the finite-dimensional setting:

**Lemma 4.1.** *TFAE for a LI pair* $\mathcal{A} = \{A, B\} \in M^2_{n,m}$*:*

(i) $\mathcal{A}$ *is GLD (i.e., LLD);*
(ii) $B \in [A]_{\text{gref}}$*;*
(iii) $A \in [B]_{\text{gref}}$*;*
(iv) $A = uv'$ *and* $B = uw'$ *for some non-trivial* $u \in F^n$ *and LI* $\{v, w\} \subset F^m$.

*Proof.* The first three items are equivalent in view of Lemma 2.16, and (iv) clearly implies (i). It remains to show that (ii) implies (iv). Assume that $B \in [A]_{\text{gref}}$. If $rank(A) > 1$ then, clearly, $A$ is 2LI, hence by Lemma 3.2(ii) $B \in [A]$. But then $\mathcal{A}$ is LD, contradicting the assumptions. So assume $\text{rank}(A) = 1$, i.e., $A = uv'$. The relation $B \in [A]_{\text{gref}}$ implies that $Bx \in [u]$ ae, hence $BX = [u]$, so $B = uw'$. Since $\mathcal{A}$ is assumed LI, $\{v, w\}$ is LI. □

We now provide a proof of Theorem 2.13.

*Proof.* To avoid trivialities we assume that $n > 1$ and $A \neq 0$. If $\text{rank}(A) \geq 2$ then $\{A\}$ is 2LI, hence $[A]_{\text{gref}} = [A]_{\text{ref}} = [A]$ by Lemma 3.2(ii). So assume $rank(A) = 1$, i.e., $A = uv'$. Every matrix of the form $B = uw'$ certainly belongs to $[A]_{\text{gref}}$. Conversely, by Lemma 4.1 every matrix in $[A]_{\text{gref}}$ is of this form. *A fortiori*, every matrix in $[A]_{\text{ref}}$ is of this form, and moreover $w \in [v]$, namely, $B \in [A]$. Indeed, if $\{v, w\}$ is LD we choose $x \in F^m$ so that $v'x = 0 \neq w'x$. Then $Bx \neq 0$ cannot depend linearly on $Ax = 0$, contradicting the assumption. □

## 5. Properties of localized span

Clearly, a canonical form for $k$-tuples in $M_{n,m}$ could be useful in analyzing localized span; unfortunately, a suitable canonical form exists only for $k \leq 2$. Nevertheless, in this section we include several useful structural observations valid for general $k$. For the sake of precision we shall work with *ordered* sets of matrices, denoted as usual by parentheses. We shall use the common conventions for operations involving ordered k-tuples. For example, if $\mathcal{A} = (A_1, \ldots, A_k)$ and $\mathcal{B} = (B_1, \ldots, B_k)$ we use $S\mathcal{A}T = (SA_1T, \ldots, SA_kT)$, $\quad \mathcal{A} \oplus \mathcal{B} = (A_1 \oplus B_1, \ldots, A_k \oplus B_k)$ etc.

**Definition 5.1.** A Gaussian operation on $\mathcal{A} = (A_1, \ldots, A_k)$ is one of the following operations: (i) permuting $A_i$ with $A_j$; (ii) replacing $A_i$ by $rA_i$; (iii) replacing $A_i$ by $A_i + rA_j$ $(0 \neq r \in F)$.
Strict equivalence is the operation which for all $1 \leq i \leq k$ replaces $A_i$ by $SA_iT$ for some $S \in GL(n, F)$ and $T \in GL(m, F)$.

**Lemma 5.2.** *The verification of the LI/LD and LLI/LLD properties, as well as the calculation of* $[\mathcal{A}], [\mathcal{A}]_{\text{ref}}, [\mathcal{A}]_{\text{gref}}$ *for finite sets* $\mathcal{A} \subset M_{n,m}$*, are preserved under Gaussian operations and strict equivalence.*

In terms of $\mathbb{S}$, we may ignore the Gaussian operations, and invariance under strict equivalence is well known (see, e.g., [1]). Next we study big blocks in terms of smaller blocks.

**Lemma 5.3.**

(i) $[\mathcal{A} \oplus \mathcal{B}] \subset [\mathcal{A}] \oplus [\mathcal{B}]$,
(ii) $[\mathcal{A} \oplus \mathcal{B}]_{\text{ref}} \subset [\mathcal{A}]_{\text{ref}} \oplus [\mathcal{B}]_{\text{ref}}$,
(iii) *If the "direct sum decomposition" $\mathcal{A} \oplus \mathcal{B}$ is LD (resp. LLD) then both $\mathcal{A}$ and $\mathcal{B}$ are LD (resp. LLD).*

*Proof.* The less trivial item is (ii). Assume that $Z = \begin{pmatrix} Z_1 & Z_2 \\ Z_3 & Z_4 \end{pmatrix} \in [\mathcal{A} \oplus \mathcal{B}]_{\text{ref}}$. Restriction to vectors of the form $(x_1', 0')'$ shows that $Z_1 \in [\mathcal{A}]_{\text{ref}}$ and $Z_3 = 0$. By symmetry of argument, $Z_4 \in [\mathcal{B}]_{\text{ref}}$ and $Z_2 = 0$. □

The $2 \times 2$ example with $A = 1 \oplus 0$ and $B = 0 \oplus 1$ shows that the equality in (ii), and the converse in (iii), do not hold in general. Some simplification occurs when one of the direct summands is essentially scalar. In that case, it may effectively be replaced by a truly scalar summand.

**Corollary 5.4.** *Define the map $\tau : F \to M_q$ by $\tau(a) = aI$. Extend $\tau$ by $\tilde{\tau}(A \oplus B) = A \oplus \tau(B)$. Let $\mathcal{B} \subset F$ be a k-tuple and $\mathcal{C} = \tau(\mathcal{B}) \subset M_q$. Then in our "direct sum notation"*

$$\text{(i)} \quad [\mathcal{A} \oplus \mathcal{C}] = \tilde{\tau}([\mathcal{A} \oplus \mathcal{B}]), \qquad \text{(ii)} \quad [\mathcal{A} \oplus \mathcal{C}]_{\text{ref}} = \tilde{\tau}([\mathcal{A} \oplus \mathcal{B}])_{\text{ref}}.$$

(Observe that $\tilde{\tau}$ defines linear isomorphisms between the corresponding spans.)

Our final result concerns some types of zero-filling not entirely covered by Lemma 5.3:

**Lemma 5.5.** *Let $\mathcal{A} = \{A_1, \dots, A_k\} \subset M_{n,m}$.*

(i) *If $Z \in [\mathcal{A}]_{\text{ref}}$ then the $j$th column of $Z$ is spanned by the $j$th columns of $A_i$.*
(ii) *If $\mathcal{A} = \begin{pmatrix} \mathcal{B} & 0 \\ 0 & 0 \end{pmatrix}$ with $\mathcal{B}$ of size $n' \times m'$ then $[\mathcal{A}] = \begin{pmatrix} [\mathcal{B}] & 0 \\ 0 & 0 \end{pmatrix}$, $[\mathcal{A}]_{\text{ref}} = \begin{pmatrix} [\mathcal{B}]_{\text{ref}} & 0 \\ 0 & 0 \end{pmatrix}$.*
(iii) *Under the same condition, $[\mathcal{A}]_{\text{gref}} \subset \begin{pmatrix} M_{n',m} \\ 0 \end{pmatrix}$, with equality iff $\mathcal{B}$ has full local rank.*

*Proof.* Item (i) follows immediately from the observation that a column in a matrix $A$ has the form $Ax$ with $x \in F^m$ an element of the canonical basis. The first statement in item (ii) is obvious, the second follows from item (i) (if $\min\{n', m'\} > 0$, also from Lemma 5.3). Inclusion in (iii) follows from the fact that if, for some vector space $V$, $Ax \subset V$ for almost all $x$ then $\Im(A) \subset V$. The case of equality follows from Lemma 3.2(i). □

As an example for item (iii) we choose $\mathcal{A} = \{A\}$ with $A = uv' \neq 0$. Here, up to strict equivalence, we may choose $n' = m' = 1$ and $\mathcal{B} = \{1\}$. Thus $\mathcal{B}$ has full local rank, which is consistent with $[\mathcal{A}]_{\text{gref}} = uM_{1,m}$ in Theorem 2.13. This also shows that the inclusion $[\mathcal{A} \oplus \mathcal{B}]_{\text{gref}} \subset [\mathcal{A}]_{\text{gref}} \oplus [\mathcal{B}]_{\text{gref}}$ in Lemma 5.3(iii) does not hold in general.

Additional results can be found in [1], especially Propositions 3.9–3.10.

## 6. Canonical forms for a matrix pencil

We concentrate on reflexivity and generic reflexivity for two-dimensional spaces. An ordered pair $(A, B)$ in $M_{n,m}$ is traditionally called a *matrix pencil.* Our proof of Theorems 2.3 and 2.14 is based on the Kronecker-Weierstrass canonical form for matrix pencils under strict equivalence, which is described in detail in Sections 5 and especially 13 of [14]. We shall very briefly review its construction and set our notation.

The building blocks of the canonical form are called *elementary cells,* each representing a *prime divisor.* The three types of prime divisors are denoted as $\mathbf{R_k}, \mathbf{L_k}, \mathbf{J_k}(\mu)$ and the corresponding blocks are resp. called *right null cells, left null cells* and *Jordan cells.* Each elementary cell is a pencil $\tilde{\mathcal{A}} = (\tilde{A}, \tilde{B})$ as described below:

| Divisor | $\tilde{\mathbf{A}}$ | $\tilde{\mathbf{B}}$ | Size | | |
|---|---|---|---|---|---|
| $\mathbf{R_k}$ | $[I, 0]$ | $[0, I]$ | $k \times (k+1)$ | | |
| $\mathbf{L_k}$ | $[I, 0]'$ | $[0, I]'$ | $(k+1) \times k$ | | (6.1) |
| $\mathbf{J_k}(\mu)$ | $I$ | $\mu I + J$ | $k \times k$ | $(\mu \in \mathbb{F})$ | |
| $\mathbf{J_k}(\infty)$ | $J$ | $I$ | $k \times k$ | $(``\mu = \infty'')$ | |

where $J$ denotes a $k \times k$ nilpotent upper Jordan block. As we see, Jordan cells may have an infinite eigenvalue, i.e., an element in the one-point compactification $\hat{F} := F \oplus \{\infty\}$ of $F$. As an example, the divisor of the $k \times k$ pencil $(aI, bI)$ is $\mathbf{J_1^k}(\mu)$ where $\mu = b/a$.

Each pencil $(A, B)$ defines the pair of integers

$$n' = \dim(AF^m + BF^m) \leq n, \quad m' = m - \dim(\operatorname{Ker}(A) \cap \operatorname{Ker}(B)) \leq m.$$

We call $\hat{\mathcal{A}} = (\hat{A}, \hat{B})$ *reduced* if $(n', m') = (n, m)$. We call $\mathcal{A} = (A, B)$ *canonical* if $\mathcal{A} = \begin{pmatrix} \hat{\mathcal{A}} & 0 \\ 0 & 0 \end{pmatrix}$ where $\hat{\mathcal{A}} \in M_{n',m'}$ is a (clearly reduced, possibly void) direct product of elementary cells. The *divisor* $\mathbf{d}$ associated with $\mathcal{A}$ is 1 if $\hat{\mathcal{A}}$ is void; otherwise, $\mathbf{d}$ is the (formally commutative) product of the elementary divisors of the cells in $\mathcal{A}$. Note that each divisor defines uniquely the reduced dimension $(n', m')$.

Let $\mathbf{d_1}, \mathbf{d_2}$ be two divisors. Let $\mathcal{A}_1, \mathcal{A}_2$ be the corresponding reduced canonical forms. We say that $\mathbf{d}_2$ *dominates* $\mathbf{d}_1$ (or $\mathbf{d_1} \leq \mathbf{d_2}$) if $\mathcal{A}_1$ is strictly equivalent to a minor of $\mathcal{A}_2$; we say that $\mathbf{d}_1$ *divides* $\mathbf{d}_2$ (or $\mathbf{d_1}|\mathbf{d_2}$) if $\mathcal{A}_2$ is strictly equivalent to $\mathcal{A}_1 \oplus \mathcal{B}$ for some $\mathcal{B}$.

*The central result of the Kronecker-Weierstrass theory is that every matrix pencil $\mathcal{A}$ is strictly equivalent, over an algebraically closed field, to a canonical pencil which is unique up to permutation of its non-trivial cells. Thus, $\mathcal{A}$ defines a unique divisor* $\mathbf{d}$.

The group of transformations on matrix pencils which includes both the Gaussian operations and strict equivalence was studied in [20]. Practically, gaussian operations do not affect the types and sizes of the various $\mathbf{L}, \mathbf{R}, \mathbf{J}$ cells, and act as a Moebius transformation on the eigenvalues of the Jordan cells, including

$\infty$. Thus, we may shift up to three eigenvalues arbitrarily, and then the remaining eigenvalues are completely determined.

## 7. Genericity-related classification of divisors

As a first step in proving Theorems 2.3 and 2.14, we classify divisors according to their genericity properties. A property of a divisor $\mathbf{d}$, by decree, is inherited from the (essentially unique) associated reduced canonical form $\hat{\mathcal{A}} = (\hat{A}, \hat{B})$ in $M_{n',m'}$. We define the following sets of divisors

$$\mathbf{\Gamma}_k(\mu) = \{\mathbf{J_k}(\mu), \mathbf{R_k}, \mathbf{L_{k-1}}\}; \qquad \mathbf{\Delta}_k(\mu_1, \dots, \mu_k) = \{\mathbf{J_1}(\mu_1)\cdots\mathbf{J_1}(\mu_\mathbf{q})\mathbf{R_1^{k-q}}\}_{q=0}^{k}$$

describing prime and linearly divisible divisors with reduced size $n' = k$.

**Lemma 7.1.** *Assume that $F$ is algebraically closed. A divisor $\mathbf{d}$ of reduced size $n' \times m'$ is:*

(i) *of full local rank iff $n' \leq 2$ and $\mathbf{d}$ is not of type $\mathbf{J_1^2}$.*
(ii) *LD iff $\mathbf{d} = \mathbf{J_1^k}$ $(k \geq 0)$.*
(iii) *LLD iff $\mathbf{d} = \mathbf{J_1^k}$ $(k \geq 0)$ or $\mathbf{d} = \mathbf{R_1}$.*
(iv) *Minimal LLI iff $\mathbf{d} \in \mathbf{\Gamma}_2(\mu) \cup \mathbf{\Delta}_2(\mu, \nu)$ for some $\mu \neq \nu$.*
(v) *2LD iff either $n' \leq 3$ or $\mathbf{d} = \mathbf{d_1 d_2}$ where $\mathbf{d_1} = \mathbf{J_1^k}(\mu)$ $(k \geq 0)$ and $\mathbf{d_2} \in \{\mathbf{1}, \mathbf{L_1}, \mathbf{J_2}(\mu), \mathbf{J_1}(\nu), \mathbf{R_1}\}$ for some $\mu \neq \nu$.*

*Proof.* First we treat items (i–iv). We use $\text{rank}_{\text{ref}}$ to denote the local rank $r$ defined in the preliminaries.

(i) Let $\hat{\mathcal{A}}$ in $M_{n',m'}$ be the corresponding reduced pencil. If $\min\{m', n'\} = 1$ then $\mathbf{d} \in \{\mathbf{L_1}, \mathbf{R_1}, \mathbf{J_1}\}$, hence $[\hat{\mathcal{A}}] = M_{n',m'}$ and $\text{rank}_{\text{ref}}(\mathcal{A}) = n'$. If $n' \geq 3$ this is clearly not the case. Finally, if $n' = 2$ then $\mathbf{d} \in \mathbf{\Gamma}_2(\mu) \cup \mathbf{\Delta}(\mu, \nu)$. Of these, $J_1^2(\mu)$ is LD and $\text{rank}_{\text{ref}}(\mathcal{A}) = 1 < 2 = n'$. The remaining cases are LI, hence (given $n' = 2$) are of full local rank.

Item (ii) is clear and item (iii) follows easily from Theorem 2.13 with $r = 1$, where $\mathbf{J_1^k}$ and $\mathbf{R_1}$ represent the LD case and the unit-rank LI case, respectively. Also, $\mathbf{J_1^k}$ is reduced to $\mathbf{J_1}$ via Corollary 5.4. Item (iv) follows easily from item (iii). □

Item (v) of the lemma is a bit tougher. We observe that most of the divisor types mentioned in this item are 2LD by Lemma 3.3(iv); the question is why the remaining types are 2LI. First we verify the case $n' = 4$:

**Lemma 7.2.** *If $\mathbf{d}$ is 2LD with $n' = 4$ then*

$$\mathbf{d} \in \{\mathbf{J_1^4}(\mu), \mathbf{J_1^3}(\mu)\mathbf{R_1}, \mathbf{J_1^3}(\mu)\mathbf{J_1}(\nu), \mathbf{J_1^2}(\mu)\mathbf{L_1}, \mathbf{J_2}(\mu)\mathbf{J_1^2}(\mu)\}$$

*for some $\mu \neq \nu$.*

*Proof.* The divisors $\mathbf{L_3}$, $\mathbf{L_2 J_1}(\mu)$, $\mathbf{R_2 J_1^2}(\mu), \mathbf{J_2}(\lambda)\mathbf{J_1^2}(\mu), \mathbf{J_3}(\mu)\mathbf{J_1}(\mu)$ $(\lambda \neq \mu)$ are 2LI. We verify this directly from Definition 3.1, i.e., we find $\xi, \eta \in F^m$ so that $d_2(\xi, \eta) = 4$. By Gaussian operations plus strict equivalence we may assume that

$\mu = 0$ and $\lambda = \infty$. In the first case ($m = 3$, $\hat{A} = [I,0]', \hat{B} = [0,I]'$) we choose $\xi = e_1, \eta = e_3$. In the second case $m = 3$, $\hat{A} = E_{11} + E_{22} + E_{43}$, $\hat{B} = E_{21} + E_{32}$. Here we choose $\xi = e_1, \eta = e_2 + e_3$. In the third case, $m = 5$ and we choose $\xi = e_2 + e_4$, $\eta = e_3 + e_5$. In the fourth case, $\xi = e_1 + e_3$, $\eta = e_2 + e_4$. In the fifth case, $\xi = e_2 + e_4$, $\eta = e_3$.

It then follows that every divisor in $\mathbf{\Gamma}_4(\mu) \cup \mathbf{\Gamma}_3(\mu)\mathbf{\Delta}_1(\nu)$ (with possibly $\mu = \nu$) is 2LI. Indeed, again we may choose $\mu = 0$ and $\nu = \infty$; and then for either subclass removal of one or two columns provides a 2LI divisor $\tilde{\mathbf{d}} \leq \mathbf{d}$. Indeed, if $\mathbf{d} \in \mathbf{\Gamma}_4$ we choose $\tilde{\mathbf{d}} = \mathbf{L_3}$ and if $\mathbf{d} \in \mathbf{\Gamma}_3(\mu)\mathbf{\Delta}_1(\nu)$ we choose $\tilde{\mathbf{d}} = \mathbf{L_2J_1}(\nu)$.

Finally, if

$$\mathbf{d} = \mathbf{d_1d_2} \text{ with } \mathbf{d_1} \in \mathbf{\Gamma}_2(\mu)\mathbf{\Delta}_2(\mu,\nu) \quad (\mu \neq \nu)$$

and

$$\mathbf{d_2} \in \mathbf{\Gamma}_2(\lambda)\mathbf{\Delta}_2(\lambda,\psi) \quad (\lambda \neq \psi)$$

then $\mathbf{d_1}, \mathbf{d_2}$ are LLI by Lemma 7.1(iv), and so $\mathbf{d}$ is 2LI by Lemma 3.3(iii).

The only divisors with $n' = 4$ not covered by the three classes described above are those mentioned in the lemma, which are clearly LLD. □

We now prove Lemma 7.1(v).

*Proof.* Let $\mathbf{d}$ be any 2LD divisor with $n' \geq 4$. Our strategy is as follows: since $\mathbf{d}$ is 2LD, by Lemma 3.3(ii–iii) we cannot have $\tilde{\mathbf{d}} \leq \mathbf{d}$ with $\tilde{\mathbf{d}}$ 2LI, nor $\mathbf{d} = \mathbf{d_1d_2}$ with $\mathbf{d_1}, \mathbf{d_2}$ LLI. We shall repeatedly use these facts to restrict $\mathbf{d}$, where the 2LI property of $\tilde{\mathbf{d}}$ will be guaranteed by Lemma 7.2 and the LLI property of $\mathbf{d_1}, \mathbf{d_2}$ by item (iii).

If $\mathbf{d}$ is LLD we are done by item (iii) of the lemma; and $\mathbf{d}$ cannot be prime, i.e., in $\mathbf{\Gamma}_k$ for some $k \geq 4$, in view of some $\tilde{\mathbf{d}} \in \mathbf{\Gamma}_4$. So we assume that $\mathbf{d}$ is LLI and composite.

Define $\mathbf{d}_1$ as a minimal LLI divisor which divides $\mathbf{d}$; define $\tilde{\mathbf{d}}_1$ as a minimal LLI divisor with is dominated by $\mathbf{d}_1$. By item (iv), $\tilde{\mathbf{d}}_1 \in \mathbf{\Gamma}_2(\mu) \cup \mathbf{\Delta}_2(\mu,\nu)$ for some pair $\mu \neq \nu$. By minimality of $\mathbf{d}_1$, both $\mathbf{d}, \mathbf{d_1}$ have the same number of prime divisors, and so it is not difficult to conclude that $\mathbf{d_1} \in \mathbf{\Gamma}_k(\mu) \cup \mathbf{\Delta}_2(\mu,\nu)$ for some $k \geq 2$.

By definition, $\mathbf{d} = \mathbf{d_1d_2}$ for some divisor $\mathbf{d_2}$ which is necessarily LLD; hence by item (iii) of the Lemma, $\mathbf{d}_2 \in \{\mathbf{R_1}\} \cup \{\mathbf{J_1^q}(\lambda) : q \geq 0\}$. If $\mathbf{d}_2 \in \{\mathbf{R_1}, \mathbf{1}\}$ the restriction $n' \geq 4$ leaves us with $\mathbf{d} \in \mathbf{\Gamma}_k \cup \mathbf{R_1}\mathbf{\Gamma}_{k-1}$ for some $k \geq 4$. But this case is impossible in view of some $\tilde{\mathbf{d}} \in \mathbf{\Gamma}_4 \cup \mathbf{R_1}\mathbf{\Gamma}_3$. So we conclude that $\mathbf{d_2} = \mathbf{J_1^q}(\lambda)$ for some $q \geq 1$, and $\mathbf{d}$ must be one of the following:

1) $\mathbf{d} = \mathbf{d_1J_1^q}(\lambda)$ with $\mathbf{d_1} \in \mathbf{\Delta}_2(\mu,\nu)$ and $q \geq 2$ : the case $\lambda \notin \{\mu,\nu\}$ is impossible in view of $\tilde{\mathbf{d}} = \mathbf{d_1J_1^2}(\lambda)$; otherwise $\mathbf{d}$ is of the desired form.
2) $\mathbf{d} = \mathbf{d_1J_1^q}(\lambda)$ with $\mathbf{d_1} \in \mathbf{\Gamma}_k(\mu)$ ($k \geq 2$): we have three possibilities:
   2.1) $\mathbf{d} = \mathbf{L_kJ_1^q}(\lambda)$ ($k \geq 1$, $q \geq 1$, $k + q \geq 3$): the case $k \geq 2$ is impossible in view of $\tilde{\mathbf{d}} = \mathbf{L_2J_1}$. Thus, $k = 1$ and $\mathbf{d}$ is of the desired form.

2.2) $\mathbf{d} = \mathbf{R_k J_1^q}(\lambda)$ $(k \geq 2,\ q \geq 1,\ k+q \geq 4)$ : this is impossible in view of $\tilde{\mathbf{d}} = \mathbf{R_3 J_1}$ or $\tilde{\mathbf{d}} = \mathbf{R_2 J_1^2}$.

2.3) $\mathbf{d} = \mathbf{J_k}(\mu)\mathbf{J_1^q}(\lambda)$ $(k \geq 2,\ q \geq 1,\ k+q \geq 4)$. The cases $\lambda \neq \mu$ and $k \geq 3$ are impossible in view of $\tilde{\mathbf{d}} = \mathbf{J_3}(\mu)\mathbf{J_1}(\lambda)$ or $\tilde{\mathbf{d}} = \mathbf{J_2}(\mu)\mathbf{J_1^2}(\lambda)$. So $\lambda = \mu$ and $k = 2$, hence $\mathbf{d}$ is of the desired form. □

## 8. Proof of Theorems 2.3, 2.10 and 2.14

As in the case of Lemma 7.1, we divide the divisors into a finite set of classes, except that now the canonical forms pertaining to each class cannot be assumed reduced. In each case we assume that $\mathcal{A} = (A, B)$ is a pencil in $M_{n,m}$ with divisor $\mathbf{d}$ of reduced dimension $n' \times m'$. We may assume that

$$\mathcal{A} = K\hat{\mathcal{A}}N, \qquad K = [I, 0]', \quad N = [I, 0], \qquad m', n' \geq 1$$

where $\hat{\mathcal{A}} = (\hat{A}, \hat{B})$ is in reduced canonical form. For each canonical form we calculate $[\hat{\mathcal{A}}]_{\mathrm{gref}}$ using genericity arguments, and then the possibly smaller set $[\hat{\mathcal{A}}]_{\mathrm{ref}}$.

For the sake of completeness, we summarize the case $r(= n') \leq 2$ which was excluded from Theorems 2.3, 2.14.

**Lemma 8.1.** *Let $\mathcal{A} = \{A, B\} \in M_{nm}$ be LI and have a divisor with reduced dimension $n' \times m'$ with $n' \leq 2$. Then $\mathcal{A}$ is g-reflexive iff $m = m'$ and reflexive iff $\mathbf{d} \neq \mathbf{J_2}$.*

*Proof.* 1. First we consider the case $\min\{n', m'\} = 1$ which includes the divisors $\mathbf{L_1}, \mathbf{R_1}, \mathbf{J_1}$. Here $[\hat{\mathcal{A}}] = M_{n',m'}$ is both reflexive and g-reflexive. By Lemma 5.5.(ii), $[\mathcal{A}]$ is reflexive. As $\hat{\mathcal{A}}$ is of maximal local rank, we also have $[\mathcal{A}]_{\mathrm{gref}} = KM_{n',m}$ according to Lemma 5.5(iii). Thus $[\mathcal{A}]$ is g-reflexive only if $m = m'$.

2. $n' = 2 \leq m'$ and $\mathbf{d}$ is not of the LD type $\mathbf{J_1^2}$. We still have maximal local rank, hence $[\mathcal{A}]_{\mathrm{gref}} = KM_{2,m'}$. Also, we have $[\mathcal{A}]_{\mathrm{ref}} = K[\hat{\mathcal{A}}]_{\mathrm{ref}}N$ by Lemmas 5.3,5.2 and it remains to analyze $[\hat{\mathcal{A}}]_{\mathrm{ref}}$. If $Z \in [\hat{\mathcal{A}}]_{\mathrm{ref}}$ is arbitrary, and if $\mathbf{d}$ contains Jordan blocks, using Lemma 5.2 we may assume that one Jordan eigenvalue is 0 and, in the case $\mathbf{J_1}(\mu)\mathbf{J_1}(\nu)$, the second eigenvalue is $\infty$. In each of the following cases, the support of $Z$ is calculated from Lemma 5.5(i). Also, define the generic set

$$\Omega := \{x \in F^{m'} : \hat{\mathcal{A}}x \text{ is } LI\} = \{x \in F^{m'} : p(x) \neq 0\}, \qquad p(x) = \det(\hat{A}x, \hat{B}x).$$

For all $x \in \Omega$ we have $[\mathcal{A}x] = F^n$ and automatically $Zx \in [\mathcal{A}x]$. In each case we provide vectors in $\Omega^C$, which we call *test vectors*, which force $Z$ to be in $[\hat{\mathcal{A}}]$. The easy verification is left for the reader.

2.1. $\mathbf{d} = \mathbf{R_2}$. Here $m' = 3$, $\hat{A} = (I, 0)$, $\hat{B} = (0, I)$, $Z_{13} = Z_{21} = 0$, $p(x, y, z) = xz - y^2$, and the test vectors are $(1, \pm 1, 1)'$.

2.2. $\mathbf{d} = \mathbf{R_1^2}$. Here $m' = 4$, $\hat{A} = E_{11} + E_{23}$, $\hat{B} = E_{12} + E_{24}$, $Z$ is supported on these four entries, $p(x, y, z, u) = xu - zy$, and the test vectors are $e_1 + e_3$, $e_2 + e_4$.

2.3. $\mathbf{d} = \mathbf{R_1 J_1(0)}$. Here $m' = 3$, $\hat{A} = E_{11} + E_{23}$, $\hat{B} = E_{12}$, $Z$ is supported on these three entries, $p(x,y,z) = yz$, and the test vectors are $e_1 \pm e_3$.

2.4. $\mathbf{d} = \mathbf{J_1(0)J_1}(\infty)$. $\hat{A} = E_{11}$, $\hat{B} = E_{22}$, hence $Z$ is diagonal so that $Z \in [\hat{\mathcal{A}}]$.

2.5. $\mathbf{d} = \mathbf{J_2(0)}$ is the only non-reflexive divisor in this group. Indeed, we have $m' = 2$, $\hat{A} = I$, $\hat{B} = E_{12}$, $Z \in U_2$ (upper triangular). Here $p(x,y) = y$, and the test vector $e_1$ shows that conversely every $Z \in U_2$ satisfies $Zx \in [\hat{A}x]$. □

We shall now prove together Theorems 2.3 and 2.14, by repeating the same type of analysis for $n' \geq 3$.

*Proof.* We have three cases:

1. If $\mathbf{d}$ is 2LI then $[\mathcal{A}]$ is g-reflexive, hence also reflexive, according to Lemma 3.2(ii).

2. If $n' = 3$, we cannot use Lemma 3.2 to avoid the direct calculation of $[\mathcal{A}]_{\text{gref}}$ since, according to Lemma 3.4, $\mathbf{d}$ is neither 2LI nor of full local rank. Let $Z \in M_{n,m}$ be arbitrary. $Z \in [\mathcal{A}]_{\text{gref}}$ iff $Z = K[Z_1, Z_2]$ and $\{A, B, Z\}$ is GLD (see Lemmas 5.5(ii),2.16), iff $Z = K[Z_1, Z_2]$ and the $3 \times 3$ matrix $R(x', y')' = [\hat{A}x, \hat{B}x, K(Z_1 x + Z_2 y)]$ is singular for almost all $(x, y) \in F^3 \times F^{m-3}$; i.e., iff the polynomial $q(x,y) := \det R(x', y')'$ vanishes for all $x \in F^{m'}$ and $y \in F^{m-m'}$. This forces all the coefficients of $q(x,y)$ to vanish, putting the needed restriction on $Z_1$ and $Z_2$. Once $[\mathcal{A}]_{\text{gref}}$ is calculated, test vectors may be required to show, in each case, that $[\mathcal{A}]$ is reflexive.

The set of divisors with $n' = 3$ consists of

$$\mathbf{\Gamma}_3(\mu), \quad \mathbf{\Delta}_3(\mu, \nu, \lambda) \quad \text{and} \quad \mathbf{\Gamma}_2(\mu)\Delta_1(\nu, \lambda).$$

For the time being we assume that $\mu, \nu, \lambda$ are distinct. Also we avoid the case $\mathbf{L_1 R_1}$. Using Lemma 5.2, we assume that $\mu = 0$, $\nu = \infty$, $\lambda = 1$. In each case, if the coefficients of $q(x,y)$ vanish, a straightforward but tedious calculation (which the reader can easily complete) shows that $Z_1 \in [\hat{\mathcal{A}}]$ and $Z_2 = 0$, hence $Z \in [\mathcal{A}]$. We conclude that for these cases $[\mathcal{A}]$ is both reflexive and g-reflexive.

3. Cases not included by items 1–2 are of two types: either $\mathbf{d}$ is 2LD and $n' \geq 4$, as described by Lemma 7.1(v); or $n' = 3$ and $\mathbf{d}$ has a repeated eigenvalue. Again, we use Lemma 5.2 to shift eigenvalues: the first to 0, the second to $\infty$.

3.1. $\mathbf{d} = \mathbf{L_1 R_1}$. Here $n' = m' = 3$, $\hat{A} = E_{11} + E_{32}$, $\hat{B} = E_{21} + E_{33}$, and analysis of $q(x,y,z)$ implies that $[\hat{\mathcal{A}}]_{\text{gref}} = [\hat{A}, \hat{B}, \hat{C}]$ where $\hat{C} = E_{13} - E_{22}$, and moreover $[\mathcal{A}]_{\text{gref}} = K[\hat{\mathcal{A}}]N$. On the other hand, it is easy to see that $\hat{C} \notin [\hat{A}, \hat{B}]_{\text{ref}}$. Thus $[\hat{\mathcal{A}}]$, hence also $[\mathcal{A}]$, is reflexive.

3.2. $\mathbf{d} = \mathbf{J_1}^q(\mu)$ with $q > 1$ : We have $[\hat{\mathcal{A}}] = [I]$. As before, if $Z = K[Z_1, Z_2] \in [\mathcal{A}]_{\text{gref}}$ then the matrices $[Z_1, Z_2]$ and $[I, 0]$ are GLD, hence LD, so $Z \in [\mathcal{A}]$.

3.3. We are left with the divisors

$$\mathbf{J_1}(\infty)\mathbf{J_1^q}(\mathbf{0}), \mathbf{L_1 J_1^{q-1}}(\mathbf{0}), \mathbf{R_1 J_1^q}(\mathbf{0}), \mathbf{J_2}(\mathbf{0})\mathbf{J_1^{q-1}}(\mathbf{0}) \tag{8.1}$$

with $q \geq 2$ and $\nu \neq \mu$. For the corresponding canonical forms, $B = E_{st}$ for some $s, t$. If $Z \in [\mathcal{A}]_{\text{gref}}$ then $A, B, Z$ are GLD. Thus the pair $\{(I - E_{ss})A, (I - E_{ss})Z)\}$

is GLD, hence LD by Theorem 2.7. It follows easily that $[\mathcal{A}]_{\text{gref}} = [A] + e_s M_{1,m}$. Next we analyze $Z \in [\mathcal{A}]_{\text{ref}} \subset [\mathcal{A}]_{\text{gref}}$ in each case. Both relations $Z \in [\mathcal{A}]$ and $Z \in [\mathcal{A}]_{\text{ref}}$ are preserved by removing the $j$th row from $Z$ and from all matrices of $[\mathcal{A}]$. Thus, to establish $Z \in [\mathcal{A}]$ we only need to treat the second and fourth cases in (8.1).

In the second case we use the block division $n = 2 + (q-1) + (n-q+1)$, $m = 1 + (q-1) + (m-q)$ obtaining $A - E_{11} \oplus I \oplus 0$, $B = E_{21} \oplus 0 \oplus 0$. By the previous part, $Z = wA + [e_2', 0', 0']'[z_1, z_2', z_3]$, and $z_2, z_3$ vanish by Lemma 5.5. So $Z = wA + z_1 B \in [\mathcal{A}]$ as claimed.

In the fourth case we use the block division $n = m = 2 + (q-1) + (m-q-1)$, obtaining $A = I \oplus I \oplus 0$, $B = E_{12} \oplus 0 \oplus 0$. By the previous part, $Z = wA + [[1,0], 0', 0']'[[z_1, z_2], z_3', z_4']$. $z_3, z_4$ vanish by Lemma 5.5. For the test vector $\xi = (x, 0, y', z')' \in F \times F \times F^{q-1} \times F^{n-q-1}$, $Z\xi = ((w+z_1)x, 0, z', 0)'$ must be spanned by $A\xi = (x, 0, z', 0')'$ and $B\xi = 0$. Thus $z_1 = 0$ and $Z = wA + z_2 B \in [\mathcal{A}]$. □

This completes the analysis of all the canonical forms, confirming Theorems 2.3 and 2.14 in all the cases. We now prove Theorem 2.10.

*Proof.* Assume that $\mathbb{L}$ is LLD. Then there exists a non-g-reflexive strict subspace $\mathbb{S}$ of $\mathbb{L}$ so that $\mathbb{L} \subset \mathbb{S}_{\text{gref}}$. According to Theorem 2.14, we have two cases. (i) $\mathbb{S}$ contains an essentially unique unit-rank matrix $C = uv'$ and $\mathbb{S}_{\text{gref}} = \mathbb{S} + uM_{1,m}$. Thus, $\mathbb{L}$ should admit at least one more matrix of the form $D = uw'$ with $v, w$ LI in order to include $\mathbb{L}$ as a strict subspace. (ii) $\mathbb{S}$ is a two-dimensional subspace of a copy of the 3-dimensional space $S_3$ and $\mathbb{L}$ is also a subspace of $S_3$. Necessarily, $\mathbb{L} = S_3$. □

## 9. Can the product $AB$ belong to $[A, B]_{\text{ref}} \setminus [A, B]$?

As an application of Theorem 2.3, we consider the description of pencils $\mathcal{A} = (A, B)$ in $M_n$ such that $AB \in [\mathcal{A}]_{\text{ref}}$, and especially when $AB$ is not in $[\mathcal{A}]$. This problem was posed as an open problem in the research note [18]. We also indicate how to solve the similar problem $AB \in [\mathcal{A}]_{\text{gref}}$ and the somewhat more general problems $CQD \in [C, D]_{\text{ref}}$ and $CQD \in [C, D]_{\text{gref}}$, where $C, D, Q' \in M_{n,m}$.

First we discuss the global analogue $AB \in [\mathcal{A}]$. A brief analysis of this situation over an arbitrary field was given in [19]. The solution given there, in terms of generalized inverses, is not very explicit and we offer an alternative description.

**Theorem 9.1.** *Assume that $ch(F) = 0$ and $A, B \in M_n$ and $AB \in [A, B]$. Then exactly one of the following conditions holds:*

(i) *$A = aX$, $B = bX$ for some $X \in M_n$ and $abX(X - cI) = 0$ for some $a, b, c \in F$;*
(ii) *$A = X + aI$, $B = Y + bI$ for some $X, Y \in M_n$, $a, b \in F$ with $ab = 0$ and $XY = 0$;*
(iii) *$A = a(I + X)$ and $B = b(I + X^{-1})$ with $X \in GL(n, F)$ and $a, b \in F \setminus \{0\}$.*

*Proof.* If $\mathcal{A}$ is LD then clearly $A = aX$ and $B = bX$. If $ab = 0$ then $AB = 0 \in \mathcal{A}$. Otherwise $AB = abX^2$ belongs to $[\mathcal{A}] = [X]$ only if $X^2 = cX$. On the other hand, if $\mathcal{A}$ is LI, there exists a unique pair $a, b \in F$ such that $AB = aA + bB$. If $b = 0$ we are in case (ii) with $X = A$; if $a = 0$ we are in case (ii) with $Y = B$; otherwise, we are in case (iii) with $X = (1/a)A - I = ((1/b)B - I)^{-1}$. □

The following result on the localized problem $AB \in [\mathcal{A}]_{\text{ref}}$ subsumes Theorem 2.4.

**Theorem 9.2.** *Let $\mathcal{A} = (A, B)$ be a pencil in $M_n$. Assume that $AB \in [\mathcal{A}]_{\text{ref}} \setminus [\mathcal{A}]$. Then there exist a left-invertible matrix $K \in M_{n,2}$ and a right-invertible matrix $N \in M_{2,n}$ such that exactly one of the following holds:*

(i) *$\{A, B\} = \{KN, KE_{12}N\}$ and $NK \notin U_2$;*

(ii) *$\{A, B\} = \{KN, K(\mu I + E_{12})N\}$ $(0 \neq \mu \in F)$ and $NK \in U_2 \setminus T_2$.*

*Recall that by $\{A, B\}$ we understand an* unordered *pair.*

A difficulty in reducing Theorem 9.2 to Theorem 2.3 is that the conditions $AB \in [A, B]$ and $AB \in [A, B]_{\text{ref}}$ are not preserved by Gaussian operations or strict equivalence (see Lemma 5.2). We therefore consider the slightly more general problem of finding triples $(C, Q, D)$ such that $CQD \in [C, D]_{\text{ref}}$. Note that in the new problem $C, D$ need not be square; and reduction of $(C, D)$ to canonical form is possible via the extended strict equivalence $(C, Q, D) \to (SCT, T^{-1}QS^{-1}, SDT)$. The new problem is still not invariant under Gaussian operations, but this minor problem can be handled and Theorem 2.3 can be invoked. A complete analysis of the new problem becomes possible; for lack of space, we shall only consider the elements needed for proving Theorem 9.2. In particular, we shall only consider triples $(C, Q, D)$ with $n = m$ with $Q$ invertible.

We now prove Theorem 9.2.

*Proof.* Let $(\dot{C}, \dot{D})$ be the reduced canonical form of $\mathcal{A} = (A, B)$. By strict equivalence we have some $S, T \in M_n$ so that

$$C := \begin{pmatrix} \hat{C} & 0 \\ 0 & 0 \end{pmatrix} = SAT, \qquad D := \begin{pmatrix} \hat{D} & 0 \\ 0 & 0 \end{pmatrix}, = SBT, \qquad Q := T^{-1}S^{-1} = \begin{pmatrix} \hat{Q} & * \\ * & * \end{pmatrix}$$

where $Q$ is block-divided conformably. The condition $AB \in [A, B]_{\text{ref}} \setminus [A, B]$ is equivalent to $CQD \in [C, D]_{\text{ref}} \setminus [C, D]$, and via Lemma 5.3(ii), also to the condition

$$Z \in [\hat{C}, \hat{D}]_{\text{ref}} \setminus [\hat{C}, \hat{D}] \qquad (Z = \hat{C}\hat{Q}\hat{D}). \tag{9.1}$$

Given that $AB \in [A, B]_{\text{ref}} \setminus [A, B]$, the set $\{A, B\}$ is not reflexive. According to Theorem 2.3, the divisor $\mathbf{d}$ of $\mathcal{A}$ must be $\mathbf{J_2}(\mu)$ for some $\mu \in \hat{F}$, hence $n' = m' = 2$ and we set accordingly $\hat{Q} = \begin{pmatrix} a & b \\ c & d \end{pmatrix}$, $K = S^{-1}\begin{pmatrix} I_2 \\ 0 \end{pmatrix}$, $N = (I_2 \;\; 0)T^{-1}$. Since the problem is not invariant under Gaussian operations, we must consider separately three cases, depending on $\mu$.

1) If $\mu = 0$ then $\hat{C} = I$, $\hat{D} = E_{12}$, $Z = \begin{pmatrix} 0 & a \\ 0 & c \end{pmatrix}$. It can be verified that $A = KN$, $B = KE_{12}N$, $\hat{Q} = NK$, $AB = KZN$. Now consider (9.1). The condition $Z \in [\hat{C}, \hat{D}]_{\text{ref}}$ becomes $Z \in [I, E_{12}]_{\text{ref}} = U_2$, i.e., $Z$ is upper triangular. In our case,

this is automatically satisfied. The condition $Z \in [\hat{C}, \hat{D}]$ implies that $Z$ is upper Toeplitz, which is possible only if $0 = c = (\hat{Q})_{21} = (NK)_{21}$. Thus, we are in case (i) of Theorem 9.2.

2) The case $\mu = \infty$ is analogous and leads again to case (i), with $(\hat{C}, \hat{D})$ switching roles.

3) If $\mu \neq 0, \infty$, with $K, N, \hat{Q}$ as before we have $\hat{C} = I$ $\hat{D} = E_{12} + \mu I$, $Z = \begin{pmatrix} \mu a & \mu b + a \\ \mu c & \mu d + c \end{pmatrix}$ $A = KN$, $B = K(E_{12} + \mu I)N$, $\hat{Q} = NK$, $AB = KZN$. The condition $Z \in [\hat{C}, \hat{D}]_{\mathrm{ref}}$ means that $Z$ is upper triangular, which now means that $c = 0$. The condition $Z \in [\hat{C}, \hat{D}]$ means that $Z$, hence $\hat{Q}$, is upper Toeplitz. Thus, we are in case (ii) of Theorem 9.2. □

The same technique provides a reduction from the problem $AB \in [A, B]_{\mathrm{gref}}$ to Theorem 2.14, and can be easily extended to discuss the full problem $CQD \in [C, D]_{\mathrm{gref}}$.

**Acknowledgment**

The author acknowledges fruitful discussions with Dan Haran (Tel Aviv University). This research was supported by a CNPq grant.

## References

[1] E.A. Azoff *On finite rank operators and preannihilators.* Mem. Amer. Math Soc. **64** (1986) 357.

[2] S.A. Amitsur, *Generalized polynomial identities and pivotal monomials.* Trans. Amer. Math. Soc. **114** (1965) 210–226.

[3] B. Aupetit, *A primer on spectral theory.* Springer 1991.

[4] J. Bračič, B. Kuzma, *Reflexive defect of spaces of linear operators.* Lin. Algebra Appl. **430** ( 2009) 344–359.

[5] M. Brešar, P. Šemrl, *On locally linearly dependent operators and derivations.* Trans. Amer. Math. Soc. **351**.3 (1999) 1257–1275.

[6] M. Chebotar, W.-F. Ke, P.-H. Lee, *On a Brešar-Šemrl conjecture and derivations of Banach algebras.* Quart. J. Math. **57** (2006) 469–478.

[7] M. Chebotar, P. Šemrl, *Minimal locally linearly dependent spaces of operators.* Lin. Algebra Appl. **429** (2008) 887–900.

[8] J.A. Deddens, P.A. Fillmore, *Reflexive linear transformations.* Lin. Algebra Appl. **10** (1975) 89–93.

[9] H. Derksen, J. Weyman, *Quiver representations.* Notices Amer. Math. Soc. **52**.2 (2005) 200–206.

[10] L. Ding, *An algebraic reflexivity result.* Houston J. Math **9** (1993) 533–540.

[11] L. Ding, *On a pattern of reflexive spaces.* Proc. AMS **124**.10 (1996) 3101–3108.

[12] L. Ding, *Separating vectors and reflexivity.* Linear Algebra Appl. **174** (1992) 37–52.

[13] P.F. Fillmore, C. Laurie, H. Radjavi, *On matrix spaces with zero determinant.* Linear Multilin. Algebra **18** (1985) 255–266.

[14] F.R. Gantmacher, *Theory of matrices I–II.* Chelsea 1959.

[15] W. Gong, D.R. Larson, W.R. Wogen, *Two results on separating vectors.* Indiana J. Math **43**.3 (1994) 1159–1165.

[16] D. Hadwin, *Algebraically reflexive linear transformations.* Linear Multilinear Alg. **14** (1983) 225–233.

[17] D. Hadwin, *A general view of reflexivity.* Trans. AMS **344** (1994) 325–360.

[18] R.E. Hartwig, P. Šemrl and H.J. Werner, Problem 19-3, Open problem section, Image (the ILAS bulletin) **19** (1997) 32.

[19] R.E. Hartwig, P. Šemrl and H.J. Werner, Solution to Problem 19-3, Open problem section, Image (the ILAS bulletin) **22** (1999) 25.

[20] J. Ja'Ja', *An addendum to Kronecker's theory of pencils.* SIAM Appl. Math **37**.3 (1979) 700–712.

[21] I. Kaplansky, *Infinite Abelian Groups.* Ann Arbor, 1954.

[22] D.R. Larson, *Reflexivity, algebraic reflexivity and linear interpolation.* Amer. J. Math **110** (1988) 283–299.

[23] D. Larson, W. Wogen, *Reflexivity properties of* $T \oplus 0$. Journal of Functional Analysis **92** (1990) 448–467.

[24] J. Li, Z. Pan, *Algebraic reflexivity of linear transformations.* Proc. AMS **135** (2007) 1695–1699.

[25] R. Meshulam, P. Šemrl, *Locally linearly dependent operators.* Pacific J. Math **203**.2 (2002) 441–459.

[26] R. Meshulam, P. Šemrl, *Locally linearly dependent operators and reflexivity of operator spaces.* Linear Algebra Appl. **383** (2004) 143–150.

[27] R. Meshulam, P. Šemrl, *Minimal rank and reflexivity of operator spaces.* Proc. Amer. Math. Soc. **135** (2007) 1839–1842.

Nir Cohen
Department of Mathematics
UFRN – Federal University of Rio Grande do Norte
Natal, Brazil
e-mail: nir@ccet.ufrn.br

Operator Theory:
Advances and Applications, Vol. 221, 241-254


# An Invitation to the $\mathcal{S}$-functional Calculus

Fabrizio Colombo and Irene Sabadini

**Abstract.** In this paper we give an overview of the $\mathcal{S}$-functional calculus which is based on the Cauchy formula for slice monogenic functions. Such a functional calculus works for $n$-tuples of noncommuting operators and it is based on the notion of $\mathcal{S}$-spectrum. There is a commutative version of the $\mathcal{S}$-functional calculus, due to the fact that the Cauchy formula for slice monogenic functions admits two representations of the Cauchy kernel. We will call $\mathcal{SC}$-functional calculus the commutative version of the $\mathcal{S}$-functional calculus. This version has the advantage that it is based on the notion of $\mathcal{F}$-spectrum, which turns out to be more simple to compute with respect to the $\mathcal{S}$-spectrum. For commuting operators the two spectra are equal, but when the operators do not commute among themselves the $\mathcal{F}$-spectrum is not well defined. We finally briefly introduce the main ideas on which the $\mathcal{F}$-functional calculus is inspired. This functional calculus is based on the integral version of the Fueter-Sce mapping theorem and on the $\mathcal{F}$-spectrum.

**Mathematics Subject Classification (2000).** Primary 47A10, 47A60.

**Keywords.** Functional calculus for $n$-tuples of commuting operators, the $\mathcal{S}$-spectrum, the $\mathcal{F}$-spectrum, $\mathcal{S}$-functional calculus, $\mathcal{SC}$ functional calculus, $\mathcal{F}$-functional calculus.

## 1. Introduction

The theory of slice monogenic functions, mainly developed in the papers [2], [3], [11], [12], [13], [15], [16] turned out to be very important because of its applications to the so-called $\mathcal{S}$-functional calculus for $n$-tuples of not necessarily commuting operators (bounded or unbounded), see [3] and [14]. In the paper [14] with D.C. Struppa we have introduced the notion of $\mathcal{S}$-spectrum for $n$-tuples of not necessarily commuting operators and we have given a first version of the $\mathcal{S}$-functional calculus, in particular we have developed all the necessary results to define the $\mathcal{S}$-functional calculus for unbounded operators. However in [14] there are restrictions on the class of slice monogenic functions to which the functional calculus can be applied. In the papers [2], [3], [6] we have proved the Cauchy formula for slice monogenic functions and we have developed the theory of the $\mathcal{S}$-functional calculus removing all the restrictions and proving some additional properties of the $\mathcal{S}$-spectrum. The Cauchy formula for slice monogenic functions (see [2] and [3]) plays a fundamental role in proving most of the properties of the $\mathcal{S}$-functional calculus. We point out that the $\mathcal{S}$-functional calculus has most of the properties that the Riesz-Dunford functional calculus for a single operator has. There exists a quaternionic version of the $\mathcal{S}$-functional calculus, which is called quaternionic functional calculus and, in its more general setting, can be found in [4] and [5]. It is crucial to note that slice monogenic functions have a Cauchy formula with slice monogenic kernel that admits two expressions. These two expressions of the Cauchy kernel are not equivalent when we want to define a functional calculus for not necessarily commuting operators. In fact, one version of the kernel is suitable for noncommuting operators and gives rise to the notion of $\mathcal{S}$-spectrum, while the second version of the Cauchy kernel gives rise to the $\mathcal{F}$-spectrum and it is well defined only in the case of commuting $n$-tuple of operators. For the analogies between the $\mathcal{S}$-functional calculus and the Riesz-Dunford functional calculus see for example the classical books [18] and [21].

The notion of $\mathcal{F}$-spectrum was introduced for the first time in the paper [9] with F. Sommen where we have defined the $\mathcal{F}$-functional calculus which is based on the integral version of the Fueter-Sce mapping theorem. Such a functional calculus associates to every slice monogenic functions a function of operators $\breve{f}(T)$ where $\breve{f}$ is a subclass of the standard monogenic functions in the sense of Dirac.

We conclude by recalling that the well-known theory of monogenic functions, see [1], [10], is the natural tool to define the monogenic functional calculus which has been studied and developed by several authors among which we mention A. McIntosh and his coauthors, see the book of B. Jefferies [20] and the literature therein for more details on the monogenic functional calculus.

## 2. Slice monogenic functions

In this section we introduce the main notions related to slice monogenic functions. The setting in which we will work is the real Clifford algebra $\mathbb{R}_n$ over $n$ imaginary units $e_1, \dots, e_n$ satisfying the relations $e_i e_j + e_j e_i = -2\delta_{ij}$. An element in the Clifford algebra will be denoted by $\sum_A e_A x_A$ where $A = i_1 \dots i_r$, $i_\ell \in \{1, 2, \dots, n\}$, $i_1 < \dots < i_r$ is a multi-index, $e_A = e_{i_1} e_{i_2} \dots e_{i_r}$ and $e_\emptyset = 1$. In the Clifford algebra $\mathbb{R}_n$, we can identify some specific elements with the vectors in the Euclidean space $\mathbb{R}^n$: an element $(x_1, x_2, \dots, x_n) \in \mathbb{R}^n$ can be identified with a so-called 1-vector in the Clifford algebra through the map $(x_1, x_2, \dots, x_n) \mapsto \underline{x} = x_1 e_1 + \dots + x_n e_n$. An element $(x_0, x_1, \dots, x_n) \in \mathbb{R}^{n+1}$ will be identified with the element $x = x_0 + \underline{x} = x_0 + \sum_{j=1}^n x_j e_j$ called paravector. The norm of $x \in \mathbb{R}^{n+1}$ is defined as $|x|^2 = x_0^2 + x_1^2 + \dots + x_n^2$. The real part $x_0$ of $x$ will be also denoted by $\mathrm{Re}[x]$. A function $f : U \subseteq \mathbb{R}^{n+1} \to \mathbb{R}_n$ is seen as a function $f(x)$ of $x$ (and similarly for a function $f(\underline{x})$ of $\underline{x} \in U \subset \mathbb{R}^{n+1}$).

**Definition 2.1.** We will denote by $\mathbb{S}$ the sphere of unit 1-vectors in $\mathbb{R}^n$, i.e.,

$$\mathbb{S} = \{\underline{x} = e_1x_1 + \cdots + e_nx_n \ : \ x_1^2 + \cdots + x_n^2 = 1\}.$$

Note that $\mathbb{S}$ is an $(n-1)$-dimensional sphere in $\mathbb{R}^{n+1}$. The vector space $\mathbb{R}+I\mathbb{R}$ passing through 1 and $I \in \mathbb{S}$ will be denoted by $\mathbb{C}_I$, while an element belonging to $\mathbb{C}_I$ will be denoted by $u + Iv$, for $u$, $v \in \mathbb{R}$. Observe that $\mathbb{C}_I$, for every $I \in \mathbb{S}$, is a 2-dimensional real subspace of $\mathbb{R}^{n+1}$ isomorphic to the complex plane. The isomorphism turns out to be an algebra isomorphism.
Given a paravector $x = x_0 + \underline{x} \in \mathbb{R}^{n+1}$ let us set

$$I_x = \begin{cases} \dfrac{\underline{x}}{|\underline{x}|} & \text{if } \underline{x} \neq 0, \\ \text{any element of } \mathbb{S} \text{ otherwise.} \end{cases}$$

By definition we have that a paravector $x$ belongs to $\mathbb{C}_{I_x}$.

**Definition 2.2.** Given an element $x \in \mathbb{R}^{n+1}$, for $I \in \mathbb{S}$, we define

$$[x] = \{y \in \mathbb{R}^{n+1} \ : \ y = x_0 + I|\underline{x}|\}.$$

*Remark* 2.3. The set $[x]$ is an $(n-1)$-dimensional sphere in $\mathbb{R}^{n+1}$. When $x \in \mathbb{R}$, then $[x]$ contains $x$ only. In this case, the $(n-1)$-dimensional sphere has radius equal to zero.

**Definition 2.4 (Slice monogenic functions, see [11]).** Let $U \subseteq \mathbb{R}^{n+1}$ be an open set and let $f : U \to \mathbb{R}_n$ be a real differentiable function. Let $I \in \mathbb{S}$ and let $f_I$ be the restriction of $f$ to the complex plane $\mathbb{C}_I$. We say that $f$ is a (left) slice monogenic function, or s-monogenic function, if for every $I \in \mathbb{S}$, we have

$$\frac{1}{2}\left(\frac{\partial}{\partial u} + I\frac{\partial}{\partial v}\right) f_I(u + Iv) = 0.$$

We denote by $\mathcal{SM}(U)$ the set of s-monogenic functions on $U$.

The natural class of domains in which we can develop the theory of s-monogenic functions are the so-called slice domains and axially symmetric domains.

**Definition 2.5 (Slice domains).** Let $U \subseteq \mathbb{R}^{n+1}$ be a domain. We say that $U$ is a slice domain (s-domain for short) if $U \cap \mathbb{R}$ is non empty and if $\mathbb{C}_I \cap U$ is a domain in $\mathbb{C}_I$ for all $I \in \mathbb{S}$.

**Definition 2.6 (Axially symmetric domains).** Let $U \subseteq \mathbb{R}^{n+1}$. We say that $U$ is axially symmetric if, for every $u + Iv \in U$, the whole $(n-1)$-sphere $[u + Iv]$ is contained in $U$.

**Definition 2.7 (Noncommutative Cauchy kernel series).** Let $x$, $s \in \mathbb{R}^{n+1}$. We call noncommutative Cauchy kernel series the following series expansion (for $|x| < |s|$)

$$S^{-1}(s, x) := \sum_{n \geq 0} x^n s^{-1-n}.$$

The definition of noncommutative Cauchy kernel series is the starting point to prove a Cauchy formula for slice monogenic functions. In fact, observe that on each complex plane $\mathbb{C}_I$, for $I \in \mathbb{S}$, a slice monogenic function $f$ admits a Cauchy formula related to such a plane. On each $\mathbb{C}_I$ the Cauchy kernel admits the usual power series expansion $\sum_{n\geq 0} x(I)^n s(I)^{-1-n}$ whose sum is $(s(I) - x(I))^{-1}$ for $x(I) := x_0 + I|\underline{x}|$ and $s(I) := s_0 + I|\underline{s}|$, where obviously $x(I)$ and $s(I)$ commute. We ask what is the sum of the series $\sum_{n\geq 0} x^n s^{-1-n}$ when $x$, $s \in \mathbb{R}^{n+1}$ do not commute. The following theorem answers such question.

**Theorem 2.8 (See [11]).** *Let $x$, $s \in \mathbb{R}^{n+1}$. Then*

$$\sum_{n\geq 0} x^n s^{-1-n} = -(x^2 - 2x\mathrm{Re}[s] + |s|^2)^{-1}(x - \overline{s})$$

*for $|x| < |s|$, where $\overline{s} = s_0 - \underline{s}$.*

So the function $-(x^2 - 2x\mathrm{Re}[s] + |s|^2)^{-1}(x - \overline{s})$ is the natural candidate to be the Cauchy kernel for slice monogenic functions. Moreover, such function can be written in an other way as the following result shows.

**Theorem 2.9 (See [3]).** *Let $x$, $s \in \mathbb{R}^{n+1}$ be such that $x \notin [s]$. Then the following identity holds:*

$$-(x^2 - 2x\mathrm{Re}[s] + |s|^2)^{-1}(x - \overline{s}) = (s - \bar{x})(s^2 - 2\mathrm{Re}[x]s + |x|^2)^{-1}. \tag{2.1}$$

We give the following definition because we now have two possible ways of writing the Cauchy kernel for the Cauchy formula of slice monogenic functions.

**Definition 2.10.** Let $x$, $s \in \mathbb{R}^{n+1}$ be such that $x \notin [s]$.

- We say that $S^{-1}(s, x)$ is written in the form I if
$$S^{-1}(s, x) := -(x^2 - 2x\mathrm{Re}[s] + |s|^2)^{-1}(x - \overline{s}).$$
- We say that $S^{-1}(s, x)$ is written in the form II if
$$S^{-1}(s, x) := (s - \bar{x})(s^2 - 2\mathrm{Re}[x]s + |x|^2)^{-1}.$$

We have used the seme symbol $S^{-1}(s, x)$ to denote the Cauchy kernel in both forms since in the sequel no confusion will arise. We now state the Cauchy formula for slice monogenic functions on axially symmetric and s-domains.

**Theorem 2.11 (Cauchy formula for axially symmetric open sets, see [2] and [3]).** *Let $W \subset \mathbb{R}^{n+1}$ be an open set and let $f \in \mathcal{SM}(W)$. Let $U$ be a bounded axially symmetric s-domain such that $\overline{U} \subset W$. Suppose that the boundary of $U \cap \mathbb{C}_I$ consists of a finite number of rectifiable Jordan curves for any $I \in \mathbb{S}$. Then, if $x \in U$, we have*

$$f(x) = \frac{1}{2\pi} \int_{\partial(U\cap\mathbb{C}_I)} S^{-1}(s, x)ds_I f(s) \tag{2.2}$$

*where $ds_I = ds/I$ and the integral does not depend on $U$ nor on the imaginary unit $I \in \mathbb{S}$.*

We finally recall the well-known notion of monogenic functions that will be used in the sequel.

**Definition 2.12 (Monogenic functions, see [1]).** Let $U \subseteq \mathbb{R}^{n+1}$ be an open set and let $f : U \to \mathbb{R}_n$ be a real differentiable function. We say that $f$ is a (left) monogenic function, or monogenic function, if $\mathcal{D}f(x) = 0$ where $\mathcal{D} = \partial_{x_0} + e_1\partial_{x_1} + \cdots + e_n\partial_{x_n}$ is the Dirac operator. We denote by $\mathcal{M}(U)$ the set of monogenic functions on $U$.

## 3. The functional setting and preliminary results

Let us now introduce the notations necessary to deal with linear operators. By $V$ we denote a Banach space over $\mathbb{R}$ with norm $\|\cdot\|$ and we set $V_n := V \otimes \mathbb{R}_n$. We recall that $V_n$ is a two-sided Banach module over $\mathbb{R}_n$ and its elements are of the type $\sum_A v_A \otimes e_A$ (where $A = i_1 \dots i_r$, $i_\ell \in \{1, 2, \dots, n\}$, $i_1 < \cdots < i_r$ is a multi-index). The multiplications (right and left) of an element $v \in V_n$ with a scalar $a \in \mathbb{R}_n$ are defined as $va = \sum_A v_A \otimes (e_A a)$ and $av = \sum_A v_A \otimes (ae_A)$. For short, in the sequel we will write $\sum_A v_A e_A$ instead of $\sum_A v_A \otimes e_A$. Moreover, we define $\|v\|^2_{V_n} = \sum_A \|v_A\|^2_V$.

Let $\mathcal{B}(V)$ be the space of bounded $\mathbb{R}$-homomorphisms of the Banach space $V$ into itself endowed with the natural norm denoted by $\|\cdot\|_{\mathcal{B}(V)}$. If $T_A \in \mathcal{B}(V)$, we can define the operator $T = \sum_A e_A T_A$ and its action on $v = \sum_B v_B e_B$ as $T(v) = \sum_{A,B} T_A(v_B) e_A e_B$. The set of all such bounded operators is denoted by $\mathcal{B}_n(V_n)$ and the norm is defined by $\|T\|_{\mathcal{B}_n(V_n)} = \sum_A \|T_A\|_{\mathcal{B}(V)}$. Note that, in the sequel, we will omit the subscript $\mathcal{B}_n(V_n)$ in the norm of an operator and note also that $\|TS\| \leq \|T\|\|S\|$. A bounded operator $T = T_0 + \sum_{j=1}^n e_j T_j$, where $T_\mu \subset \mathcal{B}(V)$ for $\mu = 0, 1, \dots, n$, will be called, with an abuse of language, an operator in paravector form. The set of such operators will be denoted by $\mathcal{B}^{0,1}_n(V_n)$. The set of bounded operators of the type $T = \sum_{j=1}^n e_j T_j$, where $T_\mu \in \mathcal{B}(V)$ for $\mu = 1, \dots, n$, will be denoted by $\mathcal{B}^1_n(V_n)$ and $T$ will be said operator in vector form. We will consider operators of the form $T = T_0 + \sum_{j=1}^n e_j T_j$ where $T_\mu \in \mathcal{B}(V)$ for $\mu = 0, 1, \dots, n$ for the sake of generality, but when dealing with $n$-tuples of operators, we will embed them into $\mathcal{B}_n(V_n)$ as operators in vector form, by setting $T_0 = 0$. The subset of those operators in $\mathcal{B}_n(V_n)$ whose components commute among themselves will be denoted by $\mathcal{BC}_n(V_n)$. In the same spirit we denote by $\mathcal{BC}^{0,1}_n(V_n)$ the set of paravector operators with commuting components. We now recall some definitions and results. Since we now want to construct a functional calculus for noncommuting operators we consider the noncommutative Cauchy kernel series in which we formally replace the paravector $x$ by the paravector operator $T$, whose components do not necessarily commute. So we define the noncommutative Cauchy kernel operator series.

**Definition 3.1.** Let $T \in \mathcal{B}^{0,1}_n(V_n)$ and $s \in \mathbb{R}^{n+1}$. We define the $\mathcal{S}$-resolvent operator series as

$$S^{-1}(s,T) := \sum_{n \geq 0} T^n s^{-1-n} \tag{3.1}$$

for $\|T\| < |s|$.

The most surprisingly fact is the following theorem that opens the way to the functional calculus for noncommuting operators.

**Theorem 3.2.** *Let $T \in \mathcal{B}_n^{0,1}(V_n)$ and $s \in \mathbb{R}^{n+1}$. Then*

$$\sum_{n\geq 0} T^n s^{-1-n} = -(T^2 - 2T\mathrm{Re}[s] + |s|^2\mathcal{I})^{-1}(T - \overline{s}\mathcal{I}), \tag{3.2}$$

*for $\|T\| < |s|$.*

In fact, Theorem 3.2 shows that even when operator $T = T_0+e_1T_1+\cdots+e_nT_n$ has components $T_j$, for $j = 0, 1, \ldots, n$, that do not necessarily commute among themselves the sum of the series remains the same as in the case $T$ is a paravector $x$. From Theorem 3.2 naturally arise the following definitions.

**Definition 3.3 (The $\mathcal{S}$-spectrum and the $\mathcal{S}$-resolvent set).** Let $T \in \mathcal{B}_n^{0,1}(V_n)$ and $s \in \mathbb{R}^{n+1}$. We define the $\mathcal{S}$-spectrum $\sigma_S(T)$ of $T$ as:

$$\sigma_S(T) = \{s \in \mathbb{R}^{n+1} \ : \ T^2 - 2\ \mathrm{Re}[s]T + |s|^2\mathcal{I} \quad \textit{is not invertible}\}.$$

The $\mathcal{S}$-resolvent set $\rho_S(T)$ is defined by

$$\rho_S(T) = \mathbb{R}^{n+1} \setminus \sigma_S(T).$$

**Definition 3.4 (The $\mathcal{S}$-resolvent operator).** Let $T \in \mathcal{B}_n^{0,1}(V_n)$ and $s \in \rho_S(T)$. We define the $\mathcal{S}$-resolvent operator as

$$S^{-1}(s,T) := -(T^2 - 2\mathrm{Re}[s]T + |s|^2\mathcal{I})^{-1}(T - \overline{s}\mathcal{I}).$$

The $\mathcal{S}$-resolvent operator satisfy the following equation which we call $\mathcal{S}$-resolvent equation.

**Theorem 3.5.** *Let $T \in \mathcal{B}_n^{0,1}(V_n)$ and $s \in \rho_S(T)$. Let $S^{-1}(s,T)$ be the $\mathcal{S}$-resolvent operator. Then $S^{-1}(s,T)$ satisfies the ( $\mathcal{S}$-resolvent) equation*

$$S^{-1}(s,T)s - TS^{-1}(s,T) = \mathcal{I}.$$

Having in mind the definition of $\sigma_S(T)$ we can state the following result:

**Theorem 3.6 (Structure of the $\mathcal{S}$-spectrum).** *Let $T \in \mathcal{B}_n^{0,1}(V_n)$ and suppose that $p = p_0 + \underline{p}$ belongs $\sigma_S(T)$ with $\underline{p} \neq 0$. Then all the elements of the $(n-1)$-sphere $[p]$ belong to $\sigma_S(T)$.*

This result implies that if $p \in \sigma_S(T)$ then either $p$ is a real point or the whole $(n-1)$-sphere $[p]$ belongs to $\sigma_S(T)$. For bounded paravector operators the $\mathcal{S}$-spectrum shares the same properties that the usual spectrum of a single operator has, that is the spectrum is a compact and nonempty set.

**Theorem 3.7 (Compactness of $\mathcal{S}$-spectrum).** *Let $T \in \mathcal{B}_n^{0,1}(V_n)$. Then the $\mathcal{S}$-spectrum $\sigma_S(T)$ is a compact nonempty set. Moreover, $\sigma_S(T)$ is contained in $\{s \in \mathbb{R}^{n+1} : |s| \leq \|T\| \}$.*

## 4. The $\mathcal{S}$-functional calculus for bounded operators

**Definition 4.1 (Admissible sets $U$).** We say that $U \subset \mathbb{R}^{n+1}$ is an admissible set if

- $U$ is axially symmetric s-domain that contains the $\mathcal{S}$-spectrum $\sigma_S(T)$ of $T \in \mathcal{B}_n^{0,1}(V_n)$,
- $\partial(U \cap \mathbb{C}_I)$ is union of a finite number of rectifiable Jordan curves for every $I \in \mathbb{S}$.

**Definition 4.2 (Locally s-monogenic functions on $\sigma_S(T)$).** Suppose that $U$ is admissible and $\overline{U}$ is contained in a domain of s-monogenicity of a function $f$. Then such a function $f$ is said to be locally s-monogenic on $\sigma_S(T)$.
We will denote by $\mathcal{SM}_{\sigma_S(T)}$ the set of locally s-monogenic functions on $\sigma_S(T)$.

The first crucial point for the definition of a functional calculus based on the Cauchy formula for slice monogenic functions is Theorem 3.2 which asserts, as we have already observed, that the Cauchy kernel for for slice monogenic functions can work as a resolvent operator for the functional calculus. But now, in order to have a good definition of the functional calculus, we also have to be sure that the integral $\int_{\partial(U\cap\mathbb{C}_I)} S^{-1}(s,T)\; ds_I\; f(s)$ is independent of $U$ and of $I \in \mathbb{S}$. Precisely we have:

**Theorem 4.3.** *Let $T \in \mathcal{B}_n^{0,1}(V_n)$ and $f \in \mathcal{SM}_{\sigma_S(T)}$. Let $U \subset \mathbb{R}^{n+1}$ be an admissible set and let $ds_I = ds/I$ for $I \in \mathbb{S}$. Then the integral*

$$\frac{1}{2\pi}\int_{\partial(U\cap\mathbb{C}_I)} S^{-1}(s,T)\; ds_I\; f(s)$$

*does not depend on the open set $U$ nor on the choice of the imaginary unit $I \in \mathbb{S}$.*

Thanks to Theorem 4.3 we can finally give the definition of the $\mathcal{S}$-functional calculus for noncommuting operators.

**Definition 4.4 ($\mathcal{S}$-functional calculus).** Let $T \in \mathcal{B}_n^{0,1}(V_n)$ and $f \in \mathcal{SM}_{\sigma_S(T)}$. Let $U \subset \mathbb{R}^{n+1}$ be an admissible set and let $ds_I = ds/I$ for $I \in \mathbb{S}$. We define

$$f(T) := \frac{1}{2\pi}\int_{\partial(U\cap\mathbb{C}_I)} S^{-1}(s,T)\; ds_I\; f(s). \tag{4.1}$$

### 4.1. Properties of the $\mathcal{S}$-functional calculus

As one will easily see there is a great analogy between the $\mathcal{S}$-functional calculus and the Riesz-Dunford functional calculus for a single operator. Since the product and the composition of slice monogenic functions is not always slice monogenic, we give some preliminary results to assure this facts. The following definition is based on the Splitting Lemma for slice monogenic functions.

**Lemma 4.5 (Splitting Lemma).** *Let $U$ be an open set in $\mathbb{R}^{n+1}$ and let $f \in \mathcal{SM}(U)$. Choose $I = I_1 \in \mathbb{S}$ and let $I_2, \ldots, I_n$ be a completion to a basis of $\mathbb{R}_n$ such that*

$I_iI_j + I_jI_i = -2\delta_{ij}$. *Denote by* $f_I$ *the restriction of* $f$ *to* $\mathbb{C}_I$. *Then we have*

$$f_I(z) = \sum_{|A|=0}^{n-1} F_A(z)I_A, \quad I_A = I_{i_1}\dots I_{i_s}, \quad z = u + Iv$$

*where* $F_A : U \cap \mathbb{C}_I \to \mathbb{C}_I$ *are holomorphic functions. The multi-index* $A = i_1 \dots i_s$ *is such that* $i_\ell \in \{2, \dots, n\}$, *with* $i_1 < \dots < i_s$, *or, when* $|A| = 0$, $I_\emptyset = 1$.

**Definition 4.6.** We denote by $\widetilde{\mathcal{SM}}(U)$ the subclass of $\mathcal{SM}(U)$ consisting of those functions $f$ such that

$$f_I(z) = \sum_{|A|=0, |A| \text{ even}}^{n-1} F_A(z)I_A, \quad I_A = I_{i_1}\dots I_{i_s}, \quad z = u + Iv.$$

**Definition 4.7.** Let $f : U \to \mathbb{R}_n$ be an s-monogenic function where $U$ is an open set in $\mathbb{R}^{n+1}$. We define

$$\mathcal{N}(U) = \{f \in \mathcal{SM}(U) \ : \ f(U \cap \mathbb{C}_I) \subseteq \mathbb{C}_I, \ \ \forall I \in \mathbb{S}\}.$$

Thanks to the definition of the subsets $\widetilde{\mathcal{SM}}(U)$ and $\mathcal{N}(U)$ of the space of slice monogenic functions $\mathcal{SM}(U)$ we have the following result:

**Theorem 4.8.** *Let* $U$, $U'$ *be two open sets in* $\mathbb{R}^{n+1}$.

- *Let* $f \in \widetilde{\mathcal{SM}}(U)$, $g \in \mathcal{SM}(U)$, *then* $fg \in \mathcal{SM}(U)$.
- *Let* $f \in \mathcal{N}(U')$ *and let* $g \in \mathcal{N}(U)$ *with* $g(U) \subseteq U'$. *Then* $f(g(x))$ *is slice monogenic for* $x \in U$.

We are now in the position to state some properties of the $\mathcal{S}$-functional calculus.

**Theorem 4.9.** *Let* $T \in \mathcal{B}_n^{0,1}(V_n)$.

(a) *Let* $f$ *and* $g \in \mathcal{SM}_{\sigma_S(T)}$. *Then we have*

$$(f+g)(T) = f(T) + g(T), \qquad (f\lambda)(T) = f(T)\lambda, \qquad \textit{for all } \ \lambda \in \mathbb{R}_n.$$

(b) *Let* $\phi \in \widetilde{\mathcal{SM}}_{\sigma_S(T)}$ *and* $g \in \mathcal{SM}_{\sigma_S(T)}$. *Then we have*

$$(\phi g)(T) = \phi(T)g(T).$$

(c) *Let* $f(s) = \sum_{n\geq 0} s^n p_n$ *where* $p_n \in \mathbb{R}_n$ *be such that* $f \in \mathcal{SM}_{\sigma_S(T)}$. *Then we have*

$$f(T) = \sum_{n\geq 0} T^n p_n.$$

**Theorem 4.10 (Spectral Mapping Theorem).** *Let* $T \in \mathcal{B}_n^{0,1}(V_n)$, $f \in \mathcal{N}_{\sigma_S(T)}$. *Then*

$$\sigma_S(f(T)) = f(\sigma_S(T)) = \{f(s) : s \in \sigma_S(T)\}.$$

**Definition 4.11 (The $\mathcal{S}$-spectral radius of $T$).** Let $T \in \mathcal{B}_n^{0,1}(V_n)$. We call $\mathcal{S}$-spectral radius of $T$ the nonnegative real number

$$r_S(T) := \sup\{\, |s| \ : \ s \in \sigma_S(T)\, \}.$$

**Theorem 4.12 (The $\mathcal{S}$-spectral radius theorem).** *Let $T \in \mathcal{B}_n^{0,1}(V_n)$ and let $r_{\mathcal{S}}(T)$ be the $\mathcal{S}$-spectral radius of $T$. Then $r_{\mathcal{S}}(T) = \lim_{m\to\infty} \|T^m\|^{1/m}$.*

**Theorem 4.13 (Perturbation of the $\mathcal{S}$-resolvent).** *Let $T, Z \in \mathcal{B}_n^{0,1}(V_n)$, $f \in \mathcal{SM}_{\sigma_{\mathcal{S}}(T)}$ and let $\varepsilon > 0$. Then there exists $\delta > 0$ such that, for $\|Z - T\| < \delta$, we have $f \in \mathcal{SM}_{\sigma_{\mathcal{S}}(Z)}$ and $\|f(Z) - f(T)\| < \varepsilon$.*

## 5. The $\mathcal{SC}$-functional calculus

We now consider what happens if we use the Cauchy kernel in form II, see Definition 2.10, to define the functional calculus. For more details see [7]. First of all we observe that $S^{-1}(s,x)$ written in the form II, that is $(s-\bar{x})(s^2 - 2\mathrm{Re}[x]s + |x|^2)^{-1}$, contains the term $|x|^2$ which can be also written as $|x|^2 = x\overline{x} = \overline{x}x$. So in terms of operators $|x|^2$ has to be interpreted as $T\overline{T}$ and we must have $T\overline{T} = \overline{T}T$. So operator $T\overline{T}$ is well defined only in the case that $T = T_0 + T_1e_1 + \cdots + T_ne_n$ has commuting components $T_j$, for $j = 0, 1, \ldots, n$. Now a natural question arise: why do we have to consider this commutative case when we already have the general case for non commuting components? The reason is that the associated spectrum, called the $\mathcal{F}$-spectrum, for commuting operators is more simple to compute with respect to the $\mathcal{S}$-spectrum. In a certain sense the $\mathcal{F}$-spectrum takes into account the fact that the components of $T$ commute and it reduces the computational difficulties. In the sequel we will show this fact by an example. Moreover the two spectra are the same for commuting operators.

**Definition 5.1 (The $\mathcal{F}$-spectrum and the $\mathcal{F}$-resolvent sets).** Let $T \in \mathcal{BC}_n^{0,1}(V_n)$. We define the $\mathcal{F}$-spectrum of $T$ as:

$$\sigma_{\mathcal{F}}(T) = \{s \in \mathbb{R}^{n+1} \quad : \quad s^2\mathcal{I} - s(T + \overline{T}) + T\overline{T} \quad \textit{is not invertible}\}.$$

The $\mathcal{F}$-resolvent set of $T$ is defined by

$$\rho_{\mathcal{F}}(T) = \mathbb{R}^{n+1} \setminus \sigma_{\mathcal{F}}(T).$$

**Theorem 5.2 (Structure of the $\mathcal{F}$-spectrum).** *Let $T \in \mathcal{BC}_n^{0,1}(V_n)$ and let $p = p_0 + p_1I \in [p_0 + p_1I] \subset \mathbb{R}^{n+1} \setminus \mathbb{R}$, such that $p \in \sigma_{\mathcal{F}}(T)$. Then all the elements of the $(n-1)$-sphere $[p_0 + p_1I]$ belong to $\sigma_{\mathcal{F}}(T)$. Thus the $\mathcal{F}$-spectrum consists of real points and/or $(n-1)$-spheres.*

**Theorem 5.3 (Compactness of $\mathcal{F}$-spectrum).** *Let $T \in \mathcal{BC}_n^{0,1}(V_n)$. Then the $\mathcal{F}$-spectrum $\sigma_{\mathcal{F}}(T)$ is a compact nonempty set. Moreover $\sigma_{\mathcal{F}}(T)$ is contained in $\{s \in \mathbb{R}^{n+1} : |s| \leq \|T\| \}$.*

**Definition 5.4.** Let $T \in \mathcal{BC}_n^{0,1}(V_n)$ and let $U \subset \mathbb{R}^{n+1}$ be an axially symmetric s-domain containing the $\mathcal{F}$-spectrum $\sigma_{\mathcal{F}}(T)$, and such that $\partial(U \cap \mathbb{C}_I)$ is union of a finite number of continuously differential Jordan curves for every $I \in \mathbb{S}$. Let $W$ be an open set in $\mathbb{R}^{n+1}$. A function $f \in \mathcal{SM}(W)$ is said to be locally s-monogenic on $\sigma_{\mathcal{F}}(T)$ if there exists a domain $U \subset \mathbb{R}^{n+1}$ as above such that $\overline{U} \subset W$. We will denote by $\mathcal{SM}_{\sigma_{\mathcal{F}}(T)}$ the set of locally s-monogenic functions on $\sigma_{\mathcal{F}}(T)$.

**Definition 5.5 (The $\mathcal{S}_\mathcal{C}$-resolvent operator).** Let $T \in \mathcal{BC}_n^{0,1}(V_n)$ and $s \in \rho_\mathcal{F}(T)$. We define the $\mathcal{S}_\mathcal{C}$-resolvent operator as

$$\mathcal{S}_\mathcal{C}{}^{-1}(s,T) := (s\mathcal{I} - \overline{T})(s^2\mathcal{I} - s(T+\overline{T}) + T\overline{T})^{-1}. \tag{5.1}$$

**Definition 5.6 (The $\mathcal{SC}$-functional calculus).** Let $T \in \mathcal{BC}_n^{0,1}(V_n)$ and $f \in \mathcal{SM}_{\sigma_\mathcal{F}(T)}$. Let $U \subset \mathbb{R}^{n+1}$ be a domain as in Definition 5.4 and set $ds_I = ds/I$ for $I \in \mathbb{S}$. We define the $\mathcal{SC}$-functional calculus as

$$f(T) = \frac{1}{2\pi}\int_{\partial(U\cap\mathbb{C}_I)} \mathcal{S}_\mathcal{C}{}^{-1}(s,T)\, ds_I\, f(s). \tag{5.2}$$

### 5.1. The $\mathcal{F}$-spectrum is easier to compute with respect to the $\mathcal{S}$-spectrum

We now treat an explicit example: consider the two commuting matrices

$$T_1 = \begin{bmatrix} 0 & 1 \\ 0 & 1 \end{bmatrix} \qquad T_2 = \begin{bmatrix} 1 & 1 \\ 0 & 2 \end{bmatrix}.$$

We consider the operator

$$T = e_1T_1 + e_2T_2 = \begin{bmatrix} e_2 & e_1+e_2 \\ 0 & e_1+2e_2 \end{bmatrix}$$

and

$$\overline{T} = \begin{bmatrix} -e_2 & -e_1-e_2 \\ 0 & -e_1-2e_2 \end{bmatrix}.$$

It is immediate that $T+\overline{T} = 0$ and

$$T\overline{T} = \begin{bmatrix} 1 & 4 \\ 0 & 5 \end{bmatrix}.$$

To compute the $\mathcal{F}$-spectrum we have to solve

$$\begin{bmatrix} s^2+1 & 4 \\ 0 & s^2+5 \end{bmatrix}\begin{bmatrix} w \\ u \end{bmatrix} = 0,$$

which gives the two equations $(s^2+1)w + 4u = 0$ and $(s^2+5)u = 0$. It is easy to see that on the complex plane $\mathbb{C}_I$ the solutions are $\{I, \sqrt{5}I\}$, thus

$$\sigma_\mathcal{F}(T) = \{I,\ \ \sqrt{5}I,\ \ \text{for all } I \in \mathbb{S}\}.$$

The same result can be obtained by solving the equation

$$(T^2 - 2s_0T + |s|^2\mathcal{I})v = 0.$$

In this case it is

$$\begin{bmatrix} -1-2s_0e_2+|s|^2 & -4-2s_0(e_1+e_2) \\ 0 & -5-2s_0(e_1+2e_2)+|s|^2 \end{bmatrix}\begin{bmatrix} w \\ u \end{bmatrix} = 0$$

which corresponds the two equations

$$(-1-2s_0e_2+|s|^2)w - (4+2s_0(e_1+e_2))u = 0, \qquad (-5-2s_0(e_1+2e_2)+|s|^2)u = 0$$

where $w = w_0+e_1w_1+e_2w_2+e_1e_2w_{12}$ and similarly for $u$. If we suppose that $s_0 = 0$ we get $s = I$, or $s = \sqrt{5}I$, when $I$ varies in $\mathbb{S}$. If $s_0 \neq 0$, long calculations show that there are no solutions, thus the $\mathcal{S}$-spectrum coincides with the $\mathcal{F}$-spectrum. Note that, in general, when working in a Clifford algebra over more than two imaginary units, we have to take into account the fact that there can be zero divisors thus to solve equations requires more attention.

## 6. The $\mathcal{F}$-functional calculus for bounded operators

The $\mathcal{F}$-spectrum also arise in the case we want to define a functional calculus for monogenic functions (see Definition 2.12) using slice monogenic functions. The $\mathcal{F}$-functional calculus is based on the Fueter-Sce mapping theorem in integral form.

In its differential form the Fueter-Sce mapping theorem states that given a slice monogenic function $f$ we can generate a monogenic function $\breve{f}$ by the formula

$$\breve{f}(x) = \Delta^{\frac{n-1}{2}} f(x)$$

where $\Delta = \partial_0^2 + \partial_1^2 + \cdots + \partial_n^2$ is the Laplace operator in dimension $n+1$. The authors and F. Sommen in [9] have proved an integral version of the Fueter-Sce mapping theorem and, based on such integral formula, it has been defined the $\mathcal{F}$-functional calculus. Precisely we state our results as follows.

**Definition 6.1 (The $\mathcal{F}_n$-kernel).** Let $n$ be an odd number. Let $x,\ s \in \mathbb{R}^{n+1}$. We define, for $s \notin [x]$, the $\mathcal{F}_n$-kernel as

$$\mathcal{F}_n(s,x) := \Delta^{\frac{n-1}{2}} S^{-1}(s,x) = \gamma_n(s-\bar{x})(s^2 - 2\mathrm{Re}[x]s + |x|^2)^{-\frac{n+1}{2}},$$

where

$$\gamma_n := (-1)^{(n-1)/2} 2^{(n-1)/2} (n-1)! \Big(\frac{n-1}{2}\Big)!. \tag{6.1}$$

**Theorem 6.2 (The Fueter-Sce mapping theorem in integral form).** *Let $n$ be an odd number. Let $W \subset \mathbb{R}^{n+1}$ be an axially symmetric open set and let $f \in \mathcal{SM}(W)$. Let $U$ be a bounded axially symmetric open set such that $\overline{U} \subset W$. Suppose that the boundary of $U \cap \mathbb{C}_I$ consists of a finite number of rectifiable Jordan curves for any $I \in \mathbb{S}$. Then, if $x \in U$, the function $\breve{f}(x)$ given by*

$$\breve{f}(x) = \Delta^{\frac{n-1}{2}} f(x)$$

*is monogenic and it admits the integral representation*

$$\breve{f}(x) = \frac{1}{2\pi} \int_{\partial(U\cap\mathbb{C}_I)} \mathcal{F}_n(s,x)\, ds_I\, f(s), \quad ds_I = ds/I, \tag{6.2}$$

*where the integral does not depend on $U$ nor on the imaginary unit $I \in \mathbb{S}$.*

*Remark* 6.3. The Fueter-Sce mapping theorem defines a map from the space $\mathcal{SM}(U)$ to $\mathcal{M}(U)$. The integral version of the Fueter-Sce mapping theorem allows the definition of a functional calculus for a subclass of the space of monogenic functions $\mathcal{M}(U)$, that is functions that are in the kernel of the Dirac operator.

**Definition 6.4 ($\mathcal{F}$-resolvent operator).** Let $n$ be an odd number and let $T \in \mathcal{BC}_n^{0,1}(V_n)$. For $s \in \rho_{\mathcal{F}}(T)$ we define the $\mathcal{F}$-resolvent operator as

$$\mathcal{F}_n^{-1}(s,T) := \gamma_n(s\mathcal{I} - \overline{T})(s^2\mathcal{I} - s(T+\overline{T}) + T\overline{T})^{-\frac{n+1}{2}},$$

where $\gamma_n$ are given by (6.1).

Now, using the kernel $\mathcal{F}_n^{-1}(s,T)$ as a resolvent operator, for $T \in \mathcal{BC}_n^{0,1}(V_n)$, we define $\breve{f}(T)$ when $\breve{f} \in \mathcal{M}(U)$ is a monogenic function which comes from a slice monogenic function $f \in \mathcal{M}(SU)$ via the Fueter-Sce theorem in integral form. The $\mathcal{F}$-functional calculus will be defined only for those monogenic functions that are of the form $\breve{f}(x) = \Delta^{\frac{n-1}{2}} f(x)$, where $f$ is a slice monogenic function. For the functional calculus associated to standard monogenic functions we mention the book [20] and the literature therein.

**Definition 6.5 (The $\mathcal{F}$-functional calculus).** Let $n$ be an odd number and let $T \in \mathcal{BC}_n^{0,1}(V_n)$. Let $U$ be an open set as in Definition 5.4. Suppose that $f \in \mathcal{SM}_{\sigma_{\mathcal{F}}(T)}$ and let $\breve{f}(x) = \Delta^{\frac{n-1}{2}} f(x)$. We define the $\mathcal{F}$-functional calculus as

$$\breve{f}(T) = \frac{1}{2\pi} \int_{\partial(U \cap \mathbb{C}_I)} \mathcal{F}_n^{-1}(s,T)\, ds_I\, f(s). \tag{6.3}$$

*Remark* 6.6. The definitions of the $\mathcal{SC}$-functional calculus and of the $\mathcal{F}$-functional calculus are well posed since the integrals in (5.2) and in (6.3) are independent of $I \in \mathbb{S}$ and of the open set $U$.

**Theorem 6.7 (Bounded perturbations of the $\mathcal{F}$-functional calculus, see [8]).** *Let $n$ be an odd number, $T, Z \in \mathcal{BC}_n^{0,1}(V_n)$, $f \in \mathcal{SM}_{\sigma_{\mathcal{F}}(T)}$ and let $\varepsilon > 0$. Then there exists $\delta > 0$ such that, for $\|Z - T\| < \delta$, we have $f \in \mathcal{SM}_{\sigma_{\mathcal{F}}(Z)}$ and*

$$\|\breve{f}(Z) - \breve{f}(T)\| < \varepsilon.$$

## 7. The $\mathcal{S}$-functional calculus for unbounded operators

We conclude this paper by pointing out that the $\mathcal{S}$-functional calculus and its commutative formulation the $\mathcal{S}_{\mathcal{C}}$-functional calculus can be extended to the case of unbounded operators using analogous strategies suggested by the Riesz-Dunford functional calculus. Here we have to take into account the fact that the composition of functions is not always well defined as it happens in the case of holomorphic functions of a complex variable. Even thought there are several difficulties due to the non commutativity all the results known for the Riesz-Dunford functional calculus have been extended to the $\mathcal{S}$-functional calculus for non commuting unbounded operators. In the following we simply give an idea of how the $\mathcal{S}$-functional calculus looks like for unbounded operators.

**Definition 7.1.** Let $k \in \mathbb{R}^{n+1}$ and define the homeomorphism $\Phi : \overline{\mathbb{R}}^{n+1} \to \overline{\mathbb{R}}^{n+1}$,

$$p = \Phi(s) = (s-k)^{-1}, \quad \Phi(\infty) = 0, \quad \Phi(k) = \infty.$$

Observe that, in the case $k \in \rho_S(T) \cap \mathbb{R}$ we have that $(T - k\mathcal{I})^{-1} = -S^{-1}(k, T)$. We use the symbol $\overline{\sigma}_S(T)$ to denote the fact that the point at infinity can be in the $\mathcal{S}$-spectral set.

**Definition 7.2 (Definition of the $\mathcal{S}$-functional calculus for unbounded operators).** Let $T : \mathcal{D}(T) \to V_n$ be a linear closed operator with $\rho_S(T) \cap \mathbb{R} \neq \emptyset$ and suppose that $f \in \mathcal{SM}_{\overline{\sigma}_S(T)}$. Let us consider $\phi(p) := f(\Phi^{-1}(p))$ and the operator $A := (T - k\mathcal{I})^{-1}$, *for some* $k \in \rho_S(T) \cap \mathbb{R}$. We define

$$f_k(T) = \phi(A). \tag{7.1}$$

The operator $f_k(T)$ is well defined since it does not depend on $k \in \rho_S(T) \cap \mathbb{R}$ as the following theorem shows.

**Theorem 7.3.** *Let $T : \mathcal{D}(T) \to V_n$ be a linear closed operator with $\rho_S(T) \cap \mathbb{R} \neq \emptyset$ and suppose that $f \in \mathcal{SM}_{\overline{\sigma}_S(T)}$. Then operator $f_k(T)$ is independent of $k \in \rho_S(T) \cap \mathbb{R}$. Let $f$ be an s-monogenic function such that its domain of s-monogenicity contains $\overline{U}$. Set $ds_I = ds/I$ for $I \in \mathbb{S}$, then we have*

$$f(T) = f(\infty)\mathcal{I} + \frac{1}{2\pi} \int_{\partial(U \cap \mathbb{C}_I)} \hat{S}^{-1}(s, T)\, ds_I\, f(s), \tag{7.2}$$

*where* $\hat{S}^{-1}(s, T) := (T^2 - 2T\mathrm{Re}[s] + |s|^2\mathcal{I})^{-1}\overline{s} - T(T^2 - 2T\mathrm{Re}[s] + |s|^2\mathcal{I})^{-1}$.

Most of the results related to the $\mathcal{S}$-functional calculus and on its function theory are now collected in the recent book [17].

## References


[1] F. Brackx, R. Delanghe, F. Sommen, *Clifford Analysis*, Pitman Res. Notes in Math., 76, 1982.

[2] F. Colombo, I. Sabadini, *A structure formula for slice monogenic functions and some of its consequences*, Hypercomplex Analysis, Trends in Mathematics, Birkhäuser, 2009, 69–99.

[3] F. Colombo, I. Sabadini, *The Cauchy formula with s-monogenic kernel and a functional calculus for noncommuting operators*, J. Math. Anal. Appl., **373** (2011), 655–679.

[4] F. Colombo, I. Sabadini, *On some properties of the quaternionic functional calculus*, J. Geom. Anal., **19** (2009), 601–627.

[5] F. Colombo, I. Sabadini, *On the formulations of the quaternionic functional calculus*, J. Geom. Phys., **60** (2010), 1490–1508.

[6] F. Colombo, I. Sabadini, *Some remarks on the $\mathcal{S}$-spectrum*, to appear in Complex Var. Elliptic Equ., (2011).

[7] F. Colombo, I. Sabadini, *The $\mathcal{F}$-spectrum and the $\mathcal{SC}$-functional calculus*, to appear in Proceedings of the Royal Society of Edinburgh, Section A.

[8] F. Colombo, I. Sabadini, *Bounded perturbations of the resolvent operators associated to the $\mathcal{F}$-spectrum*, Hypercomplex Analysis and its applications, Trends in Mathematics, Birkhäuser, (2010), 13–28.

[9] F. Colombo, I. Sabadini, F. Sommen, *The Fueter mapping theorem in integral form and the $\mathcal{F}$-functional calculus*, Math. Meth. Appl. Sci., **33** (2010), 2050–2066.

[10] F. Colombo, I. Sabadini, F. Sommen, D.C. Struppa, *Analysis of Dirac Systems and Computational Algebra*, Progress in Mathematical Physics, Vol. 39, Birkhäuser, Boston, 2004.

[11] F. Colombo, I. Sabadini, D.C. Struppa, *Slice monogenic functions*, Israel J. Math., **171** (2009), 385–403.

[12] F. Colombo, I. Sabadini, D.C. Struppa, *Extension properties for slice monogenic functions*, Israel J. Math., **177** (2010), 369–389.

[13] F. Colombo, I. Sabadini, D.C. Struppa, *The Pompeiu formula for slice hyperholomorphic functions*, Michigan Math. J., **60** (2011), 163–170.

[14] F. Colombo, I. Sabadini, D.C. Struppa, *A new functional calculus for noncommuting operators*, J. Funct. Anal., **254** (2008), 2255–2274.

[15] F. Colombo, I. Sabadini, D.C. Struppa, *Duality theorems for slice hyperholomorphic functions*, J. Reine Angew. Math., **645** (2010), 85–104.

[16] F. Colombo, I. Sabadini, D.C. Struppa, *The Runge theorem for slice hyperholomorphic functions*, Proc. Amer. Math. Soc., **139** (2011), 1787–1803.

[17] F. Colombo, I. Sabadini, D.C. Struppa, *Noncommutative functional calculus. Theory and Applications of Slice Hyperholomorphic Functions*, Progress in Mathematics, Vol. 289, Birkhäuser, 2011, VI, 222 p.

[18] N. Dunford, J. Schwartz, *Linear operators, part I: general theory* , J. Wiley and Sons (1988).

[19] G. Gentili, D.C. Struppa, A new approach to Cullen-regular functions of a quaternionic variable, C. R. Math. Acad. Sci. Paris, **342** (2006), 741–744.

[20] B. Jefferies, *Spectral properties of noncommuting operators*, Lecture Notes in Mathematics, 1843, Springer-Verlag, Berlin, 2004.

[21] W. Rudin, *Functional Analysis*, Functional analysis. McGraw-Hill Series in Higher Mathematics. McGraw-Hill Book Co., New York-Düsseldorf-Johannesburg, 1973.

Fabrizio Colombo and Irene Sabadini
Dipartimento di Matematica
Politecnico di Milano
Via Bonardi, 9
I-20133 Milano, Italy
e-mail: `fabrizio.colombo@polimi.it`
`irene.sabadini@polimi.it`

Operator Theory:
Advances and Applications, Vol. 221, 255-268


# Necessity of Parameter-ellipticity for Multi-order Systems of Differential Equations

R. Denk and M. Faierman

**Abstract.** In this paper we investigate parameter-ellipticity conditions for multi-order systems of differential equations on a bounded domain. Under suitable assumptions on smoothness and on the order structure of the system, it is shown that parameter-dependent a priori estimates imply the conditions of parameter-ellipticity, i.e., interior ellipticity, conditions of Shapiro-Lopatinskii type, and conditions of Vishik-Lyusternik type. The mixed-order systems considered here are of general form; in particular, it is not assumed that the diagonal operators are of the same order. This paper is a continuation of an article by the same authors where the sufficiency was shown, i.e., a priori estimates for the solutions of parameter-elliptic multi-order systems were established.

**Mathematics Subject Classification (2000).** Primary 35J55; Secondary 35S15.

**Keywords.** Parameter-ellipticity, multi-order systems, a priori estimates.

## 1. Introduction and main results

In this paper, we will study multi-order boundary value problems defined over a bounded domain in $\mathbb{R}^n$. Under rather general assumptions on the structure of the system, it was shown in the paper [DF] that parameter-ellipticity implies uniform a priori estimates for the solutions. Now we will show that the conditions of parameter-ellipticity are also necessary.

Parameter-elliptic boundary value problems and a priori estimates for them were treated, e.g., in [ADF] (scalar problems), [DFM] (systems of homogeneous type), and [F] (multi-order systems). The notion of parameter-ellipticity for general multi-order systems was introduced by Kozhevnikov ([K1], [K2]) and by Denk, Mennicken, and Volevich ([DMV]). The mentioned papers had restrictions on the orders of the operators which excluded, for instance, boundary conditions of Dirichlet type (see [ADN, Section 2], [G, p. 448]). In the paper [DF], these restrictions were removed.

Let us consider in a bounded domain $\Omega \subset \mathbb{R}^n$, $n \geq 2$, with boundary $\Gamma$ the boundary value problem

$$\begin{aligned} A(x,D)u(x) - \lambda u(x) &= f(x) \quad \text{in } \Omega, \\ B(x,D)u(x) &= g(x) \quad \text{on } \Gamma. \end{aligned} \tag{1.1}$$

Here $A(x,D) = \big(A_{jk}(x,D)\big)_{j,k=1,\ldots,N}$ is an $N\times N$-matrix of linear differential operators, $N \in \mathbb{N}$, $N \geq 2$, $u(x) = (u_1(x),\ldots,u_N(x))^T$ and $f(x) = (f_1(x),\ldots,f_N(x))^T$ are defined on $\Omega$ ($^T$ denoting the transpose), whereas

$$B(x,D) = \big(B_{jk}(x,D)\big)_{\substack{j=1,\ldots,\widetilde{N}\\ k=1,\ldots,N}}$$

is an $\widetilde{N} \times N$-matrix of boundary operators, and $g(x) = (g_1(x),\ldots,g_{\widetilde{N}}(x))^T$ is defined on $\Gamma$.

To describe the order structure of the boundary value problem $(A,B)$, let $\{s_j\}_{j=1}^N$ and $\{t_j\}_{j=1}^N$ denote sequences of integers satisfying $s_1 \geq \cdots \geq s_N$, $t_1 \geq \cdots \geq t_N \geq 0$, and put $m_j := s_j + t_j$ $(j = 1,\ldots,N)$. We assume

$$m_1 = \cdots = m_{k_1} > m_{k_1+1} = \cdots = m_{k_{d-1}} > m_{k_{d-1}+1} = \cdots = m_{k_d} > 0,$$

where $k_d = N$. We set $\widetilde{m}_j := m_{k_j}$ $(j = 1,\ldots,d)$, and assume that $2N_r := \sum_{j=1}^{k_r} m_j$ is even for $r = 1,\ldots,d$. We also set $k_0 := 0$ and $N_0 := 0$. Further, let $\{\sigma_j\}_{j=1}^{\widetilde{N}}$, $\widetilde{N} := N_d$, be a sequence of integers satisfying $\max_j \sigma_j < s_N$. It was shown in [DF, Section 2] that we may also assume $s_j \geq 0$ $(j = 1,\ldots,N)$ and $\sigma_j < 0$ $(j = 1,\ldots,\widetilde{N})$. Define $\kappa_0 := \max\{t_1, -\sigma_1,\ldots,-\sigma_{\widetilde{N}}\}$. Concerning $(A,B)$, we will assume that

$$\begin{aligned} &\operatorname{ord} A_{jk} \leq s_j + t_k \quad (j,k = 1,\ldots,N), \\ &\operatorname{ord} B_{jk} \leq \sigma_j + t_k \quad (j = 1,\ldots,\widetilde{N},\ k = 1,\ldots,N). \end{aligned}$$

Using the standard multi-index notation $D^\alpha = D_1^{\alpha_1}\cdots D_n^{\alpha_n}$, $D_j = -i\frac{\partial}{\partial x_j}$, we write $A_{jk}(x,D) = \sum_{|\alpha|\leq s_j+t_k} a_\alpha^{jk}(x)D^\alpha$ for $x \in \Omega$ and $j,k = 1,\ldots,N$ and $B_{jk}(x,D) = \sum_{|\alpha|\leq\sigma_j+t_k} b_\alpha^{jk}(x)D^\alpha$ for $x \in \Gamma$, $j = 1,\ldots,\widetilde{N}$, and $k = 1,\ldots,N$. With respect to the smoothness, we will suppose

**(S)** (1) $\Gamma$ is of class $C^{\kappa_0-1,1} \cap C^{s_1}$,
(2) $a_\alpha^{jk} \in C^{s_j}(\overline{\Omega})$ $(|\alpha| \leq s_j + t_k)$ if $s_j > 0$ and
$a_\alpha^{jk} \in C^0(\overline{\Omega})$ $(|\alpha| = s_j + t_k)$, $a_\alpha^{jk} \in L_\infty(\Omega)$ $(|\alpha| < s_j + t_k)$ if $s_j = 0$,
(3) $b_\alpha^{jk} \in C^{-\sigma_j-1,1}(\Gamma)$ $(|\alpha| \leq \sigma_j + t_k)$.

Let $\mathring{A}_{jk}(x,\xi)$ consist of all terms in $A_{jk}(x,\xi)$ which are exactly of order $s_j + t_k$, and set

$$\mathring{A}(x,\xi) := \big(\mathring{A}_{jk}(x,\xi)\big)_{j,k=1,\ldots,N} \quad (x \in \overline{\Omega},\ \xi \in \mathbb{R}^n).$$

Analogously, define $\mathring{B}(x,\xi) = \big(\mathring{B}_{jk}(x,\xi)\big)_{\substack{j=1,\dots,\widetilde{N}\\ k=1,\dots,N}}$ for $x \in \Gamma$, $\xi \in \mathbb{R}^n$. The operator $B(x,D)$ is said to be essentially upper triangular if $\mathring{B}_{jk}(x,D) = 0$ for $j = N_{\ell-1}+1,\dots,N_\ell$, $k = 1,\dots,k_{\ell-1}$, $\ell = 2,\dots,d$.

To formulate the ellipticity conditions, let

$$\mathcal{A}_{11}^{(r)}(x,\xi) := \big(\mathring{A}_{jk}(x,\xi)\big)_{j,k=1,\dots,k_r} \quad (r = 1,\dots,d).$$

Let $I_\ell$ denote the $\ell \times \ell$ unit matrix, $\widetilde{I}_\ell := I_{k_\ell - k_{\ell-1}}$, and $\widetilde{I}_{\ell,0} := \operatorname{diag}\big(0\cdot\widetilde{I}_1,\dots,0\cdot\widetilde{I}_{\ell-1},\widetilde{I}_\ell\big)$. In the following, let $\mathcal{L} \subset \mathbb{C}$ be a closed sector in the complex plane with vertex at the origin. The following condition is taken from [DF, Section 2] (cf. also [DMV, Section 3]).

**(E)** For each $x \in \overline{\Omega}$, $\xi \in \mathbb{R}^n \setminus \{0\}$, $\lambda \in \mathcal{L}$, and $r = 1,\dots,d$ we have

$$\det\big(\mathcal{A}_{11}^{(r)}(x,\xi) - \lambda\widetilde{I}_{r,0}\big) \neq 0.$$

If condition (E) holds, the operator $A(x,D) - \lambda I_N$ is said to be parameter-elliptic in $\mathcal{L}$. In order to formulate conditions of Shapiro-Lopatinskii type, for $x^0 \in \Gamma$ we rewrite the boundary value problem (1.1) in terms of local coordinates associated to $x^0$. In these coordinates $x^0 = 0$, and the positive $x_n$-axis coincides with the direction of the inner normal to $\Gamma$. We will keep the notation for $A$ and $B$ in the new coordinates. In local coordinates associated to $x^0 \in \Gamma$, let

$$\mathcal{B}_{r,1}^{(r,r)}(0,\xi',D_n) := \big(\mathring{B}_{jk}(0,\xi',D_n)\big)_{\substack{j=1,\dots,N_r\\ k=1,\dots,k_r}} \quad (r = 1,\dots,d),$$

$$\mathcal{B}_{r,1}^{(1,r)}(0,\xi',D_n) := \big(\mathring{B}_{jk}(0,\xi',D_n)\big)_{\substack{j=N_{r-1}+1,\dots,N_r\\ k=1,\dots,k_r}} \quad (r = 2,\dots,d),$$

The following conditions (see [DF, Section 2]) are of Shapiro-Lopatinskii type and of Vishik-Lyusternik type, respectively (cf. also [DV, Section 2.3]).

**(SL)** For each $x^0 \in \Gamma$ rewrite (1.1) in local coordinates associated to $x^0$. Then for $r = 1,\dots,d$, the boundary value problem on the half-line,

$$\begin{aligned} \mathcal{A}_{11}^{(r)}(0,\xi',D_n)v(x_n) - \lambda\widetilde{I}_{r,0}v(x_n) &= 0 \quad (x_n > 0),\\ \mathcal{B}_{r,1}^{(r,r)}(0,\xi',D_n)v(x_n) &= 0 \quad (x_n = 0),\\ |v(x_n)| &\to 0 \quad (x_n \to \infty) \end{aligned} \tag{1.2}$$

has only the trivial solution for $\xi' \in \mathbb{R}^{n-1} \setminus \{0\}$, $\lambda \in \mathcal{L}$.

**(VL)** For each $x^0 \in \Gamma$ rewrite (1.1) in local coordinates associated to $x^0$. Then for $r = 2,\dots,d$, the boundary value problem on the half-line,

$$\begin{aligned} \mathcal{A}_{11}^{(r)}(0,0,D_n)v(x_n) - \lambda\widetilde{I}_{r,0}v(x_n) &= 0 \quad (x_n > 0),\\ \mathcal{B}_{r,1}^{(1,r)}(0,0,D_n)v(x_n) &= 0 \quad (x_n = 0),\\ |v(x_n)| &\to 0 \quad (x_n \to \infty) \end{aligned} \tag{1.3}$$

has only the trivial solution for $\lambda \in \mathcal{L} \setminus \{0\}$.

We will show that (E), (SL), and (VL) are necessary for a priori estimates to hold. In order to formulate these estimates, we will introduce parameter-dependent norms. For $G \subset \mathbb{R}^\ell$ open, $\ell \in \mathbb{N}$, $s \in \mathbb{N}$ and $1 < p < \infty$, let $\|u\|_{s,p,G}$ denote the norm in the standard Sobolev space $W_p^s(G)$. For $\lambda \in \mathbb{C} \setminus \{0\}$ and $j = 1, \dots, d$ set

$$\|\!|u|\!\|_{s,p,G}^{(j)} := \|u\|_{s,p,G} + |\lambda|^{s/\widetilde{m}_j} \|u\|_{0,p,G} \quad (u \in W_p^s(G)).$$

For $s < 0$, $s \in \mathbb{Z}$, and $j = 1, \dots, d$, let $H_p^s(\mathbb{R}^n)$ be the Bessel-potential space equipped with the parameter-dependent norm $\|\!|u|\!\|_{s,p,\mathbb{R}^n}^{(j)} := \|F^{-1}\langle \xi, \lambda\rangle_j^s F u\|_{0,p,\mathbb{R}^n}$ where $F$ denotes the Fourier transform in $\mathbb{R}^n$ $(x \to \xi)$ and where $\langle \xi, \lambda\rangle_j^s := \big(|\xi|^2 + |\lambda|^{2/\widetilde{m}_j}\big)^{1/2}$. For $G \subset \mathbb{R}^n$ open, set $\|\!|u|\!\|_{s,p,G}^{(j)} := \inf\{\|\!|v|\!\|_{s,p,\mathbb{R}^n}^s : v \in H_p^s(\mathbb{R}^n),\, v|_G = u\}$. Finally, for $s \in \mathbb{N}$ we define the parameter-dependent norm on the boundary by

$$\|\!|v|\!\|_{s-1/p,p,\partial G}^{(j)} := \|v\|_{s-1/p,p,\partial G} + |\lambda|^{(s-1/p)/\widetilde{m}_j} \|v\|_{0,p,\partial G} \quad (v \in W_p^{s-1/p}(\partial G)).$$

For $j = 1, \dots, N$, let $\pi_1(j) := r$ if $k_{r-1} < j \le k_r$. Similarly, for $j = 1, \dots, N_d$ let $\pi_2(j) := r$ if $N_{r-1} < j \le N_r$. Note that, by definition, $\widetilde{m}_{\pi_1(j)} = m_j$ for $j = 1, \dots, N$.

The aim of the paper is to show the following result.

**Theorem 1.1.** *Let* (S) *hold, let* $1 < p < \infty$, *and assume that there exist constants* $C_0, C_1 > 0$ *such that for all* $\lambda \in \mathcal{L}$, $|\lambda| \ge C_0$ *and all* $u \in \prod_{j=1}^N W_p^{t_j}(\Omega)$ *the a priori estimate*

$$\sum_{j=1}^N \|\!|u_j|\!\|_{t_j,p,\Omega}^{(\pi_1(j))} \le C_1 \Big( \sum_{j=1}^N \|\!|f_j|\!\|_{-s_j,p,\Omega}^{(\pi_1(j))} + \sum_{j=1}^{\widetilde{N}} \|\!|g_j|\!\|_{-\sigma_j-1/p,p,\Gamma}^{(\pi_2(j))} \Big) \tag{1.4}$$

*holds for* $f := A(x,D)u - \lambda u$ *and* $g := B(x,D)u$. *Assume further that* $B(x,D)$ *is essentially upper triangular. Then the parameter-ellipticity conditions* (E), (SL), *and* (VL) *are satisfied.*

*Remark* 1.2. In [DF], the following result was shown, where we refer to [DF] for the definitions of properly parameter-elliptic and compatible: Let (S), (E), (SL), and (VL) hold. Assume further that $(A, B)$ is properly parameter-elliptic, that $B(x,D)$ is essentially upper triangular, and that $A(x^0, D)$ and $B(x^0, D)$ are compatible at every $x_0 \in \Gamma$. Then there exist $C_0, C_1 > 0$ such that for all $\lambda \in \mathcal{L}$, $|\lambda| \ge C_0$, the boundary value problem (1.1) has a unique solution $u \in \prod_{j=1}^N W_p^{t_j}(\Omega)$ for every $f \in \prod_{j=1}^N H_p^{-s_j}(\Omega)$ and every $g \in \prod_{j=1}^{\widetilde{N}} W_p^{-\sigma_j-1/p}(\Gamma)$, and the a priori estimate (1.4) holds.

In this sense, the sufficiency of parameter-ellipticity for the validity of the a priori estimate was shown in [DF] while Theorem 1.1 states the necessity of the conditions (E), (SL), and (VL).

## 2. Proof of the necessity

Throughout this section, we assume condition (S) to hold, and fix a closed sector $\mathcal{L} \subset \mathbb{C}$. In the following, $C$ stands for a generic constant which may vary from inequality to inequality but which is independent of the functions appearing in the inequality and independent of $\lambda$. Let $B_\delta(x^0) := \{x \in \mathbb{R}^n : |x - x^0| < \delta\}$, and let $\mathbb{R}^n_+ := \{x \in \mathbb{R}^n : x_n > 0\}$, $\mathbb{R}_+ := (0, \infty)$. We start with some useful remarks on negative-order Sobolev spaces where $C_0^\infty(\overline{\mathbb{R}^n_+})$ stands for the set of all restrictions of functions in $C_0^\infty(\mathbb{R}^n)$ to $\mathbb{R}^n_+$.

**Lemma 2.1.** *Let $s \in \mathbb{N}$, $1 < p < \infty$, $j \in \{1, \dots, N\}$. Then for all $v \in L_p(\mathbb{R}^n)$ and all $\lambda \in \mathbb{C}$, $|\lambda| \geq 1$, we have:*

a) $|\!|\!| v |\!|\!|^{(j)}_{-s,p,\mathbb{R}^n} \leq C \|v\|_{-s,p,\mathbb{R}^n}$,

b) $|\!|\!| D^\alpha v |\!|\!|^{(j)}_{-s,p,\mathbb{R}^n} \leq |\lambda|^{(|\alpha|-s)/\widetilde{m}_j} \|v\|_{0,p,\mathbb{R}^n}$ *for all* $|\alpha| \leq s$,

c) *for each $\phi \in C_0^\infty(\mathbb{R}^n)$ there exists a constant $C_\phi > 0$ independent of $v$ such that $\|v\phi\|_{-s,p,\mathbb{R}^n} \leq C_\phi \|v\|_{-s,p,\mathbb{R}^n}$ and $|\!|\!| v\phi |\!|\!|^{(j)}_{-s,p,\mathbb{R}^n} \leq C_\phi |\!|\!| v |\!|\!|^{(j)}_{-s,p,\mathbb{R}^n}$.*

*The same assertions hold if we replace $\mathbb{R}^n$ by $\mathbb{R}^n_+$ and $C_0^\infty(\mathbb{R}^n)$ in* c) *by $C_0^\infty(\overline{\mathbb{R}^n_+})$.*

*Proof.* a) We have

$$|\!|\!| v |\!|\!|^{(j)}_{-s,p,\mathbb{R}^n} = \|F^{-1}\langle \xi, \lambda\rangle_j^{-s} Fv\|_{0,p,\mathbb{R}^n} = \left\| F^{-1} \frac{\langle \xi \rangle^s}{\langle \xi, \lambda \rangle_j^s} \langle \xi \rangle^{-s} Fv \right\|_{0,p,\mathbb{R}^n}.$$

Now the assertion follows immediately from the Mikhlin-Lizorkin multiplier theorem.

b) Similarly,

$$|\lambda|^{(s-|\alpha|)/\widetilde{m}_j} |\!|\!| D^\alpha v |\!|\!|^{(j)}_{-s,p,\mathbb{R}^n} = \|F^{-1} m(\xi, \lambda) Fv\|_{0,p,\mathbb{R}^n}$$

with $m(\xi, \lambda) := |\lambda|^{(s-|\alpha|)/\widetilde{m}_j} \xi^\alpha \langle \xi, \lambda\rangle_j^{-s}$. Noting that $m$ is infinitely smooth in $\xi$ and quasi-homogeneous in $(\xi, \lambda)$ of degree 0 in the sense that $m(\rho\xi, \rho^{\widetilde{m}_j}\lambda) = m(\xi, \lambda)$ for $\rho > 0$, we see that we may apply the Mikhlin-Lizorkin theorem to obtain the statement in b).

c) We make use of the dual pairing of $H_p^{-s}(\mathbb{R}^n)$ and $W_q^s(\mathbb{R}^n)$, $\frac{1}{p} + \frac{1}{q} = 1$, and get

$$\|v\phi\|_{-s,p,\mathbb{R}^n} = \sup_\zeta |\langle v\phi, \zeta\rangle| = \sup_\zeta \left| \int v(x)\phi(x)\zeta(x)dx \right| = \sup_\zeta |\langle v, \phi\zeta\rangle|,$$

where the supremum is taken over all $\zeta \in C_0^\infty(\mathbb{R}^n)$ with $\|\zeta\|_{s,q,\mathbb{R}^n} \leq 1$. Now we make use of $\|\phi\zeta\|_{s,q,\mathbb{R}^n} \leq C_\phi \|\zeta\|_{s,q,\mathbb{R}^n}$ with $C_\phi := C_{s,q} \sup\{|D^\alpha\phi(x)| : |\alpha| \leq s,\, x \in \mathbb{R}^n\}$ where $C_{s,q}$ is a constant depending on $s$ and $q$ only. We obtain $\sup_\zeta |\langle v, \phi\zeta\rangle| \leq C_\phi \sup_\zeta |\langle v, \zeta\rangle| = C_\phi \|v\|_{-s,p,\mathbb{R}^n}$.

For the parameter-dependent norms $|\!|\!| \cdot |\!|\!|^{(j)}_{-s,p,\mathbb{R}^n}$ we again consider the dual pairing between $H_p^{-s}(\mathbb{R}^n)$ and $W_q^s(\mathbb{R}^n)$, but now with respect to the parameter-dependent norm $|\!|\!| \cdot |\!|\!|^{(j)}_{s,q,\mathbb{R}^n}$ on $W_q^s(\mathbb{R}^n)$. Then the result follows in exactly the same way, noting that

$$\|\phi\zeta\|^{(j)}_{s,q,\mathbb{R}^n} = \|\phi\zeta\|_{s,q,\mathbb{R}^n} + |\lambda|^{s/\widetilde{m}_j}\|\phi\zeta\|_{0,p,\mathbb{R}^n} \le C_\phi\Big(\|\zeta\|_{s,q,\mathbb{R}^n} + |\lambda|^{s/\widetilde{m}_j}\|\zeta\|_{0,p,\mathbb{R}^n}\Big).$$

Finally, in the case of $\mathbb{R}^n_+$ instead of $\mathbb{R}^n$ the assertions of the lemma follow easily from the results in $\mathbb{R}^n$ and the fact that there exists an extension operator $E\colon u \mapsto Eu$ which is continuous as an operator from $H^r_p(\mathbb{R}^n_+)$ to $H^r_p(\mathbb{R}^n)$ for all $|r| \le s$ (see [T, p. 218]). □

The following lemma will allow us to consider the model problem in $\mathbb{R}^n$ for the proof of the necessity.

**Lemma 2.2.** *Assume that there exist constants $C_0, C_1 > 0$ such that for all $u \in \prod_{j=1}^N W_p^{t_j}(\mathbb{R}^n)$ and all $\lambda \in \mathcal{L}$, $|\lambda| \ge C_0$, the a priori estimate* (1.4) *holds. Let $x^0 \in \overline{\Omega}$. Then there exist an $x^1 \in \Omega$, a $\delta > 0$ with $\overline{B_\delta(x^1)} \subset \Omega$, and a $\widetilde{\lambda} > 0$ such that for all $\lambda \in \mathcal{L}$ with $|\lambda| \ge \widetilde{\lambda}$ and all $u \in \prod_{j=1}^N W_p^{t_j}(\mathbb{R}^n)$ with* $\operatorname{supp} u \subset B_\delta(x^1)$, *we have*

$$\sum_{j=1}^N \|u\|^{(\pi_1(j))}_{t_j,p,\mathbb{R}^n} \le C \sum_{j=1}^N \|f_j^0\|^{(\pi_1(j))}_{-s_j,p,\mathbb{R}^n}, \tag{2.1}$$

*where we have set $f^0 := (\mathring{A}(x^0, D) - \lambda)u$.*

*Proof.* In [DF, Prop. 4.1] it was shown that for any $\varepsilon > 0$ there exist a $\delta_0 > 0$ and a $\lambda_0 > 0$ such that for $\lambda \in \mathcal{L}$, $|\lambda| \le \lambda_0$, and all $u \in \prod_{j=1}^N W_p^{t_j}(\mathbb{R}^n)$ with $\operatorname{supp} u \subset B_\delta(x^0) \cap \overline{\Omega}$ we have

$$\sum_{j=1}^N \|f_j - f_j^0\|^{(\pi_1(j))}_{-s_j,p,\Omega} \le \varepsilon \sum_{j=1}^N \|u_j\|_{t_j,p,\Omega}$$

where $f := (A(x, D) - \lambda)u$. Let $\varepsilon$ be sufficiently small. If $x^0 \in \Omega$ we choose $x^1 := x^0$ and $\delta := \frac{1}{2}\min\{\delta_0, \operatorname{dist}(x^0, \Gamma)\}$. If $x^0 \in \Gamma$ we choose $x^1 \in B_\delta(x^0) \cap \Omega$ and $\delta > 0$ sufficiently small such that $\overline{B_\delta(x^1)} \subset B_\delta(x^0) \cap \Omega$. In both cases, the statement of the lemma follows easily by arguments similar to those used in the proof of [AV, Lemma 4.2]. □

**Proposition 2.3.** *Under the assumptions of Lemma 2.2, condition* (E) *is satisfied, i.e., for $r = 1, \dots, d$, $x^0 \in \overline{\Omega}$, $\xi^0 \in \mathbb{R}^n \setminus \{0\}$, and $\lambda^0 \in \mathcal{L}$ we have*

$$\det\big(\mathcal{A}_{11}^{(r)}(x^0, \xi^0) - \lambda^0 \widetilde{I}_{r,0}\big) \ne 0.$$

*Proof.* Assume that (E) does not hold. Then there exist $r \in \{1, \dots, d\}$, $x^0 \in \overline{\Omega}$, $\xi^0 \in \mathbb{R}^n \setminus \{0\}$, $\lambda^0 \in \mathcal{L}$, and a vector $h \in \mathbb{C}^{k_r} \setminus \{0\}$ such that $(\mathcal{A}_{11}^{(r)}(x^0, \xi^0) - \lambda^0 \widetilde{I}_{r,0})h = 0$.

Let us first consider the case $\lambda^0 = 0$. We choose $x^1 \in \Omega$, $\delta > 0$ and $\widetilde{\lambda} > 0$ according to Lemma 2.1. Let $\phi \in C_0^\infty(B_\delta(x^1))$ with $\phi \not\equiv 0$, and for $\rho > 1$ set

$$u_j(x) := \begin{cases} \phi(x) e^{i\rho\xi^0\cdot x} \rho^{-t_j} h_j, & j = 1, \ldots, k_r, \\ 0, & j = k_r + 1, \ldots, N, \end{cases}$$

where $\cdot$ denotes the inner product in $\mathbb{R}^n$. We are now going to use (2.1) to arrive at a contradiction. Indeed, we easily see that for $j = 1, \ldots, k_r$,

$$\|u_j\|_{t_j,p,\mathbb{R}^n} \geq |h_j|\,|\xi_\ell^0|^{t_j} \|\phi\|_{0,p,\mathbb{R}^n} - C\rho^{-1}$$

where $\xi_\ell^0 \neq 0$. We further choose $\mu$ with

$$\widetilde{m}_{r+1} < \mu < \widetilde{m}_r \text{ if } r < d \text{ and } \widetilde{m}_d/2 < \mu < \widetilde{m}_d \text{ if } r = d, \tag{2.2}$$

and choose $\lambda \in \mathcal{L}$ with $|\lambda| = \rho^\mu$. Then it is clear that

$$|\lambda|^{t_j/\widetilde{m}_j} \|u_j\|_{0,p,\mathbb{R}^n} = \rho^{-t_j(1-\mu/\widetilde{m}_j)} |h_j|\, \|\phi\|_{0,p,\mathbb{R}^n}.$$

Thus we have shown that

$$\sum_{j=1}^N \|u_j\|_{t_j,p,\mathbb{R}^n}^{(\pi_1(j))} \geq \frac{1}{2}\Big(\sum_{j=1}^{k_r} |h_j|\,|\xi_\ell^0|^{t_j}\Big) \|\phi\|_{0,p,\mathbb{R}^n} \tag{2.3}$$

for sufficiently large $\rho$.

Turning next to the right-hand side of (2.1), let $j \in \{1, \ldots, N\}$. Then

$$\begin{aligned}
\|f_j^0\|_{-s_j,p,\mathbb{R}^n}^{(\pi_1(j))} &= \Big\| \sum_{k=1}^{k_r} \mathring{A}_{jk}(x^0, D) u_k - \delta_{jk}\lambda u_k \Big\|_{-s_j,p,\mathbb{R}^n}^{(\pi_1(j))} \\
&< \Big\| \sum_{k=1}^{k_r} \sum_{|\alpha|=s_j+t_k} a_\alpha^{jk}(x^0) \sum_\beta \binom{\alpha}{\beta} \rho^{-t_k} h_k D^\beta(e^{i\rho\xi^0\cdot x}) D^{\alpha-\beta}\phi \Big\|_{-s_j,p,\mathbb{R}^n}^{(\pi_1(j))} \\
&\quad + \sum_{k=1}^{k_r} \delta_{jk} \rho^{-r_k+\mu} |h_k|\, \|e^{i\rho\xi^0\cdot x}\phi\|_{-s_j,p,\mathbb{R}^n}^{(\pi_1(j))} =: I_1 + I_2,
\end{aligned}$$

where $\delta_{jk}$ denotes the Kronecker delta and where $\sum_\beta = \sum_{\beta<\alpha}$ if $j \leq k_r$ and $\sum_\beta = \sum_{\beta\leq\alpha}$ if $r < d$ and $j > k_r$. (Here we used the fact that $\mathcal{A}_{11}^{(r)}(x^0, \xi^0)h = 0$.)

It is clear that $I_2 \to 0$ as $\rho \to \infty$. Hence fixing our attention next upon $I_1$, we see that $I_1 \leq \sum_{k=1}^{k_r} \sum_{|\alpha|=s_j+t_k} \sum_\beta I_{1,k}^{\alpha,\beta}$ with

$$I_{1,k}^{\alpha,\beta} := \Big\| \binom{\alpha}{\beta} a_\alpha^{jk}(x^0) \rho^{-t_k} D^\beta(e^{i\rho\xi^0\cdot x}) D^{\alpha-\beta}\phi \Big\|_{-s_j,p,\mathbb{R}^n}^{(\pi_1(j))}.$$

To establish $I_{1,k}^{\alpha,\beta} \to 0$ $(\rho \to \infty)$ and, in consequence, a contradiction, it remains to show that for all appearing indices we have

$$\rho^{-t_k} \|D^\beta(e^{i\rho\xi^0\cdot x}) D^{\alpha-\beta}\phi\|_{-s_j,p,\mathbb{R}^n}^{(\pi_1(j))} \to 0 \quad (\rho \to \infty).$$

Let $j \in \{1,\dots,k_r\}$, $|\alpha| = s_j + t_k$, and $\beta < \alpha$. If $|\beta| \le t_k$, we apply Lemma 2.1 b) to obtain

$$\begin{aligned}\rho^{-t_k}\|D^\beta(e^{i\rho\xi^0\cdot x})D^{\alpha-\beta}\phi\|^{(\pi_1(j))}_{-s_j,p,\mathbb{R}^n} &\le C\rho^{-t_k-s_j\mu/m_j}\|D^\beta(e^{i\rho\xi^0\cdot x})D^{\alpha-\beta}\phi\|_{0,p,\mathbb{R}^n}\\ &\le C\rho^{-t_k+|\beta|-s_j\mu/m_j}|(\xi^0)^\beta|\,\|D^{\alpha-\beta}\phi\|_{0,p,\mathbb{R}^n}\to 0\quad(\rho\to\infty).\end{aligned}$$

Note here that $s_j = 0$ implies $|\beta| < t_k$.

If $|\beta| \ge t_k$, we write $\beta = \beta_1+\beta_2$ with $|\beta_1| = t_k$, $|\beta_2| < s_j$ and fix $\psi \in C_0^\infty(\mathbb{R}^n)$ with $\psi = 1$ on $\operatorname{supp}\phi$. Then, using Lemma 2.1 b) and c),

$$\begin{aligned}\rho^{-t_k}\|D^\beta(e^{i\rho\xi^0\cdot x})D^{\alpha-\beta}\phi\|^{(\pi_1(j))}_{-s_j,p,\mathbb{R}^n} &= \rho^{-t_k}\|D^\beta(e^{i\rho\xi^0\cdot x}\psi)D^{\alpha-\beta}\phi\|^{(\pi_1(j))}_{-s_j,p,\mathbb{R}^n}\\ &\le C\rho^{-t_k}\|D^{\beta_2}(D^{\beta_1}e^{i\rho\xi^0\cdot x}\psi)\|^{(\pi_1(j))}_{-s_j,p,\mathbb{R}^n}\\ &\le C\rho^{-t_k-(s_j-|\beta_2|)\mu/m_j}\|D^{\beta_1}e^{i\rho\xi^0\cdot x}\psi\|_{0,p,\mathbb{R}^n}\\ &\le C\rho^{-(s_j-|\beta_2|)\mu/m_j}\big(\|\psi\|_{0,p,\mathbb{R}^n}+C\rho^{-1}\big)\to 0\quad(\rho\to\infty).\end{aligned}$$

Now let $j \in \{k_r+1,\dots,N\}$. Again by Lemma 2.1 b), we have for $|\alpha| = s_j+t_k$ and $|\beta| \le |\alpha|$

$$\rho^{-t_k}\|D^\beta(e^{i\rho\xi^0\cdot x})D^{\alpha-\beta}\phi\|^{(\pi_1(j))}_{-s_j,p,\mathbb{R}^n} \le C\rho^{-s_j\mu/m_j+s_j}\|D^{\alpha-\beta}\phi\|_{0,p,\mathbb{R}^n}\to 0\quad(\rho\to\infty)$$

as $\mu/m_j > 1$.

Finally, the case $\lambda^0 \ne 0$ can be dealt with by arguing in a manner similar to that above, except now we take $\lambda = \lambda^0\rho^{\widetilde{m}_r}$. □

To prove the necessity of (SL) and (VL), we transform the problem to the half-space. For this let $x^0 \in \Gamma$ and assume that $(A,B)$ is given in local coordinates associated to $x^0$. Let $\{U,\Phi\}$ be a chart on $\Gamma$ such that $x^0 = 0 \in U$, $\Phi(0) = 0$, and $\Phi$ is a diffeomorphism of class $C^{\kappa_0-1,1}\cap C^{s_1}$ mapping $U$ onto an open set in $\mathbb{R}^n$ with $\Phi(U\cap\Omega)\subset\mathbb{R}^n_+$, $\Phi(U\cap\Gamma)\subset\mathbb{R}^{n-1}$. We denote the push-forward of the operators $A(x,D)$ and $B(x,D)$ by $\widetilde{A}(y,D)$ and $\widetilde{B}(y,D)$, respectively, where $y = \Phi(x)$.

Replacing $\Phi(x)$ by $D\Phi(0)^{-1}\Phi(x)$, it is easily seen that we may assume the Jacobian $D\Phi(0)$ to be equal to $I_n$. Then we have $\mathring{\widetilde{A}}_{jk}(0,\xi) = \mathring{A}_{jk}(0,\xi)$ and $\mathring{\widetilde{B}}_{jk}(0,\xi) = \mathring{B}_{jk}(0,\xi)$. In particular, (SL) and (VL) are satisfied for $(\widetilde{A},\widetilde{B})$ at 0 if only if this holds for $(A,B)$ at $x^0 = 0$ (see also [DHP, p. 205]).

**Lemma 2.4.** *Under the assumptions of Lemma 2.2, let $x^0\in\Gamma$ and assume $(A,B)$ to be written in coordinates associated to $x^0$. Then there exist a $\delta > 0$ and a $\widetilde{\lambda} > 0$ such that for all $u \in \prod_{j=1}^N W_p^{t_j}(\mathbb{R}^n_+)$ with $\operatorname{supp} u \subset B_\delta(0)\cap\overline{\mathbb{R}^n_+}$ and all $\lambda\in\mathcal{L}$ with $|\lambda|\ge\widetilde{\lambda}$, we have*

$$\sum_{j=1}^N \|u_j\|^{(\pi_1(j))}_{t_j,p,\mathbb{R}^n_+} \le C\Big(\sum_{j=1}^N \|f_j^0\|^{(\pi_1(j))}_{-s_j,p,\mathbb{R}^n_+} + \sum_{j=1}^{\widetilde{N}} \|g_j^0\|^{(\pi_2(j))}_{-\sigma_j-1/p,p,\mathbb{R}^{n-1}}\Big), \tag{2.4}$$

*where $f^0 := (\mathring{A}(0,D)-\lambda)u$, $g^0 := \mathring{B}(0,D)u$.*

*Proof.* Let $\Phi$ be as above, and let $\widetilde{A}(y,D)$ and $\widetilde{B}(y,D)$ be the push-forward of $A(x,D)$ and $B(x,D)$, respectively. Then

$$\Phi_*\big[(a_\alpha^{jk}(x)-a_\alpha^{jk}(0))D^\alpha u_k\big]=(\tilde{a}_\alpha^{jk}(y)-\tilde{a}_\alpha^{jk}(0))D_y^\alpha\tilde{u}_k+\sum_{|\beta|<|\alpha|}\tilde{a}_{\alpha,\beta}^{jk}(y)D_y^\beta\tilde{u}_k.$$

It was shown in the proof of [DF, Prop. 4.1], that for each $\varepsilon>0$ there exist a $\delta_0>0$ and a $\lambda_0>0$ such that for all $u\in\prod_{j=1}^N W_p^{t_j}(\Omega)$ with $\operatorname{supp} u\subset B_{\delta_0}(0)\cap\overline{\Omega}$ and all $\lambda\in\mathcal{L}$, $|\lambda|\geq\lambda_0$, we have

$$\|\Phi_*\big[(a_\alpha^{jk}(x)-a_\alpha^{jk}(0))D^\alpha u_k\big]\|_{-s_j,p,\mathbb{R}_+^n}^{(\pi_1(j))}\leq\varepsilon\|\tilde{u}_k\|_{t_k,p,\mathbb{R}_+^n}$$

for $|\alpha|=s_j+t_k$, and

$$\|\Phi_*\big[a_\alpha^{jk}(x)D^\alpha u_k\big]\|_{-s_j,p,\mathbb{R}_+^n}^{(\pi_1(j))}\leq\varepsilon\|\tilde{u}_k\|_{t_k,p,\mathbb{R}_+^n}$$

for $|\alpha|<s_j+t_k$. From this we easily obtain that for all $\varepsilon>0$ there exist $\delta_0,\lambda_0>0$ such that for all $u\in\prod_{j=1}^N W_p^{t_j}(\Omega)$ with $\operatorname{supp} u\subset B_{\delta_0}(0)$,

$$\sum_{j=1}^N\|\tilde{f}_j\|_{-s_j,p,\mathbb{R}_+^n}\leq C\sum_{j=1}^N\|\widetilde{f_j^0}\|_{-s_j,p,\mathbb{R}_+^n}+\varepsilon\sum_{j=1}^N\|u_j\|_{t_j,p,\mathbb{R}_+^n}$$

where we have set $f:=(A(x,D)-\lambda)u$, $\tilde{f}:=\Phi_*f$, $\widetilde{f^0}:=\Phi_*f^0$.

To estimate $g^0$, we first remark that we may assume $b_\alpha^{jk}$ to be defined on $\overline{\Omega}$ with $b_\alpha^{jk}\in C^{-\sigma_j-1,1}(\overline{\Omega})$. We define the function $h$ on $\Omega$ by $h_j:=\sum_{j=1}^{\widetilde{N}}b_\alpha^{jk}(x)D^\alpha u_k$, $h_j^0:=\sum_{j=1}^{\widetilde{N}}b_\alpha^{jk}(0)D^\alpha u_k$ and set $\tilde{h}:=\Phi_*h$, $\widetilde{h^0}:=\Phi_*h^0$. In the same way as above, we obtain

$$\begin{aligned}\sum_{j=1}^{\widetilde{N}}\|\tilde{g}_j\|_{-\sigma_j-1/p,p,\mathbb{R}^{n-1}}^{(\pi_2(j))}&\leq\sum_{j=1}^{\widetilde{N}}\|\tilde{h}_j\|_{-\sigma_j-1/p,p,\mathbb{R}^{n-1}}^{(\pi_2(j))}\\&\leq C\sum_{j=1}^{\widetilde{N}}\|\widetilde{h_j^0}\|_{-\sigma_j-1/p,p,\mathbb{R}^{n-1}}^{(\pi_2(j))}+\varepsilon\sum_{j=1}^{N}\|\tilde{u}\|_{t_j,p,\mathbb{R}_+^n}.\end{aligned}$$

Finally, it was shown in [DF, p. 362-363] that there exist constants $c_1,c_2>0$ such that for all $u\in\prod_{j=1}^N W_p^{t_j}(\Omega)$ with $\operatorname{supp} u\subset B_\delta(x^0)$, $B_{2\delta}(x^0)\subset U$, we have

$$\begin{aligned}c_1\|u_j\|_{t_j,p,\Omega}^{(\pi_1(j))}&\leq\|\tilde{u}_j\|_{t_j,p,\mathbb{R}_+^n}^{(\pi_1(j))}&&\leq c_2\|u_j\|_{t_j,p,\Omega}^{(\pi_1(j))},\\c_1\|f_j\|_{-s_j,p,\Omega}^{(\pi_1(j))}&\leq\|\tilde{f}_j\|_{-s_j,p,\mathbb{R}_+^n}^{(\pi_1(j))}&&\leq c_2\|f_j\|_{-s_j,p,\Omega}^{(\pi_1(j))},\\c_1\|g_j\|_{-\sigma_j-1/p,p,\Gamma}^{(\pi_1(j))}&\leq\|\tilde{g}_j\|_{-\sigma_j-1/p,p,\mathbb{R}^{n-1}}^{(\pi_1(j))}&&\leq c_2\|g_j\|_{-\sigma_j-1/p,p,\Gamma}^{(\pi_1(j))}.\end{aligned}\tag{2.5}$$

Therefore, from the a priori estimate (1.4) we obtain that for each $\varepsilon>0$ there exist $\delta,\tilde{\lambda}>0$ such that for $u\in\prod_{j=1}^N W_p^{t_j}(\Omega)$ with $\operatorname{supp} u\subset B_\delta(0)$ and $\lambda\in\mathcal{L}$,

$|\lambda| \geq \widetilde{\lambda}$, we have

$$\sum_{j=1}^{N} \|\widetilde{u}_j\|_{t_j,p,\mathbb{R}^n_+}^{(\pi_1(j))} \leq C \sum_{j=1}^{N} \|u_j\|_{t_j,p,\Omega}^{(\pi_1(j))} \leq C\Big(\sum_{j=1}^{N} \|f_j\|_{-s_j,p,\Omega}^{(\pi_1(j))} + \sum_{j=1}^{\widetilde{N}} \|g_j\|_{-\sigma_j-1/p,p,\Gamma}^{(\pi_2(j))}\Big)$$
$$\leq C\Big(\sum_{j=1}^{N} \|\widetilde{f}_j\|_{-s_j,p,\mathbb{R}^n_+}^{(\pi_1(j))} + \sum_{j=1}^{\widetilde{N}} \|\widetilde{g}_j\|_{-\sigma_j-1/p,p,\mathbb{R}^{n-1}}^{(\pi_2(j))}\Big) + \varepsilon \sum_{j=1}^{N} \|\widetilde{u}_j\|_{t_j,p,\mathbb{R}^n_+}^{(\pi_1(j))}.$$

Taking $\varepsilon$ small enough and $\lambda$ large enough and noting (2.5) and $\mathring{\widetilde{A}}(0,D) = \mathring{A}(0,D)$ and $\mathring{\widetilde{B}}(0,D) = \mathring{B}(0,D)$, we obtain the assertion of the Lemma. □

**Proposition 2.5.** *Assume that there exist constants $C_0, C_1 > 0$ such that for all $u \in \prod_{j=1}^{N} W_p^{t_j}(\mathbb{R}^n)$ and all $\lambda \in \mathcal{L}$, $|\lambda| \geq C_0$, the a priori estimate* (1.4) *holds. Further, let $x^0 \in \Gamma$, and assume that $B(x^0, D)$ is essentially upper triangular. Then condition* (SL) *holds at $x^0$.*

*Proof.* Let $(A, B)$ be written in coordinates associated to $x^0$ and assume that (SL) does not hold. Then there exist $r \in \{1, \dots, d\}$, $\lambda^0 \in \mathcal{L}$, $\xi_0' = (\xi_1^0, \dots, \xi_{n-1}^0) \in \mathbb{R}^{n-1} \setminus \{0\}$, and $v \not\equiv 0$ satisfying (1.2). By Proposition 2.3, we know that the polynomial $\det(\mathcal{A}_{11}^{(r)}(0, \xi_0', \tau) - \lambda^0 \widetilde{I}_{r,0})$ as a function of $\tau$ has no real roots. Therefore, $v = v(x_n)$ is infinitely smooth and decays exponentially for $x_n \to \infty$, in particular, $v \in L_p(\mathbb{R}_+)$.

Again, let us first consider the case $\lambda^0 = 0$. We choose $\phi' \in C_0^\infty(\mathbb{R}^{n-1})$ such that $\phi' \not\equiv 0$ and $\operatorname{supp} \phi' \subset B_\delta(0)$ with $\delta$ from Lemma 2.4, $\psi \in C_0^\infty([0, \delta))$ with $0 \leq \psi \leq 1$ and $\psi(x_n) = 1$ for $0 \leq x_n \leq \delta/2$, and $\lambda \in \mathcal{L}$ with $|\lambda| = \rho^\mu$ where $\mu$ satisfies (2.2). For $x \in \mathbb{R}^n_+$, we set $w(x) := e^{i\xi_0' \cdot x'} v(x_n)$, $\phi(x) := \phi'(x')\psi(x_n)$, and

$$u_j(x) := \begin{cases} \rho^{-t_j+1/p} w_j(\rho x)\phi(x), & j = 1, \dots, k_r, \\ 0, & j = k_r + 1, \dots, N. \end{cases} \tag{2.6}$$

We will show that (2.4) leads to a contradiction for large $\rho$. For this we first remark that for $j = 1, \dots, k_r$

$$\rho \|v_j(\rho x_n)\psi(x_n)\|_{0,p,\mathbb{R}_+}^p = \rho \int_0^\infty |v_j(\rho x_n)\psi(x_n)|^p dx_n$$
$$= \int_0^\infty |v_j(y_n)\psi(y_n/\rho)|^p dy_n \nearrow \|v_j\|_{0,p,\mathbb{R}_+}^p \quad (\rho \to \infty).$$

Therefore, for $\rho \geq \rho_0$, $\rho_0$ being sufficiently large, we have

$$\frac{1}{2}\rho^{-1/p} \|v_j\|_{0,p,\mathbb{R}_+} \leq \|v_j(\rho x_n)\psi(x_n)\|_{0,p,\mathbb{R}_+} \leq \rho^{-1/p} \|v_j\|_{0,p,\mathbb{R}_+}.$$

In the same way, we see that for any $\zeta \in C_0^\infty(\overline{\mathbb{R}^n_+})$ and $\alpha \in \mathbb{N}_0^n$ we have

$$\|D^\alpha w_j(\rho x)\zeta(x)\|_{0,p,\mathbb{R}^n_+} \leq C_\zeta \rho^{|\alpha|-1/p} \|v_j\|_{|\alpha|,p,\mathbb{R}_+}$$

with a constant $C_\zeta$ depending on $\zeta$ but not on $v$ or $\rho$.

Turning now to the left-hand side of (2.4), the above considerations show that for $\rho$ sufficiently large,

$$\|u_j\|^{(\pi_1(j))}_{t_j,p,\mathbb{R}^n_+} \geq \|u_j\|_{t_j,p,\mathbb{R}^n_+} \geq \frac{1}{2}|\xi^0_\ell|^{t_j}\|\phi'\|_{0,p,\mathbb{R}^{n-1}}\|v_j\|_{0,p,\mathbb{R}_+}. \tag{2.7}$$

On the right-hand side of (2.4), the terms $\|f^0_j\|^{(\pi_1(j))}_{-s_j,p,\mathbb{R}^n_+}$ can be estimated in the same way as in the proof of Proposition 2.3. Indeed, we have

$$f^0_j(x) = \sum_{k=1}^{k_r}\Big(\sum_{|\alpha|=s_j+t_k}\sum_{\beta} a^{jk}_\alpha(0)\binom{\alpha}{\beta}\rho^{-t_k+1/p}(D^\beta w)(\rho x)(D^{\alpha-\beta}\phi)(x) + \delta_{jk}\lambda u_k\Big)$$

where $\sum_\beta = \sum_{\beta<\alpha}$ if $j \leq k_r$ and $\sum_\beta = \sum_{\beta\leq\alpha}$ if $j > k_r$. Here we used the fact $\mathring{A}(0,D)w(x) = e^{i\xi'_0\cdot x'}\mathring{A}(0,\xi'_0,D_n)v(x_n) = 0$. From this we obtain in the same way as in the proof of Proposition 2.3

$$\sum_{j=1}^{N}\|f^0_j\|^{(\pi_1(j))}_{-s_j,p,\mathbb{R}^n_+} \to 0 \quad (\rho\to\infty). \tag{2.8}$$

To estimate $g^0_j$, we first remark that

$$\mathring{B}_{jk}(0,\rho\xi'_0,D_n)v_k(\rho x_n)\psi(x_n)\big|_{x_n=0} = \rho^{\sigma_j+t_k}\mathring{B}_{jk}(0,\zeta'_0,D_n)v_k(x_n)\big|_{x_n=0}$$

by homogeneity and as $\psi(x_n) = 1$ near $x_n = 0$. Therefore, for $j = 1,\dots,N_r$ we have

$$\begin{aligned} g_j &= \sum_{k=1}^{k_r}\rho^{-t_k+1/p}\mathring{B}_{jk}(0,D)w_k(\rho x)\phi(x)\big|_{x_n=0} \\ &\quad -\sum_{k=1}^{k_r}\sum_{|\alpha|=\sigma_j+t_k}\sum_{\beta<\alpha} b^{jk}_\alpha(0)\binom{\alpha}{\beta}\rho^{-t_k+1/p}D^\beta w_k(\rho x)D^{\alpha-\beta}\phi(x)\big|_{x_n=0} \end{aligned}$$

due to $\sum_{k=1}^{k_r}\mathring{B}_{jk}(0,\xi'_0,D_n)v_k(x_n)|_{x_n=0} = 0$. For $j = 1,\dots,N_r$, $|\alpha| = \sigma_j + t_k$, and $\beta < \alpha$ we can estimate

$$\begin{aligned} &\rho^{-t_k+1/p}\big\|D^\beta w(\rho x)D^{\alpha-\beta}\phi(x)\big|_{x_n=0}\big\|_{-\sigma_j-1/p,p,\mathbb{R}^{n-1}} \\ &\quad\leq \rho^{-t_k+1/p}\big\|D^\beta w(\rho x)D^{\alpha-\beta}\phi(x)\big\|_{-\sigma_j,p,\mathbb{R}^n_+} \\ &\quad\leq C\rho^{-t_k+1/p}\sum_{|\gamma|\leq-\sigma_j}\big\|D^\gamma\big[D^\beta w(\rho x)D^{\alpha-\beta}\phi(x)\big]\big\|_{0,p,\mathbb{R}^n_+} \\ &\quad\leq C_\phi\rho^{-t_k-\sigma_j+|\beta|}\|v\|_{0,p,\mathbb{R}_+} \to 0 \quad (\rho\to\infty). \end{aligned} \tag{2.9}$$

Further, for $|\lambda| = \rho^\mu$ we obtain

$$\begin{aligned}
&|\lambda|^{(-\sigma_j-1/p)/\widetilde{m}_{\pi_2(j)}}\|g_j\|_{0,p,\mathbb{R}^{n-1}}\\
&\quad = |\lambda|^{(-\sigma_j-1/p)/\widetilde{m}_{\pi_2(j)}}\Big\|\sum_{k=1}^{k_r}\sum_{|\alpha|=\sigma_j+t_k} b_\alpha^{jk}(0)D^\alpha u_k(x)\big|_{x_n=0}\Big\|_{0,p,\mathbb{R}^{n-1}}\\
&\quad \le C\rho^{(\sigma_j+1/p)(1-\mu/\widetilde{m}_{\pi_2(j)})} \to 0 \quad (\rho\to\infty)
\end{aligned}$$

as $\mu/\widetilde{m}_{\pi_2(j)} < 1$ and $\sigma_j \le -1$. From this and (2.9) we see that for $j = 1,\dots,N_r$

$$\|g_j^0\|^{(\pi_2(j))}_{-\sigma_j-1/p,p,\mathbb{R}^{n-1}} \to 0 \quad (\rho\to\infty). \tag{2.10}$$

Finally, for $j > N_r$ we have $g_j^0 = 0$ as $B(0,D)$ is assumed to be essentially upper triangular. From (2.7), (2.8), and (2.10) we obtain a contradiction to the a priori estimate (2.4).

In the case $\lambda^0 \ne 0$, the result follows from similar considerations where we now set $\lambda = \lambda^0\rho^{\widetilde{m}_r}$ again. □

**Proposition 2.6.** *Under the assumptions of Proposition* 2.5*, condition* (VL) *holds at* $x^0$*.*

*Proof.* The proof is similar to the proof of Proposition 2.5, and we only indicate some changes and additional remarks. Assuming $v$ to be a nontrivial solution of (1.3), define $\lambda := \rho^{\widetilde{m}_r}\lambda^0$ and $u$ as in (2.6), but now setting $\xi_0' = 0$, i.e., we set

$$u(x) := \begin{cases} \rho^{-t_j+1/p}\phi(x)v_j(\rho x_n), & j = 1,\dots,k_r,\\ 0, & j = k_r+1,\dots,N.\end{cases}$$

Now the left-hand side of (2.4) can be estimated from below by

$$\|u_j\|^{(\pi_2(j))}_{t_j,p,\mathbb{R}^n_+} \ge \|u_j\|_{t_j,p,\mathbb{R}^n_+} \ge \|D_n^{t_j}u_j\|_{0,p,\mathbb{R}^n_+} \ge \frac{1}{2}\|v_j\|_{0,p,\mathbb{R}_+}\|\phi'\|_{0,p,\mathbb{R}^{n-1}}$$

for $\rho \ge \rho_0$. To estimate $f_j^0$, note that

$$\begin{aligned}
f_j^0 &= \rho^{s_j+1/p}\phi(x)\Big[\sum_{k=1}^{k_r}\sum_{\substack{|\alpha|=s_j+t_k\\ \alpha'=0}} a_\alpha^{jk}(0)(D_n^{\alpha_n}v_k)(\rho x_n) - \delta_{jk}\rho^{\widetilde{m}_r-m_j}\lambda^0 v_k(\rho x_n)\Big]\\
&\quad + \phi(x)\sum_{k=1}^{k_r}\sum_{\substack{|\alpha|=s_j+t_k\\ \alpha'=0}}\sum_{\beta_n<\alpha_n}\rho^{-t_k+1/p}a_\alpha^{jk}(0)\binom{\alpha_n}{\beta_n}D_n^{\beta_n}v(\rho x_n)D_n^{\alpha_n-\beta_n}\psi(x_n)\\
&\quad + \sum_{k=1}^{k_r}\sum_{\substack{|\alpha|=s_j+t_k\\ \alpha'\ne 0}}\rho^{-t_k+1/p}a_\alpha^{jk}(0)D_{x'}^{\alpha'}\phi(x')D_n^{\alpha_n}(v_k(\rho x_n)\psi(x_n)).
\end{aligned}$$

Here $\alpha' = (\alpha_1,\dots,\alpha_{n-1})$. For $\pi_1(j) < r$ we have $\widetilde{m}_r - m_j < 0$, and the term in brackets tends to the $j$th row of $A(0,0,D_n)v(\rho x_n)$. For $\pi_1(j) = r$, the term equals the $j$th row of $A(0,0,D_n)v(\rho x_n) - \lambda v(\rho x_n)$. All other terms are of lower order with respect to $\rho$ and can be estimated in the same way as in the proof of Proposition 2.5. As $(\mathcal{A}_{11}^{(r)}(0,0,D_n) - \lambda^0 \widetilde{I}_{r,0})v = 0$ by assumption, we obtain (2.8) again.

By considerations similar to those above, the estimate of $g_j^0$ is reduced to the estimate of

$$\rho^{-t_k+1/p} \| D_{x'}^{\alpha'} \phi(x') D_n^{\alpha_n} v_k(\rho x_n)\big|_{x_n=0} \|_{-\sigma_j - 1/p,p,\mathbb{R}^{n-1}}^{(\pi_2(j))}. \tag{2.11}$$

Here $j = 1,\dots,N_r$, $|\alpha| = \sigma_j + t_k$, and $\alpha_n < \sigma_j + t_k$ if $\pi_2(j) = r$. Taking into account $\sigma_j < 0$ and therefore $\alpha_n \leq t_k - 1$, we may estimate

$$\begin{aligned}
&\rho^{-t_k+1/p} \| D_{x'}^{\alpha'} \phi(x') D_n^{\alpha_n} v_k(\rho x_n)\big|_{x_n=0} \|_{-\sigma_j-1/p,p,\mathbb{R}^{n-1}} \\
&\qquad \leq \rho^{-t_k+\alpha_n+1/p} \| D_{x'}^{\alpha'} \phi(x') (D_n^{\alpha_n} v_k)(0) \|_{-\sigma_j-1/p,p,\mathbb{R}^{n-1}} \\
&\qquad \leq C_{\phi'} \rho^{-t_k+\alpha_n+1/p} |(D_n^{\alpha_n} v_k)(0)| \to 0 \quad (\rho \to \infty)
\end{aligned}$$

for $j = 1,\dots,N_r$. Similarly, for $j = 1,\dots,N_{r-1}$, i.e., for $\pi_2(j) < r$, we have

$$\begin{aligned}
&|\lambda|^{(-\sigma_j-1/p)\widetilde{m}_r/\widetilde{m}_{\pi_2(j)}} \| D_{x'}^{\alpha'} \phi(x') D_n^{\alpha_n} v_k(\rho x_n)\big|_{x_n=0} \|_{0,p,\mathbb{R}^{n-1}} \\
&\qquad \leq C \rho^{(\sigma_j+1/p)(1-\widetilde{m}_r/\widetilde{m}_{\pi_2(j)})} |(D_n^{\alpha_n} v_k)(0)| \to 0 \quad (\rho \to \infty),
\end{aligned}$$

as $\sigma_j \leq -1$ and $\widetilde{m}_r/\widetilde{m}_{\pi_2(j)} < 1$. Therefore, we see that in all cases the expression in (2.11) tends to zero for $\rho \to \infty$ which finally leads to a contradiction. □

Now the proof of Theorem 1.1, i.e., of the necessity of the parameter-ellipticity conditions (E), (SL) and (VL), follows from Propositions 2.3, 2.5, and 2.6, respectively.

## References


[ADN] S. Agmon, R. Douglis, and L. Nirenberg: Estimates near the boundary for solutions of elliptic partial differential equations satisfying general boundary conditions II. *Comm. Pure Appl. Math.* **17** (1964), 35–92.

[ADF] M.S. Agranovich, R. Denk, and M. Faierman: Weakly smooth nonselfadjoint elliptic boundary problems. In: Advances in partial differential equations: Spectral theory, microlocal analysis, singular manifolds, *Math. Top.* **14** (1997), 138–199.

[AV] M.S. Agranovich and M.I. Vishik: Elliptic problems with a parameter and parabolic problems of general form. *Russ. Math. Surveys* **19** (1964), 53–157.

[DHP] R. Denk, M. Hieber, and J. Prüss: Optimal $L_p$-$L_q$-regularity for parabolic problems with inhomogeneous boundary data. *Math. Z.* **257** (2007), 193–224.

[DF] R. Denk and M. Faierman: Estimates for solutions of a parameter-elliptic multi-order system of differential equations. *Integral Equations Operator Theory* **66** (2010), 327–365.

[DFM] R. Denk, M. Faierman, and M. Möller: An elliptic boundary problem for a system involving a discontinuous weight. *Manuscripta Math.* **108** (2001), 289–317.

[DMV] R. Denk, R. Mennicken, and L. Volevich: The Newton polygon and elliptic problems with parameter. *Math. Nachr.* **192** (1998), 125–157.

[DV] R. Denk and L. Volevich: Elliptic boundary value problems with large parameter for mixed order systems. *Amer. Math. Soc. Transl.* (2) **206** (2002), 29–64.

[F] M. Faierman: Eigenvalue asymptotics for a boundary problem involving an elliptic system. *Math. Nachr.* **279** (2006), 1159–1184.

[G] G. Grubb: *Functional calculus of pseudodifferential boundary problems* 2nd edn. Birkhäuser, Boston, 1996.

[K1] A.N. Kozhevnikov: Spectral problems for pseudodifferential systems elliptic in the Douglis-Nirenberg sense, and their applications. *Math. USSR Sobornik* **21** (1973), 63–90.

[K2] A.N. Kozhevnikov: Parameter-ellipticity for mixed order systems in the sense of Petrovskii. *Commun. Appl. Anal.* **5** (2001), 277–291.

[T] H. Triebel: *Interpolation theory, function spaces, differential operators.* North-Holland, Amsterdam, 1978.

[VL] M.I. Vishik and L.A. Lyusternik: Regular degeneration and boundary layer for linear differential equations with small parameter. *Amer. Math. Soc. Transl.* (2) **20** (1962), 239–264.

[V] L.R. Volevich: Solvability of boundary value problems for general elliptic systems. *Amer. Math. Soc. Transl.* (2), **67** (1968), 182–225.

R. Denk
Department of Mathematics and Statistics
University of Konstanz
D-78457 Konstanz, Germany
e-mail: `robert.denk@uni-konstanz.de`

M. Faierman
School of Mathematics and Statistics
The University of New South Wales
UNSW Sydney
NSW 2052, Australia
e-mail: `m.faierman@unsw.edu.au`

Operator Theory:
Advances and Applications, Vol. 221, 269-291


# Canonical Eigenvalue Distribution of Multilevel Block Toeplitz Sequences with Non-Hermitian Symbols

Marco Donatelli, Maya Neytcheva and Stefano Serra-Capizzano

**Abstract.** Let $f : I_k \to \mathcal{M}_s$ be a bounded symbol with $I_k = (-\pi, \pi)^k$ and $\mathcal{M}_s$ be the linear space of the complex $s \times s$ matrices, $k, s \geq 1$. We consider the sequence of matrices $\{T_n(f)\}$, where $n = (n_1, \ldots, n_k)$, $n_j$ positive integers, $j = 1 \ldots, k$. Let $T_n(f)$ denote the multilevel block Toeplitz matrix of size $\widehat{n}\, s$, $\widehat{n} = \prod_{j=1}^{k} n_j$, constructed in the standard way by using the Fourier coefficients of the symbol $f$. If $f$ is Hermitian almost everywhere, then it is well known that $\{T_n(f)\}$ admits the canonical eigenvalue distribution with the eigenvalue symbol given exactly by $f$ that is $\{T_n(f)\} \sim_\lambda (f, I_k)$. When $s = 1$, thanks to the work of Tilli, more about the spectrum is known, independently of the regularity of $f$ and relying only on the topological features of $R(f)$, $R(f)$ being the essential range of $f$. More precisely, if $R(f)$ has empty interior and does not disconnect the complex plane, then $\{T_n(f)\} \sim_\lambda (f, I_k)$. Here we generalize the latter result for the case where the role of $R(f)$ is played by $\bigcup_{j=1}^{s} R(\lambda_j(f))$, $\lambda_j(f)$, $j = 1, \ldots, s$, being the eigenvalues of the matrix-valued symbol $f$. The result is extended to the algebra generated by Toeplitz sequences with bounded symbols. The theoretical findings are confirmed by numerical experiments, which illustrate their practical usefulness.

**Mathematics Subject Classification (2000).** Primary 47B35, Secondary 15A18.

**Keywords.** Matrix sequence, joint eigenvalue distribution, Toeplitz matrix.

The work of the first and the third authors is partly supported by MIUR grant no. 20083KLJEZ; the work of the second and the third authors is partly supported by a grant for visiting researchers of the Mathematics and Computer Science section of the Faculty of Science and Technology, Uppsala University, Sweden, 2008–2010.

## 1. Introduction and basic notations

Let $\mathcal{M}_s$ be the linear space of the complex $s \times s$ matrices and let $f$ be a $\mathcal{M}_s$-valued function of $k$ variables, integrable on the $k$-dimensional cube $I_k := (-\pi, \pi)^k$. Throughout, the symbol $\displaystyle{-}\!\!\!\!\!\!\int_{I_k}$ stands for $(2\pi)^{-k} \int_{I_k}$ and the symbol $L^p(s)$ stands for $L^p(I^k, (2\pi)^{-k} dx, \mathcal{M}_s)$, $p \in [1, \infty]$, that is $f \in L^p(s)$ if and only if every scalar function $f_{i,j} \in L^p(I^k, (2\pi)^{-k} dx, \mathbb{C})$. The Fourier coefficients of $f$, given by $\widehat{f}_j := {-}\!\!\!\!\!\!\int_{I_k} f(t) e^{-\mathbf{i}\langle j,t\rangle}\, dt \in \mathcal{M}_s$, $\mathbf{i}^2 = -1$, $j \in \mathbb{Z}^k$, $\langle j, t\rangle = \sum_{l=1}^k j_l t_l$, are the entries of the $k$-level Toeplitz matrices generated by $f$. More precisely, if $n = (n_1, \ldots, n_k)$ is a $k$-index with positive entries, then $T_n(f)$ denotes the matrix of order $\widehat{n}\, s$ given by $T_n(f) = \sum_{|j_1|<n_1} \cdots \sum_{|j_k|<n_k} \left[ J_{n_1}^{(j_1)} \otimes \cdots \otimes J_{n_k}^{(j_k)} \right] \otimes \widehat{f}_{(j_1,\ldots,j_k)}$ (throughout, we let $\widehat{n} := \prod_{i=1}^k n_i$). In this case, we say that the sequence $\{T_n(f)\}$ is *generated* by $f$. In the above equation, $\otimes$ denotes the tensor product, while $J_t^{(l)}$ denotes the matrix of order $t$ whose $(i, j)$ entry equals 1 if $j - i = l$ and equals zero otherwise. In the following, the notation $n \to \infty$ indicates that $\min_{1 \le r \le k} n_r \to \infty$ and $\mathrm{tr}(A)$ denotes the trace of a square matrix $A$, that is the sum of its eigenvalues.

In this paper we consider the global behavior of the spectrum of the sequence $\{T_n(f)\}$ and more precisely, its canonical distribution in the sense of Szegö (cf. [7], see also the rich book [3] and references therein for the standard scalar-valued case where $s = 1$). For the general case when $s$ is any positive integer, an earlier result is proved by Tilli [21], but imposing the strong restriction on the symbol $f$ to be Hermitian-valued almost everywhere (a.e.). For completeness we include the result.

**Theorem 1.1 ([21]).** *Let $f \in L^1(s)$ be Hermitian-valued a.e. and let $\Lambda_n$ be the spectrum of $T_n(f)$, the set of all $\widehat{n}s$ eigenvalues of $T_n(f)$. (Note that these are real since $f$ is Hermitian-valued and the matrix $T_n(f)$ is Hermitian.) Then, for every function $F \in \mathcal{C}_0(\mathbb{R})$ continuous with bounded support, the following asymptotic formula holds,*

$$\lim_{n\to\infty} \frac{1}{\widehat{n}s} \sum_{\lambda \in \Lambda_n} F(\lambda) = {-}\!\!\!\!\!\!\int_{I_k} \frac{1}{s} \mathrm{tr}(F(f(t)))\, dt \tag{1}$$

*that is, $\{T_n(f)\} \sim_\lambda (f, I_k)$.*

In this paper we extend Theorem 1.1 to the case where $f$ is not necessarily Hermitian-valued and prove the following result.

**Theorem 1.2.** *Let $f \in L^\infty(s)$ with eigenvalues $\lambda_j(f)$, $j = 1, \ldots, s$, and let $\Lambda_n$ be the spectrum of $T_n(f)$. Define the union of the essential ranges of the eigenvalues of $f$ as $R(f) = \bigcup_{j=1}^s R(\lambda_j(f))$. If $R(f)$ has empty interior and does not disconnect the complex plane, then for every function $F \in \mathcal{C}_0(\mathbb{C})$, which is continuous and with bounded support, the following asymptotic formula holds,*

$$\lim_{n\to\infty} \frac{1}{\widehat{n}s} \sum_{\lambda \in \Lambda_n} F(\lambda) = {-}\!\!\!\!\!\!\int_{I_k} \frac{1}{s} \mathrm{tr}(F(f(t)))\, dt \tag{2}$$

*that is, $\{T_n(f)\} \sim_\lambda (f, I_k)$.*

Theorem 1.2 generalizes a result by Tilli [22] to the matrix-valued case including the scalar-valued case ($s = 1$). We note that the same proof with obvious variations can be used for matrix sequences belonging to the algebra, generated by all Toeplitz sequences with bounded symbols (see also [8, 17]).

The paper is organized as follows. In Section 2 we introduce preliminary definitions and tools. In Section 3 we extend to the block case some technical instruments, based on the Mergelyan theorem (refer to [10]), that we employ in Section 4 in order to prove Theorem 1.2 and its generalization to the case of the block Toeplitz algebra. In Section 5 we include numerical illustrations and indicate the practical usefulness of our theoretical findings. The conclusions in Section 6, including a few open questions, finalize the paper.

## 2. Preliminary definitions and tools

We start with a general definition of an eigenvalue distribution and of weak/strong clustering for a matrix sequence, and recall the notion of *essential range*, which plays an important role in the study of the asymptotic properties of the spectrum.

For a matrix $A \in \mathbb{C}^{n\times n}$ with singular values $\sigma_1(A), \dots, \sigma_n(A)$, and $p \in [1, \infty]$ we define $\|A\|_p$, the Schatten $p$-norm of $A$, to be the $\ell^p$ norm of the vector of the singular values $\|A\|_p = [\sum_{k=1}^{n} (\sigma_k(A))^p]^{\frac{1}{p}}$. We also consider the norm $\|\cdot\|_1$ which is also known as the trace norm, the norm $\|\cdot\|_2$ also known in the numerical analysis community as the Frobenius norm, and the spectral norm $\|\cdot\|_\infty$, which is equal to the operator norm. Now, if $\lambda_j(A)$, $j = 1, \dots, n$ are the eigenvalues of $A$, and $F$ is a function defined on $\mathbb{C}$, we define the mean

$$\Sigma_\lambda(F, A) := \frac{1}{n}\sum_{j=1}^{n} F(\lambda_j(A)) = \frac{1}{n}\sum_{\lambda\in\Lambda_n} F(\lambda).$$

Analogously, we define $\Sigma_\sigma(F, A)$ with the singular values replacing the eigenvalues.

**Definition 2.1.** *Let $\mathcal{C}_0(\mathbb{C})$ be the set of continuous functions with bounded support defined over the complex field, $k$ be a positive integer, and $\theta$ be a $s \times s$ matrix-valued measurable function defined on a set $G \subset \mathbb{R}^k$ of finite and positive Lebesgue measure $m(G)$. Here $G$ is equal to $I_k$ and a function is considered to be measurable if and only if the component functions are.*

(i) *A matrix sequence $\{A_n\}$ is said to be* distributed (in the sense of the eigenvalues) as the pair $(\theta, G)$, *that is $\{A_n\} \sim_\lambda (\theta, G)$, or to* have the distribution function $\theta$, *if for any $F \in \mathcal{C}_0(\mathbb{C})$, the following limit relation holds*

$$\lim_{n\to\infty} \Sigma_\lambda(F, A_n) = \fint_G \frac{\sum_{j=1}^{s} F(\lambda_j(\theta(t)))}{s}\, dt, = \fint_G \frac{\mathrm{tr}\,(F(\theta(t)))}{s}\, dt, \qquad (3)$$

*where $\lambda_i(\theta(t))$ are the eigenvalues of the matrix $\theta(t)$ and $\fint_G = \frac{1}{m(G)}\int_G$.*

(ii) *If* (3) *holds for every* $F \in \mathcal{C}_0(\mathbb{R}_0^+)$ *in place of* $F \in \mathcal{C}_0(\mathbb{C})$, *with the singular values* $\sigma_j(A_n)$, $j = 1, \ldots, n$, *in place of the eigenvalues, and with* $|\theta(t)| = (\theta(t)^*\theta(t))^{1/2}$ *in place of* $\theta(t)$, *we say that* $\{A_n\} \sim_\sigma (\theta, G)$ *or that the matrix sequence* $\{A_n\}$ *is* distributed in the sense of the singular values *as the pair* $(\theta, G)$ : *more specifically, for every* $F \in \mathcal{C}_0(\mathbb{R}_0^+)$ *we have*

$$\lim_{n\to\infty} \Sigma_\sigma(F, A_n) = \int_G \frac{\sum_{j=1}^s F\left(\sigma_j(\theta(t))\right)}{s}\, dt = \int_G \frac{\operatorname{tr}\left(F\left(|\theta(t)|\right)\right)}{s}\, dt. \qquad (4)$$

**Definition 2.2.** *A matrix sequence* $\{A_n\}$ *is said to be* strongly clustered at $r \in \mathbb{C}$ *(in the eigenvalue sense), if for any* $\epsilon > 0$ *the number of the eigenvalues of* $A_n$ *off the disc*

$$D(r, \epsilon) := \{z : |z - r| < \epsilon\}, \qquad (5)$$

*can be bounded by a constant* $q_\epsilon$ *possibly depending on* $\epsilon$, *but not on* $n$. *In other words*

$$q_\epsilon(n, r) := \#\{j : \lambda_j(A_n) \notin D(r, \epsilon)\} = O(1), \quad n \to \infty.$$

*If each* $A_n$ *has only real eigenvalues (at least for large* $n$*) then we may assume that* $r$ *is real and that the disc* $D(r, \epsilon)$ *is the interval* $(r-\epsilon, r+\epsilon)$. *A matrix sequence* $\{A_n\}$ *is said to be* strongly clustered at a nonempty closed set $S \subset \mathbb{C}$ *(in the eigenvalue sense) if for any* $\epsilon > 0$

$$q_\epsilon(n, S) := \#\{j : \lambda_j(A_n) \notin D(S, \epsilon)\} = O(1), \quad n \to \infty, \qquad (6)$$

*where* $D(S, \epsilon) := \cup_{r \in S} D(r, \epsilon)$ *is the* $\epsilon$*-neighborhood of* $S$. *If each* $A_n$ *has only real eigenvalues, then* $S$ *is a nonempty closed subset of* $\mathbb{R}$. *We replace the term "strongly" by "weakly", if*

$$q_\epsilon(n, r) = o(n), \qquad \left(q_\epsilon(n, S) = o(n)\right), \quad n \to \infty,$$

*in the case of a point* $r$ *or a closed set* $S$. *If* $S$ *is not connected, then its disjoint parts are called sub-clusters. By replacing 'eigenvalues' with 'singular values' we obtain all the corresponding definitions for singular values.*

It is clear that $\{A_n\} \sim_\lambda (\theta, G)$, with $\theta \equiv r$ equal to a constant function if and only if $\{A_n\}$ is weakly clustered at $r \in \mathbb{C}$. For more results and relations between the notions of equal distribution, equal localization, spectral distribution, spectral clustering etc., see [14, Section 4].

**Definition 2.3.** *Given a measurable complex-valued function* $\theta$ *defined on a Lebesgue measurable set* $G$, *the* essential range of $\theta$ *is the set* $R(\theta)$ *of points* $r \in \mathbb{C}$ *such that, for every* $\epsilon > 0$, *the Lebesgue measure of the set* $\theta^{(-1)}(D(r, \epsilon)) := \{t \in G : \ \theta(t) \in D(r, \epsilon)\}$ *is positive, with* $D(r, \epsilon)$ *as in* (5). *The function* $\theta$ *is* essentially bounded *if its essential range is bounded. Furthermore, if* $\theta$ *is real-valued, then the essential supremum (infimum) is defined as the supremum (infimum) of its essential range. Finally, if the function* $\theta$ *is* $s \times s$ *matrix-valued and measurable, then the essential range of* $\theta$, *denoted again by* $R(\theta)$, *is the union of the essential ranges of the complex-valued eigenvalues* $\lambda_j(\theta)$, $j = 1, \ldots, s$, *that is* $R(\theta) = \bigcup_{j=1}^s R(\lambda_j(\theta))$.

We note that $R(\theta)$ is clearly a closed set and, thus, its complement is open. Moreover, if $\{A_n\}$ is a matrix sequence distributed as $\theta$ in the sense of eigenvalues, then $R(\theta)$ is a weak cluster for $\{A_n\}$.

## 3. Further tools for general matrix sequences and main results

The basic ideas used in this section originate in [22] and [6], where the same questions are considered in a scalar Toeplitz context and in a scalar Jacobi context, respectively. We consider now how to extend these results to the case, where the symbol is matrix-valued, that is when $s > 1$.

**Theorem 3.1.** *Let $\{A_n\}$ be a matrix sequence and $S$ be a subset of $\mathbb{C}$. Assume that the following assumptions hold:*

**(a1)** *$S$ is a compact set and $\mathbb{C}\backslash S$ is connected;*
**(a2)** *the matrix sequence $\{A_n\}$ is weakly clustered at $S$;*
**(a3)** *the spectra $\Lambda_n$ of $A_n$ are uniformly bounded, i.e., $\exists C \in \mathbb{R}^+$ such that $|\lambda| < C$, $\lambda \in \Lambda_n$, for all $n$;*
**(a4)** *there exists a $s \times s$ matrix-valued function $\theta$, which is measurable, bounded, and defined on a set $G$ of positive and finite Lebesgue measure, such that, for every positive integer $L$, we have $\lim_{n\to\infty} \frac{\operatorname{tr}(A_n^L)}{n} = \fint_G \frac{\operatorname{tr}(\theta^L(t))}{s}\, dt$, i.e., relation (3) holds with $F$ being any polynomial of an arbitrary fixed degree;*
**(a5)** *the essential range of $\theta$ is contained in $S$.*

*Then relation (3) is true for every continuous function $F$ with bounded support, which is holomorphic in the interior of $S$. If also the interior of $S$ is empty, then the sequence $\{A_n\}$ is distributed as $(\theta, G)$, in the sense of the eigenvalues.*

*Proof.* The proof follows that of Theorem 2.2 in [6]. Let $F$ be continuous over $S$ and holomorphic in its interior. By Mergelyan's Theorem [10], for every $\epsilon > 0$, we can find a polynomial $P$ such that for every $z \in S$, $|P(z) - F(z)| \le \epsilon$. Since $R(\theta)$ is contained in $S$, it is clear that $|P(\lambda_j(\theta(t))) - F(\lambda_j(\theta(t)))| \le \epsilon$, $j = 1, \ldots, s$ a.e. in its domain $G$ so that $|\operatorname{tr}(F(\theta(t))) - \operatorname{tr}(P(\theta(t)))| \le s\epsilon$. Therefore,

$$\left| \fint_G \frac{\operatorname{tr}(F(\theta(t)))}{s}\, dt - \fint_G \frac{\operatorname{tr}(P(\theta(t)))}{s}\, dt \right| \le \fint_G \epsilon\, dt = \epsilon. \tag{7}$$

Next, we consider the left-hand side of (3). By the definition of clustering, for any fixed $\epsilon' > 0$, we have

$$\#\{\lambda \in \Sigma_n,\ |\lambda - z| \ge \epsilon',\ \forall z \in S\} = \#\{\lambda \in \Sigma_n,\ \lambda \notin D(S, \epsilon')\} = o(n).$$

Moreover, by the assumption of the uniform boundedness of $\Sigma_n$, the bound $|\lambda| < C$ holds for every $\lambda \in \Sigma_n$ with a constant $C$, independent of $n$. Therefore, by extending $F$ outside $S$ in such a way that it is continuous with a bounded support, we infer

$$\left|\frac{1}{n}\sum_{\lambda\in\Sigma_n,\ \lambda\notin D(S,\epsilon')} F(\lambda)\right| \le \frac{M}{n}\#\{\lambda\in\Sigma_n,\ \lambda\notin D(S,\epsilon')\} = o(1),$$

$$\left|\frac{1}{n}\sum_{\lambda\in\Sigma_n,\ \lambda\notin D(S,\epsilon')} P(\lambda)\right| \le \frac{M}{n}\#\{\lambda\in\Sigma_n,\ \lambda\notin D(S,\epsilon')\} = o(1),$$

with $M = \max(\|F\|_\infty, \|P\|_\infty)$, and infinity norms taken over $\{z\in\mathbb{C},\ |z|\le C\}$. Consequently, by setting $\Delta = |\Sigma_\lambda(F-P,A_n)|$, we deduce that

$$\begin{aligned}
\Delta &= \left|\frac{1}{n}\sum_{\lambda\in\Sigma_n}(F(\lambda)-P(\lambda))\right| \le \frac{1}{n}\sum_{\lambda\in\Sigma_n}|F(\lambda)-P(\lambda)|\\
&= \frac{1}{n}\sum_{\lambda\in\Sigma_n,\ \lambda\in D(S,\epsilon')}|F(\lambda)-P(\lambda)| + \frac{1}{n}\sum_{\lambda\in\Sigma_n,\ \lambda\notin D(S,\epsilon')}|F(\lambda)-P(\lambda)|\\
&\le \frac{1}{n}\sum_{\lambda\in\Sigma_n,\ \lambda\in D(S,\epsilon')}|F(\lambda)-P(\lambda)| + o(1)\\
&= \frac{1}{n}\sum_{\lambda\in\Sigma_n,\ \lambda\in S}|F(\lambda)-P(\lambda)| + \frac{1}{n}\sum_{\lambda\in\Sigma_n,\ \lambda\in D(S,\epsilon')\backslash S}|F(\lambda)-P(\lambda)| + o(1).
\end{aligned}$$

For $\lambda\in S$ we use the relation $|F(\lambda)-P(\lambda)|\le\epsilon$. For $\lambda\in D(S,\epsilon')\backslash S$, we see that

$$|F(\lambda)-P(\lambda)| \le |F(\lambda)-F(\lambda')| + |F(\lambda')-P(\lambda')| + |P(\lambda')-P(\lambda)|,$$
$$|\lambda-\lambda'|<\epsilon',\quad \lambda'\in S,$$

thus, $|F(\lambda)-P(\lambda)| \le c_1(\epsilon')+\epsilon+c_2(\epsilon,\epsilon') \equiv \theta(\epsilon,\epsilon')$ with

$$\lim_{\epsilon\to 0}\lim_{\epsilon'\to 0}\theta(\epsilon,\epsilon') = 0. \tag{8}$$

Hence

$$\Delta \le \epsilon + \theta(\epsilon,\epsilon') + o(1). \tag{9}$$

Furthermore, from the hypothesis of the theorem, there holds

$$\lim_{n\to\infty}\Sigma_\lambda(P,A_n) = \int\!\!\!\!\!\!-_{G} \frac{\operatorname{tr}(P(\theta(t)))}{s}\,dt. \tag{10}$$

Since $\epsilon$ and $\epsilon'$ are arbitrary, it is clear that relations (7)–(10) imply that (3) holds for $F$ as well. Finally, when $S$ has an empty interior, we have no restriction on $F$ except for being continuous with a bounded support, and therefore what we have proved is equivalent to $\{A_n\}\sim_\lambda(\theta,G)$. □

Next, we show that the hypotheses **(a3)** and **(a4)** (or a slightly stronger form of it) imply **(a1)**, **(a2)**, and **(a3)** for the set $S$ defined by "filling in" the essential range of the function $\theta$ from **(a4)** (or its strengthened version). This shows that, when the set $R(\theta)$ has an empty interior, the matrix sequence has the desired distribution.

Here, "filling in" means taking the "Area" in the following sense:

**Definition 3.2.** *Let $K$ be a compact subset of $\mathbb{C}$. We define* Area$(K)$ *as*

$$\text{Area}(K) = \mathbb{C} \backslash U,$$

*where $U$ is the (unique) unbounded connected component of $\mathbb{C}\backslash K$.*

**Theorem 3.3.** *Let $\{A_n\}$ be a matrix sequence. If*

**(b1)** *the spectra $\Lambda_n$ of $A_n$ are uniformly bounded, i.e., $\exists C \in \mathbb{R}^+$ such that $|\lambda| < C$, $\lambda \in \Lambda_n$, for all $n$;*

**(b2)** *there exists a $s \times s$ matrix-valued function $\theta$-measurable, bounded, and defined on a set $G$ of positive and finite Lebesgue measure, such that, for all positive integers $L$ and $l$, we have $\lim_{n\to\infty} \frac{\mathrm{tr}((A_n^*)^l A_n^L)}{n} = \fint_G \frac{\mathrm{tr}\left(\overline{\theta^l(t)}\theta^L(t)\right)}{s}\,dt$;*

*then $R(\theta)$ is compact, the matrix sequence $\{A_n\}$ is weakly clustered at* Area$(R(\theta))$, *and relation (3) is true for every continuous function $F$ with bounded support, which is holomorphic in the interior of $S =$* Area$(R(\theta))$.

*If it is also true that $\mathbb{C}\backslash R(\theta)$ is connected and the interior of $R(\theta)$ is empty then the sequence $\{A_n\}$ is distributed as $(\theta, G)$, in the sense of the eigenvalues.*

*Proof.* Since $\theta$ is bounded, any of its eigenvalues $\lambda_j(\theta)$ is bounded, $j = 1, \ldots, s$ so that $R(\theta) = \bigcup_{j=1}^s R(\lambda_j(\theta))$ is bounded. Consequently, the set $R(\theta)$ is compact, since the essential range is always closed. Hence we can define $S = \text{Area}(R(\theta))$ according to Definition 3.2.

We prove that $S$ is a weak cluster for the spectra of $\{A_n\}$. First, we notice that the compact set $S_C = \{z \in \mathbb{C} : |z| \le C\}$ is a strong cluster for the spectra of $\{A_n\}$ since by **(b1)** it contains all the eigenvalues. Moreover $C$ can be chosen such that $S_C$ contains $S$. Therefore, to prove that $S$ is a weak cluster for $\{A_n\}$ it suffices to prove that, for every $\epsilon > 0$ the compact set $S_C \backslash D(S, \epsilon)$ contains at most only $o(n)$ eigenvalues, with $D(S, \epsilon)$ as in Definition 2.2. By compactness, for any $\delta > 0$, there exists a finite covering of $S_C \backslash D(S, \epsilon)$ made of balls $D(z, \delta)$, $z \in S_C \backslash S$ with $D(z, \delta) \cap S = \emptyset$, and so, it suffices to show that, for a particular $\delta$, at most $o(n)$ eigenvalues lie in $D(z, \delta)$. Let $F(t)$ be the characteristic function of the compact set $\overline{D(z, \delta)}$. Then restricting our attention to the compact set $\overline{D(z, \delta)} \bigcup S$, Mergelyan's theorem implies that for each $\epsilon > 0$ there exists a polynomial $P$ such that $|F(t) - P(t)|$ is bounded by $\epsilon$ on $\overline{D(z, \delta)} \bigcup S$.

Therefore $|P(\lambda_j(\theta(t)))| \le \epsilon$ a.e. in its domain for every $j = 1, \ldots, s$. However the latter does not guarantee

$$\|P(\theta(t))\|_2 < l(\epsilon) \tag{11}$$

a.e. for some $l(\epsilon)$ converging to zero as $\epsilon$ converges to zero (it would be trivially true if $\theta(t)$ is normal a.e. that is $\theta^*(t)\theta(t) = \theta(t)\theta^*(t)$ a.e., since, in that case $P(\theta(t))$ is also normal a.e. and $\|P(\theta(t))\|_2 = (\sum_{j=1}^s |\lambda_j(P(\theta(t)))|^2)^{1/2} =$

$(\sum_{j=1}^{s} |P(\lambda_j(\theta(t)))|^2)^{1/2} < \sqrt{s}\epsilon$ a.e.). The condition (11) on the Frobenius norm is essential in the proof as we will see later. Thus, in order to fulfill it, starting from the Schur normal form of $\theta(t)$, we define a new polynomial $P_k$ satisfying the claim and at the same time approximating $F(t)$ over $\overline{D(z,\delta)} \bigcup S$. We write $\theta(t) = U(t)T(t)U^*(t)$ with $U(t)$ unitary a.e. and $T(t) = \Lambda(t) + R(t)$ upper triangular, where $\Lambda(t) = \mathrm{diag}_{j=1,\ldots,s}\lambda_j(\theta(t))$ and $R(t)$ strictly upper triangular and bounded. Clearly, $P(\theta(t)) = U(t)(P(\Lambda(t)) + \tilde{R}(t))U^*(t)$ with $\tilde{R}(t)$ still strictly upper triangular and bounded. Let $k$ be any integer larger than $s-1$. Then $\tilde{R}^k(t) = 0$ since $\tilde{R}(t)$ is of order $s$ and strictly upper triangular (so, nilpotent). Consequently, by defining $P_k(y) = p^k(y)$, we have

$$P_k(\theta(t)) = U(t)(P(\Lambda(t)) + \tilde{R}(t))^k U^*(t)$$

so that

$$(P(\Lambda(t)) + \tilde{R}(t))^k = \sum_{j=k-s+1}^{k} \binom{k}{j} p^j(\Lambda(t))\tilde{R}^{k-j}(t). \tag{12}$$

Of course $\|P_k(\theta(t))\|_2 = \|(P(\Lambda(t)) + \tilde{R}(t))^k\|_2$ and thanks to (12) and to the boundedness of the symbol $\theta$, we can choose $k$ independent of $\epsilon$ and $t$ so that

$$\|P_k(\theta(t))\|_2^2 < s\epsilon \tag{13}$$

a.e., which is the desired relation (11) with $l(\epsilon) = s\epsilon$.

$$\begin{aligned}
(1-\epsilon)^k \gamma_n(z,\delta) &\le \sum_{i=1}^{n} F(\lambda_i)|P_k(\lambda_i)| && (14)\\
&\le \left(\sum_{i=1}^{n} F^2(\lambda_i)\right)^{1/2} \left(\sum_{i=1}^{n} |P_k(\lambda_i)|^2\right)^{1/2} && (15)\\
&= \left(\sum_{i=1}^{n} F(\lambda_i)\right)^{1/2} \left(\sum_{i=1}^{n} |P_k(\lambda_i)|^2\right)^{1/2} && (16)\\
&= (\gamma_n(z,\delta))^{1/2} \left(\sum_{i=1}^{n} |P_k(\lambda_i)|^2\right)^{1/2} && (17)\\
&\le (\gamma_n(z,\delta))^{1/2} \|P_k(A_n)\|_2 && (18)\\
&= (\gamma_n(z,\delta))^{1/2} (\mathrm{tr}(P_k^*(A_n)P_k(A_n)))^{1/2} && (19)\\
&= (\gamma_n(z,\delta))^{1/2} \left(\mathrm{tr}\left(\sum_{l,L=0}^{M} \overline{c_l} c_L (A_n^*)^l A_n^L\right)\right)^{1/2} && (20)\\
&= (\gamma_n(z,\delta))^{1/2} \left(\sum_{l,L=0}^{M} \overline{c_l} c_L \mathrm{tr}((A_n^*)^l A_n^L)\right)^{1/2}, && (21)
\end{aligned}$$

where inequality (14) follows from the definition of $F$ and from the approximation properties of $P$, relation (15) is the Cauchy-Schwartz inequality, relations (16)–(17) follow from the definitions of $F$ and $\gamma_n(z,\delta)$, (18) is a consequence of the Schur decomposition of $\theta(t)$ and of the unitary invariance of the Schatten norms, identities (19)–(21) follow from the entry-wise definition of the Schatten 2 norm (the Frobenius norm), from the monomial expansion of the polynomial $P$, and from the linearity of the trace.

Given $\epsilon_2 > 0$, we choose $\epsilon_1 > 0$ so that the inequality

$$\epsilon_1 \sum_{l,L=0}^{M} |c_l||c_L| \leq \epsilon_2,$$

holds true and then we choose $N$ so that for $n > N$, inequality

$$\left| \frac{\operatorname{tr}((A_n^*)^l A_n^L)}{n} - \int_G \frac{\operatorname{tr}\left(\overline{\theta^l(t)}\theta^L(t)\right)}{s} dt \right| < \epsilon_1,$$

is true. Hence, from (21) we obtain

$$(1-\epsilon)^k \gamma_n(z,\delta) \leq (\gamma_n(z,\delta))^{1/2} \left( n \left( \epsilon_2 + \int_G \sum_{l,L=0}^{M} \overline{c_l} c_L \left( \frac{\operatorname{tr}\left(\overline{\theta^l(t)}\theta^L(t)\right)}{s} \right) dt \right) \right)^{1/2} \tag{22}$$

$$= (\gamma_n(z,\delta))^{1/2} \left( n \left( \epsilon_2 + \int_G \frac{\|P_k(\theta(t))\|^2}{s} dt \right) \right)^{1/2} \tag{23}$$

$$\leq (\gamma_n(z,\delta))^{1/2} n^{1/2} (\epsilon^2 + \epsilon_2)^{1/2}, \tag{24}$$

where inequality (22) is assumption **(b2)**. The latter two relations are again consequences of the monomial expansion of $P$ and of the crucial inequality (13), where $\epsilon_2$ is arbitrarily small. Therefore, by choosing $\epsilon_2 = \epsilon^2$, (14)–(24) imply that, for $n$ sufficiently large,

$$\gamma_n(z,\delta) \leq 2n\epsilon^2(1-\epsilon)^{-2k},$$

which means that, since $k$ is chosen independent of $\epsilon$, $\gamma_n(z,\delta) = o(n)$.

Thus, hypotheses **(a1)**–**(a5)** of Theorem 3.1 hold with $S = \operatorname{Area}(R(\theta))$, which is necessarily compact and with connected complement. Consequently, the first conclusion in Theorem 3.1 holds. Finally, if $\mathbb{C}\backslash R(\theta)$ is connected and the interior of $R(\theta)$ is empty, then $\operatorname{Area}(R(\theta)) = R(\theta)$ and, thus, all the hypotheses of Theorem 3.1 are satisfied, we conclude that the sequence $\{A_n\}$ is distributed in the sense of the eigenvalues as $(\theta, G)$. □

Next, we present a second version of Theorem 3.3, replacing hypotheses **(a1)**–**(a5)** by only **(a3)**, **(a4)**, and a condition on the Schatten $p$ norm for a certain $p$.

**Theorem 3.4.** *Let* $\{A_n\}$ *be a matrix sequence. Assume that*

**(c1)** *the spectra* $\Lambda_n$ *of* $A_n$ *are uniformly bounded, i.e.,* $|\lambda| < C$, $\lambda \in \Lambda_n$, *for all* $n$;

**(c2)** *there exists a* $s \times s$ *matrix-valued function* $\theta$, *which is measurable, bounded and defined over* $G$ *having positive and finite Lebesgue measure, such that for every positive integer* $L$ *there holds* $\lim_{n\to\infty} \frac{\mathrm{tr}(A_n^L)}{n} = \fint_G \frac{\mathrm{tr}(\theta^L(t))}{s}\, dt$;

**(c3)** *for every* $n$ *large enough, there exist a constant* $\widehat{C}$ *and a positive real number* $p \in [1, \infty)$, *independent of* $n$, *such that* $\|P(A_n)\|_p^p \le \widehat{C} n \fint_G \|P(\theta(t))\|_p^p\, dt$ *for every fixed polynomial* $P$ *independent of* $n$.

*Then the matrix sequence* $\{A_n\}$ *is weakly clustered at* $\mathrm{Area}(R(\theta)) := \mathbb{C}\backslash U$ *(see Definition* 3.2*) and relation* (3) *is true for every continuous function* $F$ *with bounded support which is holomorphic in the interior of* $S = \mathrm{Area}(R(\theta))$.

*Moreover, if*

**(c4)** $\mathbb{C}\backslash R(\theta)$ *is connected and the interior of* $R(\theta)$ *is empty,*

*then the sequence* $\{A_n\}$ *is distributed as* $(\theta, G)$, *in the sense of the eigenvalues.*

*Proof.* The proof follows that of Theorem 3.3 until relation (14). Then, with $q$ being the conjugate of $p$, i.e., $1/q + 1/p = 1$, we have

$$
\begin{aligned}
(1-\epsilon)^k \gamma_n(z,\delta) &\le \left(\sum_{i=1}^n F^q(\lambda_i)\right)^{1/q} \left(\sum_{i=1}^n |P(\lambda_i)|^p\right)^{1/p} && (25)\\
&= \left(\sum_{i=1}^n F(\lambda_i)\right)^{1/q} \left(\sum_{i=1}^n |P(\lambda_i)|^p\right)^{1/p} && (26)\\
&= (\gamma_n(z,\delta))^{1/q} \left(\sum_{i=1}^n |P(\lambda_i)|^p\right)^{1/p} && (27)\\
&\le (\gamma_n(z,\delta))^{1/q} \|P(A_n)\|_p && (28)\\
&\le (\gamma_n(z,\delta))^{1/q} \left(\frac{\widehat{C} n}{m(G)} \int_G \|P(\theta(t))\|_p^p\, dt\right)^{1/p} && (29)\\
&\le (\gamma_n(z,\delta))^{1/q} (\widehat{C} n)^{1/p} \epsilon, && (30)
\end{aligned}
$$

where relation (25) is the Hölder inequality, relations (26)–(27) follow from the definitions of $F$ and $\gamma_n(z,\delta)$, (28) holds due to the fact that, for any square matrix, the vector with the moduli of the eigenvalues is weakly-majorized by the vector of the singular values (see [1] for more details), inequality (29) is assumption **(c3)** (which holds for any polynomial of fixed degree), and finally inequality (30) follows from the approximation properties of $P$ over the area delimited by the range of $\theta$. Therefore,

$$\gamma_n(z,\delta) \le \widehat{C} n \epsilon^p (1-\epsilon)^{-kp},$$

and since $\epsilon$ is arbitrary we have the desired result, namely, $\gamma_n(z,\delta) = o(n)$.

The rest of the proof is the same as in Theorem 3.3. □

The next result shows that the key assumption **(c3)** follows from the distribution in the singular value sense of $\{P(A_n)\}$ and that the latter is equivalent to the very same limit relation with only polynomial test functions. We mention here, that distribution results in the singular value sense (e.g., [21, 23, 20, 15, 16]) are much easier to obtain and to prove due to the higher stability of the singular values under perturbations (cf. [24]).

**Theorem 3.5.** *Using the notation of Section 2, if the sequence $\{A_n\}$ is uniformly bounded in spectral norm then,* (1), *$\{A_n\} \sim_\sigma (\theta, G)$ is true whenever condition (4) holds for all polynomial test functions. Moreover,* (2), *if $\{P(A_n)\} \sim_\sigma (P(\theta), G)$ for every polynomial $P$ then the claim* **(c3)** *is true for every value $p \in [1, \infty)$, for every $\epsilon > 0$ where $\widehat{C} = 1 + \epsilon$ and for $n$ larger than some fixed value $\bar{n}_\epsilon$.*

*Proof.* The first claim is proved by using the fact that one can approximate any continuous function, defined on a compact set contained in the (positive) real line, by polynomials. The second claim follows from taking the function $z^p$, with positive $p$, as a test function and exploiting the limit relation from the assumption $\{P(A_n)\} \sim_\sigma (P(\theta), G)$. Indeed, the sequence $\{P(A_n)\}$ is uniformly bounded since $\{A_n\}$ is, so we can use as test functions continuous functions with no restriction on the support. Therefore, by definition (see (4)), $\{P(A_n)\} \sim_\sigma (P(\theta), G)$ implies that

$$\lim_{n\to\infty} \frac{1}{n} \sum_{j=1}^{n} \sigma_j^p(P(A_n)) = \fint_G \|P(\theta(t))\|_p^p \, dt.$$

Hence, by observing that $\sum_{j=1}^{n} \sigma_j^p(P(A_n))$ is by definition $\|P(A_n)\|_p^p$ and by taking the limit, we see that, for every $\epsilon > 0$, there exists an integer $\bar{n}_\epsilon$ such that

$$\|P(A_n)\|_p^p \leq n \frac{1+\epsilon}{m(G)} \int_G \|P(\theta(t))\|_p^p \, dt, \quad \forall n \geq \bar{n}_\epsilon.$$

The latter inequality coincides with **(c3)** with $\widehat{C} = 1 + \epsilon$ and $p \in [1, \infty)$. □

We conclude this part of general tools by stating a simple but useful approximation result.

**Theorem 3.6.** *Assume that two sequences $\{A_n\}$ and $\{B_n\}$ are given, with $\{A_n\}$ satisfying conditions* **(b1)**, **(b2)** *or conditions* **(c1)**, **(c2)**, **(c3)**. *In addition, we assume that the sequence $\{B_n\}$*

**(d1)** *is uniformly bounded in spectral norm and*

**(d2)** $\|B_n - A_n\|_1 = o(n)$ *(that is $\{B_n\}$ approximates $\{A_n\}$ in trace norm).*

*Then $\{B_n\}$ satisfies the same conditions as $\{A_n\}$, that is,* **(b1)**, **(b2)** *or* **(c1)**, **(c2)**, **(c3)**.

*Proof.* The assumption **(d1)** directly implies the equivalence between **(b1)** and **(c1)**. Moreover the trace norm condition implies that any of the requirements **(b2)**, **(c2)**, **(c3)** is satisfied with $\{A_n\}$ if and only if the same requirement is satisfied for $\{B_n\}$ (we omit the details). □

## 4. Proof of the main result for the case of block Toeplitz sequences and their algebra

In this section we consider the case of Toeplitz sequences and prove Theorem 1.2, by exploiting the general tools developed above.

*Proof of Theorem* 1.2. We make use of Theorem 3.4. (An alternative proof, similar to the approach used by Tilli and based on Theorem 3.3 is also possible, but requires more effort.)

Recall, that in [16, Section 3.3.1] it is proved that the algebra, generated by block Toeplitz sequences, is a subalgebra of the (block) GLT class so that, in particular, for every $\mathcal{M}_s$-valued symbol $f$ and for every polynomial of a given degree $P$ we have

$$\{P(T_n(f))\} \sim_\sigma (P(f), I_k).$$

Therefore, by invoking Theorem 3.5, we infer that the assumption **(c3)** is fulfilled for every $p$, when ever $f \in L^\infty(s)$. Furthermore, if $f \in L^\infty(s)$ then

$$\|T_n(f)\|_\infty \leq \text{esssup}_{t\in I_k} \|f(t)\|_\infty$$

and hence assumption **(c1)** is satisfied with any constant $C > \text{esssup}_{t\in I_k} \|f(t)\|_\infty$. Finally, a simple check (see, e.g., [2, 17]) shows that also **(c2)** is fulfilled. Thus, Theorem 3.4 can be applied and from that it follows that the matrix sequence $\{T_n(f)\}$ is weakly clustered at $\text{Area}(R(f))$ (see Definition 3.2) and relation (2) is true for every continuous function $F$ with bounded support, which is holomorphic in the interior of $S = \text{Area}(R(f))$. Furthermore, if $\mathbb{C}\backslash R(f)$ is connected and the interior of $R(f)$ is empty, then $\{T_n(f)\} \sim_\lambda (f, I_k)$ which concludes the proof of Theorem 1.2. □

Finally, Theorem 3.6 allows us to generalize Theorem 1.2 to the algebra generated by block Toeplitz sequences with bounded symbols.

**Theorem 4.1.** *Let $f_{\alpha,\beta} \in L^\infty(s)$ with $\alpha = 1, \ldots, \rho$, $\beta = 1, \ldots, q_\alpha$, $\rho, q_\alpha < \infty$. Let*

$$h = \sum_{\alpha=1}^{\rho} \prod_{\beta=1}^{q_\alpha} f_{\alpha,\beta},$$

*and consider the sequence $\{B_n\}$ with $B_n = \sum_{\alpha=1}^{\rho} \prod_{\beta=1}^{q_\alpha} T_n(f_{\alpha,\beta})$. Then $\|B_n - T_n(h)\|_1 = o(\widehat{n})$, $\{A_n\}$ is weakly clustered at $\text{Area}(R(h))$ and relation*

$$\lim_{n\to\infty} \frac{1}{\widehat{n}s} \sum_{\lambda\in\Lambda_n} F(\lambda) = \fint_{I_k} \frac{1}{s}\text{tr}(F(h(t)))\, dt$$

*is true for every continuous function $F$ with bounded support, which is holomorphic in the interior of $S = \text{Area}(R(h))$. Furthermore, when $\mathbb{C}\backslash R(h)$ is connected and the interior of $R(h)$ is empty, then $\{B_n\} \sim_\lambda (h, I_k)$.*

*Proof.* The first claim, namely, $\|B_n - T_n(h)\|_1 = o(\widehat{n})$, can be shown by simple computation (see, for instance, [2, 17]). Further, the sequence $\{B_n\}$ is uniformly bounded in spectral norm since it belongs to the algebra generated by block Toeplitz sequences with bounded symbols. Therefore, Theorem 3.6 with $A_n = T_n(f)$ implies that conditions **(c1)**, **(c2)**, **(c3)** are satisfied with $\{B_n\}$, since the same conditions are satisfied by the sequence $\{A_n\}$ (see the proof of Theorem 1.2). Hence, the use of Theorem 3.4 allows to conclude the proof. □

**Remark 4.2.** Observe that the case when $f(t)$ is diagonalizable by a constant transformation independent of $t$, is special in the sense that the Szegö-type distribution result holds under the milder assumption that every eigenvalue of $f$ (now a scalar complex-valued function) shows a range with empty interior and which does not disconnect the complex plane. This leaves open the question whether this weaker requirement is sufficient in general.

Other problems remain open. For instance, it would be interesting to extend the results of this paper to the case where the involved symbols are not necessarily bounded, but only integrable. Such situations occur when constructing preconditioners for Krylov methods. As already stressed in [16], in that case, the matrix theory-based approach seems more convenient, since the corresponding Toeplitz operators are not well defined when the symbols are not bounded.

Finally, it should be observed that the conditions on the symbol reported in Theorem 1.2 for the existence of a canonical distribution corresponding to the symbol are sufficient, but not necessary. In fact, for $f(t) = e^{-\mathrm{i}t}$ the range of $f$ is the complex unit circle, disconnecting the complex plane, while the eigenvalues are all equal to zero. However, if one takes the symbol $f(t)$ in (3.24), p.80 in [3] ($f(t) = e^{2\mathrm{i}t}$, $t \in [0, \pi)$, $f(t) = e^{-2\mathrm{i}t}$, $t \in [\pi, 2\pi)$), then the range of $f$ is again the complex unit circle, which disconnects the complex plane, however the eigenvalues indeed distribute as determined by the symbol, as discussed in Example 5.39, pp. 167-169 in [3]. It would be instructive to understand how to discriminate between these two types of generating functions.

In the next section we illustrate numerically some of these issues.

## 5. Numerical experiments

The numerical experiments are divided into two parts. In the first part (Section 5.1) we consider examples, covered by the theoretical results from the previous section. As expected, the numerical results confirm the theoretical findings with the clustering, which is often of strong type.

An example of an important application, where the related problem can be formulated and analyzed using the spectral analysis of a Toeplitz sequence with non-Hermitian bounded symbol, is a signal restoration problem, where some of the sampling data are not available (cf. [12, 4]). For that problem, the numerical tests in [4], are exactly driven by the theory developed in the present paper.

In the second group of tests (Section 5.2) we consider block Toeplitz sequences with unbounded symbols, for which the Mergelyan Theorem cannot be used and our tools do not apply. The numerical tests, however, indicate that there is room for improving the theory by allowing unbounded symbols. The latter can be foreseen, provided the weaker assumption in Theorem 1.1, when dealing with Hermitian structures.

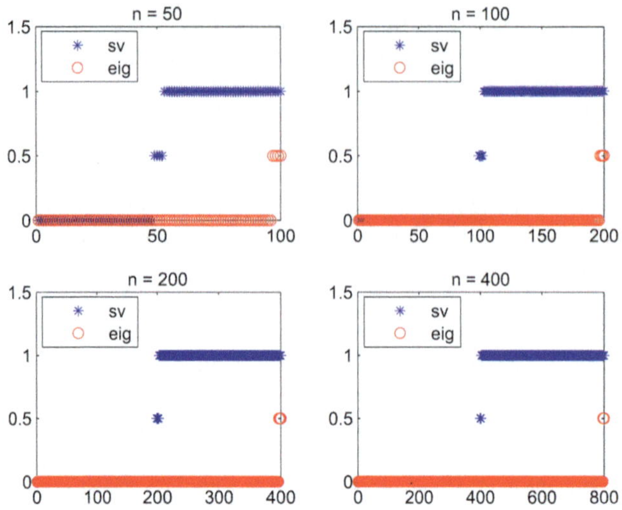


FIGURE 1. $T_n(f^{(1)})$: singular values and moduli of the eigenvalues in non decreasing order.

### 5.1. Examples covered by the theory

We considering first special classes of symbols $f$ where $s = 2$ and

$$f(t) = Q(t)B(t)Q(t)^T, \qquad t \in I_1,$$
$$Q(t) = \begin{pmatrix} \cos(t) & \sin(t) \\ -\sin(t) & \cos(t) \end{pmatrix}. \tag{31}$$

Below, $B(t)$ chosen in various ways, but having either constant eigenvalues, that is scalar functions independent of $t$, or eigenvalues with nicely behaved ranges. Thus, we expect a clustering of the eigenvalues of the corresponding block Toeplitz sequence, closely related to the shape of the spectra of the matrices.

**Example 5.1 (Nonparametrized symbols).** We choose

$$B^{(1)}(t) = \begin{pmatrix} 0 & 0 \\ 1 & 0 \end{pmatrix} \qquad \text{and} \qquad B^{(2)}(t) = \begin{pmatrix} 0 & 0 \\ 1 & 1 \end{pmatrix}.$$

In that case we find that the singular values of the matrix $T_n(f^{(1)})$ are divided into two sub-clusters of the same size (one at zero, the other at one), while its eigenvalues are clustered at zero (Figure 1). Interestingly enough, the eigenvalues clustered at zero are real, while the outliers are just four, independently of the size $n$, and lie on the imaginary axis (Figure 2 (a)). Regarding the singular values of the matrix $T_n(f^{(2)})$, we observe again two sub-clusters of the same size (one at zero, the other at $\sqrt{2}$). The eigenvalues, as predicted by our results, are distributed as the eigenvalues of the symbol, namely, half of them are around zero and half of them – around one (Figure 3). Observe, that also in this case, the range of

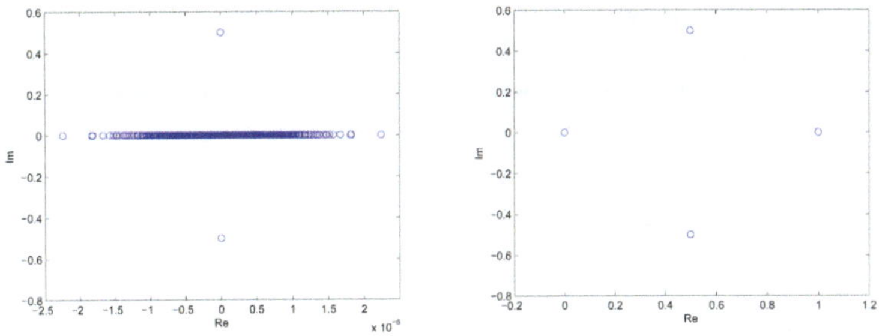


(a) (b)

FIGURE 2. Eigenvalues in the complex plane for $n = 400$: (a) eigenvalues of $T_n(f^{(1)})$, (b) eigenvalues of $T_n(f^{(2)})$.

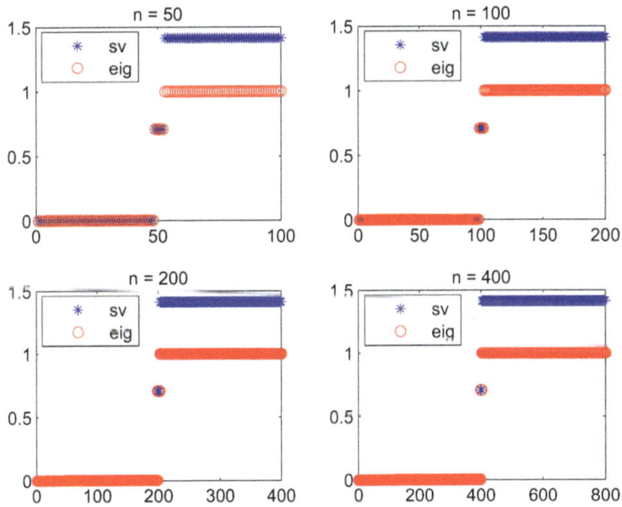


FIGURE 3. $T_n(f^{(2)})$: singular values and moduli of the eigenvalues in non decreasing order.

the symbol is a strong cluster since the outliers are only four, again lying on the imaginary axis (Figure 2 (b)).

**Example 5.2 (Parametrized symbols).** Consider

$$B^{(3)}_{(c,r)}(t) = \begin{pmatrix} 0 & 0 \\ 1 & c + r\,e^{\mathrm{i}t} \end{pmatrix} \quad \text{and} \quad B^{(4)}(t) = \begin{pmatrix} 0 & 0 \\ 1 & 2 - \cos(t) \end{pmatrix}.$$

Let $T_n(f^{(3)}_{(c,r)})$ and $T_n(f^{(4)})$ be the corresponding block Toeplitz matrices with generating functions

$$f^{(3)}_{(c,r)}(t) = Q(t)B^{(3)}_{(c,r)}(t)Q(t)^T,\ t \in I_1 \quad \text{and} \quad f^{(4)}(t) = Q(t)B^{(4)}(t)Q(t)^T.$$

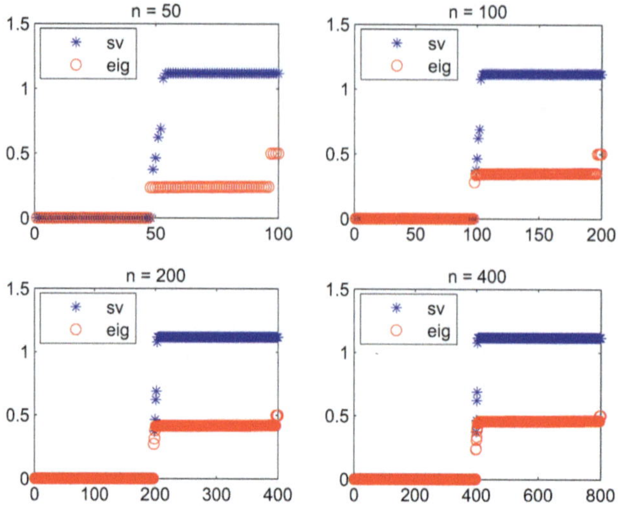


FIGURE 4. $T_n(f^{(3)}_{(0,\frac{1}{2})})$: singular values and moduli of the eigenvalues in non decreasing order.

By varying the parameters $c$ and $r$ we simulate various spectral and singular value behaviour.

The matrix $T_n(f^{(3)}_{(c,r)})$ has two sub-clusters for the singular values, one at zero and one, as expected, at the range of $\sqrt{1+|c+r\,e^{\mathbf{i}x}|^2}$ (Figure 4 and Figure 5). Its eigenvalues, are one half equal to zero and one half – residing in a disc, centered at $c$ with radius $r$ (Figure 6). More precisely, most of the eigenvalues belonging to the second sub-cluster set (in the sense of Definition 2.2) stay very close to the frontier.

For the block Toeplitz matrix $T_n(f^{(4)})$, both eigenvalues and singular values have two sub-clusters, one at zero and one in a positive interval (Figure 7). For the eigenvalues, as expected, the interval is $[1, 3]$, which represents the range of the second eigenvalue symbol, namely, $2-\cos(t)$. For the singular values the interval is $[\sqrt{2}, \sqrt{10}]$, which represents the range of the second singular value symbol - $\sqrt{1+(2-\cos(t))^2}$ (Figure 8).

**Example 5.3 (Solution of linear systems).** Consider now the solution of systems of linear equations with such matrices. Taking into account that the matrices are not Hermitian and our understanding of the spectral behavior, we use the GMRES method (cf. [11]). We test matrices with two sub-cluster points $r_1$ and $r_2$ with $|r_1| \geq |r_2|$ and pose the question how the number of iterations depends on the size of the box containing $r_1$ and $r_2$. In addition, if these values are both real and positive, it is instructive to see how the number of iterations depends on $r_1/r_2$ (spectral conditioning).

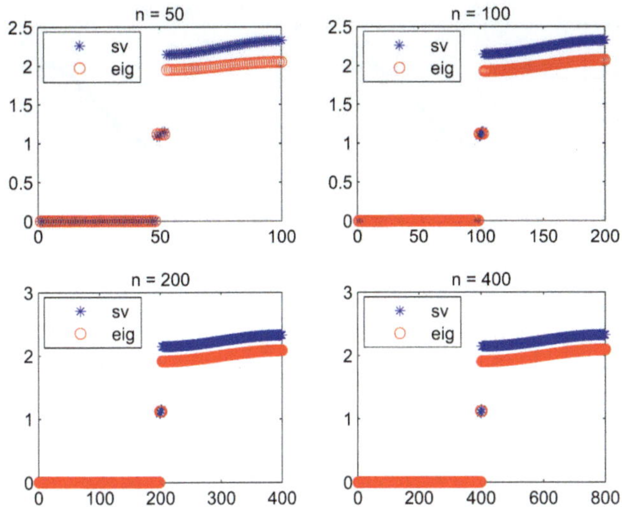


FIGURE 5. $T_n(f^{(3)}_{(2,\frac{1}{10})})$: singular values and moduli of the eigenvalues in non decreasing order.

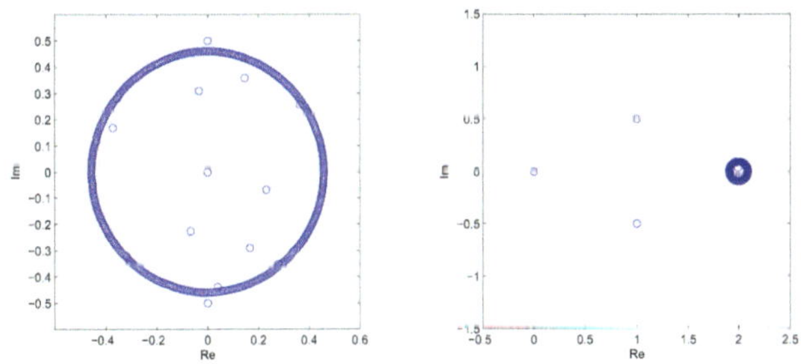


(a) (b)

FIGURE 6. Eigenvalues in the complex plane for $n = 400$: (a) eigenvalues of $T_n(f^{(3)}_{(0,\frac{1}{2})})$, (b) eigenvalues of $T_n(f^{(3)}_{(2,\frac{1}{10})})$.

We begin by considering the symbol

$$B^{(5)}_{(c_1,r_1,c_2,r_2)}(t) = \begin{pmatrix} c_1 - r_1\cos(t) & 0 \\ 1 & c_2 + r_2\, e^{\mathrm{i}t} \end{pmatrix}$$

and the corresponding matrix $T_n(B^{(5)}_{(c_1,r_1,c_2,r_2)})$. The eigenvalues of $T_n(B^{(5)}_{(c_1,r_1,c_2,r_2)})$ are divided in two sub-clusters, one in the interval $[c_1 - r_1,\ c_1 + r_1]$ and one in the disc centered at $c_2$ with radius $r_2$. We solve linear system with $T_n(B^{(5)}_{(c_1,r_1,c_2,r_2)})$ and a random right-hand side. We apply full GMRES with a relative stopping tolerance $10^{-6}$and use Matlab's built-in function gmres.

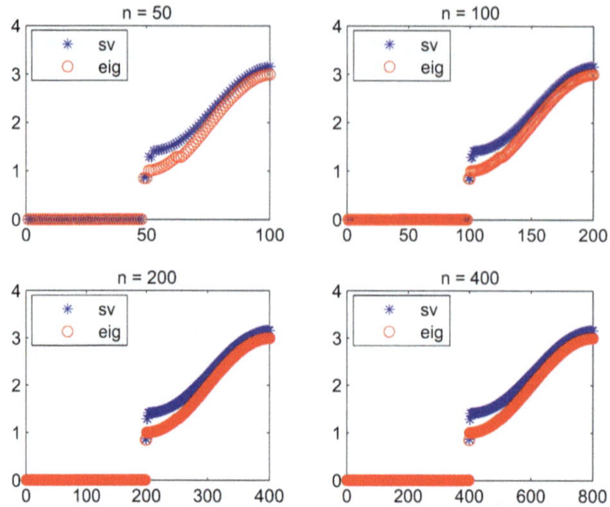


FIGURE 7. $T_n(f^{(4)})$: singular values and moduli of the eigenvalues in non decreasing order.

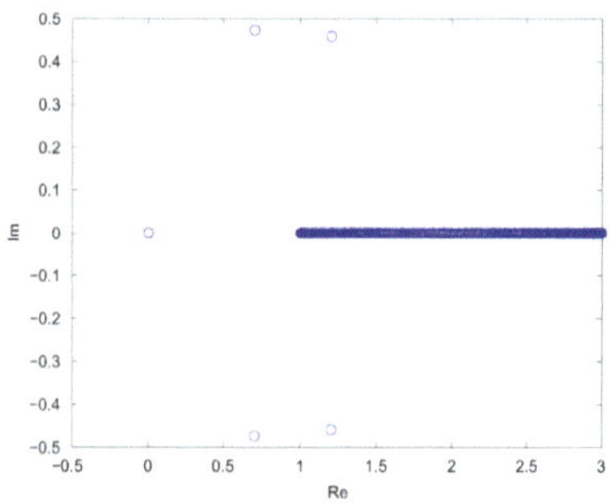


FIGURE 8. $T_n(f^{(4)})$: eigenvalues in the complex plane for $n = 400$.

Table 1 shows the number of GMRES iterations required to reach the prescribed tolerance for varying $n$ and moving the center of the disc $c_2$. The other parameters are chosen as $c_1 = 2$, $r_1 = 1$, and $r_2 = 1$. We note, that the number of iterations is independent of the size of the system, $2n$, however, it grows with increasing the spectral conditioning (the ratio $c_2/c_1$) until it reaches an asymptotic constant value (26 in this example). Table 2 reports the number of GMRES iterations for a fixed problem size, $n = 200$, and moving the disc with the center $c_2$ and the radius $r_2$. The other sub-cluster is the complex segment $[\mathbf{i}, 2+\mathbf{i},]$. We note that the number of iterations grows when increasing the radius of the disc and decreases when moving the disc away from the origin, until it reaches a constant

| $n \backslash c_2$ | 5 | 15 | 25 | 35 | 45 |
|---|---|---|---|---|---|
| 50 | 19 | 24 | 26 | 26 | 26 |
| 100 | 19 | 24 | 25 | 26 | 26 |
| 200 | 18 | 24 | 25 | 26 | 26 |
| 400 | 18 | 24 | 25 | 26 | 26 |

TABLE 1. Number of GMRES iterations for $T_n(B^{(5)}_{(2,1,c_2,1)})$ varying $n$ and $c_2$.

| $c_2 \backslash r_2$ | 1 | 2 | 3 | 4 | 5 |
|---|---|---|---|---|---|
| 10 | 30 | 35 | 40 | 45 | 53 |
| 20 | 31 | 35 | 39 | 43 | 46 |
| 30 | 31 | 35 | 38 | 41 | 43 |
| 40 | 31 | 34 | 37 | 39 | 42 |
| 50 | 32 | 35 | 37 | 39 | 41 |

TABLE 2. Number of GMRES iterations for $T_{200}(B^{(5)}_{(1+\mathrm{i},1,c_2,r_2)})$ varying $c_2$ and $r_2$.

| $r_1 \backslash r_2$ | 1 | 2 | 3 | 4 | 5 |
|---|---|---|---|---|---|
| 1 | 12 | 15 | 19 | 25 | 38 |
| 2 | 13 | 15 | 19 | 26 | 39 |
| 3 | 14 | 16 | 20 | 27 | 40 |
| 4 | 15 | 17 | 21 | 28 | 41 |
| 5 | 16 | 19 | 23 | 29 | 41 |

TABLE 3. Number of GMRES iterations for $T_{200}(B^{(5)}_{(10,r_1,5+5\mathrm{i},r_2)})$ varying $r_1$ and $r_2$.

number, depending on both parameters $c_2$ and $r_2$. In other words, when the two sub-clusters are far away from each other, the convergence of GMRES is driven by the two sub-clusters, independently. When the two sub-clusters start to approach each other, they start to act as a unique clustered area and the convergence is accelerated.

Another perspective is given by Table 3, where the centers of the two sub-clusters are fixed but we increase their radii. As expected, the number of iterations increases more noticeably when increasing the radius of the sub-cluster closer to the origin.

### 5.2. Examples with unbounded symbols

Here we consider examples with unbounded symbols. The numerical results indicate that the main distribution result, that is Theorem 1.2, holds also under these weaker assumptions. However, the tools cannot be exactly the same: the Mergelyan Theorem requires the compactness and the compactness of the range does not hold if the symbol is unbounded. Probably some approximation arguments have to be introduced.

Let $\alpha(t)$ be the unbounded function

$$\alpha(t) = \frac{1}{\sqrt{|t|}}, \qquad t \in I_1.$$

The function $\alpha(t)$ has an infinite Fourier series expansion, thus, for a fixed order $m$ it is approximated with its truncated series $\tilde{\alpha}(t) = \sum_{j=-m}^{m} \hat{\alpha}_j e^{\mathbf{i}jt}$. The Fourier coefficients $\hat{\alpha}_j$ are computed with high accuracy using the symbolic software package `Mathematica`.

Similarly to Subsection 5.1, for $s = 2$, we consider the special classes of symbols $g$ of the form

$$g(t) = Q(t)U(t)Q(t)^T, \qquad t \in I_1,$$

where $Q(t)$ is defined in (31) but the choice of $U(t)$ varies. Since the behaviour of the singular values and eigenvalues of $T_n(g)$ does not change for large enough $n$, in the following examples we fix $n = 100$.

**Example 5.4.** With

$$U^{(1)}(t) = \begin{pmatrix} \mathbf{i} & 0 \\ \alpha(t) & 1 \end{pmatrix},$$

we find that the matrix $T_n(g^{(1)})$ shows an unbounded maximal singular value, while the eigenvalues are divided in two sub-clusters – one at $\mathbf{i}$ and one at 1 (Figure 9). In other words, the unbounded character of the symbol does not play a role in the spectrum, which is determined only by the spectrum of the symbol.

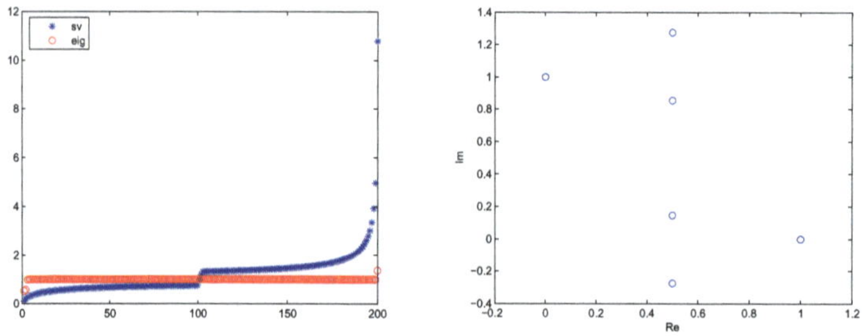


(a) (b)

FIGURE 9. $T_{100}(g^{(1)})$: (a) singular values and moduli of the eigenvalues in non decreasing order. (b) eigenvalues in the complex plane.

**Example 5.5.** Consider

$$U^{(2)}(t) = \begin{pmatrix} \alpha(t) & 0 \\ 1 & \mathbf{i} \end{pmatrix} \quad \text{and} \quad U^{(3)}(t) = \begin{pmatrix} \alpha(t)(1+\mathbf{i}\cos(t)) & 0 \\ 1 & 5\mathbf{i} \end{pmatrix}.$$

In these cases, the spectrum of the symbol is unbounded and in both cases, the spectrum and the singular values are distributed as the eigenvalues and the singular values of the symbol, correspondingly. The latter is clearly indicated in Figure 10 for the singular values of $T_n(g^{(2)})$ and $T_n(g^{(3)})$ and in Figure 11 for the eigenvalues of $T_n(g^{(2)})$ and $T_n(g^{(3)})$.

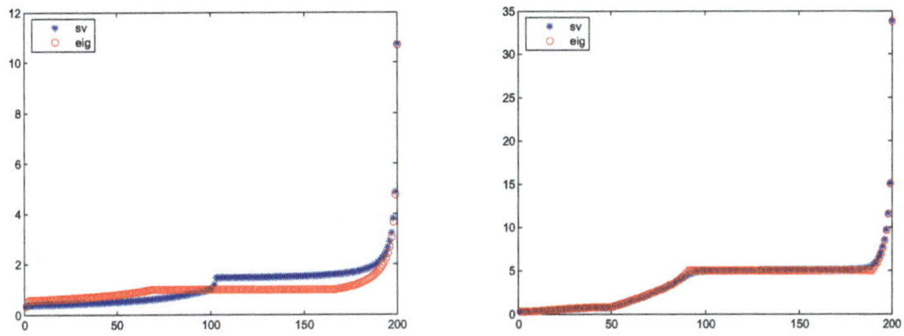


(a) (b)

FIGURE 10. Singular values and moduli of the eigenvalues in non decreasing order of $T_{100}(g^{(2)})$ in (a) and of $T_{100}(g^{(3)})$ in (b).

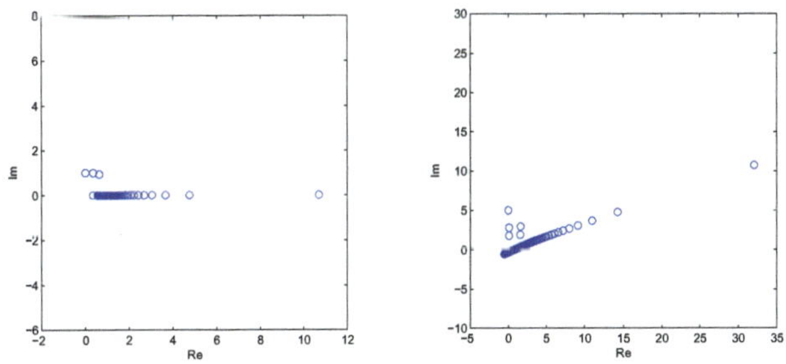


(a) (b)

FIGURE 11. Eigenvalues in the complex plane: (a) eigenvalues of $T_{100}(g^{(2)})$, (b) eigenvalues of $T_{100}(g^{(3)})$.

We finally remark that the small number of large eigenvalues is expected by virtue of the distribution results, since the measure of the set, where the symbol $\alpha(t)$ is large, is indeed very small when compared with the measure $2\pi$ of the domain $I_1$.

## 6. Concluding remarks and open problems

As a conclusion, tools from approximation theory in the complex field (Mergelyan Theorem, see [10]), combined with those from asymptotic linear algebra [21, 22, 14] are shown to be crucial when proving results on the eigenvalue distribution of non-Hermitian matrix sequences. As stated in Remark 4.2, there are still open questions, some of which have been studied numerically in Section 5. An issue, with a significant practical importance, is related to constructing structured preconditioners. Thus, given a linear system with a block Toeplitz matrix with a symbol $f$, how to find a symbol $g$ such that (1) the spectrum of $g^{-1}f$ is well localized away from zero and as clustered as possible, and (2) a generic system with the matrix $T_n(g)$ is cheap to solve. The expectation is that the eigenvalues of $T_n^{-1}(g)T_n(f)$ are described asymptotically by the range of the eigenvalues of the new symbol $h = g^{-1}f$. Numerical tests in [5] seem to confirm the latter hypothesis in the distributional sense. We recall that in the case of Hermitian symbols such a result is rigorously proven both for the distribution and the localization (see [13] and references therein) and in the case of general non-Hermitian symbols - with respect only to the localization (see [9]). A future line of research should consider the extension of Theorem 1.2 for sequences $\{A_n\}$, where $A_n = T_n^{-1}(g)T_n(f)$ and where the role of the symbol in formula (2) is played by $h = g^{-1}f$. Some preliminary results can be found in [18]. A detailed study of the above open problems is a subject of future research.

## References

[1] R. Bhatia, *Matrix Analysis*, Springer Verlag, New York, 1997.

[2] A. Böttcher, J. Gutiérrez-Gutiérrez, and P. Crespo, *Mass concentration in quasi-commutators of Toeplitz matrices*, J. Comput. Appl. Math., **205** (2007), 129–148.

[3] A. Böttcher and B. Silbermann, *Introduction to Large Truncated Toeplitz Matrices*, Springer-Verlag, New York, 1999.

[4] F. Di Benedetto, M. Donatelli, and S. Serra-Capizzano, *Symbol approach in a signal-restoration problem involving block Toeplitz matrices*, in preparation.

[5] M. Donatelli, S. Serra-Capizzano, and E. Strouse, *Spectral behavior of preconditioned non-Hermitian Toeplitz matrix sequences*, in preparation.

[6] L. Golinskii and S. Serra-Capizzano, *The asymptotic properties of the spectrum of non symmetrically perturbed Jacobi matrix sequences*, J. Approx. Theory, **144-1** (2007), 84–102.

[7] U. Grenander and G. Szegö, *Toeplitz Forms and Their Applications*, second edition, Chelsea, New York, 1984.

[8] J. Gutiérrez-Gutiérrez, P. Crespo, and A. Böttcher, *Functions of the banded Hermitian block Toeplitz matrices in signal processing*, Linear Algebra Appl, **422-2/3** (2007), 788–807.

[9] T. Huckle, S. Serra-Capizzano, and C. Tablino Possio, *Preconditioning strategies for non Hermitian Toeplitz linear systems*, Numerical Linear Algebra Appl., **12-2/3** (2005), 211–220.

[10] W. Rudin, *Real and Complex Analysis*, McGraw-Hill, New York, 1974.

[11] Y. Saad and M.H. Schultz, *GMRES: a generalized minimal residual algorithm for solving nonsymmetric linear systems*, SIAM J. Sci. Stat. Comput., **7** (1986), pp. 856–869.

[12] D.M.S. Santos and P.J.S.G. Ferreira, *Reconstruction from Missing Function and Derivative Samples and Oversampled Filter Banks*, Proceedings of the IEEE International Conference on Acoustics, Speech, and Signal Processing, ICASSP **III** (2004), 941–944.

[13] S. Serra-Capizzano, *Spectral and computational analysis of block Toeplitz matrices with nonnegative definite generating functions*, BIT, **39** (1999), 152–175.

[14] S. Serra-Capizzano, *Spectral behavior of matrix sequences and discretized boundary value problems*, Linear Algebra Appl., **337** (2001), 37–78.

[15] S. Serra-Capizzano, *Generalized Locally Toeplitz sequences: spectral analysis and applications to discretized Partial Differential Equations*, Linear Algebra Appl., **366-1** (2003), 371–402.

[16] S. Serra-Capizzano, *The GLT class as a Generalized Fourier Analysis and applications*, Linear Algebra Appl., **419-1** (2006), 180–233.

[17] S. Serra-Capizzano, D. Sesana, and E. Strouse, *The eigenvalue distribution of products of Toeplitz matrices – clustering and attraction*, Linear Algebra Appl., **432-10** (2010), 2658–2678.

[18] S. Serra-Capizzano and P. Sundqvist, *Stability of the notion of approximating class of sequences and applications*, J. Comput. Appl. Math., **219** (2008), pp. 518–536.

[19] P. Tilli, *Singular values and eigenvalues of non-Hermitian block Toeplitz matrices*, Linear Algebra Appl., **272** (1998), 59–89.

[20] P. Tilli, *Locally Toeplitz matrices: spectral theory and applications*, Linear Algebra Appl., **278** (1998), 91–120.

[21] P. Tilli, *A note on the spectral distribution of Toeplitz matrices*, Linear Multilin. Algebra, **45** (1998), pp. 147–159.

[22] P. Tilli, *Some results on complex Toeplitz eigenvalues*, J. Math. Anal. Appl., **239–2** (1999), 390–401.

[23] E. Tyrtyshnikov and N. Zamarashkin, *Spectra of multilevel Toeplitz matrices: advanced theory via simple matrix relationships*, Linear Algebra Appl., **270** (1998), 15–27.

[24] J. Wilkinson, *The Algebraic Eigenvalue Problem*, Clarendon Press, Oxford, 1965.

Marco Donatelli and Stefano Serra-Capizzano
Dipartimento di Scienza ed alta Tecnologia, Università dell Insubria
Via Valleggio 11, I-22100 Como, Italy
e-mail: `marco.donatelli@uninsubria.it`
`stefano.serrac@uninsubria.it`

Maya Neytcheva
Department of Information Technology, Uppsala University
Box 337, SE-751 05 Uppsala, Sweden
e-mail: `Maya.Neytcheva@it.uu.se`

Operator Theory:
Advances and Applications, Vol. 221, 293-304


# Curvature Invariant and Generalized Canonical Operator Models – I

Ronald G. Douglas, Yun-Su Kim, Hyun-Kyoung Kwon and Jaydeb Sarkar

**Abstract.** One can view contraction operators given by a canonical model of Sz.-Nagy and Foias as being defined by a quotient module where the basic building blocks are Hardy spaces. In this note we generalize this framework to allow the Bergman and weighted Bergman spaces as building blocks, but restricting attention to the case in which the operator obtained is in the Cowen-Douglas class and requiring the multiplicity to be one. We view the classification of such operators in the context of complex geometry and obtain a complete classification up to unitary equivalence of them in terms of their associated vector bundles and their curvatures.

**Mathematics Subject Classification (2000).** 46E22, 46M20, 47A20, 47A45, 47B32.

**Keywords.** Cowen-Douglas class, Sz.-Nagy-Foias model operator, curvature, resolutions of Hilbert modules.

## 1. Introduction

One goal of operator theory is to obtain unitary invariants, ideally, in the context of a concrete model for the operators being studied. For a multiplication operator on a space of holomorphic functions on the unit disk $\mathbb{D}$, which happens to be contractive, there are two distinct approaches to models and their associated invariants, one

The work of Douglas and Sarkar was partially supported by a grant from the National Science Foundation. The work of Kwon was supported by the Basic Science Research Program through the National Research Foundation of Korea (NRF) grant funded by the Korean government (MEST) (No. 2010-0024371), and in part, by a Young Investigator Award at the NSF-sponsored Workshop in Analysis and Probability, Texas A & M University, 2009. The research began in the summer of 2009. Sarkar was at Texas A & M University at the time and Kim and Kwon were participants of the workshop. Sarkar would also like to acknowledge the hospitality of the mathematics departments of Texas A & M University and the University of Texas at San Antonio, where part of his research was done.

due to Sz.-Nagy and Foias [12] and the other due to M. Cowen and the first author [4]. The starting point for this work was an attempt to compare the two sets of invariants and models obtained in these approaches. We will work at the simplest level of generality for which these questions make sense. Extensions of these results to more general situations are pursued later in [6].

For the Sz.-Nagy-Foias canonical model theory, the Hardy space $H^2 = H^2(\mathbb{D})$, of holomorphic functions on the unit disk $\mathbb{D}$ is central if one allows the functions to take values in some separable Hilbert space $\mathcal{E}$. In this case, we will now denote the space by $H^2 \otimes \mathcal{E}$. One can view the canonical model Hilbert space (in the case of a $C_{\cdot 0}$ contraction $T$) as given by the quotient of $H^2 \otimes \mathcal{E}_*$, for some Hilbert space $\mathcal{E}_*$, by the range of a map $M_\Theta$ defined to be multiplication by a contractive holomorphic function, $\Theta(z) \in \mathcal{L}(\mathcal{E}, \mathcal{E}_*)$, from $H^2 \otimes \mathcal{E}$ to $H^2 \otimes \mathcal{E}_*$. If one assumes that the multiplication operator associated with $\Theta(z)$ defines an isometry (or is inner) and $\Theta(z)$ is purely contractive, that is, $\|\Theta(0)\eta\| < \|\eta\|$ for all $\eta (\neq 0)$ in $\mathcal{E}$, then $\Theta(z)$ is the characteristic operator function for the operator $T$. Hence, $\Theta(z)$ provides a complete unitary invariant for the compression of multiplication by $z$ to the quotient Hilbert space of $H^2 \otimes \mathcal{E}_*$ by the range of $\Theta(z)$. In general, neither the operator $T$ nor its adjoint $T^*$ is in the $B_n(\mathbb{D})$ class of [4] but we are interested in the case in which the adjoint $T^*$ is in $B_n(\mathbb{D})$ and we study the relation between its complex geometric invariants (see [4]) and $\Theta(z)$.

We use the language of Hilbert modules [9] which we believe to be natural in this context. The Cowen-Douglas theory can also be recast in the language of Hilbert modules [3]. With this approach, the problem of the unitary equivalence of operators becomes identical to that of the isomorphism of the corresponding Hilbert modules.

Furthermore, we consider "models" obtained as quotient Hilbert modules in which the Hardy module is replaced by other Hilbert modules of holomorphic functions on $\mathbb{D}$ such as the Bergman module $A^2 = A^2(\mathbb{D})$ or the weighted Bergman modules $A^2_\alpha = A^2_\alpha(\mathbb{D})$ with weight parameter $\alpha > -1$. We require in these cases that some analogue of the corona condition holds for the multiplier $\Theta(z)$.

As previously mentioned, we concentrate on a particularly simple case of the problem. We focus on the case of $\Theta \in H^\infty_{\mathcal{L}(\mathbb{C},\mathbb{C}^2)}$, where $H^\infty_{\mathcal{L}(\mathbb{C},\mathbb{C}^2)} = H^\infty_{\mathcal{L}(\mathbb{C},\mathbb{C}^2)}(\mathbb{D})$ is the space of bounded, holomorphic $\mathcal{L}(\mathbb{C}, \mathbb{C}^2)$-valued functions on $\mathbb{D}$, so that $\Theta(z) = \theta_1(z) \otimes e_1 + \theta_2(z) \otimes e_2$ for an orthonormal basis $\{e_1, e_2\}$ for $\mathbb{C}^2$ and $\theta_i(z) \in \mathcal{L}(\mathbb{C})$, $i = 1, 2$, and $z \in \mathbb{D}$. We shall adopt the notation $\Theta = \{\theta_1, \theta_2\}$. Recall that $\Theta$ is said to satisfy the *corona condition* if there exists an $\epsilon > 0$ such that

$$|\theta_1(z)|^2 + |\theta_2(z)|^2 \geq \epsilon,$$

for all $z \in \mathbb{D}$. Moreover, we will use the notation $\mathcal{H}_\Theta$ to denote the quotient Hilbert module $(\mathcal{H} \otimes \mathbb{C}^2)/\Theta\mathcal{H}$, where $\mathcal{H}$ is the Hardy, the Bergman, or a weighted Bergman module.

Now we state the main results in this note which we will prove in Section 4. Let $\Theta = \{\theta_1, \theta_2\}$ and $\Phi = \{\varphi_1, \varphi_2\}$ both satisfy the corona condition and denote by $\bigtriangledown^2$ the Laplacian $\bigtriangledown^2 = 4\partial\bar{\partial} = 4\bar{\partial}\partial$.

**Theorem 4.4.** *The quotient Hilbert modules $\mathcal{H}_\Theta$ and $\mathcal{H}_\Phi$ are isomorphic if and only if*

$$\nabla^2 \log \frac{|\theta_1(z)|^2 + |\theta_2(z)|^2}{|\varphi_1(z)|^2 + |\varphi_2(z)|^2} = 0,$$

*for all $z \in \mathbb{D}$, where $\mathcal{H}$ is the Hardy module $H^2$, the Bergman module $A^2$, or a weighted Bergman module $A^2_\alpha$.*

**Theorem 4.5.** *The quotient Hilbert modules $(A^2_\alpha)_\Theta$ and $(A^2_\beta)_\Phi$ are isomorphic if and only if $\alpha = \beta$ and*

$$\nabla^2 \log \frac{|\theta_1(z)|^2 + |\theta_2(z)|^2}{|\varphi_1(z)|^2 + |\varphi_2(z)|^2} = 0,$$

*for all $z \in \mathbb{D}$.*

**Theorem 4.7.** *Under no circumstances can $(H^2)_\Theta$ be isomorphic to $(A^2_\alpha)_\Phi$.*

## 2. Hilbert modules

In the present section and the next, we take care of some preliminaries. We begin with the following definition.

**Definition 2.1.** Let $T$ be a linear operator on a Hilbert space $\mathcal{H}$. We say that $\mathcal{H}$ is a contractive Hilbert module over $\mathbb{C}[z]$ relative to $T$ if the module action from $\mathbb{C}[z] \times \mathcal{H}$ to $\mathcal{H}$ given by

$$p \cdot f \mapsto p(T)f,$$

for $p \in \mathbb{C}[z]$ defines bounded operators such that

$$\|p \cdot f\|_{\mathcal{H}} = \|p(T)f\|_{\mathcal{H}} \leq \|p\|_\infty \|f\|_{\mathcal{H}},$$

for all $f \in \mathcal{H}$, where $\|p\|_\infty$ is the supremum norm of $p$ on $\mathbb{D}$.

The module multiplication by the coordinate function will be denoted by $M_z$, that is,

$$M_z f = z \cdot f = Tf,$$

for all $f \in \mathcal{H}$.

**Definition 2.2.** Given two Hilbert modules $\mathcal{H}$ and $\tilde{\mathcal{H}}$ over $\mathbb{C}[z]$, we say that $X : \mathcal{H} \to \tilde{\mathcal{H}}$ is a module map if it is a bounded, linear map satisfying $X(p \cdot f) = p \cdot (Xf)$ for all $p \in \mathbb{C}[z]$ and $f \in \mathcal{H}$. Two Hilbert modules are said to be isomorphic if there exists a unitary module map between them.

Since one can extend the module action of a contractive Hilbert module $\mathcal{H}$ over $\mathbb{C}[z]$ from $\mathbb{C}[z]$ to the disk algebra $A(\mathbb{D})$ using the von Neumann inequality, a contraction operator gives rise to a contractive Hilbert module over $A(\mathbb{D})$. Recall that $A(\mathbb{D})$ denotes the disk algebra, the algebra of holomorphic functions on $\mathbb{D}$ that are continuous on the closure of $\mathbb{D}$. Thus, the unitary equivalence of contraction operators is the same as the isomorphism of the associated contractive Hilbert modules over $A(\mathbb{D})$.

Next, let us recall that the Hardy space $H^2$ consists of the holomorphic functions $f$ on $\mathbb{D}$ such that

$$\|f\|_2^2 = \sup_{0<r<1} \frac{1}{2\pi} \int_0^{2\pi} |f(re^{i\theta})|^2 d\theta < \infty.$$

Similarly, the weighted Bergman spaces $A^2_\alpha$, $-1 < \alpha < \infty$, consist of the holomorphic functions $f$ on $\mathbb{D}$ for which

$$\|f\|^2_{2,\alpha} = \frac{1}{\pi} \int_{\mathbb{D}} |f(z)|^2 dA_\alpha(z) < \infty,$$

where $dA_\alpha(z) = (\alpha+1)(1-|z|^2)^\alpha dA(z)$ and $dA(z)$ denote the weighted area measure and the area measure on $\mathbb{D}$, respectively. Note that $\alpha = 0$ gives the (unweighted) Bergman space $A^2$. We mention [14] for a comprehensive treatment of the theory of Bergman spaces. The Hardy space, the Bergman space and the weighted Bergman spaces are contractive modules under the multiplication by the coordinate function.

The Hardy, the Bergman, and the weighted Bergman modules serve as examples of *contractive reproducing kernel Hilbert modules.* A reproducing kernel Hilbert module is a Hilbert module with a function called a *positive definite kernel* whose definition we now review.

**Definition 2.3.** We say that a function $K : \mathbb{D} \times \mathbb{D} \to \mathcal{L}(\mathcal{E})$ for a Hilbert space $\mathcal{E}$, is a positive definite kernel if $\langle \sum_{i,j=1}^p K(z_i, z_j)\eta_i, \eta_j \rangle \geq 0$ for all $z_i \in \mathbb{D}$, $\eta_i \in \mathcal{E}$, and $p \in \mathbb{N}$.

Given a positive definite kernel $K$, we can construct a Hilbert space $\mathcal{H}_K$ of $\mathcal{E}$-valued functions defined to be

$$\vee_{z\in\mathbb{D}} \vee_{\eta\in\mathcal{E}} K(\cdot, z)\eta,$$

with inner product

$$\langle K(\cdot, w)\eta, K(\cdot, z)\zeta \rangle_{\mathcal{H}_K} = \langle K(z, w)\eta, \zeta \rangle_{\mathcal{E}},$$

for all $z, w \in \mathbb{D}$ and $\eta, \zeta \in \mathcal{E}$. The evaluation of $f \in \mathcal{H}_K$ at a point $z \in \mathbb{D}$ is given by the reproducing property so that

$$\langle f(z), \eta \rangle_{\mathcal{E}} = \langle f, K(\cdot, z)\eta \rangle_{\mathcal{H}_K},$$

for all $f \in \mathcal{H}_K, z \in \mathbb{D}$ and $\eta \in \mathcal{E}$. In particular, the evaluation operator $\boldsymbol{ev}_z$ : $\mathcal{H}_K \to \mathcal{E}$, $\boldsymbol{ev}_z(f) := f(z)$ is bounded for all $z \in \mathbb{D}$.

Conversely, given a Hilbert space $\mathcal{H}$ of holomorphic $\mathcal{E}$-valued functions on $\mathbb{D}$ with bounded evaluation operator $\boldsymbol{ev}_z \in \mathcal{L}(\mathcal{H}, \mathcal{E})$ for each $z \in \mathbb{D}$, we can construct a reproducing kernel

$$\boldsymbol{ev}_z \circ \boldsymbol{ev}_w^* : \mathbb{D} \times \mathbb{D} \to \mathcal{L}(\mathcal{E}),$$

for all $z, w \in \mathbb{D}$. To ensure that $\boldsymbol{ev}_z \circ \boldsymbol{ev}_w^*$ is injective, we must assume for every $z \in \mathbb{D}$ that $\overline{\{f(z) : f \in \mathcal{H}\}} = \mathcal{E}$.

A reproducing kernel Hilbert module is said to be a *contractive reproducing kernel Hilbert module* over $A(\mathbb{D})$ if the operator $M_z$ is contractive.

The kernel function for $H^2$ is $K(z,w) = (1 - \bar{w}z)^{-1}$. For $A^2_\alpha$, it is

$$K(z,w) = (1 - \bar{w}z)^{-2-\alpha} = \sum_{k=0}^{\infty} \frac{\Gamma(k+2+\alpha)}{k!\Gamma(2+\alpha)} (\bar{w}z)^k,$$

where $\Gamma$ is the gamma function.

It is well known that the multiplier algebra of $\mathcal{H}$ is $H^\infty$, that is, $M_\varphi \mathcal{H} \subseteq \mathcal{H}$, for $M_\varphi$ the operator of multiplication by $\varphi \in H^\infty$, where $H^\infty = H^\infty(\mathbb{D})$ is the algebra of bounded, analytic functions on $\mathbb{D}$ and $\mathcal{H}$ is $H^2$, $A^2$ or $A^2_\alpha$. Moreover, for all $z, w \in \mathbb{D}$, $\varphi \in H^\infty$ and $\eta \in \mathcal{E}$,

$$M_\varphi^* K(\,\cdot\,, w) = \overline{\varphi(w)} K(\,\cdot\,, w).$$

## 3. The class $B_n(\mathbb{D})$

In [4], M. Cowen and the first author introduced a class of operators $B_n(\mathbb{D})$, which includes $M_z^*$ for the operator $M_z$ defined on contractive reproducing kernel Hilbert modules of interest in this note. We now recall the notion of $B_n(\mathbb{D})$. Let $\mathcal{H}$ be a Hilbert space and $n$ a positive integer.

**Definition 3.1.** An operator $T \in \mathcal{L}(\mathcal{H})$ is in the class $B_n(\mathbb{D})$ if

(i) $\dim \ker(T - w) = n$ for all $w \in \mathbb{D}$,
(ii) $\bigvee_{w \in \mathbb{D}} \ker(T - w) = \mathcal{H}$, and
(iii) $\operatorname{ran}(T - w) = \mathcal{H}$ for all $w \in \mathbb{D}$.

**Remark 3.2.** Since it follows from (iii) that $T - w$ is semi-Fredholm for all $w \in \mathbb{D}$, (iii) actually implies (i) if we assume that $\dim \ker(T - w) < \infty$ for some $w \in \mathbb{D}$.

It is a result of Shubin [11] that for $T \in B_n(\mathbb{D})$, there exists a hermitian holomorphic rank $n$ vector bundle $E_T$ over $\mathbb{D}$ defined as the pull-back of the holomorphic map $w \mapsto \ker(T - w)$ from $\mathbb{D}$ to the Grassmannian $Gr(n, \mathcal{H})$ of the $n$-dimensional subspaces of $\mathcal{H}$. As mentioned earlier in the Introduction, in this note we consider contraction operators $T$ such that $T^* \in B_n(\mathbb{D})$. In other words, we investigate contractive Hilbert modules $\mathcal{H}$ with $M_z^* \in B_n(\mathbb{D})$. For simplicity of notation, we will write $\mathcal{H} \in B_n(\mathbb{D})$. Thus, we have an anti-holomorphic map $w \mapsto \ker(M_z - w)^*$ instead of a holomorphic one and therefore obtain a frame $\{\psi_i\}_{i=1}^n$ of anti-holomorphic $\mathcal{H}$-valued functions on $\mathbb{D}$ such that

$$\bigvee_{i=1}^n \psi_i(w) = \ker(M_z - w)^* \subseteq \mathcal{H},$$

for every $w \in \mathbb{D}$. We will use the notation $E^*_{\mathcal{H}}$ for this anti-holomorphic vector bundle since it is the dual of the natural hermitian holomorphic vector bundle $E_{\mathcal{H}}$ defined by localization.

One can show for an operator belonging to a "weaker" class than $B_n(\mathbb{D})$ that there still exists an anti-holomorphic frame. Since having such a frame is sufficient for many purposes, one can consider operators in this "weaker" class, which will be introduced after the following proposition:

**Proposition 3.3.** *Let $T \in \mathcal{L}(\mathcal{H})$ and $\tilde{T} \in \mathcal{L}(\tilde{\mathcal{H}})$. Suppose that there exist anti-holomorphic functions $\{\psi_i\}_{i=1}^n$ and $\{\tilde{\psi}_i\}_{i=1}^n$ from $\mathbb{D}$ to $\mathcal{H}$ and $\tilde{\mathcal{H}}$, respectively, satisfying*

(1) *$T\psi_i(w) = \bar{w}\psi_i(w)$ and $\tilde{T}\tilde{\psi}_i(w) = \bar{w}\tilde{\psi}_i(w)$, for all $1 \le i \le n$, $w \in \mathbb{D}$, and*
(2) *$\bigvee_{w\in\mathbb{D}} \bigvee_{i=1}^n \psi_i(w) = \mathcal{H}$ and $\bigvee_{w\in\mathbb{D}} \bigvee_{i=1}^n \tilde{\psi}_i(w) = \tilde{\mathcal{H}}$.*

*Then there is an anti-holomorphic partial isometry-valued function $V(w) : \mathcal{H} \to \tilde{\mathcal{H}}$ such that $\ker V(w) = [\vee_{i=1}^n \psi_i(w)]^\perp$ and $\operatorname{ran} V(w) = \vee_{i=1}^n \tilde{\psi}_i(w)$ if and only if there exists a unitary operator $V : \mathcal{H} \to \tilde{\mathcal{H}}$ such that $(V\psi_i)(w) = V(w)\psi_i(w)$ for every $1 \le i \le n$ and $w \in \mathbb{D}$.*

*Proof.* We refer the reader to the proof of the rigidity theorem in [4], where the language of bundles is used. □

It was pointed out by N.K. Nikolski to the first author that the basic calculation used to prove the rigidity theorem [4] appeared earlier in [10].

**Definition 3.4.** Suppose $T \in \mathcal{L}(\mathcal{H})$ is such that $\dim\ker(T - w) \geqslant n$ for all $w \in \mathbb{D}$. We say that $T$ is in the class $B_n^w(\mathbb{D})$ or weak-$B_n(\mathbb{D})$ if there exist anti-holomorphic functions $\{\psi_i\}_{i=1}^n$ from $\mathbb{D}$ to $\mathcal{H}$ such that

(i) $\{\psi_i(w)\}_{i=1}^n$ is linearly independent for all $w \in \mathbb{D}$,
(ii) $\vee_{i=1}^n \psi_i(w) \subseteq \ker(T - w)$ for all $w \in \mathbb{D}$, and
(iii) $\bigvee_{w\in\mathbb{D}} \bigvee_{i=1}^n \psi_i(w) = \mathcal{H}$.

**Remark 3.5.** The class $B_n^w(\mathbb{D})$ is closely related to the one considered by Uchiyama in [13].

Since the $\{\psi_i\}_{i=1}^n$ in Definition 3.4 frame a rank $n$ hermitian anti-holomorphic bundle, it suffices for our purpose to consider contractive Hilbert modules $\mathcal{H}$ with $M_z^* \in B_n^w(\mathbb{D})$ instead of those with $M_z^* \in B_n(\mathbb{D})$. We will write $\mathcal{H} \in B_n^w(\mathbb{D})$ to represent this case.

We continue this section with a brief discussion of some complex geometric notions. Since the anti-holomorphic vector bundle $E_{\mathcal{H}}^*$ also has hermitian structure, one can define the canonical Chern connection $\mathcal{D}_{E_{\mathcal{H}}^*}$ on $E_{\mathcal{H}}^*$ along with its associated curvature two-form $\mathcal{K}_{E_{\mathcal{H}}^*}$. For the case $n = 1$, $E_{\mathcal{H}}^*$ is a line bundle and

$$\mathcal{K}_{E_{\mathcal{H}}^*}(z) = -\frac{1}{4}\,\nabla^2 \log\|\gamma_z\|^2\, dz \wedge d\bar{z}, \tag{3.1}$$

for $z \in \mathbb{D}$, where $\gamma_z$ is an anti-holomorphic cross section of the bundle. For instance, by taking $\gamma_z$ to be the kernel functions for $H^2$ and $A_\alpha^2$, we see that

$$\mathcal{K}_{E_{H^2}^*}(z) = -\frac{1}{(1-|z|^2)^2}, \qquad \text{and} \qquad \mathcal{K}_{E_{A_\alpha^2}^*}(z) = -\frac{2+\alpha}{(1-|z|^2)^2}.$$

In [4], M. Cowen and the first author proved that the curvature is a complete unitary invariant, that is, two Hilbert modules $\mathcal{H}$ and $\tilde{\mathcal{H}}$ in $B_1(\mathbb{D})$ are isomorphic if and only if for every $z \in \mathbb{D}$,

$$\mathcal{K}_{E_{\mathcal{H}}^*}(z) = \mathcal{K}_{E_{\tilde{\mathcal{H}}}^*}(z).$$

Now that we have Proposition 3.3 available, the result can be extended to Hilbert modules in $B_1^w(\mathbb{D})$. Note that two weighted Bergman modules cannot be isomorphic to each another, that is, $A_\alpha^2$ is isomorphic to $A_\beta^2$ if and only if $\alpha = \beta$. We also conclude that the Hardy module $H^2$ cannot be isomorphic to the weighted Bergman modules $A_\alpha^2$.

## 4. Proof of the main results

Let $\Theta = \{\theta_1, \theta_2\} \in H^\infty_{\mathcal{L}(\mathbb{C},\mathbb{C}^2)}$ satisfy the corona condition. Now denote by $\mathcal{H}_\Theta$ the quotient Hilbert module $(\mathcal{H} \otimes \mathbb{C}^2)/\Theta\mathcal{H}$, where $\mathcal{H}$ is $H^2$, $A^2$, or $A_\alpha^2$. This means that we have the following short exact sequence

$$0 \longrightarrow \mathcal{H} \otimes \mathbb{C} \xrightarrow{M_\Theta} \mathcal{H} \otimes \mathbb{C}^2 \xrightarrow{\pi_\Theta} \mathcal{H}_\Theta \longrightarrow 0,$$

where the first map $M_\Theta$ is $M_\Theta f = \theta_1 f \otimes e_1 + \theta_2 f \otimes e_2$ and the second map $\pi_\Theta$ is the quotient Hilbert module map. The fact that $\Theta$ satisfies the corona condition implies that $\operatorname{ran} M_\Theta$ is closed. We denote the module multiplication $P_{\mathcal{H}_\Theta}(M_z \otimes I_{\mathbb{C}^2})|_{\mathcal{H}_\Theta}$ of the quotient Hilbert module $\mathcal{H}_\Theta$ by $N_z$. We will see later that $\mathcal{H}_\Theta \in B_1(\mathbb{D})$, but for the time being, we first show that $\mathcal{H}_\Theta \in B_1^w(\mathbb{D})$.

**Theorem 4.1.** *For $\Theta = \{\theta_1, \theta_2\}$ satisfying the corona condition, $\mathcal{H}_\Theta \in B_1^w(\mathbb{D})$.*

*Proof.* We first prove that $\dim \ker(N_z - w)^* = 1$ for all $w \in \mathbb{D}$. To this end, let $I_w := \{p(z) \in \mathbb{C}[z] : p(w) = 0\}$, a maximal ideal in $\mathbb{C}[z]$. One considers the localization of the sequence

$$0 \to \mathcal{H} \otimes \mathbb{C} \overset{M_\Theta}{\to} \mathcal{H} \otimes \mathbb{C}^2 \overset{\pi_\Theta}{\to} \mathcal{H}_\Theta \to 0,$$

to $w \in \mathbb{D}$ to obtain

$$\mathcal{H}/I_w \cdot \mathcal{H} \longrightarrow (\mathcal{H} \otimes \mathbb{C}^2)/I_w \cdot (\mathcal{H} \otimes \mathbb{C}^2) \longrightarrow \mathcal{H}_\Theta/I_w \cdot \mathcal{H}_\Theta \longrightarrow 0,$$

or equivalently,

$$\mathbb{C}_w \otimes \mathbb{C} \xrightarrow{I_{\mathbb{C}_w} \otimes \Theta(w)} \mathbb{C}_w \otimes \mathbb{C}^2 \xrightarrow{\pi_\Theta(w)} \mathcal{H}_\Theta/I_w \cdot \mathcal{H}_\Theta \longrightarrow 0.$$

Since this sequence is exact and $\dim \operatorname{ran} \Theta(w) = 1$ for all $w \in \mathbb{D}$, we have $\dim \ker \pi_\Theta(w) = 1$ (see [9]). Thus, $\dim \mathcal{H}_\Theta/I_w \cdot \mathcal{H}_\Theta = 1$, and so $\dim \ker(N_z - w)^* = 1$ for all $w \in \mathbb{D}$.

Now denote by $k_w$ a kernel function $k(\cdot, w)$ for $\mathcal{H}$, and by $\{e_1, e_2\}$ an orthonormal basis for $\mathbb{C}^2$. We prove that

$$\gamma_w := k_w \otimes (\overline{\theta_2(w)} e_1 - \overline{\theta_1(w)} e_2)$$

is a non-vanishing anti-holomorphic function from $\mathbb{D}$ to $\mathcal{H} \otimes \mathbb{C}^2$ such that

(1) $\gamma_w \in \ker(N_z - w)^*$ for all $w \in \mathbb{D}$, and
(2) $\bigvee_{w \in \mathbb{D}} \gamma_w = \mathcal{H}_\Theta$.

Since the $\theta_i$ are holomorphic and $k_w$ is anti-holomorphic, the fact that $w \mapsto \gamma_w$ is anti-holomorphic follows. Furthermore, since $\Theta$ satisfies the corona condition, the $\theta_i$ have no common zero and hence $\gamma_w \neq \mathbf{0}$ for all $w \in \mathbb{D}$. Now, for $f \in \mathcal{H}$, $M_\Theta f = \theta_1 f \otimes e_1 + \theta_2 f \otimes e_2$ and therefore for all $w \in \mathbb{D}$,

$$\begin{aligned}\langle M_\Theta f, \gamma_w \rangle &= \langle \theta_1 f, k_w \rangle \langle e_1, \overline{\theta_2(w)} e_1 \rangle - \langle \theta_2 f, k_w \rangle \langle e_2, \overline{\theta_1(w)} e_2 \rangle \\ &= \theta_1(w) f(w) \theta_2(w) - \theta_2(w) f(w) \theta_1(w) = 0.\end{aligned}$$

Hence, $\gamma_w \in (\operatorname{ran} M_\Theta)^\perp = \mathcal{H}_\Theta$. Moreover, since $M_z^* k_w = \bar{w} k_w$ for all $w \in \mathbb{D}$,

$$\begin{aligned}N_z^* \gamma_w &= (M_z \otimes I_{\mathbb{C}^2})^* \gamma_w = M_z^*(\overline{\theta_2(w)} k_w) \otimes e_1 - M_z^*(\overline{\theta_1(w)} k_w) \otimes e_2 \\ &= \overline{\theta_2(w)} \bar{w} k_w \otimes e_1 - \overline{\theta_1(w)} \bar{w} k_w \otimes e_2 \\ &= \bar{w} \gamma_w.\end{aligned}$$

Next, in order to show that (2) holds, it suffices to prove that for $h = h_1 \otimes e_1 + h_2 \otimes e_2 \in \mathcal{H} \otimes \mathbb{C}^2$ such that $h \perp \bigvee_{w \in \mathbb{D}} \gamma_w$, we have $h \in \operatorname{ran} M_\Theta$. We first claim that there exists a function $\eta$ defined on $\mathbb{D}$ such that for all $w \in \mathbb{D}$ and $i = 1, 2$,

$$h_i(w) = \theta_i(w) \eta(w).$$

Since $h \perp \gamma_w$ for every $w \in \mathbb{D}$, we have

$$\begin{aligned}\langle h, \gamma_w \rangle &= \langle h_1, k_w \rangle \langle e_1, \overline{\theta_2(w)} e_1 \rangle - \langle h_2, k_w \rangle \langle e_2, \overline{\theta_1(w)} e_2 \rangle \\ &= h_1(w) \theta_2(w) - h_2(w) \theta_1(w) = 0,\end{aligned}$$

or equivalently,

$$\det \begin{bmatrix} h_1(w) & \theta_1(w) \\ h_2(w) & \theta_2(w) \end{bmatrix} = 0, \tag{4.1}$$

for all $w \in \mathbb{D}$. Thus using the fact that rank $\begin{bmatrix} \theta_1(w) \\ \theta_2(w) \end{bmatrix} = 1$ for all $w \in \mathbb{D}$, we obtain a unique nonzero function $\eta(w)$ satisfying $h_i(w) = \theta_i(w) \eta(w)$ for $i = 1, 2$.

The proof is completed once we show that $\eta \in \mathcal{H}$. Note that by the corona theorem, we get $\psi_1, \psi_2 \in H^\infty$ such that $\psi_1(w)\theta_1(w) + \psi_2(w)\theta_2(w) = 1$ for every $w \in \mathbb{D}$. Since $\eta = (\psi_1 \theta_1 + \psi_2 \theta_2)\eta = \psi_1 h_1 + \psi_2 h_2$, and $H^\infty$ is the multiplier algebra for $\mathcal{H}$, the result follows. □

**Remark 4.2.** Observe that the above proof shows that the hermitian anti-holomorphic line bundle corresponding to the quotient Hilbert module $\mathcal{H}_\Theta$ is the twisted vector bundle obtained as the bundle tensor product of the hermitian anti-holomorphic line bundle for $\mathcal{H}$ with the anti-holomorphic dual of the line bundle $\coprod_{w \in \mathbb{D}} \mathbb{C}^2 / \Theta(w) \mathbb{C}$. This phenomenon holds in general; suppose that for Hilbert spaces $\mathcal{E}$ and $\mathcal{E}_*$, $\Theta \in H^\infty_{\mathcal{L}(\mathcal{E}, \mathcal{E}_*)}$ and $M_\Theta$ has closed range. If the quotient Hilbert module $\mathcal{H}_\Theta$,

$$0 \to \mathcal{H} \otimes \mathcal{E} \stackrel{M_\Theta}{\to} \mathcal{H} \otimes \mathcal{E}_* \to \mathcal{H}_\Theta \to 0,$$

is in $B_n(\mathbb{D})$, then the rank $n$ hermitian anti-holomorphic vector bundle $E^*_{\mathcal{H}_\Theta}$ for $\mathcal{H}_\Theta$ is the bundle tensor product of $E^*_{\mathcal{H}}$ with the anti-holomorphic dual of the rank $n$ bundle $\coprod_{w \in \mathbb{D}} \mathcal{E}_* / \Theta(w) \mathcal{E}$ (see [6]).

In order to have $\mathcal{H}_\Theta \in B_1(\mathbb{D})$, it now remains to check only one condition. We do this in the following proposition.

**Proposition 4.3.** $\operatorname{ran}(N_z - w)^* = \mathcal{H}_\Theta$ *for all* $w \in \mathbb{D}$.

*Proof.* We write

$$M_z \otimes I_{\mathbb{C}^2} \backsim \begin{bmatrix} * & * \\ 0 & N_z \end{bmatrix}$$

relative to the decomposition $\mathcal{H} \otimes \mathbb{C}^2 = \operatorname{ran} M_\Theta \oplus (\operatorname{ran} M_\Theta)^\perp$. It suffices to note that $\mathcal{H} \in B_1(\mathbb{D})$ implies that $\operatorname{ran}(M_z - w)^* = \mathcal{H}$. □

Let us now consider the curvature $\mathcal{K}_{E^*_{\mathcal{H}_\Theta}}$. By (3.1), one needs only to compute the norm of the section

$$\gamma_w = k_w \otimes (\overline{\theta_2(w)} e_1 - \overline{\theta_1(w)} e_2)$$

given in Theorem 4.1. Since

$$\|\gamma_w\|^2 = \|k_w\|^2 (|\theta_1(w)|^2 + |\theta_2(w)|^2),$$

we get the identity

$$\mathcal{K}_{E^*_{\mathcal{H}_\Theta}}(w) = \mathcal{K}_{E^*_{\mathcal{H}}}(w) - \frac{1}{4} \nabla^2 \log(|\theta_1(w)|^2 + |\theta_2(w)|^2), \tag{4.2}$$

for all $w \in \mathbb{D}$.

We are now ready to prove Theorem 4.4. For the sake of convenience, we restate it here.

**Theorem 4.4.** *Let* $\Theta = \{\theta_1, \theta_2\}$ *and* $\Phi = \{\varphi_1, \varphi_2\}$ *satisfy the corona condition. The quotient Hilbert modules* $\mathcal{H}_\Theta$ *and* $\mathcal{H}_\Phi$ *are isomorphic if and only if*

$$\nabla^2 \log \frac{|\theta_1(z)|^2 + |\theta_2(z)|^2}{|\varphi_1(z)|^2 + |\varphi_2(z)|^2} = 0,$$

*for all* $z \in \mathbb{D}$, *where* $\mathcal{H}$ *is the Hardy, the Bergman, or a weighted Bergman module.*

*Proof.* Since $\mathcal{H}_\Theta, \mathcal{H}_\Phi \in B_1^w(\mathbb{D})$, (we have seen that they actually belong to $B_1(\mathbb{D})$), they are isomorphic if and only if $\mathcal{K}_{E^*_{\mathcal{H}_\Theta}}(w) = \mathcal{K}_{E^*_{\mathcal{H}_\Phi}}(w)$ for all $w \in \mathbb{D}$. But note that (4.2) and an analogous identity for $\Phi$ hold, where the $\theta_i$ are replaced with the $\varphi_i$. Since both $\Theta$ and $\Phi$ satisfy the corona condition, the result then follows. □

We once again state Theorem 4.5.

**Theorem 4.5.** *Suppose that* $\Theta = \{\theta_1, \theta_2\}$ *and* $\Phi = \{\varphi_1, \varphi_2\}$ *satisfy the corona condition. The quotient Hilbert modules* $(A^2_\alpha)_\Theta$ *and* $(A^2_\beta)_\Phi$ *are isomorphic if and only if* $\alpha = \beta$ *and*

$$\nabla^2 \log \frac{|\theta_1(z)|^2 + |\theta_2(z)|^2}{|\varphi_1(z)|^2 + |\varphi_2(z)|^2} = 0, \tag{4.3}$$

*for all* $z \in \mathbb{D}$.

*Proof.* Since we have by (4.2),

$$\mathcal{K}_{E^*_{(A^2_\alpha)_\Theta}}(w) = -\frac{2+\alpha}{(1-|w|^2)^2} - \frac{1}{4}\nabla^2 \log(|\theta_1(w)|^2 + |\theta_2(w)|^2),$$

and

$$\mathcal{K}_{E^*_{(A^2_\beta)_\Phi}}(w) = -\frac{2+\beta}{(1-|w|^2)^2} - \frac{1}{4}\nabla^2 \log(|\varphi_1(w)|^2 + |\varphi_2(w)|^2),$$

one implication is obvious. For the other one, suppose that $(A^2_\alpha)_\Theta$ is isomorphic to $(A^2_\beta)_\Phi$ so that the curvatures coincide. Observe next that

$$\frac{4(\beta-\alpha)}{(1-|w|^2)^2} = \nabla^2 \log \frac{|\theta_1(w)|^2 + |\theta_2(w)|^2}{|\varphi_1(w)|^2 + |\varphi_2(w)|^2}.$$

Since a function $f$ with $\nabla^2 f(z) = \frac{1}{(1-|z|^2)^2}$ for all $z \in \mathbb{D}$ is necessarily unbounded, we have a contradiction unless $\alpha = \beta$ (see Lemma 4.6 below) and (4.3) holds. This is due to the assumption that the bounded functions $\Theta$ and $\Phi$ satisfy the corona condition. $\square$

**Lemma 4.6.** *There is no bounded function $f$ defined on the unit disk $\mathbb{D}$ that satisfies $\nabla^2 f(z) = \frac{1}{(1-|z|^2)^2}$ for all $z \in \mathbb{D}$.*

*Proof.* Suppose that such $f$ exists. Since $\frac{1}{4}\nabla^2[(|z|^2)^m] = \partial\bar{\partial}[(|z|^2)^m] = m^2(|z|^2)^{m-1}$ for all $m \in \mathbb{N}$, we see that for

$$g(z) := \frac{1}{4}\sum_{m=1}^{\infty} \frac{|z|^{2m}}{m} = -\frac{1}{4}\log(1-|z|^2),$$

$\nabla^2 g(z) = \frac{1}{(1-|z|^2)^2}$ for all $z \in \mathbb{D}$. Consequently, $f(z) = g(z) + h(z)$ for some harmonic function $h$. Since the assumption is that $f$ is bounded, there exists an $M > 0$ such that $|g(z) + h(z)| \leq M$ for all $z \in \mathbb{D}$. It follows that

$$\exp(h(z)) \leq \exp(-g(z) + M) = (1-|z|^2)^{\frac{1}{4}} \exp(M),$$

and letting $z = re^{i\theta}$, we have $\exp(h(re^{i\theta})) \leq (1-r^2)^{\frac{1}{4}} \exp(M)$. Thus $\exp(h(re^{i\theta})) \to 0$ uniformly as $r \to 1^-$, and hence $\exp h(z) \equiv 0$. This is due to the maximum modulus principle because $\exp h(z) = |\exp(h(z) + i\tilde{h}(z))|$, where $\tilde{h}$ is a harmonic conjugate for $h$. We then have a contradiction, and the proof is complete. $\square$

We thank E. Straube for providing us with a key idea used in the proof of Lemma 4.6.

**Theorem 4.7.** *For $\Theta = \{\theta_1, \theta_2\}$ and $\Phi = \{\varphi_1, \varphi_2\}$ satisfying the corona condition, $(H^2)_\Theta$ cannot be isomorphic to $(A^2_\alpha)_\Phi$.*

*Proof.* By identity (4.2), we conclude that $(H^2)_\Theta$ is isomorphic to $(A^2_\alpha)_\Phi$ if and only if

$$\frac{4(1+\alpha)}{(1-|w|^2)^2} = \nabla^2 \log \frac{|\varphi_1(w)|^2 + |\varphi_2(w)|^2}{|\theta_1(w)|^2 + |\theta_2(w)|^2}.$$

But according to Lemma 4.6, this is impossible unless $\alpha = -1$. $\square$

## 5. Concluding remark

Although the case of quotient modules we have been studying in this note may seem rather elementary, the class of examples obtained is not without interest. The ability to control the data in the construction, that is, the multiplier, provides one with the possibility of obtaining examples of Hilbert modules over $\mathbb{C}[z]$ and hence operators with precise and refined properties. In [1] and [2] the authors utilized this framework to exhibit operators with properties that responded to questions raised in the papers.

In particular, in [2] the authors are interested in characterizing contraction operators that are quasi-similar to the unilateral shift of multiplicity one. In the earlier part of the paper, which explores a new class of operators, a plausible conjecture presents itself but examples defined in the framework of this note, introduced in Corollary 7.9, show that it is false.

In [1], the authors study canonical models for bi-shifts; that is, for commuting pairs of pure isometries. A question arises concerning the possible structure of such pairs and again, examples built using the framework of this note answer the question.

Finally in [8], the authors determine when a contractive Hilbert module in $B_1(\mathbb{D})$ can be represented as a quotient Hilbert module of the form $\mathcal{H}_\Theta$, where $\mathcal{H}$ is the Hardy, the Bergman, or a weighted Bergman module. For the case of the Hardy module, the result is contained in the model theory of Sz.-Nagy and Foias [12].

One can consider a much larger class of quotient Hilbert modules replacing the Hardy, the Bergman and the weighted Bergman modules by a quasi-free Hilbert module [7] of rank one. In that situation, one can raise several questions relating curvature invariant, similarity and the multiplier corresponding to the given quotient Hilbert modules. These issues will be discussed in the forthcoming paper [6].

## References


[1] H. Bercovici, R.G. Douglas, and C. Foias, *Canonical models for bi-isometries*, Operator Theory: Advances and Applications, 218 (2011), 177–205.

[2] H. Bercovici, R.G. Douglas, C. Foias, and C. Pearcy, *Confluent operator algebras and the closability property*, J. Funct. Anal. **258** (2010) 4122–4153.

[3] X. Chen and R.G. Douglas, *Localization of Hilbert modules*, Mich. Math. J. **39** (1992), 443–454.

[4] M.J. Cowen and R.G. Douglas, *Complex geometry and operator theory*, Acta Math. **141** (1978), 187–261.

[5] R.G. Douglas, C. Foias, and J. Sarkar, *Resolutions of Hilbert modules and similarity*, Journal of Geometric Analysis, to appear.

[6] R.G. Douglas, Y. Kim, H. Kwon, and J. Sarkar, *Curvature invariant and generalized canonical operator models – II*, in preparation.

[7] R.G. Douglas and G. Misra, *Quasi-free resolutions of Hilbert modules*, Integral Equations Operator Theory **47** (2003), No. 4, 435–456.

[8] R.G. Douglas, G. Misra, and J. Sarkar, *Contractive Hilbert modules and their dilations over natural function algebras*, Israel Journal of Math, to appear.

[9] R.G. Douglas and V.I. Paulsen, *Hilbert Modules over Function Algebras*, Research Notes in Mathematics Series, 47, Longman, Harlow, 1989.

[10] G. Polya, *How to Solve It: A New Aspect of Mathematical Method*, Princeton University Press, Princeton, 1944.

[11] M.A. Shubin, *Factorization of parameter-dependent matrix functions in normal rings and certain related questions in the theory of Noetherian operators*, Mat. Sb. **73** (113) (1967) 610–629; Math. USSR Sb.

[12] B. Sz.-Nagy and C. Foias, *Harmonic Analysis of Operators on Hilbert Space*, North-Holland, Amsterdam, 1970.

[13] M. Uchiyama, *Curvatures and similarity of operators with holomorphic eigenvectors*, Trans. Amer. Math. Soc. **319** (1990), 405–415.

[14] K. Zhu, *Operator Theory in Function Spaces*, Mathematical Surveys and Monographs, 138, American Mathematical Society, Providence, 2007.

Ronald G. Douglas
Department of Mathematics
Texas A&M University
College Station, TX 77843, USA
e-mail: rdouglas@math.tamu.edu

Yun-Su Kim
Department of Mathematics
The University of Toledo
Toledo, OH 43606, USA
Deceased

Hyun-Kyoung Kwon
Department of Mathematical Sciences
Seoul National University
Seoul, 151-747, Republic of Korea
e-mail: hyunkwon@snu.ac.kr

Jaydeb Sarkar
Indian Statistical Institute
Statistic and Mathematics Unit
8th Mile, Mysore Road
Bangalore 560059, India
e-mail: jay@isibang.ac.in

Operator Theory:
Advances and Applications, Vol. 221, 305-321


# About Compressible Viscous Fluid Flow in a 2-dimensional Exterior Domain

Yuko Enomoto and Yoshihiro Shibata

**Abstract.** We report our results [5, 6, 7] concerning a global in time unique existence theorem of strong solutions to the equation describing the motion of compressible viscous fluid flow in a 2-dimensional exterior domain for small initial data and some decay properties of the analytic semigroup associated with Stokes operator of compressible viscous fluid flow in a 2-dimensional exterior domain. Our results are an extension of the works due to Matsumura and Nishida [13] and Kobayashi and Shibata [10] in a 3-dimensional exterior domain to the 2-dimensional case. We also discuss some analytic semigroup approach to the compressible viscous fluid flow in a bounded domain, which was first investigated by G. Strömer [20, 21, 22].

**Mathematics Subject Classification (2000).** 35Q30, 76N10.

**Keywords.** 2-dimensional exterior domain, global in time unique existence theorem, local energy decay, $L_p$-$L_q$ decay estimate .

## 1. Introduction

Let $\Omega$ be a smooth domain in the $n$-dimensional Euclidean space $\mathbb{R}^n$ and we consider a motion of compressible viscous fluid flow occupying $\Omega$. The mathematical problem is to find a solution $\rho(x,t)$, $\vec{u}(x,t) = (u_1(x,t),\dots,u_n(x,t))$ describing the mass density and the velocity field, respectively, that satisfies the initial boundary value problem:

$$\begin{cases} \rho_t + \operatorname{div}(\rho\vec{u}\,) = 0 & \text{in } \Omega\times(0,T), \\ \rho(\vec{u}_t + \vec{u}\cdot\nabla\vec{u}) - \mu\Delta\vec{u} - (\mu+\mu')\nabla\operatorname{div}\vec{u} + \nabla P(\rho) = \vec{g} & \text{in } \Omega\times(0,T), \\ \vec{u}|_{\partial\Omega} = \vec{0} = (0,\dots,0), \quad (\rho,\vec{u})|_{t=0} = (\bar{\rho}_0+\theta_0,\vec{u}_0). & \end{cases} \tag{1.1}$$

Here, $\partial\Omega$ is the boundary of $\Omega$, $\mu$ and $\mu'$ denote the viscosity coefficient and the second viscosity coefficient, respectively, $\vec{g}$ is an external force that is a given vector of functions, and $(\bar{\rho}_0+\theta_0,\vec{u}_0)$ is an initial data. We assume that $P =$

$P(\rho)$ is a smooth function of $\rho$ defined near $\rho = \bar{\rho}_0$, $\bar{\rho}_0$ being a positive constant that represents a reference mass density. Such equations from fluid dynamics with viscosity are a good model for us to use pde techniques as well as the analytic semigroup approach to solve the equations.

In fact, when $\Omega$ is a bounded domain, G. Strömer [20, 21, 22] found that the Stokes operator of compressible viscous fluid flow generates an analytic semigroup, which decays exponentially. The decay is a key property to prove the global in time unique existence theorem of strong solutions to nonlinear evolution equations in general. He proved a global in time unique existence theorem of (1.1) by using the Lagrange coordinate to eliminate the material derivative causing the hyperbolicity and the exponential decay property of the Stokes semigroup. But, if we consider the case where $\Omega$ is a unbounded domain like $\mathbb{R}^n$ or an exterior domain, 0 is in the continuous spectrum of the Stokes operator. Therefore, the hyperbolic-parabolic structure of (1.1) requires that we combine the analytic semigroup approach with energy method to show the existence of solutions to (1.1) and their decay properties.

When $\Omega = \mathbb{R}^3$ or $\Omega$ is a 3-dimensional exterior domain, that is $\Omega = \mathbb{R}^3 \setminus \mathcal{O}$, $\mathcal{O}$ being a bounded domain, Matsumura and Nishida [12, 13] proved the global in time unique existence of strong solutions to problem (1.1) with potential force $\vec{g} = \nabla\Phi$ under some smallness assumptions on $(\rho_0 - \bar{\rho}_0, \vec{u}_0)$. And also, Matsumura and Nishida [12, 13] and later on Deckelnick [3, 4] proved some convergence rate of solutions to the stationary solutions. The proofs in [12, 13, 3, 4] relied on the energy method. The optimal decay properties of solutions to problem (1.1) with $\vec{g} = 0$ were obtained by Ponce [14] when $\Omega = \mathbb{R}^3$ and by Kobayashi and Shibata [10] when $\Omega$ is a 3-dimensional exterior domain. Their proofs in [14, 10] relied on so-called $L_p$-$L_q$ decay properties of solutions to the Stokes equation of compressible viscous fluid flow which is obtained by the linearization of (1.1) at $(\bar{\rho}_0, \vec{0})$.

In the papers [12, 13, 3, 4, 14, 10], their arguments rely essentially on the fact that $\Omega$ is a 3-dimensional unbounded domain. We are interested in the 2-dimensional exterior domain case. But, in this case, so far the asymptotic behaviour of global in time unique solutions has not yet obtained although its unique existence can be proved in the same manner as in Matumura and Nishida by replacing the inequality: $\|\vec{u}\|_{L_6(\Omega)} \leq C\|\nabla\vec{u}\|_{L_2(\Omega)}$ by $\|\vec{u}\|^2_{L_4(\Omega)} \leq C\|\nabla\vec{u}\|_{L_2(\Omega)}\|\vec{u}\|_{L_2(\Omega)}$. We think that the reason why we have not yet obtained asymptotic behaviour is the following:

To prove the optimal rate of decay of global in time strong solutions to (1.1), a known method is to use $L_p$-$L_q$ decay estimate of solutions to the Stokes equations of compressible viscous fluid flow. When we prove the decay estimate of solutions to the Stokes equations in the exterior domain, it is standard to use so-called local energy decay properties of solutions, which was proved by Kobayashi [9] in the 3-dimensional exterior domain case. The proof in [9] relied on the fact that the fundamental solutions of the resolvent problem of the Stokes equation is continuous up to $\lambda = 0$, $\lambda$ being the resolvent parameter. On the other hand, in the 2-dimensional case, the fundamental solution of the resolvent problem of the Stokes equation has logarithmical singularity at $\lambda = 0$, so that we need a new idea to prove the local energy decay.

From these observations, we organize this paper as follows. In section 2, we discuss some analytic semigroup approach to (1.1). In section 3, we discuss decay properties of solutions to the linear problem. In section 4, we state our global in time unique existence theorem of strong solutions to (1.1).

Finally, we outline notation used throughout the paper. For any domain $D$, $L_p(D)$ and $W_p^m(D)$ denote the usual Lebesgue space and Sobolev space, while their norms are denoted by $\|\cdot\|_{L_p(D)}$ and $\|\cdot\|_{W_p^m(D)}$, respectively. We write $W_p^0(D) = L_p(D)$ for the notational convention. For any $\ell$-dimensional vector of functions $\vec{h} = (h_1, \ldots, h_\ell)$, we set $\|\vec{h}\|_{W_p^m(D)} = \sum_{j=1}^{\ell} \|h_j\|_{W_p^m(D)}$. We set

$$
\begin{aligned}
W_p^m(D)^n &= \{\vec{u} = (u_1, \ldots, u_n) \mid u_i \in W_p^m(D) \ (i = 1, \ldots, n)\},\\
W_p^{m,\ell}(D) &= \{\mathbf{F} = (f, \vec{g}) \mid f \in W_p^m(D),\ \vec{g} \in W_p^\ell(D)^n\},\\
\|\mathbf{F}\|_{W_p^{m,\ell}(D)} &= \|(f, \vec{g})\|_{W_p^{m,\ell}(D)} = \|f\|_{W_p^m(D)} + \|\vec{g}\|_{W_p^\ell(D)},\\
\nabla^m f &= (\partial_x^\alpha f \mid |\alpha| = m),\ \ \nabla^m \vec{g} = (\nabla^m g_1, \ldots, \nabla^m g_n).
\end{aligned}
$$

We write $\nabla^1 f = \nabla f$ and $\nabla^1 \vec{g} = \nabla \vec{g}$. For any Banach spaces $X$ and $Y$, $\mathcal{L}(X, Y)$ denotes the set of all bounded linear operators from $X$ into $Y$ and $\|\cdot\|_{\mathcal{L}(X,Y)}$ denotes its operator norm. When $X = Y$, we use the abbreviations: $\mathcal{L}(X) = \mathcal{L}(X, X)$ and $\|\cdot\|_{\mathcal{L}(X)} = \|\cdot\|_{\mathcal{L}(X,X)}$. For a complex domain $U$, $\mathrm{Anal}\,(U, X)$ denotes the set of all $X$-valued holomorphic functions defined on $U$. For $I = (a, b)$, $L_p(I, X)$ and $W_p^m(I, X)$ denote the set of all $X$-valued $L_p(I)$ functions and $W_p^m(I)$ functions, while their norms are denoted by $\|\cdot\|_{L_p(I,X)}$ and $\|\cdot\|_{W_p^m(I,X)}$. The letter $C$ stands for a generic constant and $C_{A,B,\ldots}$ means that the constant $C_{A,B,\ldots}$ depends on the quantities $A$, $B$, .... Constants $C$ and $C_{A,B,\ldots}$ may change from line to line. The projections $P_m$ and $P_v$ are defined by $P_m(f, \vec{g}) = f$ and $P_v(f, \vec{g}) = \vec{g}$.

## 2. An analytic semigroup approach to the compressible viscous fluid flow

In this section[1] $\Omega$ stands for a bounded domain or an exterior domain in $\mathbb{R}^n$ ($n \geq 2$). We assume that the boundary $\partial\Omega$ of $\Omega$ is a compact $C^{1,1}$ hypersurface. Our equation here is:

$$
\begin{cases}
\rho_t + \gamma \mathrm{div}\, \vec{u} = f & \text{in } \Omega \times (0, \infty),\\
\vec{u}_t - \alpha\Delta\vec{u} - \beta\nabla(\mathrm{div}\, \vec{u}) + \gamma\nabla\rho = \vec{g} & \text{in } \Omega \times (0, \infty),\\
\vec{u}|_{\partial\Omega} = 0 \quad (\rho, \vec{u})|_{t=0} = (\rho_0, \vec{u}_0).
\end{cases}
\tag{2.1}
$$

[1]The detailed proofs of all the theorems mentioned in this section will be given in the forthcoming paper [6].

Here, $\alpha$ and $\gamma$ are given positive constants while $\beta$ is a constant such that $\alpha+\beta > 0$. Let $1 < q < \infty$ and we define a linear operator $A$ by

$$\begin{aligned} A(\rho, \vec{u}) &= (-\gamma \mathrm{div}\, \vec{u}, \alpha\Delta\vec{u} + \beta\nabla \mathrm{div}\, \vec{u} - \gamma\nabla\rho) \quad \text{for } (\rho, \vec{u}) \in \mathcal{D}_q(A), \\ \mathcal{D}_q(A) &= \{(\rho, \vec{u}) \in W_q^{1,2}(\Omega) \mid \vec{u}|_{\partial\Omega} = \vec{0}\}. \end{aligned}$$

The underlying space for $A$ is $W_q^{1,0}(\Omega)$. In terms of $A$, the problem (2.1) is written in the form:

$$U_t = AU + F \ \ (t > 0), \quad U|_{t=0} = U_0, \tag{2.2}$$

where $U = (\rho, \vec{u})$, $F = (f, \vec{g})$ and $U_0 = (\rho_0, \vec{u}_0)$. In what follows, we consider the generation of analytic semigroup and the maximal $L_p$-$L_q$ regularity of the operator $A$ and their application to local in time and global in time existence of (1.1) via the Lagrange coordinate. To prove the maximal $L_p$-$L_q$ regularity via the Weis operator-valued Fourier multiplier theorem ([25]), we introduce the notion of $\mathcal{R}$ boundedness.

**Definition 2.1.** Let $X$ and $Y$ be two Banach spaces while $\|\cdot\|_X$ and $\|\cdot\|_Y$ denote their norms, respectively. A family of operators $\mathcal{T} \subset \mathcal{L}(X, Y)$ is called $\mathcal{R}$-bounded, if there exist $C > 0$ and $p \in [1, \infty)$ such that for each $N \in \mathbb{N}$, $\mathbb{N}$ being the set of all natural numbers, $T_j \in \mathcal{T}$, $x_j \in X$ $(j = 1, \ldots, N)$ and sequences $\{r_j(u)\}_{j=1}^N$ of independent, symmetric, $\{-1, 1\}$-valued random variables on $(0, 1)$, there holds the inequality:

$$\int_0^1 \Big\| \sum_{j=1}^N r_j(u) T_j(x_j) \Big\|_Y^p \, du \le C \int_0^1 \Big\| \sum_{j=1}^N r_j(u) x_j \Big\|_X^p \, du.$$

The smallest such $C$ is called $\mathcal{R}$-bound of $\mathcal{T}$, which is denoted by $\mathcal{R}(\mathcal{T})$.

For $0 < \epsilon < \pi/2$ and $\lambda_0 > 0$ we define a set $\Sigma_{\epsilon,\lambda_0}$ in the complex plane $\mathbb{C}$ by

$$\Sigma_{\epsilon,\lambda_0} = \{\lambda \in \mathbb{C} \mid |\arg \lambda| \le \pi - \epsilon, \ \ |\lambda| \ge \lambda_0\}.$$

For the notational simplicity, the resolvent $(\lambda I - A)^{-1}$ is denoted by $R(\lambda)$. To prove the generation of analytic semigroup and the maximal $L_p$-$L_q$ regularity, we introduce the following notion.

**Definition 2.2.**

(i) $A$ is called an admissible sectorial operator if for any $\epsilon$ $(0 < \epsilon < \pi/2)$ there exists a positive number $\lambda_0$ such that

$$\begin{aligned} |\lambda| \|R(\lambda)(f, \vec{g})\|_{W_q^{1,0}(\Omega)} &+ |\lambda|^{\frac{1}{2}} \|\nabla P_v R(\lambda)(f, \vec{g})\|_{L_q(\Omega)} \\ &+ \|\nabla^2 P_v R(\lambda)(f, \vec{g})\|_{L_q(\Omega)} \le C_{\epsilon,\lambda_0} \|(f, \vec{g})\|_{W_q^{1,0}(\Omega)} \end{aligned}$$

for any $\lambda \in \Sigma_{\epsilon,\lambda_0}$ and $(f, \vec{g}) \in W_q^{1,0}(\Omega)$.

(ii) $A$ is called an admissible $\mathcal{R}$ sectorial operator if for any $\epsilon$ $(0 < \epsilon < \pi/2)$ there exists a positive number $\lambda_0$ such that the following sets:

$$\{R(\lambda) \mid \lambda \in \Sigma_{\epsilon,\lambda_0}\}, \ \left\{\lambda \frac{\partial}{\partial \lambda} R(\lambda) \mid \lambda \in \Sigma_{\epsilon,\lambda_0}\right\},$$
$$\{\lambda R(\lambda) \mid \lambda \in \Sigma_{\epsilon,\lambda_0}\}, \ \left\{\lambda \frac{\partial}{\partial \lambda} (\lambda R(\lambda)) \mid \lambda \in \Sigma_{\epsilon,\lambda_0}\right\}$$

are $\mathcal{R}$ bounded operator families in $\mathcal{L}(W^{1,0}_q(\Omega))$, and the following sets:

$$\{|\lambda|^{\frac{1}{2}} \nabla P_v R(\lambda) \mid \lambda \in \Sigma_{\epsilon,\lambda_0}\}, \ \left\{\lambda \frac{\partial}{\partial \lambda} (|\lambda|^{\frac{1}{2}} \nabla P_v R(\lambda)) \mid \lambda \in \Sigma_{\epsilon,\lambda_0}\right\},$$
$$\{\nabla^2 P_v R(\lambda) \mid \lambda \in \Sigma_{\epsilon,\lambda_0}\}, \ \left\{\lambda \frac{\partial}{\partial \lambda} (\nabla^2 P_v R(\lambda)) \mid \lambda \in \Sigma_{\epsilon,\lambda_0}\right\}$$

are $\mathcal{R}$-bounded families in $\mathcal{L}(W^{1,0}_q(\Omega), L_q(\Omega))$.

By Shibata and Tanaka [19] we see that $A$ is an admissible sectorial operator. Therefore, we have the following theorem.

**Theorem 2.3.** *Let $1 < q < \infty$. Then $A$ generates a $C^0$ semigroup $\{T(t)\}_{t \geq 0}$ on $W^{1,0}_q(\Omega)$ that is analytic, and there exist positive constants $c$ and $M$ such that*

$$\sup_{t>0} e^{-ct} \|T(t)(f, \vec{g})\|_{W^{1,0}_q(\Omega)} + \sup_{t>0} e^{-ct} t^{\frac{1}{2}} \|\nabla P_v T(t)(f, \vec{g})\|_{L_q(\Omega)}$$
$$+ \sup_{t>0} e^{-ct} t \|\nabla^2 P_v T(t)(f, \vec{g})\|_{L_q(\Omega)} \leq M \|(f, \vec{g})\|_{W^{1,0}_q(\Omega)}$$

*for any $(f, \vec{g}) \in W^{1,0}_q(\Omega)$.*

Employing the argument due to Shibata and Shimizu [18], we also see that $A$ is an admissible $\mathcal{R}$ sectorial operator, so that by the Weis operator-valued Fourier multiplier theorem we have the following theorem.

**Theorem 2.4.** *Let $1 < p, q < \infty$. Set $E_{p,q} = [W^{1,0}_q(\Omega), \mathcal{D}_q(A)]_{1-1/p,p}$, where $[\cdot, \cdot]_{\theta,p}$ denotes the real interpolation functor. For a Banach space $X$, we define the spaces $L_{p,\gamma}((0,\infty), X)$ and $W^1_{p,\gamma}((0,\infty), X)$ by*

$$L_{p,\gamma}((0,\infty), X) = \{f \in L_{p,\mathrm{loc}}((0,\infty), X) \mid e^{-\gamma t} f \in L_p((0,\infty), X)\},$$
$$W^1_{p,\gamma}((0,\infty), X) = \{f \in L_{p,\gamma}((0,\infty), X) \mid e^{-\gamma t} f_t \in L_p((0,\infty), X)\}.$$

*Then, there exists a constant $\gamma_0 > 0$ such that for any $(\rho_0, \vec{u}_0) \in E_{p,q}$ and $(f, \vec{g}) \in L_{p,\gamma_0}((0,\infty), W^{1,0}_q(\Omega))$ the problem (2.1) admits a unique solution $(\rho, \vec{u})$ such that*

$$\rho \in W^1_{p,\gamma_0}((0,\infty), W^1_q(\Omega)),$$
$$\vec{u} \in W^1_{p,\gamma_0}((0,\infty), L_q(\Omega)^2) \cap L_{p,\gamma_0}((0,\infty), W^2_q(\Omega)^2).$$

*Moreover, there exists a constant $C_{p,q}$ such that for any $\gamma \geq \gamma_0$*

$$\begin{aligned}\|e^{-\gamma t}(\rho_t,\gamma\rho)\|_{L_p((0,\infty),W^1_q(\Omega))} + \|e^{-\gamma t}(\vec{u}_t,\gamma\vec{u})\|_{L_p((0,\infty),L_q(\Omega)^2)} \\ + \gamma^{\frac12}\|e^{-\gamma t}\nabla\vec{u}\,\|_{L_p((0,\infty),L_q(\Omega)^2)} + \|e^{-\gamma t}\nabla^2\vec{u}\,\|_{L_p((0,\infty),L_q(\Omega)^2)} \\ \leq C_{p,q}(\|(\rho_0,\vec{u}_0)\|_{E_{p,q}} + \|e^{-\gamma t}(f,\vec{g})\|_{L_p((0,\infty),W^{1,0}_q(\Omega))}).\end{aligned}$$

In order to solve (1.1) locally in time by using Theorem 2.4, we have to eliminate the material derivative causing the hyperbolic effect. By this reason, we go to the Lagrange coordinate from the Euler coordinate. Let $\vec{u}(x,t)$ be a velocity vector field in the Euler coordinate $x$ and let $x(\xi,t)$ be a solution of the Cauchy problem:

$$\frac{dx}{dt} = \vec{u}(x,t)\ \ (t>0), \quad x|_{t=0} = \xi = (\xi_1,\ldots,\xi_n).$$

If a velocity vector field $\vec{v}(\xi,t)$ is known as a function of the Lagrange coordinate $\xi$, then the connection between the Euler coordinate and the Lagrange coordinate is written in the form:

$$x = \xi + \int_0^t \vec{v}(\xi,\tau)\,d\tau = X_u(\xi,t).$$

Passing to the Lagrange coordinate in (1.1) and setting $\rho(X_u(\xi,t),t) = \bar\rho_0 + \theta(\xi,t)$, we have

$$\begin{aligned}&\theta_t + (\bar\rho_0+\theta)\mathrm{div}\,_{\vec{v}}\vec{v} = 0 && \text{in } \Omega\times(0,T),\\ &(\bar\rho_0+\theta)\vec{v}_t - \mu\Delta_{\vec{v}}\vec{v} - (\mu+\mu')\nabla_{\vec{v}}\mathrm{div}\,_{\vec{v}}\vec{v} && \\ &\quad + \nabla_{\vec{v}}[P(\bar\rho_0+\theta)] = \vec{g}(X_u(\xi,t),t) && \text{in } \Omega\times(0,T),\\ &\vec{v}|_{\partial\Omega} = 0, \quad (\theta,\vec{v})|_{t=0} = (\theta_0,\vec{u}_0). && \end{aligned} \tag{2.3}$$

Employing the argument in [16, appendix] to calculate the change of variables in (2.3), we see that the first two equations in (2.3) are rewritten in the form:

$$\begin{aligned}&\theta_t + \bar\rho_0\mathrm{div}\,\vec{v} + \bar\rho_0 V_1(\vec{v})\nabla\vec{v} + \theta(\mathrm{div}\,\vec{v} + V_1(\vec{v})\nabla\vec{v}) = 0,\\ &\bar\rho_0\vec{v}_t - \mu\Delta\vec{v} - (\mu+\mu')\nabla\mathrm{div}\,\vec{v} + P'(\bar\rho_0)\nabla\theta + \theta\vec{v}_t + V_2(\vec{v})\nabla^2\vec{v}\\ &\quad + (P'(\bar\rho_0+\theta) - P'(\bar\rho_0))\nabla\theta + P'(\bar\rho_0+\theta)V_3(\vec{v})\nabla\theta = \vec{g}(X_{\vec{v}}(\xi,t),t),\end{aligned} \tag{2.4}$$

where $V_i(\vec{v})$ $(i=1,2,3)$ have the forms: $V_i(\vec{v}) = W_i(\int_0^t \nabla\vec{v}(\xi,\tau)\,d\tau)$ with some polynomials $W_i(s)$ such that $W_i(0)=0$. Therefore, setting

$$\theta = (\bar\rho_0/\sqrt{P'(\bar\rho_0)})\tilde\theta, \quad \alpha = \mu/\bar\rho_0,\ \ \beta = (\mu+\mu')/\bar\rho_0,\ \ \gamma = \sqrt{P'(\bar\rho_0)}, \tag{2.5}$$

we have a quasi-linear system:

$$U_t - AU + Q(U) = F\ \ (t>0), \quad U|_{t=0} = U_0$$

with $U = (\tilde\theta,\vec{v})$, $U_0 = (\bar\rho_0/\sqrt{P'(\bar\rho_0)})(\rho_0-\bar\rho_0),\vec{u}_0)$. Applying Theorem 2.4 yields a local in time unique existence of solutions to (2.3). As was proved in Ströhmer [21], the map $x = X_u(\xi,t)$ is a diffeomorphism on $\Omega$, so that we have the following theorem.

**Theorem 2.5.** *Let $p$ and $q$ be indices such that $n < q < \infty$ and $2 < p < \infty$. Let $P(\rho)$ be a smooth function defined on $(\bar\rho_0/8, 8\bar\rho_0)$ and we assume that there exist positive constants $\rho_1$ and $\rho_2$ such that*

$$\rho_1 < P'(\rho) < \rho_2 \quad \text{for any } \rho \in (\bar\rho_0/8, 8\bar\rho_0). \tag{2.6}$$

*We assume that two viscosity constants $\mu$ and $\mu'$ satisfy the condition: $\mu > 0$ and $\mu + \mu' > 0$. Then, for any $M > 0$ there exists a time $T > 0$ such that for any initial data $(\rho_0 - \bar\rho_0, \vec u_0) \in E_{p,q}$ with $\|(\rho_0 - \bar\rho_0, \vec u_0)\|_{E_{p,q}} \le M$, the problem (1.1) with $\vec g = \vec 0$ admits a unique solution $(\theta, \vec v)$ such that*

$$\begin{aligned}&\theta \in W^1_p((0,T), W^1_q(\Omega)),\\ &\vec v \in W^1_q((0,T), L_q(\Omega)^2) \cap L_p((0,T), W^2_q(\Omega)^2).\end{aligned}$$

Now, we consider a global in time unique existence theorem of strong solutions to (1.1) when $\Omega$ is a $C^{1,1}$ bounded domain. To assure the uniqueness for $\lambda = 0$ in the resolvent problem, we have to assume that $\int_\Omega f\,dx = 0$ in the bounded domain case. Therefore, we consider the operator $A$ on $\dot W^{1,0}_q(\Omega)$ with $\dot W^{1,0}_q(\Omega) = \{(f,\vec g) \in W^{1,0}_q(\Omega) \mid \int_\Omega f\,dx = 0\}$ and its domain is now $\dot{\mathcal D}_q(A) = \mathcal D_q(A) \cap \dot W^{1,0}_q(\Omega)$. Applying the homotopic argument to the results due to [19], we see that the resolvent set $\rho(A)$ of $A$ contains a set $\Xi \cup \{\lambda \in \mathbb C \mid |\lambda| \le \lambda_1\}$ for some positive number $\lambda_1$, where the set $\Xi$ is defined by $\Xi = \cup_{0<\epsilon<\pi/2} \Xi_\epsilon$ and $\Xi_\epsilon$ is defined by

$$\begin{aligned}\Xi_\epsilon = {}& \{\lambda \in \mathbb C \setminus \{0\} \mid |\arg\lambda| < \pi - \epsilon\}\\ &\cap \left\{\lambda \in \mathbb C \mid \left(\operatorname{Re}\lambda + \frac{\gamma^2}{\alpha+\beta} + \epsilon\right)^2 + (\operatorname{Im}\lambda)^2 > \left(\frac{\gamma^2}{\alpha+\beta} + \epsilon\right)^2\right\}.\end{aligned} \tag{2.7}$$

Therefore, $\{T(t)\}_{t\ge0}$ decays exponentially, that is there exist positive constants $\sigma$ and $M$ such that

$$\begin{aligned}\sup_{t>0} e^{\sigma t}\|T(t)(f,\vec g)\|_{W^{1,0}_q(\Omega)} + \sup_{t>0} e^{\sigma t} t^{\frac12}\|\nabla P_v T(t)(f,\vec g)\|_{L_q(\Omega)}&\\ + \sup_{t>0} e^{\sigma t} t\|\nabla^2 P_v T(t)(f,\vec g)\|_{L_q(\Omega)} \le M\|(f,\vec g)\|_{W^{1,0}_q(\Omega)}&\end{aligned} \tag{2.8}$$

for any $(f,\vec g) \in \dot W^{1,0}_q(\Omega)$. Combining (2.8) and the localization technique due to [17], we have the following theorem.

**Theorem 2.6.** *Let $\Omega$ be a bounded $C^{1,1}$ domain. Let $1 < p, q < \infty$. Set $\dot E_{p,q} = [\dot W^{1,0}_q(\Omega), \dot{\mathcal D}_q(A)]_{1-1/p,p}$. Then, there exists a $\gamma > 0$ such that for any $(\rho_0, \vec u_0) \in \dot E_{p,q}$ and $(f, \vec g) \in L_{p,-\gamma}((0,\infty), W^{1,0}_q(\Omega))$ the problem (2.1) admits a unique solution $(\rho, \vec u)$ such that*

$$\begin{aligned}&\rho \in W^1_{p,-\gamma}((0,\infty), W^1_q(\Omega)),\\ &\vec u \in W^1_{p,-\gamma}((0,\infty), L_q(\Omega)^2) \cap L_{p,-\gamma}((0,\infty), W^2_q(\Omega)^2)\end{aligned}$$

*and there holds the estimate:*

$$\|e^{\gamma t}(\rho_t,\rho)\|_{(0,\infty,W^1_q(\Omega))} + \|e^{\gamma t}\vec{u}_t\|_{L_p((0,\infty),L_q(\Omega)^2)} + \|e^{\gamma t}\vec{u}\,\|_{L_p((0,\infty),W^2_q(\Omega)^2)} \\ \leq C_{p,q}(\|(\rho_0,\vec{u}_0)\|_{E_{p,q}} + \|e^{\gamma t}(f,\vec{g})\|_{L_p((0,\infty),W^{1,0}_q(\Omega))}).$$

Applying Theorem 2.6, we have the following global in time unique existence theorem for (1.1) with small initial data.

**Theorem 2.7.** *Let $\Omega$ be a bounded $C^{1,1}$ domain. Let $p$ and $q$ be indices such that $n<q<\infty$ and $2<p<\infty$. Let $P(\rho)$ be a smooth function defined on $(\bar\rho_0/8, 8\bar\rho_0)$ that satisfies the condition (2.6). We assume that two viscosity constants $\mu$ and $\mu'$ satisfy the condition: $\mu>0$ and $\mu+\mu'>0$. Then, there exists a small positive number $\epsilon$ such that for any initial data $(\rho_0-\bar\rho_0,\vec{u}_0)\in \dot{E}_{p,q}$ with $\|(\rho_0-\bar\rho_0,\vec{u}_0)\|_{\dot{E}_{p,q}}\leq\epsilon$, the problem (1.1) with $\vec{g}=\vec{0}$ admits a unique solution $(\theta,\vec{v})$ such that*

$$\theta\in W^1_p((0,\infty),W^1_q(\Omega)),$$
$$\vec{v}\in W^1_q((0,\infty),L_q(\Omega)^2)\cap L_p((0,\infty),W^2_q(\Omega)^2).$$

*Moreover, there exists a positive number $\gamma$ such that there holds the estimate:*

$$\|e^{\gamma t}(\theta_t,\theta)\|_{(0,\infty,W^1_q(\Omega))} + \|e^{\gamma t}\vec{v}_t\|_{L_p((0,\infty),L_q(\Omega)^2)} + \|e^{\gamma t}\vec{v}\,\|_{L_p((0,\infty),W^2_q(\Omega)^2)} \\ \leq C_{p,q}\|(\theta_0,\vec{u}_0)\|_{E_{p,q}}.$$

## 3. Some decay properties of the Stokes semigroup in a 2-dimensional exterior domain

In this section, we consider some decay properties of solutions to the linear equation (2.1) when $\Omega$ is a 2-dimensional exterior domain. If we consider the Stokes semigroup of incompressible viscous fluid flow in a 2-dimensional exterior domain, the boundedness of the semigroup in $L_p(\Omega)$ for any $1<p<\infty$ was proved by Borchers and Varnhorn [1]. And, the $L_p$-$L_q$ decay property was proved by Maremonti and Solonnikov [11] and Dan and Shibata [2] independently. In the compressible viscous fluid flow case, we have the following theorem.

**Theorem 3.1 ($L_p$-$L_q$ estimate).** *Let $\Omega$ be a 2-dimensional exterior domain of $C^{1,1}$ class and let $\{T(t)\}_{t\geq0}$ be the Stokes semigroup given in Theorem 2.3. Let $p$ and $q$ be indices such that $1\leq q\leq 2\leq p<\infty$. Then, for any $\mathbf{F}=(f,\vec{g})\in W^{1,0}_p(\Omega)\cap W^{0,0}_q(\Omega)$ and $t\geq1$ we have*

$$\|T(t)\mathbf{F}\|_{L_p(\Omega)}\leq Ct^{-\left(\frac{1}{q}-\frac{1}{p}\right)}(\|\mathbf{F}\|_{L_q(\Omega)}+\|\mathbf{F}\|_{W^{1,0}_p(\Omega)}) \qquad (1<q\leq2),$$
$$\|T(t)\mathbf{F}\|_{L_p(\Omega)}\leq Ct^{-\left(1-\frac{1}{p}\right)}(\log t)(\|\mathbf{F}\|_{L_1(\Omega)}+\|\mathbf{F}\|_{W^{1,0}_p(\Omega)}),$$
$$\|\nabla T(t)\mathbf{F}\|_{L_p(\Omega)}\leq Ct^{-\frac{1}{q}}(\|\mathbf{F}\|_{L_q(\Omega)}+\|\mathbf{F}\|_{W^{1,0}_p(\Omega)}) \qquad (2<p<\infty),$$
$$\|\nabla T(t)\mathbf{F}\|_{L_2(\Omega)}\leq Ct^{-\frac{1}{q}}(\log t)(\|\mathbf{F}\|_{L_q(\Omega)}+\|\mathbf{F}\|_{W^{1,0}_2(\Omega)}).$$

*Remark* 3.2. Combining Theorem 3.1 with $p = q = 2$ and Theorem 2.3, we have the boundedness of the semigroup, that is

$$\|T(t)(f,\vec{g})\|_{W_2^{1,0}(\Omega)} \leq M\|(f,\vec{g})\|_{W_2^{1,0}(\Omega)}$$

for any $t > 0$ and $(f,\vec{g}) \in W_2^{1,0}(\Omega)$. Unlike the incompressible viscous fluid flow case, it seems that we are unable to prove the boundedness of the semigroup except for $p = 2$, because of the hyperbolic-parabolic effect of the linear operator $A$ which is discussed a little bit more after the solution formula (3.2) to the Cauchy problem in $\mathbb{R}^2$ below.

To prove Theorem 3.1, the key step is to prove the following theorem.

**Theorem 3.3 (Local energy decay).** *Let $1 < p < \infty$ and let $b_0$ be a positive number such that $\Omega^c \subset B_{b_0}$, $B_L$ being the ball of radius $L$ with center at the origin in $\mathbb{R}^2$. For $b > b_0$, let $W_{p,b}^{1,0}(\Omega)$ denote a subset of $W_p^{1,0}(\Omega)$ defined by*

$$W_{p,b}^{1,0}(\Omega) = \{(f,\vec{g}) \in W_p^{1,0}(\Omega) \mid (f(x),\vec{g}(x)) \text{ vanishes for } |x| > b\}.$$

*Then, for any $b > b_0$ there exists a constant $C = C_{p,b}$ such that*

$$\|T(t)(f,\vec{g})\|_{W_p^{1,2}(\Omega_b)} \leq Ct^{-1}(\log t)^{-2}\|(f,\vec{g})\|_{W_p^{1,0}(\Omega)}$$

*for any $(f,\vec{g}) \in W_{p,b}^{1,0}(\Omega)$ and $t \geq 1$. Here, we have set $\Omega_b = \Omega \cap B_b$.*

To prove Theorem 3.1, we also need to know the $L_p$-$L_q$ decay estimate of solutions to the problem in $\mathbb{R}^2$:

$$\begin{cases} \theta_t + \gamma \operatorname{div} \vec{v} = 0 & \text{in } \mathbb{R}^2 \times (0,\infty), \\ \vec{v}_t - \alpha\Delta\vec{v} - \beta\nabla(\operatorname{div}\vec{v}) + \gamma\nabla\theta = \vec{0} & \text{in } \mathbb{R}^2 \times (0,\infty), \\ (\theta,\vec{v})|_{t=0} = (\theta_0,\vec{v}_0) = (\rho_0, v_{01}, v_{02}). \end{cases} \tag{3.1}$$

By taking the Fourier transform of (3.1) with respect to $x$ and solving the ordinary differential equation with respect to $t$, we have

$$\begin{aligned} \hat{\varrho}(\xi,t) &= -i\gamma\frac{e^{\lambda_+(\xi)t} - e^{\lambda_-(\xi)t}}{\lambda_+(\xi) - \lambda_-(\xi)}\sum_{j=1}^{2}\xi_j\hat{v}_{0j}(\xi) - \frac{\lambda_-(\xi)e^{\lambda_+(\xi)t} - \lambda_+(\xi)e^{\lambda_-(\xi)t}}{\lambda_+(\xi) \quad \lambda_-(\xi)}\hat{\varrho}_0(\xi), \\ \hat{v}_\ell(\xi,t) &= e^{-\alpha|\xi|^2t}\sum_{j=1}^{2}(\delta_{\ell j} - \xi_\ell\xi_j|\xi|^{-2})\hat{v}_{0j}(\xi) - i\gamma\frac{e^{\lambda_+(\xi)t} - e^{\lambda_-(\xi)t}}{\lambda_+(\xi) - \lambda_-(\xi)}\xi_\ell\hat{\varrho}_0(\xi) \\ &\quad - \frac{((\alpha+\beta)|\xi|^2 + \lambda_-(\xi))e^{\lambda_+(\xi)t} - ((\alpha+\beta)|\xi|^2 + \lambda_+(\xi))e^{\lambda_-(\xi)t}}{(\lambda_+(\xi) - \lambda_-(\xi))|\xi|^2}\sum_{j=1}^{2}\xi_\ell\xi_j\hat{v}_{0j}(\xi) \end{aligned} \tag{3.2}$$

where $\hat{\rho}(\xi,t)$ and $\hat{v}_\ell(\xi,t)$ ($\xi = (x_1,\xi_2) \in \mathbb{R}^2$) are the Fourier transform of $\rho(x,t)$ and $v_\ell(x,t)$ with respect to $x$ ($\vec{v} = (v_1,v_2)$), and

$$\lambda_\pm(\xi) = -\frac{\alpha+\beta}{2}|\xi|^2 \pm \sqrt{\left(\frac{\alpha+\beta}{2}\right)^2|\xi|^4 - \gamma^2|\xi|^2}.$$

To show a decay estimate of $(\varrho(x,t), \vec{v}\,(x,t))$, we divide the solution formulas of (3.2) into the low frequency part and the high frequency part and we use the following facts:

$$\lambda_+(\xi) = -\frac{2\gamma^2}{\alpha+\beta} + O(|\xi|^{-2}), \qquad \text{as } |\xi| \to \infty,$$

$$\lambda_-(\xi) = -(\alpha+\beta)|\xi|^2 + \frac{2\gamma^2}{\alpha+\beta} + O(|\xi|^{-2}) \quad \text{as } |\xi| \to \infty,$$

$$\lambda_\pm(\xi) = \pm i\gamma|\xi| - \frac{\alpha+\beta}{2}|\xi|^2 + O(|\xi|^3) \qquad \text{as } |\xi| \to 0.$$

Especially, the expansion formula for small $|\xi|$ shows the hyperbolic-parabolic coupled feature of solutions, that is the hyperbolic feature comes from $\pm i\gamma|\xi|$ and the parabolic one from $-2^{-1}(\alpha+\beta)|\xi|^2$. Employing the same argument as in the proof of Theorem 3.1 of Kobayashi-Shibata [10], we have the following theorem.

**Theorem 3.4.** *Let $(\varrho, \vec{v})$ be a vector of functions given in (3.2). Then, $(\varrho, \vec{v})$ solves the equation (3.1). Moreover, there exist $\varrho^0$, $\varrho^\infty$, $\vec{v}^{\,0}$ and $\vec{v}^{\,\infty}$ such that $\varrho = \varrho^0 + \varrho^\infty$, $\vec{v} = \vec{v}^0 + \vec{v}^\infty$ and the following estimates hold for non-negative integers $\ell$ and $m$:*
(i) *For all $t \geq 1$, there exists a $C = C(m, \ell, p, q) > 0$ such that*

$$\sum_{|\alpha|=\ell} \|\partial_t^m \partial_x^\alpha (\varrho^0(t), \vec{v}^0(t))\|_{L_p(\mathbb{R}^2)} \leq C t^{-\left(\frac{1}{q}-\frac{1}{p}\right)-\frac{m+\ell}{2}} \|(\varrho_0, \vec{v}_0)\|_{L_q(\mathbb{R}^2)},$$

*where $1 \leq q \leq 2 \leq p \leq \infty$.*
(ii) *Set $(k)^+ = k$ if $k \geq 0$ and $(k)^+ = 0$ if $k < 0$. Let $1 < p < \infty$. Then, there exist positive constants $C$ and $c$ such that for any $t \geq 1$*

$$\|\partial_t^m \nabla^\ell \varrho^\infty(t)\|_{L_p(\mathbb{R}^2)} \leq Ce^{-ct}\Big\{\|\nabla^{2m+\ell}\varrho_0\|_{L_p(\mathbb{R}^2)} + \|\nabla^{(2m+\ell-1)^+}\vec{v}_0\|_{L_p(\mathbb{R}^2)}\Big\},$$

$$\|\partial_t^m \nabla^\ell \vec{v}^\infty(t)\|_{L_p(\mathbb{R}^2)} \leq Ce^{-ct}\Big\{\|\nabla^{(2m+\ell-1)^+}\varrho_0\|_{L_p(\mathbb{R}^2)} + \|\nabla^{(2m+\ell-2)^+}\vec{v}_0\|_{L_p(\mathbb{R}^2)}\Big\}.$$

Combining the decay estimate near the boundary guaranteed by Theorem 3.3 and the decay estimate at far field guaranteed by Theorem 3.4 by a cut-off technique, we can prove Theorem 3.1. The argument is rather standard (cf. [10] and [2]), so that we would like to omit the detailed proof.

To prove Theorem 3.3, we use the strategy due to [2] and [8]. Our proof is very technical and rather long, so that we just give a sketch of our proof below and we would like to refer to [5] for the detailed proof. To prove Theorem 3.3, we consider the resolvent equation:

$$\lambda\rho + \gamma \operatorname{div}\vec{u} = f, \;\; \lambda\vec{u} - \alpha\Delta\vec{u} - \beta\nabla(\operatorname{div}\vec{u}) + \gamma\nabla\rho = \vec{g} \;\text{ in } \Omega, \;\; \vec{u}|_{\partial\Omega} = \vec{0}. \tag{3.3}$$

According to Shibata-Tanaka [19], we have a stronger result concerning the resolvent of $A$ as the admissible sectoriality of $A$ defined in Definition 2.2. In fact, let $\Xi$ and $\Xi_\epsilon$ be sets defined in (2.7). Then, the resolvent set $\rho(A)$ of $A$ contains $\Xi$ and for any $\epsilon$ $(0 < \epsilon < \pi/2)$ and $\lambda_0 > 0$, there exists a constant $C = C_{\epsilon,\lambda_0}$ such that there holds the resolvent estimate:

$$|\lambda|\|R(\lambda)(f,\vec{g})\|_{W_p^{1,0}(\Omega)} + |\lambda|^{\frac{1}{2}}\|\nabla P_v R(\lambda)(f,\vec{g})\|_{L_p(\Omega)} + \|\nabla^2 P_v R(\lambda)(f,\vec{g})\|_{L_p(\Omega)} \le C\|(f,\vec{g})\|_{W_p^{1,0}(\Omega)} \tag{3.4}$$

for any $(f,\vec{g}) \in W_p^{1,0}(\Omega)$ and $\lambda \in \Xi_\epsilon$ with $|\lambda| \ge \lambda_0$, where $R(\lambda) = (\lambda I - A)^{-1}$. Therefore, to prove Theorem 3.3 what we have to investigate is the behavior of $R(\lambda)$ near $\lambda = 0$. Since $\Omega$ is unbounded, $\lambda = 0$ is in the continuous spectrum of $A$, and therefore to investigate the asymptotic behavior of $R(\lambda)$ we have to shrink the domain of $A$ from $W_p^{1,0}(\Omega)$ to $W_{p,b}^{1,0}(\Omega)$ and widen the range of $A$ from $W_p^{1,2}(\Omega)$ to $W_p^{1,2}(\Omega_b)$. Namely, we use the topology of $\mathcal{L}(W_{p,b}^{1,0}(\Omega), W_p^{1,2}(\Omega_b))$ that is weaker than that of $\mathcal{L}(W_p^{1,0}(\Omega), W_p^{1,2}(\Omega))$. This is the reason why we are interested in the local energy estimate. The following lemma is the key in proving Theorem 3.3.

**Lemma 3.5.** *Let $1 < p < \infty$ and $0 < \varepsilon < \pi/2$. Then, there exist $\sigma > 0$ and $S(\lambda) \in \mathrm{Anal}\,(U^\sigma, \mathcal{L}(W_{p,b}^{1,0}(\Omega), W_p^{1,2}(\Omega_b)))$ such that*

$$S(\lambda)(f,g) = (\lambda I - A)^{-1}(f,g) \quad (f,g) \in W_{p,b}^{1,0}(\Omega) \quad (\lambda \in U^\sigma \cap \Xi),$$
$$S(\lambda) = S_0 + S_1(\log\lambda)^{-1} + \mathcal{O}\left((\log\lambda)^{-2}\right) \quad (\lambda \in U^\sigma),$$

*where $S_0, S_1 \in \mathcal{L}(W_{p,b}^{1,0}(\Omega), W_p^{1,2}(\Omega_b))$, and $U^\sigma = \{\lambda \in \mathbb{C}\backslash(-\infty,0] \mid |\lambda| < \sigma\}$.*

Once getting Lemma 3.5, we can prove Theorem 3.3 as follows. Let $\mathbf{F} = (f,\vec{g}) \in W_{p,b}^{1,0}(\Omega)$ and choose $\delta > 0$ so small that $\Gamma_1 = \cup_\pm\{\lambda \in \mathbb{C} \mid \arg\lambda = \pm((\pi/2)+\delta), |\lambda| \ge \sigma/2\} \subset \Xi$. If we set $\Gamma_2 = \Gamma_3 \cup \Gamma_4$ where

$\Gamma_3 = \cup_\pm \{\lambda = -(\sigma/2)\sin\delta \pm i\ell \mid 0 \le \ell \le (\sigma/2)\cos\delta\}$,

$\Gamma_4$ : a smooth loop jointing the points $\lambda = e^{i\pi}(\sigma/2)\sin\delta$ and $\lambda = e^{-i\pi}(\sigma/2)\sin\delta$ and going around the cut in $\mathbb{C}\backslash(-\infty,0]$,

then, by Lemma 3.5 the semigroup $\{T(t)\}_{t\ge0}$ is represented by

$$T(t)\mathbf{F} = \frac{1}{2\pi i}\int_{\Gamma_1} e^{\lambda t}(\lambda I - A)^{-1}\mathbf{F}\,d\lambda + \frac{1}{2\pi i}\int_{\Gamma_2} e^{\lambda t}S(\lambda)\mathbf{F}\,d\lambda.$$

By (3.4) and Lemma 3.5, the integrations over $\Gamma_1$ and $\Gamma_3$ decay exponentially and the integration over $\Gamma_4$ decays $t^{-1}(\log t)^{-2}$. This completes the proof of Theorem 3.3.

To prove Lemma 3.5, first we prove that the operator $S(\lambda)$ has the expansion formula:

$$S(\lambda)\mathbf{F} = \lambda^s(\log\lambda)^\tau(S_0\mathbf{F} + (\log\lambda)^{-1}S_1\mathbf{F} + \cdots) + \mathcal{O}(\lambda^{s+1}(\log\lambda)^\beta) \tag{3.5}$$

for $\lambda \in U^\sigma$ and $\mathbf{F} \in W_{p,b}^{1,0}(\Omega)$ with some integers $s$, $\tau$ and $\beta$. In fact, the inverse of $\lambda I$ + compact operator has such expansion formula in general, the idea of which goes back to Seeley [15] (and also Vainberg [24]). We used the Seeley idea to prove (3.5). Then, by the contradiction argument and the uniqueness of solution to (3.3), we can show that $s = 0$ and $\tau = 0$. In fact, we assume that there exists a $\mathbf{F} \in W^{1,0}_{p,b}(\Omega)$ such that $\mathbf{F} = (f, \vec{g}) \neq (0, \vec{0})$ and $S_0\mathbf{F} \neq (0, \vec{0})$. We plug the expansion formula (3.5) into (3.3). Then, if $s > 0$, letting $\lambda \to 0$, we see that $\mathbf{F} = (0, \vec{0})$. If $s < 0$, then we can construct a $(\rho, \vec{u}) \in W^{1,2}_{p,\mathrm{loc}}(\Omega)$ such that $(\rho, \vec{u})|_{\Omega_b} = S_0\mathbf{F}$, and $(\rho, \vec{u})$ satisfies the homogeneous equation:

$$\lambda\rho + \gamma \mathrm{div}\, \vec{u} = 0, \lambda\vec{u} - \alpha\Delta\vec{u} - \beta\nabla(\mathrm{div}\, \vec{u}) + \gamma\nabla\rho = \vec{0} \quad \text{in } \Omega, \quad \vec{u}|_{\partial\Omega} = 0,$$

and the radiation condition: $\rho(x) = O(|x|^{-1})$ and $\vec{u}(x) = O(1)$ as $|x| \to \infty$. But then, using the uniqueness theorem due to Dan-Shibata [2], we see that $(\rho, \vec{u}) = (0.\vec{0})$, which implies that $S_0\mathbf{F} = (0, \vec{0})$. This leads to a contradiction. Therefore, $s = 0$. Then, the same argument also implies that $\tau = 0$, which completes the proof of Lemma 3.5.

## 4. Global in time unique existence theorem of strong solution to (1.1) in a 2-dimensional exterior domain

In this section, we state our global in time unique existence theorem for the equation (1.1) in a 2-dimensional exterior domain[2]. We assume that the boundary of $\Omega$ is a compact $C^{4,1}$ hypersurface and that $\vec{g} = \vec{0}$ for simplicity. Following rather standard notation, we rewrite $W^{0,0}_p(\Omega) = L_p(\Omega)^3$, $W^\ell_2(\Omega) = H^\ell$, $W^\ell_2(\Omega)^2 = \mathbf{H}^\ell$, $W^{\ell,m}_2 = H^{\ell,m}$, $\|\cdot\|_{L_p(\Omega)} = \|\cdot\|_p$, $\|\cdot\|_{W^\ell_2(\Omega)} = \|\cdot\|_{H^\ell}$ and $\|\cdot\|_{W^{\ell,m}_2(\Omega)} = \|\cdot\|_{H^{\ell,m}}$ only in this section. To define $P(\rho_0)$ and $P(\rho)$, we assume that

$$\bar{\rho}_0/2 \le \rho_0(x) \le 2\bar{\rho}_0 \quad \text{for any } x \in \Omega, \tag{4.1}$$

$$\bar{\rho}_0/4 \le \rho(x,t) \le 4\bar{\rho}_0 \quad \text{for any } (x,t) \in \Omega \times (0,\infty). \tag{4.2}$$

Let $k$ stand for 1 or 2 and $X^k$ denote the class of our strong solutions to (1.1), which is defined by

$$\begin{aligned} X^k = \Big\{ &(\rho, \vec{u}) \mid \rho - \bar{\rho}_0 \in \bigcap_{j=0}^k C^j([0,\infty, H^{k+2-j}), \ \ \rho \text{ satisfies (4.2)}, \\ &\partial_\ell \rho \in L_2((0,\infty), H^{k+1}) \ (\ell = 1,2), \quad \partial_t^j \rho \in L_2((0,\infty), H^{k+2-j}) \ (j = 1,k), \\ &\vec{u} \in \bigcap_{j=0}^k C^j([0,\infty), \mathbf{H}^{k+2-2j}), \quad \partial_\ell \vec{u} \in L_2((0,\infty), \mathbf{H}^{k+2}) \ (\ell = 1,2), \\ &\partial_t^j \vec{u} \in L_2((0,\infty), \mathbf{H}^{k+3-2j}) \ (j = 1,k) \Big\}. \end{aligned}$$

And also, $N_k(t)$ denotes the norms of our solutions in $X^k$, which is defined by

$$\begin{aligned} N_1(t) = &\sup_{0<s<t} (\|(\rho(s) - \bar{\rho}_0, \vec{u}(s))\|_{H^3} + \|\partial_s(\rho(s), \vec{u}(s))\|_{H^{2,1}}) \\ &+ \Big(\int_0^t (\|\nabla(\rho(s), \vec{u}(s))\|^2_{H^{2,3}} + \|\partial_s(\rho(s), \vec{u}(s))\|^2_{H^2}\, ds\Big)^{\frac{1}{2}}, \end{aligned}$$

[2]The detailed proof is given in a forthcoming paper [7].

$$N_2(t) = \sup_{0<s<t} (\|(\rho(s) - \bar{\rho}_0, \vec{u}(t))\|_{H^4} + \|\partial_s(\rho(s), \vec{u}(s))\|_{H^{3,2}} + \|\partial_s^2(\rho(s), \vec{u}(s))\|_{H^{2,0}})$$
$$+ \Big(\int_0^t (\|\nabla(\rho(s), \vec{u}(s))\|_{H^{3,4}}^2 + \|\partial_s(\rho(s), \vec{u}(s))\|_{H^3}^2 + \|\partial_s^2(\rho(s), \vec{u}(s))\|_{H^{2,1}}^2 \, ds\Big)^{\frac{1}{2}}.$$

Let us define the compatibility condition on the initial data $(\rho_0, \vec{u}_0)$ that is a necessary condition for the existence of strong solutions in $X^k$. If the equation (1.1) admits a solution in $X^k$, then $\partial_t^j(\rho, \vec{u})|_{t=0}$ $(j = 1, k)$ are determined successively by the initial data $(\rho_0, \vec{u}_0)$ through the equation (1.1). We write $(\rho_j, \vec{u}_j) = \partial_t^j(\rho, \vec{u})|_{t=0}$ $(j = 1, k)$. In fact,

$$\rho_1 = -(\vec{u}_0 \cdot \nabla)\rho_0 - \rho_0 \operatorname{div} \vec{u}_0,$$
$$\vec{u}_1 = -(\vec{u}_0 \cdot \nabla)\vec{u}_0 + \frac{\mu}{\rho_0}\Delta\vec{u}_0 + \frac{\mu + \mu'}{\rho_0}\nabla \operatorname{div} \vec{u}_0 - \frac{\nabla P(\rho_0)}{\rho_0}.$$

To define $(\rho_2, \vec{u}_2)$, we differentiate the equations in (1.1) once with respect to t, put the resultant equations on $t = 0$ and use $(\rho_0, \vec{u}_0)$ and $(\rho_1, \vec{u}_1)$. On the other hand, the boundary condition: $\vec{u}(x, t)|_{\partial\Omega} = \vec{0}$ for any $t > 0$ requires that $\partial_t\vec{u}(x, t)|_{\partial\Omega} = 0$, because $\partial_t\vec{u}(x, t) \in \mathbf{H}^k$, so that the boundary trace $\partial_t\vec{u}(x, t)|_{\partial\Omega}$ exists. Combining this observation and the definition of $X^k$ implies that the initial data $(\rho_0, \vec{u}_0)$ should satisfy the following condition:

$$\begin{aligned} &\rho_0 - \bar{\rho}_0 \in H^{k+2}, \quad \rho_j \in H^{k+2-j} \ (j = 1, k), \\ &\vec{u}_j \in \mathbf{H}^{k+2-2j} \ (j = 0, 1, k), \quad \vec{u}_j|_{\partial\Omega} = \vec{0} \ (j = 0, 1). \end{aligned} \tag{4.3}$$

Note that when $k = 2$, $\partial_t^2\vec{u} \in L_2(\Omega)^2$, so that the boundary trace $\partial_t^2\vec{u}|_{\partial\Omega}$ does not necessary exit. Therefore, it is not necessary to assume that $\vec{u}_2|_{\partial\Omega} = 0$. The following theorem is our results concerning the global in time unique existence of strong solutions to (1.1) and their decay properties.

**Theorem 4.1.** *Let $k = 1$ or $2$. Let $P(\rho)$ be a smooth function defined on $(\bar{\rho}_0/8, 8\bar{\rho}_0)$ that satisfies the condition (2.6). We assume that two viscosity constants $\mu$ and $\mu'$ satisfy the condition: $\mu > 0$ and $\mu + \mu' \geq 0$. Then, there exists an $\epsilon > 0$ such that if $(\rho_0, \vec{u}_0)$ satisfies the conditions (4.1), (4.3) and $\|(\rho_0 - \bar{\rho}_0, \vec{u}_0)\|_{H^3} \leq \epsilon$, then the equation (1.1) with $\vec{g} = \vec{0}$ admits a global in time unique solution $(\rho, \vec{u}) \in X^k$ which satisfies the estimate*

$$N_k(t) \leq C\|(\rho - \bar{\rho}_0, \vec{u}_0)\|_{H^{k+2}} \tag{4.4}$$

*for any $t > 0$ with some constant $C$.*

*Moreover, if we assume that $(\rho_0 - \bar{\rho}_0, \vec{u}_0) \in L_1$ and that $\|(\rho_0 - \bar{\rho}_0, \vec{u}_0)\|_{H^4} \leq \epsilon$ additionally, then we have the following decay estimate:*

$$\begin{aligned} &\|(\rho(t) - \bar{\rho}_0, \vec{u}(t))\|_2 = O(t^{-\frac{1}{2}} \log t), \\ &\|\nabla(\rho(t), \vec{u}(t))\|_2 = O(t^{-1} \log t), \\ &\|\partial_t(\rho(t), \vec{u}(t))\|_{H^{2,1}} + \|(\nabla\rho(t), \nabla^2\vec{u}(t))\|_{H^1} = O(t^{-1}). \end{aligned} \tag{4.5}$$

In what follows, we give a sketch of our proof of Theorem 4.1. Let $(\cdot,\cdot)$ denote the $L_2$ inner-product on $\Omega$ and we use the abbreviation: $\|\cdot\|_2 = \|\cdot\|$. Instead of Gagliardo-Nirenberg-Sobolev inequality: $\|\vec u\|_6 \leq C\|\nabla \vec u\|$ in the 3-dimensional case, we use the Sobolev inequality:

$$\|\vec u\|_4 \leq C\|\vec u\|^{\frac12}\|\nabla\vec u\|^{\frac12} \quad \text{for } \vec u \in \mathbf{H}^1 \text{ with } \vec u|_{\partial\Omega} = \vec 0. \tag{4.6}$$

We know the existence of a local in time solutions to (1.1), so that we prove *a priori* estimates of such local in time solutions. Therefore, we assume that $(\rho, \vec u)$ is a solution to (1.1) and satisfies the range condition (4.2). Substituting $\rho = \bar\rho_0 + \theta$ into (1.1), we have

$$\theta_t + \operatorname{div}(\rho\vec u) = 0, \quad \rho(\vec u_t + \vec u\cdot\nabla\vec u) - \mu\Delta\vec u - (\mu+\mu')\nabla\operatorname{div}\vec u + P'(\rho)\nabla\theta = \vec 0. \tag{4.7}$$

Multiplying the first equation of (4.7) by $\theta$ and the second equation of (4.7) by $(\rho/P'(\rho))\vec u$, integrating the resulting formulas over $\Omega$ and summing two equalities up, by integration by parts we have

$$\begin{aligned}&\frac12\frac{d}{dt}\{\|\theta\|^2 + (\frac{\rho^2}{P'(\rho)}\vec u, \vec u)\} - \frac12\Big(\Big(\frac{\rho^2}{P'(\rho)}\Big)_t\vec u, \vec u) + (\frac{\rho^2}{P'(\rho)}\vec u\cdot\nabla\vec u, \vec u) + \mu(\frac{\rho}{P'(\rho)}\nabla\vec u, \nabla\vec u)\\ &\quad + (\mu+\mu')(\frac{\rho}{P'(\rho)}\operatorname{div}\vec u, \operatorname{div}\vec u) + (\nabla\Big(\frac{\rho}{P'(\rho)}\Big)\vec u, \nabla\vec u) + (\nabla\Big(\frac{\rho}{P'(\rho)}\Big)\cdot\vec u, \operatorname{div}\vec u).\end{aligned} \tag{4.8}$$

Using (4.6), (2.6) and (4.2), we proceed our estimates as follows:

$$\begin{aligned}&\Big(\frac{\rho^2}{P'(\rho)}\vec u, \vec u\Big) \geq \rho_1^{-1}(\bar\rho_0/4)^2\|\vec u\|^2,\\ &\Big|\Big(\Big(\frac{\rho^2}{P'(\rho)}\Big)_t\vec u, \vec u\Big)\Big| \leq C\|\theta_t\|\|\vec u\|_{L_4}^2 \leq C\|\vec u\|\|\theta_t\|\|\nabla\vec u\|,\\ &\Big|\Big(\frac{\rho^2}{P'(\rho)}\vec u\cdot\nabla\vec u, \vec u\Big)\Big| \leq C\|\vec u\|_{L_4}^2\|\nabla\vec u\| \leq C\|\vec u\|\|\nabla\vec u\|^2,\\ &\mu\Big(\frac{\rho}{P'(\rho)}\nabla\vec u, \nabla\vec u\Big) \geq \mu(\bar\rho_0/4)\rho_1^{-1}\|\nabla\vec u\|^2\\ &(\mu+\mu')\Big(\frac{\rho}{P'(\rho)}\operatorname{div}\vec u, \operatorname{div}\vec u\Big) \geq 0,\\ &\Big|\mu\Big(\nabla\Big(\frac{\rho}{P'(\rho)}\Big)\vec u, \nabla\vec u\Big)\Big| + \Big|(\mu+\mu')\Big(\nabla\Big(\frac{\rho}{P'(\rho)}\Big)\cdot\vec u, \nabla\vec u\Big)\Big|\\ &\quad \leq C|(|\nabla\rho||\vec u|, |\nabla\vec u|)| \leq C\|\vec u\|_\infty\|\nabla\theta\|\|\nabla\vec u\|.\end{aligned} \tag{4.9}$$

Integrating (4.8) on $(0,t)$ and using the estimates (4.9), we have

$$\begin{aligned}&\|(\theta(t), \vec u(t))\|^2 + \int_0^t \|\nabla\vec u(s)\|^2\,ds \leq C\{\|\rho_0 - \bar\rho_0, \vec u_0)\|^2\\ &\quad + \int_0^t (\|\vec u(s)\| + \|\vec u(s)\|_\infty)(\|\nabla\vec u(s)\|^2 + \|\nabla\theta(s)\|^2 + \|\theta_s(s)\|^2)\,ds\}.\end{aligned}$$

for some constant $C$. Differentiating the equations with respect to $t$ and using the multiplicative method as above, we also have the estimate for

$$\|\partial_t^j(\theta(t),\vec{u}(t))\|^2+\int_0^t\|\nabla\partial_s^j\vec{u}(s)\|^2\,ds.$$

Here, if necessary, we use the mollifier with respect to the time variable and the Friedrichs commutator lemma, so that we may assume that $(\theta,\vec{u})$ is smooth enough in $t$. To get the rest of required estimates of $\theta(t)$ and $\vec{u}(t)$ appearing in the definition of $N_k(t)$, we can copy the argument due to Matsumura and Nishida [13], which is very complicated to explain here, so that we refer to [13] for this argument. Finally, we arrive at the inequality:

$$N_3(t)\le C_1\|(\rho_0-\bar{\rho}_0,\vec{u}_0)\|_{H^3}+C_2N_3(t)^{\frac{3}{2}},\tag{4.10}$$

$$N_4(t)\le C_3\|(\rho_0-\bar{\rho}_0,\vec{u}_0)\|_{H^4}+C_4N_3(t)^{\frac{1}{2}}N_4(t)\tag{4.11}$$

for some positive constants $C_i$ $(i=1,2,3,4)$ as long as the solution $(\rho(t),\vec{u}(t))$ exists and satisfies suitable regularity condition and assumption (4.2). Let $\sigma>0$ be a small number such that $C_2\sigma^{\frac{1}{2}}\le 1/2$ and we choose $\epsilon>0$ so small that $C_1\|(\rho_0-\bar{\rho}_0,\vec{u}_0)\|_{H^3}\le C_1\epsilon<\sigma/4$. From (4.10) we have $N_3(t)\le\sigma/4+(1/2)N_3(t)$ whenever $N_3(t)\le\sigma$. Thus, assuming that $N_3(t)\le\sigma$, we have $N_3(t)\le\sigma/2$. If we choose $\epsilon>0$ small enough, then by the continuity of solutions up to $t=0$ we may assume that $N_3(t)\le\sigma$ in some short interval $(0,T_1)$. But then, $N_3(t)\le\sigma/2$, and therefore we can prolong solution to larger time interval $(0,T_2)$ with $N_3(t)\le\sigma$. Repeated use of this argument implies the global in time existence of strong solutions to (1.1) that possesses the estimate (4.4). If $(\rho-\bar{\rho}_0,\vec{u}_0)\in H^{4,4}$, then choosing $\sigma>0$ so small that $C_2\sigma^{\frac{1}{2}}\le 1/2$, by (4.11) we have (4.4) with $k=2$.

Since we already have (4.4) when $(\rho-\bar{\rho}_0,\vec{u}_0)\in H^{4,4}$, to prove (4.5) it suffices to consider the linear problem:

$$\begin{aligned}&\theta_t+\bar{\rho}_0\operatorname{div}\vec{u}=f && \text{in } \Omega\times(0,\infty),\\ &\bar{\rho}_0\vec{u}_t-\mu\Delta\vec{u}-(\mu+\mu')\nabla\operatorname{div}\vec{u}+P'(\bar{\rho}_0)\nabla\theta=\vec{g} && \text{in } \Omega\times(0,\infty),\\ &\vec{u}|_{\partial\Omega}=\vec{0},\quad (\theta,\vec{u})|_{t=0}=(\rho_0-\bar{\rho}_0,\vec{u}_0),\end{aligned}\tag{4.12}$$

where we have set $f=-\operatorname{div}(\theta\vec{u})$ and

$$\vec{g}=-\vec{u}\cdot\nabla\vec{u}+(\rho^{-1}-(\bar{\rho}_0)^{-1})(\mu\Delta\vec{u}+(\mu+\mu')\nabla\operatorname{div}\vec{u})-(P'(\rho)-P'(\bar{\rho}_0))\nabla\theta.$$

First we consider the decay estimate of the time derivatives of $(\theta,\vec{u})$. To this end, we differentiate the equations in (4.12) with respect to $t$ and we consider the equation:

$$\begin{aligned}&(t\theta_t)_t+\bar{\rho}_0\operatorname{div}(t\vec{u}_t)=f_t+\theta_t,\\ &\bar{\rho}_0(t\vec{u}_t)_t-\mu\Delta\vec{u}_t-(\mu+\mu')\nabla\operatorname{div}\vec{u}_t+P'(\bar{\rho}_0)\nabla(t\theta_t)=t\vec{g}_t+\vec{u}_t.\end{aligned}$$

in $\Omega\times(0,\infty)$ with boundary condition $t\vec{u}|_{\partial\Omega}=\vec{0}$ and zero initial data. Since we already know the estimates for $\theta_t$ and $\vec{u}_t$ on the right-hand side, employing the multiplicative technique and the Matsumura and Nishida argument, we have the estimate: $\|(\theta_t, \vec{u}_t)\|_{H^{2,1}} \leq Ct^{-1}$. And then, moving $\theta_t$ and $\vec{u}_t$ to the right-hand sides in (4.12) and using the elliptic estimate (so-called Cattabriga estimate after late Cattabriga), we have $\|(\nabla\theta(t), \nabla^2\vec{u}(t))\|_{H^1} \leq Ct^{-1}$. Finally, choosing $\tilde{\theta}$, $\alpha$, $\beta$ and $\gamma$ in the same way as in (2.5) and applying Theorem 3.1, we have

$$\|(\rho(t) - \bar{\rho}_0, \vec{u}(t))\| \leq Ct^{-\frac{1}{2}} \log t, \quad \|\nabla(\rho(t), \vec{u}(t))\| \leq Ct^{-1} \log t.$$

This ends a rough explanation of our approach to (1.1).

**Acknowledgment**

We are very grateful to the referee for valuable suggestions to the first draft.

## References

[1] W. Borchers and W. Varnhorn, *On the boundedness of the Stokes semigroup in two-dimensional exterior domains,* Math. Z., **213** (1993), 275–299.

[2] W. Dan and Y. Shibata, *On the $L_q$-$L_r$ estimates of the Stokes semigroup in a two-dimensional exterior domain,* J. Math. Soc. Japan, **51** (1999), no. 1, 181–207.

[3] K. Deckelnick, *Decay estimates for the compressible Navier-Stokes equations in unbounded domain*, Math. Z., **209** (1992), 115–130.

[4] K. Deckelnick, *$L^2$-decay for the compressible Navier-Stokes equations in unbounded domains*, Comm. Partial Differential Equations, **18** (1993), 1445–1476.

[5] Y. Enomoto and Y. Shibata, *On some decay properties of Stokes semigroup of compressible viscous fluid flow in a two-dimensional exterior domain*, preprint.

[6] Y. Enomoto, Y. Shibata and M. Suzuki, *On the maximal $L_p$-$L_q$ regularity of the Stokes semigroup of compressible viscous fluid flow and its application to a non-linear problem*, preprint.

[7] Y. Enomoto, Y. Shibata and M. Suzuki, *On the global in time unique existence theorem for the compressible viscous fluid flow in 2-dimensional exterior domains*, preprint.

[8] R. Kleinman and B. Vainberg, *Full low-frequency asymptotic expansion for second-order elliptic equations in two dimensions*, Math. Meth. Appl. Sci., **17** (1994), 989–1004.

[9] T. Kobayashi, *On a local energy decay of solutions for the equations of motion of compressible viscous and heat-conductive gases in an exterior domain in* $\mathbb{R}^3$, Tsukuba J. Math., **21** (1997), No. 3, 629–670.

[10] T. Kobayashi and Y. Shibata, *Decay estimates of solutions for the equations of motion of compressible viscous and heat-conductive gases in an exterior domain in* $\mathbb{R}^3$, Commun. Math. Phys., **200** (1999), 621–659.

[11] P. Maremonti and V.A. Solonnikov, *On nonstationary Stokes problem in exterior domain,* Ann. Scuola Norm. Sup. Pisa CI. Sci., **24** (1997) no. 4, 395–449.

[12] A. Matsumura and T. Nishida, *The initial value problem for the equations of motion of viscous and heat-conductive gases*, J. Math. Kyoto Univ., **20** (1980), 67–104.

[13] A. Matsumura and T. Nishida, *Initial boundary value problems for the equations of motion of compressible viscous and heat-conductive fluids*, Commun. Math. Phys., **89** (1983), 445–464.

[14] G. Ponce, *Global existence of small solutions to a class of nonlinear evolution equations*, Nonlinear Anal., **9** (1985), 339–418.

[15] R.T. Seeley, *Integral equations depending analytically on a parameter*, Indag. Math., **24** (1964), 434–443.

[16] Y. Shibata and S. Shimizu, *On a free boundary problem for the Navier-Stokes equations*, Diff. Int. Eqns., **20** (2007), 241–276.

[17] Y. Shibata and S. Shimizu, *On the $L_p$-$L_q$ maximal regularity of the Neumann problem for the Stokes equations in a bounded domain*, J. reine angew. Math., **615** (2008), 157–209 (CRELLE).

[18] Y. Shibata and S. Shimizu, *On the maximal $L_p$-$L_q$ regularity of the Stokes problem with first-order boundary condition: Model Problem*, to appear in J. Math. Soc. Japan.

[19] Y. Shibata and K. Tanaka, *On a resolvent problem for the linearized system from the dynamical system describing the compressible viscous fluid motion,* Math. Methods Appl. Sci., **27** (2004) no. 13, 1579–1606.

[20] G. Ströhmer, *About the resolvent of an operator from fluid dynamics,* Math. Z., **194** (1987), 183–191.

[21] G. Ströhmer, *About a certain class of parabolic-hyperbolic systems of differential equations*, Analysis, **9** (1989), 1–39.

[22] G. Ströhmer, *About compressible viscous fluid flow in a bounded region*, Pacific J. Math., **143** (1990) no. 2, 359–375.

[23] A. Tani, *On the first initial-boundary value problem of compressible viscous fluid motion*, Publ. RIMS Kyoto Univ., **13** (1977), 193–253.

[24] B. Vainberg, *Asymptotic Methods in Equations of Mathematical Physics, in Russian,* Moscow Univ. Press, 1982; Gordon and Breach Publishers, New York, London, Paris, Montreux, Tokyo, 1989; English translation.

[25] L. Weis, *Operator-valued Fourier multiplier theorems and maximal $L_p$-regularity,* Math. Ann., **319** (2001), 735–758.



Yuko Enomoto
Fukasaku 307
Minuma-ku
Saitama-shi
Saitama 337-8570, Japan
e-mail: `e-yuko@shibaura-it.ac.jp`

Yoshihiro Shibata
Ohkubo 3-4-1
Shinjuku-ku
Tokyo 169-8555, Japan
e-mail: `yshibata@waseda.jp`

Operator Theory:
Advances and Applications, Vol. 221, 323-372


# On Canonical Solutions of a Moment Problem for Rational Matrix-valued Functions

Bernd Fritzsche, Bernd Kirstein and Andreas Lasarow

**Abstract.** We discuss extremal solutions of a certain finite moment problem for rational matrix functions which satisfy an additional rank condition. We will see, among other things, that these solutions are molecular nonnegative Hermitian matrix-valued Borel measures on the unit circle and that these measures have a particular structure. We study the above-mentioned solutions in all generality, but later focus on the nondegenerate case. In this case, the family of these special solutions can be parametrized by the set of unitary matrices. This realization allows us to further examine the structure of these solutions. Here, the analysis of the structural properties relies, to a great extent, on the theory of orthogonal rational matrix functions on the unit circle.

**Mathematics Subject Classification (2000).** 30E05, 42C05, 44A60, 47A56.

**Keywords.** Nonnegative Hermitian matrix measures, matrix moment problem, orthogonal rational matrix functions, matricial Carathéodory functions.

## 0. Introduction

A moment problem for rational matrix-valued functions with finitely many given data, which we will call Problem (R), will serve as main focus (see Section 2, where Problem (R) is stated). This problem can be regarded as a generalization of the truncated trigonometric matrix moment problem. For an introduction to Problem (R) and related topics, we refer the reader to [23]–[25].

In many ways, we build on previous work in [30] and [31] (see also [40]), where we discussed extremal solutions of Problem (R) in the nondegenerate case. For instance, the considerations in [30] and [31] led to some entropy optimality and some maximality properties of right and left outer spectral factors with respect to the Löwner semiordering for Hermitian matrices in keeping with Arov and Kreĭn in [4] (see also [3, Chapter 10] as well as [19, Chapter 11]). However, we

The third author's research for this paper was supported by the German Research Foundation (Deutsche Forschungsgemeinschaft) on badge LA 1386/3–1.

now concern ourselves with another kind of extremal solutions for Problem (R), namely solutions which can be understood as having the highest degree of degeneracy. Because these solutions share many of the structural properties of particular solutions to the nondegenerate matricial Schur problem studied in [22, Section 5], we will call these extremal solutions the canonical solutions of Problem (R).

A matricial extension of the theory of orthogonal rational functions on the unit circle by Bultheel, González-Vera, Hendriksen, and Njåstad in [12] (see also [8]–[11]) drew our attention to Problem (R). Thus, the work here is connected to that of [26], [27], [32], and [39] as well.

The classical Gaussian quadrature formulas are exact on sets of polynomials and, in a sense, optimal (for a survey, see [33]). The Szegő quadrature formulas are the counterpart (on the unit circle) to the Gaussian formulas. This is underscored by the fact that these formulas are exact on sets of Laurent polynomials. Since Laurent polynomials are rational functions with poles at the origin and at infinity, it seems natural to next consider the more general case in which there are poles at several other fixed points in the extended complex plane. This leads to orthogonal (or, more precisely, para-orthogonal) rational functions with arbitrary, but fixed poles. For a discussion of the rational Szegő quadrature formulas from a numerical perspective, we refer the reader to [16] (see also [15] and [34]).

As explained in [8], para-orthogonal rational (complex-valued) functions can be used to obtain quadrature formulas on the unit circle. This is achieved in much the same way as in the classical polynomial case, covered in [37] by Jones, Njåstad, and Thron (see also Geronimus [35, Theorem 20.1]). The present work includes ideas which could lead to similar results for rational matrix functions. It should also be mentioned that [10] offers an alternate approach to obtaining Szegő quadrature formulas with rational (complex-valued) functions, using Hermite interpolation. Furthermore, [7] (see also [13]) demonstrates how, based on the matricial representation for orthogonal rational functions on the unit circle (recently established in [45]), the nodes and weights for rational quadrature formulas can be obtained. This method emphasizes the relationship between para-orthogonal rational functions and eigenvalue problems for special matrices.

In [23, Theorem 31], a particular solution of Problem (R) was built from given data. Using much the same idea, we construct the members of the family of extremal solutions of Problem (R), studied in [30] and [31]. We focus, mainly, on the nondegenerate case. In broad terms, this means that the given moment matrix $\mathbf{G}$ for Problem (R) must meet an additional regularity condition. We will specify this condition as part of the final two sections. Then it will become clear that the set of particular solutions we later discussed and the family of solutions introduced in [30] share many formal structural properties.

In [31] we verified, among other things, that the Riesz–Herglotz transform of a member of the family of extremal solutions is a rational matrix function which can be expressed in terms of orthogonal rational matrix functions. Here, too, there is a connection to our present work. For the nondegenerate case, the solutions we later discuss make up a family of solutions of Problem (R), where the elements of the associated family of Riesz–Herglotz transforms admit representations similar to the ones mentioned above. These representations differ in that strictly contractive matrices appear as parameters in the representations of [31], whereas those considered below have unitary matrices as parameters. That canonical solutions of Problem (R) can be characterized in this way is one of our main results.

For the special case of complex-valued functions, the family of extremal solutions studied in this paper consists of those elements which can be used to get rational Szegő quadrature formulas. For this case, these elements correspond in a certain way to molecular measures which are composed of a minimal number of Dirac measures (cf. Remark 2.15). However, we will see that the matrix case is somewhat more complicated. In particular, a canonical solution of Problem (R) in general cannot be characterized by the property that it admits a representation as a sum of Dirac measures with a minimal number of summands. The first main result of the paper clarifies this and points out an overall characterization (see Theorem 2.10). Furthermore, for the nondegenerate case, the associated weights of the mass points of canonical solutions of Problem (R) are not given by the values of related reproducing kernels (or of the Christoffel functions) in general, whereas the opposite is the case for complex-valued functions (compare, e.g., Theorem 4.3 and Remark 4.11). Nevertheless, we will present a connection between these weights and values of related reproducing kernels (see (4.5) and (4.11)). This is the most important result of this paper.

The basic approach, in the nondegenerate case, will rely heavily on the theory of orthogonal rational matrix functions with respect to nonnegative Hermitian matrix Borel measures on the unit circle. Part of the work here can thus be seen as a generalization of our previous work in [29, Section 9], where the Szegő theory of orthogonal matrix polynomials was used to describe extremal solutions of the matricial Carathéodory problem. There, the corresponding extremal properties within the solution set were related to the maximal weights assigned to points of the unit circle (see also Arov [2] and Sakhnovich [44]). Our objective was not only to obtain similar results (as in [29, Section 9] for the matricial Carathéodory problem) for Problem (R), but also to gain additional insight into the structure of particular solutions for maximal weights.

It seems, at this point, best to forgot a substantial discussion of the concrete extremal properties of maximal weights within the solution set of Problem (R) to allow for more detail elsewhere. We intend, however, to return to this topic at a later time. (In the classical case for complex-valued functions, the corresponding extremal problems can be dealt with via para-orthogonal rational functions.)

This paper is organized as follows. In anticipation of our main topic, we present some facts on canonical nonnegative definite sequences of matrices in Section 1. There, our main interest will be in characterizing what it means for a nonnegative definite sequence to be canonical (see Theorem 1.7 and Corollary 1.14). Section 2 begins with a quick review of the notation we will be using which, to a large extent, carries over from previous publications. Afterwards, we turn our attention to the main theme of this paper, namely towards canonical solutions of

Problem (R). Naturally, we begin by stating Problem (R) and then continue with basic observations on the canonical solutions associated with this problem. We will see that these solutions to Problem (R) are molecular matrix measures with a very specific structure (see Theorem 2.10 and Remark 2.14). In particular, the Fourier coefficient sequence for a canonical solution is always a canonical nonnegative definite sequences of matrices. The spectral measure belonging to a canonical nonnegative definite sequences of matrices is, conversely, a canonical solution of a suitably chosen Problem (R). We have kept Sections 1 and 2 more general, but in later sections, focus increasingly on the nondegenerate case. It is in this case that the canonical solutions of Problem (R) can be parametrized by the set of unitary matrices (of suitable size). In Section 3 we discuss two such parameterizations (see Theorems 3.6 and 3.10). Finally, in Section 4, Theorem 3.6 will allow us to further examine the structure of canonical solutions regarding Problem (R) in the nondegenerate case (see Theorem 4.3).

## 1. On canonical nonnegative definite sequences of matrices and corresponding matrix measures

First, we explain some general notation and terms which we use in this paper.

Let $\mathbb{N}_0$ and $\mathbb{N}$ be the set of all nonnegative integers and the set of all positive integers, respectively. For each $k \in \mathbb{N}_0$ and each $\tau \in \mathbb{N}_0$ or $\tau = +\infty$, let $\mathbb{N}_{k,\tau}$ be the set of all integers $n$ for which $k \leq n \leq \tau$ holds. Furthermore, let $\mathbb{D}$ and $\mathbb{T}$ be the unit disk and the unit circle of the complex plane $\mathbb{C}$, respectively, i.e., let $\mathbb{D} := \{w \in \mathbb{C} : |w| < 1\}$ and $\mathbb{T} := \{z \in \mathbb{C} : |z| = 1\}$. We will use the notation $\mathbb{C}_0$ for the extended complex plane $\mathbb{C} \cup \{\infty\}$.

Throughout this paper, let $p, q \in \mathbb{N}$. If $\mathfrak{X}$ is a nonempty set, then $\mathfrak{X}^{p\times q}$ stands for the set of all $p \times q$ matrices each entry of which belongs to $\mathfrak{X}$. If $\mathbf{A}$ belongs to $\mathbb{C}^{p\times q}$, then the null space (resp., the range) of a $\mathbf{A}$ will be denoted by $\mathcal{N}(\mathbf{A})$ (resp., $\mathcal{R}(\mathbf{A})$), the notation $\mathbf{A}^*$ means the adjoint matrix of $\mathbf{A}$, and $\mathbf{A}^+$ stands for the Moore–Penrose inverse of $\mathbf{A}$. For the null matrix which belongs to $\mathbb{C}^{p\times q}$ we will write $0_{p\times q}$. The identity matrix which belongs to $\mathbb{C}^{q\times q}$ will be denoted by $\mathbf{I}_q$. If $\mathbf{A} \in \mathbb{C}^{q\times q}$, then $\det \mathbf{A}$ is the determinant of $\mathbf{A}$ and $\operatorname{Re} \mathbf{A}$ (resp., $\operatorname{Im} \mathbf{A}$) stands for the real (resp., imaginary) part of $\mathbf{A}$, i.e., $\operatorname{Re} \mathbf{A} := \frac{1}{2}(\mathbf{A} + \mathbf{A}^*)$ (resp., $\operatorname{Im} \mathbf{A} := \frac{1}{2\mathrm{i}}(\mathbf{A} - \mathbf{A}^*)$). We will write $\mathbf{A} \geq \mathbf{B}$ (resp., $\mathbf{A} > \mathbf{B}$) when $\mathbf{A}$ and $\mathbf{B}$ are Hermitian matrices (in particular, square and of the same size) such that $\mathbf{A}-\mathbf{B}$ is a nonnegative (resp., positive) Hermitian matrix. Recall that a complex $p \times q$ matrix $\mathbf{A}$ is *contractive* (resp., *strictly contractive*) when $\mathbf{I}_q \geq \mathbf{A}^*\mathbf{A}$ (resp., $\mathbf{I}_q > \mathbf{A}^*\mathbf{A}$). If $\mathbf{A}$ is a nonnegative Hermitian matrix, then $\sqrt{\mathbf{A}}$ stands for the (unique) nonnegative Hermitian matrix $\mathbf{B}$ given by $\mathbf{B}^2 = \mathbf{A}$.

Now, we turn our attention to the main definition of this section. If $\tau \in \mathbb{N}_0$ and if $(\mathbf{c}_k)_{k=0}^{\tau}$ is a sequence of complex $q \times q$ matrices, then $(\mathbf{c}_k)_{k=0}^{\tau}$ is called *nonnegative definite* (resp., *positive definite*) when the block Toeplitz matrix

$$\mathbf{T}_\tau := \begin{pmatrix} \mathbf{c}_0 & \mathbf{c}_1^* & \dots & \mathbf{c}_\tau^* \\ \mathbf{c}_1 & \mathbf{c}_0 & \ddots & \vdots \\ \vdots & \ddots & \ddots & \mathbf{c}_1^* \\ \mathbf{c}_\tau & \dots & \mathbf{c}_1 & \mathbf{c}_0 \end{pmatrix} \tag{1.1}$$

is nonnegative (resp., positive) Hermitian. A sequence $(\mathbf{c}_k)_{k=0}^\infty$ of complex $q \times q$ matrices is *nonnegative definite* (resp., *positive definite*) if, for all $\tau \in \mathbb{N}_0$, the sequence $(\mathbf{c}_k)_{k=0}^\tau$ is nonnegative definite (resp., positive definite).

Suppose that $\tau \in \mathbb{N}$ or $\tau = +\infty$ and let $n \in \mathbb{N}_{1,\tau}$. A nonnegative definite sequence $(\mathbf{c}_k)_{k=0}^\tau$ of complex $q \times q$ matrices is called *canonical of order* $n$ when

$$\operatorname{rank} \mathbf{T}_n = \operatorname{rank} \mathbf{T}_{n-1}, \tag{1.2}$$

where $\mathbf{T}_n$ and $\mathbf{T}_{n-1}$ are the block Toeplitz matrices given via (1.1). Note that the size of $\mathbf{T}_n$ is $(n+1)q \times (n+1)q$, whereas the size of $\mathbf{T}_{n-1}$ is $nq \times nq$.

*Example* 1.1. Let $r \in \mathbb{N}$ and let $z_1, z_2, \dots, z_r \in \mathbb{T}$ be pairwise different. Furthermore, let $(\mathbf{A}_j)_{j=1}^r$ be a sequence of nonnegative Hermitian $q \times q$ matrices and let

$$\mathbf{c}_k := \sum_{s=1}^{r} z_s^{-k} \mathbf{A}_s$$

for each $k \in \mathbb{N}_0$. By [29, Remarks 9.4 and 9.5] it is proven that the sequence $(\mathbf{c}_k)_{k=0}^\infty$ of complex $q \times q$ matrices is nonnegative definite and that

$$\operatorname{rank} \mathbf{T}_m = \sum_{s=1}^{r} \operatorname{rank} \mathbf{A}_s$$

for each $m \in \mathbb{N}_0$ with $m \geq r-1$. In particular, for each $n \in \mathbb{N}$ with $n \geq r$, the nonnegative definite sequence $(\mathbf{c}_k)_{k=0}^\infty$ is canonical of order $n$.

*Remark* 1.2. Let $(\mathbf{c}_k)_{k=0}^\tau$ be (with $\tau \in \mathbb{N}$ or $\tau = +\infty$) a nonnegative definite sequence of complex $q \times q$ matrices and let $n \in \mathbb{N}_{1,\tau}$. By (1.1) and some well-known results on nonnegative definite sequence (use, e.g., [18, Remark 3.4.3] along with [18, Lemma 1.1.7]) one can see that the following statements are equivalent:

(i) The sequence $(\mathbf{c}_k)_{k=0}^\tau$ is canonical of order $n$.
(ii) For each $m \in \mathbb{N}_{n,\tau}$, the sequence $(\mathbf{c}_k)_{k=0}^\tau$ is canonical of order $m$.
(iii) For each $m \in \mathbb{N}_{n,\tau}$, the equality $\operatorname{rank} \mathbf{T}_m = \operatorname{rank} \mathbf{T}_{n-1}$ holds.

The main result in this section will be a characterization of the property that a nonnegative definite sequence $(\mathbf{c}_k)_{k=0}^\tau$ is canonical of order $n$ in terms of $\mathbf{c}_n$. As an intermediate result we will first prove that any finite nonnegative definite sequence can be extended to such a sequence which is canonical. In doing so, we will use some formulas on the extension of nonnegative definite sequences stated in [18, Section 3.4]. Furthermore, the following lemma will be useful in the proofs.

**Lemma 1.3.** *Let* $\mathbf{K}$, $\mathbf{L}$, *and* $\mathbf{R}$ *be complex* $q \times q$ *matrices. Then:*

(a) *The equality* $\mathbf{K}\mathbf{K}^* = \mathbf{L}\mathbf{L}^+$ (*resp.,* $\mathbf{K}^*\mathbf{K} = \mathbf{R}\mathbf{R}^+$) *holds if and only if there is a unitary* $q \times q$ *matrix* $\mathbf{U}$ (*resp.,* $\mathbf{V}$) *such that* $\mathbf{K} = \mathbf{L}\mathbf{L}^+\mathbf{U}$ (*resp.,* $\mathbf{K} = \mathbf{V}\mathbf{R}\mathbf{R}^+$).

(b) *Suppose that the identities* $\mathbf{K}\mathbf{K}^* = \mathbf{L}\mathbf{L}^+$ *and* $\mathbf{K}^*\mathbf{K} = \mathbf{R}\mathbf{R}^+$ *hold and let* $\mathbf{W}$ *be a unitary* $q \times q$ *matrix. Then the following statements are equivalent:*
  (i) $\mathbf{K} = \mathbf{L}\mathbf{L}^+\mathbf{W}$.
  (ii) $\mathbf{K} = \mathbf{W}\mathbf{R}\mathbf{R}^+$.

*Moreover, if* (i) *is satisfied, then* $\mathbf{K} = \mathbf{L}\mathbf{L}^+\mathbf{W}\mathbf{R}\mathbf{R}^+$ *and* $\mathbf{L}\mathbf{L}^+ = \mathbf{W}\mathbf{R}\mathbf{R}^+\mathbf{W}^*$. *In particular,* (i) *implies* $\mathbf{K} = \mathbf{K}\mathbf{K}^*\mathbf{K}$ *and* $\mathbf{K} = \mathbf{L}\mathbf{L}^+\mathbf{K}\mathbf{R}\mathbf{R}^+$.

(c) *If the equations* $\mathbf{K} = \mathbf{L}\mathbf{L}^+\mathbf{W}\mathbf{R}\mathbf{R}^+$ *and* $\mathbf{L}\mathbf{L}^+ = \mathbf{W}\mathbf{R}\mathbf{R}^+\mathbf{W}^*$ *are fulfilled for some unitary* $q \times q$ *matrix* $\mathbf{W}$, *then* $\mathbf{K}\mathbf{K}^* = \mathbf{L}\mathbf{L}^+$ *and* $\mathbf{K}^*\mathbf{K} = \mathbf{R}\mathbf{R}^+$ *hold.*

(d) *The rank identity* $\operatorname{rank}\mathbf{L} = \operatorname{rank}\mathbf{R}$ *holds if and only if there is a unitary* $q \times q$ *matrix* $\mathbf{W}$ *such that* $\mathbf{L}\mathbf{L}^+ = \mathbf{W}\mathbf{R}\mathbf{R}^+\mathbf{W}^*$.

*Proof.* (a) Obviously, if there exists a unitary $q \times q$ matrix $\mathbf{U}$ such that the identity $\mathbf{K} = \mathbf{L}\mathbf{L}^+\mathbf{U}$ holds, then we get

$$\mathbf{K}\mathbf{K}^* = \mathbf{L}\mathbf{L}^+\mathbf{U}\mathbf{U}^*(\mathbf{L}\mathbf{L}^+)^* = \mathbf{L}\mathbf{L}^+\mathbf{L}\mathbf{L}^+ = \mathbf{L}\mathbf{L}^+.$$

Conversely, we suppose now that the equality $\mathbf{K}\mathbf{K}^* = \mathbf{L}\mathbf{L}^+$ is fulfilled. Then the polar decomposition of complex matrices implies the existence of a unitary $q \times q$ matrix $\mathbf{U}$ such that $\mathbf{K} = \sqrt{\mathbf{K}\mathbf{K}^*}\mathbf{U}$ holds, where (note [18, Theorem 1.1.1])

$$\sqrt{\mathbf{K}\mathbf{K}^*} = \sqrt{\mathbf{L}\mathbf{L}^+} = \mathbf{L}\mathbf{L}^+.$$

Therefore, it follows that $\mathbf{K} = \mathbf{L}\mathbf{L}^+\mathbf{U}$. Similarly, one can verify that $\mathbf{K}^*\mathbf{K} = \mathbf{R}\mathbf{R}^+$ holds if and only if there is a unitary $q \times q$ matrix $\mathbf{V}$ such that $\mathbf{K} = \mathbf{V}\mathbf{R}\mathbf{R}^+$.
(b) Let (i) be fulfilled. Therefore, we have

$$\mathbf{K}^* = \mathbf{W}^*(\mathbf{L}\mathbf{L}^+)^* = \mathbf{W}^*\mathbf{L}\mathbf{L}^+.$$

Consequently, from $\mathbf{K}^*\mathbf{K} = \mathbf{R}\mathbf{R}^+$ it follows that

$$\mathbf{R}\mathbf{R}^+ = \mathbf{K}^*\mathbf{K} = \mathbf{W}^*\mathbf{L}\mathbf{L}^+\mathbf{L}\mathbf{L}^+\mathbf{W} = \mathbf{W}^*\mathbf{L}\mathbf{L}^+\mathbf{W},$$

i.e., that $\mathbf{L}\mathbf{L}^+ = \mathbf{W}\mathbf{R}\mathbf{R}^+\mathbf{W}^*$ holds. Furthermore, this along with (i) yields

$$\mathbf{W}\mathbf{R}\mathbf{R}^+ = \mathbf{W}\mathbf{W}^*\mathbf{L}\mathbf{L}^+\mathbf{W} = \mathbf{L}\mathbf{L}^+\mathbf{W} = \mathbf{K}.$$

Hence, (i) leads to (ii). A similar argumentation shows that (ii) implies (i). Thus, (i) is equivalent to (ii). Taking this into account, from (i) one can conclude as an intermediate step that $\mathbf{K} = \mathbf{L}\mathbf{L}^+\mathbf{K}$ and $\mathbf{K} = \mathbf{K}\mathbf{R}\mathbf{R}^+$. So, in view of (ii) and $\mathbf{K}\mathbf{K}^* = \mathbf{L}\mathbf{L}^+$, one can finally see that

$$\mathbf{K} = \mathbf{L}\mathbf{L}^+\mathbf{K} = \mathbf{L}\mathbf{L}^+\mathbf{W}\mathbf{R}\mathbf{R}^+, \qquad \mathbf{K} = \mathbf{L}\mathbf{L}^+\mathbf{K} = \mathbf{K}\mathbf{K}^*\mathbf{K},$$

and

$$\mathbf{K} = \mathbf{L}\mathbf{L}^+\mathbf{K} = \mathbf{L}\mathbf{L}^+\mathbf{K}\mathbf{R}\mathbf{R}^+.$$

(c) Suppose that $\mathbf{K} = \mathbf{L}\mathbf{L}^+\mathbf{W}\mathbf{R}\mathbf{R}^+$ and $\mathbf{L}\mathbf{L}^+ = \mathbf{W}\mathbf{R}\mathbf{R}^+\mathbf{W}^*$ are fulfilled for some unitary $q \times q$ matrix $\mathbf{W}$. Then we obtain

$$\mathbf{K}\mathbf{K}^* = \mathbf{L}\mathbf{L}^+\mathbf{W}\mathbf{R}\mathbf{R}^+(\mathbf{R}\mathbf{R}^+)^*\mathbf{W}^*(\mathbf{L}\mathbf{L}^+)^* = \mathbf{L}\mathbf{L}^+\mathbf{W}\mathbf{R}\mathbf{R}^+\mathbf{W}^*\mathbf{L}\mathbf{L}^+ = \mathbf{L}\mathbf{L}^+$$

and analogously $\mathbf{K}^*\mathbf{K} = \mathbf{R}\mathbf{R}^+$.

(d) The assertion of part (d) is an immediate consequence of the rank identities $\operatorname{rank}\mathbf{L} = \operatorname{rank}\mathbf{L}\mathbf{L}^+$ and $\operatorname{rank}\mathbf{R} = \operatorname{rank}\mathbf{R}\mathbf{R}^+$ in combination with the fact that $\mathbf{L}\mathbf{L}^+$ and $\mathbf{R}\mathbf{R}^+$ are nonnegative Hermitian $q \times q$ matrices, where only 0 and 1 can appear as eigenvalues. □

In the following, with some $n \in \mathbb{N}_0$, let $(\mathbf{c}_k)_{k=0}^n$ be a sequence of complex $q \times q$ matrices which is nonnegative definite. Using $(\mathbf{c}_k)_{k=0}^n$, then we set

$$\mathbf{L}_{n+1} := \begin{cases} \mathbf{c}_0 & \text{if } n = 0 \\ \mathbf{c}_0 - \mathbf{Z}_n \mathbf{T}_{n-1}^+ \mathbf{Z}_n^* & \text{if } n \geq 1, \end{cases} \quad \mathbf{R}_{n+1} := \begin{cases} \mathbf{c}_0 & \text{if } n = 0 \\ \mathbf{c}_0 - \mathbf{Y}_n^* \mathbf{T}_{n-1}^+ \mathbf{Y}_n & \text{if } n \geq 1, \end{cases} \tag{1.3}$$

and

$$\mathbf{M}_{n+1} := \begin{cases} 0_{q\times q} & \text{if } n = 0 \\ \mathbf{Z}_n \mathbf{T}_{n-1}^+ \mathbf{Y}_n & \text{if } n \geq 1, \end{cases} \tag{1.4}$$

where (if $n \geq 1$) the matrix $\mathbf{T}_{n-1}$ is given via (1.1) with $\tau = n - 1$ and where

$$\mathbf{Y}_n := \begin{pmatrix} \mathbf{c}_1 \\ \mathbf{c}_2 \\ \vdots \\ \mathbf{c}_n \end{pmatrix} \quad \text{and} \quad \mathbf{Z}_n := \left(\mathbf{c}_n, \mathbf{c}_{n-1}, \ldots, \mathbf{c}_1\right).$$

Since $(\mathbf{c}_k)_{k=0}^n$ is nonnegative definite, in view of (1.3) and [18, Lemma 1.1.9] one can see that $\mathbf{L}_{n+1} \geq 0_{q\times q}$ and $\mathbf{R}_{n+1} \geq 0_{q\times q}$. Furthermore, it is well known that the extended sequence $(\mathbf{c}_k)_{k=0}^{n+1}$ with some $\mathbf{c}_{n+1} \in \mathbb{C}^{q\times q}$ is also nonnegative definite if and only if there is a contractive $q \times q$ matrix $\mathbf{K}$ such that the representation $\mathbf{c}_{n+1} = \mathbf{M}_{n+1} + \sqrt{\mathbf{L}_{n+1}}\mathbf{K}\sqrt{\mathbf{R}_{n+1}}$ is satisfied (see, e.g., [18, Theorem 3.4.1]).

Later on (like in [18]), we use the notation $\mathfrak{K}(\mathbf{M}; \mathbf{A}, \mathbf{B})$ with certain complex $q \times q$ matrices $\mathbf{M}$, $\mathbf{A}$, and $\mathbf{B}$ for the set of all complex $q \times q$ matrices $\mathbf{X}$ fulfilling $\mathbf{X} = \mathbf{M} + \mathbf{A}\mathbf{K}\mathbf{B}$ with some contractive $q \times q$ matrix $\mathbf{K}$. In other words, $\mathfrak{K}(\mathbf{M}; \mathbf{A}, \mathbf{B})$ is the *matrix ball* with center $\mathbf{M}$, left semi-radius $\mathbf{A}$, and right semi-radius $\mathbf{B}$. Furthermore (taking the elementary case $q = 1$ into account), we call the set of all complex $q \times q$ matrices $\mathbf{Y}$ fulfilling the equality $\mathbf{Y} = \mathbf{M} + \mathbf{A}\mathbf{U}\mathbf{B}$ with some unitary $q \times q$ matrix $\mathbf{U}$ the *boundary* of the matrix ball $\mathfrak{K}(\mathbf{M}; \mathbf{A}, \mathbf{B})$.

*Remark* 1.4. Let $n \in \mathbb{N}$ and suppose that $(\mathbf{c}_k)_{k=0}^n$ is a sequence of complex $q \times q$ matrices which is nonnegative definite. Furthermore, let $\mathbf{L}_{n+1}$ and $\mathbf{R}_{n+1}$ be the complex $q \times q$ matrices defined as in (1.3). Based on [18, Lemmas 1.1.7 and 1.1.9] one can see that the following statements are equivalent:

(i) The nonnegative definite sequence $(\mathbf{c}_k)_{k=0}^n$ is canonical of order $n$.
(ii) $\mathbf{L}_{n+1} = 0_{q\times q}$.
(iii) $\mathbf{R}_{n+1} = 0_{q\times q}$.

The following result implies that each finite nonnegative definite sequence of matrices admits a canonical extension.

**Proposition 1.5.** *Let $n \in \mathbb{N}_0$ and let $(\mathbf{c}_k)_{k=0}^n$ be a sequence of complex $q \times q$ matrices which is nonnegative definite. Let $\mathbf{L}_{n+1}$ and $\mathbf{R}_{n+1}$ be given by* (1.3)*. Then:*

(a) *There is a unitary $q \times q$ matrix $\mathbf{W}_{n+1}$ such that*

$$\mathbf{L}_{n+1}\mathbf{L}_{n+1}^{+} = \mathbf{W}_{n+1}\mathbf{R}_{n+1}\mathbf{R}_{n+1}^{+}\mathbf{W}_{n+1}^{*}. \tag{1.5}$$

(b) *Let $\mathbf{M}_{n+1}$ be defined as in* (1.4) *and let $\mathbf{W}_{n+1}$ be a unitary $q \times q$ matrix such that* (1.5) *holds. Then, by setting*

$$\mathbf{c}_{n+1} := \mathbf{M}_{n+1} + \sqrt{\mathbf{L}_{n+1}}\mathbf{W}_{n+1}\sqrt{\mathbf{R}_{n+1}}, \tag{1.6}$$

*$(\mathbf{c}_k)_{k=0}^{n+1}$ is a nonnegative definite sequence which is canonical of order $n+1$.*

(c) *Suppose that $\mathbf{c}_{n+1} \in \mathbb{C}^{q\times q}$ such that $(\mathbf{c}_k)_{k=0}^{n+1}$ forms a nonnegative definite sequence which is canonical of order $n+1$. Furthermore, let $\mathbf{M}_{n+2}$ be defined as in* (1.4) *with $(\mathbf{c}_k)_{k=0}^{n+1}$ instead of $(\mathbf{c}_k)_{k=0}^{n}$. For a given $\mathbf{c}_{n+2} \in \mathbb{C}^{q\times q}$, then $(\mathbf{c}_k)_{k=0}^{n+2}$ is a nonnegative definite sequence if and only if $\mathbf{c}_{n+2} = \mathbf{M}_{n+2}$.*

*Proof.* (a) Since $(\mathbf{c}_k)_{k=0}^n$ is nonnegative definite, the matrix $\mathbf{T}_n$ defined as in (1.1) with $\tau = n$ is nonnegative Hermitian. Consequently, in view of (1.1) and [18, Lemmas 1.1.7 and 1.1.9] one can see that the identity

$$\operatorname{rank}\mathbf{L}_{n+1} = \operatorname{rank}\mathbf{R}_{n+1}$$

holds. Thus, by using part (d) of Lemma 1.3 one can realize that there is a unitary $q \times q$ matrix $\mathbf{W}_{n+1}$ fulfilling (1.5).
(b) Let $\mathbf{c}_{n+1}$ be defined as in (1.6) and let

$$\mathbf{K}_{n+1} := \mathbf{L}_{n+1}\mathbf{L}_{n+1}^{+}\mathbf{W}_{n+1}\mathbf{R}_{n+1}\mathbf{R}_{n+1}^{+}.$$

Recalling that $\mathbf{W}_{n+1}$ is a unitary $q \times q$ matrix, from (1.6) and [18, Theorem 3.4.1] it follows that the extended sequence $(\mathbf{c}_k)_{k=0}^{n+1}$ is nonnegative definite as well, where

$$\begin{aligned} \mathbf{c}_{n+1} &= \mathbf{M}_{n+1} + \sqrt{\mathbf{L}_{n+1}}\mathbf{W}_{n+1}\sqrt{\mathbf{R}_{n+1}} \\ &= \mathbf{M}_{n+1} + \sqrt{\mathbf{L}_{n+1}}\mathbf{L}_{n+1}\mathbf{L}_{n+1}^{+}\mathbf{W}_{n+1}\mathbf{R}_{n+1}\mathbf{R}_{n+1}^{+}\sqrt{\mathbf{R}_{n+1}} \\ &= \mathbf{M}_{n+1} + \sqrt{\mathbf{L}_{n+1}}\mathbf{K}_{n+1}\sqrt{\mathbf{R}_{n+1}} \end{aligned} \tag{1.7}$$

(note [18, Lemma 1.1.6]). Furthermore, because of part (c) of Lemma 1.3 we get $\mathbf{K}_{n+1}\mathbf{K}_{n+1}^{*} = \mathbf{L}_{n+1}\mathbf{L}_{n+1}^{+}$ (and $\mathbf{K}_{n+1}^{*}\mathbf{K}_{n+1} = \mathbf{R}_{n+1}\mathbf{R}_{n+1}^{+}$). Hence (note that the matrix $\mathbf{T}_{n+1}$ defined as in (1.1) with $\tau = n+1$ is nonnegative Hermitian since the sequence $(\mathbf{c}_k)_{k=0}^{n+1}$ is nonnegative definite), by (1.7) and (1.1) one can conclude (use, e.g., [18, Remark 3.4.3]) that the complex $q \times q$ matrices $\mathbf{L}_{n+2}$ defined via (1.3) is equal to $0_{q\times q}$. Admittedly, from Remark 1.4 it follows that the nonnegative definite sequence $(\mathbf{c}_k)_{k=0}^{n+1}$ is canonical of order $n$.
(c) Taking into account that $\mathbf{T}_{n+1} \geq 0_{q\times q}$ and that $\mathbf{L}_{n+2} = 0_{q\times q}$ (see Remark 1.4), the assertion of part (c) is an immediate consequence of [18, Theorem 3.4.1]. □

**Corollary 1.6.** *Let $\tau \in \mathbb{N}$ or $\tau = +\infty$ and let $n \in \mathbb{N}_{1,\tau}$. Suppose that $(\mathbf{c}_k)_{k=0}^{\tau}$ is a sequence of complex $q \times q$ matrices such that $(\mathbf{c}_k)_{k=0}^{n}$ is a nonnegative definite sequence which is canonical of order $n$. Furthermore, for each $m \in \mathbb{N}_{n,\tau} \setminus \{\tau\}$,* *let $\mathbf{M}_{m+1}$ be the complex $q \times q$ matrix defined as in (1.4) with $(\mathbf{c}_k)_{k=0}^m$ instead of $(\mathbf{c}_k)_{k=0}^n$. Then the following statements are equivalent:*

(i) *$(\mathbf{c}_k)_{k=0}^\tau$ is a nonnegative definite sequence which is canonical of order $n$.*
(ii) *$(\mathbf{c}_k)_{k=0}^\tau$ is a nonnegative definite sequence.*
(iii) *$\mathbf{c}_k = \mathbf{M}_k$ for each $k \in \mathbb{N}_{n+1,\tau}$.*

*Proof.* Use part (c) of Proposition 1.5 along with Remark 1.2. □

**Theorem 1.7.** *Let $\tau \in \mathbb{N}$ or $\tau = +\infty$ and let $n \in \mathbb{N}_{1,\tau}$. Suppose that $(\mathbf{c}_k)_{k=0}^\tau$ is a sequence of complex $q \times q$ matrices which is nonnegative definite. Furthermore, let $\mathbf{K}_n := \sqrt{\mathbf{L}_n}^+(\mathbf{c}_n - \mathbf{M}_n)\sqrt{\mathbf{R}_n}^+$, where $\mathbf{L}_n$ and $\mathbf{R}_n$ are given by (1.3) and where $\mathbf{M}_n$ is given by (1.4) using $(\mathbf{c}_k)_{k=0}^{n-1}$. Then the following statements are equivalent:*

(i) *The nonnegative definite sequence $(\mathbf{c}_k)_{k=0}^\tau$ is canonical of order $n$.*
(ii) *$\mathbf{K}_n\mathbf{K}_n^* = \mathbf{L}_n\mathbf{L}_n^+$.*
(iii) *$\mathbf{K}_n^*\mathbf{K}_n = \mathbf{R}_n\mathbf{R}_n^+$.*
(iv) *There is a unitary $q \times q$ matrix $\mathbf{W}$ such that $\mathbf{K}_n = \mathbf{L}_n\mathbf{L}_n^+\mathbf{W}\mathbf{R}_n\mathbf{R}_n^+$ and $\mathbf{L}_n\mathbf{L}_n^+ = \mathbf{W}\mathbf{R}_n\mathbf{R}_n^+\mathbf{W}^*$ hold.*
(v) *There is a unitary $q \times q$ matrix $\mathbf{W}$ such that $\mathbf{c}_n = \mathbf{M}_n + \sqrt{\mathbf{L}_n}\mathbf{W}\sqrt{\mathbf{R}_n}$ and $\mathbf{L}_n\mathbf{L}_n^+ = \mathbf{W}\mathbf{R}_n\mathbf{R}_n^+\mathbf{W}^*$ hold.*

*Moreover, if* (i) *is satisfied, then $\mathbf{K}_n\mathbf{K}_n^*\mathbf{K}_n = \mathbf{K}_n$ holds and $\mathbf{c}_k = \mathbf{M}_k$ for each $k \in \mathbb{N}_{n+1,\tau}$, where $\mathbf{M}_k$ is defined via (1.4) using $(\mathbf{c}_k)_{k=0}^{k-1}$.*

*Proof.* Obviously, the assumptions ensure that the sequences $(\mathbf{c}_k)_{k=0}^n$ and $(\mathbf{c}_k)_{k=0}^{n-1}$ are nonnegative definite. Let $\mathbf{L}_{n+1}$ and $\mathbf{R}_{n+1}$ be the complex $q \times q$ matrices defined as in (1.3). Suppose that (i) is fulfilled. In particular, the nonnegative definite sequence $(\mathbf{c}_k)_{k=0}^n$ is canonical of order $n$. Consequently, from Remark 1.4 we get $\mathbf{L}_{n+1} = 0_{q\times q}$. Thus, because of the choice of $\mathbf{K}_n$ and [18, Lemma 1.1.6 and Remark 3.4.3] it follows that

$$\begin{aligned} 0_{q\times q} &= \sqrt{\mathbf{L}_n}^+\mathbf{L}_{n+1}\sqrt{\mathbf{L}_n}^+ \\ &= \sqrt{\mathbf{L}_n}^+\sqrt{\mathbf{L}_n}(\mathbf{I}_q - \mathbf{K}_n\mathbf{K}_n^*)\sqrt{\mathbf{L}_n}\sqrt{\mathbf{L}_n}^+ \\ &= \mathbf{L}_n\mathbf{L}_n^+ - \mathbf{K}_n\mathbf{K}_n^*, \end{aligned}$$

i.e., that $\mathbf{K}_n\mathbf{K}_n^* = \mathbf{L}_n\mathbf{L}_n^+$ holds. Hence, (i) yields (ii). Since Remark 1.4 shows that the equation $\mathbf{L}_{n+1} = 0_{q\times q}$ holds if and only if $\mathbf{R}_{n+1} = 0_{q\times q}$, by using [18, Lemma 1.1.6 and Remark 3.4.3] one can also conclude that (ii) is equivalent to (iii). Therefore, from parts (a) and (b) of Lemma 1.3 one can realize that (ii) implies (iv) and that the identity $\mathbf{K}_n\mathbf{K}_n^*\mathbf{K}_n = \mathbf{K}_n$ is satisfied. Now, we suppose that (iv) is satisfied. Recalling the choice of the complex $q \times q$ matrix $\mathbf{K}_n$ and that $(\mathbf{c}_k)_{k=0}^n$ is a nonnegative definite sequence, based on [18, Theorem 3.4.1] in combination with the first identity in (iv) we get

$$\begin{aligned} \mathbf{c}_n &= \mathbf{M}_n + \sqrt{\mathbf{L}_n}\mathbf{K}_n\sqrt{\mathbf{R}_n} \\ &= \mathbf{M}_n + \sqrt{\mathbf{L}_n}\mathbf{L}_n\mathbf{L}_n^+\mathbf{W}\mathbf{R}_n\mathbf{R}_n^+\sqrt{\mathbf{R}_n} \\ &= \mathbf{M}_n + \sqrt{\mathbf{L}_n}\mathbf{W}\sqrt{\mathbf{R}_n}. \end{aligned}$$

Thus, (iv) results in (v). Taking into account that $(\mathbf{c}_k)_{k=0}^{n-1}$ is a nonnegative definite sequence, applying Proposition 1.5 one can see that (v) implies (i). Finally, Corollary 1.6 yields $\mathbf{c}_k = \mathbf{M}_k$ for each $k \in \mathbb{N}_{n+1,\tau}$. □

In particular, Theorem 1.7 (resp., Corollary 1.6) shows that, if $(\mathbf{c}_k)_{k=0}^{\tau}$ is a nonnegative definite sequence of matrices which is canonical of order $n$ with $\tau \in \mathbb{N}$ or $\tau = +\infty$ and $n \in \mathbb{N}_{1,\tau}$, then the elements with a greater index than $n$ are uniquely determined by $(\mathbf{c}_k)_{k=0}^{n}$. It is therefore enough to study nonnegative definite sequences $(\mathbf{c}_k)_{k=0}^{n}$ which are canonical of order $n$ with $n \in \mathbb{N}$.

**Corollary 1.8.** *Let $n \in \mathbb{N}$ and $\mathbf{c}_n \in \mathbb{C}^{q\times q}$. Suppose that $(\mathbf{c}_k)_{k=0}^{n-1}$ is a positive definite sequence of complex $q \times q$ matrices. Furthermore, let $\mathbf{L}_n$, $\mathbf{R}_n$, and $\mathbf{M}_n$ be given by* (1.3) *and* (1.4) *using $(\mathbf{c}_k)_{k=0}^{n-1}$. Then the following statements are equivalent:*

(i) *$(\mathbf{c}_k)_{k=0}^{n}$ is a nonnegative definite sequence which is canonical of order $n$.*

(ii) *$\mathbf{K} := \sqrt{\mathbf{L}_n}^{+}(\mathbf{c}_n - \mathbf{M}_n)\sqrt{\mathbf{R}_n}^{+}$ is a unitary $q \times q$ matrix such that the equality $\mathbf{c}_n = \mathbf{M}_n + \sqrt{\mathbf{L}_n}\mathbf{K}\sqrt{\mathbf{R}_n}$ holds.*

(iii) *There is a unitary $q \times q$ matrix $\mathbf{U}$ such that $\mathbf{c}_n = \mathbf{M}_n + \sqrt{\mathbf{L}_n}\mathbf{U}\sqrt{\mathbf{R}_n}$.*

*Moreover, if* (i) *is satisfied, then the matrices $\sqrt{\mathbf{L}_n}$ and $\sqrt{\mathbf{R}_n}$ are nonsingular and there exists a unique matrix $\mathbf{K} \in \mathbb{C}^{q\times q}$ such that $\mathbf{c}_n = \mathbf{M}_n + \sqrt{\mathbf{L}_n}\mathbf{K}\sqrt{\mathbf{R}_n}$ holds, namely $\mathbf{K} = \sqrt{\mathbf{L}_n}^{-1}(\mathbf{c}_n - \mathbf{M}_n)\sqrt{\mathbf{R}_n}^{-1}$.*

*Proof.* Since $(\mathbf{c}_k)_{k=0}^{n-1}$ is a positive definite sequence, the block Toeplitz matrix defined as in (1.1) with $\tau = n-1$ is positive Hermitian. Thus, the matrices $\mathbf{L}_n$ and $\mathbf{R}_n$ are positive Hermitian $q \times q$ matrices (see, e.g., [18, Remark 3.4.1]). Consequently, the assertion is an immediate consequence of Theorem 1.7. □

*Remark* 1.9. Let $n \in \mathbb{N}$ and let $(\mathbf{c}_k)_{k=0}^{n-1}$ be a positive definite sequence of complex $q \times q$ matrices. Because of Corollary 1.8 it follows that there is a bijective correspondence between the set $\mathfrak{C}[(\mathbf{c}_k)_{k=0}^{n-1}]$ of all $\mathbf{c}_n \in \mathbb{C}^{q\times q}$ such that $(\mathbf{c}_k)_{k=0}^{n}$ is a nonnegative definite sequence which is canonical of order $n$ and the set of all unitary $q \times q$ matrices. In particular, Corollary 1.8 implies that the set $\mathfrak{C}[(\mathbf{c}_k)_{k=0}^{n-1}]$ is the boundary of the matrix ball $\mathfrak{K}(\mathbf{M}_n; \sqrt{\mathbf{L}_n}, \sqrt{\mathbf{R}_n})$.

In view of Theorem 1.7 we consider a simple example (with $\tau = n = 1$ and $q = 2$) which shows that the identity $\mathbf{K}_n\mathbf{K}_n^*\mathbf{K}_n = \mathbf{K}_n$ does not in general imply that a nonnegative definite sequence $(\mathbf{c}_k)_{k=0}^{\tau}$ is canonical of order $n$.

*Example* 1.10. Let

$$\mathbf{c}_0 := \begin{pmatrix} 1 & 0 \\ 0 & 1 \end{pmatrix} \quad \text{and} \quad \mathbf{c}_1 := \begin{pmatrix} 1 & 0 \\ 0 & 0 \end{pmatrix}.$$

It is not hard to check that $(\mathbf{c}_k)_{k=0}^{1}$ is a nonnegative definite sequence such that $\mathbf{K}_1 = \mathbf{c}_1$ holds, where $\mathbf{K}_1$ is defined as in Theorem 1.7 (with $\tau = n = 1$). In particular, the identity $\mathbf{K}_1\mathbf{K}_1^*\mathbf{K}_1 = \mathbf{K}_1$ is fulfilled, but $\operatorname{rank}\mathbf{T}_1 = 3 \neq 2 = \operatorname{rank}\mathbf{T}_0$.

The following example points out that Remark 1.9 is a consequence of the fact that the underlying sequence is positive definite.

*Example* 1.11. Let

$$\mathbf{c}_0 := \begin{pmatrix} 1 & 0 \\ 0 & 0 \end{pmatrix}, \quad \mathbf{c}_1 := \begin{pmatrix} 0 & 0 \\ 0 & 0 \end{pmatrix}, \quad \text{and} \quad \mathbf{W} := \begin{pmatrix} 0 & 1 \\ 1 & 0 \end{pmatrix}.$$

By a straightforward calculation one can see that $(\mathbf{c}_k)_{k=0}^{1}$ is a nonnegative definite sequence and $\mathbf{W}$ is a unitary matrix such that the identity

$$\mathbf{c}_1 = \mathbf{M}_1 + \sqrt{\mathbf{L}_1}\mathbf{W}\sqrt{\mathbf{R}_1}$$

holds, where the matrices $\mathbf{M}_1$, $\mathbf{L}_1$, and $\mathbf{R}_1$ are given by (1.3) and (1.4), but $\operatorname{rank}\mathbf{T}_1 = 2 \neq 1 = \operatorname{rank}\mathbf{T}_0$. In particular, the set $\mathfrak{C}[(\mathbf{c}_k)_{k=0}^{0}]$ does not coincide with the boundary of the matrix ball $\mathfrak{K}(\mathbf{M}_1; \sqrt{\mathbf{L}_1}, \sqrt{\mathbf{R}_1})$.

We will use $\mathcal{M}_{\geq}^{q}(\mathbb{T}, \mathfrak{B}_{\mathbb{T}})$ to denote the set of all nonnegative Hermitian $q \times q$ measures defined on the $\sigma$-algebra $\mathfrak{B}_{\mathbb{T}}$ of all Borel subsets of the unit circle $\mathbb{T}$. Furthermore, given a matrix measure $F \in \mathcal{M}_{\geq}^{q}(\mathbb{T}, \mathfrak{B}_{\mathbb{T}})$, we set

$$\mathbf{c}_{\ell}^{(F)} := \int_{\mathbb{T}} z^{-\ell} F(\mathrm{d}z) \tag{1.8}$$

for an integer $\ell$. For each $n \in \mathbb{N}_0$, let

$$\mathbf{T}_n^{(F)} := \left(\mathbf{c}_{j-k}^{(F)}\right)_{j,k=0}^{n}. \tag{1.9}$$

In Sections 3 and 4, particular attention will be paid to the situation that some nondegeneracy condition holds. In doing so, an $F \in \mathcal{M}_{\geq}^{q}(\mathbb{T}, \mathfrak{B}_{\mathbb{T}})$ is called *nondegenerate of order* $n$ if the matrix $\mathbf{T}_n^{(F)}$ is nonsingular. We write $\mathcal{M}_{\geq}^{q,n}(\mathbb{T}, \mathfrak{B}_{\mathbb{T}})$ for the set of all $F \in \mathcal{M}_{\geq}^{q}(\mathbb{T}, \mathfrak{B}_{\mathbb{T}})$ which are nondegenerate of order $n$.

As is well known, via (1.8), there is a close relationship between nonnegative definite sequences of $q \times q$ matrices and measures belonging to $\mathcal{M}_{\geq}^{q}(\mathbb{T}, \mathfrak{B}_{\mathbb{T}})$ (see, e.g., [18, Theorems 2.2.1 and 3.4.2]). We will give next some information on the measures which are associated with canonical nonnegative definite sequences.

In what follows, the notation $\varepsilon_z$ with some $z \in \mathbb{T}$ stands for the Dirac measure defined on $\mathfrak{B}_{\mathbb{T}}$ with unit mass located at $z$. We shall show that an $F \in \mathcal{M}_{\geq}^{q}(\mathbb{T}, \mathfrak{B}_{\mathbb{T}})$ which is associated with a canonical nonnegative definite sequence of $q \times q$ matrices admits the representation

$$F = \sum_{s=1}^{r} \varepsilon_{z_s} \mathbf{A}_s \tag{1.10}$$

for some $r \in \mathbb{N}$, pairwise different points $z_1, z_2, \dots, z_r \in \mathbb{T}$, and a sequence $(\mathbf{A}_s)_{s=1}^{r}$ of nonnegative Hermitian $q \times q$ matrices.

Recall that a function $\Omega : \mathbb{D} \to \mathbb{C}^{q\times q}$ which is holomorphic in $\mathbb{D}$ and satisfies the condition $\operatorname{Re}\Omega(w) \geq 0_{q\times q}$ for each $w \in \mathbb{D}$ is a $q \times q$ *Carathéodory function* (in $\mathbb{D}$). In particular, if $F \in \mathcal{M}_{\geq}^{q}(\mathbb{T}, \mathfrak{B}_{\mathbb{T}})$, then $\Omega : \mathbb{D} \to \mathbb{C}^{q\times q}$ defined by

$$\Omega(w) := \int_{\mathbb{T}} \frac{z+w}{z-w} F(\mathrm{d}z)$$

is a $q \times q$ Carathéodory function (see, e.g., [18, Theorem 2.2.2]). We will call this matrix-valued function $\Omega$ the *Riesz–Herglotz transform of* (the Borel measure) $F$.

The following result shows that each nonnegative definite sequence which is canonical of some order has the form considered in Example 1.1.

**Proposition 1.12.** *Let $n \in \mathbb{N}$ and let $(\mathbf{c}_k)_{k=0}^n$ be a sequence of complex $q \times q$ matrices. Then the following statements are equivalent:*

(i) *$(\mathbf{c}_k)_{k=0}^n$ is a nonnegative definite sequence which is canonical of order $n$.*

(ii) *There is a unique $F \in \mathcal{M}_{\geq}^q(\mathbb{T}, \mathfrak{B}_{\mathbb{T}})$ such that $\mathbf{c}_k = \mathbf{c}_k^{(F)}$ holds for $k \in \mathbb{N}_{0,n}$.*

*If* (i) *is satisfied, then the matrix measure $F$ in* (ii) *admits* (1.10) *with some $r \in \mathbb{N}$, pairwise different points $z_1, z_2, \ldots, z_r \in \mathbb{T}$, and a sequence $(\mathbf{A}_s)_{s=1}^r$ of nonnegative Hermitian $q \times q$ matrices. In particular, if* (i) *holds, then*

$$\mathbf{c}_k = \sum\nolimits_{s=1}^{r} z_s^{-k} \mathbf{A}_s, \quad k \in \mathbb{N}_{0,n}.$$

*Proof.* Let (i) be fulfilled. In particular, the sequence $(\mathbf{c}_k)_{k=0}^n$ of complex $q \times q$ matrices is nonnegative definite. Thus, [18, Theorem 3.4.2] implies that there exists a measure $F \in \mathcal{M}_{\geq}^q(\mathbb{T}, \mathfrak{B}_{\mathbb{T}})$ such that

$$\mathbf{c}_k = \mathbf{c}_k^{(F)}, \quad k \in \mathbb{N}_{0,n}, \tag{1.11}$$

holds. Using Theorem 1.7 along with the fact that a measure $H \in \mathcal{M}_{\geq}^q(\mathbb{T}, \mathfrak{B}_{\mathbb{T}})$ is uniquely determined by its sequence $(\mathbf{c}_k^{(H)})_{k=0}^\infty$ of moments (cf. [18, Theorem 2.2.1]) one can see that the measure $F$ is uniquely determined by (1.11). Consequently, (i) implies (ii). We suppose now that (ii) holds. From [18, Theorem 2.2.1] in combination with (1.11) it follows that the sequence $(\mathbf{c}_k)_{k=0}^n$ of complex $q \times q$ matrices is nonnegative definite. Furthermore, because of [18, Theorems 3.4.1 and 3.4.2] (note also [18, Remark 3.4.3]) one can conclude that (ii) yields

$$\mathbf{L}_{n+1} = 0_{q \times q},$$

where $\mathbf{L}_{n+1}$ is the matrix defined via (1.3). Hence, Remark 1.4 shows that the nonnegative definite sequence $(\mathbf{c}_k)_{k=0}^n$ is canonical of order $n$, i.e., that (i) holds. It remains to be shown that, if (i) is satisfied, then the measure $F$ in (ii) admits (1.10) with some $r \in \mathbb{N}$, pairwise different points $z_1, z_2, \ldots, z_r \in \mathbb{T}$, and a sequence $(\mathbf{A}_s)_{s=1}^r$ of nonnegative Hermitian $q \times q$ matrices. Since (i) implies (ii), in view of the matricial version of the Riesz–Herglotz Theorem (see, e.g., [18, Theorem 2.2.2]) along with [28, Theorem 6.7] one can conclude that the Riesz–Herglotz transform $\Omega$ of $F$ admits the representation

$$\Omega(w) = \sum\nolimits_{s=1}^{r} \frac{z_s + w}{z_s - w} \mathbf{A}_s, \quad w \in \mathbb{D},$$

for some $r \in \mathbb{N}$, pairwise different points $z_1, z_2, \ldots, z_r \in \mathbb{T}$, and a sequence $(\mathbf{A}_s)_{s=1}^r$ of nonnegative Hermitian $q \times q$ matrices. But (note again [18, Theorem 2.2.2]), this leads to the asserted representation of $F$. In particular, in view of (1.8) we get

$$\mathbf{c}_k = \mathbf{c}_k^{(F)} = \sum\nolimits_{s=1}^{r} z_s^{-k} \mathbf{A}_s \tag{1.12}$$

for each $k \in \mathbb{N}_{0,n}$. □

Following the notation in [24, Section 6], Proposition 1.12 shows that a measure which is associated with a canonical nonnegative definite sequence of $q \times q$ matrices is a (special) molecular measure in $\mathcal{M}^q_{\geq}(\mathbb{T}, \mathfrak{B}_{\mathbb{T}})$.

Let $F \in \mathcal{M}^q_{\geq}(\mathbb{T}, \mathfrak{B}_{\mathbb{T}})$. If $r \in \mathbb{N}$ and if there is a subset $\Delta$ of $r$ elements of $\mathbb{T}$ such that $F(\mathbb{T} \setminus \Delta) = 0_{q\times q}$ holds, then $F$ is called *molecular of order at most* $r$. We call $F$ *molecular* when $F$ is molecular of order at most $r$ for some $r \in \mathbb{N}$. Also (as a convention), $F$ is *molecular of order at most* 0 means that $F(B) = 0_{q\times q}$ for each $B \in \mathfrak{B}_{\mathbb{T}}$ (i.e., that $F$ is the zero measure $\mathfrak{o}_q$ in $\mathcal{M}^q_{\geq}(\mathbb{T}, \mathfrak{B}_{\mathbb{T}})$).

**Corollary 1.14.** *Suppose that $(\mathbf{c}_k)_{k=0}^{\infty}$ is a sequence of complex $q\times q$ matrices. Then the following statements are equivalent:*

(i) *There is an $n \in \mathbb{N}$ such that $(\mathbf{c}_k)_{k=0}^{\infty}$ is a nonnegative definite sequence which is canonical of order $n$.*

(ii) *There exists a measure $F \in \mathcal{M}^q_{\geq}(\mathbb{T}, \mathfrak{B}_{\mathbb{T}})$ which is molecular such that the identity $\mathbf{c}_k = \mathbf{c}_k^{(F)}$ is satisfied for each $k \in \mathbb{N}_0$.*

(iii) *There are some $r \in \mathbb{N}$, pairwise different points $z_1, z_2, \ldots, z_r \in \mathbb{T}$, and a sequence $(\mathbf{A}_j)_{j=1}^{r}$ of nonnegative Hermitian $q \times q$ matrices such that*

$$\mathbf{c}_k = \sum_{s=1}^{r} z_s^{-k} \mathbf{A}_s, \quad k \in \mathbb{N}_0.$$

*Proof.* The measure $F \in \mathcal{M}^q_{\geq}(\mathbb{T}, \mathfrak{B}_{\mathbb{T}})$ is molecular means that there exist some $r \in \mathbb{N}$ and a subset $\Delta$ of $r$ elements of $\mathbb{T}$ such that $F(\mathbb{T} \setminus \Delta) = 0_{q\times q}$ holds. Therefore, the measure $F \in \mathcal{M}^q_{\geq}(\mathbb{T}, \mathfrak{B}_{\mathbb{T}})$ is molecular if and only if there are some $r \in \mathbb{N}$, pairwise different points $z_1, z_2, \ldots, z_r \in \mathbb{T}$, and a sequence $(\mathbf{A}_j)_{j=1}^{r}$ of nonnegative Hermitian $q\times q$ matrices such that $F$ admits (1.10). Thus, recalling [18, Theorem 2.2.1] and Corollary 1.6, from Proposition 1.12 one can see that (i) implies (ii). We suppose now that (ii) is fulfilled. Since the measure $F \in \mathcal{M}^q_{\geq}(\mathbb{T}, \mathfrak{B}_{\mathbb{T}})$ is molecular, there are some $r \in \mathbb{N}$, pairwise different points $z_1, z_2, \ldots, z_r \subset \mathbb{T}$, and a sequence $(\mathbf{A}_j)_{j=1}^{r}$ of nonnegative Hermitian $q \times q$ matrices such that $F$ admits (1.10). Consequently, in view of (ii) and (1.8) we get (1.12) for each $k \in \mathbb{N}_0$. Hence, (ii) yields (iii). Finally, because of Example 1.1 one can see that (iii) results in (i). □

Note that there is a relationship between the integers $n$ and $r$ occurring in Corollary 1.14, but the number $n$ does not coincide with $r$ in general (cf. [29, Example 9.11]). However, in the scalar case $q = 1$, one can always choose $n = r$. This fact will be emphasized by the following remark.

*Remark* 1.15. Let $n \in \mathbb{N}$ and suppose that $(c_k)_{k=0}^{\infty}$ is a sequence of complex numbers. Because of Corollary 1.14 and [24, Proposition 6.4 and Corollary 6.12] one can conclude that the following statements are equivalent:

(i) $(c_k)_{k=0}^{\infty}$ is a nonnegative definite sequence which is canonical of order $n$.

(ii) There exists a measure $F \in \mathcal{M}^1_{\geq}(\mathbb{T}, \mathfrak{B}_{\mathbb{T}})$ which is molecular of order at most $n$ such that $c_k = \mathbf{c}_k^{(F)}$ holds for each $k \in \mathbb{N}_0$.

(iii) There are pairwise different points $z_1, z_2, \ldots, z_n \in \mathbb{T}$ and a sequence $(a_j)_{j=1}^n$ of nonnegative real numbers such that

$$c_k = \sum_{s=1}^n a_s z_s^{-k}, \quad k \in \mathbb{N}_0.$$

Moreover (see again [24, Corollary 6.12]), if (i) holds and if $\mathbf{T}_n$ is the matrix given via (1.1), then $\operatorname{rank} \mathbf{T}_n$ is the smallest nonnegative integer $m$ such that the nonnegative definite sequence $(c_k)_{k=0}^\infty$ is canonical of order $m$.

The next sections will offer us further insight into the structure of the molecular matrix measures which are associated with canonical nonnegative definite sequences (cf. Corollaries 2.12 and 4.10). Roughly speaking, we will see that these measures can be characterized by a special structure involving the corresponding weights. In particular, the relationship between the integers $n$ and $r$ occurring in Corollary 1.14 will become more clear.

## 2. On canonical solutions of Problem (R), the general case

Let $\tau \in \mathbb{N}$ or $\tau = +\infty$. Let $(\alpha_j)_{j=1}^\tau$ be a sequence of numbers belonging to $\mathbb{C} \setminus \mathbb{T}$ and let $n \in \mathbb{N}_{0,\tau}$. If $n = 0$, then let $\pi_{\alpha,0}$ be the constant function on $\mathbb{C}_0$ with value 1 and let $\mathcal{R}_{\alpha,0}$ denote the set of all constant complex-valued functions defined on $\mathbb{C}_0$. Let $\mathbb{P}_{\alpha,0} := \emptyset$ and $\mathbb{Z}_{\alpha,0} := \emptyset$. If $n \in \mathbb{N}$, then let $\pi_{\alpha,n} : \mathbb{C} \to \mathbb{C}$ be given by

$$\pi_{\alpha,n}(u) := \prod_{j=1}^n (1 - \overline{\alpha_j} u)$$

and let $\mathcal{R}_{\alpha,n}$ denote the set of all rational functions $f$ which admit a representation

$$f = \frac{p_n}{\pi_{\alpha,n}}$$

with some polynomial $p_n : \mathbb{C} \to \mathbb{C}$ of degree not greater than $n$. Furthermore (using the convention $\frac{1}{0} := \infty$), let

$$\mathbb{P}_{\alpha,n} := \bigcup_{j=1}^n \left\{ \frac{1}{\overline{\alpha_j}} \right\} \quad \text{and} \quad \mathbb{Z}_{\alpha,n} := \bigcup_{j=1}^n \{\alpha_j\}.$$

Let $F \in \mathcal{M}_{\geq}^q(\mathbb{T}, \mathfrak{B}_{\mathbb{T}})$. Similar to [23]–[25], the right (resp., left) $\mathbb{C}^{q\times q}$-module $\mathcal{R}_{\alpha,n}^{q\times q}$ will have a matrix-valued inner product by

$$\left(X, Y\right)_{F,r} := \int_{\mathbb{T}} (X(z))^* F(\mathrm{d}z)\, Y(z)$$

$$\left(\text{resp., } \left(X, Y\right)_{F,l} := \int_{\mathbb{T}} X(z)\, F(\mathrm{d}z)\, (Y(z))^* \right)$$

for all $X, Y \in \mathcal{R}_{\alpha,n}^{q\times q}$. (For details on integration theory for nonnegative Hermitian $q \times q$ measures, we refer to Kats [38] and Rosenberg [41]–[43].) Moreover, if

$(X_k)_{k=0}^n$ is a sequence of matrix functions belonging to $\mathcal{R}_{\alpha,n}^{q\times q}$, then with $(X_k)_{k=0}^n$ we associate the nonnegative Hermitian matrix

$$\mathbf{G}_{X,n}^{(F)} := \left( \int_{\mathbb{T}} (X_j(z))^* F(\mathrm{d}z) X_k(z) \right)_{j,k=0}^n$$

$$\left( \text{resp., } \mathbf{H}_{X,n}^{(F)} := \left( \int_{\mathbb{T}} X_j(z) F(\mathrm{d}z) (X_k(z))^* \right)_{j,k=0}^n \right).$$

We now consider the following moment problem for rational matrix-valued functions, called Problem (R).

**Problem (R)**: *Let $n \in \mathbb{N}$ and $\alpha_1, \alpha_2, \ldots, \alpha_n \in \mathbb{C} \setminus \mathbb{T}$. Let $\mathbf{G} \in \mathbb{C}^{(n+1)q\times(n+1)q}$ and suppose that $X_0, X_1, \ldots, X_n$ is a basis of the right $\mathbb{C}^{q\times q}$-module $\mathcal{R}_{\alpha,n}^{q\times q}$. Describe the set $\mathcal{M}[(\alpha_j)_{j=1}^n, \mathbf{G}; (X_k)_{k=0}^n]$ of all measures $F \in \mathcal{M}_{\geq}^q(\mathbb{T}, \mathfrak{B}_{\mathbb{T}})$ such that $\mathbf{G}_{X,n}^{(F)} = \mathbf{G}$.*

If $\alpha_j = 0$ for each $j \in \mathbb{N}_{1,n}$, then $\mathcal{R}_{\alpha,n}^{q\times q}$ is the set of all complex $q \times q$ matrix polynomials of degree not greater than $n$. Thus, Problem (R) leads to the truncated trigonometric matrix moment problem, choosing $X_k$ as the complex $q \times q$ matrix polynomial $E_{k,q}$ given by

$$E_{k,q}(u) := u^k \mathbf{I}_q, \quad u \in \mathbb{C}, \tag{2.1}$$

for each $k \in \mathbb{N}_{0,n}$ (cf. [23, Section 2]). In this case (see also (1.8) and (1.9)), we write $\mathcal{M}[\mathbf{T}_n^{(F)}]$ instead of $\mathcal{M}[(\alpha_j)_{j=1}^n, \mathbf{T}_n^{(F)}; (E_{k,q})_{k=0}^n]$.

In what follows, unless otherwise indicated, let $\alpha_1, \alpha_2, \ldots, \alpha_n \in \mathbb{C} \setminus \mathbb{T}$ with some $n \in \mathbb{N}$. Furthermore, in view of Problem (R), let $\mathbf{G} \in \mathbb{C}^{(n+1)q\times(n+1)q}$ and suppose that $X_0, X_1, \ldots, X_n$ is a basis of the right $\mathbb{C}^{q\times q}$-module $\mathcal{R}_{\alpha,n}^{q\times q}$.

Similar to the notation of the previous section (cf. (1.2)), we will call a measure $F$ which belongs to $\mathcal{M}[(\alpha_j)_{j=1}^n, \mathbf{G}; (X_k)_{k=0}^n]$ a *canonical* solution when

$$\operatorname{rank} \mathbf{T}_{n+1}^{(F)} = \operatorname{rank} \mathbf{G}. \tag{2.2}$$

We point out that (see (1.8) and (1.9)) the size of the block Toeplitz matrix $\mathbf{T}_{n+1}^{(F)}$ in (2.2) is $(n+2)q \times (n+2)q$, whereas the size of $\mathbf{G}$ is $(n+1)q \times (n+1)q$.

We next consider how this definition applies to the solution set of Problem (R) and to nonnegative definite sequences.

*Remark* 2.1. Let $F \in \mathcal{M}_{\geq}^q(\mathbb{T}, \mathfrak{B}_{\mathbb{T}})$. Recalling [18, Theorems 2.2.1] and the special choice of $\mathcal{M}[\mathbf{T}_n^{(F)}]$ (see also (1.1) and (1.9) along with (1.8)), a comparison of (2.2) with (1.2) shows immediately that the following statements are equivalent:

(i) $F$ is a canonical solution in $\mathcal{M}[\mathbf{T}_n^{(F)}]$.

(ii) $(\mathbf{c}_k^{(F)})_{k=0}^\infty$ is a nonnegative definite sequence which is canonical of order $n+1$.

(iii) $(\mathbf{c}_k^{(F)})_{k=0}^{n+1}$ is a nonnegative definite sequence which is canonical of order $n+1$.

It turns out that the notion canonical solution is independent of the concrete way of looking at a problem. This will be clarified by the following.

*Remark* 2.2. Let $F \in \mathcal{M}[(\alpha_j)_{j=1}^n, \mathbf{G}; (X_k)_{k=0}^n]$. Recalling $\mathbf{G}_{X,n}^{(F)} = \mathbf{G}$, from [24, Theorem 4.4] it follows that the identity

$$\operatorname{rank} \mathbf{T}_n^{(F)} = \operatorname{rank} \mathbf{G} \tag{2.3}$$

is satisfied. Because of (2.2) and (2.3) one can see that $F$ is a canonical solution in $\mathcal{M}[(\alpha_j)_{j=1}^n, \mathbf{G}; (X_k)_{k=0}^n]$ if and only if

$$\operatorname{rank} \mathbf{T}_{n+1}^{(F)} = \operatorname{rank} \mathbf{T}_n^{(F)}.$$

Thus, Remark 2.1 shows that $F$ is a canonical solution in $\mathcal{M}[(\alpha_j)_{j=1}^n, \mathbf{G}; (X_k)_{k=0}^n]$ if and only if $(\mathbf{c}_k^{(F)})_{k=0}^\infty$ is a nonnegative definite sequence which is canonical of order $n+1$. In particular (cf. Remark 1.2), if the measure $F$ is a canonical solution in $\mathcal{M}[(\alpha_j)_{j=1}^n, \mathbf{G}; (X_k)_{k=0}^n]$ and if $m \in \mathbb{N}$ with $m \geq n$, then $\operatorname{rank} \mathbf{T}_m^{(F)} = \operatorname{rank} \mathbf{T}_n^{(F)}$.

*Remark* 2.3. Let $F \in \mathcal{M}_{\geq}^q(\mathbb{T}, \mathfrak{B}_\mathbb{T})$. In view of Remark 2.2 one can in particular see that the following statements are equivalent:

(i) $(\mathbf{c}_k^{(F)})_{k=0}^\infty$ is a nonnegative definite sequence which is canonical of order $n+1$.

(ii) There exist points $\alpha_1, \alpha_2, \ldots, \alpha_n \in \mathbb{C} \setminus \mathbb{T}$ and a basis $X_0, X_1, \ldots, X_n$ of the right $\mathbb{C}^{q\times q}$-module $\mathcal{R}_{\alpha,n}^{q\times q}$ such that the measure $F$ is a canonical solution in $\mathcal{M}[(\alpha_j)_{j=1}^n, \mathbf{G}_{X,n}^{(F)}; (X_k)_{k=0}^n]$.

(iii) The measure $F$ is a canonical solution in $\mathcal{M}[(\alpha_j)_{j=1}^n, \mathbf{G}_{X,n}^{(F)}; (X_k)_{k=0}^n]$ for every choice of points $\alpha_1, \alpha_2, \ldots, \alpha_n \in \mathbb{C} \setminus \mathbb{T}$ and each basis $X_0, X_1, \ldots, X_n$ of the right $\mathbb{C}^{q\times q}$-module $\mathcal{R}_{\alpha,n}^{q\times q}$.

*Remark* 2.4. If $F \in \mathcal{M}[(\alpha_j)_{j=1}^n, \mathbf{G}; (X_k)_{k=0}^n]$ and if $F^{(\alpha,n)} : \mathfrak{B}_\mathbb{T} \to \mathbb{C}^{q\times q}$ is given by

$$F^{(\alpha,n)}(B) := \int_B \Big(\frac{1}{\pi_{\alpha,n}(z)} \mathbf{I}_q\Big)^* F(\mathrm{d}z) \Big(\frac{1}{\pi_{\alpha,n}(z)} \mathbf{I}_q\Big)$$

(with a view to [23, Proposition 5]), then [24, Lemma 1.1, Remark 3.9, and Proposition 4.2] in combination with Remark 2.2 imply that $F$ is a canonical solution in $\mathcal{M}[(\alpha_j)_{j=1}^n, \mathbf{G}; (X_k)_{k=0}^n]$ if and only if $F^{(\alpha,n)}$ is a canonical solution in $\mathcal{M}[\mathbf{T}_n^{(F^{(\alpha,n)})}]$.

If $F \in \mathcal{M}_{\geq}^q(\mathbb{T}, \mathfrak{B}_\mathbb{T})$ and if $\Omega$ stands for the Riesz–Herglotz transform of $F$, then $\Omega^+$ is a $q \times q$ Carathéodory function as well (see, e.g., [14, Theorem 4.5]). Taking this and the matricial version of the Riesz–Herglotz Theorem (see [18, Theorem 2.2.2]) into account, the unique measure $F^\# \in \mathcal{M}_{\geq}^q(\mathbb{T}, \mathfrak{B}_\mathbb{T})$ fulfilling

$$\big(\Omega(w)\big)^+ = \int_\mathbb{T} \frac{z+w}{z-w} F^\#(\mathrm{d}z), \quad w \in \mathbb{D},$$

is called the *reciprocal measure corresponding to* $F$. (This is a generalization of [18, Definition 3.6.10].)

In view of the concept of reciprocal measures in the set $\mathcal{M}_{\geq}^q(\mathbb{T}, \mathfrak{B}_\mathbb{T})$ we mention the following (cf. [31, Proposition 6.3]).

*Remark* 2.5. Let $F^\#$ be the reciprocal measure corresponding to $F$. Based on [14, Lemma 4.3] and Remark 2.2, by a similar argumentation as for [18, Lemma 3.6.24], it follows that $F$ is a canonical solution in $\mathcal{M}[(\alpha_j)_{j=1}^n, \mathbf{G}; (X_k)_{k=0}^n]$ if and only if its reciprocal measure $F^{\#}$ is a canonical solution in $\mathcal{M}[(\alpha_j)_{j=1}^n, \mathbf{G}_{X,n}^{(F^{\#})}; (X_k)_{k=0}^n]$.

Regarding Problem (R) we assume, from now on, the following.

**Basic assumption:** Let $\alpha_1, \alpha_2, \dots, \alpha_n \in \mathbb{C} \setminus \mathbb{T}$ with some $n \in \mathbb{N}$. Furthermore, let $X_0, X_1, \dots, X_n$ be a basis of the right $\mathbb{C}^{q\times q}$-module $\mathcal{R}_{\alpha,n}^{q\times q}$ and suppose that the matrix $\mathbf{G} \in \mathbb{C}^{(n+1)q\times(n+1)q}$ is chosen such that the set $\mathcal{M}[(\alpha_j)_{j=1}^n, \mathbf{G}; (X_k)_{k=0}^n]$ is nonempty. In this case, $F$ stands for an element of $\mathcal{M}[(\alpha_j)_{j=1}^n, \mathbf{G}; (X_k)_{k=0}^n]$.

The fundamental question regarding the existence of a canonical solution for Problem (R) can be simply answered as follows.

**Theorem 2.6.** *There is (always) a canonical solution in* $\mathcal{M}[(\alpha_j)_{j=1}^n, \mathbf{G}; (X_k)_{k=0}^n]$.

*Proof.* Note that we have assumed that

$$\mathcal{M}[(\alpha_j)_{j=1}^n, \mathbf{G}; (X_k)_{k=0}^n] \neq \emptyset$$

holds. To prove the assertion, in view of [23, Proposition 5] and Remark 2.4 one can restrict the considerations without loss of generality to the special case in which $\alpha_j = 0$ for each $j \in \mathbb{N}_{1,n}$, the underlying matrix $\mathbf{G}$ is a nonnegative Hermitian $q\times q$ block Toeplitz matrix, and $X_k$ coincides with the complex $q \times q$ matrix polynomial $E_{k,q}$ defined by (2.1) for each $k \in \mathbb{N}_{0,n}$. For this case (note Remark 2.1), however, Proposition 1.5 in combination with [18, Theorem 3.4.2] implies the existence of such canonical solution. □

The next considerations are related to Proposition 1.12 (note Remark 2.1).

**Lemma 2.7.** *Let* $\alpha_{n+1} \in \mathbb{C} \setminus \mathbb{T}$ *and suppose that* $Y_0, Y_1, \dots, Y_{n+1}$ *is a basis of the right* $\mathbb{C}^{q\times q}$*-module* $\mathcal{R}_{\alpha,n+1}^{q\times q}$. *Then the measure* $F$ *is a canonical solution in the set* $\mathcal{M}[(\alpha_j)_{j=1}^n, \mathbf{G}; (X_k)_{k=0}^n]$ *if and only if*

$$\mathcal{M}[(\alpha_j)_{j=1}^{n+1}, \mathbf{G}_{Y,n+1}^{(F)}; (Y_k)_{k=0}^{n+1}] = \{F\}. \tag{2.4}$$

*Proof.* Since $F \in \mathcal{M}[(\alpha_j)_{j=1}^n, \mathbf{G}; (X_k)_{k=0}^n]$, we have

$$\mathbf{G}_{X,n}^{(F)} = \mathbf{G}. \tag{2.5}$$

By [25, Theorem 1] we know that (2.4) holds if and only if the identity

$$\operatorname{rank} \mathbf{G}_{Y,n+1}^{(F)} = \operatorname{rank} \mathbf{G}_{X,n}^{(F)}$$

is satisfied. Therefore, recalling (2.2) and (2.5), by using the fact that [24, Theorem 4.4] implies $\operatorname{rank} \mathbf{G}_{Y,n+1}^{(F)} = \operatorname{rank} \mathbf{T}_{n+1}^{(F)}$ we obtain the assertion. □

Based on Lemma 2.7 one can conclude that the values of the Riesz–Herglotz transform of a canonical solution of Problem (R) are in a certain way unique within the possible values of some Riesz–Herglotz transform associated with a solution of that problem. In fact, we get the following characterization (cf. [31, Corollary 5.9]). Here, the function $\widehat{\Omega} : \mathbb{C} \setminus \mathbb{T} \to \mathbb{C}^{q\times q}$ is defined by

$$\widehat{\Omega}(v) := \begin{cases} \Omega(v) & \text{if } v \in \mathbb{D} \\ -\left(\Omega\left(\frac{1}{\overline{v}}\right)\right)^* & \text{if } v \in \mathbb{C} \setminus (\mathbb{D} \cup \mathbb{T}) \end{cases}$$

for some holomorphic function $\Omega : \mathbb{D} \to \mathbb{C}^{q\times q}$ and $\widehat{\Omega}^{(t)}(v)$ is the value of the $t$th derivative of the function $\widehat{\Omega}$ at the point $v \in \mathbb{C} \setminus \mathbb{T}$ for $t \in \mathbb{N}_0$.

**Proposition 2.8.** *Suppose that $\overline{\alpha_j}\alpha_k \neq 1$ holds for all $j, k \in \mathbb{N}_{1,n}$. Let $m$ be the number of pairwise different points amongst $(\alpha_j)_{j=0}^n$ with $\alpha_0 := 0$ and let $\gamma_1, \gamma_2, \ldots, \gamma_m$ denote these points. Furthermore, let $\Omega$ be the Riesz–Herglotz transform of $F$ as well as let $\mathbf{\Omega}_v := \widehat{\Omega}(v)$ if $v \in \mathbb{C} \setminus (\mathbb{T} \cup \mathbb{P}_{\alpha,n} \cup \mathbb{Z}_{\alpha,n})$ and let $\mathbf{\Omega}_{\gamma_k} := \widehat{\Omega}^{(l_k)}(\gamma_k)$ if $k \in \mathbb{N}_{1,m}$, where $l_k$ stands for the number of occurrences of $\gamma_k$ in $(\alpha_j)_{j=0}^n$. Let $v \in \mathbb{C} \setminus (\mathbb{T} \cup \mathbb{P}_{\alpha,n} \cup \mathbb{Z}_{\alpha,n})$. Then the following statements are equivalent:*

(i) *$F$ is a canonical solution in $\mathcal{M}[(\alpha_j)_{j=1}^n, \mathbf{G}; (X_k)_{k=0}^n]$.*

(ii) *$F$ is the unique element in the set $\mathcal{M}[(\alpha_j)_{j=1}^n, \mathbf{G}; (X_k)_{k=0}^n]$ such that the value $\widehat{\Omega}(v)$ given by its Riesz–Herglotz transform coincides with $\mathbf{\Omega}_v$.*

(iii) *$F$ is the unique element in $\mathcal{M}[(\alpha_j)_{j=1}^n, \mathbf{G}; (X_k)_{k=0}^n]$ so that the value $\widehat{\Omega}^{(l_k)}(\gamma_k)$ related to its Riesz–Herglotz transform coincides with $\mathbf{\Omega}_{\gamma_k}$ for some $k \in \mathbb{N}_{1,m}$.*

(iv) *For all $k \in \mathbb{N}_{1,m}$, $F$ is the unique element in $\mathcal{M}[(\alpha_j)_{j=1}^n, \mathbf{G}; (X_k)_{k=0}^n]$ so that the value $\widehat{\Omega}^{(l_k)}(\gamma_k)$ related to its Riesz–Herglotz transform coincides with $\mathbf{\Omega}_{\gamma_k}$.*

*Proof.* Let $\alpha_{n+1} := v$. Furthermore, let $Y_0, Y_1, \ldots, Y_{n+1}$ be a basis of the right $\mathbb{C}^{q\times q}$-module $\mathcal{R}_{\alpha,n+1}^{q\times q}$. Recalling $F \in \mathcal{M}[(\alpha_j)_{j=1}^n, \mathbf{G}; (X_k)_{k=0}^n]$ (which implies the equality in (2.5)), from the interrelation between Problem (R) and an interpolation problem of Nevanlinna–Pick type for matrix-valued Carathéodory functions (see, e.g., [30, Proposition 2.1]) it follows that (ii) is satisfied if and only if (2.4) holds. Thus, an application of Lemma 2.7 yields that (i) is equivalent to (ii). A similar argumentation, based on the setting $\alpha_{n+1} := \gamma_k$ for some $k \in \mathbb{N}_{1,m}$, shows that (i) holds if and only if (iii) (resp., (iv)) is satisfied. □

*Remark* 2.9. Obviously, $\mathbf{G} = 0_{(n+1)q\times(n+1)q}$ holds if and only if $\operatorname{rank} \mathbf{G} = 0$. In particular, $\operatorname{rank} \mathbf{G} = 0$ holds if and only if $F(B) = 0_{q\times q}$ is satisfied for each $B \in \mathfrak{B}_{\mathbb{T}}$ (i.e., if $F \in \mathcal{M}[(\alpha_j)_{j=1}^n, \mathbf{G}; (X_k)_{k=0}^n]$ is the zero measure $\mathfrak{o}_q$ in $\mathcal{M}_{\geq}^q(\mathbb{T}, \mathfrak{B}_{\mathbb{T}})$).

In view of Remark 2.3 and Corollary 1.14 one can see that a canonical solution in $\mathcal{M}[(\alpha_j)_{j=1}^n, \mathbf{G}; (X_k)_{k=0}^n]$ is a molecular measure. In fact, canonical solutions in $\mathcal{M}[(\alpha_j)_{j=1}^n, \mathbf{G}; (X_k)_{k=0}^n]$ can be characterized by a special structure with respect to mass points and corresponding weights as follows.

**Theorem 2.10.** *The following statements are equivalent:*

(i) *$F$ is a canonical solution in $\mathcal{M}[(\alpha_j)_{j=1}^n, \mathbf{G}; (X_k)_{k=0}^n]$.*

(ii) *There is a finite subset* $\Delta$ *of* $\mathbb{T}$ *such that* $F(\mathbb{T} \setminus \Delta) = 0_{q\times q}$ *and*

$$\sum_{z\in\Delta} \operatorname{rank} F(\{z\}) = \operatorname{rank} \mathbf{G}.$$

*In particular, if* (i) *holds, then* $F$ *is molecular of order at most* $\operatorname{rank} \mathbf{G}$.

*Proof.* First, suppose that (i) is fulfilled. Because of Remark 2.3 and Proposition 1.12 there exist some $r \in \mathbb{N}$, pairwise different points $z_1, z_2, \dots, z_r \in \mathbb{T}$, and a sequence $(\mathbf{A}_s)_{s=1}^r$ of nonnegative Hermitian $q \times q$ matrices such that $F$ admits the representation (1.10). Therefore, in view of (1.8) we get

$$\mathbf{c}_k^{(F)} = \sum_{s=1}^{r} z_s^{-k} \mathbf{A}_s$$

for each $k \in \mathbb{N}_0$. Hence, recalling (1.9), from Example 1.1 it follows that

$$\operatorname{rank} \mathbf{T}_m^{(F)} = \sum_{s=1}^{r} \operatorname{rank} \mathbf{A}_s \tag{2.6}$$

for each $m \in \mathbb{N}_0$ with $m \geq r - 1$. Since the measure $F$ is a canonical solution in $\mathcal{M}[(\alpha_j)_{j=1}^n, \mathbf{G}; (X_k)_{k=0}^n]$, by (2.2) and (2.6) along with Remark 2.2 we obtain then

$$\operatorname{rank} \mathbf{G} = \operatorname{rank} \mathbf{T}_{n+1}^{(F)} = \operatorname{rank} \mathbf{T}_{n+r}^{(F)} = \sum_{s=1}^{r} \operatorname{rank} \mathbf{A}_s.$$

Thus, if we set $\Delta := \{z_1, z_2, \dots, z_r\}$, then $F(\mathbb{T} \setminus \Delta) = 0_{q\times q}$ and

$$\sum_{z\in\Delta} \operatorname{rank} F(\{z\}) = \operatorname{rank} \mathbf{G}$$

hold. In particular, (i) implies (ii). Furthermore, if (i) holds, then from the argumentation above one can see that $F$ is molecular of order at most $\operatorname{rank} \mathbf{G}$. Conversely, we suppose now (ii). Hence, there are some $r \in \mathbb{N}$, pairwise different points $z_1, z_2, \dots, z_r \in \mathbb{T}$, and a sequence $(\mathbf{A}_s)_{s=1}^r$ of nonnegative Hermitian $q \times q$ matrices such that the matrix measure $F$ admits (1.10). This leads to the inequality

$$\operatorname{rank} \mathbf{T}_{n+1}^{(F)} \leq \sum_{s=1}^{r} \operatorname{rank} \mathbf{A}_s$$

(see, e.g., [24, Theorem 6.6] or [29, Remark 9.4]). Since (1.10) and (ii) yield

$$\sum_{s=1}^{r} \operatorname{rank} \mathbf{A}_s = \sum_{z\in\Delta} \operatorname{rank} F(\{z\}) = \operatorname{rank} \mathbf{G}$$

and since the estimate $\operatorname{rank} \mathbf{T}_n^{(F)} \leq \operatorname{rank} \mathbf{T}_{n+1}^{(F)}$ is always satisfied (see, e.g., [18, Lemmas 1.1.7 and 1.1.9] as well as [24, Remarks 3.9 and 3.10]), we obtain

$$\operatorname{rank} \mathbf{T}_n^{(F)} \leq \operatorname{rank} \mathbf{T}_{n+1}^{(F)} \leq \operatorname{rank} \mathbf{G}.$$

Furthermore, from [24, Theorem 4.4] and $F \in \mathcal{M}[(\alpha_j)_{j=1}^n, \mathbf{G}; (X_k)_{k=0}^n]$ we get (2.3). Therefore, (2.2) is fulfilled, i.e., (i) holds. □

**Corollary 2.11.** *Suppose that the solution $F$ admits (1.10) with some $r \in \mathbb{N}_{1,n+1}$, pairwise different points $z_1, z_2, \dots, z_r \in \mathbb{T}$, and a sequence $(\mathbf{A}_s)_{s=1}^r$ of nonnegative Hermitian $q\times q$ matrices. Then $F$ is a canonical solution in $\mathcal{M}[(\alpha_j)_{j=1}^n, \mathbf{G}; (X_k)_{k=0}^n]$, where $\mathbf{G}$ is singular in the case of $r \neq n+1$. Moreover, if $r = n+1$, then $\mathbf{G}$ is nonsingular if and only if $\mathbf{A}_s$ is nonsingular for each $s \in \mathbb{N}_{1,r}$.*

*Proof.* Taking $r \le n+1$ into account and using Example 1.1 (cf. [29, Remark 9.4]), a similar argumentation as in the proof of Theorem 2.10 (cf. (2.6)) implies

$$\operatorname{rank} \mathbf{T}_n^{(F)} = \sum_{s=1}^r \operatorname{rank} \mathbf{A}_s.$$

Furthermore, by [24, Theorem 4.4] and $F \in \mathcal{M}[(\alpha_j)_{j=1}^n, \mathbf{G}; (X_k)_{k=0}^n]$ we have (2.3). Thus, if we set $\Delta := \{z_1, z_2, \dots, z_r\}$, then (1.10) yields $F(\mathbb{T} \setminus \Delta) = 0_{q\times q}$ and

$$\sum_{z\in\Delta} \operatorname{rank} F(\{z\}) = \sum_{s=1}^r \operatorname{rank} \mathbf{A}_s = \operatorname{rank} \mathbf{T}_n^{(F)} = \operatorname{rank} \mathbf{G}.$$

Finally, an application of Theorem 2.10 shows that $F$ is a canonical solution in $\mathcal{M}[(\alpha_j)_{j=1}^n, \mathbf{G}; (X_k)_{k=0}^n]$. Moreover, the $(n+1)q \times (n+1)q$ matrix $\mathbf{G}$ is singular if $r \neq n+1$ and in the case $r = n+1$ it follows that $\mathbf{G}$ is nonsingular if and only if $\mathbf{A}_s$ is nonsingular for each $s \in \mathbb{N}_{1,r}$. □

**Corollary 2.12.** *Let $\tau \in \mathbb{N}$ or $\tau = +\infty$ and let $m \in \mathbb{N}_{1,\tau}$. Suppose that $(\mathbf{c}_k)_{k=0}^\tau$ is a sequence of complex $q \times q$ matrices. The following statements are equivalent:*

(i) *$(\mathbf{c}_k)_{k=0}^\tau$ is a nonnegative definite sequence which is canonical of order $m$.*

(ii) *There are an $H \in \mathcal{M}_{\geq}^q(\mathbb{T}, \mathfrak{B}_\mathbb{T})$ and a finite $\Delta \subset \mathbb{T}$ so that $H(\mathbb{T}\setminus\Delta) = 0_{q\times q}$ and*
$$\sum_{z\in\Delta} \operatorname{rank} H(\{z\}) = \operatorname{rank} \mathbf{T}_{m-1}^{(H)},$$
*where $\mathbf{c}_k = \mathbf{c}_k^{(H)}$ for each $k \in \mathbb{N}_{0,\tau}$.*

(iii) *There are some $\ell \in \mathbb{N}$, pairwise different points $z_1, z_2, \dots, z_\ell \in \mathbb{T}$, and a sequence $(\mathbf{A}_j)_{j=1}^\ell$ of nonnegative Hermitian $q \times q$ matrices such that*
$$\sum_{s=1}^\ell \operatorname{rank} \mathbf{A}_s = \operatorname{rank} \mathbf{T}_{m-1} \quad \text{and} \quad \mathbf{c}_k = \sum_{s=1}^\ell z_s^{-k} \mathbf{A}_s, \quad k \in \mathbb{N}_{0,\tau},$$
*where $\mathbf{T}_{m-1}$ is the block Toeplitz matrix given by (1.1) with $\tau = m-1$.*

*In particular, if (i) is satisfied, then one can choose $\Delta$ and $z_1, z_2, \dots, z_\ell$ in (ii) and (iii) such that $\Delta = \{z_1, z_2, \dots, z_\ell\}$ and $\ell \le \max\{1, \operatorname{rank} \mathbf{T}_{m-1}\}$.*

*Proof.* Recalling Remark 2.1 and Proposition 1.12, the assertion for $n \ge 2$ follows from Theorem 2.10 along with Corollary 1.14. The case $n = 1$ is then a consequence of Remark 1.2 and [29, Remark 9.4]. □

*Remark* 2.14. Suppose that $\mathbf{G} \neq 0_{(n+1)q\times(n+1)q}$. Furthermore, let $r_1 := \operatorname{rank} \mathbf{G}$ and let $r_2$ be the smallest integer not less than $\frac{r_1}{q}$. If $F$ is a canonical solution in $\mathcal{M}[(\alpha_j)_{j=1}^n, \mathbf{G}; (X_k)_{k=0}^n]$, then Theorem 2.10 and Remark 2.9 imply that $F$ admits

(1.10) with some $r \in \mathbb{N}_{r_2,r_1}$, pairwise different points $z_1, z_2, \dots, z_r \in \mathbb{T}$, and a sequence $(\mathbf{A}_s)_{s=1}^r$ of nonnegative Hermitian $q\times q$ matrices each of which is not equal to the zero matrix. (Similarly, if (i) is fulfilled in Corollary 2.12 and if $\mathbf{c}_0 \neq 0_{q\times q}$, then the matrix measure $H$ in (ii) admits such a representation.)

In the scalar case $q = 1$, the situation is somewhat more straightforward. This fact will be emphasized by the following remark (cf. Remark 1.15).

*Remark* 2.15. Observe the special case $q = 1$, where $\mathbf{G} \in \mathbb{C}^{(n+1)\times(n+1)}$ and where $X_0, X_1, \dots, X_n$ is a basis of the linear space $\mathcal{R}_{\alpha,n}$. Then $F$ is a canonical solution in $\mathcal{M}[(\alpha_j)_{j=1}^n, \mathbf{G}; (X_k)_{k=0}^n]$ if and only if there exists a set $\Delta$ of exactly $\operatorname{rank}\mathbf{G}$ pairwise different points belonging to $\mathbb{T}$ such that $F(\mathbb{T}\setminus\Delta) = 0$ holds and that $F(\{z\}) \neq 0$ is satisfied for each $z \in \Delta$. This follows from Theorem 2.10 (note also [24, Proposition 6.4 and Corollary 6.12]). Moreover, since $\operatorname{rank}\mathbf{G} \neq n+1$ is equivalent to the condition $\mathcal{M}[(\alpha_j)_{j=1}^n, \mathbf{G}; (X_k)_{k=0}^n] = \{F\}$ (cf. [25, Theorem 1]), based on Theorem 2.6 one can see that $F$ is a canonical solution (and the unique element) in $\mathcal{M}[(\alpha_j)_{j=1}^n, \mathbf{G}; (X_k)_{k=0}^n]$ when $\operatorname{rank}\mathbf{G} \neq n+1$.

The case $n = 0$ which includes only a condition on the total weight $F(\mathbb{T})$ of some measure $F \in \mathcal{M}_{\geq}^q(\mathbb{T}, \mathfrak{B}_{\mathbb{T}})$ actually does not enter into Problem (R). However, we now present a few comments on this elementary case.

Let $X_0 \in \mathcal{R}_{\alpha,0}^{q\times q}$, i.e., suppose that $X_0$ is a constant function defined on $\mathbb{C}_0$ with some complex $q\times q$ matrix $\mathbf{X}_0$ as value. Then $X_0$ is a basis of the right $\mathbb{C}^{q\times q}$-module $\mathcal{R}_{\alpha,0}^{q\times q}$ if and only if $\det\mathbf{X}_0 \neq 0$. Moreover, if $\det\mathbf{X}_0 \neq 0$ and if $\mathbf{G} \in \mathbb{C}^{q\times q}$, then there exists a matrix measure $F \in \mathcal{M}_{\geq}^q(\mathbb{T}, \mathfrak{B}_{\mathbb{T}})$ such that

$$\int_{\mathbb{T}} \left(X_0(z)\right)^* F(\mathrm{d}z)\, X_0(z) = \mathbf{G} \tag{2.7}$$

holds if and only if $\mathbf{G} \geq 0_{q\times q}$ (see also [18, Theorem 3.4.2]).

Suppose that $\det\mathbf{X}_0 \neq 0$ and that $\mathbf{G} \geq 0_{q\times q}$. Similar to the notation introduced at the beginning of this section (see (2.2)), an $F \in \mathcal{M}_{\geq}^q(\mathbb{T}, \mathfrak{B}_{\mathbb{T}})$ fulfilling (2.7) is called a *canonical* solution of that problem when

$$\operatorname{rank}\mathbf{T}_1^{(F)} = \operatorname{rank}\mathbf{G}. \tag{2.8}$$

(If $\mathbf{X}_0 = \mathbf{I}_q$, then we will use the term canonical solution in $\mathcal{M}[\mathbf{G}]$.) In fact, the argumentations above (and below) are also applicable to this elementary case and lead to adequate results (cf. Corollary 2.12). In particular, for such a canonical solution $F$, it must not exist a set $\Delta$ of exactly $\operatorname{rank}\mathbf{G}$ pairwise different points belonging to $\mathbb{T}$ such that $F(\mathbb{T}\setminus\Delta) = 0_{q\times q}$ holds and $F(\{z\}) \neq 0$ for each $z \in \Delta$ in the case $q \geq 2$ (see also [29, Example 9.11]). This is, however, in contrast to the scalar case $q = 1$ (cf. Remark 2.15).

## 3. Descriptions of canonical solutions in the nondegenerate case

Starting with this section, we concentrate the considerations on the nondegenerate case, where the underlying complex $(n+1)q \times (n+1)q$ matrix $\mathbf{G}$ in Problem (R) is assumed to be nonsingular. Furthermore, unless otherwise indicated, we assume that $X_0, X_1, \ldots, X_n$ is a basis of the right $\mathbb{C}^{q\times q}$-module $\mathcal{R}_{\alpha,n}^{q\times q}$ and that the solution set $\mathcal{M}[(\alpha_j)_{j=1}^n, \mathbf{G}; (X_k)_{k=0}^n]$ is nonempty, where $n \in \mathbb{N}$ and where the points $\alpha_1, \alpha_2, \ldots, \alpha_n \in \mathbb{C} \setminus \mathbb{T}$ are arbitrary, but fixed.

*Remark* 3.1. Let $F \in \mathcal{M}[(\alpha_j)_{j=1}^n, \mathbf{G}; (X_k)_{k=0}^n]$. For the nondegenerate case, in view of Theorem 2.10 it follows that $F$ is a canonical solution in $\mathcal{M}[(\alpha_j)_{j=1}^n, \mathbf{G}; (X_k)_{k=0}^n]$ if and only if there exists a finite subset $\Delta$ of $\mathbb{T}$ such that $F(\mathbb{T} \setminus \Delta) = 0_{q\times q}$ and

$$\sum_{z\in\Delta} \operatorname{rank} F(\{z\}) = (n+1)q.$$

In particular, such a canonical solution $F$ is molecular of order at most $(n+1)q$.

*Remark* 3.2. Let $F \in \mathcal{M}[(\alpha_j)_{j=1}^n, \mathbf{G}; (X_k)_{k=0}^n]$ and suppose that $F$ admits (1.10) for an $r \in \mathbb{N}_{1,n+1}$, pairwise different points $z_1, z_2, \ldots, z_r \in \mathbb{T}$, and a sequence $(\mathbf{A}_s)_{s=1}^r$ of nonnegative Hermitian $q \times q$ matrices. By Corollary 2.11 (note that here $\det \mathbf{G} \neq 0$) it follows that $F$ is a canonical solution in $\mathcal{M}[(\alpha_j)_{j=1}^n, \mathbf{G}; (X_k)_{k=0}^n]$, where $r = n+1$ and the matrix $\mathbf{A}_s$ is nonsingular for each $s \in \mathbb{N}_{1,r}$.

*Remark* 3.3. Suppose that $F$ is a canonical solution in $\mathcal{M}[(\alpha_j)_{j=1}^n, \mathbf{G}; (X_k)_{k=0}^n]$. In view of Remarks 2.14 and 3.1 one can see that the matrix measure $F$ admits (1.10) with some $r \in \mathbb{N}_{n+1,(n+1)q}$, pairwise different points $z_1, z_2, \ldots, z_r \in \mathbb{T}$, and a sequence $(\mathbf{A}_s)_{s=1}^r$ of nonnegative Hermitian $q \times q$ matrices each of which is not equal to the zero matrix. In particular, $F$ admits such a representation with $r = (n+1)q$ if and only if $\operatorname{rank} \mathbf{A}_s = 1$ for each $s \in \mathbb{N}_{1,r}$.

We are now going to verify that, in the nondegenerate case, the canonical solutions of Problem (R) form a family which can be parametrized by the set of unitary matrices (cf. Corollary 1.8). Here, our considerations tie in with the necessary and sufficient condition in [31] and [40] for the fact that a measure belongs to the solution set of Problem (R). This condition is expressed in terms of orthogonal rational matrix functions. We first recall the relevant objects.

As to an application of results stated in [26] (see also [27] and [39]), we focus on the situation in which the elements of the underlying sequence $(\alpha_j)_{j=1}^n$ are in a sense well positioned with respect to $\mathbb{T}$. In doing so, the notation $\mathcal{T}_1$ stands for the set of all sequences $(\alpha_j)_{j=1}^\infty$ of complex numbers which satisfy $\overline{\alpha_j}\alpha_k \neq 1$ for all $j, k \in \mathbb{N}$. Obviously, if $(\alpha_j)_{j=1}^\infty \in \mathcal{T}_1$, then $\alpha_j \notin \mathbb{T}$ for all $j \in \mathbb{N}$.

Let $(\alpha_j)_{j=1}^\infty \in \mathcal{T}_1$ and $\alpha_0 := 0$. Furthermore, for each $k \in \mathbb{N}_0$, let

$$\eta_k := \begin{cases} -1 & \text{if } \alpha_k = 0 \\ \dfrac{\overline{\alpha_k}}{|\alpha_k|} & \text{if } \alpha_k \neq 0 \end{cases}$$

and let the function $b_{\alpha_k} : \mathbb{C}_0 \setminus \{\frac{1}{\overline{\alpha_k}}\} \to \mathbb{C}$ be defined by

$$b_{\alpha_k}(u) := \begin{cases} \eta_k \dfrac{\alpha_k - u}{1 - \overline{\alpha_k} u} & \text{if } u \in \mathbb{C} \setminus \{\frac{1}{\overline{\alpha_k}}\} \\ \dfrac{1}{|\alpha_k|} & \text{if } u = \infty . \end{cases}$$

Let $m \in \mathbb{N}_0$. If $B^{(q)}_{\alpha,0}$ stands for the constant function on $\mathbb{C}_0$ with value $\mathbf{I}_q$ and if

$$B^{(q)}_{\alpha,j} := \left( \prod_{k=1}^{j} b_{\alpha_k} \right) \mathbf{I}_q, \quad j \in \mathbb{N}_{1,m},$$

then the system $B^{(q)}_{\alpha,0}, B^{(q)}_{\alpha,1}, \ldots, B^{(q)}_{\alpha,m}$ forms a basis of the right (resp., left) $\mathbb{C}^{q\times q}$-module $\mathcal{R}^{q\times q}_{\alpha,m}$ (see, e.g., [24, Section 2]). Accordingly, if $X \in \mathcal{R}^{q\times q}_{\alpha,m}$, then there are unique matrices $\mathbf{A}_0, \mathbf{A}_1, \ldots, \mathbf{A}_m$ belonging to $\mathbb{C}^{q\times q}$ such that

$$X = \sum_{k=0}^{m} \mathbf{A}_k B^{(q)}_{\alpha,k}.$$

Based on this, the *reciprocal rational (matrix-valued) function* $X^{[\alpha,m]}$ *of* $X$ *with respect to* $(\alpha_j)_{j=1}^{\infty}$ *and* $m$ is given by

$$X^{[\alpha,m]} := \sum_{k=0}^{m} \mathbf{A}^*_{m-k} B^{(q)}_{\beta,k},$$

where $(\beta_j)_{j=1}^{\infty}$ is defined by $\beta_j := \alpha_{m+1-j}$ for each $j \in \mathbb{N}_{1,m}$ and $\beta_j := \alpha_j$ otherwise (cf. [26, Section 2]). Note that, in the special case that $\alpha_j = 0$ for each $j \in \mathbb{N}_{1,m}$, a function $X \in \mathcal{R}^{q\times q}_{\alpha,m}$ is a complex $q \times q$ matrix polynomial of degree not greater than $m$ and $X^{[\alpha,m]}$ is the reciprocal matrix polynomial $\widetilde{X}^{[m]}$ of $X$ with respect to $\mathbb{T}$ and formal degree $m$ (as used, e.g., in [17] or [18, Section 1.2]).

Let $F \in \mathcal{M}^q_{\geq}(\mathbb{T}, \mathfrak{B}_{\mathbb{T}})$. Suppose that $\tau \in \mathbb{N}_0$ or $\tau = +\infty$. A sequence $(Y_k)_{k=0}^{\tau}$ of matrix functions with $Y_k \in \mathcal{R}^{q\times q}_{\alpha,k}$ for each $k \in \mathbb{N}_{0,\tau}$ is called a *left* (resp., *right*) *orthonormal system corresponding to* $(\alpha_j)_{j=1}^{\infty}$ *and* $F$ when

$$\Big(Y_m, Y_s\Big)_{F,l} = \delta_{m,s}\mathbf{I}_q \quad \left(\text{resp., } \Big(Y_m, Y_s\Big)_{F,r} = \delta_{m,s}\mathbf{I}_q \right), \quad m, s \in \mathbb{N}_{0,\tau},$$

where $\delta_{m,s} := 1$ if $m = s$ and $\delta_{m,s} := 0$ otherwise (cf. [26, Definition 3.3]). If $(L_k)_{k=0}^{\tau}$ is a left orthonormal system and if $(R_k)_{k=0}^{\tau}$ is a right orthonormal system, respectively, corresponding to $(\alpha_j)_{j=1}^{\infty}$ and $F$, then we call $[(L_k)_{k=0}^{\tau}, (R_k)_{k=0}^{\tau}]$ a *pair of orthonormal systems corresponding to* $(\alpha_j)_{j=1}^{\infty}$ *and* $F$.

In what follows, let $\mathbf{L}_0$ and $\mathbf{R}_0$ be nonsingular complex $q\times q$ matrices fulfilling

$$\mathbf{L}_0^*\mathbf{L}_0 = \mathbf{R}_0\mathbf{R}_0^* \tag{3.1}$$

and let $(\mathbf{U}_j)_{j=1}^{\tau}$ be a sequence of complex $2q \times 2q$ matrices such that

$$\mathbf{U}_j^* \mathbf{j}_{qq} \mathbf{U}_j = \begin{cases} \mathbf{j}_{qq} & \text{if } (1-|\alpha_{j-1}|)(1-|\alpha_j|) > 0 \\ -\mathbf{j}_{qq} & \text{if } (1-|\alpha_{j-1}|)(1-|\alpha_j|) < 0 \end{cases} \tag{3.2}$$

for each $j \in \mathbb{N}_{1,\tau}$ (with $\tau \geq 1$), where $\mathbf{j}_{qq}$ is the $2q \times 2q$ signature matrix given by

$$\mathbf{j}_{qq} := \begin{pmatrix} \mathbf{I}_q & 0_{q\times q} \\ 0_{q\times q} & -\mathbf{I}_q \end{pmatrix}$$

and where we use for technical reasons the setting $\alpha_0 := 0$. Besides, let

$$\rho_j := \begin{cases} \sqrt{\dfrac{1-|\alpha_j|^2}{1-|\alpha_{j-1}|^2}} & \text{if } (1-|\alpha_{j-1}|)(1-|\alpha_j|) > 0 \\ -\sqrt{\dfrac{|\alpha_j|^2-1}{1-|\alpha_{j-1}|^2}} & \text{if } (1-|\alpha_{j-1}|)(1-|\alpha_j|) < 0 \end{cases}$$

for each $j \in \mathbb{N}_{1,\tau}$. As in [27, Section 3], we define sequences of rational matrix functions $(L_k)_{k=0}^{\tau}$ and $(R_k)_{k=0}^{\tau}$ by the initial conditions

$$L_0(u) = \mathbf{L}_0 \qquad \text{and} \qquad R_0(u) = \mathbf{R}_0 \tag{3.3}$$

for each $u \in \mathbb{C}$ and recursively by

$$\begin{pmatrix} L_j(u) \\ R_j^{[\alpha,j]}(u) \end{pmatrix} = \rho_j \frac{1-\overline{\alpha_{j-1}}u}{1-\overline{\alpha_j}u} \mathbf{U}_j \begin{pmatrix} b_{\alpha_{j-1}}(u)\mathbf{I}_q & 0_{q\times q} \\ 0_{q\times q} & \mathbf{I}_q \end{pmatrix} \begin{pmatrix} L_{j-1}(u) \\ R_{j-1}^{[\alpha,j-1]}(u) \end{pmatrix}$$

for each $j \in \mathbb{N}_{1,\tau}$ and each $u \in \mathbb{C}\backslash\mathbb{P}_{\alpha,j}$. The pair $[(L_k)_{k=0}^{\tau}, (R_k)_{k=0}^{\tau}]$ of rational matrix functions is called *the pair which is left-generated by* $[(\alpha_j)_{j=1}^{\tau}; (\mathbf{U}_j)_{j=1}^{\tau}; \mathbf{L}_0, \mathbf{R}_0]$.

Roughly speaking, there is a bijective correspondence between pairs which are left-generated and pairs of orthonormal systems of rational matrix functions (see [27] for details). Based on this relationship we will use the notation dual pair of orthonormal systems as explained below.

Let $F \in \mathcal{M}_{\geq}^{q,\tau}(\mathbb{T}, \mathfrak{B}_{\mathbb{T}})$ and let $[(L_k)_{k=0}^{\tau}, (R_k)_{k=0}^{\tau}]$ be a pair of orthonormal systems corresponding to $(\alpha_j)_{j=1}^{\tau}$ and $F$. First, we consider the case $\tau = 0$. Obviously (cf. [26, Remark 5.3]), there are nonsingular $\mathbf{L}_0, \mathbf{R}_0 \in \mathbb{C}^{q\times q}$ satisfying (3.1) and (3.3). The pair $[(L_k^{\#})_{k=0}^{0}, (R_k^{\#})_{k=0}^{0}]$ which is given, for each $u \in \mathbb{C}$, by

$$L_0^{\#}(u) = \mathbf{L}_0^{-*} \qquad \text{and} \qquad R_0^{\#}(u) = \mathbf{R}_0^{-*}$$

is called the *dual pair of orthonormal systems corresponding to* $[(L_k)_{k=0}^{0}, (R_k)_{k=0}^{0}]$. Now, let $\tau \geq 1$ and (recalling [27, Remark 3.5, Definition 3.6, Proposition 3.14, and Theorem 4.12]) let $(\mathbf{U}_j)_{j=1}^{\tau}$ be the unique sequence of complex $2q \times 2q$ matrices fulfilling (3.2) for each $j \in \mathbb{N}_{1,\tau}$ such that $[(L_k)_{k=0}^{\tau}, (R_k)_{k=0}^{\tau}]$ is the pair which is left-generated by $[(\alpha_j)_{j=1}^{\tau}; (\mathbf{U}_j)_{j=1}^{\tau}; \mathbf{L}_0, \mathbf{R}_0]$ with some nonsingular $q \times q$ matrices $\mathbf{L}_0$ and $\mathbf{R}_0$ satisfying (3.1) and (3.3). The pair $[(L_k^{\#})_{k=0}^{\tau}, (R_k^{\#})_{k=0}^{\tau}]$ which is left-generated by $[(\alpha_j)_{j=1}^{\tau}; (\mathbf{j}_{qq}\mathbf{U}_j\mathbf{j}_{qq})_{j=1}^{\tau}; \mathbf{L}_0^{-*}, \mathbf{R}_0^{-*}]$ is the *dual pair of orthonormal systems corresponding to* $[(L_k)_{k=0}^{\tau}, (R_k)_{k=0}^{\tau}]$.

In view of Problem (R), again, let $n \in \mathbb{N}$ and let $X_0, X_1, \dots, X_n$ be a basis of the right $\mathbb{C}^{q\times q}$-module $\mathcal{R}^{q\times q}_{\alpha,n}$. Furthermore, suppose that $\mathbf{G}$ is a nonsingular matrix such that $\mathcal{M}[(\alpha_j)_{j=1}^n, \mathbf{G}; (X_k)_{k=0}^n] \neq \emptyset$ holds. Taking [31, Remark 3.3] into account, a pair of orthonormal systems $[(L_k)_{k=0}^n, (R_k)_{k=0}^n]$ corresponding to $(\alpha_j)_{j=1}^\infty$ and some $F \in \mathcal{M}[(\alpha_j)_{j=1}^n, \mathbf{G}; (X_k)_{k=0}^n]$ is called a *pair of orthonormal systems corresponding to* $\mathcal{M}[(\alpha_j)_{j=1}^n, \mathbf{G}; (X_k)_{k=0}^n]$. ([31, Remark 3.3] points out that the additional assumption $\det \mathbf{G} \neq 0$ is essential for the existence of such a pair of orthonormal systems.) Keeping this in mind, we will also speak of the dual pair of orthonormal systems corresponding to this pair $[(L_k)_{k=0}^n, (R_k)_{k=0}^n]$.

From now on, let $[(L_k)_{k=0}^n, (R_k)_{k=0}^n]$ be a pair of orthonormal systems corresponding to the solution set $\mathcal{M}[(\alpha_j)_{j=1}^n, \mathbf{G}; (X_k)_{k=0}^n]$ and let $[(L_k^\#)_{k=0}^n, (R_k^\#)_{k=0}^n]$ be the dual pair of orthonormal systems corresponding to $[(L_k)_{k=0}^n, (R_k)_{k=0}^n]$.

With a view to the concept of reciprocal measures in the set $\mathcal{M}^q_{\geq}(\mathbb{T}, \mathfrak{B}_{\mathbb{T}})$, we have to note the following (see also Remark 2.5).

*Remark* 3.4. Let $F \in \mathcal{M}[(\alpha_j)_{j=1}^n, \mathbf{G}; (X_k)_{k=0}^n]$ and let $F^\#$ be the reciprocal measure corresponding to $F$. In view of [39, Theorem 4.6] and the terms introduced above it follows that $[(L_k^\#)_{k=0}^n, (R_k^\#)_{k=0}^n]$ is a pair of orthonormal systems corresponding to $\mathcal{M}[(\alpha_j)_{j=1}^n, \mathbf{G}_{X,n}^{(F^\#)}; (X_k)_{k=0}^n]$, where $[(L_k)_{k=0}^n, (R_k)_{k=0}^n]$ is obviously the dual pair of orthonormal systems corresponding to $[(L_k^\#)_{k=0}^n, (R_k^\#)_{k=0}^n]$.

In what follows, the rational matrix functions

$$\begin{aligned} &P_{n;\mathbf{U}}^{(\alpha)} := R_n^{[\alpha,n]} + b_{\alpha_n}\mathbf{U}L_n \quad \text{and} \quad P_{n;\mathbf{U}}^{(\alpha,\#)} := (R_n^\#)^{[\alpha,n]} - b_{\alpha_n}\mathbf{U}L_n^\# \\ &\Big(\text{resp., } Q_{n;\mathbf{U}}^{(\alpha)} := L_n^{[\alpha,n]} + b_{\alpha_n}R_n\mathbf{U} \quad \text{and} \quad Q_{n;\mathbf{U}}^{(\alpha,\#)} := (L_n^\#)^{[\alpha,n]} - b_{\alpha_n}R_n^\#\mathbf{U}\Big) \end{aligned} \tag{3.4}$$

with some unitary $q\times q$ matrix $\mathbf{U}$ will be of particular interest. Here, as an aside, we briefly mention that the matrix functions in (3.4) can be interpreted as elements of special para-orthogonal systems of rational matrix functions (note Remark 3.4 as well as [32, Theorem 3.8, Definition 6.1, and Proposition 6.15]).

*Remark* 3.5. Because of (3.4) and the unitarity of $\mathbf{U}$ it follows that the rational matrix function $\Psi_{n;\mathbf{U}}^{(\alpha)} := \big(P_{n;\mathbf{U}}^{(\alpha)}\big)^{-1} P_{n;\mathbf{U}}^{(\alpha,\#)}$ admits the representation

$$\Psi_{n;\mathbf{U}}^{(\alpha)} = \Big(\frac{1}{b_{\alpha_n}}\mathbf{U}^* R_n^{[\alpha,n]} + L_n\Big)^{-1}\Big(\frac{1}{b_{\alpha_n}}\mathbf{U}^*(R_n^\#)^{[\alpha,n]} - L_n^\#\Big).$$

Thus, [31, Lemma 5.7] implies that $\Psi_{n;\mathbf{U}}^{(\alpha)} = Q_{n;\mathbf{U}}^{(\alpha,\#)}\big(Q_{n;\mathbf{U}}^{(\alpha)}\big)^{-1}$ and

$$\Psi_{n;\mathbf{U}}^{(\alpha)} = \Big(\frac{1}{b_{\alpha_n}}(L_n^\#)^{[\alpha,n]}\mathbf{U}^* - R_n^\#\Big)\Big(\frac{1}{b_{\alpha_n}}L_n^{[\alpha,n]}\mathbf{U}^* + R_n\Big)^{-1}$$

as well, where the complex $q \times q$ matrices $P_{n;\mathbf{U}}^{(\alpha)}(v)$, $\frac{1}{b_{\alpha_n}(v)}\mathbf{U}^* R_n^{[\alpha,n]}(v) + L_n(v)$, $Q_{n;\mathbf{U}}^{(\alpha)}(v)$, and $\frac{1}{b_{\alpha_n}(v)}L_n^{[\alpha,n]}(v)\mathbf{U}^* + R_n(v)$ are nonsingular for each $v \in \mathbb{D} \setminus \mathbb{P}_{\alpha,n}$.

The values of the function $\Psi_{n;\mathbf{U}}^{(\alpha)}$ in Remark 3.5 are subject to a geometrical constraint. More precisely, the restriction of that function to $\mathbb{D}$ is a $q \times q$ Carathéodory function. In fact, by using the notation of Remark 3.5, we get a characterization of canonical solutions of Problem (R) for the nondegenerate case as follows.

**Theorem 3.6.** *Let $(\alpha_j)_{j=1}^{\infty} \in \mathcal{T}_1$. Furthermore, let $F \in \mathcal{M}_{\geq}^{q}(\mathbb{T}, \mathfrak{B}_{\mathbb{T}})$ and let $\Omega$ be the Riesz–Herglotz transform of $F$. Then the following statements are equivalent:*

(i) *$F$ is a canonical solution in $\mathcal{M}[(\alpha_j)_{j=1}^{n}, \mathbf{G}; (X_k)_{k=0}^{n}]$.*

(ii) *There is a unitary $q \times q$ matrix $\mathbf{U}$ such that $\Omega(v) = \Psi_{n;\mathbf{U}}^{(\alpha)}(v)$ for $v \in \mathbb{D} \setminus \mathbb{P}_{\alpha,n}$.*

*Moreover, if* (i) *holds, then the unitary $q \times q$ matrix $\mathbf{U}$ in* (ii) *is uniquely determined.*

*Proof.* Denote by $\mathcal{S}_{q\times q}(\mathbb{D})$ the set of all functions $S : \mathbb{D} \to \mathbb{C}^{q\times q}$ which are holomorphic in $\mathbb{D}$ and which have a contractive $q \times q$ matrix as their value $S(w)$ for each $w \in \mathbb{D}$. Furthermore, for each $S \in \mathcal{S}_{q\times q}(\mathbb{D})$ and $v \in \mathbb{D} \setminus \mathbb{P}_{\alpha,n}$, let

$$\Omega_S(v) := \Big((L_n^{\#})^{[\alpha,n]}(v) - b_{\alpha_n}(v) R_n^{\#}(v) S(v)\Big)\Big(L_n^{[\alpha,n]}(v) + b_{\alpha_n}(v) R_n(v) S(v)\Big)^{-1}$$

if $\alpha_n \in \mathbb{D}$ and in the case of $\alpha_n \notin \mathbb{D}$ let

$$\Omega_S(v) := \Big(\frac{1}{b_{\alpha_n}(v)}(L_n^{\#})^{[\alpha,n]}(v) S(v) - R_n^{\#}(v)\Big)\Big(\frac{1}{b_{\alpha_n}(v)} L_n^{[\alpha,n]}(v) S(v) + R_n(v)\Big)^{-1}$$

(where the inverse matrices exist according to [31, Lemma 5.7]). Let (i) be fulfilled. In particular, we have that $F$ belongs to $\mathcal{M}[(\alpha_j)_{j=1}^{n}, \mathbf{G}; (X_k)_{k=0}^{n}]$. Thus, [40, Theorem 4.6] implies that there exists a uniquely determined matrix function $S$ belonging to $\mathcal{S}_{q\times q}(\mathbb{D})$ such that

$$\Omega(v) = \Omega_S(v), \quad v \in \mathbb{D} \setminus \mathbb{P}_{\alpha,n}. \tag{3.5}$$

Recalling (2.2), due to (i) and the fact that the matrix $\mathbf{G}$ is nonsingular it follows

$$\operatorname{rank} \mathbf{T}_{n+1}^{(F)} = (n+1)q. \tag{3.6}$$

From (3.5) and (3.6) along with [40, Remark 4.7] one can conclude that $S(0)$ is a unitary $q \times q$ matrix. Because of $S \in \mathcal{S}_{q\times q}(\mathbb{D})$ this yields that $S$ is the constant function on $\mathbb{D}$ with that unitary $q \times q$ matrix $S(0)$ as value (see, e.g., [18, Corollary 2.3.2]). Therefore, in view of the definition of $\Omega_S$ and Remark 3.5 we get (ii). Moreover, since the matrix function $S \in \mathcal{S}_{q\times q}(\mathbb{D})$ in (3.5) is uniquely determined, we find that the unitary $q \times q$ matrix $\mathbf{U}$ in (ii) is uniquely determined. Conversely, we suppose now that (ii) holds. Obviously, a constant function on $\mathbb{D}$ with a unitary $q \times q$ matrix as value belongs to $\mathcal{S}_{q\times q}(\mathbb{D})$. Consequently, by using Remark 3.5 in combination with [31, Theorem 5.8] we obtain that the matrix measure $F$ belongs to the solution set $\mathcal{M}[(\alpha_j)_{j=1}^{n}, \mathbf{G}; (X_k)_{k=0}^{n}]$. Furthermore, the unitarity of $\mathbf{U}$ and [40, Remark 4.7] yield that (3.6) holds. Hence, taking (2.2) and the nonsingularity of $\mathbf{G}$ into account, it finally follows that (i) is fulfilled. □

In view of Theorem 3.6 (note also [18, Theorem 2.2.2]), if $\mathbf{U}$ is a unitary $q \times q$ matrix, then we will use the notation $F_{n;\mathbf{U}}^{(\alpha)}$ and $\Omega_{n;\mathbf{U}}^{(\alpha)}$. Here, $F_{n;\mathbf{U}}^{(\alpha)}$ stands for the (uniquely determined) matrix measure belonging to $\mathcal{M}_{\geq}^{q}(\mathbb{T}, \mathfrak{B}_{\mathbb{T}})$ such that its Riesz–Herglotz transform $\Omega_{n;\mathbf{U}}^{(\alpha)}$ satisfies $\Omega_{n;\mathbf{U}}^{(\alpha)}(v) = \Psi_{n;\mathbf{U}}^{(\alpha)}(v)$ for each $v \in \mathbb{D} \setminus \mathbb{P}_{\alpha,n}$.

**Corollary 3.7.** *Let $(\alpha_j)_{j=1}^{\infty} \in \mathcal{T}_1$. Let $m$ be the number of pairwise different points amongst $(\alpha_j)_{j=0}^{n}$ with $\alpha_0 := 0$ and let $\gamma_1, \gamma_2, \dots, \gamma_m$ denote these points, where $l_k$ stands for the number of occurrences of $\gamma_k$ in $(\alpha_j)_{j=0}^{n}$ for all $k \in \mathbb{N}_{1,m}$. Suppose that $F \in \mathcal{M}[(\alpha_j)_{j=1}^{n}, \mathbf{G}; (X_k)_{k=0}^{n}]$ and let $\Omega$ be the Riesz–Herglotz transform of $F$. If $\mathbf{U}$ is a unitary $q \times q$ matrix, then the following statements are equivalent:*

(i) $F = F_{n;\mathbf{U}}^{(\alpha)}$.

(ii) *There is some $v \in \mathbb{C} \setminus (\mathbb{T} \cup \mathbb{P}_{\alpha,n} \cup \mathbb{Z}_{\alpha,n})$ such that $\widehat{\Omega}(v) = \widehat{\Omega_{n;\mathbf{U}}^{(\alpha)}}(v)$.*

(iii) *There exists some $k \in \mathbb{N}_{1,m}$ such that $\widehat{\Omega}^{(l_k)}(\gamma_k) = \big(\widehat{\Omega_{n;\mathbf{U}}^{(\alpha)}}\big)^{(l_k)}(\gamma_k)$.*

(iv) *For each $k \in \mathbb{N}_{1,m}$, the equality $\widehat{\Omega}^{(l_k)}(\gamma_k) = \big(\widehat{\Omega_{n;\mathbf{U}}^{(\alpha)}}\big)^{(l_k)}(\gamma_k)$ holds.*

*Proof.* Use Theorem 3.6 in combination with Proposition 2.8. □

*Remark* 3.8. Let the assumptions of Corollary 3.7 be fulfilled. Furthermore, suppose that $w \in \mathbb{D} \setminus (\mathbb{P}_{\alpha,n} \cup \mathbb{Z}_{\alpha,n})$ (resp., $w = \gamma_k$ for a $k \in \mathbb{N}_{1,m}$ with $\gamma_k \in \mathbb{D}$). From [40, Proposition 5.1] (resp., [40, Proposition 5.5]) we know that the set of possible values for $\Omega(w)$ (resp., $\frac{1}{l_k!}\Omega^{(l_k)}(w)$) fills a matrix ball $\mathfrak{K}(\mathbf{M}_w; \mathbf{A}_w, \mathbf{B}_w)$. In [40], the corresponding parameters $\mathbf{M}_w$, $\mathbf{A}_w$, and $\mathbf{B}_w$ are also expressed in terms of orthogonal rational matrix functions. Based on these formulas along with Theorem 3.6 and Corollary 3.7 one can realize that the matrix measure $F$ is a canonical solution in $\mathcal{M}[(\alpha_j)_{j=1}^{n}, \mathbf{G}; (X_k)_{k=0}^{n}]$ if and only if the value $\Omega(w)$ (resp., $\frac{1}{l_k!}\Omega^{(l_k)}(w)$) belongs to the boundary of $\mathfrak{K}(\mathbf{M}_w; \mathbf{A}_w, \mathbf{B}_w)$ (cf. Remark 1.9).

The representation of the Riesz–Herglotz transform $\Omega_{n;\mathbf{U}}^{(\alpha)}$ of the matrix measure $F_{n;\mathbf{U}}^{(\alpha)}$ (defined as in Corollary 3.7 with some unitary $q \times q$ matrix $\mathbf{U}$) which appears in Theorem 3.6 depends on the choice of the pair of orthonormal systems $[(L_k)_{k=0}^{n}, (R_k)_{k=0}^{n}]$ corresponding to $\mathcal{M}[(\alpha_j)_{j=1}^{n}, \mathbf{G}; (X_k)_{k=0}^{n}]$ (with associated dual pair $[(L_k^{\#})_{k=0}^{n}, (R_k^{\#})_{k=0}^{n}]$). However, by [31, Lemma 5.6] one can see that this is not essential. If we choose another pair of orthonormal systems $[(\tilde{L}_k)_{k=0}^{n}, (\tilde{R}_k)_{k=0}^{n}]$ corresponding to the solution set $\mathcal{M}[(\alpha_j)_{j=1}^{n}, \mathbf{G}; (X_k)_{k=0}^{n}]$ (with associated dual pair $[(\tilde{L}_k^{\#})_{k=0}^{n}, (\tilde{R}_k^{\#})_{k=0}^{n}]$), then the only possible difference is that another unitary $q \times q$ matrix $\tilde{\mathbf{U}}$ occurs in that representation of $\Omega_{n;\mathbf{U}}^{(\alpha)}$.

The rational matrix functions $L_n$, $R_n$, $L_n^{\#}$, and $R_n^{\#}$ occurring in the representation of $\Omega_{n;\mathbf{U}}^{(\alpha)}$ can be constructed from the given data in different ways. In view of the recurrence relations for orthogonal rational matrix functions one needs to determine the corresponding matrices, which realize these. In keeping with that, one can apply the formulas presented in [27]. Moreover, because of [31, Remark 5.4] one can also use Szegő parameters to obtain representations of $\Omega_{n;\mathbf{U}}^{(\alpha)}$. In particular, the associated Szegő parameters can be calculated by integral formulas. In addition, the rational matrix functions $L_n^{\#}$ and $R_n^{\#}$ can be extracted directly from $L_n$ and $R_n$ by using the integral formulas in [39, Section 5] as well.

Below, we will present an alternative to gain descriptions of the Riesz–Herglotz transform of a canonical solution of Problem (R) for the nondegenerate case.

In other words, we will reformulate the above representations in terms of reproducing kernels based on the procedure already used in [31, Section 6].

Along the lines of the scalar theory of reproducing kernels, which goes back to [1] by Aronszajn, this machinery can be extended to the matrix case (see, e.g., [5], [6], [20], [21], and [36]). The reproducing kernels of the $\mathbb{C}^{q\times q}$-Hilbert modules of rational matrix functions, which are of particular interest here, were studied intensively in [23], [25], and [26] (see also [30]).

Let $m \in \mathbb{N}_0$ and $\alpha_1, \alpha_2, \ldots, \alpha_m \in \mathbb{C} \setminus \mathbb{T}$. Suppose that $F \in \mathcal{M}_{\geq}^{q,m}(\mathbb{T}, \mathfrak{B}_{\mathbb{T}})$. In view of [24, Theorem 5.8] and [23, Theorem 10] one can see that by $(\mathcal{R}_{\alpha,m}^{q\times q}, (\cdot,\cdot)_{F,r})$ (resp., $(\mathcal{R}_{\alpha,m}^{q\times q}, (\cdot,\cdot)_{F,l})$) a right (resp., left) $\mathbb{C}^{q\times q}$-Hilbert module with reproducing kernel $K_{m;r}^{(\alpha,F)}$ (resp., $K_{m;l}^{(\alpha,F)}$) is given. Here, the relevant kernel is a mapping from $(\mathbb{C}_0 \setminus \mathbb{P}_{\alpha,m}) \times (\mathbb{C}_0 \setminus \mathbb{P}_{\alpha,m})$ into $\mathbb{C}^{q\times q}$. In fact, if $X_0, X_1, \ldots, X_m$ is a basis of the right (resp., left) $\mathbb{C}^{q\times q}$-module $\mathcal{R}_{\alpha,m}^{q\times q}$, then the representation

$$K_{m;r}^{(\alpha,F)}(v,w) = \Big(X_0(v), X_1(v), \ldots, X_m(v)\Big)\Big(\mathbf{G}_{X,m}^{(F)}\Big)^{-1}\begin{pmatrix} (X_0(w))^* \\ (X_1(w))^* \\ \vdots \\ (X_m(w))^* \end{pmatrix}$$

$$\left(\text{resp., } K_{m;l}^{(\alpha,F)}(w,v) = \Big((X_0(w))^*, (X_1(w))^*, \ldots, (X_m(w))^*\Big)\Big(\mathbf{H}_{X,m}^{(F)}\Big)^{-1}\begin{pmatrix} X_0(v) \\ X_1(v) \\ \vdots \\ X_m(v) \end{pmatrix}\right)$$

holds for all $v, w \in \mathbb{C}_0 \setminus \mathbb{P}_{\alpha,m}$ (cf. [23, Remark 12]). Furthermore, for $w \in \mathbb{C}_0 \setminus \mathbb{P}_{\alpha,m}$, let $A_{m,w}^{(\alpha,F)} : \mathbb{C}_0 \setminus \mathbb{P}_{\alpha,m} \to \mathbb{C}^{q\times q}$ (resp., $C_{m,w}^{(\alpha,F)} : \mathbb{C}_0 \setminus \mathbb{P}_{\alpha,m} \to \mathbb{C}^{q\times q}$) be defined by

$$A_{m,w}^{(\alpha,F)}(v) := K_{m;r}^{(\alpha,F)}(v,w) \quad \Big(\text{resp., } C_{m,w}^{(\alpha,F)}(v) := K_{m;l}^{(\alpha,F)}(w,v)\Big).$$

Let $(\alpha_j)_{j=1}^{\infty} \in \mathcal{T}_1$ and $n \in \mathbb{N}$. Also, in view of Problem (R), let $X_0, X_1, \ldots, X_n$ be a basis of the right $\mathbb{C}^{q\times q}$-module $\mathcal{R}_{\alpha,n}^{q\times q}$ and suppose that $\mathbf{G}$ is a nonsingular matrix such that $\mathcal{M}[(\alpha_j)_{j=1}^{n}, \mathbf{G}; (X_k)_{k=0}^{n}] \neq \emptyset$. Let $v, w \in \mathbb{C}_0 \setminus \mathbb{P}_{\alpha,n}$. Then we set

$$A_{n,w}^{(\alpha)}(v) := \Big(X_0(v), X_1(v), \ldots, X_n(v)\Big)\mathbf{G}^{-1}\Big(X_0(w), X_1(w), \ldots, X_n(w)\Big)^* \tag{3.7}$$

and

$$C_{n,w}^{(\alpha)}(v) := \begin{pmatrix} X_0^{[\alpha,n]}(w) \\ X_1^{[\alpha,n]}(w) \\ \vdots \\ X_n^{[\alpha,n]}(w) \end{pmatrix}^* \mathbf{G}^{-1} \begin{pmatrix} X_0^{[\alpha,n]}(v) \\ X_1^{[\alpha,n]}(v) \\ \vdots \\ X_n^{[\alpha,n]}(v) \end{pmatrix}. \tag{3.8}$$

Because of these settings, if $F \in \mathcal{M}[(\alpha_j)_{j=1}^{n}, \mathbf{G}; (X_k)_{k=0}^{n}]$, then it follows

$$A_{n,w}^{(\alpha,F)} = A_{n,w}^{(\alpha)} \quad \text{and} \quad C_{n,w}^{(\alpha,F)} = C_{n,w}^{(\alpha)} \tag{3.9}$$

(see also [31, Remark 2.1]). In particular (cf. [25, Proposition 11]), we have

$$A_{n,w}^{(\alpha)}(w) > 0_{q\times q} \quad \text{and} \quad C_{n,w}^{(\alpha)}(w) > 0_{q\times q}, \quad w \in \mathbb{C}_0 \setminus \mathbb{P}_{\alpha,n}. \tag{3.10}$$

Based on [31, Remarks 6.1 and 6.2], similar to (3.7) and (3.8), we set

$$A_{n,w}^{(\alpha,\#)}(v) := \Big(X_0(v), X_1(v), \dots, X_n(v)\Big)\left(\mathbf{G}^{\#}\right)^{-1}\Big(X_0(w), X_1(w), \dots, X_n(w)\Big)^*$$

and

$$C_{n,w}^{(\alpha,\#)} := \begin{pmatrix} X_0^{[\alpha,n]}(w) \\ X_1^{[\alpha,n]}(w) \\ \vdots \\ X_n^{[\alpha,n]}(w) \end{pmatrix}^* \left(\mathbf{G}^{\#}\right)^{-1} \begin{pmatrix} X_0^{[\alpha,n]}(v) \\ X_1^{[\alpha,n]}(v) \\ \vdots \\ X_n^{[\alpha,n]}(v) \end{pmatrix},$$

where $\mathbf{G}^{\#} \in \mathbb{C}^{(n+1)q\times(n+1)q}$ stands for matrix $\mathbf{G}_{X,n}^{(F^{\#})}$ which is (uniquely) given by the reciprocal measure $F^{\#}$ corresponding to an $F \in \mathcal{M}[(\alpha_j)_{j=1}^n, \mathbf{G}; (X_k)_{k=0}^n]$.

*Remark* 3.9. By Remark 3.5 and [31, Lemma 6.4] (note (3.10)) one can conclude that, for some unitary $q \times q$ matrix $\mathbf{V}$, the rational matrix function

$$\begin{aligned}\Phi_{n;\mathbf{V}}^{(\alpha)} := \Big(& A_{n,\alpha_n}^{(\alpha,\#)}\,\mathbf{\Omega}_n^{-*}\sqrt{A_{n,\alpha_n}^{(\alpha)}(\alpha_n)}^{-1} - b_{\alpha_n}(C_{n,\alpha_n}^{(\alpha,\#)})^{[\alpha,n]}\,\mathbf{\Omega}_n^{-1}\sqrt{C_{n,\alpha_n}^{(\alpha)}(\alpha_n)}^{-1}\mathbf{V}\Big)\\ &\cdot\Big(A_{n,\alpha_n}^{(\alpha)}\sqrt{A_{n,\alpha_n}^{(\alpha)}(\alpha_n)}^{-1} + b_{\alpha_n}(C_{n,\alpha_n}^{(\alpha)})^{[\alpha,n]}\sqrt{C_{n,\alpha_n}^{(\alpha)}(\alpha_n)}^{-1}\mathbf{V}\Big)^{-1},\end{aligned}$$

where $\mathbf{\Omega}_n := (L_n^{\#})^{[\alpha,n]}(\alpha_n)\big(L_n^{[\alpha,n]}(\alpha_n)\big)^{-1}$, admits also the representations

$$\begin{aligned}\Phi_{n;\mathbf{V}}^{(\alpha)} = \Big(&\sqrt{C_{n,\alpha_n}^{(\alpha)}(\alpha_n)}^{-1} C_{n,\alpha_n}^{(\alpha)} + b_{\alpha_n}\mathbf{V}\sqrt{A_{n,\alpha_n}^{(\alpha)}(\alpha_n)}^{-1}(A_{n,\alpha_n}^{(\alpha)})^{[\alpha,n]}\Big)^{-1}\\ &\cdot\Big(\sqrt{C_{n,\alpha_n}^{(\alpha)}(\alpha_n)}^{-1}\mathbf{\Omega}_n^{-*}C_{n,\alpha_n}^{(\alpha,\#)} - b_{\alpha_n}\mathbf{V}\sqrt{A_{n,\alpha_n}^{(\alpha)}(\alpha_n)}^{-1}\mathbf{\Omega}_n^{-1}(A_{n,\alpha_n}^{(\alpha,\#)})^{[\alpha,n]}\Big),\end{aligned}$$

$$\begin{aligned}\Phi_{n;\mathbf{V}}^{(\alpha)} = \Big(&\frac{1}{b_{\alpha_n}}A_{n,\alpha_n}^{(\alpha,\#)}\,\mathbf{\Omega}_n^{-*}\sqrt{A_{n,\alpha_n}^{(\alpha)}(\alpha_n)}^{-1}\mathbf{V}^* - (C_{n,\alpha_n}^{(\alpha,\#)})^{[\alpha,n]}\,\mathbf{\Omega}_n^{-1}\sqrt{C_{n,\alpha_n}^{(\alpha)}(\alpha_n)}^{-1}\Big)\\ &\cdot\Big(\frac{1}{b_{\alpha_n}}A_{n,\alpha_n}^{(\alpha)}\sqrt{A_{n,\alpha_n}^{(\alpha)}(\alpha_n)}^{-1}\mathbf{V}^* + (C_{n,\alpha_n}^{(\alpha)})^{[\alpha,n]}\sqrt{C_{n,\alpha_n}^{(\alpha)}(\alpha_n)}^{-1}\Big)^{-1},\end{aligned}$$

$$\begin{aligned}\Phi_{n;\mathbf{V}}^{(\alpha)} = \Big(&\frac{1}{b_{\alpha_n}}\mathbf{V}^*\sqrt{C_{n,\alpha_n}^{(\alpha)}(\alpha_n)}^{-1}C_{n,\alpha_n}^{(\alpha)} + \sqrt{A_{n,\alpha_n}^{(\alpha)}(\alpha_n)}^{-1}(A_{n,\alpha_n}^{(\alpha)})^{[\alpha,n]}\Big)^{-1}\\ &\cdot\Big(\frac{1}{b_{\alpha_n}}\mathbf{V}^*\sqrt{C_{n,\alpha_n}^{(\alpha)}(\alpha_n)}^{-1}\mathbf{\Omega}_n^{-*}C_{n,\alpha_n}^{(\alpha,\#)} - \sqrt{A_{n,\alpha_n}^{(\alpha)}(\alpha_n)}^{-1}\mathbf{\Omega}_n^{-1}(A_{n,\alpha_n}^{(\alpha,\#)})^{[\alpha,n]}\Big).\end{aligned}$$

As an aside, we note that [31, Lemma 6.4] points out some other possibilities to calculate the nonsingular $q \times q$ matrix $\mathbf{\Omega}_n$ in Remark 3.9.

Using the notation of Remark 3.9, we get the following characterization of canonical solutions of Problem (R) for the nondegenerate case (similar to Theorem 3.6, but now) in terms of reproducing kernels of rational matrix functions.

**Theorem 3.10.** *Let $(\alpha_j)_{j=1}^{\infty} \in \mathcal{T}_1$. Furthermore, let $F \in \mathcal{M}_{\geq}^{q}(\mathbb{T}, \mathfrak{B}_{\mathbb{T}})$ and let $\Omega$ be the Riesz–Herglotz transform of $F$. Then the following statements are equivalent:*

(i) *$F$ is a canonical solution in $\mathcal{M}[(\alpha_j)_{j=1}^{n}, \mathbf{G}; (X_k)_{k=0}^{n}]$.*

(ii) *There is a unitary $q \times q$ matrix $\mathbf{V}$ such that $\Omega(v) = \Phi_{n;\mathbf{V}}^{(\alpha)}(v)$ for $v \in \mathbb{D} \setminus \mathbb{P}_{\alpha,n}$.*

*Moreover, if* (i) *holds, then the unitary $q \times q$ matrix $\mathbf{V}$ in* (ii) *is uniquely determined.*

*Proof.* Use Theorem 3.6 along with [31, Lemma 6.4]. □

As an aside, we briefly mention that similar to Corollary 3.7 one can draw a conclusion based on Theorem 3.10 instead of Theorem 3.6.

Note that, for a fixed canonical solution of Problem (R), the unitary matrices $\mathbf{U}$ and $\mathbf{V}$ occurring in Theorems 3.6 and 3.10 do not coincide in general. Furthermore, in view of Remark 2.1 (see also Proposition 1.12), Theorems 3.6 and 3.10 can be used for nonnegative definite sequences of matrices which are canonical of order $n+1$. In this context, Theorems 3.6 and 3.10 are related to [29, Theorem 6.5, Theorem 6.9, and Proposition 10.3].

In Theorem 3.6 (resp., Theorem 3.10) the condition $(\alpha_j)_{j=1}^{\infty} \in \mathcal{T}_1$ is chosen, since we want to apply the results on orthogonal rational matrix functions stated in [26] and [27]. However, for the somewhat more general case that only $\alpha_1, \alpha_2, \dots, \alpha_n \in \mathbb{C} \setminus \mathbb{T}$ is assumed, one can at least conclude the following.

**Proposition 3.11.** *Suppose that $\alpha_1, \alpha_2, \dots, \alpha_n \in \mathbb{C} \setminus \mathbb{T}$. Then there is a bijective correspondence between the set of all canonical solutions in $\mathcal{M}[(\alpha_j)_{j=1}^{n}, \mathbf{G}; (X_k)_{k=0}^{n}]$ and the set of all unitary $q \times q$ matrices. In particular, there exist uncountably infinitely many canonical solutions in $\mathcal{M}[(\alpha_j)_{j=1}^{n}, \mathbf{G}; (X_k)_{k=0}^{n}]$.*

*Proof.* Taking [24, Theorem 5.6 and Remark 5.10] into account, in view of [23, Proposition 5] and Remark 2.4 one can restrict the considerations without loss of generality to the special case in which $\alpha_j = 0$ for each $j \in \mathbb{N}_{1,n}$, the underlying matrix $\mathbf{G}$ is a nonnegative Hermitian $q \times q$ block Toeplitz matrix, and in which $X_k$ coincides with the complex $q \times q$ matrix polynomial $E_{k,q}$ defined by (2.1) for each $k \in \mathbb{N}_{0,n}$. For this case, however, the assertion follows immediately from Theorem 3.6 (resp., Theorem 3.10 or Remark 1.9 along with Proposition 1.12). □

*Remark* 3.12. Suppose that $\alpha_1, \alpha_2, \dots, \alpha_n \in \mathbb{C} \setminus \mathbb{T}$. Observe the special case $q = 1$, where $\mathbf{G}$ is a nonsingular $(n+1) \times (n+1)$ matrix and where $X_0, X_1, \dots, X_n$ is a basis of the linear space $\mathcal{R}_{\alpha,n}$. For this case, in view of Remark 2.15 (see also Remark 3.3) it follows that an $F \in \mathcal{M}[(\alpha_j)_{j=1}^{n}, \mathbf{G}; (X_k)_{k=0}^{n}]$ is a canonical solution if and only if $F$ admits (1.10) with $r = n+1$, some pairwise different points $z_1, z_2, \dots, z_r \in \mathbb{T}$, and a sequence $(\mathbf{A}_s)_{s=1}^{r}$ of positive numbers. Moreover, Proposition 3.11 shows that there is a bijective correspondence between the set of all canonical solutions in $\mathcal{M}[(\alpha_j)_{j=1}^{n}, \mathbf{G}; (X_k)_{k=0}^{n}]$ and the unit circle $\mathbb{T}$.

Finally, we now consider the elementary case $n = 0$, based on a constant function $X_0$ defined on $\mathbb{C}_0$ with a nonsingular $q \times q$ matrix $\mathbf{X}_0$ as value and a positive Hermitian $q \times q$ matrix $\mathbf{G}$. Suppose that $F \in \mathcal{M}_{\geq}^{q}(\mathbb{T}, \mathfrak{B}_{\mathbb{T}})$. Then, similar to Theorem 3.6 or Theorem 3.10 (cf. [29, Remark 6.8 and Example 9.11]), one can conclude that $F$ is a canonical solution with respect to (2.7) and (2.8) if and only if the Riesz–Herglotz transform $\Omega$ of $F$ admits the representation

$$\Omega(v) = \sqrt{\mathbf{X}_0\mathbf{G}^{-1}\mathbf{X}_0^*}^{\,-1}\mathbf{W}\begin{pmatrix} \frac{z_1+v}{z_1-v} & 0 & \cdots & 0 \\ 0 & \frac{z_2+v}{z_2-v} & \ddots & \vdots \\ \vdots & \ddots & \ddots & 0 \\ 0 & \cdots & 0 & \frac{z_q+v}{z_q-v} \end{pmatrix}\mathbf{W}^*\sqrt{\mathbf{X}_0\mathbf{G}^{-1}\mathbf{X}_0^*}^{\,-1} \tag{3.11}$$

for each $v \in \mathbb{D}$ with some unitary $q \times q$ matrix $\mathbf{W}$ and (not necessarily pairwise different) points $z_1, z_2, \dots, z_q \in \mathbb{T}$. In view of (3.11) and the matricial version of the Riesz–Herglotz Theorem (see [18, Theorem 2.2.2]) it follows that $F$ is a canonical solution with respect to (2.7) and (2.8) if and only if $F$ admits the representation

$$F = \sqrt{\mathbf{X}_0\mathbf{G}^{-1}\mathbf{X}_0^*}^{\,-1}\mathbf{W}\begin{pmatrix} \varepsilon_{z_1} & \mathfrak{o}_1 & \cdots & \mathfrak{o}_1 \\ \mathfrak{o}_1 & \varepsilon_{z_2} & \ddots & \vdots \\ \vdots & \ddots & \ddots & \mathfrak{o}_1 \\ \mathfrak{o}_1 & \cdots & \mathfrak{o}_1 & \varepsilon_{z_q} \end{pmatrix}\mathbf{W}^*\sqrt{\mathbf{X}_0\mathbf{G}^{-1}\mathbf{X}_0^*}^{\,-1} \tag{3.12}$$

(where $\mathfrak{o}_1$ is the zero measure in $\mathcal{M}^1_{\geq}(\mathbb{T}, \mathfrak{B}_{\mathbb{T}})$) with some unitary $q \times q$ matrix $\mathbf{W}$ and points $z_1, z_2, \dots, z_q \in \mathbb{T}$. In particular, one can see that for a canonical solution in this context each case of $r$ mass points with $r \in \mathbb{N}_{1,q}$ is possible.

## 4. Some conclusions from Theorem 3.6

Because of Remarks 2.9 and 2.14 (see also Theorem 2.10) we already know that canonical solutions of Problem (R) are molecular matrix measures with a special structure. In the present section, we will provide somewhat more insight into the weights corresponding to mass points which are associated with canonical solutions of Problem (R) for the nondegenerate case. The characterization of canonical solutions in Theorem 3.6 will be our starting point.

From now on, unless otherwise indicated, we act on the assumption that a basis $X_0, X_1, \dots, X_n$ of the right $\mathbb{C}^{q\times q}$-module $\mathcal{R}^{q\times q}_{\alpha,n}$ is given and that $\mathbf{G}$ is a nonsingular $(n+1)q\times(n+1)q$ matrix such that $\mathcal{M}[(\alpha_j)_{j=1}^n, \mathbf{G}; (X_k)_{k=0}^n] \neq \emptyset$ holds, where $(\alpha_j)_{j=1}^\infty \in \mathcal{T}_1$ and $n \in \mathbb{N}$ are arbitrary, but fixed. Let $[(L_k)_{k=0}^n, (R_k)_{k=0}^n]$ be a pair of orthonormal systems corresponding to the solution set $\mathcal{M}[(\alpha_j)_{j=1}^n, \mathbf{G}; (X_k)_{k=0}^n]$ and let $[(L_k^\#)_{k=0}^n, (R_k^\#)_{k=0}^n]$ be the dual pair of orthonormal systems corresponding to $[(L_k)_{k=0}^n, (R_k)_{k=0}^n]$. Furthermore, based on a unitary $q\times q$ matrix $\mathbf{U}$, let the rational matrix functions $P_{n;\mathbf{U}}^{(\alpha)}$ and $P_{n;\mathbf{U}}^{(\alpha,\#)}$ (resp., $Q_{n;\mathbf{U}}^{(\alpha)}$ and $Q_{n;\mathbf{U}}^{(\alpha,\#)}$) be given by (3.4).

In view of (3.4) and the choice of the set $\mathcal{R}^{q\times q}_{\alpha,n}$ there exist (uniquely determined) complex $q \times q$ matrix polynomials $p_{n;\mathbf{U}}^{(\alpha)}$ and $p_{n;\mathbf{U}}^{(\alpha,\#)}$ (resp., $q_{n;\mathbf{U}}^{(\alpha)}$ and $q_{n;\mathbf{U}}^{(\alpha,\#)}$)

of degree not greater than $n+1$ such that the identities

$$P_{n;\mathbf{U}}^{(\alpha)} = \frac{1}{\varpi_{\alpha,n}} p_{n;\mathbf{U}}^{(\alpha)} \quad \text{and} \quad P_{n;\mathbf{U}}^{(\alpha,\#)} = \frac{1}{\varpi_{\alpha,n}} p_{n;\mathbf{U}}^{(\alpha,\#)}$$

$$\left(\text{resp., } Q_{n;\mathbf{U}}^{(\alpha)} = \frac{1}{\varpi_{\alpha,n}} q_{n;\mathbf{U}}^{(\alpha)} \quad \text{and} \quad Q_{n;\mathbf{U}}^{(\alpha,\#)} = \frac{1}{\varpi_{\alpha,n}} q_{n;\mathbf{U}}^{(\alpha,\#)}\right)$$

are satisfied, where $\varpi_{\alpha,n} : \mathbb{C} \to \mathbb{C}$ is the polynomial defined by

$$\varpi_{\alpha,n}(u) := (1-\overline{\alpha_n}u)\prod_{j=1}^{n}(1-\overline{\alpha_j}u) \quad \Big( = (1-\overline{\alpha_n}u)\pi_{\alpha,n}(u)\Big).$$

With regard to these particular complex $q \times q$ matrix polynomials one can realize the following, where we use (for technical reasons) the notation

$$\eta := (-1)^{n+1}\cdot\eta_n\cdot\eta_1\cdot\ldots\cdot\eta_n \quad \left( = \frac{b_{\alpha_n}(0)\cdot b_{\alpha_1}(0)\cdot\ldots\cdot b_{\alpha_n}(0)}{\left(\widetilde{\varpi_{\alpha,n}}\right)^{[n+1]}(0)}\right).$$

**Lemma 4.1.** *Let $p_{n;\mathbf{U}}^{(\alpha)}$, $p_{n;\mathbf{U}}^{(\alpha,\#)}$, $q_{n;\mathbf{U}}^{(\alpha)}$, and $q_{n;\mathbf{U}}^{(\alpha,\#)}$ be the complex $q \times q$ matrix polynomials and $\eta$ be the complex number given by the expressions above. Then:*

(a) *The equalities $p_{n;\mathbf{U}}^{(\alpha)} = \eta\mathbf{U}(\widetilde{q_{n;\mathbf{U}}^{(\alpha)}})^{[n+1]}$ and $p_{n;\mathbf{U}}^{(\alpha,\#)} = -\eta\mathbf{U}(\widetilde{q_{n;\mathbf{U}}^{(\alpha,\#)}})^{[n+1]}$ hold. In particular, for each $v \in \mathbb{C}$, the relations $\mathcal{N}(p_{n;\mathbf{U}}^{(\alpha)}(v)) = \mathcal{N}((\widetilde{q_{n;\mathbf{U}}^{(\alpha)}})^{[n+1]}(v))$ and $\mathcal{N}(p_{n;\mathbf{U}}^{(\alpha,\#)}(v)) = \mathcal{N}((\widetilde{q_{n;\mathbf{U}}^{(\alpha,\#)}})^{[n+1]}(v))$ are satisfied as well as, for each $z \in \mathbb{T}$, $\mathcal{N}(p_{n;\mathbf{U}}^{(\alpha)}(z)) = \mathcal{N}((q_{n;\mathbf{U}}^{(\alpha)}(z))^*)$ and $\mathcal{N}(p_{n;\mathbf{U}}^{(\alpha,\#)}(z)) = \mathcal{N}((q_{n;\mathbf{U}}^{(\alpha,\#)}(z))^*)$ hold.*

(b) *There is some $w \in \mathbb{T}$ (resp., $\breve{w} \in \mathbb{T}$) such that $\det p_{n;\mathbf{U}}^{(\alpha)} = w\det(\widetilde{q_{n;\mathbf{U}}^{(\alpha)}})^{[n+1]}$ and $\det p_{n;\mathbf{U}}^{(\alpha,\#)} = (-1)^q w \det(\widetilde{q_{n;\mathbf{U}}^{(\alpha,\#)}})^{[n+1]}$ (resp., such that $\det p_{n;\mathbf{U}}^{(\alpha)} = \breve{w}\det q_{n;\mathbf{U}}^{(\alpha)}$ and $\det p_{n;\mathbf{U}}^{(\alpha,\#)} = \breve{w}\det q_{n;\mathbf{U}}^{(\alpha,\#)}$) hold.*

(c) *There exist at most $n+1$ pairwise different points $u_1, u_2, \ldots, u_{n+1} \in \mathbb{C}$ such that $p_{n;\mathbf{U}}^{(\alpha)}(u_s) = 0_{q\times q}$ (resp., $p_{n;\mathbf{U}}^{(\alpha,\#)}(u_s) = 0_{q\times q}$) for each $s \in \mathbb{N}_{1,n+1}$ and there exist at most $(n+1)q$ pairwise different points $z_1, z_2, \ldots, z_{(n+1)q} \in \mathbb{C}$ such that $\det p_{n;\mathbf{U}}^{(\alpha)}(z_s) = 0$ (resp., $\det p_{n;\mathbf{U}}^{(\alpha,\#)}(z_s) = 0$) for each $s \in \mathbb{N}_{1,(n+1)q}$.*

(d) *If any of the $q\times q$ matrices $p_{n;\mathbf{U}}^{(\alpha)}(z)$, $q_{n;\mathbf{U}}^{(\alpha)}(z)$, $(\widetilde{p_{n;\mathbf{U}}^{(\alpha)}})^{[n+1]}(z)$, or $(\widetilde{q_{n;\mathbf{U}}^{(\alpha)}})^{[n+1]}(z)$ (resp., $p_{n;\mathbf{U}}^{(\alpha,\#)}(z)$, $q_{n;\mathbf{U}}^{(\alpha,\#)}(z)$, $(\widetilde{p_{n;\mathbf{U}}^{(\alpha,\#)}})^{[n+1]}(z)$, or $(\widetilde{q_{n;\mathbf{U}}^{(\alpha,\#)}})^{[n+1]}(z)$) is singular for some $z \in \mathbb{C}$, then all of them are singular, where the point $z$ must belong to $\mathbb{T}$ and $p_{n;\mathbf{U}}^{(\alpha,\#)}(z) \neq 0_{q\times q}$ (resp., $p_{n;\mathbf{U}}^{(\alpha)}(z) \neq 0_{q\times q}$).*

(e) *If the value $p_{n;\mathbf{U}}^{(\alpha)}(u)$, $q_{n;\mathbf{U}}^{(\alpha)}(u)$, $(\widetilde{p_{n;\mathbf{U}}^{(\alpha)}})^{[n+1]}(u)$, or $(\widetilde{q_{n;\mathbf{U}}^{(\alpha)}})^{[n+1]}(u)$ (resp., $p_{n;\mathbf{U}}^{(\alpha,\#)}(u)$, $q_{n;\mathbf{U}}^{(\alpha,\#)}(u)$, $(\widetilde{p_{n;\mathbf{U}}^{(\alpha,\#)}})^{[n+1]}(u)$, or $(\widetilde{q_{n;\mathbf{U}}^{(\alpha,\#)}})^{[n+1]}(u)$) is $0_{q\times q}$ for some $u \in \mathbb{C}$, then all of them are $0_{q\times q}$, where $u \in \mathbb{T}$ and $\det p_{n;\mathbf{U}}^{(\alpha,\#)}(u) \neq 0$ (resp., $\det p_{n;\mathbf{U}}^{(\alpha)}(u) \neq 0$).*

(f) *The complex $q \times q$ matrix polynomials* $p_{n;\mathbf{U}}^{(\alpha)}$, $q_{n;\mathbf{U}}^{(\alpha)}$, $(\widetilde{p_{n;\mathbf{U}}^{(\alpha)}})^{[n+1]}$, $(\widetilde{q_{n;\mathbf{U}}^{(\alpha)}})^{[n+1]}$, $p_{n;\mathbf{U}}^{(\alpha,\#)}$, $q_{n;\mathbf{U}}^{(\alpha,\#)}$, $(\widetilde{p_{n;\mathbf{U}}^{(\alpha,\#)}})^{[n+1]}$, *and* $(\widetilde{q_{n;\mathbf{U}}^{(\alpha,\#)}})^{[n+1]}$ *are of (exact) degree* $n+1$ *and each of them has a nonsingular matrix as leading coefficient.*

*Proof.* (a) Let $l_n$ and $r_n$ be the (uniquely determined) complex $q \times q$ matrix polynomials of degree not greater than $n$ such that the relations

$$L_n = \frac{1}{\pi_{\alpha,n}} l_n \quad \text{and} \quad R_n = \frac{1}{\pi_{\alpha,n}} r_n$$

are satisfied. Thus, in view of [26, Proposition 2.13] we get

$$L_n^{[\alpha,n]} = \frac{\tilde{\eta}}{\pi_{\alpha,n}} \tilde{l}_n^{[n]} \quad \text{and} \quad R_n^{[\alpha,n]} = \frac{\tilde{\eta}}{\pi_{\alpha,n}} \tilde{r}_n^{[n]}$$

with $\tilde{\eta} := (-\eta_1) \cdot \ldots \cdot (-\eta_n)$. Similarly, [26, Proposition 2.13] leads to

$$(Q_{n;\mathbf{U}}^{(\alpha)})^{[\alpha,n+1]} = \frac{\eta}{\varpi_{\alpha,n}} (\widetilde{q_{n;\mathbf{U}}^{(\alpha)}})^{[n+1]},$$

where $\eta = (-\eta_1) \cdot \ldots \cdot (-\eta_n) \cdot (-\eta_n)$ holds (cf. the choice of $\eta$) and where $\alpha_{n+1} = \alpha_n$ (for technical reasons). Consequently, because of (3.4) it follows that

$$\begin{aligned} \frac{1}{\varpi_{\alpha,n}(u)} p_{n;\mathbf{U}}^{(\alpha)}(u) &= P_{n;\mathbf{U}}^{(\alpha)}(u) = \frac{\tilde{\eta}}{\pi_{\alpha,n}(u)} \tilde{r}_n^{[n]}(u) + \frac{b_{\alpha_n}(u)}{\pi_{\alpha,n}(u)} \mathbf{U} l_n(u) \\ &= \frac{1}{\varpi_{\alpha,n}(u)} \Big( \tilde{\eta}(1-\alpha_n u) \tilde{r}_n^{[n]}(u) + \eta_n(\alpha_n - u) \mathbf{U} l_n(u) \Big) \end{aligned} \tag{4.1}$$

and (using some rules for working with reciprocal rational matrix functions presented in [26, Section 2]) that

$$\begin{aligned} \frac{\eta}{\varpi_{\alpha,n}(u)} \mathbf{U} (\widetilde{q_{n;\mathbf{U}}^{(\alpha)}})^{[n+1]}(u) &= \mathbf{U} (Q_{n;\mathbf{U}}^{(\alpha)})^{[\alpha,n+1]}(u) = \mathbf{U} \Big( b_{\alpha_n}(u) L_n(u) + \mathbf{U}^* R_n^{[\alpha,n]}(u) \Big) \\ &= \frac{b_{\alpha_n}(u)}{\pi_{\alpha,n}(u)} \mathbf{U} l_n(u) + \frac{\tilde{\eta}}{\pi_{\alpha,n}(u)} \tilde{r}_n^{[n]}(u) \\ &= \frac{1}{\varpi_{\alpha,n}(u)} \Big( \eta_n(\alpha_n - u) \mathbf{U} l_n(u) + \tilde{\eta}(1-\overline{\alpha_n} u) \tilde{r}_n^{[n]}(u) \Big) \end{aligned}$$

for each $u \in \mathbb{C} \setminus \mathbb{P}_{\alpha,n}$. This implies $p_{n;\mathbf{U}}^{(\alpha)} = \eta \mathbf{U} (\widetilde{q_{n;\mathbf{U}}^{(\alpha)}})^{[n+1]}$. In particular, we get

$$\mathcal{N}(p_{n;\mathbf{U}}^{(\alpha)}(v)) = \mathcal{N}((\widetilde{q_{n;\mathbf{U}}^{(\alpha)}})^{[n+1]}(v))$$

for each $v \in \mathbb{C}$. Let $z \in \mathbb{T}$. Since $(\widetilde{q_{n;\mathbf{U}}^{(\alpha)}})^{[n+1]}(z) = z^{n+1} \big(q_{n;\mathbf{U}}^{(\alpha)}(z)\big)^*$ (see, e.g., [18, Lemma 1.2.2]), we obtain

$$\mathcal{N}(p_{n;\mathbf{U}}^{(\alpha)}(z)) = \mathcal{N}((\widetilde{q_{n;\mathbf{U}}^{(\alpha)}})^{[n+1]}(z)) = \mathcal{N}((q_{n;\mathbf{U}}^{(\alpha)}(z))^*).$$

An analogous argumentation yields the relations for $p_{n;\mathbf{U}}^{(\alpha,\#)}$ and $q_{n;\mathbf{U}}^{(\alpha,\#)}$.

(b) From the first two identities in (a) it follows immediately that

$$\det p_{n;\mathbf{U}}^{(\alpha)} = w \det(\widetilde{q_{n;\mathbf{U}}^{(\alpha)}})^{[n+1]} \quad \text{and} \quad \det p_{n;\mathbf{U}}^{(\alpha,\#)} = (-1)^q w \det(\widetilde{q_{n;\mathbf{U}}^{(\alpha,\#)}})^{[n+1]}$$

hold for some $w \in \mathbb{T}$. Furthermore, similar to (4.1) one can find that

$$\frac{1}{\varpi_{\alpha,n}(u)} q_{n;\mathbf{U}}^{(\alpha)}(u) = \frac{1}{\varpi_{\alpha,n}(u)} \Big( \tilde{\eta}(1-\overline{\alpha_n}u)\widetilde{l}_n^{[n]}(u) + \eta_n(\alpha_n - u) r_n(u)\mathbf{U} \Big)$$

for each $u \in \mathbb{C} \setminus \mathbb{P}_{\alpha,n}$. Beyond that, by [26, Proposition 2.13 and Theorem 6.10] we know that $\det \widetilde{r}_n^{[n]} = \breve{w} \det \widetilde{l}_n^{[n]}$ for some $\breve{w} \in \mathbb{T}$. Thus, recalling [26, Remark 6.2 and Lemma 6.5] and [18, Lemma 1.1.8], from (4.1) one can conclude

$$\begin{aligned}
\det p_{n;\mathbf{U}}^{(\alpha)}(u) &= \det\Big( \tilde{\eta}(1-\overline{\alpha_n}u)\mathbf{I}_q + \eta_n(\alpha_n - u)\mathbf{U} l_n(u) \big(\widetilde{r}_n^{[n]}(u)\big)^{-1} \Big) \det \widetilde{r}_n^{[n]}(u) \\
&= \breve{w} \big(\tilde{\eta}(1-\overline{\alpha_n}u)\big)^q \det\Big( \mathbf{I}_q + \frac{\eta_n(\alpha_n-u)}{\tilde{\eta}(1-\overline{\alpha_n}u)} \mathbf{U} \big(\widetilde{l}_n^{[n]}(u)\big)^{-1} r_n(u) \Big) \det \widetilde{l}_n^{[n]}(u) \\
&= \breve{w} \big(\tilde{\eta}(1-\overline{\alpha_n}u)\big)^q \det\Big( \mathbf{I}_q + \frac{b_{\alpha_n}(u)}{\tilde{\eta}} r_n(u)\mathbf{U} \big(\widetilde{l}_n^{[n]}(u)\big)^{-1} \Big) \det \widetilde{l}_n^{[n]}(u) \\
&= \breve{w} \det q_{n;\mathbf{U}}^{(\alpha)}(u)
\end{aligned}$$

for each $u \in \mathbb{C} \setminus \mathbb{P}_{\alpha,n}$ satisfying $\det \widetilde{r}_n^{[n]}(u) \neq 0$. Note that [26, Corollaries 4.4 and 4.7] and [18, Lemma 1.2.3] imply that the polynomial $\det \widetilde{r}_n^{[n]}$ has at most $nq$ pairwise different zeros. Therefore, by a continuity argument we get

$$\det p_{n;\mathbf{U}}^{(\alpha)} = \breve{w} \det q_{n;\mathbf{U}}^{(\alpha)}.$$

Since [26, Proposition 2.13 and Theorem 6.10] along with [39, Proposition 3.5 and Theorem 4.6] involving $\det \widetilde{r}_{n,\#}^{[n]} = \breve{w} \det \widetilde{l}_{n,\#}^{[n]}$, where $l_{n,\#}$ and $r_{n,\#}$ are the $q \times q$ matrix polynomials of degree not greater than $n$ such that the representations

$$L_n^{\#} = \frac{1}{\pi_{\alpha,n}} l_{n,\#} \quad \text{and} \quad R_n^{\#} = \frac{1}{\pi_{\alpha,n}} r_{n,\#}$$

are fulfilled, an analogous argumentation shows

$$\det p_{n;\mathbf{U}}^{(\alpha,\#)} = \breve{w} \det q_{n;\mathbf{U}}^{(\alpha,\#)}.$$

(d) Let any of the matrices $p_{n;\mathbf{U}}^{(\alpha)}(z)$, $q_{n;\mathbf{U}}^{(\alpha)}(z)$, $(\widetilde{p_{n;\mathbf{U}}^{(\alpha)}})^{[n+1]}(z)$, or $(\widetilde{q_{n;\mathbf{U}}^{(\alpha)}})^{[n+1]}(z)$ be singular for some $z \in \mathbb{C}$. By (b) we see that all of them are singular matrices. Let $\alpha_n \in \mathbb{D}$. With a view to [26, Corollary 4.4 and Remark 6.2] and [27, Theorem 3.10 and Lemma 3.11] we find out that $\det \widetilde{r}_n^{[n]}(v) \neq 0$ for each $v \in \mathbb{D}$. Additionally, considering [26, Corollary 4.4], from [31, Lemma 3.11] it follows that the matrix $b_{\alpha_n}(v)\, l_n(v) \big(\widetilde{r}_n^{[n]}(v)\big)^{-1}$ is strictly contractive for each $v \in \mathbb{D} \setminus \mathbb{P}_{\alpha,n}$. Accordingly, an elementary result on matricial Schur functions (see, e.g., [18, Lemma 2.1.5]) shows that $b_{\alpha_n}(v)\, l_n(v) \big(\widetilde{r}_n^{[n]}(v)\big)^{-1}$ is strictly contractive for each $v \in \mathbb{D}$. Thus (use, e.g., [18, Remark 1.1.2 and Lemma 1.1.13]), for each $v \in \mathbb{D}$, the matrix $\frac{b_{\alpha_n}(v)}{\tilde{\eta}}\mathbf{U}l_n(v)\big(\widetilde{r}_n^{[n]}(v)\big)^{-1}$ is strictly contractive as well and

$$\det p_{n;\mathbf{U}}^{(\alpha)}(v) = \big(\tilde{\eta}(1-\overline{\alpha_n}v)\big)^q \det\Big(\mathbf{I}_q + \frac{b_{\alpha_n}(v)}{\tilde{\eta}}\mathbf{U}l_n(v)\big(\widetilde{r}_n^{[n]}(v)\big)^{-1}\Big)\det \widetilde{r}_n^{[n]}(v) \neq 0.$$

Similarly, based on [26, Corollary 4.4 and Remark 6.2], [27, Theorem 3.10 and Lemma 3.11], and [31, Lemma 3.11] we have $\det l_n(v) \neq 0$ and one can verify that $\frac{\tilde{\eta}}{b_{\alpha_n}(v)}\mathbf{U}^*\widetilde{r}_n^{[n]}(v)\big(l_n(v)\big)^{-1}$ is a strictly contractive matrix and

$$\det p_{n;\mathbf{U}}^{(\alpha)}(v) \neq 0$$

for each $v \in \mathbb{D}$ in the case of $\alpha_n \in \mathbb{C}\setminus\mathbb{D}$. Combining this with (b) we obtain

$$\det\widetilde{(p_{n;\mathbf{U}}^{(\alpha)})}^{[n+1]}(v) \neq 0$$

for each $v \in \mathbb{D}$ as well. Therefore, from [18, Lemma 1.2.3] one can conclude that $z \in \mathbb{T}$. We suppose now, for a moment, that $p_{n;\mathbf{U}}^{(\alpha,\#)}(u) = 0_{q\times q}$ for some $u \in \mathbb{T}$. Thus, we have $P_{n;\mathbf{U}}^{(\alpha,\#)}(u) = 0_{q\times q}$ such that in view of (3.4) and [31, Lemma 3.11] we see that the matrix $(R_n^{\#})^{[\alpha,n]}(u)$ is nonsingular and that

$$\mathbf{U} = \overline{b_{\alpha_n}(u)}\left(L_n^{\#}(u)\big((R_n^{\#})^{[\alpha,n]}(u)\big)^{-1}\right)^*.$$

Moreover, from [39, Proposition 3.3] we can deduce that

$$R_n^{\#}(u)R_n^{[\alpha,n]}(u) + (L_n^{\#})^{[\alpha,n]}(u)L_n(u) = 2\frac{1-|\alpha_n|^2}{|1-\overline{\alpha_n}u|^2}B_{\alpha,n}^{(q)}(u).$$

In summary, recalling (3.4) and $u \in \mathbb{T}$, by using [26, Lemma 2.2] we get

$$\begin{aligned}
P_{n;\mathbf{U}}^{(\alpha)}(u) &= R_n^{[\alpha,n]}(u) + |b_{\alpha_n}(u)|^2\Big(L_n^{\#}(u)\big((R_n^{\#})^{[\alpha,n]}(u)\big)^{-1}\Big)^*L_n(u)\\
&= \big((R_n^{\#})^{[\alpha,n]}(u)\big)^{-*}\Big(\big((R_n^{\#})^{[\alpha,n]}(u)\big)^*R_n^{[\alpha,n]}(u) + \big(L_n^{\#}(u)\big)^*L_n(u)\Big)\\
&= \overline{B_{\alpha,n}^{(q)}(u)}\big((R_n^{\#})^{[\alpha,n]}(u)\big)^{-*}\Big(R_n^{\#}(u)R_n^{[\alpha,n]}(u) + (L_n^{\#})^{[\alpha,n]}(u)L_n(u)\Big)\\
&= 2\frac{1-|\alpha_n|^2}{|1-\overline{\alpha_n}u|^2}\big((R_n^{\#})^{[\alpha,n]}(u)\big)^{-*}.
\end{aligned}$$

In particular, the matrix $P_{n;\mathbf{U}}^{(\alpha)}(u)$ is nonsingular. Since $p_{n;\mathbf{U}}^{(\alpha)}(z)$ is singular, it follows that $p_{n;\mathbf{U}}^{(\alpha,\#)}(z) \neq 0_{q\times q}$ holds. Using the already proved part of (d) in combination with Remark 3.4 one can infer that, for some $z \in \mathbb{C}$, if any element of the set

$$\Big\{p_{n;\mathbf{U}}^{(\alpha,\#)}(z), q_{n;\mathbf{U}}^{(\alpha,\#)}(z), \widetilde{(p_{n;\mathbf{U}}^{(\alpha,\#)})}^{[n+1]}(z), \widetilde{(q_{n;\mathbf{U}}^{(\alpha,\#)})}^{[n+1]}(z)\Big\}$$

is a singular matrix, then all of them are singular, where $z \in \mathbb{T}$ and $p_{n;\mathbf{U}}^{(\alpha)}(z) \neq 0_{q\times q}$.

(e) Considering (a) and (b), the assertion of (e) follows from (d).

(f) By (d) one can see that the complex $q \times q$ matrices $(\widetilde{p_{n;\mathbf{U}}^{(\alpha)}})^{[n+1]}(0)$, $(\widetilde{q_{n;\mathbf{U}}^{(\alpha)}})^{[n+1]}(0)$, $p_{n;\mathbf{U}}^{(\alpha)}(0)$, and $q_{n;\mathbf{U}}^{(\alpha)}(0)$ as well as $(\widetilde{p_{n;\mathbf{U}}^{(\alpha,\#)}})^{[n+1]}(0)$, $(\widetilde{q_{n;\mathbf{U}}^{(\alpha,\#)}})^{[n+1]}(0)$, $p_{n;\mathbf{U}}^{(\alpha,\#)}(0)$, and $q_{n;\mathbf{U}}^{(\alpha,\#)}(0)$ are all nonsingular. This entails that each function of the set

$$\left\{p_{n;\mathbf{U}}^{(\alpha)}, q_{n;\mathbf{U}}^{(\alpha)}, (\widetilde{p_{n;\mathbf{U}}^{(\alpha)}})^{[n+1]}, (\widetilde{q_{n;\mathbf{U}}^{(\alpha)}})^{[n+1]}, p_{n;\mathbf{U}}^{(\alpha,\#)}, q_{n;\mathbf{U}}^{(\alpha,\#)}, (\widetilde{p_{n;\mathbf{U}}^{(\alpha,\#)}})^{[n+1]}, (\widetilde{q_{n;\mathbf{U}}^{(\alpha,\#)}})^{[n+1]}\right\}$$

is a complex $q \times q$ matrix polynomial of (exact) degree $n+1$ with a nonsingular $q \times q$ matrix as leading coefficient.
(c) This is a consequence of the Fundamental Theorem of Algebra and (f). □

**Lemma 4.2.** *Let $P_{n;\mathbf{U}}^{(\alpha)}$, $P_{n;\mathbf{U}}^{(\alpha,\#)}$, $Q_{n;\mathbf{U}}^{(\alpha)}$, and $Q_{n;\mathbf{U}}^{(\alpha,\#)}$ be the rational matrix function given by (3.4) with a unitary $q \times q$ matrix $\mathbf{U}$. Furthermore, let $\alpha_{n+1} = \alpha_n$. Then:*

(a) *The equalities $P_{n;\mathbf{U}}^{(\alpha)} = \mathbf{U}(Q_{n;\mathbf{U}}^{(\alpha)})^{[\alpha,n+1]}$ and $P_{n;\mathbf{U}}^{(\alpha,\#)} = -\mathbf{U}(Q_{n;\mathbf{U}}^{(\alpha,\#)})^{[\alpha,n+1]}$ are satisfied. In particular, the relations $\mathcal{N}(P_{n;\mathbf{U}}^{(\alpha)}(v)) = \mathcal{N}((Q_{n;\mathbf{U}}^{(\alpha)})^{[\alpha,n+1]}(v))$ and $\mathcal{N}(P_{n;\mathbf{U}}^{(\alpha,\#)}(v)) = \mathcal{N}((Q_{n;\mathbf{U}}^{(\alpha,\#)})^{[\alpha,n+1]}(v))$ for all $v \in \mathbb{C}_0 \setminus \mathbb{P}_{\alpha,n}$ holds as well as $\mathcal{N}(P_{n;\mathbf{U}}^{(\alpha)}(z)) = \mathcal{N}((Q_{n;\mathbf{U}}^{(\alpha)}(z))^*)$ and $\mathcal{N}(P_{n;\mathbf{U}}^{(\alpha,\#)}(z)) = \mathcal{N}((Q_{n;\mathbf{U}}^{(\alpha,\#)}(z))^*)$ for $z \in \mathbb{T}$.*

(b) *There is a $w \in \mathbb{T}$ (resp., a $\breve{w} \in \mathbb{T}$) such that $\det P_{n;\mathbf{U}}^{(\alpha)} = w \det(Q_{n;\mathbf{U}}^{(\alpha)})^{[\alpha,n+1]}$ and $\det P_{n;\mathbf{U}}^{(\alpha,\#)} = (-1)^q w \det(Q_{n;\mathbf{U}}^{(\alpha,\#)})^{[\alpha,n+1]}$ (resp., that $\det P_{n;\mathbf{U}}^{(\alpha)} = \breve{w} \det Q_{n;\mathbf{U}}^{(\alpha)}$ and $\det P_{n;\mathbf{U}}^{(\alpha,\#)} = \breve{w} \det Q_{n;\mathbf{U}}^{(\alpha,\#)}$) hold.*

(c) *There exist at most $n+1$ pairwise different points $u_1, u_2, \dots, u_{n+1} \in \mathbb{C}_0 \setminus \mathbb{P}_{\alpha,n}$ such that $P_{n;\mathbf{U}}^{(\alpha)}(u_s) = 0_{q\times q}$ (resp., $P_{n;\mathbf{U}}^{(\alpha,\#)}(u_s) = 0_{q\times q}$) for each $s \in \mathbb{N}_{1,n+1}$ and at most $(n+1)q$ pairwise different points $z_1, z_2, \dots, z_{(n+1)q} \in \mathbb{C}_0 \setminus \mathbb{P}_{\alpha,n}$ such that $\det P_{n;\mathbf{U}}^{(\alpha)}(z_s) = 0$ (resp., $\det P_{n;\mathbf{U}}^{(\alpha,\#)}(z_s) = 0$) for each $s \in \mathbb{N}_{1,(n+1)q}$.*

(d) *If $P_{n;\mathbf{U}}^{(\alpha)}(z)$, $Q_{n;\mathbf{U}}^{(\alpha)}(z)$, $(P_{n;\mathbf{U}}^{(\alpha)})^{[\alpha,n+1]}(z)$, or $(Q_{n;\mathbf{U}}^{(\alpha)})^{[\alpha,n+1]}(z)$ (resp., $P_{n;\mathbf{U}}^{(\alpha,\#)}(z)$, $Q_{n;\mathbf{U}}^{(\alpha,\#)}(z)$, $(Q_{n;\mathbf{U}}^{(\alpha,\#)})^{[\alpha,n+1]}(z)$, or $(Q_{n;\mathbf{U}}^{(\alpha,\#)})^{[\alpha,n+1]}(z)$) is a singular $q \times q$ matrix for some $z \in \mathbb{C}_0 \setminus \mathbb{P}_{\alpha,n}$, then all of them are singular, where $z$ must belong to $\mathbb{T}$ and $P_{n;\mathbf{U}}^{(\alpha,\#)}(z) \neq 0_{q\times q}$ (resp., $P_{n;\mathbf{U}}^{(\alpha)}(z) \neq 0_{q\times q}$).*

(e) *If any of the values $P_{n;\mathbf{U}}^{(\alpha)}(u)$, $Q_{n;\mathbf{U}}^{(\alpha)}(u)$, $(P_{n;\mathbf{U}}^{(\alpha)})^{[\alpha,n+1]}(u)$, or $(Q_{n;\mathbf{U}}^{(\alpha)})^{[\alpha,n+1]}(u)$ (resp., of $P_{n;\mathbf{U}}^{(\alpha,\#)}(u)$, $Q_{n;\mathbf{U}}^{(\alpha,\#)}(u)$, $(Q_{n;\mathbf{U}}^{(\alpha,\#)})^{[\alpha,n+1]}(u)$, or $(Q_{n;\mathbf{U}}^{(\alpha,\#)})^{[\alpha,n+1]}(u)$) is equal to $0_{q\times q}$ for some $u \in \mathbb{C}_0 \setminus \mathbb{P}_{\alpha,n}$, then all of them are equal to $0_{q\times q}$, where $u \in \mathbb{T}$ and $\det P_{n;\mathbf{U}}^{(\alpha,\#)}(u) \neq 0$ (resp., $\det P_{n;\mathbf{U}}^{(\alpha)}(u) \neq 0$).*

(f) *The matrix functions $P_{n;\mathbf{U}}^{(\alpha)}$, $Q_{n;\mathbf{U}}^{(\alpha)}$, $(P_{n;\mathbf{U}}^{(\alpha)})^{[\alpha,n+1]}$, $(Q_{n;\mathbf{U}}^{(\alpha)})^{[\alpha,n+1]}$, $P_{n;\mathbf{U}}^{(\alpha,\#)}$, $Q_{n;\mathbf{U}}^{(\alpha,\#)}$, $(P_{n;\mathbf{U}}^{(\alpha,\#)})^{[\alpha,n+1]}$, and $(Q_{n;\mathbf{U}}^{(\alpha,\#)})^{[\alpha,n+1]}$ belong to $\mathcal{R}_{\alpha,n+1}^{q\times q} \setminus \mathcal{R}_{\alpha,n}^{q\times q}$.*

*Proof.* Taking (3.4) and the choice of the complex $q \times q$ matrix polynomials $p_{n;\mathbf{U}}^{(\alpha)}$, $p_{n;\mathbf{U}}^{(\alpha,\#)}$, $q_{n;\mathbf{U}}^{(\alpha)}$, and $q_{n;\mathbf{U}}^{(\alpha,\#)}$ in Lemma 4.1 into account, the assertions of (a)–(e) are an easy consequence of Lemma 4.1 along with [26, Proposition 2.13]. Furthermore, from (d) we can conclude that the complex $q \times q$ matrices $(P_{n;\mathbf{U}}^{(\alpha)})^{[\alpha,n+1]}(\alpha_{n+1})$,

$(Q_{n;\mathbf{U}}^{(\alpha)})^{[\alpha,n+1]}(\alpha_{n+1})$, $P_{n;\mathbf{U}}^{(\alpha)}(\alpha_{n+1})$, and $Q_{n;\mathbf{U}}^{(\alpha)}(\alpha_{n+1})$ as well as $(P_{n;\mathbf{U}}^{(\alpha,\#)})^{[\alpha,n+1]}(\alpha_{n+1})$, $(Q_{n;\mathbf{U}}^{(\alpha,\#)})^{[\alpha,n+1]}(\alpha_{n+1})$, $P_{n;\mathbf{U}}^{(\alpha,\#)}(\alpha_{n+1})$, and $Q_{n;\mathbf{U}}^{(\alpha,\#)}(\alpha_{n+1})$ are all nonsingular. This leads in combination with [26, Equation (2.10)] to (f). □

Recall that, in view of [32], the rational matrix functions defined by (3.4) can be interpreted as elements of special para-orthogonal systems. In this regard, Lemma 4.2 is closely related to [32, Theorem 6.8].

Now, based on Theorem 3.6 and Lemma 4.2, we gain somewhat more insight into the structure of canonical solutions of Problem (R) for the nondegenerate case. In view of Theorem 3.6 and Corollary 3.7, having fixed a unitary $q \times q$ matrix $\mathbf{U}$, we will reapply the notation $F_{n;\mathbf{U}}^{(\alpha)}$ for the matrix measure belonging to $\mathcal{M}_{\geq}^{q}(\mathbb{T}, \mathfrak{B}_{\mathbb{T}})$ such that its Riesz–Herglotz transform $\Omega_{n;\mathbf{U}}^{(\alpha)}$ satisfies the identity

$$\Omega_{n;\mathbf{U}}^{(\alpha)}(v) = \Psi_{n;\mathbf{U}}^{(\alpha)}(v), \quad v \in \mathbb{D} \setminus \mathbb{P}_{\alpha,n}.$$

In doing so, $\Psi_{n;\mathbf{U}}^{(\alpha)}$ is the rational matrix function defined as in Remark 3.5 by

$$\Psi_{n;\mathbf{U}}^{(\alpha)} := \left(P_{n;\mathbf{U}}^{(\alpha)}\right)^{-1} P_{n;\mathbf{U}}^{(\alpha,\#)}.$$

**Theorem 4.3.** *Let $(\alpha_j)_{j=1}^{\infty} \in \mathcal{T}_1$ and let $n \in \mathbb{N}$. Let $X_0, X_1, \dots, X_n$ be a basis of the right $\mathbb{C}^{q\times q}$-module $\mathcal{R}_{\alpha,n}^{q\times q}$ and suppose that $\mathbf{G}$ is a nonsingular matrix such that $\mathcal{M}[(\alpha_j)_{j=1}^{n}, \mathbf{G}; (X_k)_{k=0}^{n}] \neq \emptyset$ holds. Furthermore, let $\mathbf{U}$ be a unitary $q \times q$ matrix and let $P_{n;\mathbf{U}}^{(\alpha)}$ and $Q_{n;\mathbf{U}}^{(\alpha)}$ be the rational matrix functions given by (3.4). Then:*

(a) *There exist some $r \in \mathbb{N}_{n+1,(n+1)q}$, pairwise different points $z_1, z_2, \dots, z_r \in \mathbb{T}$, and a sequence $(\mathbf{A}_s)_{s=1}^{r}$ of nonnegative Hermitian $q \times q$ matrices each of which is not equal to the zero matrix such that the representation*

$$F_{n;\mathbf{U}}^{(\alpha)} = \sum_{s=1}^{r} \varepsilon_{z_s} \mathbf{A}_s \tag{4.2}$$

*holds. In particular, the equality*

$$\sum_{s=1}^{r} \operatorname{rank} \mathbf{A}_s = (n+1)q \tag{4.3}$$

*is satisfied, where $\mathcal{R}(\mathbf{A}_s) = \mathcal{R}\big(F_{n;\mathbf{U}}^{(\alpha)}(\{z_s\})\big)$ for each $s \in \mathbb{N}_{1,r}$.*

(b) *If $z \in \mathbb{T}$, then the relations*

$$\mathcal{N}\big(P_{n;\mathbf{U}}^{(\alpha)}(z)\big) = \mathcal{N}\big((Q_{n;\mathbf{U}}^{(\alpha)}(z))^*\big) = \mathcal{R}\big(F_{n;\mathbf{U}}^{(\alpha)}(\{z\})\big) \tag{4.4}$$

*and*

$$F_{n;\mathbf{U}}^{(\alpha)}(\{z\}) A_{n,z}^{(\alpha)}(z)\mathbf{x} = \mathbf{x}, \quad \mathbf{x} \in \mathcal{R}\big(F_{n;\mathbf{U}}^{(\alpha)}(\{z\})\big), \tag{4.5}$$

*hold, where $A_{n,z}^{(\alpha)}(z) = C_{n,z}^{(\alpha)}(z)$ and where $A_{n,z}^{(\alpha)}$ and $C_{n,z}^{(\alpha)}$ are the rational matrix functions given by (3.7) and (3.8) with $w = z$.*

(c) *For some $z \in \mathbb{C}_0 \setminus \mathbb{P}_{\alpha,n}$, $\det P_{n;\mathbf{U}}^{(\alpha)}(z) = 0$ holds if and only if $z \in \{z_1, z_2, \dots, z_r\}$.*

*Proof.* Obviously, to prove (4.2) it is enough to show that the Riesz–Herglotz transform $\Omega_{n;\mathbf{U}}^{(\alpha)}$ of $F_{n;\mathbf{U}}^{(\alpha)}$ admits the representation

$$\Omega_{n;\mathbf{U}}^{(\alpha)}(v) = \sum_{s=1}^{r} \frac{z_s + v}{z_s - v} \mathbf{A}_s, \quad v \in \mathbb{D}, \tag{4.6}$$

with some $r \in \mathbb{N}_{n+1,(n+1)q}$, pairwise different points $z_1, z_2, \ldots, z_r \in \mathbb{T}$, and a sequence $(\mathbf{A}_s)_{s=1}^r$ of nonnegative Hermitian $q \times q$ matrices each of which is not equal to the zero matrix. From part (c) of Lemma 4.2 we know that the set $\Delta$ of points $z \in \mathbb{C}_0 \setminus \mathbb{P}_{\alpha,n}$ such that $\det P_{n;\mathbf{U}}^{(\alpha)}(z) = 0$ holds consists of at most $(n+1)q$ pairwise different elements. Let $z \in \mathbb{T} \setminus \Delta$. From [39, Proposition 3.3] we get

$$L_n(z) R_n^{\#}(z) = L_n^{\#}(z) R_n(z), \quad (R_n^{\#})^{[\alpha,n]}(z) L_n^{[\alpha,n]}(z) = R_n^{[\alpha,n]}(z) (L_n^{\#})^{[\alpha,n]}(z)$$

as well as

$$\begin{aligned} L_n(z)(L_n^{\#})^{[\alpha,n]}(z) + L_n^{\#}(z) L_n^{[\alpha,n]}(z) &= 2 \frac{1-|\alpha_n|^2}{|1-\overline{\alpha_n} z|^2} B_{\alpha,n}^{(q)}(z), \\ (R_n^{\#})^{[\alpha,n]}(z) R_n(z) + R_n^{[\alpha,n]}(z) R_n^{\#}(z) &= 2 \frac{1-|\alpha_n|^2}{|1-\overline{\alpha_n} z|^2} B_{\alpha,n}^{(q)}(z). \end{aligned}$$

Hence, if $\Psi_{n;\mathbf{U}}^{(\alpha)} := \left(P_{n;\mathbf{U}}^{(\alpha)}\right)^{-1} P_{n;\mathbf{U}}^{(\alpha,\#)}$ as in Remark 3.5 and if

$$N(z) := 2 \operatorname{Re} \Psi_{n;\mathbf{U}}^{(\alpha)}(z),$$

then by (3.4), [26, Lemma 2.2], $z \in \mathbb{T} \setminus \Delta$, and the unitarity of $\mathbf{U}$ it follows

$$\begin{aligned} N(z) &= \left(P_{n;\mathbf{U}}^{(\alpha)}(z)\right)^{-1} P_{n;\mathbf{U}}^{(\alpha,\#)}(z) + \left(\left(P_{n;\mathbf{U}}^{(\alpha)}(z)\right)^{-1} P_{n;\mathbf{U}}^{(\alpha,\#)}(z)\right)^* \\ &= \left(P_{n;\mathbf{U}}^{(\alpha)}(z)\right)^{-1} \Big(P_{n;\mathbf{U}}^{(\alpha,\#)}(z)\left(P_{n;\mathbf{U}}^{(\alpha)}(z)\right)^* + P_{n;\mathbf{U}}^{(\alpha)}(z)\left(P_{n;\mathbf{U}}^{(\alpha,\#)}(z)\right)^*\Big)\left(P_{n;\mathbf{U}}^{(\alpha)}(z)\right)^{-*} \\ &= \left(P_{n;\mathbf{U}}^{(\alpha)}(z)\right)^{-1} \Big((R_n^{\#})^{[\alpha,n]}(z)\left(R_n^{[\alpha,n]}(z)\right)^* + \overline{b_{\alpha_n}(z)}(R_n^{\#})^{[\alpha,n]}(z)\left(L_n(z)\right)^* \mathbf{U}^* \\ &\qquad - b_{\alpha_n}(z) \mathbf{U} L_n^{\#}(z)\left(R_n^{[\alpha,n]}(z)\right)^* - \mathbf{U} L_n^{\#}(z)\left(L_n(z)\right)^* \mathbf{U}^* \\ &\qquad - \overline{b_{\alpha_n}(z)} R_n^{[\alpha,n]}(z)\left(L_n^{\#}(z)\right)^* \mathbf{U}^* + b_{\alpha_n}(z) \mathbf{U} L_n(z)\left((R_n^{\#})^{[\alpha,n]}(z)\right)^* \\ &\qquad - \mathbf{U} L_n(z)\left(L_n^{\#}(z)\right)^* \mathbf{U}^* + R_n^{[\alpha,n]}(z)\left((R_n^{\#})^{[\alpha,n]}(z)\right)^*\Big)\left(P_{n;\mathbf{U}}^{(\alpha)}(z)\right)^{-*} \\ &= \overline{B_{\alpha,n}^{(q)}(z)}\left(P_{n;\mathbf{U}}^{(\alpha)}(z)\right)^{-1} \Big((R_n^{\#})^{[\alpha,n]}(z) R_n(z) + \overline{b_{\alpha_n}(z)}(R_n^{\#})^{[\alpha,n]}(z) L_n^{[\alpha,n]}(z) \mathbf{U}^* \\ &\qquad - b_{\alpha_n}(z) \mathbf{U} L_n^{\#}(z) R_n(z) - \mathbf{U} L_n^{\#}(z) L_n^{[\alpha,n]}(z) \mathbf{U}^* + R_n^{[\alpha,n]}(z) R_n^{\#}(z) \\ &\qquad - \overline{b_{\alpha_n}(z)} R_n^{[\alpha,n]}(z)(L_n^{\#})^{[\alpha,n]}(z) \mathbf{U}^* + b_{\alpha_n}(z) \mathbf{U} L_n(z) R_n^{\#}(z) \\ &\qquad - \mathbf{U} L_n(z)(L_n^{\#})^{[\alpha,n]}(z) \mathbf{U}^*\Big)\left(P_{n;\mathbf{U}}^{(\alpha)}(z)\right)^{-*} \\ &= 2 \frac{1-|\alpha_n|^2}{|1-\overline{\alpha_n} z|^2}\left(P_{n;\mathbf{U}}^{(\alpha)}(z)\right)^{-1}\left(\mathbf{I}_q - \mathbf{U}\mathbf{U}^*\right)\left(P_{n;\mathbf{U}}^{(\alpha)}(z)\right)^{-*} = 0_{q \times q}. \end{aligned}$$

Recall that the set $\Delta$ consists of at most $(n+1)q$ elements, that $\Omega_{n;\mathbf{U}}^{(\alpha)}$ is by definition the restriction of the rational matrix function $\Psi_{n;\mathbf{U}}^{(\alpha)}$ to $\mathbb{D}$, and that

$\Omega_{n;\mathbf{U}}^{(\alpha)}(0) = F_{n;\mathbf{U}}^{(\alpha)}(\mathbb{T})$. Taking this and $N(z) = 0_{q\times q}$ into account, an application of [28, Lemma 6.6] yields the representation (4.6) with some $r \in \mathbb{N}_{1,(n+1)q}$, pairwise different points $z_1, z_2, \ldots, z_r \in \mathbb{T}$, and some sequence $(\mathbf{A}_s)_{s=1}^r$ of nonnegative Hermitian $q \times q$ matrices. Since Theorem 3.6 implies that the matrix measure $F_{n;\mathbf{U}}^{(\alpha)}$ belongs to the solution set $\mathcal{M}[(\alpha_j)_{j=1}^n, \mathbf{G}; (X_k)_{k=0}^n]$ and since $\mathbf{G}$ is nonsingular, in view of [24, Theorem 5.6 and Proposition 6.1] and (4.6) one can conclude that $r \in \mathbb{N}_{n+1,(n+1)q}$ and that $\mathbf{A}_s \neq 0_{q\times q}$ for each $s \in \mathbb{N}_{1,r}$. In particular, we get that $F$ admits the asserted representation subject to (4.2) and that

$$F_{n;\mathbf{U}}^{(\alpha)}(\{z\}) = \begin{cases} 0_{q\times q} & \text{if } z \in \mathbb{T} \setminus \{z_1, z_2, \ldots, z_r\} \\ \mathbf{A}_s & \text{if } z = z_s \text{ for some } s \in \mathbb{N}_{1,r}\,. \end{cases}$$

Consequently, the relation

$$\mathcal{R}\big(F_{n;\mathbf{U}}^{(\alpha)}(\{z\})\big) = \begin{cases} 0_{q\times 1} & \text{if } z \in \mathbb{T} \setminus \{z_1, z_2, \ldots, z_r\} \\ \mathcal{R}(\mathbf{A}_s) & \text{if } z = z_s \text{ for some } s \in \mathbb{N}_{1,r} \end{cases} \tag{4.7}$$

holds. Considering (3.4), a comparison of the choice of $F_{n;\mathbf{U}}^{(\alpha)}$ with (4.6) shows that

$$P_{n;\mathbf{U}}^{(\alpha,\#)}(u) = P_{n;\mathbf{U}}^{(\alpha)}(u) \sum_{s=1}^{r} \frac{z_s + u}{z_s - u} \mathbf{A}_s \tag{4.8}$$

for each $u \in \mathbb{C} \setminus (\mathbb{P}_{\alpha,n} \cup \{z_1, z_2, \ldots, z_r\})$. In view of (4.8), part (f) of Lemma 4.2, and $\mathbb{P}_{\alpha,n} \subset \mathbb{C}_0 \setminus \mathbb{T}$ one can see that $\{z_1, z_2, \ldots, z_r\} \subseteq \Delta$ and that

$$\mathcal{R}(\mathbf{A}_s) \subseteq \mathcal{N}\big(P_{n;\mathbf{U}}^{(\alpha)}(z_s)\big)$$

for all $s \in \mathbb{N}_{1,r}$. Now, let $z \in \Delta$. Therefore, there is some $\mathbf{x} \in \mathbb{C}^{q\times 1} \setminus \{0_{q\times 1}\}$ such that $P_{n;\mathbf{U}}^{(\alpha)}(z)\mathbf{x} = 0_{q\times 1}$, where part (d) of Lemma 4.2 implies $z \in \mathbb{T}$. Thus, (3.4) yields

$$\mathbf{x}^* \big(R_n^{[\alpha,n]}(z)\big)^* = -\mathbf{x}^* \big(b_{\alpha_n}(z) L_n(z)\big)^* \mathbf{U}^*.$$

Accordingly, by (3.4) and (4.8) we get

$$\begin{aligned}
&\mathbf{x}^* \Big( \big(R_n^{[\alpha,n]}(z)\big)^* (R_n^{\#})^{[\alpha,n]}(u) + \overline{b_{\alpha_n}(z)} b_{\alpha_n}(u) \big(L_n(z)\big)^* L_n^{\#}(u) \Big) \\
&= -\mathbf{x}^* \big(b_{\alpha_n}(z) L_n(z)\big)^* \mathbf{U}^* \Big( (R_n^{\#})^{[\alpha,n]}(u) - b_{\alpha_n}(u) \mathbf{U} L_n^{\#}(u) \Big) \\
&= -\mathbf{x}^* \big(b_{\alpha_n}(z) L_n(z)\big)^* \mathbf{U}^* \Big( R_n^{[\alpha,n]}(u) + b_{\alpha_n}(u) \mathbf{U} L_n(u) \Big) \sum_{s=1}^{r} \frac{z_s + u}{z_s - u} \mathbf{A}_s \\
&= \mathbf{x}^* \Big( \big(R_n^{[\alpha,n]}(z)\big)^* R_n^{[\alpha,n]}(u) - \overline{b_{\alpha_n}(z)} b_{\alpha_n}(u) \big(L_n(z)\big)^* L_n(u) \Big) \sum_{s=1}^{r} \frac{z_s + u}{z_s - u} \mathbf{A}_s
\end{aligned}$$

for each $u \in \mathbb{C} \setminus (\mathbb{P}_{\alpha,n} \cup \{z_1, z_2, \ldots, z_r\})$. Recalling $\mathcal{M}[(\alpha_j)_{j=1}^n, \mathbf{G}; (X_k)_{k=0}^n] \neq \emptyset$ and (3.9), this leads along with the Christoffel–Darboux formulas for orthogonal rational matrix functions (see [26, Lemma 5.1 and Corollary 5.5] and [39, Proposition 3.1]) to the relation

$$\mathbf{x}^*\left(\frac{2}{1-\overline{z}u}\mathbf{I}_q - \sum_{k=0}^{n}\big(L_k(z)\big)^* L_k^{\#}(u)\right) = \mathbf{x}^*\sum_{k=0}^{n}\big(L_k(z)\big)^* L_k(u)\sum_{s=1}^{r}\frac{z_s+u}{z_s-u}\mathbf{A}_s$$
$$= \mathbf{x}^* C_{n,z}^{(\alpha)}(u)\sum_{s=1}^{r}\frac{z_s+u}{z_s-u}\mathbf{A}_s$$

for each $u \in \mathbb{C} \setminus (\mathbb{P}_{\alpha,n} \cup \{z, z_1, z_2, \dots, z_r\})$. Consequently, for some $s \in \mathbb{N}_{1,r}$, a continuity argument gives rise to $z = z_s$ and to

$$\mathbf{x}^* = \mathbf{x}^* C_{n,z_s}^{(\alpha)}(z_s)\mathbf{A}_s. \tag{4.9}$$

From (4.9) it follows $\mathbf{x} \in \mathcal{R}(\mathbf{A}_s)$. Summing up, we obtain $\Delta = \{z_1, z_2, \dots, z_r\}$ and (4.4) for each $z \in \mathbb{T}$ (note $\mathcal{R}(\mathbf{A}_s) \subseteq \mathcal{N}\big(P_{n;\mathbf{U}}^{(\alpha)}(z_s)\big)$ and (4.7) along with part (a) of Lemma 4.2). Furthermore, taking into account $\Delta = \{z_1, z_2, \dots, z_r\}$ and (4.7), in view of (4.4) and the Fundamental Theorem of Algebra one can conclude that (4.3) holds (cf. part (c) of Lemma 4.1 and part (c) of Lemma 4.2). For $z \in \mathbb{T}$, based on (3.7) and (3.8) along with some rules for working with reciprocal rational matrix functions stated in [26, Section 2] one can also see that the equality

$$A_{n,z}^{(\alpha)}(z) = C_{n,z}^{(\alpha)}(z)$$

is satisfied (besides, note (3.9) and [24, Theorem 5.6] as well as [25, Lemma 5]), where from (3.10) we know that $A_{n,z}^{(\alpha)}(z)$ is a positive Hermitian $q \times q$ matrix. Thus, by using (4.9) and $\Delta = \{z_1, z_2, \dots, z_r\}$ in combination with (4.4) and (4.7) we obtain that (4.5) holds as well. □

As an aside we will point out that, based on (4.3) and [24, Theorem 6.6], one can conclude that the matrix measure $F_{n;\mathbf{U}}^{(\alpha)}$ is a canonical solution of Problem (R) in an alternative way (to the given proof in Theorem 3.6). Furthermore, since the proof of Theorem 4.3 is performed without recourse to Theorem 2.10 (and Proposition 1.12), this leads to an alternative approach to Remarks 3.1–3.3 as well.

With a view to the functions given by (3.4) and the concept of reciprocal measures in $\mathcal{M}_{\geq}^{q}(\mathbb{T}, \mathfrak{B}_{\mathbb{T}})$, Theorem 4.3 implies the following (cf. [31, Proposition 6.3]).

**Corollary 4.4.** *Let the assumptions of* Theorem 4.3 *be fulfilled. Furthermore, let* $F_{n;\mathbf{U}}^{(\alpha,\#)}$ *be the reciprocal measure corresponding to the matrix measure* $F_{n;\mathbf{U}}^{(\alpha)}$. *Then:*

(a) *There exist some* $\ell \in \mathbb{N}_{n+1,(n+1)q}$, *pairwise different points* $u_1, u_2, \dots, u_\ell \in \mathbb{T}$, *and a sequence* $(\mathbf{A}_s^{\#})_{s=1}^{\ell}$ *of nonnegative Hermitian* $q \times q$ *matrices each of which is not equal to the zero matrix such that*

$$F_{n;\mathbf{U}}^{(\alpha,\#)} = \sum_{s=1}^{\ell} \varepsilon_{u_s}\mathbf{A}_s^{\#}.$$

*In particular, the equality*

$$\sum_{s=1}^{\ell} \operatorname{rank} \mathbf{A}_s^{\#} = (n+1)q$$

*is satisfied, where* $\mathcal{R}(\mathbf{A}_s^{\#}) = \mathcal{R}(F_{n;\mathbf{U}}^{(\alpha,\#)}(\{u_s\}))$ *for each* $s \in \mathbb{N}_{1,\ell}$.

(b) *If* $z \in \mathbb{T}$, *then the relations*

$$\mathcal{N}(P_{n;\mathbf{U}}^{(\alpha,\#)}(z)) = \mathcal{N}((Q_{n;\mathbf{U}}^{(\alpha,\#)}(z))^*) = \mathcal{R}(F_{n;\mathbf{U}}^{(\alpha,\#)}(\{z\}))$$

*and*

$$F_{n;\mathbf{U}}^{(\alpha,\#)}(\{z\}) A_{n,z}^{(\alpha,\#)}(z)\mathbf{x} = \mathbf{x}, \quad \mathbf{x} \in \mathcal{R}(F_{n;\mathbf{U}}^{(\alpha,\#)}(\{z\})),$$

*hold, where* $A_{n,z}^{(\alpha,\#)}(z) = C_{n,z}^{(\alpha,\#)}(z)$.

(c) *For a* $z \in \mathbb{C}_0 \setminus \mathbb{P}_{\alpha,n}$, $\det P_{n;\mathbf{U}}^{(\alpha,\#)}(z) = 0$ *holds if and only if* $z \in \{u_1, u_2, \dots, u_\ell\}$.

*Proof.* Recalling that the matrix $\mathbf{G}$ is nonsingular, by [31, Remarks 6.1 and 6.2] and the choice of $\mathbf{G}^{\#}$ we see that the matrix $\mathbf{G}^{\#}$ is nonsingular as well. Furthermore, in view of Theorem 3.6 we know that $F_{n;\mathbf{U}}^{(\alpha)} \in \mathcal{M}[(\alpha_j)_{j=1}^n, \mathbf{G}; (X_k)_{k=0}^n]$. Thus, for each $w \in \mathbb{C}_0 \setminus \mathbb{P}_{\alpha,n}$, similar to (3.9) we get

$$A_{n,w}^{(\alpha, F_{n;\mathbf{U}}^{(\alpha,\#)})} = A_{n,w}^{(\alpha,\#)} \quad \text{and} \quad C_{n,w}^{(\alpha, F_{n;\mathbf{U}}^{(\alpha,\#)})} = C_{n,w}^{(\alpha,\#)}.$$

Taking this into account (note also Remark 2.5 and Theorem 3.6), the assertion is a direct consequence of Theorem 4.3 along with Remarks 3.4 and 3.5. □

In view of Theorem 4.3 we emphasize the following special cases.

**Corollary 4.5.** *Let the assumptions of* Theorem 4.3 *be fulfilled. Furthermore, let* $F_{n;\mathbf{U}}^{(\alpha,\#)}$ *be the reciprocal measure corresponding to* $F_{n;\mathbf{U}}^{(\alpha)}$. *Let* $z \in \mathbb{T}$. *Then:*

(a) *The following statements are equivalent:*
   (i) $\det F_{n,\mathbf{U}}^{(\alpha)}(\{z\}) \neq 0$.
   (ii) $P_{n;\mathbf{U}}^{(\alpha)}(z) = 0_{q\times q}$ (*resp.*, $Q_{n;\mathbf{U}}^{(\alpha)}(z) = 0_{q\times q}$).
   *Moreover, if* (i) *is satisfied, then* $\det P_{n;\mathbf{U}}^{(\alpha,\#)}(z) \neq 0$ (*resp.*, $\det Q_{n;\mathbf{U}}^{(\alpha,\#)}(z) \neq 0$). *In particular, if* (i) *holds, then* $F_{n;\mathbf{U}}^{(\alpha,\#)}(\{z\}) = 0_{q\times q}$.

(b) *The following statements are equivalent:*
   (iii) $F_{n;\mathbf{U}}^{(\alpha)}(\{z\}) = 0_{q\times q}$.
   (iv) $\det P_{n;\mathbf{U}}^{(\alpha)}(z) \neq 0$ (*resp.*, $\det Q_{n;\mathbf{U}}^{(\alpha)}(z) \neq 0$).
   *Moreover, if* $P_{n;\mathbf{U}}^{(\alpha,\#)}(z) = 0_{q\times q}$ (*resp.*, $Q_{n;\mathbf{U}}^{(\alpha,\#)}(z) = 0_{q\times q}$) *is fulfilled, then* (iii) *is satisfied. In particular, if* $\det F_{n;\mathbf{U}}^{(\alpha,\#)}(\{z\}) \neq 0$, *then* (iii) *holds.*

*Proof.* The equivalence of (i) and (ii) (resp., of (iii) and (iv)) is an immediate consequence of Theorem 4.3 (see (4.4)). Taking this into account, from parts (d) and (e) of Lemma 4.2 one can see that, if (i) is satisfied, then $\det P_{n;\mathbf{U}}^{(\alpha,\#)} \neq 0$ and $\det Q_{n;\mathbf{U}}^{(\alpha,\#)} \neq 0$. Similarly, if $P_{n;\mathbf{U}}^{(\alpha,\#)} = 0_{q\times q}$ (resp., $Q_{n;\mathbf{U}}^{(\alpha,\#)} = 0_{q\times q}$) holds, then parts (d) and (e) of Lemma 4.2 imply (iii). Finally, an application of part (b) of Corollary 4.4 completes the proof. □

**Corollary 4.6.** *Let the assumptions of* Theorem 4.3 *be fulfilled. Furthermore, let* $F_{n;\mathbf{U}}^{(\alpha,\#)}$ *be the reciprocal measure corresponding to the matrix measure* $F_{n;\mathbf{U}}^{(\alpha)}$. *Then the following statements are equivalent:*

(i) *For each* $z \in \mathbb{T}$, *the value* $F_{n;\mathbf{U}}^{(\alpha)}(\{z\})$ *is either a nonsingular matrix or* $0_{q\times q}$.

(ii) *There exist exactly* $n+1$ *pairwise different points* $z_1, z_2, \ldots, z_{n+1} \in \mathbb{T}$ *such that the value* $F_{n;\mathbf{U}}^{(\alpha)}(\{z_s\})$ *is nonsingular for each* $s \in \mathbb{N}_{1,n+1}$.

(iii) *There exist exactly* $n+1$ *pairwise different points* $z_1, z_2, \ldots, z_{n+1} \in \mathbb{T}$ *such that* $F_{n;\mathbf{U}}^{(\alpha)}(\{z_s\}) \neq 0_{q\times q}$ *holds for each* $s \in \mathbb{N}_{1,n+1}$.

(iv) *For each* $z \in \mathbb{T}$, *the value* $P_{n;\mathbf{U}}^{(\alpha)}(z)$ *(resp.,* $Q_{n;\mathbf{U}}^{(\alpha)}(z)$*) is either nonsingular or* $0_{q\times q}$.

(v) *There exist exactly* $n+1$ *pairwise different points* $u_1, u_2, \ldots, u_{n+1} \in \mathbb{T}$ *such that* $P_{n;\mathbf{U}}^{(\alpha)}(u_s) = 0_{q\times q}$ *(resp.,* $Q_{n;\mathbf{U}}^{(\alpha)}(u_s) = 0_{q\times q}$*) holds for each* $s \in \mathbb{N}_{1,n+1}$.

(vi) *There exist exactly* $n+1$ *pairwise different points* $u_1, u_2, \ldots, u_{n+1} \in \mathbb{T}$ *such that* $P_{n;\mathbf{U}}^{(\alpha)}(u_s)$ *(resp.,* $Q_{n;\mathbf{U}}^{(\alpha)}(u_s)$*) is singular for each* $s \in \mathbb{N}_{1,n+1}$.

*Moreover, if* (i) *is satisfied, then* $\{z_1, z_2, \ldots, z_{n+1}\} = \{u_1, u_2, \ldots, u_{n+1}\}$ *holds in view of the points occurring in* (ii), (iii), (v), *and* (vi), *where* $\det P_{n;\mathbf{U}}^{(\alpha,\#)}(z_s) \neq 0$ *(resp.,* $\det Q_{n;\mathbf{U}}^{(\alpha,\#)}(z_s) \neq 0$*) and* $F_{n;\mathbf{U}}^{(\alpha,\#)}(\{z_s\}) = 0_{q\times q}$ *for each* $s \in \mathbb{N}_{1,n+1}$.

*Proof.* Use Theorem 4.3 in combination with Corollary 4.5. □

We briefly mention that the functions $P_{n;\mathbf{U}}^{(\alpha)}$, $P_{n;\mathbf{U}}^{(\alpha,\#)}$, $Q_{n;\mathbf{U}}^{(\alpha)}$, and $Q_{n;\mathbf{U}}^{(\alpha,\#)}$ in Corollaries 4.5 and 4.6 can also be replaced by the corresponding reciprocal functions $\big(P_{n;\mathbf{U}}^{(\alpha)}\big)^{[\alpha,n+1]}$, $\big(P_{n;\mathbf{U}}^{(\alpha,\#)}\big)^{[\alpha,n+1]}$, $\big(Q_{n;\mathbf{U}}^{(\alpha)}\big)^{[\alpha,n+1]}$, and $\big(Q_{n;\mathbf{U}}^{(\alpha,\#)}\big)^{[\alpha,n+1]}$ (cf. Lemma 4.2).

Recalling Theorem 3.6, we see that Theorem 4.3 and the resulting Corollaries offer us further insight into the weights corresponding to mass points which are associated with canonical solutions of Problem (R) for the nondegenerate case. In this regard, Corollary 4.6 is closely related to Remark 3.2 (see also Corollary 4.9). Furthermore, for the somewhat more general case that only $\alpha_1, \alpha_2, \ldots, \alpha_n \in \mathbb{C}\setminus\mathbb{T}$ (instead of $(\alpha_j)_{j=1}^{\infty} \in \mathcal{T}_1$), one can conclude the following.

**Proposition 4.7.** *Let* $n \in \mathbb{N}$ *and let* $\alpha_1, \alpha_2, \ldots, \alpha_n \in \mathbb{C}\setminus\mathbb{T}$. *Let* $X_0, X_1, \ldots, X_n$ *be a basis of the right* $\mathbb{C}^{q\times q}$*-module* $\mathcal{R}_{\alpha,n}^{q\times q}$ *and suppose that* $\mathbf{G}$ *is a nonsingular matrix such that* $\mathcal{M}[(\alpha_j)_{j=1}^{n}, \mathbf{G}; (X_k)_{k=0}^{n}] \neq \emptyset$ *holds. Furthermore, let* $F$ *be a canonical solution in* $\mathcal{M}[(\alpha_j)_{j=1}^{n}, \mathbf{G}; (X_k)_{k=0}^{n}]$. *Then the matrix measure* $F$ *admits* (1.10) *with some* $r \in \mathbb{N}_{n+1,(n+1)q}$, *pairwise different points* $z_1, z_2, \ldots, z_r \in \mathbb{T}$, *and a sequence* $(\mathbf{A}_s)_{s=1}^{r}$ *of nonnegative Hermitian* $q\times q$ *matrices each of which is not equal to the zero matrix. In particular,* (4.3) *is satisfied. Moreover, if* $z \in \mathbb{T}$ *and if*

$$A_{n,z}^{(\alpha)}(z) := \Big(X_0(z), X_1(z), \ldots, X_n(z)\Big)\mathbf{G}^{-1}\Big(X_0(z), X_1(z), \ldots, X_n(z)\Big)^{*}, \tag{4.10}$$

*then*

$$F(\{z\})A_{n,z}^{(\alpha)}(z)\mathbf{x} = \mathbf{x}, \quad \mathbf{x} \in \mathcal{R}\big(F(\{z\})\big). \tag{4.11}$$

*Proof.* Because of Remark 3.3 we already know that the matrix measure $F$ admits (1.10) with some $r \in \mathbb{N}_{n+1,(n+1)q}$, pairwise different points $z_1, z_2, \ldots, z_r \in \mathbb{T}$, and a sequence $(\mathbf{A}_s)_{s=1}^{r}$ of nonnegative Hermitian $q\times q$ matrices each of which is not equal to the zero matrix. This and the fact that $F$ is a canonical solution in the set $\mathcal{M}[(\alpha_j)_{j=1}^n, \mathbf{G}; (X_k)_{k=0}^n] \neq \emptyset$ implies along with Remark 3.1 that (4.3) is fulfilled as well. It remains to be shown that (4.11) holds. For this, let $z \in \mathbb{T}$ and (based on $F$) let $F^{(\alpha,n)}$ be the measure given as in Remark 2.4. In view of (1.10) we get

$$F^{(\alpha,n)}(\{z\}) = \frac{1}{|\pi_{\alpha,n}(z)|^2} F(\{z\}).$$

Furthermore, the choice $F \in \mathcal{M}[(\alpha_j)_{j=1}^n, \mathbf{G}; (X_k)_{k=0}^n]$ and (4.10) lead to

$$A_{n,z}^{(\alpha,F)}(z) = A_{n,z}^{(\alpha)}(z).$$

Hence, by using [23, Remarks 12 and 29] one can conclude that

$$A_{n,z}^{(\alpha)}(z) = \frac{1}{|\pi_{\alpha,n}(z)|^2} A_{n,z}(z),$$

where

$$A_{n,z}(z) = \Big(E_{0,q}(z), \dots, E_{n,q}(z)\Big)\big(\mathbf{T}_n^{(F^{(\alpha,n)})}\big)^{-1}\Big(E_{0,q}(z), \dots, E_{n,q}(z)\Big)^*$$

and where $E_{k,q}$ is the matrix polynomial defined as in (2.1) for each $k \in \mathbb{N}_{0,n}$. Note that $A_{n,z}(z)$ is the special matrix given via (3.7) in which $\alpha_j = 0$ for each $j \in \mathbb{N}_{1,n}$ and $X_k = E_{k,q}$ for each $k \in \mathbb{N}_{0,n}$ is chosen as well as $v = w = z$ and $\mathbf{G}$ is replaced by $\mathbf{T}_n^{(F^{(\alpha,n)})}$. Thus, taking Theorem 3.6 and Remark 2.4 into account, (4.11) follows from Theorem 4.3 (see (4.5)). □

**Corollary 4.8.** *Let the assumptions of* Proposition 4.7 *be fulfilled. Furthermore, let $F^{\#}$ be the reciprocal measure corresponding to a $F$. Then*

$$F^{\#} = \sum_{s=1}^{\ell} \varepsilon_{u_s} \mathbf{A}_s^{\#} \quad \text{and} \quad \sum_{s=1}^{\ell} \operatorname{rank} \mathbf{A}_s^{\#} = (n+1)q$$

*hold with some $\ell \in \mathbb{N}_{n+1,(n+1)q}$, pairwise different points $u_1, u_2, \dots, u_\ell \in \mathbb{T}$, and a sequence $(\mathbf{A}_s^{\#})_{s=1}^{\ell}$ of nonnegative Hermitian $q \times q$ matrices each of which is not equal to the zero matrix. Moreover, if $z \in \mathbb{T}$ and if*

$$A_{n,z}^{(\alpha,\#)}(z) := \Big(X_0(z), X_1(z), \dots, X_n(z)\Big)\big(\mathbf{G}^{\#}\big)^{-1}\Big(X_0(z), X_1(z), \dots, X_n(z)\Big)^*, \quad (4.12)$$

*then*

$$F^{\#}(\{z\}) A_{n,z}^{(\alpha,\#)}(z)\mathbf{x} = \mathbf{x}, \quad \mathbf{x} \in \mathcal{R}(F^{\#}(\{z\})).$$

*Proof.* By using a similar argumentation as in the proof of Corollary 4.4, the assertion follows from Proposition 4.7 along with Remark 2.5. □

**Corollary 4.9.** *Let the assumptions of* Proposition 4.7 *be fulfilled. Furthermore, let $F^{\#}$ be the reciprocal measure corresponding to $F$ and let $A_{n,z}^{(\alpha)}(z)$ and $A_{n,z}^{(\alpha,\#)}(z)$ be the complex $q \times q$ matrices given by (4.10) and (4.12) for some $z \in \mathbb{T}$.*

(a) *Let $z \in \mathbb{T}$. Then* $\det F(\{z\}) \neq 0$ *holds if and only if* $F(\{z\}) = \big(A_{n,z}^{(\alpha)}(z)\big)^{-1}$. *Moreover, if* $\det F(\{z\}) \neq 0$ *is satisfied, then $F^{\#}(\{z\}) = 0_{q\times q}$. In particular, if* $\det F(\{z\}) \neq 0$, *then $F^{\#}(\{z\}) \neq \big(A_{n,z}^{(\alpha,\#)}(z)\big)^{-1}$ holds.*

(b) *The following statements are equivalent:*

(i) *For each* $z \in \mathbb{T}$, *the value* $F(\{z\})$ *is either a nonsingular matrix or* $0_{q\times q}$.

(ii) *For each* $z \in \mathbb{T}$, *the value* $F(\{z\})$ *is either* $\big(A^{(\alpha)}_{n,z}(z)\big)^{-1}$ *or* $0_{q\times q}$.

(iii) *There exist exactly* $n+1$ *pairwise different points* $u_1, u_2, \dots, u_{n+1} \in \mathbb{T}$ *such that* $F(\{u_s\}) = \big(A^{(\alpha)}_{n,u_s}(u_s)\big)^{-1}$ *for each* $s \in \mathbb{N}_{1,n+1}$.

(iv) *There exist exactly* $n+1$ *pairwise different points* $u_1, u_2, \dots, u_{n+1} \in \mathbb{T}$ *such that the value* $F(\{u_s\})$ *is nonsingular for each* $s \in \mathbb{N}_{1,n+1}$.

(v) *There exist exactly* $n+1$ *pairwise different points* $u_1, u_2, \dots, u_{n+1} \in \mathbb{T}$ *such that* $F(\{u_s\}) \neq 0_{q\times q}$ *holds for each* $s \in \mathbb{N}_{1,n+1}$.

(vi) *For some* $r \in \mathbb{N}_{1,n+1}$, *pairwise different points* $z_1, z_2, \dots, z_r \in \mathbb{T}$, *and sequence* $(\mathbf{A}_s)_{s=1}^{r}$ *of nonnegative Hermitian* $q\times q$ *matrices,* $F$ *admits* (1.10).

*Moreover, if* (i) *is satisfied, then in view of* (ii)–(vi) *the relations* $r = n+1$ *and* $\{z_1, z_2, \dots, z_{n+1}\} = \{u_1, u_2, \dots, u_{n+1}\}$ *hold, where* $\mathbf{A}_s = \big(A^{(\alpha)}_{n,z_s}(z_s)\big)^{-1}$ *as well as* $F^{\#}(\{z_s\}) = 0_{q\times q}$ *and* $F^{\#}(\{z_s\}) \neq \big(A^{(\alpha,\#)}_{n,z_s}(z_s)\big)^{-1}$ *for* $s \in \mathbb{N}_{1,n+1}$.

*Proof.* Let $z \in \mathbb{T}$. Taking into account that (4.10) implies $A^{(\alpha)}_{n,z}(z) > 0_{q\times q}$ (cf. (3.10)), from Proposition 4.7 (see, in particular, (4.11)) it follows directly that the condition $\det F(\{z\}) \neq 0$ holds if and only if

$$F(\{z\}) = \big(A^{(\alpha)}_{n,z}(z)\big)^{-1}.$$

Moreover, since the concept of reciprocal measures in the set $\mathcal{M}^q_{\geq}(\mathbb{T}, \mathfrak{B}_{\mathbb{T}})$ is independent of the chosen points $\alpha_1, \alpha_2, \dots, \alpha_n$ belonging to $\mathbb{C} \setminus \mathbb{T}$, by using Theorem 3.6 and part (a) of Corollary 4.5 along with Remarks 2.3 and 2.5 one can conclude that $\det F(\{z\}) \neq 0$ implies the equality $F^{\#}(\{z\}) = 0_{q\times q}$. In particular, because of $A^{(\alpha,\#)}_{n,z}(z) > 0_{q\times q}$ holds, we have $F^{\#}(\{z\}) \neq \big(A^{(\alpha,\#)}_{n,z}(z)\big)^{-1}$ in the case of $\det F(\{z\}) \neq 0$. Consequently, part (a) is proven. Recalling (a) and Remark 3.2, part (b) is a simple consequence of Proposition 4.7 (note (4.3)). □

**Corollary 4.10.** *Let* $\tau \in \mathbb{N}$ *or* $\tau = +\infty$ *and let* $m \in \mathbb{N}_{1,\tau}$. *Suppose that* $(\mathbf{c}_k)_{k=0}^{\tau}$ *is a sequence of complex* $q\times q$ *matrices such that the block Toeplitz matrix* $\mathbf{T}_{m-1}$ *given via* (1.1) *is nonsingular. Then the following statements are equivalent:*

(i) $(\mathbf{c}_k)_{k=0}^{\tau}$ *is a nonnegative definite sequence which is canonical of order* $m$.

(ii) *There are an* $F \in \mathcal{M}^q_{\geq}(\mathbb{T}, \mathfrak{B}_{\mathbb{T}})$ *and a finite* $\Delta \subset \mathbb{T}$ *so that* $F(\mathbb{T}\setminus\Delta) = 0_{q\times q}$ *and*

$$\sum_{z\in\Delta} \operatorname{rank} F(\{z\}) = mq$$

*hold, where* $\mathbf{c}_k = \mathbf{c}_k^{(F)}$ *for each* $k \in \mathbb{N}_{0,\tau}$.

(iii) *There are some* $\ell \in \mathbb{N}_{m,mq}$, *pairwise different points* $z_1, z_2, \dots, z_\ell \in \mathbb{T}$, *and a sequence* $(\mathbf{A}_j)_{j=1}^{\ell}$ *of nonnegative Hermitian* $q\times q$ *matrices each of which is not equal to the zero matrix such that*

$$\sum_{s=1}^{\ell} \operatorname{rank} \mathbf{A}_s = mq \quad \text{and} \quad \mathbf{c}_k = \sum_{s=1}^{\ell} z_s^{-k} \mathbf{A}_s, \quad k \in \mathbb{N}_{0,\tau}.$$

*If* (i) *holds, then one can choose* $\Delta = \{z_1, z_2, \dots, z_\ell\}$ *regarding* (ii) *and* (iii), *where*

$$\mathbf{A}_s\left(z_s^0\mathbf{I}_q, \dots, z_s^{m-1}\mathbf{I}_q\right)\mathbf{T}_{m-1}^{-1}\left(z_s^0\mathbf{I}_q, \dots, z_s^{m-1}\mathbf{I}_q\right)^* \mathbf{x} = \mathbf{x}, \quad \mathbf{x} \in \mathcal{R}(\mathbf{A}_s),$$

*for each* $s \in \mathbb{N}_{1,\ell}$. *In particular, if* (i) *is satisfied, then the nonsingularity of the matrix* $\mathbf{A}_s$ *for some* $s \in \mathbb{N}_{1,\ell}$ *occurring in* (iii) *is equivalent to the identity*

$$\mathbf{A}_s = \left(\left(z_s^0\mathbf{I}_q, \dots, z_s^{m-1}\mathbf{I}_q\right)\mathbf{T}_{m-1}^{-1}\left(z_s^0\mathbf{I}_q, \dots, z_s^{m-1}\mathbf{I}_q\right)^*\right)^{-1}.$$

*Proof.* Taking $\operatorname{rank} \mathbf{T}_{m-1} = mq$ and Remark 2.14 into account, the equivalence of (i)–(iii) is an easy consequence of Corollary 2.12. Based on that, in the elementary case $m = 1$, the rest of the assertion is obvious (cf. (3.12) or [29, Example 9.11]). Furthermore, if $m \geq 2$, then the remaining part of the assertion follows from Proposition 4.7 along with Remark 2.1. □

At first glance, the situation studied in Corollary 4.6 (resp., in part (b) of Corollary 4.9) seems slightly artificial, but does occur (for instance, always in the scalar case $q = 1$ which we will next see).

*Remark* 4.11. Consider the case $q = 1$, where $\mathbf{G}$ is a nonsingular $(n+1) \times (n+1)$ matrix and where $X_0, X_1, \dots, X_n$ is a basis of the linear space $\mathcal{R}_{\alpha,n}$. Suppose that $F \in \mathcal{M}[(\alpha_j)_{j=1}^n, \mathbf{G}; (X_k)_{k=0}^n]$. In view of Remark 2.15 and Corollary 4.9 it follows that $F$ is a canonical solution in $\mathcal{M}[(\alpha_j)_{j=1}^n, \mathbf{G}; (X_k)_{k=0}^n]$ for that case if and only if there are pairwise different points $z_1, z_2, \dots, z_{n+1} \in \mathbb{T}$ such that $F$ admits

$$F = \sum_{s=1}^{n+1} \frac{1}{A_{n,z_s}^{(\alpha)}(z_s)} \varepsilon_{z_s}.$$

Thus, by using Remark 2.5 and Corollary 4.9 one can see that $F$ is a canonical solution in $\mathcal{M}[(\alpha_j)_{j=1}^n, \mathbf{G}; (X_k)_{k=0}^n]$ if and only if there are pairwise different points $u_1, u_2, \dots, u_{n+1} \in \mathbb{T}$ so that the reciprocal measure $F^\#$ corresponding to $F$ admits

$$F^\# = \sum_{s=1}^{n+1} \frac{1}{A_{n,u_s}^{(\alpha,\#)}(u_s)} \varepsilon_{u_s},$$

where $z_j \neq u_k$ for all $j, k \in \mathbb{N}_{1,n+1}$.

Let us again consider the elementary case $n = 0$, based on a constant function $X_0$ defined on $\mathbb{C}_0$ with a nonsingular $q \times q$ matrix $\mathbf{X}_0$ as value, a positive Hermitian $q \times q$ matrix $\mathbf{G}$, and a measure $F \in \mathcal{M}_{\geq}^q(\mathbb{T}, \mathfrak{B}_\mathbb{T})$ fulfilling (2.7). We already know that $F$ is a canonical solution of that problem if and only if the Riesz–Herglotz transform $\Omega$ of $F$ admits (3.11) for each $v \in \mathbb{D}$ with some unitary $q \times q$ matrix $\mathbf{W}$ and points $z_1, z_2, \dots, z_q \in \mathbb{T}$. Thus, it is clear that the relevant results in this section for canonical solutions are valid in the case $n = 0$ as well. Moreover, one can see that, if $\Omega$ admits (3.11), then the Riesz–Herglotz transform $\Omega^\#$ of the reciprocal measure $F^{\#}$ corresponding to $F$ satisfies

$$\Omega^{\#}(v) = \sqrt{\mathbf{X}_0\mathbf{G}^{-1}\mathbf{X}_0^*}\,\mathbf{W}\begin{pmatrix} \frac{-z_1+v}{-z_1-v} & 0 & \cdots & 0 \\ 0 & \frac{-z_2+v}{-z_2-v} & \ddots & \vdots \\ \vdots & \ddots & \ddots & 0 \\ 0 & \cdots & 0 & \frac{-z_q+v}{-z_q-v} \end{pmatrix}\mathbf{W}^*\sqrt{\mathbf{X}_0\mathbf{G}^{-1}\mathbf{X}_0^*}$$

for each $v \in \mathbb{D}$. Therefore (cf. (3.12)), we get the representation

$$F^{\#} = \sqrt{\mathbf{X}_0\mathbf{G}^{-1}\mathbf{X}_0^*}\,\mathbf{W}\begin{pmatrix} \varepsilon_{-z_1} & o_1 & \cdots & o_1 \\ o_1 & \varepsilon_{-z_2} & \ddots & \vdots \\ \vdots & \ddots & \ddots & o_1 \\ o_1 & \cdots & o_1 & \varepsilon_{-z_q} \end{pmatrix}\mathbf{W}^*\sqrt{\mathbf{X}_0\mathbf{G}^{-1}\mathbf{X}_0^*}.$$

In particular, if $F$ admits (1.10) with $r = 1$ and some $z_1 \in \mathbb{T}$, then $\mathbf{A}_1$ is a positive Hermitian $q \times q$ matrix and

$$F^{\#} = \varepsilon_{-z_1}\mathbf{A}_1^{-1}.$$

With a view to the special case of $n = 0$ (resp., of $q = 1$ in Remark 4.11), we present an example for $n > 1$ and $q > 1$ which clarifies that, if $F$ is a canonical solution of $\mathcal{M}[(\alpha_j)_{j=1}^n, \mathbf{G}; (X_k)_{k=0}^n]$ such that $F(\{z\})$ is nonsingular for some $z \in \mathbb{T}$, then it can be possible that $F^{\#}(\{u\})$ is singular for all $u \in \mathbb{T}$, where $F^{\#}$ stands for the reciprocal measure corresponding to $F$ (even under the strong condition that $F$ admits (1.10) with $r = n+1$; cf. Corollaries 4.6 and 4.9).

*Example* 4.12. Let $\alpha_1 \in \mathbb{C} \setminus \mathbb{T}$ and let $X_0, X_1$ be a basis of the right $\mathbb{C}^{2\times 2}$-module $\mathcal{R}_{\alpha,1}^{2\times 2}$. Furthermore, let $F := \varepsilon_1\mathbf{A}_1 + \varepsilon_{-1}\mathbf{A}_2$, where

$$\mathbf{A}_1 := \begin{pmatrix} 1 & 0 \\ 0 & 1 \end{pmatrix} \quad \text{and} \quad \mathbf{A}_2 := \begin{pmatrix} 1 & 1 \\ 1 & 2 \end{pmatrix}.$$

Then $F \in \mathcal{M}_{\geq}^2(\mathbb{T}, \mathfrak{B}_{\mathbb{T}})$ and the matrices $\mathbf{G}_{X,1}^{(F)}$, $\mathbf{A}_1$, and $\mathbf{A}_2$ are nonsingular, where $A_{1,1}^{(\alpha,F)}(1) = \mathbf{A}_1$ as well as $A_{1,-1}^{(\alpha,F)}(-1) = \mathbf{A}_2^{-1}$ and where

$$\mathbf{A}_2^{-1} = \begin{pmatrix} 2 & -1 \\ -1 & 1 \end{pmatrix}.$$

Moreover, $F$ is a canonical solution in $\mathcal{M}[(\alpha_j)_{j=1}^1, \mathbf{G}_{X,1}^{(F)}; (X_k)_{k=0}^1]$ and the reciprocal measure $F^{\#}$ corresponding to $F$ is given by

$$F^{\#} = \varepsilon_{z_1}\mathbf{B}_1 + \varepsilon_{\overline{z_1}}\mathbf{B}_1 + \varepsilon_{-z_1}\mathbf{B}_2 + \varepsilon_{-\overline{z_1}}\mathbf{B}_2,$$

where $z_1 := \frac{1}{\sqrt{5}}(1 + 2\mathrm{i})$ and where

$$\mathbf{B}_1 := \frac{1}{20}\begin{pmatrix} 3-\sqrt{5} & \sqrt{5}-1 \\ \sqrt{5}-1 & 2 \end{pmatrix} \quad \text{and} \quad \mathbf{B}_2 := \frac{1}{20}\begin{pmatrix} 3+\sqrt{5} & -\sqrt{5}-1 \\ -\sqrt{5}-1 & 2 \end{pmatrix}.$$

In particular, the complex $2 \times 2$ matrices $F(\{1\})$ and $F(\{-1\})$ are nonsingular, but $F^{\#}(\{u\})$ is singular for each $u \in \mathbb{T}$.

*Proof.* Obviously, 1 and $-1$ are pairwise different points belonging to $\mathbb{T}$ as well as $\mathbf{A}_1$ and $\mathbf{A}_2$ are positive Hermitian $2 \times 2$ matrices, where $\mathbf{A}_1^{-1} = \mathbf{A}_1$ and

$$\mathbf{A}_2^{-1} = \begin{pmatrix} 2 & -1 \\ -1 & 1 \end{pmatrix}.$$

In particular, the choice of $F$ implies $F \in \mathcal{M}_{\geq}^2(\mathbb{T}, \mathfrak{B}_\mathbb{T})$. Furthermore, in view of Corollary 2.11 we get that the matrix $\mathbf{G}_{X,1}^{(F)}$ is nonsingular and that $F$ is a canonical solution in $\mathcal{M}[(\alpha_j)_{j=1}^1, \mathbf{G}_{X,1}^{(F)}; (X_k)_{k=0}^1]$. Hence, by using Corollary 4.9 and (3.9) we get $A_{1,1}^{(\alpha,F)}(1) = \mathbf{A}_1$ and $A_{1,-1}^{(\alpha,F)}(-1) = \mathbf{A}_2^{-1}$. Moreover, one can check that the Riesz–Herglotz transform $\Omega$ of $F$ satisfies the representation

$$\Omega(v) = \frac{1}{(1-v)(1+v)} \begin{pmatrix} (1+v)^2 + (1-v)^2 & (1-v)^2 \\ (1-v)^2 & (1+v)^2 + 2(1-v)^2 \end{pmatrix}$$

for each $v \in \mathbb{D}$. Thus, for each $v \in \mathbb{D}$, one can conclude that $\det \Omega(v) \neq 0$ and that

$$\big(\Omega(v)\big)^{-1} = \frac{(1-v)(1+v)}{5v^4 + 6v^2 + 5} \begin{pmatrix} (1+v)^2 + 2(1-v)^2 & -(1-v)^2 \\ -(1-v)^2 & (1+v)^2 + (1-v)^2 \end{pmatrix}.$$

Consequently, by a straightforward calculation one can see that the reciprocal measure $F^\#$ corresponding to $F$ admits the representation

$$F^\# = \varepsilon_{z_1}\mathbf{B}_1 + \varepsilon_{\overline{z_1}}\mathbf{B}_1 + \varepsilon_{-z_1}\mathbf{B}_2 + \varepsilon_{-\overline{z_1}}\mathbf{B}_2.$$

Finally, since $z_1$, $\overline{z_1}$, $-z_1$, and $-\overline{z_1}$ are pairwise different points and since $\mathbf{B}_1$ and $\mathbf{B}_2$ are singular $2 \times 2$ matrices, the value $F^\#(\{u\})$ is a singular matrix for each $u \in \mathbb{T}$. There again, we have $F(\{1\}) = \mathbf{A}_1$ and $F(\{-1\}) = \mathbf{A}_2$ which shows that the values $F(\{1\})$ and $F(\{-1\})$ are nonsingular matrices. □

## Some final remarks

We discussed canonical solutions of a certain finite moment problem for rational matrix functions, i.e., of Problem (R). These solutions can be understood as having the highest degree of degeneracy (see (2.2)). Among other things, we showed that these solutions are molecular nonnegative Hermitian matrix-valued Borel measures on the unit circle which have a particular structure.

For the special case in which $q = 1$ (i.e., of complex-valued functions), the family of extremal solutions we studied in this paper consists of those elements which can be used to get rational Szegő quadrature formulas. In particular, for $q = 1$, these elements can be characterized by the property that they admit representations as sums of Dirac measures with a minimal number of terms. Furthermore, in the nondegenerate case, the associated weights of the mass points of canonical solutions of Problem (R) with $q = 1$ are given by the values of related reproducing kernels (or of the Christoffel functions). However, the case of matrix-valued functions is somewhat more complicated. To emphasize this, we presented the main results of canonical solutions of Problem (R) and stated what these imply for the special case $q = 1$ (compare Theorem 2.10 with Remark 2.15, Theorem 3.6 with Remark 3.12, and Theorem 4.3 with Remark 4.11).

Since the analysis of structural properties of canonical solutions of Problem (R) relies on the theory of orthogonal rational matrix functions on the unit circle $\mathbb{T}$, we have mostly restrict the considerations that the poles of the underlying rational matrix functions are in a sense well positioned with respect to $\mathbb{T}$ (i.e., we assumed $(\alpha_j)_{j=1}^{\infty} \in \mathcal{T}_1$). But, we might used these results to obtain similar statements for the general case (see Propositions 3.11 and 4.7).

Important results are the formulas (4.5) and (4.11) which point out some connection between the weights of canonical solutions of Problem (R) and values of related reproducing kernels of rational matrix functions. These will finally lead to profound conclusions concerning extremal properties of maximal weights within the solution set of Problem (R) for the nondegenerate case. (In particular, this will be the starting point for considering a subclass of canonical solutions which realize at least in one mass point an analogous weight as canonical solutions in the special case $q = 1$.) This will be done in a forthcoming work.

## References

[1] Aronszajn, N.: *Theory of reproducing kernels*, Trans. Amer. Math. Soc. **68** (1950), 337–404.

[2] Arov, D.Z.: *Regular J-inner matrix-functions and related continuation problems*, in: Linear Operators in Function Spaces, Operator Theory: Adv. Appl. Vol. 43, Birkhäuser, Basel, 1990, pp. 63–87.

[3] Arov, D.Z.; Dym, H.: *J-Contractive Matrix Valued Functions and Related Topics*, Encyclopedia Math. and its Appl. 116, Cambridge University Press, Cambridge 2008.

[4] Arov, D.Z.; Kreĭn, M.G.: *The problem of finding the minimum entropy in indeterminate problems of continuation* (Russian), Funct. Anal. Appl. **15** (1981), 61–64.

[5] Ben-Artzi, A.; Gohberg, I.: *Orthogonal polynomials over Hilbert modules*, in: Nonselfadjoint Operators and Related Topics, Operator Theory: Adv. Appl. Vol. 73, Birkhäuser, Basel, 1994, pp. 96–126.

[6] Bultheel, A.: *Inequalities in Hilbert modules of matrix-valued functions*, Proc. Amer. Math. Soc. **85** (1982), 369–372.

[7] Bultheel A.; Cantero, M.J.: *A matricial computation of rational quadrature formulas on the unit circle*, Numer. Algorithms **52** (2009), 47–68.

[8] Bultheel, A.; González-Vera, P.; Hendriksen, E.; Njåstad, O.: *Orthogonal rational functions and quadrature on the unit circle*, Numer. Algorithms **3** (1992), 105–116.

[9] Bultheel, A.; González-Vera, P.; Hendriksen, E.; Njåstad, O.: *Moment problems and orthogonal functions*, J. Comput. Appl. Math. **48** (1993), 49–68.

[10] Bultheel, A.; González-Vera, P.; Hendriksen, E.; Njåstad, O.: *Quadrature formulas on the unit circle based on rational functions*, J. Comput. Appl. Math. **50** (1994), 159–170.

[11] Bultheel, A.; González-Vera, P.; Hendriksen, E.; Njåstad, O.: *A rational moment problem on the unit circle*, Methods Appl. Anal. **4** (1997), 283–310.

[12] Bultheel, A.; González-Vera, P.; Hendriksen, E.; Njåstad, O.: *Orthogonal Rational Functions*, Cambridge Monographs on Applied and Comput. Math. 5, Cambridge University Press, Cambridge 1999.

[13] Cantero, M.J.; Cruz-Barroso, R.; González-Vera, P.: *A matrix approach to the computation of quadrature formulas on the unit circle*, Appl. Numer. Math. **58** (2008), 296–318.
[14] Choque Rivero, A.E.; Lasarow, A.; Rahn, A.: *On ranges and Moore–Penrose inverses related to matrix Carathéodory and Schur functions*, Complex Anal. Oper. Theory **5** (2011), 513–543.
[15] de la Calle Ysern, B.: *Error bounds for rational quadrature formulae of analytic functions*, Numer. Math. **101** (2005), 251–271.
[16] de la Calle Ysern, B.; González-Vera, P.: *Rational quadrature formulae on the unit circle with arbitrary poles*, Numer. Math. **107** (2007), 559–587.
[17] Delsarte, P.; Genin, Y.; Kamp, Y.: *Orthogonal polynomial matrices on the unit circle*, IEEE Trans. Circuits and Systems **CAS-25** (1978), 149–160.
[18] Dubovoj, V.K.; Fritzsche, B.; Kirstein, B.: *Matricial Version of the Classical Schur Problem*, Teubner-Texte zur Mathematik 129, B.G. Teubner, Stuttgart-Leipzig 1992.
[19] Dym, H.: *J Contractive Matrix Functions, Reproducing Kernel Hilbert Spaces and Interpolation*, CBMS Regional Conf. Ser. Math. 71, Providence, R.I. 1989.
[20] Ellis, R.L.; Gohberg, I.: *Extensions of matrix-valued inner products on modules and the inversion formula for block Toeplitz matrices*, in: Operator Theory and Analysis, Operator Theory: Adv. Appl. Vol. 122, Birkhäuser, Basel, 2001, pp. 191–227.
[21] Ellis, R.L.; Gohberg, I.; Lay, D.C.: *Infinite analogues of block Toeplitz matrices and related orthogonal functions*, Integral Equations Operator Theory **22** (1995), 375–419.
[22] Fritzsche, B.; Fuchs, S.; Kirstein, B.: *A Schur type matrix extension problem V*, Math. Nachr. **158** (1992), 133–159.
[23] Fritzsche, B.; Kirstein, B.; Lasarow, A.: *On a moment problem for rational matrix-valued functions*, Linear Algebra Appl. **372** (2003), 1–31.
[24] Fritzsche, B.; Kirstein, B.; Lasarow, A.: *On rank invariance of moment matrices of nonnegative Hermitian-valued Borel measures on the unit circle*, Math. Nachr. **263/264** (2004), 103–132.
[25] Fritzsche, B.; Kirstein, B.; Lasarow, A.: *On Hilbert modules of rational matrix-valued functions and related inverse problems*, J. Comput. Appl. Math. **179** (2005), 215–248.
[26] Fritzsche, B.; Kirstein, B.; Lasarow, A.: *Orthogonal rational matrix-valued functions on the unit circle*, Math. Nachr. **278** (2005), 525–553.
[27] Fritzsche, B.; Kirstein, B.; Lasarow, A.: *Orthogonal rational matrix-valued functions on the unit circle: Recurrence relations and a Favard-type theorem*, Math. Nachr. **279** (2006), 513–542.
[28] Fritzsche, B.; Kirstein, B.; Lasarow, A.: *The matricial Carathéodory problem in both nondegenerate and degenerate cases*, in: Interpolation, Schur Functions and Moment Problems, Operator Theory: Adv. Appl. Vol. 165, Birkhäuser, Basel, 2006, pp. 251–290.
[29] Fritzsche, B.; Kirstein, B.; Lasarow, A.: *On a class of extremal solutions of the nondegenerate matricial Carathéodory problem*, Analysis **27** (2007), 109–164.
[30] Fritzsche, B.; Kirstein, B.; Lasarow, A.: *On a class of extremal solutions of a moment problem for rational matrix-valued functions in the nondegenerate case I*, Math. Nachr. **283** (2010), 1706–1735.
[31] Fritzsche, B.; Kirstein, B.; Lasarow, A.: *On a class of extremal solutions of a moment problem for rational matrix-valued functions in the nondegenerate case II*, J. Comput. Appl. Math. **235** (2010), 1008–1041.
[32] Fritzsche, B.; Kirstein, B.; Lasarow, A.: *Para-orthogonal rational matrix-valued functions on the unit circle*, Oper. Matrices (in press), OaM-0412.

[33] Gautschi, W.: *A survey of Gauss-Christoffel quadrature formulae*, E.B. Christoffel, The Influence of His Work on Mathematical and Physical Sciences, Birkhäuser, Basel, 1981, pp. 72–147.
[34] Gautschi, W.; Gori, L.; Lo Cascio, M.L.: *Quadrature rules for rational functions*, Numer. Math. **86** (2000), 617–633.
[35] Geronimus, Ja.L.: *Polynomials orthogonal on a circle and their applications* (Russian), Zap. Naučno-Issled. Inst. Mat. Meh. Har'kov. Mat. Obšč. **19** (1948), 35–120.
[36] Itoh, S.: *Reproducing kernels in modules over $C^*$-algebras and their applications*, Bull. Kyushu Inst. Tech. Math. Natur. Sci. **37** (1990), 1–20.
[37] Jones, W.B.; Njåstad, O.; Thron, W.J.: *Moment theory, orthogonal polynomials, quadrature, and continued fractions associated with the unit circle*, Bull. London Math. Soc. **21** (1989), 113–152.
[38] Kats, I.S.: *On Hilbert spaces generated by monotone Hermitian matrix-functions* (Russian), Zap. Mat. Otd. Fiz.-Mat. Fak. i Har'kov. Mat. Obšč. **22** (1950), 95–113.
[39] Lasarow, A.: *Dual pairs of orthogonal systems of rational matrix-valued functions on the unit circle*, Analysis **26** (2006), 209–244.
[40] Lasarow, A.: *More on a class of extremal solutions of a moment problem for rational matrix-valued functions in the nondegenerate case*, J. Approx. Theory **163** (2011), 864–887.
[41] Rosenberg, M.: *The square integrability of matrix-valued functions with respect to a non-negative Hermitian measure*, Duke Math. J. **31** (1964), 291–298.
[42] Rosenberg, M.: *Operators as spectral integrals of operator-valued functions from the study of multivariate stationary stochastic processes*, J. Mult. Anal. **4** (1974), 166–209.
[43] Rosenberg, M.: *Spectral integrals of operator-valued functions – II. From the study of stationary processes*, J. Mult. Anal. **6** (1976), 538–571.
[44] Sakhnovich, A.L.: *On a class of extremal problems* (Russian), Izv. Akad. Nauk SSSR, Ser. Mat. **51** (1987), 436–443.
[45] Velázquez, L.: *Spectral methods for orthogonal rational functions*, J. Funct. Anal. **254** (2008), 954–986.

B. Fritzsche and B. Kirstein
Fakultät für Mathematik und Informatik
Universität Leipzig
Postfach: 10 09 20
D-04009 Leipzig, Germany
e-mail: fritzsche@mathematik.uni-leipzig.de
kirstein@mathematik.uni-leipzig.de

Andreas Lasarow
Departement Computerwetenschappen
Katholieke Universiteit Leuven
Celestijnenlaan 200a – postbus: 02402
B-3001 Leuven, Belgium
e-mail: Andreas.Lasarow@cs.kuleuven.be

Operator Theory:
Advances and Applications, Vol. 221, 373-384


# Maximal $L^p$-regularity for a 2D Fluid-Solid Interaction Problem

Karoline Götze

**Abstract.** We study a coupled system of equations which appears as a suitable linearization of the model for the free motion of a rigid body in a Newtonian fluid in two space dimensions. For this problem, we show maximal $L^p$-regularity estimates. The method rests on a suitable reformulation of the problem as the question of invertibility of a bounded operator on $W^{1,p}(0,T;\mathbb{R}^3)$.

**Mathematics Subject Classification (2000).** Primary 35Q35; Secondary 74F10.

**Keywords.** Rigid body in a fluid, maximal regularity, Stokes equations.

## 1. Introduction

The free motion of a rigid body in a Newtonian fluid is a classical problem in fluid mechanics and it has been studied extensively, especially during the last 15 years. It is known that weak solutions exist in two and three space dimensions, see, e.g., [4, 7]. The existence of strong solutions was shown in the $L^2$-setting, in [9] and [15]. However, several interesting open problems remain, as for example the attainability of steady falls when gravity is applied, the problem of collision of several bodies or in the study of a non-Newtonian coupled flow. For a summary of results and a discussion of open problems, we refer to [8].

The aim of this paper is to show maximal $L^p$-regularity estimates for a system of equations which appears as a suitable linearization for this problem in two dimensions. For the proof, we use a versatile method which can also be applied to generalized Newtonian fluids and, similarly, in three space dimensions, [10]. The main difficulty lies in the strong, non-local coupling between fluid and body movements. To deal with this, Hilbert space techniques are available, where the geometric properties of the rigid body can be encoded in the underlying function

This work was supported by the international research training group 1529: mathematical fluid dynamics.

space, [9], [15]. We introduce a different approach, putting these properties into the operator, based on the idea in [8] of how to decompose the coupling forces.

The paper is organized as follows. In the next section, we introduce the linearized model and the main result, as well as some definitions and notations. Section 3 is devoted to several preliminary estimates for the fluid velocity and pressure. The main part is Section 4, where we construct a suitable reduction of the problem to the question of invertibility of a bounded operator on $W^{1,p}(0,T;\mathbb{R}^3)$. In the last Section 5, we prove this invertibility and the main result.

## 2. Model and main result

We consider the equations

$$\left\{\begin{aligned} u_t - \mu\Delta u + \nabla\pi &= f_0, && \text{in } J_{T_0}\times\mathcal{D},\\ \operatorname{div} u &= 0, && \text{in } J_{T_0}\times\mathcal{D},\\ u(t,y) - v_{\mathcal{B}}(t,y) &= 0, && t\in J_{T_0}, y\in\Gamma,\\ u(0) &= u_0, && \text{in } \mathcal{D},\\ \mathrm{m}\xi' + \textstyle\int_\Gamma \mathbf{T}(u,\pi)N\,d\sigma &= f_1, && \text{in } J_{T_0},\\ I\Omega' + \textstyle\int_\Gamma y^\perp\mathbf{T}(u,\pi)N\,d\sigma &= f_2, && \text{in } J_{T_0},\\ \xi(0) &= \xi_0, &&\\ \Omega(0) &= \Omega_0, && \end{aligned}\right. \tag{2.1}$$

for an arbitrary time $T_0 > 0$, $J_{T_0} := (0,T_0)$. In the first four lines, we have the Stokes equations for the fluid velocity field $u$ and scalar pressure $\pi$ in an exterior domain $\mathcal{D}\subset\mathbb{R}^2$. The body occupies the bounded domain $\mathcal{B} = \mathbb{R}^2\backslash\mathcal{D}$ and body and fluid meet at the boundary $\Gamma$ of $\mathcal{B}$ with normal $N$. The body moves with velocity

$$v_{\mathcal{B}}(t,y) = \xi(t) + \Omega(t)y^\perp,$$

where $\xi(t)\in\mathbb{R}^2$ is its translational velocity and $\Omega(t)\in\mathbb{R}$ its angular velocity at time $t$ and we write $\begin{pmatrix} x_1\\ x_2\end{pmatrix}^\perp = \begin{pmatrix} x_2\\ -x_1\end{pmatrix}$ for all $(x_1,x_2)^T\in\mathbb{R}^2$. We assume 1 for the density of the fluid and a density $\rho_{\mathcal{B}}(y) > 0$, a mass $\mathrm{m} = \int_{\mathcal{B}}\rho_{\mathcal{B}}(x)\,dx$ and a moment of inertia $I = \int_{\mathcal{B}}\rho_{\mathcal{B}}(x)|x|^2\,dx$ for the rigid body.

The movement of body and fluid is coupled in both ways. The equations on the body velocity include the Newtonian stress tensor

$$\mathbf{T}(u,p) = 2\mu D(u) - \pi\mathrm{Id},$$

where $\mu$ is the kinematic viscosity and

$$D(u) := \frac{1}{2}(\nabla u + (\nabla u)^T)$$

the deformation tensor. The fluid exerts a drag or lift force $-\int_\Gamma \mathbf{T}(u,\pi)N\,d\sigma$ and a torque $-\int_\Gamma y^\perp\mathbf{T}(u,\pi)N\,d\sigma$ on the body. On the other hand, the full velocity $v_{\mathcal{B}}$ of the body influences the fluid flow via Dirichlet no-slip conditions on the interface $\Gamma$.

The aim of the paper is to show the following main result, yielding an isomorphism between the data $f_0, f_1, f_2, u_0, \xi_0$ and $\Omega_0$ and the solution $u, \pi, \xi, \Omega$ of (2.1) in the natural spaces of maximal $L^p$-regularity.

For $1 \le p, q \le \infty$, $0 < \alpha \le 2$ and a domain $\mathcal{D}$ of class $C^{2,1}$, we denote by $B^\alpha_{p,q}(\mathcal{D})$, $H^{\alpha,p}(\mathcal{D})$ and $\dot{H}^{1,p}(\mathcal{D})$ the Besov spaces, Bessel potential spaces and the homogeneous Sobolev space of order one, respectively. They are defined on these domains by interpolation or restriction, see [16, Section III.4.3]. The usual Sobolev spaces of order $m \in \mathbb{N}$ are denoted by $W^{m,p}(\mathcal{D})$. Especially if we denote the norms of these spaces, we may omit stating whether they are scalar- or vector-valued in $\mathbb{R}$.

**Theorem 1.** *Let $\mathcal{D}$ be a domain of class $C^{2,1}$ in $\mathbb{R}^2$ as described above. We assume that $1 < p, q < \infty$, $\frac{1}{2q} + \frac{1}{p} \neq 1$, $\xi_0 \in \mathbb{R}^2$, $\Omega_0 \in \mathbb{R}$ and that $u_0 \in B^{2-2/p}_{q,p}(\mathcal{D})$ satisfies the compatibility conditions $\operatorname{div} u_0 = 0$ and $u_0|_\Gamma(y) = \xi_0 + \Omega_0 y^\perp$, if $\frac{1}{2q} + \frac{1}{p} < 1$ and $(u_0|_\Gamma \cdot N)(y) = (\xi_0 + \Omega_0 y^\perp) \cdot N(y)$, $x \in \Gamma$, if $\frac{1}{2q} + \frac{1}{p} > 1$. We assume that $f_0 \in L^p(J_{T_0}; L^q(\mathcal{D}))$, $f_1 \in L^p(J_{T_0}; \mathbb{R}^2)$ and that $f_2 \in L^p(J_{T_0}; \mathbb{R})$. Then there exists a unique strong solution*

$$
\begin{aligned}
u &\in L^p(J_{T_0}; W^{2,q}(\mathcal{D})) \cap W^{1,p}(J_{T_0}; L^q(\mathcal{D})) =: X^{T_0}_{p,q},\\
\pi &\in L^p(J_{T_0}; \dot{H}^{1,q}(\mathcal{D})) =: Y^{T_0}_{p,q},\\
\xi &\in W^{1,p}(J_{T_0}; \mathbb{R}^2),\\
\Omega &\in W^{1,p}(J_{T_0}),
\end{aligned}
$$

*which satisfies the estimate*

$$
\begin{aligned}
&\|u\|_{X^{T_0}_{p,q}} + \|\pi\|_{Y^{T_0}_{p,q}} + \|\xi\|_{W^{1,p}(J_{T_0})} + \|\Omega\|_{W^{1,p}(J_{T_0})}\\
&\quad \le C(\|f_0\|_{L^p(J_{T_0}, L^q(\mathcal{D}))} + \|(f_1, f_2)\|_{L^p(J_{T_0})} + \|u_0\|_{B^{2-2/p}_{q,p}(\mathcal{D})} + |(\xi_0, \Omega_0)|),
\end{aligned}
$$

*where the constant $C$ depends only on the geometry of the body and on $T_0$.*

*Remark* 2. The compatibility conditions on $u_0, \xi_0$ and $\Omega_0$ are a consequence of our construction of the solution below and of the characterization of the time-trace space $Z_{p,q} := (L^q_\sigma, W^{2,q} \cap W^{1,q}_0 \cap L^q_\sigma)_{1-1/p,p}$ given in [2, Theorem 3.4].

To prove the theorem, we first give preliminary classical estimates on the Stokes problem with inhomogeneous boundary conditions and the corresponding local pressure estimates in the next section. The most important argument is done in Section 4, where we give a reformulation of the full system (2.1) in terms of a linear equation in $W^{1,p}(0, T_0; \mathbb{R}^{2+1})$.

## 3. Preliminary results

The following proposition is a classical result due to Solonnikov [14], see also [12, Theorem 2.7] for the case $p \neq q$.

**Proposition 3.** *Let $\mathcal{D} \subset \mathbb{R}^n$, $n \ge 2$, be a bounded or exterior domain of class $C^{2,1}$ and $1 < p, q < \infty$, $0 < T < T_0$, $f \in L^p(J_T; L^q_\sigma(\Omega))$ and $u_0 \in Z_{p,q}$. Then there* *exists a unique solution* $u \in X^T_{p,q}$, $\pi \in Y^T_{p,q}$ *to the Stokes problem*

$$\begin{cases} \partial_t u - \Delta u + \nabla\pi &= f, \quad \text{in } J_T \times \mathcal{D}, \\ \operatorname{div} u &= 0, \quad \text{in } J_T \times \mathcal{D}, \\ u|_{\partial\mathcal{D}} &= 0, \quad \text{on } J_T \times \Gamma, \\ u(0) &= u_0 \end{cases} \tag{3.1}$$

*and there exists a constant* $C > 0$ *independent of* $T, u_0$ *and* $f$, *such that*

$$\|u\|_{X^T_{p,q}} + \|\pi\|_{Y^T_{p,q}} \le C(\|f\|_{L^p(L^q)} + \|u_0\|_{Z_{p,q}}).$$

We now consider the following special situation of inhomogeneous Dirichlet boundary data: Let $h \in W^{1,p}(J_T; C^2(\partial\mathcal{D}; \mathbb{R}^2))$ be a function on the boundary of an exterior domain $\mathcal{D}$ of class $C^{2,1}$ such that there exists an extension $H$ of $h$ to $\mathcal{D}$ satisfying

$$H|_{\partial\mathcal{D}} = h, \quad H \in W^{1,p}(J_T; C^2(\mathcal{D})), \quad \text{and} \quad \operatorname{div} H = 0. \tag{3.2}$$

In particular, $h$ and $H$ could be given by a rigid motion $\xi + \Omega y^\perp$ on $\mathbb{R}^2$, where $\xi \in W^{1,p}(J_T; \mathbb{R}^2)$, $\Omega \in W^{1,p}(J_T; \mathbb{R})$. We choose open balls $B_1, B_2 \subset \mathbb{R}^2$ such that $\mathcal{D}^c \subset B_1 \subset \overline{B_1} \subset B_2$ and define a cut-off function $\chi \in C^\infty(\mathbb{R}^2; [0,1])$ satisfying

$$\chi(y) := \begin{cases} 1 & \text{if } y \in \overline{B_1}, \\ 0 & \text{if } y \in \mathcal{D} \setminus B_2. \end{cases} \tag{3.3}$$

Then we define

$$b_h := \chi H - B_K((\nabla\chi)H), \tag{3.4}$$

where $K \subset \mathbb{R}^2$ is a bounded open set which contains the annulus $B_1 \backslash \overline{B_2}$ and $B_K$ denotes the Bogovskiĭ operator with respect to the domain $K$, cf. [3]. It follows that $b_h \in W^{1,p}(0,T; C^2_{c,\sigma}(\mathbb{R}^n))$ due to $\operatorname{div} b_h = \nabla\chi H + \chi \operatorname{div} H - \nabla\chi H = 0$ and by setting $u := u_b + b_h$, we can solve the Stokes problem

$$\begin{cases} u_t - \Delta u + \nabla\pi &= f \quad \text{in } J_T \times \mathcal{D}, \\ \operatorname{div} u &= 0 \quad \text{in } J_T \times \mathcal{D}, \\ u|_{\partial\Omega} &= h \quad \text{on } J_T \times \partial\mathcal{D}, \\ u(0) &= u_0 \quad \text{in } \mathcal{D}, \end{cases} \tag{3.5}$$

and get the estimate

$$\|u\|_{X^T_{p,q}} + \|\pi\|_{Y^T_{p,q}} \le C(\|f\|_{L^p(J_{T_0}, L^q(\mathcal{D}))} + \|u_0 - b_h(0)\|_{Z_{p,q}} + \|b_h\|_{X^T_{p,q}}) \tag{3.6}$$

by Proposition 3. Corresponding to this result, we define solution operators

$$\mathcal{U}(f, h, u_0) \in X^T_{p,q}, \qquad \mathcal{P}(f, h, u_0) \in Y^T_{p,q} \tag{3.7}$$

for the inhomogeneous problem (3.5).

We also need the following embedding property of $X^T_{p,q}$, which follows from the mixed derivatives theorem, see, e.g., [5, Lemma 4.1] and which can be proved as in [5, Lemma 4.3] or as in [6, Theorem 1.7.2].

**Proposition 4.** *Let $\mathcal{D} \subset \mathbb{R}^n$ be a $C^{1,1}$-domain with compact boundary, assume $p, q \in (1, \infty)$, $\alpha \in (0,1)$ and fix $T_0 > 0$. Then*

$$X_{p,q}^{T_0} \hookrightarrow H^{\alpha,p}(J_{T_0}; H^{2(1-\alpha),q}(\mathcal{D})).$$

At this point we also note the elementary embedding constants

$$\|f\|_p \le T^{1/p-1/q} \|f\|_q \qquad \text{for all } f \in L^q(J_T), \infty \ge q > p \ge 1, \tag{3.8}$$

and

$$\|f\|_\infty \le T^{1/p'} \|f\|_{W^{1,p}(J_T)} \qquad \text{for all } f \in {}_0W^{1,p}(J_T), \frac{1}{p} + \frac{1}{p'} = 1, \tag{3.9}$$

where we define

$${}_0W^{1,p}(J_T) = \{f \in W^{1,p}(J_T) : \lim_{t\to 0} f(t) = 0\}.$$

The last preliminary result gives locally improved time regularity for the pressure in (3.1) for suitable $f$. For a proof, see [13] and [10].

**Lemma 5.** *Let $\mathcal{D} \subset \mathbb{R}^n$, $n \ge 2$, be an exterior domain of class $C^{2,1}$, $1 < p, q < \infty$, $0 < T < T_0$, and $f \in L^p(J_T; L^q_\sigma(\Omega))$. Then if the pressure part $\pi = \mathcal{P}(f, 0, 0) \in Y_{p,q}^T$ of the solution of (3.5) with zero initial and boundary conditions is chosen in a way that $\pi \in L^p(J_T; L^q(\mathcal{D}_R))$ and $\int_{\mathcal{D}_R} \pi = 0$ for some $R > 0$, $\mathcal{D}_R := \mathcal{D} \cap B_R$, it satisfies the estimate*

$$\|\pi\|_{L^p(J_T;L^q(\mathcal{D}_R))} \lesssim CT^{\alpha/p} \|f\|_{L^p(J_T,L^q(\mathcal{D}))} \qquad \textit{for all } 0 \le \alpha < \frac{1}{2q}.$$

## 4. A reformulation of the problem

The reformulation procedure for problem (2.1) can be split into three smaller steps. First we obtain homogeneous initial conditions by subtracting $u^* = \mathcal{U}(f_0, \xi_0 + \Omega_0 y^\perp, u_0)$ and $\pi^* = \mathcal{P}(f_0, \xi_0 + \Omega_0 y^\perp, u_0)$ from $u, \pi$ and $\xi_0, \Omega_0$ from $\xi, \Omega$, respectively, with the estimate

$$\|u^*\|_{X_{p,q}^T} + \|\pi^*\|_{Y_{p,q}^T} \le C(\|f_0\|_{p,q} + \|u_0\|_{B_{q,p}^{2-2/p}(\mathcal{D})} + |(\xi_0, \Omega_0)|).$$

Secondly, we consider the functions $\overline{v}^{(i)}$ and $\overline{V}$ solving the weak Neumann problems

$$\begin{cases} \Delta \overline{v}^{(i)} &= 0 \quad \text{in } \mathcal{D}, \\ \frac{\partial \overline{v}^{(i)}}{\partial N}|_\Gamma &= N_i \quad \text{on } \Gamma, \end{cases} \qquad \text{and} \qquad \begin{cases} \Delta \overline{V} &= 0 \quad \text{in } \mathcal{D}, \\ \frac{\partial \overline{V}}{\partial N}|_\Gamma &= N \cdot y^\perp \quad \text{on } \Gamma. \end{cases}$$

By [11], we obtain strong regularity and the estimate

$$\|\nabla \overline{v}^{(i)}\|_{W^{2,r}(\mathcal{D})}, \|\nabla \overline{V}\|_{W^{2,r}(\mathcal{D})} \le C \tag{4.1}$$

for every $1 < r < \infty$.

This gives rise to the following construction. Let $0 < T \le T_0$. For any given $\xi \in {}_0W^{1,p}(J_T;\mathbb{R}^2)$, $\Omega \in {}_0W^{1,p}(J_T;\mathbb{R})$, we define

$$v_{\xi,\Omega}(t) := \sum_i \xi_i(t)\bar{v}^{(i)} + \Omega(t)\bar{V} \qquad \text{for all } t \in J_T, \tag{4.2}$$

which implies that

$$\begin{cases} \Delta v_{\xi,\Omega}(t) = 0, & \text{in } \mathcal{D}, \\ \dfrac{\partial v_{\xi,\Omega}(t)}{\partial N}|_\Gamma = (\xi(t) + y^\perp\Omega(t)) \cdot N, & \text{on } \Gamma, \end{cases}$$

is satisfied. It follows immediately from (4.2) that

$$\|\nabla v_{\xi,\Omega}\|_{W^{1,p}(J_T;W^{2,q}(\mathcal{D}))} + \|\partial_t v_{\xi,\Omega}\|_{Y^T_{p,q}} \le C\,\|(\xi,\Omega)\|_{W^{1,p}(J_T)}. \tag{4.3}$$

In the following, we define new unknown functions $\hat{u}, \hat{\pi}, \hat{\xi}, \hat{\Omega}$ by

$$\begin{aligned} u &:= u^* + \hat{u} + \nabla v_{\hat{\xi},\hat{\Omega}}, \\ \pi &:= \pi^* + \hat{\pi} - \partial_t v_{\hat{\xi},\hat{\Omega}}, \\ \xi &:= \xi_0 + \hat{\xi}, \\ \Omega &:= \Omega_0 + \hat{\Omega}. \end{aligned}$$

Instead of (2.1), we then get the equivalent problem

$$\begin{cases} \hat{u}_t - \Delta\hat{u} + \nabla\hat{\pi} = 0 & \text{in } J_T \times \mathcal{D}, \\ \operatorname{div}\hat{u} = 0 & \text{in } J_T \times \mathcal{D}, \\ \hat{u}|_\Gamma = h(\hat{\xi},\hat{\Omega}) & \text{on } J_T \times \Gamma, \\ \hat{u}(0) = 0 & \text{in } \mathcal{D}, \\ \mathrm{m}\hat{\xi}' + \int_\Gamma \mathbf{T}(\hat{u}, \hat{\pi} - \partial_t v_{\hat{\xi},\hat{\Omega}})N\,\mathrm{d}\sigma = f_1 - \int_\Gamma \mathbf{T}(u^*,\pi^*)N\,\mathrm{d}\sigma & \text{in } J_T, \\ I\hat{\Omega}' + \int_\Gamma y^\perp \mathbf{T}(\hat{u}, \hat{\pi} - \partial_t v_{\hat{\xi},\hat{\Omega}})N\,\mathrm{d}\sigma = f_2 - \int_\Gamma y^\perp\mathbf{T}(u^*,\pi^*)N\,\mathrm{d}\sigma & \text{in } J_T, \\ \hat{\xi}(0) = 0, & \\ \hat{\Omega}(0) = 0, & \end{cases} \tag{4.4}$$

where

$$h(\hat{\xi},\hat{\Omega})(t,y) := \hat{\xi}(t) + y^\perp\hat{\Omega}(t) - \nabla v_{\hat{\xi},\hat{\Omega}}|_\Gamma(t,y). \tag{4.5}$$

Due to the correction by $v_{\hat{\xi},\hat{\Omega}}$, we get the additional condition $\hat{u}|_\Gamma N = 0$ on the boundary. As $v_{\hat{\xi},\hat{\Omega}}$ was defined to absorb the normal component of the interface velocity into the pressure, the correction $\nabla v_{\hat{\xi},\hat{\Omega}}$ applied to $u$ does not affect the rigid body. This can be seen from the following calculations. It holds that

$$\begin{aligned} \Big(\int_\Gamma D(\nabla v_{\hat{\xi},\hat{\Omega}}))N\,\mathrm{d}\sigma\Big)_i &= \int_\Gamma (\partial_i \nabla v_{\hat{\xi},\hat{\Omega}}) \cdot N\,\mathrm{d}\sigma \\ &= \int_{\mathcal{D}} \operatorname{div}(\partial_i \nabla v_{\hat{\xi},\hat{\Omega}}) = \int_{\mathcal{D}} \partial_i \Delta v_{\hat{\xi},\hat{\Omega}} = 0 \end{aligned}$$

and that

$$\begin{aligned}\int_\Gamma y^\perp D(\nabla v_{\hat{\xi},\hat{\Omega}}))N\,\mathrm{d}\sigma &= \int_{\mathcal{D}} \operatorname{div}\big(\sum_{i=1}^2 y_i^\perp \partial_i\partial_1 v_{\hat{\xi},\hat{\Omega}}, \sum_{i=1}^2 y_i^\perp \partial_i\partial_2 v_{\hat{\xi},\hat{\Omega}}\big)\\ &= \int_{\mathcal{D}} (\partial_1\partial_2 - \partial_2\partial_1)v_{\hat{\xi},\hat{\Omega}} + (-y_2\partial_1 + y_1\partial_2)\Delta v_{\hat{\xi},\hat{\Omega}}\\ &= 0.\end{aligned}$$

For convenience of notation and in order to reformulate the problem, let

$$\mathbb{I} := \begin{pmatrix} \mathrm{mId}_{\mathbb{R}^2} & 0 \\ 0 & I \end{pmatrix} \tag{4.6}$$

the constant momentum matrix for the body and let $\mathcal{J} : (W^{1,q}_{\mathrm{loc}}(\mathcal{D}))^{2\times 2} \to \mathbb{R}^3$ the integral operator given by

$$\mathcal{J}(M) = \begin{pmatrix} \int_\Gamma MN\,\mathrm{d}\sigma \\ \int_\Gamma y^\perp MN\,\mathrm{d}\sigma \end{pmatrix}. \tag{4.7}$$

For all $\varepsilon > 0$, by the boundedness of the trace operator $\gamma : H^{\varepsilon+1/q,q}(\mathcal{D}) \to L^q(\Gamma)$,we get

$$|\mathcal{J}(M)| \le C\,\|\gamma(M)\|_{L^1(\Gamma;\mathbb{R}^4)} \le C\,\|\gamma(M)\|_{L^q(\Gamma);\mathbb{R}^4)} \le C\,\|M\|_{H^{\varepsilon+1/q,q}(\mathcal{D};\mathbb{R}^4)}, \tag{4.8}$$

for all $M \in H^{\varepsilon+1/q,q}(\mathcal{D};\mathbb{R}^{2\times 2})$. Furthermore, we define an *added mass*

$$\mathbb{M} = \begin{pmatrix} \int_\Gamma \overline{v}^{(1)} N_1\,\mathrm{d}\sigma & \int_\Gamma \overline{v}^{(2)} N_1\,\mathrm{d}\sigma & \int_\Gamma \overline{V} N_1\,\mathrm{d}\sigma \\ \int_\Gamma \overline{v}^{(1)} N_2\,\mathrm{d}\sigma & \int_\Gamma \overline{v}^{(2)} N_2\,\mathrm{d}\sigma & \int_\Gamma \overline{V} N_2\,\mathrm{d}\sigma \\ \int_\Gamma \overline{v}^{(1)} y^\perp\cdot N\,\mathrm{d}\sigma & \int_\Gamma \overline{v}^{(2)} y^\perp\cdot N\,\mathrm{d}\sigma & \int_\Gamma \overline{V} y^\perp\cdot N\,\mathrm{d}\sigma \end{pmatrix}$$

of the body, cf. [8, p. 685]. The point of this definition is that we get

$$\mathcal{J}(\partial_t v_{\hat{\xi},\hat{\Omega}}\mathrm{Id}_{\mathbb{R}^2}) = \mathbb{M}\begin{pmatrix} \hat{\xi}' \\ \hat{\Omega}' \end{pmatrix},$$

so that the body equations in (4.4) can be rewritten as

$$(\mathbb{I}+\mathbb{M})\begin{pmatrix} \hat{\xi}' \\ \hat{\Omega}' \end{pmatrix} = \begin{pmatrix} f_1 \\ f_2 \end{pmatrix} - \mathcal{J}\mathbf{T}(\mathcal{U}(0,h(\hat{\xi},\hat{\Omega}),0),\mathcal{P}(0,h(\hat{\xi},\hat{\Omega}),0)) - \mathcal{J}\mathbf{T}(u^*,\pi^*). \tag{4.9}$$

Furthermore, we obtain the following properties of the added mass matrix.

**Lemma 6.** *The matrix* $\mathbb{M}$ *is symmetric and semi positive-definite.*

*Proof.* The matrix $\mathbb{M}$ is symmetric because by the Gauss theorem and the properties of $\overline{v}^{(i)}, \overline{V}$,

$$\int_\Gamma \overline{v}^{(i)} N_j\,\mathrm{d}\sigma = \int_\Gamma \overline{v}^{(i)} \frac{\partial \overline{v}^{(j)}}{\partial N}\,\mathrm{d}\sigma = \int_{\mathcal{D}} \nabla\overline{v}^{(i)}\cdot\nabla\overline{v}^{(j)} = \sum_{l=1}^3 \int_{\mathcal{D}} \partial_l\overline{v}^{(i)}\partial_l\overline{v}^{(j)},$$

and similarly,

$$\int_\Gamma \overline{v}^{(i)} y^\perp \cdot N \,\mathrm{d}\sigma = \int_\Gamma \overline{v}^{(i)} \frac{\partial \overline{V}}{\partial N} \,\mathrm{d}\sigma = \int_{\mathcal{D}} \nabla \overline{v}^{(i)} \cdot \nabla \overline{V} = \int_\Gamma \overline{V} N_i \,\mathrm{d}\sigma.$$

The existence of these integrals follows from (4.1) for $r = 2$. Furthermore, for all $z = (x_1, x_2, y)^T \in \mathbb{R}^3$,

$$\begin{aligned} z^T \overline{\mathbb{M}} z &= \sum_{i=1}^2 \int_{\mathcal{D}} \big[(x_1^2 (\partial_i \overline{v}^{(1)})^2 + x_2^2 (\partial_i \overline{v}^{(2)})^2 + y^2 (\partial_i \overline{V})^2 \\ &\qquad + 2x_1 x_2 (\partial_i \overline{v}^{(1)})(\partial_i \overline{v}^{(2)}) + 2x_1 y (\partial_i \overline{v}^{(1)})(\partial_i \overline{V}) + 2x_2 y (\partial_i \overline{v}^{(2)})(\partial_i \overline{V})\big] \\ &= \sum_{i=1}^2 \int_{\mathcal{D}} \big(x_1(\partial_i \overline{v}^{(1)}) + x_2(\partial_i \overline{v}^{(2)}) + y(\partial_i \overline{V})\big)^2 \\ &\geq 0 \end{aligned}$$

by the Gauss Theorem. □

For every choice of the body's density $\rho_{\mathcal{B}} > 0$, $\mathbb{I}$ is strictly positive, so Lemma 6 yields that $\mathbb{I} + \mathbb{M}$ is invertible.

Thus, as a third step in our reformulation, instead of problem (4.4) we can write the equation

$$\begin{pmatrix} \hat{\xi} \\ \hat{\Omega} \end{pmatrix} = \mathcal{R} \begin{pmatrix} \hat{\xi} \\ \hat{\Omega} \end{pmatrix} + f^*, \tag{4.10}$$

where $\mathcal{R} : {}_0W^{1,p}(J_T; \mathbb{R}^3) \to {}_0W^{1,p}(J_T; \mathbb{R}^3)$ is given by

$$\mathcal{R}(\hat{\xi}, \hat{\Omega})(t) := \int_0^t (\mathbb{I} + \mathbb{M})^{-1} \mathcal{J} \left[\mathbf{T}(\mathcal{U}(0, h(\hat{\xi}, \hat{\Omega}), 0), \mathcal{P}(0, h(\hat{\xi}, \hat{\Omega}), 0))\right](s) \,\mathrm{d}s$$

and

$$f^*(t) := \int_0^t (\mathbb{I} + \mathbb{M})^{-1} \left[\begin{pmatrix} f_1 \\ f_2 \end{pmatrix} - \mathcal{J}\mathbf{T}(u^*, \pi^*)\right](s) \,\mathrm{d}s.$$

Note that despite its appearance, (4.9) is not an ODE, but we can consider it as a linear fixed point problem in this way.

## 5. Proof of Theorem 1

In this section, we show that for sufficiently small $T$, $0 < T \leq T_0$, $\mathrm{Id} - \mathcal{R}$ is invertible and thus gives a solution for (2.1). By iteration, the solution extends to $J_{T_0}$.

**Lemma 7.** *The map $\mathcal{R}$ is bounded linear and $\|\mathcal{R}\|_{\mathcal{L}({}_0W^{1,p}(J_T;\mathbb{R}^3))} < 1$ for sufficiently small $0 < T \leq T_0$.*

*Proof.* Let $(\hat{\xi}, \hat{\Omega}) \in {}_0W^{1,p}(J_T; \mathbb{R}^3)$. The functions

$$H(t, y) = \hat{\xi}(t) + \hat{\Omega}(t) y^\perp - \nabla v_{\hat{\xi},\hat{\Omega}}(t, y)$$

and $h(\hat{\xi},\hat{\Omega}) = H|_\Gamma$ satisfy the conditions (3.2), so that we get

$$\begin{aligned} \|\mathcal{U}(0,h(\hat{\xi},\hat{\Omega}),0)\|_{X^T_{p,q}} + \|\mathcal{P}(0,h(\hat{\xi},\hat{\Omega}),0)\|_{Y^T_{p,q}} &\le C\|b_{h(\hat{\xi},\hat{\Omega})}\|_{X^T_{p,q}} \\ &\le C\|(\hat{\xi},\hat{\Omega})\|_{W^{1,p}(J_T)} \end{aligned} \tag{5.1}$$

by (3.6). In the following, we abbreviate

$$\begin{aligned} \hat{u} &= \mathcal{U}(0,h(\hat{\xi},\hat{\Omega}),0)), \\ \hat{\pi} &= \mathcal{P}(0,h(\hat{\xi},\hat{\Omega}),0)). \end{aligned}$$

We can apply Lemma 5 to show

$$\begin{aligned} \|\hat{\pi}\|_{L^p(J_T;L^q(\mathcal{D}_R))} &\le CT^{\alpha/p}\|b_{h(\hat{\xi},\hat{\Omega})}\|_{X^T_{p,q}} \\ &\le CT^{\alpha/p}\|(\hat{\xi},\hat{\Omega})\|_{W^{1,p}(J_T)} \end{aligned} \tag{5.2}$$

for a suitable choice of $R > 0$ and $0 \le \alpha < \frac{1}{2q}$. Here, $b_{h(\hat{\xi},\hat{\Omega})} = \chi h(\hat{\xi},\hat{\Omega}) - B_K(\nabla\chi h(\hat{\xi},\hat{\Omega})) \in W^{1,p}(J_T;C^\infty_{c,\sigma}(\mathbb{R}^3))$ is the auxiliary function from (3.4), which moves the boundary condition $h(\hat{\xi},\hat{\Omega})$ to the right-hand side of the Stokes equation. By construction, it is solenoidal and it satisfies

$$\partial_t b_{h(\hat{\xi},\hat{\Omega})}|_\Gamma \cdot N = \partial_t h(\hat{\xi},\hat{\Omega}) \cdot N = 0$$

and $\Delta b|_\Gamma = 0$ on the boundary, so that the right-hand side $\Delta b_{h(\hat{\xi},\hat{\Omega})} - \partial_t b_{h(\hat{\xi},\hat{\Omega})} \in L^p(J_T;L^q_\sigma(\mathcal{D}))$ satisfies the assumption in Lemma 5. If we choose $\varepsilon > 0$ such that $1/q + \varepsilon < 1$,

$$\frac{1}{r} := -\alpha + \frac{1}{p} > -\frac{1}{2q} + \frac{1}{p}$$

and set $\beta := \frac{1}{p} - \frac{1}{r}$, it follows that by (4.8), (3.8), Proposition 4, and (5.1), we have

$$\begin{aligned} \|\mathcal{J}(D(\hat{u}))\|_{L^p(J_T)} &\le C\|\hat{u}\|_{L^p(J_T;H^{1+1/q+\varepsilon,q}(\mathcal{D}))} \\ &\le CT^\beta\|\hat{u}\|_{L^r(J_T;H^{1+1/q+\varepsilon,q}(\mathcal{D}))} \\ &\le CT^\beta\|\hat{u}\|_{H^{\alpha,p}(J_T;H^{2-2\alpha,q}(\mathcal{D}))} \\ &\le CT^\beta\|\hat{u}\|_{X^T_{p,q}} \\ &\le CT^\beta\|(\hat{\xi},\hat{\Omega})\|_{W^{1,p}(J_T)}, \end{aligned} \tag{5.3}$$

where the constant $C$ does not depend on $T$, $0 < T \le T_0$, as $\hat{u}(0) = 0$. Let now $c := \frac{1}{q} + \varepsilon < 1$. Similarly, by interpolation, by the Poincaré inequality, by (5.2) and by (5.1),

$$\begin{aligned} \|\mathcal{J}(\hat{\pi}\mathrm{Id}_{\mathbb{R}^2})\|_{L^p(J_T)} &\le C\|\hat{\pi}\|_{L^p(J_T;H^{1/q+\varepsilon,q}(\mathcal{D}_R))} \\ &\le C\|\hat{\pi}\|^c_{L^p(J_T;L^q(\mathcal{D}_R))}\|\hat{\pi}\|^{1-c}_{L^p(J_T;W^{1,q}(\mathcal{D}_R))} \\ &\le CT^{c\alpha/p}\|(\hat{\xi},\hat{\Omega})\|_{W^{1,p}(J_T)}\|\hat{\pi}\|^{1-c}_{Y^T_{p,q}} \\ &\le CT^{c\alpha/p}\|(\hat{\xi},\hat{\Omega})\|_{W^{1,p}(J_T)} \end{aligned}$$

for all $0 \le \alpha < \frac{1}{2q}$. In conclusion, by (4.3) and (3.8) and (3.9),

$$\begin{aligned}\|\mathcal{R}(\hat{\xi},\hat{\Omega})\|_{W^{1,p}(J_T)} &\le C\|\mathcal{J}\mathbf{T}(\hat{u}+\nabla v_{\hat{\xi},\hat{\Omega}},\hat{\pi})\|_{L^p(J_T)}\\ &\le C\|\nabla v_{\hat{\xi},\hat{\Omega}}\|_{L^p(J_T;W^{2,q}(\mathcal{D}))}+C(T^\beta+T^{c\alpha/p})\|(\hat{\xi},\hat{\Omega})\|_{W^{1,p}(J_T)}\\ &\le C(T+T^\beta+T^{c\alpha/p})\|(\hat{\xi},\hat{\Omega})\|_{W^{1,p}(J_T)}\end{aligned}$$

and $\mathcal{R}(\hat{\xi},\hat{\Omega})(0)=0$ by definition, so that

$$L:=\|\mathcal{R}\|_{\mathcal{L}({}_0W^{1,p}(J_T))}<1 \tag{5.4}$$

for small $T$. □

Clearly, by (3.6), we have

$$\begin{aligned}\|f^*\|_{W^{1,p}(J_T)} &\le C\left\|\begin{pmatrix} f_1\\ f_2\end{pmatrix}-\mathcal{J}\mathbf{T}(u^*,\pi^*)\right\|_{L^p(J_T)}\\ &\le C(\|f_0\|_{L^p(J_T,L^q(\mathcal{D}))}+\|(f_1,f_2)\|_{L^p(J_T)}\\ &\qquad+\|u_0\|_{B^{2-2/p}_{q,p}(\mathcal{D})}+|(\xi_0,\Omega_0)|),\end{aligned}$$

and thus $f^*\in {}_0W^{1,p}(J_T;\mathbb{R}^3)$.

The lemma shows that for some $T>0$, which depends on the geometry of the body but not on the initial data, the operator $\mathrm{Id}-\mathcal{R}$ is invertible. Thus, a unique solution $(\hat{\xi},\hat{\Omega})$ to (4.10) exists on this time interval. Furthermore, we obtain the estimate

$$\|(\hat{\xi},\hat{\Omega})\|_{W^{1,p}(J_T)}\le(1-L)C(\|f_0\|_{p,q}+\|(f_1,f_2)\|_p+\|u_0\|_{B^{2-2/p}_{q,p}(\mathcal{D})}+|(\xi_0,\Omega_0)|).$$

Plugging the solution $\hat{\xi},\hat{\Omega}$ into (4.2) and setting

$$\begin{aligned}\hat{u}&=\mathcal{U}(0,h(\hat{\xi},\hat{\Omega}),0)),\\ \hat{\pi}&=\mathcal{P}(0,h(\hat{\xi},\hat{\Omega}),0)),\end{aligned}$$

yields solutions

$$\begin{aligned}u&:=\hat{u}+\nabla v_{\hat{\xi}+\xi_0,\hat{\Omega}+\Omega_0}+u^*\in X^T_{p,q},\\ \pi&:=\hat{\pi}+\partial_t v_{\hat{\xi}+\xi_0,\hat{\Omega}+\Omega_0}-\pi^*\in Y^T_{p,q},\\ \xi&:=\hat{\xi}+\xi_0\in W^{1,p}(J_T),\\ \Omega&:=\hat{\Omega}+\Omega_0\in W^{1,p}(J_T)\end{aligned}$$

of (2.1) and the estimate

$$\begin{aligned}&\|u\|_{X^T_{p,q}}+\|\pi\|_{Y^T_{p,q}}+\|(\xi,\Omega)\|_{W^{1,p}(J_T)}\\ &\quad\le C(\|f_0\|_{L^p(L^q)}+\|(f_1,f_2)\|_{L^p}+\|u_0\|_{B^{2-2/p}_{q,p}(\mathcal{D})}+|(\xi_0,\Omega_0)|).\end{aligned}$$

The uniqueness of the solution for this linear problem immediately follows from the estimate. Since the length $T$ of our time interval arises from condition (5.4) in the proof of Lemma 7, it is unaffected by the initial data and external forces and since

$$X_{p,q}^T \hookrightarrow C([0,T]; \{u \in B_{q,p}^{2-2/p}(\mathcal{D}) : \operatorname{div} u = 0\})$$

see [1, Theorem III.4.10.2], and $W^{1,p}(J_T) \hookrightarrow C([0,T])$, we can take $u(T)$, $\xi(T)$, $\Omega(T)$ as initial values for solving the problem up to time $2T$. Iterating this procedure and gluing together the solutions on $(kT,(k+1)T)$ $k = 0,1,2,\ldots$ yields a solution on $J_{T_0}$. We can see that it satisfies the above estimate on every subinterval of $J_{T_0}$, as we can choose any starting time $T_s \in [0, T_0 - T)$ to find a solution by the above procedure on $[T_s, T_s + T)$. This proves Theorem 1.

## References


[1] H. Amann, *Linear and Quasilinear Parabolic Problems. Vol. I*, Birkhäuser, 1995.

[2] H. Amann, *On the Strong Solvability of the Navier-Stokes equations.* J. Math. Fluid Mech. **2** (2000), 16–98.

[3] M.E. Bogovskiĭ, *Solution of the first boundary value problem for an equation of continuity of an incompressible medium.* Dokl. Akad. Nauk SSSR **248** (1979), 1037–1040.

[4] C. Conca, J. San Martín, and M. Tucsnak, *Existence of solutions for the equations modeling the motion of a rigid body in a viscous fluid.* Comm. Partial Differential Equations **25** (2000) 1019–1042.

[5] R. Denk, J. Saal, and J. Seiler, *Inhomogeneous symbols, the Newton polygon, and maximal $L^p$-regularity.* Russ. J. Math. Phys. **15** (2008), 171–191.

[6] E. Dintelmann, *Fluids in the exterior domain of several moving obstacles.* PhD thesis, Technische Universität Darmstadt, 2007.

[7] E. Feireisl, *On the motion of rigid bodies in a viscous compressible fluid.* Arch. Ration. Mech. Anal. **167** (2003), 281–308.

[8] G.P. Galdi, *On the motion of a rigid body in a viscous liquid: a mathematical analysis with applications.* Handbook of mathematical fluid dynamics, Vol. I, North-Holland, 2002, 653–791.

[9] G.P. Galdi and A.L. Silvestre, *Strong solutions to the problem of motion of a rigid body in a Navier-Stokes liquid under the action of prescribed forces and torques.* Nonlinear Problems in Mathematical Physics and Related Topics I, Kluwer Academic/Plenum Publishers, 2002, 121–144.

[10] M. Geißert, K. Götze, and M. Hieber, *$L^p$-theory for strong solutions to fluid rigid-body interaction in Newtonian and generalized Newtonian fluids.* Trans. Amer. Math. Soc., to appear.

[11] M. Geißert, H. Heck, M. Hieber, and O. Sawada, *Weak Neumann implies Stokes.* J. Reine Angew. Math., to appear.

[12] Y. Giga and H. Sohr, *Abstract $L^p$ estimates for the Cauchy problem with applications to the Navier-Stokes equations in exterior domains.* J. Funct. Anal. **102** (1991), 72–94.

[13] A. Noll and J. Saal, *$H^\infty$-calculus for the Stokes operator on $L^q$-spaces.* Math. Z. **244** (2003), 651–688.

[14] V.A. Solonnikov, *Estimates for solutions of nonstationary Navier-Stokes equations.* J. Sov. Math. **8** (1977), 467–529.

[15] T. Takahashi. *Analysis of strong solutions for the equations modeling the motion of a rigid-fluid system in a bounded domain.* Adv. Differential Equations **8** (2003), 1499–1532.

[16] H. Triebel. *Interpolation Theory, Function Spaces, Differential Operators.* Johann Ambrosius Barth, 1995.

Karoline Götze
Weierstrass Institute
Mohrenstraße 39
D-10117 Berlin, Germany
e-mail: `karoline.goetze@wias-berlin.de`

Operator Theory:
Advances and Applications, Vol. 221, 385-398


# Transfer Functions for Pairs of Wandering Subspaces

Rolf Gohm

**Abstract.** To a pair of subspaces wandering with respect to a row isometry we associate a transfer function which in general is multi-Toeplitz and in interesting special cases is multi-analytic. Then we describe in an expository way how characteristic functions from operator theory as well as transfer functions from noncommutative Markov chains fit into this scheme.

**Mathematics Subject Classification (2000).** Primary 47A13; Secondary 46L53.

**Keywords.** Row isometry, multi-Toeplitz, multi-analytic, wandering subspace, transfer function, characteristic function, noncommutative Markov chain.

## Introduction

It is evident to all workers in these fields that the relationship between operator theory and the theory of analytic functions is the source of many deep results. In recent work [6] of the author a transfer function, which is in fact a multi-analytic operator, has been introduced in the context of noncommutative Markov chains. These can be thought of as toy models for interaction processes in quantum physics. The theory of multi-analytic operators, pioneered by Popescu [8, 9] and others in the late 1980's, has developed into a very successful generalization of the relationship mentioned above. Hence it is natural to expect that noncommutative Markov chains and their transfer functions open up a possibility to apply these tools in the study of models in quantum physics.

This paper is the result of an effort to discover the common geometric underpinning which ties together these at first sight rather different settings. It is found in the tree-like structure of wandering subspaces of row isometries, more precisely: the transfer function describes the relative position of two such trees. This is worked out in Section 1 below. One of the main results in Section 1 is a geometric characterization of pairs of subspaces with a multi-analytic operator as their transfer function.

With this work done we are in a position to discuss the existing applications in a new light which highlights common features. In Section 2 we give, from this point of view, an expository treatment of characteristic functions, both the well-known characteristic function of a contraction in the sense of Sz.-Nagy and Foias [13] and the less well-known characteristic function of a lifting introduced by Dey and Gohm [4]. In Section 3 we explain in the same short but expository style the transfer function of a noncommutative Markov chain from [6] and sketch a generalization which is natural in the setting of this paper. We hope and expect that this presentation is helpful for operator theorists to find their way into an area of potentially interesting applications.

## 1. Pairs of subspaces

Let $\hat{\mathcal{H}}$ be a Hilbert space and $V = (V_1, \ldots, V_d)$ a row isometry on $\hat{\mathcal{H}}$. Recall that this means that the $V_k : \hat{\mathcal{H}} \to \hat{\mathcal{H}}$ are isometries with orthogonal ranges. Here $d \in \mathbb{N}$ and additionally we also include the possibility of a sequence $(V_1, V_2, \ldots)$ of such isometries, writing symbolically $d = \infty$ in this case.

Let $F_d^+$ be the free semigroup with $d$ generators (which we denote $1, \ldots, d$). Its elements are (finite) words in the generators, including the empty word (which we denote by 0). The binary operation is concatenation of words. Let $\alpha = \alpha_1 \ldots \alpha_r$, with the $\alpha_\ell \in \{1, \ldots, d\}$, be such a word. We denote by $|\alpha| = r$ the length of the word $\alpha$. Further we define

$$V_\alpha := V_{\alpha_1} \ldots V_{\alpha_r}$$

($V_0$ is the identity operator). By $V_\alpha^*$ we mean $(V_\alpha)^* = V_{\alpha_r}^* \ldots V_{\alpha_1}^*$. We refer to [8, 9, 2, 3, 4] for further background about this type of multi-variable operator theory.

We want to establish an efficient description of the relative position of pairs of subspaces and their translates under a row isometry $V = (V_1, \ldots, V_d)$ on $\hat{\mathcal{H}}$. Suppose $\mathcal{U}$ and $\mathcal{Y}$ are Hilbert spaces and $i_0 : \mathcal{U} \to \hat{\mathcal{H}}$ and $j_0 : \mathcal{Y} \to \hat{\mathcal{H}}$ are isometric embeddings into $\hat{\mathcal{H}}$. Further we write $i_\omega := V_\omega i_0$ and $i_\omega(\mathcal{U}) =: \mathcal{U}_\omega$, similarly $j_\sigma := V_\sigma j_0$ and $j_\sigma(\mathcal{Y}) =: \mathcal{Y}_\sigma$, where $\omega, \sigma \in F_d^+$. To describe the relative position of $\mathcal{U}_\omega$ and $\mathcal{Y}_\sigma$ we consider the contraction

$$K(\sigma, \omega) := j_\sigma^* \, i_\omega : \mathcal{U} \to \mathcal{Y}.$$

Note that

$$j_\sigma \, K(\sigma, \omega) \, i_\omega^* : \hat{\mathcal{H}} \to \hat{\mathcal{H}}$$

is nothing but the orthogonal projection onto $\mathcal{Y}_\sigma$ restricted to $\mathcal{U}_\omega$. The embeddings introduced above allow us to represent these contractions for varying $\sigma$ and $\omega$ on common Hilbert spaces $\mathcal{U}$ and $\mathcal{Y}$.

**Lemma 1.1.** *$K(\sigma,\omega)$ for varying $\sigma$ and $\omega$ is a multi-Toeplitz kernel, i.e.,*

$$K : F_d^+ \times F_d^+ \to \mathcal{B}(\mathcal{U},\mathcal{Y})$$

*such that*

$$K(\sigma,\omega) = \begin{cases} K(\alpha,0) & \text{if } \sigma = \omega\alpha \\ K(0,\alpha) & \text{if } \omega = \sigma\alpha \\ 0 & \text{otherwise.} \end{cases}$$

*Proof.* If $\sigma = \omega\alpha$ then

$$K(\sigma,\omega) = j_\sigma^* i_\omega = j_0^* V_{\omega\alpha}^* V_\omega i_0 = j_0^* V_\alpha^* V_\omega^* V_\omega i_0 = j_0^* V_\alpha^* i_0 = K(\alpha,0)\,.$$

Similarly if $\omega = \sigma\alpha$ then

$$K(\sigma,\omega) = j_0^* V_\alpha i_0 = K(0,\alpha)\,.$$

Otherwise the orthogonality of the ranges of the $V_k$ forces $K(\sigma,\omega)$ to be 0. □

Multi-Toeplitz kernels, in the positive definite case, have been investigated by Popescu, cf. [11]. For more recent developments see also [2, 3]. Our focus will be on the analytic case, see Theorem 1.2 below.

Let us introduce further notation and terminology. We define

$$\mathcal{U}_+ := \overline{\text{span}}\,\{\mathcal{U}_\alpha \colon \alpha \in F_d^+\}$$

$$\mathcal{H} := \hat{\mathcal{H}} \ominus \mathcal{U}_+ .$$

$\mathcal{U}_+$ is the smallest closed subspace invariant for all $V_k$ containing $\mathcal{U}_0$, and $\mathcal{H}$ is invariant for all $V_k^*$.

A subspace $\mathcal{W} \subset \hat{\mathcal{H}}$ is called *wandering* if $V_\alpha \mathcal{W} \perp V_\beta \mathcal{W}$ for $\alpha \neq \beta$ ($\alpha, \beta \in F_d^+$). We suppose from now on that $\mathcal{U}_0$ is wandering. Then $\mathcal{U}_+ = \bigoplus_{\alpha\in F_d^+} \mathcal{U}_\alpha$ (orthogonal direct sum), $V_k \mathcal{H} \subset \mathcal{H} \oplus \mathcal{U}_0$ for all $k = 1,\ldots,d$ and $V_\alpha^* \mathcal{U}_0 \subset \mathcal{H}$ for all $\alpha \neq 0$.

We can identify the space $\mathcal{U}_+$ with $\ell^2(F_d^+,\mathcal{U})$, the $\mathcal{U}$-valued square-summable functions on $F_d^+$, in the natural way. If $\mathcal{Y}_0$ is also wandering then we can associate a multi-Toeplitz operator

$$M \colon \ell^2(F_d^+,\mathcal{U}) \to \ell^2(F_d^+,\mathcal{Y})$$

with a matrix given by the multi-Toeplitz kernel $K$ from Lemma 1.1. In fact, using the identifications of $\mathcal{U}_+ = \bigoplus_{\alpha\in F_d^+} \mathcal{U}_\alpha$ with $\ell^2(F_d^+,\mathcal{U})$ and of $\mathcal{Y}_+ = \bigoplus_{\alpha\in F_d^+} \mathcal{Y}_\alpha$ with $\ell^2(F_d^+,\mathcal{Y})$ we see that $M$ is nothing but the orthogonal projection onto $\mathcal{Y}_+$ restricted to $\mathcal{U}_+$. Hence $M$ is a contraction which describes the relative position of $\mathcal{U}_+$ and $\mathcal{Y}_+$.

We are interested in the case where the multi-Toeplitz kernel $K$ (resp. the multi-Toeplitz operator $M$) is *multi-analytic*, i.e., $K(0,\alpha) = 0$ for all $\alpha \neq 0$. We note that the notion of multi-analytic operators has been studied in great detail by Popescu, cf. for example [10].

The following theorem gives several characterizations of multi-analyticity in our setting. The notation $P_\mathcal{X}$ for the orthogonal projection onto a subspace $\mathcal{X}$ is used without further comments.

**Theorem 1.2.** *Suppose that $\mathcal{U}_0$ is wandering for the row isometry $V$ on $\hat{\mathcal{H}}$ and let $\mathcal{Y}_0$ be any subspace of $\hat{\mathcal{H}}$. Then the following assertions are equivalent:*

(1) *$K$ is multi-analytic.*
(2) $\mathcal{U}_0 \perp V_\alpha^* \mathcal{Y}_0$ *for all* $\alpha \neq 0$.
(3) $\mathcal{Y}_0 \subset \mathcal{H} \oplus \mathcal{U}_0$.
(4) $V_k^* \mathcal{Y}_0 \subset \mathcal{H}$ *for all* $k = 1, \ldots, d$.
(5) $V_\alpha^* \mathcal{Y}_0 \subset \mathcal{H}$ *for all* $\alpha \neq 0$.

*Assertions* (1)–(5) *imply the following assertion:*

(6) $P_{\mathcal{Y}_+} V_\alpha x = V_\alpha P_{\mathcal{Y}_+} x$ *for all* $\alpha \in F_d^+$ *and* $x \in \mathcal{U}_+$.

*If in addition $\mathcal{Y}_0$ is also wandering for $V$ then* (6) *is equivalent to* (1)–(5) *and can be rewritten as*

(6′) $M\, S_\alpha^{\mathcal{U}} = S_\alpha^{\mathcal{Y}}\, M$ *for all* $\alpha \in F_d^+$, *where* $S^{\mathcal{U}}$ *and* $S^{\mathcal{Y}}$ *are the row shifts obtained by restricting $V$ to $\mathcal{U}_+$ and $\mathcal{Y}_+$ and* $M = P_{\mathcal{Y}_+}|_{\mathcal{U}_+}$ *is the multi-Toeplitz operator introduced above.*

Let us describe the relative position of the embedded subspaces $\mathcal{U}$ and $\mathcal{Y}$ characterized in Theorem 1.2 by saying that there is an *orthogonal $\mathcal{Y}$-past.* This terminology is suggested by (5) and some additional motivation for it is given at the end of this section.

*Proof.* (1) $\Leftrightarrow$ (2). In fact,

$$0 = K(0, \alpha) = j_0^* V_\alpha i_0$$

means exactly that $V_\alpha \mathcal{U}_0$ is orthogonal to $\mathcal{Y}_0$ or, equivalently, that $\mathcal{U}_0$ is orthogonal to $V_\alpha^* \mathcal{Y}_0$.
(2) $\Rightarrow$ (3). If (3) is not satisfied then there exists $y \in \mathcal{Y}_0$ and $\alpha \neq 0$ such that $P_{\mathcal{U}_\alpha} y \neq 0$. But then $P_{\mathcal{U}_0} V_\alpha^* y \neq 0$ contradicting (2).
(3) $\Rightarrow$ (4). Because for $k = 1, \ldots, d$

$$V_k \bigoplus_{\alpha \in F_d^+} \mathcal{U}_\alpha \subset \bigoplus_{\alpha \neq 0} \mathcal{U}_\alpha \perp \mathcal{H} \oplus \mathcal{U}_0$$

we conclude from $\mathcal{Y}_0 \subset \mathcal{H} \oplus \mathcal{U}_0$ that $\mathcal{U}_+ \perp V_k^* \mathcal{Y}_0$, hence $V_k^* \mathcal{Y}_0 \subset \mathcal{H}$.
(4) $\Rightarrow$ (5) follows from the fact that $\mathcal{H}$ is invariant for the $V_k^*$ and
(5) $\Rightarrow$ (2) is obvious.
(3) $\Rightarrow$ (6). It is elementary that $P_{V_\alpha \mathcal{Y}_+} V_\alpha = V_\alpha P_{\mathcal{Y}_+}$ for all $\alpha \in F_d^+$. To get (6), that is $P_{\mathcal{Y}_+} V_\alpha x = V_\alpha P_{\mathcal{Y}_+} x$ for all $\alpha \in F_d^+$ and $x \in \mathcal{U}_+$, it is therefore enough to consider all vectors of the form $V_\beta y$ where $y \in \mathcal{Y}_0$ and $\beta \in F_d^+$ is a word which does not begin with $\alpha$ and to show that such vectors are always orthogonal to $V_\alpha x$ where $x \in \mathcal{U}_+$. By (3) we have $\mathcal{Y}_0 \subset \mathcal{H} \oplus \mathcal{U}_0$ which implies, because $V_k \mathcal{H} \subset \mathcal{H} \oplus \mathcal{U}_0$ for all $k = 1, \ldots, d$, that $V_\beta y$ is contained in the span of $\mathcal{H}$ and of all $V_\gamma \mathcal{U}_0$ where the word $\gamma \in F_d^+$ does not begin with $\alpha$. This is indeed orthogonal to $V_\alpha x$ because $\mathcal{U}_0$ is wandering.

Conversely we prove, under the additional assumption that $\mathcal{Y}_0$ is wandering, the implication $(6') \Rightarrow (2)$. If (2) is not satisfied then there exists $u \in \mathcal{U}_0$ and $\alpha \neq 0$ such that $V_\alpha u$ is not orthogonal to $\mathcal{Y}_0$. Hence

$$P_{\mathcal{Y}_0} M S^{\mathcal{U}}_\alpha u = P_{\mathcal{Y}_0} V_\alpha u \neq 0 \,.$$

On the other hand, from $\mathcal{Y}_0$ wandering, we get

$$P_{\mathcal{Y}_0} S^{\mathcal{Y}}_\alpha M u = 0$$

and hence $M S^{\mathcal{U}}_\alpha \neq S^{\mathcal{Y}}_\alpha M$. □

Note that if $\mathcal{Y}_0$ is not wandering then in general (6) does not imply (1)–(5), in other words (6) may be true without $K$ being multi-analytic. Choose $\mathcal{Y}_0 = \hat{\mathcal{H}}$ for example. Though in this paper we are mainly interested in pairs of wandering subspaces it is very useful to observe that all the other implications in Theorem 1.2 hold more general. For example it can be convenient in applications to start with a bigger subspace $\mathcal{Y}_0$ and to restrict only later to a suitable wandering subspace.

Now consider the following operators:

$$\begin{aligned} A_k &:= V_k^*|_{\mathcal{H}} \colon \mathcal{H} \to \mathcal{H}, \quad & B_k &:= V_k^*\, i_0 \colon \mathcal{U} \to \mathcal{H}, \qquad k = 1, \dots, d \\ C &:= j_0^*|_{\mathcal{H}} \colon \mathcal{H} \to \mathcal{Y}, & D &:= j_0^*\, i_0 \colon \mathcal{U} \to \mathcal{Y}. \end{aligned}$$

Note that the assumption that $\mathcal{U}_0$ is wandering is needed to show that the $B_k$ map $\mathcal{U}$ into $\mathcal{H}$. If $K$ is multi-analytic then it is determined by these operators. In fact, it is elementary to check that

$$K(\alpha, 0) = j_0^* V_\alpha^* i_0 = \begin{cases} D & \text{if } \alpha = 0 \\ C B_\alpha & \text{if } |\alpha| = 1 \\ C A_{\alpha_r} \dots A_{\alpha_2} B_{\alpha_1} & \text{if } \alpha = \alpha_1 \dots \alpha_r,\ r = |\alpha| \geq 2. \end{cases}$$

These formulas suggest an interpretation from the point of view of linear system theory.

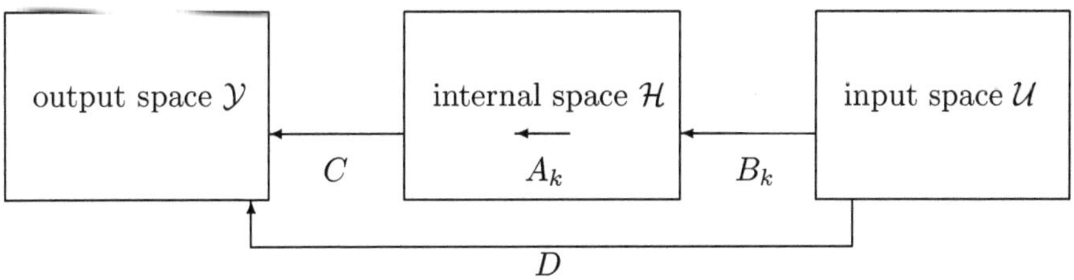


In fact, if we interpret $u \in \mathcal{U}$ as an input then we can think of $CA_\beta B_k u$ as a family of outputs originating from it, stored in suitable copies of $\mathcal{Y}$. Motivated by these observations we say, in the case of an orthogonal $\mathcal{Y}$-past, that the associated multi-analytic kernel $K$ (or the multi-analytic operator $M$ if available) is a *transfer function* (for the embedded spaces $\mathcal{U}$ and $\mathcal{Y}$).

We remark that the scheme is close to the formalism of Ball-Vinnikov in [3], compare for example formula (3.3.2) therein. Essentially the same construction, but in a commutative-variable setting, appears in [1]. In the later section on Markov chains in this paper we describe another reappearance of this structure which has been observed in [6]. For the moment, to make our terminology even more plausible, let us consider the simplest case where $\mathcal{U}_0$ and $\mathcal{Y}_0$ are both wandering and $d = 1$ (i.e., $V$ is a single isometry). Let $H^2(\mathcal{U})$ resp. $H^2(\mathcal{Y})$ denote the $\mathcal{U}$-valued resp. $\mathcal{Y}$-valued Hardy space on the complex unit disc $\mathbb{D}$. For example a function in $H^2(\mathcal{U})$ has the form

$$\mathbb{D} \ni z \mapsto \sum_{n=0}^{\infty} a_n z^n \quad \textit{with } a_n \in \mathcal{U}\,.$$

There is a natural unitary from $\bigoplus_{n=0}^{\infty} \mathcal{U}_n$ onto $H^2(\mathcal{U})$, taking the summands as coefficients (similar for $\mathcal{Y}$). It can be used to move operators from one Hilbert space to the other. For more details see for example [5], Chapter IX. This allows us to summarize the previous discussions in this special case as follows.

**Corollary 1.3.** *If $\mathcal{U}_0$ and $\mathcal{Y}_0$ are a pair of wandering subspaces (for an isometry $V$) with orthogonal $\mathcal{Y}$-past then $M := P_{\mathcal{Y}_+}|_{\mathcal{U}_+}$ moved to the Hardy spaces becomes a contractive multiplication operator $M_\Theta$ with*

$$\Theta(z) = D + \sum_{n=1}^{\infty} C A^{n-1} B z^n = D + C(I_{\mathcal{H}} - zA)^{-1} z B\,.$$

*Here $A := A_1 = V^*|_{\mathcal{H}}$, $B := B_1 = V^* i_0$ and $\Theta \in H_1^\infty(\mathcal{U},\mathcal{Y})$, the unit ball of the algebra of bounded analytic functions on $\mathbb{D}$ with values in $\mathcal{B}(\mathcal{U},\mathcal{Y})$, the bounded operators from $\mathcal{U}$ to $\mathcal{Y}$.*

This means that in this case $M$ is an analytic operator in the sense of [12] (except for the insignificant fact that it operates between different Hilbert spaces).

In the general noncommutative case we can similarly encode all the entries $K(\alpha, 0)$ (as described above) into a formal power series which fully describes a multi-analytic operator $M$.

**Corollary 1.4.** *If $\mathcal{U}_0$ is a wandering subspace for a row isometry $V = (V_1, \dots, V_d)$ and $\mathcal{Y}_0$ is another subspace then, with indeterminates $z_1, \dots, z_d$ which are freely noncommuting among each other but commuting with the operators,*

$$\Theta(z_1, \dots, z_d) := \sum_{\alpha \in F_d^+} K(\alpha, 0)\, z_\alpha = D + C \sum_{r=1}^{\infty} (ZA)^{r-1} ZB = D + C(I_{\mathcal{H}} - ZA)^{-1} ZB$$

*where $Z = (z_1 I_{\mathcal{H}}, \dots, z_d I_{\mathcal{H}})$, $A = (A_1, \dots, A_d)^T$, $B = (B_1, \dots, B_d)^T$, the transpose indicating that $A$ and $B$ should be interpreted as (operator-valued) column vectors.*

Such a formalism is explained in more detail and used systematically in [3]. Using the language of system theory we have all the relevant information in the so-called *system matrix*

$$\Sigma = \begin{pmatrix} A & B \\ C & D \end{pmatrix}.$$

Let us put these results into the context of other work already done in operator theory and focus on the case $d = 1$ again. We could have extended the isometry $V$ to a unitary $\tilde{V}$ on a larger Hilbert space. If we now define $\mathcal{Y}_k = \tilde{V}^k \mathcal{Y}_0$ also for $k < 0$ then it is natural to call $\bigoplus_{k<0} \mathcal{Y}_k$ the $\mathcal{Y}$-past. In this extended setting the fact that we have orthogonal $\mathcal{Y}$-past ensures that $\tilde{V}$ is a coupling in the sense of [5], Chapter VII.7, between the right shifts on the orthogonal spaces $\bigoplus_{k\geq 0}^{\infty} \mathcal{U}_n$ and $\bigoplus_{k<0} \mathcal{Y}_k$. Further our operator $M$ can now be interpreted as the contractive intertwining lifting of the zero intertwiner between the two shifts which is canonically associated to the coupling $\tilde{V}$. See [5], Chapter VII.8, for this construction. We don't go into this here, the book [5] contains detailed discussions how analytic functions arise in the classification of such structures.

We remark that in the case $d > 1$ it is more complicated to develop the analogue of such a 'two-sided' setting but this has been worked out in [2, 3] within a theory of Haplitz kernels and Cuntz weights. For the purposes of this paper it turns out that the simpler 'one-sided' setting, as presented in this section and in particular in Theorem 1.2, is sufficient.

## 2. Examples: Characteristic functions

The examples in this section are well known and the main emphasis is therefore to show that they fit naturally into the scheme developed in the previous section and that thinking about them in this way simplifies the constructions. For further simplification we only work through the details of the case $d = 1$, i.e., a single isometry $V : \mathcal{H} \to \hat{\mathcal{H}}$.

Suppose that $\mathcal{U}_0$ and $\mathcal{Y}_0$ are a pair of wandering subspaces with orthogonal $\mathcal{Y}$-past and with system matrix

$$\Sigma = \begin{pmatrix} A & B \\ C & D \end{pmatrix} : \mathcal{H} \oplus \mathcal{U} \to \mathcal{H} \oplus \mathcal{Y}.$$

For the adjoint $\Sigma^*$ we obtain from the definition of $A, B, C, D$:

$$\Sigma^* = \begin{pmatrix} A^* & C^* \\ B^* & D^* \end{pmatrix} : h \oplus y \mapsto P_H\big[Vh + j_0(y)\big] \oplus i_0^* P_{\mathcal{U}_0}\big[Vh + j_0(y)\big].$$

### 2.1. Example

Let us consider a special case of the previous setting where $V\mathcal{H} \perp j_0(\mathcal{Y})$. Then $\Sigma^*$ is isometric, i.e., $\Sigma$ is a coisometry.

Conversely, for any Hilbert spaces $\mathcal{H}, \mathcal{U}, \mathcal{Y}$ let $\Sigma = \begin{pmatrix} A & B \\ C & D \end{pmatrix} : \mathcal{H} \oplus \mathcal{U} \to \mathcal{H} \oplus \mathcal{Y}$ be any coisometry. Now define the Hilbert space $\hat{\mathcal{H}} := \mathcal{H} \oplus \bigoplus_{n=0}^{\infty} \mathcal{U}_n$ with the $\mathcal{U}_n$ copies of $\mathcal{U}$, the embeddings

$$i_0(\mathcal{U}) := \mathcal{U}_0, \quad j_0 := (I_{\mathcal{H}} \oplus i_0)\Sigma^*|_{\mathcal{Y}}$$

and an isometry $V$ by $V|_{\mathcal{H}} := (I_{\mathcal{H}} \oplus i_0)\Sigma^*|_{\mathcal{H}}$ and acting as a right shift on $\mathcal{U}_+ = \bigoplus_{n=0}^{\infty} \mathcal{U}_n$. Then $\mathcal{U}_0$ and $\mathcal{Y}_0$ are a pair of wandering subspaces with orthogonal

$\mathcal{Y}$-past and with system matrix $\Sigma$. In fact, orthogonal $\mathcal{Y}$-past is clear from $\mathcal{Y}_0 \subset \mathcal{H} \oplus \mathcal{U}_0$ and Theorem 1.2 and then $\mathcal{Y}_0$ wandering is an immediate consequence of $V\mathcal{H} \perp j_0(\mathcal{Y})$ and the specific form of $V$.

This situation occurs in the Sz.-Nagy-Foias theory of characteristic functions for contractions. Let us sketch briefly how this fits in. Let $T \in \mathcal{B}(\mathcal{H})$ be a contraction. Then we have defect operators $D_T = \sqrt{I - T^*T}$ and $D_{T^*}$ with defect spaces $\mathcal{D}_T$ and $\mathcal{D}_{T^*}$ defined as the closure of their ranges. The reader can easily check that the construction above applies with $\mathcal{U} = \mathcal{D}_T$, $\mathcal{Y} = \mathcal{D}_{T^*}$ and with the unitary rotation matrix

$$\Sigma = \begin{pmatrix} A & B \\ C & D \end{pmatrix} = \begin{pmatrix} T^* & D_T \\ D_{T^*} & -T \end{pmatrix} : \mathcal{H} \oplus \mathcal{U} \to \mathcal{H} \oplus \mathcal{Y}.$$

Then $V$ is the minimal isometric dilation of $T$ and the transfer function for the pair $\mathcal{U}_0$ and $\mathcal{Y}_0$ given by

$$\Theta(z) = -T + D_{T^*}(I_{\mathcal{H}} - zT^*)^{-1} z\, D_T$$

is nothing but the well-known Sz.-Nagy-Foias *characteristic function* of $T$. In fact it is characteristic in the sense that it characterizes $T$ up to unitary equivalence only if $T$ is completely non-unitary (cf. [13] or [5]). So in the general case it may be better to refer to $\Theta$ as the transfer function associated to $T$.

It is possible to handle the multi-variable case ($d > 1$), first studied by Popescu in [9], in a very similar way and the result, if expressed in the notation explained for Corollary 1.4, is very similar: The transfer function associated to a row contraction $T = (T_1, \dots, T_d)\colon \bigoplus_1^d \mathcal{H} \to \mathcal{H}$ is

$$\Theta(z_1, \dots, z_d) = -T + D_{T^*}(I_{\mathcal{H}} - ZT^*)^{-1} Z D_T$$

where $Z = (z_1 I_{\mathcal{H}}, \dots, z_d I_{\mathcal{H}})$. It is shown in ([9], 5.4) that $\Theta$ is characteristic in the sense of being a complete unitary invariant if $T$ is completely non-coisometric. It is further shown in ([3], 5.3.3) that to get a complete unitary invariant in the class of completely non-unitary row contractions one can consider a characteristic pair $(\Theta, L)$ where $L$ is a Cuntz weight.

### 2.2. Example

But there are other possibilities to obtain a pair of wandering subspaces $\mathcal{U}_0$ and $\mathcal{Y}_0$ with orthogonal $\mathcal{Y}$-past than the scheme explained in the previous example. We go back to the case $d = 1$ and assume again that $\hat{\mathcal{H}} := \mathcal{H} \oplus \bigoplus_{n=0}^\infty \mathcal{U}_n$ and that an isometry $V$ is given on $\hat{\mathcal{H}}$ which acts as a right shift on $\bigoplus_{n=0}^\infty \mathcal{U}_n$. Now suppose further that $\mathcal{H}_S$ is a subspace of $\mathcal{H}$ which is $V^*$-invariant. Then for any subspace $\mathcal{Y}_0$ satisfying

$$\mathcal{Y}_0 \subset \overline{\mathrm{span}}\{\mathcal{H}_S, V\mathcal{H}_S\} \ominus \mathcal{H}_S$$

it follows that $\mathcal{U}_0$ and $\mathcal{Y}_0$ are a pair of wandering subspaces with orthogonal $\mathcal{Y}$-past. In fact, because $\mathcal{Y}_0 \perp \mathcal{H}_S$ we have for $k \geq 1$ that $V^k\mathcal{Y}_0 \perp \mathcal{H}_S$, but also $V^{k-1}\mathcal{Y}_0 \perp \mathcal{H}_S$ so that $V^k\mathcal{Y}_0 \perp V\mathcal{H}_S$. Hence $V^k\mathcal{Y}_0 \perp \mathcal{Y}_0$ for all $k \geq 1$, i.e., $\mathcal{Y}_0$ is a wandering subspace. Together with $V\mathcal{H} \subset \mathcal{H} \oplus \mathcal{U}_0$ and Theorem 1.2 this establishes the claim.

This situation occurs in the theory of characteristic functions for liftings (cf. [4]). As this is less well known than the Sz.-Nagy-Foias theory in the previous example and the presentation in [4] gives the general case $d \geq 1$ using a different approach and a different notation we think it is instructive to work out explicitly some details of this transfer function in the case $d = 1$ with the methods of this paper.

As in the previous subsection let $T \in \mathcal{B}(\mathcal{H})$ be a contraction, $\mathcal{U} := \mathcal{D}_T$, $\hat{\mathcal{H}} := \mathcal{H} \oplus \bigoplus_{n=0}^{\infty} \mathcal{U}_n$, $i_0(\mathcal{U}) = \mathcal{U}_0$, $V$ the minimal isometric dilation and we still have $A = V^*|_{\mathcal{H}} = T^*$ and $B = V^* i_0 = D_T$. But now suppose that $\mathcal{H} = \mathcal{H}_S \oplus \mathcal{H}_R$ such that $\mathcal{H}_S$ is invariant for $T^*$, in other words $T$ is a block matrix

$$T = \begin{pmatrix} S & 0 \\ Q & R \end{pmatrix}$$

with respect to $\mathcal{H} = \mathcal{H}_S \oplus \mathcal{H}_R$. We also say that $T \in \mathcal{B}(\mathcal{H})$ is a *lifting* of $S \in \mathcal{B}(\mathcal{H}_S)$. Then $V$ is also an isometric dilation of $S$, i.e., $P_{\mathcal{H}_S} V^n|_{\mathcal{H}_S} = S^n$ for all $n \in \mathbb{N}$, and it restricts to the minimal isometric dilation $V_S$ of $S$ on a reducing subspace. The subspace $\mathcal{H}_S$ is invariant for $V^*$ and we obtain a situation as described in the beginning of this subsection by putting $\mathcal{Y} := \mathcal{D}_S$ and for $h_S \in \mathcal{H}_S$

$$j_0(D_S h_S) := (V_S - S) h_S = (V - S) h_S = Q h_S \oplus i_0(D_T h_S) \in \mathcal{H}_R \oplus \mathcal{U}_0 \,.$$

Hence we have orthogonal $\mathcal{Y}$-past and $\mathcal{U}_0$ and $\mathcal{Y}_0$ are both wandering.

It is well known about contractive liftings such as $T$ that we always have

$$Q = D_{R^*} \gamma^* D_S : \mathcal{H}_S \to \mathcal{H}_R$$

with a contraction $\gamma : \mathcal{D}_{R^*} \to \mathcal{D}_S$ (cf. [5], Chapter IV, Lemma 2.1). We conclude that

$$C = j_0^*|_{\mathcal{H}} = \gamma\, D_{R^*} P_{\mathcal{H}_R} \,.$$

To compute $D = j_0^* i_0$ more explicitly note that for $h_S \in \mathcal{H}_S$, $h_R \in \mathcal{H}_R$

$$j_0^* V h_S = j_0^* (S h_S \oplus j_0(D_S h_S)) = D_S h_S, \qquad j_0^* V h_R = 0,$$

[the latter because $V\mathcal{H}_R \perp \overline{\operatorname{span}}\{V\mathcal{H}_S, \mathcal{H}_S\} \supset j_0(\mathcal{D}_S)$].

With $\mathcal{H} \ni h = h_S \oplus h_R \in \mathcal{H}_S \oplus \mathcal{H}_R$ we can compute $D$ as follows:

$$\begin{aligned} D(D_T h) &= j_0^* \big( (V - T) h \big) \\ &= j_0^* V h - j_0^* T h = j_0^* (V h_S + V h_R) - C T h \\ &= D_S h_S - \gamma D_{R^*} (T h_S + R h_R) \\ &= (D_S - \gamma D_{R^*} Q) h_S - \gamma D_{R^*} R h_R \\ &= (D_S - \gamma D_{R^*} Q) h_S - \gamma R D_R h_R \end{aligned}$$

(using $D_{R^*} R = R D_R$ in the last line). Hence we get a transfer function

$$\begin{aligned} \Theta(z) &= D + \sum_{n \geq 1} \gamma D_{R^*} P_{\mathcal{H}_R} (z T^*)^{n-1} z D_T = D + \gamma D_{R^*} P_{\mathcal{H}_R} (I_{\mathcal{H}} - z T^*)^{-1} z D_T \\ &= D + \sum_{n \geq 1} \gamma D_{R^*} (z R^*)^{n-1} P_{\mathcal{H}_R} z D_T = D + \gamma D_{R^*} (I_{\mathcal{H}_R} - z R^*)^{-1} P_{\mathcal{H}_R} z D_T . \end{aligned}$$

The domain of $\Theta(z)$ (for each $z$) is $\mathcal{U} = \mathcal{D}_T$. We gain additional insights by evaluating $\Theta(z)$ on $D_T h_R = D_R h_R$ with $h_R \in \mathcal{H}_R$ and on $D_T h_S$ with $h_S \in \mathcal{H}_S$.

$$\begin{aligned}\Theta(z)(D_T h_R) &= D(D_T h_R) + \sum_{n\geq 1} \gamma D_{R^*}(zR^*)^{n-1} P_{\mathcal{H}_R} z D_T (D_T h_R)\\ &= \gamma\big[-R + D_{R^*}(I - zR^*)^{-1} z D_R\big](D_R h_R)\end{aligned}$$

which shows that $\Theta(z)$ restricted to $D_T\mathcal{H}_R = D_R\mathcal{H}_R$ is nothing but $\gamma$ times the transfer function associated to $R$ in the sense of Sz.-Nagy and Foias, as discussed in the previous subsection. Its presence can be explained by the fact that $V$ restricted to $\hat{\mathcal{H}} \ominus \mathcal{H}_S$ also provides an isometric dilation of $R$. For the other restriction $\Theta(z)|_{D_T\mathcal{H}_S}$ we find, using $P_{\mathcal{H}_R} z D_T^2 h_S = P_{\mathcal{H}_R} z(I - T^*T)h_S = -zR^*Q\, h_S$,

$$\begin{aligned}\Theta(z)(D_T h_S) &= D(D_T h_S) + \sum_{n\geq 1} \gamma D_{R^*}(zR^*)^{n-1} P_{\mathcal{H}_R} z D_T (D_T h_S)\\ &= \left[D_S - \gamma D_{R^*} Q - \sum_{n\geq 1} \gamma D_{R^*}(zR^*)^n Q\right](h_S)\\ &= \left[D_S - \gamma D_{R^*} \sum_{n\geq 0} (zR^*)^n Q\right](h_S)\\ &= \big[I - \gamma D_{R^*}(I - zR^*)^{-1} D_{R^*}\gamma^*\big] D_S h_S.\end{aligned}$$

Again the multi-variable case ($d > 1$) can be handled similarly and yields similar results. Here we investigate a row contraction $T = (T_1, \dots, T_d)$ which is a lifting in the sense that

$$T_k = \begin{pmatrix} S_k & 0 \\ Q_k & R_k \end{pmatrix}$$

for all $k = 1, \dots, d$. All the formulas for transfer functions derived above have been written in a form which makes sense and which is still valid for the multi-variable case if we simply replace the variable $z$ by $Z = (z_1 I, \dots, z_d I)$ (as in Corollary 1.4) and the vectors $h_S$, $h_R$ by $d$-tuples of vectors.

These transfer functions have been introduced in [4]; see Section 4 therein, in particular formulas (4.6) and (4.5), for an alternative approach to the facts sketched above. Among other things it is further investigated in [4] in which cases such transfer functions are characteristic for the lifting, i.e., characterize the lifting $T$ given $S$ up to unitary equivalence.

## 3. Examples: Noncommutative Markov chains

There is another way how transfer functions as described in Section 1 appear in applications, namely in the theory of noncommutative Markov chains. This has been observed in [6] and to work out a common framework in order to facilitate the discussion of similarities has been a major motivation for this paper.

We quickly review the setting of [6] as far as it is needed to make our point, referring to that paper for more details. An *interaction* is defined as a unitary

$$U\colon \mathcal{H}\otimes\mathcal{K}\to\mathcal{H}\otimes\mathcal{P}$$

where $\mathcal{H},\mathcal{K},\mathcal{P}$ are Hilbert spaces. In quantum physics it is common to describe the aggregation of different parts by a tensor product of Hilbert spaces and in this case we may think of $U$ as one step of a discretized interacting dynamics. (For such an interpretation we may take $\mathcal{K}=\mathcal{P}$ and think of $\mathcal{K}$ and $\mathcal{P}$ as describing the same part before and after the interaction. But mathematically it is more transparent to treat them as two distinguishable spaces.) If $\mathcal{H}$ represents a fixed quantum system, say an atom, and interactions take place with a wave passing by, say a light beam, then it is natural, at least as a toy model, to represent repeated interactions ($n$ steps) by

$$U(n):=U_n\dots U_2U_1\colon \mathcal{H}\otimes\bigotimes_{\ell=1}^{n}\mathcal{K}_\ell\mapsto\mathcal{H}\otimes\bigotimes_{\ell=1}^{n}\mathcal{P}_\ell$$

where the $\mathcal{K}_\ell$ (resp. $\mathcal{P}_\ell$) are copies of $\mathcal{K}$ (resp. $\mathcal{P}$), and $U_\ell$ acts as $U$ from $\mathcal{H}\otimes\mathcal{K}_\ell$ to $\mathcal{H}\otimes\mathcal{P}_\ell$, identical at the other parts.

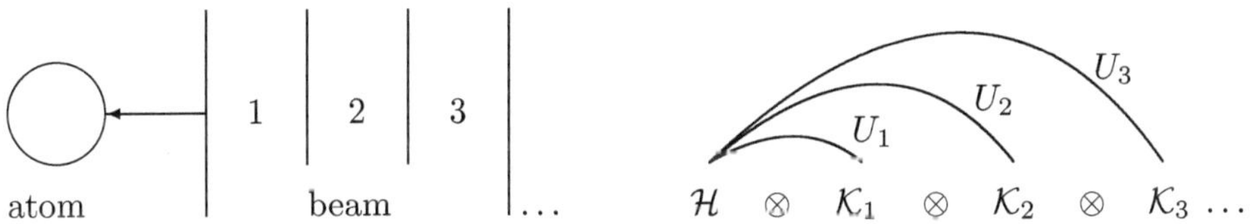


Choosing unit vectors $\Omega_\mathcal{K}\in\mathcal{K}$ and $\Omega_\mathcal{P}\in\mathcal{P}$ we can also form infinite tensor products along these unit vectors and obtain $U(n)$ for every $n\in\mathbb{N}$ on a common Hilbert space. Such a toy model of quantum repeated interactions can mathematically be thought of as a noncommutative Markov chain. We refer to [6] for some motivation for this terminology by analogies with classical Markov chains.

It is proved in [6] (in a slightly different language) that if we have another unit vector $\Omega_\mathcal{H}\in\mathcal{H}$ such that $U(\Omega_\mathcal{H}\otimes\Omega_\mathcal{K})=\Omega_\mathcal{H}\otimes\Omega_\mathcal{P}$ (we call these unit vectors *vacuum vectors* in this case) then we obtain a pair of wandering subspaces $\mathcal{U}_0$ and $\mathcal{Y}_0$ with orthogonal $\mathcal{Y}$-past for a row isometry $V$, notation consistent with Section 1, as follows:

$$\hat{\mathcal{H}}:=\mathcal{H}\otimes\bigotimes_{\ell=1}^{\infty}\mathcal{K}_\ell\quad\supset\mathcal{H}\otimes\bigotimes_{\ell=1}^{\infty}\Omega_{\mathcal{K}_\ell}\simeq\mathcal{H},$$

i.e., the latter subspace of $\hat{\mathcal{H}}$ is identified with $\mathcal{H}$. The row isometry $V$ on $\hat{\mathcal{H}}$ is of the form

$$V:=(V_1,\dots,V_d),\quad d=\dim\,\mathcal{P},$$

where dim $\mathcal{P}$ is the number of elements in an orthonormal basis of $\mathcal{P}$. Let $(\epsilon_k)_{k=1}^{d}$ be such an orthonormal basis of $\mathcal{P}=\mathcal{P}_1$, fixed from now on.

Then for $\xi \in \mathcal{H}$ and $\eta \in \bigotimes_{\ell=1}^{\infty} \mathcal{K}_\ell$

$$V_k(\xi \otimes \eta) := U_1^*(\xi \otimes \epsilon_k \otimes \eta) \in (\mathcal{H} \otimes \mathcal{K}_1) \otimes \bigotimes_{\ell=2}^{\infty} \mathcal{K}_\ell .$$

Note that $\eta$ is shifted to the right in the tensor product and appears as $\eta \in \bigotimes_{\ell=2}^{\infty} \mathcal{K}_\ell$ on the right-hand side. It is immediate that $V$ is a row isometry. To get used to this definition the reader is invited to verify the formula

$$V_\alpha(\xi \otimes \eta) = U(r)^*(\xi \otimes \epsilon_{\alpha_1} \otimes \cdots \otimes \epsilon_{\alpha_r} \otimes \eta) \in (\mathcal{H} \otimes \mathcal{K}_1 \otimes \cdots \mathcal{K}_r) \otimes \bigotimes_{\ell=r+1}^{\infty} \mathcal{K}_\ell$$

where $\alpha = \alpha_1 \ldots \alpha_r \in F_d^+$ with $|\alpha| = r$ and on the right-hand side $\eta$ now appears as $\eta \in \bigotimes_{\ell=r+1}^{\infty} \mathcal{K}_\ell$. It becomes clear that the properties of the repeated interaction are encoded into properties of the row isometry $V$.

Finally we define the pair of embedded subspaces:

$$\mathcal{U} := \mathcal{H} \otimes (\Omega_\mathcal{K})^\perp \subset \mathcal{H} \otimes \mathcal{K},$$

$$\mathcal{U}_0 = i_0(\mathcal{U}) := \mathcal{H} \otimes (\Omega_{\mathcal{K}_1})^\perp \otimes \bigotimes_{\ell=2}^{\infty} \Omega_{\mathcal{K}_\ell},$$

$$\mathcal{Y} := (\Omega_\mathcal{P})^\perp \subset \mathcal{P},$$

$$\mathcal{Y}_0 = j_0(\mathcal{Y}) := U_1^* \left( \Omega_\mathcal{H} \otimes (\Omega_{\mathcal{P}_1})^\perp \otimes \bigotimes_{\ell=2}^{\infty} \Omega_{\mathcal{K}_\ell} \right).$$

From the specific form of the isometries $V_k$ it is easy to check that $\mathcal{U}_0$ is wandering and that $\hat{\mathcal{H}} = \mathcal{H} \oplus \mathcal{U}_+$. The proof that $\mathcal{Y}_0$ is wandering can be found in [6] or deduced from Proposition 3.1 below (which covers a more general situation). From Theorem 1.2 we have an associated transfer function which can be made explicit as a multi-analytic kernel $K$ or as a (contractive) multi-analytic operator $M$. It may be called a transfer function of the noncommutative Markov process. With $h \oplus u \in \mathcal{H} \oplus \mathcal{U} = \mathcal{H} \otimes \mathcal{K}$ (here we identify $\mathcal{H}$ with $\mathcal{H} \otimes \Omega_\mathcal{K}$) we find the operators $A_k, B_k, C, D$ appearing in the system matrix $\Sigma$ to be related to the interaction $U$ by

$$U(h \oplus u) = \sum_{k=1}^{d} (A_k h + B_k u) \otimes \epsilon_k \ \in \mathcal{H} \otimes \mathcal{P}$$

$$P_{\Omega_\mathcal{H} \otimes \mathcal{Y}} U(h \oplus u) = Ch + Du \ \in \mathcal{Y}$$

(where we have to identify $\Omega_\mathcal{H} \otimes \mathcal{Y}$ and $\mathcal{Y}$ for the last equation). It is further discussed in [6] how for models in quantum physics these operators and the coefficients of the transfer function built from them can be interpreted, and it is shown that the transfer function can be used to study questions about observability and about scattering theory (outgoing Cuntz scattering systems [3] and scattering theory for Markov chains [7]).

Let us finally indicate that the additional ideas introduced in this paper provide a flexible setting for generalizations. Let us consider the situation above but without assuming the existence of vacuum vectors. With $\Omega_{\mathcal{K}} \in \mathcal{K}$ being an arbitrary unit vector we can easily check that $\mathcal{U}_0$ as defined above is still a wandering subspace for $V$. Hence for an arbitrary subspace $\mathcal{Y}_0$ of

$$\mathcal{H} \oplus \mathcal{U}_0 = \mathcal{H} \otimes \mathcal{K}_1 \otimes \bigotimes_{\ell=2}^{\infty} \Omega_{\mathcal{K}_\ell}$$

we conclude, by Theorem 1.2, that we have orthogonal $\mathcal{Y}$-past and there exists a corresponding transfer function corresponding to a multi-analytic kernel $K$. When is $\mathcal{Y}_0$ wandering? A sufficient criterion generalizing the situation with vacuum vectors is provided by the following

**Proposition 3.1.** *Let $\mathcal{H}_S$ be a subspace of $\mathcal{H}$ such that $U(\mathcal{H}_S \otimes \Omega_{\mathcal{K}}) \subset \mathcal{H}_S \otimes \mathcal{P}$. Then any subspace*

$$\mathcal{Y}_0 \subset U_1^*(\mathcal{H}_S \otimes \mathcal{P}_1) \ominus (\mathcal{H}_S \otimes \Omega_{\mathcal{K}_1})$$

*is wandering.*
*(Here we adapt the convention to omit a tensoring with $\bigotimes_{\ell=2}^{\infty} \Omega_{\mathcal{K}_\ell}$ in the notation.)*

*Proof.* Let $\zeta \in \hat{\mathcal{H}}$ be any vector orthogonal to $\mathcal{H}_S \otimes \Omega_{\mathcal{K}_1}$. Our first observation is that for all $k = 1, \dots, d$ the vectors $V_k\zeta$ are orthogonal to $\mathcal{H}_S \otimes \Omega_{\mathcal{K}_1}$ too. In fact, we can write $\zeta = \zeta_1 \oplus \zeta_2$ where $\zeta_1 = \xi_0 \otimes \eta$ with $\xi_0 \in \mathcal{H}_S$ and with $\eta \in \bigotimes_{\ell=1}^{\infty} \mathcal{K}_\ell$ orthogonal to $\bigotimes_{\ell=1}^{\infty} \Omega_{\mathcal{K}_\ell}$ and $\zeta_2 \in (\mathcal{H} \ominus \mathcal{H}_S) \otimes \bigotimes_{\ell=1}^{\infty} \mathcal{K}_\ell$. Using the specific form of $V_k$ it follows immediately that $V_k\zeta_1$ is orthogonal to $\mathcal{H}_S \otimes \Omega_{\mathcal{K}_1}$ and the same is also true for $V_k\zeta_2$ taking into account the assumption $U(\mathcal{H}_S \otimes \Omega_{\mathcal{K}}) \subset \mathcal{H}_S \otimes \mathcal{P}$, in the form: $U_1^*\big((\mathcal{H} \ominus \mathcal{H}_S) \otimes \mathcal{P}_1\big)$ is orthogonal to $\mathcal{H}_S \otimes \Omega_{\mathcal{K}_1}$.

The second observation is that for all $k = 1, \dots, d$ the vectors $V_k\zeta$ are orthogonal to $U_1^*(\mathcal{H}_S \otimes \mathcal{P}_1)$. As $\zeta$ can be approximated by a finite sum $\sum_j \xi_j \otimes \eta_j$ with $\xi_j \in \mathcal{H}$ and $\eta_j \in \bigotimes_{\ell=1}^{\infty} \mathcal{K}_\ell$ we may assume for simplicity that $\zeta$ is of this form. But then $\sum_j \xi_j \otimes \epsilon_k \otimes \eta_j$ is orthogonal to $\mathcal{H}_S \otimes \mathcal{P}_1$ and now an application of $U_1^*$ gives us the result.

Applying these observations repeatedly to elements of $\mathcal{Y}_0$ we conclude that $\mathcal{Y}_0$ is orthogonal to $V_\alpha \mathcal{Y}_0$ for all $\alpha \neq 0$. This implies that $\mathcal{Y}_0$ is wandering. □

**Acknowledgment**

During my visit to Mumbai in 2010 results closely related to Proposition 3.1 were communicated to me by Santanu Dey. These were in the back of my mind when I worked out a version fitting into the setting of this paper. Further I want to thank the referee for constructive comments leading to a better presentation and for pointing out additional links to the existing literature.

## References

[1] J. Ball, C. Sadosky, V. Vinnikov, *Scattering Systems with Several Evolutions and Multidimensional Input/State/Output Systems.* Integral Equations and Operator Theory **52** (2005), 323–393.

[2] J. Ball, V. Vinnikov, *Functional Models for Representations of the Cuntz Algebra.* In: Operator Theory, System Theory and Scattering Theory: Multidimensional Generalizations. Operator Theory, Advances and Applications, vol. **157**, Birkhäuser, 2005, 1–60.

[3] J. Ball, V. Vinnikov, *Lax-Phillips Scattering and Conservative Linear Systems: A Cuntz-Algebra Multidimensional Setting.* Memoirs of the AMS, vol. 178, no. **837** (2005).

[4] S. Dey, R. Gohm, *Characteristic Functions of Liftings.* Journal of Operator Theory **65** (2011), 17–45.

[5] C. Foias, A.E. Frazho, *The Commutant Lifting Approach to Interpolation Problems.* Operator Theory, Advances and Applications, vol. **44**, Birkhäuser, 1990.

[6] R. Gohm, *Noncommutative Markov Chains and Multi-Analytic Operators.* J. Math. Anal. Appl., vol. **364**(1) (2009), 275–288.

[7] B. Kümmerer, H. Maassen, *A Scattering Theory for Markov Chains.* Inf. Dim. Analysis, Quantum Prob. and Related Topics, vol.**3** (2000), 161–176.

[8] G. Popescu, *Isometric Dilations for Infinite Sequences of Noncommuting Operators.* Trans. Amer. Math. Soc. **316** (1989), 523–536.

[9] G. Popescu, *Characteristic Functions for Infinite Sequences of Noncommuting Operators.* J. Operator Theory **22** (1989), 51–71.

[10] G. Popescu, *Multi-Analytic Operators on Fock Spaces.* Math. Ann. **303** (1995), no. 1, 31–46.

[11] G. Popescu, *Structure and Entropy for Toeplitz Kernels.* C.R. Acad. Sci. Paris Ser. I Math. **329** (1999), 129–134.

[12] M. Rosenblum, J. Rovnyak, *Hardy Classes and Operator Theory.* Oxford University Press, 1995.

[13] B. Sz.-Nagy, C. Foias, *Harmonic Analysis of Operators.* North-Holland, 1970.

Rolf Gohm
Institute of Mathematics and Physics
Aberystwyth University
Aberystwyth SY23 3BZ
United Kingdom
e-mail: rog@aber.ac.uk

Operator Theory:
Advances and Applications, Vol. 221, 399-412


# Non-negativity Analysis for Exponential-Polynomial-Trigonometric Functions on $[0, \infty)$

Bernard Hanzon and Finbarr Holland

**Abstract.** This note concerns the class of functions that are solutions of homogeneous linear differential equations with constant real coefficients. This class, which is ubiquitous in the mathematical sciences, is denoted throughout the paper by $EPT$, and members of it can be written in the form

$$\sum_{i=0}^{d} q_i(t)e^{\lambda_i t}\cos(\theta_i t + \tau_i),$$

where the $q_i$ are real polynomials, and $\lambda_i$, $\theta_i$ and $\tau_i$ are real numbers. The subclass of these functions, for which all the $\theta_i$ are zero, is denoted by $EP$. In this paper, we address the characterization of those members of $EPT$ that are non-negative on the half-line $[0, \infty)$. We present necessary conditions, some of which are known, and a new sufficient condition, and describe methods for the verification of this sufficient condition. The main idea is to represent an $EPT$ function as the product of a row vector of $EP$ functions, and a column vector of multivariate polynomials with unimodular exponential functions $e^{i\theta_k t}$, $k = 1, 2, \ldots, m$, as arguments, where $\{\theta_k : k = 1, 2, \ldots, m\}$ is a set of real numbers that is linearly independent over the set of rational numbers $\mathbf{Q}$. From this we deduce necessary conditions for an $EPT$ function to be non-negative on an unbounded subinterval of $[0, \infty)$. The completion of the analysis is reliant on a generalized Budan Fourier sequence technique, devised by the authors, to examine the non-negativity of the function on the complementary interval.

**Mathematics Subject Classification (2000).** Primary 34H05; Secondary 33B10.

**Keywords.** Non-negativity, polynomials, exponential functions, trigonometric polynomials, generalized Budan-Fourier sequence, Kronecker's approximation theorem, Lipschitz continuity.

---

Research of the first-named author is supported by the Science Foundation Ireland under grant numbers RFP2007-MATF802 and 07/MI/008.

## 1. Introduction

We consider a class of functions on $[0,\infty)$ that can be described in various ways:

- As the *Euler-d'Alembert*[1] class of infinitely differentiable functions $y:[0,\infty)\to \mathbf{R}$ that satisfy a homogeneous linear differential equation with real constant coefficients:
$$y^{(n)}+a_1y^{(n-1)}+a_2y^{(n-2)}+\cdots+a_ny=0,$$
with real initial conditions:
$$y(0)=b_1, y^{(1)}(0)=b_2,\dots,y^{(n-1)}(0)=b_n.$$
- As the *matrix-exponential* class of functions of the form
$$y(t)=ce^{At}b,\;\; A\in\mathbf{R}^{n\times n}, c\in\mathbf{R}^{1\times n}, b\in\mathbf{R}^{n\times 1},$$
with $t\geq 0$, where $A,b,c$ are independent of $t$. While the triple $(A,b,c)$ need not be unique, one such triple can always be chosen so that, for some choice of the positive integer $n$, the sets of vectors $\{b,Ab,\dots,A^{n-1}b\}$, $\{c,cA,\dots,cA^{n-1}\}$ are bases for $\mathbf{R}^n$. Such triples are said to provide a *minimal realization* of the function $y$. For the theory of minimal realization we refer to the theory of linear state space systems, (cf., e.g., [3] or Chapter 10 of [4] for a more algebraic approach).
- As the class $EPT$ of real exponential-polynomial-trigonometric functions $y:[0,\infty)\to\mathbf{R}$ of the form
$$y(t)=\Re\left(\sum_{k=1}^{K}p_k(t)e^{\mu_k t}\right)=\frac{1}{2}\sum_{k=1}^{K}p_k(t)e^{\mu_k t}+\frac{1}{2}\sum_{k=1}^{K}\overline{p_k}(t)e^{\overline{\mu_k}t},$$
where $p_k\in\mathbf{C}[t]$, $\mu_k\in\mathbf{C}, k=1,2,\dots,K,\ t\geq 0$, and the bar denotes complex conjugation. Note that this can also be written in the form $y(t)=\sum_{i=0}^{d}q_i(t)e^{\lambda_i t}\cos(\theta_i t+\tau_i)$, where the $q_i$ are real polynomials, $\lambda_i$, $\theta_i$ and $\tau_i$ are real numbers, and $K\leq d\leq 2K$.

The fact that these classes are equal is well known. For instance, using the companion matrix associated with the characteristic polynomial $z^n+a_1z^{n-1}+\cdots+a_n$, it can be seen that the second class contains the first. The Cayley-Hamilton theorem ensures that the first class is contained in the second. That the second class is contained in the third follows from the theory of the Jordan normal form. Since, finally, for every number $\lambda$, it's easy to see that
$$\left(\frac{d}{dt}-\lambda\right)^{k+1}(t^ke^{\lambda t})=0,\;\; k=0,1,2,\dots,$$
and the Euler-d'Alembert class forms a ring of functions, every $EPT$ function satisfies a homogeneous linear differential equation with constant coefficients. Thus, the third class is a subset of the first.

[1]cf. Euler (1743), d'Alembert (1748)

A further useful characterization of this class can be given as the class of continuous functions on $[0,\infty)$ whose Laplace transform exists on a right half-plane of the complex plane, and is a proper rational function there. In fact, the Laplace transform of $t \mapsto ce^{At}b$ is equal to $r(z) = c(zI - A)^{-1}b$ for all $z \in \mathbf{C}$ with $\Re z > \max\{\Re\lambda : \det(A - \lambda I) = 0\}$. (This is well known in linear systems theory where $r$ is called the *transfer function*, and the corresponding $EPT$ function is called the *impulse response function.*) Conversely, for any strictly proper rational function $r$, one can construct a triple $(A, b, c)$ such that $r(z) = c(zI - A)^{-1}b$, and the corresponding $EPT$ function is $ce^{At}b$.

Important subclasses can be characterized by the location of the eigenvalues of the matrix $A$ in a minimal realization $(A, b, c)$ of an $EPT$ function. The subclass $P$ of polynomials coincides with the subclass for which all eigenvalues of $A$ are zero. The subclass $E$ of *real exponential sums*, i.e., linear combinations with constant coefficients of real exponential functions of the form $t \mapsto e^{\lambda t}$ coincides with the subclass for which the eigenvalues are real and distinct. The subclass $T$ is the subclass for which the eigenvalues are purely imaginary and distinct.

Clearly, $EPT$ functions are ubiquitous in the mathematical sciences! Here we want to mention some places where they are required to be non-negative on $[0,\infty)$.

- In *financial mathematics* they appear as *forward rate curves*, e.g., the Nelson-Siegel forward rate curves (cf. [5]):
$$t \mapsto z_0 + z_1 e^{-\lambda t} + z_2 t e^{-\lambda t},$$
or the Svensson forward rate curves (cf. [6]). As the function values of these curves denote interest rates, we want them to be non-negative!
- In *probability theory* they appear as *probability density functions*; such functions must be non-negative and integrate to one over $[0,\infty)$. For instance, they occur in the form of Gamma densities with positive integer shape parameter $k$:
$$f(t; k, \beta) = \big(\beta^k/(k-1)!\big)t^{k-1}e^{-t\beta}, \ t \geq 0, \ k \in \mathbf{N}, \ \beta > 0.$$
- In *systems theory* they appear as impulse response functions of linear systems. In the case of so-called positive (respectively, non-negative) systems it is a requirement that the impulse response functions be positive (respectively, non-negative).

Given the importance of non-negative $EPT$ functions in these and other areas, the question arises of how to analyze whether a given $EPT$ function is non-negative or not, and how to characterize classes of non-negative $EPT$ functions. The present paper addresses these questions. We provide (i) a necessary condition and (ii) a sufficient condition for the non-negativity of the tail of an $EPT$ function, and (iii) a method that can be used to determine whether the sufficient condition is satisfied. Use will be made of an earlier method, that was found by the authors, to determine whether an $EPT$ function is non-negative on any finite closed subinterval of $[0,\infty)$. In the next section, a summary of that method will be given, as it will be used in the sequel.

## 2. Non-negativity analysis on a finite interval – a summary

In [10], a method is presented to determine non-negativity of an $EP$ function $y$ on an interval $[0, T]$, where $0 < T < \infty$. This is based on the construction of a generalized Budan-Fourier sequence ($BF$-sequence) given by $y$. Such a function can be represented as $y(t) := ce^{At}b$, for some minimal triple $(A, b, c) \in \mathbf{R}^{n\times n} \times \mathbf{R}^{n\times 1} \times \mathbf{R}^{1\times n}$, where the eigenvalues of $A$ are real numbers, and can be ordered in decreasing order as $\lambda_1 \geq \lambda_2 \geq \cdots \geq \lambda_n$. For such an $EP$ function $y$, a $BF$-sequence is generated on $[0, T]$ by:

$$\begin{aligned} y_0(t) &:= y(t) = ce^{At}b, \\ y_1(t) &:= c(\lambda_1 I - A)e^{At}b, \\ y_2(t) &:= c(\lambda_1 I - A)(\lambda_2 I - A)e^{At}b, \\ &\ \ \vdots \\ y_n(t) &:= c(\lambda_1 I - A)(\lambda_2 I - A)\dots(\lambda_n I - A)e^{At}b \equiv 0. \end{aligned}$$

This sequence has the property that $y_k$ has at most one sign-changing zero in between any two consecutive sign-changing zeros or boundary points of $y_{k+1}$, $k = 0, 1, \dots, n-1$ on $[0, T]$. Using this, and applying a bisection technique one can find all sign-changing zeros of $y_n, y_{n-1}, \dots, y_0$, and hence determine whether or not $y$ is non-negative on $[0, T]$.

Also in [10], a $BF$-sequence is presented for any $EPT$ function $y$ on $[0, T]$. This allows us to determine whether $y(t) \geq 0$, $\forall t \in [0, T]$. Based on this we now want to study the possible tail behaviour of $EPT$ functions. Obviously, if we can show that an $EPT$ function $y$ is non-negative for all $t \geq T_0$ for some $T_0 > 0$, and we can verify that $y$ is non-negative on $[0, T_0]$ using the $BF$-sequence method, then non-negativity of $y$ on $[0, \infty)$ follows.

## 3. A special representation of *EPT* functions

The following representation of an arbitrary $EPT$ function will play an important role in the remainder of the paper.

**Theorem 3.1.** *Any $EPT$ function $y$ can be written in the form*

$$y(t) = \sum_{k=0}^{N} b_k(t)\Re f_k(e^{i\theta_1 t}, e^{i\theta_2 t}, \dots, e^{i\theta_m t}), \tag{3.1}$$

*where each $b_k$ is an $EP$ function of the form*

$$b_k(t) = (t + T_1)^{d_k} e^{\lambda_k (t+T_1)}, \ k = 0, 1, \dots, N,$$

*for some $T_1 \geq 0$, such that $b_0(t) > b_1(t) > \cdots > b_N(t) > 0$, $\forall t > 0$, each $f_k$ is a multivariate trigonometric polynomial in $m$ complex variables, and $\{\theta_1, \theta_2, \dots, \theta_m\}$ is a subset of $\mathbf{R}$ that is linearly independent over $\mathbf{Q}$.*

*Remark.* A function $f$ on the $m$-dimensional unit torus

$$\mathbf{T}^m = \{z = (z_1, z_2, \ldots, z_m) : |z_j| = 1,\ j = 1, 2, \ldots, m\},$$

will be defined to be a multivariate trigonometric polynomial if it is of the form

$$f(z) = \sum_{\alpha \in I^m} c_\alpha z^\alpha,\ z \in \mathbf{T}^m,$$

where $c_\alpha \in \mathbf{C}$, for each multi-index $\alpha = (\alpha_1, \alpha_2, \ldots, \alpha_m) \in I^m$, $z^\alpha := z_1^{\alpha_1} z_2^{\alpha_2} \cdots z_m^{\alpha_m}$ and $I$ is a finite subset of the integers $\mathbf{Z}$. So, in particular, negative powers are allowed. Note that, for $z \in \mathbf{T}^m$, $\Re f(z) = \frac{1}{2} f(z_1, z_2, \ldots, z_m) + \frac{1}{2} \bar{f}(z_1^{-1}, z_2^{-1}, \ldots, z_m^{-1})$ where $\bar{f}(z) = \sum_{\alpha \in I^m} \overline{c_\alpha} z^\alpha$, and $\overline{c_\alpha}$ is the complex conjugate of $c_\alpha$.

*Proof.* From the representation of a *complex EPT* function as a sum of products of polynomials with complex exponential functions, it follows, by treating each of the monomials in the polynomials separately, that an *EPT* function $y$ can be written as

$$y(t) = \Re\left(\sum_{k=0}^{N} \tilde{b}_k(t) \tilde{\ell}_k(e^{i\eta_1 t}, e^{i\eta_2 t}, \ldots, e^{i\eta_K t})\right)$$

where each $\tilde{b}_k$ is an *EP* function of the form

$$\tilde{b}_k(t) = t^{d_k} e^{\lambda_k t},\ k = 0, 1, \ldots, N,$$

such that $\tilde{b}_0(t) > \tilde{b}_1(t) > \cdots > \tilde{b}_N(t)$, $\forall t > T_1$ for some sufficiently large number $T_1 > 0$; and, for each $k = 0, 1, 2, \ldots, N$, $\tilde{\ell}_k(w_1, w_2, \ldots, w_K)$ is a degree-one (linear) polynomial defined on $\mathbf{T}^K$. Here, the $\lambda_k$ are the real parts of the eigenvalues of the matrix $A$ and $\eta_j$ are the imaginary parts of the non-real eigenvalues of $A$, where $A$ is the square matrix in a minimal $(A, b, c)$ representation of $y$, so that $y(t) = ce^{At}b$. Because of the nature of *EP* functions, for each $k = 0, 1, 2, \ldots, N-1$,

$$\lim_{t \to \infty} \frac{\tilde{b}_{k+1}(t)}{\tilde{b}_k(t)} = 0.$$

Hence, $\tilde{b}_k(t) = o(\tilde{b}_0(t))$ as $t \to \infty$, for each $k = 1, 2, \ldots, N$.

Now consider the auxiliary function $g(t) := ce^{-AT_1} e^{At} b$. Note that $g(t+T_1) = y(t)$ for all $t \geq 0$. The function $g$ belongs to $EPT$, and hence it has a representation of the same form as $y$. In fact, since minimal representations of both $g$ and $y$ involve the *same* matrix $A$, the same exponential functions appear in both, so that

$$g(t) = \Re\left(\sum_{k=0}^{N} \tilde{b}_k(t) \hat{\ell}_k(e^{i\eta_1 t}, e^{i\eta_2 t}, \ldots, e^{i\eta_K t})\right),$$

with $\hat{l}_k$ again a linear polynomial on the $K$-dimensional torus, for each $k = 0, 1, \ldots, N$. It follows from $y(t) = g(t + T_1)$ that

$$y(t) = \Re\left(\sum_{k=0}^{N} b_k(t) \ell_k(e^{i\eta_1 t}, e^{i\eta_2 t}, \ldots, e^{i\eta_K t})\right),$$

where $\ell_k$ is another linear polynomial on $\mathbf{T}^K$, for each $k = 0, 1, \ldots, N$. Now consider the vector space $V$ over the field of rational numbers $\mathbf{Q}$ spanned by the set of the real numbers $\{\eta_1, \eta_2, \ldots, \eta_K\}$. Let $m$ be the dimension of this vector space $V$, and consider a $V$-basis $\{\theta_1, \theta_2, \ldots, \theta_m\}$ which has the special property that each of the elements $\eta_1, \eta_2, \ldots, \eta_K$ can be expressed as a linear combination of the basis elements with *integer* coefficients. Such a basis can be obtained from an arbitrary basis by an appropriate basis transformation. To see this, let $\{\tilde{\theta}_1, \tilde{\theta}_2, \ldots, \tilde{\theta}_m\}$ be an arbitrary $V$-basis, and suppose that $\eta = \tilde{M}\tilde{\theta}$, where $\eta = (\eta_1, \eta_2, \ldots, \eta_K)'$, $\tilde{\theta} = (\tilde{\theta}_1, \tilde{\theta}_2, \ldots, \tilde{\theta}_m)'$, and $\tilde{M} \in \mathbf{Q}^{K\times m}$. Let $\delta_i \in \mathbf{N}$ denote the least common denominator of the $i$th column of $\tilde{M}$, for each $i = 1, 2, \ldots, m$. Then take $\theta_j := \frac{\tilde{\theta}_j}{\delta_j}$ and let $D$ be the diagonal matrix with $\delta_i$ as its $i$th diagonal element. Then the entries of $M = \tilde{M}D$ are integers, and $\{\theta_1, \theta_2, \ldots, \theta_m\}$ is a basis with the required property. Let the $(i, j)$th element of $M$ be denoted by $m_{i,j} \in \mathbf{Z}$. We can now replace each $\eta_j$ in the expression $\ell_k(e^{i\eta_1 t}, e^{i\eta_2 t}, \ldots, e^{i\eta_K t})$ by $\eta_j = \sum_{i=1}^m m_{j,i}\theta_i$. Once that is done for each $\eta_j$ then we obtain a trigonometric polynomial $f_k$ on $\mathbf{T}^m$, with

$$f_k(e^{i\theta_1 t}, e^{i\theta_2 t}, \ldots, e^{i\theta_m t}) = \ell_k(e^{i\eta_1 t}, e^{i\eta_2 t}, \ldots, e^{i\eta_K t}).$$

If we denote by $m_j$ the multi-index given by the $j$th row of $M$, then we can write

$$f_k(z) = \ell_k(z^{m_1}, z^{m_2}, \ldots, z^{m_K}),\ k = 1, 2, \ldots, N.$$

By construction, the set $\{\theta_1, \theta_2, \ldots, \theta_m\}$ is linearly independent over $\mathbf{Q}$. □

Let us say that an *EPT* function $y$ is non-negative eventually if there exists a non-negative number $T$ such that $y(t) \geq 0$ for all $t \geq T$. So, if an *EPT* function is non-negative eventually, and hence non-negative on $[T, \infty)$ for some $T \geq 0$, then one can apply the *BF*-sequence approach to determine whether it is non-negative on $[0, T)$ as well, and hence on all of $[0, \infty)$.

**Theorem 3.2.** *Suppose an EPT function $y$ has the representation* (3.1). *Assume without loss of generality that the real part of the polynomial $f_0$ is not the zero function, i.e., $\Re f_0 \not\equiv 0$. Necessary conditions for $y$ to be non-negative eventually are that*

(i) $\Re f_0(z) \geq 0,\ \forall z \in \mathbf{T}^m$,
(ii) $\Re f_0$ *has a positive constant term, $\lambda_0$ is a real eigenvalue of $A$, and*

$$\lambda_0 = \max\{\lambda_k : k = 0, 1, 2 \ldots, N\}.$$

*Proof.* ad (i) Since $\{\theta_1, \theta_2, \ldots, \theta_m\}$ is linearly independent over $\mathbf{Q}$, it follows from Kronecker's approximation theorem [7] that, for any number $T > 0$, $\{(f_0(z), f_1(z), \ldots, f_N(z)) : z \in \mathbf{T}^m\}$ is the closure of the set

$$\{(f_0(e^{i\theta_1 t}, \ldots, e^{i\theta_m t}), \ldots, f_N(e^{i\theta_1 t}, \ldots, e^{i\theta_m t})) : t \geq T\}.$$

This implies that if there exists a point $z \in \mathbf{T}^m$ with the property that $\Re f_0(z) =: -\epsilon < 0$, then, due to the continuity of $f_0$, for each $T > 0$, there will be a number

$t > T$ such that $\Re f_0(e^{i\theta_1 t}, \ldots, e^{i\theta_m t}) < -\epsilon/2$. Because they are continuous functions on a compact set, the functions $f_1, f_2, \ldots, f_N$ are bounded on $\mathbf{T}^m$, and so $b_k(t) f_k(e^{i\theta_1 t}, e^{i\theta_2 t}, \ldots, e^{i\theta_m t})$, $k = 1, 2, \ldots, N$, are all $o(b_0(t))$ as $t \to \infty$. It follows that $y$ has negative values for some $t > T$, for any positive number $T$. Since this conflicts with the hypothesis, it follows that $\Re f_0(z) \geq 0$ for all $z$ on the unit torus $\mathbf{T}^m$.

ad (ii) As is well known, the constant term of any trigonometric polynomial in $m$ variables can be obtained by integrating it over the unit torus (viewed as a subset of Euclidean space), and dividing by $(2\pi)^m$. Applying this to the non-negative and not identically zero function

$$z \mapsto \frac{1}{2} f_0(z_1, z_2, \ldots, z_m) + \frac{1}{2} \bar{f}_0(z_1^{-1}, z_2^{-1}, \ldots, z_m^{-1}), \; z = (z_1, z_2, \ldots, z_m) \in \mathbf{T}^m,$$

we obtain that the real part of the constant term of $f_0$ is positive. This implies that the real part of the constant term of the corresponding linear polynomial $\ell_0$ is also positive (the constant terms in both $f_0$ and $\ell_0$ are the same, since only the non-constant terms can change when going from $\ell_0$ to $f_0$), and hence that the real number $\lambda_0$ is an eigenvalue of $A$, with multiplicity $d_0 + 1$. That $\lambda_0$ is at least as big as the real part of the other eigenvalues follows from the fact that $b_0$ is at least as large as the $b_k$, $k = 1, 2, \ldots, N$. □

*Remark.* The fact that if $t \mapsto ce^{At}b$ is a minimal representation of a function $y \in EPT$, and is eventually non-negative, then $A$ must have a dominant real eigenvalue is also shown in [11], using the classical Pringsheim theorem about power series with nonnegative coefficients. The discrete analogue of this result is dealt with in [9].

The question of determining criteria for the non-negativity of a trigonometrical polynomial on $\mathbf{T}^m$ arises naturally from this theorem. While the well-known L. Fejér and F. Riesz characterization of non-negative trigonometric polynomials of one real variable settles the issue when $m = 1$ [12], the question is more challenging when $m \geq 2$. However, it's easy to resolve it for a polynomial that is linear in each variable separately.

**Theorem 3.3.** *Let*

$$f(z) = c_0 + \sum_{k=1}^{m} c_k z_k, \; z = (z_1, z_2, \ldots, z_m) \in \mathbf{T}^m.$$

*Then the real part of $f$ is non-negative on $\mathbf{T}^m$ if and only if*

$$\sum_{k=1}^{m} |c_k| \leq \Re c_0.$$

*Proof.* If the displayed condition holds, and $z \in \mathbf{T}^m$, then

$$\Re f(z) = \Re c_0 + \sum_{k=1}^{m} \Re\{c_k z_k\} \geq \Re c_0 - \sum_{k=1}^{m} |c_k z_k| = \Re c_0 - \sum_{k=1}^{m} |c_k| \geq 0,$$

whence the sufficiency part follows. Conversely, if $\Re f \geq 0$ on $\mathbf{T}^m$, and $1 \leq j \leq m$, consider $c_j$: if $c_j \neq 0$, select $w_j$ so that $w_j c_j = -|c_j|$; and otherwise select $w_j = 1$. Then, $w = (w_1, w_2, \dots, w_m) \in \mathbf{T}^m$ and so

$$0 \leq \Re f(w) = \Re c_0 - \sum_{k=1}^{m} |c_k|,$$

whence the necessity part follows. □

**Theorem 3.4.** *Suppose $y \in EPT$. Then, in the notation of Theorem 3.1, a sufficient condition for $y$ to be non-negative on $[T, \infty)$, $T \geq 0$ is that*

$$\forall z \in \mathbf{T}^m : \forall t \in [T, \infty) : \sum_{k=0}^{N} b_k(t) \Re f_k(z) \geq 0. \tag{3.2}$$

*Proof.* This is self-evident. □

At this point, the following examples may help to elucidate matters.

*Example* 1. Let

$$g(t) = \frac{1}{2} + \frac{1}{2}\cos 2t - 2(\cos t)e^{-t} + e^{-2t},\ 0 \leq t < \infty.$$

Then $g$ belongs to $EPT$, and is non-negative on $[0, \infty)$. Also,

$$g(t) = \Re\big(b_0(t) f_0(e^{it}) + b_1(t) f_1(e^{it}) + b_2(t) f_2(e^{it})\big),$$

where $b_0(t) = 1$, $b_1(t) = e^{-t}$, $b_2(t) = e^{-2t}$, and

$$f_0(z) = \frac{1 + z^2}{2},\ f_1(z) = -2z,\ f_2(z) = 1,\ \forall z \in \mathbf{C}.$$

*Proof.* That $g \in EPT$ is clear, and its non-negativity can be easily verified directly. But this also follows from Theorem 3.4 because, if $z \in \mathbf{C}$, with $|z| = 1$, and $t \in [0, \infty)$, then

$$\begin{aligned}
\Re\big(b_0(t) f_0(z) + b_1(t) f_1(z) + b_2(t) f_2(z)\big) &= \Re\Big(\frac{1 + z^2}{2} - 2ze^{-t} + e^{-2t}\Big) \\
&= \frac{2z\bar{z} + z^2 + \bar{z}^2}{4} - 2\Re z e^{-t} + e^{-2t} \\
&= (\Re z)^2 - 2\Re z e^{-t} + e^{-2t} \\
&= \Big(\Re z - e^{-t}\Big)^2 \geq 0.
\end{aligned}$$

□

*Example* 2. Let

$$h(t) = 2 + \cos \pi t + \cos t - e^{-t},\ 0 \leq t < \infty.$$

Then $h$ belongs to $EPT$, and is eventually non-negative on $[0, \infty)$. However, it fails to satisfy the sufficient condition (3.2).

*Proof.* That $h \in EPT$ is clear. Moreover,

$$h(t) = \Re\big(b_0(t)f_0(e^{it}, e^{i\pi t}) + b_1(t)f_1(e^{it}, e^{i\pi t})\big),$$

where $b_0(t) = 1$, $b_1(t) = e^{-t}$, and

$$f_0(z_1, z_2) = 2 + z_1 + z_2,\ f_1(z_1, z_2) = -1.$$

Hence,

$$\Re\big(b_0(t)f_0(-1,-1) + b_1(t)f_1(-1,-1)\big) = -e^{-t} < 0,\ \forall t \geq 0.$$

Nevertheless, $h$ is eventually non-negative. The verification of this fact is due to Pat McCarthy, the essentials of whose proof we reproduce here with his permission. So, suppose $h(x) = 0$ for some positive $x$, so that

$$2 + \cos \pi x + \cos x = e^{-x},$$

whence

$$2\sin^2\left(\frac{x-\pi}{2}\right) = 1 + \cos x \leq e^{-x}, \text{ and } 2\sin^2\left(\frac{\pi x - \pi}{2}\right) = 1 + \cos \pi x \leq e^{-x}.$$

But there are non-negative integers $p, q$ such that

$$2p \leq x \leq 2p + 2, \text{ and } 2q\pi \leq x \leq (2q+2)\pi.$$

Moreover, it is easy to easy to see that $\sin s \geq 2s/\pi$ if $0 \leq s \leq \pi/2$, so that

$$\sin^2\left(\frac{t-\pi}{2}\right) \geq \frac{(t-\pi)^2}{\pi^2}, \text{ if } 0 \leq t \leq 2\pi.$$

Hence, by periodicity,

$$\frac{2(x-(2q+1)\pi)^2}{\pi^2} \leq 2\sin^2\left(\frac{(x-2q\pi)-\pi}{2}\right) = 1 + \cos x \leq e^{-x} \leq e^{-2q\pi},$$

and so

$$|x - (2q+1)\pi| \leq \frac{\pi e^{-q\pi}}{\sqrt{2}},$$

Similarly,

$$|x - (2p+1)| \leq \frac{e^{-x/2}}{\sqrt{2}} \leq \frac{e^{-q\pi}}{\sqrt{2}}.$$

By the triangle inequality it now follows that

$$\left|\pi - \frac{2p+1}{2q+1}\right| \leq \frac{(\pi+1)e^{\pi/2}e^{-\frac{\pi}{2}(2q+1)}}{(2q+1)\sqrt{2}}.$$

However, it is well known that such an inequality can hold for at most finitely many pairs of non-negative integers $p, q$. Otherwise, it conflicts, for instance, with Kurt Mahler's remarkable result that, if $a, b$ are positive integers, then

$$\left|\pi - \frac{a}{b}\right| > \frac{1}{b^{42}}.$$

(For improvements of this, see [8].) In the light of this fact, it now follows that $h$ has at most a finite number of positive real zeros. Since $h(t) > 1$ whenever $t$ is a multiple of $2\pi$, it follows that $h$ is eventually non-negative, as claimed. □

*Remarks.*

- The sufficient condition (3.2) also applies to functions for which a representation like (3.1) holds, even when $\{\theta_1, \theta_2, \ldots, \theta_m\}$ is not linearly independent.
- A special case in which the sufficient condition for eventual non-negativity holds is when $\Re f_0$ has a positive minimum on the unit torus; the sufficient condition (3.2) is satisfied in that case for sufficiently large $T$, because $b_k(t) = o(b_0(t)),\ k = 1, 2, \ldots, N$, as $t \to \infty$.
- A further special case of this occurs when the set of eigenvalues with maximum real part consists just of one point (higher multiplicity would be no problem in this case, only the location of the points matters here). According to Theorem 3.2, such an eigenvalue would actually have to be real. Then $\Re f_0$ is a positive constant, and hence the sufficient condition (3.2) is satisfied for sufficiently large $T$.
- And a further special case of this is when the function is actually an *EP* function, because then all eigenvalues are real, hence, also, the one with largest real part.
- The class of *EPT* functions satisfying (3.2) is likely to be important in practice; for this class, one can determine non-negativity on $[0, \infty)$ by determining an appropriate $T$ such that the function is non-negative for all $t > T$, while for $t \in [0, T]$ one can use the *BF*-sequence approach mentioned in Section 2 to determine non-negativity (assuming it is possible to determine the sign of the *EPT* functions involved in the *BF*-sequence at each of the points at which we need to know it).

## 4. A projection on the boundary of a set of non-negative functions

The question now arises as to how one can ascertain that a given *EPT* function satisfies the sufficient condition for some sufficiently large value of $T$. As was noted in the previous section, if $\Re f_0$ has a positive minimum over the unit torus, then the sufficient condition is indeed satisfied for sufficiently large values of $T$. If this function has minimum zero (recall that non-negativity of this function is a necessary condition) and $\Re f_1$ has a positive minimum, then again the sufficient condition is satisfied for large enough $T$. In fact, more generally, if there exists a value $0 < \lambda \leq 1$ such that $\Re(f_0 + \lambda f_1)$ has positive minimum, then there exists a value of $T$ such that the sufficient condition is satisfied (use $b_0(t) > b_1(t)/\lambda$ for sufficiently large $T$ and replace $b_0$ by $b_1/\lambda$; as $\Re f_0 \geq 0$ the result follows). Of course, for any given *EPT* function, one could try to analyze whether it satisfies the sufficient condition using these or other "ad hoc" arguments. In what follows, we will present a more systematic method to analyze the problem.

We first introduce some notation. For $z = (z_1, z_2, \ldots, z_m) \in \mathbf{T}^m$, and $k = 0, 1, 2, \ldots, N$, let

$$\Phi_k(z) = \frac{1}{2} f_k(z_1, z_2, \ldots, z_m) + \frac{1}{2} \bar{f}_k(z_1^{-1}, z_2^{-1}, \ldots, z_m^{-1}),$$

and define

$$\Phi(z) := (\Phi_0(z), \Phi_1(z), \dots, \Phi_N(z))',$$

so that $\Phi$ maps the unit torus $\mathbf{T}^m$ into $\mathbf{R}^{N+1}$. Let $F = \Phi(\mathbf{T}^m)$ denote the subset of $\mathbf{R}^{N+1}$ of points in the image of $\Phi$.

Let $b = (b_0, b_1, \dots, b_N)$, and let $T > 0$ be given. Define the following subsets of (column vectors) of $\mathbf{R}^{N+1}$ that relate, respectively, to non-negativity and positivity of linear combinations of the $EP$ functions $b_0, b_1, \dots, b_N$, for $t \geq T$ :

$$Q := \{\phi \in \mathbf{R}^{N+1} : b(t)\phi \geq 0, \forall t \in [T, \infty)\},$$

$$P := \{\phi \in \mathbf{R}^{N+1} : b(t)\phi > 0, \forall t \in [T, \infty)\}.$$

Let $\partial Q$ be the boundary of $Q$ in $\mathbf{R}^{N+1}$.

*Remark.* Note that $Q \setminus P \subseteq \partial Q,\ Q \setminus P \neq \partial Q$. To see this, consider, for instance, the case $N = 1$ and $T = 0$, when $b_0(t) = e^{-\alpha_0 t}$, $b_1(t) = e^{-\alpha_1 t}$, where $0 < \alpha_0 < \alpha_1$. The set $Q$ consists of vectors $\phi = (\phi_0, \phi_1)$ such that

$$e^{-\alpha_0 t}\phi_0 + e^{-\alpha_1 t}\phi_1 \geq 0,\ \forall t \geq 0.$$

Then $(\phi_0, \phi_1) = (0, 1) \in \partial Q \setminus (Q \setminus P)$, as $(-\epsilon, 1) \notin Q, \forall \epsilon > 0; (0, 1) \in P \subset Q$. □

Define the projection

$$\mathbf{R}^{N+1} \to \mathbf{R}^{N+1}, \phi = \begin{pmatrix} \phi_0 \\ \phi_1 \\ \vdots \\ \phi_N \end{pmatrix} \mapsto R(\phi) - \begin{pmatrix} \rho_0 \\ \phi_1 \\ \vdots \\ \phi_N \end{pmatrix}$$

by $\rho_0 = \min \left\{ \hat{\phi}_0 : \begin{pmatrix} \hat{\phi}_0 \\ \phi_1 \\ \vdots \\ \phi_N \end{pmatrix} \in Q \right\}$

Note that $\rho_0 \geq 0$ (due to the dominance of $b_0$ over $b_1, \dots, b_N$). Let $R_0(\phi) :- e_1' R(\phi) = \rho_0$ and let $L(\phi) := \phi_0 - \rho_0 = e_1'\phi - R_0(\phi))$, $\forall \phi \in \mathbf{R}^{N+1}$, where $e_1, e_2, \dots, e_{N+1}$ form the standard basis for $\mathbf{R}^{N+1}$.

This function has some interesting properties.

**Proposition 4.1.**

(i) *The function $L$ is Lipschitz continuous with respect to the $\ell_1$ norm on $\mathbf{R}^{N+1}$, and $|L(\phi) - L(\tilde{\phi})| \leq ||\phi - \tilde{\phi}||_1$ for any $\phi, \tilde{\phi} \in \mathbf{R}^{N+1}$.*

(ii) *$L(\phi) \geq 0 \Longleftrightarrow \phi \in Q$.*

*Proof.* ad (i) Consider a pair of vectors $\phi, \tilde{\phi} \in \mathbf{R}^{N+1}$. As before, let $\rho_0$ denote $R_0(\phi)$. Similarly, let $\tilde{\rho}_0$ denote $R_0(\tilde{\phi})$. Then $L(\phi) - L(\tilde{\phi}) = \phi_0 - \tilde{\phi}_0 + \tilde{\rho}_0 - \rho_0$. Using the fact that

$$e_1 - e_j = \begin{pmatrix} 1 \\ 0 \\ \vdots \\ 0 \\ -1 \\ 0 \\ \vdots \\ 0 \end{pmatrix} \in Q, \ \forall j \in \{2, 3, \dots, N+1\},$$

it can be seen that:

$$\tilde{\rho}_0 - \rho_0 \leq |\phi_1 - \tilde{\phi}_1| + |\phi_2 - \tilde{\phi}_2| + \cdots + |\phi_N - \tilde{\phi}_N|.$$

Similarly, it can be seen that

$$\rho_0 - \tilde{\rho}_0 \leq |\phi_1 - \tilde{\phi}_1| + |\phi_2 - \tilde{\phi}_2| + \cdots + |\phi_N - \tilde{\phi}_N|,$$

and hence we obtain

$$|\rho_0 - \tilde{\rho}_0| \leq |\phi_1 - \tilde{\phi}_1| + |\phi_2 - \tilde{\phi}_2| + \cdots + |\phi_N - \tilde{\phi}_N|.$$

It follows that

$$|L(\phi) - L(\tilde{\phi})| \leq ||\phi - \tilde{\phi}||_1.$$

ad (ii) For any $\phi \in Q$, increasing the first entry gives another vector in $Q$, because it corresponds to adding a positive multiple of $b_0$ to a non-negative $EP$ function, which gives another non-negative $EP$ function. It follows, by construction of $L$, that if $L(\phi) \geq 0$, then $\phi \in Q$. Also, if $\phi \in Q$, then, by construction, $L(\phi) \geq 0$, hence the statement in the proposition follows. □

This result can now be applied as follows.

Let $T$ be fixed. We can parametrize the torus $\mathbf{T}^m$ by a map $Z : (\mathbf{R}/\mathbf{Z})^m \to \mathbf{T}^m$ that is given by $\omega = (\omega_1, \omega_2, \dots, \omega_m) \mapsto Z(\omega) = (e^{2\pi i \omega_1}, e^{2\pi i \omega_2}, \dots, e^{2\pi i \omega_m})$. Here, the $\omega_j$ are taken to be real numbers between zero and one. Composing this with $\Phi$ one can parametrize $F$ by $F = \Phi(Z([0,1)^m)) = \Phi(Z([0,1]^m))$. Now note that $\min_{\phi \in F} L(\phi)$ is non-negative if and only if $F \subset Q$ if and only if the sufficient condition (3.2) holds. As $F$ is compact ($\mathbf{T}^m$ is compact and $L$ is continuous), this minimum exists. We can build a grid in $[0,1]^m$ with a prescribed mesh size $\mu > 0$, i.e., every point in the set is at a distance of at most $\mu$ to a point in the grid. Because the composition $\Phi \circ Z$ is real analytic, and $L$ is Lipschitz continuous, the composition $L(\Phi \circ Z)$ is Lipschitz continuous as well, and a uniform Lipschitz constant can be found for this mapping using the previous proposition, together with the fact that the derivatives of $\omega \mapsto \Phi(Z(\omega))$ consist of trigonometric polynomials each of which can be bounded by the sum of the absolute values of their coefficients. Therefore, one can construct a lower bound for the minimum within distance $\epsilon > 0$ of the minimum, for arbitrarily small positive $\epsilon$ by calculating the values of $\omega \mapsto L(\Phi(Z(\omega)))$ on the grid with sufficiently small mesh size $\mu$. That leaves the question on how $L(\phi)$ can actually be calculated for a given vector

$\phi = \Phi(Z(\omega))$. Here, we can use the fact that, for any $EP$ function $y(t) := b(t)\phi$, non-negativity can be verified by determining a $T_1$ such that $y(t) > 0$ for all $t > T_1$ if such a number $T_1$ exists and if so, determining the minimum of $y$ on $[T, T_1]$ using the $BF$-sequence algorithm. So, we can check for each point in $\mathbf{R}^{N+1}$ to see whether it is in $Q$ or not. The idea is now to apply the bisection technique to determine $R_0(\phi)$ and hence $L(\phi)$. Recall that $R_0(\phi) \geq 0$. In the bisection algorithm we will create vectors which are obtained from $\phi$ by replacing the first entry by another number, say $r = r(n)$, where $n = 0, 1, 2, \ldots$ denote the stages in the algorithm. Note that $r < 0$ gives a vector outside $Q$. If taking $r = 0$ gives a vector in $Q$, then $R_0(\phi) = 0$, and we can stop. Now, assume taking $r = 0$ produces a vector outside $Q$ and let $r(0) := 0$. If $\phi \in Q$, we can take $r(1) = \phi_0$; otherwise, one can take $r(1) = |\phi_1| + |\phi_2| + \cdots + |\phi_N|$ to be sure that we obtain a vector in $Q$ (this follows from the fact that $b_0 \geq b_1 \geq \cdots \geq b_N$). Now do bisection to create a sequence $r(n)$, $n = 0, 1, 2, \ldots$, where each next value in the sequence is a mid-point between two earlier values in the sequence depending on whether the various corresponding vectors are inside or outside $Q$. At each stage, the pair $r(n)$ and $r(n+1)$ corresponds to a pair of vectors one of which is inside $Q$ and the other one is outside $Q$. This sequence will converge, and the limit will lie in $Q$, because $Q$ is closed! So, this gives us $R_0(\phi)$, and hence also $L(\phi)$.

Note that if $\Re f_0$ has minimum equal to zero, and $\Re f_0(\hat{z}) = 0$, then $L(\Phi(\hat{z})) = R_0(\Phi(\hat{z})) = 0$. Therefore, in that case the minimum of $L$ on $F$ is at most zero. Therefore, this method will not guarantee that the sufficient condition (3.2) is satisfied. On the other hand, if the sufficient condition is not satisfied, then, for a sufficiently refined mesh, the grid calculations will reveal this.

## 5. Conclusions and further research

Further research is needed in order to obtain systematic methods that can tell us whether a given $EPT$ function satisfies the sufficient condition (3.2) presented here. It may be possible to obtain this using algebraic methods (exploiting the ring structure of the class of $EP$ functions).

Also, it would be interesting to study the properties of the class of $EPT$ functions that satisfy condition (3.2), such as the possible invariance under certain operations (addition, multiplication etc.). Gaining a better understanding in the class of $EPT$ functions that are non-negative, but do not satisfy the sufficient conditions is also of interest.

To verify the necessary condition one needs to verify whether a certain multivariable trigonometric polynomial is non-negative. Non-negativity of polynomials including trigonometric polynomials is a topic of active research at the moment, where techniques from constructive algebra, real algebraic geometry and numerical optimization (such as interior point methods for convex optimization) are combined. See, e.g., [13] and references given there.

For practical applications it will be important to see how one can deal with questions of linear dependence and independence of certain numbers over the rational numbers. If these numbers are replaced by approximations (e.g., through round-off procedures) the dependency structure could be changed inadvertently. However, if the $EPT$ functions are obtained by operating on other $EPT$ functions (addition, multiplication, convolution etc.) rational dependencies may well show up explicitly.

**Acknowledgement.** We record our indebtedness to an eagle-eyed referee who drew our attention to several points that needed to be clarified.

## References

[1] L. Euler (1743): *De integratione aequationum differentialium altiorum gradum*, Miscellanea Berolinensia vol. 7, pp. 193–242; Opera Omnia vol. XXII, pp. 108–149.

[2] J. d'Alembert, Hist. Acad. Berlin 1748, p. 283.

[3] T. Kailath, *Linear Systems*, Prentice Hall, Englewood Cliffs, NJ, 1980.

[4] R.E. Kalman, P. Falb, M. Arbib, *Topics in mathematical system theory*, McGraw-Hill, New York, 1969.

[5] C.R. Nelson, A.F. Siegel, *Parsimonious modeling of yield curves*, Journal of Business, vol. 60, nr. 4, pp. 473–489, 1987.

[6] L. Svensson, *Estimating and interpreting forward interest rates: Sweden* 1992-4, Discussion paper, Centre for Economic Policy Research (1051), 1994.

[7] G.H. Hardy, E.M. Wright, *The Theory of Numbers*, Third edition, Oxford University Press, Amen House, London, 1954.

[8] M. Hata, *Rational approximations to $\pi$ and some other numbers*, Acta. Arith., 63 (1993), pp. 335–369.

[9] B. Hanzon, F. Holland, *The long-term behaviour of Markov sequences*, Mathematical Proceedings of the Royal Irish Academy, 110A (1), pp. 163–185 (2010)

[10] B. Hanzon, F. Holland, *Non-negativity of Exponential Polynomial Trigonometric Functions-a Budan Fourier sequence approach*, Poster 429 presented at the Bachelier Finance Society Congress, Toronto, 2010; available on-line at web-address http://euclid.ucc.ie/staff/hanzon/BFSTorontoPosterHanzonHollandFinal.pdf

[11] B. Hanzon, F. Holland, *On a Perron-Frobenius type result for non-negative impulse response functions*, internal report, School of Mathematical Sciences, UCC, 20 March 2010; available on-line at web-address http://euclid.ucc.ie/staff/hanzon/HanzonHollandDominantPoleLisbonPaperisn.pdf

[12] G. Pólya, G. Szegö, *Problems and Theorems in Analysis*, Vol II, Springer-Verlag, Heidelberg, Berlin, 1976

[13] B. Dumitrescu, *Positive Trigonometric Polynomials and Signal Processing Applications*, Springer-Verlag, New York, 2007.

Bernard Hanzon
Edgeworth Centre for Financial Mathematics
Department of Mathematics
University College Cork, Ireland
e-mail: `b.hanzon@ucc.ie`

Finbarr Holland
Department of Mathematics
University College Cork, Ireland
e-mail: `f.holland@ucc.ie`

Operator Theory:
Advances and Applications, Vol. 221, 413-420


# Compactness of the $\overline{\partial}$-Neumann Operator on Weighted $(0,q)$-forms

Friedrich Haslinger

**Abstract.** As an application of a characterization of compactness of the $\overline{\partial}$-Neumann operator we derive a sufficient condition for compactness of the $\overline{\partial}$-Neumann operator on $(0,q)$-forms in weighted $L^2$-spaces on $\mathbb{C}^n$.

**Mathematics Subject Classification (2000).** Primary 32W05; Secondary 32A36. 35J10.

**Keywords.** $\overline{\partial}$-Neumann problem, Sobolev spaces, compactness.

## 1. Introduction

In this paper we continue the investigations of [12] and [11] concerning existence and compactness of the canonical solution operator to $\overline{\partial}$ on weighted $L^2$-spaces over $\mathbb{C}^n$.

Let $\varphi : \mathbb{C}^n \longrightarrow \mathbb{R}^+$ be a plurisubharmonic $\mathcal{C}^2$-weight function and define the space

$$L^2(\mathbb{C}^n,\varphi) = \left\{ f : \mathbb{C}^n \longrightarrow \mathbb{C} \ : \ \int_{\mathbb{C}^n} |f|^2 \, e^{-\varphi} \, d\lambda < \infty \right\},$$

where $\lambda$ denotes the Lebesgue measure, the space $L^2_{(0,q)}(\mathbb{C}^n,\varphi)$ of $(0,q)$-forms with coefficients in $L^2(\mathbb{C}^n,\varphi)$, for $1 \leq q \leq n$. Let

$$(f,g)_\varphi = \int_{\mathbb{C}^n} f \, \overline{g} e^{-\varphi} \, d\lambda$$

denote the inner product and

$$\|f\|^2_\varphi = \int_{\mathbb{C}^n} |f|^2 e^{-\varphi} \, d\lambda$$

the norm in $L^2(\mathbb{C}^n,\varphi)$.

We consider the weighted $\overline{\partial}$-complex

$$L^2_{(0,q-1)}(\mathbb{C}^n,\varphi) \underset{\overline{\partial}^*_\varphi}{\overset{\overline{\partial}}{\rightleftarrows}} L^2_{(0,q)}(\mathbb{C}^n,\varphi) \underset{\overline{\partial}^*_\varphi}{\overset{\overline{\partial}}{\rightleftarrows}} L^2_{(0,q+1)}(\mathbb{C}^n,\varphi),$$

where for $(0,q)$-forms $u=\sum'_{|J|=q} u_J\, d\overline{z}_J$ with coefficients in $\mathcal{C}_0^\infty(\mathbb{C}^n)$ we have

$$\overline{\partial} u = \sum_{|J|=q}{}' \sum_{j=1}^n \frac{\partial u_J}{\partial \overline{z}_j}\, d\overline{z}_j \wedge d\overline{z}_J,$$

and

$$\overline{\partial}^*_\varphi u = -\sum_{|K|=q-1}{}' \sum_{k=1}^n \delta_k u_{kK}\, d\overline{z}_K,$$

where $\delta_k = \frac{\partial}{\partial z_k} - \frac{\partial \varphi}{\partial z_k}$.

There is an interesting connection between $\overline{\partial}$ and the theory of Schrödinger operators with magnetic fields, see for example [5], [2], [8] and [6] for recent contributions exploiting this point of view.

The complex Laplacian on $(0,q)$-forms is defined as

$$\Box_\varphi := \overline{\partial}\,\overline{\partial}^*_\varphi + \overline{\partial}^*_\varphi \overline{\partial},$$

where the symbol $\Box_\varphi$ is to be understood as the maximal closure of the operator initially defined on forms with coefficients in $\mathcal{C}_0^\infty$, i.e., the space of smooth functions with compact support.

$\Box_\varphi$ is a selfadjoint and positive operator, which means that

$$(\Box_\varphi f, f)_\varphi \geq 0\ ,\ \text{ for } f \in \mathrm{dom}(\Box_\varphi).$$

The associated Dirichlet form is denoted by

$$Q_\varphi(f,g) = (\overline{\partial} f, \overline{\partial} g)_\varphi + (\overline{\partial}^*_\varphi f, \overline{\partial}^*_\varphi g)_\varphi, \tag{1.1}$$

for $f,g \in \mathrm{dom}(\overline{\partial}) \cap \mathrm{dom}(\overline{\partial}^*_\varphi)$. The weighted $\overline{\partial}$-Neumann operator $N_{\varphi,q}$ is – if it exists – the bounded inverse of $\Box_\varphi$.

We indicate that a $(0,1)$-form $f=\sum_{j=1}^n f_j\, d\overline{z}_j$ belongs to $\mathrm{dom}(\overline{\partial}^*_\varphi)$ if and only if

$$\sum_{j=1}^n \left( \frac{\partial f_j}{\partial z_j} - \frac{\partial \varphi}{\partial z_j} f_j \right) \in L^2(\mathbb{C}^n,\varphi)$$

and that forms with coefficients in $\mathcal{C}_0^\infty(\mathbb{C}^n)$ are dense in $\mathrm{dom}(\overline{\partial}) \cap \mathrm{dom}(\overline{\partial}^*_\varphi)$ in the graph norm $f \mapsto (\|\overline{\partial} f\|^2_\varphi + \|\overline{\partial}^*_\varphi f\|^2_\varphi)^{\frac{1}{2}}$ (see [10]).

We consider the Levi-matrix

$$M_\varphi = \left( \frac{\partial^2 \varphi}{\partial z_j \partial \overline{z}_k} \right)_{jk}$$

of $\varphi$ and suppose that the sum $s_q$ of any $q$ (equivalently: the smallest $q$) eigenvalues of $M_\varphi$ satisfies

$$\liminf_{|z|\to\infty} s_q(z) > 0. \tag{1.2}$$

We show that (1.2) implies that there exists a continuous linear operator

$$N_{\varphi,q} : L^2_{(0,q)}(\mathbb{C}^n,\varphi) \longrightarrow L^2_{(0,q)}(\mathbb{C}^n,\varphi),$$

such that $\Box_\varphi \circ N_{\varphi,q} u = u$, for any $u \in L^2_{(0,q)}(\mathbb{C}^n,\varphi)$.

If we suppose that the sum $s_q$ of any $q$ (equivalently: the smallest $q$) eigenvalues of $M_\varphi$ satisfies

$$\lim_{|z|\to\infty} s_q(z) = \infty. \tag{1.3}$$

Then the $\overline{\partial}$-Neumann operator $N_{\varphi,q} : L^2_{(0,q)}(\mathbb{C}^n,\varphi) \longrightarrow L^2_{(0,q)}(\mathbb{C}^n,\varphi)$ is compact.

This generalizes results from [12] and [11], where the case of $q = 1$ was handled.

Finally we discuss some examples in $\mathbb{C}^2$.

## 2. The weighted Kohn-Morrey formula

First we compute

$$(\Box_\varphi u, u)_\varphi = \|\overline{\partial} u\|^2_\varphi + \|\overline{\partial}^*_\varphi u\|^2_\varphi$$

for $u \in \mathrm{dom}(\Box_\varphi)$.

We obtain

$$\|\overline{\partial} u\|^2_\varphi + \|\overline{\partial}^*_\varphi u\|^2_\varphi = \sum_{|J|=|M|=q}{}' \sum_{j,k=1}^n \epsilon^{kM}_{jJ} \int_{\mathbb{C}^n} \frac{\partial u_J}{\partial \overline{z}_j} \overline{\frac{\partial u_M}{\partial \overline{z}_k}}\, e^{-\varphi}\, d\lambda$$
$$+ \sum_{|K|=q-1}{}' \sum_{j,k=1}^n \int_{\mathbb{C}^n} \delta_j u_{jK} \overline{\delta_k u_{kK}}\, e^{-\varphi}\, d\lambda,$$

where $\epsilon^{kM}_{jJ} = 0$ if $j \in J$ or $k \in M$ or if $k \cup M \neq j \cup J$, and equals the sign of the permutation $\binom{kM}{jJ}$ otherwise. The right-hand side of the last formula can be rewritten as

$$\sum_{|J|=q}{}' \sum_{j=1}^n \left\| \frac{\partial u_J}{\partial \overline{z}_j} \right\|^2_\varphi + \sum_{|K|=q-1}{}' \sum_{j,k=1}^n \int_{\mathbb{C}^n} \left( \delta_j u_{jK} \overline{\delta_k u_{kK}} - \frac{\partial u_{jK}}{\partial \overline{z}_k} \overline{\frac{\partial u_{kK}}{\partial \overline{z}_j}} \right) e^{-\varphi}\, d\lambda,$$

see [18] Proposition 2.4 for the details. Now we mention that for $f, g \in \mathcal{C}^\infty_0(\mathbb{C}^n)$ we have

$$\left( \frac{\partial f}{\partial \overline{z}_k}, g \right)_\varphi = -(f, \delta_k g)_\varphi$$

and hence

$$\left( \left[ \delta_j, \frac{\partial}{\partial \overline{z}_k} \right] u_{jK}, u_{kK} \right)_\varphi = -\left( \frac{\partial u_{jK}}{\partial \overline{z}_k}, \frac{\partial u_{kK}}{\partial \overline{z}_j} \right)_\varphi + (\delta_j u_{jK}, \delta_k u_{kK})_\varphi.$$

Since

$$\left[\delta_j, \frac{\partial}{\partial \overline{z}_k}\right] = \frac{\partial^2 \varphi}{\partial z_j \partial \overline{z}_k},$$

we get

$$\|\overline{\partial} u\|_\varphi^2 + \|\overline{\partial}_\varphi^* u\|_\varphi^2 = \sum_{|J|=q}{}' \sum_{j=1}^n \left\| \frac{\partial u_J}{\partial \overline{z}_j} \right\|_\varphi^2 + \sum_{|K|=q-1}{}' \sum_{j,k=1}^n \int_{\mathbb{C}^n} \frac{\partial^2 \varphi}{\partial z_j \partial \overline{z}_k} u_{jK} \overline{u}_{kK} \, e^{-\varphi} \, d\lambda. \tag{2.1}$$

Formula (2.1) is a version of the Kohn-Morrey formula, compare [18] or [16].

**Proposition 2.1.** *Let $1 \leq q \leq n$ and suppose that the sum $s_q$ of any $q$ (equivalently: the smallest $q$) eigenvalues of $M_\varphi$ satisfies*

$$\liminf_{|z| \to \infty} s_q(z) > 0. \tag{2.2}$$

*Then there exists a uniquely determined bounded linear operator*

$$N_{\varphi,q} : L^2_{(0,q)}(\mathbb{C}^n, \varphi) \longrightarrow L^2_{(0,q)}(\mathbb{C}^n, \varphi),$$

*such that $\Box_\varphi \circ N_{\varphi,q} u = u$, for any $u \in L^2_{(0,q)}(\mathbb{C}^n, \varphi)$.*

*Proof.* Let $\mu_{\varphi,1} \leq \mu_{\varphi,2} \leq \cdots \leq \mu_{\varphi,n}$ denote the eigenvalues of $M_\varphi$ and suppose that $M_\varphi$ is diagonalized. Then, in a suitable basis,

$$\begin{aligned}\sum_{|K|=q-1}{}' \sum_{j,k=1}^n \frac{\partial^2 \varphi}{\partial z_j \partial \overline{z}_k} u_{jK} \overline{u}_{kK} &= \sum_{|K|=q-1}{}' \sum_{j=1}^n \mu_{\varphi,j} |u_{jK}|^2 \\ &= \sum_{J=(j_1,\dots,j_q)}{}' (\mu_{\varphi,j_1} + \cdots + \mu_{\varphi,j_q}) |u_J|^2 \\ &\geq s_q |u|^2\end{aligned}$$

It follows from (2.1) that there exists a constant $C > 0$ such that

$$\|u\|_\varphi^2 \leq C(\|\overline{\partial} u\|_\varphi^2 + \|\overline{\partial}_\varphi^* u\|_\varphi^2) \tag{2.3}$$

for each $(0,q)$-form $u \in$dom $(\overline{\partial}) \cap$ dom $(\overline{\partial}_\varphi^*)$. For a given $v \in L^2_{(0,q)}(\mathbb{C}^n, \varphi)$ consider the linear functional $L$ on dom $(\overline{\partial}) \cap$ dom $(\overline{\partial}_\varphi^*)$ given by $L(u) = (u, v)_\varphi$. Notice that dom $(\overline{\partial}) \cap$ dom $(\overline{\partial}_\varphi^*)$ is a Hilbert space in the inner product $Q_\varphi$. Since we have by (2.3)

$$|L(u)| = |(u, v)_\varphi| \leq \|u\|_\varphi \, \|v\|_\varphi \leq C Q_\varphi(u,u)^{1/2} \, \|v\|_\varphi.$$

Hence by the Riesz representation theorem there exists a uniquely determined $(0,q)$-form $N_{\varphi,q} v$ such that

$$(u, v)_\varphi = Q_\varphi(u, N_{\varphi,q} v) = (\overline{\partial} u, \overline{\partial} N_{\varphi,q} v)_\varphi + (\overline{\partial}_\varphi^* u, \overline{\partial}_\varphi^* N_{\varphi,q} v)_\varphi,$$

from which we immediately get that $\Box_\varphi \circ N_{\varphi,q} v = v$, for any $v \in L^2_{(0,q)}(\mathbb{C}^n, \varphi)$. If we set $u = N_{\varphi,q}v$ we get again from 2.3

$$\begin{aligned}\|\overline{\partial} N_{\varphi,q}v\|^2_\varphi + \|\overline{\partial}^*_\varphi N_{\varphi,q}v\|^2_\varphi &= Q_\varphi(N_{\varphi,q}v, N_{\varphi,q}v) = (N_{\varphi,q}v, v)_\varphi \le \|N_{\varphi,q}v\|_\varphi \, \|v\|_\varphi \\ &\le C_1(\|\overline{\partial} N_{\varphi,q}v\|^2_\varphi + \|\overline{\partial}^*_\varphi N_{\varphi,q}v\|^2_\varphi)^{1/2} \, \|v\|_\varphi,\end{aligned}$$

hence

$$(\|\overline{\partial} N_{\varphi,q}v\|^2_\varphi + \|\overline{\partial}^*_\varphi N_{\varphi,q}v\|^2_\varphi)^{1/2} \le C_2\|v\|_\varphi$$

and finally again by (2.3)

$$\|N_{\varphi,q}v\|_\varphi \le C_3(\|\overline{\partial} N_{\varphi,q}v\|^2_\varphi + \|\overline{\partial}^*_\varphi N_{\varphi,q}v\|^2_\varphi)^{1/2} \le C_4\|v\|_\varphi,$$

where $C_1, C_2, C_3, C_4 > 0$ are constants. Hence we get that $N_{\varphi,q}$ is a continuous linear operator from $L^2_{(0,q)}(\mathbb{C}^n, \varphi)$ into itself (see also [13] or [4]). □

## 3. Compactness of $N_{\varphi,q}$

We use a characterization of precompact subsets of $L^2$-spaces, see [1]:

A bounded subset $\mathcal{A}$ of $L^2(\Omega)$ is precompact in $L^2(\Omega)$ if and only if for every $\epsilon > 0$ there exists a number $\delta > 0$ and a subset $\omega \subset\subset \Omega$ such that for every $u \in \mathcal{A}$ and $h \in \mathbb{R}^n$ with $|h| < \delta$ both of the following inequalities hold:

$$\text{(i)} \int_\Omega |\tilde{u}(x+h) - \tilde{u}(x)|^2 \, dx < \epsilon^2 \quad , \quad \text{(ii)} \int_{\Omega\setminus\overline{\omega}} |u(x)|^2 \, dx < \epsilon^2, \tag{3.1}$$

where $\tilde{u}$ denotes the extension by zero of $u$ outside of $\Omega$.

In addition we define an appropriate Sobolev space and prove compactness of the corresponding embedding, for related settings see [3], [14], [15] .

**Definition 3.1.** Let

$$\mathcal{W}^{Q_\varphi}_q = \{u \in L^2_{(0,q)}(\mathbb{C}^n, \varphi) \ : \ \|\overline{\partial} u\|^2_\varphi + \|\overline{\partial}^*_\varphi u\|^2_\varphi < \infty\}$$

with norm

$$\|u\|_{Q_\varphi} = (\|\overline{\partial} u\|^2_\varphi + \|\overline{\partial}^*_\varphi u\|^2_\varphi)^{1/2}.$$

*Remark* 3.2. $\mathcal{W}^{Q_\varphi}_q$ coincides with the form domain $\mathrm{dom}(\overline{\partial}) \cap \mathrm{dom}(\overline{\partial}^*_\varphi)$ of $Q_\varphi$ (see [9], [10]).

**Proposition 3.3.** *Let $\varphi$ be a plurisubharmonic $\mathcal{C}^2$- weight function. Let $1 \le q \le n$ and suppose that the sum $s_q$ of any $q$ (equivalently: the smallest $q$) eigenvalues of $M_\varphi$ satisfies*

$$\lim_{|z|\to\infty} s_q(z) = \infty. \tag{3.2}$$

*Then $N_{\varphi,q} : L^2_{(0,q)}(\mathbb{C}^n, \varphi) \longrightarrow L^2_{(0,q)}(\mathbb{C}^n, \varphi)$ is compact.*

*Proof.* For $(0,q)$ forms one has by (2.1) and Proposition 2.1 that

$$\|\overline{\partial}u\|_\varphi^2 + \|\overline{\partial}_\varphi^* u\|_\varphi^2 \geq \int_{\mathbb{C}^n} s_q(z)\,|u(z)|^2\, e^{-\varphi(z)}\, d\lambda(z). \tag{3.3}$$

We indicate that the embedding

$$j_{\varphi,q} : \mathcal{W}_q^{Q_\varphi} \hookrightarrow L^2_{(0,q)}(\mathbb{C}^n, \varphi)$$

is compact by showing that the unit ball of $\mathcal{W}_q^{Q_\varphi}$ is a precompact subset of $L^2_{(0,q)}(\mathbb{C}^n, \varphi)$, which follows by the above-mentioned characterization of precompact subsets in $L^2$-spaces with the help of Gårding's inequality to verify (3.1) (i)(see for instance [7] or [4]) and to verify (3.1) (ii): we have

$$\int_{\mathbb{C}^n \setminus \mathbb{B}_R} |u(z)|^2 e^{-\varphi(z)}\, d\lambda(z) \leq \int_{\mathbb{C}^n \setminus \mathbb{B}_R} \frac{s_q(z)|u(z)|^2}{\inf\{s_q(z) : |z| \geq R\}} e^{-\varphi(z)} d\lambda(z),$$

which implies by (3.3) that

$$\int_{\mathbb{C}^n \setminus \mathbb{B}_R} |u(z)|^2 e^{-\varphi(z)}\, d\lambda(z) \leq \frac{\|u\|^2_{Q_\varphi}}{\inf\{s_q(z) \,:\, |z| \geq R\}} < \epsilon,$$

if $R$ is big enough, see [11] for the details.

This together with the fact that $N_{\varphi,q} = j_{\varphi,q} \circ j^*_{\varphi,q}$, (see [18]) gives the desired result. □

*Remark* 3.4. If $q = 1$ condition (3.2) means that the lowest eigenvalue $\mu_{\varphi,1}$ of $M_\varphi$ satisfies

$$\lim_{|z|\to\infty} \mu_{\varphi,1}(z) = \infty. \tag{3.4}$$

This implies compactness of $N_{\varphi,1}$ (see [11]).

**Examples:** a) We consider the plurisubharmonic weight function

$$\varphi(z,w) = |z|^2|w|^2 + |w|^4$$

on $\mathbb{C}^2$. The Levi matrix of $\varphi$ has the form

$$\begin{pmatrix} |w|^2 & \overline{z}w \\ \overline{w}z & |z|^2 + 4|w|^2 \end{pmatrix}$$

and the eigenvalues are

$$\mu_{\varphi,1}(z,w) = \frac{1}{2}\left(5|w|^2 + |z|^2 - \sqrt{9|w|^4 + 10|z|^2|w|^2 + |z|^4}\right)$$

$$= \frac{16|w|^4}{2\left(5|w|^2 + |z|^2 + \sqrt{9|w|^4 + 10|z|^2|w|^2 + |z|^4}\right)},$$

and

$$\mu_{\varphi,2}(z,w) = \frac{1}{2}\left(5|w|^2 + |z|^2 + \sqrt{9|w|^4 + 10|z|^2|w|^2 + |z|^4}\right).$$

It follows that (3.4) fails, since even

$$\lim_{|z|\to\infty} |z|^2\mu_{\varphi,1}(z,0) = 0,$$

but

$$s_2(z,w) = \frac{1}{4}\,\Delta\varphi(z,w) = |z|^2 + 5|w|^2,$$

hence (3.2) is satisfied for $q = 2$.

b) In the next example we consider decoupled weights. Let $n \geq 2$ and

$$\varphi(z_1, z_2, \ldots, z_n) = \varphi(z_1) + \varphi(z_2) + \cdots + \varphi(z_n)$$

be a plurisubharmonic decoupled weight function and suppose that $|z_\ell|^2\Delta\varphi_\ell(z_\ell) \to +\infty$, as $|z_\ell| \to \infty$ for some $\ell \in \{1,\ldots,n\}$. Then the $\overline{\partial}$-Neumann operator$N_{\varphi,1}$ acting on $L^2_{(0,1)}(\mathbb{C}^n,\varphi)$ fails to be compact (see [12], [9], [17]).

Finally we discuss two examples in $\mathbb{C}^2$ : for $\varphi(z_1,z_2) = |z_1|^2 + |z_2|^2$ all eigenvalues of the Levi matrix are 1 and $N_{\varphi,1}$ fails to be compact by the above result on decoupled weights, for the weight function $\varphi(z_1,z_2) = |z_1|^4 + |z_2|^4$ the eigenvalues are $4|z_1|^2$ and $4|z_2|^2$ and $N_{\varphi,1}$ fails to be compact again by the above result, whereas $N_{\varphi,2}$ is compact by 3.3.

# References


[1] R.A. Adams and J.J.F. Fournier, *Sobolev spaces.* Pure and Applied Math. Vol. 140, Academic Press, 2006.

[2] B. Berndtsson, *$\overline{\partial}$ and Schrödinger operators.* Math. Z. **221** (1996), 401–413.

[3] P. Bolley, M. Dauge and B. Helffer, *Conditions suffisantes pour l'injection compacte d'espace de Sobolev à poids.* Séminaire équation aux dérivées partielles (France), Université de Nantes **1** (1989), 1–14.

[4] So-Chin Chen and Mei-Chi Shaw, *Partial differential equations in several complex variables.* Studies in Advanced Mathematics, Vol. 19, Amer. Math. Soc., 2001.

[5] M. Christ, *On the $\overline{\partial}$ equation in weighted $L^2$ norms in $\mathbb{C}^1$.* J. of Geometric Analysis **1** (1991), 193–230.

[6] M. Christ and S. Fu, *Compactness in the $\overline{\partial}$ Neumann problem, magnetic Schrödinger operators, and the Aharonov-Bohm effect.* Adv. Math. **197** (2005), 1–40.

[7] G.B. Folland, *Introduction to partial differential equations.* Princeton University Press, Princeton, 1995.

[8] S. Fu and E.J. Straube, *Semi-classical analysis of Schrödinger operators and compactness in the $\overline{\partial}$ Neumann problem.* J. Math. Anal. Appl. **271** (2002), 267–282.

[9] K. Gansberger, *Compactness of the $\overline{\partial}$-Neumann operator.* Dissertation, University of Vienna, 2009.

[10] K. Gansberger and F. Haslinger, *Compactness estimates for the $\overline{\partial}$-Neumann problem in weighted $L^2$ - spaces.* Complex Analysis (P. Ebenfelt, N. Hungerbühler, J.J. Kohn, N. Mok, E.J. Straube, eds.), Trends in Mathematics, Birkhäuser (2010), 159–174.

[11] F. Haslinger, *Compactness for the $\overline{\partial}$-Neumann problem – a functional analysis approach.* Collectanea Mathematica **62** (2011), 121–129.

[12] F. Haslinger and B. Helffer, *Compactness of the solution operator to $\overline{\partial}$ in weighted $L^2$-spaces.* J. of Functional Analysis **255** (2008), 13–24.

[13] L. Hörmander, *An introduction to complex analysis in several variables.* North-Holland, 1990.

[14] J. Johnsen, *On the spectral properties of Witten Laplacians, their range projections and Brascamp-Lieb's inequality.* Integral Equations Operator Theory **36** (2000), 288–324.

[15] J.-M. Kneib and F. Mignot, *Equation de Schmoluchowski généralisée.* Ann. Math. Pura Appl. (IV) **167** (1994), 257–298.

[16] J.D. McNeal, *$L^2$ estimates on twisted Cauchy-Riemann complexes.* 150 years of mathematics at Washington University in St. Louis. Sesquicentennial of mathematics at Washington University, St. Louis, MO, USA, October 3–5, 2003. Providence, RI: American Mathematical Society (AMS). Contemporary Mathematics **395** (2006), 83–103.

[17] G. Schneider, *Non-compactness of the solution operator to $\overline{\partial}$ on the Fock-space in several dimensions.* Math. Nachr. **278** (2005), 312–317.

[18] E. Straube, *The $L^2$-Sobolev theory of the $\overline{\partial}$-Neumann problem.* ESI Lectures in Mathematics and Physics, EMS, 2010.

Friedrich Haslinger
Fakultät für Mathematik
Universität Wien
Nordbergstr. 15
A-1090 Wien, Austria
e-mail: friedrich.haslinger@univie.ac.at

Operator Theory:
Advances and Applications, Vol. 221, 421-432


# Dislocation Problems for Periodic Schrödinger Operators and Mathematical Aspects of Small Angle Grain Boundaries

Rainer Hempel and Martin Kohlmann

**Abstract.** We discuss two types of defects in two-dimensional lattices, namely (1) translational dislocations and (2) defects produced by a rotation of the lattice in a half-space.

For Lipschitz-continuous and $\mathbb{Z}^2$-periodic potentials, we first show that translational dislocations produce spectrum inside the gaps of the periodic problem; we also give estimates for the (integrated) density of the associated surface states. We then study lattices with a small angle defect where we find that the gaps of the periodic problem fill with spectrum as the defect angle goes to zero. To introduce our methods, we begin with the study of dislocation problems on the real line and on an infinite strip. Finally, we consider examples of muffin tin type. Our overview refers to results in [HK1, HK2].

**Mathematics Subject Classification (2000).** Primary 35J10, 35P20, 81Q10.

**Keywords.** Schrödinger operators, eigenvalues, spectral gaps.

## 1. Introduction

In solid state physics, pure matter in a crystallized form is usually described by a periodic Schrödinger operator $-\Delta + V(x)$ in $\mathbb{R}^3$, where the potential $V$ is a periodic function. In reality, however, crystals are not perfectly periodic since the periodic pattern of atomic arrangement is disturbed by various types of crystal defects, most notably:

- point defects where single atoms are removed (vacancies) or replaced by foreign atoms (impurities),
- large scale defects that produce a surface at which two portions of the lattice (or two different half-lattices) face each other (line defects, grain boundaries).

For the modeling of point defects, random Schrödinger operators are the appropriate setting (cf., e.g., [PF] or [V]). Here we present a deterministic approach to some two-dimensional models with defects from the second class.

Let $V\colon \mathbb{R}^2 \to \mathbb{R}$ be (bounded and) periodic with respect to the lattice $\mathbb{Z}^2$ and consider the family of potentials

$$W_t(x,y) := \begin{cases} V(x,y), & x \geq 0, \\ V(x+t,y), & x < 0, \end{cases} \qquad t \in [0,1]. \tag{1.1}$$

We then let $D_t := -\Delta + W_t$ denote the associated (self-adjoint) Schrödinger operators, acting in $L_2(\mathbb{R}^2)$. The operators $D_t$ are the Hamiltonians for a two-dimensional lattice where the potential equals the $\mathbb{Z}^2$-periodic function $V$ on $\{x \geq 0\}$ and a shifted copy of $V$ for $\{x < 0\}$, i.e., we study a *dislocation problem.* We call $W_t$ the *dislocation potential*, $t$ the *dislocation parameter* and $D_t$ the *dislocation operators.* The spectrum of $D_0 = D_1$ is purely absolutely continuous and has a band-gap structure,

$$\sigma(D_0) = \sigma_{\mathrm{ess}}(D_0) = \cup_{k=1}^{\infty}[a_k, b_k], \quad a_k < b_k \leq a_{k+1}. \tag{1.2}$$

The spectral gaps $(b_k, a_{k+1})$ are denoted as $\Gamma_k$. For simplicity, we will sometimes write $a$ and $b$ for the edges of a given $\Gamma_k$ with $\Gamma_k \neq \emptyset$. We shall say that a gap $\Gamma_k = (a,b)$ is *non-trivial* if $a < b$ and $a$ is above the infimum of the essential spectrum of the given self-adjoint operator.

We will show that the operators $D_t$ possess *surface states* (i.e., spectrum produced by the interface) in the gaps $\Gamma_k$ of $D_0$, for suitable values of $t \in (0,1)$. More strongly, we have a positivity result for the (integrated) density of the surface states associated with the above spectrum in the gaps. Here we first have to choose an appropriate scaling which permits to distinguish the bulk from the surface density of states. To this end, we consider the operators $-\Delta + W_t$ on squares $Q_n = (-n,n)^2$ with Dirichlet boundary conditions, for $n$ large, count the number of eigenvalues inside a compact subset of a non-degenerate spectral gap of $D_0$ and scale with $n^{-2}$ for the bulk and with $n^{-1}$ for the surface states. Taking the limit $n \to \infty$ (which exists as explained in [KS, EKSchrS]), we obtain the (integrated) density of states measures $\varrho_{\mathrm{bulk}}(D_t, I)$ for the bulk and $\varrho_{\mathrm{surf}}(D_t, J)$ for the surface states of this model; here $I \subset \mathbb{R}$ and $J \subset \mathbb{R}\backslash\sigma(D_0)$ are open intervals and $\overline{J} \subset \mathbb{R}\backslash\sigma(D_0)$. (The fact that an integrated surface density of states exists does not necessarily mean it is non-zero and there are only rare examples where we know $\varrho_{\mathrm{surf}}$ to be non-trivial.) Our first main result can be described as follows:

**1.1. Theorem.** *If $(a,b)$ is a non-trivial spectral gap of the periodic operator $-\Delta+V$, acting in $L_2(\mathbb{R}^2)$ with $V$ Lipschitz-continuous, then for any interval $(\alpha,\beta)$ with $a < \alpha < \beta < b$ there is a $t \in (0,1)$ such that $\varrho_{\mathrm{surf}}(D_t,(\alpha,\beta)) > 0$.*

We also explain how to obtain upper bounds (as in [HK2]) for the surface density of states. In Section 2, we will outline a proof of Theorem 1.1 starting from dislocation problems on $\mathbb{R}$ and on the strip $\Sigma := \mathbb{R} \times [0,1]$. The one-dimensional dislocation problem has been studied extensively by Korotyaev [K1, K2], and we use the 1D model mainly for testing our methods in the simplest possible case.

The techniques and results connected with Theorem 1.1 are mainly presented as a preparation for the study of rotational defects where we consider the potential

$$V_\vartheta(x,y) := \begin{cases} V(x,y), & x \geq 0, \\ V(M_{-\vartheta}(x,y)), & x < 0, \end{cases} \tag{1.3}$$

where $M_\vartheta \in \mathbb{R}^{2\times 2}$ is the usual orthogonal matrix associated with rotation through the angle $\vartheta$. The self-adjoint operators $R_\vartheta := -\Delta + V_\vartheta$ in $L_2(\mathbb{R}^2)$ are the Hamiltonians for two half-lattices given by the potential $V$ in $\{x \geq 0\}$ and a rotated copy of $V$ for $\{x < 0\}$; we obtain an interface at $x = 0$ where the two copies meet under the defect angle $\vartheta$. Our main assumption is that the periodic operator $H := R_0$ has a non-trivial gap $(a, b)$. We then have $R_\vartheta \to R_{\vartheta_0}$ in the strong resolvent sense as $\vartheta \to \vartheta_0 \in [0, \pi/2)$; in particular $R_\vartheta$ converges to $H$ in the strong resolvent sense as $\vartheta \to 0$. Our main result, Theorem 1.2 below, shows that the spectrum of $R_\vartheta$ is discontinuous at $\vartheta = 0$; in particular, $R_\vartheta$ cannot converge to $H$ in the norm resolvent sense as $\vartheta \to 0$.

**1.2. Theorem.** *Let $H$, $R_\vartheta$ and $(a, b)$ as above with a Lipschitz-continuous potential $V$. Then, for any $\varepsilon > 0$, there exists $0 < \vartheta_\varepsilon < \pi/2$ such that for any $E \in (a, b)$ we have*

$$\sigma(R_\vartheta) \cap (E - \varepsilon, E + \varepsilon) \neq \emptyset, \quad \forall 0 < \vartheta < \vartheta_\varepsilon. \tag{1.4}$$

As an illustration, we will consider potentials of *muffin tin type* which can be specified by fixing a radius $0 < r < 1/2$ for the discs where the potential vanishes, and the center $P_0 = (x_0, y_0) \in [0, 1)^2$ for the generic disc. In other words, we consider the periodic sets

$$\Omega_{r,P_0} := \cup_{(i,j)\in\mathbb{Z}^2} B_r(P_0 + (i,j)), \tag{1.5}$$

and we let $V = V_{r,P_0}$ be zero on $\Omega_{r,P_0}$ while we assume that $V$ is infinite on $\mathbb{R}^2 \setminus \Omega_{r,P_0}$. If $H_{i,j}$ is the Dirichlet Laplacian of the disc $B_r(P_0 + (i,j))$, then the form sum of $-\Delta$ and $V_{r,P_0}$ is $\oplus_{(i,j)\in\mathbb{Z}^2} H_{i,j}$. In our examples, we can see the behavior of surface states in the dislocation problem and the rotation problem for $-\Delta + V_{r,P_0}$ directly.

The paper is organized as follows: Section 2 is devoted to the dislocation problem on $\mathbb{R}$, on $\Sigma$ and in $\mathbb{R}^2$. Section 3 is about a small angle defect model in 2D and explains some details of the proof of Theorem 1.2. Finally, in Section 4, we turn to dislocations and rotations for muffin tin potentials where results analogous to Theorem 1.1 and Theorem 1.2 can be obtained. For further reading, we refer to [HK1, HK2].

## 2. Dislocation problems on the real line, on the strip $\mathbb{R} \times [0, 1]$, and in the plane

In this section, we study Schrödinger operators in one and two dimensions where the potential is obtained from a periodic potential by a coordinate shift on $\{x < 0\}$. We begin with a brief overview of the one-dimensional dislocation problem. In a second step, we study the dislocation problem on the strip $\Sigma = \mathbb{R} \times [0, 1]$ which provides a connection between the dislocation problems in one and two dimensions. Finally, we deal with dislocations in $\mathbb{R}^2$. Some of the results obtained in this section will be used in our treatment of rotational defects in the following section.

Let $h_0$ denote the (unique) self-adjoint extension of $-\frac{\mathrm{d}^2}{\mathrm{d}x^2}$ defined on $C_c^\infty(\mathbb{R})$. Our basic class of potentials is given by

$$\mathcal{P} := \{V \in L_{1,\mathrm{loc}}(\mathbb{R}, \mathbb{R}) \mid \forall x \in \mathbb{R} : V(x+1) = V(x)\}. \tag{2.1}$$

Potentials $V \in \mathcal{P}$ belong to the class $L_{1,\,\mathrm{loc},\,\mathrm{unif}}(\mathbb{R})$ which coincides with the Kato-class on the real line; in particular, any $V \in \mathcal{P}$ has relative form-bound zero with respect to $h_0$ and thus the form sum $H$ of $h_0$ and $V \in \mathcal{P}$ is well defined (cf. [CFrKS]).

For $V \in \mathcal{P}$ and $t \in [0, 1]$, we define the dislocation potentials $W_t$ by $W_t(x) := V(x)$, for $x \geq 0$, and $W_t(x) := V(x + t)$, for $x < 0$. As before, the form-sum $H_t$ of $h_0$ and $W_t$ is well defined.

We begin with some well-known results pertaining to the spectrum of $H = H_0$. As explained in [E, RS-IV], we have

$$\sigma(H) = \sigma_{\mathrm{ess}}(H) = \cup_{k=1}^\infty [\gamma_k, \gamma_k'], \tag{2.2}$$

where the numbers $\gamma_k$ and $\gamma_k'$ satisfy $\gamma_k < \gamma_k' \leq \gamma_{k+1}$, for all $k \in \mathbb{N}$, and $\gamma_k \to \infty$ as $k \to \infty$. Moreover, the spectrum of $H$ is purely absolutely continuous. The intervals $[\gamma_k, \gamma_k']$ are called the *spectral bands* of $H$. The open intervals $\Gamma_k := (\gamma_k', \gamma_{k+1})$ are the *spectral gaps* of $H$; we say the $k$th gap is *open* or *non-degenerate* if $\gamma_{k+1} > \gamma_k'$. It is easy to see ([HK1]) that

$$\sigma_{\mathrm{ess}}(H_t) = \sigma_{\mathrm{ess}}(H), \quad 0 \leq t \leq 1, \tag{2.3}$$

since inserting a Dirichlet boundary condition at a finite number of points means a finite rank perturbation of the resolvent, as is well known. Hence each non-trivial gap $(a, b)$ of $H$ is a gap in the essential spectrum of $H_t$, for all $t$. However, the dislocation may produce discrete (and simple) eigenvalues inside the spectral gaps of $H$: for any $(a, b)$ with $\inf \sigma_{\mathrm{ess}}(H) < a < b$ and $(a, b) \cap \sigma(H) = \emptyset$ there exists $t \in (0, 1)$ such that

$$\sigma(H_t) \cap (a, b) \neq \emptyset. \tag{2.4}$$

We thus have the following picture: while the essential spectrum remains unchanged under the perturbation, eigenvalues of $H_t$ cross the (non-trivial) gaps of $H$ as $t$ ranges through $(0, 1)$. These eigenvalues of $H_t$ can be described by continuous functions of $t$ (cf. [K1, K2] and Lemma 2.1 below). Lemma 2.1 states the (more or less obvious) fact that the eigenvalues of $H_t$ inside a given gap $\Gamma_k$ of

$H$ can be described by an (at most) countable, locally finite family of continuous functions, defined on suitable subintervals of $[0, 1]$. The proof of Lemma 2.1 uses a straight-forward compactness argument (cf. [HK1]). The result stated in Lemma 2.1 is presumably far from optimal if one assumes periodicity of the potential. On the other hand, the lemma and its proof in [HK1] allow for a generalization to non-periodic situations.

**2.1. Lemma.** *Let $V \in \mathcal{P}$ and $k \in \mathbb{N}$ and suppose that the gap $\Gamma_k$ of $H$ is open. Then there is a (finite or countable) family of continuous functions $f_j\colon (\alpha_j, \beta_j) \to \Gamma_k$, where $0 \le \alpha_j < \beta_j \le 1$, with the following properties:*

(i) *For all $j$ and for all $\alpha_j < t < \beta_j$, $f_j(t)$ is an eigenvalue of $H_t$. Conversely, for any $t \in (0,1)$ and any eigenvalue $E \in \Gamma_k$ of $H_t$ there is a unique index $j$ such that $f_j(t) = E$.*

(ii) *As $t \downarrow \alpha_j$ (or $t \uparrow \beta_j$), the limit of $f_j(t)$ exists and belongs to the set $\{a, b\}$.*

(iii) *For all but a finite number of indices $j$ the range of $f_j$ does not intersect a given compact subinterval of $\Gamma_k$.*

Under stronger assumptions on $V$ one can show that the eigenvalue branches are Hölder- or Lipschitz-continuous, or even analytic (cf. [K1]): we consider potentials from the classes

$$\mathcal{P}_\alpha := \left\{ V \in \mathcal{P} \,\middle|\, \exists C \ge 0\colon \int_0^1 |V(x+s) - V(x)|\,\mathrm{d}x \le C s^\alpha,\ \forall 0 < s \le 1 \right\}, \tag{2.5}$$

where $0 < \alpha \le 1$. The class $\mathcal{P}_\alpha$ consists of all periodic functions $V \in \mathcal{P}$ which are (locally) $\alpha$-Hölder-continuous in the $L_1$-mean; for $\alpha = 1$ this is a Lipschitz-condition in the $L_1$-mean. The class $\mathcal{P}_1$ is of particular practical importance since it contains the periodic step functions. As shown by J. Voigt, $\mathcal{P}_1$ coincides with the class of periodic functions on the real line which are locally of bounded variation (cf. [HK1]).

**2.2. Proposition.** *For $V \in \mathcal{P}_1$, let $(a, b)$ denote any of the gaps $\Gamma_k$ of $H$ and let $f_j\colon (\alpha_j, \beta_j) \to (a, b)$ be as in Lemma 2.1. Then the functions $f_j$ are uniformly Lipschitz-continuous. More precisely, there exists a constant $C \ge 0$ such that for all $j$*

$$|f_j(t) - f_j(t')| \le C|t - t'|, \qquad \alpha_j \le t, t' \le \beta_j. \tag{2.6}$$

*If $0 < \alpha < 1$ and $V \in \mathcal{P}_\alpha$, then each of the functions $f_j\colon (\alpha_j, \beta_j) \to (a, b)$ is locally uniformly Hölder-continuous, i.e., for any compact subset $[\alpha_j', \beta_j'] \subset (\alpha_j, \beta_j)$ there is a constant $C = C(j, \alpha_j', \beta_j')$ such that $|f_j(t) - f_j(t')| \le C|t - t'|^\alpha$, for all $t, t' \in [\alpha_j', \beta_j']$.*

Our basic result in the study of the one-dimensional dislocation problem says that at least $k$ eigenvalues move from the upper to the lower edge of the $k$th gap as the dislocation parameter ranges from 0 to 1. Using the notation of Lemma 2.1 and writing $f_i(\alpha_i) := \lim_{t \downarrow \alpha_i} f_i(t)$, $f_i(\beta_i) := \lim_{t \uparrow \beta_i} f_i(t)$, we define

$$\mathcal{N}_k := \#\{i \mid f_i(\alpha_i) = b,\ f_i(\beta_i) = a\} - \#\{i \mid f_i(\alpha_i) = a,\ f_i(\beta_i) = b\} \tag{2.7}$$

(note that both terms on the RHS of eqn. (2.7) are finite by Lemma 2.1 (iii)). Thus $\mathcal{N}_k$ is precisely the number of eigenvalue branches of $H_t$ that cross the $k$th gap moving from the upper to the lower edge minus the number crossing from the lower to the upper edge. Put differently, $\mathcal{N}_k$ is the spectral multiplicity which *effectively* crosses the gap $\Gamma_k$ in downwards direction as $t$ increases from 0 to 1. We then have the following result.

**2.3. Theorem.** (cf. [K1, HK1]) *Let $V \in \mathcal{P}$ and let $k \in \mathbb{N}$ be such that the $k$th spectral gap of $H$ is open, i.e., $\gamma_k' < \gamma_{k+1}$. Then $\mathcal{N}_k = k$.*

In fact, the results obtained by Korotyaev in [K1, K2] are more detailed; e.g., Korotyaev shows that the dislocation operator produces at most two states (an eigenvalue and a resonance) in a gap of the periodic problem. On the other hand, our variational arguments are more flexible and allow an extension to higher dimensions, as we will see in the sequel. The main idea of our proof – somewhat reminiscent of [DH, ADH] – goes as follows: consider a sequence of approximations on intervals $(-n-t, n)$ with associated operators $H_{n,t} = -\frac{\mathrm{d}^2}{\mathrm{d}x^2} + W_t$ with periodic boundary conditions. We first observe that the gap $\Gamma_k$ is free of eigenvalues of $H_{n,0}$ and $H_{n,1}$ since both operators are obtained by restricting a periodic operator on the real line to some interval of length equal to an entire multiple of the period, with periodic boundary conditions. Second, the operators $H_{n,t}$ have purely discrete spectrum and it follows from Floquet theory (cf. [E, RS-IV]) that $H_{n,0}$ has precisely $2n$ eigenvalues in each band while $H_{n,1}$ has precisely $2n+1$ eigenvalues in each band. As a consequence, effectively $k$ eigenvalues of $H_{n,t}$ must cross any fixed $E \in \Gamma_k$ as $t$ increases from 0 to 1. To obtain the result of Theorem 2.3 we only have to take the limit $n \to \infty$; cf. [HK1] for the technical arguments. In [HK1], we also discuss a one-dimensional periodic step potential and perform some explicit (and also numerical) computations resulting in a plot of an eigenvalue branch for the associated dislocation problem.

We now turn to the dislocation problem on the infinite strip $\Sigma = \mathbb{R} \times [0,1]$. Let $V\colon \mathbb{R}^2 \to \mathbb{R}$ be $\mathbb{Z}^2$-periodic and Lipschitz continuous. We denote by $S_t$ the (self-adjoint) operator $-\Delta + W_t$, acting in $L_2(\Sigma)$, with periodic boundary conditions in the $y$-variable and with $W_t$ defined as in eqn. (1.1); again, the parameter $t$ ranges between 0 and 1. Since $S_0$ is periodic in the $x$-variable, its spectrum has a band-gap structure. To see that the essential spectrum of the family $S_t$ does not depend on the parameter $t$, i.e., $\sigma_{\mathrm{ess}}(S_t) = \sigma_{\mathrm{ess}}(S_0)$ for all $t \in [0,1]$, it suffices to prove compactness of the resolvent difference $(S_t - c)^{-1} - (S_{t,D} - c)^{-1}$, where $S_{t,D}$ is $S_t$ with an additional Dirichlet boundary condition at $x = 0$, say. (While, in one dimension, adding in a Dirichlet boundary condition at a single point causes a rank-one perturbation of the resolvent, the resolvent difference is now Hilbert-Schmidt, which can be seen from the following well-known line of argument: If $-\Delta_\Sigma$ denotes the (negative) Laplacian in $L_2(\Sigma)$ and $-\Delta_{\Sigma;D}$ is the (negative) Laplacian in $L_2(\Sigma)$ with an additional Dirichlet boundary condition at $x = 0$, then $(-\Delta_\Sigma + 1)^{-1} - (-\Delta_{\Sigma;D} + 1)^{-1}$ has an integral kernel which can be written down explicitly using the Green's function for $-\Delta_\Sigma$ and the reflection principle.)

While the band gap structure of the essential spectrum of $S_t$ is independent of $t \in [0,1]$, $S_t$ will have discrete eigenvalues in the spectral gaps of $S_0$ for appropriate values of $t$. We have the following result.

**2.4. Theorem.** *Assume that $V$ is Lipschitz-continuous. Let $(a,b)$ denote a non-trivial spectral gap of $S_0$ and let $E \in (a,b)$. Then there exists $t = t_E \in (0,1)$ such that $E$ is a discrete eigenvalue of $S_t$.*

As on the real line, we work with approximating problems on finite size sections of the infinite strip $\Sigma$. Let $\Sigma_{n,t} := (-n-t, n) \times (0,1)$ for $n \in \mathbb{N}$, and consider $S_{n,t} := -\Delta + W_t$ acting in $L_2(\Sigma_{n,t})$ with periodic boundary conditions in both coordinates. The operator $S_{n,t}$ has compact resolvent and purely discrete spectrum accumulating only at $+\infty$. The rectangles $\Sigma_{n,0}$ (respectively, $\Sigma_{n,1}$) consist of $2n$ (respectively, $2n+1$) period cells. By routine arguments (see, e.g., [RS-IV, E]), the number of eigenvalues below the gap $(a,b)$ is an integer multiple of the number of cells in these rectangles; we conclude that eigenvalues of $S_{n,t}$ must cross the gap as $t$ increases from 0 to 1. Thus for any $n \in \mathbb{N}$ we can find $t_n \in (0,1)$ such that $E \in \sigma_{\text{disc}}(S_{n,t_n})$; furthermore, there are eigenfunctions $u_n \in D(S_{n,t_n})$ satisfying $S_{n,t_n} u_n = E u_n$, $\|u_n\| = 1$, and $\|\nabla u_n\| \le C$ for some constant $C \ge 0$. Multiplying $u_n$ with a suitable cut-off function, we obtain (after extracting a suitable subsequence) functions $v_n \in D(S_t)$ and $t \in (0,1)$ satisfying

$$\|(S_t - E)v_n\| \to 0 \quad \text{and} \quad \|v_n\| \to 1, \tag{2.8}$$

as $n \to \infty$, which implies $E \in \sigma(S_t)$, cf. [HK1].

Finally, we consider the dislocation problem on the plane $\mathbb{R}^2$ where we study the operators

$$D_t = -\Delta + W_t, \quad 0 \le t \le 1. \tag{2.9}$$

Denote by $S_t(\vartheta)$ the operator $S_t$ on the strip $\Sigma$ with $\vartheta$-periodic boundary conditions in the $y$-variable. Since $W_t$ is periodic with respect to $y$, we have

$$D_t \sim \int_{[0,2\pi]}^{\oplus} S_t(\vartheta) \frac{d\vartheta}{2\pi}; \tag{2.10}$$

in particular, $D_t$ has no singular continuous part, cf. [DS]. As for the spectrum of $S_t$ inside the gaps of $S_0$, Theorem 2.4 yields the following result.

**2.5. Theorem.** *Assume that $V$ is Lipschitz-continuous. Let $(a,b)$ denote a non-trivial spectral gap of $D_0$ and let $E \in (a,b)$. Then there exists $t = t_E \in (0,1)$ with $E \in \sigma(D_t)$.*

*Proof.* Let $v_n \in D(S_t)$ denote an approximate solution of the eigenvalue problem for $S_t$ and $E$; see (2.8). We extend $v_n$ to a function $\tilde{v}_n(x,y)$ on $\mathbb{R}^2$ which is periodic in $y$. By multiplying $\tilde{v}_n$ by smooth cut-off functions $\Phi_n(x,y)$, we obtain functions

$$w_n = w_n(x,y) := \frac{1}{\|\Phi_n \tilde{v}_n\|} \Phi_n \tilde{v}_n \tag{2.11}$$

belonging to the domain of $D_t$ and satisfying $\|w_n\| = 1$, supp $w_n \subset [-n, n]^2$, and

$$(D_t - E)w_n \to 0, \qquad n \to \infty; \tag{2.12}$$

this implies the desired result. □

The stronger statement in Theorem 1.1. follows by a very similar line of argument. The upshot is that the dislocation moves enough states through the gap to have a non-trivial (integrated) surface density of states, for suitable parameters $t$.

The lower estimate established in Theorem 1.1. is complemented by an upper bound which is of the expected order (up to a logarithmic factor) in [HK2]. Note that the situation treated in [HK2] is far more general than the rotation or dislocation problems studied so far. In fact, here we allow for different potentials $V_1$ on the left and $V_2$ on the right which are only linked by the assumption that there is a common spectral gap; neither $V_1$ nor $V_2$ are required to be periodic. The proof uses technology which is fairly standard and is based on exponential decay estimates for resolvents, cf. [S].

**2.6. Theorem.** *Let $V_1$, $V_2 \in L_\infty(\mathbb{R}^2, \mathbb{R})$ and suppose that the interval $(a, b) \subset \mathbb{R}$ does not intersect the spectra of the self-adjoint operators $H_k := -\Delta + V_k$, $k = 1, 2$, both acting in the Hilbert space $L_2(\mathbb{R}^2)$. Let*

$$W := \chi_{\{x<0\}} \cdot V_1 + \chi_{\{x \geq 0\}} \cdot V_2 \tag{2.13}$$

*and define $H := -\Delta + W$, a self-adjoint operator in $L_2(\mathbb{R}^2)$. Finally, we let $H^{(n)}$ denote the self-adjoint operator $-\Delta + W$ acting in $L_2(Q_n)$ with Dirichlet boundary conditions. Then, for any interval $[a', b'] \subset (a, b)$, we have*

$$\limsup_{n\to\infty} (1/(n \log n)) N_{[a',b']}(H^{(n)}) < \infty, \tag{2.14}$$

*where $N_{[a',b']}(H^{(n)})$ denotes the number of eigenvalues of $H^{(n)}$ in $[a', b']$.*

We note that the factor $\log n$ in eqn. (2.14) can presumably be dropped under appropriate assumptions (H. Cornean, private communication); however, this seems to require substantially different, and less elementary, methods.

## 3. Rotational defect in a two-dimensional lattice

In this section, we will use our results on the translational problem to obtain spectral information about rotational problems in the limit of small angles. Our main theorem deals with the following situation. Let $V \colon \mathbb{R}^2 \to \mathbb{R}$ be a Lipschitz-continuous function which is periodic w.r.t. the lattice $\mathbb{Z}^2$. For $\vartheta \in (0, \pi/2)$, let

$$M_\vartheta := \begin{pmatrix} \cos\vartheta & -\sin\vartheta \\ \sin\vartheta & \cos\vartheta \end{pmatrix} \in \mathbb{R}^{2\times 2}, \tag{3.1}$$

and $V_\vartheta$ as in (1.3). We then let $H_0$ denote the (unique) self-adjoint extension of $-\Delta \restriction C_c^\infty(\mathbb{R}^2)$, acting in the Hilbert space $L_2(\mathbb{R}^2)$, and

$$R_\vartheta := H_0 + V_\vartheta, \qquad D(R_\vartheta) = D(H_0). \tag{3.2}$$

Then $R_\vartheta$ is essentially self-adjoint on $C_c^\infty(\mathbb{R}^2)$ and semi-bounded from below.

Now our key observation consists in the following: for any $t \in (0,1)$ given, any $\varepsilon > 0$, and any $n \in \mathbb{N}$, we can find points $(0,\eta)$ on the $y$-axis with $\eta \in \mathbb{N}$ such that

$$|V_\vartheta(x,y) - W_t(x,y)| < \varepsilon, \qquad (x,y) \in Q_n(0,\eta), \tag{3.3}$$

where $Q_n(0,\eta) = (-n,n) \times (\eta - n, \eta + n)$, provided $\vartheta > 0$ is small enough and satisfies a condition which ensures an appropriate alignment of the period cells on the $y$-axis. Put differently: for very small angles, the rotated potential $V_\vartheta$ will almost look like a dislocation potential $W_t$, on suitable squares $Q_n(0,\eta)$.

To prove Theorem 1.2 we proceed as follows: Fix an arbitrary $E$ in a gap of $H = H_0 + V = R_0$. Knowing that there exists $t \in (0,1)$ such that $E \in \sigma_{\text{disc}}(D_t)$, we choose an associated approximate eigenfunction as constructed in the proof of Theorem 2.5 and shift it along the $y$ axis until its support is contained in $Q_n(0,\eta)$. In this way we obtain an approximate eigenfunction for $R_\vartheta$ and $E$. It is easy to see that, in view of the Lipschitz continuity and the $\mathbb{Z}^2$-periodicity of $V$, the geometric conditions for the estimate (3.3) are

$$|k \tan\vartheta - [k\tan\vartheta] - t| < \varepsilon, \quad |k/\cos\vartheta - \eta| < \varepsilon \tag{3.4}$$

for some $k \in \mathbb{N}$. It thus remains to prove the existence of natural numbers $k$ satisfying the conditions in (3.4), for given $\vartheta \in (0,\pi/2)$. Actually, the existence of numbers $k$ as desired can only be established for a dense set of angles.

**3.1. Lemma.** *Let $\mathbb{T}^2 = \mathbb{R}^2/\mathbb{Z}^2$ be the flat two-dimensional torus and let $T_\vartheta\colon \mathbb{T}^2 \to \mathbb{T}^2$ be defined by*

$$T_\vartheta(x,y) := (x + \tan\vartheta, y + 1/\cos\vartheta). \tag{3.5}$$

*Then there is a set $\Theta \subset (0,\pi/2)$ with countable complement such that the transformation $T_\vartheta$ in (3.5) is ergodic for all $\vartheta \in \Theta$.*

The assertion of Theorem 1.2 now follows from Birkhoff's ergodic theorem, cf. [CFS, HK2]: Let us first assume that $\vartheta \in \Theta$. Let $\varepsilon > 0$ and let us denote by $\chi = \chi_Q$ the characteristic function of the set $Q := (t-\varepsilon, t+\varepsilon) \times (-\varepsilon,\varepsilon) \subset \mathbb{T}^2$. Then

$$\lim_{n\to\infty} \frac{1}{n} \sum_{m=0}^{n-1} \chi(T_\vartheta^m(0,0)) = \int_Q \mathrm{d}x\,\mathrm{d}y = 4\varepsilon^2 > 0. \tag{3.6}$$

By a simple approximation argument the statement of Theorem 1.2 also holds for angles $\vartheta \notin \Theta$. Altogether, this completes the proof of Theorem 1.2.

Recall that strong resolvent convergence implies upper semi-continuity of the spectrum while the spectrum may contract considerably when the limit is reached. In the present section, we are dealing with a situation where the spectrum in fact behaves discontinuously at $\vartheta = 0$ since, counter to first intuition, the spectrum of $R_\vartheta$ "fills" the gap $(a,b)$ as $0 \neq \vartheta \to 0$. This implies, in particular, that $R_\vartheta$ cannot converge to $H$ in the norm resolvent sense, as $\vartheta \to 0$.

## 4. Muffin tin potentials

In this section, we present a class of examples where one can arrive at rather precise statements that illustrate some of the phenomena described before. Our potentials $V = V_{r,P_0}$ are of muffin tin type, as defined in the introduction.

**(1) The dislocation problem.** In the simplest case we would take $x_0 = 1/2$ and $y_0 = 0$ so that the disks $B_r(1/2 + i, j)$, for $i, j \in \mathbb{Z}$, will not intersect or touch the interface $\{(x, y) \mid x = 0\}$, for $0 < r < 1/2$. Defining the dislocation potential $W_t$ as in (1.1), we see that there are bulk states given by the Dirichlet eigenvalues of all the discs that do not meet the interface, and there may be surface states given as the Dirichlet eigenvalues of the sets $B_r(1/2 - t, j) \cap \{x < 0\}$ for $j \in \mathbb{Z}$ and $1/2 - r < t < 1/2 + r$.

More precisely, let $\mu_k = \mu_k(r)$ denote the Dirichlet eigenvalues of the Laplacian on the disc of radius $r$, ordered by min-max and repeated according to their respective multiplicities. The Dirichlet eigenvalues of the domains $B_r(1/2 - t, 0) \cap \{x < 0\}$, for $1/2 - r < t < 1/2 + r$, are denoted as $\lambda_k(t) = \lambda_k(t, r)$; they are continuous, monotonically decreasing functions of $t$ and converge to $\mu_k$ as $t \uparrow 1/2 + r$ and to $+\infty$ as $t \downarrow 1/2 - r$. In this simple model, the eigenvalues $\mu_k$ correspond to the bands of a periodic operator. We see that the gaps are crossed by surface states as $t$ increases from 0 to 1, in agreement with Theorem 1.1. In [HK1] we also discuss muffin tin potentials with dislocation in the $y$ direction.

**(2) The rotation problem.** In [HK2], we look at three types of muffin tin potentials and discuss the effect of the "filling up" of the gaps at small angles of rotation. We begin with muffin tins with walls of infinite height, then approximate by muffin tin potentials of height $n$, for $n \in \mathbb{N}$ large. By another approximation step, one may obtain examples with Lipschitz-continuous potentials. These examples show, among other things, that Schrödinger operators of the form $R_\vartheta$ may in fact have spectral gaps for some $\vartheta > 0$. For the sake of brevity, we only state our main results and refer to [HK2] for further details.

We write (in the notation of (1.5)) $\Omega_r = \Omega_{r,(1/2,1/2)}$ and

$$\Omega_{r,\vartheta} := \Omega_r \cap \{x \geq 0\} \cup (M_\vartheta \Omega_r) \cap \{x < 0\}, \tag{4.1}$$

and let $H_{r,\vartheta}$ denote the Dirichlet Laplacian on $\Omega_{r,\vartheta}$ for $0 < r < 1/2$ and $0 \leq \vartheta \leq \pi/4$. Denote the Dirichlet eigenvalues of the Laplacian $H_r$ of $\Omega_r$ by $(\tilde{\mu}_j(r))_{j \in \mathbb{N}}$, with $\tilde{\mu}_j(r) \to \infty$ as $j \to \infty$ and $\tilde{\mu}_j(r) < \tilde{\mu}_{j+1}(r)$ for all $j \in \mathbb{N}$; note that the eigenvalues $\tilde{\mu}_j$ may have multiplicity $> 1$.

**4.1. Proposition.** *Let $(a, b)$ be one of the gaps $(\tilde{\mu}_j, \tilde{\mu}_{j+1})$ and let $0 < r < 1/2$ be fixed.*

(a) *Each $\tilde{\mu}_j(r)$, $j = 1, 2, \ldots$, is an eigenvalue of infinite multiplicity of $H_{r,\vartheta}$, for all $0 \leq \vartheta \leq \pi/2$. The spectrum of $H_{r,\vartheta}$ is pure point, for all $0 \leq \vartheta \leq \pi/2$.*

(b) *For any $\varepsilon > 0$ there is a $\vartheta_\varepsilon = \vartheta_\varepsilon(r) > 0$ such that any interval $(\alpha, \beta) \subset (a, b)$ with $\beta - \alpha \geq \varepsilon$ contains an eigenvalue of $H_{r,\vartheta}$ for any $0 < \vartheta < \vartheta_\varepsilon$.*

(c) *There exists a set* $\Theta \subset (0, \pi/2)$ *of full measure such that* $\sigma(H_{r,\vartheta}) = [\tilde{\mu}_1(r), \infty)$. *The eigenvalues different from the* $\tilde{\mu}_j(r)$ *are of finite multiplicity for* $\vartheta \in \Theta$.

**4.2. Remark.** If $\tan\vartheta$ is rational, the grid $M_\vartheta \mathbb{Z}^2$ is periodic in the $x$- and $y$-directions with $\vartheta$-dependent periods $p, q \in \mathbb{N}$. As a consequence, $H_{r,\vartheta}$ has at most a finite number of eigenvalues in $(a, b)$ for $\tan\vartheta$ rational, each of them of infinite multiplicity. Hence we see a drastic change in the spectrum for $\tan\vartheta \in \mathbb{Q}$ as compared with $\vartheta \in \Theta$. Furthermore, if $\tan\vartheta$ is rational with $\tan\vartheta \notin \{1/(2k+1) \mid k \in \mathbb{N}\}$, then there is some $r_\vartheta > 0$ such that $\sigma(H_{r,\vartheta}) = \sigma(H_r)$ for all $0 < r < r_\vartheta$.

We next turn to muffin tin potentials of finite height. Here we define the potential $V_{r,\vartheta}$ to be zero on $\Omega_{r,\vartheta}$ and $V_{r,\vartheta} = 1$ on the complement of $\Omega_{r,\vartheta}$, where $0 < r < 1/2$ and $0 \le \vartheta \le \pi/4$; we also let $H_{r,n,\vartheta} := H_0 + nV_{r,\vartheta}$. The periodic operators $H_{r,n,0}$ have purely absolutely continuous spectrum and $H_{r,n,\vartheta} \to H_{r,\vartheta}$ in the sense of norm resolvent convergence, uniformly for $\vartheta \in [0, \pi/4]$.

**4.3. Proposition.** *Let* $(a, b)$ *be one of the gaps* $(\tilde{\mu}_j, \tilde{\mu}_{j+1})$. *We then have:*

(a) *For* $\tan\vartheta \in \mathbb{Q}$ *the spectrum of* $H_{r,n,\vartheta}$ *has gaps inside the interval* $(a, b)$ *for* $n$ *large. More precisely, if* $H_{r,\vartheta}$ *has a gap* $(a', b') \subset (a, b)$, *then, for* $\varepsilon > 0$ *given, the interval* $(a' + \varepsilon, b' - \varepsilon)$ *will be free of spectrum of* $H_{r,n,\vartheta}$ *for* $n$ *large.*

(b) *For any* $\varepsilon > 0$ *there are* $\vartheta_0 > 0$ *and* $n_0 > 0$ *such that any interval* $(c - \varepsilon, c + \varepsilon) \subset (a, b)$ *contains spectrum of* $H_{r,n,\vartheta}$ *for all* $0 < \vartheta < \vartheta_0$ *and* $n \ge n_0$.

By similar arguments, we can approximate $V_{r,\vartheta}$ by Lipschitz-continuous muffin tin potentials that converge monotonically (from below) to $V_{r,\vartheta}$ in such a way that norm resolvent convergence holds for the associated Schrödinger operators (again uniformly in $\vartheta \in [0, \pi/4]$). The spectral properties obtained are analogous to the ones stated in Proposition 4.3. Note, however, that the statement corresponding to part (*b*) in Proposition 4.3 is weaker than the result of our main Theorem 1.1.

**Acknowledgement**

The authors thank E. Korotyaev (St. Petersburg) and J. Voigt (Dresden) for useful discussions.

## References

[ADH] S. Alama, P.A. Deift, and R. Hempel, *Eigenvalue branches of the Schrödinger operator* $H - \lambda W$ *in a gap of* $\sigma(H)$, Commun. Math. Phys. **121** (1989), 291–321.

[CFS] I.P. Cornfield, S.V. Fomin, and Y.G. Sinai, *Ergodic theory*, Springer, New York, 1982.

[CFrKS] H.L. Cycon, R.G. Froese, W. Kirsch, and B. Simon, *Schrödinger Operators with Applications to Quantum Mechanics and Global Geometry*, Springer, New York, 1987.

[DH] P.A. Deift and R. Hempel, *On the existence of eigenvalues of the Schrödinger operator $H - \lambda W$ in a gap of $\sigma(H)$*, Commun. Math. Phys. **103** (1986), 461–490.

[DS] E.B. Davies and B. Simon, *Scattering theory for systems with different spatial asymptotics on the left and right*, Commun. Math. Phys. **63** (1978), 277–301.

[E] M.S.P. Eastham, *The Spectral Theory of Periodic Differential Equations*, Scottish Academic Press, Edinburgh, London, 1973.

[EKSchrS] H. Englisch, W. Kirsch, M. Schröder, and B. Simon, *Random Hamiltonians ergodic in all but one direction*, Commun. Math. Phys. **128** (1990), 613–625.

[HK1] R. Hempel and M. Kohlmann, *A variational approach to dislocation problems for periodic Schrödinger operators*, J. Math. Anal. Appl., to appear.

[HK2] ______, *Spectral properties of grain boundaries at small angles of rotation*, J. Spect. Th., to appear.

[K1] E. Korotyaev, *Lattice dislocations in a 1-dimensional model*, Commun. Math. Phys. **213** (2000), 471–489.

[K2] ______, *Schrödinger operators with a junction of two* 1*-dimensional periodic potentials*, Asymptotic Anal. **45** (2005), 73–97.

[KS] V. Kostrykin and R. Schrader, *Regularity of the surface density of states*, J. Funct. Anal. **187** (2001), 227–246.

[PF] L. Pastur and A. Figotin, *Spectra of Random and almost-periodic Operators*, Springer, New York, 1991.

[RS-IV] M. Reed and B. Simon, *Methods of Modern Mathematical Physics, Vol. IV*, Analysis of Operators, Academic Press, New York, 1978.

[S] B. Simon, *Schrödinger semigroups*, Bull. Amer. Math. Soc. **7** (1982), 447–526.

[V] I. Veselic, *Existence and regularity properties of the integrated density of states of random Schrödinger operators*, Springer Lecture Notes in Mathematics, vol. 1917, Springer, New York, 2008.

Rainer Hempel
Institute for Computational Mathematics
Technische Universität Braunschweig
Pockelsstraße 14
D-38106 Braunschweig, Germany
e-mail: r.hempel@tu-bs.de

Martin Kohlmann
Institute for Applied Mathematics
Leibniz Universität Hannover
Welfengarten 1
D-30167 Hannover, Germany
e-mail: kohlmann@ifam.uni-hannover.de

Operator Theory:
Advances and Applications, Vol. 221, 433-441


# The Riemann Boundary Value Problem on Non-rectifiable Arcs and the Cauchy Transform

Boris A. Kats

**Abstract.** In this paper we introduce an alternative way of defining the curvilinear Cauchy integral over non-rectifiable arcs on the complex plane. We construct this integral as the convolution of the distribution $(2\pi iz)^{-1}$ with a certain distribution such that its support is a non-rectifiable arc. These convolutions are called Cauchy transforms. As an application, solvability conditions of the Riemann boundary value problem are derived under very weak conditions on the boundary.

**Mathematics Subject Classification (2000).** Primary 30E25; secondary 30E20.

**Keywords.** Non-rectifiable arc, metric dimension, Cauchy transform, Riemann boundary value problem.

## Introduction

Let $\Gamma$ be a non-rectifiable Jordan arc on the complex plane. In the present paper we construct analogs of the curvilinear integral $\int_\Gamma \phi(z)dz$ and the Cauchy integral $(2\pi i)^{-1}\int_\Gamma \phi(t)(t-z)^{-1}dt$ for non-rectifiable arcs and apply these constructions to solve the Riemann boundary value problem on such arcs.

We put $\Gamma^\circ := \Gamma \setminus \{a_1, a_2\}$, where $a_1$ and $a_2$ are the endpoints of the arc $\Gamma$. Below we consider non-rectifiable arcs $\Gamma$ under the following additional restriction:

- *we can find a finite domain $\Delta$ with piecewise-smooth boundary such that $\Gamma^\circ \subset \Delta$ and $\Delta \setminus \Gamma$ consists of two connected components $\Delta^+$ and $\Delta^-$.*

If $\Gamma$ satisfies this restriction then we call it a $G$-arc. We assume that the arc $\Gamma$ is directed from the point $a_1$ to the point $a_2$, and that the domains $\Delta^+$ and $\Delta^-$ are located on the left and on the right from $\Gamma$ correspondingly.

This work was partially supported by Russian Fund for Basic Researches.

Let a function $F(z)$ be locally integrable in the domain $\Delta$. We identify it with the distribution

$$F : C_0^\infty(\Delta) \ni \phi \mapsto \iint_\Delta F(\zeta)\phi(\zeta)d\zeta d\overline{\zeta}.$$

As usual, $C_0^\infty(\Delta)$ is the test space consisting of all infinitely smooth functions with compact supports in $\Delta$.

We assume that a function $F$ is holomorphic in $\Delta \setminus \Gamma$ and has boundary values $F^+(t)$ and $F^-(t)$ from the left and from the right at each point $t \in \Gamma^\circ$. If the arc $\Gamma$ is rectifiable then by virtue of the Green formula we have

$$\langle \overline{\partial} F, \phi\rangle := -\iint_\Delta F(\zeta)\frac{\partial \phi}{\partial \overline{\zeta}}d\zeta d\overline{\zeta} = -\left(\iint_{\Delta^+} + \iint_{\Delta^-}\right)\frac{\partial F\phi}{\partial \overline{\zeta}}d\zeta d\overline{\zeta}$$
$$= \int_\Gamma (F^+(\zeta) - F^-(\zeta))\phi(\zeta)d\zeta.$$

Thus, the distribution

$$\langle \overline{\partial} F, \phi\rangle = \int_\Gamma j_F(\zeta)\phi(\zeta)d\zeta$$

is weighted integration over $\Gamma$, and its weight

$$j_F(\zeta) = F^+(\zeta) - F^-(\zeta), \zeta \in \Gamma,$$

is the jump of function $F$ on $\Gamma$.

If $\Gamma$ is not rectifiable, then we consider the distribution $\overline{\partial} F$ as generalized integration. In the present paper we study certain properties of the generalized integrations (Section 1) and its Cauchy transforms (Section 2). In the final Section 3 we apply these results to solve the Riemann boundary value problem on non-rectifiable arcs.

## 1. Integrations and dimensions

Let us put $D := \overline{\mathbb{C}}\setminus\Gamma$. We assume that a function $F(z)$ is holomorphic in $D$, it has boundary values $F^+(t)$ and $F^-(t)$ from the left and from the right at each point $t \in \Gamma^\circ$, $F(\infty) = 0$, and $F$ is integrable in a neighborhood of each of the endpoints $a_1$ and $a_2$. Then the distributional derivative $\overline{\partial} F$ is defined on $C_0^\infty(\mathbb{C}) \equiv C_0^\infty$. We call $\overline{\partial} F$ *primary integration* and use notation $\int[F]$ instead of $\overline{\partial} F$.

Let $\mathfrak{X}$ be a functional space such that $C^\infty(\mathbf{C})$ is dense in $\mathfrak{X}$ and $\mathfrak{X}C^\infty = \mathfrak{X}$, i.e., $f\phi \in \mathfrak{X}$ for any $\phi \in C^\infty, f \in \mathfrak{X}$. If a primary integration $\int[F]$ is continuous in $\mathfrak{X}$, then it is continuable up to functional $\int[F]$ on $\mathfrak{X}$ and generates a family of distributions

$$\left\langle \int [F]f, \phi \right\rangle := \int [F]f(\zeta)\phi(\zeta)d\zeta. \tag{1}$$

We call them *integrations* and write $\int[F]f\phi d\zeta$ instead of $\langle \int[F]f, \phi\rangle$.

In the present paper we define the space $\mathfrak{X}$ in terms of the Hölder condition

$$h_\nu(f,A) := \sup\left\{\frac{|f(t')-f(t'')|}{|t'-t''|^\nu} : t',t'' \in A, t' \neq t''\right\} < \infty, \tag{2}$$

where $A$ is a compact set on the complex plane, $\nu \in (0,1]$. Let $H_\nu(A)$ stand for the set of all functions $f$ defined on $A$ which satisfy condition (2). It is Banach space with norm $\|f\|_{C(A)} + h_\nu(f,A)$, where $\|f\|_{C(A)} = \sup\{|f(\zeta)| : \zeta \in A\}$. But $C^\infty$ is not dense in this space. We denote $H^*(A,\nu) := \bigcup_{\mu>\nu} H_\mu(A)$ and fix a sequence of exponents $\{\nu_j\}$ such that $1 > \nu_1 > \nu_2 > \cdots > \nu_j > \nu_{j+1} > \cdots$ and $\lim_{j\to\infty} \nu_j = \nu$. The values $\{h_{\nu_j}(\cdot, A)\}$, $j = 1,2,\dots$ and $\|f\|_{C(A)}$ are semi-norms, and the space $H^*(A,\nu)$ is countably normed. Obviously, $C^\infty$ is dense there and $H^*(A,\nu)C^\infty = H^*(A,\nu)$. Thus, we can use $H^*(A,\nu)$ as the space $\mathfrak{X}$.

In accordance with the Whitney theorem (see, for instance, [1]) any function $f \in H_\nu(A)$ is extendable to a function $f^w$ which is defined in the whole complex plane and satisfies there the Hölder condition with the same exponent $\nu$.

We will describe the continuity of primary integrations in $H^*(\Gamma,\nu)$ in terms of chain dimension of the $G$-arc $\Gamma$. The analogous characteristics for closed curves is introduced in [2].

A rectifiable chain $\mathfrak{C}$ is a sequence of pairs $(\Delta_j, s_j)$, where $\Delta_j$ is domain with rectifiable boundary, $s_j$ is either 1 or $-1$, $j = 0,1,2,\dots$, and $s_0 = 1$. Given a chain $\mathfrak{C}$, we associate the sequence of domains $B_j$, $j = 0,1,2,\dots$, such that $B_0 = \Delta_0$, $B_j = B_{j-1} \cup \Delta_j$ for $s_j = 1$, and $B_j = B_{j-1} \setminus \Delta_j$ for $s_j = -1$, $j = 1,2,\dots$. We say that the chain $\mathfrak{C}$ converges to the $G$-arc $\Gamma$ if it satisfies the following conditions:

i. if $s_j = -1$, then $\Delta_j \subset B_{j-1}$, $j = 1,2,\dots$;
ii. if $s_j = 1$, then $\overline{\Delta_j} \cap \overline{B_{j-1}} \subset \partial\Delta_j$, $j = 1,2,\dots$;
iii. $\lim_{j\to\infty} \operatorname{dist}(\Gamma, \Delta_j) = 0$;
iv. the set $\bigcup_{n\geq 0}\bigcap_{j\geq n} B_j$ is one of the components $\Delta^+$, $\Delta^-$.

Let $d$ be a value from the segment $[1,2]$. We define the $d$-mass of the chain $\mathfrak{C}$ by the equality

$$M_d(\mathfrak{C}) := \sum_{j>0} \Lambda(\Delta_j) w^{d-1}(\Delta_j),$$

where $\Lambda(\Delta_j)$ is the length of boundary of domain $\Delta_j$, and $w(\Delta_j)$ is the width of this domain, i.e., it is equal to the diameter of the largest disk lying in $\Delta_j$.

If $M_d(\mathfrak{C}) < \infty$ for some rectifiable chain $\mathfrak{C}$ which is convergent to $\Gamma$, then we associate the value $d$ to the set $\mathfrak{F}(\Gamma)$.

**Definition 1.1.** The chain dimension of the $G$-arc $\Gamma$ is the greatest lower bound of the set $\mathfrak{F}(\Gamma)$:

$$\operatorname{dmc}\Gamma := \inf \mathfrak{F}(\Gamma).$$

Let us compare this notion of dimension with the following well-known version of fractal dimension

$$\operatorname{dmb}\Gamma = \limsup_{\varepsilon\to 0} \frac{\log N(\varepsilon,\Gamma)}{-\log\varepsilon},$$

where $N(\varepsilon,\Gamma)$ is the least number of disks of diameter $\varepsilon$ covering $\Gamma$. It is called upper metric dimension [3], box dimension, or Minkowski dimension [4]. In the same way as in the paper [2], we see that $1 \le \operatorname{dmc}\Gamma \le \operatorname{dmb}\Gamma \le 2$ for any $G$-arc $\Gamma$. Moreover, for any value $d \in (1,2]$ we can construct a $G$-arc $\Gamma$ such that $d = \operatorname{dmb}\Gamma > \operatorname{dmc}\Gamma$.

The main result of this section is the following.

**Theorem 1.2.** *Assume that $\Gamma$ is a $G$-arc, $F(z)$ is a holomorphic in $\overline{\mathbb{C}}\setminus\Gamma$ function such that its boundary values $F^+(t)$ and $F^-(t)$ from the left and from the right exist at any point $t \in \Gamma^\circ$, $F$ is integrable with any degree $p > 1$ in a neighborhood of endpoints $a_1$ and $a_2$, and $F(\infty) = 0$.*

*If $\operatorname{dmc}\Gamma < 2$, then the primary integration $\int[F]$ is continuous in the space $H^*(\overline{A}, \operatorname{dmc}\Gamma - 1)$.*

The proof is analogous to the proof of Theorem 1 in [5].

Thus, the primary integration $\int[F]$ extends to a functional on the space $H^*(\Gamma,\operatorname{dmc}\Gamma-1)$ and determines the family of integrations $\int[F]f$, $f\in H^*(\Gamma,\operatorname{dmc}\Gamma-1)$, by formula (1). Let us describe the explicit construction of these integrations. If $f \in H^*(\Gamma, \operatorname{dmc}\Gamma - 1)$, then $f \in H_\nu(\Gamma)$ for any $\nu > \operatorname{dmc}\Gamma - 1$. We fix real values $\nu$ and $d$ such that $1 > \nu > d-1 > \operatorname{dmc}\Gamma - 1$. By definition of the chain dimension we find a rectifiable chain $\mathfrak{C} = \{(\Delta_0, s_0), (\Delta_1, s_1), (\Delta_2, s_2), \dots\}$ with finite $d$-mass convergent to $\Gamma$. We extend $f$ from $\Gamma$ onto $\mathbb{C}$ by means of the Whitney extension operator (see [1]), restrict the extension on the compact set $\partial\mathfrak{C} := \overline{\bigcup_{j\ge 0}\partial\Delta_j}$ and again extend this restriction onto $\mathbb{C}$ by the Whitney extension operator. Let $f^*$ be the result of that double Whitney extension. Then the following Lemma holds.

**Lemma 1.3.** *The first partial derivatives of $f^*$ belong to $L^p_{\mathrm{loc}}(\mathbb{C})$ for any*

$$p < \frac{2 - \operatorname{dmc}\Gamma}{1-\nu}.$$

The proof of this lemma is analogous to the proof of the parallel result in the paper [2]. The right side of the last inequality exceeds 1 under the assumptions of Theorem 1.2. Therefore, if $\omega$ is smooth function with compact support equaling 1 in a neighborhood of $\Gamma$, then the first partial derivatives of the product $\omega f^*$ belong to $L^p(\mathbb{C})$ for certain $p > 1$. This fact allows one to write the integrations as follows:

$$\int [F]f(\zeta)\phi(\zeta)d\zeta = -\iint_{\mathbb{C}} F(\zeta)\frac{\partial\omega\phi f^*}{\partial\overline{\zeta}}d\zeta d\overline{\zeta}. \tag{3}$$

## 2. The Cauchy transform of integrations

A number of recent publications (see, for instance, [6, 7, 8]) deal with various properties of the Cauchy transforms of measures. The Cauchy transform of a finite measure $\mu$ on the complex plane is the integral

$$\mathsf{C}\mu := \frac{1}{2\pi i}\int \frac{d\mu(\zeta)}{\zeta - z}.$$

If the support $S$ of the measure $\mu$ is rectifiable curve and $d\mu = f(\zeta)d\zeta$, then we have the Cauchy type integral

$$\mathsf{C}(f(\zeta)d\zeta) = \frac{1}{2\pi i}\int_{\Gamma} \frac{f(\zeta)d\zeta}{\zeta - z}. \tag{4}$$

The Cauchy transform of a distribution $\varphi$ with compact support $S$ on the complex plane is defined by the equality

$$\mathsf{C}\varphi := \frac{1}{2\pi i}\left\langle \varphi, \frac{1}{\zeta - z}\right\rangle,$$

where $z \notin S$, and $\varphi$ is applied to the Cauchy kernel $E(\zeta - z) := \frac{1}{2\pi i(\zeta - z)}$ viewed as a function of the variable $\zeta$. In other words, it is the convolution $\varphi * E$ where $E$ is the distribution $\frac{1}{2\pi i \zeta}$. As $E$ is the fundamental solution of differential operator $\overline{\partial}$ (i.e., $\overline{\partial}E = \delta_0$, see [1]), since $\overline{\partial}\mathsf{C}\varphi = \varphi$, and the function $\mathsf{C}\varphi(z)$ is holomorphic in $\overline{\mathbf{C}} \setminus S$. Obviously, it vanishes at the point $\infty$.

Let us consider the Cauchy transforms of the integrations $\int[F]f$. We put

$$\Phi(z) := \mathsf{C}\int[F]f = \frac{1}{2\pi i}\left\langle \int[F]f, \frac{1}{\zeta - z}\right\rangle, z \in \mathbb{C}\setminus\Gamma. \tag{5}$$

More precisely, we must replace here the function $\frac{1}{\zeta - z}$ by $\frac{\omega_z(\zeta)}{\zeta - z}$, where the function $\omega_z(\zeta) \in C_0^\infty(\mathbb{C})$ is equal to 1 in a neighborhood of $\Gamma$ and vanishes in a neighborhood of the point $z$. By virtue of the equality (3) we obtain the representation

$$\Phi(z) = F(z)f^*(z)\omega(z) - \frac{1}{2\pi i}\iint_{\mathbb{C}} F(\zeta)\frac{\partial f^*\omega}{\partial\overline{\zeta}}\frac{d\zeta d\overline{\zeta}}{\zeta - z}. \tag{6}$$

The function $\omega \in C_0^\infty(\mathbb{C})$ is equal to 1 in a neighborhood of $\Gamma$.

The properties of the integral term of (6) are well known (see, for instance, [1]). If $F(\zeta)\frac{\partial f^*\omega}{\partial\overline{\zeta}} \in L^p(\mathbb{C})$, $p > 2$, then it is continuous in the whole complex plane and satisfies the Hölder condition with exponent $1 - \frac{2}{p}$. By virtue of Lemma 1.3 the exponent $p$ exceeds 2 under restriction $\nu > \frac{1}{2}\operatorname{dmc}\Gamma$. Then the integral term of (6) satisfies the Hölder condition with exponent $\mu - \varepsilon$ in the whole complex plane, where

$$\mu = \frac{2\nu - \operatorname{dmc}\Gamma}{2 - \operatorname{dmc}\Gamma} \tag{7}$$

and $\varepsilon$ is an arbitrarily small positive number.

The first term of the right side of (6) has jump of $j_F(t)f(t)$ on $\Gamma^\circ$. The factor $f^*(z)\omega(z)$ satisfies the Hölder condition with exponent $\nu > \mu$ in sufficiently small closed half-neighborhoods of points $t \in \Gamma^\circ$. The smoothness of the whole product $F(z)f^*(z)\omega(z)$ and its growth at the points $a_{1,2}$ depend on $F(z)$. We restrict $F$ near $\Gamma$ as follows.

**Definition 2.1.** We say that a function $F(z)$ holomorphic in $\mathbb{C} \setminus \Gamma$ is in the class $\mathsf{H}^\circ_\alpha(\Gamma)$ if it satisfies the Hölder condition with exponent $\alpha$ in a sufficiently small closed half-neighborhood of any point $t \in \Gamma^\circ$.

As a result, we obtain

**Theorem 2.2.** *Assume that $\Gamma$ is a G-arc, that $F(z)$ is a function holomorphic in $\overline{\mathbb{C}} \setminus \Gamma$ such that its boundary values $F^+(t)$ and $F^-(t)$ exist both from the left and from the right at any point $t \in \Gamma^\circ$, that $F$ is integrable with any degree $p > 1$ in certain neighborhoods of endpoints $a_1$ and $a_2$, and that $F(\infty) = 0$.*

*If $f \in H^*(\Gamma, \frac{1}{2}\,\mathrm{dmc}\,\Gamma)$, then the Cauchy transform $\Phi(z)$ of integration $\int[F]f$ has continuous boundary values $\Phi^+(t)$ and $\Phi^-(t)$ from the left and from the right at any point $t \in \Gamma^\circ$,*

$$\Phi^+(t) - \Phi^-(t) = j_F(t)f(t), t \in \Gamma^\circ, \tag{8}$$

*and near the endpoints of $\Gamma$ the function $\Phi$ satisfies estimates*

$$\Phi(z) = f(a_j)F(z) + o(1), \quad z \to a_j, \quad j = 1, 2. \tag{9}$$

*In addition, for $F \in \mathsf{H}^\circ_\alpha(\Gamma)$ we have $\Phi(z) \in \mathsf{H}^\circ_\beta(\Gamma)$, where $\beta = \min\{\alpha, \mu - \varepsilon\}$, $\mu$ is defined by equality (7) and $\varepsilon > 0$ is arbitrarily small.*

## 3. The Riemann boundary value problem on non-rectifiable arcs

Let us consider the Riemann boundary value problem on non-rectifiable $G$-arc, i.e., the problem of finding a function $\Phi(z)$ holomorphic in $\overline{\mathbb{C}} \setminus \Gamma$ such that

$$\Phi^+(t) = G(t)\Phi^-(t) + g(t), t \in \Gamma^\circ, \tag{10}$$

and

$$\Phi(z) = O(|z - a_j|^{-\gamma}), \quad z \to a_j, \quad j = 1, 2, \quad \gamma = \gamma(\Phi) < 1. \tag{11}$$

The solution of this problem is well known for piecewise smooth arcs $\Gamma$ (see, for instance, [9, 10]). In this case the solution is expressed in terms of Cauchy-type integrals over $\Gamma$. In the present paper we show that for non-rectifiable arcs the solutions are representable by Cauchy transforms of integrations as introduced above.

In the simplest case $G \equiv 1$ the Riemann boundary-value problem turns into a so-called jump problem:

$$\Phi^+(t) - \Phi^-(t) = f(t), t \in \Gamma^\circ. \tag{12}$$

If $j_F(t) \equiv 1$ then by virtue of Theorem 2.2 the Cauchy transform of integration $\int[F]f$ gives a solution of this problem. We consider function

$$k_\Gamma(z) = \frac{1}{2\pi i} \log \frac{z - a_2}{z - a_1},$$

where the branch of logarithm is determined by means of the cut along $\Gamma$ and condition $k_\Gamma(\infty) = 0$. Obviously, $k_\Gamma \in \mathsf{H}_1^\circ(\Gamma)$, and the jump of this function on $\Gamma$ equals to 1. If $\Gamma$ is $G$-arc, then $k_\Gamma(z) = \frac{(-1)^j}{2\pi i}\log|z-a_j| + O(1)$ for $z \to a_j$, $j = 1, 2$. Thus, we obtain

**Theorem 3.1.** *If $\Gamma$ is $G$-arc and $f \in H^*(\Gamma, \frac{1}{2}\,\mathrm{dmc}\,\Gamma)$, then the Cauchy transform*

$$\Phi_0(z) = \mathsf{C}\int[k_\Gamma]f(z) \tag{13}$$

*is a solution of the jump problem* (12). *In addition, it satisfies the estimate*

$$\Phi_0(z) = \frac{(-1)^j}{2\pi i}f(a_j)\log|z-a_j| + O(1), z \to a_j, j = 1, 2, \tag{14}$$

$\Phi_0 \in \mathsf{H}^\circ_{\mu-\varepsilon}(\Gamma)$ *for any sufficiently small $\varepsilon > 0$, and $\Phi_0(\infty) = 0$.*

Let us denote by $\mathrm{dmh}\,\Gamma$ the Hausdorff dimension of arc $\Gamma$. If $\mathrm{dmh}\,\Gamma > 1$, then we can find a non-trivial function $\Phi_1(z)$ holomorphic in $\overline{\mathbb{C}}\setminus\Gamma$ which is continuous in $\overline{\mathbb{C}}$ and vanishes at the point at infinity (see, for instance, [11]). Then the sum $\Phi_0(z) + \sum_{k=1}^n c_k\Phi_1^k(z)$ satisfies the equality (12) for any $n$ and $c_k, k = 1, 2, \ldots$. Thus, the set of solutions of the jump problem on the non-rectifiable arc is infinite in general. On the other hand, if a holomorphic function in $\mathbb{C}\setminus\Gamma$ $\Phi(z)$ satisfies the Hölder condition with exponent $\lambda > \mathrm{dmh}\,\Gamma - 1$ in a neighborhood of $\Gamma$, then $\Phi(z)$ is holomorphic in the whole plane $\mathbb{C}$ (see [11]). Thus, for

$$\mathrm{dmh}\,\Gamma - 1 < \lambda < \mu \tag{15}$$

the function $\Phi_0(z)$ is a unique solution of the jump problem in the class $\mathsf{H}^\circ_\lambda(\Gamma)$ which vanishes at the point at infinity and satisfies the condition (11). The restriction (15) has meaning only if

$$\mathrm{dmh}\,\Gamma - 1 < \mu = \frac{2\nu - \mathrm{dmc}\,\Gamma}{2 - \mathrm{dmc}\,\Gamma}. \tag{16}$$

Then we consider the Riemann boundary value problem (10) in the class of functions satisfying conditions (11) and $\Phi(\infty) = 0$. We assume that the coefficients $G$ and $g$ of the problem (10) belong to the space $H^*(\Gamma, \frac{1}{2}\,\mathrm{dmc}\,\Gamma)$ and $G(t) \neq 0$ for $t \in \Gamma$. Then $G(t) = \exp f(t)$, $f \in H^*(\Gamma, \frac{1}{2}\,\mathrm{dmc}\,\Gamma)$, and the problem (12) with jump $f$ has a solution $\Phi_0(z) := \mathsf{C}[k_\Gamma]f(z)$ satisfying relation (14). We put $v_j = \Re(-1)^j(2\pi i)^{-1}f(a_j)$, $\kappa_j = ]v_j[+1$, where $]x[:= \max\{n \in \mathbb{N} : n < x\}$, $j = 1, 2$, $\kappa := \kappa_1 + \kappa_2$, and

$$X(z) := (z-a_1)^{-\kappa_1}(z-a_2)^{-\kappa_2}\exp\Phi_0(z).$$

The function $X$ satisfies estimates (11), and $X^{-1}(z)$ is bounded near the points $a_1$ and $a_2$. Obviously, $X^+(t) = G(t)X^-(t)$. We apply customary factorization and reduce the Riemann boundary value problem to the jump problem

$$\frac{\Phi^+(t)}{X^+(t)} - \frac{\Phi^-(t)}{X^-(t)} = \frac{g(t)}{X^+(t)}, t \in \Gamma^\circ. \tag{17}$$

We consider the function

$$\Psi(z) := \mathsf{C}\int\left[\frac{1}{X}\right]h(z),$$

where $h \in H^*(\Gamma, \frac{1}{2}\operatorname{dmc}\Gamma)$ will be defined later. Clearly,

$$\Psi^+(t) - \Psi^-(t) = \frac{h(t)}{X^+(t)} - \frac{h(t)}{X^-(t)} = \frac{(1-G(t))h(t)}{X^+(t)}.$$

Hence, the function $\Psi$ satisfies the equality

$$\Psi^+(t) - \Psi^-(t) = \frac{g(t)}{X^+(t)}$$

for

$$h(t) = \frac{g(t)}{1-G(t)}.$$

Thus, for $G(t) \neq 1$ the product

$$\Phi(z) := X(z)\mathsf{C}\int\left[\frac{1}{X}\right]\left(\frac{g}{1-G}\right)(z)$$

satisfies the boundary value condition (10). As a result, we obtain

**Theorem 3.2.** *Assume that $\Gamma$ is a $G$-arc satisfying restriction* (16)*, $G$ and $g$ belong to $H^*(\Gamma, \frac{1}{2}\operatorname{dmc}\Gamma)$ and $G(t) \neq 0, 1$ on $\Gamma$. Then the family of solutions of the problem* (10)*,* (11) *in the class $\mathsf{H}^\circ_\lambda(\Gamma)$ with $\lambda \in (\operatorname{dmh}\Gamma - 1, \mu]$ has the same structure as in the classical case of piecewise smooth arc (see* [10, 9]*), i.e., for $\kappa > 0$ this family is affine $\kappa$-dimensional over $\mathbb{C}$ and for $\kappa \leq 0$ it consists of unique solution existing under $-\kappa$ conditions.*

Apparently, the detailed study of the jump problem (17) enables one to remove the restriction that $G(t) \neq 1$.

Let us note that the similarity of solvability of the Riemann boundary value problem on non-rectifiable $G$-arcs and on piecewise smooth arcs is rather formal. In the case of non-rectifiable arc we need additional conditions both for existence (the condition $\nu > \frac{1}{2}\operatorname{dmc}\Gamma$) and for uniqueness (the condition $\Phi(z) \in \mathsf{H}^\circ_\lambda(\Gamma)$, $\lambda \in (\operatorname{dmh}\Gamma - 1, \mu]$) of the solutions.

The first solution of the Riemann boundary value problem on non-rectifiable arcs was constructed in the paper [12] for $\nu > \frac{1}{2}\operatorname{dmb}\Gamma$. In the present paper we replace the dimension $\operatorname{dmb}\Gamma$ by the lower value $\operatorname{dmc}\Gamma$. In addition, we obtain explicit solutions in terms of the Cauchy transforms.

## References

[1] L. Hörmander, *The Analysis of Linear Partial Differential Operators I. Distribution theory and Fourier Analysis*, Springer Verlag, 1983.

[2] B.A. Kats, *The refined metric dimension with applications*, Computational Methods and Function Theory 7 (2007), No. 1, 77–89

[3] A.N. Kolmogorov, V.M. Tikhomirov, *ε-entropy and capacity of set in functional spaces*, Uspekhi Math. Nauk, **14** (1959), 3–86.

[4] I. Feder, *Fractals*, Mir Publishers, Moscow, 1991.

[5] B.A. Kats, *The Cauchy transform of certain distributions with application*, Complex Analysis and Operator Theory, to appear.

[6] P. Mattila and M.S. Melnikov, *Existence and weak type inequalities for Cauchy integrals of general measure on rectifiable curves and sets*, Proc. Am. Math. Soc. 120(1994), pp. 143–149.

[7] X. Tolsa, *Bilipschitz maps, analytic capacity and the Cauchy integral*, Ann. of Math. (2) 162(2005), pp. 1243–1304.

[8] J. Verdera, *A weak type inequality for Cauchy transform of finite measure*, Publ. Mat. 36(1992), pp. 1029–1034.

[9] N.I. Muskhelishvili, *Singular integral equations*, Nauka publishers, Moscow, 1962.

[10] F.D. Gakhov, *Boundary value problems*, Nauka publishers, Moscow, 1977.

[11] E.P. Dolzhenko, *On "erasing" of singularities of analytical functions*, Uspekhi Mathem. Nauk, **18** (1963), No. 4, 135–142.

[12] B.A. Kats, *The Riemann boundary value problem on open Jordan arc*, Izv. vuzov, Mathem., **12** 1983, 30–38.

Boris A. Kats
Kazan Federal University
Zelenaya str. 1
Kazan, 420043, Russia
e-mail: katsboris877@gmail.com

Operator Theory:
Advances and Applications, Vol. 221, 443-451


# Decay Estimates for Fourier Transforms of Densities Defined on Surfaces with Biplanar Singularities

Otto Liess and Claudio Melotti

**Abstract.** In this note we describe results on decay estimates for Fourier transforms of densities which live on surfaces $S$ in three space dimensions with isolated biplanar singularities and we also describe briefly how such results are related to the theory of crystal elasticity for tetragonal crystals.

**Mathematics Subject Classification (2000).** Primary 42B10; Secondary 35Q72.

**Keywords.** Decay estimates, crystal acoustics.

## 1. Decay of Fourier transforms

Let $U$ be open in $\mathbb{R}^3$ and consider a surface $S \subset U$, which is $\mathcal{C}^1$ except a finite number of isolated points. We assume that the singularities at these points are "biplanar", in a sense which will be recalled in a moment. In particular our assumptions on the singularities will imply that compactly supported continuous functions defined on $S$ are integrable, when we integrate with respect to the measure given by the surface element on $S$. When $f$ is a bounded compactly supported function defined on $S$ we can then associate a distribution $u$ with it by the map $C_0^\infty(U) \ni \varphi \to u(\varphi) = \int_S \bar{\varphi}\hat{f} d\sigma$, where $d\sigma$ denotes the surface element on $S$. (The notations are as in standard distribution theory.) Our main result will be that the Fourier transform $\lambda \to \hat{u}(\lambda)$ of $u$ decays at infinity of order $|\lambda|^{-1/2} \ln(1 + |\lambda|)$, provided $f$ satisfies some natural regularity assumptions, which we will describe in Proposition 1.3. In Section 2 below we will then show how this result relates to the problem of long time behavior of solutions to the system of crystal acoustics in a case when biplanar singularities appear.

We next describe the setting of our problem in detail. We assume that $S$ is given by an equation $h(x, y, z) = 0$, where $h : U \to \mathbb{R}$ is a $\mathcal{C}^1$-function. The conditions which we mention at this moment are not related to the three-dimensional case, so for a short while, we shall denote the variables by "$x$" (and may think that all conditions refer to some surfaces in $\mathbb{R}^n$).

We assume that $0 \in U$ and that $h(0) = 0$, $\operatorname{grad}_x h(0) = 0$. In principle 0 could then be a singular point of $S$ and we assume that $S$ does not have other singular points in an open neighborhood $U'$ of 0 in $U$. More precisely, we assume that $h(x) = 0$ implies $\operatorname{grad}_x h(x) \neq 0$ when $x \neq 0$, $x \in U'$. We define the "tangent set" to $S$ at 0 to consist of all vectors $v \in \mathbb{R}^3$ for which we can find a $\mathcal{C}^1$-curve $\gamma : (-c, c) \to S$ with $\gamma(0) = 0$ and $\dot{\gamma}(0) = v$. We say that 0 is a biplanarly singular point if the tangent set is the union of two distinct planes.

It may be worthwhile to describe a simple situation in which such biplanar singularities appear. We first recall that if $h$ is a $\mathcal{C}^\infty$-function defined in a neighborhood of the origin in $\mathbb{R}^n$ and if the derivatives of $h$ up to inclusively order $s-1$ vanish, then we call "localization polynomial" of $h$ at 0 the polynomial $J_s h(v) = \sum_{|\alpha|=s} \partial_x^\alpha h(0) v^\alpha / \alpha!$. (It is just the Taylor polynomial of order $s$ of $h$ at 0.) Clearly, tangent vectors $v$ to $S$ at 0 must satisfy $J_s h(v) = 0$, but the converse is also almost true: if $v$ satisfies $J_s h(v) = 0$ and $\operatorname{grad}_v J_s h(v) \neq 0$, then $v$ must be a tangent vector to $S$ at 0 (cf., e.g., [5]). An example is when $J_s h(v) = v_1^2 - v_2^2$.

From this moment on we return definitively to the case of three variables and the variables will be denoted again by $(x, y, z)$. The Fourier dual variables will be denoted by $\lambda = (\xi, \eta, \tau)$.

An example of a surface with a biplanar singularity is when $U = \mathbb{R}^3$ and $S$ is given by $h(x, y, z) = z^2 - 2(x^2 + y^2)z + x^4 + 2x^2y^2 + y^4 - \delta^2 x^2 - \delta^2 y^4$, with $\delta$ a constant. In this case, $S$ consists of the two sheets $z = x^2 + y^2 \pm \delta\sqrt{x^2 + y^4}$, but the equation $z - x^2 - y^2 - \delta\sqrt{x^2 + y^4} = 0$, already defines a surface with a biplanar singularity in the sense above at 0. Indeed, in our main result we shall work with precisely such a sheet in a situation which is only slightly more general. More precisely, we assume that for some neighborhood $V$ of the origin in $\mathbb{R}^2$

$$S = \{(x, y, z); z = g(x, y), (x, y) \in V\},$$

where $g : V \to \mathbb{R}$ is for some constants $A, B, C, A_1, B_1, C_1, \delta$, a function of form

$$Ax^2 + 2Bxy + Cy^2 + g_1(x, y) + \delta\sqrt{A_1 x^2 + 2B_1 xy^2 + C_1 y^4 + g_2(x, y)}. \tag{1.1}$$

Here $g_1, g_2 : V \to \mathbb{R}$ are two functions in $\mathcal{C}^\infty(V)$. In addition we suppose that:

(i) $g_1(x, y) = O(|(x, y)|^3)$, for $(x, y) \to 0$,
(ii) $g_2(x, y) = o(|(x^2 + y^4)|)$, for $(x, y) \to 0$,
(iii) the functions $(x, y) \to Ax^2 + 2Bxy + Cy^2$, $(x, t) \to A_1 x^2 + 2B_1 xt + C_1 t^2$ are strictly positive for $(x, y) \neq 0$, $(x, t) \neq 0$,
(iv) $\max(|A|, |B|, |C|) \leq 1$, $\max(|A_1|, |B_1|, |C_1|) \leq 1$ and $\delta$ is small and positive.

Thus, $S$ has a singular point at $0 \in \mathbb{R}^3$ and the singularity is biplanar there.

*Remark* 1.1. The condition (iv) just means that if we denote by $A_1' = \delta^2 A_1, B_1' = \delta^2 B_1, C_1' = \delta^2 C_1$, then the constants $A_1', B_1', C_1'$ are small compared with the constants $A, B, C$. As for the constant $\delta$ we shall assume it to be small. Indeed, what we need is that the second derivatives of the function

$$\delta\sqrt{A_1x^2 + 2B_1xy^2 + C_1y^4 + g_2(x,y)}$$

must be small when compared with the second derivatives of the term $Ax^2 + 2Bxy + Cy^2 + g_1(x,y)$.

Furthermore the estimates which we obtain later on will depend on $\delta$, but must not depend on $A_1, B_1, C_1$.

Finally, the assumption (iii) implies in particular that $Ax^2 + 2Bxy + Cy^2 \geq c_1|(x,y)|^2$ and $A_1x^2 + 2B_1xt + C_1t^2 \geq c_2|(x,t)|^2$, for some constants $c_1$, $c_2$.

We mentioned already that $S$ is not smooth at 0. The quantity which measures the strength of the singularity is in some sense

$$\Delta = A_1x^2 + 2B_1xy^2 + C_1y^4 + g_2(x,y),$$

but in a neighborhood of the origin, the main contribution will come from the part $\Delta' = A_1x^2 + 2B_1xy^2 + C_1y^4$ of $\Delta$. Indeed, from the assumptions on $\Delta'$ and the fact that $g_2 = o(|(x^2+y^4)|)$, for $(x,y) \to 0$ it follows that we have $\Delta(x,y) \geq c(x^2+y^4)$ in a neighborhood of 0 for some constant $c > 0$. We also recall that the surface element $d\sigma$ on $S$ is given by $\sqrt{1 + g_x^2(x,y) + g_y^2(x,y)}\ d(x,y)$. The following easy estimates for derivatives of $g$ and the function $(1 + g_x^2(x,y) + g_y^2(x,y))^{1/2}$ is needed in the arguments:

*Remark* 1.2. There are constants $c, \varepsilon$ such that

$$|\operatorname{grad}_{x,y}\sqrt{\Delta(x,y)}| < c,\ |\operatorname{grad}_{x,y} g(x,y)| \leq c,$$
$$|H_{x,y}\sqrt{\Delta(x,y)}| \leq \frac{c}{\sqrt{\Delta(x,y)}},\ |H_{x,y}g(x,y)| \leq \frac{c}{\sqrt{\Delta(x,y)}}, \tag{1.2}$$

provided $0 \neq |(x,y)| \leq \varepsilon$. ($H_{x,y}f$ is for a given function $f$ the Hessian matrix of $f$ in the variables $(x,y)$.)

Note that it follows from these estimates also that we have for the same $(x,y)$

$$(1 + g_x^2(x,y) + g_y^2(x,y))^{1/2} \leq c,\ |\operatorname{grad}_{x,y}(1 + g_x^2(x,y) + g_y^2(x,y))^{1/2}| \leq c. \tag{1.3}$$

If $F$ is now a function which is defined on $S$, the Fourier transform of the density $F d\sigma$ can be written as

$$I(\xi,\eta,\tau) = \int_S \exp\left[ix\xi + iy\eta + ig(x,y)\tau\right] F(x,y,g(x,y))d\sigma, (\xi,\eta,\tau) \in \mathbb{R}^3. \tag{1.4}$$

We have not yet specified explicit conditions under which this integral converges. Actually, since $(1 + g_x^2(x,y) + g_y^2(x,y))^{1/2}$ is bounded, local integrability near 0 holds if we assume for example that $F$ is bounded in a neighborhood of the origin. However, application of results from the theory of stationary phase requires at least that $F$ be also differentiable in the variables in which we want to apply this method. It seems reasonable to ask for conditions which are modeled on the regularity properties and estimates which we have obtained for the surface element in Remark 1.2. We can now state the main result

**Proposition 1.3.** *Assume that $S$ is as above and let $F : S \to \mathbb{C}$ be a continuous function which is bounded in a neighborhood of the origin which is such that the function $f(x,y) = F(x,y,g(x,y))$ is $C^1$ for $(x,y) \neq 0$ small, and for which there is a constant $c$ such that*

$$|\operatorname{grad}_{(x,y)} F(x,y,g(x,y))| \leq c/\sqrt{\Delta(x,y)} \text{ for } 0 \neq |(x,y)| \text{ small.}$$

*If $\delta$ and $\kappa$ are small enough, we can find a constant $c'$, such that*

$$I(\xi,\eta,\tau) = \int_S \exp\left[i\xi x + i\eta y + i\tau z\right] F(x,y,z) d\sigma,$$

*satisfies the estimate*

$$|I(\xi,\eta,\tau)| \leq c'(1 + |(\xi,\eta,\tau)|)^{-1/2} \ln(1 + |(\xi,\eta,\tau)|), \tag{1.5}$$

*provided $F(x,y,g(x,y))$ vanishes for $|(x,y)| \geq \kappa$.*

The proof of proposition 1.3 will be given in a forthcoming paper. (For a related result see [1].) We only mention here a result from the method of stationary phase, due to E. Stein, [12], which is used in the argument and a lemma which collects two intermediate estimates needed when we want to apply the result of Stein to the situation at hand.

**Lemma 1.4 (Stein).** *Let $\varphi$ be a real-valued function on the interval $[a,b]$ which is $k$ times differentiable. Assume that $k \geq 2$ and that $|\varphi^{(k)}(r)| \geq 1$ for all $r$. Also consider $\psi \in C^1[a,b]$. Then it follows that*

$$\left| \int_a^b e^{it\varphi(r)} \psi(r) dr \right| \leq c_1 t^{-1/k} \left[ |\psi(b)| + \int_a^b |\psi'(r)| dr \right], \text{ for } t > 0,$$

*for some constant $c_1$ which does not depend on $\varphi$, $\psi$, $a$ and $b$.*

**Lemma 1.5.** *There are constants $c$, $c'$, $\kappa$, such that $\sup_\alpha |\frac{\partial}{\partial r}(rf(r\cos\alpha, r\sin\alpha))| \leq cr^{-1}$ for $r \leq \kappa$, $\left|\frac{d^2}{dr^2}[\tau g(r\cos\alpha, r\sin\alpha) + r\xi\cos\alpha + r\eta\sin\alpha]\right| \geq c'(|\xi| + |\eta| + |\tau|)$, uniformly in $\alpha$, and for $|\tau| \geq (|\xi| + |\eta|)$, $|(x,y)| \leq \kappa$.*

## 2. Applications to crystal theory

We recall the (homogeneous) system of crystal acoustics in the specific case of tetragonal crystals (see [9]). It is

$$\partial_t^2 \begin{pmatrix} u_1(t,x) \\ u_2(t,x) \\ u_3(t,x) \end{pmatrix} = A(\partial_x) \begin{pmatrix} u_1(t,x) \\ u_2(t,x) \\ u_3(t,x) \end{pmatrix}, \tag{2.1}$$

where $A(\partial_x)$ is the differential matrix

$$\begin{pmatrix} c_{11}\partial_{11}^2 + c_{66}\partial_{22}^2 + c_{44}\partial_{33}^2 & (c_{12}+c_{66})\partial_{12}^2 & (c_{13}+c_{44})\partial_{13}^2 \\ (c_{12}+c_{66})\partial_{12}^2 & c_{66}\partial_{11}^2 + c_{11}\partial_{22}^2 + c_{44}\partial_{33}^2 & (c_{13}+c_{44})\partial_{23}^2 \\ (c_{13}+c_{44})\partial_{13}^2 & (c_{13}+c_{44})\partial_{23}^2 & c_{44}\partial_{11}^2 + c_{44}\partial_{22}^2 + c_{33}\partial_{33}^2 \end{pmatrix}$$

(Here $\partial_{ij} = \partial_{x_i}\partial_{x_j}$, etc.) Note that variables in the physical space $\mathbb{R}^3$ are now denoted by $x = (x_1, x_2, x_3)$, $t$ corresponds to the time variable, and the Fourier dual variables to $x$ are denoted by $\xi = (\xi_1, \xi_2, \xi_3)$. (The notation has thus changed with respect to Section 1.) We are interested in decay properties of global (say $\mathcal{C}^\infty$) solutions $u(t,x)$ of the systems for $t \to \infty$. By this we mean that the $(u_1(t,x), u_2(t,x), u_3(t,x))$ are defined on all of $\mathbb{R}^4$ and we are interested in estimates of type $|u_i(t,x)| \leq c(1+|t|)^{-\chi}$ (or similar) for some positive constant $\chi > 0$, $\forall t$. For such "pointwise" estimates to hold we must of course assume that the support of the initial conditions $u(0,x)$ and $\partial_t u(0,x)$ be compact in $x$.

Apart from their intrinsic interest, such estimates can be interesting when one wants to study long-time existence of small nonlinear perturbations to the system. Similar studies in the case of isotropic wave type equations are by now classical, see, e.g., [3], [4], [10], [11].

The constants $c_{ij}$ above are subject to a number of restrictions, and we mention some of them which come from the fact that, due to physical considerations, the system must be hyperbolic.

*Remark* 2.1. The conditions on constants which we assume are

$$c_{ii} > 0, \text{ for } i = 1,3,4,6, \quad c_{66} - c_{12} > 0, \quad c_{44} \neq c_{13},$$

$$c_{11} - c_{66} - c_{12} > 0, \quad c_{33} - \frac{(c_{13}+c_{44})^2}{c_{12}+c_{66}} > 0.$$

Thus, when $c_{11} = c_{66}$, we must have $c_{12} < 0$ and we shall assume that $c_{44} > c_{66}$. Cubic crystals are a particular case of tetragonal crystals. They are obtained when we ask for the conditions $c_{11} = c_{33}$, $c_{44} = c_{66}$, $c_{12} = c_{13}$. We observe that tetragonal crystals depend on 6 constants, whereas cubic crystals depend only on 3 constants.

The way by which we want to obtain results on decay for solutions of the system (2.1) is to represent them as parametric Fourier-type integrals which live on the so-called "slowness surface" of the system. We briefly recall the relevant terminology. The first thing is to recall that the "characteristic polynomial" associated with the system is the polynomial $p(\tau,\xi) = \det(\tau^2 I - A(\xi))$, where $A(\xi)$ is the $3 \times 3$ matrix which is obtained by formally replacing $\partial_{ij}^2$ by $\xi_i\xi_j$ in the matrix $A(\partial_x)$. The characteristic surface of the system is given by the condition $\{(\tau,\xi) \in \mathbb{R}^4; p(\tau,\xi) = 0\}$. Hyperbolicity means here that the polynomial equation $p(\tau,\xi) = 0$ has 6 real solutions $\tau$ if $\xi$ is real. Finally, we call "slowness surface" the set $\{\xi \in \mathbb{R}^3; p(1,\xi) = 0\}$. Most of the analysis is done on the slowness surface. This is a compact algebraic surface of degree 6 in $\mathbb{R}^3$ and its equation can be understood best when one puts it into Kelvin's form. To obtain this form, we introduce the following notations:

$$n_1(\xi_1) = (c_{12}+c_{66})\xi_1^2, n_2(\xi_2) = (c_{12}+c_{66})\xi_2^2, n_3(\xi_3) = \frac{(c_{13}+c_{44})^2}{c_{12}+c_{66}}\xi_3^2,$$

and

$$d_1(\tau,\xi) = \tau^2 - d_1'(\xi), \quad d_1'(\xi) = c_{11}\xi_1^2 + c_{66}\xi_2^2 + c_{44}\xi_3^2 - (c_{12}+c_{66})\xi_1^2,$$

$$d_2(\tau,\xi) = \tau^2 - d_2'(\xi), \quad d_2'(\xi) = c_{66}\xi_1^2 + c_{11}\xi_2^2 + c_{44}\xi_3^2 - (c_{12}+c_{66})\xi_2^2,$$
$$d_3(\tau,\xi) = \tau^2 - d_3'(\xi), \quad d_3'(\xi) = c_{44}\xi_1^2 + c_{44}\xi_2^2 + c_{33}\xi_3^2 - \frac{(c_{13}+c_{44})^2}{c_{12}+c_{66}}\xi_3^2.$$

"Kelvin's form" of the condition $p(1,\xi)=0$ is then (see [2]):

$$\frac{n_1(\xi_1)}{1-d_1'(\xi)} + \frac{n_2(\xi_2)}{1-d_2'(\xi)} + \frac{n_3(\xi_3)}{1-d_3'(\xi)} = 1. \tag{2.2}$$

Hyperbolicity then implies that the slowness surface is a three-sheeted surface. Indeed, in terms of the positive roots $\tau_j(\xi)$, $j=1,2,3$ of the characteristic polynomial $\tau \to p(\tau,\xi)$, ordered, e.g., such that $\tau_1(\xi) \le \tau_2(\xi) \le \tau_3(\xi)$, they can be written as $\{\xi \in \mathbb{R}^3; \tau_j(\xi) = 1\}$. For later purpose, we also denote for $i=4,5,6$, by $\tau_4(\xi) = -\tau_1(\xi)$, $\tau_5(\xi) = -\tau_2(\xi)$, $\tau_6(\xi) = -\tau_3(\xi)$ the negative roots of the characteristic polynomial. (Observe that $p(-\tau,\xi) = p(\tau,\xi)$, so if $p(\tau,\xi)=0$, also $p(-\tau,\xi)=0$.)

We next denote by $\varphi = u(0,x)$ and by $\psi = \partial_t u(0,x)$ the initial data of some solution $u$ of the system (2.1). We mentioned already that we will assume $\varphi$ and $\psi$ to have compact support. The solution $u(t,x)$ will then have (by "finite propagation speed of signals") compact support in the variable $x$, for all $t$, and we can take the partial Fourier transform in the variable $x$. Denoting these partial Fourier transform of $u$ by $\hat{u}(t,\xi)$, we see that $\hat{u}$ satisfies the following $3\times 3$ system of ordinary differential equations in the variable $t$ with $\xi$ as a parameter:

$$\partial_t^2 \hat{u}(t,\xi) = A(\xi)\hat{u}(t,\xi).$$

The initial conditions are now $\hat{u}(0,\xi) = \hat{\varphi}(\xi)$, $\partial_t\hat{u}(0,\xi) = \hat{\psi}(\xi)$, where we have used "hats" also to denote the Fourier transform for functions which depend only on $x$. This gives the possibility to write down $u$ explicitly in terms of Fourier type integrals. Indeed, G.F.D. Duff gives explicit formulas for $u$ in terms of the roots $\tau_j(\xi)$, $j=1,2,3,4,5,6$, of the characteristic polynomial $\tau \to p(\tau,\xi)$. He introduces the notations

$$T_{ipj} = \frac{(n_i(\xi)n_j(\xi))^{1/2}}{2i\tau_p(\xi)} \cdot \frac{(\tau_p^2(\xi) - d_{j+1}'(\xi))(\tau_p^2(\xi) - d_{j+2}'(\xi))}{(\tau_p^2(\xi) - d_i'(\xi))(\tau_p^2(\xi) - \tau_{p+1}^2(\xi))(\tau_p^2(\xi) - \tau_{p+2}^2(\xi))}, \tag{2.3}$$

and shows, assuming that $\varphi \equiv 0$ (as is often done when one studies a Cauchy problem), that

$$u_i(t,x) = \int_{\mathbb{R}^3} \sum_{p=1}^{6}\sum_{j=1}^{3} e^{it\tau_p(\xi) + i\langle x,\xi\rangle} T_{ipj}(\xi)\hat{\psi}_j(\xi)d\xi, \qquad i=1,2,3. \tag{2.4}$$

The problem with this representation is that it is singular at $\xi = 0$ and also when for some $\xi^0 \ne 0$ we have multiple roots. Note that, at a first glance, e.g., double roots can create problems in the denominators of the expressions which define the $T_{ipj}$, since then one or the other of the factors $(\tau_p^2(\xi)-\tau_{p+1}^2(\xi))(\tau_p^2(\xi)-\tau_{p+2}^2(\xi))$ may vanish on the line $\nu\xi^0$, $\nu \in \mathbb{R}$. (A more careful analysis shows moreover that also the factors $\tau_p^2(\xi) - d_{j+2}'(\xi)$ can vanish on that line.) The intersection of the line $\{\nu\xi^0,\ \nu \in \mathbb{R}\}$, with the slowness surface in such a situation, corresponds to a multiple point on that surface. However, if we assume that $c_{44} > c_{66}$, then we can have at most "double" points on the surface, and for this reason in the following we will refer only to double points. Moreover, it is not difficult to show that the double points on the slowness surface are isolated (and therefore, since the slowness surface is compact, in finite number) and their location is easily calculated, albeit the expressions which give them are quite involved. We will come back to this problem in a forthcoming paper (also see [8]), and we only want here to describe how we can deal with the difficulties related to the singularities in (2.3). The first remark is that one can show that the singularities can only be of three types: conical, uniplanar or biplanar, with biplanarity understood in the sense it was defined above. The definitions for conical and uniplanar singularities are perhaps better known, but for completeness we mention briefly that for a surface $S \subset \mathbb{R}^3$ given in a neighborhood of 0 by an equation $h(x) = 0$, $h(0) = 0$, $\operatorname{grad} h(0) = 0$, $Hh(0) \neq 0$, the point 0 is a called "conical", respectively "uniplanar", if for a suitable choice of linear coordinates $y = (y_1, y_2, y_3)$, we have $J_2 h = y_1^2 - y_2^2 - y_3^2$, respectively $J_2 h = y_1^2$, and if the surface can be written near 0 as $y_1^2 + A(y_2, y_3)y_1 + B(y_1, y_3) = 0$, with $A(0) = 0, B(0) = 0, \nabla_y A(0) = 0$ for some $C^\infty$- functions $A$, $B$. $c_1(|y_2|^4 + |y_3|^4) \leq A^2(y_1, y_2) - 4B(y_1, y_2) \leq c_2(|y_2|^4 + |y_3|^4)$, in a neighborhood of 0 for two constants $c_i > 0$. $Hh$ is again the Hessian of $h$. Geometrically speaking, $S$ is thus, if 0 is a uniplanarly singular point, the union of the two sheets $S^\pm = \{y; y_1 = \frac{1}{2}(-A(y_2, y_3) \pm \sqrt{\Delta(y_2, y_2)}\,)\}$ and these sheets have $y_1 = 0$ as a common tangent plane at 0. Note that "biplanarity" is somehow "intermediate" between "conicity" and "uniplanarity".)

It is known that for cubic crystals only conical and uniplanar singularities can appear: see, e.g., [2], [5]. Also in the tetragonal case this is the "generic" situation. However, it was discovered by one of us (C.M.) that in the special case when $c_{11} = c_{66}$ (and only then), biplanar singularities will appear precisely on the intersection of the slowness surface with the $\xi_1$ and the $\xi_2$ axis. It is in fact not difficult to find the double points on the axes and they have the following expressions:

$$(\pm\frac{1}{\sqrt{c_{44}}}, 0, 0), (\pm\frac{1}{\sqrt{c_{66}}}, 0, 0), (\pm\frac{1}{\sqrt{c_{11}}}, 0, 0),$$
$$(0, \pm\frac{1}{\sqrt{c_{44}}}, 0), (0, \pm\frac{1}{\sqrt{c_{66}}}, 0), (0, \pm\frac{1}{\sqrt{c_{11}}}, 0),$$
$$(0, 0, \pm\frac{1}{\sqrt{c_{44}}}), (0, 0, \pm\frac{1}{\sqrt{c_{44}}}), (0, 0, \pm\frac{1}{\sqrt{c_{33}}}).$$

It is clear from these expressions that when $c_{11} = c_{66}$, then we have double points on the $\xi_1$ and on the $\xi_2$ axis. There will always be a double point on the $\xi_3$ axis and in view of our assumption $c_{44} > c_{66}$ we never have triple points. This latter assumption will also imply that when $c_{11} = c_{66}$, then the double points lie on the outer sheets of the slowness surface.

We conclude this note with some remarks on how one can study the integrals which appear in (2.4) by reducing them to parametric integrals on the slowness surface. The first thing is that for some constant $c$ we have $|T_{ipj}(\xi)| \le c|\xi|^{-1}$, for $\xi \in \mathbb{R}^3$. This is not difficult to show and we refer to [2] and [6] for similar results. In particular this means that the integrands in (2.4) are absolutely integrable in $\mathbb{R}^3$. The contribution of a small neighborhood of the origin in $\xi$-space to (2.4) will therefore be small, and we are left with the integrals there for $|\xi| \ge 1/|x|$, say. (The point why we are interested in $|\xi| \ge 1/|x|$, is that our argument in fact refers to the case when we assume that $|t|$ dominates $|x|$.) We also observe that the functions $T_{ipj}$ are homogeneous of degree $-1$ and that the exponent $it\tau_p(\xi) + i\langle x, \xi\rangle$ is homogeneous of degree 1. We want to show how one can reduce estimates for (2.4) to estimates on the slowness surface of the system. To be specific, we fix a small open conic neighborhood $G$ of the point $(1,0,0)$ and shall study the term in (2.4) for $p = 1$ and some $i, j$. We also denote by $\rho : \{|(\xi_2, \xi_3)| \le \varepsilon\} \to \mathbb{R}$ the continuous function determined by the conditions $\tau_1(\rho(\xi'), \xi') \equiv 1$, $\rho(1,0,0) = 1/c_{66}^{1/2}$, and by $K$ the map $K(\nu, \xi') = \nu(\rho(\xi'), \xi') = \xi$. The determinant of the Jacobian of this map is easily calculated and is for $\xi' \neq 0$ equal to $\nu^2(\rho(\xi') - \langle \xi', \mathrm{grad}_{\xi'} \rho(\xi')\rangle)$. (See [7].) We now change coordinates in (2.4) for the term corresponding to $p = 1$ and some $i, j$. The exponent $it\tau_p(\xi) + i\langle x, \xi\rangle$ is in the new coordinates $it\nu + i\nu\rho(\xi')x_1 + i\nu\langle x', \xi'\rangle$ and the term $T_{i1j}(\xi)\hat{\psi}_j(\xi)$ transforms to $\nu^{-1}T_{i1j}(\rho(\xi'), \xi')\nu^2(\rho(\xi') - \langle \xi', \mathrm{grad}_{\xi'} \rho(\xi')\rangle)\hat{\psi}_j(\nu(\rho(\xi'), \xi'))$. We see in this way that (2.4) has transformed into an integral in the variables $(\nu, \xi')$. When $\nu$ is fixed we obtain an integral in $\xi'$ which we may regard as an integral on the sheet of the slowness surface corresponding to $\tau_1(\xi) = 1$ (which we have parametrized by $\xi' \to (\rho(\xi), \xi')$). It is a technical matter to show that the assumptions of proposition 1.3 hold when $\nu$ is fixed and one can also see how constants depend on $\nu$. By arguing as in [7] we can obtain starting from this the following result:

**Proposition 2.2.** *Let $\chi : \mathbb{R}^3 \to \mathbb{R}$ be a function which is positively homogeneous of degree 0, which vanishes identically outside $\Gamma = \{\xi; |\xi'| < d\xi_1\}$, which is identically one in a conic neighborhood of $(1,0,0)$ and is $C^\infty$ outside the origin. If $d > 0$ is small enough, there are constants $c$ and $k$ such that the functions $v_i$ defined by*

$$v_i(t,x) = \int_{\mathbb{R}^3} \sum_{p=1}^{6} \sum_{j=1}^{3} e^{it\tau_p(\xi)+i\langle x,\xi\rangle} T_{ipj}(\xi)\hat{\psi}_j(\xi)\chi(\xi)d\xi, \qquad i = 1,2,3.$$

*satisfy the estimate $|v_i(t,x)| \le c(1+|t|)^{-1/2}\ln(1+|t|)\sum_{j=1}^{3}\sum_{|\alpha|\le k} ||\partial_x^\alpha \psi_j(x)||_1$, where $||g||_1$ is for some integrable function $g$ defined on $\mathbb{R}^3$ its $L^1$ norm $||g||_1 = \int_{\mathbb{R}^3} |g(x)|dx$.*

## References

[1] A. Bannini, O. Liess: *Estimates for Fourier transforms of surface carried densities on surfaces with singular points. II.* Ann. Univ. Ferrara, Sez. VII, Sci. Mat. **52**, No. 2, (2006), 211–232.

[2] G.F.D. Duff: *The Cauchy problem for elastic waves in an anisotropic medium,* Phil. Transactions Royal Soc. London, Ser. A Nr. 1010, Vol. **252**, (1960), 249–273.

[3] S. Klainerman: *Global existence for nonlinear wave equations,* Comm. Pure Appl. Math. 33 (1980), 43–101.

[4] S. Klainerman, G. Ponce: *Global, small amplitude solutions to nonlinear evolution equations.* Commun. Pure Appl. Math. 36, 133–141 (1983).

[5] O. Liess: *Conical refraction and higher microlocalization.* Springer Lecture Notes in Mathematics, vol. **1555**, (1993).

[6] O. Liess: *Curvature properties of the slowness surface of the system of crystal acoustics for cubic crystals.* Osaka J. Math., vol.**45** (2008), 173–210.

[7] O. Liess: *Decay estimates for the solutions of the system of crystal acoustics for cubic crystals.* Asympt. Anal. **64** (2009), 1–27.

[8] C. Melotti: *Ph D thesis at Bologna University.* In preparation.

[9] M.J.P. Musgrave, Crystal acoustics, Holden and Day, San Francisco 1979.

[10] R. Racke, *Lectures on nonlinear evolution equations: initial value problems,* Vieweg Verlag, Wiesbaden, Aspects of mathematics, E 19, 1992.

[11] T.C. Sideris, *The null condition and global existence of nonlinear elastic waves.* Invent. Math. 123, No. 2, 323–342 (1996).

[12] E.M. Stein: *Harmonic analysis: real variable methods, orthogonality, and oscillatory integrals,* Princeton University Press, Princeton, 1993.

Otto Liess and Claudio Melotti
Department of Mathematics
Bologna University
Bologna, Italy
e-mail: liess@dm.unibo.it
melotti@dm.unibo.it

Operator Theory:
Advances and Applications, Vol. 221, 453-468


# Schatten-von Neumann Estimates for Resolvent Differences of Robin Laplacians on a Half-space

Vladimir Lotoreichik and Jonathan Rohleder

**Abstract.** The difference of the resolvents of two Laplacians on a half-space subject to Robin boundary conditions is studied. In general this difference is not compact, but it will be shown that it is compact and even belongs to some Schatten-von Neumann class, if the coefficients in the boundary condition are sufficiently close to each other in a proper sense. In certain cases the resolvent difference is shown to belong even to the same Schatten-von Neumann class as it is known for the resolvent difference of two Robin Laplacians on a domain with a compact boundary.

**Mathematics Subject Classification (2000).** Primary 47B10; Secondary 35P20.

**Keywords.** Robin Laplacian, Schatten-von Neumann class, non-selfadjoint operator, quasi-boundary triple.

## 1. Introduction

Schatten-von Neumann properties for resolvent differences of elliptic operators on domains have been studied basically since M.Sh. Birman's famous paper [6], which appeared fifty years ago and was followed by important contributions of M.Sh. Birman and M.Z. Solomjak as well as of G. Grubb, see [7, 16, 17]; moreover, the topic has attracted new interest very recently, see [4, 5, 20, 21, 26]. Recall that a compact operator belongs to the Schatten-von Neumann class $\mathfrak{S}_p$ (weak Schatten-von Neumann class $\mathfrak{S}_{p,\infty}$) of order $p > 0$ if the sequence of its singular values is $p$-summable (is $O(k^{-1/p})$ as $k \to \infty$); see Section 3 for more details. The objective of the present paper is to study the resolvent difference of two (in general non-selfadjoint) Robin Laplacians on the half-space $\mathbb{R}^{n+1}_+ = \{(x', x_{n+1})^{\mathrm{T}} : x' \in \mathbb{R}^n, x_{n+1} > 0\}$, $n \geq 1$, of the form

$$A_\alpha f = -\Delta f, \quad \operatorname{dom}(A_\alpha) = \left\{ f \in H^2(\mathbb{R}^{n+1}_+) : \partial_\nu f|_{\mathbb{R}^n} = \alpha f|_{\mathbb{R}^n} \right\}, \tag{1.1}$$

in $L^2(\mathbb{R}^{n+1}_+)$ with a function $\alpha : \mathbb{R}^n \to \mathbb{C}$ belonging to the Sobolev space $W^{1,\infty}(\mathbb{R}^n)$, i.e., $\alpha$ is bounded and has bounded partial derivatives of first order; here $f|_{\mathbb{R}^n}$ is the trace of a function $f$ at the boundary $\mathbb{R}^n$ of $\mathbb{R}^{n+1}_+$ and $\partial_\nu f|_{\mathbb{R}^n}$ is the trace of the normal derivative of $f$ with the normal pointing outwards of $\mathbb{R}^{n+1}_+$. We emphasize that, as a special case, our discussion contains the resolvent difference of the self-adjoint operator with a Neumann boundary condition and a Robin Laplacian. If the half-space $\mathbb{R}^{n+1}_+$ is replaced by a domain with a compact, smooth boundary, it is known that for real-valued $\alpha_1$ and $\alpha_2$ the operators $A_{\alpha_1}$ and $A_{\alpha_2}$ are selfadjoint and the difference of their resolvents

$$(A_{\alpha_2} - \lambda)^{-1} - (A_{\alpha_1} - \lambda)^{-1} \tag{1.2}$$

belongs to the class $\mathfrak{S}_{\frac{n}{3},\infty}$; see [5] and [4, 21], where also more general elliptic differential expressions and certain non-selfadjoint cases are discussed.

On the half-space $\mathbb{R}^{n+1}_+$ the situation is fundamentally different. Here, in general, the resolvent difference (1.2) is not even compact. For instance, if $\alpha_1 \neq \alpha_2$ are real, positive constants, the essential spectra of $A_{\alpha_1}$ and $A_{\alpha_2}$ are given by $[-\alpha_1^2, \infty)$ and $[-\alpha_2^2, \infty)$, respectively. Consequently, in this case the difference (1.2) cannot be compact. Nevertheless, the main results of the present paper show that under the assumption of a certain decay of the difference $\alpha_2(x) - \alpha_1(x)$ for $|x| \to \infty$, compactness of the resolvent difference in (1.2) can be guaranteed, and that this difference belongs to $\mathfrak{S}_p$ or $\mathfrak{S}_{p,\infty}$ for certain $p$, if $\alpha_2 - \alpha_1$ has a compact support or belongs to $L^q(\mathbb{R}^n)$ for some $q$. It is a question of special interest under which assumptions on $\alpha_2 - \alpha_1$ the difference (1.2) belongs to $\mathfrak{S}_{\frac{n}{3},\infty}$, i.e., to the same class as in the case of a domain with a compact boundary. Our results show that if $\alpha_2 - \alpha_1$ has a compact support this is always true, and that in dimensions $n > 3$ a sufficient condition is

$$\alpha_2 - \alpha_1 \in L^{n/3}(\mathbb{R}^n).$$

If $\alpha_2 - \alpha_1 \in L^p(\mathbb{R}^n)$ with $p \geq 1$ and $p > n/3$, we show that the resolvent difference in (1.2) belongs to the larger class $\mathfrak{S}_p \supsetneq \mathfrak{S}_{\frac{n}{3},\infty}$. In dimensions $n = 1, 2$ for $\alpha_2 - \alpha_1 \in L^1(\mathbb{R}^n)$ the difference (1.2) belongs to the trace class $\mathfrak{S}_1$. As an immediate consequence, for $n = 1, 2$ and real-valued $\alpha_1$, $\alpha_2$ with $\alpha_2 - \alpha_1 \in L^1(\mathbb{R}^n)$ wave operators for the pair $\{A_{\alpha_1}, A_{\alpha_2}\}$ exist and are complete, which is of importance in scattering theory. Two further corollaries of our results concern the case that $A_{\alpha_1}$ is the Neumann operator, i.e., $\alpha_1 = 0$: on the one hand, if $\alpha_2$ is real-valued and satisfies some decay condition, then the Neumann operator and the absolutely continuous part of $A_{\alpha_2}$ are unitarily equivalent, cf. [27]; on the other hand, with the help of recent results from perturbation theory for non-selfadjoint operators, see [9, 22], we conclude some statements on the accumulation of the (in general non-real) eigenvalues in the discrete spectrum of $A_{\alpha_2}$.

Our results complement and extend the result by M.Sh. Birman in [6]. He considers a realization of a symmetric second-order elliptic differential expression on an unbounded domain with combined boundary conditions, a Robin boundary condition on a compact part and a Dirichlet boundary condition on the remaining non-compact part of the boundary, and showed that the resolvent difference of the described realization and the realization with a Dirichlet boundary condition on the whole boundary belongs to the class $\mathfrak{S}_{\frac{n}{2},\infty}$. It is remarkable that in our situation in some cases the singular values converge faster than in the situation Birman considers. This phenomenon is already known for domains with compact boundaries, when a Neumann boundary condition instead of a Dirichlet boundary condition is considered; see [4].

It is worth mentioning that all results in this paper on compactness and Schatten-von Neumann estimates remain valid for $-\Delta$ replaced by a Schrödinger differential expression $-\Delta + V$ with a real-valued, bounded potential $V$ and the proofs are completely analogous.

Our considerations are based on an abstract concept from the extension theory of symmetric operators, namely, the notion of quasi-boundary triples, which was introduced by J. Behrndt and M. Langer in [2] and has been developed further by them together with the first author of the present paper in [5]. The key tool provided by the theory of quasi-boundary triples is a convenient factorization of the resolvent difference in (1.2). For the proof of our main theorem we combine this factorization with results on the compactness of the embedding of $H^1(\Omega)$ into $L^2(\Omega)$ for $\Omega$ being a (possibly unbounded) domain of finite measure and with $\mathfrak{S}_p$- and $\mathfrak{S}_{p,\infty}$-properties of the operator $\sqrt{|\alpha_2 - \alpha_1|}(I - \Delta_{\mathbb{R}^n})^{-3/4}$; the proof of the most optimal $\mathfrak{S}_{\frac{n}{3},\infty}$-estimate is based on an asymptotic result proved by M. Cwikel in [8], conjectured earlier by B. Simon in [28].

A short outline of this paper looks as follows. In Section 2 we give an overview of some known results on quasi-boundary triples which are used in the further analysis and provide a quasi-boundary triple for the Laplacian on the half-space; furthermore we prove that for each two coefficients $\alpha_1, \alpha_2$ the operators $A_{\alpha_1}$ and $A_{\alpha_2}$ have joint points in their resolvent sets and that $A_\alpha$ is selfadjoint if and only if $\alpha$ is real-valued. In Section 3 we establish sufficient conditions for the resolvent difference (1.2) to be compact or even to belong to certain Schatten-von Neumann classes. The paper concludes with some corollaries of the main results.

Let us fix some notation. If $T$ is a linear operator from a Hilbert space $\mathcal{H}$ into a Hilbert space $\mathcal{G}$ we denote by $\operatorname{dom} T$, $\operatorname{ran} T$, and $\ker T$ its domain, range, and kernel, respectively. If $T$ is densely defined, we write $T^*$ for the adjoint operator of $T$. If $\Theta$ and $\Lambda$ are linear relations from $\mathcal{H}$ to $\mathcal{G}$, i.e., linear subspaces of $\mathcal{H} \times \mathcal{G}$, we define their sum to be

$$\Theta + \Lambda = \big\{\{f, g_\Theta + g_\Lambda\} : \{f, g_\Theta\} \in \Theta, \{f, g_\Lambda\} \in \Lambda\big\}.$$

We write $T \in \mathcal{B}(\mathcal{H}, \mathcal{G})$, if $T$ is a bounded, everywhere defined operator from $\mathcal{H}$ into $\mathcal{G}$; if $\mathcal{G} = \mathcal{H}$ we simply write $T \in \mathcal{B}(\mathcal{G})$. For a closed operator $T$ in $\mathcal{H}$ we denote by $\rho(T)$ and $\sigma(T)$ its resolvent set and spectrum, respectively. Moreover, $\sigma_{\mathrm{d}}(T)$ denotes the discrete spectrum of $T$, i.e., the set of all eigenvalues of $T$ which are isolated in $\sigma(T)$ and have finite algebraic multiplicity, and $\sigma_{\mathrm{ess}}(T)$ is the essential spectrum of $T$, which consists of all points $\lambda \in \mathbb{C}$ such that $T - \lambda$ is not a semi-Fredholm operator. Finally, for a bounded, measurable function $\alpha : \mathbb{R}^n \to \mathbb{C}$ we denote its norm by $\|\alpha\|_\infty = \sup_{x \in \mathbb{R}^n} |\alpha(x)|$. Furthermore, for simplicity we identify $\alpha$ with the corresponding multiplication operator in $L^2(\mathbb{R}^n)$.

## 2. Quasi-boundary triples and Robin Laplacians on a half-space

In this section we provide some general facts on quasi-boundary triples as introduced in [2]. Afterwards we apply the theory to the Robin Laplacian in (1.1). Let us start with the basic definition.

**Definition 2.1.** Let $A$ be a closed, densely defined, symmetric operator in a Hilbert space $(\mathcal{H}, (\cdot,\cdot)_{\mathcal{H}})$. We say that $\{\mathcal{G}, \Gamma_0, \Gamma_1\}$ is a *quasi-boundary triple* for $A^*$, if $T \subset A^*$ is an operator satisfying $\overline{T} = A^*$ and $\Gamma_0$ and $\Gamma_1$ are linear mappings defined on $\operatorname{dom} T$ with values in the Hilbert space $(\mathcal{G}, (\cdot,\cdot)_{\mathcal{G}})$ such that the following conditions are satisfied.

(i) $\Gamma := \binom{\Gamma_0}{\Gamma_1} : \operatorname{dom} T \to \mathcal{G} \times \mathcal{G}$ has a dense range.

(ii) The *abstract Green identity*

$$(Tf, g)_{\mathcal{H}} - (f, Tg)_{\mathcal{H}} = (\Gamma_1 f, \Gamma_0 g)_{\mathcal{G}} - (\Gamma_0 f, \Gamma_1 g)_{\mathcal{G}} \tag{2.1}$$

holds for all $f, g \in \operatorname{dom} T$.

(iii) $A_0 := T \upharpoonright \ker \Gamma_0 = A^* \upharpoonright \ker \Gamma_0$ is selfadjoint.

We set $\mathcal{G}_i = \operatorname{ran} \Gamma_i$, $i = 0, 1$. Note that the definition of a quasi-boundary triple as given above is only a special case of the original one given in [2] for the adjoint of a closed, symmetric linear relation $A$. We remark that if $\{\mathcal{G}, \Gamma_0, \Gamma_1\}$ is a quasi-boundary triple with the additional property $\mathcal{G}_0 = \mathcal{G}$, then $\{\mathcal{G}, \Gamma_0, \Gamma_1\}$ is a generalized boundary triple in the sense of [11]. Let us also mention that a quasi-boundary triple for $A^*$ exists if and only if the deficiency indices $\dim \ker(A^* \mp i)$ of $A$ coincide. If $\{\mathcal{G}, \Gamma_0, \Gamma_1\}$ is a quasi-boundary triple for $A^*$ with $T$ as in Definition 2.1, then $A$ coincides with $T \upharpoonright \ker \Gamma$.

The next proposition contains a sufficient condition for a triple $\{\mathcal{G}, \Gamma_0, \Gamma_1\}$ to be a quasi-boundary triple. For a proof see [2, Thm. 2.3]; cf. also [3, Thm. 2.3].

**Proposition 2.2.** *Let $\mathcal{H}$ and $\mathcal{G}$ be Hilbert spaces and let $T$ be a linear operator in $\mathcal{H}$. Assume that $\Gamma_0, \Gamma_1 : \operatorname{dom} T \to \mathcal{G}$ are linear mappings such that the following conditions are satisfied:*

(a) $\Gamma := \binom{\Gamma_0}{\Gamma_1} : \operatorname{dom} T \to \mathcal{G} \times \mathcal{G}$ *has a dense range.*

(b) *The identity* (2.1) *holds for all* $f, g \in \operatorname{dom} T$.

(c) $T \upharpoonright \ker \Gamma_0$ *contains a selfadjoint operator and* $\ker \Gamma_0 \cap \ker \Gamma_1$ *is dense in* $\mathcal{H}$.

*Then $A := T \upharpoonright \ker \Gamma$ is a closed, densely defined, symmetric operator in $\mathcal{H}$ and $\{\mathcal{G}, \Gamma_0, \Gamma_1\}$ is a quasi-boundary triple for $A^*$.*

Let us recall next the definition of two related analytic objects, the $\gamma$-field and the Weyl function associated with a quasi-boundary triple.

**Definition 2.3.** Let $A$ be a closed, densely defined, symmetric operator in $\mathcal{H}$ and let $\{\mathcal{G}, \Gamma_0, \Gamma_1\}$ be a quasi-boundary triple for $A^*$ with $T$ as in Definition 2.1 and $A_0 = T \upharpoonright \ker \Gamma_0$. Then the operator-valued functions $\gamma$ and $M$ defined by

$$\gamma(\lambda) := \big(\Gamma_0 \upharpoonright \ker(T - \lambda)\big)^{-1} \quad \text{and} \quad M(\lambda) := \Gamma_1 \gamma(\lambda), \quad \lambda \in \rho(A_0),$$

are called the *$\gamma$-field* and the *Weyl function*, respectively, corresponding to the triple $\{\mathcal{G}, \Gamma_0, \Gamma_1\}$.

These definitions coincide with the definitions of the $\gamma$-field and the Weyl function in the case that $\{\mathcal{G},\Gamma_0,\Gamma_1\}$ is an ordinary boundary triple, see [10]. It is an immediate consequence of the decomposition

$$\operatorname{dom} T = \operatorname{dom} A_0 \dotplus \ker(T-\lambda) = \ker\Gamma_0 \dotplus \ker(T-\lambda), \quad \lambda\in\rho(A_0), \tag{2.2}$$

that the mappings $\gamma(\lambda)$ and $M(\lambda)$ are well defined. Note that for each $\lambda\in\rho(A_0)$, $\gamma(\lambda)$ maps $\mathcal{G}_0$ onto $\ker(T-\lambda)\subset\mathcal{H}$ and $M(\lambda)$ maps $\mathcal{G}_0$ into $\mathcal{G}_1$. Furthermore, it follows immediately from the definitions of $\gamma(\lambda)$ and $M(\lambda)$ that

$$\gamma(\lambda)\Gamma_0 f_\lambda = f_\lambda \quad\text{and}\quad M(\lambda)\Gamma_0 f_\lambda = \Gamma_1 f_\lambda, \quad f_\lambda\in\ker(T-\lambda), \tag{2.3}$$

holds for all $\lambda\in\rho(A_0)$.

In the next proposition we collect some properties of the $\gamma$-field and the Weyl function; all statements can be found in [2, Proposition 2.6].

**Proposition 2.4.** *Let $A$ be a closed, densely defined, symmetric operator in a Hilbert space $\mathcal{H}$ and let $\{\mathcal{G},\Gamma_0,\Gamma_1\}$ be a quasi-boundary triple for $A^*$ with $\gamma$-field $\gamma$ and Weyl function $M$. Denote by $A_0$ the restriction of $A^*$ to* $\ker\Gamma_0$. *Then for $\lambda\in\rho(A_0)$ the following assertions hold.*

(i) *$\gamma(\lambda)$ is a bounded, densely defined operator from $\mathcal{G}$ into $\mathcal{H}$.*

(ii) *The adjoint of $\gamma(\overline{\lambda})$ can be expressed as*

$$\gamma(\overline{\lambda})^* = \Gamma_1(A_0-\lambda)^{-1}\in\mathcal{B}(\mathcal{H},\mathcal{G}).$$

(iii) *$M(\lambda)$ is a densely defined, in general unbounded operator in $\mathcal{G}$, whose range is contained in $\mathcal{G}_1$ and which satisfies $M(\overline{\lambda})\subset M(\lambda)^*$.*

A quasi-boundary triple provides a parametrization for a class of extensions of a closed, densely defined, symmetric operator $A$. If $\{\mathcal{G},\Gamma_0,\Gamma_1\}$ is a quasi-boundary triple for $A^*$ with $T$ as in Definition 2.1 and $\Theta$ is a linear relation in $\mathcal{G}$, we denote by $A_\Theta$ the restriction of $T$ given by

$$A_\Theta f = Tf, \quad \operatorname{dom} A_\Theta = \left\{ f\in\operatorname{dom}T\colon \left(\begin{smallmatrix}\Gamma_0 f\\ \Gamma_1 f\end{smallmatrix}\right)\in\Theta\right\}. \tag{2.4}$$

In contrast to the case of an ordinary boundary triple, this parametrization does not cover all extensions of $A$ which are contained in $A^*$, and selfadjointness of $\Theta$ does not imply selfadjointness or essential selfadjointness of $A_\Theta$; cf. [2, Proposition 4.11] for a counterexample and [2, Proposition 2.4]. The following proposition shows that under certain conditions $\rho(A_\Theta)$ is non-empty, and it implies a sufficient condition for selfadjointness of $A_\Theta$. It also provides a formula of Krein type for the resolvent difference of $A_\Theta$ and $A_0$. In the present form the proposition is a special case of [2, Theorem 2.8].

**Proposition 2.5.** *Let $A$ be a closed, densely defined, symmetric operator in $\mathcal{H}$ and let $\{\mathcal{G},\Gamma_0,\Gamma_1\}$ be a quasi-boundary triple for $A^*$ with $A_0 = A^*\upharpoonright\ker\Gamma_0$. Let $\gamma$ be the corresponding $\gamma$-field and $M$ the corresponding Weyl function. Furthermore, let $\Theta$ be a linear relation in $\mathcal{G}$ and assume that $(\Theta-M(\lambda))^{-1}\in\mathcal{B}(\mathcal{G})$ is satisfied for some $\lambda\in\rho(A_0)$. Then $\lambda\in\rho(A_\Theta)$ and*

$$(A_\Theta-\lambda)^{-1}-(A_0-\lambda)^{-1} = \gamma(\lambda)\big(\Theta-M(\lambda)\big)^{-1}\gamma(\overline{\lambda})^* \quad \textit{holds.}$$

In order to construct a specific quasi-boundary triple for $-\Delta$ on the half-space $\mathbb{R}^{n+1}_+$, $n \geq 1$, let us recall some basic facts on traces of functions from Sobolev spaces. For proofs and further details see, e.g., [1, 19, 25]. We denote by $H^s(\mathbb{R}^{n+1}_+)$ and $H^s(\mathbb{R}^n)$ the $L^2$-based Sobolev spaces of order $s \geq 0$ on $\mathbb{R}^{n+1}_+$ and its boundary $\mathbb{R}^n$, respectively. The closure in $H^s(\mathbb{R}^{n+1}_+)$ of the space of infinitely-differentiable functions with a compact support is denoted by $H^s_0(\mathbb{R}^{n+1}_+)$. The trace map $C_0^\infty(\overline{\mathbb{R}^{n+1}_+}) \ni f \mapsto f|_{\mathbb{R}^n} \in C^\infty(\mathbb{R}^n)$ and the trace of the derivative $C_0^\infty(\overline{\mathbb{R}^{n+1}_+}) \ni f \mapsto \partial_\nu f|_{\mathbb{R}^n} = -\frac{\partial f}{\partial x_{n+1}}|_{\mathbb{R}^n} \in C^\infty(\mathbb{R}^n)$ in the direction of the normal vector field pointing outwards of $\mathbb{R}^{n+1}_+$ extend by continuity to $H^s(\mathbb{R}^{n+1}_+)$, $s > 3/2$, such that the mapping

$$H^s(\mathbb{R}^{n+1}_+) \ni f \mapsto \begin{pmatrix} f|_{\mathbb{R}^n} \\ \partial_\nu f|_{\mathbb{R}^n} \end{pmatrix} \in H^{s-1/2}(\mathbb{R}^n) \times H^{s-3/2}(\mathbb{R}^n)$$

is well defined and surjective onto $H^{s-1/2}(\mathbb{R}^n) \times H^{s-3/2}(\mathbb{R}^n)$. Moreover, this mapping can be extended to the spaces

$$H^s_\Delta(\mathbb{R}^{n+1}_+) := \{f \in H^s(\mathbb{R}^{n+1}_+) \colon \Delta f \in L^2(\mathbb{R}^{n+1}_+)\}, \quad s \geq 0;$$

cf. [14, 18, 19, 25]. We remark that for $s \geq 2$ the latter space coincides with the usual Sobolev space $H^s(\mathbb{R}^n)$. In contrast to the case $s \geq 2$, the mapping

$$H^s_\Delta(\mathbb{R}^{n+1}_+) \ni f \mapsto \begin{pmatrix} f|_{\mathbb{R}^n} \\ \partial_\nu f|_{\mathbb{R}^n} \end{pmatrix} \in H^{s-1/2}(\mathbb{R}^n) \times H^{s-3/2}(\mathbb{R}^n), \quad s \in [0, 2),$$

is not surjective onto the product $H^{s-1/2}(\mathbb{R}^n) \times H^{s-3/2}(\mathbb{R}^n)$, but the separate mappings

$$H^s_\Delta(\mathbb{R}^{n+1}_+) \ni f \mapsto f|_{\mathbb{R}^n} \in H^{s-1/2}(\mathbb{R}^n), \quad s \in [0, 2),$$

and

$$H^s_\Delta(\mathbb{R}^{n+1}_+) \ni f \mapsto \partial_\nu f|_{\mathbb{R}^n} \in H^{s-3/2}(\mathbb{R}^n), \quad s \in [0, 2), \tag{2.5}$$

are surjective onto $H^{s-1/2}(\mathbb{R}^n)$ and $H^{s-3/2}(\mathbb{R}^n)$, respectively.

Let us introduce the operator realizations of $-\Delta$ in $L^2(\mathbb{R}^{n+1}_+)$ given by

$$Af = -\Delta f, \quad \operatorname{dom} A = H^2_0(\mathbb{R}^{n+1}_+), \tag{2.6}$$

and

$$Tf = -\Delta f, \quad \operatorname{dom} T = H^{3/2}_\Delta(\mathbb{R}^{n+1}_+),$$

and the boundary mappings $\Gamma_0$ and $\Gamma_1$ defined by

$$\Gamma_0 f = \partial_\nu f|_{\mathbb{R}^n}, \quad \Gamma_1 f = f|_{\mathbb{R}^n}, \qquad f \in \operatorname{dom} T. \tag{2.7}$$

Furthermore, let us mention that the *Neumann operator*

$$A_{\mathrm{N}} f = -\Delta f, \quad \operatorname{dom} A_{\mathrm{N}} = \{f \in H^2(\mathbb{R}^{n+1}_+) \colon \partial_\nu f\big|_{\mathbb{R}^n} = 0\}, \tag{2.8}$$

is selfadjoint in $L^2(\mathbb{R}^{n+1}_+)$ and its spectrum is given by $\sigma(A_N) = [0,\infty)$; see, e.g., [19, Chapter 9]. We prove now that the mappings $\Gamma_0$ and $\Gamma_1$ in (2.7) provide a quasi-boundary triple for the operator $A^*$ with $A_0 := A^* \upharpoonright \ker\Gamma_0 = A_N$.

**Proposition 2.6.** *The operator $A$ in* (2.6) *is closed, densely defined, and symmetric, and the triple $\{\mathcal{G},\Gamma_0,\Gamma_1\}$ with $\mathcal{G} = L^2(\mathbb{R}^n)$ and $\Gamma_0,\Gamma_1$ defined in* (2.7) *is a quasi-boundary triple for $A^*$. Moreover, $A^* \upharpoonright \ker\Gamma_0 = A_N$ holds. For $\lambda \in \rho(A_N)$ the associated $\gamma$-field is given by the* Poisson operator

$$\gamma(\lambda)\partial_\nu f_\lambda|_{\mathbb{R}^n} = f_\lambda, \quad f_\lambda \in \ker(T-\lambda), \tag{2.9}$$

*and the associated Weyl function is given by the* Neumann-to-Dirichlet map

$$M(\lambda)\partial_\nu f_\lambda|_{\mathbb{R}^n} = f_\lambda|_{\mathbb{R}^n}, \quad f_\lambda \in \ker(T-\lambda), \tag{2.10}$$

*and satisfies $M(\lambda) \in \mathcal{B}(L^2(\mathbb{R}^n))$.*

*Proof.* We verify the conditions (a)–(c) of Proposition 2.2. The mapping

$$H^2(\mathbb{R}^{n+1}_+) \ni f \mapsto \begin{pmatrix} \partial_\nu f|_{\mathbb{R}^n} \\ f|_{\mathbb{R}^n} \end{pmatrix} \in H^{1/2}(\mathbb{R}^n) \times H^{3/2}(\mathbb{R}^n)$$

is surjective, see above. Since it is a restriction of the mapping $\Gamma = \binom{\Gamma_0}{\Gamma_1}$, the density of $H^{1/2}(\mathbb{R}^n) \times H^{3/2}(\mathbb{R}^n)$ in $L^2(\mathbb{R}^n) \times L^2(\mathbb{R}^n)$ yields (a). Condition (b) is just the usual second Green identity,

$$(-\Delta f, g) - (f, -\Delta g) = (f|_{\mathbb{R}^n}, \partial_\nu g|_{\mathbb{R}^n}) \quad (\partial_\nu f|_{\mathbb{R}^n}, g|_{\mathbb{R}^n}),$$

for $f,g \in H^{3/2}_\Delta(\mathbb{R}^{n+1}_+)$, which can be found in, e.g., [14, Theorem 5.5]; here the inner products in $L^2(\mathbb{R}^{n+1}_+)$ and in $L^2(\mathbb{R}^n)$ both are denoted by $(\cdot,\cdot)$. In order to verify (c) we observe that the operator $T \upharpoonright \ker\Gamma_0$ is $-\Delta$ on the domain

$$\left\{ f \in H^{3/2}_\Delta(\mathbb{R}^{n+1}_+) : \partial_\nu f|_{\mathbb{R}^n} = 0 \right\}$$

in $L^2(\mathbb{R}^{n+1}_+)$, which contains the domain of the selfadjoint Neumann operator $A_N$ in (2.8). Moreover, $\ker\Gamma_0 \cap \ker\Gamma_1 = H^2_0(\mathbb{R}^{n+1}_+)$ is dense in $L^2(\mathbb{R}^{n+1}_+)$. Thus by Proposition 2.2 $\{L^2(\mathbb{R}^n),\Gamma_0,\Gamma_1\}$ is a quasi-boundary triple for $A^*$ and the statements on $A$ are true. In particular, $T \upharpoonright \ker\Gamma_0$ coincides with the Neumann operator $A_N$. The representations (2.9) and (2.10) follow immediately from (2.3) and the definition of the boundary mappings $\Gamma_0$ and $\Gamma_1$. It remains to show that $M(\lambda)$ is bounded and everywhere defined. Since by Proposition 2.4 (iii) $M(\lambda) \subset M(\overline{\lambda})^*$ holds for each $\lambda \in \rho(A_N)$ and the latter operator is closed, $M(\lambda)$ is closable. It follows from $\operatorname{dom} M(\lambda) = \operatorname{ran}\Gamma_0 = L^2(\mathbb{R}^n)$, see (2.5), that $M(\lambda)$ is even closed and, hence, $M(\lambda) \in \mathcal{B}(L^2(\mathbb{R}^n))$ by the closed graph theorem. □

For the sake of completeness we remark that the adjoint of $A$ is given by

$$A^* f = -\Delta f, \quad \operatorname{dom} A^* = \left\{ f \in L^2(\mathbb{R}^{n+1}_+) : \Delta f \in L^2(\mathbb{R}^{n+1}_+) \right\},$$

but this will not play a role in our further considerations.

We are now able to provide some information on the operator $A_\alpha$ in (1.1).

**Theorem 2.7.** *Let $\alpha \in W^{1,\infty}(\mathbb{R}^n)$. Then each $\lambda < -\|\alpha\|_\infty^2$ belongs to $\rho(A_\alpha)$. Moreover, $A_\alpha$ is selfadjoint if and only if $\alpha$ is real valued. In particular, in this case $A_\alpha$ is semibounded from below by $-\|\alpha\|_\infty^2$.*

*Remark* 2.8. We emphasize that in certain cases the estimate for the spectrum of $A_\alpha$ given in Theorem 2.7 is very rough. For example, if $\alpha$ is a real, nonpositive function, the first Green identity implies that $A_\alpha$ is even nonnegative.

*Proof of Theorem* 2.7. Let $\{L^2(\mathbb{R}^n), \Gamma_0, \Gamma_1\}$ be the quasi-boundary triple for $A^*$ in Proposition 2.6, $\gamma$ the corresponding $\gamma$-field, and $M$ the corresponding Weyl function. We verify first that with respect to the quasi-boundary triple $\{L^2(\mathbb{R}^n), \Gamma_0, \Gamma_1\}$ in Proposition 2.6 the operator $A_\alpha$ admits a representation $A_\alpha = A_\Theta$ in the sense of (2.4) with

$$\Theta = \left\{ \begin{pmatrix} \alpha f \\ f \end{pmatrix} : f \in L^2(\mathbb{R}^n) \right\}. \tag{2.11}$$

In fact, it is obvious from the definitions that $A_\alpha \subset A_\Theta$ holds, and it remains to show $\operatorname{dom} A_\Theta \subset H^2(\mathbb{R}^{n+1}_+)$. Let $f \in \operatorname{dom} A_\Theta \subset \operatorname{dom} T$ and $\eta \in \rho(A_{\mathrm{N}})$. By (2.2) there exist $f_{\mathrm{N}} \in \operatorname{dom} A_{\mathrm{N}}$ and $f_\eta \in \ker(T - \eta)$ with $f = f_{\mathrm{N}} + f_\eta$. Clearly, $f_{\mathrm{N}}$ belongs to $H^2(\mathbb{R}^{n+1}_+)$. Moreover, $f$ satisfies the boundary condition

$$\alpha \Gamma_1 f = \Gamma_0 f = \Gamma_0 f_\eta;$$

in particular, $\operatorname{ran} \Gamma_1 = H^1(\mathbb{R}^n)$ and the regularity assumption on $\alpha$ imply $\Gamma_0 f_\eta \in H^1(\mathbb{R}^n) \subset H^{1/2}(\mathbb{R}^n)$. Since the mapping $f \mapsto \partial_\nu f|_{\mathbb{R}^n}$ provides a bijection between $H^2(\mathbb{R}^{n+1}_+) \cap \ker(T - \eta)$ and $H^{1/2}(\mathbb{R}^n)$, see, e.g., [18, Section 3], it follows $f_\eta \in H^2(\mathbb{R}^{n+1}_+)$. This shows $f \in H^2(\mathbb{R}^{n+1}_+)$ and, hence, $A_\Theta = A_\alpha$.

Let $\lambda < -\|\alpha\|_\infty^2$ be fixed. Then $\lambda \in \rho(A_{\mathrm{N}})$ holds and by Proposition 2.5 in order to verify $\lambda \in \rho(A_\alpha)$ it is sufficient to show $0 \in \rho(\Theta - M(\lambda))$. Note first that $\Theta$ is injective; hence we can write

$$(\Theta - M(\lambda))^{-1} = \Theta^{-1}\big(I - M(\lambda)\Theta^{-1}\big)^{-1}, \tag{2.12}$$

where the equality has first to be understood in the sense of linear relations. Since $\Theta^{-1} = \alpha \in \mathcal{B}(L^2(\mathbb{R}^n))$, we only need to show that $I - M(\lambda)\Theta^{-1}$ has a bounded, everywhere defined inverse. In fact, the Neumann-to-Dirichlet map is given by

$$M(\lambda) = (-\Delta_{\mathbb{R}^n} - \lambda)^{-1/2},$$

where $\Delta_{\mathbb{R}^n}$ denotes the Laplacian in $L^2(\mathbb{R}^n)$, defined on $H^2(\mathbb{R}^n)$; cf., e.g., [19, Chapter 9]. In particular, $\|M(\lambda)\| = 1/\sqrt{-\lambda}$ holds. This implies $\|M(\lambda)\Theta^{-1}\| < 1$. Now (2.12) yields that $(\Theta - M(\lambda))^{-1}$ is a bounded operator, which is everywhere defined. This implies $\lambda \in \rho(A_\alpha)$.

It follows immediately from the Green identity that $A_\alpha$ is symmetric if and only if $\alpha$ is real valued. In this case $A_\alpha$ is even selfadjoint as $\rho(A_\alpha) \cap \mathbb{R}$ is nonempty. Since we have shown that each $\lambda < -\|\alpha\|_\infty^2$ belongs to $\rho(A_\alpha)$, the statement on the semiboundedness of $A_\alpha$ follows immediately. This completes the proof. $\square$

## 3. Compactness and Schatten-von Neumann estimates for resolvent differences of Robin Laplacians

The present section is devoted to our main results on compactness and Schatten-von Neumann properties of the resolvent difference

$$(A_{\alpha_2} - \lambda)^{-1} - (A_{\alpha_1} - \lambda)^{-1} \tag{3.1}$$

of two Robin Laplacians as in (1.1) with boundary coefficients $\alpha_1$ and $\alpha_2$ in dependence of the asymptotic behavior of $\alpha_2 - \alpha_1$. Let us shortly recall the definition of the Schatten-von Neumann classes and some of their basic properties. For more details see [15, Chapter II and III] and [29]. Let $\mathfrak{S}_\infty(\mathcal{G}, \mathcal{H})$ denote the linear space of all compact linear operators mapping the Hilbert space $\mathcal{G}$ into the Hilbert space $\mathcal{H}$. Usually the spaces $\mathcal{G}$ and $\mathcal{H}$ are clear from the context and we simply write $\mathfrak{S}_\infty$. For $K \in \mathfrak{S}_\infty$ we denote by $s_k(K)$, $k = 1, 2, \ldots$, the *singular values* (or *s-numbers*) of $K$, i.e., the eigenvalues of the compact, selfadjoint, nonnegative operator $(K^*K)^{1/2}$, enumerated in decreasing order and counted according to their multiplicities. Note that for a selfadjoint, nonnegative operator $K \in \mathfrak{S}_\infty$ the singular values are precisely the eigenvalues of $K$.

**Definition 3.1.** An operator $K \in \mathfrak{S}_\infty$ is said to belong to the *Schatten-von Neumann class* $\mathfrak{S}_p$ of order $p > 0$, if its singular values satisfy

$$\sum_{k=1}^{\infty} \big(s_k(K)\big)^p < \infty.$$

An operator $K$ is said to belong to the *weak Schatten-von Neumann class* $\mathfrak{S}_{p,\infty}$ of order $p > 0$, if

$$s_k(K) = O(k^{-1/p}), \quad k \to \infty,$$

holds.

Some well-known properties of the Schatten-von Neumann classes are collected in the following lemma; for proofs see the above-mentioned references and [5, Lemma 2.3].

**Lemma 3.2.** *For $p, q, r > 0$ the following assertions hold.*

(i) *Let $\frac{1}{p} + \frac{1}{q} = \frac{1}{r}$. If $K \in \mathfrak{S}_p$ and $L \in \mathfrak{S}_q$, then $KL \in \mathfrak{S}_r$; if $K \in \mathfrak{S}_{p,\infty}$ and $L \in \mathfrak{S}_{q,\infty}$, then $KL \in \mathfrak{S}_{r,\infty}$;*

(ii) *$K \in \mathfrak{S}_p \iff K^* \in \mathfrak{S}_p$ and $K \in \mathfrak{S}_{p,\infty} \iff K^* \in \mathfrak{S}_{p,\infty}$;*

(iii) *$\mathfrak{S}_p \subset \mathfrak{S}_{p,\infty}$ and $\mathfrak{S}_{p,\infty} \subset \mathfrak{S}_q$ for all $q > p$, but $\mathfrak{S}_{p,\infty} \not\subset \mathfrak{S}_p$.*

Let us now come to the investigation of compactness and Schatten-von Neumann properties of (3.1). The condition

$$\mu\left(\{x \in \mathbb{R}^n : |\alpha(x)| \geq \varepsilon\}\right) < \infty \quad \text{for all } \varepsilon > 0 \tag{3.2}$$

for $\alpha = \alpha_2 - \alpha_1$ turns out to be sufficient for the compactness of the resolvent difference (3.1), see Theorem 3.6 below; here $\mu$ denotes the Lebesgue measure on

$\mathbb{R}^n$. We remark that the condition (3.2) includes, e.g., the case that $\alpha$ belongs to $L^q(\mathbb{R}^n)$ for some $q \geq 1$, and the case that $\sup_{|x|\geq r} |\alpha(x)| \to 0$ as $r \to \infty$.

The following lemma contains the main ingredients of the proof of Theorem 3.6 and Theorem 3.7 below.

**Lemma 3.3.** *Let $\mathcal{K}$ be a Hilbert space and let $K \in \mathcal{B}(\mathcal{K}, L^2(\mathbb{R}^n))$ be an operator with* $\operatorname{ran} K \subset H^{3/2}(\mathbb{R}^n)$. *Assume $\alpha \in L^\infty(\mathbb{R}^n)$.*

(i) *If $\alpha$ satisfies the condition (3.2), then $\alpha K \in \mathfrak{S}_\infty$.*
(ii) *If $\alpha$ has a compact support or if $n > 3$ and $\alpha \in L^{2n/3}(\mathbb{R}^n)$, then*

$$\alpha K \in \mathfrak{S}_{\frac{2n}{3},\infty}.$$

(iii) *If $\alpha \in L^2(\mathbb{R}^n)$ and $n \geq 3$, then*

$$\alpha K \in \mathfrak{S}_r \quad \textit{for all } r > 2n/3.$$

(iv) *If $\alpha \in L^p(\mathbb{R}^n)$ for $p \geq 2$ and $p > \frac{2n}{3}$, then*

$$\alpha K \in \mathfrak{S}_p.$$

*Proof.* Assume first that $\alpha$ satisfies (3.2). Then there exists a sequence $\Omega_1 \subset \Omega_2 \subset \cdots$ of smooth domains of finite measure whose union is all of $\mathbb{R}^n$ such that for each $m \in \mathbb{N}$ we have $|\alpha(x)| < \frac{1}{m}$ for all $x \in \mathbb{R}^n \setminus \Omega_m$. For each $m \in \mathbb{N}$ let $\chi_m$ be the characteristic function of the set $\Omega_m$. Denote by $P_m$ the canonical projection from $L^2(\mathbb{R}^n)$ to $L^2(\Omega_m)$ and by $J_m$ the canonical embedding of $L^2(\Omega_m)$ into $L^2(\mathbb{R}^n)$. Then $\operatorname{ran}(P_m\chi_m K) \subset H^{3/2}(\Omega_m) \subset H^1(\Omega_m)$ and, by embedding statements, $P_m\chi_m K : \mathcal{K} \to L^2(\Omega_m)$ is compact; see [12, Theorem 3.4 and Theorem 4.11] and [13, Chapter V]. Since $\alpha J_m$ is bounded, it turns out that $\alpha\chi_m K = \alpha J_m P_m \chi_m K$ is compact. From the assumption (3.2) on $\alpha$ it follows easily that the sequence of operators $\alpha\chi_m K$ converges to $\alpha K$ in the operator-norm topology. Thus also $\alpha K$ is compact, which is the assertion of item (i).

Let us assume that $\alpha$ has a compact support and that $\Omega \subset \mathbb{R}^n$ is a bounded, smooth domain with $\Omega \supset \operatorname{supp}\alpha$. Let $P$ be the canonical projection in $L^2(\mathbb{R}^n)$ onto $L^2(\Omega)$ and let $J$ be the canonical embedding of $L^2(\Omega)$ into $L^2(\mathbb{R}^n)$, and let $\widetilde{\alpha} := \alpha|_\Omega$. Since $\operatorname{ran}(PK) \subset H^{3/2}(\Omega)$ and $\Omega$ is a bounded, smooth domain, the embedding operator from $H^{3/2}(\Omega)$ into $L^2(\Omega)$ is contained in the class $\mathfrak{S}_{\frac{2n}{3},\infty}$, see [23, Theorem 7.8]. It follows $PK \in \mathfrak{S}_{\frac{2n}{3},\infty}$ as a mapping from $\mathcal{K}$ into $L^2(\Omega)$. Since $J\widetilde{\alpha}$ is bounded, we obtain $\alpha K = J\widetilde{\alpha}PK \in \mathfrak{S}_{\frac{2n}{3},\infty}$.

The proofs of the remaining statements make use of spectral estimates for the operator $\alpha D$ in $L^2(\mathbb{R}^n)$ with

$$D = (I - \Delta_{\mathbb{R}^n})^{-3/4} = g(-i\nabla), \quad g(x) = (1 + |x|^2)^{-3/4}, \; x \in \mathbb{R}^n, \tag{3.3}$$

where the formal notation $g(-i\nabla)$ can be made precise with the help of the Fourier transformation. We remark that $D$, regarded as an operator from $L^2(\mathbb{R}^n)$ into $H^{3/2}(\mathbb{R}^n)$, is an isometric isomorphism. Recall that a function $f$ is said to belong to the *weak Lebesgue space* $L^{p,\infty}(\mathbb{R}^n)$ for some $p > 0$, if the condition

$$\sup_{t>0} \left( t^p \mu(\{x \in \mathbb{R}^n : |f(x)| > t\}) \right) < \infty$$

is satisfied, where $\mu$ denotes the Lebesgue measure on $\mathbb{R}^n$. The function $g$ in (3.3) belongs to $L^{2n/3,\infty}(\mathbb{R}^n)$. In fact, one easily verifies that the set $\{x \in \mathbb{R}^n : |g(x)| > t\}$ is contained in the ball of radius $t^{-2/3}$ centered at the origin, and the formula for the volume of a ball leads to the claim. Let now $n > 3$ and $\alpha \in L^{2n/3}(\mathbb{R}^n)$. Then a result by M. Cwikel in [8] yields

$$\alpha D \in \mathfrak{S}_{\frac{2n}{3},\infty};$$

see also [29, Theorem 4.2]. We conclude

$$\alpha K = \alpha D D^{-1} K \in \mathfrak{S}_{\frac{2n}{3},\infty}.$$

Thus we have proved (ii).

In order to show (iii) let us assume $\alpha \in L^2(\mathbb{R}^n)$ and $n \geq 3$. Since $\alpha$ is bounded, $\alpha \in L^p(\mathbb{R}^n)$ for each $p > 2$. It is easy to check that $g$ in (3.3) belongs to $L^p(\mathbb{R}^n)$ for each $p > 2n/3$. The standard result [29, Theorem 4.1] and $\alpha, g \in L^r(\mathbb{R}^n)$ for all $r > 2n/3 \geq 2$ imply

$$\alpha D \in \mathfrak{S}_r \quad \text{for all } r > 2n/3.$$

It follows

$$\alpha K = \alpha D D^{-1} K \in \mathfrak{S}_r \quad \text{for all } r > 2n/3,$$

which is the assertion of (iii).

Let now $\alpha \in L^p(\mathbb{R}^n)$ for $p \geq 2$ and $p > 2n/3$. As above, $g \in L^p(\mathbb{R}^n)$ and [29, Theorem 4.1] yields $\alpha D \in \mathfrak{S}_p$. Hence, $\alpha K = \alpha D D^{-1} K \in \mathfrak{S}_p$, which completes the proof of (iv). □

*Remark* 3.4. The condition in Lemma 3.3 (i) can still be slightly weakened using the optimal prerequisites on a domain $\Omega$ which imply compactness of the embedding of $H^1(\Omega)$ into $L^2(\Omega)$; see, e.g., [13, Chapter VIII]. To avoid too inconvenient and technical assumptions, we restrict ourselves to the above condition.

We continue with giving a factorization of the resolvent difference of two Robin Laplacians. It is based on the formula of Krein type in Proposition 2.5 and will be crucial for the proofs of our main results. We remark that an analogous formula as below is well known for ordinary boundary triples and abstract boundary conditions, see [10, Proof of Theorem 2].

**Lemma 3.5.** *Let $\alpha_1, \alpha_2 \in W^{1,\infty}(\mathbb{R}^n)$ and let $A_{\alpha_1}, A_{\alpha_2}$ be the corresponding Robin Laplacians as in* (1.1). *Then*

$$\begin{aligned}(A_{\alpha_2} - \lambda)^{-1} &- (A_{\alpha_1} - \lambda)^{-1} \\ &= \gamma(\lambda) \left(I - \alpha_1 M(\lambda)\right)^{-1} (\alpha_2 - \alpha_1) \left(I - M(\lambda)\alpha_2\right)^{-1} \gamma(\lambda)^*\end{aligned}$$

*holds for each $\lambda < -\max\{\|\alpha_1\|_\infty^2, \|\alpha_2\|_\infty^2\}$, where $\gamma(\lambda)$ is the Poisson operator in* (2.9) *and $M(\lambda)$ is the Neumann-to-Dirichlet map in* (2.10).

*Proof.* Let $A$ be given as in (2.6) and let $\{L^2(\mathbb{R}^n), \Gamma_0, \Gamma_1\}$ be the quasi-boundary triple for $A^*$ in Proposition 2.6, so that $\gamma$ is the corresponding $\gamma$-field and $M$ is the corresponding Weyl function. Let us fix $\lambda$ as in the proposition. Then $\lambda$ belongs to $\rho(A_{\alpha_1}) \cap \rho(A_{\alpha_2})$ by Theorem 2.7. Moreover, if $\Theta_1$ and $\Theta_2$ denote the linear relations corresponding to $\alpha_1$ and $\alpha_2$, respectively, as in (2.11), then we have

$$\begin{aligned}&\big(\Theta_2 - M(\lambda)\big)^{-1} - \big(\Theta_1 - M(\lambda)\big)^{-1}\\&\quad = \alpha_2\big(I - M(\lambda)\alpha_2\big)^{-1} - \big(I - \alpha_1 M(\lambda)\big)^{-1}\alpha_1\\&\quad = \big(I - \alpha_1 M(\lambda)\big)^{-1}\Big(\big(I - \alpha_1 M(\lambda)\big)\alpha_2 - \alpha_1\big(I - M(\lambda)\alpha_2\big)\Big)\big(I - M(\lambda)\alpha_2\big)^{-1},\end{aligned}$$

which, together with Proposition 2.5, completes the proof. □

The following two theorems contain the main results of the present paper. Since their proofs have similar structures, we give a joint proof below. The first of the two main theorems states that under the condition (3.2) on $\alpha = \alpha_2 - \alpha_1$ the resolvent difference (3.1) is compact.

**Theorem 3.6.** *Let $\alpha_1, \alpha_2 \in W^{1,\infty}(\mathbb{R}^n)$, let $A_{\alpha_1}$, $A_{\alpha_2}$ be the corresponding operators as in* (1.1)*, and let $\alpha := \alpha_2 - \alpha_1$ satisfy* (3.2)*. Then*

$$(A_{\alpha_2} - \lambda)^{-1} - (A_{\alpha_1} - \lambda)^{-1} \in \mathfrak{S}_\infty$$

*holds for each $\lambda \in \rho(A_{\alpha_1}) \cap \rho(A_{\alpha_2})$, and, in particular, $\sigma_{\rm ess}(A_{\alpha_1}) = \sigma_{\rm ess}(A_{\alpha_2})$.*

As mentioned before, the condition (3.2) covers the case that $\alpha$ belongs to $L^p(\mathbb{R}^n)$ for an arbitrary $p > 0$. For certain $p$, if $\alpha \in L^p(\mathbb{R}^n)$, the result of Theorem 3.6 can be improved as follows.

**Theorem 3.7.** *Let $\alpha_1, \alpha_2 \in W^{1,\infty}(\mathbb{R}^n)$, let $A_{\alpha_1}$, $A_{\alpha_2}$ be the corresponding operators as in* (1.1)*, and let $\alpha := \alpha_2 - \alpha_1$. Then for $\lambda \in \rho(A_{\alpha_1}) \cap \rho(A_{\alpha_2})$ the following assertions hold.*

(i) *If $\alpha$ has a compact support or if $n > 3$ and $\alpha \in L^{n/3}(\mathbb{R}^n)$, then*
$$(A_{\alpha_2} - \lambda)^{-1} - (A_{\alpha_1} - \lambda)^{-1} \in \mathfrak{S}_{\frac{n}{3},\infty}.$$

(ii) *If $\alpha \in L^1(\mathbb{R}^n)$ and $n = 3$, then*
$$(A_{\alpha_2} - \lambda)^{-1} - (A_{\alpha_1} - \lambda)^{-1} \in \mathfrak{S}_r \quad \text{for all } r > 1.$$

(iii) *If $\alpha \in L^p(\mathbb{R}^n)$ for $p \geq 1$ and $p > n/3$, then*
$$(A_{\alpha_2} - \lambda)^{-1} - (A_{\alpha_1} - \lambda)^{-1} \in \mathfrak{S}_p.$$

*Proof of Theorem* 3.6 *and Theorem* 3.7. Let us fix $\lambda < -\max\{\|\alpha_1\|_\infty^2, \|\alpha_2\|_\infty^2\}$. We first observe that

$$\operatorname{ran}\big((I - M(\lambda)\alpha_2)^{-1}\gamma(\lambda)^*\big) \subset H^{3/2}(\mathbb{R}^n). \tag{3.4}$$

Note first that $\operatorname{dom} A_{\rm N} \subset H^2(\mathbb{R}^{n+1}_+)$ and Proposition 2.4 (ii) imply $\operatorname{ran}\gamma(\lambda)^* \subset H^{3/2}(\mathbb{R}^n)$. Thus, for $\varphi \in \operatorname{ran}(I - M(\lambda)\alpha_2)^{-1}\gamma(\lambda)^*$ we have $\varphi - M(\lambda)\alpha_2\varphi \in H^{3/2}(\mathbb{R}^n)$, and $\operatorname{ran} M(\lambda) \subset H^1(\mathbb{R}^n)$ implies $\varphi \in H^1(\mathbb{R}^n)$. Since $\alpha_2\varphi$ belongs to

$H^1(\mathbb{R}^n)$, $M(\lambda)\alpha_2\varphi$ automatically belongs to $H^2(\mathbb{R}^n)$; this can be seen as in the proof of Theorem 2.7, see also [18, Section 3]. This proves (3.4). Analogously also

$$\operatorname{ran}\left((I - M(\lambda)\overline{\alpha_1})^{-1}\gamma(\lambda)^*\right) \subset H^{3/2}(\mathbb{R}^n) \tag{3.5}$$

holds. The factorization given in Lemma 3.5 can be written as

$$\begin{aligned}(A_{\alpha_2} - \lambda)^{-1} &- (A_{\alpha_1} - \lambda)^{-1} \\ &= \gamma(\lambda)\left(I - \alpha_1 M(\lambda)\right)^{-1}\sqrt{|\alpha|}\widetilde{\alpha}\sqrt{|\alpha|}\left(I - M(\lambda)\alpha_2\right)^{-1}\gamma(\lambda)^*,\end{aligned} \tag{3.6}$$

where $\widetilde{\alpha}(x)$ is given by 0 if $\alpha(x) = 0$ and by $\alpha(x)/|\alpha(x)|$ if $\alpha(x) \neq 0$.

If $\alpha$ satisfies (3.2), then the same holds for $\alpha$ replaced by $\sqrt{|\alpha|}$. Now (3.4) and Lemma 3.3 (i) imply

$$\sqrt{|\alpha|}\left(I - M(\lambda)\alpha_2\right)^{-1}\gamma(\lambda)^* \in \mathfrak{S}_\infty.$$

Since $\gamma(\lambda)\left(I - \alpha_1 M(\lambda)\right)^{-1}\sqrt{|\alpha|}\widetilde{\alpha} \in \mathcal{B}(L^2(\mathbb{R}^n))$, the assertion of Theorem 3.6 follows from (3.6).

If $\alpha$ has a compact support or if $n > 3$ and $\alpha$ belongs to $L^{n/3}(\mathbb{R}^n)$, then $\sqrt{|\alpha|}$ has a compact support or belongs to $L^{2n/3}(\mathbb{R}^n)$, respectively; thus (3.4) and (3.5) together with Lemma 3.3 (ii) imply

$$\sqrt{|\alpha|}(I - M(\lambda)\alpha_2)^{-1}\gamma(\lambda)^* \in \mathfrak{S}_{\frac{2n}{3},\infty} \quad \text{and} \quad \sqrt{|\alpha|}(I - M(\lambda)\overline{\alpha_1})^{-1}\gamma(\lambda)^* \in \mathfrak{S}_{\frac{2n}{3},\infty}.$$

Taking the adjoint of the latter operator, Lemma (3.2) (i) and (ii) and (3.6) yield Theorem 3.7 (i).

The proofs of Theorem 3.7 (ii) and (iii) are completely analogous; one uses Lemma 3.3 (iii) and (iv), respectively, instead of item (ii). □

As an immediate consequence of Theorem 3.7 we obtain the following result concerning scattering theory. Note that in the case $n < 3$ and $\alpha_2 - \alpha_1 \in L^1(\mathbb{R}^n)$ Theorem 3.7 (iii) implies that the difference (3.1) is contained in the trace class $\mathfrak{S}_1$. Now basic statements from scattering theory yield the following corollary, see, e.g., [24, Theorem X.4.12]; we remark that for $n = 1$ this result can already be found in [10, Section 9].

**Corollary 3.8.** *Let $n < 3$ and let $\alpha_1, \alpha_2 \in W^{1,\infty}(\mathbb{R}^n)$ be real valued with $\alpha_2 - \alpha_1 \in L^1(\mathbb{R}^n)$. Then wave operators for the pair of selfadjoint operators $\{A_{\alpha_1}, A_{\alpha_2}\}$ exist and are complete. Moreover, the absolutely continuous spectra of $A_{\alpha_1}$ and $A_{\alpha_2}$ coincide and their absolutely continuous parts are unitarily equivalent.*

We would like to put some emphasis on the important special case $\alpha_1 = 0$, in which $A_{\alpha_1}$ is the selfadjoint Neumann operator $A_{\mathrm{N}}$ in (2.8). In this situation Theorem 3.6 and Theorem 3.7 read as follows. Recall that the spectrum of $A_{\mathrm{N}}$ has the simple structure $\sigma(A_{\mathrm{N}}) = \sigma_{\mathrm{ess}}(A_{\mathrm{N}}) = [0, \infty)$.

**Corollary 3.9.** *Let $\alpha \in W^{1,\infty}(\mathbb{R}^n)$ satisfy (3.2) and let $A_\alpha$ be the operator in (1.1). Then*

$$(A_\alpha - \lambda)^{-1} - (A_{\mathrm{N}} - \lambda)^{-1} \in \mathfrak{S}_\infty$$

*holds for each $\lambda \in \rho(A_{\alpha_1}) \cap \rho(A_{\alpha_2})$, and, in particular, $\sigma_{\mathrm{ess}}(A_\alpha) = [0, \infty)$.*

**Corollary 3.10.** *Let $\alpha \in W^{1,\infty}(\mathbb{R}^n)$ and let $A_\alpha$ be the operator in (1.1). Then for $\lambda \in \rho(A_\alpha) \cap \rho(A_{\rm N})$ the following assertions hold.*

(i) *If $\alpha$ has a compact support or if $n > 3$ and $\alpha \in L^{n/3}(\mathbb{R}^n)$, then*

$$(A_\alpha - \lambda)^{-1} - (A_{\rm N} - \lambda)^{-1} \in \mathfrak{S}_{\frac{n}{3},\infty}.$$

(ii) *If $\alpha \in L^1(\mathbb{R}^n)$ and $n = 3$, then*

$$(A_\alpha - \lambda)^{-1} - (A_{\rm N} - \lambda)^{-1} \in \mathfrak{S}_r \quad \textit{for all } r > 1.$$

(iii) *If $\alpha \in L^p(\mathbb{R}^n)$ for $p \geq 1$ and $p > n/3$, then*

$$(A_\alpha - \lambda)^{-1} - (A_{\rm N} - \lambda)^{-1} \in \mathfrak{S}_p.$$

As a consequence of Corollary 3.9, applying [27, Proposition 5.11 (v) and (vii)] we obtain the following statement on the absolutely continuous part of $A_\alpha$, if $\alpha$ is real valued.

**Corollary 3.11.** *Let $\alpha \in W^{1,\infty}(\mathbb{R}^n)$ be real valued satisfying (3.2) and let $A_\alpha$ be the selfadjoint operator in (1.1). Then $A_{\rm N}$ and the absolutely continuous part of $A_\alpha$ are unitarily equivalent.*

We conclude our paper with an observation connected with the speed of accumulation of the discrete spectrum of the operator $A_\alpha$, where $\alpha$ is a complex-valued function subject to the condition (3.2). As Corollary 3.9 shows, the essential spectrum of $A_\alpha$ in this case is given by $[0,\infty)$ and, additionally, discrete, (in general) non-real eigenvalues may appear. The following statement combines our main result with some recent advances in the theory of non-selfadjoint perturbations of selfadjoint operators; it is based on [22, Theorem 2.1].

**Corollary 3.12.** *Let $\alpha \in W^{1,\infty}(\mathbb{R}^n)$ and let $A_\alpha$ be the operator in (1.1). Then for all $a > \|\alpha\|_\infty^2$ the following assertions hold.*

(i) *If $\alpha \in L^p(\mathbb{R}^n)$ for $p \geq 1$ and $p > n/3$, then*

$$\sum_{\lambda \in \sigma_{\rm d}(A_\alpha)} \operatorname{dist}\big((\lambda + a)^{-1}, [0, a^{-1}]\big)^p < \infty.$$

(ii) *If $n \geq 3$ and $\alpha \in L^{n/3}(\mathbb{R}^n)$, then*

$$\sum_{\lambda \in \sigma_{\rm d}(A_\alpha)} \operatorname{dist}\big((\lambda + a)^{-1}, [0, a^{-1}]\big)^{\frac{n}{3}+\varepsilon} < \infty \quad \textit{for all} \quad \varepsilon > 0.$$

*Above the eigenvalues in the discrete spectrum are counted according to their algebraic multiplicities, and $\operatorname{dist}(\cdot,\cdot)$ denotes the usual distance in the complex plane.*

For the proof recall that the numerical range of a bounded operator $A$ in a Hilbert space $\mathcal{H}$ is defined as

$$\operatorname{Num}(A) := \{(Af,f)_{\mathcal{H}} \colon f \in \mathcal{H}, \|f\|_{\mathcal{H}} = 1\}.$$

*Proof.* Let us assume first $\alpha \in L^p(\mathbb{R}^n)$ for some $p \geq 1$ with $p > n/3$. Clearly $-a \in \rho(A_{\rm N})$ and by Theorem 2.7 also $-a \in \rho(A_\alpha)$. In view of the assumptions on

$\alpha$ it follows from Corollary 3.10 (iii) that

$$(a + A_\alpha)^{-1} - (a + A_{\mathrm{N}})^{-1} \in \mathfrak{S}_p. \tag{3.7}$$

The operator $(a + A_{\mathrm{N}})^{-1}$ is bounded and selfadjoint and it has a purely essential spectrum given by $[0, a^{-1}]$. Since

$$\overline{\mathrm{Num}\big((a + A_{\mathrm{N}})^{-1}\big)} = \sigma\left((a + A_{\mathrm{N}})^{-1}\right) = [0, a^{-1}]$$

and, trivially,

$$\lambda \in \sigma_{\mathrm{d}}(A_\alpha) \quad \Longleftrightarrow \quad (\lambda + a)^{-1} \in \sigma_{\mathrm{d}}\big((a + A_\alpha)^{-1}\big),$$

the claim of (i) follows from [22, Theorem 2.1].

The proof of (ii) uses Corollary 3.10 (i) and (ii) instead of item (iii) and is completely analogous. □

## References


[1] R.A. Adams and J.J.F. Fournier, Sobolev Spaces, 2nd edition, Pure and Applied Mathematics, vol. 140, Elsevier/Academic Press, Amsterdam, 2003.

[2] J. Behrndt and M. Langer, *Boundary value problems for elliptic partial differential operators on bounded domains*, J. Funct. Anal. 243 (2007), 536–565.

[3] J. Behrndt and M. Langer, *Elliptic operators, Dirichlet-to-Neumann maps and quasi boundary triples*, to appear in London Math. Soc. Lecture Note Series.

[4] J. Behrndt, M. Langer, I. Lobanov, V. Lotoreichik, and I.Yu. Popov, *A remark on Schatten-von Neumann properties of resolvent differences of generalized Robin Laplacians on bounded domains*, J. Math. Anal. Appl. 371 (2010), 750–758.

[5] J. Behrndt, M. Langer, and V. Lotoreichik, *Spectral estimates for resolvent differences of self-adjoint elliptic operators*, submitted, preprint, arXiv:1012.4596.

[6] M.Sh. Birman, *Perturbations of the continuous spectrum of a singular elliptic operator by varying the boundary and the boundary conditions*, Vestnik Leningrad. Univ. 17 (1962), 22–55 (in Russian); translated in Amer. Math. Soc. Transl. 225 (2008), 19–53.

[7] M.Sh. Birman and M.Z. Solomjak, *Asymptotic behavior of the spectrum of variational problems on solutions of elliptic equations in unbounded domains*, Funktsional. Anal. i Prilozhen. 14 (1980), 27–35 (in Russian); translated in Funct. Anal. Appl. 14 (1981), 267–274.

[8] M. Cwikel, *Weak type estimates for singular values and the number of bound states of Schrödinger operators*, Ann. of Math. 106 (1977), 93–100.

[9] M. Demuth, M. Hansmann, and G. Katriel, *On the discrete spectrum of non-selfadjoint operators,* J. Funct. Anal. 257 (2009), 2742–2759.

[10] V.A. Derkach and M.M. Malamud, *Generalized resolvents and the boundary value problems for Hermitian operators with gaps*, J. Funct. Anal. 95 (1991), 1–95.

[11] V.A. Derkach and M.M. Malamud, *The extension theory of Hermitian operators and the moment problem*, J. Math. Sci. 73 (1995), 141–242.

[12] D.E. Edmunds and W.D. Evans, *Orlicz and Sobolev spaces on unbounded domains*, Proc. R. Soc. Lond. A. 342 (1975), 373–400.

[13] D.E. Edmunds and W.D. Evans, Spectral Theory and Differential Operators, Clarendon Press, Oxford, 1987.

[14] R.S. Freeman, *Closed operators and their adjoints associated with elliptic differential operators*, Pac. J. Math. 22 (1967), 71–97.

[15] I.C. Gohberg and M.G. Kreĭn, Introduction to the Theory of Linear Nonselfadjoint Operators, Transl. Math. Monogr., vol. 18., Amer. Math. Soc., Providence, RI, 1969.

[16] G. Grubb, *Remarks on trace estimates for exterior boundary problems*, Comm. Partial Differential Equations 9 (1984), 231–270.

[17] G. Grubb, *Singular Green operators and their spectral asymptotics*, Duke Math. J. 51 (1984), 477–528.

[18] G. Grubb, *Krein resolvent formulas for elliptic boundary problems in nonsmooth domains*, Rend. Semin. Mat. Univ. Politec. Torino 66 (2008), 271–297.

[19] G. Grubb, Distributions and Operators, Springer, 2009.

[20] G. Grubb, *Perturbation of essential spectra of exterior elliptic problems*, Appl. Anal. 90 (2011), 103–123.

[21] G. Grubb, *Spectral asymptotics for Robin problems with a discontinuous coefficient*, J. Spectr. Theory 1 (2011), 155–177.

[22] M. Hansmann, *An eigenvalue estimate and its application to non-selfadjoint Jacobi and Schrödinger operators,* to appear in Lett. Math. Phys., doi: 10.1007/s11005-011-0494-9.

[23] D. Haroske and H. Triebel, Distributions, Sobolev Spaces, Elliptic Equations, EMS Textbooks in Mathematics, European Mathematical Society, Zürich, 2008.

[24] T. Kato, Perturbation Theory for Linear Operators, Springer-Verlag, Berlin, 1995.

[25] J. Lions and E. Magenes, Non-Homogeneous Boundary Value Problems and Applications I, Springer-Verlag, Berlin–Heidelberg–New York, 1972.

[26] M.M. Malamud, *Spectral theory of elliptic operators in exterior domains*, Russ. J. Math. Phys. 17 (2010), 96–125.

[27] M.M. Malamud and H. Neidhardt, *Sturm-Liouville boundary value problems with operator potentials and unitary equivalence*, preprint, arXiv:1102.3849.

[28] B. Simon, *Analysis with weak trace ideals and the number of bound states of Schrödinger operators*, Trans. Amer. Math. Soc. 224 (1976), 367–380.

[29] B. Simon, Trace Ideals and their Applications, Second Edition, Mathematical Surveys and Monographs 120. Providence, RI: American Mathematical Society (AMS), 2005.

Vladimir Lotoreichik and Jonathan Rohleder
Technische Universität Graz
Institut für Numerische Mathematik
Steyrergasse 30
A-8010 Graz, Austria
e-mail: `rohleder@math.tugraz.at`
`lotoreichik@math.tugraz.at`

Operator Theory:
Advances and Applications, Vol. 221, 469-479


# Smoothness of Hill's Potential and Lengths of Spectral Gaps

Vladimir Mikhailets and Volodymyr Molyboga

**Abstract.** The paper studies the Hill–Schrödinger operators with potentials in the space $H^\omega \subset L^2(\mathbb{T},\mathbb{R})$. Explicit description for the classes of sequences being the lengths of spectral gaps of these operators is found. The functions $\omega$ may be nonmonotonic. The space $H^\omega$ coincides with the Hörmander space $H_2^\omega(\mathbb{T},\mathbb{R})$ with the weight function $\omega(\sqrt{1+\xi^2})$ if $\omega$ is in Avakumovich's class OR.

**Mathematics Subject Classification (2000).** Primary 34L40; Secondary 47A10, 47A75.

**Keywords.** Hill–Schrödinger operators, spectral gaps, Hörmander spaces.

## 1. Introduction

Let us consider the Hill–Schrödinger operators

$$S(q)u := -u'' + q(x)u, \quad x \in \mathbb{R}, \tag{1}$$

with 1-periodic real-valued potentials

$$q(x) = \sum_{k\in\mathbb{Z}} \widehat{q}(k)e^{ik2\pi x} \in L^2(\mathbb{T},\mathbb{R}), \quad \mathbb{T} := \mathbb{R}/\mathbb{Z}.$$

This condition means that

$$\sum_{k\in\mathbb{Z}} |\widehat{q}(k)|^2 < \infty \quad \text{and} \quad \widehat{q}(k) = \overline{\widehat{q}(-k)}, \quad k \in \mathbb{Z}.$$

It is well known that the operators $S(q)$ are lower semibounded and self-adjoint in the Hilbert space $L^2(\mathbb{R})$. Their spectra are absolutely continuous and have a zone structure [22].

Spectra of the operators $S(q)$ are completely defined by the location of the endpoints of spectral gaps $\{\lambda_0(q), \lambda_n^{\pm}(q)\}_{n=1}^{\infty}$, which satisfy the inequalities:

$$-\infty < \lambda_0(q) < \lambda_1^-(q) \le \lambda_1^+(q) < \lambda_2^-(q) \le \lambda_2^+(q) < \cdots . \tag{2}$$

Some gaps can degenerate, then corresponding bands merge. For even/odd numbers $n \in \mathbb{Z}_+$ the endpoints of spectral gaps $\{\lambda_0(q), \lambda_n^{\pm}(q)\}_{n=1}^{\infty}$ are eigenvalues of the periodic/semiperiodic problems on the interval $(0,1)$:

$$S_{\pm}(q)u := -u'' + q(x)u = \lambda u,$$
$$\operatorname{Dom}(S_{\pm}(q)) := \left\{ u \in H^2[0,1] \,\middle|\, u^{(j)}(0) = \pm u^{(j)}(1),\ j = 0,1 \right\}.$$

The interiors of spectral bands (stability zones)

$$\mathcal{B}_0(q) := (\lambda_0(q), \lambda_1^-(q)), \qquad \mathcal{B}_n(q) := (\lambda_n^+(q), \lambda_{n+1}^-(q)), \quad n \in \mathbb{N},$$

together with the collapsed gaps,

$$\lambda = \lambda_{n_i}^- = \lambda_{n_i}^+,$$

are characterized as the set of those $\lambda \in \mathbb{R}$, for which all solutions of the equation

$$-u'' + q(x)u = \lambda u \tag{3}$$

are bounded. The open spectral gaps (instability zones)

$$\mathcal{G}_0(q) := (-\infty, \lambda_0(q)), \qquad \mathcal{G}_n(q) := (\lambda_n^-(q), \lambda_n^+(q)) \neq \emptyset, \quad n \in \mathbb{N},$$

form the set of those $\lambda \in \mathbb{R}$ for which any nontrivial solution of the equation (3) is unbounded.

We study the behaviour of the lengths of spectral gaps

$$\gamma_q(n) := \lambda_n^+(q) - \lambda_n^-(q), \quad n \in \mathbb{N},$$

of the operators $S(q)$ in terms of behaviour of the Fourier coefficients $\{\widehat{q}(n)\}_{n\in\mathbb{N}}$ of the potentials $q$ with respect to test sequence spaces, that is in terms of potential regularity.

Hochstadt [5, 6], Marchenko and Ostrovskii [14], McKean and Trubowitz [12, 23] proved that the potential $q(x)$ is an infinitely differentiable function if and only if the lengths of spectral gaps $\{\gamma_q(n)\}_{n=1}^{\infty}$ decrease faster than an arbitrary power of $1/n$:

$$q(x) \in C^{\infty}(\mathbb{T}, \mathbb{R}) \Leftrightarrow \gamma_q(n) = O(n^{-k}),\ n \to \infty, \quad k \in \mathbb{Z}_+.$$

However, the scale of spaces $\left\{C^k(\mathbb{T}, \mathbb{R})\right\}_{k\in\mathbb{N}}$ turned out unusable to obtain precise quantitative results. Marchenko and Ostrovskii [14] (see also [13]) found that

$$q \in H^s(\mathbb{T}, \mathbb{R}) \Leftrightarrow \sum_{n\in\mathbb{N}} (1+2n)^{2s} \gamma_q^2(n) < \infty, \qquad s \in \mathbb{Z}_+, \tag{4}$$

where $H^s(\mathbb{T}, \mathbb{R})$, $s \in \mathbb{Z}_+$, is the Sobolev space on the circle $\mathbb{T}$.

Djakov and Mityagin [2], Pöschel [21] extended the Marchenko–Ostrovskii Theorem (4) to a very general class of weights $\omega = \{\omega(k)\}_{k\in\mathbb{N}}$ satisfying the following conditions:

(i) $\omega(k) \uparrow \infty$, $k \in \mathbb{N}$; (monotonicity)

(ii) $\omega(k+m) \leq \omega(k)\omega(m)$, $k, m \in \mathbb{N}$; (submultiplicity)

(iii) $\dfrac{\log \omega(k)}{k} \downarrow 0$, $k \to \infty$, (subexponentiality).

For such weights they proved that

$$q \in H^{\omega}(\mathbb{T}, \mathbb{R}) \Leftrightarrow \{\gamma_q(\cdot)\} \in h^{\omega}(\mathbb{N}). \tag{5}$$

Here

$$H^{\omega} = \left\{ \sum_{k\in\mathbb{Z}} \widehat{f}(k) e^{ik2\pi x} \in L^2(\mathbb{T}) \,\middle|\, \sum_{k\in\mathbb{N}} \omega^2(k) |\widehat{f}(k)|^2 < \infty,\ \widehat{f}(k) = \overline{\widehat{f}(-k)},\ k \in \mathbb{Z} \right\},$$

and $h^{\omega}(\mathbb{N})$ is the Hilbert space of weighted sequences generated by the weight $\omega(\cdot)$,

$$h^{\omega}(\mathbb{N}) := \left\{ a = \{a(k)_{k\in\mathbb{N}}\} \,\middle|\, \sum_{k\in\mathbb{N}} \omega^2(k) |a(k)|^2 < \infty \right\}.$$

To characterize regularity of potentials in the finer way, in this paper we apply the real function spaces $H^{\omega}(\mathbb{T}, \mathbb{R})$ where $\omega(\cdot)$ is a positive, in general, non-monotonic weight. This extension is essential (see Remark 1.1). The space $H^{\omega}$ coincides with the Hörmander space $H_2^{\omega}(\mathbb{T}, \mathbb{R})$ with the weight function $\omega(\sqrt{1+\xi^2})$ if $\omega \in \mathrm{OR}$ (see Appendix A). In the case of the power weight $H_2^{\omega}(\mathbb{T}, \mathbb{R})$ is a Sobolev space.

## 2. Main result

As is well known, the sequence of the lengths of spectral gaps $\{\gamma_q(n)\}_{n\in\mathbb{N}}$ of the Hill–Schrödinger operators $S(q)$ with $L^2(\mathbb{T}, \mathbb{R})$-potentials $q$ belongs to the sequence space $h_+^0(\mathbb{N})$,

$$h_+^0(\mathbb{N}) := \left\{ a = \{a(k)\}_{k\in\mathbb{N}} \in l^2(\mathbb{N}) \mid a(k) \geq 0,\ k \in \mathbb{N} \right\}.$$

Let us consider the map

$$\gamma : L^2(\mathbb{T}, \mathbb{R}) \ni q \mapsto \{\gamma_q(n)\}_{n\in\mathbb{N}} \in h_+^0(\mathbb{N}).$$

Garnett and Trubowitz [4] established that for any sequence $\{\gamma(n)\}_{n\in\mathbb{N}} \in h_+^0(\mathbb{N})$ we can place the open intervals $I_n$ of the lengths $\gamma(n)$ on the positive semi-axis $(0, \infty)$ in a such single way that there exists a potential $q \in L^2(\mathbb{T}, \mathbb{R})$ for which the sequence $\{\gamma(n)\}_{n\in\mathbb{N}}$ is a sequence of the lengths of spectral gaps of the Hill–Schrödinger operator $S(q)$. Thus the map $\gamma$ maps the space $L^2(\mathbb{T}, \mathbb{R})$ *onto* the sequence space $h_+^0(\mathbb{N})$:

$$\gamma(L^2(\mathbb{T}, \mathbb{R})) = h_+^0(\mathbb{N}). \tag{6}$$

Note that Korotyaev [10] proved a more general result than (6): the mapping $L^2(\mathbb{T},\mathbb{R}) \ni q \mapsto g(q) = \{(g_{sn}, g_{cn})\}_{n\in\mathbb{N}} \in h^0(\mathbb{N}) \oplus h^0(\mathbb{N})$ is a real analytic isomorphism, where $g_n = (g_{sn}, g_{cn})$ are some spectral data such that $|g_n| = \frac{1}{2}\gamma_q(n)$. For more detailed exposition of this problem see [10] and the references therein.

We use the notations:

$$h^\omega_+(\mathbb{N}) := \{a = \{a(k)\}_{k\in\mathbb{N}} \in h^\omega(\mathbb{N}) \mid a(k) \geq 0,\ k \in \mathbb{N}\},$$

and

$$b_k \ll a_k \ll c_k, \qquad k \in \mathbb{N}.$$

It means that there exist the positive constants $C_1$ and $C_2$ such that the inequalities

$$C_1 b_k \leq a_k \leq C_2 c_k, \qquad k \in \mathbb{N},$$

hold.

The main purpose of this paper is to prove the following result.

**Theorem 1.** *Let $q \in L^2(\mathbb{T},\mathbb{R})$ and the weight $\omega = \{\omega(k)\}_{k\in\mathbb{N}}$ satisfy conditions:*

$$k^s \ll \omega(k) \ll k^{1+s}, \qquad s \in [0,\infty).$$

*Then the map $\gamma : q \mapsto \{\gamma_q(n)\}_{n\in\mathbb{N}}$ satisfies the equalities:*

(i) $\gamma\left(H^\omega(\mathbb{T},\mathbb{R})\right) = h^\omega_+(\mathbb{N})$,

(ii) $\gamma^{-1}\left(h^\omega_+(\mathbb{N})\right) = H^\omega(\mathbb{T},\mathbb{R})$.

Theorem 1 immediately implies the following statement.

**Corollary 1.1.** *Let for the weight $\omega = \{\omega(k)\}_{k\in\mathbb{N}}$ there exists the limit*

$$\lim_{k\to\infty} \frac{\log \omega(k)}{\log k} = s \in [0,\infty),$$

*called an order of the weight sequence $\omega = \{\omega(k)\}_{k\in\mathbb{N}}$, and let for $s = 0$ the values of the weight $\omega = \{\omega(k)\}_{k\in\mathbb{N}}$ be separated from zero. Then*

$$q \in H^\omega(\mathbb{T},\mathbb{R}) \Leftrightarrow \{\gamma_q(\cdot)\} \in h^\omega(\mathbb{N}).$$

Let us remind that a measurable function $f > 0$ satisfying the relationship

$$f(\lambda x)/f(x) \to \lambda^s, \quad x \to \infty, \qquad (\forall \lambda > 0)$$

is called *regular varying in Karamata's sense of the index s*. Its restriction to $\mathbb{N}$ we call *a regular varying sequence in Karamata's sense of the index s*, for more details see, for example, [1].

From Corollary 1.1 we obtain the following result.

**Corollary 1.2 ([15]).** *Let the weight $\omega = \{\omega(k)\}_{k\in\mathbb{N}}$ be a regular varying sequence in the Karamata sense of the index $s \in [0,\infty)$, and let for $s = 0$ its values be separated from zero. Then*

$$q \in H^\omega(\mathbb{T},\mathbb{R}) \Leftrightarrow \{\gamma_q(\cdot)\} \in h^\omega(\mathbb{N}).$$

Note that the assumption of Corollary 1.2 holds for instance for the weight

$$\omega(k) = (1+2k)^s (\log(1+k))^{r_1} (\log\log(1+k))^{r_2} \dots (\log\log\dots\log(1+k))^{r_p},$$
$$s \in (0,\infty),\ \{r_1,\dots,r_p\} \subset \mathbb{R},\ p \in \mathbb{N}.$$

The following example shows that statement (5) does not cover Corollary 1.1 and all the more Theorem 1.

*Example.* Let $s \in [0,\infty)$. Set

$$\omega(k) := \begin{cases} k^s \log(1+k) & \text{if} \quad k \in 2\mathbb{N}; \\ k^s & \text{if} \quad k \in (2\mathbb{N}-1). \end{cases}$$

Then the weight $\omega = \{\omega(k)\}_{k\in\mathbb{N}}$ satisfies conditions of Corollary 1.1. But one can prove that it is not equivalent to any monotonic weight.

*Remark* 1.1. Theorem 1 shows that if the sequence $\{|\widehat{q}(n_k)|\}_{k=1}^{\infty}$ decreases particularly fast on a certain subsequence $\{n_k\}_{k=1}^{\infty} \subset \mathbb{N}$, then so does sequence $\{\gamma_q(n_k)\}_{k=1}^{\infty}$ on the *same subsequence*. The converse statement is also true.

*Remark* 1.2. In the papers [9, 17, 11, 15, 3, 16] (see also the references therein) the case of singular periodic potentials $q \in H^{-1}(\mathbb{T})$ was studied.

## 3. Preliminarics

Here we define Hilbert spaces of weighted two-sided sequences and formulate the Convolution Lemma.

For every positive sequence $\omega = \{\omega(k)\}_{k\in\mathbb{N}}$, its unique extension on $\mathbb{Z}$ exists being a two-sided sequence satisfying the conditions:

(i) $\omega(0) = 1$; (ii) $\omega(-k) = \omega(k), \quad k \in \mathbb{N}$; (iii) $\omega(k) > 0, \quad k \in \mathbb{Z}$.

Let $h^\omega(\mathbb{Z}) \equiv h^\omega(\mathbb{Z},\mathbb{C})$ be the Hilbert space of two-sided sequences:

$$h^\omega(\mathbb{Z}) := \left\{ a = \{a(k)\}_{k\in\mathbb{Z}} \,\middle|\, \sum_{k\in\mathbb{Z}} \omega^2(k)|a(k)|^2 < \infty \right\},$$

$$(a,b)_{h^\omega(\mathbb{Z})} := \sum_{k\in\mathbb{Z}} \omega^2(k) a(k)\overline{b(k)}, \quad a,b \in h^\omega(\mathbb{Z}),$$

$$\|a\|_{h^\omega(\mathbb{Z})} := (a,a)^{1/2}_{h^\omega(\mathbb{Z})}, \quad a \in h^\omega(\mathbb{Z}).$$

For convenience we will denote by $h^\omega(n)$ the $n$th element of a sequence $a = \{a(k)\}_{k\in\mathbb{Z}}$ in $h^\omega(\mathbb{Z})$.

The basic weights we apply are the power ones:

$$w_s(k) = (1+2|k|)^s, \quad s \in \mathbb{R}.$$

In this case it is convenient to use shorter notations:

$$h^{\omega_s}(\mathbb{Z}) \equiv h^s(\mathbb{Z}), \quad s \in \mathbb{R}.$$

The operation of convolution for the two-sided sequences

$$a = \{a(k)\}_{k\in\mathbb{Z}} \quad \text{and} \quad b = \{b(k)\}_{k\in\mathbb{Z}}$$

is formally defined as follows:

$$(a,b) \mapsto a * b, \quad (a*b)(k) := \sum_{j\in\mathbb{Z}} a(k-j)\, b(j), \quad k \in \mathbb{Z}.$$

Sufficient conditions for the convolution to exist as a continuous map are given by the following known lemma, see for example [9, 17].

**Lemma 2 (The Convolution Lemma).** *Let $s, r \geq 0$, and $t \leq \min(s,r)$, $t \in \mathbb{R}$. If $s + r - t > 1/2$, then the convolution $(a,b) \mapsto a * b$ is well defined as a continuous map acting between the spaces:*

$$\text{(a)}\ h^s(\mathbb{Z}) \times h^r(\mathbb{Z}) \to h^t(\mathbb{Z}); \qquad \text{(b)}\ h^{-t}(\mathbb{Z}) \times h^s(\mathbb{Z}) \to h^{-r}(\mathbb{Z}).$$

*In the case $s + r - t < 1/2$ this statement fails to hold.*

## 4. The proofs

The basic point of our proof of Theorem 1 is a sharp asymptotic formula for the lengths of spectral gaps $\{\gamma_q(n)\}_{n\in\mathbb{N}}$ of the operators $S(q)$ and the fundamental result of [4, Theorem 1].

**Lemma 3.** *The lengths of spectral gaps $\{\gamma_q(n)\}_{n\in\mathbb{N}}$ of the operators $S(q)$ with $q \in H^s(\mathbb{T},\mathbb{R})$, $s \in [0,\infty)$, uniformly on the bounded sets of potentials $q$ in the corresponding Sobolev spaces $H^s(\mathbb{T})$ for $n \geq n_0\left(\|q\|_{H^s(\mathbb{T})}\right)$ satisfy the asymptotic formula*

$$\gamma_q(n) = 2|\widehat{q}(n)| + h^{1+s}(n). \tag{7}$$

*Proof of Lemma* 3. To prove the asymptotic formula (7), we apply [8, Theorem 1.2] and the Convolution Lemma.

Indeed, applying [8, Theorem 1.2] with $q \in H^s(\mathbb{T},\mathbb{R})$, $s \in [0,\infty)$, we get

$$\sum_{n\in\mathbb{N}} (1+2n)^{2(1+s)} \left(\min_{\pm} \left|\gamma_q(n) \pm 2\sqrt{(\widehat{q}+\varrho)(-n)(\widehat{q}+\varrho)(n)}\right|\right)^2 \leq C\left(\|q\|_{H^s(\mathbb{T})}\right), \tag{8}$$

where

$$\varrho(n) := \frac{1}{\pi^2} \sum_{j\in\mathbb{Z}\setminus\{\pm n\}} \frac{\widehat{q}(n-j)\widehat{q}(n+j)}{(n-j)(n+j)}.$$

Without losing generality we assume that

$$\widehat{q}(0) = 0. \tag{9}$$

Taking into account that the potentials $q$ are real valued we have

$$\widehat{q}(k) = \overline{\widehat{q}(-k)}, \ \varrho(k) = \overline{\varrho(-k)}, \quad k \in \mathbb{Z}.$$

Therefore from inequality (8) we get the estimates

$$\{\gamma_n(q) - 2\,|\widehat{q}(n) + \varrho(n)|\}_{n\in\mathbb{N}} \in h^{1+s}(\mathbb{N}). \tag{10}$$

Further, since by assumption $q \in H^s(\mathbb{T},\mathbb{R})$, that is $\{\widehat{q}(k)\}_{k\in\mathbb{Z}} \in h^s(\mathbb{Z})$, then taking into account assumption (9)

$$\left\{\frac{\widehat{q}(k)}{k}\right\}_{k\in\mathbb{Z}} \in h^{1+s}(\mathbb{Z}), \quad s \in [0,\infty).$$

Applying the Convolution Lemma we obtain

$$\varrho(n) = \frac{1}{\pi^2}\sum_{j\in\mathbb{Z}} \frac{\widehat{q}(n-j)\widehat{q}(n+j)}{(n-j)(n+j)} = \frac{1}{\pi^2}\sum_{j\in\mathbb{Z}} \frac{\widehat{q}(2n-j)}{2n-j}\cdot\frac{\widehat{q}(j)}{j} \tag{11}$$

$$= \left(\left\{\frac{\widehat{q}(k)}{k}\right\}_{k\in\mathbb{Z}} * \left\{\frac{\widehat{q}(k)}{k}\right\}_{k\in\mathbb{Z}}\right)(2n) \in h^{1+s}(\mathbb{N}).$$

Finally, from (10) and (11) we get the necessary estimates (7).

The proof of Lemma 3 is complete. □

### 4.1. Proof of Theorem 1

Let $q \in L^2(\mathbb{T},\mathbb{R})$ and $\omega = \{\omega(k)\}_{k\in\mathbb{N}}$ be a given weight satisfying the conditions of Theorem 1:

$$k^s \ll \omega(k) \ll k^{1+s}, \qquad s \in [0,\infty). \tag{12}$$

At first, we need to prove the statement

$$q \in H^\omega(\mathbb{T},\mathbb{R}) \Leftrightarrow \{\gamma_q(\cdot)\} \in h^\omega(\mathbb{N}). \tag{13}$$

Due to the condition (12) the embeddings

$$H^{1+s}(\mathbb{T}) \hookrightarrow H^\omega(\mathbb{T}) \hookrightarrow H^s(\mathbb{T}), \tag{14}$$

$$h^{1+s}(\mathbb{N}) \hookrightarrow h^\omega(\mathbb{N}) \hookrightarrow h^s(\mathbb{N}), \qquad s \in [0,\infty), \tag{15}$$

are valid since

$$H^{\omega_1}(\mathbb{T}) \hookrightarrow H^{\omega_2}(\mathbb{T}), \quad h^{\omega_1}(\mathbb{N}) \hookrightarrow h^{\omega_2}(\mathbb{N}) \quad \text{if} \quad \omega_1 \gg \omega_2. \tag{16}$$

Let $q \subset H^\omega(\mathbb{T},\mathbb{R})$, then from (14) we get $q \in H^s(\mathbb{T},\mathbb{R})$. Due to Lemma 3, we find that

$$\gamma_q(n) = 2|\widehat{q}(n)| + h^{1+s}(n).$$

Applying (15) from the latter we derive

$$\gamma_q(n) = 2|\widehat{q}(n)| + h^\omega(n).$$

As a consequence, we obtain that $\{\gamma_q(\cdot)\} \in h^\omega(\mathbb{N})$.

The direct implication in statement (13) has been proved.

Let $\{\gamma_q(\cdot)\} \in h^\omega(\mathbb{N})$. Applying (15) we get $\{\gamma_q(\cdot)\} \in h^s(\mathbb{N})$. Further, from (5) with $\omega(k) = (1+2k)^s$, $s \in [0,\infty)$, we obtain $q \in H^s(\mathbb{T},\mathbb{R})$.

We have already proved the implication

$$q \in H^s(\mathbb{T},\mathbb{R}) \Rightarrow \gamma_q(n) = 2|\widehat{q}(n)| + h^\omega(n).$$

Hence $\{\widehat{q}(\cdot)\} \in h^\omega(\mathbb{N})$ and $q \in H^\omega(\mathbb{T},\mathbb{R})$.

The inverse implication in statement (13) has been proved. Now we are ready to prove the statement of Theorem 1.

From statement (13) we get

$$\gamma\left(H^{\omega}(\mathbb{T},\mathbb{R})\right) \subset h_{+}^{\omega}(\mathbb{N}). \tag{17}$$

To establish the equality (i) of Theorem 1 it is necessary to prove the inverse inclusion to formula (17). So, let $\{\gamma(n)\}_{n\in\mathbb{N}}$ be an arbitrary sequence in the space $h_{+}^{\omega}(\mathbb{N})$. Then $\{\gamma(n)\}_{n\in\mathbb{N}} \in h_{+}^{0}(\mathbb{N})$. Due to [4, Theorem 1], a potential $q \in L^2(\mathbb{T},\mathbb{R})$ exists, such that the sequence $\{\gamma(n)\}_{n\in\mathbb{N}} \in h_{+}^{0}(\mathbb{N})$ is its sequence of the lengths of spectral gaps. Since by assumption $\{\gamma(n)\}_{n\in\mathbb{N}} \in h_{+}^{\omega}(\mathbb{N})$ due to (13), we conclude that $q \in H^{\omega}(\mathbb{T},\mathbb{R})$. Therefore the inclusion

$$\gamma\left(H^{\omega}(\mathbb{T},\mathbb{R})\right) \supset h_{+}^{\omega}(\mathbb{N}) \tag{18}$$

holds.

Inclusions (17) and (18) give the equality (i).

Now, let us prove the equality (ii) of Theorem 1. Let $\{\gamma(n)\}_{n\in\mathbb{N}}$ be an arbitrary sequence from the space $h_{+}^{\omega}(\mathbb{N})$ and $q \in L^2(\mathbb{T},\mathbb{R})$ is such that $\gamma(q) = \{\gamma(n)\}_{n\in\mathbb{N}}$. Then according to (13) we have $q \in H^{\omega}(\mathbb{T},\mathbb{R})$. That is

$$\gamma^{-1}\left(h_{+}^{\omega}(\mathbb{N})\right) \subset H^{\omega}(\mathbb{T},\mathbb{R}). \tag{19}$$

Conversely, let $q$ be an arbitrary function in the space $H^{\omega}(\mathbb{T},\mathbb{R})$. Then, due to statement (13), we have $\gamma_q = \{\gamma_q(n)\}_{n\in\mathbb{N}} \in h_{+}^{\omega}(\mathbb{N})$. Therefore

$$\gamma^{-1}\left(h_{+}^{\omega}(\mathbb{N})\right) \supset H^{\omega}(\mathbb{T},\mathbb{R}). \tag{20}$$

Inclusions (19) and (20) give the equality (ii) of Theorem 1.

The proof of Theorem 1 is complete.

## Appendix A. Hörmander spaces on the circle

Let OR be a class of all Borel measurable functions $\omega:\ (0,\infty) \to (0,\infty)$, for which real numbers $a, c > 1$ exist, such that

$$c^{-1} \le \frac{\omega(\lambda t)}{\omega(t)} \le c, \qquad t \ge 1,\ \lambda \in [1,a].$$

The space $H_2^{\omega}(\mathbb{R}^n)$, $n \in \mathbb{N}$, consists of all complex-valued distributions $u \in \mathcal{S}'(\mathbb{R}^n)$ such that their Fourier transformations $\widehat{u}$ are locally Lebesgue integrable on $\mathbb{R}^n$ and $\omega(\langle\xi\rangle)|\widehat{u}(\xi)| \in L^2(\mathbb{R}^n)$ with $\langle\xi\rangle := (1+\xi^2)^{1/2}$. This space is a Hilbert space with respect to the inner product

$$(u_1,u_2)_{H_2^{\omega}(\mathbb{R}^n)} := \int_{\mathbb{R}^n} \omega^2(\langle\xi\rangle)\widehat{u}_1(\xi)\overline{\widehat{u}_2(\xi)}\,d\xi.$$

It is a special case of the isotropic Hörmander spaces [7]. If $\Omega$ is a domain in $\mathbb{R}^n$ with a smooth boundary, then the spaces $H_2^{\omega}(\Omega)$ are defined in a standard way.

Let $\Gamma$ be an infinitely smooth closed oriented manifold of dimension $n \ge 1$ with density $dx$ given on it. Let $\mathfrak{D}'(\Gamma)$ be a topological vector space of distributions on $\Gamma$ dual to $C^\infty(\Gamma)$ with respect to the extension by continuity of the inner product in the space $L^2(\Gamma) := L^2(\Gamma, dx)$.

Now, let us define the Hörmander spaces on the manifold $\Gamma$. Choose a finite atlas from the $C^\infty$-structure on $\Gamma$ formed by the local charts $\alpha_j : \mathbb{R}^n \leftrightarrow U_j$, $j = 1, \ldots, r$, where the open sets $U_j$ form a finite covering of the manifold $\Gamma$. Let functions $\chi_j \in C^\infty(\Gamma)$, $j = 1, \ldots, r$, satisfying the condition $\operatorname{supp} \chi_j \subset U_j$ form a partition of unity on $\Gamma$. By definition, the linear space $H_2^\omega(\Gamma)$ consists of all distributions $f \in \mathfrak{D}'(\Gamma)$ such that $(\chi_j f) \circ \alpha_j \in H_2^\omega(\mathbb{R}^n)$ for every $j$, where $(\chi_j f) \circ \alpha_j$ is a representation of the distribution $\chi_j f$ in the local chart $\alpha_j$. In the space $H_2^\omega(\Gamma)$ the inner product is defined by the formula

$$(f_1, f_2)_{H_2^\omega(\Gamma)} := \sum_{j=1}^{r} ((\chi_j f_1) \circ \alpha_j, (\chi_j f_2) \circ \alpha_j)_{H_2^\omega(\mathbb{R}^n)},$$

and induces the norm $\|f\|_{H_2^\omega(\Gamma)} := (f, f)^{1/2}_{H_2^\omega(\Gamma)}$.

There exists an alternative definition of the space $H_2^\omega(\Gamma)$ which shows that this space does not depend (up to equivalence of norms) on the choice of the local charts, the partition of unity and that it is a Hilbert space.

Let a $\Psi$DO $A$ of order $m > 0$ be elliptic on $\Gamma$, and let it be a positive unbounded operator on the space $L^2(\Gamma)$. For instance, we can set $A := (1 - \Delta_\Gamma)^{1/2}$, where $\Delta_\Gamma$ is the Beltrami–Laplace operator on the Riemannian manifold $\Gamma$. Redefine the function $\omega \in \mathrm{OR}$ on the interval $0 < t < 1$ by the equality $\omega(t) := \omega(1)$ and introduce the norm

$$f \mapsto \|\omega(A^{1/m}) f\|_{L^2(\Gamma)}, \qquad f \in C^\infty(\Gamma). \tag{A.1}$$

**Theorem A.1.** *If $\omega \in \mathrm{OR}$, then the space $H_2^\omega(\Gamma)$ coincides up to the equivalence of norms with the completion of the linear space $C^\infty(\Gamma)$ by the norm* (A.1).

Since the operator $A$ has a discrete spectrum, the space $H_2^\omega(\Gamma)$ can be described by means of the Fourier series. Let $\{\lambda_k\}_{k\in\mathbb{N}}$ be a monotonically nondecreasing, positive sequence of all eigenvalues of the operator $A$, enumerated according to their multiplicity. Let $\{h_k\}_{k\in\mathbb{N}}$ be an orthonormal basis in the space $L^2(\Gamma)$ formed by the corresponding eigenfunctions of the operator $A$: $Ah_k = \lambda_k h_k$. Then for any distribution, the following expansion into the Fourier series converging in the linear space $\mathfrak{D}'(\Gamma)$ holds:

$$f = \sum_{k=1}^{\infty} c_k(f) h_k, \quad f \in \mathfrak{D}'(\Gamma),\ c_k(f) := (f, h_k). \tag{A.2}$$

**Theorem A.2.** *The following formulae are fulfilled:*

$$H_2^\omega(\Gamma) = \left\{ f = \sum_{k=1}^{\infty} c_k(f) h_k \in \mathfrak{D}'(\Gamma) \;\middle|\; \sum_{k=1}^{\infty} \omega^2(k^{1/n}) |c_k(f)|^2 < \infty \right\},$$

$$\|f\|^2_{H_2^\omega(\Gamma)} \asymp \sum_{k=1}^{\infty} \omega^2(k^{1/n}) |c_k(f)|^2.$$

Note that for every distribution $f \in H_2^{\omega}(\Gamma)$, series (A.2) converges by the norm of the space $H_2^{\omega}(\Gamma)$. If values of the function $\omega$ are separated from zero, then $H_2^{\omega}(\Gamma) \subseteq L^2(\Gamma)$, and everywhere above we may replace the space $\mathfrak{D}'(\Gamma)$ by the space $L^2(\Gamma)$. For more details, see [18, 19] and [20].

*Example.* Let $\Gamma = \mathbb{T}$. Then we choose $A = \left(1 - d^2/dx^2\right)^{1/2}$, where we denote by $x$ the natural parametrization on $\mathbb{T}$. The eigenfunctions $h_k = e^{ik2\pi x}$, $k \in \mathbb{Z}$, of the operator $A$ form an orthonormal basis in the space $L^2(\mathbb{T})$. For $\omega \in \mathrm{OR}$ we have

$$f \in H_2^{\omega}(\mathbb{T}) \Leftrightarrow f = \sum_{k\in\mathbb{Z}} \widehat{f}(k) e^{ik2\pi x}, \qquad \sum_{k\in\mathbb{Z}\setminus\{0\}} |\widehat{f}(k)|^2 \omega^2(|k|) < \infty.$$

In this case the function $f$ is real valued if and only if $\widehat{f}(k) = \overline{\widehat{f}(-k)}$, $k \in \mathbb{Z}$. Therefore *the class $H^{\omega}$ coincides with the Hörmander space $H_2^{\omega}(\mathbb{T}, \mathbb{R})$* with the weight function $\omega(\sqrt{1+\xi^2})$ if $\omega \in \mathrm{OR}$.

## References


[1] N. Bingham, C. Goldie, J. Teugels, *Regular Variation*. Encyclopedia of Math. and its Appl., vol. 27, Cambridge University Press, Cambridge, etc., 1989.

[2] P. Djakov, B. Mityagin, *Instability zones of one-dimentional periodic Schrödinger and Dirac operators*. Uspekhi Mat. Nauk **61** (2006), no. 4, 77–182. (Russian); English Transl. in Russian Math. Surveys **61** (2006), no. 4, 663–766.

[3] P. Djakov, B. Mityagin, *Spectral gaps of Schrödinger operators with periodic singular potentials*. Dynamics of PDE **6** (2009), no. 2, 95–165.

[4] J. Garnett, E. Trubowitz, *Gaps and bands of one-dimensional periodic Schrödinger operators*. Comm. Math. Helv., **59** (1984), 258–312.

[5] H. Hochstadt, *Estimates of the stability intervals for Hill's equation*. Proc. Amer. Math. Soc. **14** (1963), 930–932.

[6] H. Hochstadt, *On the determination of Hill's equation from its spectrum*. Arch. Rat. Mech. Anal. **19** (1965), 353–362.

[7] L. Hörmander, *Linear Partial Differential Operators*. Springer-Verlag, Berlin, 1963.

[8] T. Kappeler, B. Mityagin, *Estimates for periodic and Dirichlet eigenvalues of the Schrödinger operator*. SIAM J. Math. Anal. **33** (2001), no. 1, 113–152.

[9] T. Kappeler, C. Möhr, *Estimates for periodic and Dirichlet eigenvalues of the Schrödinger operator with singular potentials*. J. Funct. Anal. **186** (2001), 62–91.

[10] E. Korotyaev, *Inverse problem and the trace formula for the Hill operator*. II. Math. Z. **231** (1999), no. 2, 345–368.

[11] E. Korotyaev, *Characterization of the spectrum of Schrödinger operators with periodic distributions*. Int. Math. Res. Not. **37** (2003), no. 2, 2019–2031.

[12] H. McKean, E. Trubowitz, *Hill's operators and hyperelliptic function theory in the presence of infinitely many branch points*. Comm. Pure Appl. Math. **29** (1976), no. 2, 143–226.

[13] V. Marchenko, *Sturm–Liouville Operators and Applications*. Birkhäuser Verlag, Basel, 1986. (Russian edition: Naukova Dumka, Kiev, 1977)

[14] V. Marchenko, I. Ostrovskii, *A characterization of the spectrum of Hill's operator.* Matem. Sbornik **97** (1975), no. 4, 540–606. (Russian); English Transl. in Math. USSR-Sb. **26** (1975), no. 4, 493–554.

[15] V. Mikhailets, V. Molyboga, *Spectral gaps of the one-dimensional Schrödinger operators with singular periodic potentials.* Methods Funct. Anal. Topology **15** (2009), no. 1, 31–40.

[16] V. Mikhailets, V. Molyboga, *Hill's potentials in Hörmander spaces and their spectral gaps.* Methods Funct. Anal. Topology **17** (2011), no. 3, 235–243.

[17] C. Möhr, *Schrödinger Operators with Singular Potentials on the Circle: Spectral Analysis and Applications.* Thesis at the University of Zürich, 2001, 134 p.

[18] V. Mikhailets, A. Murach, *Interpolation with a function parameter and refined scale of spaces.* Methods Funct. Anal. Topology **14** (2008), no. 1, 81–100.

[19] V. Mikhailets, A. Murach, *On the elliptic operators on a closed compact manifold.* Reports of NAS of Ukraine (2009), no. 3, 29–35. (Russian)

[20] V. Mikhailets, A. Murach, *Hörmander Spaces, Interpolation, and Elliptic Problems.* Institute of Mathematics of NAS of Ukraine, Kyiv, 2010. (Russian)

[21] J. Pöschel, *Hill's potentials in weighted Sobolev spaces and their spectral gaps.* In: W. Craig (ed), Hamiltonian Systems and Applications, Springer, 2008, 421–430.

[22] M. Reed, B. Simon, *Methods of Modern Mathematical Physics*: Vols 1-4. Academic Press, New York, etc., 1972–1978, V. 4: *Analysis of Operators.* 1972.

[23] E. Trubowitz, *The inverse problem for periodic potentials.* Comm. Pure Appl. Math. **30** (1977), 321–337.

Vladimir Mikhailets and Volodymyr Molyboga
Institute of Mathematics
National Academy of Science of Ukraine
3 Tereshchenkivs'ka Str.
01601 Kyiv-4, Ukraine
e-mail: mikhailets@imath.kiev.ua
molyboga@imath.kiev.ua

Operator Theory:
Advances and Applications, Vol. 221, 481-490


# A Frucht Theorem for Quantum Graphs

Delio Mugnolo

**Abstract.** A celebrated theorem due to R. Frucht states that, roughly speaking, each group is abstractly isomorphic to the symmetry group of some graph. By "symmetry group" the group of all graph automorphisms is meant. We provide an analogue of this result for *quantum* graphs, i.e., for Schrödinger equations on a metric graph, after suitably defining the notion of symmetry.

**Mathematics Subject Classification (2000).** Primary 05C25; Secondary 35B06, 81Q35.

**Keywords.** Symmetries for evolution equations; quantum graphs; algebraic graph theory.

## 1. Introduction

Beginning with the second half of the XIX century, symmetries have played an important rôle in analysis. A key observation that paved the road to the seminal work of Lie, Klein and Noether is the group structure typical of symmetries. Ever since, the notion of symmetry has been very important in the theory of differential equations, but has also appeared in further contexts eventually leading to the typical general question:

> *Let* $\Gamma$ *be a group and* $C$ *be a category. Is there an object in* $C$ *whose group of symmetries is isomorphic to* $\Gamma$*?*

Possibly, the earliest example of mathematical problem related to the above question is the following, concerning the category of sets. Clearly, the notion of symmetry is not univocal and strongly depends from the considered category. In the case of sets, the symmetry group is by definition simply the group of all permutations of the set's elements.

- *Let* $\Gamma$ *be a group and* $C$ *be the category of sets. Is there an object of* $C$ *such that (a subgroup of) its symmetry group* $S_n$ *is isomorphic to* $\Gamma$*?*

Yes, there is. Indeed, by Cayley's theorem every group $\Gamma$ is isomorphic to a subgroup of the symmetric group on $\Gamma$ (defined as the group of all bijections on $\Gamma$).

Further examples include the following ones.

- *Let* $\Gamma$ *be a group and* $C$ *be the category of topological spaces. Is there an object* $M$ *of* $C$ *whose symmetry group (i.e., the group of all homeomorphisms on* $M$*) is isomorphic to* $\Gamma$*?*

Again, the answer is positive. Actually, such a topological space can be chosen to be a complete, connected, locally connected, 1-dimensional metric space: this has been proved by de Groot [9].

Symmetries also play an important rôle in graph theory. By definition, a **symmetry** (or **automorphism**) of a graph $G$ is a permutation of nodes of $G$ that preserves adjacency. Equivalently, a permutation is a symmetry of $G$ if and only if it commutes with the adjacency matrix of $G$. With this definition, the following can be formulated (here $A(G)$ denotes the group of all symmetries of a graph $G$).

- *Let* $\Gamma$ *be a* finite *group and* $C$ *the category of (simple) connected graphs. Is there an object* $G$ *of* $C$ *whose symmetry group* $A(G)$ *is (abstractly) isomorphic to* $\Gamma$*?*

Yes, there is. This affirmative answer is the statement of Frucht's classical theorem [10] (in fact, there exist infinitely many, pairwise non-isomorphic, finite graphs $G$ such that $A(G) \cong \Gamma$). This assertion has been significantly strengthened by a later work of Sabidussi [21], who has shown that these graphs can be constructed to be $k$-regular for any $k \geq 3$, to have arbitrary connectivity, or arbitrary chromatic number.[1] We also mention the following related results.

- Let $\Gamma$ be a finite group. Then for any $k \in \{3, 4, 5\}$ there exist uncountably many, pairwise non-isomorphic, $k$-regular connected infinite (simple) graphs $G$ such that $A(G) \cong \Gamma$ [15].
- Let $\Gamma$ be an infinite group. Then there exists uncountably many, pairwise non-isomorphic (simple) connected infinite graphs $G$ such that $A(G) \cong \Gamma$ [9, 22].

An account of several further results related to Frucht's theorem obtained over the last decades can be found in [2, § 4].

Aim of this note is to discuss a question similar to the above ones with respect to the category of *quantum graphs*. Quantum graphs arise as differential models on quasi-1-dimensional quantum systems. The great attention such models have received in the mathematical and physical communities is reflected by hundreds of articles that have been published in the field of quantum graphs over the last 10 years. Two concise but excellent overviews on this topic have been provided in [19, 16].

---

[1]Both Frucht and Sabidussi begin with the construction of a basic graph $G$ related to the Cayley graph of the group $\Gamma$ and then extend this construction to infinitely many further graphs by suitably decorating $G$. However, already the basic graph $G$ is in general highly redundant: e.g., according to Frucht's construction in [11] the 3-regular graph constructed in order to realize the symmetric group $S_n$ has $8 \cdot n!$ nodes, whereas the Petersen graph' symmetry group is abstractly isomorphic to $S_5$ but the Pertersen graph has only 10 nodes.

## 2. Quantum graphs

Let $H$ be a separable complex Hilbert space and $A$ a self-adjoint operator on $H$. By Stone's theorem, the abstract Cauchy problem of Schrödinger type

$$iu'(t) = Au(t), \quad t \in \mathbb{R}, \qquad u(0) = u_0 \in H, \tag{1}$$

is well posed and the solution $u$ is given by $u(t) := e^{itA}u_0$, where $(e^{itA})_{t\in\mathbb{R}}$ denotes the $C_0$-group of unitary operators on $H$ generated by $A$.

In the following we will consider closed operators $\Sigma : D(\Sigma) \to H$ such that

$$\Sigma e^{itA} f = e^{itA} \Sigma f \qquad \text{for all } t \in \mathbb{R} \text{ and } f \in D(\Sigma) \tag{2}$$

In mathematical physics, a unitary operator $\Sigma$ satisfying (2) is said to be a **symmetry** of the system described by (1).

It has been observed in [7] that if $A$ is self-adjoint and dissipative (and hence it generates both a $C_0$-group $(e^{itA})_{t\in\mathbb{R}}$ of unitary linear operators on $H$ and a $C_0$-semigroup $(e^{tA})_{t\geq 0}$ of linear contractive operators on $H$), then a closed subspace of $H$ is invariant under $(e^{itA})_{t\in\mathbb{R}}$ if and only if it is invariant under $(e^{tA})_{t\geq 0}$. Observe that self-adjoint dissipative operators are always associated with a (symmetric, $H$-elliptic, continuous) sesquilinear form, cf. [18]. The following criterion holds.

**Lemma 2.1.** *Let $a$ be a sesquilinear, symmetric, $H$-elliptic, continuous form with dense domain $D(a)$ associated with an operator $A$ on $H$. Consider a closed operator $\Sigma$ on $H$. Then $\Sigma$ satisfies (2) if and only if*

- *both $\Sigma, \Sigma^*$ leave $D(a)$ invariant and moreover*
- *for all $f, g \in D(a)$*

$$a(Lf + \Sigma^* Rg, \Sigma^*\Sigma Lf - \Sigma^* Rg) - a(\Sigma Lf + \Sigma\Sigma^* Rg, \Sigma Lf - Rg),$$

*where $L := (I + \Sigma^*\Sigma)^{-1}$. $R := (I + \Sigma\Sigma^*)^{-1}$ and hence $I - R = \Sigma\Sigma^* R$ and $I - L = \Sigma^*\Sigma L$.*

We deduce as a special case the well-known characterization of unitary operators that are symmetries: If $\Sigma$ is unitary, then it is a symmetry of the system described by (1) if and only if $\Sigma f \in D(a)$ and $a(\Sigma f, \Sigma f) = a(f, f)$ for all $f \in D(a)$.

*Proof.* The proof of (1) is based on the observation that

$$\Sigma e^{itA} = e^{itA}\Sigma \qquad \text{for all } t \in \mathbb{R}$$

if and only if the graph of $\Sigma$, i.e., the closed subspace

$$\mathrm{Graph}(\Sigma) := \left\{ \begin{pmatrix} x \\ \Sigma x \end{pmatrix} \in D(\Sigma) \times H \right\}$$

is invariant under the matrix group

$$\begin{pmatrix} e^{itA} & 0 \\ 0 & e^{itA} \end{pmatrix}, \qquad t \in \mathbb{R},$$

on the Hilbert space $H \times H$, or equivalently under the matrix semigroup

$$\begin{pmatrix} e^{tA} & 0 \\ 0 & e^{tA} \end{pmatrix}, \qquad t \geq 0,$$

associated with the sesquilinear form $\mathbf{a} = a \oplus a$ with dense domain $D(\mathbf{a}) := D(a) \times D(a)$. A classical formula due to von Neumann yields that the orthogonal projection of $H \times H$ onto Graph($\Sigma$) is given by

$$P_{\mathrm{Graph}(\Sigma)} = \begin{pmatrix} (I+\Sigma^*\Sigma)^{-1} & \Sigma^*(I+\Sigma\Sigma^*)^{-1} \\ \Sigma(I+\Sigma^*\Sigma)^{-1} & I-(I+\Sigma\Sigma^*)^{-1} \end{pmatrix} := \begin{pmatrix} L & \Sigma^* R \\ \Sigma L & I-R \end{pmatrix},$$

cf. [17, Thm. 23]. The remainder of the proof is based on a known criterion by Ouhabaz, see [18, §2.1], stating that a closed subspace $Y$ of a Hilbert space is invariant under a semigroup associated with a form $b$ with domain $D(b)$ if and only if

- the orthogonal projection $P_Y$ onto $Y$ leaves $D(b)$ invariant and
- $b(P_Y f, f - P_Y f) = 0$ for all $f \in D(b)$.

Clearly

$$P_{\mathrm{Graph}(\Sigma)} D(\mathbf{a}) \subset D(\mathbf{a})$$

if and only if each of the four entries of $P_{\mathrm{Graph}(\Sigma)}$ leave $D(a)$ invariant. In particular, the upper-left entry leaves $D(a)$ invariant if and only if $\Sigma^*\Sigma$ leaves $D(a)$ invariant, but then the lower-left entry leaves $D(a)$ invariant if and only if additionally $\Sigma$ leaves $D(a)$ invariant, too. Similarly, the lower-right entry leaves $D(a)$ invariant if and only if $\Sigma\Sigma^*$ leaves $D(a)$ invariant, but then the upper-right (resp., lower-left) entry leaves $D(a)$ invariant if and only if additionally $\Sigma^*$ (resp., $\Sigma$) leaves $D(a)$ invariant, too. Since however invariance of $D(a)$ under $\Sigma, \Sigma^*$ already implies invariance of $D(a)$ under $\Sigma\Sigma^*, \Sigma^*\Sigma$, the claim follows – the second condition is in fact just a plain reformulation of Ouhabaz's second condition. □

A special class of Cauchy problems is given by so-called quantum graphs. In its easiest form (to which we restrict ourselves for the sake of simplicity), a **quantum graph** $\mathcal{G}$ is a pair $(G, L)$, where $G = (V, E)$ is a (possibly infinite) simple connected metric graph and $L$ is a Hamiltonian (i.e., a self-adjoint operator) on $\bigoplus_{e\in E} L^2(e)$. For technical reasons, edges have to be directed (in an arbitrary way which is not further relevant for the problem) and given a metric structure. Hence, we identify each edge $e = (v, w)$ with the interval $[0, 1]$ and write $\psi(v) := \psi(0)$ and $\psi(w) := \psi(1)$ whenever we consider a function $\psi : (v, w) \equiv [0, 1] \to \mathbb{C}$.

In this note, we consider for the sake of simplicity the easiest possible choice of Hamiltonian: the second derivative. To each quantum graph is naturally associated a system of Schrödinger type equations

$$i\frac{\partial \psi_e}{\partial t}(t,x) = \frac{\partial^2 \psi_e}{\partial x^2}(t,x), \qquad t \in \mathbb{R},\ x \in (0,1),\ e \in E,$$

where $\psi_e : (v, w) \equiv e \equiv [0, 1] \to \mathbb{C}$, i.e., $\psi$ are vector-valued wavefunctions from $[0, 1] \to \ell^2(E)$. The natural operator theoretical setting of this problem includes the Hilbert space

$$H := L^2(0,1;\ell^2(E)) \cong L^2(0,1) \otimes \ell^2(E) \cong \bigoplus_{e\in E} L^2(0,1;\mathbb{C})$$

and the Hamiltonian defined by the diagonal operator matrix

$$L\psi := \Delta\psi := \operatorname{diag}\left(\frac{\partial^2\psi_e}{\partial x^2}\right)_{e\in E}.$$

Naturally, some compatibility conditions have to be satisfied in the boundary, i.e., in the nodes of the graph. These are typically given by so-called *continuity/Kirchhoff coupling conditions*

$$\psi_e(t,v) = \psi_f(t,v) \qquad \text{for all } t\in\mathbb{R},\ v\in V,\ \text{whenever } e,f\sim v \tag{3}$$

(here and in the following we write $e\sim v$ if the edge $e$ is incident in the node $v$) and moreover

$$\sum_{e\sim v}\frac{\partial\psi_e}{\partial n}(t,v) = 0 \qquad \text{for all } t\in\mathbb{R},\ v\in V.$$

Here $\frac{\partial\psi_e}{\partial n}$ denotes the outer normal derivative of $\psi_e$ at 0 or 1. Alternatively, we may formulate the above boundary conditions by

$$\underline{\psi} := \begin{pmatrix}\psi(0)\\ \psi(1)\end{pmatrix} \in Y \qquad \text{and} \qquad \underline{\psi}' := \begin{pmatrix}-\psi'(0)\\ \psi'(1)\end{pmatrix} \in Y^\perp, \tag{4}$$

where $Y := \operatorname{Range}\tilde{I}$ is a closed subspace of $\ell^2(E)\times\ell^2(E)$. Here $I$ is the $n\times m$ (signed) incidence matrix $G$, and $I^+, I^-$ are the matrices whose entries are the positive and negative parts of the entries of $I$ and

$$\tilde{I} := \begin{pmatrix}(I^+)^T\\ (I^-)^T\end{pmatrix}. \tag{5}$$

It can be easily shown that $\Delta$ is associated with the sesquilinear, symmetric, $H$-elliptic, continuous form $a$ defined by

$$a(\psi,\phi) := \int_0^1 (\psi'(x)|\phi'(x))_{\ell^2(E)}dx$$

with form domain

$$D(a) := \{\psi\in H^1(0,1;\ell^2(E)) : \underline{\psi}\in Y\}.$$

Consistently with the general definition, a symmetry of a quantum graph $\mathcal{G}$ is a unitary operator on $H$ that commutes with the unitary group generated by $i\Delta$. Symmetries of $\mathcal{G}$ define a group, which we denote by $\mathfrak{A}(\mathcal{G})$.

## 3. Symmetries of quantum graphs

Following the pattern suggested in Section 1, one can address the following.

- *Let $\Gamma$ be a group and $C$ the category of quantum graphs. Is there an object $\mathcal{G}$ of $C$ whose symmetry group $\mathfrak{A}(\mathcal{G})$ is abstractly isomorphic to $G$?*

The above question can be easily answered in the negative. Since by linearity of the Schrödinger equation $U(1)$ is always abstractly isomorphic to a subgroup of the symmetry group $\mathfrak{A}(\mathcal{G})$ of a quantum graph[2], no finite group $\Gamma$ can be abstractly isomorphic to $\mathfrak{A}(\mathcal{G})$. The above isomorphy condition can be relaxed, though.

In the proof of our main theorem we will need the notion of **edge symmetry** of a graph: by definition, this is a permutation of edges of $G$ that preserves edge adjacency (or equivalently, a permutation of edges that commutes with the adjacency matrix of the line graph of $G$). In other words, by definition a permutation $\tilde{\pi}$ on $E$ is an edge symmetry if $\tilde{\pi}(e), \tilde{\pi}(f)$ have a common endpoint whenever $e, f \in E$ do. Edge symmetries form a group which is usually denoted by $A^*(G)$.

Now, observe that each symmetry $\pi \in A(G)$ naturally induces an edge symmetry $\tilde{\pi} \in A^*(G)$: simply define

$$\tilde{\pi}(e) := (\pi(v), \pi(w)) \qquad \text{whenever } e = (v, w).$$

While clearly

$$A'(G) := \{\tilde{\pi} : \pi \in A(G)\}$$

(whose elements we call **induced edge symmetries**) is a group, it can be strictly smaller than $A(G)$: simply think of the graph $G$ defined by

for which $A(G) = C_2 \times C_2$ (independent switching of the adjacent nodes and/or of the isolated nodes) but $A'(G)$ is trivial. However, this is an exceptional case. The following has been proved by Sabidussi in the case of a finite group, but its proof (see [14, Thm. 1]) carries over verbatim to the case of an infinite graph.

**Lemma 3.1.** *Let $G$ be a simple graph. Then the groups $A(G)$ and $A'(G)$ are isomorphic provided that $G$ contains at most one isolated node and no isolated edge.*

Hence, $A(G) \cong A'(G)$ in any connected graph with at least 3 nodes, and in particular in any connected, 3-regular graph.

**Theorem 3.2.** *Let $\Gamma$ be a (possibly infinite) group. Then there exists a quantum graph $\mathcal{G}$ such that $\Gamma$ is abstractly isomorphic to a subgroup of $\mathfrak{A}(\mathcal{G})$.*

[2] While usual Schrödinger equations in $\mathbb{R}^3$ are invariant under the Galilean transformations, this is in general not true for quantum graphs. Indeed, a quantum graph constructed over a cycle is invariant under space translation (that is, rotations), but this is not true in case of general ramifications.

*Proof.* To begin with, apply Frucht's theorem or its infinite generalization and consider some graph $G$ such that $A(G)$ is abstractly isomorphic to $\Gamma$: If $\Gamma$ is infinite consider an infinite connected graph yielded by the results of de Groot [9] and Sabidussi [22], whereas if $\Gamma$ is finite consider a 3-regular connected graph given by Frucht's theorem [11]. In any case, $G$ is connected and has more than 3 nodes, hence by Lemma 3.1 $A(G) \cong A'(G)$.

Now, any $\tilde{\pi} \in A'(G)$ can be associated with a bounded linear operator $\Pi$ on $H = L^2(0,1;\ell^2(E))$ defined by

$$(\Pi\psi)_e := \psi_{\tilde{\pi}(e)}, \qquad \psi \in L^2(e),\ e \in E. \tag{6}$$

Such an operator is clearly unitary, since $\tilde{\pi}$ is a permutation, and the identifications

$$\pi \mapsto \tilde{\pi} \mapsto \Pi$$

define a group

$$\{\Pi \in \mathcal{L}(H) : \pi \in A(G)\} \cong A(G)$$

of unitary operators on $H$.

It remains to prove that each such $\Pi$ commutes with the unitary group $(e^{it\Delta})_{t\geq 0}$, or rather with its generator $\Delta$. In order to apply Lemma 2.1, it suffices to observe that $\Pi$ is unitary and not dependent on the space variable, so that the second condition is trivially satisfied.

Finally, observe that if $\psi \in D(a)$, then clearly $\Pi\psi \in H^1(0,1;\ell^2(E))$ and moreover for all $v \in V$ and all $e, f \sim v$ one has

$$\Pi\psi_e(v) = \psi_{\tilde{\pi}(e)}(\pi(v)) = \psi_{\tilde{\pi}(f)}(\pi(v)) = \Pi\psi_f(v),$$

by definition of edge symmetry induced by $\pi$ and by (3). This yields invariance of $D(a)$ under $\Pi$ and concludes the proof. □

**Remark 3.3.** In recent years, quantum systems over graphs under more general boundary conditions have become popular. The easiest case is obtained whenever the boundary conditions in (4) are replaced by

$$\underline{\psi} \in Y^\perp \qquad \text{and} \qquad \underline{\psi}' \in Y, \tag{7}$$

i.e., whenever a continuity condition is imposed on the normal derivatives and a Kirchhoff one on the trace of the wavefunction. These are sometimes referred to as *anti-Kirchhoff boundary conditions* or *$\delta'$ coupling* in the literature. Interesting results are available for quantum graphs with such boundary conditions, see, e.g., [12, 20].

Then, Theorem 3.2 also holds if instead, given a group, we look for a quantum graph with boundary conditions (7) whose automorphism group contains a subgroup abstractly isomorphic to $A(G)$. To see this, observe that the corresponding Hamiltonian $\Delta^\perp$ is associated with the form $a^\perp := a$, but now defined on

$$D(a^\perp) := \{\psi \in H^1(0,1;\ell^2(E)) : \underline{\psi} \in Y^\perp\}.$$

We have proved above that $\Pi D(a) \subset D(a)$. As observed in the proof of Lemma 2.1, this is equivalent to asking that

$$P_{\mathrm{Graph}(\Pi)} D(\mathbf{a}) \subset D(\mathbf{a}),$$

or rather

$$P_{\mathrm{Graph}(\tilde{\pi})} Y \subset Y.$$

In other words, the orthogonal projections $P_{\mathrm{Graph}(\tilde{\pi})}$ and $P_Y$ onto the closed subspaces $\mathrm{Graph}(\tilde{\pi})$ and $Y$ of $\ell^2(E) \times \ell^2(E)$ commute. But then also the orthogonal projections $P_{\mathrm{Graph}(\tilde{\pi})}$ and $P_{Y^\perp} = \mathrm{Id} - P_Y$ onto $\mathrm{Graph}(\tilde{\pi})$ and $Y^\perp$ commute, i.e., $\Pi D(a^\perp) \subset D(a^\perp)$. The claim follows.

**Remark 3.4.** Clearly, any edge permutation $\tilde{\pi}$ induces a unitary operator $\Pi$ on $H$ defined as in (6), but it generally ignores the adjacency structure of a graph. One could imagine that a general *edge symmetry* $\tilde{\pi} \in A^*(G)$ may then suffice in the last part of the proof of Theorem 3.2, in order to deduce that $\Pi$ is a symmetry of $\mathfrak{A}(\mathcal{G})$. This is tempting, because on the one hand edge symmetries seem to be more general than induced edge symmetries (i.e., than elements of $A'(G)$), on the other hand they still preserve adjacency. However, general edge symmetries are not fit for our framework, as the following example shows: the permutation $\tilde{\pi} \equiv (e_1\ e_4)$ (which is not induced by any of the two (node) symmetries) is clearly an edge symmetry of the graph

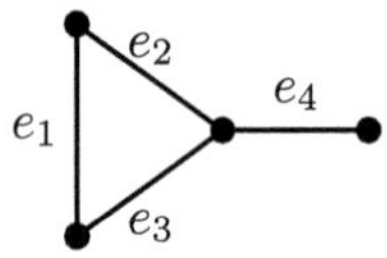


but the induced unitary operator on $H$ does not preserve the boundary conditions (3): in fact, both $e_1$ and $e_4$ are adjacent to (say) $e_2$, but their node which is common to $e_2$ is different.

Is there room for a generalization of Theorem 3.2 invoking edge permutations that are more general than those induced by (node) symmetries but less general than edge symmetries? In the absolute majority of cases the answer is negative: a classical result going back to Whitney (see [4, Cor. 9.5b]) states that in a connected simple graph $G$ with at least three nodes the three groups $A(G), A'(G), A^*(G)$ are pairwise isomorphic if and only if $G$ is different from each of the following graphs:

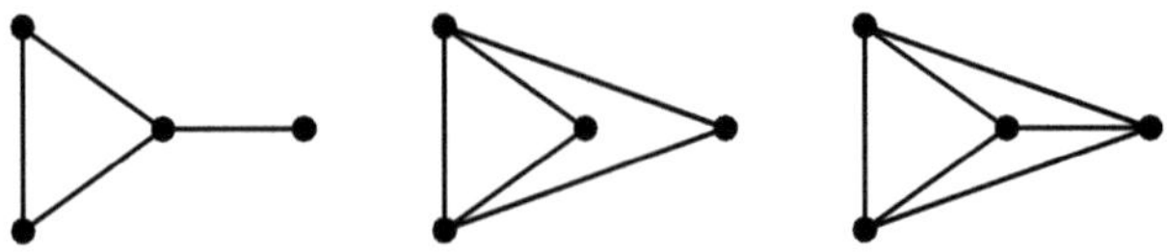

## 4. Isospectral non-isomorphic graphs

"Can one hear the shape of a drum?": this has become a popular question in several mathematical fields, since Kac first addressed it in 1966. We conclude this note by commenting on a simple application of the above results.

In the context of quantum graphs we do not have drums but networks (of strings), and the corresponding question has been discussed in [6, 13]. In particular, it has been observed in [6] that there actually exist pairs of isospectral quantum graphs (that is, isospectral Hamiltonians) constructed on graphs that are not mutually isomorphic. This is a direct consequence of a useful description of the spectrum of $\Delta$ obtained in [5, §6] in the case of regular graphs; and of the well-known existence of pairs of isospectral[3], non-isomorphic, connected regular graphs, see, e.g., [8, § 6.1][4]. Von Below's result can be extended as follows.

**Proposition 4.1.** *There exist arbitrarily long sequences of isospectral, pairwise non isomorphic quantum graphs $\mathcal{G}_1, \dots, \mathcal{G}_N$ whose automorphism groups $\mathfrak{A}(\mathcal{G}_1), \dots, \mathfrak{A}(\mathcal{G}_N)$ contain subgroups that are abstractly isomorphic to arbitrary finite groups.*

*Proof.* The main ingredient of the proof is a beautiful theorem that Babai has obtained in [1], see also [8, § 5.4]:

*Given a finite family of finite groups $\Gamma_1, \dots, \Gamma_N$, there exist pairwise non-isomorphic finite graphs $G_1, \dots, G_N$ such that all graphs $G_1, \dots, G_N$ are isospectral and $A(G_i)$ is abstractly isomorphic to $\Gamma_i$ for all $i = 1, \dots, N$.*

Carefully checking Babai's proof shows that the graphs $G_1, \dots, G_N$ can even be taken to be 3-regular [3]. Now, in order to complete the proof it suffices to take arbitrarily many groups $\Gamma_1, \dots, \Gamma_N$. Then, Babai's theorem yields pairwise non-isomorphic, isospectral 3-regular graphs $G_1, \dots, G_N$ (with $A(G_i)$ abstractly isomorphic to $\Gamma_i$). By the above-mentioned result by von Below, isospectral regular graphs induce isospectral quantum graphs. By the proof of Theorem 3.2, $\Gamma_i$ is also abstractly isomorphic to a subgroup of $\mathfrak{A}(\mathcal{G}_i)$. □

## References

[1] L. Babai. Automorphism group and category of cospectral graphs. *Acta Math. Acad. Sci. Hung.*, 31:295–306, 1978.

[2] L. Babai. Automorphism groups, isomorphism, reconstruction. In R.L. Graham, M. Grötschel, and L. Lovász, editors, *Handbook of Combinatorics – Vol.* 2, pages 1447–1540. North-Holland, Amsterdam, 1995.

[3] L. Babai. Private communication, 2010.

[4] M. Behzad, G. Chartrand, and L. Lesniak-Foster. *Graphs & Digraphs.* Prindle, Weber & Schmidt, Boston, 1979.

[3] Two graphs are called isospectral if their adjacency matrices have the same spectrum.

[4] According to [8, § 6.1], the earliest example of pairs of isospectral, non-isomorphic, connected 4-regular graphs has been obtained by Hoffman and Ray-Chaudhuri in 1965, whereas Sunada's approach to the original Kac's question, upon which the construction of Gutkin and Smilansky relies, has been developed 20 years later.

[5] J. von Below. A characteristic equation associated with an eigenvalue problem on $C^2$-networks. *Lin. Algebra Appl.*, 71:309–325, 1985.

[6] J. von Below. Can one hear the shape of a network? In F. Ali Mehmeti, J. von Below, and S. Nicaise, editors, *Partial Differential Equations on Multistructures* (*Proc. Luminy* 1999), volume 219 of *Lect. Notes Pure Appl. Math.*, pages 19–36, New York, 2001. Marcel Dekker.

[7] S. Cardanobile, D. Mugnolo, and R. Nittka. Well-posedness and symmetries of strongly coupled network equations. *J. Phys. A*, 41:055102, 2008.

[8] D.M. Cvetković, M. Doob, and H. Sachs. *Spectra of Graphs – Theory and Applications.* Pure Appl. Math. Academic Press, New York, 1979.

[9] J. De Groot. Groups represented by homeomorphism groups I. *Math. Ann.*, 138:80–102, 1959.

[10] R. Frucht. Herstellung von Graphen mit vorgegebener abstrakter Gruppe. *Compositio Math*, 6:239–250, 1938.

[11] R. Frucht. Graphs of degree three with a given abstract group. *Canadian J. Math*, 1:365–378, 1949.

[12] S.A. Fulling, P. Kuchment, and J.H. Wilson. Index theorems for quantum graphs. *J. Phys. A*, 40:14165–14180, 2007.

[13] B. Gutkin and U. Smilansky. Can one hear the shape of a graph? *J. Phys. A*, 34:6061–6068, 2001.

[14] F. Harary and E.M. Palmer. On the point-group and line-group of a graph. *Acta Math. Acad. Sci. Hung.*, 19:263–269, 1968.

[15] H. Izbicki. Unendliche Graphen endlichen Grades mit vorgegebenen Eigenschaften. *Monats. Math.*, 63:298–301, 1959.

[16] P. Kuchment. Quantum graphs: an introduction and a brief survey. In P. Exner, J. Keating, P. Kuchment, T. Sunada, and A. Teplyaev, editors, *Analysis on Graphs and its Applications*, volume 77 of *Proc. Symp. Pure Math.*, pages 291–314, Providence, RI, 2008. Amer. Math. Soc.

[17] J.W. Neuberger. *Sobolev Gradients and Differential Equations*, volume 1670 of *Lect. Notes Math.* Springer-Verlag, Berlin, 1997.

[18] E.M. Ouhabaz. *Analysis of Heat Equations on Domains*, volume 30 of *Lond. Math. Soc. Monograph Series.* Princeton Univ. Press, Princeton, 2005.

[19] Y.V. Pokornyi and A.V. Borovskikh. Differential equations on networks (geometric graphs). *J. Math. Sci.*, 119:691–718, 2004.

[20] O. Post. First-order approach and index theorems for discrete and metric graphs. *Ann. Henri Poincaré*, 10:823–866, 2009.

[21] G. Sabidussi. Graphs with given group and given graph-theoretical properties. *Canad. J. Math*, 9:515–525, 1957.

[22] G. Sabidussi. Graphs with given infinite group. *Monats. Math.*, 64:64–67, 1960.

Delio Mugnolo
Institut für Analysis, Universität Ulm
Helmholtzstraße 18
D-89081 Ulm, Germany
e-mail: delio.mugnolo@uni-ulm.de

Operator Theory:
Advances and Applications, Vol. 221, 491-496


# Note on Characterizations of the Harmonic Bergman Space

Kyesook Nam, Kyunguk Na and Eun Sun Choi

**Abstract.** In this paper, we solve the problem which was stated in [6]. Actually we obtain a characterization of harmonic Bergman space in terms of Lipschitz type condition with pseudo-hyperbolic metric on the unit ball in $\mathbf{R}^n$.

**Mathematics Subject Classification (2000).** Primary 31B05; Secondary 46E30.

**Keywords.** Harmonic Bergman space, hyperbolic metric, Lipschitz condition.

## 1. Introduction

For a fixed positive integer $n \geq 2$, let $B$ be the open unit ball in $\mathbf{R}^n$. Given $\alpha > -1$ and $1 \leq p < \infty$, let $L^p_\alpha = L^p(B, dV_\alpha)$ denote the weighted Lebesgue spaces on $B$ where $dV_\alpha$ denotes the weighted measure given by

$$dV_\alpha(x) = (1 - |x|^2)^\alpha \, dV(x).$$

Here $dV$ is the Lebesgue volume measure on $B$. For simplicity, we use the notation $dx = dV(x)$.

Given $\alpha > -1$ and $1 \leq p < \infty$, the weighted harmonic Bergman space $b^p_\alpha$ is the space of all harmonic functions $f$ on $B$ such that

$$\|f\|_{b^p_\alpha} := \left( \int_B |f|^p \, dV_\alpha \right)^{1/p} < \infty.$$

We now recall the pseudo-hyperbolic on $B$. Let $\rho$ be the pseudo-hyperbolic distance between two points $x, y \in B$ defined by

$$\rho(x,y) = \frac{|x-y|}{[x,y]}$$

where

$$[x,y] = \sqrt{1 - 2x \cdot y + |x|^2|y|^2}.$$

The next theorem is our main result in this paper.

**Theorem 1.1.** *Suppose $\alpha > -1$, $1 \le p < \infty$ and $f$ is harmonic in $B$. Then $f \in b^p_\alpha$ if and only if there exists a continuous function $g \in L^p_\alpha$ such that*

$$|f(x) - f(y)| \le \rho(x, y)[g(x) + g(y)]$$

*for all $x, y \in B$.*

In Section 2 we collect some lemmas to be used later. In Section 3 we prove norm equivalences in terms of radial and gradient norms. In the last section, our main theorem is proved.

*Constants.* In the rest of the paper we use the same letter $C$, often depending on the allowed parameters, to denote various constants which may change at each occurrence. For nonnegative quantities $X$ and $Y$, we often write $X \lesssim Y$ if $X$ is dominated by $Y$ times some *inessential* positive constant. Also, we write $X \approx Y$ if $X \lesssim Y \lesssim X$.

## 2. Preliminaries

For $x \in B$ and $r \in (0, 1)$, let $E_r(x)$ denote the pseudo-hyperbolic ball of radius $r$ centered at $x$. A straightforward calculation shows that

$$E_r(x) = B\left(\frac{1-r^2}{1-r^2|x|^2}x,\ \frac{1-|x|^2}{1-r^2|x|^2}r\right) \tag{2.1}$$

where $B(a, t)$ is an Euclidean ball of radius $t$ centered at $a$.

The following lemma comes from [3].

**Lemma 2.1.** *$\rho$ is a distance function on $B$.*

The following two lemmas come from [2].

**Lemma 2.2.** *The inequality*

$$\frac{1-\rho(x,y)}{1+\rho(x,y)} \le \frac{1-|x|}{1-|y|} \le \frac{1+\rho(x,y)}{1-\rho(x,y)}$$

*holds for $x, y \in B$.*

**Lemma 2.3.** *The inequality*

$$\frac{1-\rho(x,y)}{1+\rho(x,y)} \le \frac{[x,a]}{[y,a]} \le \frac{1+\rho(x,y)}{1-\rho(x,y)}$$

*holds for $x, y, a \in B$.*

**Lemma 2.4.** *Let $\alpha > -1$, $1 \le p < \infty$ and $r \in (0, 1)$. Then there exists a positive constant $C = C(p, r, \alpha)$ such that*

$$|\nabla f(x)|^p \le \frac{C}{(1-|x|^2)^{n+\alpha+p}} \int_{E_r(x)} |f|^p \, dV_\alpha, \qquad x \in B$$

*for all function $f$ harmonic on $B$.*

*Proof.* Let $r \in (0,1)$ and $x \in B$. Taking $\Omega = E_r(x)$ in Corollary 8.2 of [1], there exists a constant $C = C(p) > 0$ such that

$$\left|\frac{\partial f}{\partial x_i}(x)\right|^p \leq \frac{C}{d(x, \partial E_r(x))^{n+p}} \int_{E_r(x)} |f|^p \, dV \tag{2.2}$$

where $d(x, \partial E_r(x))$ denotes the Euclidean distance from a point $x$ to $\partial E_r(x)$. Note that from (2.1)

$$d(x, \partial E_r(x)) = \frac{(1-|x|^2)r}{1+r|x|}.$$

Thus, by Lemma 2.2 and (2.2), we have

$$\left|\frac{\partial f}{\partial x_i}(x)\right|^p \leq \frac{C}{(1-|x|^2)^{n+\alpha+p}} \int_{E_r(x)} |f|^p \, dV_\alpha$$

where the constant $C$ depends only on $p, r$ and $\alpha$. Consequently, we obtain that

$$|\nabla f(x)|^p \leq \frac{C}{(1-|x|^2)^{n+\alpha+p}} \int_{E_r(x)} |f|^p \, dV_\alpha$$

as desired. □

We use the notation $\mathcal{D}$ for the radial differentiation defined by

$$\mathcal{D}f(x) = \sum_{i=1}^n x_i \frac{\partial f}{\partial x_i}(x), \qquad x \in B$$

for functions $f \in C^1(B)$.

**Proposition 2.5.** *Suppose $\alpha > -1$, $1 \leq p < \infty$ and $f$ is harmonic on $B$. Then the following conditions are equivalent.*

(a) $f \in b^p_\alpha$.
(b) $(1-|x|^2)\mathcal{D}f(x) \in L^p_\alpha$.
(c) $(1-|x|^2)|\nabla f(x)| \in L^p_\alpha$.

*Proof.* The implication $(c) \Longrightarrow (b) \Longrightarrow (a)$ has been proved by Theorem 5.1 of [5]. Also, see [4] for a proof. Thus we only need to prove the implication $(a) \Longrightarrow (c)$.

Assume $f \in b^p_\alpha$ and fix $r \in (0,1)$. Then Lemma 2.4 and Fubini's theorem give us that

$$\begin{aligned}\int_B (1-|x|^2)^p |\nabla f(x)|^p \, dV_\alpha(x) &\lesssim \int_B \frac{1}{(1-|x|^2)^n} \int_{E_r(x)} |f(y)|^p \, dV_\alpha(y) \, dx \\ &= \int_B |f(y)|^p \int_B \frac{\chi_{E_r(x)}(y)}{(1-|x|^2)^n} \, dx \, dV_\alpha(y).\end{aligned}$$

Here $\chi_{E_r(x)}$ denotes the characteristic function of the set $E_r(x)$. Using Lemma 2.2 and (2.1), we obtain

$$\int_B \frac{\chi_{E_r(x)}(y)}{(1-|x|^2)^n} \, dx = \int_{E_r(y)} \frac{dx}{(1-|x|^2)^n} \approx 1 \tag{2.3}$$

so that

$$\int_B (1-|x|^2)^p |\nabla f(x)|^p \, dV_\alpha(x) \lesssim \int_B |f(y)|^p \, dV_\alpha(y).$$

This completes the proof. □

## 3. Proof of the main theorem

Now we are ready to our main theorem.

We write $\beta$ for the hyperbolic distance between two points $x, y \in B$ defined by

$$\beta(x,y) = \frac{1}{2} \log \frac{1+\rho(x,y)}{1-\rho(x,y)}.$$

**Theorem 3.1.** *Suppose $\alpha > -1$, $1 \le p < \infty$ and $f$ is harmonic in $B$. Then the following conditions are equivalent.*

(a) $f \in b^p_\alpha$.

(b) *There exists a continuous function $g \in L^p_\alpha$ such that*

$$|f(x) - f(y)| \le \rho(x,y)[g(x) + g(y)]$$

*for all $x, y \in B$.*

(c) *There exists a continuous function $g \in L^p_\alpha$ such that*

$$|f(x) - f(y)| \le \beta(x,y)[g(x) + g(y)].$$

*for all $x, y \in B$.*

(d) *There exists a continuous function $g \in L^p_{p+\alpha}$ such that*

$$|f(x) - f(y)| \le |x-y|[g(x) + g(y))].$$

*for all $x, y \in B$.*

*Proof.* The implications (b) $\Longrightarrow$ (c) $\Longrightarrow$ (a) and (b) $\Longrightarrow$ (d) $\Longrightarrow$ (a) have been proved in [4]. Thus we only need to prove the implication (a) $\Longrightarrow$ (b).

Let $f \in b^p_\alpha$. Fix $r \in (0, 1/2)$ and let $x \in E_r(y)$. Then we have

$$\begin{aligned} |f(x) - f(y)| &\le \int_0^1 \left| \frac{d}{dt} [f(t(x-y)+y)] \right| dt \\ &\le \int_0^1 |x-y||\nabla f(t(x-y)+y)| \, dt. \end{aligned} \tag{3.1}$$

Since $[x,x] = 1 - |x|^2$, Lemma 2.3 implies

$$|x-y| \approx \rho(x,y)(1-|x|^2) \approx \rho(x,y)(1-|z|^2)$$

for all $y, z \in E_r(x)$. So, letting

$$h(x) = \sup_{z \in E_r(x)} (1-|z|^2)|\nabla f(z)|,$$

we have from (3.1)

$$|f(x) - f(y)| \lesssim \rho(x,y)h(x).$$

If $x \in E_r(y)^c$, then

$$|f(x) - f(y)| \le \rho(x,y)\left\{\frac{|f(x)|}{r} + \frac{|f(y)|}{r}\right\}.$$

Let

$$g(x) = \frac{|f(x)|}{r} + h(x).$$

Then $g$ is continuous on $B$ and

$$|f(x) - f(y)| \le \rho(x,y)[g(x) + g(y)]$$

for all $x, y \in B$.

Now, we need to show that $g \in L^p_\alpha$. By Lemma 2.1, $\rho$ satisfies the triangle inequality. Thus we can see that $E_r(z) \subset E_{2r}(x)$ for every $z \in E_r(x)$. So Lemma 2.4 and Lemma 2.2 give us that

$$\begin{aligned} h(x)^p &\lesssim \sup_{z\in E_r(x)} \frac{1}{(1-|z|^2)^{n+\alpha}} \int_{E_r(z)} |f(y)|^p \, dV_\alpha(y) \\ &\lesssim \frac{1}{(1-|x|^2)^{n+\alpha}} \int_{E_{2r}(x)} |f(y)|^p \, dV_\alpha(y) \end{aligned}$$

for all $x \in B$. Consequently, we obtain by Fubini's theorem,

$$\begin{aligned} \int_B h(x)^p \, dV_\alpha(x) &\lesssim \int_B \frac{1}{(1-|x|^2)^{n+\alpha}} \int_{E_{2r}(x)} |f(y)|^p \, dV_\alpha(y)\, dV_\alpha(x) \\ &= \int_B |f(y)|^p \int_{E_{2r}(y)} \frac{1}{(1-|x|^2)^n} \, dx \, dV_\alpha(y) \\ &\lesssim \int_B |f|^p \, dV_\alpha \end{aligned}$$

where the last inequality comes from (2.3).

The proof is complete. □

## References


[1] S. Axler, P. Bourdon and W. Ramey, *Harmonic function theory*, Springer-Verlag, New York, 1992.

[2] B.R. Choe, H. Koo and Y.J. Lee, *Positive Schatten(-Herz) class Toeplitz operators on the ball*, Studia Math, 189(2008), No. 1, 65–90.

[3] B.R. Choe and Y.J. Lee, *Note on atomic decompositions of harmonic Bergman functions*, Proceedings of the 15th ICFIDCAA; Complex Analysis and its Applications, Osaka Municipal Universities Press, OCAMI Studies Series 2(2007), 11–24.

[4] E.S. Choi and K. Na, *Characterizations of the harmonic Bergman space on the ball*, J. Math. Anal. Appl. 353 (2009), 375–385.

[5] B.R. Choe, H. Koo and H. Yi, *Derivatives of harmonic Bergman and Bloch functions on the ball*, J. Math. Anal. Appl. 260(2001), 100–123.

[6] K. Nam, K. Na and E.S. Choi, *Corrigendum of "Characterizations of the harmonic Bergman space on the ball"* [*J. Math. Anal. Appl.* 353 (2009), 375–385], in press.

[7] H. Wulan and K. Zhu, *Lipschitz type characterizations for Bergman spaces*, to appear in Canadian Math. Bull.

Kyesook Nam
Department of Mathematical Sciences
BK21-Mathematical Sciences Division
Seoul National University
Seoul 151-742, Republic of Korea
e-mail: `ksnam@snu.ac.kr`

Kyunguk Na
General Education, Mathematics
Hanshin University
Gyeonggi 447-791, Republic of Korea
e-mail: `nakyunguk@hanshin.ac.kr`

Eun Sun Choi
Department of Mathematics
Korea University
Seoul 136-701, Republic of Korea
e-mail: `eschoi93@korea.ac.kr`

Operator Theory:
Advances and Applications, Vol. 221, 497-513


# On Some Boundary Value Problems for the Helmholtz Equation in a Cone of 240°

A.P. Nolasco and F.-O. Speck

*Dedicated to Professor Erhard Meister on the occasion of his 80th anniversary*

**Abstract.** In this paper we present the explicit solution in closed analytic form of Dirichlet and Neumann problems for the Helmholtz equation in the non-convex and non-rectangular cone $\Omega_{0,\alpha}$ with $\alpha = 4\pi/3$. Actually, these problems are the only known cases of exterior (i.e., $\alpha > \pi$) wedge diffraction problems explicitly solvable in closed analytic form with the present method. To accomplish that, we reduce the BVPs in $\Omega_{0,\alpha}$ each to a pair of BVPs with symmetry in the same cone and each BVP with symmetry to a pair of semi-homogeneous BVPs in the convex half-cones. Since $\alpha/2$ is an (odd) integer part of $2\pi$, we obtain the explicit solution of the semi-homogeneous BVPs for half-cones by so-called Sommerfeld potentials (resulting from special Sommerfeld problems which are explicitly solvable).

**Mathematics Subject Classification (2000).** Primary 35J25; Secondary 30E25, 35J05, 45E10, 47B35, 47G30.

**Keywords.** Wedge diffraction problem; Helmholtz equation; boundary value problem; half-line potential; pseudodifferential operator; Sommerfeld potential.

## 1. Introduction

The present work originates from diffraction theory, the time-harmonic scattering of waves from infinite wedges in a so-called canonical formulation [15], [16], [21], [27] where the basic problems are modeled by Dirichlet or Neumann boundary value problems for the two-dimensional Helmholtz equation in a cone $\Omega$ with an angle $\alpha \in ]0, 2\pi]$. With most of the existing methods the "interior problems" where $\alpha < \pi$ are somehow easier to tackle than the "exterior problems" where $\pi < \alpha < 2\pi$, see for instance [12], [13], [21], [24], [27], [28]. Sometimes a structural difference between the interior and the exterior problem ($\alpha$ vs. $\pi - \alpha$) becomes apparent and most interesting in what concerns the question of explicit solution, e.g., see the case of $\alpha = \pi/2$ vs. $\alpha = 3\pi/2$ in [1], [3], [4] and [17]. This observation holds for the present boundary value problems (BVPs), as well. However, the solution technique for the exterior problems with $\alpha = 4\pi/3$ is surprising and absolutely exceptional in view of the fact that it can be obtained from explicit solution of the corresponding interior problems ($\alpha = 2\pi/3$) studied before, see [6]. The case $\alpha = 4\pi/3$ represents the only non-convex cone (with $\alpha \in ]\pi, 2\pi[$) such that $\alpha/2$ has the form $\alpha/2 = 2\pi/n$ with $n \in \mathbb{N}$.

In that previous paper, we solved explicitly and analytically the following boundary value problems. Let

$$\Omega_{0,\alpha} = \{(x_1, x_2) \in \mathbb{R}^2 : 0 < \arg(x_1 + ix_2) < \alpha\}$$

be a convex cone with $\alpha = 2\pi/n$, $n \in \mathbb{N}_2 = 2, 3, 4, \ldots$, bordered by

$$\Gamma_1 = \{(x_1, x_2) \in \mathbb{R}^2 : x_1 > 0, x_2 = 0\},$$
$$\Gamma_2 = \{(x_1, x_2) \in \mathbb{R}^2 : \arg(x_1 + ix_2) = \alpha\},$$

and the origin. We were looking for all weak solutions $u \in H^{1+\varepsilon}(\Omega_{0,\alpha})$ ($\varepsilon \in [0, 1/2[$) of the Helmholtz equation

$$(\Delta + k^2)u = 0 \text{ in } \Omega_{0,\alpha},$$

where the wave number $k$ is always complex with $\mathcal{I}m\, k > 0$, briefly denoted by $u \in \mathcal{H}^{1+\varepsilon}(\Omega_{0,\alpha})$, which satisfy Dirichlet or Neumann conditions on the two half-lines $\Gamma_1$ and $\Gamma_2$ bordering $\Omega_{0,\alpha}$, admitting the mixed type as well:

$$\text{(DD)}\quad T_{0,\Gamma_1} u = g_1 \text{ on } \Gamma_1,\ T_{0,\Gamma_2} u = g_2 \text{ on } \Gamma_2 \text{ or} \tag{1.1}$$
$$\text{(NN)}\quad T_{1,\Gamma_1} u = g_1 \text{ on } \Gamma_1,\ T_{1,\Gamma_2} u = g_2 \text{ on } \Gamma_2 \text{ or} \tag{1.2}$$
$$\text{(DN)}\quad T_{0,\Gamma_1} u = g_1 \text{ on } \Gamma_1,\ T_{1,\Gamma_2} u = g_2 \text{ on } \Gamma_2,$$

respectively. Here and in what follows, $T_{0,\Gamma_j}$ stand for the trace operators due to the corresponding parts $\Gamma_j$ of the boundary, $T_{1,\Gamma_j} = T_{0,\Gamma_j} \frac{\partial}{\partial n_j}$, the normal derivative $n_j$ on $\Gamma_j$ is directed into the interior of $\Omega$, and the boundary data $g_1$ and $g_2$ are given in the corresponding trace space $H^{1/2+\varepsilon}(\mathbb{R}_+)$ or $H^{-1/2+\varepsilon}(\mathbb{R}_+)$, where $\Gamma_j$ are "copied onto $\mathbb{R}_+$". The range $[0, 1/2[$ of the *regularity parameter* $\varepsilon$ is chosen for convenience, see [18] and [25]. In some cases (connected with DD conditions), it may be extended to $|\varepsilon| < 1/2$, and in some other cases (connected with mixed DN conditions) it must be restricted to $|\varepsilon| < 1/4$ or to $\varepsilon \in [0, 1/4[$.

The extension of the results to cones $\Omega_{\beta,\gamma}$ with $\gamma - \beta = \alpha$ is simply obtained by rotation. For that purpose, consider the $\alpha$-rotation $\mathcal{R}_\alpha$ given by

$$y = \begin{pmatrix} y_1 \\ y_2 \end{pmatrix} = \mathcal{R}_\alpha\, x = \begin{pmatrix} \cos\alpha & -\sin\alpha \\ \sin\alpha & \cos\alpha \end{pmatrix} \begin{pmatrix} x_1 \\ x_2 \end{pmatrix}.$$

The backward rotation is given by $\mathcal{R}_\alpha^{-1} = \mathcal{R}_{-\alpha} : \Omega_\alpha^\pm \to \Omega_0^\pm = \Omega^\pm$, and we shall denote the $\alpha$-rotation operator acting on suitable functions (or distributions) by

$$(\mathcal{J}_\alpha f)(x) = f(\mathcal{R}_\alpha^{-1} x), \quad x \in \mathbb{R}^2. \tag{1.3}$$

Here and in what follows, $\Omega_\alpha^\pm = \mathcal{R}_\alpha \Omega^\pm = \{\mathcal{R}_\alpha x : x = (x_1, x_2) \in \Omega^\pm\}$, where $\Omega^\pm = \{(x_1, x_2) \in \mathbb{R}^2 : x_2 \gtrless 0\}$ denotes the upper/lower half-planes, respectively.

The Dirichlet problems (DD) are uniquely solvable and well posed in corresponding topologies under certain compatibility conditions for the Dirichlet data

$$g_1 - g_2 \in \widetilde{H}^{1/2+\varepsilon}(\mathbb{R}_+), \tag{1.4}$$

i.e., this function is extendible by zero onto the full line $\mathbb{R}$ such that the zero extension $\ell_0(g_1 - g_2)$ belongs to $H^{1/2+\varepsilon}(\mathbb{R})$. For the study of "tilde spaces" we refer to [5], [9], [11], [20]. The Neumann problems (NN) need a compatibility condition

$$g_1 + g_2 \in \widetilde{H}^{-1/2+\varepsilon}(\mathbb{R}_+), \tag{1.5}$$

if and only if $\varepsilon = 0$ (as a rule, normal derivatives are directed into the interior of $\Omega_{0,\alpha}$ if nothing else is said). The mixed problems (DN) are well posed without any additional condition.

The spaces of functionals $(g_1, g_2)$ which satisfy the compatibility conditions (1.4) and (1.5) will be denoted by $H^{1/2+\varepsilon}(\mathbb{R}_+)^2_\sim$ for $\varepsilon \in [0, 1/2[$ and $H^{-1/2+\varepsilon}(\mathbb{R}_+)^2_\sim$ (relevant for $\varepsilon = 0$), respectively, equipped with the norms

$$\|(g_1, g_2)\|_{H^{1/2+\varepsilon}(\mathbb{R}_+)^2_\sim} = \|g_1\|_{H^{1/2+\varepsilon}(\mathbb{R}_+)} + \|g_1 - g_2\|_{\widetilde{H}^{1/2+\varepsilon}(\mathbb{R}_+)},$$
$$\|(g_1, g_2)\|_{H^{-1/2+\varepsilon}(\mathbb{R}_+)^2_\sim} = \|g_1\|_{H^{-1/2+\varepsilon}(\mathbb{R}_+)} + \|g_1 + g_2\|_{\widetilde{H}^{-1/2+\varepsilon}(\mathbb{R}_+)}.$$

The resolvent operators $(g_1, g_2) \mapsto u \in \mathcal{H}^{1+\varepsilon}(\Omega_{0,\alpha})$ due to the DD and NN problem were seen to be linear homeomorphisms, if $\alpha = 2\pi/n$, $n \in \mathbb{N}_1 = 1, 2, 3, \ldots$, [6].

In particular cases, the solution is immediate, e.g., for $\alpha = \pi$, $\Omega_{0,\alpha} = \Omega^+$ being the upper half-plane, cases DD and NN, respectively, we have the double and simple layer potentials [11] in its simplest form [26]

$$u(x_1, x_2) = \mathcal{K}_{D,\Omega^+} \iota g(x_1, x_2) = \mathcal{F}^{-1}_{\xi \mapsto x_1} e^{-t(\xi) x_2} \widehat{\iota g}(\xi),$$
$$u(x_1, x_2) = \mathcal{K}_{N,\Omega^+} \iota g(x_1, x_2) = -\mathcal{F}^{-1}_{\xi \mapsto x_1} e^{-t(\xi) x_2} t^{-1}(\xi) \widehat{\iota g}(\xi),$$

which are called *line potentials* (LIPs) with density $\iota g$. Similarly, the solution of the DD and NN problems in the lower half-plane $\Omega^-$ are given, respectively, by

$$u(x_1, x_2) = \mathcal{K}_{D,\Omega^-} \iota g(x_1, x_2) = \mathcal{F}^{-1}_{\xi \mapsto x_1} e^{t(\xi) x_2} \widehat{\iota g}(\xi),$$
$$u(x_1, x_2) = \mathcal{K}_{N,\Omega^-} \iota g(x_1, x_2) = -\mathcal{F}^{-1}_{\xi \mapsto x_1} e^{t(\xi) x_2} t^{-1}(\xi) \widehat{\iota g}(\xi).$$

Here $\mathcal{F}$ denotes de Fourier transformation, $\widehat{f} = \mathcal{F} f$, and $t(\xi) = (\xi^2 - k^2)^{1/2}$ with vertical branch cut from $k$ to $-k$ via $\infty$, not crossing the real line. Considering $g = (g_1, g_2) \in H^s(\mathbb{R}_+)^2$, $\iota g$ is the "natural composition" of the two boundary data

$$\iota g(x) = \begin{cases} g_1(x), & x > 0 \\ g_2(-x), & x < 0 \end{cases},$$

(where the data are given on $\mathbb{R}_+$ as a copy of $\Gamma_j$), taking any value in $x = 0$ if $s \in [0, 1/2[$. For $s \in ]-1/2, 1/2[$ the formula might be replaced by $\iota g = \ell_0 g_1 + \mathcal{J} \ell_0 g_2$ (using the reflection operator $\mathcal{J} f(x) = f(-x)$, $x \in \mathbb{R}$), i.e., by a continuous extension of $\iota g$ to negative values of $s$. On one hand, if $(g_1, g_2) \in H^{1/2+\varepsilon}(\mathbb{R}_+)^2_\sim$, we consider $\iota g$ as a function of $H^{1/2+\varepsilon}(\mathbb{R})$ since the value in a single point does not matter. On the other hand, if $(g_1, g_2) \in H^{-1/2+\varepsilon}(\mathbb{R}_+)^2_\sim$, the functional $\iota g \in H^{-1/2+\varepsilon}(\mathbb{R})$ is understood in a distributional sense, cf. [6, Section 4], for details.

The solution of the mixed DN problem for the upper half-plane $\Omega^+$ needs already more sophisticated methods such as the Wiener-Hopf technique but can be presented explicitly, as well, by an *analytic formula*

$$\begin{aligned} u(x_1,x_2) &= \mathcal{K}_{DN,\Omega^+}\,(g_1,g_2)(x_1,x_2) \\ &= \mathcal{F}^{-1}_{\xi\mapsto x_1} e^{-t(\xi)x_2} t_-^{-1/2}(\xi)\left\{\widehat{P_+} t_-^{1/2}\widehat{\ell g_1} - \widehat{P_-} t_+^{-1/2}\widehat{\mathcal{J}\ell g_2}\right\}(\xi) \\ &= \mathcal{K}_{D,\Omega^+}\, A_{t_-^{-1/2}}\left\{P_+ A_{t_-^{1/2}}\ell g_1 - P_- A_{t_+^{-1/2}}\mathcal{J}\ell g_2\right\}(x_1,x_2) \end{aligned}$$

where $t_\pm(\xi) = \xi \pm k$ , $A_\phi = \mathcal{F}^{-1}\phi \cdot \mathcal{F}$ is referred to as a (distributional) convolution operator with Fourier symbol $\phi$, $P_\pm = \ell_0 r_\pm$ are projectors in $H^\varepsilon(\mathbb{R})$, $\widehat{P_\pm} = \mathcal{F}\ell_0 r_\pm \mathcal{F}^{-1}$, and $\ell g_1, \ell g_2$ denote any extensions from $\mathbb{R}_+$ to $\mathbb{R}$ such that $\ell g_1 \in H^{1/2+\varepsilon}(\mathbb{R})$ and $\ell g_2 \in H^{-1/2+\varepsilon}(\mathbb{R})$; the operator does not depend on the particular choice of extension.

The method presented in [6] consists of a combination of our knowledge about the analytical solution of Sommerfeld and rectangular wedge diffraction problems, see [3], [4], [17] and [19], with new symmetry arguments that relate the present to previously solved problems and yield the explicit analytical solution in a great number of cases. For this purpose we introduce the so-called "Sommerfeld potentials" (explicit solutions to special Sommerfeld problems) whose use turns out to be most efficient, see Section 3. It is surprising that the case where the angle is an *integer part* of $2\pi$ (where the cone is convex) can be solved completely whilst the case of "rational" angles $\alpha = 2\pi m/n$ for $m \geq 2$ appears much harder. An exception is the rectangular exterior wedge problem ($\alpha = 3\pi/2$) which includes the study of Hankel operators and can be solved in closed analytical form, cf. [1] and [17].

The second exception, that will be studied in this paper, is the case of Dirichlet and Neumann exterior wedge problems in $\Omega_{0,4\pi/3}$, for which we present the explicit solution in closed analytic form.

In order to obtain the solution of these problems we first show that the solution of a DD or NN problem for the cone $\Omega_{0,\alpha}$ ($\alpha = 4\pi/3$) is given by the sum of the solutions of two BVPs with symmetry in the same cone $\Omega_{0,\alpha}$ and then show that the solution of each BVP with symmetry in $\Omega_{0,\alpha}$ is given by the sum of the solutions of two semi-homogeneous BVPs for the half-cones $\Omega_{0,\alpha/2}$ and $\Omega_{\alpha/2,\alpha}$. Since $\alpha/2 = 2\pi/3$, [6] gives us the explicit form of the solution of the semi-homogeneous BVPs in the convex half-cones $\Omega_{0,\alpha/2}$ and $\Omega_{\alpha/2,\alpha}$, in terms of the sum of Sommerfeld potentials, and consequently the explicit solution of the Dirichlet and Neumann problems in $\Omega_{0,4\pi/3}$. Hence the present work is a direct extension of [6] which particularly shows the explicit formulas in the special case.

Unfortunately the mixed Dirichlet/Neumann problem ($DN$) is not reducible in this way and needs a different method of explicit solution, so far only possible by series expansion, [7], and a rigorous approach based upon the treatise of BVPs in conical Riemann surfaces that will be published elsewhere.

## 2. Reduction to a pair of problems in convex cones (half-angle)

Let $\alpha = 4\pi/3$. By superposition we first reduce the $DD$ and $NN$ problems each to a pair of *boundary value problems with symmetry* in $\Omega = \Omega_{0,\alpha}$. As far as we know this idea is new, at least in the present context.

**Theorem 2.1.** *Let* $\varepsilon \in ]-1/2, 1/2[$ *and* $(g_1, g_2) \in H^{1/2+\varepsilon}(\mathbb{R}_+)^2_\sim$. *Then the following assertions are equivalent:*

(i) $u \in \mathcal{H}^{1+\varepsilon}(\Omega)$ *solves the DD problem in* $\Omega$.
(ii) $u = u^e + u^o$, $u^{e,o} \in \mathcal{H}^{1+\varepsilon}(\Omega)$ *and*

$$(DD1) \quad T_{0,\Gamma_1} u^e = \quad T_{0,\Gamma_2} u^e = g^e = \frac{1}{2}(g_1 + g_2),$$

$$(DD2) \quad T_{0,\Gamma_1} u^o = -T_{0,\Gamma_2} u^o = g^o = \frac{1}{2}(g_1 - g_2).$$

*Proof.* The step from (i) to (ii) results from a decomposition of $u$ into the even and the odd part with respect to the ray $\Gamma = \{(x_1, x_2) \in \mathbb{R}^2 : \arg(x_1 + ix_2) = \alpha/2\}$, noting that the reflected function solves the Helmholtz equation, as well. The inverse conclusion is evident. Note that the case $\varepsilon < 0$ needs a separate investigation as carried out in [6], at the end of Section 3. Roughly speaking one needs to admit distributional solutions and to show that the involved operators have continuous extensions. ☐

Figure 1 illustrates Theorem 2.1.

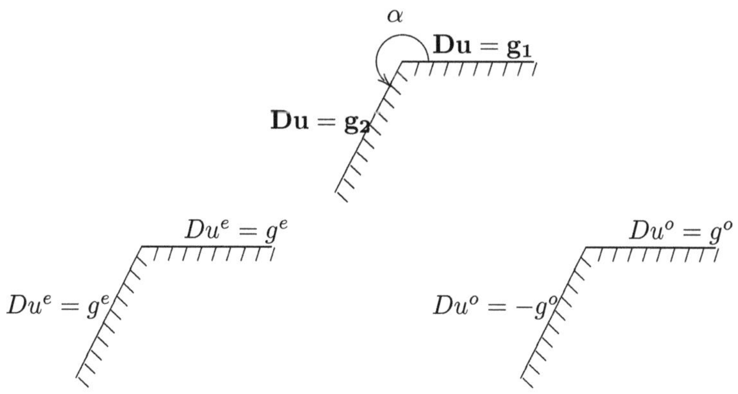


FIGURE 1. Reduction of the DD problem for $u$ to DD problems for $u^e$ and for $u^o$

**Theorem 2.2.** *Let $\varepsilon \in [0, 1/2[$ and $(g_1, g_2) \in H^{-1/2+\varepsilon}(\mathbb{R}_+)^2_\sim$ (the tilde being relevant only for $\varepsilon = 0$). Then the following assertions are equivalent:*

(i) *$u \in \mathcal{H}^{1+\varepsilon}(\Omega)$ solves the NN problem in $\Omega$.*

(ii) *$u = u^e + u^o$, $u^{e,o} \in \mathcal{H}^{1+\varepsilon}(\Omega)$ and*

$$(NN1) \quad T_{1,\Gamma_1} u^e = \quad T_{1,\Gamma_2} u^e = g^e = \frac{1}{2}(g_1 + g_2),$$

$$(NN2) \quad T_{1,\Gamma_1} u^o = -T_{1,\Gamma_2} u^o = g^o = \frac{1}{2}(g_1 - g_2).$$

*Proof.* It is similar to the previous. Negative $\varepsilon$ is avoided, see [6, Section 4]. □

*Remark* 2.3. The mixed Dirichlet/Neumann problem (DN) is not reducible in this way. However, mixed Dirichlet/Neumann problems will become important later on. We speak of equivalent (linear) problems, if they result from each other by one-to-one substitutions in the solution and data spaces, and those are linear homeomorphisms, cf. [2].

In a second step, we show that the BVPs considered in Theorems 2.1 and 2.2 in $\Omega_{0,\alpha}$ are each one equivalent to a pair of semi-homogeneous BVPs in the convex half-cones $\Omega_{0,\alpha/2}$ and $\Omega_{\alpha/2,\alpha}$. For that purpose, consider

$$\Gamma = \left\{(x_1, x_2) \in \mathbb{R}^2 : \arg(x_1 + ix_2) = \frac{\alpha}{2}\right\}.$$

**Theorem 2.4.**

(i) *Let $\varepsilon \in \left]-\frac{1}{4}, \frac{1}{4}\right[$ and $g^e \in H^{1/2+\varepsilon}(\mathbb{R}_+)$. The solution of the* BVP *with symmetry* (DD1)*, $u^e \in \mathcal{H}^{1+\varepsilon}(\Omega)$, is given by*

$$u^e = \begin{cases} u_1^e & in\ \Omega_{0,\alpha/2} \\ u_2^e & in\ \Omega_{\alpha/2,\alpha} \end{cases},$$

*where $u_1^e$ and $u_2^e$ are, respectively, the solutions of the semi-homogeneous* BVPs

(a) *$T_{0,\Gamma_1} u_1^e = g^e$, $\quad T_{1,\Gamma} u_1^e = 0$, with $u_1^e \in \mathcal{H}^{1+\varepsilon}(\Omega_{0,\alpha/2})$,*

(b) *$T_{1,\Gamma} u_2^e = 0$, $\quad T_{0,\Gamma_2} u_2^e = g^e$, with $u_2^e \in \mathcal{H}^{1+\varepsilon}(\Omega_{\alpha/2,\alpha})$.*

(ii) *Let $\varepsilon \in \left]-\frac{1}{2}, \frac{1}{2}\right[$ and $g^o \in \widetilde{H}^{1/2+\varepsilon}(\mathbb{R}_+)$. The solution of the* BVP *with symmetry* (DD2)*, $u^o \in \mathcal{H}^{1+\varepsilon}(\Omega)$, is given by*

$$u^o = \begin{cases} u_1^o & in\ \Omega_{0,\alpha/2} \\ u_2^o & in\ \Omega_{\alpha/2,\alpha} \end{cases},$$

*where $u_1^o$ and $u_2^o$ are, respectively, the solutions of the semi-homogeneous* BVPs

(a) *$T_{0,\Gamma_1} u_1^o = g^o$, $\quad T_{0,\Gamma} u_1^o = 0$, with $u_1^o \in \mathcal{H}^{1+\varepsilon}(\Omega_{0,\alpha/2})$,*

(b) *$T_{0,\Gamma} u_2^o = 0$, $\quad T_{0,\Gamma_2} u_2^o = -g^o$, with $u_2^o \in \mathcal{H}^{1+\varepsilon}(\Omega_{\alpha/2,\alpha})$.*

*Proof.* According to the boundary conditions on $\Gamma$ ($T_{1,\Gamma} u_{1,2}^e = 0$ or $T_{0,\Gamma} u_{1,2}^o = 0$, respectively), the jumps of $\partial u/\partial n$ and $u$ across $\Gamma$ are zero, which means that the functions $u^e$ and $u^o$ satisfy the Helmholtz equation throughout $\Gamma$. Finally, the Dirichlet conditions on $\Gamma_1$ and $\Gamma_2$ are obviously satisfied. □

Figures 2 and 3 illustrate Theorem 2.4.

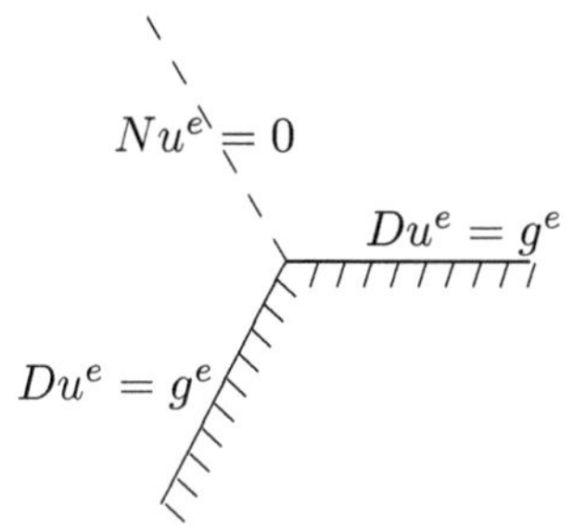


FIGURE 2. Pair of semi-homogeneous BVPs connected with the DD problem for $u^e$

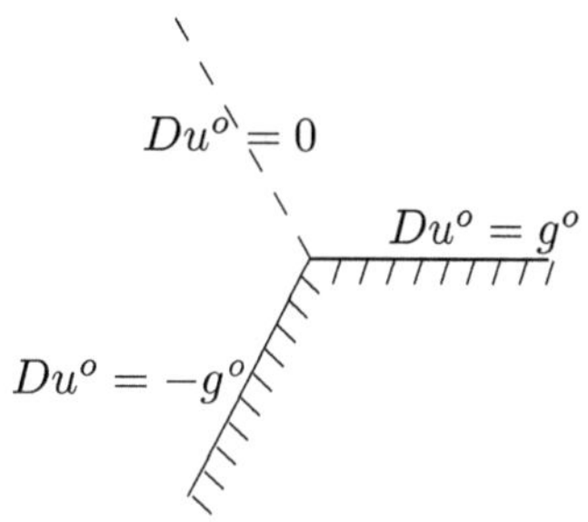


FIGURE 3. Pair of semi-homogeneous BVPs connected with the DD problem for $u^o$

*Remark* 2.5. In the first part of Theorem 2.4, the restriction in the regularity parameter $\varepsilon$ to $|\varepsilon| < 1/4$ is due to the semi-homogeneous DN and ND problems (i) (a) and (b), respectively.

Similarly as before, we can prove the following.

**Theorem 2.6.** (i) *Let* $\varepsilon \in [0, 1/2[$ *and* $g^e \in \widetilde{H}^{-1/2+\varepsilon}(\mathbb{R}_+)$ *(the tilde being relevant only for* $\varepsilon = 0$*). The solution of the* BVP *with symmetry* (NN1), $u^e \in \mathcal{H}^{1+\varepsilon}(\Omega)$, *is given by*

$$u^e = \begin{cases} u_1^e & in\ \Omega_{0,\alpha/2} \\ u_2^e & in\ \Omega_{\alpha/2,\alpha} \end{cases},$$

*where* $u_1^e$ *and* $u_2^e$ *are, respectively, the solutions of the semi-homogeneous* BVPs

(a) $T_{1,\Gamma_1} u_1^e = g^e, \quad T_{1,\Gamma} u_1^e = 0$, *with* $u_1^e \in \mathcal{H}^{1+\varepsilon}(\Omega_{0,\alpha/2})$,

(b) $T_{1,\Gamma} u_2^e = 0, \quad T_{1,\Gamma_2} u_2^e = g^e$, *with* $u_2^e \in \mathcal{H}^{1+\varepsilon}(\Omega_{\alpha/2,\alpha})$.

(ii) *Let* $\varepsilon \in [0, 1/4[$ *and* $g^o \in H^{-1/2+\varepsilon}(\mathbb{R}_+)$. *The solution of the* BVP *with symmetry* (NN2), $u^o \in \mathcal{H}^{1+\varepsilon}(\Omega)$, *is given by*

$$u^o = \begin{cases} u_1^o & in\ \Omega_{0,\alpha/2} \\ u_2^o & in\ \Omega_{\alpha/2,\alpha} \end{cases},$$

*where* $u_1^o$ *and* $u_2^o$ *are, respectively, the solutions of the semi-homogeneous* BVPs

(a) $T_{1,\Gamma_1} u_1^o = g^o, \quad T_{0,\Gamma} u_1^o = 0$, *with* $u_1^o \in \mathcal{H}^{1+\varepsilon}(\Omega_{0,\alpha/2})$,

(b) $T_{0,\Gamma} u_2^o = 0, \quad T_{1,\Gamma_2} u_2^o = -g^o$, *with* $u_2^o \in \mathcal{H}^{1+\varepsilon}(\Omega_{\alpha/2,\alpha})$.

Figures 4 and 5 illustrate Theorem 2.6.

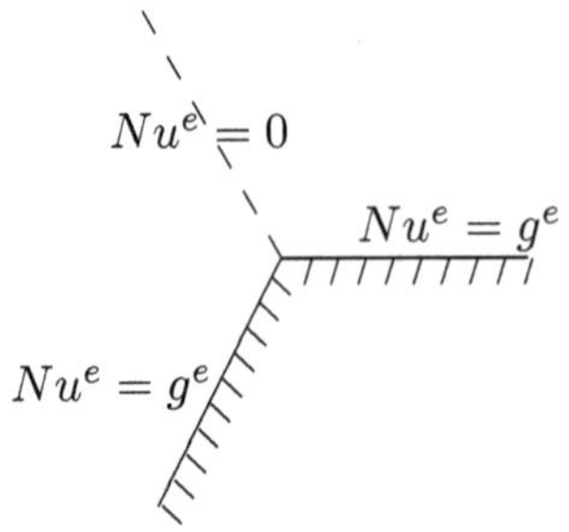


FIGURE 4. Pair of semi-homogeneous BVPs connected with the NN problem for $u^e$

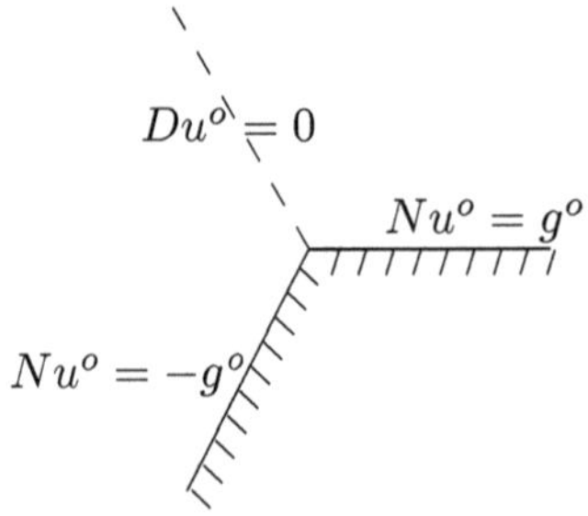


FIGURE 5. Pair of semi-homogeneous BVPs connected with the NN problem for $u^o$

*Remark* 2.7. The method described in this section, relating solutions of BVPs in a cone with solutions of BVPs in half-angled cones, works for any $\alpha \in ]0, 2\pi]$ and even in conical Riemann surfaces [7] (with $\alpha > 2\pi$) where it gains further interest. Here we refined ourselves to $\alpha = 4\pi/3$ for the sake of clarity.

## 3. Sommerfeld potentials

In the previous section, we proved that the BVPs with symmetry (DD1), (DD2), (NN1) and (NN2) are equivalent to semi-homogeneous BVPs in cones with an angle $2\pi/3$. In order to find the solution of these semi-homogeneous BVPs, we will use the *Sommerfeld potentials* (SOPs) introduced in [6] according to the angle $\alpha = 2\pi$.

The Sommerfeld potentials are the solutions of the Sommerfeld diffraction problems in the slit-plane

$$\mathbb{R}^2 \backslash \overline{\Sigma} = \Omega_{0,2\pi}$$

where boundary data $g_1, g_2$ are given on the upper and lower banks $\Sigma^\pm$ of $\Sigma = \mathbb{R}_+ \times \{0\}$ (we write $\Gamma_1 = \Sigma^+$ and $\Gamma_2 = \Sigma^-$ here, both copied to $\mathbb{R}_+$ again). The uniqueness of the Sommerfeld diffraction problem in the slit-plane with Dirichlet, Neumann or mixed boundary conditions is well known [5], [11].

The Dirichlet problem for the Helmholtz equation in $\mathbb{R}^2 \backslash \overline{\Sigma}$, provided $(g_1, g_2) \in H^{1/2+\varepsilon}(\mathbb{R}_+)^2_\sim$, $\varepsilon \in [0, 1/2[$, is uniquely solved by the Sommerfeld potential

$$\begin{aligned} u = \mathcal{K}_{DD,\mathbb{R}^2\backslash\overline{\Sigma}}(g_1, g_2) &= \begin{cases} \mathcal{K}_{D,\Omega^+}\, u_0^+ & \text{in } \Omega^+ \\ \mathcal{K}_{D,\Omega^-}\, u_0^- & \text{in } \Omega^- \end{cases} \\ &= \mathcal{K}_{D,\Omega^+}\, u_0^+ + \mathcal{K}_{D,\Omega^-}\, u_0^-, \end{aligned}$$

$$\begin{pmatrix} u_0^+ \\ u_0^- \end{pmatrix} = \Upsilon_D^{-1} \begin{pmatrix} I & 0 \\ 0 & \Pi_{1/2} \end{pmatrix} \begin{pmatrix} \ell_0 & 0 \\ 0 & \ell \end{pmatrix} \Upsilon_D \begin{pmatrix} g_1 \\ g_2 \end{pmatrix}, \tag{3.1}$$

where we used the notation

$$\Upsilon_D = \begin{pmatrix} I & -I \\ I & I \end{pmatrix},$$

$$\Pi_s = A_{t_-^{-s}} \ell_0 r_+ A_{t_-^s} : H^{s+\varepsilon} \to H^{s+\varepsilon},\ s \in \mathbb{R},\ |\varepsilon| < 1/2$$

and, by convention, the two terms $\mathcal{K}_{D,\Omega^\pm} u_0^\pm$ are extended by zero from $\Omega^\pm$ to $\Omega_{0,2\pi}$. The solution space consists of all $H^{1+\varepsilon}$ functions which satisfy the Helmholtz equation in any proper sub-cone of $\Omega_{0,2\pi}$ and can be written as

$$\mathcal{H}^{1+\varepsilon}(\mathbb{R}^2\backslash\overline{\Sigma}) = \Big\{ u \in L^2(\mathbb{R}^2) : u|_{\Omega^\pm} \in H^{1+\varepsilon}(\Omega^\pm), (\Delta + k^2)u = 0 \text{ in } \Omega^+ \cup \Omega^-,$$
$$u_0^+ - u_0^- \in \widetilde{H}_+^{1/2+\varepsilon}\ ,\ u_1^+ - u_1^- \in \widetilde{H}_+^{-1/2+\varepsilon} \Big\}.$$

The two differences in the last line of the formula denote the jumps of the traces $u_0^+ - u_0^- = u(x_1, 0+0) - u(x_1, 0-0)$ or of the $x_2$-derivatives of $u$, namely $u_1^+ - u_1^- = \frac{\partial u}{\partial x_2}(x_1, 0+0) - \frac{\partial u}{\partial x_2}(x_1, 0-0)$, respectively, across the line $x_2 = 0$. We use the notation $\widetilde{H}_\pm^s = \{ f \in H^s(\mathbb{R}) : \operatorname{supp} f \subset \overline{\mathbb{R}_\pm} \}$ [8].

As for the rotated slit-planes $\mathbb{R}^2\backslash\overline{\Sigma}_\alpha = \Omega_{\alpha,\alpha+2\pi}$, where $\Sigma_\alpha = \mathcal{R}_\alpha \Sigma$ and $\alpha \in \mathbb{R}$, the Dirichlet problem for the Helmholtz equation in $\mathbb{R}^2\backslash\overline{\Sigma}_\alpha$, provided $(g_1, g_2) \in H^{1/2+\varepsilon}(\mathbb{R}_+)^2_\sim$, $\varepsilon \in [0, 1/2[$, is uniquely solved by the Sommerfeld potential

$$u = \mathcal{J}_\alpha\, \mathcal{K}_{DD,\mathbb{R}^2\backslash\overline{\Sigma}}(g_1, g_2) = \mathcal{K}_{DD,\mathbb{R}^2\backslash\overline{\Sigma}_\alpha}(g_1, g_2) = \begin{cases} \mathcal{K}_{D,\Omega_\alpha^+}\, u_0^+ & \text{in } \Omega_\alpha^+ \\ \mathcal{K}_{D,\Omega_\alpha^-}\, u_0^- & \text{in } \Omega_\alpha^- \end{cases},$$

where $u_0^\pm$ are given by (3.1). In this case, the solution space is defined by

$$\mathcal{H}^{1+\varepsilon}(\mathbb{R}^2\backslash\overline{\Sigma}_\alpha) = \mathcal{J}_\alpha \mathcal{H}^{1+\varepsilon}(\mathbb{R}^2\backslash\overline{\Sigma}).$$

The Neumann problem for the Helmholtz equation in $\mathbb{R}^2\backslash\overline{\Sigma}_\alpha$, with $(g_1, g_2) \in H^{-1/2+\varepsilon}(\mathbb{R}_+)^2_\sim$ (in case $\varepsilon = 0$ only, superfluous for $\varepsilon \in ]0, 1/2[$), is uniquely solved by the Sommerfeld potential

$$u = \mathcal{K}_{NN,\mathbb{R}^2\backslash\overline{\Sigma}_\alpha}(g_1, g_2) = \begin{cases} \mathcal{K}_{N,\Omega_\alpha^+}\, u_1^+ & \text{in } \Omega_\alpha^+ \\ \mathcal{K}_{N,\Omega_\alpha^-}\, u_1^- & \text{in } \Omega_\alpha^- \end{cases},$$

$$\begin{pmatrix} u_1^+ \\ u_1^- \end{pmatrix} = \Upsilon_N^{-1} \begin{pmatrix} I & 0 \\ 0 & \Pi_{-1/2} \end{pmatrix} \begin{pmatrix} \ell_0 & 0 \\ 0 & \ell \end{pmatrix} \Upsilon_N \begin{pmatrix} g_1 \\ -g_2 \end{pmatrix},$$

$$\Upsilon_N = \begin{pmatrix} I & I \\ I & -I \end{pmatrix}.$$

Note that in some publications (such as [19]) the normal derivative was taken in the positive $x_2$-direction in both banks $\Sigma^\pm$ of the screen $\Sigma = \partial\Omega$ (i.e., the interior derivative $g_2$ on the lower bank has here to be replaced by $-g_2$ in the cited formulas, such that $g_1 + g_2 = r_+(u_1^+ - u_1^-)$ satisfies (1.5) with identification of $\Sigma$ and $\mathbb{R}_+$).

Finally, the solution of the mixed Dirichlet/Neumann problem for the Helmholtz equation in $\mathbb{R}^2\backslash\overline{\Sigma}$, provided $(g_1, g_2) \in H^{1/2+\varepsilon}(\mathbb{R}_+) \times H^{-1/2+\varepsilon}(\mathbb{R}_+)$, $\varepsilon \in ]-1/2, 1/2[$, (considered by Meister in 1977, [14]) was given by a celebrated matrix factorization due to Rawlins in 1981, [22], see also [10], [15], [23], and its operator theoretical interpretation [19]:

$$
\begin{aligned}
u = \mathcal{K}_{DN,\mathbb{R}^2\backslash\overline{\Sigma}}(g_1, g_2) &= \begin{cases} \mathcal{K}_{D,\Omega^+} u_0^+ & \text{in} \quad \Omega^+ \\ \mathcal{K}_{N,\Omega^-} u_1^- & \text{in} \quad \Omega^- \end{cases} \\
&= \mathcal{K}_{D,\Omega^+}\, u_0^+ + \mathcal{K}_{N,\Omega^-}\, u_1^-,
\end{aligned}
$$

$$
\begin{pmatrix} u_0^+ \\ u_1^- \end{pmatrix} = \mathcal{A} W_{DN}^{-1} \begin{pmatrix} g_1 \\ g_2 \end{pmatrix},
$$

$$
W_{DN}^{-1} = (r_+\mathcal{A})^{-1} = \mathcal{A}_+^{-1}\ell_0 r_+ \mathcal{A}_-^{-1}\ell,
$$

$$
\begin{aligned}
\mathcal{A} &= \mathcal{F}^{-1}\begin{pmatrix} 1 & -t^{-1} \\ -t & -1 \end{pmatrix}\mathcal{F} = \mathcal{A}_-\, \mathcal{A}_+ \\
&= -\frac{1}{\sqrt{4k}}\,\mathcal{F}^{-1}\begin{pmatrix} -t_{+-} & t^{-1}t_{--} \\ -t_{--} & -t_{+-} \end{pmatrix}\begin{pmatrix} t_{++} & -t^{-1}t_{-+} \\ t_{-+} & t_{++} \end{pmatrix}\mathcal{F},
\end{aligned}
$$

where $t_{\pm\pm}(\xi) = \left(\sqrt{2k} \pm \sqrt{k \pm \xi}\right)^{1/2}$ and the first/second index corresponds to the first/second sign, respectively. In some papers the factors are written in terms of $\sqrt{\xi - k}$ instead of $\sqrt{k - \xi}$ and one has to substitute $\sqrt{k - \xi} = i\sqrt{\xi - k}$, $\sqrt{k^2 - \xi^2} = i\sqrt{\xi^2 - k^2}$, due to the vertical branch cut from $k$ to $\infty$ in the upper half-plane and from $\infty$ to $-k$ in the lower half-plane.

Analogously, the mixed Dirichlet/Neumann problem for the Helmholtz equation in $\mathbb{R}^2\backslash\overline{\Sigma}_\alpha$, provided $(g_1, g_2) \in H^{1/2+\varepsilon}(\mathbb{R}_+) \times H^{-1/2+\varepsilon}(\mathbb{R}_+)$, $\varepsilon \in ] - 1/2, 1/2[$, is uniquely solved by the Sommerfeld potential

$$
u = \mathcal{J}_\alpha\, \mathcal{K}_{DN,\mathbb{R}^2\backslash\overline{\Sigma}}\,(g_1, g_2).
$$

Additionally, for the same boundary data $g_1, g_2$ given on the upper and lower banks $\Sigma_\alpha^\pm$, we obtain by reflection the solution of the mixed Neumann/Dirichlet problem for the Helmholtz equation in $\mathbb{R}^2\backslash\overline{\Sigma}_\alpha$,

$$
u = \mathcal{K}_{ND,\mathbb{R}^2\backslash\overline{\Sigma}_\alpha}\,(g_2, g_1) = R\,\mathcal{K}_{DN,\mathbb{R}^2\backslash\overline{\Sigma}_\alpha}\,(g_1, g_2),
$$

where the two conditions on $\Sigma_\alpha^\pm$ are exchanged and $R$ is the reflection operator in $x_2$-direction given by

$$
Rf(x_1, x_2) = f(x_1, -x_2), \quad (x_1, x_2) \in \mathbb{R}^2. \tag{3.2}
$$

## 4. Solution of the reduced semi-homogeneous BVPs

Firstly we present the solutions of the semi-homogeneous BVPs in the convex cone $\Omega_{0,\alpha/2}$ considered in Theorems 2.4 and 2.6.

**Proposition 4.1.**

(i) *Let* $\varepsilon\in\left]-\frac{1}{2},\frac{1}{2}\right[$. *The solution of the semi-homogeneous problem* $\mathcal{P}(DD,\Omega_{0,\alpha/2})$

$$T_{0,\Gamma_1}u_1^o = g^o,\ T_{0,\Gamma}u_1^o = 0 \text{ with } g^o \in \widetilde{H}^{1/2+\varepsilon}(\mathbb{R}_+)$$

*is given by*

$$u_1^o = \mathcal{K}_{D,\mathbb{R}^2\backslash\overline{\Sigma}_0}(g^o,0) + \mathcal{K}_{D,\mathbb{R}^2\backslash\overline{\Sigma}_{\alpha/2}}(g^o,0) + \mathcal{K}_{D,\mathbb{R}^2\backslash\overline{\Sigma}_\alpha}(0,-g^o).$$

(ii) *Let* $\varepsilon\in\left[0,\frac{1}{2}\right[$. *The solution of the semi-homogeneous problem* $\mathcal{P}(NN,\Omega_{0,\alpha/2})$

$$T_{1,\Gamma_1}u_1^e = g^e,\ T_{1,\Gamma}u_1^e = 0 \text{ with } g^e \in \widetilde{H}^{-1/2+\varepsilon}(\mathbb{R}_+),$$

*the tilde being relevant only for* $\varepsilon = 0$, *is given by*

$$u_1^e = \mathcal{K}_{N,\mathbb{R}^2\backslash\overline{\Sigma}_0}(g^e,0) + \mathcal{K}_{N,\mathbb{R}^2\backslash\overline{\Sigma}_{\alpha/2}}(g^e,0) + \mathcal{K}_{N,\mathbb{R}^2\backslash\overline{\Sigma}_\alpha}(0,g^e).$$

(iii) *Let* $\varepsilon\in\left]-\frac{1}{4},\frac{1}{4}\right[$. *The solution of the semi-homogeneous problem* $\mathcal{P}(DN,\Omega_{0,\alpha/2})$

$$T_{0,\Gamma_1}u_1^e = g^e,\ T_{1,\Gamma}u_1^e = 0 \text{ with } g^e \in H^{1/2+\varepsilon}(\mathbb{R}_+)$$

*is given by*

$$u_1^e = \mathcal{K}_{DN,\mathbb{R}^2\backslash\overline{\Sigma}_0}(g^e,0) + \mathcal{K}_{DN,\mathbb{R}^2\backslash\overline{\Sigma}_{\alpha/2}}(-g^e,0) + \mathcal{K}_{ND,\mathbb{R}^2\backslash\overline{\Sigma}_\alpha}(0,g^e).$$

(iv) *Let* $\varepsilon\in\left]-\frac{1}{4},\frac{1}{4}\right[$. *The solution of the semi-homogeneous problem* $\mathcal{P}(ND,\Omega_{0,\alpha/2})$

$$T_{1,\Gamma_1}u_1^o = g^o,\ T_{0,\Gamma}u_1^o = 0 \text{ with } g^o \in H^{-1/2+\varepsilon}(\mathbb{R}_+)$$

*is given by*

$$u_1^o = \mathcal{K}_{ND,\mathbb{R}^2\backslash\overline{\Sigma}_0}(g^o,0) + \mathcal{K}_{ND,\mathbb{R}^2\backslash\overline{\Sigma}_{\alpha/2}}(-g^o,0) + \mathcal{K}_{DN,\mathbb{R}^2\backslash\overline{\Sigma}_\alpha}(0,-g^o).$$

*All of them are unique for the indicated parameters.*

*Proof.* These results follow from Theorems 3.9, 4.5 and 5.10, respectively, of [6]. Roughly speaking, uniqueness is shown by arguments based upon Green's theorem, cf. [1], [5], [11] for instance. A verification of the formulas is possible because of symmetry arguments [7] incorporating the compatibility conditions [17], [20] by help of the "tilde spaces" [9], [11]. □

Figures 6–9 illustrate the fact that the solutions of the semi-homogeneous problems mentioned in Proposition 4.1 are composed of three terms defined in the slit-planes $\mathbb{R}^2\backslash\overline{\Sigma}_0$, $\mathbb{R}^2\backslash\overline{\Sigma}_{\alpha/2}$ and $\mathbb{R}^2\backslash\overline{\Sigma}_\alpha$. The "E" in "E-stars" stands for Torsten Ehrhardt who presented the geometrical idea to us in 2003 during his visit at Lisbon.

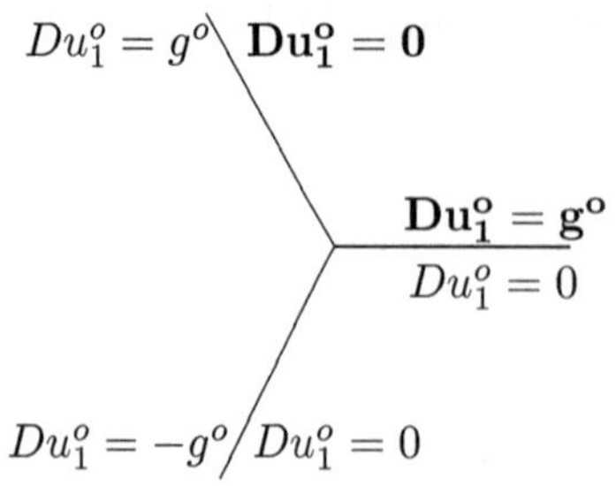


FIGURE 6. E-star for the $\mathcal{P}(DD, \Omega_{0,\alpha/2})$

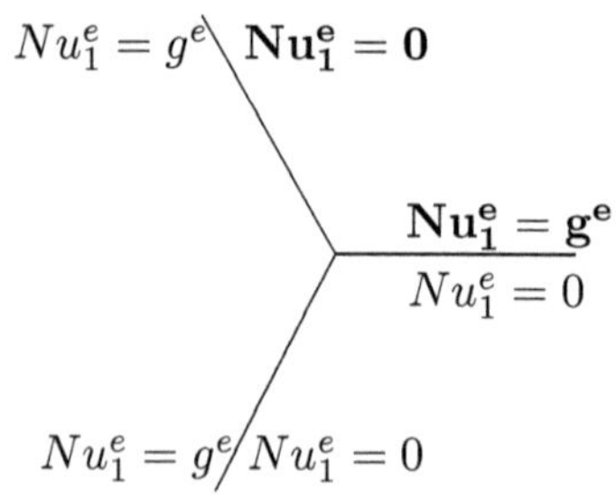


FIGURE 7. E-star for the $\mathcal{P}(NN, \Omega_{0,\alpha/2})$

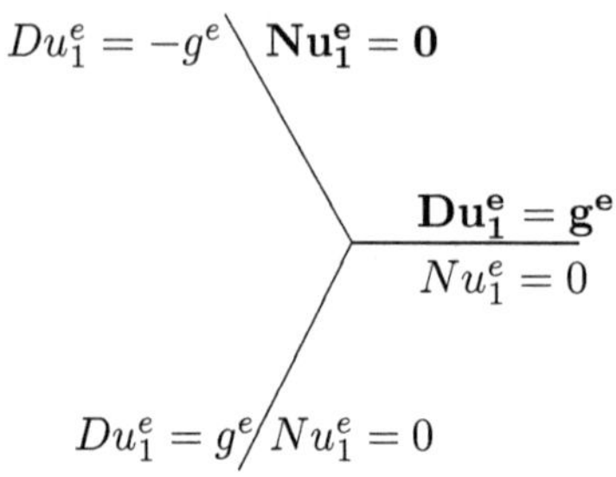


FIGURE 8. E-star for the $\mathcal{P}(DN, \Omega_{0,\alpha/2})$

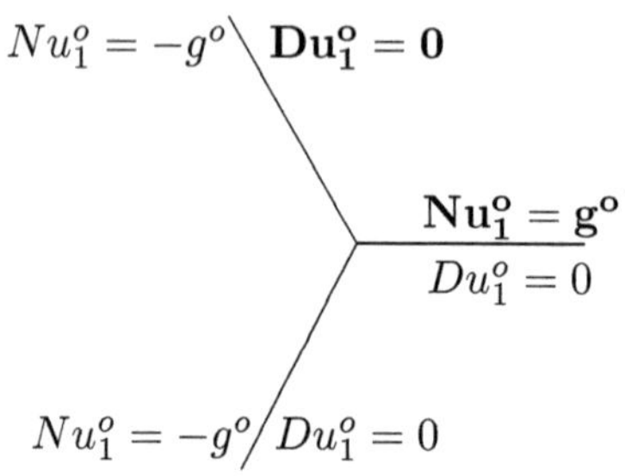


FIGURE 9. E-star for the $\mathcal{P}(ND, \Omega_{0,\alpha/2})$

Using the operator $\mathcal{J}_\alpha$ of $\alpha$-rotation around zero, cf. (1.3), and the operator $R$ of reflection in $x_2$-direction, cf. (3.2), we are able to relate the solutions of the semi-homogeneous problems in $\Omega_{\alpha/2,\alpha}$ (mentioned in Theorems 2.4 and 2.6) with the solutions of the semi-homogeneous problems in $\Omega_{0,\alpha/2}$, see Figure 10.

**Corollary 4.2.**

(i) *Let* $\varepsilon \in \left]-\frac{1}{2}, \frac{1}{2}\right[$. *The solution of the semi-homogeneous problem* $\mathcal{P}(DD, \Omega_{\alpha/2,\alpha})$

$$T_{0,\Gamma} u_2^o = 0, \; T_{0,\Gamma_2} u_2^o = -g^o \text{ with } g^o \in \widetilde{H}^{1/2+\varepsilon}(\mathbb{R}_+)$$

*is given by*

$$u_2^o = -\mathcal{J}_{\alpha/2} R \mathcal{J}_{-\alpha/2} u_1^o,$$

*where* $u_1^o$ *is the solution of the corresponding semi-homogeneous problem* $\mathcal{P}(DD, \Omega_{0,\alpha/2})$.

(ii) *Let* $\varepsilon \in \left[0, \frac{1}{2}\right[$. *The solution of the semi-homogeneous problem* $\mathcal{P}(NN, \Omega_{\alpha/2,\alpha})$

$$T_{1,\Gamma} u_2^e = 0, \; T_{1,\Gamma_2} u_2^e = g^e \text{ with } g^e \in \widetilde{H}^{-1/2+\varepsilon}(\mathbb{R}_+),$$

*the tilde being relevant only for* $\varepsilon = 0$, *is given by*

$$u_2^e = \mathcal{J}_{\alpha/2} R \mathcal{J}_{-\alpha/2} u_1^e,$$

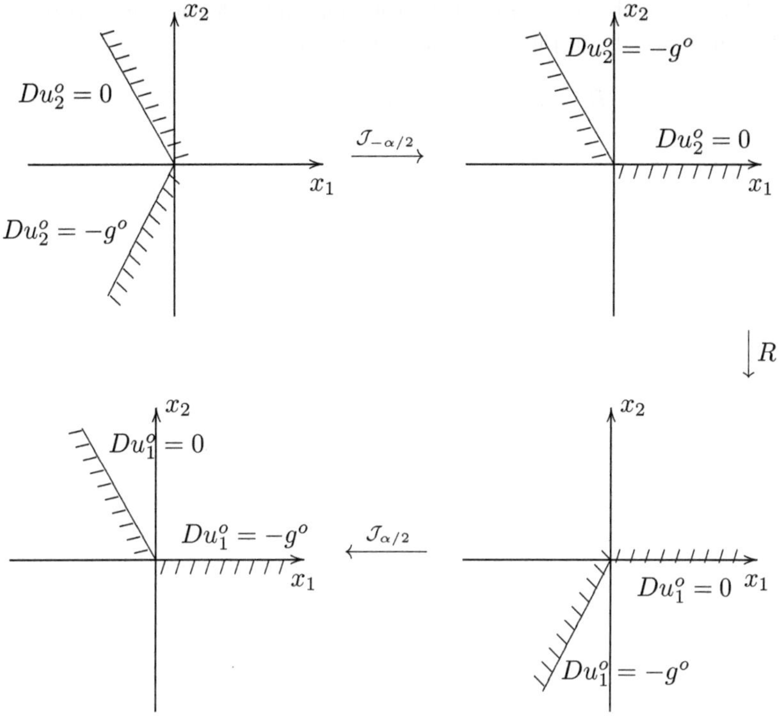


FIGURE 10. Relation between the solution of the problems $\mathcal{P}(DD, \Omega_{\alpha/2,\alpha})$ and $\mathcal{P}(DD, \Omega_{0,\alpha/2})$

*where* $u_1^e$ *is the solution of the corresponding semi-homogeneous problem* $\mathcal{P}(NN, \Omega_{0,\alpha/2})$.

(iii) *Let* $\varepsilon \in \left]-\frac{1}{4}, \frac{1}{4}\right[$ *The solution of the semi-homogeneous problem* $\mathcal{P}(DN,\Omega_{\alpha/2,\alpha})$

$$T_{0,\Gamma} u_2^o = 0,\ T_{1,\Gamma_2} u_2^o = -g^o \ \textit{with}\ g^o \in H^{-1/2+\varepsilon}(\mathbb{R}_+)$$

*is given by*

$$u_2^o = -\mathcal{J}_{\alpha/2} R \mathcal{J}_{-\alpha/2} u_1^o,$$

*where* $u_1^o$ *is the solution of the corresponding semi-homogeneous problem* $\mathcal{P}(ND, \Omega_{0,\alpha/2})$.

(iv) *Let* $\varepsilon \in \left]-\frac{1}{4}, \frac{1}{4}\right[$. *The solution of the semi-homogeneous problem* $\mathcal{P}(ND,\Omega_{\alpha/2,\alpha})$

$$T_{1,\Gamma} u_2^e = 0,\ T_{0,\Gamma_2} u_2^e = g^e \ \textit{with}\ g^e \in H^{1/2+\varepsilon}(\mathbb{R}_+)$$

*is given by*

$$u_2^e = \mathcal{J}_{\alpha/2} R \mathcal{J}_{-\alpha/2} u_1^e,$$

*where* $u_1^e$ *is the solution of the corresponding semi-homogeneous problem* $\mathcal{P}(DN, \Omega_{0,\alpha/2})$.

*All of them are unique for the indicated parameters.*

*Remark* 4.3. Using the definition of the rotation and reflection operators and after some straightforward computations one shows that

$$u_2^{e,o}(x_1,x_2) = \left(\mathcal{J}_{\alpha/2} R \mathcal{J}_{-\alpha/2} u_1^{e,o}\right)(x_1,x_2) = u_1^{e,o}\left(-\tfrac{1}{2}x_1 + \tfrac{\sqrt{3}}{2}x_2, \tfrac{\sqrt{3}}{2}x_1 - \tfrac{1}{2}x_2\right),$$

for all $(x_1,x_2) \in \Omega_{\alpha/2,\alpha}$.

## 5. Solution of the BVPs with symmetry in $\Omega = \Omega_{0,\alpha}$

Now we come back to the BVPs with symmetry presented earlier in Section 2.

**Theorem 5.1.**

(i) *Let* $\varepsilon \in \left]-\frac{1}{4}, \frac{1}{4}\right[$. *The* BVP *with symmetry* (DD1) *is uniquely solved by* $u^e \in \mathcal{H}^{1+\varepsilon}(\Omega)$ *given by*

$$u^e = \begin{cases} u_1^e & \text{in } \Omega_{0,\alpha/2} \\ u_2^e & \text{in } \Omega_{\alpha/2,\alpha} \end{cases} = \begin{cases} u_1^e & \text{in } \Omega_{0,\alpha/2} \\ \mathcal{J}_{\alpha/2} R \mathcal{J}_{-\alpha/2} u_1^e & \text{in } \Omega_{\alpha/2,\alpha} \end{cases},$$

*where* $u_1^e = \mathcal{K}_{DN,\mathbb{R}^2\setminus\overline{\Sigma}_0}(g^e,0) + \mathcal{K}_{DN,\mathbb{R}^2\setminus\overline{\Sigma}_{\alpha/2}}(-g^e,0) + \mathcal{K}_{ND,\mathbb{R}^2\setminus\overline{\Sigma}_\alpha}(0,g^e)$.

(ii) *Let* $\varepsilon \in \left]-\frac{1}{2}, \frac{1}{2}\right[$. *The* BVP *with symmetry* (DD2) *is uniquely solved by* $u^o \in \mathcal{H}^{1+\varepsilon}(\Omega)$ *given by*

$$u^o = \begin{cases} u_1^o & \text{in } \Omega_{0,\alpha/2} \\ u_2^o & \text{in } \Omega_{\alpha/2,\alpha} \end{cases} = \begin{cases} u_1^o & \text{in } \Omega_{0,\alpha/2} \\ -\mathcal{J}_{\alpha/2} R \mathcal{J}_{-\alpha/2} u_1^o & \text{in } \Omega_{\alpha/2,\alpha} \end{cases},$$

*where* $u_1^o = \mathcal{K}_{D,\mathbb{R}^2\setminus\overline{\Sigma}_0}(g^o,0) + \mathcal{K}_{D,\mathbb{R}^2\setminus\overline{\Sigma}_{\alpha/2}}(g^o,0) + \mathcal{K}_{D,\mathbb{R}^2\setminus\overline{\Sigma}_\alpha}(0,-g^o)$.

*Proof.* The result follows from Theorem 2.4, Proposition 4.1 and Corollary 4.2. □

**Theorem 5.2.** (i) *Let* $\varepsilon \in \left[0, \frac{1}{4}\right[$. *The* BVP *with symmetry* (NN1) *is uniquely solved by* $u^e \in \mathcal{H}^{1+\varepsilon}(\Omega)$ *given by*

$$u^e = \begin{cases} u_1^e & \text{in } \Omega_{0,\alpha/2} \\ u_2^e & \text{in } \Omega_{\alpha/2,\alpha} \end{cases} = \begin{cases} u_1^e & \text{in } \Omega_{0,\alpha/2} \\ \mathcal{J}_{\alpha/2} R \mathcal{J}_{-\alpha/2} u_1^e & \text{in } \Omega_{\alpha/2,\alpha} \end{cases},$$

*where* $u_1^e = \mathcal{K}_{N,\mathbb{R}^2\setminus\overline{\Sigma}_0}(g^e,0) + \mathcal{K}_{N,\mathbb{R}^2\setminus\overline{\Sigma}_{\alpha/2}}(g^e,0) + \mathcal{K}_{N,\mathbb{R}^2\setminus\overline{\Sigma}_\alpha}(0,g^e)$.

(ii) *Let* $\varepsilon \in \left]-\frac{1}{4}, \frac{1}{4}\right[$. *The* BVP *with symmetry* (NN2) *is uniquely solved by* $u^o \in \mathcal{H}^{1+\varepsilon}(\Omega)$ *given by*

$$u^o = \begin{cases} u_1^o & \text{in } \Omega_{0,\alpha/2} \\ u_2^o & \text{in } \Omega_{\alpha/2,\alpha} \end{cases} = \begin{cases} u_1^o & \text{in } \Omega_{0,\alpha/2} \\ -\mathcal{J}_{\alpha/2} R \mathcal{J}_{-\alpha/2} u_1^o & \text{in } \Omega_{\alpha/2,\alpha} \end{cases},$$

*where* $u_1^o = \mathcal{K}_{ND,\mathbb{R}^2\setminus\overline{\Sigma}_0}(g^o,0) + \mathcal{K}_{ND,\mathbb{R}^2\setminus\overline{\Sigma}_{\alpha/2}}(-g^o,0) + \mathcal{K}_{DN,\mathbb{R}^2\setminus\overline{\Sigma}_\alpha}(0,-g^o)$.

*Proof.* It follows from Theorem 2.6, Proposition 4.1 and Corollary 4.2. □

## 6. Analytic solution of DD and NN problems in $\Omega = \Omega_{0,\alpha}$

Now we are in the position to present the final results: well-posedness and explicit solution of the DD and NN problems in $\Omega_{0,\alpha}$ in closed analytical form.

**Theorem 6.1.** *Let $\varepsilon \in \left]-\frac{1}{4}, \frac{1}{4}\right[$. The Dirichlet problem for the Helmholtz equation in $\Omega = \Omega_{0,\alpha}$ in weak formulation with Dirichlet data $g = (g_1, g_2) \in H^{1/2+\varepsilon}(\mathbb{R}_+)^2_\sim$ is uniquely solved by*

$$u = \begin{cases} u_1^e + u_1^o & \text{in} \quad \Omega_{0,\alpha/2} \\ \mathcal{J}_{\alpha/2} R \mathcal{J}_{-\alpha/2}(u_1^e - u_1^o) & \text{in} \quad \Omega_{\alpha/2,\alpha} \end{cases},$$

*where*

$$u_1^e = \mathcal{K}_{DN,\mathbb{R}^2\setminus\overline{\Sigma}_0}(g^e, 0) + \mathcal{K}_{DN,\mathbb{R}^2\setminus\overline{\Sigma}_{\alpha/2}}(-g^e, 0) + \mathcal{K}_{ND,\mathbb{R}^2\setminus\overline{\Sigma}_\alpha}(0, g^e)$$

*and*

$$u_1^o = \mathcal{K}_{D,\mathbb{R}^2\setminus\overline{\Sigma}_0}(g^o, 0) + \mathcal{K}_{D,\mathbb{R}^2\setminus\overline{\Sigma}_{\alpha/2}}(g^o, 0) + \mathcal{K}_{D,\mathbb{R}^2\setminus\overline{\Sigma}_\alpha}(0, -g^o).$$

*Proof.* We just assemble the previous results: Theorem 2.1 shows that the DD problem in $\Omega_{0,\alpha}$ is equivalent to a pair of symmetrized problems by decomposition $u = u^e + u^o$. Theorem 2.4 yields the equivalence of each of them to a pair of semi-homogeneous BVPs in $\Omega_{0,\alpha/2}$ and $\Omega_{\alpha/2,\alpha}$, respectively. The solution of these problems was given in Proposition 4.1 and Corollary 4.2. Hence we obtain the explicit formulas, which represent the solution in dependence of the data, where the resolvent operator is a linear homeomorphism in the above-mentioned spaces. □

**Theorem 6.2.** *Let $\varepsilon \in \left[0, \frac{1}{4}\right[$. The Neumann problem for the Helmholtz equation in $\Omega = \Omega_{0,\alpha}$ in weak formulation with Neumann data $g = (g_1, g_2) \in H^{-1/2+\varepsilon}(\mathbb{R}_+)^2_\sim$ is uniquely solved by*

$$u = \begin{cases} u_1^e + u_1^o & \text{in} \quad \Omega_{0,\alpha/2} \\ \mathcal{J}_{\alpha/2} R \mathcal{J}_{-\alpha/2}(u_1^e - u_1^o) & \text{in} \quad \Omega_{\alpha/2,\alpha} \end{cases},$$

*where*

$$u_1^e = \mathcal{K}_{N,\mathbb{R}^2\setminus\overline{\Sigma}_0}(g^e, 0) + \mathcal{K}_{N,\mathbb{R}^2\setminus\overline{\Sigma}_{\alpha/2}}(g^e, 0) + \mathcal{K}_{N,\mathbb{R}^2\setminus\overline{\Sigma}_\alpha}(0, g^e)$$

*and*

$$u_1^o = \mathcal{K}_{ND,\mathbb{R}^2\setminus\overline{\Sigma}_0}(g^o, 0) + \mathcal{K}_{ND,\mathbb{R}^2\setminus\overline{\Sigma}_{\alpha/2}}(-g^o, 0) + \mathcal{K}_{DN,\mathbb{R}^2\setminus\overline{\Sigma}_\alpha}(0, -g^o).$$

*Proof.* This is analogous to the proof of Theorem 6.1 and based on Theorem 2.2, Theorem 2.6, Proposition 4.1 and Corollary 4.2, as well. □

*Remark* 6.3. If we consider $g_1 = g_2$ or $g_1 = -g_2$ in the DD problem (1.1) or in the NN problem (1.2), then we obtain a BVP with symmetry, (DD1) or (NN2), respectively, with solution given by Theorems 5.1 and 5.2.

**Acknowledgment**

The present work was partially supported by *Centro de Matemática e Aplicações* of Instituto Superior Técnico and *Centro de Investigação e Desenvolvimento em Matemática e Aplicações* of Universidade de Aveiro, through the Portuguese Science Foundation (*FCT – Fundação para a Ciência e a Tecnologia*), co-financed by the European Community.

A.P. Nolasco acknowledges the support of *Fundação para a Ciência e a Tecnologia* (grant number SFRH/BPD/38273/2007).

## References

[1] L.P. Castro and D. Kapanadze, *Exterior wedge diffraction problems with Dirichlet, Neumann and impedance boundary conditions.* Acta Appl. Math. **110** (2010), 289–311.

[2] L.P. Castro and F.-O. Speck, *Regularity properties and generalized inverses of delta-related operators.* Z. Anal. Anwend. **17** (1998), 577–598.

[3] L.P. Castro, F.-O. Speck and F.S. Teixeira, *On a class of wedge diffraction problems posted by Erhard Meister.* Oper. Theory Adv. Appl. **147** (2004), 211–238.

[4] L.P. Castro, F.-O. Speck and F.S. Teixeira, *Mixed boundary value problems for the Helmholtz equation in a quadrant.* Integr. Equ. Oper. Theory **56** (2006), 1–44.

[5] M. Costabel and E. Stephan, *A direct boundary integral equation method for transmission problems.* J. Math. Anal. Appl. **106** (1985), 367–413.

[6] T. Ehrhardt, A.P. Nolasco and F.-O. Speck, *Boundary integral methods for wedge diffraction problems: the angle $2\pi/n$, Dirichlet and Neumann conditions*, Operators and Matrices, **5** (2011), 1–40.

[7] T. Ehrhardt, A.P. Nolasco and F.-O. Speck, *A Riemannn surface approach for diffraction from rational wedges*, to appear.

[8] G.I. Èskin, *Boundary Value Problems for Elliptic Pseudodifferential Equations* Translations of Mathematical Monographs **52**, AMS, 1981.

[9] P. Grisvard, *Elliptic Problems in Nonsmooth Domains.* Monographs and Studies in Mathematics **24**, Pitman, 1985.

[10] A.E. Heins, *The Sommerfeld half-plane problem revisited, II, the factoring of a matrix of analytic functions.* Math. Meth. Appl. Sci. **5** (1983), 14–21.

[11] G.C. Hsiao and W.L. Wendland, *Boundary Integral Equations.* Applied Mathematical Sciences Series **164**, Springer-Verlag, 2008.

[12] A.I. Komech, N.J. Mauser and A.E. Merzon, *On Sommerfeld representation and uniqueness in scattering by wedges.* Math. Meth. Appl. Sci. **28** (2005), 147–183.

[13] G.D. Malujinetz, *Excitation, reflection and emission of the surface waves on a wedge with given impedances of the sides.* Dokl. Acad. Nauk SSSR **121** (1958), 436–439.

[14] E. Meister, *Ein Überblick über analytische Methoden zur Lösung singulärer Integralgleichungen.* Proceedings of the GAMM Conference in Graz in 1976, Z. Angew. Math. Mech. **57** (1977), T23–T35.

[15] E. Meister, *Some multiple-part Wiener-Hopf problems in Mathematical Physics.* Mathematical Models and Methods in Mechanics, Banach Center Publications **15**, PWN – Polish Scientific Publishers, Warsaw (1985), 359–407.

[16] E. Meister, *Some solved and unsolved canonical problems of diffraction theory.* Lecture Notes in Math. **1285**, (1987) 320–336.

[17] E. Meister, F. Penzel, F.-O. Speck and F.S. Teixeira, *Some interior and exterior boundary-value problems for the Helmholtz equation in a quadrant.* Proc. R. Soc. Edinb., Sect. A **123** (1993), 193–237.

[18] E. Meister, P.A. Santos and F.S. Teixeira, *A Sommerfeld-type diffraction problem with second-order boundary conditions.* Z. Angew. Math. Mech. **72** (1992), 621–630.

[19] E. Meister and F.-O. Speck, *Modern Wiener-Hopf methods in diffraction theory.* Pitman Res. Notes Math. Ser. **216** (1989), 130–171.

[20] A. Moura Santos, F.-O. Speck and F.S. Teixeira, *Minimal normalization of Wiener-Hopf operators in spaces of Bessel potentials.* J. Math. Anal. Appl. **225** (1998), 501–531.

[21] A.V. Osipov and A.N. Norris, *The Malyuzhinets theory for scattering from wedge boundaries: a review.* Wave Motion **29** (1999), 313–340.

[22] A.D. Rawlins, *The explicit Wiener-Hopf factorization of a special matrix.* Z. Angew. Math. Mech. **61** (1981), 527–528.

[23] A.D. Rawlins, *The solution of a mixed boundary value problem in the theory of diffraction.* J. Eng. Math. **18** (1984), 37–62.

[24] A.D. Rawlins, *Plane-wave diffraction by a rational angle.* Proc. R. Soc. Lond. **A 411** (1987), 265–283.

[25] A.F. dos Santos, A.B. Lebre and F.S. Teixeira, *The diffraction problem for a half-plane with different face impedances revisited.* J. Math. Anal. Appl. **140** (1989), 485–509.

[26] F.-O. Speck, *Mixed boundary value problems of the type of Sommerfeld's half-plane problem.* Proc. R. Soc. Edinb., Sect. A **104** (1986), 261–277.

[27] P.Ya. Ufimtsev, *Theory of Edge Diffraction in Electromagnetics.* Tech Science Press, 2003.

[28] P. Zhevandrov and A. Merzon, *On the Neumann problem for the Helmholtz equation in a plane angle.* Math. Meth. Appl. Sci. **23** (2000), 1401–1446.



A.P. Nolasco
Department of Mathematics
University of Aveiro
3810-193 Aveiro, Portugal
e-mail: anolasco@ua.pt

F.-O. Speck
Department of Mathematics, I.S.T.
Technical University of Lisbon
1049-001 Lisbon, Portugal
e-mail: fspeck@math.ist.utl.pt

Operator Theory:
Advances and Applications, Vol. 221, 515-531


# The Infinite-dimensional Sylvester Differential Equation and Periodic Output Regulation

Lassi Paunonen

**Abstract.** In this paper the solvability of the infinite-dimensional Sylvester differential equation is considered. This is an operator differential equation on a Banach space. Conditions for the existence of a unique classical solution to the equation are presented. In addition, a periodic version of the equation is studied and conditions for the existence of a unique periodic solution are given. These results are applied to generalize a theorem which characterizes the controllers achieving output regulation of a distributed parameter system with a nonautonomous signal generator.

**Mathematics Subject Classification (2000).** Primary 47A62; Secondary 93C25.

**Keywords.** Sylvester differential equation, strongly continuous evolution family, output regulation.

## 1. Introduction

In this paper we consider the solvability of a Sylvester differential equation on a Banach space. This is an operator differential equation of form

$$\dot{\Sigma}(t) = A(t)\Sigma(t) - \Sigma(t)B(t) + C(t), \qquad \Sigma(0) = \Sigma_0, \tag{1.1}$$

where $(A(t), \mathcal{D}(A(t)))$ and $(B(t), \mathcal{D}(B(t)))$ are families of unbounded operators on Banach spaces $X$ and $Y$, respectively, $C(\cdot)$ is an operator-valued function and $\Sigma_0$ is a bounded linear operator. The equations of this type have an application in the output regulation of linear distributed parameter systems when the reference signals are generated with a periodic exosystem of form

$$\dot{v}(t) = S(t)v(t), \qquad v(0) = v_0 \in \mathbb{C}^q, \tag{1.2a}$$

$$y_{ref}(t) = F(t)v(t). \tag{1.2b}$$

By the periodicity of the exosystem we mean that $S(\cdot)$ and $F(\cdot)$ are periodic functions with the same period, i.e., there exists $\tau > 0$ such that $S(t+\tau) = S(t)$ and $F(t+\tau) = F(t)$ for all $t \in \mathbb{R}$. Paunonen and Pohjolainen [9] have shown that the solvability of the output regulation problem related to this type of exosystem can be characterized using the properties of the solution to a certain Sylvester differential equation.

The results in [9] generalize the theory of periodic output regulation of linear finite-dimensional systems presented by Zhang and Serrani [13]. In the finite-dimensional theory the Sylvester differential equations are ordinary matrix differential equations and $A(\cdot)$, $B(\cdot)$ and $C(\cdot)$ are smooth matrix-valued functions. However, if we want to consider output regulation of infinite-dimensional linear systems, the matrix-valued function $A(\cdot)$ becomes a family $(A(t), \mathcal{D}(A(t)))$ of unbounded operators associated to the closed-loop system consisting of the distributed parameter system to be controlled and the controller.

The treatment of the Sylvester differential equation presented in this paper generalizes the results on the solvability of finite-dimensional equations of this type [13, 7]. Also the infinite-dimensional equation has been studied in the case where $A(t) \equiv A$ and $B(t) \equiv B$ are generators of strongly continuous semigroups [5, 4]. In the case of time-dependent families of operators some results are also known for time-dependent Riccati equations [2]. On the other hand, in the time-invariant case the equation becomes an infinite-dimensional Sylvester equation [1, 11, 12]. Our approach in solving the Sylvester differential equation (1.1) generalizes the methods used in [11].

We have in [9] studied a Sylvester differential equation of form (1.1), where $(A(t), \mathcal{D}(A(t)))$ is a family of unbounded operators and $B(t) \equiv B$ is a matrix. In this paper we consider the solvability of the equation in a more general case where also $B(t)$ are allowed to be unbounded operators. We restrict ourselves to a situation where the domains of the unbounded operators are independent of time. The main tools in our analysis are the strongly continuous evolution families associated to families of unbounded operators and nonautonomous abstract Cauchy problems [10, Ch. 5], [6, Sec. VI.9].

We apply the theoretic results on the solvability of (1.1) to the output regulation of infinite-dimensional systems. In particular we present a characterization of the controllers achieving output regulation of a linear distributed parameter system to the signals generated by a nonautonomous periodic signal generator.

The paper is organized as follows. In Section 2 we introduce notation, recall the definition of a strongly continuous evolution family and state the basic assumptions on the families of operators. The solvability of the Sylvester differential equation is considered in Section 3. The main results of the paper are Theorems 3.2 and 3.3. In Section 4 we apply these results to output regulation. Section 5 contains concluding remarks.

## 2. Notation and definitions

If $X$ and $Y$ are Banach spaces and $A : X \to Y$ is a linear operator, we denote by $\mathcal{D}(A)$ and $\mathcal{R}(A)$ the domain and the range of $A$, respectively. The space of bounded linear operators from $X$ to $Y$ is denoted by $\mathcal{L}(X, Y)$. If $A : X \to X$, then $\sigma(A)$ and $\rho(A)$ denote the spectrum and the resolvent set of $A$, respectively. For $\lambda \in \rho(A)$ the resolvent operator is given by $R(\lambda, A) = (\lambda I - A)^{-1}$. The space of continuous functions $f : I \subset \mathbb{R} \to X$ is denoted by $C(I, X)$ and the space of continuously differentiable functions by $C^1(I, X)$. Finally, we denote by $C(I, \mathcal{L}_s(X, Y))$ the space of strongly continuous $\mathcal{L}(X, Y)$-valued functions.

In dealing with families of unbounded operators we use the theory of strongly continuous evolution families [10, Ch. 5], [6, Sec. VI.9].

**Definition 2.1 (A Strongly Continuous Evolution Family).** A family of bounded operators $(U(t, s))_{t \geq s} \subset \mathcal{L}(X)$ is called a *strongly continuous evolution family* if

(a) $U(s, s) = I$ for $s \in \mathbb{R}$.

(b) $U(t, s) = U(t, r)U(r, s)$ for $t \geq r \geq s$.

(c) $\{ (t, s) \in \mathbb{R}^2 \mid t \geq s \} \ni (t, s) \mapsto U(t, s)$ is a strongly continuous mapping.

A strongly continuous evolution family is called *exponentially bounded* if there exist constants $M \geq 1$ and $\omega \in \mathbb{R}$ such that

$$\|U(t, s)\| \leq M e^{\omega(t-s)}$$

for all $t \geq s$. The evolution family is called *periodic* (with period $\tau > 0$) if

$$U(t + \tau, s + \tau) = U(t, s)$$

for all $t \geq s$.

Strongly continuous evolution families are related to nonautonomous abstract Cauchy problems. If we consider an equation

$$\begin{aligned} \dot{x}(t) &= A(t)x(t) + f(t), \\ x(s) &= x_s \in X \end{aligned}$$

and if $U(t, s)$ is a strongly continuous evolution family associated to the family $(A(t), \mathcal{D}(A(t)))$ of operators, then if for every $s \in \mathbb{R}$ this equation has a classical solution $x(\cdot) \in C^1([s, \infty), X)$ such that $x(t) \in \mathcal{D}(A(t))$ for all $t \geq s$, this solution is given by

$$x(t) = U(t, s)x_s + \int_s^t U(t, r)f(r)ds \tag{2.1}$$

for all $t \geq s$. If the family $(A(t), \mathcal{D}(A(t)))$ of operators is periodic with period $\tau > 0$, then also the associated evolution family is periodic with the same period.

Throughout this paper we consider a case where the domains of the unbounded operators are independent of time, i.e.,

$$A(t) : \mathcal{D}(A) \subset X \to X, \qquad B(t) : \mathcal{D}(B) \subset Y \to Y$$

for all $t$. We assume that there exist exponentially bounded strongly continuous evolution families $U_A(t,s)$ and $U_B(t,s)$ related to the families $(A(t),\mathcal{D}(A))$ and $(-B(t),\mathcal{D}(B))$ of operators, respectively, and that the evolution family $U_B(t,s)$ satisfies Definition 2.1 for all $t,s,r\in\mathbb{R}$. This means that the nonautonomous abstract Cauchy problem associated to this family of operators can be solved forward and backwards in time and the minus sign in the family of operators corresponds to the reversal of time in the equation. Because of this, we can also think of the situation in such a way that the evolution family related to the family $(B(t),\mathcal{D}(B))$ of operators satisfies Definition 2.1 for all $t\le r\le s$. Motivated by this, we denote this evolution family by $U_B(s,t)$ for $t\ge s$.

## 3. The infinite-dimensional Sylvester differential equation

In this section we consider the infinite-dimensional Sylvester differential equation

$$\dot{\Sigma}(t)=A(t)\Sigma(t)-\Sigma(t)B(t)+C(t),\qquad \Sigma(0)=\Sigma_0 \tag{3.1}$$

on an interval $[0,T]$. The equation is considered in the strong sense for $y\in\mathcal{D}(B)$.

The main result of this paper is Theorem 3.2 which states sufficient conditions for the existence of a classical solution to the Sylvester differential equation. As we are motivated by the periodic output regulation problem for distributed parameter systems [9], we will also show that if the families of operators $(A(t),\mathcal{D}(A))$ and $(B(t),\mathcal{D}(B))$ and the function $C(\cdot)$ are periodic with the same period, then under suitable additional assumptions on the growths of the evolution families $U_A(t,s)$ and $U_B(s,t)$ the Sylvester differential equation has a unique periodic solution. This result is presented in Theorem 3.3.

We begin by defining the classical solution of the Sylvester differential equation on the interval $[0,T]$.

**Definition 3.1.** A strongly continuous function $\Sigma(\cdot)\in C([0,T],\mathcal{L}_s(Y,X))$ satisfying $\Sigma(\cdot)y\in C^1([0,T],X)$ and $\Sigma(t)y\in\mathcal{D}(A)$ for all $y\in\mathcal{D}(B)$ and $t\in[0,T]$ is called the *classical solution* of the Sylvester differential equation (3.1) if it satisfies the equation on $[0,T]$.

The next theorem is the main result of the paper. It states sufficient conditions for the solvability of the Sylvester differential equation on the interval $[0,T]$. The *parabolic conditions* [10, Sec. 5.6] appearing in the theorem essentially require that the operators $A(t)$ for $t\in[0,T]$ are generators of analytic semigroups on $X$.

**Theorem 3.2.** *Assume the following are satisfied.*

1. *There exists $\mu\in\mathbb{R}$ such that $U_A(t,s)$ satisfies the* parabolic conditions:
   (P$_1$) *The domain $\mathcal{D}(A)$ is dense in $X$.*
   (P$_2$) *We have $\{\lambda\in\mathbb{C}\mid \operatorname{Re}\lambda\ge\mu\}\subset\rho(A(t))$ for every $t\in[0,T]$ and there exists a constant $M\ge 1$ such that*

$$\|R(\lambda,A(t))\|\le\frac{M}{|\lambda-\mu|+1},\qquad \operatorname{Re}\lambda\ge\mu,\ t\in[0,T].$$

(P$_3$) *There exists a constant $L \geq 0$ such that for $t, s, r \in [0,T]$*

$$\|(A(t) - A(s))R(\mu, A(r))\| \leq L|t-s|.$$

2. *The domain $\mathcal{D}(A(t)^*) =: \mathcal{D}(A^*)$ is independent of $t \in [0,T]$ and dense in $X^*$. For all $x \in X$ and $x^* \in \mathcal{D}(A^*)$ the mapping*

$$t \mapsto \langle x, A(t)^* x^* \rangle$$

*is continuous on $[0,T]$.*

3. *The domain $\mathcal{D}(B)$ is dense in $Y$. For every $y \in \mathcal{D}(B)$ the function $B(\cdot)y$ is continuous, we have $U_B(s,t)y \in \mathcal{D}(B)$ and the evolution family $U_B(s,t)$ satisfies the differentiation rules*

$$\frac{\partial}{\partial t} U_B(s,t)y = -U_B(s,t)B(t)y, \qquad \frac{\partial}{\partial s} U_B(s,t)y = B(s)U_B(s,t)y.$$

*for all $t, s \in [0,T]$.*

4. *For every $y \in Y$ the function $C(\cdot)y$ is Hölder continuous on $[0,T]$.*
5. *$\Sigma_0(\mathcal{D}(B)) \subset \mathcal{D}(A)$.*

*The infinite-dimensional Sylvester differential equation* (3.1) *has a unique classical solution $\Sigma(\cdot)$ on $[0,T]$ given by the formula*

$$\Sigma(t)y = U_A(t,0)\Sigma_0 U_B(0,t)y + \int_0^t U_A(t,s)C(s)U_B(s,t)y\,ds \tag{3.2}$$

*for all $y \in Y$.*

*Proof.* Since $U_A(t,s)$ satisfies the parabolic conditions, we have from [10, Sec. 5.6] that for all $x \in X$, $x' \in \mathcal{D}(A)$, and $t > s$

$$\frac{\partial}{\partial t} U_A(t,s)x = A(t)U_A(t,s)x, \qquad \frac{\partial}{\partial s} U_A(t,s)x' = -U_A(t,s)A(s)x'.$$

Let $y \in \mathcal{D}(B)$, $x^* \in \mathcal{D}(A^*)$, and $s \in [0,T]$. Using the differentiation rules for $U_A(t,s)$ and $U_B(s,t)$ we see that for any $t \in (s,T]$

$$\begin{aligned}
&\frac{\partial}{\partial t}\langle U_A(t,s)C(s)U_B(s,t)y, x^*\rangle \\
&= \langle A(t)U_A(t,s)C(s)U_B(s,t)y, x^*\rangle - \langle U_A(t,s)C(s)U_B(s,t)B(t)y, x^*\rangle \\
&= \langle U_A(t,s)C(s)U_B(s,t)y, A(t)^*x^*\rangle - \langle U_A(t,s)C(s)U_B(s,t)B(t)y, x^*\rangle
\end{aligned}$$

$$\begin{aligned}
&\frac{\partial}{\partial t}\langle U_A(t,0)\Sigma_0 U_B(0,t)y, x^*\rangle \\
&= \langle A(t)U_A(t,0)\Sigma_0 U_B(0,t)y, x^*\rangle - \langle U_A(t,0)\Sigma_0 U_B(0,t)B(t)y, x^*\rangle \\
&= \langle U_A(t,0)\Sigma_0 U_B(0,t)y, A(t)^*x^*\rangle - \langle U_A(t,0)\Sigma_0 U_B(0,t)B(t)y, x^*\rangle.
\end{aligned}$$

To show that (3.2) is a solution of the Sylvester differential equation we will use the Leibniz integral rule [8, Lem. VIII.2.2]. This result states that if the function $f : \{ (t,s) \mid 0 \le s \le t \le T \} \to \mathbb{C}$ is continuous, and if $\frac{\partial}{\partial t} f(t,s)$ exists and is continuous and uniformly bounded on $\{ (t,s) \mid 0 \le s < t \le T \}$, then the mapping $t \mapsto \int_0^t f(t,s)ds$ is differentiable on $(0,T)$ and

$$\frac{d}{dt}\int_0^t f(t,s)ds = f(t,t) + \int_0^t \frac{\partial}{\partial t} f(t,s)ds.$$

Our assumptions imply that the function

$$(t,s) \to f(t,s) = \langle U_A(t,s)C(s)U_B(s,t)y, x^* \rangle$$

is continuous for $0 \le s \le t \le T$ and the computation above shows that its derivative with respect to $t$ is continuous. It thus remains to show that this derivative is uniformly bounded. Since the mappings $(t,s) \to U_A(t,s)$ and $(t,s) \to U_B(s,t)$ are strongly continuous, there exist constants $M_A, M_B > 0$ such that

$$\max_{0\le s\le t\le T} \|U_A(t,s)\| \le M_A, \qquad \max_{0\le s\le t\le T} \|U_B(s,t)\| \le M_B.$$

Using these estimates we see that

$$\begin{aligned}\left|\frac{\partial}{\partial t} f(t,s)\right| &\le \|U_A(t,s)C(s)U_B(s,t)y\| \cdot \|A(t)^* x^*\| \\ &\quad + \|U_A(t,s)C(s)U_B(s,t)B(t)y\| \cdot \|x^*\| \\ &\le \|U_A(t,s)\| \cdot \|C(s)\| \cdot \|U_B(s,t)\| \left(\|y\| \cdot \|A(t)^*x^*\| + \|B(t)y\| \cdot \|x^*\|\right) \\ &\le M_A M_B \max_{r\in[0,T]} \|C(r)\| \left( \|y\| \max_{r\in[0,T]} \|A(r)^*x^*\| + \|x^*\| \max_{r\in[0,T]} \|B(r)y\| \right) \\ &< \infty.\end{aligned}$$

This concludes that we can use the Leibniz integral rule.

For the function $\Sigma(\cdot)$ defined in (3.2) we now have

$$\begin{aligned}&\frac{d}{dt}\langle \Sigma(t)y, x^*\rangle = \frac{d}{dt}\langle U_A(t,0)\Sigma_0 U_B(0,t)y, x^*\rangle \\ &\quad + \frac{d}{dt}\int_0^t \langle U_A(t,s)C(s)U_B(s,t)y, x^*\rangle ds \\ &= \langle U_A(t,0)\Sigma_0 U_B(0,t)y, A(t)^*x^*\rangle - \langle U_A(t,0)\Sigma_0 U_B(0,t)B(t)y, x^*\rangle \\ &\quad + \int_0^t \left(\langle U_A(t,s)C(s)U_B(s,t)y, A(t)^*x^*\rangle - \langle U_A(t,s)C(s)U_B(s,t)B(t)y, x^*\rangle\right) ds \\ &\quad + \langle U_A(t,t)C(t)U_B(t,t)y, x^*\rangle \\ &= \langle \Sigma(t)y, A(t)^*x^*\rangle - \langle \Sigma(t)B(t)y, x^*\rangle + \langle C(t)y, x^*\rangle. \qquad (3.3)\end{aligned}$$

We will next show that the mapping $t \mapsto \Sigma(t)y$ is continuously differentiable on $(0,T)$ and that $\Sigma(t)y \in \mathcal{D}(A)$ for all $t \in [0,T]$. We will do this by first considering the nonautonomous Cauchy problem

$$\dot{x}(t) = A(t)x(t) + C(t)U_B(t,0)v, \qquad x(0) = \Sigma_0 v,$$

where $v \in \mathcal{D}(B)$. Since $x(0) \in \mathcal{D}(A)$ and since $t \mapsto C(t)U_B(t,0)v$ is Hölder continuous on $[0,T]$ we have from [10, Thm. 5.7.1] that this equation has a unique classical solution given by

$$x(t) = U_A(t,0)\Sigma_0 v + \int_0^t U_A(t,s)C(s)U_B(s,0)v ds$$

such that $x(\cdot)$ is continuously differentiable on $(0,T)$ and $x(t) \in \mathcal{D}(A)$ for all $t \in [0,T]$. If we denote by $H(\cdot) : [0,T] \to \mathcal{L}(Y,X)$ the strongly continuous mapping $x(t) = H(t)v$, then for all $v \in \mathcal{D}(B)$ the function $t \mapsto H(t)v$ is continuously differentiable on $(0,T)$ and $H(t)v \in \mathcal{D}(A)$. Since $t \mapsto U_B(0,t)y$ is strongly continuously differentiable, the choice $v = U_B(0,t)y \in \mathcal{D}(B)$ and a straight-forward computation finally show that the function

$$\begin{aligned} t \mapsto H(t)U_B(0,t)y &= U_A(t,0)\Sigma_0 U_B(0,t)y + \int_0^t U_A(t,s)C(s)U_B(s,0)U_B(0,t)y ds \\ &= U_A(t,0)\Sigma_0 U_B(0,t)y + \int_0^t U_A(t,s)C(s)U_B(s,t)y ds \\ &= \Sigma(t)y \end{aligned}$$

is continuously differentiable on $(0,T)$ and $\Sigma(t)y \in \mathcal{D}(A)$ for all $[0,T]$. Now equation (3.3) becomes

$$\langle \frac{d}{dt}\Sigma(t)y, x^* \rangle = \langle A(t)\Sigma(t)y, x^* \rangle - \langle \Sigma(t)B(t)y, x^* \rangle + \langle C(t)y, x^* \rangle.$$

Since $x^* \in \mathcal{D}(A^*)$ was arbitrary and since $\mathcal{D}(A^*)$ is dense in $X^*$, this implies

$$\frac{d}{dt}\Sigma(t)y = A(t)\Sigma(t)y - \Sigma(t)B(t)y + C(t)y.$$

This concludes that $\Sigma(\cdot)$ is a classical solution of the Sylvester differential equation.

To prove the uniqueness of the solution, let $\Sigma_1(\cdot) \in C([0,T], \mathcal{L}_s(Y,X))$ be a classical solution of the Sylvester differential equation (3.2). Letting $y \in \mathcal{D}(B)$ and applying both sides of the equation to $U_B(s,t)y \in \mathcal{D}(B)$ for $t > s$ we obtain

$$\begin{aligned} &\dot{\Sigma}_1(s)U_B(s,t)y = A(s)\Sigma_1(s)U_B(s,t)y - \Sigma_1(s)B(s)U_B(s,t)y + C(s)U_B(s,t)y \\ \Rightarrow\quad & U_A(t,s)\dot{\Sigma}_1(s)U_B(s,t)y = U_A(t,s)A(s)\Sigma_1(s)U_B(s,t)y \\ & - U_A(t,s)\Sigma_1(s)B(s)U_B(s,t)y + U_A(t,s)C(s)U_B(s,t)y \\ \Rightarrow\quad & \frac{d}{ds}\left(U_A(t,s)\Sigma_1(s)U_B(s,t)y\right) = U_A(t,s)C(s)U_B(s,t)y \end{aligned}$$

Integrating both sides of the last equation from 0 to $t$ and using $\Sigma_1(0) = \Sigma_0$ gives

$$\int_0^t U_A(t,s)C(s)U_B(s,t)y\,ds = U_A(t,t)\Sigma_1(t)U_B(t,t)y - U_A(t,0)\Sigma_1(0)U_B(0,t)y$$
$$= \Sigma_1(t)y - U_A(t,0)\Sigma_0 U_B(0,t)y$$

and thus $\Sigma_1(\cdot) = \Sigma(\cdot)$. □

As already mentioned, the conditions imposed on the evolution family $U_A(t,s)$ in Theorem 3.2 require that for $t \in [0,T]$ the operators $A(t)$ generate analytic semigroups on $X$. If these conditions are not satisfied, the solution (3.2) can under weaker conditions be seen as a *mild solution* of the Sylvester differential equation (3.1).

To illustrate the parabolic conditions we will present an example of a family of unbounded operators satisfying these conditions.

*Example.* Let $\alpha(\cdot), \gamma(\cdot) \in C([0,T],\mathbb{R})$ be Lipschitz continuous functions such that $\alpha(t) > 0$ for all $t \in [0,T]$. Consider a one-dimensional heat equation with time-varying coefficients

$$\begin{aligned} \frac{\partial x}{\partial t}(z,t) &= \alpha(t)\frac{\partial^2 x}{\partial t^2}(z,t) + \gamma(t)x(z,t), \\ x(z,0) &= x_0(z) \\ x(0,t) &= x(1,t) = 0 \end{aligned}$$

on the interval $[0,1]$. This can be written as a nonautonomous Cauchy problem

$$\dot{x} = A(t)x(t), \qquad x(t) = x_0 \in X$$

on the space $X = L^2(0,1)$ where the family of operators $(A(t),\mathcal{D}(A))$ is given by

$$\begin{aligned} A(t)x &= \alpha(t)x'' + \gamma(t)x, \\ \mathcal{D}(A) &= \left\{\, x \in X \;\middle|\; x, x' \text{ abs. cont. } x'' \in L^2(0,1),\ x(0) = x(1) = 0 \,\right\}. \end{aligned}$$

Furthermore, the operators $A(t)$ have spectral decompositions [3, Ex. A.4.26]

$$A(t)x = \sum_{n=1}^{\infty} \lambda_n(t)\langle x, \phi_n\rangle\phi_n, \qquad x \in \mathcal{D}(A) = \left\{ x \in X \,\middle|\, \sum_{n=1}^{\infty} n^4 |\langle x, \phi_n\rangle|^2 < \infty \right\}$$

where the eigenvalues are given by $\lambda_n(t) = -\alpha(t)n^2\pi^2 + \gamma(t)$ and the corresponding eigenvectors $\phi_n = \sqrt{2}\sin(n\pi\cdot)$ form an orthonormal basis of $X$. These decompositions and the fact that $\alpha(\cdot)$ and $\gamma(\cdot)$ are Lipschitz continuous functions can be used to verify that the parabolic conditions are satisfied.

We can also show that the second condition in Theorem 3.2 is satisfied for this family of operators. The operators $A(t)$ are self-adjoint and thus we can achieve this by showing that the mapping $t \mapsto A(t)x$ is continuous for all $x \in \mathcal{D}(A)$. If we define the operator $A_0 : \mathcal{D}(A) \to X$ by $A_0 x = x''$ we can write

$$A(t)x = \alpha(t)A_0 x + \gamma(t)x, \qquad x \in \mathcal{D}(A).$$

Since the functions $\alpha(\cdot)$ and $\gamma(\cdot)$ are continuous, we can conclude that the second condition in Theorem 3.2 is satisfied.

Families of operators satisfying the conditions concerning $(B(t), \mathcal{D}(B))$ include, for example, all functions $B(\cdot) \in C([0,T], \mathcal{L}_s(Y))$ and the case where $B(t) \equiv B$ is a generator of a strongly continuous group on $Y$.

We conclude this section by considering the periodic Sylvester differential equation. By this we mean the equation

$$\dot{\Sigma}(t) = A(t)\Sigma(t) - \Sigma(t)B(t) + C(t) \tag{3.4}$$

for $t \in \mathbb{R}$ when the families of unbounded operators and the function $C(\cdot)$ are periodic with the same period $\tau > 0$. The periodic solution of this equation is a periodic function $\Sigma(\cdot) \in C(\mathbb{R}, \mathcal{L}_s(Y,X))$ which is a classical solution of the Sylvester differential equation (3.1) with some initial condition $\Sigma(0) = \Sigma_0 \in \mathcal{L}(Y,X)$ on an interval $[0,T]$. The following theorem states that if the exponential growths of the evolution families $U_A(t,s)$ and $U_B(s,t)$ satisfy a certain condition, then under the assumptions of Theorem 3.2 the periodic Sylvester differential equation (3.4) has a unique periodic solution and that this solution has period $\tau$.

**Theorem 3.3.** *Assume the conditions of Theorem 3.2 are satisfied and that the evolution families $(A(t), \mathcal{D}(A))$ and $(B(t), \mathcal{D}(B))$ and the function $C(\cdot)$ are periodic with period $\tau > 0$. If there exist constants $M_A, M_B \geq 1$ and $\omega_A, \omega_B \in \mathbb{R}$ such that $\omega_A + \omega_B < 0$ and such that for all $t \geq s$*

$$\|U_A(t,s)\| \leq M_A e^{\omega_A(t-s)}, \qquad \|U_B(s,t)\| \leq M_B e^{\omega_B(t-s)},$$

*then the periodic Sylvester differential equation (3.4) has a unique periodic solution $\Sigma_\infty(\cdot) \in C(\mathbb{R}, \mathcal{L}_s(Y,X))$ such that $\Sigma_\infty(\cdot)y \in C^1(\mathbb{R}, X)$ and $\Sigma(t)y \in \mathcal{D}(A)$ for all $y \in \mathcal{D}(B)$ and $t \in \mathbb{R}$. The function $\Sigma_\infty(\cdot)$ has period $\tau$ and is given by the formula*

$$\Sigma_\infty(t)y = \int_{-\infty}^{t} U_A(t,s)C(s)U_B(s,t)y\,ds, \qquad y \in Y.$$

*Proof.* We will first show that $\Sigma_\infty(\cdot)$ is a classical solution of the Sylvester differential equation (3.1) on the interval $[0, 2\tau]$. Since for every $y \in Y$ we have

$$\begin{aligned}\Sigma_\infty(t)y = U_A(t,0) &\int_{-\infty}^{0} U_A(0,s)C(s)U_B(s,0)U_B(0,t)y\,ds \\ &+ \int_0^t U_A(t,s)C(s)U_B(s,t)y\,ds,\end{aligned}$$

it suffices to show that the linear operator $\Sigma_\infty(0) : Y \to X$ defined by

$$\Sigma_\infty(0)y = \int_{-\infty}^{0} U_A(0,s)C(s)U_B(s,0)y\,ds, \qquad y \in Y$$

is bounded and $\Sigma_\infty(0)(\mathcal{D}(B)) \subset \mathcal{D}(A)$. Our assumptions imply that for all $y \in Y$ we have

$$\int_{-\infty}^0 \|U_A(0,s)C(s)U_B(s,0)y\|ds \le M_A M_B \max_{r\in[0,\tau]} \|C(r)\| \int_{-\infty}^0 e^{-(\omega_A+\omega_B)s} ds \cdot \|y\| =: M\|y\|,$$

where $M < \infty$. This concludes that $\Sigma_\infty(0) : Y \to X$ is a well-defined linear operator and since

$$\left\| \int_{-\infty}^0 U_A(0,s)C(s)U_B(s,0)yds \right\| \le \int_{-\infty}^0 \|U_A(0,s)C(s)U_B(s,0)y\|ds \le M\|y\|,$$

we have $\Sigma_\infty(0) \in \mathcal{L}(Y,X)$. To show that $\Sigma_\infty(0)(\mathcal{D}(B)) \subset \mathcal{D}(A)$, let $y \in \mathcal{D}(B)$ and write

$$\Sigma_\infty(0)y = \int_{-\infty}^{-1} U_A(0,s)C(s)U_B(s,0)yds + \int_{-1}^0 U_A(0,s)C(s)U_B(s,0)yds =: v_0 + v_1.$$

If we denote $f(s) = U_A(0,s)C(s)U_B(s,0)y$, then $f(s) \in \mathcal{D}(A)$ for all $s < 0$ and from the previous estimate we have $f \in L^1((-\infty,-1),X)$. We have from [10, Thm. 5.6.1] that $A(0)U_A(0,-1) \in \mathcal{L}(X)$ and thus

$$\begin{aligned}
&\int_{-\infty}^{-1} \|A(0)U_A(0,s)C(s)U_B(s,0)y\|ds \\
&\le \|A(0)U_A(0,-1)\| \int_{-\infty}^{-1} \|U_A(-1,s)C(s)U_B(s,0)y\|ds \\
&\le M_A M_B \max_{r\in[0,\tau]} \|C(r)\| \cdot \|A(0)U_A(0,-1)\| \cdot \|y\| \cdot e^{-\omega_A} \int_{-\infty}^{-1} e^{-(\omega_A+\omega_B)s} ds < \infty.
\end{aligned}$$

This shows that $A(0)f \in L^1((-\infty,-1),X)$ and since $A(0)$ is a closed linear operator we have that $v_0 \in \mathcal{D}(A(0)) = \mathcal{D}(A)$. As in the proof of Theorem 3.2 we have that since the mapping $t \mapsto C(t)U_B(t,0)y$ is Hölder continuous on $[-1,0]$, the nonautonomous abstract Cauchy problem

$$\dot{x}(t) = A(t)x(t) + C(t)U_B(t,0)y, \qquad x(-1) = 0$$

has a unique classical solution

$$x(t) = \int_{-1}^t U_A(t,s)C(s)U_B(s,0)yds$$

on $[-1,0]$. Thus we also have $v_1 = x(0) \in \mathcal{D}(A)$. Combining these results shows that we have $\Sigma_\infty(0)y = v_0 + v_1 \in \mathcal{D}(A)$ and thus $\Sigma_\infty(0)$ is the unique classical solution of the Sylvester differential equation on $[0,2\tau]$ associated to the initial condition $\Sigma_\infty(0)$.

To prove the periodicity of $\Sigma_\infty(\cdot)$, let $t \in \mathbb{R}$. For all $y \in Y$ we then have

$$\begin{aligned}\Sigma_\infty(t+\tau)y &= \int_{-\infty}^{t+\tau} U_A(t+\tau,s)C(s)U_B(s,t+\tau)yds \\ &= \int_{-\infty}^{t} U_A(t+\tau,s+\tau)C(s+\tau)U_B(s+\tau,t+\tau)yds \\ &= \int_{-\infty}^{t} U_A(t,s)C(s)U_B(s,t)yds = \Sigma_\infty(t)y.\end{aligned}$$

This shows that $\Sigma_\infty(\cdot)$ is periodic with period $\tau$. This and the fact that $\Sigma_\infty(\cdot)$ is the classical solution of the Sylvester differential equation (3.1) on the interval $[0,2\tau]$ imply that $\Sigma_\infty(\cdot)y \in C^1(\mathbb{R},X)$ and $\Sigma_\infty(t)y \in \mathcal{D}(A)$ for all $t \in \mathbb{R}$. This concludes that $\Sigma_\infty(\cdot)$ is a periodic solution of the periodic Sylvester differential equation.

It remains to prove that the periodic Sylvester differential equation (3.4) has no other periodic solutions. To this end, let $\Sigma(\cdot)$ be any periodic solution of the equation corresponding to an arbitrary initial condition $\Sigma(0) = \Sigma_0 \in \mathcal{L}(W,X)$. Let $y \in Y$. We have

$$\Sigma(t)y = U_A(t,0)\Sigma_0 U_B(0,t)y + \int_0^t U_A(t,s)C(s)U_B(s,t)yds$$

and the difference $\Delta(t)y = \Sigma_\infty(t)y - \Sigma(t)y$ satisfies

$$\begin{aligned}\Delta(t)y = &\int_{-\infty}^{t} U_A(t,s)C(s)U_B(s,t)yds - U_A(t,0)\Sigma_0 U_B(0,t)y \\ &- \int_0^t U_A(t,s)C(s)U_B(s,t)yds \\ = &\int_{-\infty}^{0} U_A(t,s)C(s)U_B(s,t)yds - U_A(t,0)\Sigma_0 U_B(0,t)y \\ = &\, U_A(t,0)\Sigma_\infty(0)U_B(0,t) - U_A(t,0)\Sigma_0 U_B(0,t)y = U_A(t,0)\Delta(0)U_B(0,t)y.\end{aligned}$$

Thus

$$\|\Delta(t)\| \le M_A M_B e^{(\omega_A+\omega_B)t}\|\Delta(0)\|$$

and the assumption $\omega_A + \omega_B < 0$ implies $\lim_{t\to\infty} \Delta(t) = 0$. Since $\Sigma(\cdot)$ and $\Sigma_\infty(\cdot)$ are periodic and since $\lim_{t\to\infty}\|\Sigma(t) - \Sigma_\infty(t)\| = 0$, we must have $\Sigma(t) \equiv \Sigma_\infty(t)$. This concludes that no other periodic solutions than $\Sigma_\infty(\cdot)$ may exist. □

## 4. Periodic Output Regulation

In this section we finally apply the results on the solvability of the Sylvester differential equation to obtain a characterization for the controllers solving the output regulation problem related to a distributed parameter system and a nonautonomous periodic signal generator. We will use notation typical to mathematical systems theory and because of this the choices of symbols differ from the ones used in the earlier sections.

We consider the output regulation of an infinite-dimensional linear system in a situation where the reference and disturbance signals are generated by an exosystem

$$\dot{v}(t) = S(t)v(t), \qquad v(0) = v_0 \in W \tag{4.1}$$

on a finite-dimensional space $W = \mathbb{C}^q$. We assume the input and output spaces $U$ and $Y$, respectively, are Hilbert spaces and that the plant can be written in a standard form as

$$\begin{aligned} \dot{x}(t) &= Ax(t) + Bu(t) + E(t)v(t), \qquad x(0) = x_0 \in X \\ e(t) &= Cx(t) + Du(t) + F(t)v(t) \end{aligned}$$

on a Banach space $X$. Here $e(t) \in Y$ is the regulation error, $u(t) \in U$ the input, $E(t)v(t)$ is the disturbance signal to the state and $F(t)v(t)$ contains the disturbance signal to the output and the reference signal. The operator $A : \mathcal{D}(A) \subset X \to X$ is assumed to generate an analytic semigroup on $X$ and the rest of the operators are bounded. We consider a dynamic error feedback controller of form

$$\begin{aligned} \dot{z}(t) &= \mathcal{G}_1(t)z(t) + \mathcal{G}_2(t)e(t), \qquad z(0) = z_0 \in Z \\ u(t) &= K(t)z(t) \end{aligned}$$

on a Banach space $Z$. Here $(\mathcal{G}_1(t), \mathcal{D}(\mathcal{G}_1))$ is a family of unbounded operators, and $\mathcal{G}_2(t) \in \mathcal{L}(Y, Z)$ and $K(t) \in \mathcal{L}(Z, U)$ for all $t \geq 0$. The plant and the controller can be written as a closed-loop system

$$\dot{x}_e(t) = A_e(t)x_e(t) + B_e(t)v(t) \qquad x_e(0) = x_{e0} \in X_e \tag{4.2a}$$

$$e(t) = C_e(t)x_e(t) + D_e(t)v(t) \tag{4.2b}$$

on the Banach space $X_e = X \times Z$ by choosing

$$A_e(t) = \begin{pmatrix} A & BK(t) \\ \mathcal{G}_2(t)C & \mathcal{G}_1(t) + \mathcal{G}_2(t)DK(t) \end{pmatrix}, \qquad B_e(t) = \begin{pmatrix} E(t) \\ \mathcal{G}_2(t)F(t) \end{pmatrix}$$

$C_e(t) = \big(C,\ DK(t)\big)$ and $D_e(t) = F(t)$. We assume the family $(A_e(t), \mathcal{D}(A_e(t)))$ of unbounded operators and the operator-valued functions $S(\cdot)$, $B_e(\cdot)$, $C_e(\cdot)$ and $D_e(\cdot)$ are periodic with the same period $\tau > 0$. The Periodic Output Regulation Problem is defined as follows.

**Definition 4.1 (Periodic Output Regulation Problem).** Choose the parameters $(\mathcal{G}_1(\cdot), \mathcal{G}_2(\cdot), K(\cdot))$ of the dynamic error feedback controller in such a way that

1. The evolution family $U_e(t,s)$ associated to the family $(A_e(t), \mathcal{D}(A_e(t)))$ is exponentially stable, i.e., there exist $M_e, \omega_e > 0$ such that for all $t \geq s$
$$\|U_e(t,s)\| \leq M_e e^{-\omega_e(t-s)}.$$
2. For all initial values $x_{e0} \in X_e$ and $v_0 \in W$ of the closed-loop system and the exosystem, respectively, the regulation error $e(t)$ goes to zero asymptotically, i.e., $e(t) \to 0$ as $t \to \infty$.

It has been shown in [9] that under suitable assumptions the solvability of the Periodic Output Regulation Problem can be characterized using the periodic Sylvester differential equation

$$\dot{\Sigma}(t) = A_e(t)\Sigma(t) - \Sigma(t)S(t) + B_e(t). \tag{4.3}$$

Using Theorem 3.3 we can weaken the assumptions required for this characterization and thus extend the results presented in [9] for more general classes of systems and exosystems. We will first state the required assumptions. Since the space $W = \mathbb{C}^q$ is finite-dimensional, the strong continuity of the operator-valued functions $S(\cdot)$ and $B_e(\cdot)$ coincide with the continuity with respect to the uniform operator topology.

1. The family $(A_e(t), \mathcal{D}(A_e(t)))$ satisfies the parabolic conditions.
2. The domain $\mathcal{D}(A_e(t)^*) =: \mathcal{D}(A_e^*)$ is independent of $t \in \mathbb{R}$ and dense in $X_e^*$. For all $x \in X_e$ and $x^* \in X_e^*$ the mapping $t \mapsto \langle x, A_e(t)^* x^* \rangle$ is continuous.
3. The matrix-valued function $S(\cdot)$ is continuous, we have $|\lambda| = 1$ for all eigenvalues $\lambda$ of $U_S(\tau, 0)$ and there exists $M_S \geq 1$ such that $\|U_S(t,s)\| \leq M_S$ for all $t, s \in \mathbb{R}$.
4. The function $B_e(\cdot)$ is Hölder continuous.
5. The functions $C_e(\cdot)$ and $D_e(\cdot)$ are strongly continuous.

The following theorem characterizes the controllers solving the Periodic Output Regulation Problem using the properties of the Sylvester differential equation (4.3).

**Theorem 4.2.** *Assume that the above conditions are satisfied. If the controller stabilizes the closed-loop system exponentially, then the periodic Sylvester differential equation* (4.3) *has a unique periodic classical solution* $\Sigma_\infty(\cdot)$ *and the controller solves the Periodic Output Regulation Problem if and only if this solution satisfies*

$$C_e(t)\Sigma_\infty(t) + D_e(t) = 0 \tag{4.4}$$

*for all* $t \in [0, \tau]$.

*Proof.* Since the conditions of Theorem 3.3 are satisfied, the Sylvester differential equation (4.3) has a unique periodic classical solution $\Sigma_\infty(\cdot)$ with period $\tau$.

Since the space $W$ is finite-dimensional we have $\Sigma_\infty(\cdot) \in C^1(\mathbb{R}, \mathcal{L}(W, X_e))$ and $\mathcal{R}(\Sigma_\infty(t)) \subset \mathcal{D}(A_e)$ for all $t \in \mathbb{R}$.

We will first study the asymptotic behaviour of the regulation error. For any initial conditions $x_{e0} \in X_e$ and $v_0 \in W$ and for any $t \geq 0$ the state of the closed-loop system is given by

$$x_e(t) = U_e(t,0)x_{e0} + \int_0^t U_e(t,s)B_e(s)U_S(s,0)v_0 ds.$$

Using the Sylvester differential equation we see that

$$\begin{aligned} U_e(t,s)B_e(s)U_S(t,0)v_0 &= U_e(t,s)(\dot{\Sigma}_\infty(s) + \Sigma_\infty(s)S(s) - A_e(s)\Sigma_\infty(s))U_S(s,0)v_0 \\ &= U_e(t,s)\dot{\Sigma}_\infty(s)U_S(s,0)v_0 + U_e(t,s)\Sigma_\infty(s)S(s)U_S(s,0)v_0 \\ &\quad - U_e(t,s)A_e(s)\Sigma_\infty(s)U_S(s,0)v_0 \\ &= \frac{d}{ds}U_e(t,s)\Sigma_\infty(s)U_S(s,0)v_0. \end{aligned}$$

The state of the closed-loop system can thus be expressed using a formula

$$\begin{aligned} x_e(t) &= U_e(t,0)x_{e0} + \int_0^t U_e(t,s)B_e(s)U_S(s,0)v_0 ds \\ &= U_e(t,0)x_{e0} + \Sigma_\infty(t)U_S(t,0)v_0 - U_e(t,0)\Sigma_\infty(0)v_0 \\ &= U_e(t,0)(x_{e0} - \Sigma_\infty(0)v_0) + \Sigma_\infty(t)v(t) \end{aligned}$$

and the regulation error corresponding to these initial states is given by

$$\begin{aligned} e(t) &= C_e(t)x_e(t) + D_e v(t) \\ &= C_e(t)U_e(t,0)(x_{e0} - \Sigma_\infty(0)v_0) + (C_e(t)\Sigma_\infty(t) + D_e(t))\, v(t). \end{aligned}$$

Since the closed-loop system is stable there exist constants $M_e \geq 1$ and $\omega_e > 0$ such that for all $t \geq s$ we have $\|U_e(t,s)\| \leq M_e e^{-\omega_e(t-s)}$. Using the formula for the regulation error we have

$$\begin{aligned} &\|e(t) - (C_e(t)\Sigma_\infty(t) + D_e(t))v(t)\| = \|C_e(t)U_e(t,0)(x_{e0} - \Sigma_\infty(0)v_0)\| \\ &\leq M_e e^{-\omega_e t} \max_{s\in[0,T]} \|C_e(s)\| \cdot \|x_{e0} - \Sigma_\infty(0)v_0\| \longrightarrow 0 \end{aligned}$$

as $t \to \infty$ since $\omega_e > 0$. This property describing the asymptotic behaviour of the regulation error allows us to prove the theorem.

Assume first that (4.4) is satisfied for all $t \in [0, \tau]$. The periodicity of the functions implies that it is satisfied for all $t \in \mathbb{R}$ and thus for all initial states $x_{e0} \in X_e$ and $v_0 \in W$ the regulation error satisfies

$$\|e(t)\| = \|e(t) - (C_e(t)\Sigma_\infty(t) + D_e(t))v(t)\| \longrightarrow 0$$

as $t \to \infty$. This concludes that the controller solves the Periodic Output Regulation Problem.

To prove the converse implication assume that the controller solves the Periodic Output Regulation Problem. Let $t_0 \in [0, \tau)$ and $n \in \mathbb{N}_0$ and denote $t = n\tau + t_0$.

Using the periodicity of the functions and the above property of the regulation error we have that for any initial state $v_0 \in W$ of the exosystem and any $x_{e0} \in X_e$

$$
\begin{aligned}
&\|(C_e(t_0)\Sigma_\infty(t_0) + D_e(t_0))v(t)\| = \|(C_e(t)\Sigma_\infty(t) + D_e(t))v(t)\| \\
&= \|e(t) - (C_e(t)\Sigma_\infty(t) + D_e(t))v(t)\| + \|e(t)\| \longrightarrow 0
\end{aligned}
$$

as $n \to \infty$. Let $\lambda \in \sigma(U_S(\tau,0))$ and let $\{\phi_k\}_{k=1}^m$ be a Jordan chain associated to this eigenvalue. We will use the above limit to show that for all $k \in \{1,\dots,m\}$ we have $(C_e(t_0)\Sigma_\infty(t_0) + D_e(t_0))\phi_k = 0$. By assumption we have $|\lambda| = 1$ and

$$
U_S(\tau,0)\phi_1 = \lambda\phi_1, \qquad U_S(\tau,0)\phi_k = \lambda\phi_k + \phi_{k-1}, \quad k \in \{2,\dots,m\}. \tag{4.5}
$$

The periodicity of the evolution family $U_S(t,s)$ implies

$$
\begin{aligned}
U_S(t,0) &= U_S(n\tau + t_0, 0) = U_S(n\tau + t_0, n\tau)U_S(n\tau,(n-1)\tau)\cdots U_S(\tau,0) \\
&= U_S(t_0,0)U_S(\tau,0)^n
\end{aligned}
$$

and thus

$$
\begin{aligned}
0 &= \lim_{n\to\infty} \|(C_e(t_0)\Sigma_\infty(t_0) + D_e(t_0))\, U_S(t,0)\phi_1\| \\
&= \|(C_e(t_0)\Sigma_\infty(t_0) + D_e(t_0))\, U_S(t_0,0)\phi_1\| \cdot \left(\lim_{n\to\infty} |\lambda|^n\right).
\end{aligned}
$$

This implies $(C_e(t_0)\Sigma_\infty(t_0) + D_e(t_0))\, U_S(t_0,0)\phi_1 = 0$ since $|\lambda| = 1$. Using this and (4.5) we get

$$
\begin{aligned}
0 &= \lim_{n\to\infty} \|(C_e(t_0)\Sigma_\infty(t_0) + D_e(t_0))\, U_S(t,0)\phi_2\| \\
&= \|(C_e(t_0)\Sigma_\infty(t_0) + D_e(t_0))\, U_S(t_0,0)\phi_2\| \cdot \left(\lim_{n\to\infty} |\lambda|^n\right)
\end{aligned}
$$

and thus also $(C_e(t_0)\Sigma_\infty(t_0) + D_e(t_0))\, U_S(t_0,0)\phi_2 = 0$. Continuing this we finally obtain

$$
\begin{aligned}
0 &= \lim_{n\to\infty} \|(C_e(t_0)\Sigma_\infty(t_0) + D_e(t_0))\, U_S(t,0)\phi_m\| \\
&= \|(C_e(t_0)\Sigma_\infty(t_0) + D_e(t_0))\, U_S(t_0,0)\phi_m\| \cdot \left(\lim_{n\to\infty} |\lambda|^n\right)
\end{aligned}
$$

which implies $(C_e(t_0)\Sigma_\infty(t_0) + D_e(t_0))\, U_S(t_0,0)\phi_m = 0$. Since $\lambda \in \sigma(U_S(\tau,0))$ and the associated Jordan chain were arbitrary, we must have

$$
(C_e(t_0)\Sigma_\infty(t_0) + D_e(t_0))U_S(t_0,0) = 0.
$$

The invertibility of $U_S(t_0,0)$ further concludes that $C_e(t_0)\Sigma_\infty(t_0) + D_e(t_0) = 0$. Since $t_0 \in [0,\tau)$ was arbitrary, this finally shows that $C_e(t)\Sigma_\infty(t) + D_e(t) = 0$ for every $t \in [0,\tau]$. □

It should also be noted that Theorem 4.2 is independent of the form of the controller in the sense that if the closed-loop system can be written in the form (4.2), then this result implies that the output $e(t)$ of the closed-loop system driven by the nonautonomous exosystem (4.1) decays to zero asymptotically if and only if the solution of the Sylvester differential equation satisfies the constraint (4.4). This makes it possible to study the Periodic Output Regulation Problems with different types of controllers simultaneously. The general results obtained this way can subsequently be used to derive separate conditions for the solvability of the problem using different controller types.

## 5. Conclusions

In this paper we have considered the solvability of the infinite-dimensional Sylvester differential equation. We have introduced conditions under which the equation has a unique classical solution. We have also considered the periodic version of the equation and shown that if a certain condition on the growth of the evolution families associated to the equation is satisfied, then the periodic Sylvester differential equation has a unique periodic solution.

We applied the results on the solvability of the equation to the output regulation of a distributed parameter system with a time-dependent exosystem. In particular we showed that the controllers solving the output regulation problem can be characterized using the properties of the solution of the Sylvester differential equation. Developing the results for the solvability of these types of equations is crucial to the generalization of the theory of output regulation for more general classes of infinite-dimensional systems and exogeneous signals.

## References

[1] W. Arendt, F. Räbiger, and A. Sourour, *Spectral properties of the operator equation* $AX + XB = Y$. Q. J. Math **45** (1994), 133–149.

[2] A. Bensoussan, G. Da Prato, M.C. Delfour, and S.K. Mitter, *Representation and Control of Infinite Dimensional Systems.* 2nd Edition, Birkhäuser Boston, 2007.

[3] R.F. Curtain and H.J. Zwart, *An Introduction to Infinite-Dimensional Linear Systems Theory.* Springer-Verlag New York, 1995.

[4] Z. Emirsajlow and S. Townley, *On application of the implemented semigroup to a problem arising in optimal control.* Internat. J. Control, **78** (2005), 298–310.

[5] Z. Emirsajlow, *A composite semigroup for the infinite-dimensional differential Sylvester equation.* In Proceedings of the 7th Mediterranean Conference on Control and Automation, Haifa, Israel (1999), 1723–1726.

[6] K.-J. Engel and R. Nagel, *One-Parameter Semigroups for Linear Evolution Equations.* Springer-Verlag New York, 2000.

[7] H. Kano and T. Nishimura, *A note on periodic Lyapunov equations.* In Proceedings of the 35th Conference on Decision and Control, Kobe, Japan (1996).

[8] S. Lang, *Real and Functional Analysis.* 3rd Edition, Springer-Verlag New York, 1993.

[9] L. Paunonen and S. Pohjolainen, *Output regulation of distributed parameter systems with time-periodic exosystems.* In Proceedings of the Mathematical Theory of Networks and Systems, Budapest, Hungary (2010), 1637–1644.

[10] A. Pazy, *Semigroups of Linear Operators and Applications to Partial Differential Equations.* Springer-Verlag New York, 1983.

[11] Q.P. Vũ, *The operator equation $AX - XB = C$ with unbounded operators $A$ and $B$ and related abstract Cauchy problems.* Math. Z., **208** (1991), 567–588.

[12] Q.P. Vũ and E. Schüler, *The operator equation $AX - XB = C$, admissibility and asymptotic behaviour of differential equations.* J. Differential Equations, **145** (1998), 394–419.

[13] Z. Zhen and A. Serrani, *The linear periodic output regulation problem.* Systems Control Lett., **55** (2006), 518–529.

Lassi Paunonen
Department of Mathematics
Tampere University of Technology
PO. Box 553
SF-33101 Tampere, Finland
e-mail: `lassi.paunonen@tut.fi`

Operator Theory:
Advances and Applications, Vol. 221, 533-548


# A Class of Evolutionary Problems with an Application to Acoustic Waves with Impedance Type Boundary Conditions

Rainer Picard

**Dedicated to Prof. Dr. Rolf Leis on the occasion of his 80th birthday**

**Abstract.** A class of evolutionary operator equations is studied. As an application the equations of linear acoustics are considered with complex material laws. A dynamic boundary condition is imposed which in the time-harmonic case corresponds to an impedance or Robin boundary condition. Memory and delay effects in the interior and also on the boundary are built into the problem class.

**Mathematics Subject Classification (2000).** Primary 35F05; Secondary 35L40, 35F10, 76Q05.

**Keywords.** Evolution equations, partial differential equations, causality, acoustic waves, impedance type boundary condition, memory, delay.

## 0. Introduction

In [2] and [5], Chapter 5, a theoretical framework has been presented to discuss typical linear evolutionary problems as they arise in various fields of applications. The suitability of the problem class described for such applications has been demonstrated by numerous examples of varying complexity. The problem class can be heuristically described as: finding $U, V$ satisfying

$$\partial_0 V + AU = f,$$

where $V$ is linked to $U$ by a linear material law

$$V = \mathbb{M}U.$$

Here $\partial_0$ denotes the time derivative, $\mathbb{M}$ is a bounded linear operator commuting with $\partial_0$ and $A$ is a (usually) unbounded translation invariant linear operator. The material law operator $\mathbb{M}$ is given in terms of a suitable operator-valued function of the time derivative $\partial_0$ in the sense of a functional calculus associated with $\partial_0$ realized as a normal operator. The focus in the quoted references is on the case where $A$ is skew-selfadjoint.

The aim of this paper is to extend the general theory to encompass an even larger class of problems by allowing $A$ to be more general. The presentation will rest on a conceptually more elementary version of this theory as presented in [3], where the above problem is discussed as establishing the continuous invertibility of the unbounded operator sum $\partial_0\mathbb{M} + A$, i.e., we shall develop a solution theory for a class of operator equations of the form $\overline{\partial_0\mathbb{M} + A}\, U = f$. For suggestiveness of notation we shall simply write $\partial_0\mathbb{M} + A$ instead of $\overline{\partial_0\mathbb{M} + A}$, which can indeed be made rigorous in a suitable distributional sense.

To exemplify the utility of the generalization we shall apply the ideas developed here to impedance type boundary conditions in linear acoustics.

After briefly describing the corner stones of a general solution theory in section 1 we shall discuss in section 2 the particular issue of causality, which is a characteristic feature of problems we may rightfully call *evolutionary.*

The general findings will be illustrated by an application to acoustic equations with a dynamic boundary condition allowing for additional memory effects on the boundary of the underlying domain. We refer to the boundary condition as of *impedance type* due to its form after Fourier-Laplace transformation with respect to time. The reasoning in [4] finds its generalization in the arguments presented here in so far as here evolutionary boundary conditions modelling a separate dynamics on the boundary are included.

## 1. General solution theory

First we specify the space and the class of material law operators we want to consider.

**Assumptions on the material law operator**: Let $M = (M(z))_{z\in B_{\mathbb{C}}(r,r)}$ be a family of uniformly bounded linear operators in a Hilbert space $H$ with inner product $\langle\,\cdot\,|\,\cdot\,\rangle_H$, assumed to be linear in the second factor, with $z \mapsto M(z)$ holomorphic in the ball $B_{\mathbb{C}}(r,r)$ of radius $r$ centered at $r$. Then, for $\varrho > \frac{1}{2r}$, we define

$$\mathbb{M} := M\left(\partial_0^{-1}\right),$$

where

$$M\left(\partial_0^{-1}\right) := \mathbb{L}_\varrho^* M\left(\frac{1}{\mathrm{i}\, m_0 + \varrho}\right)\mathbb{L}_\varrho.$$

Here $\mathbb{L}_\varrho : H_\varrho(\mathbb{R}, H) \to H_0(\mathbb{R}, H)$, $\varrho \in \mathbb{R}_{\geq 0}$, denotes the unitary extension of the Fourier-Laplace transform to $H_\varrho(\mathbb{R}, H)$, the space of $H$-valued $L^{2,\mathrm{loc}}$-functions $f$ on $\mathbb{R}$ with

$$|f|_\varrho := \sqrt{\int_{\mathbb{R}} |f(t)|_H^2 \exp(-2\varrho t)\, dt} < \infty.$$

The Fourier-Laplace transform is given by

$$(\mathbb{L}_\varrho\varphi)(s) = \frac{1}{\sqrt{2\pi}} \int_{\mathbb{R}} \exp\left(-\mathrm{i}\left(s - \mathrm{i}\varrho\right)t\right) \varphi(t)\, dt,\ s \in \mathbb{R},$$

for $\varphi \in \mathring{C}_\infty(\mathbb{R}, H)$, i.e., for smooth $H$-valued functions $\varphi$ with compact support. The multiplicatively applied operator $M\left(\frac{1}{\mathrm{i}m_0+\varrho}\right) : H_0(\mathbb{R}, H) \to H_0(\mathbb{R}, H)$ is given by

$$\left(M\left(\frac{1}{\mathrm{i}m_0 + \varrho}\right)\varphi\right)(s) = M\left(\frac{1}{\mathrm{i}s + \varrho}\right)\varphi(s),\ s \in \mathbb{R},$$

for $\varphi \in \mathring{C}_\infty(\mathbb{R}, H)$, (so that $m_0$ simply denotes the multiplication by the argument operator).

$H_\varrho(\mathbb{R}, H)$ is a Hilbert space with inner product $\langle\,\cdot\,|\,\cdot\,\rangle_\varrho$ given by

$$\langle f|g\rangle_\varrho = \int_{\mathbb{R}} \langle f(t)\,|g(t)\rangle_H \exp(-2\varrho t)\; dt$$

and the associated norm will be denoted by $|\cdot|_\varrho$. In the case $\varrho = 0$ the space $H_\varrho(\mathbb{R}, H)$ is simply the space $L^2(\mathbb{R}, H)$ of $H$-valued $L^2$-functions on $\mathbb{R}$. In our general framework there is, however, a bias to consider large $\varrho \in \mathbb{R}_{>0}$.

Note that for $r \in \mathbb{R}_{>0}$

$$\begin{aligned} B_{\mathbb{C}}(r,r) &\to [\mathrm{i}\,\mathbb{R}] + \left[\mathbb{R}_{>1/(2r)}\right] \\ z &\mapsto z^{-1} \end{aligned}$$

is a bijection. In $H_\varrho(\mathbb{R}, H)$ the closure $\partial_0$ of the derivative on $\mathring{C}_\infty(\mathbb{R}, H)$ turns out to be a normal operator, see, e.g., [3], with $\mathfrak{Re}(\partial_0) = \varrho$.

With the time translation operator $\tau_h : H_\varrho(\mathbb{R}, H) \to H_\varrho(\mathbb{R}, H)$, $h \in \mathbb{R}$, given by

$$(\tau_h\varphi)(s) = \varphi(s+h),\ s \in \mathbb{R},$$

for $\varphi \in \mathring{C}_\infty(\mathbb{R}, H)$ it is easy to see that $M\left(\partial_0^{-1}\right)$ is translation invariant, i.e.,

$$\tau_h M\left(\partial_0^{-1}\right) = M\left(\partial_0^{-1}\right)\tau_h\,,\ h \in \mathbb{R}.$$

This is indeed clear since $M\left(\partial_0^{-1}\right)$ commutes by construction with $\partial_0^{-1}$ and

$$\tau_h = \exp\left(h\left(\partial_0^{-1}\right)^{-1}\right).$$

**Assumptions on $A$:** For the densely defined, closed linear operator $A$ in $H_\varrho(\mathbb{R},H)$ we also assume commutativity with $\partial_0^{-1}$ which implies commutativity with bounded Borel functions of $\partial_0^{-1}$, in particular, translation invariance

$$\tau_h A = A\tau_h\,,\ h \in \mathbb{R}.$$

We shall use $M\left(\partial_0^{-1}\right)$ and $A$ without denoting a reference to $\varrho \in \mathbb{R}_{>0}$. Thus we are led to solving

$$\left(\partial_0 M\left(\partial_0^{-1}\right) + A\right)U = f$$

in $H_\varrho(\mathbb{R},H)$ with $f\in H_\varrho(\mathbb{R},H)$ given. Recall that we have chosen to write $\partial_0 M\left(\partial_0^{-1}\right)+A$ for the closure of the sum of the two discontinuous operators involved. Note that initial data are not explicitly prescribed. It is assumed that they are built into the source term $f$ so that vanishing initial data may be assumed.

**Condition (Positivity 1):** *For all $U\in D(\partial_0)\cap D(A)$ and $V\in D(\partial_0)\cap D(A^*)$ we have, uniformly for all sufficiently large $\varrho\in\mathbb{R}_{>0}$, some $\beta_0\in\mathbb{R}_{>0}$ such that*

$$\mathfrak{Re}\left\langle\chi_{\mathbb{R}_{\leq 0}}(m_0)U\middle|\left(\partial_0 M\left(\partial_0^{-1}\right)+A\right)U\right\rangle_\varrho\geq\beta_0\left\langle\chi_{\mathbb{R}_{\leq 0}}(m_0)U\middle|U\right\rangle_\varrho,$$
$$\mathfrak{Re}\left\langle V\middle|\left(\partial_0^* M^*\left(\left(\partial_0^{-1}\right)^*\right)+A^*\right)V\right\rangle_\varrho\geq\beta_0\left\langle V|V\right\rangle_\varrho.$$

Here we have used the notation

$$M^*(z):=M(z^*)^*$$

with which we have

$$M\left(\partial_0^{-1}\right)^*=M^*\left(\left(\partial_0^{-1}\right)^*\right).$$

Note that due to translation invariance **Condition (Positivity 1)** is equivalent to

**Condition (Positivity 2)**[1]: *For all $a\in\mathbb{R}$ and all $U\in D(\partial_0)\cap D(A)$, $V\in D(\partial_0)\cap D(A^*)$ we have, uniformly for all sufficiently large $\varrho\in\mathbb{R}_{>0}$, some $\beta_0\in\mathbb{R}_{>0}$ such that*

$$\mathfrak{Re}\left\langle\chi_{\mathbb{R}_{\leq a}}(m_0)U\middle|\left(\partial_0 M\left(\partial_0^{-1}\right)+A\right)U\right\rangle_\varrho\geq\beta_0\left\langle\chi_{\mathbb{R}_{\leq a}}(m_0)U\middle|U\right\rangle_\varrho,$$
$$\mathfrak{Re}\left\langle V\middle|\left(\partial_0^* M^*\left(\left(\partial_0^{-1}\right)^*\right)+A^*\right)V\right\rangle_\varrho\geq\beta_0\left\langle V|V\right\rangle_\varrho.$$

Here $\chi_M$ denotes the characteristic function of the set $M$. As a general notation we introduce for every measurable function $\psi$ the associated multiplication operator

$$\psi(m_0):H_\varrho(\mathbb{R},H)\to H_\varrho(\mathbb{R},H)$$

determined by

$$(\psi(m_0)f)(t):=\psi(t)\,f(t),\ t\in\mathbb{R},$$

for $f\in\mathring{C}_\infty(\mathbb{R},H)$. Letting $a$ go to $\infty$ in this leads to

**Condition (Positivity 3)**: *For all $U\in D(\partial_0)\cap D(A)$ and $V\in D(\partial_0)\cap D(A^*)$ we have, uniformly for all sufficiently large $\varrho\in\mathbb{R}_{>0}$, some $\beta_0\in\mathbb{R}_{>0}$ such that*

$$\mathfrak{Re}\left\langle U\middle|\left(\partial_0 M\left(\partial_0^{-1}\right)+A\right)U\right\rangle_\varrho\geq\beta_0\left\langle U|U\right\rangle_\varrho,$$
$$\mathfrak{Re}\left\langle V\middle|\left(\partial_0^* M^*\left(\left(\partial_0^{-1}\right)^*\right)+A^*\right)V\right\rangle_\varrho\geq\beta_0\left\langle V|V\right\rangle_\varrho.$$

[1]This is the assumption employed in [4]. The second inequality in the corresponding assumption in [4] erroneously also contains the cut-off multiplier $\chi_{\mathbb{R}_{\leq a}}(m_0)$ which should not be there and is indeed never utilized in the arguments.

This is the constraint we have used in previous work. In the earlier considered cases the seemingly stronger **Condition (Positivity 2)** can be shown to hold. It turns out, however, that in the translation invariant case **Condition (Positivity 1)** adds flexibility to the solution theory (to include more general cases) and makes the issue of causality more easily accessible.

In preparation of our well-posedness result we need the following lemma. Note that for sake of clarity here we do distinguish between the natural sum $A+B$ and its closure.

**Lemma 1.1.** *Let $A, B$ be densely defined, closed linear operators in a Hilbert space $H$ such that $A+B$ and $A^*+B^*$ are densely defined and let $(P_n)_{n\in\mathbb{N}}$ be a monotone sequence of orthogonal projectors commuting with $A$ and $B$ with $P_n \overset{n\to\infty}{\to} 1$ strongly, such that $(P_nBP_n)_{n\in\mathbb{N}}$ is a sequence of continuous linear operators. Then*

$$(P_nAP_n)^* = P_nA^*P_n,\ (P_nBP_n)^* = P_nB^*P_n$$

*for every $n \in \mathbb{N}$. Moreover,*

$$\overline{A^*+B^*} = (A+B)^* = \text{s-}\lim_{n\to\infty} (P_n(A+B)P_n)^*.$$

*Proof.* We have for $y \in D(A^*)$ that $\langle Ax|y\rangle_H = \langle x|A^*y\rangle_H$ for all $x \in D(A)$ and so also for $n \in \mathbb{N}$

$$\langle Ax|P_ny\rangle_H = \langle P_nAx|y\rangle_H = \langle AP_nx|y\rangle_H = \langle P_nx|A^*y\rangle_H = \langle x|P_nA^*y\rangle_H$$

for all $x \in D(A)$. Thus, we find $P_ny \in D(A^*)$ and

$$A^*P_ny = P_nA^*y.$$

This shows that $P_n$ commutes with $A^*$ and similarly we find $P_n$ commuting with $B^*$.

From $P_nBx = BP_nx$ for $x \in D(B)$ we get

$$\langle x|(P_nBP_n)^*y\rangle_H = \langle P_nBP_nx|y\rangle_H = \langle BP_nx|P_ny\rangle_H = \langle Bx|P_ny\rangle_H$$

for all $x \in D(B)$, $y \in H$ and so

$$\langle x|(P_nBP_n)^*y\rangle_H = \langle P_nx|B^*P_ny\rangle_H = \langle x|P_nB^*P_ny\rangle_H$$

proving

$$(P_nBP_n)^* = P_nB^*P_n$$

for every $n \in \mathbb{N}$.

Now let $y \in D(A^*+B^*)$. Then $\langle (A+B)x|y\rangle_H = \langle x|(A+B)^*y\rangle_H$ for all $x \in D(A+B)$. Since $P_n(A+B) \subseteq (A+B)P_n$ we have

$$\begin{aligned}\langle x|P_n(A+B)^*y\rangle_H &= \langle P_nx|(A+B)^*y\rangle_H\\ &= \langle (A+B)P_nx|y\rangle_H\\ &= \langle P_n(A+B)P_nx|y\rangle_H\\ &= \langle P_nAP_nx + P_nBP_nx|y\rangle_H\\ &= \langle P_nAP_nx|y\rangle_H + \langle x|P_nB^*P_ny\rangle_H\end{aligned}$$

implying $y \in D\left((P_nAP_n)^*\right)$ and $(P_nAP_n)^*\, y = P_n\,(A+B)^*\, y - P_nB^*P_ny$. Thus, we see that

$$P_n\,(A+B)^* \subseteq (P_nAP_n)^* + P_nB^*P_n$$

and so, since clearly we have

$$A^* + B^* \subseteq (A+B)^*\,,$$

we have the relations

$$\begin{aligned} P_nA^*P_n + P_nB^*P_n &= P_n\,(A^*+B^*)\,P_n \\ &\subseteq P_n\,(A+B)^*\,P_n \\ &\subseteq (P_nAP_n)^* + P_nB^*P_n. \end{aligned} \tag{1.1}$$

In particular, this yields

$$P_nA^*P_n \subseteq (P_nAP_n)^*$$

for every $n \in \mathbb{N}$.

Let now $y \in D\left((P_nAP_n)^*\right)$ then

$$\langle P_nAP_nx|y\rangle_H = \left\langle x|\,(P_nAP_n)^*\, y\right\rangle_H$$

for all $x \in D\,(P_nAP_n)$. In particular, for $x \in D\,(A)$ we get

$$\langle Ax|P_ny\rangle_H = \langle P_nAx|y\rangle_H = \langle P_nAP_nx|y\rangle_H = \left\langle x|\,(P_nAP_n)^*\, y\right\rangle_H$$

showing that $P_ny \in D\,(A^*)$ and

$$A^*P_ny = (P_nAP_n)^*\, y.$$

Moreover,

$$\left\langle x|\,(P_nAP_n)^*\, y\right\rangle_H = \langle P_nAP_nx|y\rangle_H = \langle P_nx|A^*P_ny\rangle_H = \langle x|P_nA^*P_ny\rangle_H$$

for all $x \in D\,(A)$ proving that

$$(P_nAP_n)^* \subseteq P_nA^*P_n.$$

Thus, we also have

$$(P_nAP_n)^* = P_nA^*P_n.$$

From (1.1) we obtain now

$$\begin{aligned} P_n\,(A+B)^* \subseteq P_nA^*P_n + P_nB^*P_n &= P_n\,(A^*+B^*)\,P_n \\ &= P_n\,(A+B)^*\,P_n = (P_nAP_n)^* + P_nB^*P_n. \end{aligned}$$

I.e., for $z \in (A+B)^*$ we have

$$P_n\,(A+B)^*\, z = P_n\,(A^*+B^*)\,P_nz\,, = (A^*+B^*)\,P_nz.$$

Since $P_n \overset{n\to\infty}{\to} 1$ and from the closability of $A^*+B^*$ follows $z \in D\left(\overline{A^*+B^*}\right)$ and

$$\overline{A^*+B^*}z = (A+B)^*\, z.$$

Thus we have indeed shown that

$$(A+B)^* = \overline{A^*+B^*}. \qquad \square$$

This lemma will be crucial in the proof of our solution theorem.

**Theorem 1.2 (Solution Theory).** *Let $A$ and $M$ be as above and satisfy* **Condition (Positivity 1)**. *Then for every $f \in H_\varrho(\mathbb{R}, H)$, $\varrho \in \mathbb{R}_{>0}$ sufficiently large, there is a unique solution $U \in H_\varrho(\mathbb{R}, H)$ of*

$$\left(\partial_0 M\left(\partial_0^{-1}\right) + A\right) U = f.$$

*The solution depends continuously on the data in the sense that*

$$|U|_\varrho \le \beta_0^{-1} |f|_\varrho$$

*uniformly for all $f \in H_\varrho(\mathbb{R}, H)$ and $\varrho \in \mathbb{R}_{>0}$ sufficiently large.*

*Proof.* From **Condition (Positivity 3)** we see that the operators $\left(\partial_0 M\left(\partial_0^{-1}\right) + A\right)$ and $\left(\partial_0^* M^*\left(\left(\partial_0^{-1}\right)^*\right) + A^*\right)$ both have inverses bounded by $\beta_0^{-1}$. The invertibility of the operator $\left(\partial_0 M\left(\partial_0^{-1}\right) + A\right)$ already confirms the continuous dependence estimate. Moreover, we know that the null spaces of these operators are trivial. In particular,

$$N\left(\overline{\partial_0^* M^*\left(\left(\partial_0^{-1}\right)^*\right) + A^*}\right) = \{0\}. \tag{1.2}$$

It remains to be seen that the range $\left(\partial_0 M\left(\partial_0^{-1}\right) + A\right)\left[H_\varrho(\mathbb{R}, H)\right]$ of the operator $\left(\partial_0 M\left(\partial_0^{-1}\right) + A\right)$ is dense in $H_\rho(\mathbb{R}, H)$. Then the result follows (recall that $\left(\partial_0 M\left(\partial_0^{-1}\right) + A\right)$ is used in the above as a suggestive notation for $\overline{\partial_0 M\left(\partial_0^{-1}\right) + A}$).

The previous lemma applied with[2]

$$P_n := \chi_{[-n,n]}\left(\mathfrak{Im}\left(\partial_0\right)\right),\ n \in \mathbb{N},$$

and $B := \partial_0 M\left(\partial_0^{-1}\right)$ yields

$$\left(\partial_0 M\left(\partial_0^{-1}\right) + A\right)^* = \overline{\partial_0^* M^*\left(\left(\partial_0^{-1}\right)^*\right) + A^*}.$$

Since from the projection theorem we have the orthogonal decomposition

$$H_\varrho(\mathbb{R}, H) = N\left(\left(\partial_0 M\left(\partial_0^{-1}\right) + A\right)^*\right) \oplus \overline{\left(\partial_0 M\left(\partial_0^{-1}\right) + A\right)\left[H_\varrho(\mathbb{R}, H)\right]}$$

and from (1.2) we see that

$$N\left(\left(\partial_0 M\left(\partial_0^{-1}\right) + A\right)^*\right) = \{0\},$$

it follows indeed that

$$H_\varrho(\mathbb{R}, H) = \overline{\left(\partial_0 M\left(\partial_0^{-1}\right) + A\right)\left[H_\varrho(\mathbb{R}, H)\right]}.$$ □

---

[2]Recall that $\left(\chi_{]-\infty,\lambda]}\left(\mathfrak{Im}\left(\partial_0\right)\right)\right)_{\lambda\in\mathbb{R}}$ is the spectral family associated with the selfadjoint operator $\mathfrak{Im}\left(\partial_0\right)$.

This well-posedness result is, however, not all we would like to have. Note that so far we have only used **Condition (Positivity 3)** which was a simple consequence of **Condition (Positivity 1)**. For an actual *evolution* to take place we also need additionally a property securing causality for the solution. This is where the **Condition (Positivity 1)** comes into play.

## 2. Causality

We first need to specify what we mean by causality. This can here be done in a more elementary way than in [2].

**Definition 2.1.** A mapping $F : H_\varrho(\mathbb{R}, H) \to H_\varrho(\mathbb{R}, H)$ is called causal if for every $a \in \mathbb{R}$ and every $u, v \in H_\varrho(\mathbb{R}, H)$ we have

$$\chi_{]-\infty,a]}(m_0)(u - v) = 0 \implies \chi_{]-\infty,a]}(m_0)(F(u) - F(v)) = 0.$$

*Remark* 2.2. Note that if $F$ is translation invariant, then $a \in \mathbb{R}$ in this statement can be fixed (for example to $a = 0$). If $F$ is linear we may fix $v = 0$ to simplify the requirement.

It is known that, by construction from an analytic, bounded $M$, the operator $M(\partial_0^{-1})$ is causal, see [3].

**Theorem 2.3 (Causality of Solution Operator).** *Let $A$ and $M$ be as above and satisfy* **Condition (Positivity 1)**. *Then the solution operator*

$$\left(\partial_0 M\left(\partial_0^{-1}\right) + A\right)^{-1} : H_\varrho(\mathbb{R}, H) \to H_\varrho(\mathbb{R}, H)$$

*is causal for all sufficiently large $\varrho \in \mathbb{R}_{>0}$.*

*Proof.* By translation invariance we may base our arguments on **Condition (Positivity 2)**. We estimate

$$\begin{aligned}
&\left|\chi_{\mathbb{R}_{\leq a}}(m_0) U\right|_\varrho \left|\chi_{\mathbb{R}_{\leq a}}(m_0)\left(\partial_0 M\left(\partial_0^{-1}\right) + A\right) U\right|_\varrho \\
&\quad \geq \mathfrak{Re}\left\langle \chi_{\mathbb{R}_{\leq a}}(m_0) U \middle| \left(\partial_0 M\left(\partial_0^{-1}\right) + A\right) U\right\rangle_\varrho, \\
&\quad \geq \beta_0 \left|\chi_{\mathbb{R}_{\leq a}}(m_0) U\right|_\varrho^2,
\end{aligned}$$

yielding

$$\beta_0 \left|\chi_{\mathbb{R}_{\leq a}}(m_0) U\right|_\varrho \leq \left|\chi_{\mathbb{R}_{\leq a}}(m_0)\left(\partial_0 M\left(\partial_0^{-1}\right) + A\right) U\right|_\varrho$$

for every $a \in \mathbb{R}$. Substituting $\left(\partial_0 M\left(\partial_0^{-1}\right) + A\right)^{-1} f$ for $U$ this gives

$$\beta_0 \left|\chi_{\mathbb{R}_{\leq a}}(m_0)\left(\partial_0 M\left(\partial_0^{-1}\right) + A\right)^{-1} f\right|_\varrho \leq \left|\chi_{\mathbb{R}_{\leq a}}(m_0) f\right|_\varrho, \ a \in \mathbb{R}. \tag{2.1}$$

We read off that if $\chi_{\mathbb{R}_{\leq a}}(m_0) f = 0$ then $\chi_{\mathbb{R}_{\leq a}}(m_0)\left(\partial_0 M\left(\partial_0^{-1}\right) + A\right)^{-1} f = 0$, $a \in \mathbb{R}$. This is the desired causality of $\left(\partial_0 M\left(\partial_0^{-1}\right) + A\right)^{-1}$. □

## 3. An application: Acoustic waves with impedance type boundary condition

We want to conclude our discussion with a more substantial utilization of the theory presented. We assume that the material law is of the form

$$M\left(\partial_0^{-1}\right) = M_0 + \partial_0^{-1} M_1\left(\partial_0^{-1}\right)$$

with $M_0$ selfadjoint and strictly positive definite. The underlying Hilbert space is $H = L^2(\Omega) \oplus L^2(\Omega)^3$ and so we consider the material law as an operator in the spaces $H_\varrho(\mathbb{R}, L^2(\Omega) \oplus L^2(\Omega)^3)$, $\varrho \in \mathbb{R}_{>0}$.

The Hilbert space $H(\mathring{\mathrm{div}}, \Omega)$ is (equipped with the graph norm) the domain of the closure of the classical divergence operator on vector fields with $\mathring{C}_\infty(\Omega, \mathbb{C})$ components considered as a mapping from $L^2(\Omega)^3$ to $L^2(\Omega)$. To be an element of $H(\mathring{\mathrm{div}}, \Omega)$ generalizes the classical boundary condition of vanishing normal component on the boundary of $\Omega$ to cases with non-smooth boundary. Analogously we define the Hilbert space $H(\mathring{\mathrm{grad}}, \Omega)$ as the domain of the closure of the classical gradient operator on functions in $\mathring{C}_\infty(\Omega, \mathbb{C})$ (equipped with the graph norm).

Moreover, we let

$$A = \begin{pmatrix} 0 & \mathrm{div} \\ \mathrm{grad} & 0 \end{pmatrix}$$

with

$$D(A) := \left\{ \begin{pmatrix} p \\ v \end{pmatrix} \in D\left(\begin{pmatrix} 0 & \mathrm{div} \\ \mathrm{grad} & 0 \end{pmatrix}\right) \middle| a\left(\partial_0^{-1}\right) p - \partial_0^{-1} v \in H_\varrho\left(\mathbb{R}, H\left(\mathring{\mathrm{div}}, \Omega\right)\right) \right\}.$$

Here we assume that $a = (a(z))_{z \in B_{\mathbb{C}}(r,r)}$ is analytic and bounded and that $a\left(\partial_0^{-1}\right)$ is multiplicative in the sense that it is of the form

$$a(z) = \sum_{k=0}^{\infty} a_{k,r}(m)\,(z-r)^k,$$

where $a_{k,r}$ are $L^\infty(\Omega)$-vector fields and $a_{k,r}(m)$ is the associated multiplication operator

$$(a_{k,r}(m)\,\varphi)(t,x) := a_{k,r}(x)\,\varphi(t,x)$$

for $\varphi \in \mathring{C}_\infty(\mathbb{R} \times \Omega, \mathbb{C})$. Moreover, we assume that

$$(\mathrm{div}\, a)(z) := \sum_{k=0}^{\infty} (\mathrm{div}\, a_{k,r})(m)\,(z-r)^k$$

is analytic and bounded with $\mathrm{div}\, a_{k,r} \in L^\infty(\Omega)$. Then, in particular, the product rule holds

$$\mathrm{div}\left(a\left(\partial_0^{-1}\right) p\right) = \left(\mathrm{div}\, a\left(\partial_0^{-1}\right)\right) p + a\left(\partial_0^{-1}\right) \cdot \mathrm{grad}\, p. \tag{3.1}$$

As a consequence, we have

$$a\left(\partial_0^{-1}\right) : H_\varrho(\mathbb{R}, H(\mathrm{grad}, \Omega)) \to H_\varrho(\mathbb{R}, H(\mathrm{div}, \Omega)) \tag{3.2}$$

uniformly bounded for all sufficiently large $\varrho \in \mathbb{R}_{>0}$.

The boundary condition given in the description of the domain of $A$ is in classical terms[3]

$$n \cdot a\left(\partial_0^{-1}\right) \partial_0 p\,(t) - n \cdot v\,(t) \;= 0 \text{ on } \partial\Omega,\ t \in \mathbb{R},$$

in the case that the boundary $\partial\Omega$ of $\Omega$ and the solution are smooth and $\partial\Omega$ has $n$ as exterior unit normal field.

We also will impose a sign requirement on $a$:

$$\mathfrak{Re} \int_{-\infty}^{0} \left(\left\langle \operatorname{grad} p | \partial_0 a\left(\partial_0^{-1}\right) p\right\rangle_0 (t) + \left\langle p \right| \operatorname{div} \partial_0 a\left(\partial_0^{-1}\right) p\right\rangle_0 (t)\right) \exp\left(-2\varrho t\right)\, dt \geq 0 \tag{3.3}$$

for all sufficiently large $\varrho \in \mathbb{R}_{>0}$. This is the appropriate generalization of the condition:

$$\mathfrak{Re} \int_{\partial\Omega} p\,(t)^* \left(\partial_0 n \cdot a\left(\partial_0^{-1}\right) p\right)(t)\; do \geq 0,\ t \in \mathbb{R},$$

in the case that $\partial\Omega$ and $p$ are smooth and $n$ is the exterior unit normal field.

*Remark* 3.1. We could require instead that the quadratic functional $Q_{\Omega, a\left(\partial_0^{-1}\right)}$ given by

$$p \mapsto \left\langle \operatorname{grad} p \right| \mathfrak{Re}\left(\partial_0 a\left(\partial_0^{-1}\right)\right) p\right\rangle_\varrho + \left\langle \mathfrak{Re}\left(\partial_0 a\left(\partial_0^{-1}\right)\right) p \right| \operatorname{grad} p\right\rangle_\varrho$$
$$+ \left\langle \operatorname{div}\left(\mathfrak{Re}\left(\partial_0 a\left(\partial_0^{-1}\right)\right)\right) p | p\right\rangle_\varrho$$

is non-negative on $H\left(\operatorname{grad}, \Omega\right)$.

Note that this functional vanishes on $H\left(\mathring{\operatorname{grad}}, \Omega\right)$ and therefore the positivity condition constitutes a boundary constraint on $a\left(\partial_0^{-1}\right)$ and on the underlying domain $\Omega$. The constraint on $\Omega$ is that the requirement $Q_{\Omega, a\left(\partial_0^{-1}\right)}\left[H\left(\operatorname{grad}, \Omega\right)\right] \subseteq \mathbb{R}_{\geq 0}$ must be non-trivial, i.e., there must be an $a\left(\partial_0^{-1}\right)$ for which this does not hold. For this surely we must have $H\left(\mathring{\operatorname{grad}}, \Omega\right) \neq H\left(\operatorname{grad}, \Omega\right)$.

**Proposition 3.2.** *Let $A$ be as given above. Then $A$ is closed, densely defined and*

$$\mathfrak{Re}\left\langle \chi_{\mathbb{R}_{<0}}\left(m_0\right) U | AU\right\rangle_\varrho \geq 0$$

*for all sufficiently large $\varrho \in \mathbb{R}_{>0}$ and all $U \in D\left(A\right)$.*

---

[3]This includes as highly special cases boundary conditions of the form $kp\,(t) - n \cdot v\,(t) = 0$ (Robin boundary condition), $k\partial_0 p\,(t) - n \cdot v\,(t) = 0$ or $kp\,(t) - n \cdot \partial_0 v\,(t) = 0$, on $\partial\Omega$, $t \in \mathbb{R}$, $k \in \mathbb{R}_{>0}$. It should be noted that in the time-independent case the above sign constraints become void since causality is not an issue anymore and in simple cases the problem is elliptic, which can be dealt with by sesqui-linear form methods, compare, e.g., [1, Section 2.4].

The general class of boundary conditions considered here in the time-dependent, time-translation invariant case covers for example cases of additional temporal convolution terms also on the boundary.

*Proof.* Any $U$ with components in $\mathring{C}_\infty(\mathbb{R}\times\Omega)$ is in $D(A)$. Note that

$$U \in D(A)$$

is equivalent to

$$\begin{pmatrix} 1 & 0 \\ -a\left(\partial_0^{-1}\right) & \partial_0^{-1}\end{pmatrix} U \in H_\varrho\left(\mathbb{R}, H(\operatorname{grad},\Omega)\oplus H\left(\mathring{\operatorname{div}},\Omega\right)\right).$$

According to (3.2) we have that

$$\begin{pmatrix} 1 & 0 \\ -a\left(\partial_0^{-1}\right) & \partial_0^{-1}\end{pmatrix} : H_\varrho(\mathbb{R}, H(\operatorname{grad},\Omega)\oplus H(\operatorname{div},\Omega)) \to H_\varrho(\mathbb{R}, H(\operatorname{grad},\Omega)\oplus H(\operatorname{div},\Omega)) \tag{3.4}$$

is a well-defined continuous linear mapping. Moreover, since multiplication by an $L^\infty(\Omega)$-multiplier and application by $\partial_0^{-1}$ does not increase the support, we have that if $\Phi$ has support in $\mathbb{R}\times K$ for some compact set $K\subseteq\Omega$ then $\begin{pmatrix} 1 & 0 \\ -a\left(\partial_0^{-1}\right) & \partial_0^{-1}\end{pmatrix}\Phi$ also has support in $\mathbb{R}\times K$. This confirms that $\mathring{C}_\infty(\mathbb{R}\times\Omega)\subseteq D(A)$ and since $\mathring{C}_\infty(\mathbb{R}\times\Omega)$ is dense in $H_\varrho\left(\mathbb{R}, L^2(\Omega)\oplus L^2(\Omega)\right)$ the operator $A$ is densely defined.

Now let $\Phi_k \overset{k\to\infty}{\to} \Phi_\infty$ and $A\Phi_k \overset{k\to\infty}{\to} \Psi_\infty$. We have first, due to the closedness of $\begin{pmatrix} 0 & \operatorname{div} \\ \operatorname{grad} & 0\end{pmatrix}$ that $\Psi_\infty = \begin{pmatrix} 0 & \operatorname{div} \\ \operatorname{grad} & 0\end{pmatrix}\Phi_\infty$. Moreover, we have from (3.4)

$$\begin{pmatrix} 1 & 0 \\ -a\left(\partial_0^{-1}\right) & \partial_0^{-1}\end{pmatrix}\Phi_k \overset{k\to\infty}{\to} \begin{pmatrix} 1 & 0 \\ -a\left(\partial_0^{-1}\right) & \partial_0^{-1}\end{pmatrix}\Phi_\infty.$$

Using (3.1), a straightforward calculation yields on $D(A)$

$$\begin{aligned}
&\begin{pmatrix} 0 & \mathring{\operatorname{div}} \\ \operatorname{grad} & 0\end{pmatrix}\begin{pmatrix} 1 & 0 \\ -a\left(\partial_0^{-1}\right) & \partial_0^{-1}\end{pmatrix} \\
&= \begin{pmatrix} 0 & \operatorname{div} \\ \operatorname{grad} & 0\end{pmatrix}\begin{pmatrix} 1 & 0 \\ -a\left(\partial_0^{-1}\right) & \partial_0^{-1}\end{pmatrix} \\
&= \begin{pmatrix} \partial_0^{-1} & 0 \\ 0 & 1\end{pmatrix}\begin{pmatrix} 0 & \operatorname{div} \\ \operatorname{grad} & 0\end{pmatrix} + \begin{pmatrix} 0 & \operatorname{div} \\ 0 & 0\end{pmatrix}\begin{pmatrix} 0 & 0 \\ -a\left(\partial_0^{-1}\right) & 0\end{pmatrix} \\
&= \begin{pmatrix} \partial_0^{-1} & 0 \\ 0 & 1\end{pmatrix}\begin{pmatrix} 0 & \operatorname{div} \\ \operatorname{grad} & 0\end{pmatrix} + \begin{pmatrix} -\operatorname{div} a\left(\partial_0^{-1}\right) & 0 \\ 0 & 0\end{pmatrix} \\
&= \begin{pmatrix} 0 & \partial_0^{-1}\operatorname{div} \\ \operatorname{grad} & 0\end{pmatrix} - \begin{pmatrix} (\operatorname{div} a)\left(\partial_0^{-1}\right) & 0 \\ 0 & 0\end{pmatrix} - \begin{pmatrix} a\left(\partial_0^{-1}\right)\cdot\operatorname{grad} & 0 \\ 0 & 0\end{pmatrix} \\
&= \begin{pmatrix} \partial_0^{-1} & -a\left(\partial_0^{-1}\right)\cdot \\ 0 & 1\end{pmatrix}\begin{pmatrix} 0 & \operatorname{div} \\ \operatorname{grad} & 0\end{pmatrix} - \begin{pmatrix} (\operatorname{div} a)\left(\partial_0^{-1}\right) & 0 \\ 0 & 0\end{pmatrix}.
\end{aligned}$$

Thus, we have

$$\begin{aligned}
\partial_0^{-1}\begin{pmatrix} 0 & \operatorname{div} \\ \operatorname{grad} & 0\end{pmatrix} U = &\begin{pmatrix} 1 & a\left(\partial_0^{-1}\right)\cdot \\ 0 & \partial_0^{-1}\end{pmatrix}\begin{pmatrix} 0 & \mathring{\operatorname{div}} \\ \operatorname{grad} & 0\end{pmatrix}\begin{pmatrix} 1 & 0 \\ -a\left(\partial_0^{-1}\right) & \partial_0^{-1}\end{pmatrix} U + \\
&+\begin{pmatrix} (\operatorname{div} a)\left(\partial_0^{-1}\right) & 0 \\ 0 & 0\end{pmatrix} U.
\end{aligned} \tag{3.5}$$

Here we have used that

$$\begin{pmatrix} \partial_0^{-1} & -a\left(\partial_0^{-1}\right)\cdot \\ 0 & 1 \end{pmatrix}^{-1} = \partial_0 \begin{pmatrix} 1 & a\left(\partial_0^{-1}\right)\cdot \\ 0 & \partial_0^{-1} \end{pmatrix}.$$

Consequently,

$$\begin{aligned}
&\begin{pmatrix} 0 & \mathring{\operatorname{div}} \\ \operatorname{grad} & 0 \end{pmatrix} \begin{pmatrix} 1 & 0 \\ -a\left(\partial_0^{-1}\right) & \partial_0^{-1} \end{pmatrix} \Phi_k \\
&= \begin{pmatrix} \partial_0^{-1} & -a\left(\partial_0^{-1}\right)\cdot \\ 0 & 1 \end{pmatrix} \begin{pmatrix} 0 & \operatorname{div} \\ \operatorname{grad} & 0 \end{pmatrix} \Phi_k - \begin{pmatrix} (\operatorname{div} a)\left(\partial_0^{-1}\right) & 0 \\ 0 & 0 \end{pmatrix} \Phi_k \\
&\overset{k\to\infty}{\to} \begin{pmatrix} \partial_0^{-1} & -a\left(\partial_0^{-1}\right)\cdot \\ 0 & 1 \end{pmatrix} \Psi_\infty - \begin{pmatrix} (\operatorname{div} a)\left(\partial_0^{-1}\right) & 0 \\ 0 & 0 \end{pmatrix} \Phi_\infty
\end{aligned}$$

and so, by the closedness of $\begin{pmatrix} 0 & \mathring{\operatorname{div}} \\ \operatorname{grad} & 0 \end{pmatrix}$

$$\begin{pmatrix} 1 & 0 \\ -a\left(\partial_0^{-1}\right) & \partial_0^{-1} \end{pmatrix} \Phi_\infty \in H_\varrho\left(\mathbb{R}, H(\operatorname{grad}, \Omega) \oplus H\left(\mathring{\operatorname{div}}, \Omega\right)\right)$$

and so

$$\Phi_\infty \in D(A).$$

Moreover, we have from (3.5)

$$\begin{aligned}
\partial_0^{-1} A \Phi_\infty = &\begin{pmatrix} 1 & a\left(\partial_0^{-1}\right)\cdot \\ 0 & \partial_0^{-1} \end{pmatrix} \begin{pmatrix} 0 & \mathring{\operatorname{div}} \\ \operatorname{grad} & 0 \end{pmatrix} \begin{pmatrix} 1 & 0 \\ -a\left(\partial_0^{-1}\right) & \partial_0^{-1} \end{pmatrix} \Phi_\infty \\
&+ \begin{pmatrix} (\operatorname{div} a)\left(\partial_0^{-1}\right) & 0 \\ 0 & 0 \end{pmatrix} \Phi_\infty.
\end{aligned} \tag{3.6}$$

We shall now show the (real) non-negativity of $A$. Assume that

$$U = \begin{pmatrix} p \\ v \end{pmatrix} \in \bigcup_{n\in\mathbb{N}} \chi_{[-n,n]}\left(\mathfrak{Im}\left(\partial_0\right)\right)[D(A)].$$

Then $U \in D(\partial_0) \cap D(A)$ and we may calculate

$$\begin{aligned}
&\mathfrak{Re}\,\langle U|AU\rangle_0 \\
&= \frac{1}{2}\left(\langle U|AU\rangle_0 + \langle AU|U\rangle_0\right) \\
&= \frac{1}{2}\left(\left(\langle p|\operatorname{div} v\rangle_0 + \langle \operatorname{grad} p|v\rangle_0\right) + \left(\langle \operatorname{div} v|p\rangle_0 + \langle v|\operatorname{grad} p\rangle_0\right)\right) \\
&= \frac{1}{2}\left(\left\langle p\middle|\mathring{\operatorname{div}}\left(v - \partial_0 a\left(\partial_0^{-1}\right)p\right)\right\rangle_0 + \left\langle p\middle|\operatorname{div} \partial_0 a\left(\partial_0^{-1}\right)p\right\rangle_0 + \langle \operatorname{grad} p|v\rangle_0\right) \\
&\quad + \frac{1}{2}\left(\left\langle \mathring{\operatorname{div}}\left(v - \partial_0 a\left(\partial_0^{-1}\right)p\right)\middle|p\right\rangle_0 + \left\langle \operatorname{div} \partial_0 a\left(\partial_0^{-1}\right)p\middle|p\right\rangle_0 + \langle v|\operatorname{grad} p\rangle_0\right) \\
&= \frac{1}{2}\left(-\left\langle \operatorname{grad} p\middle|\left(v - \partial_0 a\left(\partial_0^{-1}\right)p\right)\right\rangle_0 + \left\langle p\middle|\operatorname{div} \partial_0 a\left(\partial_0^{-1}\right)p\right\rangle_0 + \langle \operatorname{grad} p|v\rangle_0\right) \\
&\quad + \frac{1}{2}\left(-\left\langle \left(v + \partial_0 a\left(\partial_0^{-1}\right)p\right)\middle|\operatorname{grad} p\right\rangle_0 + \left\langle \operatorname{div} \partial_0 a\left(\partial_0^{-1}\right)p\middle|p\right\rangle_0 + \langle v|\operatorname{grad} p\rangle_0\right)
\end{aligned}$$

$$= \frac{1}{2}\left(\left\langle \operatorname{grad} p|\partial_0 a\left(\partial_0^{-1}\right)p\right\rangle_0 + \left\langle p|\operatorname{div}\partial_0 a\left(\partial_0^{-1}\right)p\right\rangle_0\right)$$
$$+\frac{1}{2}\left(\left\langle \partial_0 a\left(\partial_0^{-1}\right)p|\operatorname{grad} p\right\rangle_0 + \left\langle \operatorname{div}\partial_0 a\left(\partial_0^{-1}\right)p|p\right\rangle_0\right).$$

Integrating this over $\mathbb{R}_{<0}$ yields with requirement (3.3)

$$\mathfrak{Re}\left\langle \chi_{\mathbb{R}_{<0}}(m_0)\, U|AU\right\rangle_\varrho \geq 0$$

for $U \in \bigcup_{n\in\mathbb{N}} \chi_{[-n,n]}(\mathfrak{Im}(\partial_0))[D(A)]$ and so by density for every $U \in D(A)$ (and all sufficiently large $\varrho \in \mathbb{R}_{>0}$). □

We need to find the adjoint of $A$, which must be a restriction of

$$-\begin{pmatrix} 0 & \operatorname{div} \\ \operatorname{grad} & 0 \end{pmatrix}$$

and an extension of

$$-\begin{pmatrix} 0 & \mathring{\operatorname{div}} \\ \mathring{\operatorname{grad}} & 0 \end{pmatrix}.$$

We suspect it is

$$D(A^*) := \left\{\begin{pmatrix} p \\ v \end{pmatrix} \in D\left(\begin{pmatrix} 0 & \operatorname{div} \\ \operatorname{grad} & 0 \end{pmatrix}\right) \middle|\, a\left(\partial_0^{-1}\right)^* p + \left(\partial_0^{-1}\right)^* v \in H_\varrho\left(\mathbb{R}, H\left(\mathring{\operatorname{div}},\Omega\right)\right)\right\}.$$

Using (3.5) and letting $W := \partial_0 \begin{pmatrix} 1 & 0 \\ -a\left(\partial_0^{-1}\right) & \partial_0^{-1} \end{pmatrix} U = \begin{pmatrix} \partial_0^{-1} & 0 \\ a\left(\partial_0^{-1}\right) & 1 \end{pmatrix}^{-1} U$ for

$$U \in \bigcup_{n\in\mathbb{N}} \chi_{[-n,n]}(\mathfrak{Im}(\partial_0))[D(A)]$$

we have $V \in D(A^*)$ if and only if for all

$$W \in \bigcup_{n\in\mathbb{N}} \chi_{[-n,n]}(\mathfrak{Im}(\partial_0))\left[D\left(\begin{pmatrix} 0 & \mathring{\operatorname{div}} \\ \operatorname{grad} & 0 \end{pmatrix}\right)\right]$$

$$0 = \left\langle \begin{pmatrix} 1 & a\left(\partial_0^{-1}\right)\cdot \\ 0 & \partial_0^{-1} \end{pmatrix}\begin{pmatrix} 0 & \mathring{\operatorname{div}} \\ \operatorname{grad} & 0 \end{pmatrix} W|V\right\rangle_{\varrho,0,0}$$
$$+\left\langle \begin{pmatrix} (\operatorname{div} a)\left(\partial_0^{-1}\right) & 0 \\ 0 & 0 \end{pmatrix}\partial_0\begin{pmatrix} \partial_0^{-1} & 0 \\ a\left(\partial_0^{-1}\right) & 1 \end{pmatrix} W|V\right\rangle_{\varrho,0,0}$$
$$+\left\langle \begin{pmatrix} \partial_0^{-1} & 0 \\ a\left(\partial_0^{-1}\right) & 1 \end{pmatrix} W|\begin{pmatrix} 0 & \operatorname{div} \\ \operatorname{grad} & 0 \end{pmatrix} V\right\rangle_{\varrho,0,0},$$
$$= \left\langle \begin{pmatrix} 1 & a\left(\partial_0^{-1}\right)\cdot \\ 0 & \partial_0^{-1} \end{pmatrix}\begin{pmatrix} 0 & \mathring{\operatorname{div}} \\ \operatorname{grad} & 0 \end{pmatrix} W|V\right\rangle_{\varrho,0,0} + \left\langle \begin{pmatrix} (\operatorname{div} a)\left(\partial_0^{-1}\right) & 0 \\ 0 & 0 \end{pmatrix} W|V\right\rangle_{\varrho,0,0}$$
$$+\left\langle \begin{pmatrix} \partial_0^{-1} & 0 \\ a\left(\partial_0^{-1}\right) & 1 \end{pmatrix} W|\begin{pmatrix} 0 & \operatorname{div} \\ \operatorname{grad} & 0 \end{pmatrix} V\right\rangle_{\varrho,0,0},$$

$$= \left\langle \begin{pmatrix} 0 & \mathring{\operatorname{div}} \\ \operatorname{grad} & 0 \end{pmatrix} W \middle| \begin{pmatrix} 1 & 0 \\ a\left(\partial_0^{-1}\right)^* & \left(\partial_0^{-1}\right)^* \end{pmatrix} V \right\rangle_{\varrho,0,0}$$

$$+ \left\langle W \middle| \begin{pmatrix} \operatorname{div}\left(a\left(\partial_0^{-1}\right)^*\right) & 0 \\ 0 & 0 \end{pmatrix} V \right\rangle_{\varrho,0,0} +$$

$$+ \left\langle W \middle| \begin{pmatrix} \left(\partial_0^{-1}\right)^* & a\left(\partial_0^{-1}\right)^* \\ 0 & 1 \end{pmatrix} \begin{pmatrix} 0 & \operatorname{div} \\ \operatorname{grad} & 0 \end{pmatrix} V \right\rangle_{\varrho,0,0}.$$

This implies that

$$\begin{pmatrix} 1 & 0 \\ a\left(\partial_0^{-1}\right)^* & \left(\partial_0^{-1}\right)^* \end{pmatrix} V \in D\left(\begin{pmatrix} 0 & \mathring{\operatorname{div}} \\ \operatorname{grad} & 0 \end{pmatrix}\right),$$

which is the above characterization. Moreover,

$$\begin{pmatrix} 0 & \mathring{\operatorname{div}} \\ \operatorname{grad} & 0 \end{pmatrix} \begin{pmatrix} 1 & 0 \\ a\left(\partial_0^{-1}\right)^* & \left(\partial_0^{-1}\right)^* \end{pmatrix} V = \begin{pmatrix} (\operatorname{div} a)\left(\partial_0^{-1}\right)^* & 0 \\ 0 & 0 \end{pmatrix} V$$
$$+ \begin{pmatrix} 1 & a\left(\partial_0^{-1}\right)^* \\ 0 & \left(\partial_0^{-1}\right)^* \end{pmatrix} \begin{pmatrix} 0 & \operatorname{div} \\ \operatorname{grad} & 0 \end{pmatrix} V$$

which yields the analogous formula for $A^*$ as (3.5) for $A$.

**Proposition 3.3.** *Let $A$ be as given above. Then $A^*$ is closed, densely defined and*

$$\mathfrak{Re}\left\langle \chi_{\mathbb{R}_{\leq 0}}(m_0)\, V | A^* V\right\rangle_\varrho \geq 0$$

*for all sufficiently large $\varrho \in \mathbb{R}_{>0}$ and all $V \in D(A^*)$.*

*Proof.* The proof is analogous to the proof of Proposition 3.2, since $A$ and $A^*$ share a similar structure. □

**Theorem 3.4.** *Let $A$ and $M$ be as specified in this section, such that the propagation of acoustic waves is governed by the equation*

$$\left(\partial_0 M\left(\partial_0^{-1}\right) + A\right) U = f.$$

*Then, there is a $\varrho_0 \in \mathbb{R}_{>0}$ such that for every $\varrho \in \mathbb{R}_{\geq \varrho_0}$ and every given data $f \in H_\varrho\left(\mathbb{R}, L^2(\Omega) \oplus L^2(\Omega)^3\right)$ there is a unique solution $U \in H_\varrho\left(\mathbb{R}, L^2(\Omega) \oplus L^2(\Omega)^3\right)$. Moreover, the solution operator*

$$\left(\partial_0 M\left(\partial_0^{-1}\right) + A\right)^{-1} : H_\varrho\left(\mathbb{R}, L^2(\Omega) \oplus L^2(\Omega)^3\right) \to H_\varrho\left(\mathbb{R}, L^2(\Omega) \oplus L^2(\Omega)^3\right)$$

*is a causal, continuous linear operator.*
*Furthermore, the operator norm $\left\|\left(\partial_0 M\left(\partial_0^{-1}\right) + A\right)^{-1}\right\|$ is uniformly bounded with respect to $\varrho \in \mathbb{R}_{\geq \varrho_0}$.*

*Proof.* The result is immediate if we are able to show that **Condition (Positivity 1)** is satisfied. Due to Propositions 3.2 and 3.3 we only need to show that for some $\beta_0 \in \mathbb{R}_{>0}$ we have

$$\mathfrak{Re}\left\langle \chi_{\mathbb{R}_{\leq 0}}(m_0)\, U|\partial_0 M\left(\partial_0^{-1}\right) U\right\rangle_\varrho \geq \beta_0 \left\langle \chi_{\mathbb{R}_{\leq 0}}(m_0)\, U|U\right\rangle_\varrho,$$
$$\mathfrak{Re}\left\langle U|\partial_0^* M^*\left(\left(\partial_0^{-1}\right)^*\right) U\right\rangle_\varrho \geq \beta_0 \left\langle U|U\right\rangle_\varrho$$

for all $U \in D(\partial_0) = D(\partial_0^*)$.

Let us consider the first estimate:

$$\begin{aligned}
&\mathfrak{Re}\left\langle \chi_{\mathbb{R}_{\leq 0}}(m_0)\, U|\partial_0 M\left(\partial_0^{-1}\right) U\right\rangle_\varrho \\
&= \mathfrak{Re}\left\langle \chi_{\mathbb{R}_{\leq 0}}(m_0)\, U|\partial_0 M_0 U\right\rangle_\varrho + \mathfrak{Re}\left\langle \chi_{\mathbb{R}_{\leq 0}}(m_0)\, U|M_1\left(\partial_0^{-1}\right) U\right\rangle_\varrho, \\
&= \mathfrak{Re}\left\langle \chi_{\mathbb{R}_{\leq 0}}(m_0)\, \sqrt{M_0}U|\partial_0 \sqrt{M_0}U\right\rangle_\varrho \\
&\quad + \mathfrak{Re}\left\langle \chi_{\mathbb{R}_{\leq 0}}(m_0)\, U|M_1\left(\partial_0^{-1}\right)\chi_{\mathbb{R}_{\leq 0}}(m_0)\, U\right\rangle_\varrho, \\
&\geq \frac{1}{2}\int_{\mathbb{R}_{\leq 0}} \left(\partial_0 \left|\sqrt{M_0}U\right|_0^2\right)(t)\, \exp(-2\varrho t)\, dt - \mu_0 \left|\chi_{\mathbb{R}_{\leq 0}}(m_0)\, U\right|_\varrho^2, \\
&\geq \frac{1}{2}\int_{\mathbb{R}_{\leq 0}} \left(\partial_0 \exp(-2\rho m_0) \left|\sqrt{M_0}U\right|_0^2\right)(t)\, dt \\
&\quad + \varrho\int_{\mathbb{R}_{\leq 0}} \left|\sqrt{M_0}U\right|_0^2(t)\, \exp(-2\varrho t)\, dt - \mu_0 \left|\chi_{\mathbb{R}_{\leq 0}}(m_0)\, U\right|_\varrho^2, \\
&\geq \frac{1}{2}\left|\sqrt{M_0}U(0)\right|_0^2 + (\varrho\gamma_0 - \mu_0)\left|\chi_{\mathbb{R}_{\leq 0}}(m_0)\, U\right|_\varrho^2, \\
&\geq (\varrho\gamma_0 - \mu_0)\left|\chi_{\mathbb{R}_{\leq 0}}(m_0)\, U\right|_\varrho^2.
\end{aligned}$$

Similarly we obtain, using $\mathfrak{Re}\,\partial_0^* = \mathfrak{Re}\,\partial_0 = \varrho$,

$$\begin{aligned}
\mathfrak{Re}\left\langle U|\partial_0^* M^*\left(\left(\partial_0^{-1}\right)^*\right) U\right\rangle_\varrho &= \mathfrak{Re}\left\langle \sqrt{M_0}U|\partial_0^* \sqrt{M_0}U\right\rangle_\varrho \\
&\quad + \mathfrak{Re}\left\langle U|M_1^*\left(\left(\partial_0^{-1}\right)^*\right) U\right\rangle_\varrho \\
&\geq \varrho\left\langle \sqrt{M_0}U|\partial_0 \sqrt{M_0}U\right\rangle_\varrho - \mu_0 \left|\chi_{\mathbb{R}_{\leq 0}}(m_0)\, U\right|_\varrho^2 \\
&\geq (\varrho\gamma_0 - \mu_0)\, |U|_\varrho^2.
\end{aligned}$$

Thus **Condition (Positivity 1)** is satisfied with $\beta_0 := \varrho_0\gamma_0 - \mu_0$ where $\varrho_0 \in \mathbb{R}_{>0}$ is chosen such that $\varrho_0 > \frac{\mu_0}{\gamma_0}$. □

## 4. Conclusion

We have presented an extension of a Hilbert space approach to a class of evolutionary problems. As an illustration we have applied the general theory to a particular problem concerning the propagation of acoustic waves in complex media with a dynamic boundary condition.

## References

[1] R. Leis. *Initial boundary value problems in mathematical physics.* John Wiley & Sons Ltd. and B.G. Teubner; Stuttgart, 1986.

[2] R. Picard. A Structural Observation for Linear Material Laws in Classical Mathematical Physics. *Math. Methods Appl. Sci.*, 32(14):1768–1803, 2009.

[3] R. Picard. On a Class of Linear Material Laws in Classical Mathematical Physics. *Int. J. Pure Appl. Math.*, 50(2):283–288, 2009.

[4] R. Picard. An Elementary Hilbert Space Approach to Evolutionary Partial Differential Equations. *Rend. Istit. Mat. Univ. Trieste*, 42 suppl.:185–204, 2010.

[5] R. Picard and D.F. McGhee. *Partial Differential Equations: A unified Hilbert Space Approach*, volume 55 of *De Gruyter Expositions in Mathematics.* De Gruyter. Berlin, New York. 518 p., 2011. To appear.

Rainer Picard
Institut für Analysis, Fachrichtung Mathematik
Technische Universität Dresden, Germany
e-mail: `rainer.picard@tu-dresden.de`

Operator Theory:
Advances and Applications, Vol. 221, 549-570


# Maximal Semidefinite Invariant Subspaces for $J$-dissipative Operators

S.G. Pyatkov

**Abstract.** We describe some sufficient conditions for a $J$-dissipative operator in a Krein space to have maximal semidefinite invariant subspaces. The semigroup properties of the restrictions of an operator to these subspaces are studied. Applications are given to the case when an operator admits matrix representation with respect to the canonical decomposition of the space and to some singular differential operators. The main conditions are given in the terms of the interpolation theory of Banach spaces.

**Mathematics Subject Classification (2000).** Primary 47B50; Secondary 46C20; 47D06.

**Keywords.** Dissipative operator, Pontryagin space, Krein space, invariant subspace, analytic semigroup.

## 1. Introduction

In this article we consider the question of existence of invariant semidefinite invariant subspaces for $J$-dissipative operators defined in a Krein space. Recall that a Krein space (see [8]) is a Hilbert space $H$ with an inner product $(\cdot,\cdot)$ in addition endowed with an indefinite inner product of the form $[x, y] = (Jx, y)$, where $J = P^+ - P^-$ ($P^\pm$ are orthoprojections in $H$, $P^+ + P^- = I$). We put $H^\pm = R(P^\pm)$. In what follows, the symbol $I$ stands for the identity and the symbols $D(A)$ and $R(A)$ designate the domain and the range of an operator $A$. The operator $J$ is called a fundamental symmetry. The Krein space is called a Pontryagin space if $\dim R(P^+) < \infty$ or $\dim R(P^-) < \infty$ and it is denoted by $\Pi_\kappa$, where $\kappa = \min(\dim R(P^+), \dim R(P^+))$. A subspace $M$ in $H$ is said to be nonnegative (positive, uniformly positive) if the inequality $[x, x] \geq 0$ ($[x, x] > 0$, $[x, x] \geq \delta\|x\|^2$ ($\delta > 0$)) holds for all $x \in M$. Nonpositive, negative, uniformly negative subspaces in $H$ are defined in a similar way. If a nonnegative subspace $M$ admits no nontrivial nonnegative extensions, then it is called a maximal nonnegative subspace. Maximal nonpositive (positive, negative, nonnegative, etc.) subspaces in $H$ are defined by analogy. A densely defined operator $A$ is said to be dissipative (strictly dissipative, uniformly dissipative) in $H$ if $-\mathrm{Re}\,(Ax,x)\geq 0$ for all $x\in D(A)$ ($-\mathrm{Re}\,(Ax,x)>0$ for all $x\in D(A)$ or $-\mathrm{Re}\,(Ax,x)\geq\delta\|u\|^2$ ($\delta>0$) for all $x\in D(A)$). Similarly, a densely defined operator $A$ is called a $J$-dissipative (strictly $J$-dissipative or uniformly $J$-dissipative) whenever the operator $JA$ is dissipative (strictly dissipative or uniformly dissipative). A dissipative ($J$-dissipative) operator is said to be maximal dissipative (maximal $J$-dissipative) if it admits no nontrivial dissipative ($J$-dissipative) extensions. Let $A:H\to H$ be a $J$-dissipative operator. We say that a subspace $M\subset H$ is invariant under $A$ if $D(A)\cap M$ is dense in $M$ and $Ax\in M$ for all $x\in D(A)\cap M$.

The main question under consideration here is the question on existence of semidefinite (i.e., of a definite sign) invariant subspaces for a given $J$-dissipative operator in a Krein space.

The first results in this direction were obtained in the Pontryagin article [1], where is was proven that every $J$-selfadjoint operator in a Pontryagin space (let $\dim H^+=\kappa<\infty$) has a maximal nonnegative invariant subspace $M$ ($\dim \mathrm{M}=\kappa$) such that the spectrum of the restriction $L|_M$ lies in the closed upper half-plane.

After this paper, the problem on existence of invariant maximal semidefinite subspaces turned out to be a focus of an attention in the theory of operators in Pontryagin and Krein spaces. The Pontryagin results are generalized for different classes of operators in [2]–[14]. A sufficiently complete bibliography and some results are presented in [8]. Among the recent articles we note the articles [11]–[14], where the most general results were obtained. In these articles the whole space $H$ is identified with the Cartesian product $H^+\times H^-$ ($H^\pm=R(P^\pm)$) and an operator $A$ with the matrix operator $A:H^+\times H^-\to H^+\times H^-$ of the form

$$A=\begin{pmatrix} A_{11} & A_{12}\\ A_{21} & A_{22}\end{pmatrix}, \tag{1.1}$$
$$A_{11}=P^+LP^+,\ A_{12}=P^+LP^-,\ A_{21}=P^-LP^+,\ A_{22}=P^-LP^-.$$

In this case the fundamental symmetry is as follows $J=\left[\begin{smallmatrix} I & 0\\ 0 & -I\end{smallmatrix}\right]$. The basic condition of existence of a maximal nonnegative invariant subspace for an operator $L$ in [14] is the condition of compactness of the operator $A_{12}(A_{22}-\mu)^{-1}$ for some $\mu$ from the left half-plane.

This article can serve as some supplement to the article [17] (see also [16, 21, 20]) in which sufficient and necessary conditions of existence of invariant subspaces are given in the regular case when $i\mathbb{R}\subset\rho(L)$. Here we study the case when the point 0 can lie in a continuous spectrum of $L$. This case has rather general applications to the operator theory in Krein spaces. In particular, even in the case of $J$-positive $J$-selfadjoint operator there are no good sufficient conditions ensuring the existence of the semidefinite invariant subspaces of $L$. In this case the operator $L$ is definitizable and has a spectral resolution similar to that for a selfadoint operator but the spectral function has the so-called critical points (0 and $\infty$ in our case). The complete analogy is achieved in the case of regular critical points. In this case the operator is similar to a selfadjoint operator. Moreover, the existence of semidefinite invariant subspaces whose sum is the whole space is equivalent to the regularity of 0 and $\infty$ and to similarity of $L$ to a selfadjoint operator [34, Prop. 2.2]. The fundamentals of the theory of definitizable operators, the definitions, and results can be found in [28]. The first results devoted to the problem of regularity of the points 0 and $\infty$ are the articles [35, 27]. The generalizations and new results can be found in [29]–[34], [40]. However it is worth noting that the positive results were obtained for very narrow class of weight functions and differential operators. In contrast to the above-mentioned articles, our main conditions are stated in terms of the interpolation theory of Banach spaces which allows to refine many previous results. We apply the results obtained to the study of the operators represented in the form (1.1) and under certain constraints on operators weaken the condition of compactness of the operator $A_{12}(A_{22}-\mu)^{-1}$ replacing it with the conditions that the operators $A_{12}$ and $A_{21}$ are subordinate in some sense to the operators $A_{11}$ and $A_{22}$. Moreover, we exhibit some examples proving under some conditions on the functions $\omega$ and $q$ that the operator $Lu = \frac{\operatorname{sgn} x}{\omega(x)}(u_{xx} - q(x)u)$ $(x \in \mathbb{R})$ is similar to a selfadjoint operator.

## 2. Preliminaries

Given Banach spaces $X, Y$, the symbol $L(X,Y)$ denotes the space of linear continuous operators defined on $X$ with values in $Y$. If $X = Y$ then we write $L(X)$ rather than $L(X,X)$. We denote by $\sigma(A)$ and $\rho(A)$ the spectrum and the resolvent set of $A$, respectively. If $M \subset X$ is a subspace then by the restriction of $L$ to $M$ we mean the operator $L|_M : M \to X$ with the domain $D(L|_M) = D(L) \cap M$ coinciding with $L$ on $D(L|_M)$. An operator $A$ such that $-A$ is dissipative (maximal dissipative) is called accretive (maximal accretive). Hence, taking the sign into account we can say that all statements valid for an accretive operator are true for a dissipative operator as well. In what follows, we replace the word "maximal" with the letter $m$ and thus we write $m$-dissipative rather than maximal dissipative. If $A$ is an operator in a Krein space $H$ then we denote by $A^*$ and $A^c$ the adjoint operators with respect to the inner product and the indefinite inner product in $H$, respectively. The latter operator possesses the usual properties of an adjoint operator (see [8]). Let $A_0$ and $A_1$ be two Banach spaces continuously embedded into a topological linear space $E$: $A_0 \subset E$, $A_1 \subset E$. Such a pair $\{A_0, A_1\}$ is called compatible or an interpolation couple. The definition of the interpolation space $(A_0, A_1)_{\theta,q}$ can be found in [23]. Denote by $N(L)$ the kernel of an operator $L$.

We now present some facts used below.

**Proposition 2.1.** *Let $H$ be a Hilbert space (a Krein space).*

1. *A maximal dissipative ($J$-dissipative) operator $A$ is always closed and a closed operator $A$ is $m$-dissipative if and only if $\rho(A) \cap \{z : Re\, z \geq 0\} \neq \emptyset$, in this case $\mathbb{C}^+ = \{z \in \mathbb{C} : \operatorname{Re} z > 0\} \subset \rho(A)$ (see [8, Lemma 2.8]).*
2. *If $A$ is $m$-$J$-dissipative and $N(A) = \{0\}$ then $N(A^c) = \{0\}$ (see [8, Chap. 2, Sect. 2, Corollary 2.17]). Note that in [8] the authors use a different definition*

*of a $J$-dissipative operator. In this definition it is required that* $\operatorname{Im}[Au, u] \geq 0$ *for all* $u \in D(A)$.

3. *If $A$ is a maximal uniformly dissipative ($J$-dissipative) operator then* $i\mathbb{R} \in \rho(A)$ (*see* [8, Chap. 2, Sect. 2, Prop. 2.32]).
4. *If $A$ is $m$-dissipative ($m$-$J$-dissipative) then the operator $A$ is injective if and only if the subspace $R(A)$ is dense in $H$* (*see* [15, Prop. 7.0.1]).
5. *If $A$ is $m$-dissipative ($m$-$J$-dissipative) then so is the operator* $A^*$ ($A^c$) ([15, Prop. C.7.2] *or* [8, Chap. 2, Sect. 2, Prop. 2.7]).
6. *An operator $A$ is $m$-$J$-dissipative if and only if both operators $JA$ and $AJ$ are $m$-dissipative* [8, Chap. 2, Sect. 2, Prop. 2.3].
7. *If $A$ is a maximal dissipative ($J$-dissipative) operator then $A + i\omega I$ ($\omega \in \mathbb{R}$) is also maximal dissipative ($J$-dissipative) (a consequence of the definition).*

Let $H$ be a complex Hilbert space with the norm $\|\cdot\|$ and the inner product $(\cdot,\cdot)$ and let $L : H \to H$ be a closed densely defined operator. Assign $S_\theta = \{z : |\arg z| < \theta\}$ for $\theta \in (0, \pi]$ and $S_\theta = (0, \infty)$ for $\theta = 0$. Recall that $L : H \to H$ is a sectorial operator if there exists $\theta \in [0, \pi)$ such that $\sigma(L) \subset \overline{S_\theta}$, $\mathbb{C} \setminus \overline{S_\theta} \subset \rho(L)$, and, for every $\omega > \theta$, there exists a constant $c(\omega)$ such that

$$\|(L - \lambda I)^{-1}\| \leq c/|\lambda| \quad \forall \lambda \in \mathbb{C} \setminus S_\omega. \tag{2.1}$$

The quantity $\theta$ is called the sectoriality angle of $L$. Let $L$ be sectorial and injective (we do not require that $0 \in \rho(L)$). In this case completing the subspaces $D(L^k)$ and $R(L^k)$ ($k \in \mathbb{N}$, $\mathbb{N}$ is the set of positive integers) with respect to the norms $\|u\|_{D_L} = \|L^k u\|$ and $\|u\|_{R_{L^k}} = \|L^{-k}u\|$ we obtain new spaces denoted by $D_{L^k}$ and $R_{L^k}$, respectively (see [15], [24], [26]). Interpolation properties of these spaces can be found in [15]. In the case of $0 \in \rho(L)$, these spaces and their interpolation properties are described in [22] (see also [23, Sect. 1.14.3]).

**Proposition 2.2.** *Let $L : H \to H$ ($H$ is a Hilbert space) be a sectorial operator.*

1. *If $L$ is injective then so is the operator $L^{-1}$ with the same sectoriality angle and $D(L^n) \cap R(L^n)$ is dense in $D(L)$, $R(L)$, and $H$ (the claim results from Prop.* 2.3.13 *and* 2.1.1 *in* [15]).
2. *The operators $L, L^*$ are sectorial simultaneously* (*Prop.* 2.1.1 *in* [15]).
3. *The space $H$ is representable as the direct sum $H = N(L) + Y$ ($Y = \overline{R(L)}$), the operator $L|_Y : Y \to Y$ is sectorial, $(\lambda I - L)^{-1}(x+y) = \frac{1}{\lambda}x + (\lambda I - L|_Y)^{-1}y$ ($x \in N(L), y \in Y$), and $N(L) = N(L^k)$ ($\forall k \in \mathbb{N}$)* (*see Prop.* 2.1.1 *and Sect.* 2.1 *in* [15]).
4. *If $L$ is an injective sectorial operator then so is the operator $L^{-1}$ with the same sectoriality angle and $\lambda(\lambda + L^{-1})^{-1} = I - \frac{1}{\lambda}(\frac{1}{\lambda} + L)^{-1}$ for $-1/\lambda \in \rho(L)$* (*Prop.* 2.1.1 *in* [15]).

**Proposition 2.3.** *Let $L$ be an injective $m$-dissipative operator. Then $-L$ is sectorial with $\theta = \pi/2$ and*

$$(D_L, R_L)_{1/2,2} = H. \tag{2.2}$$

The sectoriality with $\theta = \pi/2$ results from [15, Sect. 7.1.1]). The equality (2.2) follows from Theorems 2.2 and 4.2 in [24] and the arguments after Theorem 2.2 (see also Sect. 7.3.1, Theorem 7.3.1, and the equality (7.18) in [15]).

Let $H_1, H$ be a pair of compatible Hilbert spaces and $H_1 \cap H$ is densely embedded into $H$ and $H_1$. The symbol $(\cdot,\cdot)$ stands for the inner product in $H$. Define the negative space $H_1'$ constructed on this pair as the completion of $H \cap H_1$ with respect to the norm

$$\|u\|_{H_1'} = \sup_{v \in H_1 \cap H} |(u,v)|/\|v\|_{H_1}.$$

**Proposition 2.4.** *The spaces $H_1$, $H_1'$ are dual to each other with respect to the pairing $(\cdot,\cdot)$ and*

$$(H_1, H_1')_{1/2,2} = H. \tag{2.3}$$

*Thus the space of antilinear continuous functionals over $H_1$ can be identified with $H_1'$ and the norm in $H_1$ is equivalent to the norm $\sup_{v \in H_1'} |(v,u)|/\|v\|_{H_1'}$.*

*Proof.* Note that this proposition is well known in the case of $H_1 \subset H$ (see, for instance, [25]). Under our conditions, there exists a positive selfadjoint operator $S : H \to H$ such that $D_S = H_1$ (see the proof of Theorem 6.1 in [24]). In this case $R_S = H_1'$ and the claim easily follows from Proposition 2.3. □

If $L : H \to H$ is a sectorial operator then we can construct a functional calculus for $L$ ([15], [36]). What we need is a functional calculus a little bit different from that in [15]. This generalization of the conventional calculus is presented in [24, Sect. 8]. Assign $S_\theta^- = \{z : |\arg z| > \pi - \theta\}$, $S_\theta^+ = \{z : |\arg z| < \theta\}$ and $S_\theta^0 = S_\theta^+ \cup S_\theta^-$ ($\theta \in [0,\pi/2)$). An operator $L : H \to H$ is referred to as an operator of type $S_\theta^0$ if there exists $\theta \in [0,\pi/2)$ such that $\sigma(L) \subset \overline{S_\theta^0}$, $\mathbb{C} \setminus \overline{S_\theta^0} \subset \rho(L)$, and, for every $\omega \in (\theta, \pi/2)$, there exists a constant $c(\omega)$ such that

$$\|(L - \lambda I)^{-1}\| \le c/|\lambda| \quad \forall \lambda \in \mathbb{C} \setminus S_\omega^0. \tag{2.4}$$

**Remark 2.1.** Observe that if an operator $L$ is of type $S_\omega^0$ then the operators $\pm iL$ are sectorial with the sectoriality angle $\pi - \omega$ and $L^2$ is sectorial with the sectoriality angle $2\omega$ (Prop. 8.1 in [24]).

Fix an operator $L$ of type $S_\theta^0$. Given $\omega > \theta$, the classes $\Psi(S_\omega^0)$ and $F(S_\omega^0)$ comprise functions $f(z)$ analytic on $S_\omega^0$ and such that, for some $\alpha > 0$ and a constant $c > 0$,

$$|f(z)| \le c \min(|z|^\alpha, |z|^{-\alpha}) \quad \forall z \in S_\omega^0 \tag{2.5}$$

and

$$|f(z)| \le c(|z|^\alpha + |z|^{-\alpha}) \qquad \forall z \in S_\omega^0, \tag{2.6}$$

respectively. The class $H(S_\omega^0)$ consists of the functions $f(z)$ analytic and bounded on $S_\omega^0$. Given $f \in \Psi(S_\omega^0)$, for some $\omega' \in (\theta, \omega)$, we put

$$f(L) = \frac{1}{2\pi i} \int_{\Gamma_{\omega'}} f(z)(L - zI)^{-1}\, dz, \ \ \Gamma_{\omega'} = \Gamma_{\omega'}^+ \cup \Gamma_{\omega'}^-, \ \Gamma_{\omega'}^\pm = \partial S_{\omega'}^\pm,$$

where the direction of integration is clockwise on both boundaries $\partial S^+_{\omega'}$ and $\partial S^-_{\omega'}$. It is easy to check the operator $f(L): H \to H$ is bounded. Let now $f(z) \in F(S^0_\omega)$. Put $e(\lambda) = \lambda^2(1+\lambda^2)^{-2}$. Obviously, in this case there exists $k \in \mathbb{N}$ such that $e^k(\lambda)f(\lambda) \in \Psi(S^0_\omega)$. By definition (see [36], [24]),

$$D(f(L)) = \{x \in H: \ (e^k f)(L)x \in D(e^{-k}(L)) = D(L^{2k}) \cap R(L^{2k})\}$$

and $f(L)x = e^{-k}(L)(e^k f(L)x)$ for $x \in D(f(L))$. As is noted in [24], this functional calculus possesses the conventional properties. We note that for $x \in D(L^{2k}) \cap R(L^{2k})$ we have $f(L)x = (e^k f)(L)e^{-k}(L)x$. If $f \in H(S^0_\omega)$ then we can take $k=1$ and $e^{-1}(L) = L^2 + 2I + L^{-2}$. Thus, for $u \in D(L^2) \cap R(L^2)$, we have

$$f(L) = \frac{1}{2\pi i}\int_{\Gamma_{\omega'}} \frac{z^2 f(z)}{(1+z^2)^2}(L - zI)^{-1}(L+L^{-1})^2 u\,dz, \quad \Gamma_{\omega'} = \Gamma^+_{\omega'} \cup \Gamma^-_{\omega'}. \tag{2.7}$$

If $f(L) \in L(H)$ for every $f \in H(S^0_\omega)$ then we say that $L$ has a bounded $H_\infty$ calculus. In this case $f(L)$ is an extension of the above integral operator defined for $u \in D(L^2) \cap R(L^2)$. In particular, we can define the operators $\operatorname{sgn} L$, $\chi^\pm(L)$ for functions

$$f(z) = \begin{cases} 1, & z \in S^+_\omega \\ -1, & z \in S^-_\omega \end{cases}, \ \chi^+(z) = (1+f(z))/2, \ \chi^-(z) = (1-f(z))/2. \tag{2.8}$$

**Proposition 2.5.** ([24, Sect. 8]) *Let $L$ be an operator of type $S^0_\theta$ ($\theta \in [0,\pi/2)$. Then $L$ has a bounded $H_\infty$ calculus if and only if the equality* (2.2) *holds.*

Observe that the operators $\chi^\pm(L)$ possess the property $(\chi^\pm(L))^2 u = \chi^\pm(L)u$ ($u \in D(L^4) \cap R(L^4)$) which is established with the help of the residue theorem.

Let $L: H \to H$ be a strictly $m$-$J$-dissipative operator in the Krein space $H$ with the indefinite inner product $[\cdot,\cdot] = (J\cdot,\cdot)$ and the norm $\|\cdot\|$, where $J$ is the fundamental symmetry and the symbol $(\cdot,\cdot)$ designates the inner product in $H$. We use these notations below in all statements of Section 2. Define the quantities

$$\|u\|^2_{F_1} = -\mathrm{Re}\,[Lu,u] \ \ (u \in D(L)), \ \ \|u\|^2_{F_{-1}} = -\mathrm{Re}\,[L^{-1}u,u] \ \ (u \in R(L)).$$

Suppose also that

$$\exists c > 0: \ |[Lu,v]| \le c\|u\|_{F_1}\|v\|_{F_1} \ \ \forall u,v \in D(L). \tag{2.9}$$

Obviously, the quantity $\|u\|_{F_1}$ is a norm on $D(L)$.

**Proposition 2.6.** *Let $L: H \to H$ be a strictly $m$-$J$-dissipative operator satisfying* (2.9) *with a nonempty resolvent set. Then so is the operator $L^{-1}$ and $D(L^m) \cap R(L^m)$ is dense in $H$, $D(L)$, and $R(L)$ for every $m = 1,2,\dots$. Thus we have that*

$$\exists c > 0: \ \ |[L^{-1}u,v]| \le c\|u\|_{F_{-1}}\|v\|_{F_{-1}}. \tag{2.10}$$

*Proof.* An operator $L$ is $m$-$J$-dissipative and thereby the operators $JL$ and $LJ$ are $m$-dissipative (Prop. 2.1). Obviously, if $L$ is closed then so is $L^{-1}$. In this case, $\rho(JL) \neq \emptyset$ and $\rho(LJ) \neq \emptyset$ (Prop. 2.1) and they are both sectorial with the sectoriality angle $\pi/2$ (see Prop. 2.3). In this case $\rho(L^{-1}J) \neq \emptyset$ and $\rho(JL^{-1}) \neq \emptyset$ (see the claim 4 of Prop. 2.2). Hence, both these operators are $m$-dissipative. In this case, $L^{-1}$ is $m$-$J$-dissipative (see the claim 6 of Prop. 2.1). To prove density in $H$, we assume the contrary. In this case, there exists an element $u \in H$ such that $[u, v] = 0$ for all $v \in D(L^m) \cap R(L^m)$. Take $v = L^m(L-\lambda)^{-2m}\psi$, with $\psi \in H$ and $\lambda \in \rho(L)$. Since an element $\psi$ is arbitrary, we obtain that $(L^c)^m(L^c - \overline{\lambda})^{-2m}u = 0$ and, therefore (see Prop. 2.1) $u = 0$. To prove the density in $D(L)$ (or $R(L)$), we argue as follows. Given $f = (L - \lambda I)u$ $(u \in D(L))$, we can construct an approximation $f_n \in D(L^m) \cap R(L^m)$ of $f$ and assign $u_n = (L-\lambda)^{-1}f_n \to u \in D(L)$. It is immediate that $u_n \in D(L^m) \cap R(L^m)$. Similar arguments are used in the case of $R(L)$. Now we check (2.9) for $L^{-1}$. Indeed, let $u, v \in R(L)$. There exist $u_0, v_0 \in D(L)$ such that $u = Lu_0, v = Lv_0$. In this case,

$$\begin{aligned} |[-L^{-1}u, v]|^2 &= |[-u_0, Lv_0]|^2 \le c\mathrm{Re}\,[-Lu_0, u_0]\mathrm{Re}\,[-Lv_0, v_0]) \\ &= c\mathrm{Re}\,[-L^{-1}u, u]\mathrm{Re}\,[-L^{-1}v, v]). \end{aligned}$$ □

Define the spaces $F_1$ and $F_{-1}$ as completions of $D(L) \cap R(L)$ with respect to the norms $\|\cdot\|_{F_1}$ and $\|\cdot\|_{F_{-1}}$, respectively.

**Proposition 2.7.** *Let $L : H \to H$ be a strictly $m$-$J$-dissipative operator satisfying (2.9) with a nonempty resolvent set. Then the spaces $F_{-1}$ and $F_1$ are compatible, the subspaces $F_1 \cap H$ and $F_{-1} \cap H$ are dense in $H$ and $F_1$, in $H$ and $F_{-1}$, respectively. The operators $J : H \cap F_{\mp 1} \to H$ admit extensions to isomorphisms of $F_{-1}$ onto $F_1'$ and $F_1$ onto $F_{-1}'$, respectively, where $F_1'$ and $F_{-1}'$ are negative spaces constructed on the pairs $(F_1, H)$ and $(F_{-1}, H)$.*

*Proof.* Define the space $G_1$ as completion of $D(L) \cap R(L)$ with respect to the norm

$$\|u\|_{G_1}^2 = \mathrm{Re}\,[-Lu, u] + \mathrm{Re}\,[-L^{-1}u, u] + \|u\|^2.$$

Conventionally (see the proof of the second claim of Prop. 7.3.4 in [15]), using (2.9) and (2.10) we can prove that $G_1$ can be identified with a dense subspace of $H$. Construct the space $G_{-1}$ as completion of $D(L) \cap R(L)$ with respect to the norm

$$\|u\|_{G_{-1}} = \sup_{v \in G_1} |[u, v]| / \|v\|_{G_1}.$$

Due to the definition, we have the natural imbedding $G_1 \subset F_1 \cap F_{-1}$. Demonstrate the estimates $\|u\|_{G_{-1}} \le c\|u\|_{F_1}$ and $\|u\|_{G_{-1}} \le c\|u\|_{F_{-1}}$ for $u \in D(L) \cap R(L)$. Indeed, we have that

$$\sup_{v \in G_1} \frac{|[u, v]|}{\|v\|_{G_1}} = \sup_{v \in D(L) \cap R(L)} \frac{|[Lg, v]|}{\|v\|_{G_1}} \le c \sup_{v \in D(L) \cap R(L)} \frac{\|g\|_{F_1}\|v\|_{F_1}}{\|v\|_{G_1}} \le c\|u\|_{F_{-1}},$$

where $g = L^{-1}u$. Similarly, we conclude that

$$\sup_{v \in G_1} |[u, v]| / \|v\|_{G_1} \le c\|u\|_{F_1}.$$

These two estimates and the definitions imply that $F_1 + F_{-1} \subset G_{-1}$. The statements about the density in the formulation are obvious in view of fact that the subspace $D(L)\cap R(L)$ is dense in $F_1$, $F_{-1}$, and $H$. We now prove that the norm $\|Ju\|_{F_1'}$ is equivalent to the norm $\|u\|_{F_{-1}}$ for $u\in D(L)\cap R(L)$. Indeed, we have

$$\|u\|_{F_{-1}}=\frac{\operatorname{Re}[u,-L^{-1}u]}{\|u\|_{F_{-1}}}\le\sup_{v\in D(L)\cap R(L)}\frac{|[u,L^{-1}v]|}{\|v\|_{F_{-1}}}=\sup_{g\in F_1}\frac{|[u,g]|}{\|g\|_{F_1}}=\|Ju\|_{F_1'}.$$

Similarly,

$$\|Ju\|_{F_1'}=\sup_{v\in D(L^2)\cap R(L^2)}\frac{|[u,v]|}{\|v\|_{F_1}}\le\sup_{g\in D(L)\cap R(L)}\frac{|[u,L^{-1}g]|}{\|g\|_{F_{-1}}}\le c\|u\|_{F_{-1}}.$$

The same arguments are used to prove the equivalence of the norms $\|u\|_{F_1}$ and $\|Ju\|_{F_{-1}'}$. □

**Remark 2.2.** In view of the above arguments, we can define in $F_{-1}$ and $F_1$ the following equivalent norms

$$\|u\|_{F_{-1}}=\sup_{v\in D(L)\cap R(L)}|[u,v]|/\|v\|_{F_1},\quad \|u\|_{F_1}=\sup_{v\in D(L)\cap R(L)}|[u,v]|/\|v\|_{F_{-1}}.$$

Moreover, the expression $[u,v]$ is defined for $u\in F_1$ and $v\in F_{-1}$ and we have the inequality

$$|[u,v]|\le c_1\|u\|_{F_{-1}}\|v\|_{F_1},$$

with $c_1$ a constant independent of $u,v$.

Given a strictly $m$-$J$-dissipative operator $L$ satisfying the condition (2.9), we can construct the spaces $D_L$ and $R_L$ using the same definition as in the case of a sectorial operator. Using the same ideas as those in the proof of the previous proposition, we can establish that these spaces are compatible (they are subspaces of $(D(L)\cap R(L))'$).

**Lemma 2.1.** *Let $L:H\to H$ be a strictly $m$-$J$-dissipative operator satisfying* (2.9) *with a nonempty resolvent set and such that*

$$(F_1,F_{-1})_{1/2,2}=H. \tag{2.11}$$

*Then $L$ is of type $S^0_\omega$ for some $\omega\in(0,\pi/2)$.*

*Proof.* In view of the definition we have that $\|Lu\|_{F_{-1}}=\|u\|_{F_1}$. Therefore, the operator $L$ admits an extension to an isomorphism of $F_1$ onto $F_{-1}$. Denote this extension by $\tilde L$. We can pass to the limit in the expression $[Lu,v]$ (see Remark 2.2) and thereby it is defined for $u,v\in F_1$. Consider the operator $\tilde L|_H:H\to H$ with the domain $D(\tilde L|_H)=\{u\in F_1\cap H:\ \tilde Lu\in H\}$. This operator is obviously an extension of the operator $L$ and we have $-\operatorname{Re}[\tilde L|_Hu,u]=\|u\|^2_{F_1}\ge 0$ for all $u\in D(\tilde L|_H)$. Therefore, $\tilde L|_H$ is $J$-dissipative. In view of the maximality of $L$ we have $L=\tilde L|_H$ and, in particular,

$$D(L)=\{u\in F_1\cap H:\ \tilde Lu\in H\}. \tag{2.12}$$

Given $u\in D(L)\cap R(L)$, we put

$$Lu-i\omega u=f,\quad \omega\in\mathbb{R},\ \omega\neq 0. \tag{2.13}$$

From the definition it follows that $f \in R(L)$. As a consequence of (2.13), we have

$$-\mathrm{Re}\,[Lu, u] = -\mathrm{Re}\,[f, u] \leq c\|f\|_{F_{-1}}\|u\|_{F_1}. \tag{2.14}$$

In view of (2.13), we obtain the estimate

$$|\omega|^2\|u\|^2_{F_{-1}} \leq 2\|f\|^2_{F_{-1}} + 2\|Lu\|^2_{F_{-1}} \leq 2\|f\|^2_{F_{-1}} + 2\mathrm{Re}\,[-Lu, u]. \tag{2.15}$$

Multiplying this inequality by 1/4 and summing it with (2.14) we infer

$$\|u\|^2_{F_1} + |\omega|^2\|u\|^2_{F_{-1}} \leq 2c\|f\|_{F_{-1}}\|u\|_{F_1} + \|f\|^2_{F_{-1}}/2. \tag{2.16}$$

Using the inequality $|ab| \leq |a|^2/4 + |b|^2$ on the right-hand side of (2.16) we conclude

$$-\mathrm{Re}\,[Lu, u] + |\omega|^2\|u\|^2_{F_{-1}} \leq (4c^2 + 1)\|f\|^2_{F_{-1}} + \|u\|^2_{F_1}/2$$

and thereby

$$-\mathrm{Re}\,[Lu, u] + |\omega|^2\|u\|^2_{F_{-1}} \leq (8c^2 + 2)\|f\|^2_{F_{-1}}. \tag{2.17}$$

This estimate implies that $N(L - i\omega I) = \{0\}$, since $N(L - i\omega I) \subset D(L) \cap R(L)$. Therefore (see Prop. 2.1), the subspace $R(L - i\omega I)$ is dense in $H$. In this case the subspace $R((L - i\omega I)|_{R(L)\cap D(L)}) \subset H \cap F_{-1}$ is dense in $F_{-1}$. Moreover, we have the obvious estimate $\|(L - i\omega I)u\|_{F_{-1}} \leq (\|u\|_{F_1} + |\omega|\|u\|_{F_{-1}})$. This estimate and the estimate (2.17) ensure that the extension $\tilde{L} - i\omega I : F_1 \cap F_{-1} \to F_{-1}$ of the operator $L - i\omega I|_{D(L)\cap R(L)}$ has a a bounded inverse, i.e., this extension $\tilde{L} - i\omega I$ is an isomorphism of $F_1 \cap F_{-1}$ onto $F_{-1}$. Note that the equality (2.11) yields $F_1 \cap F_{-1} \subset H$. We can consider the operator $\tilde{L}$ as an unbounded operator from $F_{-1}$ into $F_{-1}$ with the domain $D(\tilde{L}) = F_1 \cap F_{-1}$. As we have proven, $i\mathbb{R}\setminus\{0\} \subset \rho(\tilde{L})$ and in view of (2.17)

$$|\omega|\|(\tilde{L} - i\omega I)^{-1}f\|_{F_{-1}} \leq c_1\|f\|_{F_{-1}} \quad \forall f \in F_{-1}. \tag{2.18}$$

In this case for $f \in D(\tilde{L})$, we obtain

$$|\omega|\|(\tilde{L} - i\omega I)^{-1}f\|_{F_1} = |\omega|\|\tilde{L}(\tilde{L} - i\omega I)^{-1}f\|_{F_{-1}} \leq c_1\|\tilde{L}f\|_{F_{-1}} = c_1\|f\|_{F_1}. \tag{2.19}$$

Since $F_1 \cap F_{-1}$ is dense in $F_1$ we can conclude that the operator $\omega(\tilde{L} - i\omega I)^{-1}$ is extensible to an operator of the class $L(F_1)$. In accord with (2.11), $\omega(\tilde{L} - i\omega I)^{-1}|_H \in L(H)$, i.e., there exists a constant $c > 0$ such that

$$|\omega|\|(\tilde{L} - i\omega I)^{-1}|_H f\|_H \leq c_1\|f\|_H \quad \forall f \in F_1 \cap F_{-1}. \tag{2.20}$$

Let $f \in H$. Approximating $f$ by a sequences $f_n \in F_1 \cap F_{-1}$ in the norm of $H$ we can construct a sequence of solutions $u_n \in F_1 \cap F_{-1}$ to the equation

$$\tilde{L}u_n - i\omega u_n = f_n \in F_1 \cap F_{-1} \subset H.$$

The estimate (2.20) implies that $u_n \to u \in H$ in the norm of $H$. Since $\mathrm{Re}\,[\tilde{L}(u_n - u_m), u_n - u_m] = \mathrm{Re}\,[f, u_n - u_m]$, we can conclude that $u_n \to u \in F_1$ in the space $F_1$ and

$$\tilde{L}u - i\omega u = f \in H.$$

In view of (2.12) we derive that $u \in D(L)$. We have proven that $i\mathbb{R} \setminus \{0\} \subset \rho(L)$ and the inequality (2.20) holds. The conventional arguments (see the claim a) of

Prop. 2.1.1 in [15]) validate the existence of a constant $\omega \in (0,\pi/2)$ such that the estimate
$$|\lambda|\|(L-\lambda I)^{-1}f\| \le c_1\|f\|$$
is valid for some constant $c_1$ and $\lambda \in \mathbb{C}\setminus S^0_\omega$. □

**Lemma 2.2.** *If $L$ is a strictly m-J dissipative operator satisfying* (2.9) *with a nonempty resolvent set then the conditions* (2.2) *and* (2.11) *are equivalent.*

*Proof.* Let (2.2) hold. We have $\|u\|^2_{F_1} = \operatorname{Re}[-Lu,u] \le \|Lu\|\|u\|$. The last inequality means that $F_1 \subset J(1/2, D_L, H)$ (see the definition of this class in [23, Sect. 1.10.1]). Similarly, we derive that $F_{-1} \subset J(1/2, R_L, H)$, i.e., $\|u\|^2_{F_{-1}} \le \|L^{-1}u\|\|u\|$. But (2.2) yields the inequality
$$\|u\|_H \le c\|u\|^{1/2}_{D_L}\|u\|^{1/2}_{R_L},\ u \in D_L\cap R_L.$$
Combining this inequality with the previous inequalities, we infer
$$\|u\|_{F_1} \le c\|u\|^{3/4}_{D_L}\|u\|^{1/4}_{R_L},\quad \|u\|_{F_{-1}} \le c\|u\|^{1/4}_{D_L}\|u\|^{3/4}_{R_L}.$$
Therefore, $F_1 \subset J(1/4, D_L, R_L)$ and $F_{-1} \subset J(3/4, D_L, R_L)$. Applying the reiteration theorem [23, Sect. 1.10.2] (see also Remark 1 after the theorem), we infer
$$H = (D_L, R_L)_{1/2,2} \subset (F_1, F_{-1})_{1/2,2}.$$
By the duality theorem [23, Sect. 1.11.2],
$$(F_1', F_{-1}')_{1/2,2} \subset H' = H,$$
where the duality is defined by the inner product in $H$. Proposition 2.7 ensures that $F_1' = JF_{-1}$ and $F_{-1}' = JF_1$, i.e.,
$$(JF_{-1}, JF_1)_{1/2,2} \subset H.$$
As a consequence,
$$(F_{-1}, F_1)_{1/2,2} \subset JH = H.$$
We arrive at the inverse containment and thereby the equality (2.11) holds.

Assume now that the equality (2.11) holds. By Lemma 2.1, the operator $L$ is of type $S^0_\omega$ for some $\omega \in (0,\pi/2)$. The equality (2.11) and the reiteration theorem [23] imply that
$$(F_1, H)_{1-\theta,2} = (F_1, F_{-1})_{\theta_1,2},\quad (H, F_{-1})_{\theta,2} = (F_1, F_{-1})_{\theta_2,2},$$
where $\theta_1 = (1-\theta)/2$ and $\theta_2 = (1+\theta)/2$. As a consequence, we infer
$$((F_1, H)_{1-\theta,2}, (H, F_{-1})_{\theta,2})_{1/2,2} = H. \tag{2.21}$$
As in the proof of Lemma 2.2, we have $F_1 \subset J(1/2, D_L, H)$ and $F_{-1} \subset J(1/2, R_L, H)$. Involving the reiteration theorem, we infer
$$(D_L, H)_{1-\frac{\theta}{2},2} \subset (F_1, H)_{1-\theta,2},\quad (H, R_L)_{\frac{\theta}{2},2} \subset (H, F_{-1})_{\theta,2}. \tag{2.22}$$
On the other hand, Theorem 4.2 and the remark after this theorem in [24] ensure the equalities
$$(D_L, H)_{\theta,2} = (D_L, R_L)_{\theta_1},\quad (H, R_L)_{\theta,2} = (D_L, R_L)_{\theta_2},\ \theta_1 = \theta/2,\quad \theta_2 = (1-\theta)/2.$$

The reiteration theorem yields the equality

$$((D_L, H)_{1-\frac{\theta}{2},2}, (H, R_L)_{\frac{\theta}{2},2})_{1/2,2} = (D_L, R_L)_{1/2,2}.$$

In this case, the relations (2.22) and (2.21) imply that $(D_L, R_L)_{1/2,2} \subset H$. Consider the operator $L^c$ with the same properties. Similar arguments validate the embedding $(D_{L^c}, R_{L^c})_{1/2,2} \subset H$. But $D_{L^c} = JD_{L^*}$ and $R_{L^c} = JR_{L^*}$ and thereby

$$(D_{L^*}, R_{L^*})_{1/2,2} \subset JH = H.$$

Next, we can refer to Theorem 2.2 in [24], where we take $L^*$ rather than $L'$. In accord with this theorem

$$(D_L, R_L)_{1/2,2} = H. \qquad \square$$

Next we present some sufficient conditions ensuring (2.2) and (2.11).

**Lemma 2.3 ([24, Corollary 5.5]).** *Suppose that $L$ is an injective operator of type $S^0_\omega$ and there exist $s \in (0,1)$ such that*

$$(H, D_L)_{s,2} = (H, D_{L^*})_{s,2}. \tag{2.23}$$

*Then the equality* (2.2) *holds.*

**Lemma 2.4.** *Suppose that $L$ is a strictly m-$J$-dissipative operator satisfying* (2.9) *with a nonempty resolvent set in a Krein space $H$ and there exists $s_0 \in (0,1)$ such that $J \in L(F_{s_0})$ $(F_s = (F_1, H)_{1-s,2})$. Then the equalities* (2.11) *and* (2.2) *hold.*

*Proof.* Proposition 2.7 ensures that $F_1' = JF_{-1}$ and $F_{-1}' = JF_1$. Thus $J \in L(F_{-1}, F_1') \cap L(H)$ and $J \in L(F_1, F_{-1}') \cap L(H)$. Given $s \in (0,1)$, assign $F_s = (F_1, H)_{1-s,2}$, $F_{-s} = (F_{-1}, H)_{1-s,2}$, $\tilde{F}_s = (F_{-1}', H)_{1-s,2}$, and $\tilde{F}_{-s} = (F_1', H)_{1-s,2}$. The duality theorem [23, Sect. 1.11.2] yields the equalities $F_s' = \tilde{F}_{-s}$ and $F_{-s}' = \tilde{F}_s$. The above properties of $J$ imply that $J \in L(F_s, \tilde{F}_s)$ and $J \in L(F_{-s}, \tilde{F}_{-s})$. It is immediate from the condition of the lemma that $F_{s_0} = \tilde{F}_{s_0}$. By the duality theorem [23, Sect. 1.11.2], $F_{-s_0} = \tilde{F}_{-s_0}$. In view of Proposition 2.4, we have

$$H = (F_{s_0}, F_{s_0}')_{1/2,2} = (F_{s_0}, \tilde{F}_{-s_0})_{1/2,2} = (F_{s_0}, F_{-s_0})_{1/2,2}.$$

From this equality and the definitions of the spaces $F_{\pm s}$ it follows that

$$\|u\| \le c\|u\|_{F_{s_0}}^{1/2}\|u\|_{F_{-s_0}}^{1/2} \le c_1\|u\|_{F_1}^{s/2}\|u\|_{F_{-1}}^{s/2}\|u\|^{1-s}, \quad \forall u \in F_1 \cap F_{-1}.$$

Dividing both parts by $\|u\|^{1-s}$ we arrive at the inequality

$$\|u\| \le c_2\|u\|_{F_1}^{1/2}\|u\|_{F_{-1}}^{1/2}.$$

Using this inequality, we obtain

$$\|u\|_{F_{s_0}} \le c\|u\|_{F_1}^{s_0}\|u\|^{1-s_0} \le c_1\|u\|_{F_1}^{(1+s_0)/2}\|u\|_{F_{-1}}^{(1-s_0)/2}, \ u \in F_1 \cap F_{-1}.$$

Similarly, we conclude that

$$\|u\|_{F_{-s_0}} \le c \le \|u\|_{F_1}^{(1-s_0)/2}\|u\|_{F_{-1}}^{(1+s_0)/2}.$$

Appealing to the reiteration theorem again, we infer

$$(F_1, F_{-1})_{1/2,2} \subset (F_{s_0}, F_{-s_0})_{1/2,2} = H.$$

By duality, we establish that

$$H \subset (F_1, F_{-1})'_{1/2,2} = (F'_1, F'_{-1})_{1/2,2}.$$

Therefore,

$$H \subset (JF_{-1}, JF_1)_{1/2,2}.$$

As a consequence, $H = JH \subset (F_{-1}, F_1)_{1/2,2}$. We arrive at the converse containment and thereby $H = (F_1, F_{-1})_{1/2,2}$. □

**Corollary 2.1.** *Any of the embeddings*

$$H \subset (F_1, F_{-1})_{1/2,2}, \quad (F_1, F_{-1})_{1/2,2} \subset H$$

*imply the equality* (2.11).

This statement was actually proven at the ends of the proofs of the Lemmas 2.4 and 2.2.

Suppose that $L$ is a strictly $m$-$J$-dissipative operator in a Krein space $H$ which satisfies (2.9). Define the space $\tilde{G}_1$ as the completion of $D(L)$ with respect to the norm $\|u\|^2_{\tilde{G}_1} = -\mathrm{Re}[Lu, u] + \|u\|^2_H$ and the space $\tilde{G}_{-1}$ as the completion of $H$ with respect to the norm $\|u\|_{\tilde{G}_{-1}} = \sup_{v\in\tilde{G}_1} |[u, v]|/\|v\|_{\tilde{G}_1}$. Assign $\tilde{G}_s = (G_1, H)_{1-s,2}$.

**Lemma 2.5.** *Suppose that $L : H \to H$ ($H$ is a Krein space) is a strictly $m$-$J$-dissipative operator satisfying* (2.9) *and there exists a constant $m > 0$ such that*

$$\|u\|^2 \le m(-\mathrm{Re}[Lu, u] + \|u\|^2_{\tilde{G}_{-1}}) \quad \forall u \in D(L). \tag{2.24}$$

*Then there exists a number $\omega_0 \ge 0$ such that $I_{\omega_0} = \{i\omega : |\omega| \ge \omega_0\} \subset \rho(L)\}$. The inequality* (2.24) *certainly holds whenever*

$$(\tilde{G}_1, \tilde{G}_{-1})_{1/2,2} = H. \tag{2.25}$$

*If there exists $s \in (0, 1)$ such that $J \in L(\tilde{G}_s)$ then the equality* (2.25) *holds.*

The first part of the lemma is proven in [17]. Its proof is quite similar to the proof of Lemma 2.1. So we omit it. The proof of the last claim can be found also in [19, Theorem 5] (see also [21, 20]). Moreover, its proof is quite similar to the proof of Lemma 2.4.

## 3. Main results

The theorem below supplements the results in [17, 16] (see also [20, Chap. 1, Theorem 4.1] and [21]. In view of claim 3 of Proposition 2.2, we can reduce the case of $N(L) \ne \{0\}$ to the case $N(L) = \{0\}$. So we assume below that the operator $L$ is injective.

**Theorem 3.1.** *Let $L : H \to H$ be a strictly $m$-$J$-dissipative operator in a Krein space $H$ with a nonempty resolvent set. If the equality* (2.2) *holds and either $L$*

*is of type $S^0_\theta$ for some $\theta \in (0, \pi/2)$ or $L$ satisfies the condition* (2.9) *then there exist maximal nonnegative and maximal nonpositive or maximal uniformly positive and uniformly negative, respectively, $L$-invariant subspaces $H^+$ and $H^-$ such that $H = H^+ + H^-$, $D(L) = D(L) \cap H^+ + D(L) \cap H^-$ (both sums are direct), $\sigma(L|_{H^\pm}) \subset \mathbb{C}^\mp$, and the operators $\pm L|_{H^\pm}$ are generators of analytic semigroups. If there exist uniformly positive and uniformly negative $L$-invariant subspaces $H^+$ and $H^-$ such that $H = H^+ + H^-$ and $D(L) = D(L) \cap H^+ + D(L) \cap H^-$ (the sums are direct) then the equality* (2.2) *holds.*

*Proof.* In both cases the operator $L$ is of type $S^0_\theta$ for some $\theta \in (0, \pi/2)$. Let $\theta < \omega' < \omega < \pi/2$. By Proposition 2.5, $L$ has a bounded $H_\infty$ calculus and thereby the operators $\chi^\pm(L)$ defined by the equalities (2.7) and (2.8) are bounded. For $u \in D(L^2) \cap R(L^2)$ we have

$$\chi^\pm u = \chi^\pm(L)u = -\frac{1}{2\pi i}\int_{\Gamma^\pm_{\omega'}} \frac{z^2}{(1+z^2)^2}(L - zI)^{-1}(L + L^{-1})^2 u\, dz, \tag{3.1}$$

where the integration over $\Gamma^\pm_{\omega'}$ is counterclockwise. As it is easy to see, the integrals are normally convergent for $u \in D(L^2) \cap R(L^2)$ and thus the quantities $\chi^\pm u$ are defined correctly. The operators $\chi^\pm$ are extensible to operators of the class $L(H)$ and it is not difficult to establish that $(\chi^\pm)^2 = \chi^\pm$. We put $P^\pm = \chi^\mp$. Using the definitions, we establish that $(P^+ + P^-)u = u$ for all $u \in D(L^4) \cap R(L^4)$ and thus for all $u \in H$. Thereby the space $H$ is representable as the direct sum $H = H^+ + H^-$, with $H^\pm = \{u \in H : P^\pm u = u\}$. Demonstrate that $H^+$ and $H^-$ are nonnegative and nonpositive subspaces, respectively. For example, we consider $H^+$. For every $u \in H^+$, there exists a sequence $u_n \in D(L^4) \cap R(L^4) \cap H^+$ such that $\|u_n - u\| \to 0$ as $n \to \infty$. Indeed, find a sequence $v_n \in D(L^6) \cap R(L^6)$ (see Prop. 2.2, Prop. 2.6, and Remark 2.1) such that $\|v_n - u\| \to 0$ as $n \to \infty$. Put $u_n = P^+ v_n \in D(L^4) \cap R(L^4) \cap H^+$. In this case $\|u_n - u\| = \|P^+(v_n - u)\| \to 0$ as $n \to \infty$. Define an operator

$$Pu = -\frac{1}{2\pi i}\int_{\Gamma^-_{\omega'}} e^{zt}\frac{z^2}{(1+z^2)^2}(L - zI)^{-1}(L + L^{-1})^2 u\, dz, \quad t > 0. \tag{3.2}$$

Take $v_n(t) = Pu_n$. Employing the normal convergence of the integrals obtained from (3.2) by the formal differentiation with respect to $t$ and the formal application of $L$, we can say that $v_n(t) \in L_2(0, \infty; D(L))$ and the distributional derivative $v'(t)$ possesses the property $v_n'(t) \in L_2(0, \infty; H)$. Hence, after a possible modification on a set of zero measure $v_n(t) \in C([0, \infty); H)$, i.e., $v_n(t)$ is a continuous function with values in $H$. It is immediate that

$$v_n(0) = u_n, \tag{3.3}$$

$$Sv_n = v_n' - Lv_n = 0. \tag{3.4}$$

Using the equalities (3.3) and (3.4) and integrating by parts, we infer

$$0 = \operatorname{Re}\int_0^\infty [Sv_n(t), v_n(t)]\, dt = -[v_n(0), v_n(0)]_0 - \int_0^\infty \operatorname{Re}\,[Lv_n, v_n](\tau)\, d\tau. \tag{3.5}$$

Therefore, we conclude that

$$[u_n, u_n]_0 = -\int_0^\infty \mathrm{Re}\,[Lv_n, v_n](\tau)\,d\tau, \tag{3.6}$$

i.e., $[u_n, u_n] \geq 0$. Passing to the limit on $n$ in this inequality, we derive $[u, u] \geq 0$, i.e., $H^+$ is a nonnegative subspace. By analogy, we can prove that $H^-$ is a nonpositive subspace. The maximality of $H^+$ and $H^-$ follows from Proposition 1.25 in [8]. Now we assume that the condition 2 holds. Demonstrate that the subspaces $H^\pm$ are uniformly definite. Consider the case $H^+$. In view of (3.4) and (3.6), we conclude that

$$[u_n, u_n] \geq \delta_0(\|v_n'\|^2_{L_2(0,\infty;F_{-1})} + \|v_n\|^2_{L_2(0,\infty;F_1)}), \tag{3.7}$$

where $\delta_0$ is a positive constant independent of $n$. Applying (3.7) to the difference $u_n - u_m$ and using the fact that $u_n$ is a Cauchy sequence in $H$, we derive that the sequence $v_n$ is a Cauchy sequence in $L_2(0,\infty;F_1)$ and thereby it converges to some function $v(t) \in L_2(0,\infty;F_1)$ and also $v_n'(t) \to v'(t)$ in $L_2(0,\infty;F_{-1})$. The trace theorem (see [23, Theorem 1.8.3]) and the inequality (2.11) (see Lemma 2.2) imply that $v_n(0) \to v(0)$ in $H$ and there exists a constant $\delta > 0$ such that

$$\|v_n'\|^2_{L_2(0,\infty;F_{-1})} + \|v_n\|^2_{L_2(0,\infty;F_1)} \geq \delta\|u_n\|^2_H.$$

This inequality and (3.7) imply the estimate $[u_n, u_n] \geq \delta_1\|u_n\|^2_H/2$, $\delta_1 > 0$. Passing to the limit on $n$ in this estimate we arrive at the inequality $[u, u] \geq \delta_1\|u\|^2_H/2$ valid for all $u \in H^+$ which ensures the uniform positivity of $H^+$. Similarly, we can prove that $H^-$ is a uniformly negative subspace. We now prove the last claim. Let $\mathrm{Re}\,\lambda \leq 0$ and $\lambda \neq 0$. Assume that $u \in H^-$. As before we can approximate this function by a sequence $u^n \in D(L^2) \cap R(L^2) \cap H^-$ such that $\|u^n - u\| \to 0$ as $n \to \infty$. Consider the sequence

$$v^n = -\frac{1}{2\pi i}\int_{\Gamma^+_{\omega'}} \frac{z^2\lambda}{(1+z^2)^2(z-\lambda)}(L - zI)^{-1}(L + L^{-1})^2 u^n\,dz, \quad t > 0. \tag{3.8}$$

The function $\lambda\chi^+(z)/(z-\lambda) \in H(S^0_\omega)$. Applying the operator $\chi^+(L)$ to (3.8), we can verify that $\chi^+ v^n = v^n$, i.e., $v^n \in H^-$. Since the operator $L$ has a bounded $H_\infty$ calculus, we have the estimates

$$\|v^n\| \leq c\|u^n\|, \quad \|v^n - v^m\| \leq c\|u^n - u^m\|$$

and thereby the sequence $v^n$ has a limit $v \in H^-$ in $H$. Applying the operator $L - \lambda I$ to (3.8) and using the Cauchy theorem, we infer

$$(L - \lambda I)v^n = \lambda\chi^+(L)u^n = \lambda u^n.$$

This equality yields the convergence $v^n \to v$ in $D(L)$ (with the graph norm). Passing to the limit we arrive at the equality $(L - \lambda I)v = \lambda u$, $\mathrm{Re}\,\lambda \leq 0$, $\lambda \neq 0$. Thus, for every $u \in H^-$, there exists $v \in D(L) \cap H^-$ such that $(L - \lambda I)v = \lambda u$ and we have the estimate

$$|\lambda|\|(L - \lambda I)^{-1}u\| \leq c\|u\| \quad \forall u \in H^-. \tag{3.9}$$

Moreover, $N(L|_{H^-} - \lambda I) = \{0\}$. Indeed, it suffices to consider the case of $\operatorname{Re}\lambda < 0$. If $(L - \lambda I)v = 0$ then $-[Lv, v] = -\lambda[v, v]$ and thereby $[v, v] > 0$. But $v \in H^-$, a contradiction. Therefore, $\mathbb{C}^- \subset \rho(L|_{H^-} - \lambda I)$ and (3.9) holds. This means that $-L|_{H^-}$ is a generator of an analytic semigroup [26]. Similar arguments can be applied to the operator $L|_{H^+}$.

Define the projections $P^\pm$ onto $H^\pm$, respectively, corresponding to the decomposition $H = H^+ + H^-$. Thus, $P^+ + P^- = I$, $P^+P^- = P^-P^+ = 0$, $P^\pm u = u$ for all $u \in H^\pm$. Let $\lambda \in \rho(L)$. In view of the equality $D(L) = D(L) \cap H^+ + D(L) \cap H^-$, it is easy to find that $(L - \lambda I)^{-1}\varphi \in H^\pm \cap D(L)$ for $\varphi \in H^\pm$. As a consequence, we have that the operators $\pm L|_{H^\pm}$ are closed operators with nonempty resolvent sets. Since the subspaces $H^\pm$ are uniformly definite, the expressions $(u, v)_\pm = \pm[u, v]$ are equivalent inner products on $H^\pm$, respectively. Using Proposition 2.1 we can easily find that the operators $\pm L|_{H^\pm}$ are $m$-dissipative with respect to these inner products. Involving the decomposition of $D(L)$ once more, we infer

$$P^\pm Lu = LP^\pm u \ (u \in D(L)), \quad P^\pm L^{-1}u = L^{-1}P^\pm u \ (u \in R(L)). \tag{3.10}$$

Put $H_1 = D_L$, $H_{-1} = R_L$, $H_1^\pm = D_{L|_{H^\pm}}$ and $H_{-1}^\pm = R_{L|_{H^\pm}}$. As a consequence of (3.10), $P^\pm$ are extensible to operators of the class $L(H_{\pm 1})$ and we have the equalities $H_1 = H_1^+ + H_1^-$, $H_{-1} = H_{-1}^+ + H_{-1}^-$. In this case, we infer

$$P^\pm \in L(H_{1-2\theta}) \ \forall \theta \in (0, 1), \ H_{1-2\theta} = (H_1, H_{-1})_{\theta,2}$$

and Theorem 1.17.1 in [23] implies that

$$H_{1-2\theta}^\pm = (H_1^\pm, H_{-1}^\pm)_{\theta,2} = \{u \in H_{1-2\theta} : \ P^\pm u = u\} = P^\pm H_{1-2\theta}.$$

In particular, we have that

$$H_{1-2\theta} = P^+ H_{1-2\theta} + P^- H_{1-2\theta} = H_{1-2\theta}^+ + H_{1-2\theta}^- \ \text{(the sum is direct)}. \tag{3.11}$$

Consider the operators $L|_{H^\pm} : H^\pm \to H^\pm$. The operators $L|_{H^\pm}$ are injective and $m$-dissipative. By Proposition 2.3, $H^+ = (H_1^+, H_{-1}^+)_{1/2,2}$ and $H^- = (H_1^-, H_{-1}^-)_{1/2,2}$. In this case the equality (2.2) results from (3.11) with $\theta = 1/2$. □

**Remark 3.1.** Assume that $L : H \to H$ is a $J$-selfadjoint operator in a Krein space $H$ with a nonempty resolvent set. Let $L$ be $J$-nonpositive, i.e., $-[Lu, u] \geq 0$ for all $u \in D(L)$. In this case it is $m$-$J$-dissipative and the condition (2.9) holds. First, we assume that $L$ is $J$-negative. In this case it is strictly $J$-dissipative. We put $H_1 = D(L)$, $G_1 = R(L)$ with $\|u\|_{H_1} = \|Lu\| + \|u\|$ and $\|u\|_{G_1} = \|L^{-1}u\| + \|u\|$. Given $\lambda \in \rho(L)$ and $\mu \in \rho(L^{-1})$, denote be $H_{-1}$ and $G_{-1}$ the completions of $H$ with respect to the norms

$$\|u\|_{H_{-1}} = \|(L - \lambda I)^{-1}u\|, \quad \|u\|_{G_{-1}} = \|(L^{-1} - \mu I)^{-1}u\|.$$

In this situation the regularity of the critical point infinity is connected with the condition $(H_1, H_{-1})_{1/2,2} = H$ and the regularity of the critical point 0 with the condition $(G_1, G_{-1})_{1/2,2} = H$. The condition (2.2) ensures both of them as it follows from Proposition 11.1 in [24]. The case of $N(L) \neq \{0\}$ can be reduced to the case of $N(L) = \{0\}$ and $L$ is $J$-negative. We should consider the operator $L$

as an operator from $\overline{R(L)}$ into $\overline{R(L)}$. If the subspace $N(L)$ is nondegenerate in $H$ (see the definitions in [8]) and finite-dimensional then $\overline{R(L)} = N(L)^{[\perp]}$, where $N(L)^{[\perp]}$ is the $J$-orthogonal complement to $N(L)$.

We now present some corollaries in the case when an operator $L : H \to H$ ($H$ is a Krein space with the fundamental symmetry $J = P^+ - P^-$) is representable in the form (1.1). In this case, the whole space $H$ with an inner product $(\cdot,\cdot)$ and the norm $\|\cdot\|$ is identified with the Cartesian product $H^+ \times H^-$ ($H^\pm = R(P^\pm)$) and the operator $L$ with a matrix operator $L : H^+ \times H^- \to H^+ \times H^-$ of the form

$$L = \begin{pmatrix} A_{11} & A_{12} \\ A_{21} & A_{22} \end{pmatrix}, \tag{3.12}$$
$$A_{11} = P^+LP^+,\ A_{12} = P^+LP^-,\ A_{21} = P^-LP^+,\ A_{22} = P^-LP^-.$$

The fundamental symmetry can be written in the form $J = \left[\begin{smallmatrix} I & 0 \\ 0 & -I \end{smallmatrix}\right]$. Let $u = (u^+, u^-) \in H$, $v = (v^+, v^-) \in H$. The inner product $(u, v)$ in $H$ is written as $(u^+, v^+) + (u^-, v^-)$ ($u^\pm, v^\pm \in H^\pm$) and an indefinite inner product as $[u, v] = (Ju, v) = (u^+, v^+) - (u^-, v^-)$. Describe the conditions of Theorem 3.1 in this case. Put $L_0 = \left(\begin{smallmatrix} A_{11} & 0 \\ 0 & A_{22} \end{smallmatrix}\right)$.

**Theorem 3.2.** *Let $L : H \to H$ be a strictly $m$-$J$-dissipative operator of type $S_\theta^0$ for some $\theta \in (0, \pi/2)$ in a Krein space $H$ such that $D(L) = D(L_0)$, $R(L) = R(L_0)$ and there exists constants $m, M > 0$ such that, for all $u \in D(L) \cap R(L)$,*

$$m\|Lu\| \le \|L_0 u\| \le M\|Lu\|,\quad m\|L^{-1}u\| \le \|L_0^{-1}u\| \le M\|L^{-1}u\|.$$

*Then there exist maximal nonnegative and maximal nonpositive subspaces $M^\pm$ invariant under $L$. The whole space $H$ is representable as the direct sum $H = M^+ + M^-$, $\sigma(L|_{M^\pm}) \subset \mathbb{C}^\mp$, and the operators $\pm L|_{M^\pm}$ are generators of analytic semigroups.*

*Proof.* Note that the condition of the theorem implies that the operators $A_{11} : H^+ \to H^+$ and $-A_{22} : H^- \to H^-$ are strictly $m$-dissipative. To refer to Theorem 3.1, it suffices to prove that the interpolation equality (2.2) holds. By condition, $D_L = D_{L_0} = H_1 = H_1^+ \times H_1^-$, where $H_1^+ = D_{A_{11}}$ and $H_1^- = D_{A_{22}}$. Similarly we have that the space $H_{-1} = R_L$ coincides with $R_{L_0}$, i.e., with the space $H_{-1}^+ \times H_{-1}^-$, where $H_{-1}^+ = R_{A_{11}}$ and $H_{-1}^- = R_{A_{22}}$. Proposition 2.3 implies that $(H_1^+, H_{-1}^+)_{1/2,2} = H^+$ and $(H_1^-, H_{-1}^-)_{1/2,2} = H^-$. Applying Theorem 1.17.1 of [23] yields $(H_1, H_{-1})_{1/2,2} = (H_1^+, H_{-1}^+)_{1/2,2} \times (H_1^-, H_{-1}^-)_{1/2,2}$ and thereby $(H_1, H_{-1})_{1/2,2} = H^+ \times H^- = H$. □

In the next theorem we assume that $A_{11}, -A_{22}$ are strictly $m$-dissipative operators satisfying the conditions

$$\begin{aligned} \exists c > 0 :\ &|(A_{11}u^+, v^+)| \le c\|u^+\|_{F_1^+}\|v^+\|_{F_1^+}, \\ &|(A_{22}u^-, v^-)| \le c\|u^-\|_{F_1^-}\|v^-\|_{F_1^-} \end{aligned} \tag{3.13}$$

for all $u^{\pm} \in H_1^+ = D(A_{11}) \cap R(A_{11})$, $v^{\pm} \in H_1^- = D(A_{22}) \cap R(A_{22})$, where

$$\|u\|^2_{F_1^+} = -\mathrm{Re}\,(A_{11}u, u),\ \|u\|^2_{F_1^-} = \mathrm{Re}\,(A_{22}u, u).$$

We denote by $F_1^{\pm}$ the completions of $H_1^+$ and $H_1^-$ with respect to the norms $\|\cdot\|_{F_1^+}$ and $\|\cdot\|_{F_1^-}$, respectively. By $F_{-1}^{\pm}$, we mean the negative spaces constructed on the pairs $F_1^{\pm}, H^{\pm}$, where the norms can be defined as

$$\|u\|^2_{F_{-1}^+} = -\mathrm{Re}\,((A_{11})^{-1}u, u),\ \|u\|^2_{F_{-1}^-} = \mathrm{Re}\,((A_{22})^{-1}u, u).$$

We also assume that the operators $A_{12}$ and $A_{21}$ are subordinate to the operators $A_{11}$ and $A_{22}$ in the following sense: $H_1^- \subset D(A_{12})$, $H_1^+ \subset D(A_{21})$, and

$$\exists c > 0:\ \|A_{12}u^-\|_{F_{-1}^+} \le c\|u^-\|_{F_1^-},\ \ \|A_{21}u^+\|_{F_{-1}^-} \le c\|u^+\|_{F_1^+}; \tag{3.14}$$

$$\exists c_0 > 0:\ \|u^+\|^2_{F_1^+} + \|u^-\|^2_{F_1^-} \le c_0\,\mathrm{Re}[-L\vec{u}, \vec{u}] \tag{3.15}$$

for all $\vec{u} = (u^+, u^-) \in H_1 = H_1^+ \times H_1^-$.

Let $\|u\|_{\tilde{G}_1} = \mathrm{Re}[-L\vec{u}, \vec{u}] + \|u^+\|^2 + \|u^-\|^2$. The space $\tilde{G}_{-1}$ is defined as the completion of $H$ with respect to the norm $\|u\|_{\tilde{G}_{-1}} = \sup_{v \in \tilde{G}_1} |[u, v]|/\|v\|_{\tilde{G}_1}$ (see Lemma 2.5).

**Theorem 3.3.** *Let $L : H \to H$ be a strictly m-J-dissipative operator in a Krein space $H$ satisfying the conditions* (3.13)–(3.15). *Then there exist maximal uniformly positive and maximal uniformly negative subspaces $M^{\pm}$ invariant under $L$. The whole space $H$ is representable as the direct sum $H = M^+ + M^-$, $\sigma(L|_{M^{\mp}}) \subset \mathbb{C}^{\pm}$, and the operators $\pm L|_{M^{\pm}}$ are generators of analytic semigroups.*

*Proof.* The conditions (3.15), (3.14) imply that the norm in the space $F_1$ ($\|u\|^2_{F_1} = -\mathrm{Re}\,[Lu, u]$) is equivalent to the norm $\|u\|^2_{F_1} = \|u^+\|^2_{F_1^+} + \|u^-\|^2_{F_1^-}$ ($u = (u^+, u^-)$). The definitions of the space $F_{-1}$ and the indefinite inner product in $H$ ensure that $F_{-1} = F_{-1}^+ \times F_{-1}^-$. Proposition 2.4 and Theorem 1.17.1 in [23] yields $(F_1, F_{-1})_{1/2,2} = (F_1^+, F_{-1}^+)_{1/2,2} \times (F_1^-, F_{-1}^-)_{1/2,2} = H^+ \times H^- = H$. Similarly, using the conditions (3.15), (3.14) we can prove that $(\tilde{G}_1, \tilde{G}_{-1})_{1/2,2} = H$. Now the claim results from Theorem 3.1 and Lemma 2.5. □

We now present some simple applications of Theorem 3.1. We take

$$Lu = \frac{\mathrm{sgn}\, x}{\omega(x)}(u_{xx} - q(x)u),\ x \in \mathbb{R}.$$

This operator was treated, for example, in [33] for some special functions $\omega$ and $q$. We can refer also to [29]–[34], [40], where similar problem were examined. We consider this operator imposing rather general constraints on these functions which are assumed to be real valued.

First we suppose that

(A) $\omega, q \in L_{1,\mathrm{loc}}(\mathbb{R})$, $\omega > 0$ and $q \ge 0$ almost everywhere on $\mathbb{R}$, and $\omega \notin L_1(\mathbb{R})$.
(B) there exists a constant $\delta > 0$ such that $q(x) + \omega(x) \ge \delta/(1 + x^2)$.

We take $H = L_{2,\omega}(\mathbb{R})$. This space is endowed with the inner product $(u,v) = \int_{\mathbb{R}} \omega(x)u(x)\overline{v(x)}\,dx$ and the corresponding norm. Furthermore, let

$$D(L) = \{u \in L_{2,\omega+q}(\mathbb{R}) :\ u_x \in L_2(\mathbb{R}),\ u_{xx} \in L_{1,\mathrm{loc}}(\mathbb{R}),\ Lu \in L_{2,\omega}(\mathbb{R})\},$$

where the symbols $u_x, u_{xx}$ stand for the generalized derivatives in the Sobolev sense. The space $H$ with the fundamental symmetry $Ju = \operatorname{sgn} x u$ and the indefinite inner product $[u,v] = (Ju,v)$ is a Krein space.

**Lemma 3.1.** *Under the conditions* (A), (B), *the operator* $L : H \to H$ *is* $J$-*selfadjoint and* $J$-*negative.*

*Proof.* First we prove that $\mathbb{R}^+ = \{\lambda :\ \lambda > 0\} \subset \rho(JL)$. Consider the equation

$$JLu - \lambda u = f \in L_{2,\omega}(\mathbb{R}). \tag{3.16}$$

To prove its solvability we examine the equality

$$a(u,v) = \int_{\mathbb{R}} u_x\overline{v_x} + (q(x) + \lambda\omega(x))u(x)\overline{v(x)}\,dx = -\int_{\mathbb{R}} \omega(x)f(x)\overline{v(x)}\,dx. \tag{3.17}$$

By definition, the space $\tilde{F}_1$ comprises the functions $u \in L_{2,q+\omega}(\mathbb{R})$ such that $u_x \in L_2(\mathbb{R})$ and it is endowed with the norm $\|u\|^2_{\tilde{F}_1} = \|u_x\|^2_{L_2(\mathbb{R})} + \|u\|^2_{L_{2,q+\omega}(\mathbb{R})}$. The form $a(u,v)$ possesses the following property: there exist constants $m, M > 0$ such that $m\|u\|^2_{\tilde{F}_1} \le \operatorname{Re} a(u,v) \le M\|u\|^2_{\tilde{F}_1}$. The Lax-Milgram theorem (see [15, Theorem C.5.3]) implies that there exists a function $u \in \tilde{F}_1$ such that $a(u,v) = -(f,v)$ for all $v \in \tilde{F}_1$. This equality and the definition of a generalized derivative ensures the existence of the generalized derivative $u_{xx} \in L_{1,loc}(\mathbb{R})$ and the equality

$$-u_{xx} + qu + \lambda\omega u = -f\omega.$$

Hence, $Lu \in H$ and $u \in D(L)$. Prove that $JL$ is symmetric. It suffices to establish that

$$\int_{\mathbb{R}} u_{xx}(x)\overline{v(x)}\,dx = \int_{\mathbb{R}} u\overline{v_{xx}(x)}\,dx \quad \forall u,v \in D(L). \tag{3.18}$$

Since $u_{xx} = \omega \operatorname{sgn} x Lu + qu$, we have $u_{xx}\overline{v} \in L_1(\mathbb{R})$. Let $\varphi \in C_0^\infty(\mathbb{R})$, $\varphi(x) = 1$ for $x \in (-1,1)$, and $\varphi(x) = 0$ for $|x| \ge 2$. Consider the expression

$$I = \int_{\mathbb{R}} \varphi(x/R)u_{xx}(x)\overline{v(x)}\,dx \quad (R > 0).$$

Integrating by parts, we infer

$$I = \int_{\mathbb{R}} \varphi(x/R)u\overline{v_{xx}} + 2\frac{1}{R}\varphi'(x/R)u\overline{v_x} + \frac{1}{R^2}\varphi''(x/R)u\overline{v}\,dx. \tag{3.19}$$

The second and third integrals vanish as $R \to \infty$. Consider, for example, the second integral. We have

$$\Big|\int_{\mathbb{R}} 2\frac{1}{R}\varphi'(x/R)u\overline{v_x}\,dx\Big| \le c\Big(\int_{R\le|x|\le 2R} \frac{1}{R^2}|u|^2\,dx\Big)^{1/2}\Big(\int_{R\le|x|\le 2R} |v_x|^2\,dx\Big)^{1/2}$$

In view of (B), for all $x$ such that $R \le |x| \le 2R$ we have the inequality $1/R^2 \le c(q(x)+\omega(x))$, where $c$ is a constant independent of $R$. Using this inequality we obtain that

$$\int_{R\le|x|\le 2R} \frac{1}{R^2}|u|^2\,dx \le c\int_{R\le|x|\le 2R} (q+\omega)|u|^2\,dx \to 0$$

as $R\to\infty$ due to the inclusion $u \in D(L)$. We also have the convergence

$$\int_{R\le|x|\le 2R} |v_x|^2\,dx \to 0 \text{ as } R\to\infty$$

because of the inclusion $v \in D(L)$. The first integral in (3.19) tends to $\int_{\mathbb{R}} u\overline{v_{xx}}\,dx$, since $u\overline{v_{xx}} \in L_1(\mathbb{R})$. Passing to the limit in (3.19), we arrive at the claim. We have proven that the operator $JL$ is selfadjont. As a consequence $L$ is $J$-selfadjoint. □

The following lemma follows from results of the articles [37, 38, 39] (see also [18]).

**Lemma 3.2.** *Let $f(\eta) = \int_0^\eta \omega(\tau)\,d\tau$. Assume that one of the following conditions holds:*

(a) $\exists\beta\in(0,1)$ $\exists\omega\in(0,1)$ *such that* $\forall\varepsilon\in(0,1)$ $f(\omega\varepsilon)\le\beta f(\varepsilon)$;
(b) *there exist constants* $c,d>0$ *such that*

$$f(\eta) \le c\Big(\frac{\eta}{\xi}\Big)^d f(\xi) \quad \forall \eta,\xi: \quad 0<\eta\le\xi<1.$$

*Then there exist* $\theta\in(0,1)$ *such that*

$$(\overset{\circ}{W}{}_2^1(0,1), L_{2,\omega}(0,1))_{1-\theta,2} = (\tilde{W}_2^1(0,1), L_{2,\omega}(0,1))_{1-\theta,2},$$

*where* $\tilde{W}_2^1(0,1) = \{u\in W_2^1(0,1):\ u(1)=0\}$.

Now we can state the main result. For simplicity, we assume in addition that

$$\text{(C)} \quad \mu(\{x\in\mathbb{R}: q(x)\ne 0\})>0,$$

where the symbol $\mu$ stands for the Lebesgue measure. The case of $q(x)\equiv 0$ requires a separate discussion.

**Theorem 3.4.** *Let the conditions* (A)–(C) *hold. Assume also that the conditions of Lemma* 3.2 *are fulfilled either for the function $\omega(x)$ or for the function $\omega(-x)$. Then $L: H\to H$ is similar to a selfadjoint operator.*

*Proof.* To validate the claim we can apply Lemma 2.4. As in the proof of Lemma 3.1 we can justify the equality

$$-[Lu,u] = \int_{\mathbb{R}} |u_x|^2 + q|u|^2\,dx, \quad u\in D(L).$$

The expression on the right-hand side is exactly the norm in the space $F_1$ (see the definition before Lemma 2.6). It is not difficult to establish that the space $F_1$ comprises the functions in $W^1_{2,\mathrm{loc}}(\mathbb{R})$ such that $\int_{\mathbb{R}} q|u|^2\,dx<\infty$. In view of the conditions (A) and (C), $N(L)=0$. Next, we can construct a function $\varphi(x)\in C^\infty(\mathbb{R})$

such that $\varphi(x) = 1$ for $x \in (-\infty, 1/2)$, $\varphi(x) = 0$ for $x \in (3/4, \infty)$, and $0 \le \varphi(x) \le 1$ for all $x$. Put $\psi(x) = 1 - \varphi(x)$. Define maps $R_0 : F_1 \to \tilde{W}_2^1(0,1)$ and $R_1 : F_1 \to F_1$ by the equalities $R_0 u = \varphi u$ and $R_1 u = \psi u$. Obviously, $R_1 \in L(F_1) \cap L(H)$ and thereby $R_1 \in L(F_s)$ for every $s \in [0,1]$ (recall that $F_s = (F_1, H)_{1-s,2}$). Similarly, we obtain that $R_0 \in L(F_s, \tilde{W}_2^s(0,1))$, with $\tilde{W}_2^s(0,1) = (\tilde{W}_2^1(0,1), L_2(0,1))_{1-s,2}$. Define an operator $S$ by the equality $Su = u$ for $x \in (0,1)$ and $Su = 0$ for $x \notin (0,1)$. Obviously, $S \in L(\overset{\circ}{W}{}_2^1(0,1), F_1)$ and $S \in L(L_{2,\omega}(0,1), H)$. Therefore, $S \in L(F_s^0, F_s)$, with $F_s^0 = (\overset{\circ}{W}{}_2^1(0,1), L_{2,\omega}(0,1))_{1-\theta,2}$. By Lemma 3.2, for $s = \theta$ we have that the operator $SR_0 + R_1$ belongs to $L(F_\theta)$. On the other hand, $(SR_0 + R_1)u = u$ for $x > 0$ and $(SR_0 + R_1)u = 0$ for $x \le 0$. Thus the operator of multiplication by the characteristic function $\chi_{\mathbb{R}^+}$ is continuous in $F_\theta$ which fact implies that $J \in L(F_\theta)$. Applying Lemma 2.4 validates the equalities (2.11) and (2.2). It remains to establish that $\rho(L) \neq \emptyset$. We use Lemma 2.5. The space $\tilde{G}_1$ coincides with the space $\tilde{F}_1$ from Lemma 3.1. By the same arguments as those above we can establish that there exists $s \in (0,1)$ such that $J \in \tilde{G}_s$ and thus the statement of Lemma 2.5 holds. Therefore, the conditions of Theorem 3.1 are fulfilled and there exist maximal uniformly positive and uniformly negative invariant subspaces of $L$ whose direct sum is the whole space $H$. Thereby, the critical points 0 and infinity of $L$ are both regular and the operator $L$ is similar to a selfadjoint operator in $H$ (see [34, Prop. 2.2]). □

## References


[1] Pontryagin, L.S. Hermitian operators in spaces with indefinite metric. Izv. Akad. Nauk SSSR, Ser. Mat. **8**, 243–280 (1944).

[2] Krein, M.G. On an application of the fixed point principle in the theory of linear transformations of spaces with an indefinite metric. Uspekhi Mat. Nauk. **50**, no. 2(36), 180–190 (1950).

[3] Krein, M.G. A new application of the fixed-point principle in the theory of operators in a space with indefinite metric. Dokl. Akad. Nauk SSSR. **154**, no. 5, 1023–1026 (1964).

[4] Krein, M.G. and Langer, G.K. The defect subspaces and generalized resolvents of a Hermitian operator in the space $\Pi_\kappa$. Functional Anal. Appl. **5**, no. 2, 126–146; no. 3, 217–228 (1971).

[5] Langer, G.K. On $J$-Hermitian operators. Dokl. Akad. Nauk SSSR **134**, no. 2, 263–266 (1960).

[6] Langer, H. Eine Verallgemeinerung eines Satzes von L.S. Pontrjagin. Math. Ann. **152**, no. 5, 434–436 (1963).

[7] Langer, H. Invariant subspaces for a class of operators in spaces with indefinite metric. J. Funct. Anal. **19**, no. 3, 232–241 (1975).

[8] Azizov, T.Ya. and Iokhvidov, I.S. Foundations of the Theory of Linear Operators in Spaces with Indefinite Metric (in Russian). Nauka, Moscow (1986).

[9] Azizov, T.Ya. Invariant subspaces and criteria for the completeness of the system of root vectors of $J$-dissipative operators in the Pontrjagin space. Soviet Math. Dokl. **200**, no. 5, 1513–1516 (1971).

[10] Azizov, T.Ya. Dissipative operators in a Hilbert space with indefinite metric. Izv. Akad. Nauk SSSR, Ser. Mat. **37**, no. 3, 639–662 (1973).

[11] Azizov, T.Ya. and Khatskevich, V.A. A theorem on existence of invariant subspaces for $J$-binoncontractive operators. Recent Advances in Operator Theory in Hilbert and Krein Spaces. Birkhäuser Verlag, Basel-Boston-Berlin, 41–49 (2009).

[12] Azizov, T.Ya. and Gridneva, I.V. On invariant subspaces of J-dissipative operators. Ukrainian Math. Bulletin. **6**, no. 1, 1–13 (2009).

[13] Shkalikov, A.A. On invariant subspaces of dissipative operators in a space with an indefinite metric. Proc. Steklov Inst. Math. **248**, no. 1, 287–296 (2005).

[14] Shkalikov, A.A. Dissipative Operators in the Krein Space. Invariant Subspaces and Properties of Restrictions. Functional Analysis and Its Applications. **41**, no. 2, 154–167 (2007).

[15] Haase, M. The Functional Calculus for Sectorial Operators. Operator Theory: Adv. and Appl. **169**, Birkhäuser Verlag, Basel-Boston-Berlin (2006).

[16] Pyatkov, S.G. Maximal semidefinite invariant subspaces for some classes of operators. Conditionally Well-Posed Problems. TVP/TSP, Utrecht, 336–338 (1993).

[17] Pyatkov, S.G. On existence of maximal semidefinite invariant subspaces for $J$-dissipative operators http://arxiv.org/abs/1007.4131 (2010).

[18] Pyatkov S.G. Some properties of eigenfunctions and associated functions of indefinite Sturm-Liouville problems. In book: Nonclassical Equations of Mathematical Physics. Novosibirsk. Sobolev Institute of Mathematics, pp. 240–251 (2005).

[19] Pyatkov S.G. Riesz completeness of the eigenelements and associated elements of linear selfadjont pencils. Russian Acad. Sci. Sb. Math. **81**, no. 2, pp. 343–361 (1995).

[20] Egorov, I.E., Pyatkov, S.G., and Popov, S.V. Nonclassical Operator-Differential Equations. Nauka, Novosibirsk (2000).

[21] Pyatkov, S.G. Operator theory. Nonclassical problems. VSP, Utrecht (2002)

[22] Grisvard, P. Commutatlvé de deux foncteurs d'interpolation et applications. J. Math. Pures et Appliq. **45**, no. 2, 143–206 (1966).

[23] Triebel, H. Interpolation Theory, Function spaces, Differential operators. VEB Deutscher Verlag Wiss., Berlin (1977).

[24] Ausher, P., McIntosh, A., and Nahmod, A. Holomorphic functional calculi of operators, quadratic estimates and interpolation. Ind. Univ. Math. J. **46**, no. 2, 375–403 (1997).

[25] Grisvard, P. An approach to the singular solutions of elliptic problems via the theory of differential equations in Banach Spaces. Differential Equations in Banach Spaces. Lect. Notes Math. **1223**, 131–156 (1986).

[26] Engel, K.-J. and Nagel, R. One-Parameter Semigroups for Linear Evolution Equations. Springer. Graduate Texts in Mathematics. **194**, Berlin (2000).

[27] Fleige, A. and Najman, B. Nonsingularity of critical points of some differential and difference operators, Operators Theory: Adv. and Appl., **106**, 147–155 (1998).

[28] Langer H. Spectral function of definitizable operators in Krein spaces. In: Functional Analysis. Proceedings (Dubrovnik 1981), Lect. Notes Math., **948**, 1–46 (1982).

[29] Faddeev, M.M. and Shterenberg, R.G. On the Similarity of Some Differential Operators to Self-Adjoint Ones. Math. Notes, **72**, no. 2, 261–270 (2002).

[30] Faddeev, M.M. and Shterenberg, R.G. Similarity of Some singular operators to self-adjoint ones. J. of Math. Sci., **115**, no. 2, 2279–2286 (2003).

[31] Karabash, I.M. On J-self-adjoint differential operators similar to self-adjoint operators. Mat. Zametki [Math. Notes], **68**, no. 6, 943–944 (2000).

[32] Karabash, I. and Kostenko, A. Spectral Analysis of Differential Operators with Indefinite Weights and a Local Point Interaction. Operator Theory: Advances and Applications, **175**, 169–191 (2007).

[33] Karabash, I.M., Kostenko, A.S. On the Similarity of a J-Nonnegative Sturm-Liouville Operator to a Self-Adjoint Operator. Functional Analysis and Its Applications, **43**, no. 1, pp. 65–68 (2009).

[34] Karabash, I.M. Abstract Kinetic Equations with Positive Collision Operators. Spectral Theory in Inner Product Spaces and Applications. Operator Theory: Advances and Applications, **188**, 175–196 (2008).

[35] Curgus, B. and Najman, B. The operator $-(\operatorname{sgn} x)\frac{d^2}{dx^2}$ is similar to selfadjoint operator in $L_2(\mathbb{R})$. Proc. Amer. Math. Soc. **123**, 1125–1128 (1995).

[36] Weis, L. $H^\infty$ holomorphic functional calculus for sectorial operators – a survey. Operator Theory: Advances and Applications, **168**, 263–294 (2006).

[37] Parfenov, A.I. On an embedding criterion of interpolation spaces and its application to indefinite spectral problems. Sib. Math. J., **44**, no. 4, pp. 638–644 (2003).

[38] Parfenov, A.I., On the Curgus condition in indefinite Sturm-Liouville problems. Sib. Adv. Math. **15**, no. 2, pp. 68–103 (2005).

[39] Parfenov A.I., A contracting operator and boundary values. Novosibirsk, Sobolev Institute of Math., Omega Print, preprint no. 155 (2005).

[40] Karabash, I.M., Malamud M.M. Indefinite Sturm-Liouville operators $\operatorname{sgn} x(-\frac{d^2}{dx^2} + q(x))$ with finite zone potentials. Operators and Matrices, **1**, no. 3, pp. 301–368 (2007).

S.G. Pyatkov
Yugra State University
Chekhov st. 16, 628012
Hanty-Mansiisk, Russia

fax:+7 3467357734
phone: +7 9129010471
e-mail: pyatkov@math.nsc.ru
s_pyatkov@ugrasu.ru

Operator Theory:
Advances and Applications, Vol. 221, 571-585


# The Riemann–Hilbert Boundary Value Problem with a Countable Set of Coefficient Discontinuities and Two-side Curling at Infinity of the Order Less Than 1/2

R.B. Salimov and P.L. Shabalin

**Abstract.** The Riemann–Hilbert boundary value problem is one of the oldest boundary value problems of theory of analytic functions. Its complete solution (for the case of a finite index and continuous coefficients) was given by Hilbert in 1905. In the present paper we study the inhomogeneous Riemann–Hilbert boundary value problem in the upper half of complex plane with strong singularities of boundary data. We obtain general solution for the case where coefficients of the problem have a countable set of finite discontinuity points and two-side curling of order less than 1/2 at the infinity point. We investigate also the solvability conditions.

**Keywords.** Riemann–Hilbert boundary value problem, infinite index, curling, entire functions.

## 1. Introduction

The Riemann–Hilbert boundary value problem is one of the oldest boundary value problems of theory of analytic functions. Its first explicit statement can be found in the work [1] (1883). The Hilbert problem with two-side curling at infinity for a half-plane is solved in the papers [2], [3] by reduction to the corresponding Riemann boundary value problem (so-called Muskhelishvili technique, see [4], pp. 139–150). It should be noted that the papers [2], [3] are based on N.V. Govorov's results (see [5, 6] and [7]) on the Riemann boundary value problem with infinite index.

In the paper [8] the authors solve homogeneous Hilbert problem of the same type as in [2] directly, i.e., without reduction to the corresponding Riemann bound-

The research was partially supported by Russian Foundation for Basic Research (grant no. 08-01-00381) and Federal Agency for Science and Innovations (project no. 02.740.11.0193).

ary value problem with infinite index. In other words, we apply the so-called Gahov technique, see [9], p. 273. This technique is developed by a number of authors for the Hilbert boundary value problem with infinite index (see references in [8]). The authors [8] use to this end a nontrivial modification of regularizing multiplier.

The technique of paper [8] allows us to describe the solvability conditions in details (for instance, the work [2] does not contain a theorem analogous to Theorem 3 of [8]). On the other hand, it turns out to be useful for solving of other complicated problems with infinite index and a countable set of discontinuity points. In particular, we solve the Hilbert boundary problem on the half-plane with infinite set of discontinuities of coefficients. In our paper [10] we consider two cases: in the first case the series of jumps of the arguments of coefficients converges, and in the second one it diverges. As a result, we obtain the Hilbert problems with finite and infinite indices correspondingly. In the second case we study the Hilbert problem with infinite index of order $\rho < 1$. In the present paper we obtain a general solution of the inhomogeneous problem in a half-plane for the case of a countable set of coefficient discontinuities and two-side curling at infinity point of order less than $1/2$. We investigate also the solvability conditions.

## 2. Statement of the problem

Let $L$ stand for the real axis on the complex plane, $z = x + iy$, $D = \{z : \mathrm{Im}z > 0\}$, $t \in L$. We seek an analytic function $F(z)$ in the domain $D$ satisfying boundary condition

$$a(t)\mathrm{Re}\Phi(t) - b(t)\mathrm{Im}\Phi(t) = c(t), \quad t \in L, \tag{1}$$

where $a(t)$, $b(t)$, $c(t)$ are given real functions on $L$. These functions are continuous everywhere on $L$ except of discontinuity points of jump type $t_k$, $k = \pm 1, \pm 2, \ldots$, $0 < t_1 < \cdots < t_k < t_{k+1} < \cdots$, $\lim\limits_{k\to\infty} t_k = \infty$, $0 > t_{-1} > \cdots > t_{-k} > t_{-k-1} > \cdots$, $\lim\limits_{k\to\infty} t_{-k} = -\infty$. The boundary condition is assumed to be fulfilled everywhere on $L$ except the points $t_k, t_{-k}$, $k = \overline{1,\infty}$. We put $G(t) = a(t) - ib(t)$. Let $\nu(t) = \arg G(t)$ be a continuous on intervals of continuity $\arg G$ branch of argument such that its jumps $\delta_j = \nu(t_j + 0) - \nu(t_j - 0)$ satisfy inequalities $0 \le \delta_j < 2\pi$, $j = \pm 1, \pm 2, \ldots$.

We choose integer numbers $\beta_k$ such that $0 \le \varphi(t_k + 0) - \varphi(t_k - 0) < \pi$ and $0 \le \varphi(t_{-k} + 0) - \varphi(t_{-k} - 0) < \pi$ for $k = 1, 2, \ldots$. Then we introduce function $\varphi(t) = \nu(t) - \beta(t)\pi$, where $\beta(t)$ is an integer-valued function equaling to $\beta_k$, $\beta_{-k}$ on the intervals $(t_k, t_{k+1})$ and $(t_{-k}, t_{-k-1})$ correspondingly, $k = \overline{1,\infty}$. It equals to $\beta_0 = 0$ on the range $(t_{-1}, t_1)$.

Then we put

$$\kappa_j = \frac{\varphi(t_j + 0) - \varphi(t_j - 0)}{\pi}, \; j = \pm 1, \pm 2, \ldots,$$

where $0 \le \kappa_k < 1$, $0 \le \kappa_{-k} < 1$, $k = \overline{1,\infty}$.

We assume that the discontinuity points satisfy the conditions

$$\sum_{k=1}^{\infty} \frac{1}{t_k} < \infty, \qquad \sum_{k=1}^{\infty} \frac{1}{-t_{-k}} < \infty, \tag{2}$$

and consider infinite products

$$P_+(z) = \prod_{k=1}^{\infty} \left(1 - \frac{z}{t_k}\right)^{\kappa_k}, \qquad P_-(z) = \prod_{k=1}^{\infty} \left(1 - \frac{z}{t_{-k}}\right)^{\kappa_{-k}}.$$

Here the branch of $\arg(1 - z/t_j)$ vanishes for $z = 0$, and it is continuous in the plane $z$ with cut along certain ray of the real axis. This ray connects the points $t = t_j$, $t = +\infty$ for $j > 0$ and the points $t = -\infty$, $t = t_j$ for $j < 0$. The functions $P_+(z)$, $P_-(z)$ are analytic in the whole plane $z$ excepting the real rays $t > t_1$, $t < t_{-1}$, and on the sides of these rays excepting the points $t_k$, $t_{-k}$, $k = \overline{1, \infty}$. Thus, on the upper side of the cut we have

$$\arg P_+(t) = \begin{cases} 0, & t < t_1, \\ -\sum_{j=1}^{k} \kappa_j \pi, & t_k < t < t_{k+1}, \quad k = \overline{1, \infty}, \end{cases} \tag{3}$$

and

$$\arg P_-(t) = \begin{cases} 0, & t > t_{-1} \\ \sum_{j=1}^{k} \kappa_{-j} \pi, & t_{-k-1} < t < t_{-k}. \end{cases} \tag{4}$$

We denote

$$n_-^*(x) = \begin{cases} 0, & 0 \le x < -t_{-1}, \\ \sum_{j=1}^{k-1} \kappa_{-j}, & -t_{-k+1} \le x < -t_{-k}, \quad k = \overline{2, \infty}. \end{cases} \tag{5}$$

Analogously, we put

$$n_+^*(x) = 0, \ 0 < x < t_1, \quad n_+^*(x) = \sum_{j=1}^{k-1} \kappa_j, \ t_{k-1} \le x < t_k, \quad k = \overline{2, \infty}. \tag{6}$$

Furthermore, we assume that the numbers $t_j$, $\kappa_j$, $j = \pm 1, \pm 2, \dots$ are such that for certain positive constants $\kappa_+$, $\kappa_-$ the limits

$$\lim_{x \to +\infty} \frac{n_+^*(x)}{x^{\kappa_+}} = \Delta_+, \qquad \lim_{x \to +\infty} \frac{n_-^*(x)}{x^{\kappa_-}} = \Delta_- \tag{7}$$

exist and do not vanish. In addition, we assume that $\kappa_- = \kappa_+ = \kappa < 1/2$. The relations (7), imply representations (see [12])

$$\ln P_+(z) = \frac{\pi \Delta_+ e^{-i\pi\kappa}}{\sin \pi\kappa} z^{\kappa} + I_+(z), \qquad I_+(z) = -z \int_0^{+\infty} \frac{n_+^*(x) - \Delta_+ x^{\kappa}}{x(x - z)} dx, \tag{8}$$

$0 < \arg z < 2\pi$, and, analogously,

$$\ln P_-(z) = \frac{\pi\Delta_-}{\sin \pi\kappa} z^\kappa + I_-(z), \qquad I_-(z) = z\int_0^{+\infty} \frac{n_-^*(x) - \Delta_- x^\kappa}{x(x+z)} dx, \tag{9}$$

$-\pi < \arg z < \pi$. In just the same way as in ([7], pp. 127, 128) we conclude that under assumption (7)

$$n_-^*(-t_{-k}) - \Delta_-(-t_{-k})^\kappa = p_{-k}, \tag{10}$$

$$n_+^*(t_k) - \Delta_+(t_k)^\kappa = p_k, \tag{11}$$

where the numbers $p_{-k}$, $p_k$, $t_{-k}$, $t_k$, $k = \overline{1,\infty}$ are chosen so that the inequalities

$$p_{-k} = -n_-^*(-t_{-k}) + \Delta_-(-t_{-(k+1)})^\kappa, \tag{12}$$

$$p_k = -n_+^*(t_k) + \Delta_+(t_{k+1})^\kappa \tag{13}$$

are valid.

The functions $n_-^*(x) - \Delta_- x^\kappa$, $n_+^*(x) - \Delta_+ x^\kappa$ decrease. Hence, the restrictions $p_{-k} > 0$, $p_k > 0$, $k = \overline{1,\infty}$, are necessary for validity of the inequalities $-t_{-k} < -t_{-(k+1)}$, $t_k < t_{k+1}$.

The relations (10), (5) and (12) imply equality

$$p_{-(k+1)} = \kappa_{-(k+1)} - p_{-k}. \tag{14}$$

Analogously, the conditions (11), (6) allow to rewrite the formula (13) as follows:

$$p_{k+1} = \kappa_{k+1} - p_k. \tag{15}$$

As $p_{-k} > 0$, $p_k > 0$, then owing to (14), (15) we have

$$0 < p_{-k} < \kappa_{-(k+1)}, \quad 0 < p_k < \kappa_{k+1}.$$

In what follows we assume that

$$\inf\{\kappa_{-k}\} = \kappa_0^- > 0, \quad \inf\{\kappa_k\} = \kappa_0^+ > 0, \tag{16}$$

$$\inf\{p_{-k}\} = p_0^- > 0, \quad \inf\{p_k\} = p_0^+ > 0. \tag{17}$$

Let us consider the function

$$\varphi_1(t) = \varphi(t) + \arg P_+(t) + \arg P_-(t),$$

where $\arg P_+(t)$, $\arg P_-(t)$ are defined by formulas (3), (4). Here we assume that the function $\varphi_1(t)$ satisfies conditions

$$\varphi_1(t) = \begin{cases} \nu^- t^\rho + \tilde{\nu}(t), & t > 0, \\ \nu^+ |t|^\rho + \tilde{\nu}(t), & t < 0, \end{cases}$$

$(\nu^-)^2 + (\nu^+)^2 \neq 0$, where $\nu^-$, $\nu^+$, $\rho$ are constants, $0 < \rho < 1/2$, the function $\tilde{\nu}(t)$ is continuous on $L$ including the infinity point, and satisfies the Hölder condition with exponent $\mu$, $0 < \mu \leq 1$, everywhere on $L$. According to N.V. Govorov ([7], p. 113), we denote the class of that functions by $H_L(\mu)$. We have

$$|\tilde{\nu}(t_2) - \tilde{\nu}(t_1)| \leq K \left| \frac{1}{t_1} - \frac{1}{t_2} \right|^\mu,$$

where $|t_1|$, $|t_2|$ are sufficiently large.

Let symbol $H(\mu)$ stand for the class of all functions satisfying the Hölder condition on interval $(-\infty, +\infty)$.

Below we obtain a formula for general solution of the problem with described above singularities. The class of solution will be specified later.

## 3. Solvability of the homogeneous Riemann–Hilbert boundary value problem

We rewrite the boundary condition (1) in the form

$$\mathrm{Re}[e^{-i\varphi_1(t)}\Phi(t)P_+(t)P_-(t)] = c_1(t), \tag{18}$$

where

$$c_1(t) = c(t)|G(t)|^{-1}|P_+(t)||P_-(t)|\cos(\beta(t)\pi). \tag{19}$$

We consider first the homogeneous problem (18), i.e., the case $c(t) \equiv 0$. According [8], we introduce the function

$$P(z) + iQ(z) = le^{i\alpha}z^{\rho}, \tag{20}$$

where $l$, $\alpha$ are real constants, $l > 0$, $0 \le \alpha < 2\pi$, $P(z) = \mathrm{Re}[le^{i\alpha}z^{\rho}]$. We choose the numbers $l$, $\alpha$ such that

$$P(t) = \begin{cases} \nu^- t^{\rho}, & t > 0, \\ \nu^+ |t|^{\rho}, & t < 0. \end{cases}$$

Then we have

$$Q(re^{i\theta}) = r^{\rho}\frac{\nu^- \cos((\pi-\theta)\rho) - \nu^+\cos(\theta\rho)}{\sin(\pi\rho)}. \tag{21}$$

Furthermore, the function

$$\Gamma(z) = \frac{1}{\pi}\int\limits_{-\infty}^{+\infty} \tilde{\nu}(t)\frac{dt}{t-z}$$

is analytic and bounded in the domain $D$. The boundary values of imaginary part of this function are $\varphi_1(t) - P(t) = \tilde{\nu}(t)$.

On the contour $L$ the function takes the values $\Gamma^+(t) = \Gamma(t) + i\tilde{\nu}(t)$, where

$$\Gamma(t) = \frac{1}{\pi}\int\limits_{-\infty}^{+\infty} \tilde{\nu}(t_1)\frac{dt_1}{t_1 - t}.$$

The boundary condition (18) can be written as follows:

$$\mathrm{Re}\left\{e^{-\Gamma^+(t)}e^{-iP(t)+Q(t)}\Phi(t)P_+(t)P_-(t)\right\} = \frac{c(t)|P_+(t)|P_-(t)|e^{Q(t)}}{\cos(\beta(t)\pi)|G(t)|e^{\Gamma(t)}}. \tag{22}$$

We put $c(t) \equiv 0$ in the last formulas, and obtain the boundary condition for the homogeneous problem

$$\mathrm{Im}\left\{ie^{-\Gamma^+(t)}e^{-iP(t)+Q(t)}\Phi(t)P_+(t)P_-(t)\right\} = 0. \tag{23}$$

Here the braces contain the boundary value of analytic in the domain $D$ function

$$ie^{-\Gamma^+(z)}e^{-iP(z)+Q(z)}\Phi(z)P_+(z)P_-(z) = F(z). \tag{24}$$

Under assumptions (9), (8) it can be rewritten in the following expedient form:

$$F(z) = ie^{-\Gamma^+(z)}e^{-iP(z)+Q(z)}\Phi(z)e^{I_+(z)}e^{I_-(z)}\exp\left\{\pi\frac{\Delta_+e^{-i\pi\kappa}+\Delta_-}{\sin(\pi\kappa)}z^\kappa\right\}. \tag{25}$$

According to (23) we obtain the following relation for the boundary values of the function $F(z)$:

$$\mathrm{Im}F_+(t) = 0,\ t \in L. \tag{26}$$

It allows us to extend the function $F(z)$ analytically in the lower half-plane $\mathrm{Im}z < 0$ by the rule

$$F(z) = \overline{F(\overline{z})},\ \mathrm{Im}z < 0.$$

Let $\tilde{B}$ be class of solutions $\Phi(z)$ of the problem (23) such that the product $|\Phi(z)||z-t_j|^{\kappa_j}$ is bounded in a vicinity of the point $t_j$ for $j = \pm1, \pm2, \dots$.

**Theorem 3.1.** *The homogeneous boundary value problem* (23) *has a solution* $\Phi(z)$ *in the class* $\tilde{B}$ *if and only if the product* (24) *coincides in the half-plane with restriction of an entire function satisfying the condition* (26).

*Proof.* The condition (26) must be valid at the points $t_k$, $t_{-k}$, $k = \overline{1, +\infty}$. Indeed, the extension of the function in the half-plane $\mathrm{Im}z < 0$ through the intervals $(t_{k-1}, t_k)$, $(t_k, t_{k+1})$, gives the same function by virtue of (26). If a solution $\Phi(z)$ of the problem (23) belongs to the class $\tilde{B}$, then $|F(z)|$ is bounded in a vicinity of $t_k$. Thus, at the point $t_k$ the function $F(z)$ has a removable singularity. Hence, $\mathrm{Im}F^+(t_k) = 0$. We conclude that $F(z)$ is entire function. □

The following proposition is valid (see [12]).

**Lemma 3.2.** *If the relations* (11), (13) *are valid, then the function* $P_+(z)$ *is representable in the form*

$$P_+(z) = \exp\{\pi\Delta_+e^{-i\pi\kappa}z^\kappa/\sin\pi\kappa\}\exp\{I_+(z)\},$$

*where* $|\exp\{I_+(z)\}|$ *is bounded on* $D$. *The boundary values of this function satisfy the Hölder condition on any finite segment of the real axis.*

Now we seek the solution of the homogeneous problem (23) in the class $B_*$ consisting of all functions $\Phi(z)$ such that the product

$$|\Phi(z)|e^{\mathrm{Re}I_+(z)}e^{\mathrm{Re}I_-(z)}$$

is bounded in the domain $D$. Clearly, $B_* \subset \tilde{B}$.

By means of Lemma 3.2 we conclude that if $\Phi(z) \in B_*$, then the absolute value $|\Phi(z)|$ is bounded for $r \to \infty$.

It is evident by virtue of symmetry of the entire function $F(z)$ that the following equality holds

$$M(r) = \max_{0\le\theta\le\pi} |F(re^{i\theta})|,$$

and due to (25) we have

$$\log M(r) \le \log C + lr^{\rho} + \pi \frac{\Delta_+ + \Delta_-}{\sin(\pi\kappa)} r^{\kappa}. \tag{27}$$

We discern the following cases. If $\rho > \kappa$, then we obtain by means of (27)

$$\overline{\lim_{r\to\infty}} \frac{\log\log M(r)}{\log r} \le \overline{\lim_{r\to\infty}} \left\{ \rho + \frac{1}{\log r} \log\left[ l + \frac{\log C}{r^{\rho}} + \pi \frac{\Delta_+ + \Delta_-}{r^{\rho-\kappa}\sin(\pi\kappa)} \right] \right\} = \rho.$$

Therefore, the order $\rho_F$ of the entire function $F(z)$, which is determined by formulas (25), (26), does not exceed $\rho$. In this case the following theorem holds.

**Theorem 3.3.** *Let $\rho > \kappa$. Then the homogeneous boundary value problem* (23) *has a solution $\Phi(z)$ in the class $B_*$ if and only if the product* (24) *coincides in the half-plane with restriction of some entire function $F(z)$ of order $\rho_F \le \rho$ such that $F(z)$ satisfies the condition* (26) *and*

$$|F(t)| \le C \exp\left\{ \frac{\nu^- \cos(\pi\rho) - \nu^+}{\sin(\pi\rho)} t^{\rho} + \pi \frac{\Delta_+ \cos(\pi\kappa) + \Delta_-}{\sin(\pi\kappa)} t^{\kappa} \right\}, \ t > 0, \tag{28}$$

$$|F(t)| \le C \exp\left\{ \frac{\nu^- - \nu^+ \cos(\pi\rho)}{\sin(\pi\rho)} |t|^{\rho} + \pi \frac{\Delta_+ + \Delta_- \cos(\pi\kappa)}{\sin(\pi\kappa)} |t|^{\kappa} \right\}, \ t < 0. \tag{29}$$

*on the axis $L$ for sufficiently large $|t|$.*

*The general solution of problem* (23) *is determined by formula*

$$\Phi(z) = \frac{ic^{\Gamma(z)} e^{i[P(z)+iQ(z)]} F(z)}{P_+(z)P_-(z)}. \tag{30}$$

Theorem 3.3 gives a representation for general solution of the problem under consideration. The solution contains an entire function $F(z)$ satisfying a number of restrictions. The solvability follows from the fact that any entire function of order less than $\rho$ with real values on $L$ satisfies the conditions (26), (28),(29). We can build that functions by the Weierstrass scheme; see, for instance, in [10].

Theorem 3.3 differs essentially from the corresponding theorem for the problem with finite index, where the general solution contains a polynomial with arbitrary coefficients satisfying certain additional conditions.

*Proof.* Let $\Phi(z)$ be a solution of the boundary value problem (23) in the class $B_*$. Then the relations (24), (26) are fulfilled. Here $F(z)$ is entire function of order $\rho_F \le \rho$. The validity of the inequalities (28), (29) is obvious.

Assume that the formula (24) fulfils for the function $\Phi(z)$, where $F(z)$ is an entire function of order $\rho_F \le \rho$ satisfying the condition (26) and the inequalities (28), (29) (then the function $\Phi(z)$ is a solution of the problem). We estimate the analytical in domain $D$ function $\Phi(z)e^{I_+(z)+I_-(z)}$ by means of the formulas (25), (21). As $|\mathrm{Re}\Gamma(re^{i\theta})| < q$, since

$$|\Phi(t)e^{I_+(t)+I_-(t)}| < Ce^{q}$$

everywhere on $L$. Hence, we obtain by terms of (25) and (20)

$$\max_{0\le\theta\le\pi} |\Phi(re^{i\theta})e^{I_+(re^{i\theta})+I_-(re^{i\theta})}| \le M(r)\exp\left\{lr^\rho + q + \pi\frac{\Delta_+ + \Delta_-}{\sin(\pi\kappa)}r^\kappa\right\}.$$

The relation $M(r) < \exp\{r^{\rho_F+\epsilon}\}$ is valid for any $\epsilon > 0$ and all $r > r_\epsilon$. Therefore, we can choose the numbers $\epsilon$, $\rho_1$ such that $\rho < \rho_1 < 1$, $\rho_F + \epsilon < \rho_1$ and the inequality

$$\max_{0\le\theta\le\pi} |\Phi(re^{i\theta})e^{I_+(re^{i\theta})+I_-(re^{i\theta})}| < e^{r^{\rho_1}}$$

is valid for all sufficiently large $r$.

Consequently, the order of function $\Phi(z)e^{I_+(z)+I_-(z)}$ inside the angle $0 \le \theta \le \pi$ does not exceed $\rho_1$ (see, for example , [13], p. 69). Then by virtue of the Phragmen-Lindelöf principle the relation $|\Phi(z)e^{I_+(z)+I_-(z)}| < Ce^q$ fulfils everywhere in $D$. Thus, $\Phi(z) \in B_*$. □

Sometimes the homogeneous boundary value problem has only zero solution in the class $B_*$. The following theorem gives a criterion of non-trivial solvability of the problem for the case $\kappa < \rho < 1/2$.

**Theorem 3.4.** *Let $\rho > \kappa$, $\rho < 1/2$. Then the following propositions are valid.*

a) *If $\nu^-\cos(\pi\rho) - \nu^+ < 0$, or $\nu^- - \nu^+\cos(\pi\rho) < 0$, then the homogeneous boundary value problem has only zero solution in the class $B_*$.*

b) *If* $\begin{cases} \nu^-\cos(\pi\rho) - \nu^+ = 0, \\ \nu^- - \nu^+\cos(\pi\rho) > 0, \end{cases}$ *or* $\begin{cases} \nu^- - \nu^+\cos(\pi\rho) = 0, \\ \nu^-\cos(\pi\rho) - \nu^+ > 0, \end{cases}$

*then the homogeneous boundary value problem has an infinite set of non-trivial solutions in the class $B_*$ given by the formula (30). Here $F(z)$ is any real on $L$ entire function of order $\rho_F \le \rho$, which satisfies either inequalities*

$$|F(t)| \le \begin{cases} C\exp\left\{\pi\frac{\Delta_+\cos(\kappa\pi) + \Delta_-}{\sin(\pi\kappa)}t^\kappa\right\}, & t > 0, \\ C\exp\left\{\frac{\nu^- - \nu^+\cos(\pi\rho)}{\sin(\pi\rho)}|t|^\rho + \pi\frac{\Delta_+ + \Delta_-\cos(\kappa\pi)}{\sin(\pi\kappa)}|t|^\kappa\right\}, & t < 0, \end{cases} \tag{31}$$

*or inequalities*

$$|F(t)| \le \begin{cases} C\exp\left\{\frac{\nu^-\cos(\pi\rho) - \nu^+}{\sin(\pi\rho)}t^\rho + \pi\frac{\Delta_+\cos(\kappa\pi) + \Delta_-}{\sin(\pi\kappa)}t^\kappa\right\}, & t > 0, \\ C\exp\left\{\pi\frac{\Delta_+ + \Delta_-\cos(\kappa\pi)}{\sin(\pi\kappa)}|t|^\kappa\right\}, & t < 0. \end{cases}$$

c) *If $\nu^-\cos(\pi\rho) - \nu^+ > 0$, $\nu^- - \nu^+\cos(\pi\rho) > 0$, then the homogeneous boundary value problem has an infinite set of non-trivial solutions in the class $B_*$ given by the formula (30). Here $F(z)$ is any real on $L$ entire function of order $\rho_F \le \rho$. For $\rho_F = \rho$ it satisfies, in addition, the inequalities (28), (29).*

*Proof.* a) Let $\kappa < \rho < 1/2$ and $\nu^- \cos(\pi\rho) - \nu^+ < 0$. By virtue of (28) we have $\lim_{t\to+\infty} |F(t)| = 0$, and by the Phragmen-Lindelöf principle $F(z) \equiv 0$ in the whole plane and, consequently, $\Phi(z) \equiv 0$ in $D$. The case $\nu^- - \nu^+ \cos(\pi\rho) < 0$ can be investigated in analogous way.

b) Let

$$\begin{cases} \nu^- \cos(\pi\rho) - \nu^+ = 0, \\ \nu^- - \nu^+ \cos(\pi\rho) > 0. \end{cases} \tag{32}$$

In accordance with Theorem 3.3 the solvability is connected with existences of entire functions of order $\rho_F \le \rho$ satisfying (26), (28), (29). By means of (32) we rewrite the conditions (28), (29) in the form (31). Therefore, we can build an appropriate entire function by the Weierstrass scheme (see, for instance, [10]). Let us consider the entire function of order $\kappa_0$

$$F_0(z) = \prod_{k=1}^{\infty} \left(1 - \frac{z}{r_k e^{i\theta_0}}\right)\left(1 - \frac{z}{r_k e^{-i\theta_0}}\right), \tag{33}$$

where

$$r_k = \left(\frac{2k-1}{2\Delta_0}\right)^{1/\kappa_0}, \quad \Delta_0 > 0, \quad 0 \le \theta_o \le \pi. \tag{34}$$

The authors [10] obtain representation

$$\ln F_0(z) = I_0(z,\theta_0) + I_0(z,-\theta_0)$$
$$+ \begin{cases} \Delta_0\pi(re^{i\theta})^{\kappa_0} 2\cos((\theta_0 - \pi)\kappa_0)/\sin(\pi\kappa_0), & 0 \le \theta < \theta_0, \\ \Delta_0\pi e^{-i\kappa_0\pi}(re^{i\theta})^{\kappa_0} 2\cos(\theta_0\kappa_0)/\sin(\pi\kappa_0), & \theta_0 \le \theta < 2\pi - \theta_0, \\ \Delta_0\pi e^{-2i\kappa_0\pi}(re^{i\theta})^{\kappa_0} 2\cos((\theta_0 - \pi)\kappa_0)/\sin(\pi\kappa_0), & 2\pi - \theta_0 < \theta < 2\pi, \end{cases}$$

where $z = re^{i\theta}$, $0 \le \theta \le 2\pi$,

$$I_0(z,\theta_0) = -ze^{i\theta_0}\int_0^{\infty} \frac{n(\tau) - \tau^{\kappa_0}\Delta_0}{\tau(\tau - ze^{i\theta_0})}d\tau, \quad n(\tau) = \begin{cases} k, & r_k \le \tau < r_{(k+1)}, \\ 0, & 0 \le \tau < r_0, \end{cases}$$

and $I_0(z,\theta_0) = o(r^{\kappa_0})$ for $\to \infty$. Hence, the function $F_0(z)$ has order $\rho_{F_0} = \kappa_0$. On the real axis the values of function $\ln F_0(z)$ are real:

$$\ln F_0(t) = 2\int_0^{\infty} \frac{(n(\tau) - \tau^{\kappa_0}\Delta_0)(t^2 - t\tau\cos\theta_0)}{\tau(t^2 - 2t\tau\cos(\theta_0) + \tau^2)}d\tau$$
$$+ \begin{cases} \dfrac{\Delta_0\pi 2\cos\Big((\theta_0 - \pi)\kappa_0\Big)}{\sin(\pi\kappa_0)} t^{\kappa_0}, \ t > 0, \\ \dfrac{\Delta_0\pi 2\cos(\theta_0\kappa_0)}{\sin(\pi\kappa_0)}|t|^{\kappa_0}, \quad t < 0. \end{cases}$$

Therefore, the condition (26) is valid. If $\kappa_0 < \kappa$, then the condition (31) fulfils for entire function $F_0(z)$ for any $\Delta_0$, $\theta_0$. If $\kappa_0 = \kappa$, then the numbers $\Delta_0$, $\theta_0$, $\Delta_0 > 0$, $0 \le \theta_0 \le \pi$, must satisfy inequality

$$\Delta_0 2\cos\Big((\theta_0 - \pi)\kappa_0\Big) \le \Delta_+ \cos(\pi\kappa).$$

In the case c) we consider the function (33), (34) under assumption $\kappa_0 \le \rho$. If $\kappa_0 \le \rho$ in the formula (34), then $\Delta_0$ and $\theta_0$ are arbitrary values. If $\kappa_0 = \rho$, then the values $\Delta_0$ and $\theta_0$ must satisfy the system of inequalities

$$\begin{cases} \Delta_0 2\cos\Big((\theta_0 - \pi)\rho\Big) \le \nu^- \cos(\pi\rho) - \nu^+, \\ \Delta_0 2\cos(\theta_0\rho) \le \nu^- - \nu^+ \cos(\pi\rho). \end{cases}$$

Obviously, the system is compatible. □

**Theorem 3.5.** *Let*

$$\rho = \kappa < 1/2. \tag{35}$$

*Then the following propositions are valid.*

a) *If*

$$(\nu^- + \pi\Delta_+)\cos(\pi\rho) - (\nu^+ - \pi\Delta_-) < 0,$$

*or*

$$(\nu^- + \pi\Delta_+) - (\nu^+ - \pi\Delta_-)\cos(\pi\rho) < 0,$$

*then the homogeneous boundary value problem* (23) *has only zero solution in the class* $B_*$.

b) *If either*

$$\begin{cases} (\nu^- + \pi\Delta_+)\cos(\pi\rho) - (\nu^+ - \pi\Delta_-) = 0, \\ \nu^- + \pi\Delta_+ - (\nu^+ - \pi\Delta_-)\cos(\pi\rho) > 0, \end{cases}$$

*or*

$$\begin{cases} (\nu^- + \pi\Delta_+)\cos(\pi\rho) - (\nu^+ - \pi\Delta_-) > 0, \\ \nu^- + \pi\Delta_+ - (\nu^+ - \pi\Delta_-)\cos(\pi\rho) = 0 \end{cases}$$

*then it has only solutions of the form*

$$\Phi(z) = A\frac{e^{\Gamma(z)}e^{i[P(z)+iQ(z)]}}{P_+(z)P_-(z)},$$

*where $A$ is arbitrary constant.*

c) *If*

$$\begin{cases} (\nu^- + \pi\Delta_+)\cos(\pi\rho) - (\nu^+ - \pi\Delta_-) > 0, \\ \nu^- + \pi\Delta_+ - (\nu^+ - \pi\Delta_-)\cos(\pi\rho) > 0, \end{cases}$$

*then the problem has infinite set of linearly independent solutions in the class $B_*$ given by the formula* (30), *where $F(z)$ is arbitrary real on $L$ entire function of order $\rho_F \le \rho$; for $\rho_F = \rho$ the function $F$ satisfies the inequalities* (28), (29).

*Proof.* Let us prove the proposition b). Assume that the conditions

$$\begin{cases} (\nu^- + \pi\Delta_+)\cos(\pi\rho) - (\nu^+ - \pi\Delta_-) = 0, \\ \nu^- + \pi\Delta_+ - (\nu^+ - \pi\Delta_-)\cos(\pi\rho) > 0, \end{cases} \tag{36}$$

are valid. Then (see Theorem 3.3) the general solution of the problem (23) is given by the formula (30),where $F(z)$ is an entire function of order $\rho_F \le \rho < 1/2$. The function $F(z)$ must satisfy the condition (26) and inequalities (28), (29). By means of (28), (35) and (36) we obtain $|F(t)| < C$, $t > 0$, $C = \text{const}$. By virtue of the Phragmen-Lindelöf theorem we conclude for $\rho_F < 1/2$ that $|F(z)| < C$ in the whole plane. Consequently, $|F(z)| = \text{const}$ in $D$.

The cases a), c) can be considered in just the same way as in the Theorem 3.4. □

Let $\rho < \kappa < 1/2$. According to (27), we have

$$\log\log M(r) \le \kappa \log r + \log\left[\frac{\pi(\Delta_+ + \Delta_-)}{\sin(\pi\kappa)} + \frac{\log C}{r^\kappa} + \frac{l}{r^{\kappa-\rho}}\right],$$

hence,

$$\rho_F = \overline{\lim_{r\to\infty}} \frac{\log\log M(r)}{\log r} \le \kappa,$$

whence, the order of the entire function $F(z)$ defined by formulas (25), (26) does not exceed $\kappa$.

If $\rho < \kappa$ then the inequalities (28), (29) fulfil automatically.

Thus, the following theorems hold.

**Theorem 3.6.** *If $\rho < \kappa$, then all solutions $\Phi(z)$ of the homogeneous boundary value problem* (23) *in the class $B_*$ are given by the formula* (24)*, where $F(z)$ is an entire function of order $\rho_F \le \max\{\kappa, \kappa\}$ satisfying the conditions* (26), (28), (29).

**Theorem 3.7.** *If $\rho \le \kappa < 1/2$, then the general solution of the homogeneous boundary value problem* (23) *in the class $B_*$ is defined by the formula* (30)*, where $F(z)$ is arbitrary real on $L$ entire function of order $\rho_F \le \kappa$; for $\rho_F = \rho$ it satisfies the inequalities* (28), (29).

## 4. The inhomogeneous problem

We seek a solution of inhomogeneous problem (22) in the class $B_*$ under restrictions of Section 2. In addition, let

$$c(t) = \frac{\tilde{c}(t)}{1+t^2},$$

where function $\tilde{c}(t)$ is bounded and satisfies the Hölder condition on $L$.

We consider the case $\kappa < 1/2$, under additional assumption

$$\begin{cases} \nu^-\cos(\pi\rho) - \nu^+ \ge 0, \\ \nu^- - \nu^+\cos(\pi\rho) > 0 \end{cases} \text{or} \begin{cases} \nu^-\cos(\pi\rho) - \nu^+ > 0, \\ \nu^- - \nu^+\cos(\pi\rho) \ge 0. \end{cases} \tag{37}$$

According to (8), (9) we obtain

$$|P_+(t)P_-(t)| = \exp\left\{\pi\frac{\Delta_+\cos(\pi\kappa)+\Delta_-}{\sin(\pi\kappa)}t^\kappa + I_+(t)+I_-(t)\right\}, \quad \text{if } t>0,$$

$$|P_+(t)P_-(t)| = \exp\left\{\pi\frac{\Delta_+ + \Delta_-\cos(\pi\kappa)}{\sin(\pi\kappa)}t^\kappa + I_+(t)+I_-(t)\right\}, \quad \text{if } t<0.$$

Let us find a particular solution. We consider entire function

$$F_j(z) = \prod_{k=0}^{\infty}\left(1-\frac{z}{r_{jk}e^{i\theta_j}}\right)\left(1-\frac{z}{r_{jk}e^{-i\theta_j}}\right),$$

of order $\tilde{\rho_j}$, where

$$r_{j,k} = \left(\frac{2k-1}{2\Delta_j}\right)^{1/\tilde{\rho_j}}, \ j=1,2, \ \tilde{\rho_1}=\rho, \ \tilde{\rho_2}=\kappa.$$

It is easy to check validity of formulas (see [10])

$$\log F_j(t) = 2I_j(t,\theta_j) + \begin{cases} \Delta_j\pi 2\cos((\theta_j-\pi)\tilde{\rho_j})t^{\tilde{\rho_j}}\,/\sin(\pi\tilde{\rho_j}) & t>0, \\ \Delta_j\pi 2\cos(\theta_j\tilde{\rho_j})|t|^{\tilde{\rho_j}}\,/\sin(\pi\tilde{\rho_j}) & t<0, \end{cases} \tag{38}$$

where

$$I_j(t,\theta_j) = \int\limits_0^\infty (n_j(x)-x^{\tilde{\rho_j}}\Delta_j)\frac{t^2-tx\cos(\theta_j)}{x(t^2-2tx\cos(\theta_j)+x^2)}dx,$$

$$n_j(x) = \begin{cases} k, & r_{jk}\le x< r_{j(k+1)}, \ j=1,2, \\ 0, & 0<x<r_{j1}. \end{cases}$$

By virtue of (37) the values $\Delta_1>0$, $0<\theta_1<\pi$ are uniquely determined by the system

$$\begin{cases} \Delta_1\pi 2\cos((\theta_1-\pi)\rho) = \nu^-\cos(\pi\rho)-\nu^+, \\ \Delta_1\pi 2\cos(\theta_1\rho) = \nu^- - \nu^+\cos(\pi\rho), \end{cases}$$

what ensures the boundedness of ratio $e^{Q(t)}/F_1(t)$ on $L$. This system has a unique solution

$$\Delta_1 = \frac{1}{2\pi}\sqrt{(\nu^-)^2-2\nu^-\nu^+\cos(\pi\rho)=(\nu^+)^2},$$

$$\theta_1 = \frac{1}{\rho}\arccos\frac{\nu^- - \nu^+\cos(\pi\rho)}{2\pi\Delta_1}.$$

Using (38) we obtain

$$\frac{e^{Q(t)}}{F_1(t)} = e^{-2I_1(t,\theta_1)}.$$

In [10] we have proved inclusion $I_1(t,\theta_1)\in H(\rho)$.

To estimate the smoothness of $|P_+(t)P_-(t)|/F_2(t)$ we choose the values $\Delta_2 > 0$, $0 < \theta_2 < \pi$ such that

$$\begin{cases} \Delta_2 2\cos((\theta_2 - \pi)\kappa) = \Delta_+ \cos(\pi\kappa) - \Delta_-, \\ \Delta_2 2\cos(\theta_2\kappa) = \Delta_- \cos(\pi\kappa) + \Delta_+. \end{cases}$$

As above, we have

$$\Delta_2 = \frac{1}{2}\sqrt{\Delta_+^2 + 2\Delta_-\Delta_+\cos(\pi\kappa) + \Delta_-^2},$$

$$\theta_2 = \frac{1}{\kappa}\arccos\frac{\Delta_+ + \Delta_-\cos(\pi\kappa)}{2\pi\Delta_2}.$$

Therefore,

$$\frac{|P_+(t)P_-(t)|}{F_2(t)} = e^{I_+(t)+I_-(t)-2I_2(t,\theta_2)}, \quad I_2(t,\theta_2) \in H(\kappa).$$

Now we rewrite the boundary conditions (18), (19) in the form

$$\operatorname{Re}\left\{e^{-\Gamma^+(t)}\frac{e^{-i[P(t)+iQ(t)]}}{F_1(t)}\Phi(t)\frac{P_+(t)P_-(t)}{F_2(t)}\right\} = c_2(t), \tag{39}$$

where

$$c_2(t) = \frac{\tilde{c}(t)}{|G(t)|}\frac{P_+(t)P_-(t)}{F_2(t)(1+t^2)}\cos(\beta(t)\pi)e^{-\Gamma(t)}\frac{e^{Q(t)}}{F_1(t)}.$$

We need below the following result from the paper [12].

**Lemma 4.1.** *Assume that a function $f(x)$ is bounded on $(d, +\infty)$, $d > 1$, and satisfies the condition*

$$|f(x_2) - f(x_1)| < K|x_2 - x_1|^\alpha$$

*for any points $x_1$ and $x_2$ of this interval. Then the function $f_1(x) = (1+x^2)^{-\alpha_1} f(x)$, $\alpha_1 \geq \alpha$, satisfies the inequality*

$$|f_1(x_2) - f_1(x_1)| < K_1|1/x_2 - 1/x_1|^\alpha,$$

*for sufficiently large values of the arguments.*

By Lemma 4.1 we have $c_2(t) \in H_L(\mu^*), 0 < \mu^* < 1$. The inclusion follows from the Hölder condition and from the behavior of $c(t)$ nearby $t = \infty$.

We seek a particular solution $\Phi(z)$ of the inhomogeneous problem (39) in the class $B_*$ under assumption that the braced expression in the left-hand side of (39) is analytic bounded function in the domain $D$. By virtue of the properties of $c_2(t)$ the mentioned expression is representable by the Schwarz formula for half-plane (see [4], p. 155):

$$\Phi(z) = e^{\Gamma(z)}\frac{F_1(z)}{e^{-i[P(z)+iQ(z)]}} \cdot \frac{F_2(z)}{P_+(z)P_-(z)} \cdot \frac{1}{\pi i}\int_L \frac{c_2(t)}{t-z}dt. \tag{40}$$

It is easy to show that this solution belongs to the class $B_*$. Indeed, we have

$$\left|e^{\Gamma(z)}\frac{1}{\pi i}\int_L \frac{c_2(t)}{t-z}dt\right| < C,\ z \in D, \tag{41}$$

and, consequently, the formulas (40), (9), (8) imply bound

$$|\Phi(z)e^{\mathrm{Re}(I_+(z)+I_-(z))}| < C\frac{|F_1(z)|}{e^{Q(z)}} \cdot \frac{|F_2(z)|}{\exp\{\pi[\Delta_-\mathrm{Re}z^\kappa + \Delta_+\mathrm{Re}(e^{-i\kappa\pi}z^\kappa)]/\sin(\pi\kappa)\}}. \tag{42}$$

As $F_1(z)$ and $F_2(z)$ are entire functions of orders $\rho$ and $\kappa$ correspondingly, then we conclude that the order of the function in the left-hand side of the previous formula in the angle $0 \le \theta \le \pi$ is less than unity.

It is obvious for $t \in L$ that

$$\frac{|F_1(t)|}{e^{Q(t)}} = \frac{F_1(t)}{e^{Q(t)}} < C_1,$$

and

$$\frac{|F_2(t)|}{\exp\{\pi[\Delta_-\mathrm{Re}t^\kappa + \Delta_+\mathrm{Re}(e^{-i\kappa\pi}t^\kappa)]/\sin(\pi\kappa)\}} = \frac{F_2(t)e^{I_+(t)+I_-(t)}}{|P_+(t)P_-(t)|} < C_2.$$

Therefore, passing to the limit for $z \to t$ in the relations (42), (41), we obtain

$$|\Phi(t)e^{I_+(t)+I_-(t)}| \le \tilde{C},$$

and according to the Phragmen-Lindelöf principle we have that

$$|\Phi(z)e^{\mathrm{Re}(I_+(z)+I_-(z))}| \le \tilde{C},\ z \in D,$$

i.e., the solution (40) belongs to the class $B_*$.

Thus, we proved

**Theorem 4.2.** *The general solution of the inhomogeneous boundary value problem* (22) *in the class* $B_*$ *is representable as a sum of the particular solution* (40) *of this problem and the general solution of homogeneous problem* (23) *in the class* $B_*$.

## References


[1] V. Volterra, *Sopra alkune condizioni caratteristische per functioni di variabile complessa.* Ann. Mat. (2), T. 11, (1883).

[2] I.E. Sandrigaylo, *On the Hilbert boundary value problem with infinite index for half-plane.* Izvestiya akademii nayk Belorusskoy SSSR, Seriya Fis.-Mat. Nauk, No. 6, (1974), 16–23.

[3] P.Yu. Alekna, *The Hilbert boundary value problem with infinite index of logarithmic order for half-plane.* Litovskiy Matematicheskiy Sbornik. No. 1, (1977), 5–12.

[4] N.I. Musheleshvili, *Singular integral equations.* Nauka, Moscow, 1968.

[5] N.V. Govorov, *On the Riemann boundary value problem with infinite index.* Dokl. AN USSR 154, No. 6, (1964), 1247–1249.

[6] N.V. Govorov, *Inhomogeneous Riemann boundary value problem with infinite index.* Dokl. AN USSR 159, No. 5, (1964), 961–964.

[7] N.V. Govorov, *The Riemann boundary value problem with infinite index.* Nauka, Moscow, 1986.

[8] R.B. Salimov and P.L. Shabalin, *Method of regularizing multiplier for solving of the uniform Hilbert problem with infinite index.* Izvestiya vuzov, Matematika. No. 4, (2001), 76–79.

[9] F.D. Gahov, *Boundary value problems.* Nauka, Moscow, 1977.

[10] R.B. Salimov and P.L. Shabalin, *On solving of the Hilbert problem with infinite index.* Mathematical notes. 73 (5), (2003), 724–734.

[11] A.I. Markushevich, *Theory of analytic functions.* V. 2. Nauka, Moscow, 2008.

[12] R.B. Salimov and P.L. Shabalin, *The Hilbert problem: the case of infinitely many discontinuity points of coefficients.* Siberian Mathematical Journal. Vol. 49, No. 4, (2008), 898–915.

[13] B.Ya. Levin, *Distribution of the roots of entire functions* Gostehizdat, Moscow, 1956.

R.B. Salimov and P.L. Shabalin
Kazan State University of Architecture and Engineering
Chair of higher mathematics
Zelenaya str. 1
Kazan 420043, Russian Federation
e-mail: Pavel.Shabalin@mail.ru

Operator Theory:
Advances and Applications, Vol. 221, 587-605


# Galerkin Method with Graded Meshes for Wiener-Hopf Operators with PC Symbols in $L^p$ Spaces

Pedro A. Santos

**Abstract.** This paper is concerned with the applicability of maximum defect polynomial (Galerkin) spline approximation methods with graded meshes to Wiener-Hopf operators with matrix-valued piecewise continuous generating function defined on $\dot{\mathbb{R}}$. For this, an algebra of sequences is introduced, which contains the approximating sequences we are interested in. There is a direct relationship between the stability of the approximation method for a given operator and invertibility of the corresponding sequence in this algebra. Exploring this relationship, the methods of essentialization, localization and identification of the local algebras are used in order to derive stability criteria for the approximation sequences.

**Mathematics Subject Classification (2000).** Primary 65R20; Secondary 45E10, 47B35, 47C15.

**Keywords.** Galerkin method, graded meshes, Wiener Hopf operators.

## 1. Introduction

Wiener-Hopf operators with matrix-valued piecewise continuous symbol (or generating function) appear frequently in applications, namely in Sommerfeld diffraction problems (see, for instance, [1, 4, 9, 14]). Given a matrix Wiener-Hopf operator $A : L^p_N(\mathbb{R}^+) \to L^p_N(\mathbb{R}^+)$, in order to find an approximate solution to the system of equations

$$Au = v, \quad u,\, v \in L^p_N(\mathbb{R}^+),$$

by solving a finite dimension linear system of equations, one considers a discretization of the real line by defining meshes.

---

The author wishes to thank Steffen Roch, who carefully read the manuscript and gave many useful suggestions for its improvement.

Let $(\triangle^n)_{n\in\mathbb{N}}$ denote a sequence of meshes (or partitions) in $\mathbb{R}^+$, $\triangle^n := \{x_k^{(n)} : 0 \le k \le n\}$ such that $0 = x_0^{(n)} \le x_1^{(n)} \le \cdots \le x_n^{(n)} = \infty$. Define the length of the interval between two mesh points as

$$h_{j,n} := x_j^{(n)} - x_{j-1}^{(n)}.$$

We say that the mesh (sequence) is of class $\mathcal{M}$ if $x_{n-1}^{(n)} \to \infty$ and $\max_{1\le j<n} h_{j,n} \to 0$ as $n \to \infty$. Meshes of this type are called graded meshes.

For notation simplification, consider the scalar case $N = 1$. To the sequence of meshes $(\triangle^n)_{n\in\mathbb{N}}$ we associate a sequence of piecewise polynomial spline spaces

$$S^n := \{u : u|_{]x_{j-1}^{(n)}, x_j^{(n)}[} \in \mathbb{P}_d, u|_{]x_{n-1}^{(n)}, \infty[} = 0\},$$

where $j$ runs from 1 to $n-1$ and $\mathbb{P}_d$ represents the set of polynomials of degree less than or equal to $d$ ($d$ fixed). Note that $S^n$ does not depend on $p$. Let $P^n$ be the orthogonal projection from $L^2(\mathbb{R}^+)$ onto $S^n$, which is a closed subspace of $L^p(\mathbb{R}^+)$. This projection is well defined also for $p \neq 2$ (see Proposition 4.2) and is known as the Galerkin projection. Denote by $Q^n$ the complementary projection, that is, $Q^n = I - P^n$.

Consider the approximation method

$$P^n A P^n u_n = P^n v \tag{1}$$

with $u_n \in \mathrm{Im}(P^n)$.

The purpose of this paper is then to study the stability of approximation sequences using spline Galerkin methods, that is, of the stability of sequences of the form $(P^n A P^n)_{n\in\mathbb{N}}$. The continuous symbol case for $L^p$ was already studied by Elschner in [5] (also in [10, Chapter 5]). The $L^2$ piecewise continuous case was studied in [12] using algebraic techniques. In [13], B. Silbermann and the author proposed a general theory for approximation methods of operators in C*-algebras generated by piecewise continuous functions of shifts, which also includes the $L^2$ piecewise continuous case. Numerical methods for singular integral operators in $L^p$ spaces with uniform meshes were studied using algebraic techniques for example by Hagen, Roch and Silbermann in [6]. The non-continuous case in $L^p$ presents substantial difficulties which need to be solved, namely the inverse-closedness of the Banach subalgebras and the appearance of massive local spectra. These difficulties, which are common to other settings like the finite section method, have received some attention in the last years [7, 11]. The results presented here use some techniques inspired from those works and it was possible to solve most of the difficulties, except the one arising from the discontinuity at infinity in the symbol of the Wiener-Hopf operator.

## 2. Spaces definitions

Let $\dot{\mathbb{R}} := \mathbb{R} \cup \{\infty\}$ be the one-point compactification of the real line and $\overline{\mathbb{R}} := [-\infty, +\infty]$ be its two-point compactification. For a domain $\Omega$ contained in the real line, we denote by $L^p(\Omega)$ $(1 \le p < +\infty)$ the Banach (Hilbert for $p = 2$) space of (classes of) Lebesgue measurable complex-valued functions $u$ defined in $\Omega$ such that the norm

$$\|u\|_p := \left( \int_\Omega |u(x)|^p \, dx \right)^{\frac{1}{p}}$$

is finite. In what follows, the set $\Omega$ will be the real line itself or the positive half-axis $\mathbb{R}^+ = \{x \in \mathbb{R} : x > 0\}$. The set of continuous functions defined on $\dot{\mathbb{R}}$ is denoted by $C(\dot{\mathbb{R}})$ and the set of piecewise continuous functions $a$ on $\dot{\mathbb{R}}$, that is, functions with well-defined one-sided limits $a(x_0^{\pm}) = \lim_{x \to x_0^{\pm}} a(x)$ at all points $x_0$ of $\mathbb{R}$ and at $\pm\infty$, by $PC(\dot{\mathbb{R}})$. These sets of functions are considered as subalgebras of $L^\infty(\mathbb{R})$, the algebra of essentially bounded functions on the real line. We will also work with vector-valued functions and thus consider the Banach space $L^p_N(\Omega)$ of the $N$-tuples of functions in $L^p(\Omega)$ with the norm

$$\|u\|_{p,N} := \max_{j=1\ldots N} \|u_j\|_p.$$

From now on, when it is consistent, we will use the same notation for a scalar operator acting in $L^p(\Omega)$ and its trivial extension to a diagonal matrix operator acting in $L^p_N(\Omega)$.

Let $V_1(\mathbb{R})$ be the set of functions $a : \overline{\mathbb{R}} \to \mathbb{C}$ with the finite total variation

$$V_1(a) := \sup \left\{ \sum_{i=1}^{n} |a(x_i) - a(x_{i-1})| \, : \, -\infty \le x_0 < x_1 < \cdots < x_n \le +\infty, n \in \mathbb{N} \right\}.$$

where the supremum is taken over all finite decompositions of the real line $\mathbb{R}$. It is well known that $V_1(\mathbb{R})$ is a Banach algebra under the norm

$$\|a\|_{V_1(\mathbb{R})} := \|a\|_\infty + V_1(a).$$

Let $\mathcal{B} := \mathcal{B}(L^p_N(\mathbb{R}))$ be the Banach algebra of all bounded linear operators on $L^p_N(\mathbb{R})$ and $\mathcal{K} := \mathcal{K}(L^p_N(\mathbb{R}))$ be the closed two-sided ideal of all compact operators on $L^p_N(\mathbb{R})$.

For $a \in PC(\dot{\mathbb{R}})^{N \times N}$ define the multiplication operator in $L^p_N(\mathbb{R})$ by

$$aI : u \mapsto a\, u \;\text{ with }\; (a\, u)(t) = a(t)u(t).$$

By $\chi_+$ and $\chi_-$ we denote the (diagonal) operator of multiplication by the characteristic function of $\mathbb{R}^+$ and $\mathbb{R}^- = \{x \in \mathbb{R} : x < 0\}$, respectively.

Denote by $F$ the Fourier transform defined on the Schwartz space by

$$(Fu)(y) = \frac{1}{\sqrt{2\pi}} \int_{-\infty}^{+\infty} e^{iyx} u(x) \, dx, \; y \in \mathbb{R}$$

and by $F^{-1}$ its inverse

$$(F^{-1}v)(x) = \frac{1}{\sqrt{2\pi}} \int_{-\infty}^{+\infty} e^{-ixy} v(y)\, dy,\ x \in \mathbb{R}.$$

The operators $F$ and $F^{-1}$ can be extended continuously to bounded and unitary operators acting in $L^2(\mathbb{R})$, and these extensions will be denoted by the same symbols.

We define the operator $W^0(a)$ on $L^2(\mathbb{R}) \cap L^p(\mathbb{R})$ by

$$(W^0(a)u)(x) := (F^{-1}aFu)(x), \quad \varphi \in L^2(\mathbb{R}) \cap L^p(\mathbb{R}). \tag{2}$$

A function $a \in L^\infty(\mathbb{R})$ is called a *Fourier multiplier on* $L^p(\mathbb{R})$ if the operator $W^0(a)$ given by (2) can be extended to a bounded linear operator on $L^p(\mathbb{R})$, which will again be denoted by $W^0(a)$. If $a \in L^\infty_{N\times N}(\mathbb{R})$ one can in a natural way define $W^0(a)$ as a bounded operator on $L^p_N(\mathbb{R}^+)$.

The set $\mathcal{M}_p$ of all Fourier multipliers on $L^p(\mathbb{R})$ is defined as

$$\mathcal{M}_p := \left\{ a \in L^\infty(\mathbb{R}) \ : \ W^0(a) \in \mathcal{B}(L^p(\mathbb{R})) \right\}.$$

It is well known that $\mathcal{M}_p$ is a Banach algebra with the norm

$$\|a\|_{\mathcal{M}_p} := \|W^0(a)\|_{\mathcal{B}(L^p(\mathbb{R}))},$$

and

$$\mathcal{M}_p \subset \mathcal{M}_2 = L^\infty(\mathbb{R}) \quad \text{for} \quad 1 < p < \infty. \tag{3}$$

As a consequence of the Stechkin's inequality (see, e.g., [2, Theorem 17.1]), $\mathcal{M}_p$ contains all functions $a \in PC$ of finite total variation and

$$\|a\|_{\mathcal{M}_p} \le \|S_{\mathbb{R}}\|_{\mathcal{B}}\big(\|a\|_\infty + V_1(a)\big),$$

with $S_{\mathbb{R}}$ denoting the usual singular integral operator

$$(S_{\mathbb{R}}u)(x) := \frac{1}{\pi \mathbf{i}} \int_{\mathbb{R}} \frac{u(y)}{y-x}\, dy.$$

Let $C_p(\dot{\mathbb{R}})$ and $C_p(\overline{\mathbb{R}})$ stand for the closure in $\mathcal{M}_p$ of the set of all functions with finite total variation in $C(\dot{\mathbb{R}})$ and $C(\overline{\mathbb{R}})$, respectively. Denote by $PC_p$ the closure in $\mathcal{M}_p$ of the set of all piecewise constant functions on $\mathbb{R}$ which have a finite number of jumps.

We can finally define the Wiener-Hopf operator

$$W(a) = \chi_+ W^0(a) \chi_+$$

acting on $L^p_N(\mathbb{R}^+)$. The Hankel operator

$$H(a) = \chi_+ W^0(a) J \chi_+$$

with $(Ju)(x) = u(-x)$ will also play a role. Here we identify the operator $A$ acting on the range of $\chi_+$ with the "extended" $A + \chi_-$ operator acting on the whole $L^p_N(\mathbb{R})$.

## 3. Projections and related operators

Let $\{\phi^l\}_{l=0}^d, \phi^l : ]0,1[ \to \mathbb{R}$ be an orthonormal polynomial basis for $\mathbb{P}_d$. For each $n$ and $j$, define $J_{j,n} := ]x_{j-1}^{(n)}, x_j^{(n)}[$ and the orthonormal basis (in the $L^2$-induced norm) functions for $S^n$ as

$$\phi_{j,n}^l(x) := \begin{cases} h_{j,n}^{-\frac{1}{2}} \phi^l \left( (x - x_{j-1}^{(n)}) h_{j,n}^{-1} \right) & x \in J_{j,n}, \\ 0 & \text{otherwise.} \end{cases} \tag{4}$$

The Galerkin projection $P^n$ of a function $u \in L^p(\mathbb{R}^+)$ can thus be written as

$$(P^n u)(x) = \sum_{j=1}^{n-1} \sum_{l=0}^{d} \int_{J_{j,n}} u(s) \phi_{j,n}^l(s)\, ds\, \phi_{j,n}^l(x), \tag{5}$$

and the extension to the case $L_N^p(\mathbb{R}^+)$ as a diagonal operator is straightforward.

We define here some operators related with $P^n$ which are necessary in the theory that is going to be applied. Let $\tau$ be a positive real number, let $I$ represent the identity operator and define the following (diagonal) operators acting on $L_N^p(\mathbb{R}^+)$:

$$(P_\tau u)(t) := \begin{cases} u(t) & 0 < t < \tau \\ 0 & t > \tau \end{cases}, \qquad Q_\tau := I - P_\tau,$$

$$(R_\tau u)(t) := \begin{cases} u(\tau - t) & t < \tau \\ 0 & t > \tau \end{cases},$$

$$(V_\tau u)(t) := \begin{cases} 0 & t < \tau \\ u(t - \tau) & t > \tau \end{cases}, \qquad (V_{-\tau} u)(t) := u(t + \tau).$$

The following relations between these operators are very important and easy to show:

**Lemma 3.1.**

- $P^n P_{\tau_n} = P_{\tau_n} P^n = P^n$, $Q^n Q_{\tau_n} = Q_{\tau_n} Q^n = Q_{\tau_n}$, *for* $\tau_n = x_{n-1}^{(n)}$;
- $P_\tau = R_\tau^2$ *and* $R_\tau \to 0$ *weakly as* $\tau \to \infty$;
- $V_\tau V_{-\tau} = Q_\tau$, $V_{-\tau} V_\tau = I$, $(V_\tau)^* = V_{-\tau}$.

In the above lemma, $(V_\tau)^*$ represents the adjoint operator, acting on the dual space $L_N^q(\mathbb{R}^+)$ of $L_N^p(\mathbb{R}^+)$, with $1/p + 1/q = 1$.

Define the oscillation of a function $f$ as

$$\omega(f, \epsilon) := \sup\{|f(s) - f(t)| : |s - t| < \epsilon\}. \tag{6}$$

where the points $s, t$ belong to the domain.

**Proposition 3.2.** *Let* $f \in C(\mathbb{R}^+)$ *be a continuous function and* $u \in L^p(\mathbb{R}^+)$. *Then there exists a constant* $c$ *independent of* $n$ *such that*

$$\|(f P^n - P^n f) u\|_p \le c\,\omega(f, \max_{1 \le j \le n-1} h_{j,n}) \|u\|_p.$$

*Proof.* Let $\{\phi^l_{j,n},\ 0 \le l \le d,\ 0 < j < n-1\}$ be the functions defined in (4), let $f$ be real and let $u$ be a positive function. Then, using (5), the Mean Value Theorem and writing $\phi^l_{j,n} = \phi^{l+}_{j,n} - \phi^{l-}_{j,n}$, with $\phi^{l\pm}_{j,n} \ge 0$, we obtain

$$\begin{aligned}
&\|(fP^n - P^n f)u\|_p^p \\
&= \sum_{j=1}^{n-1} \int_{J_{j,n}} \Bigg| f(s) \left( \sum_{l=0}^{d} \int_{J_{j,n}} u(x)\phi^l_{j,n}(x)\,dx\, \phi^l_{j,n}(s) \right) \\
&\quad - \sum_{l=0}^{d} \int_{J_{j,n}} f(x)u(x)\phi^l_{j,n}(x)\,dx\, \phi^l_{j,n}(s) \Bigg|^p ds \\
&= \sum_{j=1}^{n-1} \Bigg( \int_{J_{j,n}} \Bigg| (f(s)-f(s_{j+})) \left( \sum_{l=0}^{d} \int_{J_{j,n}} u(x)\phi^{l+}_{j,n}(x)\,dx\, \phi^l_{j,n}(s) \right) \\
&\quad -(f(s)-f(s_{j-})) \left( \sum_{l=0}^{d} \int_{J_{j,n}} u(x)\phi^{l-}_{j,n}(x)\,dx\, \phi^l_{j,n}(s) \right) \Bigg|^p ds \Bigg)
\end{aligned}$$

with $s_{j\pm} \in J_{j,n}$. The above is less than or equal to

$$\begin{aligned}
&\sum_{j=1}^{n-1} \Bigg( \int_{J_{j,n}} \Bigg| |f(s)-f(s_{j+})| \left( \sum_{l=0}^{d} \int_{J_{j,n}} u(x)\phi^{l+}_{j,n}(x)\,dx\, |\phi^l_{j,n}(s)| \right) \\
&\quad +|f(s)-f(s_{j-})| \left( \sum_{l=0}^{d} \int_{J_{j,n}} u(x)\phi^{l-}_{j,n}(x)\,dx\, |\phi^l_{j,n}(s)| \right) \Bigg|^p ds \Bigg) \\
&\le \sum_{j=1}^{n-1} \int_{J_{j,n}} \Bigg| \omega(f,h_{j,n}) \sum_{l=0}^{d} \left( \int_{J_{j,n}} |u(x)|^p\,dx \right)^{\frac{1}{p}} \left( \int_{J_{j,n}} |\phi^{l+}_{j,n}(x)|^q\,dx \right)^{\frac{1}{q}} |\phi^l_{j,n}(s)| \\
&\quad + \omega(f,h_{j,n}) \sum_{l=0}^{d} \left( \int_{J_{j,n}} |u(x)|^p\,dx \right)^{\frac{1}{p}} \left( \int_{J_{j,n}} |\phi^{l-}_{j,n}(x)|^q\,dx \right)^{\frac{1}{q}} |\phi^l_{j,n}(s)| \Bigg|^p ds,
\end{aligned} \tag{7}$$

with $q := p/(p-1)$. From (4), making the change of variables $y = (x - x^{(n)}_{j-1})h^{-1}_{j,n}$ one obtains

$$\left( \int_{J_{j,n}} |\phi^{l\pm}_{j,n}(x)|^q\,dx \right)^{\frac{1}{q}} \le \left( \int_{J_{j,n}} |\phi^l_{j,n}(x)|^q\,dx \right)^{\frac{1}{q}} = h_{j,n}^{-\frac{1}{2}+\frac{1}{q}} \left( \int_0^1 |\phi^l(y)|^q\,dy \right)^{\frac{1}{q}},$$

with the last integral being less than a constant $c_q$ due to the equivalence of the norms in the finite-dimensional spline space on $]0,1[$. Using this fact we obtain that (7) is less than or equal to

$$\left(2c_q\omega(f,\max_{1\le j\le n-1}(h_{j,n}))\right)^p\sum_{j=1}^{n-1}h_{j,n}^{-\frac{p}{2}+\frac{p}{q}}\left(\int_{J_{j,n}}|u(x)|^p\,dx\right)\int_{J_{j,n}}\left|\sum_{l=0}^{d}|\phi_{j,n}^l(s)|\right|^p ds. \tag{8}$$

Applying the inequality $(\sum_0^d|a_l|)^p\le(d+1)^{p-1}\sum_0^d|a_l|^p$ and a similar reasoning to the one above we have

$$\begin{aligned}\int_{J_{j,n}}\left|\sum_{l=0}^{d}|\phi_{j,n}^l(s)|\right|^p ds &\le (d+1)^{p-1}\sum_{l=0}^{d}\int_{J_{j,n}}|\phi_{j,n}^l(s)|^p\,ds\\ &\le (d+1)^p h_{j,n}^{-\frac{p}{2}+1}c_p^p.\end{aligned}$$

As $h_{j,n}^{-\frac{p}{2}+\frac{p}{q}}h_{j,n}^{-\frac{p}{2}+1}=1^p=1$, (8) is less than or equal to

$$\left(2c_pc_q(d+1)\omega(f,\max_{1\le j\le n-1}h_{j,n})\right)^p\|u\|_p^p$$

which yields the result. To end the proof we remark that any function in $L^p(\mathbb{R})$ can be written as a difference of two positive functions and that any continuous function $f$ is $f_1+if_2$, with $f_{1,2}$ real. □

## 4. Approximation methods

The results obtained so far regarding the meshes and methods under study in the $L^p$ setting were obtained by Elschner and are resumed in the next theorem (see [10, Theorem 5.37])

**Theorem 4.1.** *Let $b\in L^1(\mathbb{R})$ and let $a=1-Fb\in C(\dot{\mathbb{R}})$. If the operator $A=W(a)$ is invertible in $L^p(\mathbb{R}^+)$, then the approximation method (1) is stable, that is, there is an integer $n_0$ such that for all $n>n_0$ the operators $P^nAP^n$ are invertible in $S^n$ and the norms of the inverses are uniformly bounded.*

The result below is in [10, Theorem 5.25] with a very summary proof. A more detailed proof is presented here.

**Proposition 4.2.** *The operators $P^n$ are uniformly bounded in $L^p_N(\mathbb{R}^+)$ and $P^n\to I$, $(P^n)^*\to I$ strongly as $n\to\infty$.*

*Proof.* It is sufficient to show the results for the scalar case. To prove the uniform boundedness consider the interval $]0,1[$, and the finite-dimensional space of the polynomials of degree less than or equal to $d$ defined on the interval, with the orthogonal basis $\{\phi^l\}_{l=0}^d$ (for the $L^2$-induced norm). Let now $P$ be the orthogonal (in $L^2(]0,1[)$) projection onto the finite-dimensional space. One has

$$\|Pu\|_{p,]0,1[} = \left\| \sum_{l=0}^{d} \langle u, \phi^l \rangle \phi^l \right\|_{p,]0,1[} \leq (d+1) \max_l \|\langle u, \phi^l \rangle \phi^l\|_{p,]0,1[}$$
$$\leq (d+1) \max_l \|\phi^l\|_\infty^2 \|u\|_{p,]0,1[},$$

which means there is a constant $c$, such that

$$\|Pu\|_{p,]0,1[} \leq c\|u\|_{p,]0,1[}.$$

Given a function $v \in L^p(\mathbb{R}^+)$ and an interval $J_{j,n} \subset \mathbb{R}$, define $u(x) := v(x_{j-1} + xh_{j,n})$. Then

$$(P^n v)(x) = (Pu)\left((x - x_{j-1}^{(n)})h_{j,n}^{-1}\right) \quad \text{for } x \in J_{j,n}$$

and it is possible to conclude that $\|P^n v\|_{p,J_{j,n}} \leq c\|v\|_{p,J_{j,n}}$. Thus

$$\|P^n v\|_p \leq c\|v\|_p$$

with $c$ independent of $n$, which proves the uniform boundedness.

To prove the strong convergence, suppose first that the function $u$ is the characteristic function of an interval $[x_a, x_b] \subset \mathbb{R}^+$. Then $u - P^n u = 0$ on all intervals $J_j$ except those containing the points $x_a$ and $x_b$. Denote them by $J_a$ and $J_b$, respectively. Write $u_n$ for the characteristic function of the interval $[x_a, x_b] \setminus (J_a \cup J_b)$. Then

$$\|(I - P^n)u\|_p \leq \|u - u_n\|_p + \|u_n - P^n u_n\|_p + \|P^n u_n - P^n u\|_p.$$

Because $u_n = P^n u_n$ the middle term on the right-hand side of the expression above is zero. We have also that $\|P^n u_n - P^n u\|_p \leq \|P^n\| \|u - u_n\|_p \leq c\|u - u_n\|_p$ due to the uniform boundedness of $P^n$. All that remains to verify is that $\|u - u_n\|_p$ tends to zero, which is obviously true because the function is piecewise constant and the length of the intervals in which it is not null decreases to zero, as $n \to \infty$.

The convergence is then true for any function $u$ which is a linear combination of characteristic functions of finite intervals, and any function of $L^p(\mathbb{R}^+)$ can be uniformly approximated by such piecewise constant functions. □

We say that a sequence $(A_n)_{n\in\mathbb{N}}$ converges uniformly to an operator $A$ if for any $\epsilon > 0$ there exists a $n_0 \in \mathbb{N}$ such that for all $n > n_0$ the inequality $\|A_n - A\| < \epsilon$ holds.

The formal concept of stability is defined now. We say that a sequence $(A_n)_{n\in\mathbb{N}}$ with uniformly bounded $A_n : L_N^p(\mathbb{R}^+) \to L_N^p(\mathbb{R}^+)$ is stable if there exists a $n_0$ such that, for all $n > n_0$, $A_n$ is invertible and the norms $\|A_n^{-1}\|$ are uniformly bounded.

## 5. Algebraization

Define $\mathcal{E}$ as the Banach algebra of all sequences $\mathbf{A} := (A_n)_{n\in\mathbb{N}}$ with $A_n : S^n \to S^n$ such that $\sup_{n\in\mathbb{N}} \|A_n\|_{L^p(\mathbb{R})} < \infty$ (with pointwise defined operations, for example, $(A_n)(B_n) = (A_nB_n)$). The algebra $\mathcal{E}$ is unital with unit $\mathbf{I} = (P^n)$. Denote by $\mathcal{G} \subset \mathcal{E}$ the closed ideal of the sequences which tend uniformly to zero.

The following result goes back to Kozak [8]. For a proof, see, e.g., [11, Section 6.1].

**Theorem 5.1.** *A sequence $(A_n) \in \mathcal{E}$ is stable if and only if the coset $(A_n) + \mathcal{G}$ is invertible in the quotient algebra $\mathcal{E}/\mathcal{G}$.*

## 6. Essentialization

We will say that a sequence of operators on a Banach space converges **-strongly* if it converges strongly and the sequence of the adjoint operators converges strongly on the dual space.

Write $\tau_n$ for the "last" finite point of the mesh, $x_{n-1}^{(n)}$.

Let $\mathcal{F}$ denote the set of all sequences $\mathbf{A} := (A_n) \in \mathcal{E}$ such that the sequences $(A_n)$ and $(R_{\tau_n} A_n R_{\tau_n})$ are *-strongly convergent as $n \to +\infty$.

**Lemma 6.1.**

(i) *The set $\mathcal{F}$ is a closed unital subalgebra of the algebra $\mathcal{E}$.*

(ii) *Let $i \in \{0, 1\}$. The mappings $O_i : \mathcal{F} \to \mathcal{B}$ given by*

$$O_0(\mathbf{A}) := \operatorname*{s-lim}_{n\to\infty} A_n,$$

$$O_1(\mathbf{A}) := \operatorname*{s-lim}_{n\to\infty} R_{\tau_n} A_n R_{\tau_n}$$

*for $\mathbf{A} = (A_n) \in \mathcal{F}$ are bounded unital homomorphisms with norm 1.*

(iii) *The set $\mathcal{G}$ is a closed two-sided ideal of the algebra $\mathcal{F}$.*

(iv) *The ideal $\mathcal{G}$ lies in the kernel of the homomorphisms $O_0$ and $O_1$.*

(v) *The algebra $\mathcal{F}$ is inverse-closed in the algebra $\mathcal{E}$ and the algebra $\mathcal{F}/\mathcal{G}$ is inverse-closed in the algebra $\mathcal{E}/\mathcal{G}$.*

*Proof.* The proofs of (i) and (v) follow the proof of [11, Proposition 6.5.1], taking into account Lemma 3.1 and the different unit sequence $\mathbf{I} = (P^n)$ in $\mathcal{E}$. The proofs of (ii)–(iv) are immediate. □

The algebra $\mathcal{F}/\mathcal{G}$ is still too large to obtain a verifiable invertibility condition for its elements. It is thus necessary to apply a Lifting Theorem (see [11, Theorem 6.2.7]). Let $\mathcal{A}$ and $\mathcal{B}$ be unital algebras and $\mathsf{W} : \mathcal{A} \to \mathcal{B}$ a unital homomorphism. We say that an ideal $\mathcal{I}$ of $\mathcal{A}$ is lifted by the homomorphism $\mathsf{W}$ if $\mathrm{Ker}\,\mathsf{W} \cap \mathcal{I}$ lies in the radical of $\mathcal{A}$.

**Theorem 6.2 (Lifting theorem).** *Let $\mathcal{A}$ be a unital algebra and, for every element $k$ of a certain set $T$, let $\mathcal{J}_k$ be an ideal of $\mathcal{A}$ which is lifted by a unital homomorphism $\mathsf{W}_k : \mathcal{A} \to \mathcal{B}_k$. Suppose furthermore that $\mathsf{W}_t(\mathcal{J}_k)$ is an ideal of $\mathcal{B}_k$. Let $\mathcal{J}$ stand for the smallest ideal of $\mathcal{A}$ which contains all ideals $\mathcal{J}_k$. Then an element $a \in \mathcal{A}$*

*is invertible if and only if the coset* $a+\mathcal{J}$ *is invertible in* $\mathcal{A}/\mathcal{J}$ *and if all elements* $\mathsf{W}_k(a)$ *are invertible in* $\mathcal{B}_k$.

The natural candidates for the homomorphisms $\mathsf{W}_t$ above, in the case under study, are $O_0$ and $O_1$. Let $\mathcal{K}\subset\mathcal{L}(L^p_N(\mathbb{R}^+))$ denote the ideal of compact operators and define $\mathcal{J}_0$ and $\mathcal{J}_1$ to be the sets

$$\mathcal{J}_0 := \{(P^nKP^n)+(G_n),\ \ K\in\mathcal{K},\ (G_n)\in\mathcal{G}\},$$
$$\mathcal{J}_1 := \{(P^nR_{\tau_n}KR_{\tau_n}P^n)+(G_n),\ \ K\in\mathcal{K},\ (G_n)\in\mathcal{G}\},$$

in $\mathcal{F}$ and $\mathcal{J}_t^{\mathcal{G}} := \mathcal{J}_t/\mathcal{G}$ in $\mathcal{F}/\mathcal{G}$. Because $O_{0,1}(\mathcal{G})=\{0\}$, the homomorphisms are well defined in the quotient algebra $\mathcal{F}/\mathcal{G}$, and

$$O_0((P^nK_1P^n+P^nR_{\tau_n}K_2R_{\tau_n}P^n+G_n)) = K_1, \tag{9}$$
$$O_1((P^nK_1P^n+P^nR_{\tau_n}K_2R_{\tau_n}P^n+G_n)) = K_2. \tag{10}$$

The proof of the next result is standard (see for instance [11, Proposition 6.5.2] or [12, Proposition 4.2]).

**Proposition 6.3.** $\mathcal{J}_0$ *is a closed two-sided ideal of* $\mathcal{F}$.

To prove the same result for $\mathcal{J}_1$ we need the following auxiliary result:

**Lemma 6.4.** *Let* $K\in\mathcal{K}$. *Then* $(Q^nR_{\tau_n}K)$ *tends uniformly to* $0$ *as* $n\to\infty$.

*Proof.* We will begin by proving that $\|(I-P^n)R_{\tau_n}u\|_p\to 0$ as $n\to\infty$, for any $u\in L^p$. Suppose that $u$ is $\chi_{[x_a,x_b]}$, the characteristic function of the interval $[x_a,x_b]\subset\mathbb{R}^+$. Define $u^\tau := R_\tau u = \chi_{[\tau-x_b,\tau-x_a]}$. As in the proof of Proposition 4.2, $u^\tau - P^nu^\tau = 0$ for all intervals $J_j$ except those containing the points $\tau-x_a$ and $\tau-x_b$, which we denote by $J_a$ and $J_b$. Write $u_n^\tau$ for the characteristic function of the interval $[\tau-x_b,\tau-x_a]\setminus(J_a\cup J_b)$.

Thus $\|u^\tau-u_n^\tau\|_p$ tends to zero. The result now follows because multiplication on the right by a compact operator transforms strong convergence into uniform convergence (see, for instance [10, 1.1.(h)]). □

**Proposition 6.5.** $\mathcal{J}_1$ *is a closed two-sided ideal of* $\mathcal{F}$.

*Proof.* From (9) and (10) we have $\mathcal{J}_1\subset\mathcal{F}$. To prove that $\mathcal{J}_1$ is a left ideal denote by $\tilde{A}$ the limit $\text{s-lim}_{n\to\infty}R_{\tau_n}A_nR_{\tau_n}$. Now, $(A_n)(P^nR_{\tau_n}KR_{\tau_n}P^n)$ can be written as

$$\begin{aligned}(P^nR_{\tau_n}\tilde{A}KR_{\tau_n}P^n)&+(P^nR_{\tau_n}(R_{\tau_n}A_nR_{\tau_n}-\tilde{A})KR_{\tau_n}P^n)\\ &-(P^nA_n(I-P^n)R_{\tau_n}KR_{\tau_n}P^n)\end{aligned}$$

and note that the second term is in $\mathcal{G}$ because $\text{s-lim}_{n\to\infty}R_{\tau_n}A_nR_{\tau_n}-\tilde{A}=0$, and the third term also, due to Lemma 6.4. The proof that $\mathcal{J}_1$ is right ideal is the same, taking into account that we must apply adjoints to Lemma 6.4 so that we get $\|KR_{\tau_n}(I-P^n)\| = \|(KR_{\tau_n}(I-P^n))^*\| = \|(I-P^n)^*R_{\tau_n}^*K^*\| = \|(I-P^n)R_{\tau_n}K^*\|$. The closedness proof is again standard. □

Define $\mathcal{J}$ as the smallest closed two-sided ideal of $\mathcal{F}$ containing both $\mathcal{J}_0$ and $\mathcal{J}_1$. We can now particularize Theorem 6.2 to the algebra $\mathcal{F}/\mathcal{G}$ and obtain:

**Corollary 6.6.** *Let $\mathbf{A} \in \mathcal{F}$. Then the sequence $\mathbf{A}$ is stable if and only if the limits $O_0(\mathbf{A})$ and $O_1(\mathbf{A})$ are invertible operators in $L^p_N(\mathbb{R}^+)$ and the coset $\mathbf{A} + \mathcal{J}$ is invertible in $\mathcal{F}/\mathcal{J}$.*

We now turn to the question of what interesting sequences are in $\mathcal{F}$, besides the ones corresponding to the ideals $\mathcal{J}_0$ and $\mathcal{J}_1$.

**Proposition 6.7.** *Let $m \in \mathbb{N}$, $A_n := P^n A^{(0)} \left(\prod_{k=1}^m P_{\tau_n} A^{(k)}\right) P^n$, where $A^{(k)}$ are operators of the form $W(a)$ with $a \in PC_p^{N\times N}$. Then $\mathbf{A} := (A_n) \in \mathcal{F}$.*

*Proof.* Indeed, it is easy to verify that the limit $O_0(\mathbf{A})$ exists, because all operators in the finite product are either independent of $n$ or have well-defined strong limits. Regarding $O_1$ we have

$$R_{\tau_n} P^n R_{\tau_n} \to I,$$
$$R_{\tau_n} P_{\tau_n} R_{\tau_n} \to I,$$

and

$$\begin{aligned} O_1((P^n W(a) P^n)) &= \operatorname*{s-lim}_{n\to\infty} R_{\tau_n} P^n W(a) P^n R_{\tau_n} \\ &= \operatorname*{s-lim}_{n\to\infty} R_{\tau_n} P^n R_{\tau_n} R_{\tau_n} W(a) R_{\tau_n} R_{\tau_n} P^n R_{\tau_n}. \end{aligned}$$

Because $R_{\tau_n} P^n R_{\tau_n} \to I$, it is only necessary to know the strong limit of $R_{\tau_n} W(a) R_{\tau_n}$. It is not difficult to see that $R_\tau W(a) R_\tau = P_\tau W(\tilde{a}) P_\tau$, with $\tilde{a}(x) = a(-x)$, and so the strong limit is $W(\tilde{a})$.

One can now show that $O_1(\mathbf{A})$ exists. Due to $P_{\tau_n} = R^2_{\tau_n}$, we can write

$$\begin{aligned} O_1(\mathbf{A}) &= \operatorname*{s-lim}_{n\to\infty} R_{\tau_n} P^n A^{(0)} \left(\prod_{k=1}^m P_{\tau_n} A^{(k)}\right) P^n R_{\tau_n} \\ &= \operatorname*{s-lim}_{n\to\infty} P^n R_{\tau_n} A^{(0)} R_{\tau_n} \left(\prod_{k=1}^m R_{\tau_n} A^{(k)} R_{\tau_n}\right) P^n \end{aligned}$$

which is a product of $m+1$ factors of the form $R_{\tau_n} W(a) R_{\tau_n}$, each with a well-defined strong limit, with two sequences of operators tending strongly to the identity as the first and last factors. □

It is thus possible to write:

**Theorem 6.8.** *Let $A$ be the Wiener-Hopf operator $W(a)$ with $a \in PC_p$. The method (1) applied to $A$ is stable if and only if the operators $W(a)$ and $W(\tilde{a})$ are invertible and $(P^n A P^n) + \mathcal{J}$ is invertible in $\mathcal{F}/\mathcal{J}$.*

## 7. Localization

Recall that the *center* of an algebra $\mathcal{A}$ is the set

$$\{c \in \mathcal{A} : ac = ca \text{ for all } a \in \mathcal{A}\}.$$

Let $A$ be a Banach algebra with identity. The following result is well known. A proof can be found for instance in [11, Theorem 2.2.2 (a)].

**Theorem 7.1 (Allan's local principle).** *Let $\mathcal{A}$ be a Banach algebra with identity $e$ and let $\mathcal{B}$ be closed subalgebra of the center of $\mathcal{A}$ containing $e$. Let $M_{\mathcal{B}}$ be the maximal ideal space of $\mathcal{B}$, and for $x \in M_{\mathcal{B}}$, let $\mathcal{I}_x$ refer to the smallest closed two-sided ideal of $\mathcal{A}$ containing the ideal $x$. Then an element $a \in \mathcal{A}$ is invertible in $\mathcal{A}$ if and only if $a + \mathcal{I}_x$ is invertible in the quotient algebra $\mathcal{A}/\mathcal{I}_x$ for all $x \in M_{\mathcal{B}}$.*

There are several techniques to obtain an algebra with a large center, in order to apply localization techniques. Here we follow a similar one to that in [7]. Denote by $\mathcal{L}$ the set of all sequences $\mathbf{A}$ in $\mathcal{F}$ such that

$$\mathbf{AC} - \mathbf{CA} \in \mathcal{J}, \tag{11}$$

for all sequences $\mathbf{C}$ belonging to the set $\mathcal{C} := \{(P^n W(fI_N)P^n) : f \in C_p(\dot{\mathbb{R}})\}$, where $W(fI_N)$ represents the trivial diagonal extension of the scalar operator $W(f)$.

**Lemma 7.2.** (i) *The set $\mathcal{L}$ is a closed unital subalgebra of the algebra $\mathcal{F}$.*
(ii) *The set $\mathcal{J}$ is a closed two-sided ideal of the algebra $\mathcal{L}$.*
(iii) *The algebra $\mathcal{L}/\mathcal{J}$ is inverse-closed in the algebra $\mathcal{F}/\mathcal{J}$.*

*Proof.* By definition, $\mathcal{L} \subset \mathcal{F}$. It easy to see that $\mathcal{L}$ contains the identity of $\mathcal{F}$ and is closed for the algebra operations. To prove that $\mathcal{L}$ is topologically closed in $\mathcal{F}$, suppose $(\mathbf{A}_j)_{j\in\mathbb{N}}$ with $\mathbf{A}_j = (A_n^{(j)}) \in \mathcal{L}$ is a Cauchy sequence in $\mathcal{L}$. It has a limit $\mathbf{A} = (A_n) \in \mathcal{F}$. We will show that $\mathbf{A} \in \mathcal{L}$. From (11) we get that the sequence $\mathbf{J}_k = \mathbf{A}_j\mathbf{C} - \mathbf{CA}_j \in \mathcal{J}$, and it has the limit $\mathbf{AC} - \mathbf{CA}$. Because $\mathcal{J}$ is closed, $\mathbf{AC} - \mathbf{CA}$ must also belong to $\mathcal{J}$, and the result follows. Part (ii) is immediate. Regarding part (iii), let $\mathbf{A} \in \mathcal{L}$ and $\mathbf{A} + \mathcal{J}$ be invertible in $\mathcal{F}/\mathcal{J}$. Then there exist sequences $\mathbf{B}, \mathbf{J}_{1,2} \in \mathcal{F}$ such that $\mathbf{J}_1$, $\mathbf{J}_2$ belongs to $\mathcal{J}$ and

$$\mathbf{I} = \mathbf{AB} + \mathbf{J}_1, \quad \mathbf{I} = \mathbf{BA} + \mathbf{J}_2.$$

Let $\mathbf{C} \in \mathcal{C}$. We have

$$\begin{aligned}\mathbf{BC} - \mathbf{CB} &= \mathbf{BC}(\mathbf{AB} + \mathbf{J}_1) - (\mathbf{BA} + \mathbf{J}_2)\mathbf{CB} \\ &= \mathbf{BCAB} - \mathbf{BACB} + \mathbf{J}' \\ &= \mathbf{B}(\mathbf{CA} - \mathbf{AC})\mathbf{B} + \mathbf{J}',\end{aligned}$$

with $\mathbf{J}' \in \mathcal{J}$. Because $\mathbf{CA} - \mathbf{AC} \in \mathcal{J}$, $\mathbf{B} \in \mathcal{L}$ and therefore (iii) is proved. □

It is possible to apply Allan's local principle to the algebra $\mathcal{L}/\mathcal{J}$, due to its large center. It is necessary now to show that $\mathcal{L}$ contains the sequences that interest us, namely, the ones corresponding to Wiener-Hopf operators with piecewise continuous symbols. For this we use the following two results.

**Proposition 7.3.** *If $f \in C_p(\dot{\mathbb{R}})$ and $f(\infty) = 0$ then*

$$\|P_{\tau_n} Q^n W(f)\| \to 0, \quad \|W(f)P_{\tau_n} Q^n\| \to 0 \text{ as } n \to \infty.$$

*Proof.* Since it is possible to approximate $f$ on the $\mathcal{M}_p$ norm by functions $f_j$ such that $k_j = F^{-1} f_j$ are functions in $L^1(\mathbb{R})$ with derivatives also in $L^1(\mathbb{R})$, we may assume without loss of generality that $f$ itself has this property. In this way we obtain the continuity of $W(f)$ from $L^p(\mathbb{R}^+)$ to $L^{p,1}(\mathbb{R}^+)$ (the subspace of $L^p(\mathbb{R}^+)$ containing the functions also with derivative in $L^p(\mathbb{R}^+)$ – see [10, Remark 5.6(ii)]). Using [10, Lemma 5.25] we then get that there exists a constant $c$ such that

$$\|Q^n W(f)u\|_{J_j} \leq c h_{j,n} \|DW(f)u\|_{J_j},$$

with $h_{j,n} = x_j^{(n)} - x_{j-1}^{(n)}$ and $D$ representing the operator of differentiation. So we obtain

$$\begin{aligned} \|P_{\tau_n} Q^n W(f)u\|_{L^p_N} &= \|Q^n W(f)u\|_{[0,\tau_n]} \\ &\leq c_1 (\max_{1\leq j\leq n-1} h_{j,n}) \|DW(f)u\|_{[0,\tau_n]} \\ &\leq c_2 (\max_{1\leq j\leq n-1} h_{j,n}) \|u\|_{[0,\tau_n]}, \end{aligned}$$

for some constants $c_1, c_2$. Now as $\max_{1\leq j\leq n-1} h_{j,n}$ goes to zero as $n \to \infty$ the first result is proved. To prove the second, one uses a duality argument. □

**Proposition 7.4.** *Let $f \in C_p(\dot{\mathbb{R}})$ with $f(\infty) = 0$ and $K$ represent any compact operator. Then the following pairs of sequences differ only by a sequence which tends in the norm to zero. We represent this relation by the sign $\sim$.*

$$P^n W(f) \simeq P_{\tau_n} W(f), \quad W(f)P^n \simeq W(f)P_{\tau_n}; \tag{12}$$

$$P^n K \simeq P_{\tau_n} K, \quad KP^n \simeq KP_{\tau_n}; \tag{13}$$

$$P^n R_{\tau_n} K \simeq P_{\tau_n} R_{\tau_n} K, \quad KR_{\tau_n} P^n \simeq KR_{\tau_n} P_{\tau_n}. \tag{14}$$

*The same relations hold with $P_{\tau_n}$ and $P^n$ substituted by $Q_{\tau_n}$ and $Q^n$, respectively.*

*Proof.* The relations (12) are due to Proposition 7.3. Regarding (13) the result is evident because $P_{\tau_n} = P^n + P_{\tau_n} Q^n$ and $P_{\tau_n} Q^n$ tends strongly to zero, together with their adjoints. To prove the relations (14) note that $R_{\tau_n} Q^n R_{\tau_n} = P_{\tau_n} Q'^n$, where $Q'^n$ is a projection, similar to $Q^n$ but with the "rotated" mesh $\{\tau_n - x_k^{(n)} : 0 \leq k \leq n\}$, which has the same properties of the original mesh. □

Let $I_N$ represent the $N \times N$ identity matrix.

**Proposition 7.5.** *Let $f \in C_p(\dot{\mathbb{R}})$ and $A_n$ be the operators defined in Proposition 6.7. Then we have $(P^n W(fI_N)P^n)(A_n) - (A_n)(P^n W(fI_N)P^n) \in \mathcal{J}$.*

*Proof.* First write $W(fI_N) = W\big(f - f(\infty)\big)I_N\big) + f(\infty)I$. It is thus sufficient to prove the result for continuous functions $f$ which take the value 0 at infinity.

The proof will be done by induction with respect to the number of factors of $A_n$. We start with one factor. Using the well-known equality

$$W(ab) = W(a)W(b) + H(a)H(\tilde{b}) \tag{15}$$

where $\tilde{b}$ is the function defined by $\tilde{b}(x) := b(-x)$, we get

$$\begin{aligned} &P^n W(fa) P^n \\ &\quad = P^n W(fI_N) P^n W(a) P^n + P^n W(fI_N) Q^n W(a) P^n + P^n H(fI_N) H(\tilde{a}) P^n. \end{aligned}$$

As $H(fI_N)$ is compact, the third term is in $\mathcal{J}$. The second can be written as

$$\begin{aligned} P^n W(fI_N) Q^n W(a) P^n &\simeq P^n W(fI_N) Q_{\tau_n} W(a) P^n \\ &= P^n R_{\tau_n} R_{\tau_n} W(fI_N) V_{\tau_n} V_{-\tau_n} W(a) R_{\tau_n} R_{\tau_n} P^n. \end{aligned}$$

For $0 < x < \tau$ we have

$$\begin{aligned} (R_\tau W(f) V_\tau u)(x) &= \frac{1}{2\pi} \int_{-\infty}^{+\infty} e^{-i(\tau - x)\xi} f(\xi) \int_\tau^\infty e^{i\xi y} u(y - \tau)\, dy\, d\xi \\ &= \frac{1}{2\pi} \int_{-\infty}^{+\infty} e^{-ix(-\xi)} f(\xi) \int_0^\infty e^{i\xi y'} u(y')\, dy'\, d\xi = P_\tau H(\tilde{f}), \end{aligned}$$

and with the same arguments it can be proved that $V_{-\tau} W(a) R_\tau = H(a) P_\tau$, consequently,

$$P^n W(fI_N) Q_{\tau_n} W(a) P^n = P^n R_{\tau_n} H(\tilde{f} I_N) H(a) R_{\tau_n} P^n,$$

and $H(\tilde{f} I)$ is compact because $\tilde{f}$ is continuous. One concludes then that

$$(P^n W(fI_N) P^n W(a) P^n) - (P^n W(fa) P^n) \in \mathcal{J}.$$

Similarly, one has

$$(P^n W(a) P^n W(fI_N) P^n) - (P^n W(fa) P^n) \in \mathcal{J}.$$

Now we turn to the induction step. For notational simplicity, we will only explain this step for the passage from one factor to two factors. A little thought shows that the same argument works for the general induction step. We will use the relations (12)–(14), which allow the substitution of the projections $P^n$ by the projection $P_{\tau_n}$ under certain conditions. Some care is in order because formally those sequences by themselves do not belong to $\mathcal{E}$, since they do not act on the spline space $S^n$.

So, let $A_n = P^n W(a) P_{\tau_n} W(b) P^n$. We have:

$$\begin{aligned} P^n W(fI_N) P^n A_n \simeq\ & P^n W(fI_N) W(a) P_{\tau_n} W(b) P^n && (16) \\ & - P^n W(fI_N) Q_{\tau_n} W(a) P_{\tau_n} W(b) P^n && (17) \end{aligned}$$

where with a similar reasoning as in the first part of the proof we obtain that (17) is equal to

$$P^n R_{\tau_n} H(\tilde{f} I_N) H(a) R_{\tau_n} P_{\tau_n} W(b) P^n$$

with $H(\tilde{f} I_N)$ compact and thus belongs to $\mathcal{J}$. Regarding the right-hand side of (16), one can write

$$\begin{aligned} P^n W(fI_N) W(a) P_{\tau_n} W(b) P^n \simeq\ & P^n W(a) W(fI_N) P^n W(b) P^n && (18) \\ & + P^n K P^n W(b) P^n, && (19) \end{aligned}$$

with a compact operator $K$. The sequence in (19) is in $\mathcal{J}$. It is possible now to repeat the arguments to show that

$$P^nW(a)W(fI_N)P^nW(b)P^n = P^nW(a)P^nW(fI_N)P^nW(b)P^n + J_n$$

where $(J_n) \in \mathcal{J}$. Continuing to use the same reasoning as before one finally obtains

$$(P^nW(fI_N)P^n)(A_n) - (A_n)(P^nW(fI_N)) \in \mathcal{J}. \qquad \square$$

The last proposition shows that the sequences $(A_n)$ formed by the operators defined in Proposition 6.7 belong to the algebra $\mathcal{L}$. Thus the algebra $\mathcal{L}/\mathcal{J}$ has the cosets of the sequences of which we want to study stability, and has a large enough center that allows invertibility criteria via localization. Write $\mathcal{L}^{\mathcal{J}}$ for $\mathcal{L}/\mathcal{J}$. It is also not difficult to see that the set $\mathcal{C}^{\mathcal{J}} := \mathcal{C}/\mathcal{J}$ forms a closed commutative subalgebra of $\mathcal{L}^{\mathcal{J}}$, and is contained in its center. The algebra $\mathcal{C}^{\mathcal{J}}$ is isomorphic to the algebra $C_p$. The maximal ideal space is formed by the cosets $(P^nW(f_xI)P^n) + \mathcal{J}$ with $f_x(x) = 0$ and $x \in \dot{\mathbb{R}}$, and it is topologically isomorphic to $\dot{\mathbb{R}}$.

In order to apply Allan's localization principle, let $\mathcal{I}_\eta$, $\eta \in \dot{\mathbb{R}}$, be the smallest closed two-sided ideal of $\mathcal{L}^{\mathcal{J}}$ which contains the ideal $\eta$ of $\mathcal{C}^{\mathcal{J}}$. We denote the canonical homomorphism $\mathcal{L}^{\mathcal{J}} \to \mathcal{L}^{\mathcal{J}}/\mathcal{I}_\eta =: \mathcal{L}^{\mathcal{J}}_\eta$ by $\Phi^{\mathcal{J}}_\eta$.

After localization, Theorem 6.8 takes the form:

**Theorem 7.6.** *Let $\mathbf{A} \in \mathcal{L}$. The sequence $\mathbf{A}$ is stable if and only if $O_0(\mathbf{A})$ and $O_1(\mathbf{A})$ are invertible, and for all $\eta \in \dot{\mathbb{R}}$, $\Phi^{\mathcal{J}}_\eta(\mathbf{A})$ is invertible in $\mathcal{L}^{\mathcal{J}}_\eta$.*

## 8. Identification of the local algebras

Proposition 7.5 shows that the sequences $\mathbf{A} := (P^nW(a)P^n)$ with $a \in PC_p^{N\times N}(\dot{\mathbb{R}})$ belong to the algebra $\mathcal{L}$. It is thus possible to study the local invertibility of the coset $\Phi^{\mathcal{J}}_x(\mathbf{A})$.

Given any positive real number $\tau$, define the operator

$$Z_\tau : L^p_N(\mathbb{R}^+) \to L^p_N(\mathbb{R}^+), \quad (Z_\tau u)(x) = \tau^{-1/p}u(x/\tau).$$

Obviously $\|Z_\tau\|_{\mathcal{B}} = 1$. It is also clear that $Z_\tau$ is invertible and $Z_\tau^{-1} = Z_{1/\tau}$.

For $\eta \in \mathbb{R}$, put

$$U_\eta : L^p_N(\mathbb{R}^+) \to L^p_N(\mathbb{R}^+), \quad (U_\eta u)(x) = e^{i\eta x}u(x).$$

It is clear that $U_\eta^{-1} = U_{-\eta}$ and $\|U_\eta^{\pm 1}\|_{\mathcal{B}} = 1$.

Let $\mathcal{H}_\eta$ denote the set of all sequences $\mathbf{A} = (A_n) \in \mathcal{E}$ such that the sequence $(Z_{\tau_n}^{-1}U_\eta A_n U_\eta^{-1}Z_{\tau_n})$ is *-strongly convergent as $n \to +\infty$.

**Lemma 8.1.** *Let $\eta \in \mathbb{R}$.*

(i) *The set $\mathcal{H}_\eta$ is a closed unital subalgebra of the algebra $\mathcal{E}$.*
(ii) *The mapping $\mathsf{H}_\eta : \mathcal{H}_\eta \to \mathcal{B}$ given by*

$$\mathsf{H}_\eta(\mathbf{A}) := \operatorname*{s-lim}_{n\to+\infty} Z_{\tau_n}^{-1}U_\eta A_n U_\eta^{-1}Z_{\tau_n}$$

*for* $\mathbf{A} = (A_n) \in \mathcal{H}_\eta$ *is a bounded unital homomorphism with the norm*

$$\|\mathsf{H}_\eta\| = 1.$$

(iii) *The set* $\mathcal{G}$ *is a closed two-sided ideal of the algebra* $\mathcal{H}_\eta$.
(iv) *The ideal* $\mathcal{G}$ *lies in the kernel of the homomorphism* $\mathsf{H}_\eta$.
(v) *The algebra* $\mathcal{H}_\eta/\mathcal{G}$ *is inverse closed in the algebra* $\mathcal{E}/\mathcal{G}$.

The proof is again analogous to the proof of [11, Proposition 6.5.1].

Denote by $\mathcal{A}$ be the smallest closed subalgebra of $\mathcal{E}$ containing the sequences $(P^nW(a)P^n)$ with $a \in PC_p^{N\times N}(\dot{\mathbb{R}})$.

**Lemma 8.2.** *Let* $f \in C(\mathbb{R})$ *be a uniformly continuous function. Then* $\|fP^n - P^n fI\|_{\mathcal{L}(L^p(\mathbb{R}^+))} \to 0$.

*Proof.* It is just necessary to remark that if $f$ is a uniformly continuous function, then $\omega(f, \max_{1\le j\le n-1} h_{j,n})$ tends to zero and apply Proposition 3.2. □

**Proposition 8.3.** *If* $(A_n) \in \mathcal{A}$ *and* $\eta \in \mathbb{R}$, *then the limit* $\mathsf{H}_\eta(A_n)$ *exists. In particular,*

(i) $\mathsf{H}_\eta(P^n) = P_1$;
(ii) $\mathsf{H}_\eta(W(a)) = a(\eta^-)W(\chi_- I_N) + a(\eta^+)W(\chi_+ I_N)$ *for* $a \in PC^{N\times N}(\dot{\mathbb{R}})$;
(iii) *if* $(j_\tau) \in \mathcal{J}_0$ *or* $(j_\tau) \in \mathcal{J}_1$, *then* $\mathsf{H}_\eta(j_\tau) = 0$.

The proof of the result above is the same as the one in [12, Lemma 7.4] taking into account Corollary 8.2.

We are now in a position to identify the local algebras. We will start by proving that these local algebras are singly generated (without counting the matrix multipliers). Let $I_\eta$ represent the identity in the local algebra $\mathcal{L}_\eta^{\mathcal{J}}$.

**Proposition 8.4.** *If* $f \in C_p(\dot{\mathbb{R}})$ *and* $\eta \in \mathbb{R}$ *then* $\Phi_\eta^{\mathcal{J}}(P^nW(fI_N)P^n) = f(\eta)I_\eta$.

*Proof.* If $f(\eta) = 0$, the result is clear because $\Phi_\eta^{\mathcal{J}}(P^nW(fI_N)P^n)$ is in the coset 0. If $f(\eta) \neq 0$ then we write

$$\Phi_\eta^{\mathcal{J}}(P^nW(f)P^n) = f(\eta)\Phi_\eta^{\mathcal{J}}(P^nW(I_N)P^n) + \Phi_\eta^{\mathcal{J}}(P^nW((f - f(\eta))I_N)P^n)$$

with $f - f(\eta)$ a function in $C(\dot{\mathbb{R}})$ that takes the value zero at the point $\eta$, and the result is proved. □

It is also easy to prove that some other cosets are in the local ideals $\mathcal{I}_\eta$.

**Proposition 8.5.** *Let* $a \in PC(\dot{\mathbb{R}})^{N\times N}$ *be continuous at the point* $\eta \in \dot{\mathbb{R}}$ *and take the value* 0 *at that point. Then* $\Phi_\eta^{\mathcal{J}}(P^nW(a)P^n) = 0$.

*Proof.* Given $\epsilon > 0$, we can approximate $a$ by a matrix function $b \in PC(\dot{\mathbb{R}})^{N\times N}$ such that $b$ is zero in a neighborhood $U$ of $\eta$ and $\|W(a - b)\| < \epsilon$. Then define a continuous function $f_\epsilon$ such that its support is contained in $U$ and $f_\epsilon(\eta) = 1$. Due to the last proposition, $\Phi_\eta^{\mathcal{J}}(P^n W(f_\epsilon I_N)P^n)$ is the identity in the local algebra and so it is possible to write

$$\|\Phi_\eta^{\mathcal{J}}(P^n W(a)P^n)\| \leq \epsilon + \|\Phi_\eta^{\mathcal{J}}(P^n W(b)P^n)\|$$

and

$$\begin{aligned}\Phi_\eta^{\mathcal{J}}(P^n W(b)P^n) &= \Phi_\eta^{\mathcal{J}}(P^n W(b)P^n)\Phi_\eta^{\mathcal{J}}(P^n W(f_\epsilon I_N)P^n)\\ &= \Phi_\eta^{\mathcal{J}}(P^n W(bf I_N)P^n) = 0.\end{aligned}$$

Because we can choose $\epsilon$ as small as desired, the result follows. □

Denote by $\chi_\pm^\eta$ the characteristic function of the infinite interval $]\eta, +\infty[$ (resp. $]-\infty, \eta[$).

**Proposition 8.6.** *The algebras $\mathcal{A}_\eta^{\mathcal{J}}$, $\eta \in \dot{\mathbb{R}}$, are generated by the cosets $\Phi_\eta^{\mathcal{J}}(aP^n)$ with $a \in \mathbb{C}^{N\times N}$ and $\Phi_\eta^{\mathcal{J}}(P^n W(\chi_+^\eta I_N)P^n)$.*

*Proof.* Let $\Phi_\eta^{\mathcal{J}}(P^n W(a)P^n)$ be a generator element of $\mathcal{A}_\eta^{\mathcal{J}}$. If $\eta \in \mathbb{R}$, define the piecewise constant matrix function $a_\eta = a(\eta^-)\chi_-^\eta + a(\eta^+)\chi_+^\eta$, if $\eta = \infty$, define $a_\eta = a(-\infty)\chi_- + a(+\infty)\chi_+$. Because $a - a_\eta$ is continuous at the point $\eta$ and takes the value 0 there, Proposition 8.5 enables us to write

$$\Phi_\eta^{\mathcal{J}}(P^n W(a)P^n) = \Phi_\eta^{\mathcal{J}}(P^n W(a_\eta)P^n).$$

As $W(a_\eta)$ can be written as $a(\eta^-)W(I_N - \chi_+^\eta I_N) + a(\eta^+)W(\chi_+^\eta I_N)$ (or in an equivalent form for $\eta = \infty$) the result follows. □

Having proved that the local algebras are singly generated (times matrix multipliers), it is only necessary to know the local spectra of $\Phi_\eta^{\mathcal{J}}(P^n W(\chi_+)P^n)$ in order to obtain invertibility criteria for any element of the algebra. Denote the local spectrum of an element $\mathbf{A} + \mathcal{J}$ at $\eta \in \dot{\mathbb{R}}$ by $\sigma_\eta(\mathbf{A})$. For $\alpha > 0$, set

$$\mathfrak{A}_\alpha := \{(1 + \coth((z + \mathrm{i}\alpha)\pi))/2 : z \in \mathbb{R}\} \cup \{0, 1\}$$

and define the lentiform domain

$$\mathfrak{L}_p := \{z \in \mathfrak{A}_s : \ \min\{p, q\} \leq s \leq \max\{p, q\}\}. \tag{20}$$

where $q = p/(p-1)$.

**Theorem 8.7.** *Let $\eta \in \mathbb{R}$. Then $\sigma_\eta(P^n W(\chi_+)P^n) = \mathfrak{L}_p$. Moreover, the local algebra $\mathcal{A}_\eta^{\mathcal{J}}$ is isomorphic to the matrix algebra $[P_1 W(\chi_+)P_1]^{N\times N}$.*

*Proof.* Consider first $\eta \in \mathbb{R}$. Due to the unital homomorphism $\mathsf{H}_\eta$, which is well defined in the local algebras because of Lemma 8.3, and the known results for singular integral operators on bounded curves (see for instance [7, Theorem 3.2]), it is easy to see that the lentiform domain $\mathfrak{L}_p$ must be contained in the local spectrum.

To prove the reverse inclusion consider first $\eta \in \mathbb{R}$. Suppose $\mathbf{A} \in \mathcal{A}$. Then the sequence $\mathbf{A}_\eta$ given by

$$\mathbf{A}_\eta := (P^n U_\eta^{-1} Z_{\tau_n} \mathsf{H}_\eta(\mathbf{A}) Z_{\tau_n}^{-1} U_\eta P^n)$$

belongs to $\mathcal{L}$. In fact, the elements of $\mathsf{H}_\eta(\mathbf{A})$ are generated by $P_1$ and $P_1W(\chi_+)P_1$. We have

$$(P^nU_\eta^{-1}Z_{\tau_n}\prod_{k=1}^{m}(P_1W(\chi_+)P_1)\,Z_{\tau_n}^{-1}U_\eta P^n) = (P^n\prod_{k=1}^{m}\left(P_{\tau_n}W(\chi_+^\eta)P_{\tau_n}\right)P^n)$$

which belongs to $\mathcal{L}$ by Proposition 7.5 Define now

$$\mathsf{H}'_\eta : \mathsf{H}(\mathcal{A}) \to \Phi_\eta^{\mathcal{J}}(\mathcal{A}), \qquad A \mapsto \Phi_\eta^{\mathcal{J}}(P^nU_\eta^{-1}Z_{\tau_n}AZ_{\tau_n}^{-1}U_\eta P^n).$$

We claim that $\mathsf{H}'_\eta$ is an homomorphism. To show this, write $A := P_1W(\chi_+)P_1$ and let $f_\eta$ be a continuous function vanishing at infinity and such that $f_\eta(\eta) = 1$.

$$\begin{aligned}
\mathsf{H}'_\eta(A^2) &= \Phi_\eta^{\mathcal{J}}(P^nU_\eta^{-1}Z_{\tau_n}AZ_{\tau_n}^{-1}U_\eta U_\eta^{-1}Z_{\tau_n}AZ_{\tau_n}^{-1}U_\eta P^n)\\
&= \Phi_\eta^{\mathcal{J}}(P^nW(\chi_+^\eta)P_{\tau_n}P_{\tau_n}W(\chi_+^\eta)P^n)\\
&= \Phi_\eta^{\mathcal{J}}(P^nW(f_\eta)P^n)\Phi_\eta^{\mathcal{J}}(P^nW(\chi_+^\eta)P_{\tau_n}P_{\tau_n}W(\chi_+^\eta)P^n)\\
&= \Phi_\eta^{\mathcal{J}}(P^nW(\chi_+^\eta)W(f_\eta)P_{\tau_n}W(\chi_+^\eta)P^n)\\
&= \Phi_\eta^{\mathcal{J}}(P^nW(\chi_+^\eta)W(f_\eta)P^nW(\chi_+^\eta)P^n) \qquad\qquad \text{(by (12))}\\
&= \Phi_\eta^{\mathcal{J}}(P^nW(\chi_+^\eta)P^nW(\chi_+^\eta)P^n)\\
&= \Phi_\eta^{\mathcal{J}}(P^nW(\chi_+^\eta)P^n)\Phi_\eta^{\mathcal{J}}(P^nW(\chi_+^\eta)P^n) = \mathsf{H}'_\eta(A)\mathsf{H}'_\eta(A).
\end{aligned}$$

The claim is proved.

The homomorphism $\mathsf{H}'_\eta$ is in fact the inverse of $\mathsf{H}_\eta$, and thus the local spectrum of the element $\Phi_\eta^{\mathcal{J}}(P^nW(\chi_+^\eta)P^n)$ contains the spectrum of the operator $P_1W(\chi_+)P_1$ which is well known to be $\mathfrak{L}_p$ (see, for instance, [3, Proposition 9.15]).

Finally one can see that the local algebra $\mathcal{A}_\eta^{\mathcal{J}}$ is isomorphic to the matrix algebra $[P_1W(\chi_+)P_1]^{N\times N}$ through the isomorphism $\mathsf{H}_\eta$. □

These last observations finish the identification of the local algebras. We can enunciate the result

**Theorem 8.8.** *If $A = W(a)$ with $a \in PC^{N\times N}(\dot{\mathbb{R}})$ and $a$ continuous at infinity, then the method* (1) *is stable if and only if $W(a)$ and $W(\tilde{a})$ are invertible and $P_1\left(a(\eta^-)W(\chi_-I_N) + a(\eta^+)W(\chi_+I_N)\right)P_1$ is invertible in $[\mathrm{Im}\ P_1]^N$ for every point of discontinuity of $a$.*

What the above result means is that, in some sense, for meshes of the class $\mathcal{M}$, the specific mesh is unessential for the stability property of the sequence, as long as the generating function of the Wiener-Hopf operator is continuous at infinity. It is still an open question if the discontinuity at infinity will change this picture. That question will be object of further research.

## References

[1] M.A. Bastos, P.A. Lopes, and A. Moura Santos. The two straight line approach for periodic diffraction boundary-value problems. *J. Math. Anal. Appl.*, 338(1):330–349, 2008.

[2] A. Böttcher, Yu.I. Karlovich, and I. Spitkovsky. *Convolution operators and factorization of almost periodic matrix functions*, volume 131 of *Oper. Theory Adv. Appl.* Birkhäuser, Basel, 2002.

[3] A. Böttcher and B. Silbermann. *Analysis of Toeplitz operators*. Springer-Verlag, Berlin, second edition, 2006.

[4] L. Castro, F.-O. Speck, and F.S. Teixeira. On a class of wedge diffraction problems posted by Erhard Meister. In *Operator theoretical methods and applications to mathematical physics*, volume 147 of *Oper. Theory Adv. Appl.*, pages 213–240. Birkhäuser, Basel, 2004.

[5] J. Elschner. On spline approximation for a class of non-compact integral equations. *Math. Nachr.*, 146:271–321, 1990.

[6] R. Hagen, S. Roch, and B. Silbermann. *Spectral theory of approximation methods for convolution equations*. Birkhäuser, Basel, 1995.

[7] A. Karlovich, H. Mascarenhas, and P.A. Santos. Finite section method for a Banach algebra of convolution type operators on Lp(R) with symbols generated by PC and SO. *Integral Equations Operator Theory*, 67(4):559–600, 2010.

[8] A.V. Kozak. A local principle in the theory of projection methods. *Dokl. Akad. Nauk SSSR*, 212:1287–1289, 1973. (in Russian; English translation in *Soviet Math. Dokl.* 14:1580–1583, 1974).

[9] E. Meister, P.A. Santos, and F.S. Teixeira. A Sommerfeld-type diffraction problem with second-order boundary conditions. *Z. Angew. Math. Mech.*, 72(12):621–630, 1992.

[10] S. Prössdorf and B. Silbermann. *Numerical analysis for integral and related operator equations*. Birkhäuser Verlag, Basel, 1991.

[11] S. Roch, P.A. Santos, and B. Silbermann. *Non-commutative Gelfand theories*. Springer, 2010.

[12] P.A. Santos and B. Silbermann. Galerkin method for Wiener-Hopf operators with piecewise continuous symbol. *Integral Equations Operator Theory*, 38(1):66–80, 2000.

[13] P.A. Santos and B. Silbermann. An approximation theory for operators generated by shifts. *Numer. Funct. Anal. Optim.*, 27(3-4):451–484, 2006.

[14] P.A. Santos and F.S. Teixeira. Sommerfeld half-plane problems with higher-order boundary conditions. *Math. Nachr.*, 171:269–282, 1995.

Pedro A. Santos
Departamento de Matemática
Instituto Superior Técnico
Universidade Técnica de Lisboa
Av. Rovisco Pais, 1
1049-001 Lisboa, Portugal
e-mail: pasantos@math.ist.utl.pt

Operator Theory:
Advances and Applications, Vol. 221, 607-616


# On the Spectrum of Some Class of Jacobi Operators in a Krein Space

I.A. Sheipak

**Abstract.** A Jacobi matrix with exponential growth of its elements and a corresponding symmetric operator are considered. It is proved that the eigenvalue problem of some self-adjoint extension of the operator in some Hilbert space is equivalent to the eigenvalue problem of a Sturm–Liouville operator with discrete self-similar weight. An asymptotic formula for the eigenvalues distribution is obtained. The case of an indefinite metric and self-adjoint extension of the operator in a Krein space is also considered.

**Mathematics Subject Classification (2000).** 47B36, 47B50.

**Keywords.** Jacobi matrix, self-adjoint extension of the symmetric operators, eigenvalues asymptotic, self-similar function, Krein space.

## 1. Introduction

We consider the eigenvalue problem for three-diagonal Jacobi matrix of the following type:

$$\begin{pmatrix} \alpha & \beta & 0 & 0 & \dots & 0 & \dots \\ \gamma & \alpha q & \beta q & 0 & \dots & 0 & \dots \\ 0 & \gamma q & \alpha q^2 & \beta q^2 & \dots & 0 & \dots \\ \dots & & & & & & \\ 0 & 0 & 0 & \gamma q^{n-1} & \alpha q^n & \beta q^n & 0 \\ \dots & \dots & \dots & \dots & \dots & \dots & \dots \end{pmatrix}, \tag{1.1}$$

with parameter $q > 1$.

The spectral properties of a Jacobi matrix depend on the behaviour of its elements. The asymptotic formulas for eigenvalues are known for some narrow classes of such matrices. We consider the matrix with exponential growth of the elements. Jacobi operators with exponential and hyper-exponential varying of the elements arise in different applications but they are studied only for some simple

The work is supported by RFFI, project N 10-01-00423.

cases when elements decrease and the operators are bounded. In [1] the spectral properties of two-diagonal symmetric matrix are studied: all the main diagonal elements are zero and the off-diagonal elements $b_n$ are given by the formula $b_n = q^{n^s}$, where $0 < q < 1$ and $s$ is an arbitrary natural number. By the technique based on properties of transcendental functions the author obtains the following asymptotic estimations for the eigenvalues $\lambda_n$ of this class of a Jacobi operator: $b_{2n+1} < \lambda_n < b_{2n-1}$.

In [2] a wider class of two-diagonal Jacobi symmetric matrixes is considered: the off-diagonal elements $b_n$ are supposed to be positive and to satisfy the condition

$$\lim_{n\to\infty} \frac{b_{n+1}}{b_n} = 0.$$

In this case the asymptotic of the eigenvalues is $\lambda_n = b_{2n-1}(1 + o(1))$, $n \to \infty$. The result is obtained by variational methods.

In the cited papers the operators are self-adjoint and compact. In contrast to them in our case the elements of the matrix (1.1) grow since the parameter $q$ is greater than 1, therefore the matrix generates an unbounded operator. This raises the question how to describe the domain of the corresponding operator.

Under the additional assumption $\beta\gamma > 0$ the operator can be symmetrized in some Hilbert space. Then it can have defect numbers $(0;0)$ or $(1;1)$. In the former case the operator is self-adjoint. In the latter case we should describe the self-adjoint extensions of the operator. It should be noted that the condition of the symmetrization is often fulfilled for matrixes in different applications. For the finite order matrix this assumption is related to the notion of *normal matrix* (see, for example, [3]). If the inequality $\beta\gamma < 0$ holds the operator can be symmetrized in some Krein space.

The paper is organized as follows. Section 2 contains basic information on self-similar functions and discrete self-similar measures. In Section 3 we transform the Sturm–Liouville problem with discrete self-similar weight to the eigenvalue problem for the matrix (1.1) and describe the suitable self-adjoint extension of the corresponding operator. By the equivalence of the two spectral problems the asymptotic formulas for the Jacobi operator eigenvalues are obtained. The asymptotic formulas for eigenvalues are different for self-adjoint extensions in a Hilbert space and in a Krein space.

## 2. Self-similar functions of zero spectral order

For the solution of our spectral problem of some class of Jacobi operators we need to give some information on self-similar functions.

Let the numbers $a \in (0,1)$ and $d$ satisfy the condition

$$a|d|^2 < 1. \tag{2.1}$$

Let us define an affine transformation $G$ in $L_2[0,1]$ by the formula

$$G(f)(x) = \beta_1 \cdot \chi_{[0,1-a)}(x) + \left(d \cdot f\left(\frac{x-1+a}{a}\right) + \beta_2\right) \cdot \chi_{(1-a,1]}(x), \tag{2.2}$$

where $\beta_1$, $\beta_2$ are arbitrary real numbers.

**Statement 2.1.** *If the condition* (2.1) *holds then the operator* $G$ *is a contraction in* $L_2[0,1]$.

*Proof.* For arbitrary functions $f, g \in L_2[0,1]$ we have

$$\begin{aligned}\|G(f) - G(g)\|^2_{L_2[0,1]} &= \int_{1-a}^{1} |d|^2 \left(f\left(\frac{x-1+a}{a}\right) - g\left(\frac{x-1+a}{a}\right)\right) dx \\ &= a|d|^2 \|f-g\|^2_{L_2[0,1]}.\end{aligned}$$

□

Consequently there is the unique function $P \in L_2[0,1]$ such that the equation $G(P) = P$ holds. Such function $P$ belongs to a class of *self-similar functions of zero spectral order*. The numbers $a$, $d$, $\beta_1$ and $\beta_2$ are called *self-similarity parameters* of the function $P$.

From the definition it follows that $P$ is a piecewise constant function and it can be written as

$$P(x) = \beta_1, \quad x \in [0, 1-a), \tag{2.3}$$

$$P(x) = d^k\beta_1 + \beta_2(1 + d + \cdots + d^{k-1}), \quad x \in (1-a^k, 1-a^{k+1}), \quad k = 1, 2, \ldots \tag{2.4}$$

More detail information on self-similar functions in $L_p[0,1]$ can be found in [5]. For the general construction of self-similar functions with zero spectral order see [6] and [8].

## 3. Sturm–Liouville problem with discrete self-similar weight

The purpose of this section is to establish the equivalence of the eigenvalue problem for the matrix (1.1) and the following boundary problem

$$-y'' - \lambda\rho y = 0, \tag{3.1}$$

$$y(0) = y(1) = 0, \tag{3.2}$$

where the derivative $\rho = P'$ is understood in the distributional sense. The function $P$ is a fixed point of the operator (2.2). It is simple to get from (2.3)–(2.4) that

$$\rho := \sum_{k=1}^{\infty} m_k \delta(x - (1-a^k)), \tag{3.3}$$

where $\delta(x)$ is a Dirac delta function. Here $m_k = d^{k-1}(d\beta_1 + \beta_2 - \beta_1)$. The condition $P \in L_2[0,1]$ implies $\rho \in \overset{\circ}{W}{}_2^{-1}[0,1]$.

Via $\mathfrak{H}$ we denote a space $\overset{\circ}{W}{}_2^1[0,1]$ with a scalar product

$$\langle y, z\rangle = \int_0^1 y'\overline{z'}dx.$$

By $\mathfrak{H}'$ we denote the dual space to $\mathfrak{H}$ with respect to $L_2[0,1]$, i.e., the completion of $L_2[0,1]$ with respect to the norm

$$\|y\|_{\mathfrak{H}'} = \sup_{\|z\|_{\mathfrak{H}}=1} \left|\int_0^1 y\overline{z}dx\right|.$$

Let us consider the embedding operator $J : \mathfrak{H} \to L_2[0,1]$. The adjoint operator $J^* : L_2[0,1] \to \mathfrak{H}$ can be continuously extended to the isometry $J^+ : \mathfrak{H}' \to \mathfrak{H}$ by the definition of the space $\mathfrak{H}'$.

We associate with the problem (3.1)–(3.2) the linear pencil of the bounded operators $T_\rho : \mathfrak{H} \to \mathfrak{H}'$, which is defined by the identity

$$\forall\lambda, \forall y \in \mathfrak{H} \qquad \langle J^+T_\rho(\lambda)y, y\rangle = \int_0^1 \left(|y'|^2 + \lambda P \cdot (|y|^2)'\right) dx. \tag{3.4}$$

From Theorem 4.1 (see [4]) it follows that the problem (3.1)–(3.2) has purely discrete spectrum for any weight $\rho \in \overset{\circ}{W}{}_2^{-1}[0,1]$. The asymptotic of eigenvalues depends on a spectral order of the self-similar function $P$.

We consider the problem (3.1)–(3.2) under an assumption that the weight $\rho$ is a generalized derivative of the self-similar function $P$ with zero spectral order. The weight $\rho$ is defined by the formula (3.3).

### 3.1. Discretization of eigenfunctions of Sturm–Liouville problem with discrete self-similar weight

As far as $P$ is a piecewise constant function we can search the eigenfunction of the problem (3.1)–(3.2) as a piecewise linear function $y \in \overset{\circ}{W}{}_2^1[0,1]$, which can be defined by the formulas

$$y(x) = s_1 x, \quad x \in [0, 1-a), \tag{3.5}$$

$$y(x) = s_k x + t_k, \quad x \in (1-a^{k-1}, 1-a^k), \quad k = 2,3,\dots \tag{3.6}$$

The function $y$ is continuous in the end points $x_k := 1 - a^k$, $k = 1,2,\dots$ of intervals $(1-a^{k-1}, 1-a^k)$, therefore

$$s_{k-1}(1-a^{k-1}) + t_{k-1} = s_k(1-a^{k-1}) + t_k, \quad k = 2,3,\dots \tag{3.7}$$

Substituting the formulas (3.5)–(3.6) into the equation (3.1) and from (3.7) we get the following system of equations for the numbers $s_k$:

$$s_1 - s_2 = \lambda(d\beta_1 + \beta_2 - \beta_1)(1-a)s_1,$$
$$s_2 - s_3 = \lambda\left(d^2\beta_1 + d(\beta_2 - \beta_1)\right)(1-a)(s_1 + as_2),$$
$$s_3 - s_4 = \lambda\left(d^3\beta_1 + d^2(\beta_2 - \beta_1)\right)(1-a)(s_1 + as_2 + a^2 s_3),$$
$$\dots,$$
$$s_n - s_{n+1} = \lambda\left(d^n\beta_1 + d^{n-1}(\beta_2 - \beta_1)\right)(1-a)(s_1 + as_2 + \dots + a^{n-1}s_n),$$
$$\dots$$

We shall also denote for brevity

$$r := (1-a)(d\beta_1 + \beta_2 - \beta_1), \quad q := \frac{1}{ad}.$$

The condition (2.1) implies that

$$|q| > 1.$$

We assume that self-similarity parameters $d$, $\beta_1$ and $\beta_2$ satisfy the condition $d\beta_1 + \beta_2 - \beta_1$, hence $r \neq 0$. To the end of this section we consider only $d > 0$, consequently $q > 1$.

A boundary condition at the point $x = 0$ is fulfilled due to (3.5). Taking into account the boundary condition at the point $x = 1$ we obtain from the formula (3.6) that

$$y(1) = \lim_{n\to\infty}(s_n + t_n).$$

Then it is simple to get from (3.7) that

$$t_n = \sum_{k=1}^{n-1} s_k(x_k - x_{k-1}) - s_n x_{n-1}.$$

Finally we receive that

$$y(1) = (1-a)\sum_{k=1}^{\infty} a^{k-1}s_k = 0,$$

i.e., the sequence $\{s_k\}_{k=1}^{\infty}$ satisfies the condition

$$\sum_{k=1}^{\infty} a^{k-1}s_k = 0. \tag{3.8}$$

Let us note that norm

$$\|y\|_{\mathfrak{H}}^2 = \sum_{k=1}^{\infty}(1 - a^k - (1 - a^{k-1}))s_k^2 = (1-a)\sum_{k=1}^{\infty} a^{k-1}s_k^2.$$

So the problem (3.1)–(3.2) with the discrete self-similar weight $\rho$ is equivalent to the following problem:

$$\begin{pmatrix} 1 & -1 & 0 & 0 & \dots & 0 & \dots \\ 0 & 1 & -1 & 0 & \dots & 0 & \dots \\ 0 & 0 & 1 & -1 & \dots & 0 & \dots \\ \dots \\ 0 & 0 & 0 & 0 & 1 & -1 & 0 \\ \dots & \dots & \dots & \dots & \dots & \dots & \dots \end{pmatrix} \begin{pmatrix} s_1 \\ s_2 \\ s_3 \\ \dots \\ s_k \\ \dots \end{pmatrix} =$$

$$= \lambda r \begin{pmatrix} 1 & 0 & 0 & 0 & \dots & 0 & \dots \\ d & da & 0 & 0 & \dots & 0 & \dots \\ d^2 & d^2 a & (da)^2 & 0 & \dots & 0 & \dots \\ \dots \\ d^{k-1} & d^{k-1}a & d^{k-1}a^2 & \dots & (da)^{k-1} & 0 & \dots \\ \dots & \dots & \dots & \dots & \dots & \dots & \dots \end{pmatrix} \begin{pmatrix} s_1 \\ s_2 \\ s_3 \\ \dots \\ s_k \\ \dots \end{pmatrix}, \quad (3.9)$$

where the sequence $\{s_k\}_{k=1}^{\infty}$ additionally satisfy the condition (3.8) and

$$\sum_{k=1}^{\infty} a^{k-1} s_k^2 < \infty.$$

Let us consider operators defined by the following matrices:

$$A = \begin{pmatrix} 1 & -1 & 0 & 0 & \dots & 0 & \dots \\ 0 & 1 & -1 & 0 & \dots & 0 & \dots \\ 0 & 0 & 1 & -1 & \dots & 0 & \dots \\ \dots \\ 0 & 0 & 0 & 0 & 1 & -1 & 0 \\ \dots & \dots & \dots & \dots & \dots & \dots & \dots \end{pmatrix}$$

$$B = \begin{pmatrix} 1 & 0 & 0 & 0 & \dots & 0 & \dots \\ d & da & 0 & 0 & \dots & 0 & \dots \\ d^2 & d^2 a & (da)^2 & 0 & \dots & 0 & \dots \\ \dots \\ d^{k-1} & d^{k-1}a & d^{k-1}a^2 & \dots & (da)^{k-1} & 0 & \dots \\ \dots & \dots & \dots & \dots & \dots & \dots & \dots \end{pmatrix}.$$

Then (3.9) can be written in the reduced form

$$As = \lambda r B s.$$

It is simple to check that an inverse operator to $B$ is given by the matrix

$$B^{-1} = \begin{pmatrix} 1 & 0 & 0 & 0 & \dots \\ -dq & q & 0 & 0 & \dots \\ 0 & -dq^2 & q^2 & 0 & \dots \\ 0 & 0 & -dq^3 & q^3 & \dots \\ \dots & \dots & \dots & \dots & \dots \end{pmatrix}.$$

The natural question arises: which of the two following problems

$$B^{-1}As = \lambda rs \qquad \text{or} \qquad AB^{-1}u = \lambda ru \quad (u := Bs)$$

is equivalent to the spectral problem (3.9) in some space of sequences?

Let $w \neq 0$ be an arbitrary real number and the space of sequence $\{v_k\}_{k=1}^{\infty}$, satisfying the condition

$$\sum_{k=1}^{\infty} w^{k-1} v_k^2 < \infty,$$

is denoted by $l_{2,w}$. The scalar product in this space is defined as follows

$$\langle u, v \rangle_w = \sum_{k=1}^{\infty} w^{k-1} u_k v_k.$$

In case $w > 0$ the space $l_{2,w}$ is a Hilbert one, in case $w < 0$ the scalar product is indefinite and $l_{2,w}$ is a Krein space.

**Statement 3.1.** *The operator $AB^{-1}$ is symmetric in the space $l_{2,1/d}$.*

*Proof.* By straightforward calculation we obtain that

$$\begin{aligned}\langle AB^{-1}u, v\rangle_{1/d} &= \langle u, AB^{-1}v\rangle_{1/d} \\ &= (1+dq)\sum_{k=1}^{\infty}\left(\frac{q}{d}\right)^{k-1} u_k v_k - q\sum_{k=1}^{\infty}\left(\frac{q}{d}\right)^{k-1}(u_{k+1}v_k + u_k v_{k+1}).\end{aligned}$$

□

**Statement 3.2.** *The domain of an adjoint operator to $AB^{-1}$ consists of the sequences $u \in l_{2,1/d}$ such that*

$$\sum_{k=2}^{\infty} \frac{1}{d^{k-1}} \left(-dq^{k-1}u_{k-1} + (1+dq)q^{k-1}u_k - q^k u_{k+1}\right)^2 < \infty. \tag{3.10}$$

*Proof.* The matrix of operator $AB^{-1}$ has the form

$$AB^{-1} = \begin{pmatrix} 1+dq & -q & 0 & 0 & \dots \\ -dq & (1+dq)q & -q^2 & 0 & \dots \\ 0 & -dq^2 & (1+dq)q^2 & -q^3 & \dots \\ \dots & \dots & \dots & \dots & \dots \end{pmatrix}. \tag{3.11}$$

Applying Proposition 1.1 from [9] (p. 174) we prove the statement. □

Now we are able to find the defect numbers of the operator $AB^{-1}$. The problem (3.1)–(3.2) is self-adjoint and has two boundary conditions. In the procedure of discretization we take into consideration only one boundary condition at the point $x = 0$ by (3.5). The boundary condition at the point $x = 1$ for the function $y \in \mathfrak{H}$ is not taken into account for the sequence $\{s_k\}_{k=1}^{\infty}$ and for the sequence $\{u_k\}_{k=1}^{\infty}$ yet. Consequently, defect numbers of the operator $AB^{-1}$ equal to $(1, 1)$. The formula (3.8) for the sequence $\{s_k\}_{k=1}^{\infty}$ is equivalent to the boundary condition $y(1) = 0$

for the function $y \in \mathfrak{H}$. From the definition of the operator $B$ it follows that the condition (3.8) for the sequence $s$ can be rewritten for the sequence $u = Bs$ as

$$\lim_{n\to\infty} \frac{u_n}{d^{n-1}} = 0 \tag{3.12}$$

in the space $l_{2,1/d}$.

**Theorem 3.1.** *The problem* (3.4) *is equivalent to the problem*

$$AB^{-1}u = \lambda r u$$

*in the space* $l_{2,1/d}$ *with conditions* (3.10), (3.12) *on the sequence* $u = (u_1, u_2, \dots)$, *where* $u = Bs$.

*Proof.* On the one hand the quadratic form of the problem (3.4) looks like

$$\int_0^1 |y'(x)|^2 dx = \lambda < \rho y, y >= \lambda \sum_{k=1}^{\infty} m_k y^2 (1 - a^k),$$

where $y \in \mathfrak{H}$. Substituting the representation of $y$ (3.5)–(3.6) into this formula we can write down the quadratic form of the problem in terms of the sequence $\{s_k\}_{k=1}^{\infty}$:

$$(1-a)\sum_{k=1}^{\infty} a^{k-1}s_k^2 = \lambda r \sum_{k=1}^{\infty} \frac{d^{k-1}}{1-a}(s_k(1-a^k) + t_k)^2.$$

From (3.7) by the method of mathematical induction we obtain that

$$t_k = (1-a)\sum_{j=1}^{k-1} a^{j-1}s_j - (1-a^{k-1})s_k.$$

This representation of $t_k$ leads to the following expression for the quadratic form of the problem (3.4)

$$\sum_{k=1}^{\infty} a^{k-1}s_k^2 = \lambda r \sum_{k=1}^{\infty} d^{k-1}\left(\sum_{j=1}^{k} a^{j-1}s_j\right)^2. \tag{3.13}$$

On the other hand we can write the quadratic form

$$\langle AB^{-1}u, u\rangle_{1/d} = \lambda r \langle u, u\rangle_{1/d}$$

of the eigenvalue problem $AB^{-1}u = \lambda r u$ as

$$\langle As, Bs\rangle_{1/d} = \lambda r \langle Bs, Bs\rangle_{1/d}.$$

Due to the definitions of $A$ and $B$:

$$\begin{aligned} As &= (s_1 - s_2, s_2 - s_3, \dots, s_k - s_{k+1}, \dots), \\ Bs &= (s_1, d(s_1 + as_2), \dots, d^{k-1}\sum_{j=1}^{k} a^{j-1}s_j, \dots) \end{aligned}$$

we get that the quadratic form for the spectral problem of the operator $AB^{-1}$ is expressed by (3.13) exactly. The concordance of the domains of the problem

(3.1)–(3.2) and of the operator $AB^{-1}$ follows from (3.10) and (3.12). Substitution formulas for $u$ ($u_k = d^{k-1}\sum_{j=1}^{k-1} a^{j-1}s_j$, $k \geq 2$) in (3.10) leads to the correctness of the problem (3.4) in the space of sequences $\{s_k\}_{k=1}^{\infty}$ with the conditions $\sum_{k=1}^{\infty} a^{k-1}s_k^2 < \infty$ and (3.8). □

**Remark 3.1.** *The matrix* (3.11) *with the domain* $u \in l_{2,1/d}$, *satisfying the condition* (3.10), (3.12) *define a self-adjoint extension of the symmetric operator* $AB^{-1}$ *in the space* $l_{2,1/d}$. *The eigenvalue problem for this self-adjoint extension is equivalent to the problem* (3.1)–(3.2) (*or to the problem* (3.4)).

In the next theorem we denote this self-adjoint extension by $L$. The Jacobi matrix $AB^{-1}$ belongs to the class of matrix (1.1) $\left(\alpha = 1 + dq = 1 + \frac{1}{a}, \beta = -q,\right.$ $\left.\gamma = -dq = -\frac{1}{a}\right)$ (see (3.11)).

**Theorem 3.2.** *There exists a positive number* $c$ *such that for the eigenvalues of the operator* $L$ *numerated in increasing order the following asymptotic formula holds as* $k \to \infty$

$$\lambda_k = cq^k(1 + o(1)). \tag{3.14}$$

*Proof.* This statement directly comes from the previous theorem 3.1 and the results of the paper ([8], Theorem 4.1). □

### 3.2. Indefinite case

We can investigate the problem (3.1)–(3.2) in the case $d < 0$. Then the operator $L$ with the domain (3.10), (3.12) is self-adjoint in the space with indefinite metric.

Indeed, let us define orthogonal projections in $l_{2,1/d}$:

$$\begin{aligned} &P_+ : e_k \to e_k, \quad k = 1, 3, \ldots, 2n-1, \ldots, \\ &P_+ : e_k \to 0, \quad k = 2, 4, \ldots, 2n, \ldots; \\ &P_- : e_k \to 0, \quad k = 1, 3, \ldots, 2n-1, \ldots, \\ &P_- : e_k \to e_k, \quad k = 2, 4, \ldots, 2n, \ldots \quad n \in \mathbb{N}, \end{aligned}$$

and the operator $J = P_+ - P_-$.

The operator $L$ is $J$-self-adjoint. Then it has both positive and negative eigenvalues. Further we enumerate its positive eigenvalues $\{\lambda_k\}_{k=1}^{\infty}$ in increasing order. The negative eigenvalues $\{\lambda_{-k}\}_{k=1}^{\infty}$ are enumerated in increasing order of the absolute values.

From the statement of the paper [8] (Theorem 4.3 ) the following proposition follows.

**Theorem 3.3.** *If* $d < 0$ *then there is a number* $c > 0$ *such that the positive* $\{\lambda_k\}_{k=1}^{\infty}$ *and the negative* $\{\lambda_{-k}\}_{k=1}^{\infty}$ *eigenvalues of the operator* $L$ *have the following asymptotics as* $k \to +\infty$

$$\begin{aligned} \lambda_{k+1} &= cq^{2k}(1 + o(1)), \\ \lambda_{-(k+1)} &= -cq^{2k+1}(1 + o(1)). \end{aligned}$$

## References

[1] E.A. Tur *Eigenvalue asymptotic for one class of Jacobi matrix with limit point spectra//* Matem. zametki, 2003, **74**:3, 449–462 (in Russian); English transl.: Mathem. Notes, 2003, **74**:3, 425–437.

[2] R.V. Kozhan *Asymptotics of the eigenvalues of two-diagonal Jacobi matrices //* Matem. zametki, 2005 **77**:2, p. 313–316 (in Russian); English transl.: Mathem. Notes, 2005 **77**:2, p. 283–287.

[3] F.R. Gantmakher, M.G. Krein *Oscillation matrix and kernels and small vibrations of mechanical systems//*GITTL, Moscow, Leningrad, 1950 (in Russian).

[4] A.A. Vladimirov, I.A. Sheipak, *Self-similar functions in space* $L_2[0,1]$ *and Sturm-Liouville problem with singular weight*, Matem. sbornik, 2006, **197**:11, 13–30 (in Russian); English transl.: Sbornik: Mathematics, 2006, **197**:11, 1569–1586.

[5] I.A. Sheipak, *On the construction and some properties of self-similar functions in the spaces* $L_p[0,1]$, 2007, Matem. zametki, **81**:6, 924–938 (in Russian); English transl.: Mathem. Notes, 2007, **81**:5-6, 827–839.

[6] I.A. Sheipak, *Singular Points of a Self-Similar Function of Spectral Order Zero: Self-Similar Stieltjes String*, Matem. zametki, 2010, **88**:2, 303–316 (in Russian); English transl.: Mathem. Notes, 2010 **88**:2, 275–286.

[7] A.A. Vladimirov, I.A. Sheipak, *Indefinite Sturm–Liouville problem for some classes of self-similar singular weights*, Trudy MIRAN, 2006, **255**, 88–98 (in Russian); English transl.: Proceedings of the Steklov Institute of Mathematics, 2006, **255**, 1–10.

[8] A.A. Vladimirov, I.A. Sheipak, *Eigenvalue asymptotics for Sturm–Liouville problem with discrete self-similar weight//* Matem. zametki, 2010, **88**:5, 662–672 (in Russian); English transl.: Mathem. Notes, 2010 **88**:5, 3–12.

[9] N.I. Ahiezer *Classic moment problems and some connected calculus problems//* Moscow, Fizmatgiz., 1961 (in Russian).

I.A. Sheipak
Moscow Lomonosov State University
e-mail: `iasheip@mech.math.msu.su`

Operator Theory:
Advances and Applications, Vol. 221, 617-634


# Holomorphy in Multicomplex Spaces

D.C. Struppa, A. Vajiac and M.B. Vajiac

**Abstract.** In this paper we extend our research on bicomplex holomorphy and on the existence of bicomplex differential operators to the nested space of multicomplex numbers. Specifically, we will discuss the different sets of idempotent representations for multicomplex numbers and we will use them to develop a theory of holomorphic and multicomplex analytic functions.

**Mathematics Subject Classification (2000).** Primary 30G35; Secondary 16E05, 13P10, 35N05.

**Keywords.** Bicomplex algebra, PDE systems, computational algebra, resolutions, syzygy.

## 1. Introduction

This paper sets the analytic background for the theory of hyperholomorphic functions of multicomplex variables, and it demonstrates how much of the known theory for (bi)holomorphic functions of bicomplex variables can be extended to this setting. For simplicity and consistency of our presentation, we will replace the terms "hyperholomorphic" and "biholomorphic" with simply "holomorphic", whenever there is no danger of confusion.

Without pretense of completeness, we note that one of the first important references for multicomplex numbers is due to Price [11], who resurrected some ideas from the Italian school of Segre [17] and Scorza-Dragoni [18], and set the stage for a modern treatment of bicomplex and multicomplex numbers. Further literature on this topic contains seminal work of Shapiro, Rochon, Ryan, et all [2, 6, 12, 15]. Important results have also been obtained in [7, 8] on a generalization of the Fatou-Julia theorem in the context of multicomplex numbers. Moreover, we mention here the results of Rochon [13], which studies a relation of bicomplex pseudoanalytic function theory to a special case of the Schrödinger equation.

The starting point of this study consists in considering the space $\mathbb{C}$ of complex numbers as a real bidimensional algebra, and then complexifying it. With this process one obtains a four-dimensional algebra which is usually denoted by $\mathbb{BC}$, or, as in the next few sections, by $\mathbb{BC}_2$, in order to differentiate it from higher-dimensional multicomplex algebras. The key point of the theory of functions on this algebra is that classical holomorphic functions can be extended from one complex variable to this algebra, and one can therefore develop a new theory of (hyper)holomorphic functions.

The algebra $\mathbb{BC}$ is four-dimensional over the reals, just like the skew-field of quaternions, but while in the space of quaternions we have three anti-commutative imaginary units, $\mathbf{i}, \mathbf{j}, \mathbf{k}$, where $\mathbf{k}$ is introduced as $\mathbf{k} = \mathbf{ij}$, in the case of bicomplex numbers one considers two imaginary units $\mathbf{i}_1, \mathbf{i}_2$ which commute, and so the third unit $\mathbf{i}_3 = \mathbf{i}_1\mathbf{i}_2$ ends up being a "new" root of 1; such units are usually called *hyperbolic*. Indeed, every bicomplex number $Z$ can be written as $Z = z_1 + \mathbf{i}_2 z_2$, where $z_1$ and $z_2$ are complex numbers of the form $z_1 = x_1 + \mathbf{i}_1 y_1$, and $z_2 = x_2 + \mathbf{i}_1 y_2$, with $\mathbf{i}_1, \mathbf{i}_2$ commuting imaginary units.

In this paper, the space $\mathbb{BC}_n$ of multicomplex numbers refers to the space generated over the reals by $n$ commuting imaginary units. We will investigate algebraic properties of this space and analytic properties of multi-complex valued functions defined on $\mathbb{BC}_n$.

It is worth noting that the algebraic properties of the space $\mathbb{BC}_2$ and the properties of its holomorphic functions have been discussed before in [4, 5, 21], using computational algebra techniques. Other important references include [2, 6, 12, 15, 16, 18, 19, 20].

Let us offer a brief overview of the paper. In Section two, we recall the basic properties and results of bicomplex numbers. This section sets the stage for Section three, where we discuss the general theory of multicomplex spaces. In Section four we discuss the holomorphy on multicomplex spaces in the special case of $n = 3$. Though, technically, this case has nothing special, we thought that its detailed analysis could help the reader's intuition when the case of general $n$ becomes computationally cumbersome. Section five is a technical section, which introduces the recursive relations which are necessary to treat the general case. Finally, in Section six, we discuss the algebraic properties of the sheaf of holomorphic function on multicomplex spaces.

## 2. The 2-dimensional bicomplex space

The first trivial case, of course, is when we only have one imaginary unit, say $\mathbf{i}_1$, in which case the space $\mathbb{BC}_1$ is the usual complex plane $\mathbb{C}$. Since, in what follows, we will have to work with different complex planes, generated by different imaginary units, we will denote such a space also by $\mathbb{C}_{\mathbf{i}_1}$, in order to clarify which is the imaginary unit used in the space itself.

Thus, the first interesting case occurs when we have two commuting imaginary units $\mathbf{i}_1$ and $\mathbf{i}_2$. This yields the bicomplex space $\mathbb{BC}_2$, which we have quickly introduced in the previous section. This space has extensively been studied in [2, 6, 11, 12], and more recently by the authors in [4, 5, 21], where it is referred to as $\mathbb{BC}$.

Given that we will study the case of several imaginary units, we will consistently use the notation which explicitly indicates how many units are involved.

Because of the various units in $\mathbb{BC}_2$, we have several different conjugations that can be defined naturally. Let therefore consider a bicomplex number $Z$, and let us write it both in terms of its complex coordinates, $Z = z_1 + \mathbf{i}_2 z_2$ with $z_1, z_2 \in \mathbb{C}_{i_1}$, and in terms of its real coordinates $Z = x_1 + \mathbf{i}_1 x_2 + \mathbf{i}_2 x_3 + \mathbf{i}_1\mathbf{i}_2 x_4$ with $x_\ell \in \mathbb{R}$. If we use the traditional notation $\overline{z}$ for complex conjugation in $\mathbb{C}_{\mathbf{i}_1}$, we obtain the following bicomplex conjugations:

$$\begin{aligned} \overline{Z}^{\mathbf{i}_2} &= z_1 - \mathbf{i}_2 z_2 = x_1 + \mathbf{i}_1 x_2 - \mathbf{i}_2 x_3 - \mathbf{i}_1\mathbf{i}_2 x_4 \,, \\ \overline{Z}^{\mathbf{i}_1} &= \overline{z_1} + \mathbf{i}_2 \overline{z_2} = x_1 - \mathbf{i}_1 x_2 + \mathbf{i}_2 x_3 - \mathbf{i}_1\mathbf{i}_2 x_4 \,, \\ \overline{Z}^{\mathbf{i}_1\mathbf{i}_2} &= \overline{z_1} - \mathbf{i}_2 \overline{z_2} = x_1 - \mathbf{i}_1 x_2 - \mathbf{i}_2 x_3 + \mathbf{i}_1\mathbf{i}_2 x_4 \,. \end{aligned} \tag{1}$$

Note the following relationships between these conjugations:

$$\begin{aligned} Z \cdot \overline{Z}^{\mathbf{i}_1\mathbf{i}_2} &= z_1\overline{z_1} + z_2\overline{z_2} + 2\mathbf{i}_1\mathbf{i}_2(x_1x_4 - x_2x_3) \,, \\ \overline{Z}^{\mathbf{i}_1} \cdot \overline{Z}^{\mathbf{i}_2} &= z_1\overline{z_1} + z_2\overline{z_2} - 2\mathbf{i}_1\mathbf{i}_2(x_1x_4 - x_2x_3) \,, \\ Z \cdot \overline{Z}^{\mathbf{i}_2} &= z_1^2 + z_2^2 \,, \qquad \overline{Z}^{\mathbf{i}_1} \cdot \overline{Z}^{\mathbf{i}_1\mathbf{i}_2} = \overline{z_1}^2 + \overline{z_2}^2 \,, \end{aligned}$$

and most importantly

$$\overline{\overline{Z}^{\mathbf{i}_1}}^{\mathbf{i}_2} = \overline{Z}^{\mathbf{i}_1\mathbf{i}_2} \,,$$

which will be very significant when we will define recursively the multicomplex spaces.

A function $F : \mathbb{BC}_2 \to \mathbb{BC}_2$ can obviously be written as $F = u + \mathbf{i}_2 v$, where $u, v : \mathbb{BC}_2 \to \mathbb{BC}_1 = \mathbb{C}_{\mathbf{i}_1}$. If we use the shorthand $\partial_t := \frac{\partial}{\partial t}$ for any real variable $t$, the usual complex differential operators are defined by

$$\begin{aligned} \partial_{z_1} &:= \partial_{x_1} - \mathbf{i}_1 \partial_{x_2} \,, \qquad & \partial_{z_2} &:= \partial_{x_3} - \mathbf{i}_1 \partial_{x_4} \,, \\ \partial_{\overline{z_1}} &:= \partial_{x_1} + \mathbf{i}_1 \partial_{x_2} \,, \qquad & \partial_{\overline{z_2}} &:= \partial_{x_3} + \mathbf{i}_1 \partial_{x_4} \,, \end{aligned}$$

but, following the notations from [4], we can also introduce the bicomplex differential operators

$$\begin{aligned} \partial_{\overline{Z}^{\mathbf{i}_1\mathbf{i}_2}} := \partial_{Z^*} &:= \partial_{\overline{z_1}} + \mathbf{i}_2 \partial_{\overline{z_2}} = \partial_{x_1} + \mathbf{i}_1 \partial_{x_2} + \mathbf{i}_2 \partial_{x_3} + \mathbf{i}_1\mathbf{i}_2 \partial_{x_4} \,, \\ \partial_{\overline{Z}^{\mathbf{i}_1}} := \partial_{\tilde{Z}} &:= \partial_{\overline{z_1}} - \mathbf{i}_2 \partial_{\overline{z_2}} = \partial_{x_1} + \mathbf{i}_1 \partial_{x_2} - \mathbf{i}_2 \partial_{x_3} - \mathbf{i}_1\mathbf{i}_2 \partial_{x_4} \,, \\ \partial_{\overline{Z}^{\mathbf{i}_2}} := \partial_{Z^\dagger} &:= \partial_{z_1} + \mathbf{i}_2 \partial_{z_2} = \partial_{x_1} - \mathbf{i}_1 \partial_{x_2} + \mathbf{i}_2 \partial_{x_3} - \mathbf{i}_1\mathbf{i}_2 \partial_{x_4} \,, \\ \partial_Z &:= \partial_{z_1} - \mathbf{i}_2 \partial_{z_2} = \partial_{x_1} - \mathbf{i}_1 \partial_{x_2} - \mathbf{i}_2 \partial_{x_3} + \mathbf{i}_1\mathbf{i}_2 \partial_{x_4} \,. \end{aligned} \tag{2}$$

Define now the complex variables:

$$\begin{aligned} \zeta &:= z_1 + \mathbf{i}_1 z_2, \qquad & \zeta^\dagger &:= z_1 - \mathbf{i}_1 z_2 \,, \\ \overline{\zeta}^{\mathbf{i}_1} &:= \overline{z_1} - \mathbf{i}_1 \overline{z_2}, \qquad & \overline{\zeta^\dagger}^{\mathbf{i}_1} &:= \overline{z_1} + \mathbf{i}_1 \overline{z_2} \,. \end{aligned}$$

$\mathbb{BC}_2$ is not a division algebra, and it has two distinguished zero divisors, $\mathbf{e}_{12}$ and $\overline{\mathbf{e}_{12}}$, which are idempotent, linearly independent over the reals, and mutually orthogonal:

$$\mathbf{e}_{12} := \frac{1+\mathbf{i}_1\mathbf{i}_2}{2}, \qquad \overline{\mathbf{e}}_{12} := \frac{1-\mathbf{i}_1\mathbf{i}_2}{2}.$$

Just like $\{1, \mathbf{i}_2\}$, they form a basis of the complex algebra $\mathbb{BC}_2$, which is called the *idempotent basis.* In this basis, the *idempotent representations* for $Z = z_1 + \mathbf{i}_2 z_2$ and its conjugates are

$$Z = \zeta^\dagger \mathbf{e}_{12} + \zeta \overline{\mathbf{e}}_{12}, \qquad \overline{Z}^{\mathbf{i}_1} = \overline{\zeta}^{\mathbf{i}_1} \mathbf{e}_{12} + \overline{\zeta^\dagger}^{\mathbf{i}_1} \overline{\mathbf{e}}_{12},$$

$$\overline{Z}^{\mathbf{i}_2} = \zeta \mathbf{e}_{12} + \zeta^\dagger \overline{\mathbf{e}}_{12}, \qquad \overline{Z}^{\mathbf{i}_1\mathbf{i}_2} = \overline{\zeta^\dagger}^{\mathbf{i}_1} \mathbf{e}_{12} + \overline{\zeta}^{\mathbf{i}_1} \overline{\mathbf{e}}_{12}.$$

If we use the notations

$$\partial_{\zeta^\dagger} = \partial_{z_1} + \mathbf{i}_1 \partial_{z_2}, \qquad \partial_{\zeta} = \partial_{z_1} - \mathbf{i}_1 \partial_{z_2},$$

$$\partial_{\overline{\zeta^\dagger}^{\mathbf{i}_1}} = \partial_{\overline{z_1}} - \mathbf{i}_1 \partial_{\overline{z_2}}, \qquad \partial_{\overline{\zeta}^{\mathbf{i}_1}} = \partial_{\overline{z_1}} + \mathbf{i}_1 \partial_{\overline{z_2}},$$

we can rewrite the bicomplex differentials as follows:

$$\partial_Z = \partial_{\zeta^\dagger} \mathbf{e}_{12} + \partial_\zeta \overline{\mathbf{e}}_{12}, \qquad \partial_{\overline{Z}^{\mathbf{i}_1\mathbf{i}_2}} = \partial_{\overline{\zeta^\dagger}^{\mathbf{i}_1}} \mathbf{e}_{12} + \partial_{\overline{\zeta}^{\mathbf{i}_1}} \overline{\mathbf{e}}_{12},$$

$$\partial_{\overline{Z}^{\mathbf{i}_1}} = \partial_{\overline{\zeta}^{\mathbf{i}_1}} \mathbf{e}_{12} + \partial_{\overline{\zeta^\dagger}^{\mathbf{i}_1}} \overline{\mathbf{e}}_{12}, \qquad \partial_{\overline{Z}^{\mathbf{i}_2}} = \partial_\zeta \mathbf{e}_{12} + \partial_{\zeta^\dagger} \overline{\mathbf{e}}_{12}.$$

In [2, 11, 13] the authors introduce the following notion of bicomplex derivative:

**Definition 1.** Let $U$ be an open set in $\mathbb{BC}_2$ and let $Z_0 \in U$. A function $F : U \to \mathbb{BC}_2$ has a derivative at $Z_0$ if the limit

$$F'(Z_0) := \lim_{Z \to Z_0} (Z - Z_0)^{-1} (F(Z) - F(Z_0))$$

exists, for all $Z \in U$ such that $Z - Z_0$ is an invertible bicomplex number in $U$ (i.e., not a divisor of zero).

Note that the limit in the definition above avoids the divisors of zero in $\mathbb{BC}_2$, which are in the union of the two ideals generated by $\mathbf{e}_{12}$ and $\overline{\mathbf{e}_{12}}$, the so-called *cone of singularities.*

Functions which admit bicomplex derivative at each point in their domain are called bicomplex holomorphic, and it can be shown that this is equivalent to require that they admit a power series expansion in $Z$ [11, Definition 15.2]. There is however a third equivalent notion which is more suitable to our purposes (see [2, 13]):

**Theorem 2.** *Let $U$ be an open set in $\mathbb{BC}_2$ and let $F = u + \mathbf{i}_2 v : U \to \mathbb{BC}_2$ be of class $C^1$ in $U$. Then $F$ is bicomplex holomorphic if and only if:*

1. *$u$ and $v$ are complex holomorphic in both complex $\mathbb{C}_{\mathbf{i}_1}$ variables $z_1$ and $z_2$.*
2. *$\partial_{z_1} u = \partial_{z_2} v$ and $\partial_{z_1} v = -\partial_{z_2} u$ on $U$, i.e., $u$ and $v$ verify the* complex *Cauchy-Riemann equations.*

*Moreover, $F' = \frac{1}{2}\partial_Z F = \partial_{z_1} u + \mathbf{i}_2 \partial_{z_1} v = \partial_{z_2} v - \mathbf{i}_2 \partial_{z_2} u$, and $F'(Z)$ is invertible if and only if the corresponding Jacobian is non-zero.*

As mentioned in [14], the condition $F \in \mathcal{C}^1(U)$ can be dropped via the Hartogs' lemma, on the holomorphy of functions of several complex variables as being equivalent to their holomorphy in each variable separately (in other words, no continuity assumption is required). Note here a major difference between quaternionic and bicomplex analysis: as it is well known, in $\mathbb{H}$ the only functions who have (one-dimensional) quaternionic derivative in the usual sense are quaternionic linear functions (e.g., [22]). We mention here the important work of [9, 10], authors which define and study a two-dimensional directional derivative of a quaternionic function along a two-dimensional plane. In the bicomplex setting the class of functions admitting a usual derivative is non-trivial and consists on functions admitting power series expansion.

In an algebraic interpretation, the conditions in Theorem 2 can be translated as follows: let $F = u + \mathbf{i}_2 v$ a $\mathcal{C}^1(U)$ function, and set $\vec{F} = [u\ v]^t$; then $F$ is a bicomplex holomorphic function if and only if

$$\begin{bmatrix} \partial_{\overline{z}_1} & 0 \\ \partial_{\overline{z}_2} & 0 \\ 0 & \partial_{\overline{z}_1} \\ 0 & \partial_{\overline{z}_2} \\ \partial_{z_1} & -\partial_{z_2} \\ \partial_{z_2} & \partial_{z_1} \end{bmatrix} \vec{F} = \vec{0}. \tag{3}$$

A further interesting characterization of holomorphy in $\mathbb{BC}_2$ is the following result from [4].

**Theorem 3.** *Let $U \subseteq \mathbb{BC}_2$ be an open set and let $F : U \to \mathbb{BC}_2$ be of class $\mathcal{C}^1$ on $U$. Then $F$ is bicomplex holomorphic if and only if $F$ is $\overline{Z}^{\mathbf{i}_1}$, $\overline{Z}^{\mathbf{i}_2}$, $\overline{Z}^{\mathbf{i}_1\mathbf{i}_2}$-regular, i.e.,*

$$\frac{\partial F}{\partial \overline{Z}^{\mathbf{i}_1}} = \frac{\partial F}{\partial \overline{Z}^{\mathbf{i}_2}} = \frac{\partial F}{\partial \overline{Z}^{\mathbf{i}_1\mathbf{i}_2}} = 0.$$

We mention here that a stronger proof of this theorem can be found in [13, Lemma 1, p. 8], where the $\mathcal{C}^1$ condition is used in only one direction.

Note that both of these last results indicate that holomorphic functions on bicomplex variables can be seen as solutions of overdetermined systems of differential equations with constant coefficients. We have exploited this particularity in our papers [4, 5, 21], and we will show how similar arguments can be made for holomorphy on multicomplex spaces.

## 3. The $n$th multicomplex space

We now turn to the definition of the multicomplex spaces, $\mathbb{BC}_n$, for values of $n \geq 2$. These spaces are defined by taking $n$ commuting imaginary units $\mathbf{i}_1, \mathbf{i}_2, \ldots, \mathbf{i}_n$, i.e., $\mathbf{i}_a^2 = -1$, and $\mathbf{i}_a \mathbf{i}_b = \mathbf{i}_b \mathbf{i}_a$ for all $a, b$. Since the product of two commuting imaginary units is a hyperbolic unit, and since the product of an imaginary unit and a hyperbolic unit is an imaginary unit, we see that these units will generate a group $\mathfrak{A}_n$ of $2^n$ elements: one is the identity 1, $2^{n-1}$ elements are imaginary units and $2^{n-1}-1$ are hyperbolic units. Then the algebra generated over the real numbers by $\mathfrak{A}_n$ is the multicomplex space $\mathbb{BC}_n$ which forms a ring under the usual addition and multiplication operations. As in the $n=2$ case, the ring $\mathbb{BC}_n$ generates a real algebra, and each of its elements can be written as $Z=\sum_{I\in\mathfrak{A}_n} Z_I I$, where $Z_I$ are real numbers. This last representation of $Z\in\mathbb{BC}_n$ is quite messy, so we will build a more refined one that will show that these spaces are "nested" spaces.

In particular, following [11], it is natural to define the $n$-dimensional multicomplex space as follows, for $n\geq 1$:

$$\mathbb{BC}_n := \{Z_n = Z_{n-1,1} + \mathbf{i}_n Z_{n-1,2} \,\big|\, Z_{n-1,1}, Z_{n-1,2} \in \mathbb{BC}_{n-1}\}$$

with the natural operations of addition and multiplication. Note that $\mathbb{BC}_0 := \mathbb{R}$ and $\mathbb{BC}_1 = \mathbb{C}_{\mathbf{i}_1}$. For $n\geq 2$, since $\mathbb{BC}_{n-1}$ can be defined in a similar way, we recursively obtain, at the $k$th level:

$$Z_n = \sum_{|I|=n-k} \prod_{t=k+1}^{n} (\mathbf{i}_t)^{\alpha_t-1} Z_{k,I}$$

where $Z_{k,I}\in\mathbb{BC}_k$, $I=(\alpha_{k+1},\dots,\alpha_n)$, and $\alpha_t\in\{1,2\}$.

Because of the existence of $n$ imaginary units, we can define multiple types of conjugations as follows:

$$\overline{Z_n}^{\mathbf{i}_l} = \begin{cases} \displaystyle\sum_{|I|=n-k} \prod_{\substack{t=k+1\\ t\neq l}}^{n} (\mathbf{i}_t)^{\alpha_t-1}(-\mathbf{i}_l)^{\alpha_l-1} Z_{k,I} & \text{if } l>k \\ \displaystyle\sum_{|I|=n-k} \prod_{\substack{t=k+1\\ t\neq l}}^{n} (\mathbf{i}_t)^{\alpha_t-1}\overline{Z_{k,I}}^{\mathbf{i}_l} & \text{if } l<k \end{cases}$$

and

$$\overline{Z_n}^{\mathbf{i}_1\dots\mathbf{i}_l} = \overline{\overline{\overline{Z_n^{\mathbf{i}_1}}}^{\mathbf{i}_2\dots}}^{\mathbf{i}_l}$$

It turns out that

$$\overline{Z_n}^{\mathbf{i}_l} = \sum_{|I|=n-k}\sum_{i=k+1}^{n} \delta_{l,i}(-1)^{\alpha_l-1} c_k Z_{k,I} + \sum_{|I|=n-k}\sum_{i=1}^{k} \delta_{l,i} c_k \overline{Z_{k,I}}^{\mathbf{i}_l}.$$

These conjugations have been also defined in [7, 8], where the authors prove the following results:

**Theorem 4.** *The composition of the conjugates in the set of multicomplex numbers $\mathbb{BC}_n$ forms a commutative group of cardinality $2^n$ isomorphic with $\mathbb{Z}_2^n$.*

**Theorem 5.** *The multicomplex numbers of order n are a subalgebra of the Clifford algebra* $Cl_{\mathbb{R}}(0, 2n)$.

These results shed light on the structure of the conjugations on multicomplex spaces. However, the structure of the multicomplex space $\mathbb{BC}_n$ that will allow us to simplify the notions of differentiability is the one defined by the idempotent representations, as follows.

Just as in the case of $\mathbb{BC}_2$ we have idempotent bases in $\mathbb{BC}_n$, that will be organized at each "nested" level $\mathbb{BC}_k$ inside $\mathbb{BC}_n$ as follows. Denote by

$$\mathbf{e}_{kl} := \frac{1 + \mathbf{i}_k \mathbf{i}_l}{2},$$
$$\overline{\mathbf{e}}_{kl} := \frac{1 - \mathbf{i}_k \mathbf{i}_l}{2}.$$

Consider the following sets:

$$S_1 := \{\mathbf{e}_{n-1,n}, \overline{\mathbf{e}}_{n-1,n}\},$$
$$S_2 := \{\mathbf{e}_{n-2,n-1} \cdot S_1, \overline{\mathbf{e}}_{n-2,n-1} \cdot S_1\},$$
$$\vdots$$
$$S_{n-1} := \{\mathbf{e}_{12} \cdot S_{n-2}, \overline{\mathbf{e}}_{12} \cdot S_{n-2}\}.$$

At each stage $k$, the set $S_k$ has $2^k$ idempotents. Note the following *peculiar* identities:

$$\mathbf{i}_k \cdot \mathbf{e}_{kl} = \frac{\mathbf{i}_k - \mathbf{i}_l}{2} = -\frac{\mathbf{i}_l - \mathbf{i}_k}{2} = -\mathbf{i}_l \cdot \mathbf{e}_{kl},$$
$$\mathbf{i}_k \cdot \overline{\mathbf{e}}_{kl} = \frac{\mathbf{i}_k + \mathbf{i}_l}{2} = \mathbf{i}_l \cdot \overline{\mathbf{e}}_{kl}.$$

It is possible to immediately verify the following

**Proposition 6.** *In each set* $S_k$*, the product of any two idempotents is zero.*

We have several idempotent representations of $Z_n \in \mathbb{BC}_n$, as follows.

**Theorem 7.** *Any* $Z_n \in \mathbb{BC}_n$ *can be written as:*

$$Z_n = \sum_{j=1}^{2^k} Z_{n-k,j} \mathbf{e}_j,$$

*where* $Z_{n-k,j} \in \mathbb{BC}_{n-k}$ *and* $\mathbf{e}_j \in S_k$.

*Proof.* Let $Z_n = Z_{n-1,1} + \mathbf{i}_n Z_{n-1,2}$, where $Z_{n-1,1}, Z_{n-1,2} \in \mathbb{BC}_{n-1}$. Then the following computations are immediate:

$$\begin{aligned}
Z_n &= [Z_{n-1,1} - \mathbf{i}_{n-1} Z_{n-1,2}]\mathbf{e}_{n-1,n} + [Z_{n-1,1} + \mathbf{i}_{n-1} Z_{n-1,2}]\overline{\mathbf{e}}_{n-1,n} \\
&= [(Z_{n-2,1} - \mathbf{i}_{n-2} Z_{n-2,2})\mathbf{e}_{n-2,n-1} + (Z_{n-2,1} + \mathbf{i}_{n-2} Z_{n-2,2})\overline{\mathbf{e}}_{n-2,n-1}) \\
&\quad - \mathbf{i}_{n-1}((Z_{n-2,3} - \mathbf{i}_{n-2} Z_{n-2,4})\mathbf{e}_{n-2,n-1} + (Z_{n-2,3} + \mathbf{i}_{n-2} Z_{n-2,4})\overline{\mathbf{e}}_{n-2,n-1})]\mathbf{e}_{n-1,n} \\
&\quad + [(Z_{n-2,1} - \mathbf{i}_{n-2} Z_{n-2,2})\mathbf{e}_{n-2,n-1} + (Z_{n-2,1} + \mathbf{i}_{n-2} Z_{n-2,2})\overline{\mathbf{e}}_{n-2,n-1}) \\
&\quad + \mathbf{i}_{n-1}((Z_{n-2,3} - \mathbf{i}_{n-2} Z_{n-2,4})\mathbf{e}_{n-2,n-1} + (Z_{n-2,3} + \mathbf{i}_{n-2} Z_{n-2,4})\overline{\mathbf{e}}_{n-2,n-1})]\overline{\mathbf{e}}_{n-1,n} \\
&= [(Z_{n-2,1} - \mathbf{i}_{n-2} Z_{n-2,2})\mathbf{e}_{n-2,n-1} + (Z_{n-2,1} + \mathbf{i}_{n-2} Z_{n-2,2})\overline{\mathbf{e}}_{n-2,n-1}) \\
&\quad + \mathbf{i}_{n-2}((Z_{n-2,3} - \mathbf{i}_{n-2} Z_{n-2,4})\mathbf{e}_{n-2,n-1} + (Z_{n-2,3} + \mathbf{i}_{n-2} Z_{n-2,4})\overline{\mathbf{e}}_{n-2,n-1})]\mathbf{e}_{n-1,n} \\
&\quad + [(Z_{n-2,1} - \mathbf{i}_{n-2} Z_{n-2,2})\mathbf{e}_{n-2,n-1} + (Z_{n-2,1} + \mathbf{i}_{n-2} Z_{n-2,2})\overline{\mathbf{e}}_{n-2,n-1}) \\
&\quad - \mathbf{i}_{n-2}((Z_{n-2,3} - \mathbf{i}_{n-2} Z_{n-2,4})\mathbf{e}_{n-2,n-1} - (Z_{n-2,3} + \mathbf{i}_{n-2} Z_{n-2,4})\overline{\mathbf{e}}_{n-2,n-1})]\overline{\mathbf{e}}_{n-1,n} \\
&= [(Z_{n-2,1} + Z_{n-2,4}) - \mathbf{i}_{n-2}(Z_{n-2,2} - Z_{n-2,3})]\mathbf{e}_{n-2,n-1}\mathbf{e}_{n-1,n} \\
&\quad + [(Z_{n-2,1} - Z_{n-2,4}) + \mathbf{i}_{n-2}(Z_{n-2,2} + Z_{n-2,3})]\overline{\mathbf{e}}_{n-2,n-1}\mathbf{e}_{n-1,n} \\
&\quad + [(Z_{n-2,1} - Z_{n-2,4}) - \mathbf{i}_{n-2}(Z_{n-2,2} + Z_{n-2,3})]\mathbf{e}_{n-2,n-1}\overline{\mathbf{e}}_{n-1,n} \\
&\quad + [(Z_{n-2,1} + Z_{n-2,4}) + \mathbf{i}_{n-2}(Z_{n-2,2} - Z_{n-2,3})]\overline{\mathbf{e}}_{n-2,n-1}\overline{\mathbf{e}}_{n-1,n} \\
&= \cdots
\end{aligned}$$

The decomposition continues until the last step, where all last coefficients are in terms of $\mathbf{i}_1$. That concludes the proof. □

Due to the fact that the product of two idempotents is 0 at each level $S_k$, we will have many zero divisors in $\mathbb{BC}_n$ organized in "singular cones". The topology of the space is difficult, but just like in the case of $\mathbb{BC}_2$ we can circumvent this by avoiding the zero divisors to define the derivative of a multicomplex function as follows.

**Definition 8.** Let $\Omega$ be an open set of $\mathbb{BC}_n$ and let $Z_{n,0} \in \Omega$. A function $F : \Omega \to \mathbb{BC}_n$ has a multicomplex derivative at $Z_{n,0}$ if

$$\lim_{Z_n \to Z_{n,0}} (Z_n - Z_{n,0})^{-1} (F(Z_n) - F(Z_{n,0})) =: F'(Z_{n,0}),$$

exists for any $Z_n \in \Omega$, such that $Z_n - Z_{n,0}$ is invertible in $\mathbb{BC}_n$.

Just as in the case of $\mathbb{BC}_2$, functions which admit a multicomplex derivative at each point in their domain are called multicomplex holomorphic, and it can be shown that this is equivalent to require that they admit a power series expansion in $Z_n$ [11, Section 47].

We will denote by $\mathcal{O}(\mathbb{BC}_n)$ the space of multicomplex holomorphic functions. A multicomplex holomorphic function $F \in \mathcal{O}(\mathbb{BC}_n)$ can be split into $F = U + \mathbf{i}_n V$, where $U, V$ are holomorphic functions of two $\mathbb{BC}_{n-1}$ variables, i.e., they admit $\mathbb{BC}_{n-1}$ (partial) derivatives in both variables. As in the case $n = 2$, there is an equivalent notion of multicomplex holomorphy, which is more suitable to our computational algebraic purposes, and the following theorem can be proved in a similar fashion as its correspondent for the case $n = 2$ (the differential operators that appear in the statement are defined in detail in the next few sections, but it is not difficult to imagine their actual definition).

**Theorem 9.** *Let $\Omega$ be an open set in $\mathbb{BC}_n$ and $F : \Omega \to \mathbb{BC}_n$ such that $F = U + \mathbf{i}_n V \in \mathcal{C}^1(\Omega)$. Then $F$ is multicomplex holomorphic if and only if:*

1. *$U$ and $V$ are multicomplex holomorphic in both multicomplex $\mathbb{BC}_{n-1}$ variables $Z_{n-1,1}$ and $Z_{n-1,2}$.*
2. *$\partial_{Z_{n-1,1}} U = \partial_{Z_{n-1,2}} V$ and $\partial_{Z_{n-1,1}} V = -\partial_{Z_{n-1,2}} U$ on $\Omega$, i.e., $U$ and $V$ verify the* multicomplex *Cauchy-Riemann equations.*

## 4. Holomorphy in $\mathbb{BC}_3$

Before we delve into the general case, we will explicitly describe holomorphic functions of the space of "tricomplex" numbers $\mathbb{BC}_3$. Let $Z_3 \in \mathbb{BC}_3$ be written as:

$$\begin{aligned} Z_3 &= Z_{21} + \mathbf{i}_3 Z_{22} = Z_{111} + \mathbf{i}_2 Z_{121} + \mathbf{i}_3 Z_{112} + \mathbf{i}_2\mathbf{i}_3 Z_{122} \\ &= x_1 + \mathbf{i}_1 x_2 + \mathbf{i}_2 x_3 + \mathbf{i}_1\mathbf{i}_2 x_4 + \mathbf{i}_3 x_5 + \mathbf{i}_1\mathbf{i}_3 x_6 + \mathbf{i}_2\mathbf{i}_3 x_7 + \mathbf{i}_1\mathbf{i}_2\mathbf{i}_3 x_8 \,. \end{aligned}$$

As we indicated earlier, we can define seven different conjugations, depending on which imaginary unit we want to convert to its opposite. We write explicitly a few of them, and leave to the reader to complete the description with the remaining cases:

$$\begin{aligned} \overline{Z_3}^{\mathbf{i}_1} &= \overline{Z_{21}}^{\mathbf{i}_1} + \mathbf{i}_3 \overline{Z_{22}}^{\mathbf{i}_1} = \overline{Z_{111}} + \mathbf{i}_2 \overline{Z_{121}} + \mathbf{i}_3 \overline{Z_{112}} + \mathbf{i}_2\mathbf{i}_3 \overline{Z_{122}} \\ &= x_1 - \mathbf{i}_1 x_2 + \mathbf{i}_2 x_3 - \mathbf{i}_1\mathbf{i}_2 x_4 + \mathbf{i}_3 x_5 - \mathbf{i}_1\mathbf{i}_3 x_6 + \mathbf{i}_2\mathbf{i}_3 x_7 - \mathbf{i}_1\mathbf{i}_2\mathbf{i}_3 x_8 \,, \end{aligned}$$

$$\begin{aligned} \overline{Z_3}^{\mathbf{i}_1\mathbf{i}_2} &= \overline{Z_{21}}^{\mathbf{i}_1\mathbf{i}_2} + \mathbf{i}_3 \overline{Z_{22}}^{\mathbf{i}_1\mathbf{i}_2} = \overline{Z_{111}} - \mathbf{i}_2 \overline{Z_{121}} + \mathbf{i}_3 \overline{Z_{112}} - \mathbf{i}_2\mathbf{i}_3 \overline{Z_{122}} \\ &= x_1 - \mathbf{i}_1 x_2 - \mathbf{i}_2 x_3 + \mathbf{i}_1\mathbf{i}_2 x_4 + \mathbf{i}_3 x_5 - \mathbf{i}_1\mathbf{i}_3 x_6 - \mathbf{i}_2\mathbf{i}_3 x_7 + \mathbf{i}_1\mathbf{i}_2\mathbf{i}_3 x_8 \,, \end{aligned}$$

$$\begin{aligned} \overline{Z_3}^{\mathbf{i}_1\mathbf{i}_2\mathbf{i}_3} &= \overline{Z_{21}}^{\mathbf{i}_1\mathbf{i}_2} - \mathbf{i}_3 \overline{Z_{22}}^{\mathbf{i}_1\mathbf{i}_2} = \overline{Z_{111}} - \mathbf{i}_2 \overline{Z_{121}} - \mathbf{i}_3 \overline{Z_{112}} + \mathbf{i}_2\mathbf{i}_3 \overline{Z_{122}} \\ &= x_1 - \mathbf{i}_1 x_2 - \mathbf{i}_2 x_3 + \mathbf{i}_1\mathbf{i}_2 x_4 - \mathbf{i}_3 x_5 + \mathbf{i}_1\mathbf{i}_3 x_6 + \mathbf{i}_2\mathbf{i}_3 x_7 - \mathbf{i}_1\mathbf{i}_2\mathbf{i}_3 x_8 \,. \end{aligned}$$

In the idempotent representations (there are two of them), consider $Z_3 = Z_{21} + \mathbf{i}_3 Z_{22}$, where:

$$Z_{21} = \zeta_{21}^{\dagger} \mathbf{e}_{12} + \zeta_{21} \overline{\mathbf{e}_{12}} \,, \qquad Z_{22} = \zeta_{22}^{\dagger} \mathbf{e}_{12} + \zeta_{22} \overline{\mathbf{e}_{12}} \,.$$

Then we can write

$$\begin{aligned} Z_3 &= Z_{21} + \mathbf{i}_3 Z_{22} = (Z_{21} - \mathbf{i}_2 Z_{22}) \mathbf{e}_{23} + (Z_{21} + \mathbf{i}_2 Z_{22}) \overline{\mathbf{e}_{23}} \\ &= (\zeta_{21}^{\dagger} + \mathbf{i}_1 \zeta_{22}^{\dagger}) \mathbf{e}_{12}\mathbf{e}_{23} + (\zeta_{21} - \mathbf{i}_1 \zeta_{22}) \overline{\mathbf{e}_{12}} \mathbf{e}_{23} \\ &\quad + (\zeta_{21}^{\dagger} - \mathbf{i}_1 \zeta_{22}^{\dagger}) \mathbf{e}_{12} \overline{\mathbf{e}_{23}} + (\zeta_{21} + \mathbf{i}_1 \zeta_{22}) \overline{\mathbf{e}_{12}\mathbf{e}_{23}} \,. \end{aligned}$$

Therefore, for the conjugates introduced above, we have the following idempotent representations:

$$\begin{aligned}\overline{Z_3}^{\mathbf{i}_1} &= \overline{Z_{21}}^{\mathbf{i}_1} + \mathbf{i}_3\overline{Z_{22}}^{\mathbf{i}_1} = (\overline{Z_{21}}^{\mathbf{i}_1} - \mathbf{i}_2\overline{Z_{22}}^{\mathbf{i}_1})\mathbf{e}_{23} + (\overline{Z_{21}}^{\mathbf{i}_1} + \mathbf{i}_2\overline{Z_{22}}^{\mathbf{i}_1})\overline{\mathbf{e}_{23}}\\ &= (\overline{\zeta_{21}}^{\mathbf{i}_1} + \mathbf{i}_1\overline{\zeta_{22}}^{\mathbf{i}_1})\mathbf{e}_{12}\mathbf{e}_{23} + (\overline{\zeta^\dagger_{21}}^{\mathbf{i}_1} - \mathbf{i}_1\overline{\zeta^\dagger_{22}}^{\mathbf{i}_1})\overline{\mathbf{e}_{12}}\mathbf{e}_{23}\\ &\quad + (\overline{\zeta_{21}}^{\mathbf{i}_1} - \mathbf{i}_1\overline{\zeta_{22}}^{\mathbf{i}_1})\mathbf{e}_{12}\overline{\mathbf{e}_{23}} + (\overline{\zeta^\dagger_{21}}^{\mathbf{i}_1} + \mathbf{i}_1\overline{\zeta^\dagger_{22}}^{\mathbf{i}_1})\overline{\mathbf{e}_{12}\mathbf{e}_{23}}\,,\end{aligned}$$

$$\begin{aligned}\overline{Z_3}^{\mathbf{i}_1\mathbf{i}_2} &= \overline{Z_{21}}^{\mathbf{i}_1\mathbf{i}_2} + \mathbf{i}_3\overline{Z_{22}}^{\mathbf{i}_1\mathbf{i}_2} = (\overline{Z_{21}}^{\mathbf{i}_1\mathbf{i}_2} - \mathbf{i}_2\overline{Z_{22}}^{\mathbf{i}_1\mathbf{i}_2})\mathbf{e}_{23} + (\overline{Z_{21}}^{\mathbf{i}_1\mathbf{i}_2} + \mathbf{i}_2\overline{Z_{22}}^{\mathbf{i}_1\mathbf{i}_2})\overline{\mathbf{e}_{23}}\\ &= (\overline{\zeta^\dagger_{21}}^{\mathbf{i}_1} + \mathbf{i}_1\overline{\zeta^\dagger_{22}}^{\mathbf{i}_1})\mathbf{e}_{12}\mathbf{e}_{23} + (\overline{\zeta_{21}}^{\mathbf{i}_1} - \mathbf{i}_1\overline{\zeta_{22}}^{\mathbf{i}_1})\overline{\mathbf{e}_{12}}\mathbf{e}_{23}\\ &\quad + (\overline{\zeta^\dagger_{21}}^{\mathbf{i}_1} - \mathbf{i}_1\overline{\zeta^\dagger_{22}}^{\mathbf{i}_1})\mathbf{e}_{12}\overline{\mathbf{e}_{23}} + (\overline{\zeta_{21}}^{\mathbf{i}_1} + \mathbf{i}_1\overline{\zeta_{22}}^{\mathbf{i}_1})\overline{\mathbf{e}_{12}\mathbf{e}_{23}}\,,\end{aligned}$$

$$\begin{aligned}\overline{Z_3}^{\mathbf{i}_1\mathbf{i}_2\mathbf{i}_3} &= \overline{Z_{21}}^{\mathbf{i}_1\mathbf{i}_2} - \mathbf{i}_3\overline{Z_{22}}^{\mathbf{i}_1\mathbf{i}_2} = (\overline{Z_{21}}^{\mathbf{i}_1\mathbf{i}_2} + \mathbf{i}_2\overline{Z_{22}}^{\mathbf{i}_1\mathbf{i}_2})\mathbf{e}_{23} + (\overline{Z_{21}}^{\mathbf{i}_1\mathbf{i}_2} - \mathbf{i}_2\overline{Z_{22}}^{\mathbf{i}_1\mathbf{i}_2})\overline{\mathbf{e}_{23}}\\ &= (\overline{\zeta^\dagger_{21}}^{\mathbf{i}_1} - \mathbf{i}_1\overline{\zeta^\dagger_{22}}^{\mathbf{i}_1})\mathbf{e}_{12}\mathbf{e}_{23} + (\overline{\zeta_{21}}^{\mathbf{i}_1} + \mathbf{i}_1\overline{\zeta_{22}}^{\mathbf{i}_1})\overline{\mathbf{e}_{12}}\mathbf{e}_{23}\\ &\quad + (\overline{\zeta^\dagger_{21}}^{\mathbf{i}_1} + \mathbf{i}_1\overline{\zeta^\dagger_{22}}^{\mathbf{i}_1})\mathbf{e}_{12}\overline{\mathbf{e}_{23}} + (\overline{\zeta_{21}}^{\mathbf{i}_1} - \mathbf{i}_1\overline{\zeta_{22}}^{\mathbf{i}_1})\overline{\mathbf{e}_{12}\mathbf{e}_{23}}\,,\end{aligned}$$

and similarly for the other four conjugates $\overline{Z_3}^{\mathbf{i}_2}, \overline{Z_3}^{\mathbf{i}_3}, \overline{Z_3}^{\mathbf{i}_1\mathbf{i}_3}, \overline{Z_3}^{\mathbf{i}_2\mathbf{i}_3}$. The tricomplex differential operators are defined as follows:

$$\begin{aligned}\partial^{(3)}_{Z_3} &= \partial^{(2)}_{Z_{21}} - \mathbf{i}_3\partial^{(2)}_{Z_{22}} = \partial_{Z_{111}} - \mathbf{i}_2\partial_{Z_{121}} - \mathbf{i}_3\partial_{Z_{112}} + \mathbf{i}_2\mathbf{i}_3\partial_{Z_{122}}\\ &= \partial_1 - \mathbf{i}_1\partial_2 - \mathbf{i}_2\partial_3 + \mathbf{i}_1\mathbf{i}_2\partial_4 - \mathbf{i}_3\partial_5 + \mathbf{i}_1\mathbf{i}_3\partial_6 + \mathbf{i}_2\mathbf{i}_3\partial_7 - \mathbf{i}_1\mathbf{i}_2\mathbf{i}_3\partial_8\,,\end{aligned}$$

$$\begin{aligned}\partial^{(3)}_{\overline{Z_3}^{\mathbf{i}_1}} &= \partial^{(2)}_{\overline{Z_{21}}^{\mathbf{i}_1}} - \mathbf{i}_3\partial^{(2)}_{\overline{Z_{22}}^{\mathbf{i}_1}} = \partial_{\overline{Z_{111}}} - \mathbf{i}_2\partial_{\overline{Z_{121}}} - \mathbf{i}_3\partial_{\overline{Z_{112}}} + \mathbf{i}_2\mathbf{i}_3\partial_{\overline{Z_{122}}}\\ &= \partial_1 + \mathbf{i}_1\partial_2 - \mathbf{i}_2\partial_3 - \mathbf{i}_1\mathbf{i}_2\partial_4 - \mathbf{i}_3\partial_5 - \mathbf{i}_1\mathbf{i}_3\partial_6 + \mathbf{i}_2\mathbf{i}_3\partial_7 + \mathbf{i}_1\mathbf{i}_2\mathbf{i}_3\partial_8\,,\end{aligned}$$

$$\begin{aligned}\partial^{(3)}_{\overline{Z_3}^{\mathbf{i}_1\mathbf{i}_2}} &= \partial^{(2)}_{\overline{Z_{21}}^{\mathbf{i}_1\mathbf{i}_2}} - \mathbf{i}_3\partial^{(2)}_{\overline{Z_{22}}^{\mathbf{i}_1\mathbf{i}_2}} = \partial_{\overline{Z_{111}}} + \mathbf{i}_2\partial_{\overline{Z_{121}}} - \mathbf{i}_3\partial_{\overline{Z_{112}}} - \mathbf{i}_2\mathbf{i}_3\partial_{\overline{Z_{122}}}\\ &= \partial_1 + \mathbf{i}_1\partial_2 + \mathbf{i}_2\partial_3 + \mathbf{i}_1\mathbf{i}_2\partial_4 - \mathbf{i}_3\partial_5 - \mathbf{i}_1\mathbf{i}_3\partial_6 - \mathbf{i}_2\mathbf{i}_3\partial_7 - \mathbf{i}_1\mathbf{i}_2\mathbf{i}_3\partial_8\,,\end{aligned}$$

$$\begin{aligned}\partial^{(3)}_{\overline{Z_3}^{\mathbf{i}_1\mathbf{i}_2\mathbf{i}_3}} &= \partial^{(2)}_{\overline{Z_{21}}^{\mathbf{i}_1\mathbf{i}_2}} + \mathbf{i}_3\partial^{(2)}_{\overline{Z_{22}}^{\mathbf{i}_1\mathbf{i}_2}} = \partial_{\overline{Z_{111}}} + \mathbf{i}_2\partial_{\overline{Z_{121}}} + \mathbf{i}_3\partial_{\overline{Z_{112}}} + \mathbf{i}_2\mathbf{i}_3\partial_{\overline{Z_{122}}}\\ &= \partial_1 + \mathbf{i}_1\partial_2 + \mathbf{i}_2\partial_3 + \mathbf{i}_1\mathbf{i}_2\partial_4 + \mathbf{i}_3\partial_5 + \mathbf{i}_1\mathbf{i}_3\partial_6 + \mathbf{i}_2\mathbf{i}_3\partial_7 + \mathbf{i}_1\mathbf{i}_2\mathbf{i}_3\partial_8\,,\end{aligned}$$

with similar forms for the other four differential operators

$$\partial^{(3)}_{\overline{Z_3}^{\mathbf{i}_2}},\quad \partial^{(3)}_{\overline{Z_3}^{\mathbf{i}_3}},\quad \partial^{(3)}_{\overline{Z_3}^{\mathbf{i}_2\mathbf{i}_3}},\quad \partial^{(3)}_{\overline{Z_3}^{\mathbf{i}_1\mathbf{i}_3}}\,.$$

Let now $F : \Omega \subset \mathbb{BC}_3 \to \mathbb{BC}_3$ be a function that we write as usual as

$$\begin{aligned} F &= U + \mathbf{i}_3 V = u_1 + \mathbf{i}_2 v_1 + \mathbf{i}_3 u_2 + \mathbf{i}_2\mathbf{i}_3 v_2 \\ &= f_1 + \mathbf{i}_1 f_2 + \mathbf{i}_2 f_3 + \mathbf{i}_1\mathbf{i}_2 f_4 + \mathbf{i}_3 f_5 + \mathbf{i}_1\mathbf{i}_3 f_6 + \mathbf{i}_2\mathbf{i}_3 f_7 + \mathbf{i}_1\mathbf{i}_2\mathbf{i}_3 f_8 \end{aligned}$$

For simplicity of notation we will write $Z$ instead of $Z_{21}$ and $W$ instead of $Z_{22}$. Now we study the system formed by all seven differential operators applied to $F$ equal to 0. The equation $\partial^{(3)}_{\overline{Z_3}^{\mathbf{i}_1}} F = 0$ is equivalent to:

$$(\partial_{\overline{Z}^{\mathbf{i}_1}} - \mathbf{i}_3 \partial_{\overline{W}^{\mathbf{i}_1}})(U + \mathbf{i}_3 V) = 0$$

which is equivalent to the Cauchy-Riemann type system

$$\begin{aligned} \partial_{\overline{Z}^{\mathbf{i}_1}} U + \partial_{\overline{W}^{\mathbf{i}_1}} V &= 0\,, \\ \partial_{\overline{W}^{\mathbf{i}_1}} U - \partial_{\overline{Z}^{\mathbf{i}_1}} V &= 0\,. \end{aligned}$$

In real coordinates this system is equivalent to

$$\begin{aligned} &(\partial_1 + \mathbf{i}_1\partial_2 - \mathbf{i}_2\partial_3 - \mathbf{i}_1\mathbf{i}_2\partial_4 - \mathbf{i}_3\partial_5 - \mathbf{i}_1\mathbf{i}_3\partial_6 + \mathbf{i}_2\mathbf{i}_3\partial_7 + \mathbf{i}_1\mathbf{i}_2\mathbf{i}_3\partial_8) \\ &(f_1 + \mathbf{i}_1 f_2 + \mathbf{i}_2 f_3 + \mathbf{i}_1\mathbf{i}_2 f_4 + \mathbf{i}_3 f_5 + \mathbf{i}_1\mathbf{i}_3 f_6 + \mathbf{i}_2\mathbf{i}_3 f_7 + \mathbf{i}_1\mathbf{i}_2\mathbf{i}_3 f_8) = 0\,. \end{aligned}$$

The equation $\partial^{(3)}_{\overline{Z_3}^{\mathbf{i}_2}} F = 0$ is equivalent to

$$(\partial_{\overline{Z}^{\mathbf{i}_2}} - \mathbf{i}_3 \partial_{\overline{W}^{\mathbf{i}_2}})(U + \mathbf{i}_3 V) = 0\,,$$

which is equivalent to the Cauchy-Riemann type system:

$$\begin{aligned} \partial_{\overline{Z}^{\mathbf{i}_2}} U + \partial_{\overline{W}^{\mathbf{i}_2}} V &= 0, \\ \partial_{\overline{W}^{\mathbf{i}_2}} U - \partial_{\overline{Z}^{\mathbf{i}_2}} V &= 0. \end{aligned}$$

In real coordinates this can be expressed by

$$\begin{aligned} &(\partial_1 - \mathbf{i}_1\partial_2 + \mathbf{i}_2\partial_3 - \mathbf{i}_1\mathbf{i}_2\partial_4 - \mathbf{i}_3\partial_5 + \mathbf{i}_1\mathbf{i}_3\partial_6 - \mathbf{i}_2\mathbf{i}_3\partial_7 + \mathbf{i}_1\mathbf{i}_2\mathbf{i}_3\partial_8) \\ &(f_1 + \mathbf{i}_1 f_2 + \mathbf{i}_2 f_3 + \mathbf{i}_1\mathbf{i}_2 f_4 + \mathbf{i}_3 f_5 + \mathbf{i}_1\mathbf{i}_3 f_6 + \mathbf{i}_2\mathbf{i}_3 f_7 + \mathbf{i}_1\mathbf{i}_2\mathbf{i}_3 f_8) = 0. \end{aligned}$$

One can then argue in exactly the same way for the remaining five differential operators, and one obtains the following important result.

**Theorem 10.** *A function $F = U + \mathbf{i}_3 V : \Omega \subset \mathbb{BC}_3 \to \mathbb{BC}_3$ is in the kernels of all 7 differential operators generated by the seven conjugations, if and only if $U$ and $V$ are $\mathbb{BC}_2$-holomorphic functions satisfying the* bicomplex *Cauchy-Riemann conditions.*

*Proof.* The matrix corresponding to $P(D)F = 0$ is the following:

$$\begin{bmatrix} \partial_{\overline{Z}^{\mathbf{i}_1}} & \partial_{\overline{W}^{\mathbf{i}_1}} \\ \partial_{\overline{W}^{\mathbf{i}_1}} & -\partial_{\overline{Z}^{\mathbf{i}_1}} \\ \partial_{\overline{Z}^{\mathbf{i}_2}} & \partial_{\overline{W}^{\mathbf{i}_2}} \\ \partial_{\overline{W}^{\mathbf{i}_2}} & -\partial_{\overline{Z}^{\mathbf{i}_2}} \\ \partial_{\overline{Z}^{\mathbf{i}_1\mathbf{i}_2}} & \partial_{\overline{W}^{\mathbf{i}_1\mathbf{i}_2}} \\ \partial_{\overline{W}^{\mathbf{i}_1\mathbf{i}_2}} & -\partial_{\overline{Z}^{\mathbf{i}_1\mathbf{i}_2}} \\ \partial_{\overline{Z}^{\mathbf{i}_2}} & -\partial_{\overline{W}^{\mathbf{i}_2}} \\ \partial_{\overline{W}^{\mathbf{i}_2}} & \partial_{\overline{Z}^{\mathbf{i}_2}} \\ \partial_{\overline{Z}^{\mathbf{i}_1}} & -\partial_{\overline{W}^{\mathbf{i}_1}} \\ \partial_{\overline{W}^{\mathbf{i}_1}} & \partial_{\overline{Z}^{\mathbf{i}_1}} \\ \partial_{\overline{Z}^{\mathbf{i}_1\mathbf{i}_2}} & -\partial_{\overline{W}^{\mathbf{i}_1\mathbf{i}_2}} \\ \partial_{\overline{W}^{\mathbf{i}_1\mathbf{i}_2}} & \partial_{\overline{Z}^{\mathbf{i}_1\mathbf{i}_2}} \\ \partial_Z & -\partial_W \\ \partial_W & \partial_Z \end{bmatrix} \begin{bmatrix} U \\ V \end{bmatrix} = 0$$

which simplifies to

$$\begin{bmatrix} \partial_{\overline{Z}^{\mathbf{i}_1}} & 0 \\ \partial_{\overline{Z}^{\mathbf{i}_2}} & 0 \\ \partial_{\overline{Z}^{\mathbf{i}_1\mathbf{i}_2}} & 0 \\ 0 & \partial_{\overline{W}^{\mathbf{i}_1}} \\ 0 & \partial_{\overline{W}^{\mathbf{i}_2}} \\ 0 & \partial_{\overline{W}^{\mathbf{i}_1\mathbf{i}_2}} \\ \partial_{\overline{W}^{\mathbf{i}_1}} & 0 \\ \partial_{\overline{W}^{\mathbf{i}_2}} & 0 \\ \partial_{\overline{W}^{\mathbf{i}_1\mathbf{i}_2}} & 0 \\ 0 & \partial_{\overline{Z}^{\mathbf{i}_1}} \\ 0 & \partial_{\overline{Z}^{\mathbf{i}_2}} \\ 0 & \partial_{\overline{Z}^{\mathbf{i}_1\mathbf{i}_2}} \\ \partial_Z & -\partial_W \\ \partial_W & \partial_Z \end{bmatrix} \begin{bmatrix} U \\ V \end{bmatrix} = 0.$$

The theorem follows then immediately. □

## 5. Recursive formulas

We devote this short section to some recursive formulas that will allow us to readily define differential operators on $\mathbb{BC}_n$. The formulas that follow indicate how to represent every element $Z_n \in \mathbb{BC}_n$ in terms of $(n-2)$-complex numbers. Specifically we have

$$Z_n = Z_{n-1,1} + \mathbf{i}_n Z_{n-1,2} = Z_{n-2,11} + \mathbf{i}_{n-1} Z_{n-2,21} + \mathbf{i}_n Z_{n-2,12} + \mathbf{i}_{n-1}\mathbf{i}_n Z_{n-2,22}.$$

The conjugates are then given by

$$\begin{aligned}\overline{Z_n}^{\mathbf{i}_l} &= \overline{Z_{n-1,1}}^{\mathbf{i}_l} + \mathbf{i}_n \overline{Z_{n-1,2}}^{\mathbf{i}_l}\\ &= \overline{Z_{n-2,11}}^{\mathbf{i}_l} + \mathbf{i}_{n-1}\overline{Z_{n-2,21}}^{\mathbf{i}_l} + \mathbf{i}_n\overline{Z_{n-2,12}}^{\mathbf{i}_l} + \mathbf{i}_{n-1}\mathbf{i}_n\overline{Z_{n-2,22}}^{\mathbf{i}_l},\end{aligned}$$

for $1 \le l \le n-2$, and

$$\begin{aligned}\overline{Z_n}^{\mathbf{i}_n} &= Z_{n-1,1} - \mathbf{i}_n Z_{n-1,2}\\ &= Z_{n-2,11} + \mathbf{i}_{n-1}Z_{n-2,21} - \mathbf{i}_n Z_{n-2,12} - \mathbf{i}_{n-1}\mathbf{i}_n Z_{n-2,22}\,,\\ \overline{Z_n}^{\mathbf{i}_{n-1}} &= \overline{Z_{n-1,1}}^{\mathbf{i}_{n-1}} + \mathbf{i}_n \overline{Z_{n-1,2}}^{\mathbf{i}_{n-1}}\\ &= Z_{n-2,11} - \mathbf{i}_{n-1}Z_{n-2,21} + \mathbf{i}_n Z_{n-2,12} - \mathbf{i}_{n-1}\mathbf{i}_n Z_{n-2,22}\,,\\ \overline{Z_n}^{\mathbf{i}_{n-1}\mathbf{i}_n} &= \overline{Z_{n-1,1}}^{\mathbf{i}_{n-1}} - \mathbf{i}_n \overline{Z_{n-1,2}}^{\mathbf{i}_{n-1}}\\ &= Z_{n-2,11} - \mathbf{i}_{n-1}Z_{n-2,21} - \mathbf{i}_n Z_{n-2,12} + \mathbf{i}_{n-1}\mathbf{i}_n Z_{n-2,22}\,.\end{aligned}$$

For $J$ a set of imaginary units such that $\mathbf{i}_{n-1}, \mathbf{i}_n \notin J$, we have the following formulas:

$$\begin{aligned}\overline{Z_n}^{J} &= \overline{Z_{n-2,11}}^{J} + \mathbf{i}_{n-1}\overline{Z_{n-2,21}}^{J} + \mathbf{i}_n\overline{Z_{n-2,12}}^{J} + \mathbf{i}_{n-1}\mathbf{i}_n\overline{Z_{n-2,22}}^{J},\\ \overline{Z_n}^{\{\mathbf{i}_{n-1}\}\cup J} &= \overline{Z_{n-2,11}}^{J} - \mathbf{i}_{n-1}\overline{Z_{n-2,21}}^{J} + \mathbf{i}_n\overline{Z_{n-2,12}}^{J} - \mathbf{i}_{n-1}\mathbf{i}_n\overline{Z_{n-2,22}}^{J},\\ \overline{Z_n}^{\{\mathbf{i}_n\}\cup J} &= \overline{Z_{n-2,11}}^{J} + \mathbf{i}_{n-1}\overline{Z_{n-2,21}}^{J} - \mathbf{i}_n\overline{Z_{n-2,12}}^{J} - \mathbf{i}_{n-1}\mathbf{i}_n\overline{Z_{n-2,22}}^{J},\\ \overline{Z_n}^{\{\mathbf{i}_{n-1},\mathbf{i}_n\}\cup J} &= \overline{Z_{n-2,11}}^{J} - \mathbf{i}_{n-1}\overline{Z_{n-2,21}}^{J} - \mathbf{i}_n\overline{Z_{n-2,12}}^{J} + \mathbf{i}_{n-1}\mathbf{i}_n\overline{Z_{n-2,22}}^{J}.\end{aligned}$$

We can now use these formulas to define recursively the differential operators whose kernels consist of holomorphic functions in $\mathbb{BC}_n$.

$$\begin{aligned}\partial^{(n)}_{Z_n} &= \partial^{(n-2)}_{Z_{n-2,11}} - \mathbf{i}_{n-1}\partial^{(n-2)}_{Z_{n-2,21}} - \mathbf{i}_n\partial^{(n-2)}_{Z_{n-2,12}} + \mathbf{i}_{n-1}\mathbf{i}_n\partial^{(n-2)}_{Z_{n-2,22}}\,,\\ \partial^{(n)}_{\overline{Z_n}^{\mathbf{i}_n}} &= \partial^{(n-2)}_{Z_{n-2,11}} - \mathbf{i}_{n-1}\partial^{(n-2)}_{Z_{n-2,21}} + \mathbf{i}_n\partial^{(n-2)}_{Z_{n-2,12}} - \mathbf{i}_{n-1}\mathbf{i}_n\partial^{(n-2)}_{Z_{n-2,22}}\,,\\ \partial^{(n)}_{\overline{Z_n}^{\mathbf{i}_{n-1}}} &= \partial^{(n-2)}_{Z_{n-2,11}} + \mathbf{i}_{n-1}\partial^{(n-2)}_{Z_{n-2,21}} - \mathbf{i}_n\partial^{(n-2)}_{Z_{n-2,12}} - \mathbf{i}_{n-1}\mathbf{i}_n\partial^{(n-2)}_{Z_{n-2,22}}\,,\\ \partial^{(n)}_{\overline{Z_n}^{\mathbf{i}_{n-1}\mathbf{i}_n}} &= \partial^{(n-2)}_{Z_{n-2,11}} + \mathbf{i}_{n-1}\partial^{(n-2)}_{Z_{n-2,21}} + \mathbf{i}_n\partial^{(n-2)}_{Z_{n-2,12}} + \mathbf{i}_{n-1}\mathbf{i}_n\partial^{(n-2)}_{Z_{n-2,22}}\,.\end{aligned}$$

Similarly, for $J$ a set of imaginary units such that $\mathbf{i}_{n-1}, \mathbf{i}_n \notin J$, we have the following formulas:

$$\begin{aligned}\partial^{(n)}_{\overline{Z_n}^{J}} &= \partial^{(n-2)}_{\overline{Z_{n-2,11}}^{J}} - \mathbf{i}_{n-1}\partial^{(n-2)}_{\overline{Z_{n-2,21}}^{J}} - \mathbf{i}_n\partial^{(n-2)}_{\overline{Z_{n-2,12}}^{J}} + \mathbf{i}_{n-1}\mathbf{i}_n\partial^{(n-2)}_{\overline{Z_{n-2,22}}^{J}}\,,\\ \partial_{\overline{Z_n}^{\{\mathbf{i}_{n-1}\}\cup J}} &= \partial^{(n-2)}_{\overline{Z_{n-2,11}}^{J}} + \mathbf{i}_{n-1}\partial^{(n-2)}_{\overline{Z_{n-2,21}}^{J}} - \mathbf{i}_n\partial^{(n-2)}_{\overline{Z_{n-2,12}}^{J}} - \mathbf{i}_{n-1}\mathbf{i}_n\partial^{(n-2)}_{\overline{Z_{n-2,22}}^{J}}\,,\\ \partial_{\overline{Z_n}^{\{\mathbf{i}_n\}\cup J}} &= \partial^{(n-2)}_{\overline{Z_{n-2,11}}^{J}} - \mathbf{i}_{n-1}\partial^{(n-2)}_{\overline{Z_{n-2,21}}^{J}} + \mathbf{i}_n\partial^{(n-2)}_{\overline{Z_{n-2,12}}^{J}} - \mathbf{i}_{n-1}\mathbf{i}_n\partial^{(n-2)}_{\overline{Z_{n-2,22}}^{J}}\,,\\ \partial_{\overline{Z_n}^{\{\mathbf{i}_{n-1},\mathbf{i}_n\}\cup J}} &= \partial^{(n-2)}_{\overline{Z_{n-2,11}}^{J}} + \mathbf{i}_{n-1}\partial^{(n-2)}_{\overline{Z_{n-2,21}}^{J}} + \mathbf{i}_n\partial^{(n-2)}_{\overline{Z_{n-2,12}}^{J}} + \mathbf{i}_{n-1}\mathbf{i}_n\partial^{(n-2)}_{\overline{Z_{n-2,22}}^{J}}\,.\end{aligned}$$

## 6. Algebraic properties of the sheaf of holomorphic functions on $\mathbb{BC}_n$

We start with an analogous theorem for $\mathbb{BC}_n$ which, just as in the case of $\mathbb{BC}_2$ and $\mathbb{BC}_3$, will give a description of the holomorphy of a multicomplex map in analytic terms. It is worth mentioning here the work of Baird and Wood [1], authors which study properties of bicomplex holomorphic functions (and much more) in the context of bicomplex manifolds.

The following theorem is an immediate consequence of the nesting behavior of $\mathbb{BC}_n$ and it is easily proven by induction.

**Theorem 11.** *A $\mathcal{C}^1$-function $F : \Omega \subset \mathbb{BC}_n \to \mathbb{BC}_n$ is holomorphic on $\Omega$ if and only if*

$$\frac{\partial F}{\partial \overline{Z}^I} = 0$$

*for all sets $I$ of imaginary units such that* $\operatorname{card}(I) \leq n$ *and $I$ is in increasing order of indices of complex units.*

This theorem can be given an interesting interpretation, similarly to what is proved by Rochon in [13], as follows. Consider $F : \Omega \subset \mathbb{BC}_n \to \mathbb{BC}_n$, defined recursively by the formulas developed in Section 5, and following (recursively) the argument we used in the case of $\mathbb{BC}_3$, we obtain the fact that the multicomplex function $F$ can be written as:

$$F(Z_n) = \sum_{|I| \leq n-1} f_I(z_1, \dots, z_{2^{n-1}}) I, \tag{4}$$

for all $Z_n \in \Omega$, where $f_I$ are complex-valued functions defined on an open subset of $\mathbb{C}^{2^{n-1}}$. With this convention and notation we obtain the following theorem, in which the $\mathcal{C}^1$-condition for $F$ can be eliminated:

**Theorem 12.** *A function $F : \Omega \subset \mathbb{BC}_n \to \mathbb{BC}_n$ (written as in (4)) is holomorphic on $\Omega$ if and only if $f_I$ are complex holomorphic functions and they respect the corresponding Cauchy-Riemann type conditions.*

*Proof.* Clearly, if $F$ is holomorphic on $\mathbb{BC}_n$ then it is in $\mathcal{C}^1(\Omega)$ and Theorem 11 above yields that each $f_I$ is complex holomorphic in each of the $2^{n-1}$ variables, and, moreover, all $f_I$ satisfy the Cauchy-Riemann type conditions that correspond to $\frac{\partial F}{\partial \overline{Z}^I} = 0$.

Conversely, if each $f_I$ is holomorphic in each of the $2^{n-1}$ variables, then by Hartogs' Lemma the $\mathcal{C}^1$ condition can be removed. □

Theorem 11 indicate that multicomplex holomorphic functions are solutions of certain overdetermined systems of linear constant coefficients differential equations. Let us denote, as customary in this field, by $P(D)$ the matrix of differential operators that act on functions, and whose kernels consist of holomorphic functions of multicomplex variables.

It is worth mentioning that an analogous theorem can be proven for a decomposition of $F$ in the idempotent representation; write $Z_n = \zeta_1 e_1 + \cdots + \zeta_{2^{n-1}} e_{2^{n-1}}$ and we have:

$$F(Z_n) = \sum_{l=1}^{2^{n-1}} U_l(\zeta_1, \ldots, \zeta_{2^{n-1}}) e_l.$$

Then $F : \Omega \subset \mathbb{BC}_n \to \mathbb{BC}_n$ is holomorphic on $\Omega$ if and only if $U_l$ complex holomorphic in $\zeta_l$ and depend on $\zeta_l$ only, respectively for all $1 \le l \le 2^{n-1}$.

In the case of $\mathbb{BC}_2$, the matrix $P(D)$ is given as follows: for $Z_2 = Z_1 + \mathbf{i}_2 W_1$, where $Z_1, W_1 \in \mathbb{BC}_1 = \mathbb{C}(\mathbf{i}_1)$, we have:

$$P(D;2) = \begin{bmatrix} \partial_{\overline{Z_1}^{\mathbf{i}_1}} & 0 \\ 0 & \partial_{\overline{W_1}^{\mathbf{i}_1}} \\ \partial_{\overline{W_1}^{\mathbf{i}_1}} & 0 \\ 0 & \partial_{\overline{Z_1}^{\mathbf{i}_1}} \\ \partial_{Z_1} & -\partial_{W_1} \\ \partial_{W_1} & \partial_{Z_1} \end{bmatrix} = \begin{bmatrix} P(\mathbf{i}_1) \\ M_1 \end{bmatrix}$$

where $P(\mathbf{i}_1)$ indicates the $4 \times 2$ matrix associated to differential operators in $W_1$ and $Z_1$, while $M_1$ is the matrix representing the complex Cauchy-Riemann system.

In $\mathbb{BC}_3$ for $Z_3 = Z_2 + \mathbf{i}_3 W_2$, where $Z_2, W_2 \in \mathbb{BC}_2$, we have, with obvious meaning of the symbols,

$$P(D;3) = \begin{bmatrix} \partial_{\overline{Z_2}^{\mathbf{i}_1}} & 0 \\ 0 & \partial_{\overline{W_2}^{\mathbf{i}_1}} \\ \partial_{\overline{Z_2}^{\mathbf{i}_2}} & 0 \\ 0 & \partial_{\overline{W_2}^{\mathbf{i}_2}} \\ \partial_{\overline{Z_2}^{\mathbf{i}_1\mathbf{i}_2}} & 0 \\ 0 & \partial_{\overline{W_2}^{\mathbf{i}_1\mathbf{i}_2}} \\ \partial_{\overline{W_2}^{\mathbf{i}_1}} & 0 \\ 0 & \partial_{\overline{Z_2}^{\mathbf{i}_1}} \\ \partial_{\overline{W_2}^{\mathbf{i}_2}} & 0 \\ 0 & \partial_{\overline{Z_2}^{\mathbf{i}_2}} \\ \partial_{\overline{W_2}^{\mathbf{i}_1\mathbf{i}_2}} & 0 \\ 0 & \partial_{\overline{Z_2}^{\mathbf{i}_1\mathbf{i}_2}} \\ \partial_{Z_2} & -\partial_{W_2} \\ \partial_{W_2} & \partial_{Z_2} \end{bmatrix} = \begin{bmatrix} P(\mathbf{i}_1) \\ P(\mathbf{i}_2) \\ P(\mathbf{i}_1\mathbf{i}_2) \\ M_2 \end{bmatrix}$$

and so on.

Since holomorphic functions of multicomplex variables are vectors of differentiable functions satisfying a homogeneous system of constant coefficients differential equations, we immediately obtain the following result (which was stated for $\mathbb{BC}_2$ in [4]).

**Proposition 13.** *Holomorphic functions on $\mathbb{BC}_n$ form a sheaf $\mathcal{H}_n$ of rings.*

The cohomological properties of the sheaf of holomorphic functions on $\mathbb{BC}_2$ were derived in [4], by using some standard tools described in detail in [3]. We can easily show that those same ideas apply to holomorphic functions on $\mathbb{BC}_n$, for any value of $n$. With the customary abuse of language, we begin by considering the "Fourier transform" $P$ of the matrix $P(D,3)$. The entries of $P$ belong to the ring $R = \mathbb{C}[Z, \overline{Z}^{\mathbf{i}_1}, \overline{Z}^{\mathbf{i}_2}, \overline{Z}^{\mathbf{i}_1\mathbf{i}_2}, W, \overline{W}^{\mathbf{i}_1}, \overline{W}^{\mathbf{i}_2}, \overline{W}^{\mathbf{i}_1\mathbf{i}_2}]$ and the cokernel of the map $P^t$, i.e., $M := R^2/\langle P^t\rangle$, where $\langle P^t\rangle$ denotes the module generated by the columns of $P^t$, is the module associated to the system of differential operators which define holomorphy in $\mathbb{BC}_3$. We have the following result, which mimics what we have already proved for $\mathbb{BC}_2$, and which is proved with the same arguments.

**Theorem 14.** *The minimal free resolution of the module $M = R^2/\langle P^t\rangle$ is:*

$$0 \to R^2(-7) \overset{P_6^t}{\to} R^{14}(-6) \overset{P_5^t}{\to} R^{42}(-5) \overset{P_4^t}{\to} R^{70}(-4) \overset{P_3^t}{\to}$$
$$\overset{P_3^t}{\to} R^{70}(-3) \overset{P_2^t}{\to} R^{42}(-2) \overset{P_1^t}{\to} R^{14}(-1) \overset{P^t}{\to} R^2 \to M \to 0.$$

*Moreover* $\operatorname{Ext}^i(M,R) = 0$, $i = 0\ldots 6$ *and* $\operatorname{Ext}^7(M,R) \neq 0$.

**Corollary 15.** *The characteristic variety of $M$ has dimension* 1.

The same approach can be taken for $\mathbb{BC}_4$, and in this case the entries of the matrix $P$ belong to the ring

$$R = \mathbb{C}\Big[Z, \overline{Z}^{\mathbf{i}_1}, \overline{Z}^{\mathbf{i}_2}, \overline{Z}^{\mathbf{i}_3}, \overline{Z}^{\mathbf{i}_1\mathbf{i}_2}, \overline{Z}^{\mathbf{i}_2\mathbf{i}_3}, \overline{Z}^{\mathbf{i}_1\mathbf{i}_3}, \overline{Z}^{\mathbf{i}_1\mathbf{i}_2\mathbf{i}_3},$$
$$W, \overline{W}^{\mathbf{i}_1}, \overline{W}^{\mathbf{i}_2}, \overline{W}^{\mathbf{i}_3}, \overline{W}^{\mathbf{i}_1\mathbf{i}_2}, \overline{W}^{\mathbf{i}_2\mathbf{i}_3}, \overline{W}^{\mathbf{i}_1\mathbf{i}_3}, \overline{W}^{\mathbf{i}_1\mathbf{i}_2\mathbf{i}_3}\Big].$$

In this case we obtain the following result.

**Theorem 16.** *The minimal free resolution of the module $M = R^2/\langle P^t\rangle$ is:*

$$0 \to R^2(-15) \overset{P_{14}^t}{\to} R^{30}(-14) \overset{P_{13}^t}{\to} R^{210}(-13) \overset{P_{12}^t}{\to} R^{910}(-12) \overset{P_{11}^t}{\to}$$
$$\overset{P_{11}^t}{\to} R^{2730}(-11) \overset{P_{10}^t}{\to} R^{6006}(-10) \overset{P_9^t}{\to} R^{10010}(-9) \overset{P_8^t}{\to} R^{12870}(-8) \overset{P_7^t}{\to}$$
$$\overset{P_7^t}{\to} R^{12870}(-7) \overset{P_6^t}{\to} R^{10010}(-6) \overset{P_5^t}{\to} R^{6006}(-5) \overset{P_4^t}{\to} R^{2730}(-4) \overset{P_3^t}{\to}$$
$$\overset{P_3^t}{\to} R^{910}(-3) \overset{P_2^t}{\to} R^{210}(-2) \overset{P_1^t}{\to} R^{30}(-1) \overset{P^t}{\to} R^2 \to M \to 0.$$

*Moreover* $\operatorname{Ext}^i(M,R) = 0$, $i = 0\ldots 14$ *and* $\operatorname{Ext}^{15}(M,R) \neq 0$.

**Corollary 17.** *The characteristic variety of $M$ has dimension* 1.

We can finally address the more general case of holomorphic functions on $\mathbb{BC}_n$. In this case we have $2^n - 1$ differential operators (associated to the various conjugates), which we can relabel as

$$\partial_{\overline{Z}^1}, \ldots, \partial_{\overline{Z}^{2^n-1}}.$$

It is not difficult to show that the resolution one obtains in this general case is still Koszul-like, and more precisely one has:

$$0 \to R^2 \to R^{2(2^n-1)} \to R^{2\binom{2^n-1}{2}} \to R^{2\binom{2^n-1}{3}} \to \cdots$$
$$\cdots \to R^{2\binom{2^n-1}{2}} \to R^{2(2^n-1)} \to R^2 \to M \to 0,$$

In addition, one obtains the following precise result:

**Theorem 18.** *The first Betti number of the resolution is*

$$b_1 = 2\binom{2(2^{n-1}-1)}{2} + 4\left(2^{n-1}-1\right) = (2^n-2)(2^n-1).$$

**Acknowledgment**

The authors are grateful to the two anonymous referees for the many helpful comments that have improved and clarified significantly the paper. We are particularly grateful for the references they have pointed out to us.

## References

[1] Baird, P., Wood, J.C., Harmonic morphisms and bicomplex manifolds, *J. Geom. Phys.*, **61**, no. 1, (2011), 46–61.

[2] Charak, K.S., Rochon, D., Sharma, N., Normal families of bicomplex holomorphic functions, *Fractals*, **17**, No. 3 (2009), 257–268.

[3] Colombo, F., Sabadini, I., Sommen, F., Struppa, D.C., Analysis of Dirac Systems and Computational Algebra, Birkhäuser, Boston, (2004).

[4] Colombo, F., Sabadini, I., Struppa, D.C., Vajiac, A., Vajiac M.B., Singularities of functions of one and several bicomplex variables. *Arkiv for matematik*, (Institut Mittag-Leffler), (2011).

[5] Colombo, F., Sabadini, I., Struppa, D.C., Vajiac, A., Vajiac M.B., Bicomplex hyperfunctions. *Ann. Mat. Pura Appl.*, (2010), 1–15.

[6] Luna-Elizarraras M.E., Shapiro, M., On modules over bicomplex and hyperbolic numbers, in Applied Complex and Quaternionic Approximation, editors: R.K. Kovacheva, J. Lawrynowicz, and S. Marchiafava, Edizioni Nuova Cultura, Roma, 2009, pp. 76–92.

[7] Garant-Pelletier, V., Rochon, D., On a generalized Fatou-Julia theorem in multicomplex spaces, *Fractals*, **17**, no. 2, (2009), 241–255.

[8] Garant-Pelletier, Ensembles de Mandelbrot et de Julia remplis classiques, généralisés aux espaces multicomplexes et théorème de Fatou-Julia généralisé, *Master Thesis*, (2011).

[9] Luna-Elizarrarás, M.E., Macas-Cedeño, M.A., Shapiro, M., On some relations between the derivative and the two-dimensional directional derivatives of a quaternionic function, *AIP Conference Proceedings*, v. **936**, (2007), 758–760.

[10] Luna-Elizarrarás, M.E., Macas-Cedeño, M.A., Shapiro, M., On the derivatives of quaternionic functions along two-dimensional planes, *Adv. Appl. Clifford Algebr.*, **19**, (2009), no. 2, 375–390.

[11] Price, G.B., An Introduction to Multicomplex Spaces and Functions, Monographs and Textbooks in Pure and Applied Mathematics, **140**, Marcel Dekker, Inc., New York, 1991.

[12] Rochon, D., Shapiro, M., On algebraic properties of bicomplex and hyperbolic numbers, *An. Univ. Oradea Fasc. Mat.* **11** (2004), 71–110.

[13] Rochon, D., On a relation of bicomplex pseudoanalytic function theory to the complexified stationary Schrödinger equation, *Complex Var. Elliptic Equ.* **53**, no. 6, (2008), 501–521.

[14] Rochon, D., A Bloch constant for hyperholomorphic functions, *Complex Var. Elliptic Equ.* **44**, (2001), 85–101.

[15] Ryan, J., Complexified Clifford analysis, *Complex Variables and Elliptic Equations* **1** (1982), 119–149.

[16] Ryan, J., $\mathbb{C}^2$ extensions of analytic functions defined in the complex plane, *Adv. in Applied Clifford Algebras* **11** (2001), 137–145.

[17] Segre, C., Le rappresentazioni reali delle forme complesse e gli enti iperalgebrici, *Math. Ann.* **40** (1892), 413–467.

[18] Scorza Dragoni, G., Sulle funzioni olomorfe di una variabile bicomplessa, *Reale Accad. d'Italia, Mem. Classe Sci. Nat. Fis. Mat.* **5** (1934), 597–665.

[19] Spampinato, N., Estensione nel campo bicomplesso di due teoremi, del Levi-Civita e del Severi, per le funzioni olomorfe di due variabili bicomplesse I, II, *Reale Accad. Naz. Lincei* **22** (1935), 38–43, 96–102.

[20] Spampinato, N., Sulla rappresentazione di funzioni di variabile bicomplessa totalmente derivabili, *Ann. Mat. Pura Appli.* **14** (1936), 305–325.

[21] Struppa, D.C., Vajiac, A., Vajiac M.B., Remarks on holomorphicity in three settings: complex, quaternionic, and bicomplex. In *Hypercomplex Analysis and Applications*, Trends in Mathematics, 261–274, Birkhäuser, 2010.

[22] Sudbery, A., Quaternionic Analysis, *Math. Proc. Camb. Phil. Soc.* **85** (1979), 199–225.

D.C. Struppa, A. Vajiac and M.B. Vajiac
Chapman University
Schmid College of Science and Technology
Orange, CA 92866, USA
e-mail: `struppa@chapman.edu`
`avajiac@chapman.edu`
`mbvajiac@chapman.edu`

Operator Theory:
Advances and Applications, Vol. 221, 635-651


# On Some Class of Self-adjoint Boundary Value Problems with the Spectral Parameter in the Equations and the Boundary Conditions

Victor Voytitsky

**Abstract.** The aim of this work is to study the spectral properties of some class of self-adjoint linear boundary value and transmission problems (in domains with Lipschitz boundaries) where equations and boundary conditions depend linearly on the eigenparameter $\lambda$. We consider some abstract general problem that can be formulated on the basis of the abstract Green's formula for a triple of Hilbert spaces and a trace operator. We prove that the spectrum of the abstract general problem consists of real normal eigenvalues with unique limit point $\infty$, and the system of corresponding eigenelements forms an orthonormal basis in some Hilbert space. We find also some asymptotic formulas for positive or positive and negative branches of eigenvalues. As examples, we consider three multicomponent transmission problems arising in mathematical physics and some abstract general transmission problem.

**Mathematics Subject Classification (2000).** Primary 35P05; Secondary 35P10.

**Keywords.** Spectral problem, transmission problem, abstract Green's formula, Hilbert space, embedding of spaces, compact self-adjoint operator, positive discrete spectrum, asymptotic behavior of eigenvalues.

## 1. Introduction

Spectral boundary value problems with eigenparameter in the boundary conditions arise in different problems of mathematical physics where we have a time derivative of a function acting on the boundary or some parts of the boundary. Many different statements of such problems appeared in applications since the 60ies. These are problems of scattering theory, diffraction theory, theory of dynamic systems, thermal control, problems with moving boundaries, Stefan problem and others. As a rule, spectral problems correspond to linearizations of initial nonlinear boundary value problems. The unknown functions can be defined in a single domain or in several domains with common parts of their boundaries. If the functions from different domains are connected with each other only by their values on the common parts of the boundaries, then we have so-called transmission problems.

A great number of different statements of spectral problems with parameter in boundary conditions have been considered since the middle of the XXth century till the present time. Since the 70ies many authors have studied different problems for ordinary differential expressions with spectral parameter in boundary conditions. The general theory of such problems was built by E.M. Russakovsky and A.A. Shkalikov (see [1] and [2]). Important results on spectral properties of common elliptic formally self-adjoint equations in smooth domains with systems of covered boundary conditions were obtained in the 60ies in works of American mathematicians J. Ercolano and M. Schechter (see [3]). Similar results were obtained independently in Ukraine (see [4] and [5]). Eigenvalue asymptotics for such problems were established by N.A. Kozhevnikov (see [6] and [7]). Different problems in smooth domains with spectral parameter in the equations and the $\lambda$-dependent (not only linearly) boundary conditions have been considered systematically since the 80ies (see, e.g., works [8]–[16]).

Boundary value and transmission problems in domains with Lipschitz boundary have been considered since the 90ies by M.S. Agranovich and his coauthors (see [17], [18] and others). Important results on eigenfunctions expansions of transmission problems with parameter in the boundary conditions were obtained by A.N. Komarenko (see [19], [20]). An abstract operator approach to the transmission problems with two spectral parameters was studied by P.A. Starkov (see [21]). For the first time in his PhD-work an abstract boundary value problem was considered on the basis of abstract Green's formula for a triple of Hilbert spaces and a trace operator proved by N.D. Kopachevsky and S.G. Krein.

It should be mentioned that so-called abstract (generalized) spectral problems have been considered since the 70ies. As a rule, such problems are formulated by using operators from some abstract Green's formula. There are several different abstract Green's formulas (generalized well-known first or second Green's formula for integrals) which generate different abstract classes of boundary value problems. In the papers [22]–[29] the (second) Green's formula for symmetric relation and boundary triplets is used. This approach was applied to problems with nonlinear $\lambda$-dependent boundary conditions in papers of A. Etkin (see [25]), P. Binding with coauthors (see [9], [11], [12]), J. Behrndt (see [15], [16] and [29]) and in the PhD thesis of W. Code (see [13]). In the works [30]–[33] we can find (first) abstract Green's formula for embedded Hilbert spaces, a trace operator and a sesquilinear coercive form. An almost analogous formula was proved independently by S.G. Krein in the end of the 80ies. This formula was generalized by N.D. Kopachevsky in the works [34]–[39]. With some additional assumptions it can be applied to problems of hydrodynamics and transmission problems in Lipschitz domains. All mentioned formulas are similar but not equivalent. So, different forms of abstract problems do not describe the same classes of boundary value problems, although as special cases they consist of elliptic differential expressions.

An abstract operator approach to the transmission problems was studied by N.D. Kopachevsky, P.A. Starkov and V.I. Voytitsky in [38] (see also the English translation [39]). The author of the article used results of [38], [35] and [36] for studying the spectral Stefan problems (see [41]) and other problems of mathematical physics in his PhD-work [40]. Here we consider the abstract general problem with spectral parameter in the equation and the boundary condition that generalize statements from [38], [42] and [43]. We prove a basis property of eigenfunctions, reality and discreteness of the spectrum. For the first time we obtain the Weyl's asymptotic formulas for the positive branch of eigenvalues. The given results are applied to three transmission problems of mathematical physics and to the abstract transmission problem that contains the spectral parameter in the equations and the boundary conditions.

## 2. Formulation of the Abstract General Problem (AGP)

Let us consider one simple boundary value problem as a model:

$$-\Delta u = \lambda a u \quad (\text{in } \Omega), \tag{2.1}$$

$$\frac{\partial u}{\partial n} = \lambda V u \quad (\text{on } \Gamma), \tag{2.2}$$

$$u = 0 \quad (\text{on } S). \tag{2.3}$$

Let $\Omega \subset \mathbb{R}^m$ be a bounded domain with piecewise-smooth boundary $\partial\Omega = \Gamma \cup S$ $(\Gamma \cap S = \emptyset)$. Suppose that $a$ and $V$ are given bounded self-adjoint operators acting correspondingly in $L_2(\Omega)$ and $L_2(\Gamma)$. To study this problem we can use well-known first Green's formula

$$\int_\Omega \eta(-\Delta u)\, d\Omega = \int_\Omega \nabla\eta \cdot \nabla u\, d\Omega - \int_\Gamma \eta \frac{\partial u}{\partial n}\, d\Gamma, \tag{2.4}$$

that is truth for all functions $\eta, u \in C^2_{0,S}(\Omega) := \{u \in C^2(\Omega) : \ u|_S = 0\}$.

But if we want to find generalized eigenfunctions or to study the problem in Lipschitz domain we cannot use this formula anymore. For this situation we can apply generalized Green's formula

$$\langle \eta, -\Delta u \rangle_{L_2(\Omega)} = (\eta, u)_{H^1_{0,S}(\Omega)} - \left\langle \gamma\eta, \frac{\partial u}{\partial n} \right\rangle_{L_2(\Gamma)}, \quad \forall \eta, u \in H^1_{0,S}(\Omega), \tag{2.5}$$

where $H^1_{0,S}(\Omega) := \{u \in H^1(\Omega) : \ u|_S = 0\}$. Here integrals are replaced by functionals from dual spaces $H^1_{0,S}(\Omega)$ and $(H^1_{0,S}(\Omega))^*$, and from $H^{1/2}(\Gamma)$ and $(H^{1/2}(\Gamma))^*$; $\gamma u := u|_\Gamma$ is a trace operator. Similar formulas one can find in, e.g., [44]–[47]. Formula (2.5) is a special case of more general abstract Green's formula for a triple of Hilbert spaces proved by N.D. Kopachevsky in articles [34]–[36]:

$$\langle \eta, Lu \rangle_E = (\eta, u)_F - \langle \gamma\eta, \partial u \rangle_G, \quad \forall \eta, u \in F. \tag{2.6}$$

This identity is determined by Hilbert spaces $E, F, G$ and linear operators $L, \partial$ and $\gamma$. The main result of the articles [35] and [36] is the following theorem.

**Theorem 2.1 (N.D. Kopachevsky).** *Let $F, E, G$ be given arbitrary separable Hilbert spaces, and $F$ is connected with $G$ by given bounded abstract trace operator $\gamma : F \to \mathcal{R}(\gamma) := G_+ \subset G$, where $\operatorname{Ker}\gamma$ is dense subspace of the space $E$. If spaces $F$ and $G_+$ are boundedly embedded into $E$ and $G$ correspondingly, then there exist unique operators*

$$L : \mathcal{D}(L) = F \to F^* \supset E;$$
$$\partial : \mathcal{D}(\partial) = F \to (G_+)^* =: (G_-) \supset G,$$

*such that abstract Green's formula* (2.6) *holds.*

*Remark* 2.2. Such kind of abstract Green's formula was obtained originally by J.-P. Aubin in [30]. The main difference of abstract Green's formula from works [30]–[33] is presence of a sesquilinear coercive form instead of scalar product $(\eta, u)_F$ and using the scalar products in $E$ and $G$ instead of corresponding functionals.

Essentially, Theorem 2.1 establishes one-to-one correspondence between positive definite self-adjoint operator $Au := Lu$, $\mathcal{D}(A) := \{u \in F : \ \partial u = 0 \ \ (\text{in } G)\}$ acting in the space $E$, and the triple of Hilbert spaces $F, E, G$ with trace operator $\gamma$. Here $F$ is the energetic space of the operator $A$. For example, problem (2.1)–(2.3) corresponds to the set: $E = L_2(\Omega)$, $F = H^1_{0,S}(\Omega)$, $G = L_2(\Gamma)$, $\gamma u := u|_\Gamma \in G_+ = H^{1/2}(\Gamma)$. This collection satisfies Theorem 2.1 on the basis of embedding theorems of S.L. Sobolev (see, e.g., [33]) and Gagliardo's theorem (see [48]). We should assume here that the space $H^1_{0,S}(\Omega)$ is equipped by Dirichlet's norm, that is equivalent to the standard norm of $H^1(\Omega)$.

Let Hilbert spaces $F, E, G$ and an operator $\gamma : \ F \to G$ be given and condition of Theorem 2.1 holds for them. Moreover, suppose that $F$ and $G_+$ are compactly embedded into $E$ and $G$ correspondingly (it is truth for many problems of mathematical physics, particularly, for problem (2.1)–(2.3)). Then we can build some differential expressions $L$ and $\partial$ (generalizations of $-\Delta$ and $\partial/\partial n$), such that formula (2.6) is realized, and problem (2.1)–(2.3) admits the generalization:

$$Lu = \lambda a u \qquad (\text{in } E), \tag{2.7}$$
$$\partial u = \lambda V \gamma u \quad (\text{in } G). \tag{2.8}$$

Let us call it Abstract General Problem (AGP). Assume again that operators $a$ and $V$ are given bounded self-adjoint operators acting correspondingly in $E$ and $G$. Suppose also that operator $a$ is nonnegative in $E$ with infinite-dimensional range; $V$ has $\kappa_V$-dimensional negative subspace of $\mathcal{R}(\gamma)$, i.e., there exist exactly $\kappa_V$ $(0 \le \kappa_V \le \infty)$ linearly independent elements $\varphi_k \in G_+$, such that

$$(V\varphi_k, \varphi_k)_G < 0. \tag{2.9}$$

In the case $a = I$ and $V$ is positive or negative operator this problem was considered before in [39], [42] and [43].

*Remark* 2.3. In statement (2.7)–(2.8) we can replace expression $\partial u$ to the $\widetilde{\partial} u := \partial u + b\gamma u$, where operator $b$ is an arbitrary nonnegative bounded operator acting in $G$. This substitution corresponds to introduction the new norm $|u| := \|u\|_F + \|b^{1/2}\gamma u\|_G$ in the space $F$. This norm is an equivalent with prior one, so, in this case we obtain the same abstract problem (2.7)–(2.8). But if operator $b$ is not nonnegative or it is unbounded, then we have another type of problem which can be non self-adjoint.

## 3. Properties of the abstract boundary value problems

Now we shall use approaches from [43] and [39] for studying AGP (2.7)–(2.8). At first, let us find weak solutions of auxiliary boundary value problems:

$$Lv = \lambda av, \qquad \partial v = 0, \tag{3.1}$$

$$Lw = 0, \qquad \partial w = \lambda V\gamma w. \tag{3.2}$$

Using formula (2.6) we see that element $v$ satisfies the identity

$$(\eta, \lambda av)_E = (\eta, v)_F, \quad \forall \eta \in F. \tag{3.3}$$

On the other hand $(\eta, \lambda av)_E = (\eta, A^{-1}(\lambda av))_F$, where $A^{-1}$ is the inverse operator to the operator of Hilbert pair $(F; E)$. Since $F$ is compactly embedded in $E$, operator $A^{-1}$ is positive and compact. Therefore, identity (3.3) implies the connection $v = \lambda A^{-1}av$. Since $\mathcal{D}(A^{1/2}) = F$ and $v \in F$, there exists a unique element $\vartheta \in E$, such that $\vartheta = A^{1/2}v$. So, we have

$$\vartheta = \lambda A^{-1/2}aA^{-1/2}\vartheta, \quad \vartheta \in E.$$

This is the problem of characteristic numbers of compact nonnegative operator $\mathcal{A} := A^{-1/2}aA^{-1/2}$ acting in Hilbert space $E$. Since $\operatorname{Ker}\mathcal{A}$ can be nontrivial, theorem of Hilbert-Schmidt implies that the spectrum of this problem consists of the branch of positive normal eigenvalues with limit point $+\infty$ and, probably, eigenvalue $\lambda = \infty$ (its multiplicity is equal to dimension of $\operatorname{Ker} a$). As a rule, in applications problem (3.1) has a regular Weyl's asymptotic behavior of positive eigenvalues:

$$\lambda_n^{(1)} = (\lambda_n(\mathcal{A}))^{-1} = c_1 n^\alpha[1 + o(1)] \quad (c_1,\ \alpha > 0), \quad n \to \infty. \tag{3.4}$$

Suppose that this formula holds for abstract problem (3.1).

For problem (3.2) formula (2.6) gives the following identity:

$$(\eta, w)_F = (\gamma\eta, \lambda V\gamma w)_G = (\eta, \lambda TV\gamma w)_F, \quad \forall \eta \in F. \tag{3.5}$$

Here we denote adjoint operator to $\gamma: F \to G$ as $T: G \to M \subset F$. In [35], [36] it was proved that $M$ is such subspace of $F$ that $M \oplus N = F$, and $N := \operatorname{Ker}\gamma$. Identity (3.5) implies the connection $w = \lambda TV\gamma w$. Since $w \in F$, there exists a unique element $\omega \in E$, such that $\omega = A^{1/2}w$. Introduce the operators

$$Q := \gamma A^{-1/2}: E \to G, \quad Q^* := A^{1/2}T: G \to E.$$

They are mutually adjoint and compact (on the basis of compact embedding $G_+$ in $G$). Using this operators, problem (3.2) reduces to eigenvalue problem

$$\omega = \lambda Q^* V Q \omega, \quad \omega \in E.$$

Here we find characteristic numbers of compact self-adjoint operator $\mathcal{B} := Q^*VQ$ acting in Hilbert space $E$. By Hilbert-Schmidt theorem we conclude that the spectrum of this problem consists of $\kappa_V$ negative eigenvalues, eigenvalue $\lambda = \infty$ with infinite multiplicity ($\operatorname{Ker}\mathcal{B} = A^{1/2}N$) and, probably, positive discrete eigenvalues with unique limit point $\infty$. As a rule, in applications operator $Q^*Q$ has a regular Weyl's asymptotic of positive eigenvalues. For the further considerations it is sufficient to suppose that problem (3.2) with $V = I$ has one-sided estimate of positive eigenvalues:

$$\lambda_n^{(2)} \geq c_2 n^{\beta_0} \quad (c_2 > 0,\ \alpha \geq \beta_0 > 0). \tag{3.6}$$

Consider problem (2.7)–(2.8) again. If we replace $v$ and $w$ in right-hand sides of (3.1) and (3.2) on solution $u = v + w$ of problem (2.7)–(2.8) then it is easy to show that element $\eta = A^{1/2}u \in E$ will satisfy the relation

$$\eta = \vartheta + \omega = \lambda(A^{-1/2}aA^{-1/2} + Q^*VQ)\eta = \lambda(\mathcal{A} + \mathcal{B})\eta := \lambda\mathcal{C}\eta, \quad \eta \in E. \tag{3.7}$$

One can prove that this problem is equivalent to AGP (2.7)–(2.8) (see similar result in [36] and [43]). This is the problem of characteristic numbers of compact self-adjoint operator $\mathcal{C} := \mathcal{A} + \mathcal{B}$ acting in Hilbert space $E$. In general case, this operator can have nontrivial kernel (condition $a > 0$ is sufficient for $\operatorname{Ker}\mathcal{C} = \{0\}$), then problem (3.7) has eigenvalue $\lambda = \infty$. To find another eigenvalues of problem (3.7) consider the quadratic form of the operator $\mathcal{C}$:

$$\begin{aligned}(\mathcal{C}\eta, \eta)_E &= (A^{-1/2}aA^{-1/2}\eta, \eta)_E + (Q^*VQ\eta, \eta)_E = (au, u)_E + (TV\gamma u, u)_F \\ &= (au, u)_E + (V\gamma u, \gamma u)_G, \quad \forall \eta = A^{1/2}u \in E \quad (u \in F).\end{aligned} \tag{3.8}$$

The number of positive and negative eigenvalues of the operator $\mathcal{C}$ is equal to dimensions of positive and negative subspaces corresponding to the sign of quadratic form (3.8). These dimensions are determined by the following result.

**Lemma 3.1.** *If $N = \operatorname{Ker}\gamma$ is dense subspace of the Hilbert space E then quadratic form*

$$\Phi(u) := (au, u)_E + (V\gamma u, \gamma u)_G, \quad u \in F,$$

*takes positive values on infinite number of linearly independent elements of $F$, and there exist exactly $\kappa_V$ linearly independent elements $u$ of $F$, such that $\Phi(u)$ takes negative values on them.*

*Proof.* Since the operator $a$ is self-adjoint in $E$, then $E = \operatorname{Ker} a \oplus \overline{\mathcal{R}(a)}$. As $\mathcal{R}(a)$ is infinite-dimensional and $\overline{N} = E$, there exist infinite number of linearly independent elements $u_n \in N \subset F$ not from $\operatorname{Ker} a$. For such elements we obtain $\Phi(u_n) = (au_n, u_n)_E > 0$.

By condition (2.9) we have exactly $\kappa_V$ linearly independent elements $\varphi_k \in G_+$, such that $(V\varphi_k, \varphi_k)_G < 0$. It is known, that operator $\gamma$ is an isomorphism between the spaces $M = F \ominus N$ and $G_+$. So, for all $\varphi_k$ there exist unique elements $u_k \in M$, such that $\gamma u_k = \varphi_k$. The elements $u_k$ are linearly independent and $(V\gamma u_k, \gamma u_k)_G =: -a_k < 0$.

Since $\{u_k\}_{k=1}^{\kappa_V} \subset F \subset E$ and $\overline{N} = E$, then for all numbers $\{\varepsilon_k\}_{k=1}^{\kappa_V}$ such that $0 < \varepsilon_k < \frac{a_k}{\|a\|}$, there exist elements $\{v_k\}_{k=1}^{\kappa_V} \subset N$: $\|u_k - v_k\|_E^2 < \varepsilon_k$. Hence, according to relation $\gamma v_k = 0$, we obtain

$$\Phi(u_k - v_k) \leq \|a\| \|u_k - v_k\|_E^2 + (\gamma u_k, V\gamma u_k)_G < \|a\|\varepsilon_k - a_k < 0, \quad k = 1, \dots, \kappa_V.$$

The elements $\{u_k - v_k\}_{k=1}^{\kappa_V}$ are from $F$ and they are linearly independent. Indeed, let $\sum_{k=1}^{\kappa_V} c_k(u_k - v_k) = 0$. Then property $M \cap N = \{0\}$ implies $\sum_{k=1}^{\kappa_V} c_k u_k = \sum_{k=1}^{\kappa_V} c_k v_k = 0$. Since elements $\{u_k\}_{k=1}^{\kappa_V}$ are linearly independent, all $c_k = 0$. □

Using Lemma 3.1 and the theorem of Hilbert-Schmidt it is easy to describe the spectral properties of problem (3.7). Making inverse substitutions, we get the spectral theorem for AGP (2.7)–(2.8).

**Theorem 3.2.** *Let $\overline{\operatorname{Ker}\gamma} = E$ and embeddings $F$ in $E$ and $\mathcal{R}(\gamma) = G_+$ in $G$ are compact. Then spectrum of AGP consists of characteristic numbers of compact self-adjoint operator $\mathcal{C}$ acting in the space $E$ (spectrum of problem (3.7)). These numbers include the branch of positive eigenvalues $\lambda_k \to +\infty$ $(k \to \infty)$ and, probably, $\kappa_V$ of negative eigenvalues $\lambda_k^-$ with unique possible limit point $\infty$. The number $\lambda = \infty$ can be also the eigenvalue with multiplicity* $\dim \operatorname{Ker}(\mathcal{C})$. *The system of eigenelements $u_n$ corresponding to $|\lambda| < \infty$ forms orthonormal basis in subspace $\widetilde{F} := A^{-1/2}(E \ominus \operatorname{Ker}(\mathcal{C})) \subset F$ $(\widetilde{F} = F$ whenever $a > 0)$ with respect to formulas*

$$(u_p, u_q)_F = \lambda_p \left[(au_p, u_q)_E + (V\gamma u_p, \gamma u_q)_G\right] = \delta_{pq}.$$

*Remark* 3.3. It is clear that this theorem is valid for model problem (2.1)–(2.3). Here embeddings are compact by embedding theorems of S.L. Sobolev and theorem of Gagliardo. The property $\overline{\operatorname{Ker}\gamma} = E$ is fulfilled as $\operatorname{Ker}\gamma = H_0^1(\Omega) \supset C_0^\infty(\Omega)$, and the set of infinitely differentiable finite functions is dense in $L_2(\Omega) = E$.

## 4. Asymptotic formulas for the branch of positive eigenvalues

In previous section we establish that eigenvalues of AGP are characteristic numbers of compact self-adjoint operator $\mathcal{C} = \mathcal{A} + \mathcal{B} = A^{-1/2}aA^{-1/2} + Q^*VQ$. Now we will study eigenvalues asymptotics of operator $\mathcal{C}$, using theorem of Ky Fan and some properties of classes of compact operators. Earlier, we supposed formulas (3.4) and (3.6) for eigenvalues of auxiliary boundary value problems. These formulas imply the following

$$\lambda_n(\mathcal{A}) = c_\alpha n^{-\alpha}[1 + o(1)] \quad (c_\alpha = c_1^{-1} > 0,\ \alpha > 0), \tag{4.1}$$

$$\lambda_n(Q^*Q) \leq c_{\beta_0} n^{-\beta_0} \quad (c_{\beta_0} = c_2^{-1} > 0,\ \beta_0 > 0). \tag{4.2}$$

Eigenvalues asymptotics of operator $\mathcal{C}$ is like (4.1) or (4.2) in two main situations, when operator $\mathcal{A}$ is stronger than $\mathcal{B}$ or otherwise. As a rule, in problems of mathematical physics we have $\alpha \geq \beta_0$. According to this assumption, operator $\mathcal{A}$ can be stronger than $\mathcal{B}$ only when $V$ is compact. To obtain exact result we use theorem of Fan Ky (see [49] or [50]):

**Theorem 4.1 (Fan Ky).** *Let $A$ and $B$ be compact operators, such that the following properties of s-numbers are valid:*

$$\lim_{n\to\infty} n^r s_n(A) = a, \qquad \lim_{n\to\infty} n^r s_n(B) = 0 \quad (r > 0).$$

*Then*

$$\lim_{n\to\infty} n^r s_n(A + B) = a.$$

*Remark* 4.2. Symbol $s_n(A)$ denotes singular numbers (or $s$-numbers) of an operator $A$ numbered with respect to its decrease. By the definition we have

$$s_n(A) := \lambda_n((A^*A)^{1/2}), \quad n = 1, 2, \dots.$$

Introduce the classes of compact operators $\Sigma_p$ and $\Sigma_p^0$ $(p \geq 1)$ (see [50] and [51]). Denote $A \in \Sigma_p$, $B \in \Sigma_p^0$ whenever

$$s_n(A) = O(n^{-1/p}), \quad n \to \infty,$$

$$s_n(B) = o(n^{-1/p}), \quad n \to \infty.$$

Hence formulas (4.1), (4.2) imply that $\mathcal{A} \in \Sigma_{\alpha^{-1}}$, $Q^*Q \in \Sigma_{\beta_0^{-1}}$.

**Lemma 4.3.** *Let $V$ be compact operator from class $\Sigma_\gamma^0$, where $\gamma^{-1} = \alpha - \beta_0$. Then positive eigenvalues $\lambda_n^+(\mathcal{C})$ of the operator $\mathcal{C}$ have the asymptotic behavior*

$$\lambda_n^+(\mathcal{C}) \sim \lambda_n(\mathcal{A}) = c_\alpha n^{-\alpha}[1 + o(1)], \quad n \to \infty.$$

*Proof.* At first, let us obtain the property $\mathcal{B} = Q^*VQ \in \Sigma_{\alpha^{-1}}^0$, i.e.,

$$\lim_{n\to\infty} n^\alpha s_n(Q^*VQ) = 0. \tag{4.3}$$

Using "minimaximal principle" of R. Kurant (see [52]) one can prove the inequality:

$$s_{m+n-1}(Q^*VQ) \leq s_m(Q^*Q) \cdot s_n(V), \quad \forall n, m \in \mathbb{N}. \tag{4.4}$$

It is known that analogous inequality is valid for product of two operators acting in common Hilbert space. From (4.4) for all $n = 2m + j$ $(j = 0, 1)$ we have $s_n(Q^*VQ) = s_{(m+j)+(m+1)-1}(Q^*VQ) \leq s_{m+j}(Q^*Q) \cdot s_{m+1}(V)$. By the data $V \in \Sigma_\gamma^0$, therefore

$$s_{m+1}(V) = o((m + 1)^{-\gamma^{-1}}) = o(m^{\beta_0 - \alpha}).$$

This relation and formula (4.2) imply the following:

$$\begin{aligned} 0 &\leq \lim_{n\to\infty} n^\alpha s_n(Q^*VQ) \leq \lim_{m\to\infty} (2m + j)^\alpha s_{m+j}(Q^*Q) \cdot s_{m+1}^+(V) \\ &\leq \lim_{m\to\infty} (2m + j)^\alpha c_{\beta_0}(m + j)^{-\beta_0} s_{m+1}(V) = c_{\beta_0} 2^\alpha \lim_{m\to\infty} \frac{s_{m+1}(V)}{m^{\beta_0 - \alpha}} = 0. \end{aligned} \tag{4.5}$$

Inequalities (4.5) do not depend from $j = 0, 1$, so, formula (4.3) is proved.

According to theorem of Fan Ky and relation (4.3), we obtain the property $s_n(\mathcal{C}) \sim s_n(\mathcal{A}) = \lambda_n(\mathcal{A})$ $(n \to \infty)$. This property implies the statement of the theorem in the case of finite number of negative eigenvalues $\lambda_n^-(\mathcal{C})$. If operator $\mathcal{C}$ has infinite number of positive and negative eigenvalues then we can decompose $\mathcal{C}$ as a sum $\mathcal{C}_+ + \mathcal{C}_-$, where operator $\mathcal{C}_+$ has all nonnegative eigenvalues of the operator $\mathcal{C}$ and $\mathcal{C}_-$ has all negative eigenvalues of the operator $\mathcal{C}$. For all integers $n$ we have $|\lambda_n(\mathcal{C}_-)| = |\lambda_n^-(\mathcal{C})| = |\lambda_n^-(\mathcal{A}+\mathcal{B})| \leq |\lambda_n^-(\mathcal{B})| \leq s_n(\mathcal{B}) = o(n^{-\alpha})$. Hence

$$\lim_{n\to\infty} n^\alpha|\lambda_n(\mathcal{C}_-)| = \lim_{n\to\infty} n^\alpha s_n(\mathcal{C}_-) = 0.$$

So, the theorem of Fan Ky implies that

$$\lim_{n\to\infty} n^\alpha s_n(\mathcal{C}_+) = \lim_{n\to\infty} n^\alpha \lambda_n(\mathcal{C}_+) = \lim_{n\to\infty} n^\alpha \lambda_n^+(\mathcal{C}) = c_\alpha. \qquad \square$$

*Remark* 4.4. It seems that inequality (4.3) can be obtained easily. We know that $Q^*Q \in \Sigma_{\beta_0^{-1}}$ and $V \in \Sigma_\gamma^0$. If $Q^*Q$ and $V$ act in the same Hilbert space then, according with properties of $\Sigma_p$-classes, their product would be operator from $\Sigma_p^0$, where $p = (\beta_0 + \gamma^{-1})^{-1} = \alpha^{-1}$. But this two operators act in different spaces ($E$ and $G$), so property $Q^*VQ \in \Sigma_{\alpha^{-1}}^0$ requires its original proof.

Consider another situation, when operator $\mathcal{B} = Q^*VQ$ is stronger than operator $\mathcal{A}$. Then the following result is valid.

**Lemma 4.5.** *Let $V$ be such operator that*

$$\lambda_n^+(\mathcal{B}) = c_\beta n^{-\beta}[1+o(1)] \quad (c_\beta > 0,\ \alpha > \beta > 0), \quad n \to \infty,$$

*and operator $\mathcal{B}$ has finite number of negative eigenvalues or*

$$\lambda_n^-(\mathcal{B}) = o(n^{-\beta}), \quad n \to \infty.$$

*Then positive eigenvalues $\lambda_n^+(\mathcal{C})$ of the operator $\mathcal{C}$ have the asymptotic behavior*

$$\lambda_n^+(\mathcal{C}) \sim \lambda_n^+(\mathcal{B}) = c_\beta n^{-\beta}[1+o(1)], \quad n \to \infty. \tag{4.6}$$

*Proof.* Decomposing operator $\mathcal{B}$ as a sum $\mathcal{B}_+ + \mathcal{B}_-$ of its positive and nonnegative parts and using theorem of Fan Ky, we obtain the property $s_n(\mathcal{B}) = c_\beta n^{-\beta}[1+o(1)]$. So, asymptotics (4.1) and theorem of Fan Ky imply $s_n(\mathcal{C}) = c_\beta n^{-\beta}[1+o(1)]$. Using now decomposition of the operator $\mathcal{C}$ as a sum $\mathcal{C}_+ + \mathcal{C}_-$ of its positive and nonnegative parts and inequality $|\lambda_n(\mathcal{C}_-)| = |\lambda_n^-(\mathcal{C})| \leq |\lambda_n^-(\mathcal{B})| = o(n^{-\beta})$, we obtain

$$\lim_{n\to\infty} n^\beta|\lambda_n(\mathcal{C}_-)| = \lim_{n\to\infty} n^\beta s_n(\mathcal{C}_-) = 0.$$

So, theorem of Fan Ky implies that

$$\lim_{n\to\infty} n^\beta s_n(\mathcal{C}_+) = \lim_{n\to\infty} n^\beta \lambda_n(\mathcal{C}_+) = \lim_{n\to\infty} n^\beta \lambda_n^+(\mathcal{C}) = c_\beta. \qquad \square$$

*Remark* 4.6. It is clear that asymptotics of negative eigenvalues of operator $\mathcal{C}$ is like (4.6) whenever $\lambda_n^+(Q^*VQ) = o(n^{-\beta})$ and $-\lambda_n^-(Q^*VQ) = c_\beta n^{-\beta}[1+o(1)]$.

*Remark* 4.7. Statement of lemma (4.5) seems to be true also for $\alpha = \beta$. But theorem of Fan Ky is not sufficient for proving this result. It is easy to obtain double-sided estimate for sufficiently large integers $n$:

$$a \leq \lambda_n^+(\mathcal{C})n^\beta \leq b \quad (0 < a < b < \infty). \tag{4.7}$$

With the help of Lemmas 4.3 and 4.5 we obtain some result on asymptotic behavior of positive eigenvalues of AGP (2.7)–(2.8).

**Theorem 4.8.** *Let auxiliary spectral problems* (3.1) *and* (3.2) *have the estimates of eigenvalues* (3.4) *and* (3.6). *Then formula* (3.4) *describes asymptotic behavior of positive eigenvalues of AGP whenever $V$ is compact operator from the class $\Sigma^0_\gamma$, where $\gamma^{-1} = \alpha - \beta_0$. If $V$ is such operator that*

$$\lambda_n^+(Q^*VQ) = c_\beta n^{-\beta}[1 + o(1)] \quad (\alpha > \beta), \quad n \to \infty,$$

*then positive eigenvalues of AGP have the asymptotic behavior*

$$\lambda_n^+ = c_\beta^{-1} n^\beta [1 + o(1)], \quad n \to \infty.$$

**Corollary 4.9.** *If we have $a = \alpha I$ $(\alpha > 0)$ and $V = I$ in problem* (2.1)–(2.3), *then auxiliary spectral problems* (3.1) *and* (3.2) *have the asymptotics of eigenvalues depending on dimension $m$ of domain $\Omega \subset \mathbb{R}^m$:*

$$\lambda_n^{(1)} = c_1 n^{2/m}[1 + o(1)] \qquad (c_1 > 0), \tag{4.8}$$

$$\lambda_n^{(2)} = c_2 n^{1/(m-1)}[1 + o(1)] \quad (c_2 > 0). \tag{4.9}$$

*In this case Theorem* 4.8 *implies that formula* (4.8) *describes the asymptotic behavior of positive eigenvalues of problem* (2.1)–(2.3) *whenever $V$ is arbitrary compact operator (for $m = 2$) or $V \in \Sigma^0_6$ (for $m = 3$). If $m = 3$ and $V = \beta I + K$, where constant $\beta > 0$ and $K$ is arbitrary compact operator, then positive eigenvalues of problem* (2.1)–(2.3) *have the asymptotic behavior like* (4.9).

*The same result is valid for problem* (2.7)–(2.8) *where we have a strongly elliptic second-order system and corresponding conormal derivative instead of $L$ and $\partial$. Using results of* [18] *and* [33] *one can prove that such problem is a special case of AGP.*

## 5. Multicomponent transmission spectral problems

Abstract problem (2.7)–(2.8) not only generalizes many classical statements of spectral boundary value problems. It also consists of the transmission spectral problems. Let us consider three such problems arising in mathematical physics. The first one is the multicomponent spectral Stefan problem with Hibbs-Thomson conditions:

$$\begin{aligned} -\Delta u_j + u_j &= \lambda u_j && (\text{in } \Omega_j), \\ \frac{\partial u_j}{\partial n_k} + \frac{\partial u_k}{\partial n_j} &= \lambda V_{jk} u_j && (\text{on } \Gamma_{jk}), \\ u_j &= u_k && (\text{on } \Gamma_{jk}), \\ u_j &= 0 && (\text{on } S_j). \end{aligned} \tag{5.1}$$

This problem arises in linear model of phase transitions when pure substance situates in domains $\Omega_j \subset \mathbb{R}^m$ which are in different aggregate states (see 2-dimensional situation in the left part of Figure 1). Suppose that there is a melting process on a moving phase transition boundaries $\Gamma_{jk}$; boundaries $S_j$ are external surfaces of the substance; given operators $V_{jk}$ are positive and compact in $L_2(\Gamma_{jk})$. Similar problems were considered in [38] and [39].

The second problem is generated by the normal motions problem for the hydro-system of capillary ideal fluids (see [53] and [34]):

$$\begin{aligned} &\Delta\Phi_j = 0 \quad (\text{in } \Omega_j), \quad -\Delta\widetilde{\Phi}_j = \lambda c_j^{-2}\widetilde{\Phi}_j \quad (\text{in } \widetilde{\Omega}_j), \\ &\frac{\partial\Phi_j}{\partial n_j} = 0 \quad (\text{on } S_j), \quad \int_{\Gamma_j} \Phi_j \, d\Gamma_j = 0, \quad \int_{\widetilde{\Omega}_j} \widetilde{\Phi}_j \, d\widetilde{\Omega}_j = 0, \\ &\frac{\partial\Phi_j}{\partial n_j} = \frac{\partial\Phi_{j+1}}{\partial n_j} = \lambda V_j(\rho_j\Phi_j - \rho_{j+1}\Phi_{j+1}) \quad (\text{on } \Gamma_j). \end{aligned} \tag{5.2}$$

Suppose that given hydro-system is in a close state to equilibrium position. It fills some container situated in a low-gravity field (see the right part of Figure 1).

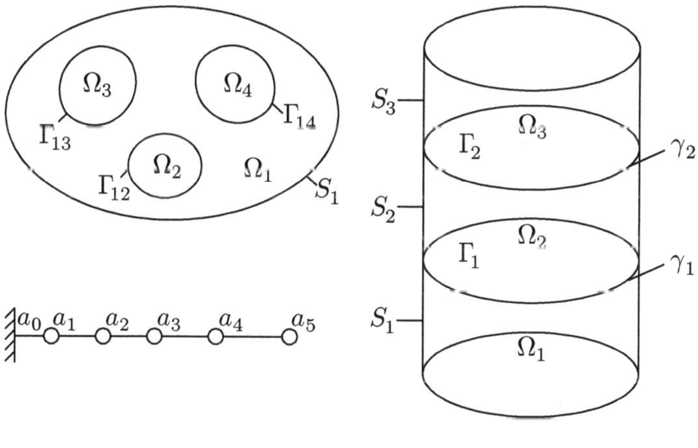


FIGURE 1

Assume also that $\Phi_j$ are potentials of fields of velocities in every layer of ideal fluids $\Omega_j$ or barotropic gas $\widetilde{\Omega}_j$. Given operators $V_j$ are positive and compact in $L_2(\Gamma_j)$.

The third problem is the problem of normal transverse oscillations of an elastic beam with a system of loads fixed on it:

$$\begin{aligned} &\frac{1}{\rho_i(x)S_i(x)}\frac{\partial^2}{\partial x^2}\left(E_i(x)J_i(x)\frac{\partial^2 u_i}{\partial x^2}\right) = \lambda u_i(x), \; x \in (a_i, a_{i+1}), \\ &\frac{d}{dx}\left(E_{i+1}J_{i+1}\frac{\partial^2 u_{i+1}}{\partial x^2}\right)(a_i) - \frac{d}{dx}\left(E_i J_i\frac{\partial^2 u_i}{\partial x^2}\right)(a_i) = \lambda m_i u_i(a_i), \end{aligned} \tag{5.3}$$

$$u_i(a_i) = u_{i+1}(a_i), \quad -E_i(a_i)J_i(a_i)\frac{\partial^2 u_i}{\partial x^2}(a_i) = -E_{i+1}(a_i)J_{i+1}(a_i)\frac{\partial^2 u_{i+1}}{\partial x^2}(a_i) = 0,$$

$$u_0(0) = u_0'(0) = 0, \quad -E_n(l)J_n(l)\frac{\partial^2 u_n}{\partial x^2}(l) = 0, \quad \frac{d}{dx}\left(E_n J_n \frac{\partial^2 u_n}{\partial x^2}\right)(l) = \lambda m_n u(l).$$

Here we suppose that the beam is situated on the segment $[0;l]$ in equilibrium position. There are $n$ loads with corresponding masses $m_i$, $i = \overline{1,n}$, situated at the points $a_0 = 0 \leq a_1 < a_2 < \cdots < a_{n-1} \leq l = a_n$ (see Figure 1). The functions $\rho_i(x), S_i(x) \in C[a_i, a_{i+1}]$, $E_i(x), J_i(x) \in C^2[a_i, a_{i+1}]$ are given and they describe physical characteristics of the beam. Such statement one can find, e.g., in [54].

In order to show that problems (5.1)–(5.3) are special cases of AGP let us construct some abstract spectral transmission problem that generalizes these three problems. If we study a boundary value problem with mixed boundary conditions or a problem of transmission then Green's formulas (2.4) and (2.5) are not useful, as trace operators and normal derivatives are defined on parts of boundaries $\Gamma_{jk}$. In this case we can use the following formulas:

$$\langle \eta_j, u_j - \Delta u_j \rangle_{L_2(\Omega_j)} \tag{5.4}$$
$$= (\eta_j, u_j)_{H^1(\Omega_j)} - \sum_{k=1}^{q} \left\langle \gamma_{jk}\eta_j, \frac{\partial u_j}{\partial n_k} \right\rangle_{L_2(\Gamma_{jk})}, \quad \forall \eta_j, u_j \in \widetilde{H}^1(\Omega_j).$$

Here spaces $\widetilde{H}^1(\Omega_j)$ are subspaces of $H^1(\Omega_j)$ which correspond to solutions of transmission problems. Formulas (5.4) were proved by N.D. Kopachevsky in [37], [38] and [39]. In these articles it was also proved the abstract generalization of these formulas:

$$\langle \eta_j, L_j u_j \rangle_{E_j} = (\eta_j, u_j)_{F_j} - \sum_{k=1}^{q} \langle \gamma_{jk}\eta_j, \partial_{jk} u_j \rangle_{G_{jk}}, \quad \forall \eta_j, u_j \in F_j, \ j = \overline{1,q}. \tag{5.5}$$

Here $\gamma_{jk} : F_j \to (G_+)_{jk} \subset G_{jk}$ are abstract bounded operators that generalize the trace operators, acting to the parts $\Gamma_{jk}$ of the boundary $\partial\Omega_j$; $\partial_{jk}$ generalizes the normal derivatives $\partial u_j/\partial n_k$ on $\Gamma_{jk}$.

Using operators of formula (5.5) we can consider some general abstract transmission problem. Let us have $q$ abstract equations (for $q$ jointed domains):

$$L_j u_j = \lambda a_j u_j, \quad 0 \leq a_j = a_j^* \in \mathcal{L}(E_j), \quad \forall u_j \in F_j, \ j = \overline{1,q}. \tag{5.6}$$

On the transmission boundaries (on $\Gamma_{jkl}$) we consider four types of conditions:

$$\begin{aligned}
&1^\circ. \ \gamma_{jk1}u_j = \gamma_{kj1}u_k, \ \partial_{jk1}u_j + \partial_{kj1}u_k + \delta_{jk1}\gamma_{jk1}u_j = \lambda\alpha_{jk1}\gamma_{jk1}.\\
&2^\circ. \ \gamma_{jk2}u_j = \gamma_{kj2}u_k, \ \partial_{jk2}u_j + \partial_{kj2}u_k + \delta_{jk2}\gamma_{jk2}u_j = 0.\\
&3^\circ. \ \partial_{jk3}u_j = -\partial_{kj3}u_k = -\delta_{jk3}(\gamma_{jk3}u_j - \gamma_{kj3}u_k) + \lambda\alpha_{jk3}(\gamma_{jk3}u_j - \gamma_{kj3}u_k).\\
&4^\circ. \ \partial_{jk4}u_j = -\partial_{kj4}u_k = -\delta_{jk4}(\gamma_{jk4}u_j - \gamma_{kj4}u_k).
\end{aligned} \tag{5.7}$$

On free parts of boundaries (on $\Gamma_{jjl}$) we consider three types of conditions:

$$\begin{aligned} &1^\circ. \quad \partial_{jj1}u_j + \delta_{jj1}\gamma_{jj1}u_j = \lambda\alpha_{jj1}\gamma_{jj1}u_j. \\ &2^\circ. \quad \partial_{jj2}u_j + \delta_{jj2}\gamma_{jj2}u_j = 0. \\ &3^\circ. \quad \gamma_{jj3}u_j = 0. \end{aligned} \tag{5.8}$$

Suppose also that $\alpha_{jkl}$ and $\delta_{jkl} \geq 0$ ($l = 1,2,3,4$) are given bounded self-adjoint operators from $\mathcal{L}(G_{jkl})$.

It is not hard to prove that problems (5.1)–(5.3) are special cases of abstract problem (5.6)–(5.8). In [39] and [38] we proved that problem (5.6)–(5.8) is a special case of AGP where operators $L$ and $\partial$ are expressions from left-hand sides of (5.6) and (5.7):

$$\begin{aligned} Lu &:= (L_1u_1, \ldots, L_qu_q), \\ \partial u &:= \Big(\partial_{jkl}u_j + \partial_{kjl}u_k + \delta_{jkl}\gamma_{jkl}u_j,\ l = 1,2,\ k > j,\ j = \overline{1,q}; \\ &\quad \partial_{jkl}u_j + \delta_{jkl}(\gamma_{jkl}u_j - \gamma_{kjl}u_k),\ l = 3,4,\ k > j,\ j = \overline{1,q}; \\ &\quad \partial_{kjl}u_k - \delta_{jkl}(\gamma_{jkl}u_j - \gamma_{kjl}u_k),\ l = 3,4,\ k > j,\ j = \overline{1,q}; \\ &\quad \partial_{jjl}u_j + \delta_{jjl}\gamma_{jjl}u_j,\ l = 1,2\Big). \end{aligned}$$

One can prove that they are operators from the abstract Green's formula

$$\langle \eta, Lu\rangle_E = (\eta, u)_{F_0} - \langle \gamma\eta, \partial u\rangle_G, \quad \forall \eta, u \in F_0,$$

which can be built by the following Hilbert spaces and the trace operator:

$$\begin{aligned} E &= \bigoplus_{j=1}^{q} E_j, \quad (\eta, u)_E := \sum_{j=1}^{q}(\eta_j, u_j)_{E_j}, \\ F_0 &:= \big\{u = (u_1, \ldots, u_q) \in F = \oplus_{j=1}^{q} F_j : \\ &\qquad \gamma_{jk1}u_j = \gamma_{kj1}u_k,\ \gamma_{jk2}u_j = \gamma_{kj2}u_k\ (k > j);\ \gamma_{jj3}u_j = 0,\ j = \overline{1,q}\big\}, \\ (\eta, u)_{F_0} &:= \sum_{j=1}^{q}(\eta_j, u_j)_{F_j} + \sum_{j=1}^{q}\sum_{k>j}\sum_{l=1}^{2}\langle \gamma_{jkl}\eta_j, \delta_{jkl}\gamma_{jkl}u_j\rangle_{G_{jkl}} \\ &\quad + \sum_{j=1}^{q}\sum_{k>j}\sum_{l=3}^{4}\langle(\gamma_{jkl}\eta_j - \gamma_{kjl}\eta_k), \delta_{jkl}(\gamma_{jkl}u_j - \gamma_{kjl}u_k)\rangle_{G_{jkl}} \\ &\quad + \sum_{j=1}^{q}\sum_{l=1}^{2}\langle \gamma_{jjl}\eta_j, \delta_{jjl}\gamma_{jjl}u_j\rangle_{G_{jjl}}, \\ \gamma\eta &:= \big(\gamma_{jkl}\eta_j,\ l = 1,2,\ k > j,\ j = \overline{1,q};\ \ \gamma_{jkl}\eta_j,\ l = 3,4,\ k > j,\ j = \overline{1,q}; \\ &\qquad \gamma_{kjl}\eta_k,\ l = 3,4,\ k > j,\ j = \overline{1,q};\ \ \gamma_{jjl}\eta_j,\ l = 1,2,\ j = \overline{1,q}\big) \in \end{aligned}$$

$$G := \left(\sum_{j=1}^{q}\sum_{k>j}\sum_{l=1}^{2} \oplus G_{jkl}\right) \oplus \left(\sum_{j=1}^{q}\sum_{k>j}\sum_{l=3}^{4} \oplus (G_{jkl} \oplus G_{jkl})\right) \oplus \left(\sum_{j=1}^{q}\sum_{l=1}^{2} \oplus G_{jjl}\right).$$

In decomposition of the space $G$ we have $au := (a_1u_1, \ldots, a_qu_q)$ and

$$V\gamma u := \big(\alpha_{jk1}\gamma_{jk1}u_j,\ 0,\ \alpha_{jk3}(\gamma_{jk3}u_j - \gamma_{kj3}u_k),\ 0,\\ -\alpha_{jk3}(\gamma_{jk3}u_j - \gamma_{kj3}u_k),\ 0\ (k > j),\ \alpha_{jj1}\gamma_{jj1}u_j,\ 0\ (j = \overline{1,q})\big).$$

So, spectral properties of problem (5.6)–(5.8) are described by Theorems 3.2 and 4.8. Therefore, we immediately obtain the property of positiveness and discreteness of the spectrum for three problems of mathematical physics considered above. Moreover, for the first and the second problem we can apply Theorem 4.8. Actually, one can prove that corresponding auxiliary boundary value problems (with homogeneous equations and boundary conditions) satisfy the estimates (3.4) and (3.6), more precisely, $\lambda_n^{(1)} = c_1n^{2/m}[1+o(1)]\ (n \to \infty)$, $\lambda_n^{(2)} \geq c_2n^{1/(m-1)}$. By physical statements operators $V_{jk}$ and $V_j$ are from the class $\Sigma_6^0$ (for $m = 2$ and $m = 3$). So, Theorem 4.8 implies that problems (5.1) and (5.2) have the following asymptotics of positive eigenvalues: $\lambda_n^+ = c_1n^{2/m}[1+o(1)]\ (n \to \infty)$. For the third problem we obtained only double-sided estimate (4.7) with $\alpha = \beta = 1$ (see Remark 4.7).

**Acknowledgment**

The author thanks his scientific adviser Prof. N.D. Kopachevsky for the statement of the problem and the help with writing the article.

## References

[1] Russakovsky E.M. *Operator Treatment of Boundary Value Problem with Spectral Parameter Entered Polynomially in Boundary Conditions.* Func. Analysis and its Appl., **9**, no. 4, 1975, 91–92 (in Russian).

[2] Shkalikov A.A. *Boundary Value Problems for Ordinary Differential Equations with Parameter in Boundary Conditions.* Matherials of I.G. Petrovsky's seminar, **9**, 1983, 140–166 (in Russian).

[3] Ercolano J., Schechter M. *Spectral Theory for Operators Generated by Elliptic Boundary Problems with Eigenvalue Parameter in Boundary Conditions I, II.* Comm. Pure and Appl. Math, **18**, 1965, 83–105, 397–414.

[4] Komarenko A.N., Lukovsky I.A., Feshenko S.F. *To the Eigenvalues Problems with Parameter in Boundary Conditions.* Ukrainian Mathematical Journal, **17**, no. 6, 1965, 22–30 (in Russian).

[5] Barkovsky V.V. *Eigenfunction's Expansion of Self-adjoint Operators Corresponding to Elliptic Problems with Parameter in Boundary Conditions.* Ukrainian mathematical journal, **19**, no. 1, 1967, 9–24 (in Russian).

[6] Kozhevnikov A.N. *On the Asymptotics of Eigenvalues and Completeness of Principal Vectors of Operator Generated by the Boundary Value Problem with Parameter in Boundary Condition.* DAS of USSR, **200**, no. 6, 1971, 1273–1276 (in Russian).

[7] Kozhevnikov A.N. *Spectral Problems for Pseudodifferential Systems with Duglis-Nirenberg Elliptic Property and Its Applications.* Math. Sbornik, **92(134)**, no. 1(9), 1973, 60–88 (in Russian).

[8] Dijksma A., Langer H. and de Snoo H.S.V. *Symmetric Sturm-Liouville Operators with Eigenvalue Depending Boundary Conditions.* CMS Conf. Proc., **8**, 1987, 87–116.

[9] Binding P., Browne P., Seddighi K. *Sturm-Liouville Problems with Eigenparameter Dependent Boundary Conditions.* Proc. Edinburgh Math. Soc., **37**, no. 2., 1993, 57–72.

[10] Binding P., Hryniv R., Langer H. and Najman B. *Elliptic Eigenvalue Problems with Eigenparameter Dependent Boundary Conditions.* J. Differential Equations, **174**, 2001, 30–54.

[11] Binding P., Browne P. and Watson B. *Sturm-Liouville Problems with Boundary Conditions Rationally Dependent on the Eigenparameter-I.* Proc. Edinb. Math. Soc., **2** (45), 2002, no. 3, 631–645.

[12] Binding P., Browne P. and Watson B. *Sturm-Liouville Problems with Boundary Conditions Rationally Dependent on the Eigenparameter-II.* J. Comput. Appl. Math., **148**, no. 1, 2002, 147–168.

[13] Code W. *Sturm-Liouville Problems with Eigenparameter-dependent Boundary Conditions.* Phd thesis, College of Graduate Studies and Research, University of Saskatchewan, Saskatoon (Canada), 2003.

[14] Çoskun H., Bayram N. *Asymptotics of Eigenvalues for Regular Sturm-Liouville Problems with Eigenvalue Parameter in the Boundary Condition.* J. of Math. Analysis and Applications, **306**, no. 2, 2005, 548–566.

[15] Behrndt J. and Langer M. *Boundary Value Problems for Elliptic Partial Differential Operators on Bounded Domains.* J. Funct. Anal., **243**, 2007, 536–565.

[16] Behrndt J. *Elliptic Boundary Value Problems with $\lambda$-dependent Boundary Conditions.* J. Differential Equations, **249**, 2010, 2663–2687.

[17] Agranovich M.S., Katsenelenbaum B.Z., Sivov A.N., Voitovich N.N. *Generalized Method of Eigenoscillations in Diffraction Theory.* Berlin: Wiley–VCH, 1999.

[18] Agranovich M.S. *Strongly Elliptic Second Order Systems with Spectral Parameter in Transmission Conditions on a Nonclosed surface.* Birkhäuser Verlag, Basel, Boston, Berlin, 2006, 1–21. (Operator Theory: Advances and Applications, Vol. 164.)

[19] Komarenko O.N. *Operators Generated by Transmission Problems with Spectral Parameter in Equations and Boundary Conditions.* Dopovidi of NAS of Ukraine, no. 1, 2002, 37–41 (in Ukrainian).

[20] Komarenko O.N. *Eigenfunctions Expansion of Self-adjoint Operators, Generated by General Transmission Problem.* Collected Papers of Mathematical Institute of NAS of Ukraine, **2**, no. 1, 2005, 127–157 (in Ukrainian).

[21] Starkov P.A. *Operator Approach to the Transmission Problems.* PhD thesis, Institute of Applied Mathematics and Mechanics, Donetsk, Ukraine, 2004 (in Russian).

[22] Bruk V.M. *On One Class of the Boundary Value Problems with the Eigenvalue Parameter in the Boundary Condition.* Mat. Sbornik, **100**, 1976, 210–216 (in Russian).

[23] Dijksma A., Langer H. and de Snoo H.S.V. *Selfadjoint $\pi_\kappa$ – Extensions of Symmetric Subspaces: an Abstract Approach to Boundary Problems with Spectral Parameter in the Boundary Conditions.* Int. Equat. Oper. Theory, **7**, 1984, 459–515.

[24] Dijksma A., Langer H. *Operator Theory and Ordinary Differential Operators.* Lectures on Operator Theory and its Applications, Amer. Math. Soc., Fields Inst. Monogr., **3**, 1996, 73–139.

[25] Etkin A. *On an Abstract Boundary Value Problem with the Eigenvalue Parameter in the Boundary Condition.* Fields Inst. Commun., **25**, 2000, 257–266.

[26] Ćurgus B., Dijksma A. and Read T. *The Linearization of Boundary Eigenvalue Problems and Reproducing Kernel Hilbert Spaces.* Linear Algebra Appl., **329**, 2001, 97–136.

[27] Derkach V.A., Hassi S., Malamud M.M., and de Snoo H.S.V. *Generalized Resolvents of Symmetric Operators and Admissibility.* Methods Funct. Anal. Topology, **6**, 2000, 24–53.

[28] Derkach V.A., Hassi S., Malamud M.M., and de Snoo H.S.V. *Boundary Relations and Generalized Resolvents of Symmetric Operators.* Russ. J. Math. Phys., **16**, no. 1, 2009, 17–60.

[29] Behrndt J. *Boundary Value Problems with Eigenvalue Depending Boundary Conditions.* Math. Nachr., **282**, 2009, 659–689.

[30] Aubin J.-P. *Approximation of Elliptic Boundary Value Problems.* New York: Wiley-Interscience, 1972.

[31] Showalter R. *Hilbert Space Methods for Partial Differential Equations.* Electronic Journal of Differential Equations, 1994.

[32] Bourland M., Cambrésis H. *Abstract Green Formula and Applications to Boundary Integral Equations.* Nummer. Funct. Anal. and Optimiz., **18**, no. 7, 8, 1997, 667–689.

[33] McLean W. *Strongly Elliptic Systems and Boundary Integral Equations.* Cambridge University Press, 2000.

[34] Kopachevsky, N.D., Krein, S.G. *Operator Approach to Linear Problems of Hydrodynamics. Vol. 1: Self-adjoint Problems for an Ideal Fluid.* Birkhäuser Verlag, Basel, Boston, Berlin, 2001. (Operator Theory: Advances and Applications, Vol. 128.)

[35] Kopachevsky N.D., Krein S.G. *The Abstract Green's Formula for a Triple of Hilbert Spaces, Abstract Boundary Value and Spectral Problems.* Ukr. Math. Bulletin, **1**, no. 1, 2004, 69–97 (in Russian).

[36] Kopachevsky N.D. *On the Abstract Green's Formula for a Triple of Hilbert spaces and its Applications to the Stokes Problem.* Taurida Bulletin of Inform. and Math. **2**, 2004, 52–80 (in Russian).

[37] Kopachevsky N.D. *The Abstract Green's Formula for Mixed Boundary Value Problems.* Scientific Notes of Taurida National University. Series "Mathematics. Mechanics. Informatics and Cybernetics", **20(59)**, no. 2, 2007, 3–12 (in Russian).

[38] Kopachevsky N.D., Voytitsky V.I., Starkov P.A. *Multicomponent Transmission Problems and Auxiliary Abstract Boundary Value Problems.* Modern Mathematics. Contemporary directions, **34**, 2009, 5–44 (in Russian).

[39] Voytitsky V.I., Kopachevsky N.D., Starkov P.A. *Multicomponent Transmission Problems and Auxiliary Abstract Boundary Value Problems.* Journal of Math Sciences, Springer, **170**, no. 2, 2010, 131–172.

[40] Voytitsky V.I. *Boundary Value Problems with Spectral Parameter in Equations and Boundary Conditions.* PhD thesis, Institute of Applied Mathematics and Mechanics, Donetsk, Ukraine, 2010 (in Russian).

[41] Kopachevsky N.D., Voytitsky V.I. *On the Modified Spectral Stefan Problem and Its Abstract Generalizations.* Birkhäuser Verlag, Basel, Boston, Berlin, 2009, 373–386. (Operator Theory: Advances and Applications, Vol. 191.)

[42] Voytitsky V.I. *The Abstract Spectral Stefan Problem.* Scientific Notes of Taurida National University. Series "Mathematics. Mechanics. Informatics and Cybernetics", **19(58)**, no. 2, 2006, 20–28 (in Russian).

[43] Voytitsky V.I. *On the Spectral Problems Generated by the Linearized Stefan Problem with Hibbs-Thomson Law.* Nonlinear Boundary Value Problems, **17**, 2007, 31–49 (in Russian).

[44] Grubb G. *A Characterization of the Non-local Boundary Value Problems Associated with an Elliptic Operator.* Ann. Scuola Norm. Sup. Pisa **22**(3), 1968, 425–513.

[45] Grubb G. *On Coerciveness and Semiboundedness of General Boundary Problems.* Israel J. Math. **10**, 1971, 32–95.

[46] Lions J., Magenes E. *Nonhomogeneous Boundary Value Problems and Applications I.* Springer Verlag, New York – Heidelberg, 1972.

[47] Brown B.M., Grubb G. and Wood I.G. *M-functions for Closed Extensions of Adjoint Pairs of Operators with Applications to Elliptic Boundary Problems.* Math. Nachr., **282**, 2009, 314–347.

[48] Gagliargo E. *Caratterizzazioni delle trace sulla frontiera relative ad alaine classi di funzioni in n variabili.* Rendiconti Sem. Mat. Univ. Padova, 1957, 284–305 (in Italian).

[49] Ky Fan. *Maximum Properties and Inequalities for the Eigenvalues of Completely Continuous Operators.* Proc. Nat. Acad. Sci. USA, **37**, 1951, 760–766.

[50] Gohberg I., Krein M. *Introduction in Theory of Non Self-adjoint Operators Acting in Hilbert Space.* Moscow, "Nauka", 1965 (in Russian).

[51] Birman M.S., Solomjak M.Z. *Spectral Theory of Self-Adjoint Operators in Hilbert Space.* Dordrecht: D. Reidel Publishing Company, 1987.

[52] Courant R., Hilbert D. *Methods of Mathematical Physics. Vol.* 1. Wiley, 1989.

[53] Myshkis A.D., Babskii V.G., Kopachevskii N.D. and others. *Low-Gravity Fluid Mechanics.* Springer, 1987.

[54] Özkaya E. *Linear Transverse Vibrations of a Simply Supported Beam Carrying Concentrated Masses.* Math. and Comp. Appl., **6**, no. 2, 2001, 147–151.

Victor Voytitsky
Taurida National University
Prosp. of Acad. V.I. Vernadsky, 4
Simferopol, 95007, Ukraine
e-mail: victor.voytitsky@gmail.com

Operator Theory:
Advances and Applications, Vol. 221, 653-666


# On the Well-posedness of Evolutionary Equations on Infinite Graphs

Marcus Waurick and Michael Kaliske

**Abstract.** The notion of systems with integration by parts is introduced. With this notion, the spatial operator of the transport equation and the spatial operator of the wave or heat equations on graphs can be defined. The graphs, which we consider, can consist of arbitrarily many edges and vertices. The respective adjoints of the operators on those graphs can be calculated and skew-selfadjoint operators can be classified via boundary values. Using the work of R. Picard (Math. Meth. App. Sci. 32: 1768–1803 [2009]), we can therefore show well-posedness results for the respective evolutionary problems.

**Mathematics Subject Classification (2000).** 47B25, 58D25, 34B45.

**Keywords.** Evolutionary equations on graphs, (skew-)self-adjoint operators.

## 1. Introduction

Evolutionary equations on graphs have wide applications and are therefore a deeply studied subject, see, e.g., [1, 10]. A survey is given in [9], where also applications to physics, chemistry and engineering sciences are discussed. In order to obtain well-posedness results for evolutionary equations on graphs, many authors use a semi-group approach and/or a form approach, see, e.g., [3, 5, 7]. Whereas the form approach in [7] treats the case of the heat equation, the semi-group approach in [5] is tailored for the transport equation. In this note, we present a different way to show well-posedness results for evolutionary equations on graphs in on one hand more restrictive Hilbert space setting. On the other hand, with this approach it is possible to treat evolutionary equations such as the transport equation, heat equation and wave equation within a unified more elementary framework. Moreover, it should be noted that the evolutionary problems discussed here include delay terms without needing additional work. Our approach uses a result given in [13], where a general structural feature of evolutionary equations in mathematical physics is observed. We show that this structure carries over to evolutionary equations on graphs. The core issue is the computation of adjoints of operators on graphs. The latter is also done in [2]. Dealing with more general operators, the author of [2] has to impose additional conditions on the graph structure. Restricting ourselves to the study of a particular type of operators, we do not need to impose additional constraints on the graph structure. Adjoints of differential operators on graphs are also studied in [4]. However, the authors of [4] focus on the second-order case. As the common structure of evolutionary equations becomes obvious by transforming the respective equations into a first-order system, we bypass the second-order case.

Our main concern will be establishing Theorem 4.1. In order to show that this theorem is indeed sufficient for obtaining well-posedness results for evolutionary equations on graphs, we briefly summarize the definitions and the main result from [13]. Later, we present our abstract model of graphs by means of tensor product constructions. Moreover, we discuss the concept of systems with integration by parts. The latter will help us to characterize skew-selfadjoint realizations of particular spatial operators belonging to evolutionary equations on graphs. In the last section, we will give some illustrating examples.

## 2. Evolutionary equations in mathematical physics

In order to state the main result of [13], we need some definitions.

At first, we need a time-derivative realized as an operator in a Hilbert space setting. We define the distributional derivative[1]

$$\partial : W_2^1(\mathbb{R}) \subseteq L_2(\mathbb{R}) \to L_2(\mathbb{R}) : f \mapsto f'.$$

It is well known that $\partial$ has an explicit spectral representation as a multiplication operator in $L_2(\mathbb{R})$ given by the unitary Fourier transform $\mathcal{F} : L_2(\mathbb{R}) \to L_2(\mathbb{R})$, i.e.,

$$\partial = \mathcal{F}^* i m \mathcal{F},$$

where

$$m : \{f \in L_2(\mathbb{R}); (x \mapsto xf(x)) \in L_2(\mathbb{R})\} \subseteq L_2(\mathbb{R}) \to L_2(\mathbb{R}) : f \mapsto (x \mapsto xf(x)).$$

From the latter representation, we deduce that $\partial$ is skew-selfadjoint. However, we prefer to establish the time-derivative as a continuously invertible operator. A possible way for doing so is to introduce a weighted $L_2$-type space ([11]). Let $\nu > 0$. We define

$$H_{\nu,0} := \{f \in L_{1,\mathrm{loc}}(\mathbb{R}); (x \mapsto \exp(-\nu x)f(x)) \in L_2(\mathbb{R})\}.$$

Endowing $H_{\nu,0}$ with the scalar-product

$$\langle \cdot, \cdot \rangle_{\nu,0} : (f, g) \mapsto \int_{\mathbb{R}} f(x)^* g(x) \exp(-2\nu x)\, \mathrm{d}x,$$

[1]For an open subset $\Omega \subseteq \mathbb{R}^n$, the space $W_2^1(\Omega)$ is the Sobolev space of weakly differentiable functions with distributional derivative representable as an $L_2(\Omega)$-function.

we deduce that

$$\exp(-\nu m) : H_{\nu,0} \to L_2(\mathbb{R}) : f \mapsto (x \mapsto \exp(-\nu x) f(x))$$

is unitary. By unitary equivalence the operator[2]

$$\partial_\nu := \exp(-\nu m)^* \partial \exp(-\nu m)$$

is skew-selfadjoint in $H_{\nu,0}$. Therefore, the real line except 0 is contained in the resolvent set of $\partial_\nu$. Hence, $\partial_{0,\nu} := \partial_\nu + \nu$ is continuously invertible in $H_{\nu,0}$. Moreover, a standard argument shows that $\left\|\partial_{0,\nu}^{-1}\right\| \leq \frac{1}{\nu}$. We choose $\partial_{0,\nu}$ to be our realization of the time-derivative. This is justified by the relation $\partial_{0,\nu}\phi = \partial\phi$ for all $\phi \in C_c^\infty(\mathbb{R})$. The Fourier-Laplace transform $\mathcal{L}_\nu := \mathcal{F}\exp(-\nu m)$ yields a representation as a multiplication operator for $\partial_{0,\nu}$, i.e.,

$$\partial_{0,\nu} = \mathcal{L}_\nu^*(im + \nu)\mathcal{L}_\nu.$$

Consequently, we have

$$\partial_{0,\nu}^{-1} = \mathcal{L}_\nu^* \left(\frac{1}{im + \nu}\right) \mathcal{L}_\nu.$$

The above representation carries over to analytic functions of $\partial_{0,\nu}^{-1}$ with values in the space $L(H)$ of continuous linear operators in some Hilbert space $H$.

**Definition 2.1.** Let $r > 0$, $H$ a Hilbert space. Define $B_\mathbb{C}(r,r) := \{z \in \mathbb{C}; |z - r| < r\}$. For $\nu > \frac{1}{2r}$ and $M : B_\mathbb{C}(r,r) \to L(H)$ bounded and analytic, we set

$$M(\partial_{0,\nu}^{-1}) := \mathbb{L}_\nu^* M\left(\frac{1}{im + \nu}\right) \mathbb{L}_\nu,$$

where[3] $\mathbb{L}_\nu := \mathcal{L}_\nu \otimes I_H$ is the tensor product of the operators $\mathcal{L}_\nu$ and the identity $I_H$ in $H$, and $M\left(\frac{1}{im+\nu}\right)\phi := \left(x \mapsto M\left(\frac{1}{ix+\nu}\right)\phi(x)\right)$ for all $\phi \in L_2(\mathbb{R}) \otimes H \cong L_2(\mathbb{R}; H)$.

For tensor products of Hilbert spaces and linear operators, we refer to [14, 15]. We use the notation given in [14]. The main result in [13] is the following.

**Theorem 2.2.** *Let $r > 0$, $H$ a Hilbert space and let $M : B_\mathbb{C}(r,r) \to L(H)$ be bounded and analytic. Assume that there is $c > 0$ such that*

$$\forall z \in B_\mathbb{C}(r,r) : \operatorname{Re}(z^{-1}M(z)) \geq c.$$

[2]For any densely defined, linear operator $A$, which maps from the Hilbert space $H_1$ into the Hilbert space $H_2$

$$A^* = (-A^{-1})^{\perp_{H_2 \oplus H_1}}$$

denotes the adjoint of $A$, where $A$ is identified with its graph, i.e., $A \subseteq H_1 \oplus H_2$.

[3]Recall that the tensor product of $H_{\nu,0}$ and $H$ is isomorphic to the space of $H$-valued functions, which are square integrable with respect to the weighted Lebesgue measure $\exp(-2\nu x)\,\mathrm{d}x$. Moreover, for two linear operators $A$, $B$ defined in the Hilbert spaces $H_1$ and $H_2$ respectively, the algebraic tensor product $A \overset{a}{\otimes} B$ is the linear extension of the mapping $\phi \otimes \psi \mapsto A\phi \otimes B\psi$, where $(\phi, \psi) \in D(A) \times D(B)$, to the linear span of simple tensors $D(A) \overset{a}{\otimes} D(B) = \operatorname{lin}\{\phi \otimes \psi; (\phi, \psi) \in D(A) \times D(B)\}$. If $A \overset{a}{\otimes} B$ is closable, then $A \otimes B := \overline{A \overset{a}{\otimes} B}$.

*Let $A : D(A) \subseteq H \to H$ be a skew-selfadjoint operator. Then for $\nu > \frac{1}{2r}$ the operator*

$$(\partial_{0,\nu} \otimes I_H) M(\partial_{0,\nu}^{-1}) + I_{H_{\nu,0}} \otimes A : D(\partial_{0,\nu} \otimes I_H) \cap D(I_{H_{\nu,0}} \otimes A) \subseteq H_{\nu,0} \otimes H \to H_{\nu,0} \otimes H$$

*has dense range and a continuous inverse.*

*Remark* 2.3. For better readability we neglect the tensor product symbols and observe that the above theorem states that the equation

$$(\partial_{0,\nu} M(\partial_{0,\nu}^{-1}) + A)u = f$$

has a unique solution $u \in H_{\nu,0} \otimes H$ for $f$ contained in a dense subspace of $H_{\nu,0} \otimes H$ and the solution $u$ depends continuously on $f$. Examples for these types of equations are discussed in [13, 14]. The latter reference also contains examples for operators which are representable as a bounded, analytic function of $\partial_{0,\nu}^{-1}$. It should be noted that the time-shift operator is included in this framework. Indeed, defining for $h \geq 0, u \in H_{\nu,0} \otimes H$ the operator $\tau_{-h} : H_{\nu,0} \otimes H \to H_{\nu,0} \otimes H$ via $\tau_{-h} u := u(\cdot - h)$, we realize $\tau_{-h} = \exp(-h(\partial_{\nu,0}^{-1})^{-1})$. Time convolution operators and fractions of $\partial_{0,\nu}$ may also be considered in this context. The reason for assuming $M$ to be an analytic operator-valued function is to enforce causality of the solution operator in the sense of [13].

## 3. Systems with integration by parts

An essential fact of the above theorem is the observation that a sufficient condition for well-posedness results for a large class of integro-differential type evolutionary equations is skew-selfadjointness of the operator $A$, modeling the spatial derivatives. The philosophy behind these considerations is to write the considered partial differential equation as a first-order system. In the particular case of graphs, we only consider tensor product constructions of derivatives on a bounded interval. Therefore, the building blocks of the spatial operator on graphs are a particular form, which we discuss as a detailed example.

*Example* 3.1. Define

$$\partial_{1,c} : C_c^\infty(0,1) \subseteq L_2(0,1) \to L_2(0,1) : f \mapsto f',$$

where $C_c^\infty(0,1)$ denotes the set of all arbitrarily often differentiable functions with compact support in $(0,1)$. It is known that $\partial_{1,c}$ is closable and densely defined. Set $\partial_{1,0} := \overline{\partial_{1,c}}$ and $\partial_1 := -\partial_{1,0}^*$. Then $\partial_{1,0} \subseteq \partial_1$ and we have

$$\partial_1 : W_2^1(0,1) \subseteq L_2(0,1) \to L_2(0,1) : f \mapsto f'.$$

With the Sobolev embedding theorem, we deduce the continuity of the mapping $W_2^1(0,1) \hookrightarrow C[0,1] : f \mapsto f$. Hence, the mapping

$$R_1 : W_2^1(0,1) \to \mathbb{C}^2 : f \mapsto (f(1-), f(0+))$$

is continuous. The formula of integration by parts, i.e.,

$$\langle \partial_1 f, g\rangle = f(1-)^* g(1-) - f(0+)^* g(0+) - \langle f, \partial_1 g\rangle,$$

for weakly differentiable functions $f, g \in W_2^1(0,1)$ is also well known. Using a method presented in [8], we note that the following decomposition holds:[4]

$$\begin{aligned} D(\partial_1) &= D(\partial_{1,0}) \oplus_{D_{\partial_1}} D(\partial_{1,0})^{\perp_{D_{\partial_1}}} \\ &= D(\partial_{1,0}) \oplus_{D_{\partial_1}} N(-\partial_1^2 + 1). \end{aligned}$$

The eigenspace corresponding to the eigenvalue 1 of the one-dimensional Laplacian $\partial_1^2$, i.e., the space $N(-\partial_1^2+1)$ is two-dimensional, namely the linear span of sinh and $t \mapsto \sinh(1-t)$. Therefore, $\Psi_1 := R_1|_{N(-\partial_1^2+1)}$ is a linear bijective mapping between two two-dimensional Banach spaces and, hence, a linear homeomorphism. With the operator $\Psi_1$, it is possible to describe the boundary conditions of any operator which is an extension of $\partial_{1,0}$ and a restriction of $\partial_1$. The latter observation will help us to generalize the operator $\partial_1$.

To underline the versatility of the above concepts and the applicability of Theorem 2.2, we give several examples. The main point is that they all fit in one scheme. As evolutionary processes on graphs are combinations of evolutionary processes on the edges with suitable boundary conditions, we focus on the one-dimensional case.

*Example* 3.2.

$$(\partial_{0,\nu} - \partial_1)u = f \qquad \text{(Transport equation)},$$

$$\left(\partial_{0,\nu}\begin{pmatrix}1 & 0\\ 0 & 1\end{pmatrix} - \begin{pmatrix}0 & \partial_1\\ \partial_1 & 0\end{pmatrix}\right)\begin{pmatrix}u\\ v\end{pmatrix} = \begin{pmatrix}f\\ 0\end{pmatrix} \qquad \text{(Wave equation)},$$

$$\left(\partial_{0,\nu}\begin{pmatrix}1 & 0\\ 0 & 0\end{pmatrix} + \begin{pmatrix}0 & 0\\ 0 & 1\end{pmatrix} - \begin{pmatrix}0 & \partial_1\\ \partial_1 & 0\end{pmatrix}\right)\begin{pmatrix}u\\ v\end{pmatrix} = \begin{pmatrix}f\\ 0\end{pmatrix} \qquad \text{(Heat equation)},$$

where $f$ is given and $u$ and $\begin{pmatrix}u\\ v\end{pmatrix}$ are solutions of the respective equations. The latter two equations may be generalized to the following

$$\left(\partial_{0,\nu}P + (I-P) - \begin{pmatrix}0 & \partial_1\\ \partial_1 & 0\end{pmatrix}\right)\begin{pmatrix}u\\ v\end{pmatrix} = \begin{pmatrix}f\\ 0\end{pmatrix} \quad \text{(Wave/Heat equation)},$$

where $P \in L(L_2(0,1)^2)$ is an orthogonal projection. For $P = \left(\begin{smallmatrix}1 & 0\\ 0 & 0\end{smallmatrix}\right)$ we recover the heat equation and for $P = \left(\begin{smallmatrix}1 & 0\\ 0 & 1\end{smallmatrix}\right)$, the above corresponds to the wave equation.

[4]For any closed operator $B : D(B) \subseteq H \to H$ on some Hilbert space $H$, we denote the Hilbert space $(D(B), \langle B\cdot, B\cdot\rangle_H + \langle\cdot,\cdot\rangle_H)$ by $D_B$, its norm will be denoted by $|\cdot|_B$.

Moreover, note that also operators of the following type can be considered:

$$(\partial_{0,\nu} + B\tau_{-h} - \partial_1)u = f \qquad \text{(Transport equation with delay)},$$

$$\left(\partial_{0,\nu}P + (I-P) + C\tau_{-h} - \begin{pmatrix} 0 & \partial_1 \\ \partial_1 & 0 \end{pmatrix}\right)\begin{pmatrix} u \\ v \end{pmatrix} = \begin{pmatrix} f \\ 0 \end{pmatrix}$$
$$\text{(Wave/Heat equation with delay)},$$

for $h > 0$ and some bounded linear operators $B \in L(L_2(0,1))$ and $C \in L(L_2(0,1)^2)$. One may also multiply the leading term $\partial_{0,\nu}$ by some continuous, linear, strictly positive definite operator $D$ in the respective spaces $L_2(0,1)$ or $L_2(0,1)^2$. Such $D$ may result from variable velocities. In order to show well-posedness of the above equations, we have to obtain skew-selfadjointness of the spatial operators $\partial_1$ and $\begin{pmatrix} 0 & \partial_1 \\ \partial_1 & 0 \end{pmatrix}$, respectively. Thus, appropriate boundary conditions have to be imposed.

*Remark* 3.3. Note that initial value problems can also be formulated in this context. The latter is carried out in [12].

We recognize that in Example 3.2, the building blocks of the spatial operator are of the form $\partial_1$ introduced in Example 3.1. The properties of that operator can be generalized in various directions. In [3], a Banach space setting combined with a higher-dimensional case is treated. However, in order to cover more general situations, the assumptions made in [3] differ from our assumptions, e.g., the boundary operator is only assumed to be surjective, whereas in Condition (iii) in Definition 3.4 we assume that the (abstract) boundary value operator restricted to a particular subspace is even a Banach space isomorphism. The considerations in [2] give another possible way. In the latter reference, the author deals with operators of higher order with variable coefficients. That generalization comes along with additional structure on the graph, which we do not want to assume here. Restricting ourselves to the Hilbert space context and the one-dimensional case, we define the following generalization of $\partial_1$:

**Definition 3.4.** A quintuple $(H, h, A_0, R, S)$ is called a *system with integration by parts* (*SWIP*) if the following conditions are satisfied:

(i) $H, h$ are Hilbert spaces,
(ii) $A_0$ is a closed, skew-symmetric operator in $H$, denote $A := -A_0^*$,
(iii) $R : D_A \to h \oplus h$ is continuous and such that $\Psi_R := R|_{N(-A^2+1)}$ is a Banach space isomorphism and $R[D(A_0)] = \{0\}$,
(iv) $S \in L(h \oplus h)$, $\|S\| \leq 1$,
(v) for all $f, g \in D(A)$ the equation

$$\langle Af, g\rangle_H = \langle SRf, Rg\rangle_{h\oplus h} - \langle f, Ag\rangle_H$$

holds.

Example 3.1 is our first example of systems with integration by parts. Indeed, the quintuple

$$(H, h, A_0, R, S) = \left(L_2(0,1), \mathbb{C}, \partial_{1,0}, R_1, \left(\begin{smallmatrix} 1 & 0 \\ 0 & -1 \end{smallmatrix}\right)\right)$$

is a system with integration by parts. Having shown that $\Psi_1 = R_1|_{N(-\partial_1^2+1)}$ is an isomorphism in Example 3.1, we only show the formula in Condition (v) of Definition 3.4. For $f, g \in D(\partial_{1,0}^*) = D(\partial_1) = W_2^1(0,1)$ we have

$$\begin{aligned}
\langle \partial_1 f, g\rangle_{L_2(0,1)} &= f(1-)^* g(1-) - f(0-)^* g(0-) - \langle f, \partial_1 g\rangle_{L_2(0,1)} \\
&= \left\langle \begin{pmatrix} f(1-) \\ -f(0-) \end{pmatrix}, \begin{pmatrix} g(1-) \\ g(0-) \end{pmatrix} \right\rangle_{\mathbb{C}\oplus\mathbb{C}} - \langle f, \partial_1 g\rangle \\
&= \left\langle \begin{pmatrix} 1 & 0 \\ 0 & -1 \end{pmatrix} \begin{pmatrix} f(1-) \\ f(0-) \end{pmatrix}, \begin{pmatrix} g(1-) \\ g(0-) \end{pmatrix} \right\rangle_{\mathbb{C}\oplus\mathbb{C}} - \langle f, \partial_1 g\rangle \\
&= \left\langle \begin{pmatrix} 1 & 0 \\ 0 & -1 \end{pmatrix} R_1(f), R_1(g) \right\rangle_{\mathbb{C}\oplus\mathbb{C}} - \langle f, \partial_1 g\rangle.
\end{aligned}$$

The advantages of introducing SWIPs become clear by Proposition 3.5 and Theorem 3.6. The first one will help us to deduce well-posedness results for the wave and heat equations on graphs. The second one describes how to build up new SWIPs out of given ones by means of tensor products. This will form a model for evolutionary equations on graphs.

**Proposition 3.5.** *Let $(H, h, A_0, R, S)$ be a SWIP. Then the quintuple*

$$\left(H \oplus H, h \oplus h, \begin{pmatrix} 0 & A_0 \\ A_0 & 0 \end{pmatrix}, \begin{pmatrix} 0 & R \\ R & 0 \end{pmatrix}, \begin{pmatrix} 0 & S \\ S & 0 \end{pmatrix}\right)$$

*is a SWIP.*

*Proof.* Clear. □

**Theorem 3.6.** *Let $(\Omega, \Sigma, \mu)$ be a measure space. Let $(H, h, A_0, R, S)$ be a SWIP. Then so is*

$$(L_2(\mu) \otimes H, L_2(\mu) \otimes h, I_{L_2(\mu)} \otimes A_0, I_{L_2(\mu)} \otimes R, I_{L_2(\mu)} \otimes S).$$

*Remark* 3.7. Since $L_2(\mu) \otimes H \cong L_2(\mu; H)$, the tensor product is in some sense the $\mu$-weighted, $\Omega$-fold copy of $H$. Note that if $\mu$ is a counting measure, the above construction coincides with an orthogonal sum construction as in [5, 7, 9]. Thus, the classical way of viewing the spatial operators on graphs as an orthogonal sum of operators is a special case of the above. If the measure space $(\Omega, \Sigma, \mu)$ is the Lebesgue-measure space on the interval $(0,1)$, we may describe a torus-like structure.

*Proof of Theorem* 3.6. It is well known that the quintuple $(L_2(\mu) \otimes H, L_2(\mu) \otimes h, I_{L_2(\mu)} \otimes A_0, I_{L_2(\mu)} \otimes R, I_{L_2(\mu)} \otimes S)$ satisfies the Conditions (i), (ii) and (iv) in Definition 3.4, see, e.g., [14, 15]. Therefore we only show (iii) and formula (v). In order to show (iii), we consider the orthogonal decomposition in the Hilbert space $D_{I_{L_2(\mu)} \otimes A_0^*}$:

$$D(I_{L_2(\mu)} \otimes A_0^*) = D(I_{L_2(\mu)} \otimes A_0) \oplus D(I_{L_2(\mu)} \otimes A_0)^{\perp}.$$

The space $D(I_{L_2(\mu)} \overset{a}{\otimes} A_0)$ is $|\cdot|_{I_{L_2(\mu)} \otimes A_0^*}$-dense in $D_{I_{L_2(\mu)} \otimes A_0}$. Hence,

$$(I_{L_2(\mu)} \otimes R)[D(I_{L_2(\mu)} \otimes A_0)] = \{0\}$$

follows from

$$(I_{L_2(\mu)} \overset{a}{\otimes} R)[D(I_{L_2(\mu)} \otimes A_0)] = \{0\}.$$

Furthermore, we have

$$\begin{aligned}(I_{L_2(\mu)} \otimes R)[D(I_{L_2(\mu)} \overset{a}{\otimes} A_0^*)] &= L_2(\mu) \overset{a}{\otimes} (h \oplus h) \\ &= (L_2(\mu) \overset{a}{\otimes} h) \oplus (L_2(\mu) \overset{a}{\otimes} h).\end{aligned}$$

The mapping $(I_{L_2(\mu)} \otimes R)|_{N(-(I_{L_2(\mu)} \overset{a}{\otimes} A_0^*)^2+1)}$ is one-to-one and bi-continuous and has dense range. Moreover, the vector space $N(-(I_{L_2(\mu)} \overset{a}{\otimes} A_0^*)^2+1)$ is $|\cdot|_{I_{L_2(\mu)} \otimes A_0^*}$-dense in $N(-(I_{L_2(\mu)} \otimes A_0^*)^2+1)$, whence $(I_{L_2(\mu)} \otimes R)|_{N(-(I_{L_2(\mu)} \otimes A_0^*)^2+1)}$ is a Banach space isomorphism.

Now we show the formula. Let $\Omega_1, \Omega_2 \subseteq \Omega$ be sets of finite measure. Let $x, y \in D(A_0^*)$. Denoting by $\chi_{\Omega_1}, \chi_{\Omega_2}$ the characteristic functions of the sets $\Omega_1$ and $\Omega_2$, respectively, we compute

$$\begin{aligned}\langle (I_{L_2(\mu)} \otimes A_0^*)(\chi_{\Omega_1} \otimes x), \chi_{\Omega_2} \otimes y \rangle_{L_2(\mu) \otimes H} &= \int_\Omega \chi_{\Omega_1}\, \chi_{\Omega_2} \langle A_0^* x, y\rangle_H \, \mathrm{d}\mu \\ = \int_\Omega \chi_{\Omega_1}\, \chi_{\Omega_2} (\langle SRx, Ry\rangle_{h\oplus h} - \langle x, A_0^* y\rangle_H)\, \mathrm{d}\mu \\ = \langle \chi_{\Omega_1} \otimes SRx, \chi_{\Omega_2} \otimes Ry\rangle_{(L_2(\mu)\otimes h)\oplus(L_2(\mu)\otimes h)} \\ \quad - \langle \chi_{\Omega_1} \otimes x, (I_{L_2(\mu)} \otimes A_0^*)(\chi_{\Omega_2} \otimes y)\rangle_{L_2(\mu)\otimes H} \\ = \langle (I_{L_2(\mu)} \otimes S)(I_{L_2(\mu)} \otimes R)(\chi_{\Omega_1} \otimes x), (I_{L_2(\mu)} \otimes R)(\chi_{\Omega_2} \otimes y)\rangle_{(L_2(\mu)\otimes h)\oplus(L_2(\mu)\otimes h)} \\ \quad - \langle \chi_{\Omega_1} \otimes x, (I_{L_2(\mu)} \otimes A_0^*)(\chi_{\Omega_2} \otimes y)\rangle_{L_2(\mu)\otimes H}.\end{aligned}$$

Due to denseness of $D(A_0^*)$-valued step functions and continuity of all considered mappings with respect to the graph-norm of $I_{L_2(\mu)} \otimes A_0^*$, the formula holds for all $f, g \in D(I_{L_2(\mu)} \otimes A_0^*)$. □

Adjacency relations on graphs come along with boundary conditions on the respective spatial operators, see, e.g., [7, 9]. In order to employ Theorem 2.2 for well-posedness results for evolutionary equations on graphs, it is favorable to characterize all boundary conditions leading to a skew-selfadjoint operator. Having the abstract structure of SWIPs at hand, we characterize skew-selfadjoint extensions of $A_0$ in the SWIP-setting. Before doing so, we need the following lemma.

**Lemma 3.8.** *Let $(H, h, A_0, R, S)$ be a SWIP. Let $K$ be a closed subspace of $D_A$ such that $K \supseteq D(A_0)$. Then $R[K] \subseteq h \oplus h$ is a closed subspace. Conversely, for each closed subspace $k \subseteq h \oplus h$ there is a uniquely determined closed subspace $K_k$ of $D_A$ such that $K_k \supseteq D(A_0)$ and $R[K_k] = k$. Furthermore, we have*

$$K_k = D(A_0) \oplus_{D_A} \Psi_R^{-1}[k].$$

*Proof.* In the space $D_A$, we have the decomposition

$$D(A) = D(A_0) \oplus D(A_0)^{\perp}.$$

An elementary calculation shows that we have $D(A_0)^{\perp_{D_A}} = N(-A^2 + 1)$. Hence,

$$K = D(A_0) \oplus (K \cap N(-A^2 + 1)).$$

Using $R[D(A_0)] = \{0\}$ and the operator $\Psi_R$ in Definition 3.4(iii), we observe that

$$\begin{aligned} R[K] &= R[D(A_0) \oplus K \cap N(-A^2 + 1)] \\ &= R[D(A_0)] \oplus R[K \cap N(-A^2 + 1)] = \Psi_R[K \cap N(-A^2 + 1)]. \end{aligned}$$

By the continuity of $\Psi_R^{-1}$ and the closedness of $K \cap N(-A^2 + 1)$ the space $R[K]$ is closed. From the above calculation, we get

$$K = D(A_0) \oplus \Psi_R^{-1}[R[K]]. \tag{1}$$

Let now $k \subseteq h \oplus h$ be a closed subspace. Define $K_k := D(A_0) \oplus \Psi_R^{-1}[k]$. The same reasoning as above yields

$$R[K_k] = \Psi_R[\Psi_R^{-1}[k]] = k.$$

$K_k$ is a closed subspace as an orthogonal sum of two closed subspaces (keep in mind that $\Psi_R$ is bicontinuous). Let $L$ be a closed subspace of $D_A$ such that $L \supseteq D(A_0)$ and $R[L] = k$. Using (1), we get

$$\begin{aligned} K_k &= D(A_0) \oplus \Psi_R^{-1}[R[K_k]] \\ &= D(A_0) \oplus \Psi_R^{-1}[k] \\ &= D(A_0) \oplus \Psi_R^{-1}[R[L]] = L. \end{aligned}$$

□

With the help of the above lemma, we are in the position to compute the adjoint of closed, linear operators restricting $A$ and extending $A_0$.

**Theorem 3.9.** *Let $(H, h, A_0, R, S)$ be a SWIP. Let $B : D(B) \subseteq H \to H$ be a densely defined, closed, linear operator such that $A_0 \subseteq B \subseteq A := -A_0^*$. Then the adjoint of $B$ is given by*

$$B^* : D(B^*) \subseteq H \to H : f \mapsto -Af$$

*where $D(B^*) = D(A_0) \oplus_{D_A} \Psi_R^{-1}[(SR[D(B)])^{\perp}]$. In particular, the following statements are equivalent:*

(i) *$B$ is skew-selfadjoint,*

(ii) *$R[D(B)] = (SR[D(B)])^{\perp}$.*

*Proof.* By $A_0 \subseteq B \subseteq A$, we have $-A_0 \subseteq B^* \subseteq -A$. In particular, we have $D(A_0) \subseteq D(B^*) \subseteq D(A)$. Thus, for $f \in D(A)$, we observe

$$\begin{aligned} f \in D(B^*) &\iff \forall g \in D(B) : \langle Bg, f\rangle_H = \langle g, -Af\rangle_H \\ &\iff \forall g \in D(B) : \langle Ag, f\rangle_H = -\langle g, Af\rangle_H \\ &\iff \forall g \in D(B) : \langle SRg, Rf\rangle_{h\oplus h} = 0 \\ &\iff Rf \in (SR[D(B)])^{\perp}. \end{aligned}$$

From "$\Rightarrow$" we read off that $R[D(B^*)] \subseteq (SR[D(B)])^{\perp}$. Let $x \in (SR[D(B)])^{\perp}$. Then $\Psi_R^{-1}(x) \in D(A)$ and $\Psi_R^{-1}(x) \in D(B^*)$, by "$\Leftarrow$". We deduce that $R[D(B^*)] = (SR[D(B)])^{\perp}$. Now, $D(B^*) \subseteq D_A$ is a closed subspace with $D(A_0) \subseteq D(B^*)$. Thus, using the formula in Lemma 3.8, we conclude that

$$D(B^*) = D(A_0) \oplus \Psi_R^{-1}[R[D(B^*)]] = D(A_0) \oplus \Psi_R^{-1}[(SR[D(B)])^{\perp}].$$

In particular, using the uniqueness result in Lemma 3.8, we have $D(B^*) = D(B)$ if and only if $(SR[D(B)])^{\perp} = R[D(B)]$. $\square$

We have the following immediate corollary of the Theorems 2.2 and 3.9, which comprises a criterion for well-posedness of evolutionary equations with a spatial operator coming from a SWIP.

**Corollary 3.10 (Solution theory).** *Let $r, c > 0$, $(H, h, A_0, R, S)$ be a SWIP and $B : D(B) \subseteq H \to H$ be a densely defined, closed, linear operator such that $A_0 \subseteq B \subseteq -A_0^*$ with $R[D(B)] = (SR[D(B)])^{\perp}$. Let $M : B_{\mathbb{C}}(r, r) \to L(H)$ be bounded and analytic such that $\operatorname{Re}(z^{-1}M(z)) \geq c$ for all $z \in B_{\mathbb{C}}(r, r)$. Then the operator*

$$(\partial_{0,\nu} M(\partial_{0,\nu}^{-1}) + B) : D(\partial_{0,\nu} \otimes I_H) \cap D(I_{H_{\nu,0}} \otimes B) \subseteq H_{\nu,0} \otimes H \to H_{\nu,0} \otimes H$$

*has dense range and a continuous inverse.*

## 4. Applications

We may now apply the above results to the examples in 3.2. Due to the construction principles of SWIPs given in the previous section, we derive all our examples from the rather simple Example 3.1. With Theorem 3.9, we are in the position to classify all skew-selfadjoint operators via boundary-values. Applying Proposition 3.5 to Example 3.1, we get a possible way to study well-posedness of the wave equation or the heat equation, as Example 3.2 indicates.

Let $(\Omega, \Sigma, \mu)$ be a measure space. We recall that

$$\left(L_2(0,1), \mathbb{C}, \partial_{1,0}, R_1, \begin{pmatrix} 1 & 0 \\ 0 & -1 \end{pmatrix}\right)$$

is a SWIP. According to Theorem 3.6 the quintuple

$$\left(L_2(\mu) \otimes L_2(0,1), L_2(\mu) \otimes \mathbb{C}, I_{L_2(\mu)} \otimes \partial_{1,0}, I_{L_2(\mu)} \otimes R_1, \begin{pmatrix} I_{L_2(\mu)} & 0 \\ 0 & -I_{L_2(\mu)} \end{pmatrix}\right)$$

is a SWIP as well. For any closed $U \subseteq L_2(\mu;\mathbb{C}) \oplus L_2(\mu;\mathbb{C})$, there is a uniquely determined subspace $D_U \subseteq L_2(\mu; L_2(0,1)) \oplus L_2(\mu; L_2(0,1))$ such that $(I_{L_2(\mu)} \otimes R_1)[D_U] = U$, by Lemma 3.8. The *associated transport operator* is defined as

$$\partial_U : D_U \subseteq L_2(\mu; L_2(0,1)) \to L_2(\mu; L_2(0,1)) : f \mapsto (I_{L_2(\mu)} \otimes \partial_1)f.$$

Due to Proposition 3.5 and Theorem 3.6 we can study the spatial operator of the wave/heat equation, i.e., the quintuple

$$\left( L_2(\mu) \otimes (L_2(0,1) \oplus L_2(0,1)), L_2(\mu) \otimes \mathbb{C}^2, \begin{pmatrix} 0 & I_{L_2(\mu)} \otimes \partial_{1,0} \\ I_{L_2(\mu)} \otimes \partial_{1,0} & 0 \end{pmatrix}, \right.$$
$$\left. \begin{pmatrix} 0 & I_{L_2(\mu)} \otimes R_1 \\ I_{L_2(\mu)} \otimes R_1 & 0 \end{pmatrix}, \begin{pmatrix} 0 & \begin{pmatrix} I_{L_2(\mu)} & 0 \\ 0 & -I_{L_2(\mu)} \end{pmatrix} \\ \begin{pmatrix} I_{L_2(\mu)} & 0 \\ 0 & -I_{L_2(\mu)} \end{pmatrix} & 0 \end{pmatrix} \right)$$

is a SWIP. For any closed subspace $W \subseteq (L_2(\mu;\mathbb{C}^2))^2$, there is a unique subspace $D_W$ such that

$$\begin{pmatrix} 0 & I_{L_2(\mu)} \otimes R_1 \\ I_{L_2(\mu)} \otimes R_1 & 0 \end{pmatrix} [D_W] = W,$$

by Lemma 3.8. Therefore, we may define the *associated wave operator*

$$\partial_W : D_W \subseteq L_2(\mu; L_2(0,1)^2) \to L_2(\mu; L_2(0,1)^2) : f \mapsto \begin{pmatrix} 0 & I_{L_2(\mu)} \otimes \partial_1 \\ I_{L_2(\mu)} \otimes \partial_1 & 0 \end{pmatrix} f.$$

We summarize the previous reasoning in the following theorem:

**Theorem 4.1.** *Let $U \subseteq L_2(\mu;\mathbb{C}) \oplus L_2(\mu;\mathbb{C})$ and $W \subseteq (L_2(\mu;\mathbb{C}^2))^2$ be closed subspaces. Let $\partial_U$ and $\partial_W$ be the associated transport and wave operator, respectively.*

(i) *The following conditions are equivalent:*
- $\partial_U$ *is skew-selfadjoint,*
- $U \subseteq L_2(\mu;\mathbb{C}) \oplus L_2(\mu;\mathbb{C})$ *is a unitary operator.*

(ii) *The following conditions are equivalent:*
- $\partial_W$ *is skew-selfadjoint,*
- $\left[ \begin{pmatrix} 0 & \begin{pmatrix} 1 & 0 \\ 0 & -1 \end{pmatrix} \\ \begin{pmatrix} 1 & 0 \\ 0 & -1 \end{pmatrix} & 0 \end{pmatrix} W \right]^{\perp} = W.$

*Proof.* In order to prove (i), we recall that

$$\left( L_2(\mu) \otimes L_2(0,1), L_2(\mu) \otimes \mathbb{C}, I_{L_2(\mu)} \otimes \partial_{1,0}, I_{L_2(\mu)} \otimes R_1, \begin{pmatrix} I_{L_2(\mu)} & 0 \\ 0 & -I_{L_2(\mu)} \end{pmatrix} \right)$$

is a SWIP by Example 3.1 and Theorem 3.6. The characterization of the skew-selfadjoint realizations follows from Theorem 3.9 and the particular form of $S := \begin{pmatrix} I_{L_2(\mu)} & 0 \\ 0 & -I_{L_2(\mu)} \end{pmatrix}$. Indeed, by Theorem 3.9 $\partial_U$ is skew-selfadjoint if and only if

$$U = R[D(\partial_U)] = (SR[D(\partial_U)])^{\perp} = (SU)^{\perp} = \left( \begin{pmatrix} 1 & 0 \\ 0 & -1 \end{pmatrix} U \right)^{\perp} = (U^*)^{-1},$$

where $U^{*-1}$ means "flipping the pairs" in the relation $U^* \subseteq L_2(\mu;\mathbb{C}) \oplus L_2(\mu;\mathbb{C})$. The latter equation shows that $U^{-1} = U^*$. An easy argument shows that $U$ is indeed a unitary operator from $L_2(\mu;\mathbb{C})$ into itself.

Using Proposition 3.5, we deduce that assertion (ii) holds, by following the same ideas as in the proof of (i). □

*Remark* 4.2. Combining Corollary 3.10 and Theorem 4.1, we presented appropriate boundary conditions on graphs in order to show well-posedness of the equations of the form in Example 3.2.

Now, we will give some examples for the above subspaces leading to skew-selfadjoint realizations of the respective operators.

*Example* 4.3 (Transport equation). Take a graph with 3 edges $\{1,2,3\}$ and the counting measure, denoted by $\mu$, thereon. The space of all boundary values is $\ell_2(\{1,2,3\})^2$. Let $A, B \in \mathbb{C}^{3\times 3}$ be invertible matrices. Consider the following space

$$U = \left\{ (x_1,x_2,x_3,y_1,y_2,y_3) \in \ell_2(\{1,2,3\})^2; A \begin{pmatrix} x_1 \\ x_2 \\ x_3 \end{pmatrix} = B \begin{pmatrix} y_1 \\ y_2 \\ y_3 \end{pmatrix} \right\}.$$

We may express $U$ in a slightly other fashion

$$U = \left\{ (x_1,x_2,x_3,y_1,y_2,y_3) \in \ell_2(\{1,2,3\})^2; B^{-1}A \begin{pmatrix} x_1 \\ x_2 \\ x_3 \end{pmatrix} = \begin{pmatrix} y_1 \\ y_2 \\ y_3 \end{pmatrix} \right\}.$$

Interpreting $(x_1,x_2,x_3)$ and $(y_1,y_2,y_3)$ as the right-hand sides and the left-hand sides of the edges $\{1,2,3\}$, respectively, we deduce that $B^{-1}A$ relates all the right-hand sides to the left-hand sides. In order to obtain skew-selfadjointness of the associated spatial operator, we have to require that $B^{-1}A$ is unitary.

In [6, Remark after Proposition 9.1], the authors obtained the result of the above example for graphs with finitely many edges. An example for the spatial operator of the wave/heat equation on graphs, we consider next.

*Example* 4.4 (Wave/heat equation). In order to describe the different types of boundary conditions that are included in the above characterization result, we only consider a graph with one edge, i.e., the space of boundary values is $\mathbb{C}^4$. Thus, we consider subspaces $W \subseteq \mathbb{C}^4$. The associated spatial operator $\partial_W$ lies in between $\begin{pmatrix} 0 & \partial_{1,0} \\ \partial_{1,0} & 0 \end{pmatrix}$ and $\begin{pmatrix} 0 & \partial_1 \\ \partial_1 & 0 \end{pmatrix}$.

- Dirichlet boundary conditions are easily obtained by $\begin{pmatrix} 0 & \partial_1 \\ \partial_{1,0} & 0 \end{pmatrix}$. Thus, $W = \{(x_1,x_2,x_3,x_4) \in \mathbb{C}^4; x_1 = x_2 = 0\}$. $W$ leads to a skew-selfadjoint $\partial_W$.

Indeed,

$$\left[\begin{pmatrix} 0 & \left(\begin{smallmatrix} 1 & 0 \\ 0 & -1 \end{smallmatrix}\right) \\ \left(\begin{smallmatrix} 1 & 0 \\ 0 & -1 \end{smallmatrix}\right) & 0 \end{pmatrix} W\right]^{\perp} = \{(x_1, x_2, x_3, x_4) \in \mathbb{C}^4; x_3 = x_4 = 0\}^{\perp}$$
$$= W.$$

- In order to consider Neumann-type boundary conditions, we consider $W = \{(x_1, x_2, x_3, x_4) \in \mathbb{C}^4; x_3 = x_4 = 0\}$. Similarly, as for the Dirichlet boundary conditions, we show that the associated spatial operator is skew-selfadjoint.

**Acknowledgment**

The authors thank the referee for useful comments as well as Jürgen Voigt and Rainer Picard for helpful discussions.

## References

[1] Felix Ali Mehmeti. *Nonlinear waves in networks.* Mathematical Research. 80. Berlin: Akademie Verlag. 171 p., 1994.

[2] Robert Carlson. Adjoint and self-adjoint differential operators on graphs. *Electronic Journal of Differential Equations*, 06:1–10, 1998.

[3] Valentina Casarino, Klaus-Jochen Engel, Rainer Nagel, and Gregor Nickel. A semigroup approach to boundary feedback systems. *Integral Equations Oper. Theory*, 47(3):289–306, 2003.

[4] Sonja Currie and Bruce A. Watson. Eigenvalue asymptotics for differential operators on graphs. *Journal of Computational and Applied Mathematics*, 182(1):13–31, 2005.

[5] Britta Dorn. Semigroup for flows in infinite networks. *Semigroup Forum*, 76:341–356, 2008.

[6] Sebastian Endres and Frank Steiner. The Berry-Keating operator on $L^2(\mathbb{R}_>, dx)$ and on compact quantum graphs with general self-adjoint realizations. *J. Phys. A*, 43, 2010.

[7] Ulrike Kant, Tobias Klauss, Jürgen Voigt, and Matthias Weber. Dirichlet forms for singular one-dimensional operators and on graphs. *Journal of Evolution Equations*, 9:637–659, 2009.

[8] Takashi Kasuga. On Sobolev-Friedrichs' Generalisation of Derivatives. *Proceedings of the Japan Academy*, 33(33):596–599, 1957.

[9] Peter Kuchment. Quantum graphs: an introduction and a brief survey. Exner, Pavel (ed.) et al., Analysis on graphs and its applications. Selected papers based on the Isaac Newton Institute for Mathematical Sciences programme, Cambridge, UK, January 8–June 29, 2007. Providence, RI: American Mathematical Society (AMS). Proceedings of Symposia in Pure Mathematics 77, 291–312 (2008)., 2008.

[10] Felix Ali Mehmeti, Joachim von Below, and Serge Nicaise. *Partial Differential Equations on Multistructures.* Marcel Dekker, 2001.

[11] Rainer Picard. *Hilbert space approach to some classical transforms.* Pitman Research Notes in Mathematics Series. 196., 1989.

[12] Rainer Picard. Evolution Equations as operator equations in lattices of Hilbert spaces. *Glasnik Matematicki Series III*, 35(1):111–136, 2000.

[13] Rainer Picard. A structural observation for linear material laws in classical mathematical physics. *Mathematical Methods in the Applied Sciences*, 32:1768–1803, 2009.

[14] Rainer Picard and Des McGhee. *Partial Differential Equations: A unified Hilbert Space Approach.* De Gruyter, 2011.

[15] Joachim Weidmann. *Linear Operators in Hilbert Spaces.* Springer Verlag, New York, 1980.

Marcus Waurick and Michael Kaliske
Institut für Statik und Dynamik der Tragwerke
TU Dresden
Fakultät Bauingenieurwesen
D-01062 Dresden
e-mail: `marcus.waurick@tu-dresden.de`
`michael.kaliske@tu-dresden.de`

Operator Theory:
Advances and Applications, Vol. 221, 667-690


# Reparametrizations of Non Trace-normed Hamiltonians

Henrik Winkler and Harald Woracek

**Abstract.** We consider a Hamiltonian system of the form $y'(x) = JH(x)y(x)$, with a locally integrable and nonnegative $2 \times 2$-matrix-valued Hamiltonian $H(x)$. In the literature dealing with the operator theory of such equations, it is often required in addition that the Hamiltonian $H$ is trace-normed, i.e., satisfies $\operatorname{tr} H(x) \equiv 1$. However, in many examples this property does not hold. The general idea is that one can reduce to the trace-normed case by applying a suitable change of scale (reparametrization). In this paper we justify this idea and work out the notion of reparametrization in detail.

**Mathematics Subject Classification (2000).** Primary 34B05; Secondary 34L40, 47E05.

**Keywords.** Hamiltonian system, reparametrization, trace-normed.

## 1. Introduction

Consider a Hamiltonian system of the form

$$y'(x) = zJH(x)y(x), \quad x \in I, \tag{1.1}$$

where $I$ is a (finite or infinite) open interval on the real line, $z \in \mathbb{C}$, $J := \begin{pmatrix} 0 & -1 \\ 1 & 0 \end{pmatrix}$, and $H : I \to \mathbb{R}^{2\times 2}$ is a function which does not vanish identically on $I$ a.e., and has the following properties:

**(Ham1)** Each entry of $H$ is (Lebesgue-to-Borel) measurable and locally integrable on $I$.

**(Ham2)** We have $H(x) \geq 0$ almost everywhere on $I$.

We call a function $H$ satisfying (Ham1) and (Ham2) a *Hamiltonian.*

In the literature dealing with systems of the form (1.1), their operator theory, and their spectral properties, it is often assumed that $H$ is *trace-normed*, i.e., that

**(Ham3)** We have $\operatorname{tr} H(x) = 1$ almost everywhere on $I$.

For example, in [HSW], where the operator model associated with (1.1) is introduced from an up-to-date viewpoint, the property (Ham3) is required from the start, in [K] trace-normed Hamiltonians are considered, and also in [GK] it is very soon required that the Hamiltonian under consideration satisfies (Ham3). Contrasting this, in [dB] no normalization conditions are required. However, that work does not deal with the operator theoretic viewpoint on the equation (1.1). In [KW/IV] boundary triples were studied which arise from Hamiltonian functions $H$ which are only assumed to be *non-vanishing*, i.e., have the property that

**(Ham3′)** The function $H$ does not vanish on any set of positive measure.

Let us now list some examples of Hamiltonian systems, where the Hamiltonian is not necessarily trace-normed, or not even non-vanishing, and which have motivated our present work.

1°. When investigating the inverse spectral problem for semibounded spectral measures $\mu$, equations (1.1) with $H$ being of the form

$$H(x) = \begin{pmatrix} v(x)^2 & v(x) \\ v(x) & 1 \end{pmatrix}$$

appear naturally, cf. [W2]. Clearly, Hamiltonians of this kind are non-vanishing but not trace-normed. The function $v$ has intrinsic meaning. For example, when $\mu$ is associated with a Kreĭn string $S[L, \mathfrak{m}]$, the function $v$ is the mass function of the dual string of $S[L, \mathfrak{m}]$, cf. [KWW2, §4].

2°. When identifying a Sturm–Liouville equation without potential term as a Hamiltonian system, one obtains an equation (1.1) with $H$ being of the form

$$H(x) = \begin{pmatrix} p(x) & 0 \\ 0 & \rho(x) \end{pmatrix}.$$

Often the functions $p$ and $\rho$ have physical meaning. For example, consider the propagation of waves in an elastic medium, and assume that the equations of isotropic elasticity hold and that the density of the medium depends only on the depth measured from the surface. Then one arrives at a hyperbolic system whose associated linear spectral problem is of the form

$$-\big(p(x)y'(x)\big)' = \omega^2 \rho(x) y(x), \quad x \geq 0,$$

where $x$ measures the depth from the surface, $\rho(x)$ is the density of the media, and $p(x) = \lambda(x) + 2\mu(x)$ with the Lamé parameters $\lambda, \mu$, cf. [BB], [McL]. Apparently, Hamiltonians of this kind are in general not trace-normed. When the medium under consideration contains layers of vacuum, they will not even be non-vanishing.

3°. Dropping normalization assumptions often leads to significant simplification. For example, transformation of Hamiltonians and their corresponding Weyl coefficients, like those given in [W1], can be treated with much more ease when the requirement that all Hamiltonians are trace-normed is dropped. Also, the natural action of such transformations on the associated chain of de Branges spaces becomes much more apparent.

For example, in our recent investigation of symmetry in the class of Hamiltonians, cf. [WW], it is much more suitable to work with Hamiltonians which may vanish on sets of positive measure. When working with transformation formulas like those introduced in [KWW1], dropping the requirement that Hamiltonians are non-vanishing is very helpful.

One obvious reason why a Hamiltonian $H$ may fail to satisfy (Ham3′), is that there exist whole intervals $(\alpha, \beta)$ with $H|_{(\alpha,\beta)} = 0$ a.e.; remember the situations described in 2°. Of course such intervals are somewhat trivial pieces of $H$. Hence, it is interesting to note that (Ham3′) may also fail for a more subtle reason.

1.1. *Example.* Choose a compact subset $K$ of the unit interval $[0, 1]$ whose Lebesgue measure $m$ is positive and less than 1, and which does not contain any open intervals. A typical example of such a set is the Smith-Volterra-Cantor set being obtained by the usual construction of the Cantor set but removing intervals of length $\frac{1}{4^n}$ instead of $\frac{1}{3^n}$ at the $n$th step of the process. For more details, see, e.g., [AB, p140 f.]. Set $I := (0, 1)$ and

$$H(x) := \begin{cases} \mathrm{id}_{2\times 2}, & x \in I \setminus K \\ 0 \quad, & x \in K \cap I \end{cases},$$

then $H$ is a Hamiltonian. It vanishes on a set of positive measure, namely on $K$. However, if $J$ is any open interval, then $J \setminus K$ is open and nonempty. Hence, there exists no interval where $H$ vanishes almost everywhere.

When dealing with Hamiltonian functions which are not normalized by (Ham3), the notion of reparametrization is (and has always been) present. The idea is:

*If two Hamiltonian functions differ only by a change of scale, they will share their operator theoretic properties.*

Reparametrizations for non-vanishing Hamiltonians were investigated in [KW/IV, §2.1.f], in the context of generalized strings reparameterizations appeared in [LW].

Our aim in this paper is to provide a rigorous fundament for the theory of (not necessarily non-vanishing) Hamiltonian, the notion of a reparametrization, and the above-quoted intuitive statement. We set up the proper environment to deal with Hamiltonians without further normalization or restriction, and provide the practical tool of reparametrization in this general setting. The definition of the associated boundary triple is in essence the same as known from the trace–normed case. The main effort is to thoroughly understand the notion of a reparametrization. As one can guess already from the above Example 1.1, the difficulties which have to be overcome are of measure theoretic nature.

To close this introduction, let us briefly describe the content of the present paper. We define a boundary triple associated with a Hamiltonian in a way which is convenient for the general situation (Section 2); we define and discuss absolutely continuous reparametrizations (Section 3); we show that for a given Hamiltonian $H$ there always exist reparametrizations which relate $H$ with a trace-normed

Hamiltonian, and that the presently defined notion of reparametrization coincides with the previously introduced one in the case of non-vanishing Hamiltonians (Section 4).

## 2. Hamiltonians and their operator models

Throughout this paper measure theoretic notions like 'integrability', 'almost everywhere', 'measurable set', 'zero set', are understood with respect to the Lebesgue measure unless explicitly stated differently.

Intervals where the Hamiltonian is of a particularly simple form play a special role.

**2.1. Definition.** Let $H$ be a Hamiltonian on $I$, and let $(\alpha, \beta) \subseteq I$ be a nonempty open interval.

(i) We call $(\alpha, \beta)$ $H$-immaterial, if $H(x) = 0$, $x \in (\alpha, \beta)$ a.e.

(ii) For $\phi \in \mathbb{R}$ set $\xi_\phi := (\cos\phi, \sin\phi)^T$. We call $(\alpha, \beta)$ $H$-indivisible of type $\phi \in \mathbb{R}$, if $H|_{(\alpha,\beta)}$ is of the form

$$H(x) = h(x)\xi_\phi \xi_\phi^T, \; x \in (\alpha, \beta) \text{ a.e.}, \tag{2.1}$$

with some scalar function $h$, and if no interval $(\alpha, \gamma)$ or $(\gamma, \beta)$ with $\gamma \in (\alpha, \beta)$ is $H$-immaterial.

(iii) We denote by $I_{\text{ind}}$ the union of all $H$-indivisible and $H$-immaterial intervals.

(iv) We say that $H$ has a heavy left endpoint, if it does not start with an immaterial interval. Analogously, $H$ has a heavy right endpoint, if it does not end with an immaterial interval. If both endpoints of $H$ are heavy, we just say that $H$ has heavy endpoints.

If no confusion is possible, we will drop the prefix '$H$-' in these notations.

Note that the type of an indivisible interval is uniquely determined up to multiples of $\pi$, and that the function $h$ in (2.1) coincides a.e. with $\operatorname{tr} H$.

For later use, let us list some simple properties of immaterial and indivisible intervals.

2.2. *Remark.*

(i) Let $(\alpha, \beta)$ and $(\alpha', \beta')$ be immaterial. If the closures of these intervals have nonempty intersection, then the interior of the union of their closures is immaterial.

(ii) Each immaterial interval is contained in a maximal immaterial interval.
Let $(\alpha, \beta)$ be maximal immaterial and let $(\alpha', \beta')$ be immaterial, then either $(\alpha', \beta') \subseteq (\alpha, \beta)$ or $[\alpha', \beta'] \cap [\alpha, \beta] = \emptyset$.
There exist at most countably many maximal immaterial intervals.

(iii) Let $(\alpha, \beta)$ be indivisible of type $\phi$, and let $(\alpha', \beta')$ be an interval which has nonempty intersection with $(\alpha, \beta)$. If $(\alpha', \beta')$ is immaterial, then $[\alpha', \beta'] \subseteq (\alpha, \beta)$. If $(\alpha', \beta')$ is indivisible of type $\phi'$, then $\phi = \phi' \bmod \pi$ and the union $(\alpha, \beta) \cup (\alpha', \beta')$ is indivisible of type $\phi$.

(iv) Each indivisible interval of type $\phi$ is contained in a maximal indivisible interval of type $\phi$.
Let $(\alpha,\beta)$ be maximal indivisible of type $\phi$ and let $(\alpha',\beta')$ be indivisible of type $\phi'$. Then either $\phi=\phi' \bmod \pi$ and $(\alpha',\beta')\subseteq(\alpha,\beta)$, or $(\alpha',\beta')\cap(\alpha,\beta)=\emptyset$.
There exist at most countably many maximal indivisible intervals.
(v) The set $I_{\text{ind}}$ is the disjoint union of all maximal indivisible intervals, and all maximal immaterial intervals which are not contained in an indivisible interval.
(vi) The following statements are equivalent:
   - The interval $(\alpha,\beta)$ is indivisible of type $\phi$.
   - We have $\xi_{\phi+\frac{\pi}{2}}\in\ker H(x)$, $x\in(\alpha,\beta)$ a.e. Neither $H$ vanishes a.e. on an interval of the form $(\alpha,\gamma)$ with $\gamma\in(\alpha,\beta)$, nor on an interval of the form $(\gamma,\beta)$.
   - We have
$$\int_{(\alpha,\beta)}\xi^*_{\phi+\frac{\pi}{2}}H(x)\xi_{\phi+\frac{\pi}{2}}\,dx=0\,.$$
Neither $H$ vanishes a.e. on an interval of the form $(\alpha,\gamma)$ with $\gamma\in(\alpha,\beta)$, nor on an interval of the form $(\gamma,\beta)$.

The first step towards the definition of the operator model associated with a Hamiltonian is to define the space of $H$-measurable functions.

**2.3. Definition.** Let $H$ be a Hamiltonian defined on $I$. Then we denote by $\mathcal{M}(H)$ the set of all $\mathbb{C}^2$-valued functions $f$ on $I$, such that:
(i) The function $Hf:I\to\mathbb{C}^2$ is (Lebesgue-to-Borel) measurable.
(ii) If $(\alpha,\beta)\subseteq I$ is immaterial, then $f$ is constant on $[\alpha,\beta]\cap I$.
(iii) If $(\alpha,\beta)\subseteq I$ is indivisible of type $\phi$, then $\xi_\phi^T f$ is constant on $(\alpha,\beta)$.

We define a relation '$=_H$' on $\mathcal{M}(H)$ by
$$f=_H g \qquad \text{if} \qquad H(f-g)=0 \text{ a.e. on } I$$

Let us point out explicitly that in the conditions (ii) and (iii) the respective functions are required to be constant, and not only constant almost everywhere. Apparently, (ii) and (iii) are a restriction only on the closure of $I_{\text{ind}}$. For example, each measurable function whose support does not intersect this closure certainly belongs to $\mathcal{M}(H)$. Also, note that the set $\mathcal{M}(H)$ does not change when $H$ is changed on a set of measure zero, and that $=_H$ is an equivalence relation.

Usually, in the literature, only measurable functions $f$ are considered. However, it turns out practical to weaken this requirement to (i) of Definition 2.3.

The next statement says that each equivalence class modulo $=_H$ in fact contains measurable functions. In particular, this implies that when factorizing modulo '$=_H$' it makes no difference whether we require $Hf$ or $f$ to be measurable.

**2.4. Lemma.** *Let $H$ be a Hamiltonian defined on $I$, and let $f\in\mathcal{M}(H)$. Then there exists a measurable function $g\in\mathcal{M}(H)$, such that $f=_H g$.*

*Proof.* Write $H := \begin{pmatrix} h_1 & h_3 \\ h_3 & h_2 \end{pmatrix}$. We divide the interval $I$ into six disjoint parts, namely

$$\begin{aligned} J_1 &:= \bigcup \{L : L \text{ maximal indivisible}\} \\ J_2 &:= \bigcup \{I \cap \overline{L} : L \text{ maximal immaterial}, L \cap J_1 = \emptyset\} \\ J_3 &:= \{x \in I : H(x) = 0\} \setminus (J_1 \cup J_2) \\ J_4 &:= \{x \in I : H(x) \neq 0, \det H(x) = 0,\ h_2(x) = 0\} \setminus (J_1 \cup J_2) \\ J_5 &:= \{x \in I : H(x) \neq 0, \det H(x) = 0,\ h_2(x) \neq 0\} \setminus (J_1 \cup J_2) \\ J_6 &:= \{x \in I : \det H(x) \neq 0\} \setminus (J_1 \cup J_2) \end{aligned}$$

Since $J_1$ is open, and $J_2$ is a countable union of (relatively) closed sets, both are measurable. Since each entry of $H$ is measurable, each of the subsets $J_3, \dots, J_6$ is measurable. If two open intervals $L_1$ and $L_2$ have empty intersection, also $L_1 \cap \overline{L}_2 = \emptyset$. Thus, $J_1 \cap J_2 = \emptyset$. The other sets $J_3, \dots, J_6$ are trivially pairwise disjoint and disjoint from $J_1$and $J_2$. We are going to define the required function $g$ on each of the sets $J_i$, $i = 1, \dots, 6$, separately.

*Definition on $J_1$:* Let $L$ be a maximal indivisible interval, say of type $\phi$. Then $\xi_\phi^T f(x)$ is constant on $L$. We set

$$g(x) := \left[\xi_\phi^T f(x)\right] \cdot \xi_\phi, \quad x \in L,$$

then

$$\begin{aligned} H(x)\big(f(x) - g(x)\big) &= h(x) \cdot \xi_\phi \xi_\phi^T \big(f(x) - g(x)\big) \\ &= h(x) \cdot \xi_\phi \Big[\xi_\phi^T f(x) - \xi_\phi^T \left[\xi_\phi^T f(x)\right] \xi_\phi\Big] = 0, \quad x \in L \text{ a.e.} \end{aligned}$$

The function $g|_L$ itself, in particular also $\xi_\phi^T g|_L$, is constant. Hence, no matter how we define $g$ on the remaining parts $J_2, \dots, J_6$, the condition (iii) of Definition 2.3 will be satisfied for $g$.

By the above procedure, $g$ is defined on all of $J_1$. Since $J_1$ is a countable union of disjoint open sets where $g$ is constant, $g|_{J_1}$ is measurable.

*Definition on $J_2$:* Let $L$ be a maximal immaterial interval which does not intersect any indivisible interval. Then $f$ is constant on $I \cap \overline{L}$. We set

$$g(x) := f(x), \quad x \in I \cap \overline{L},$$

then

$$H(x)\big(f(x) - g(x)\big) = 0, \quad x \in I \cap \overline{L}.$$

No matter how we define $g$ on the remaining parts $J_3, \dots, J_6$, the condition (ii) of Definition 2.3 will hold true for $g$: Assume that $(\alpha, \beta)$ is immaterial. Then $[\alpha, \beta] \cap I$ is either contained in some maximal indivisible interval or in some maximal immaterial interval which does not intersect any indivisible interval. In both cases, the function $g$ is constant on $[\alpha, \beta] \cap I$. Since $J_2$ is a countable disjoint union of closed sets where $g$ is constant, $g|_{J_2}$ is measurable.

*Definition on $J_3$:* We set $g(x) := 0$, $x \in J_3$, then $g|_{J_3}$ is measurable and $H(x)(f(x)-g(x)) = 0$, $x \in J_3$.

*Definition on $J_4$:* For $x \in J_4$ we have $H(x) = h(x)\xi_0\xi_0^T$ with the measurable and positive function $h(x) := \operatorname{tr} H(x)$. Write $f$ as $f(x) = f_1(x)\xi_0 + f_2(x)\xi_{\frac{\pi}{2}}$, then

$$H(x)f(x) = h(x)f_1(x)\xi_0, \quad x \in J_4 .$$

Since $h(x)$ is positive, it follows that the function

$$g(x) := f_1(x)\xi_0, \quad x \in J_4 ,$$

is measurable. Also, it satisfies

$$H(x)\big(f(x) - g(x)\big) = h(x)\xi_0\xi_0^T \cdot f_2(x)\xi_{\frac{\pi}{2}} = 0, \quad x \in J_4 .$$

*Definition on $J_5$:* We argue similar as for $J_4$. For $x \in J_5$ we have $H(x) = h(x)\xi_{\phi(x)}\xi_{\phi(x)}^T$ with the measurable and positive function $h(x) := \operatorname{tr} H(x)$ and the measurable function $\phi(x) := \operatorname{Arccot} \frac{h_3(x)}{h_2(x)}$. Write $f$ as $f(x) = f_1(x)\xi_{\phi(x)} + f_2(x)\xi_{\phi(x)+\frac{\pi}{2}}$, $x \in J_5$, then

$$H(x)f(x) = h(x)f_1(x)\xi_{\phi(x)}, \quad x \in J_5 .$$

Since $h(x)$ is positive, the function $g(x) := f_1(x)\xi_{\phi(x)}$, $x \in J_5$, is measurable. It satisfies, $H(x)(f(x) - g(x)) = 0$, $x \in J_5$.

*Definition on $J_6$:* If $\det H(x) \neq 0$, we can write $f(x) = H(x)^{-1} \cdot H(x)f(x)$, and hence $f|_{J_6}$ is measurable. Set $g(x) := f(x)$, $x \in J_6$, then $H(x)(f(x) - g(x)) = 0$, $x \in J_6$. □

**2.5. Corollary.** *Let $H$ be a Hamiltonian defined on $I$. If $f_1, f_2 \in \mathcal{M}(H)$, then the function $f_2^*Hf_1$ is measurable. For each $f \in \mathcal{M}(H)$ the function $f^*Hf$ is measurable and almost everywhere nonnegative.*

*Proof.* Choose measurable functions $g_1, g_2 \in \mathcal{M}(H)$ according to Lemma 2.4. Then

$$f_2^*Hf_1 = g_2^*Hg_1 + (f_2 - g_2)^*Hg_1 + f_2^*H(f_1 - g_1) = g_2^*Hg_1 \quad \text{a.e.}.$$

Since $H(x)$ is a.e. a nonnegative matrix, each function $f^*Hf$, $f \in \mathcal{M}(H)$, is a.e. nonnegative. □

Now we can write down the definition of the operator model associated with a Hamiltonian. It reads almost the same as in the trace-normed case.

Denote by $\operatorname{Ac}(H)$ the subset of $\mathcal{M}(H)$, which consists of all locally absolutely continuous functions in $\mathcal{M}(H)$. Moreover, call $H$ regular at the endpoint $s_- := \inf I$, if for one (and hence for all) $s \in I$

$$\int_{s_-}^{s} \operatorname{tr} H(x)\, dx < \infty .$$

If this integral is infinite, call $H$ singular at $s_-$. The terms regular/singular at the endpoint $s_+ := \sup I$ are defined analogously[†].

**2.6. Definition.** Let $H$ be a Hamiltonian defined on $I = (s_-, s_+)$. Set

$$\begin{aligned} \sigma_- &:= \sup\big\{x \in I : (s_-, x) \text{ immaterial}\big\}, \\ \sigma_+ &:= \inf\big\{x \in I : (x, s_+) \text{ immaterial}\big\}, \\ \tilde{I} &:= (\sigma_-, \sigma_+), \quad \tilde{H} := H|_{\tilde{I}} \end{aligned} \tag{2.2}$$

**2.7. Definition.** Let $H$ be a Hamiltonian defined on $I$.

(i) We define the model space $L^2(\tilde{H}) \subseteq \mathcal{M}(\tilde{H})/_{=_{\tilde{H}}}$ as

$$L^2(H) := \Big\{\hat{f}/_{=_{\tilde{H}}} : \hat{f} \in \mathcal{M}(\tilde{H}),\ \int_{\tilde{I}} \hat{f}(x)^* H(x) \hat{f}(x)\, dx < \infty\Big\}.$$

For $f_1, f_2 \in L^2(H)$ we define an inner product as ($f_1 = \hat{f}_1/_{=_H}$, $f_2 = \hat{f}_2/_{=_H}$)

$$(f_1, f_2)_H := \int_{\tilde{I}} \hat{f}_2(x)^* H(x) \hat{f}_1(x)\, dx .$$

(ii) We define the model relation $T_{\max}(H) \subseteq L^2(H) \times L^2(H)$ as

$$T_{\max}(H) := \big\{(f;g) \in L^2(H) \times L^2(H) :\ \exists \hat{f} \in \mathrm{Ac}(\tilde{H}), \hat{g} \in \mathcal{M}(\tilde{H}) \text{ with } f = \hat{f}/_{=_{\tilde{H}}}, g = \hat{g}/_{=_{\tilde{H}}} \text{ and } \hat{f}' = J\tilde{H}\hat{g} \text{ a.e.}\big\}.$$

(iii) We define the model boundary relation $\Gamma(H) \subseteq T_{\max}(H) \times (\mathbb{C}^2 \times \mathbb{C}^2)$ as the set of all elements $((f;g);(a;b))$ such that there exist representants $\hat{f} \in \mathrm{Ac}(H)$ of $f$ and $\hat{g} \in \mathcal{M}(H)$ of $g$ with $\hat{f}' = JH\hat{g}$ and ($s_- := \inf I$, $s_+ := \sup I$)

$$a = \begin{cases} \lim_{t \searrow s_-} \hat{f}(t), & \text{regular at } s_- \\ 0 \quad , & \text{singular at } s_- \end{cases}$$

$$b = \begin{cases} \lim_{t \nearrow s_+} \hat{f}(t), & \text{regular at } s_+ \\ 0 \quad , & \text{singular at } s_+ \end{cases}$$

Unless it is necessary, the equivalence relation '$=_H$' will not be mentioned explicitly and equivalence classes and their representants will not be distinguished explicitly.

The operator theoretic properties of these objects, for example the fact that $(L^2(H), T_{\max}(H), \Gamma(H))$ is a Hilbert space boundary triple, could be proved by following the known path. This, however, would be unnecessary labour. As we will see later, it is always possible to reduce to the trace-normed case by means of a reparametrization, cf. Corollary 4.4.

[†]Instead of regular and singular, one also speaks of Weyl's limit circle case or Weyl's limit point case.

For later reference let us explicitly state the obvious fact that a pair $(f;g)$ belongs to $T_{\max}(H)$ if and only if there exist representants $\hat{f}$ and $\hat{g}$ of $f$ and $g$, respectively, with

$$\hat{f}(y) = \hat{f}(x) + \int_x^y JH\hat{g}, \quad x, y \in I. \tag{2.3}$$

## 3. Absolutely continuous reparametrizations

Let us define rigorously what we understand by a reparametrization (i.e., a 'change of scale').

**3.1. Definition.** Let $H_1$ and $H_2$ be Hamiltonians defined on intervals $I_1$ and $I_2$, respectively.

(i) We say that $H_2$ is a basic reparametrization of $H_1$, and write $H_1 \rightsquigarrow H_2$, if there exists a nondecreasing, locally absolutely continuous, and surjective map $\lambda$ of $I_1$ onto $I_2$, such that

$$H_1(x) = H_2(\lambda(x)) \cdot \lambda'(x), \quad x \in I_1 \text{ a.e.} \tag{3.1}$$

Here $\lambda'$ denotes a nonnegative function which coincides a.e. with the derivative of $\lambda$.

(ii) Let numbers $\sigma_1^-, \sigma_1^+$ and $\sigma_{2,-}, \sigma_{2,+}$ be defined by (2.2) for $H_1$ and $H_2$, respectively. Then we write $H_1 ✂ H_2$, if

$$H_1|_{(\sigma_1^-,\sigma_1^+)} = H_2|_{(\sigma_{2,-},\sigma_{2,+})}\,.$$

(iii) We denote by '$\sim$' the smallest equivalence relation containing both relations '$\rightsquigarrow$' and '✂'. If $H \sim \tilde{H}$, we say that $H$ and $\tilde{H}$ are reparametrizations of each other.

First of all note that '✂' is an equivalence relation, and that '$\rightsquigarrow$' is reflexive and transitive; for transitivity apply the chain rule. However, '$\rightsquigarrow$' fails to be symmetric, see the below Example 3.2. This properties of '$\rightsquigarrow$' imply that $H \sim \tilde{H}$ if and only if there exist finitely many Hamiltonians $L_0, \dots, L_m$, such that

$$H = L_0 \approx_1 L_1 \approx_2 L_2 \approx_3 \cdots \approx_{m-1} L_{m-1} \approx_m L_m = \tilde{H} \tag{3.2}$$

where $\approx_i \in \{✂, \rightsquigarrow, \rightsquigarrow^{-1}\}$, $i = 1, \dots, m$.

3.2. *Example.* Let us show by an example that '$\rightsquigarrow$' is not symmetric. One obvious obstacle for symmetry is that a function $\lambda$ establishing a basic reparametrization by means of (3.1) need not be injective. However, if $\lambda(x_1) = \lambda(x_2)$ for some $x_1 < x_2$, then $\lambda$ is constant on the interval $(x_1, x_2)$, and hence $\lambda' = 0$ a.e. on $(x_1, x_2)$. Thus $(x_1, x_2)$ must be a $H_1$-immaterial interval; a somewhat trivial piece of the Hamiltonian.

A more subtle example is obtained from the Hamiltonian $H$ introduced in Example 1.1. Using the notation from this example, set

$$\tilde{I} := (0, 1-m), \quad \tilde{H}(y) := \mathrm{id}_{2\times 2}, \; y \in \tilde{I},$$

and consider the map

$$\lambda(x) := \int_0^x \chi_{I\setminus K}(t)\,dt, \quad x \in I\,.$$

Then, $\lambda$ is nondecreasing, absolutely continuous, and $\lambda' = \chi_{I\setminus K}$ a.e. Since $K$ does not contain any open interval, $\lambda$ is in fact an increasing bijection of $I$ onto $\tilde{I}$. Let us show that

$$H(x) = \tilde{H}(\lambda(x))\lambda'(x), \quad x \in I \text{ a.e.}$$

If $x \in I \setminus K$ and $\lambda'(x) = \chi_{I\setminus K}(x)$, both sides equal $\mathrm{id}_{2\times 2}$. If $x \in K$ and $\lambda'(x) = \chi_{I\setminus K}(x)$, both sides equal 0. We see that $H \rightsquigarrow \tilde{H}$ via $\lambda$.

Assume on the contrary that $\tilde{H} \rightsquigarrow H$ via some nondecreasing, locally absolutely continuous, and surjective map $\tau$ of $\tilde{I}$ onto $I$, so that

$$\tilde{H}(y) = H(\tau(y))\tau'(y), \quad y \in \tilde{I} \text{ a.e.}$$

For $y \in \tau^{-1}(K)$, the left side of this relation equals $\mathrm{id}_{2\times 2}$ and right side equals 0. Thus $\tau^{-1}(K)$ must be a zero set. Since $\tau$ is locally absolutely continuous and surjective, this implies that $K = \tau(\tau^{-1}(K))$ is a zero set. We have reached a contradiction, and conclude that $\tilde{H} \not\rightsquigarrow H$.

Our aim in this section is to show that Hamiltonians which are reparametrizations of each other give rise to isomorphic operator models, for the precise formulation see Theorem 3.8 below. The main effort is to understand basic reparametrizations; and this is our task in the next couple of statements.

3.3. *Remark.* Let $I_1$ and $I_2$ be nonempty open intervals on the real line, and let $\lambda : I_1 \to I_2$ be a nondecreasing, locally absolutely continuous, and surjective map.

(i) The function $\lambda$ cannot be constant on any interval of the form $(\inf I, \gamma)$ or $(\gamma, \sup I)$ with $\gamma \in I$. This is immediate from the fact that the image of $\lambda$ is an open interval.

(ii) There exists a nonnegative function $\lambda'$ which coincides almost everywhere with the derivative of $\lambda$, and which has the following property:

For each nonempty interval $(\alpha, \beta) \subseteq I$ such that $\lambda|_{(\alpha,\beta)}$ is constant, we have $\lambda'|_{[\alpha,\beta]} = 0$. (3.3)

Note here that, due to (i), always $[\alpha, \beta] \subseteq I$.

Let us show that $\lambda'$ can indeed be assumed to satisfy (3.3). Each interval $(\alpha, \beta)$ where $\lambda$ is constant is contained in a maximal interval having this property. Each two maximal intervals where $\lambda$ is constant are either equal or disjoint. Hence, there can exist at most countably many such. Let $(\alpha, \beta)$ be one of them. Then the derivative of $\lambda$ exists and is equal to zero on all of $(\alpha, \beta)$. Choose any function $\lambda'$ which coincides almost everywhere with the derivative of $\lambda$. By redefining this function on a set of measure zero, we can thus achieve that $\lambda'(x) = 0$, $x \in [\alpha, \beta]$.

We will, throughout the following, always assume that the function $\lambda'$ in Definition 3.1, (i), has the additional property (3.3). By the just said, this is no loss in generality.

**3.4. Proposition.** *Let $H_1$ and $H_2$ be Hamiltonians defined on intervals $I_1$ and $I_2$, respectively. Assume that $H_2$ is a basic reparametrization of $H_1$, and let $\lambda$ be a map which establishes this reparametrization. Moreover, let $\tilde{\lambda}$ be a right inverse of $\lambda$*[†]*.*

*Then the maps $\circ\tilde{\lambda} : f_1 \mapsto f_1 \circ \tilde{\lambda}$ and $\circ\lambda : f_2 \mapsto f_2 \circ \lambda$ induce mutually inverse linear bijections between $\mathcal{M}(H_1)$ and $\mathcal{M}(H_2)$.*

$$I_1 \underset{\tilde{\lambda}}{\overset{\lambda}{\rightleftarrows}} I_2 \qquad \lambda \circ \tilde{\lambda} = \mathrm{id} \qquad\qquad \mathcal{M}(H_1) \underset{\circ\tilde{\lambda}}{\overset{\circ\lambda}{\leftrightarrows}} \mathcal{M}(H_2)$$

*They respect the equivalence relations $=_{H_1}$ and $=_{H_2}$ in the sense that, for each two elements $f_2, g_2 \in \mathcal{M}(H_2)$,*

$$f_2 =_{H_2} g_2 \quad \Longleftrightarrow \quad (f_2 \circ \lambda) =_{H_1} (g_2 \circ \lambda) \tag{3.4}$$

*and for each two elements $f_1, g_1 \in \mathcal{M}(H_1)$,*

$$f_1 =_{H_1} g_1 \quad \Longleftrightarrow \quad (f_1 \circ \tilde{\lambda}) =_{H_2} (g_1 \circ \tilde{\lambda})$$

In the proof of this proposition there arise some difficulties of measure theoretic nature. Let us state the necessary facts separately.

**3.5. Lemma.** *Let $I_1$ and $I_2$ be nonempty open intervals on the real line, let $\lambda : I_1 \to I_2$ be a nondecreasing, locally absolutely continuous, and surjective map, and let $\tilde{\lambda}$ be a right inverse of $\lambda$. Moreover, assume that $\lambda'$ is a function which coincides almost everywhere with the derivative of $\lambda$ (and has the property (3.3)), and set*

$$L_0 := \{x \in I : \lambda'(x) = 0\}.$$

*Then the following hold:*

(i) *If $E \subseteq I_1$ is a zero set, so is $\tilde{\lambda}^{-1}(E)$.*
(ii) *The function $\tilde{\lambda}$ is Lebesgue-to-Lebesgue measurable.*
(iii) *The set $\lambda(L_0)$ is measurable and has measure zero.*
(iv) *The function $\lambda' \circ \tilde{\lambda}$ is almost everywhere positive. In fact,*

$$\{y \in I_2 : (\lambda' \circ \tilde{\lambda})(y) = 0\} = \lambda(L_0).$$

(v) *If $E \subseteq I_2$ is a measurable set, so is $\lambda^{-1}(E) \setminus L_0 \subseteq I_1$. If $E$ is a zero set, also $\lambda^{-1}(E) \setminus L_0$ has measure zero.*

---

[†]For example, one could choose $\tilde{\lambda}(y) := \min\{x \in I_1 : \lambda(x) = y\}$. Due to continuity of $\lambda$ and Remark 3.3, (i), this minimum exists and belongs to $I_1$

*Proof.*
*Item* (i): Since $\tilde{\lambda}$ is a right inverse of $\lambda$, we have $\tilde{\lambda}^{-1}(E) \subseteq \lambda(E)$. Since $\lambda$ is locally absolutely continuous, $E$ being a zero set implies that $\lambda(E)$ is a zero set. Thus $\tilde{\lambda}^{-1}(E)$ is measurable and has measure zero.
*Item* (ii): The function $\tilde{\lambda}$ is nondecreasing, and hence Borel-to-Borel measurable. Let a Lebesgue measurable set $M \subseteq I_1$ be given, and choose Borel sets $A, B$ with $A \subseteq M \subseteq B$ such that the Lebesgue measure of $B \setminus A$ equals zero. Then $\tilde{\lambda}^{-1}(A)$ and $\tilde{\lambda}^{-1}(B)$ are Borel sets,

$$\tilde{\lambda}^{-1}(A) \subseteq \tilde{\lambda}^{-1}(M) \subseteq \tilde{\lambda}^{-1}(B), \quad \tilde{\lambda}^{-1}(B) \setminus \tilde{\lambda}^{-1}(A) = \tilde{\lambda}^{-1}(B \setminus A).$$

However, by (i), $\tilde{\lambda}^{-1}(B \setminus A)$ has measure zero, and it follows that $\tilde{\lambda}^{-1}(M)$ is Lebesgue measurable.
*Item* (iii): The crucial observation is the following: If two points $x, y \in I$, $x < y$, have the same image under $\lambda$, then $\lambda$ is constant on $[x, y]$, and by (3.3) thus $x, y \in L_0$. In particular, the set $L_0$ is saturated with respect to the equivalence relation $\ker \lambda$[‡]. This implies that

$$\lambda(L_0) = \tilde{\lambda}^{-1}(L_0), \quad L_0 = \lambda^{-1}(\lambda(L_0)). \tag{3.5}$$

By (ii), the first equality already shows that $\lambda(L_0)$ is measurable. To compute the measure of $\lambda(L_0)$, we use the second equality and evaluate

$$\int_{I_2} \chi_{\lambda(L_0)}(y)\, dy = \int_{I_1} \underbrace{\big(\chi_{\lambda(L_0)} \circ \lambda\big)}_{\substack{\| \\ \chi_{\lambda^{-1}(\lambda(L_0))} = \chi_{L_0}}}(x) \cdot \lambda'(x)\, dx = 0\,.$$

*Item* (iv): Consider the function $\lambda' \circ \tilde{\lambda}$. Clearly, it is nonnegative. Let $y \in I_2$ be given. Then $(\lambda' \circ \tilde{\lambda})(y) = 0$ if and only if $\tilde{\lambda}(y) \in L_0$, and in turn, by (3.5), if and only if $y \in \lambda(L_0)$.
*Item* (v): The function $(\chi_E \circ \lambda) \cdot \lambda'$ is measurable, and hence

$$\lambda^{-1}(E) \setminus L_0 = \big\{x \in I_1 : [(\chi_E \circ \lambda) \cdot \lambda'](x) \neq 0\big\}$$

is measurable. Moreover, if $E$ is a zero set,

$$0 = \int_{I_2} \chi_E(x)\, dx = \int_{I_1} (\chi_E \circ \lambda)(x) \cdot \lambda'(x)\, dx\,,$$

and hence the (nonnegative) function $(\chi_E \circ \lambda) \cdot \lambda'$ must vanish almost everywhere. □

Next, we have to make clear how immaterial and indivisible intervals behave when performing the transformation $\lambda$.

---

[‡]Here we understand by $\ker \lambda$ the equivalence relation

$$\ker \lambda := \{(x_1; x_2) \in I_1 \times I_1 : \lambda(x_1) = \lambda(x_2)\}\,,$$

and call a subset of $I_1$ saturated with respect to this equivalence relation, if it is a union of equivalence classes.

**3.6. Lemma.** *Consider the situation described in Proposition* 3.4.

(i) *If* $(\alpha,\beta)\subseteq I_1$ *and* $\lambda$ *is constant on this interval, then* $(\alpha,\beta)$ *is* $H_1$*-immaterial.*
(ii) *If* $(\alpha,\beta)\subseteq I_1$ *is* $H_1$*-immaterial, then the set of inner points of the interval* $\lambda\big([\alpha,\beta]\cap I_1\big)$ *is either empty or* $H_2$*-immaterial.*
(iii) *If* $(\alpha,\beta)\subseteq I_2$ *is* $H_2$*-immaterial, then the set of inner points of the interval* $\lambda^{-1}\big([\alpha,\beta]\cap I_2\big)$ *is* $H_1$*-immaterial.*
(iv) *If* $(\alpha,\beta)\subseteq I_1$ *is* $H_1$*-indivisible of type* $\phi$*, then the interval* $\lambda\big((\alpha,\beta)\big)$ *is* $H_2$*-indivisible of type* $\phi$.
(v) *If* $(\alpha,\beta)\subseteq I_2$ *is* $H_2$*-indivisible of type* $\phi$*, then the interval* $\lambda^{-1}\big((\alpha,\beta)\big)$ *is* $H_1$*-indivisible of type* $\phi$.

*Proof.*
*Item* (i): This has already been noted in the first paragraph of Example 3.2.

*Item* (ii): If the set of inner points of the interval $\lambda\big([\alpha,\beta]\cap I_1\big)$ is empty, there is nothing to prove. Hence, assume that it is nonempty.

Consider first the case that $[\alpha,\beta]\cap I_1$ is saturated with respect to the equivalence relation $\ker\lambda$. Choose a zero set $E\subseteq I_1$, such that $H_1(x)=0$, $x\in\big([\alpha,\beta]\cap I_1\big)\setminus E$. Since $H_1(\tilde\lambda(y))=H_2(y)\cdot(\lambda'\circ\tilde\lambda)(y)$ a.e., we obtain

$$H_2(y)=0,\quad y\in\tilde\lambda^{-1}\big(\big([\alpha,\beta]\cap I_1\big)\setminus E\big)\setminus\lambda(L_0)\text{ a.e.}$$

Since $[\alpha,\beta]\cap I_1$ is saturated with respect to $\ker\lambda$, we have $\tilde\lambda^{-1}\big([\alpha,\beta]\cap I_1\big)=\lambda\big([\alpha,\beta]\cap I_1\big)$, and it follows that

$$\lambda^{-1}\big(\big([\alpha,\beta]\cap I_1\big)\setminus E\big)\setminus\lambda(L_0)=\lambda\big([\alpha,\beta]\cap I_1\big)\setminus\big(\tilde\lambda^{-1}(E)\cup\lambda(L_0)\big)\,.$$

In particular, $H_2$ vanishes almost everywhere on the set of inner points of the interval $\lambda\big([\alpha,\beta]\cap I_1\big)$.

Assume next that $(\alpha,\beta)$ is an arbitrary $H_1$-immaterial interval. The union of all equivalence classes of elements $x\in(\alpha,\beta)$ modulo $\ker\lambda$ is a (relatively) closed interval, say $[\alpha_0,\beta_0]\cap I_1$. Since

$$(\alpha_0,\beta_0)=(\alpha_0,\alpha]\cup(\alpha,\beta)\cup[\beta,\beta_0)\,,$$

and $\lambda$ is certainly constant on $(\alpha_0,\alpha]$ and $[\beta,\beta_0)$, it follows that $(\alpha_0,\beta_0)$ is $H_1$-immaterial. Moreover, $[\alpha_0,\beta_0]\cap I_1$ is saturated with respect to $\ker\lambda$. Applying what we have proved in the above paragraph, gives that the set of inner points of $\lambda\big([\alpha_0,\beta_0]\cap I_1\big)$ is $H_2$-immaterial. Since $[\alpha,\beta]\cap I_1\subseteq[\alpha_0,\beta_0]\cap I_1$, the required assertion follows.

*Item* (iii): Choose a zero set $E\subseteq I_2$, such that $H_2(y)=0$, $y\in\big([\alpha,\beta]\cap I_2\big)\setminus E$. Since $H_1(x)=H_2(\lambda(x))\lambda'(x)$ a.e., it follows that

$$H_1(x)=0,\quad x\in\lambda^{-1}\big(\big([\alpha,\beta]\cap I_2\big)\setminus E\big)\cup L_0\text{ a.e.}$$

However,

$$\begin{aligned}\lambda^{-1}\big([\alpha,\beta]\cap I_2\big)\setminus\big(\lambda^{-1}(E)\setminus L_0\big)&\subseteq\big[\big(\lambda^{-1}\big([\alpha,\beta]\cap I_2\big)\big)\setminus\lambda^{-1}(E)\big]\cup L_0\\&=\lambda^{-1}\big(\big([\alpha,\beta]\cap I_2\big)\setminus E\big)\cup L_0\end{aligned}$$

and we conclude that $H_1$ vanishes on $\lambda^{-1}\big([\alpha,\beta]\cap I_2\big)$ with possible exception of a zero set.

*Item* (iv): The function $\lambda$ is not constant on any interval of the form $(\alpha,\alpha+\varepsilon)$ or $(\beta-\varepsilon,\beta)$. Hence, the interval $(\alpha,\beta)$ is saturated with respect to $\ker\lambda$, and $\lambda\big((\alpha,\beta)\big)$ is open.

Choose a zero set $E\subseteq I_1$, such that $H_1(x)=h_1(x)\cdot\xi_\phi\xi_\phi^T$, $x\in(\alpha,\beta)\setminus E$. Then

$$H_2(y)=\frac{h_1(\tilde\lambda(y))}{(\lambda'\circ\tilde\lambda)(y)}\cdot\xi_\phi\xi_\phi^T,\quad y\in\tilde\lambda^{-1}\big((\alpha,\beta)\setminus E\big)\setminus\lambda(L_0)\text{ a.e.}$$

However,

$$\tilde\lambda^{-1}\big((\alpha,\beta)\setminus E\big)\setminus\lambda(L_0)=\lambda\big((\alpha,\beta)\big)\setminus\big(\tilde\lambda^{-1}(E)\cup\lambda(L_0)\big)\,.$$

Hence, $H_2$ has the required form.

Set $(\alpha',\beta'):=\lambda\big((\alpha,\beta)\big)$, and assume that for some $\gamma'>\alpha'$ the interval $(\alpha',\gamma')$ is $H_2$-immaterial. Then the interval $\lambda^{-1}\big((\alpha',\gamma')\big)$ is $H_1$-immaterial. Since $\lambda$ is continuous and $(\alpha,\beta)$ is saturated with respect to $\ker\lambda$, we have $\lambda^{-1}\big((\alpha',\gamma')\big)=(\alpha,\gamma)$ with some $\gamma\in(\alpha,\beta)$. We have reached a contradiction. The same argument shows that no interval of the form $(\gamma',\beta')$ can be $H_2$-immaterial.

*Item* (v): Choose a zero set $E\subseteq I_2$, such that

$$H_2(y)=h_2(y)\cdot\xi_\phi\xi_\phi^T,\quad y\in(\alpha,\beta)\setminus E\,.$$

Moreover, set $\lambda^{-1}\big((\alpha,\beta)\big)=:(\alpha',\beta')\subseteq I_1$.

First, we have

$$H_1(x)=h_2(\lambda(x))\lambda'(x)\cdot\xi_\phi\xi_\phi^T,\quad x\in\lambda^{-1}\big((\alpha,\beta)\setminus E\big)\text{ a.e.}$$

On the set $L_0$ this equality trivially remains true a.e. We conclude that $H_1(x)$ is of the form $h_1(x)\cdot\xi_\phi\xi_\phi^T$ for all $x\in\lambda^{-1}\big((\alpha,\beta)\big)\setminus\big(\lambda^{-1}(E)\setminus L_0\big)$ a.e.

Second, assume that for some $\gamma'\in(\alpha',\beta')$ the interval $(\alpha',\gamma')$ is $H_1$-immaterial. Then the set of inner points of $\lambda\big((\alpha',\gamma')\big)$ is $H_2$-immaterial. However, since $\lambda\big((\alpha',\beta')\big)=(\alpha,\beta)$, the function $\lambda$ cannot be constant on any interval $(\alpha',\alpha'+\varepsilon)$, and hence $\lambda\big((\alpha',\gamma')\big)\supseteq(\alpha,\gamma)$ for some $\gamma>\alpha$. We have reached a contradiction, and conclude that $(\alpha',\beta')$ cannot start with an immaterial interval. The fact that it cannot end with such an interval is seen in the same way. □

After these preparations, we turn to the proof of Proposition 3.4.

*Proof of Proposition* 3.4.
*Step* 1*:* Let $f_2\in\mathcal{M}(H_2)$ be given, and consider the function $f_1:=f_2\circ\lambda$. We have

$$H_1f_1=H_1(f_2\circ\lambda)=(H_2\circ\lambda)\lambda'\cdot(f_2\circ\lambda)=\big[(H_2f_2)\circ\lambda\big]\cdot\lambda'\text{ a.e.},\tag{3.6}$$

and hence $H_1f_1$ is measurable.

Let $(\alpha,\beta)\subseteq I_1$ be an immaterial interval. Then the set of inner points of $\lambda\big([\alpha,\beta]\cap I_1\big)$ is either empty or $H_2$-immaterial. In the first case, $\lambda$ is constant on $[\alpha,\beta]\cap I_1$, and hence also $f_1$ is constant on this interval. In the second case, $f_2$ is constant on $\lambda\big([\alpha,\beta]\cap I_1\big)$, and it follows that $f_1$ is constant on $[\alpha,\beta]\cap I_1$.

If $(\alpha,\beta)$ is $H_1$-indivisible of type $\phi$, then $\lambda\big((\alpha,\beta)\big)$ is $H_2$-indivisible of type $\phi$. Hence $\xi_\phi^T f_2$ is constant on $\lambda\big((\alpha,\beta)\big)$, and thus $\xi_\phi^T f_1$ is constant on $(\alpha,\beta)$. It follows that $f_1 \in \mathcal{M}(H_1)$, and we have shown that $\circ\lambda$ maps $\mathcal{M}(H_2)$ into $\mathcal{M}(H_1)$.

*Step* 2*:* Let $f_1 \in \mathcal{M}(H_1)$ be given, and set $f_2 := f_1 \circ \tilde{\lambda}$. First note that $(x_1;x_2) \in \ker\lambda$, $x_1 < x_2$, implies that the interval $(x_1,x_2)$ is $H_1$-immaterial, and hence that $f_1(x_1) = f_1(x_2)$. Using this fact, it follows that

$$f_2 \circ \lambda = (f_1 \circ \tilde{\lambda}) \circ \lambda = f_1 \,. \tag{3.7}$$

Next, we compute (a.e.)

$$(H_1 f_1)\circ\tilde{\lambda} = \big[H_1\cdot(f_2\circ\lambda)\big]\circ\tilde{\lambda} = \big[(H_2\circ\lambda)\lambda'\cdot(f_2\circ\lambda)\big]\circ\tilde{\lambda} = (H_2 f_2)\cdot(\lambda'\circ\tilde{\lambda})\,. \tag{3.8}$$

Since $\tilde{\lambda}$ is Lebesgue-to-Lebesgue measurable, the function $(H_1 f_1)\circ\tilde{\lambda}$ is measurable. Since $\lambda'\circ\tilde{\lambda}$ is almost everywhere positive, this implies that $H_2 f_2$ is measurable.

Let $(\alpha,\beta) \subseteq I_2$ be $H_2$-immaterial, then $f_1$ is constant on $\lambda^{-1}\big([\alpha,\beta]\cap I_2\big)$. Since $\tilde{\lambda}\big([\alpha,\beta]\cap I_2\big) \subseteq \lambda^{-1}\big([\alpha,\beta]\cap I_2\big)$, it follows that $f_1\circ\tilde{\lambda}$ is constant on $[\alpha,\beta]\cap I_2$.

If $(\alpha,\beta) \subseteq I_2$ is $H_2$-indivisible of type $\phi$, then $\xi_\phi^T f_1$ is constant on $\lambda^{-1}((\alpha,\beta))$, and in turn $\xi_\phi^T f_2$ is constant on $(\alpha,\beta)$. It follows that $f_2 \in \mathcal{M}(H_2)$, and we have shown that $\circ\tilde{\lambda}$ maps $\mathcal{M}(H_1)$ into $\mathcal{M}(H_2)$.

*Step* 3*:* Since $\tilde{\lambda}$ is a right inverse of $\lambda$, we have $(f_2\circ\lambda)\circ\tilde{\lambda} = f_2$ for any function defined on $I_2$. The fact that $(f_1\circ\tilde{\lambda})\circ\lambda = f_1$ whenever $f_1 \in \mathcal{M}(H_1)$, was shown in (3.7). We conclude that the maps $\circ\lambda$ and $\circ\tilde{\lambda}$ are mutually inverse bijections between $\mathcal{M}(H_1)$ and $\mathcal{M}(H_2)$.

*Step* 4*:* To show (3.4), it is clearly enough to consider the case that $g_2 = 0$. Let $f_2 \in \mathcal{M}(H_2)$ be given. Assume first that there exists a set $E \subseteq I_1$ of measure zero, such that $H_1(x)(f_2\circ\lambda)(x) = 0$, $x \in I_1\setminus E$. Then, by (3.8) and the fact that $H_1(f_2\circ\lambda) = [(H_2 f_2)\circ\lambda]\cdot\lambda'$, we have

$$(H_2 f_2)(y)\cdot(\lambda'\circ\lambda)(y) = 0, \quad y \in I_2 \setminus \tilde{\lambda}^{-1}(E)\,.$$

Since $\tilde{\lambda}^{-1}(E)$ is a zero set, and $(\lambda'\circ\tilde{\lambda})$ is positive a.e., this implies that $H_2 f_2 = 0$ a.e. on $I_2$. Conversely, assume that $H_2(y)f_2(y) = 0$, $y \in I_2\setminus E$, with some set $E \subseteq I_2$ of measure zero. Then, by (3.6), we have

$$H_1(x)(f_2\circ\lambda)(x) = 0, \quad x \in \big(I_1\setminus\lambda^{-1}(E)\big)\cup L_0 = I_1\setminus(\lambda^{-1}(E)\setminus L_0)\,.$$

However, we know that $\lambda^{-1}(E)\setminus L_0$ is a zero set.

Since we already know that $\circ\tilde{\lambda}$ is the inverse of $\circ\lambda$, the last equivalence follows from (3.4). □

Continuing the argument, we obtain that the model boundary triples of $H_1$ and $H_2$ are isomorphic.

**3.7. Proposition.** *Consider the situation described in Proposition* 3.4. *Then the maps* $\circ\lambda$ *and* $\circ\tilde{\lambda}$ *induce mutually inverse isometric isomorphisms between* $L^2(H_1)$ *and* $L^2(H_2)$,

$$L^2(H_1) \underset{\circ\tilde{\lambda}}{\overset{\circ\lambda}{\rightleftarrows}} L^2(H_2)$$

*which satisfy*

$$[\circ\lambda \times \circ\lambda]\big(T_{\max}(H_2)\big) = T_{\max}(H_1), \quad \Gamma(H_1) \circ [\circ\lambda \times \circ\lambda] = \Gamma(H_2).$$

*Proof.*
*Step* 1. *Mapping* $L^2$*:* Let $f_2 \in \mathcal{M}(H_2)$. Then

$$\int_{I_2} f_2^* H_2 f_2 = \int_{I_1} ([f_2^* H_2 f_2] \circ \lambda) \cdot \lambda' = \int_{I_1} (f_2 \circ \lambda)^* \cdot (H_2 \circ \lambda)\lambda' \cdot (f_2 \circ \lambda).$$

Remembering that $\circ\lambda$ maps $\mathcal{M}(H_2)$ bijectively onto $\mathcal{M}(H_1)$ and respects the equivalence relations $=_{H_1}$ and $=_{H_2}$, this relation implies that $\circ\lambda$ induces an isometric isomorphism of $L^2(H_2)$ onto $L^2(H_1)$.

*Step* 2. *Image of* $T_{\max}$*:* Let $f_2, g_2 \in L^2(H_2)$, and let $\hat{f}_2, \hat{g}_2$ be some respective representants. Then we have

$$\hat{f}_2(\lambda(x)) + \int_{\lambda(x)}^{\lambda(y)} JH_2\hat{g}_2 = (\hat{f}_2 \circ \lambda)(x) + \int_x^y JH_1(\hat{g}_2 \circ \lambda), \quad x, y \in I_1. \tag{3.9}$$

If $(f_2; g_2) \in T_{\max}(H_2)$, choose representants $\hat{f}_2, \hat{g}_2$ as in (2.3). If $x, y \in I_1$, then the left side of (3.9) is equal to $\hat{f}_2(\lambda(y))$. Hence also the right side takes this value. We see that $\hat{f}_2 \circ \lambda$ and $\hat{g}_2 \circ \lambda$ are representants as required in (2.3) to conclude that $(f_2 \circ \lambda, g_2 \circ \lambda) \in T_{\max}(H_1)$.

Conversely, assume that $f_2, g_2 \in L^2(H_2)$ with $(f_1; g_1) := (f_2 \circ \lambda; g_2 \circ \lambda) \in T_{\max}(H_1)$, let $\hat{f}_1, \hat{g}_1$ be representants as in (2.3), and set $\hat{f}_2 := \hat{f}_1 \circ \tilde{\lambda}$ and $\hat{g}_2 := \hat{g}_1 \circ \tilde{\lambda}$. First of all notice that $\hat{f}_2$ and $\hat{g}_2$ are representants of $f_2$ and $g_2$, respectively, and remember that $\hat{f}_2 \circ \lambda = \hat{f}_1$ and $\hat{g}_2 \circ \lambda = \hat{g}_1$, cf. (3.7). The right-hand side of (3.9), and thus also the left-hand side, is equal to $\hat{f}_1(y) = (\hat{f}_2 \circ \lambda)(y)$. Since $\lambda$ is surjective, it follows that

$$\hat{f}_2(\tilde{y}) = \hat{f}_2(\tilde{x}) + \int_{\tilde{x}}^{\tilde{y}} JH_2\hat{g}_2, \quad \tilde{x}, \tilde{y} \in I_2.$$

It follows that $\hat{f}_2$ is absolutely continuous, and satisfies the relation required in (2.3) to conclude that $(f_2; g_2) \in T_{\max}(H_2)$.

*Step* 3. *Boundary values:* As we have seen in the previous part of this proof, the map $\circ\lambda \times \circ\lambda$ is not only a bijection of $T_{\max}(H)$ onto $T_{\max}(H)$, but actually between the sets of all possible representants which can be used in (2.3). This implies that also $\Gamma(H_1) \circ [\circ\lambda \times \circ\lambda] = \Gamma(H_2)$. □

Now it is easy to reach our aim, and treat arbitrary reparametrizations.

**3.8. Theorem.** *Let $H$ and $\tilde{H}$ be Hamiltonians which are reparametrizations of each other. Then there exists a linear and isometric bijection $\Phi$ of $L^2(H)$ onto $L^2(\tilde{H})$ such that*

$$(\Phi \times \Phi)\big(T_{\max}(H)\big) = T_{\max}(\tilde{H}), \quad \Gamma(\tilde{H}) \circ (\Phi \times \Phi) = \Gamma(H).$$

*Proof.* Assume that $H \sim \tilde{H}$, and choose $L_0, \dots, L_m$ as in (3.2). Then there exist isometric isomorphisms $\Phi_i : L^2(L_i) \to L^2(L_{i+1})$, $i = 0, \dots, m-1$, with $(\Phi_i \times \Phi_i)(T_{\max}(H_i)) = T_{\max}(H_{i+1})$ and $\Gamma(H_{i+1}) \circ (\Phi_i \times \Phi_i) = \Gamma(H_i)$. The composition

$$\Phi := \Phi_{m-1} \circ \cdots \circ \Phi_0$$

hence does the job. □

## 4. Trace-normed and non-vanishing Hamiltonians

In this section we show that indeed it is often no loss in generality to work with trace-normed Hamiltonians. Moreover, we show that the presently introduced notion of reparametrization is consistent with what was used previously.

### a. Existence of trace-norming reparametrizations

The fact that each equivalence class of Hamiltonians modulo reparametrization contains trace-normed elements, is a consequence of the following lemma.

**4.1. Lemma.** *Let $I_1$ and $I_2$ be nonempty open intervals on the real line, and let $\lambda : I_1 \to I_2$ be a nondecreasing, locally absolutely continuous, and surjective map. Moreover, let $H_1$ be a Hamiltonian on $I_1$. Then there exists a Hamiltonian $H_2$ on $I_2$, such that $H_1 \rightsquigarrow H_2$ via the map $\lambda$.*

*Proof.* Choose a right inverse $\tilde{\lambda}$ of $\lambda$, and a function $\lambda'$ which coincides almost everywhere with the derivative of $\lambda$ (and satisfies (3.3)). Moreover, set again $L_0 := \{x \in I_1 : \lambda'(x) = 0\}$.

Then we define

$$H_2(y) := \begin{cases} \frac{1}{(\lambda' \circ \tilde{\lambda})(y)} (H_1 \circ \tilde{\lambda})(y), & y \in I_2 \setminus \lambda(L_0) \\ 0 \hfill , & y \in \lambda(L_0) \end{cases}$$

Then $H_2$ is a measurable function, and $H_2(y) \geq 0$ a.e. If $x_1, x_2 \in I_1$, $x_1 < x_2$, and $(x_1; x_2) \in \ker \lambda$, then $x_1, x_2 \in L_0$. Hence,

$$(\tilde{\lambda} \circ \lambda)(x) = x, \quad x \in I_1 \setminus L_0 .$$

Thus $H_1$ and $H_2$ are related by (3.1).

Let $\alpha, \beta \in I_1$, $\alpha < \beta$. Then

$$\int_{\lambda(\alpha)}^{\lambda(\beta)} \operatorname{tr} H_2 = \int_\alpha^\beta \big([\operatorname{tr} H_2] \circ \lambda\big) \cdot \lambda' = \int_\alpha^\beta \operatorname{tr} H_1 < \infty .$$

Whenever $K$ is a compact subset of $I_2$, we can choose $\alpha, \beta$ such that $K \subseteq \lambda\big((\alpha, \beta)\big)$. Thus $\operatorname{tr} H_2$, and hence also each entry of $H_2$, is locally integrable. □

**4.2. Proposition.** *Let $H$ be a Hamiltonian, then there exists a trace-normed reparametrization of $H$.*

*Proof.* Since we are only interested in the equivalence class modulo reparametrization which contains $H$ as a representant, we may assume without loss of generality that $H$ has heavy endpoints.

Write the domain of $H$ as $I = (s_-, s_+)$, fix $s \in (s_-, s_+)$, and set

$$\mathfrak{t}(x) := \int_s^x \operatorname{tr} H(t)\, dt, \quad x \in I, \tag{4.1}$$

$$\sigma_- := \lim_{x \searrow s_-} \mathfrak{t}(x), \quad \sigma_+ := \lim_{x \nearrow s_+} \mathfrak{t}(x).$$

Then $\mathfrak{t}$ is an absolutely continuous and nondecreasing function which maps $I$ surjectively onto the open interval $\tilde{I} := (\sigma_-, \sigma_+)$. By Lemma 4.1, there exists a basic reparametrization $\tilde{H}$ of $H$ via the map $\mathfrak{t}$.

It remains to compute ($\tilde{\mathfrak{t}}$, $\mathfrak{t}'$, and $L_0$, are as in Lemma 4.1 for $\lambda := \mathfrak{t}$)

$$\operatorname{tr} \tilde{H}(y) = \frac{1}{(\operatorname{tr} H \circ \tilde{\mathfrak{t}})(y)} \operatorname{tr}(H \circ \tilde{\mathfrak{t}})(y) = 1, \quad y \in \tilde{I} \setminus \mathfrak{t}(L_0),$$

and to remember that $\mathfrak{t}(L_0)$ is a zero set. □

4.3. *Remark.* We would like to note that the reparameterization used in Proposition 4.2 could also be obtained as a three-step result: First, we may assume that $I$ is a finite interval and that $H$ has heavy endpoints. This can be achieved by an affine reparameterization and applying the 'scissors'-operation respectively.

In the second step apply Lemma 4.1 with the map

$$\lambda(t) := \int_{\inf I}^t \chi_J(t)\, dt$$

where

$$J := \{t \in I : H(t) \neq 0\}.$$

This yields a reparameterization to a non-vanishing Hamiltonian. Note here that, since $H$ has heavy endpoints, the image of $\lambda$ is an open interval.

Finally, apply the reparameterization via the map (4.1). For this step we need not anymore use Lemma 4.1, but can refer to the classical theory (or Lemma 4.7 below).

Now we obtain without any further effort that the operator model defined in Section 2 indeed has all the properties known from the trace-normed case. For example:

**4.4. Corollary.** *Let $H$ be a Hamiltonian. Then $(L^2(H), T_{\max}(H), \Gamma(H))$ is a boundary triple with defect* 1 *or* 2 *in the sense of* [KW/IV, §2.2.a]. □

### b. Description of '∼' for non-vanishing Hamiltonians

Our last aim in this paper is to show that the restriction of the relation '∼' to the subclass of non-vanishing Hamiltonians can be described in a simple way, namely in exactly the way 'reparametrizations' were defined in [KW/IV], compare Proposition 4.9 below with [KW/IV, §2.1.f]. In particular, this tells us that the present notion of reparametrization is consistent with the one introduced earlier.

To achieve this aim, we provide some lemmata.

**4.5. Lemma.** *Let $H_1$ and $H_2$ be Hamiltonians defined on $I_1 = (s_{1,-}, s_{1,+})$ and $I_2 = (s_{2,-}, s_{2,+})$, respectively, and let $H_i' := H_i|_{(\sigma_{i,-},\sigma_{i,+})}$ where $\sigma_i^\pm$ is defined as in (2.2). Then the following are equivalent:*

(i) *We have $H_1 \rightsquigarrow H_2$.*
(ii) *We have $H_1' \rightsquigarrow H_2'$. Moreover, $H_1$ and $H_2$ together do or do not have a heavy left endpoint, and together do or do not have a heavy right endpoint.*

*Proof.* Assume that $H_1 \rightsquigarrow H_2$, and let $\lambda$ be a nondecreasing, locally absolutely continuous surjection of $I_1$ onto $I_2$ which establishes this basic reparametrization. First we show that $H_1$ does not have a heavy left endpoint if and only if $H_2$ does not have a heavy left endpoint, and that, in this case,

$$\lambda(\sigma_1^-) = \sigma_{2,-}\,. \tag{4.2}$$

Assume that $s_{1,-} < \sigma_1^-$. Then, by Lemma 3.6, the set of inner points of the interval $\lambda\big((s_{1,-}, \sigma_1^-]\big)$ is either empty or $H_2$-immaterial. However, this set is nothing but the open interval $(s_{1,-}, \lambda(\sigma_1^-))$. We conclude that $\lambda(\sigma_1^-) \leq \sigma_{2,-}$, in particular, $s_{2,-} < \sigma_{2,-}$. For the converse, assume that $s_{2,-} < \sigma_{2,-}$. Then, again by Lemma 3.6, the set of inner points of $\lambda^{-1}\big((s_{2,-}, \sigma_{2,-}]\big)$ is $H_1$-immaterial. This set is an open interval of the form $(s_{1,-}, x_0)$ with some $x_0 \in I_1$. It already follows that $s_{1,-} < \sigma_1^-$. Assume that $\lambda(\sigma_1^-) < \sigma_{2,-}$. Then there exists a point $x \in (s_{1,-}, x_0)$ with $\lambda(\sigma_1^-) < \lambda(x)$. This implies that $\sigma_1^- < x$, and we have reached a contradiction. Thus the equality (4.2) must hold.

The fact that $H_1$ and $H_2$ together do or do not have a heavy right endpoint is seen in exactly the same way. Moreover, we also obtain that $\lambda(\sigma_1^+) = \sigma_{2,+}$, in case $\sigma_{2,-} < s_{2,-}$.

Consider the restriction $\Lambda := \lambda|_{(\sigma_1^-, \sigma_1^+)}$. Then $\Lambda$ is a nondecreasing and locally absolutely continuous map. Since $\lambda$ cannot be constant on any interval having $\sigma_1^-$ as its left endpoint or $\sigma_1^+$ as its right endpoint, we have

$$\Lambda\big((\sigma_1^-, \sigma_1^+)\big) = (\sigma_{2,-}, \sigma_{2,+})\,.$$

Hence $\Lambda$ establishes a basic reparametrization of $H_1'$ to $H_2'$. We have shown that $H_1 \rightsquigarrow H_2$ implies that the stated conditions hold true.

For the converse implication, assume that the stated conditions are satisfied, and let $\Lambda$ be a nondecreasing, locally absolutely continuous surjection of $(\sigma_1^-, \sigma_1^+)$ onto $(\sigma_{2,-}, \sigma_{2,+})$ which establishes the basic reparametrization of $H_1'$ to $H_2'$. If $s_{1,-} < \sigma_1^-$, then also $s_{2,-} < \sigma_{2,-}$, and hence we can choose a linear and increasing bijection $\Lambda_-$ of $[s_{1,-}, \sigma_1^-]$ onto $[s_{2,-}, \sigma_{2,-}]$. If $s_{1,+} < \sigma_1^+$ choose analogously a linear and increasing bijection $\Lambda_+$ of $[\sigma_1^+, s_{1,+}]$ onto $[\sigma_{2,+}, s_{2,+}]$. Then the map

$\lambda : I_1 \to I_2$ defined as

$$\lambda(x) := \begin{cases} \Lambda_-(x) , & x \in (s_{1,-}, \sigma_1^-] \text{ if } s_{1,-} < \sigma_1^- \\ \Lambda(x) \quad , & x \in (\sigma_1^-, \sigma_1^+) \\ \Lambda_+(x) , & x \in [\sigma_1^+, s_{1,+}) \text{ if } \sigma_1^+ < s_{1,+} \end{cases}$$

establishes a basic reparametrization of $H_1$ to $H_2$. □

**4.6. Lemma.** *Let $H$ and $\tilde{H}$ be Hamiltonians. Then $H \sim \tilde{H}$ if and only if there exist finitely many Hamiltonians $H_1, \dots, H_n$ with heavy endpoints, such that*

$$H \text{✂} H_1 \rightsquigarrow H_2 \rightsquigarrow^{-1} H_3 \rightsquigarrow \cdots \rightsquigarrow H_{n-1} \rightsquigarrow^{-1} H_n \text{✂} \tilde{H}$$

*Proof.* First we show that

$$\text{✂} \circ \rightsquigarrow \; = \; \rightsquigarrow \circ \text{✂} \tag{4.3}$$

Assume that $(H_1; H_2) \in \text{✂} \circ \rightsquigarrow$. Then there exists a Hamiltonian $L$, such that

$$H_1 \text{✂} L \rightsquigarrow H_2$$

Let $H_1', H_2', L'$ be the Hamiltonians with heavy endpoints, such that

$$H_1' \text{✂} H_1, \;\; H_2' \text{✂} H_2, \;\; L' \text{✂} L$$

By Lemma 4.5, we have $H_1' = L' \rightsquigarrow H_2'$. Define a Hamiltonian $L''$ by appending (if necessary) immaterial intervals to $L'$ in such a way that $L'' \text{✂} L'$, and $L''$ and $H_1$ together do or do not heavy left or right endpoints. Analogously, define $H_2''$, such that $H_2'' \text{✂} H_2'$, and $H_2''$ and $H_1$ together do or do not have heavy left or right endpoints. Then $H_2'' \text{✂} H_2$ and, by Lemma 4.5,

$$H_1 \rightsquigarrow L'' \rightsquigarrow H_2''$$

Altogether it follows that

$$H_1 \rightsquigarrow H_2'' \text{✂} H_2,$$

i.e., $(H_1; H_2) \in \rightsquigarrow \circ \text{✂}$. We have established the inclusion '$\subseteq$' in (4.3). The reverse inclusion is seen in the same way.

Assume now that $H \sim \tilde{H}$, and let $L_0, \dots, L_m$ be as in (3.2). By (4.3), reflexivity, and transitivity, there exist Hamiltonians $L_0', \dots, L_n'$ with

$$H = L_0' \text{✂} L_1' \rightsquigarrow L_2' \rightsquigarrow^{-1} \cdots \rightsquigarrow L_{n-1}' \rightsquigarrow^{-1} L_n' = \tilde{H}$$

Let $H_i$, $i = 0, \dots, n$, be the Hamiltonians with heavy endpoints and $H_i \text{✂} L_i'$. Then, by Lemma 4.5,

$$H \text{✂} H_0 = H_1 \rightsquigarrow H_2 \rightsquigarrow^{-1} \cdots \rightsquigarrow H_{n-1} \rightsquigarrow^{-1} H_n \text{✂} \tilde{H}$$ □

**4.7. Lemma.** *Let $H_1$ and $H_2$ be Hamiltonians defined on intervals $I_1$ and $I_2$, respectively. Assume that $H_1 \rightsquigarrow H_2$, and let $\lambda : I_1 \to I_2$ be a nondecreasing, locally absolutely continuous, and surjective map such that (3.1) holds. If $H_1$ is non-vanishing, then $\lambda$ is bijective, $\lambda'$ is almost everywhere positive, $\lambda^{-1}$ is locally absolutely continuous, and $H_2$ is non-vanishing.*

*Proof.* Assume that $H_1$ is non-vanishing. Then the function $\lambda'$ cannot vanish on any set of positive measure, i.e., it is almost everywhere positive. In particular, $\lambda$ cannot be constant on any nonempty interval. Hence, $\lambda$ is strictly increasing, and thus also bijective.

Let $E \subseteq I_2$ be a zero set, then

$$0 = \int_{I_2} \chi_E(y)\, dy = \int_{I_1} (\chi_E \circ \lambda)(x)\lambda'(x)\, dx\,.$$

This implies that the (nonnegative) function $\chi_E \circ \lambda$ must vanish almost everywhere. However, $\chi_E \circ \lambda = \chi_{\lambda^{-1}(E)}$, i.e., $\lambda^{-1}(E)$ is a zero set.

It remains to show that $H_2$ is non-vanishing. Let $E \subseteq I_2$ be measurable. Then

$$\int_E \operatorname{tr} H_2 = \int_{\lambda^{-1}(E)} (\operatorname{tr} H_2 \circ \lambda)\lambda' = \int_{\lambda^{-1}(E)} \operatorname{tr} H_1\,.$$

If $\operatorname{tr} H_2$ vanishes on $E$, then $\operatorname{tr} H_1$ must vanish on $\lambda^{-1}(E)$. Hence, $\lambda^{-1}(E)$ is a zero set, and thus also $E$ is a zero set. □

**4.8. Lemma.** *Let $H, H_1, H_2$ be Hamiltonians with heavy endpoints, being defined on respective intervals $I, I_1, I_2$. Assume that $H_1$ and $H_2$ are non-vanishing, and that $H \rightsquigarrow H_1$ and $H \rightsquigarrow H_2$ via maps $\lambda_1 : I \to I_1$ and $\lambda_2 : I \to I_2$. Then there exists a bijective increasing map $\mu : I_1 \to I_2$ such that $\mu$ and $\mu^{-1}$ are locally absolutely continuous and*

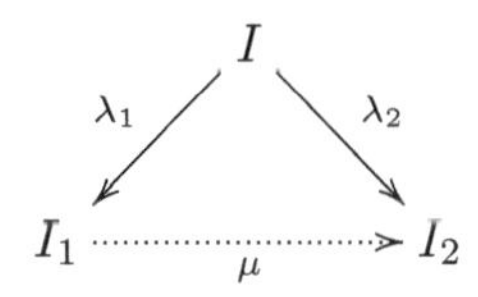


*Proof.*

*Step 1:* We start with a preliminary remark. Denote

$$L_0^j := \{x \in I : \lambda_j'(x) = 0\}, \quad j = 1, 2\,.$$

If $x \in L_0^1 \setminus L_0^2$, i.e., $\lambda_1'(x) = 0$ but $\lambda_2'(x) \neq 0$, then

$$H_2(\lambda_2(x)) = \frac{1}{\lambda_2'(x)} H(x) = \frac{1}{\lambda_2'(x)} \cdot H_1(\lambda_1(x))\lambda_1'(x) = 0\,.$$

Since $H_2$ is non-vanishing, it follows that $\lambda_2(L_0^1 \setminus L_0^2)$ is a zero set. This implies that also

$$\lambda_2^{-1}\big(\lambda_2(L_0^1 \setminus L_0^2)\big) \setminus L_0^2$$

has measure zero. However,

$$L_0^1 \setminus L_0^2 = \big(L_0^1 \setminus L_0^2\big) \setminus L_0^2 \subseteq \lambda_2^{-1}\big(\lambda_2(L_0^1 \setminus L_0^2)\big) \setminus L_0^2\,,$$

and hence also $L_0^1 \setminus L_0^2$ is a zero set. In the same way it follows that $L_0^2 \setminus L_0^1$ is a zero set.

*Step* 2: We turn to the proof of the lemma. Let $\tilde{\lambda}_1$ be a right inverse of $\lambda_1$, and set

$$\mu := \lambda_2 \circ \tilde{\lambda}_1 .$$

Then $\mu$ is a nondecreasing map of $I_1$ onto $I_2$.

First, we show that $\mu$ is surjective. Let $y \in I_2$ be given, and set $x := \lambda_1(\tilde{\lambda}_2(y))$ where $\tilde{\lambda}_2$ is a right inverse of $\lambda_2$. If $\tilde{\lambda}_1(x) = \tilde{\lambda}_2(y)$, we have

$$\mu(x) = \lambda_2(\tilde{\lambda}_1(x)) = \lambda_2(\tilde{\lambda}_2(y)) = y .$$

Assume that $\tilde{\lambda}_1(x) < \tilde{\lambda}_2(y)$. We have

$$\lambda_1\big(\tilde{\lambda}_1(x)\big) = x = \lambda_1\big(\tilde{\lambda}_2(y)\big) ,$$

and hence the interval $(\tilde{\lambda}_1(x), \tilde{\lambda}_2(y))$ is $H$-immaterial. Thus the set of inner points of $\lambda_2\big([\tilde{\lambda}_1(x), \tilde{\lambda}_2(y)] \cap I\big)$ is either empty or $H_2$-immaterial. Since $H_2$ is non-vanishing, the second possibility cannot occur. We conclude that $\lambda_2(\tilde{\lambda}_1(x)) = \lambda_2(\tilde{\lambda}_2(y))$, and hence again $\mu(x) = y$. The case that $\tilde{\lambda}_1(x) > \tilde{\lambda}_2(y)$ is treated in the same way. In any case, the given point $y$ belongs to the image of $\mu$.

Since $\mu$ is nondecreasing and surjective, $\mu$ must be continuous. To show that $\mu$ is locally absolutely continuous, let a set $E \subseteq I_1$ with measure zero be given. Denote by $A$ the union of all equivalence classes modulo $\ker \lambda_2$ which intersect $\lambda_1^{-1}(E)$. Then we have

$$\mu(E) = \lambda_2\big(\tilde{\lambda}_1(E)\big) \subseteq \lambda_2\big(\lambda_1^{-1}(E)\big) = \lambda_2(A) .$$

Hence, it suffices to show that $\lambda_2(A)$ has measure zero.

We know that the set $\lambda_1^{-1}(E) \setminus L_0^1$ has measure zero. By what we showed in Step 1, thus also $\lambda_1^{-1}(E) \setminus L_0^2$ has this property. Since $\lambda_2$ is absolutely continuous, it follows that also the set

$$\lambda_2\big(\lambda_1^{-1}(E) \setminus L_0^2\big) = \lambda_2\big(A \setminus L_0^2\big)$$

has measure zero. We can rewrite

$$\lambda_2(A) \setminus \lambda_2(L_0^2) = \lambda_2\big[\lambda_2^{-1}(\lambda_2(A) \setminus \lambda_2(L_0^2))\big] = \lambda_2\big[\underbrace{\lambda_2^{-1}(\lambda_2(A)) \setminus \lambda_2^{-1}(\lambda_2(L_0^2))}_{=A \setminus L_0^2}\big] ,$$

and conclude that

$$\lambda_2(A) \subseteq \lambda_2(A \setminus L_0^2) \cup \lambda_2(L_0^2) .$$

Thus $\lambda_2(A)$ is a zero set.

We conclude that $H_1 \rightsquigarrow H_2$ via $\mu$. The proof of the lemma is completed by applying Lemma 4.7. $\square$

Now we are ready for the proof of the following simple description of '$\sim$' for non-vanishing Hamiltonians.

**4.9. Proposition.** *Let $H$ and $\tilde{H}$ be non-vanishing Hamiltonians defined on intervals $I$ and $\tilde{I}$, respectively. Then we have $H \sim \tilde{H}$ if and only if there exists an increasing* *bijection* $\lambda$ *of* $I$ *onto* $\tilde{I}$, *such that* $\lambda$ *and* $\lambda^{-1}$ *are both locally absolutely continuous, and*

$$H(x) = \tilde{H}(\lambda(x))\lambda'(x), \quad x \in I_1 \ a.e.$$

*Proof.* Let $H_1, \dots, H_n$ be Hamiltonians with heavy endpoints as in Lemma 4.6. Since $H$ and $\tilde{H}$ are non-vanishing, they certainly have heavy endpoints. Thus $H = H_1$ and $\tilde{H} = H_n$.

Let $\lambda_i$, $i = 1, \dots, n-1$, be maps which establish the basic reparametrizations

$$\begin{cases} H_i \rightsquigarrow H_{i+1}, & i = 1, 3, \dots, n-2 \\ H_{i+1} \rightsquigarrow H_i, & i = 2, 4, \dots, n-1. \end{cases}$$

Lemma 4.8 furnishes us with maps $\mu_i$, $i = 1, \dots, n-1$, which establish basic reparametrizations

$$\begin{cases} H_i' \rightsquigarrow H_{i+1}', & i = 1, 3, \dots, n-2 \\ H_{i+1}' \rightsquigarrow H_i', & i = 2, 4, \dots, n-1 \end{cases}$$

where $H_i'$ is trace-normed basic reparametrizations of $H_i$, e.g., $H_i \rightsquigarrow H_i'$ via the map $\mathfrak{t}_i = \operatorname{tr} H_i$ as in Proposition 4.2:

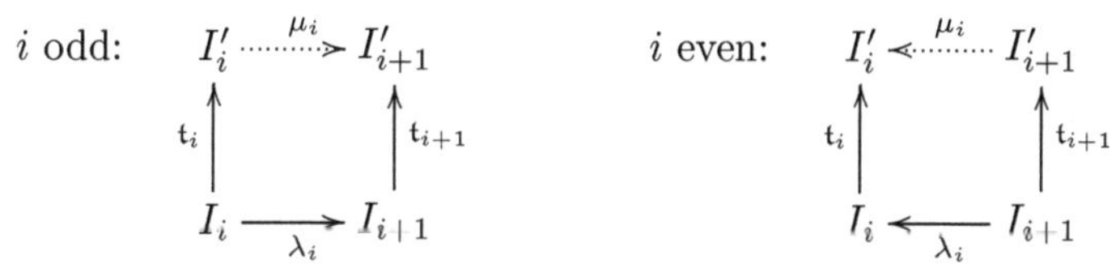


The maps $\mu_i$ are bijective and have the property that $\mu_i^{-1}$ is locally absolutely continuous. Set $\mu_0 := \mathfrak{t}_1$ and $\mu_n := \mathfrak{t}_n$. Since $H_1 = H$ and $H_n = \tilde{H}$ are non-vanishing, by Lemma 4.7, also $\mu_0$ and $\mu_n$ are bijective, and their inverses are locally absolutely continuous.

We see that the composition

$$\lambda := \mu_n^{-1} \circ \mu_{n-1}^{-1} \circ \mu_{n-2} \circ \cdots \circ \mu_3 \circ \mu_2^{-1} \circ \mu_1 \circ \mu_0$$

has the required properties:

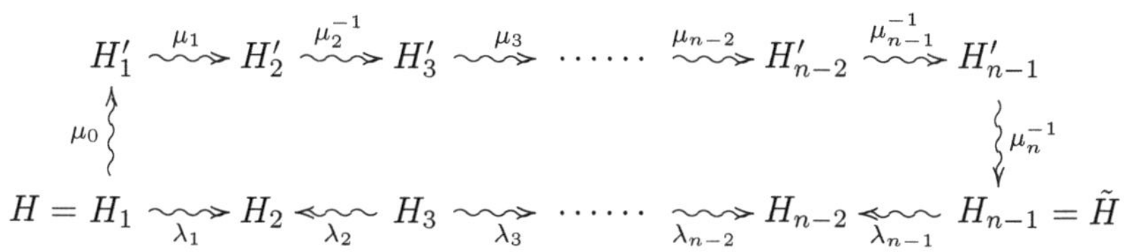


□

4.10. *Remark.* As an immediate corollary of Proposition 4.9, we obtain that the relation '$\rightsquigarrow$' restricted to the set of all non-vanishing Hamiltonians is an equivalence relation. Let us note that the proof of this fact does not require the full strength of Proposition 4.9; it follows immediately from Lemma 4.6 and the chain rule for differentiation.

## References


[AB] C. Aliprantis, O. Burkinshaw: *Principles of Real Analysis,* Academic Press, third edition, San Diego, 1998.

[dB] L. de Branges: *Hilbert spaces of entire functions,* Prentice-Hall, London 1968.

[BB] K. Bube, R. Burridge: *The one-dimensional inverse problem of reflection seismology,* SIAM Rev. 25(4) (1983), 497–559.

[GK] I. Gohberg, M.G. Kreĭn: *Theory and applications of Volterra operators in Hilbert space,* Translations of Mathematical Monographs, Amer. Math. Soc., Providence, Rhode Island, 1970.

[HSW] S. Hassi, H. de Snoo, H. Winkler: *Boundary-value problems for two-dimensional canonical systems,* Integral Equations Operator Theory 36(4) (2000), 445–479.

[K] I.S. Kac: *Linear relations, generated by a canonical differential equation on an interval with a regular endpoint, and expansibility in eigenfunctions, (Russian),* Deposited in Ukr NIINTI, No. 1453, 1984. (VINITI Deponirovannye Nauchnye Raboty, No. 1 (195), b.o. 720, 1985).

[KWW1] M. Kaltenbäck, H. Winkler, H. Woracek: *Singularities of generalized strings,* Oper. Theory Adv. Appl. 163 (2006), 191–248.

[KWW2] M. Kaltenbäck, H. Winkler, H. Woracek: *Strings, dual strings and related canonical systems,* Math. Nachr. 280 (13-14) (2007), 1518–1536.

[KW/IV] M. Kaltenbäck, H. Woracek: *Pontryagin spaces of entire functions IV,* Acta Sci. Math. (Szeged) (2006), 791–917.

[LW] H. Langer, H. Winkler: *Direct and inverse spectral problems for generalized strings,* Integral Equations Oper. Theory 30 (1998), 409–431.

[McL] J. McLaughlin: *Analytic methods for recovering coefficients in differential equations from spectral data,* SIAM Rev. 28(1) (1986), 53–72.

[R] W. Rudin: *Real and Complex Analysis,* International Edition, 3rd Edition, McGraw-Hill 1987.

[W1] H. Winkler: *On transformations of canonical systems,* Oper. Theory Adv. Appl. 80 (1995), 276–288.

[W2] H. Winkler: *Canonical systems with a semibounded spectrum,* Oper. Theory Adv. Appl. 106 (1998), 397–417.

[WW] H. Winkler, H. Woracek: *Symmetry in some classes related with Hamiltonian systems,* manuscript in preparation.



Henrik Winkler
Institut für Mathematik
Technische Universität Ilmenau
Curiebau, Weimarer Straße 25
D-98693 Illmenau, Germany
e-mail: henrik.winkler@tu-ilmenau.de

Harald Woracek
Institut für Analysis
und Scientific Computing
Technische Universität Wien
Wiedner Hauptstraße 8–10/101
A-1040 Wien, Austria
e-mail: harald.woracek@tuwien.ac.at

MIX
Papier aus verantwortungsvollen Quellen
Paper from responsible sources
FSC® C105338
FSC
www.fsc.org

If you have any concerns about our products,
you can contact us on
**ProductSafety@springernature.com**

In case Publisher is established outside the EU,
the EU authorized representative is:
**Springer Nature Customer Service Center GmbH**
**Europaplatz 3, 69115 Heidelberg, Germany**

Printed by Libri Plureos GmbH
in Hamburg, Germany